FE Review Manual

Rapid Preparation for the General Fundamentals of Engineering Exam

Second Edition

Michael R. Lindeburg, PE

The Power to Pass™
www.ppi2pass.com

Professional Publications, Inc. • Belmont, California

How to Locate and Report Errata for This Book

At PPI, we do our best to bring you error-free books. But when errors do occur, we want to make sure you can view corrections and report any potential errors you find, so the errors cause as little confusion as possible.

A current list of known errata and other updates for this book is available on the PPI website at **www.ppi2pass.com/errata**. We update the errata page as often as necessary, so check in regularly. You will also find instructions for submitting suspected errata. We are grateful to every reader who takes the time to help us improve the quality of our books by pointing out an error.

FE Review Manual
Previously published as the *EIT Review Manual.*
Second Edition

Current printing of this edition: 3

Printing History

edition number	printing number	update
2	1	New topic and chapters added. Major revisions. Copyright update.
2	2	Minor corrections.
2	3	Minor corrections.

Printed in the United States of America

PPI
1250 Fifth Avenue, Belmont, CA 94002
(650) 593-9119
www.ppi2pass.com

Library of Congress Cataloging-in-Publication Data
Lindeburg, Michael R.
    FE review manual: rapid preparation for the general fundamentals of engineering
        p. cm.
    ISBN: 978-1-59126-072-1
    1. Engineering--United States--Examinations--Study guides. 2. Engineering mathematics--Study guides. 3. Engineering--Problems, exercises, etc. 4. Engineers--Certification--United States. I. Title.

TA159.L5733   2006
620.0076--dc22
                                                        2006044868

PPI Money-Back Guarantee

The *FE Review Manual*, second edition, is your best choice to prepare for the general Fundamentals of Engineering examination. It is the only review book that covers every exam topic, provides realistic practice problems written by recent examinees, and includes a practice exam accurately simulating the number and scope of problems you will find on the actual test. The author, Michael R. Lindeburg, PE, is acknowledged by the engineering community as the leading authority on licensing exam preparation.

We are confident that if you use the *FE Review Manual* conscientiously to prepare for the general FE examination, following the guidelines described in the "How to Use This Book" section, you'll pass the exam—or we will refund the purchase price of the book.

- To qualify for your refund, send us the following within six months of your FE exam:

  (1) Your name, mailing address, and email address

  (2) Your original packing slip or store sales receipt with the price you paid for the *FE Review Manual*

  (3) Proof from your state board that you took the FE exam and did not pass

  (4) Your copy of the *FE Review Manual*

- Mail these to: PPI
  FERM Refund
  1250 Fifth Ave.
  Belmont, CA 94002

- Upon receipt, we will refund the price you paid for the book, up to our published price.

The Power to Pass™
www.ppi2pass.com

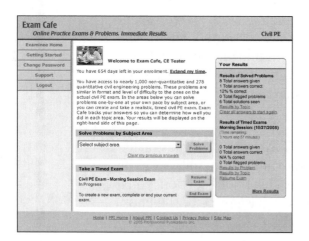
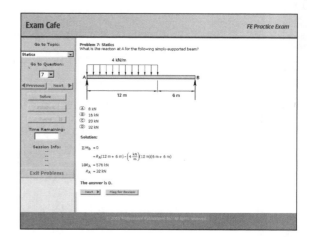

In memory of

Joanne Bergeson

1958–1995

Faithful friend, dedicated general manager, technical wizard, all-out athlete, first employee of Professional Publications.

Thanks for showing us what it means to go all out for something you believe in.

You are missed.

Topics

Topic I   Units and Fundamental Constants
Topic II   Conversion Factors
Topic III   Mathematics
Topic IV   Statics
Topic V   Dynamics
Topic VI   Mechanics of Materials
Topic VII   Fluid Mechanics
Topic VIII   Thermodynamics
Topic IX   Heat Transfer
Topic X   Transport Phenomena
Topic XI   Biology
Topic XII   Chemistry
Topic XIII   Materials Science/Structure of Matter
Topic XIV   Electric Circuits
Topic XV   Computers, Measurement, and Controls
Topic XVI   Engineering Economics
Topic XVII   Ethics

Hint: For the most current information about the exam, visit **www.ppi2pass.com** regularly. Use the Exam Forum to compare notes with other FE examinees.

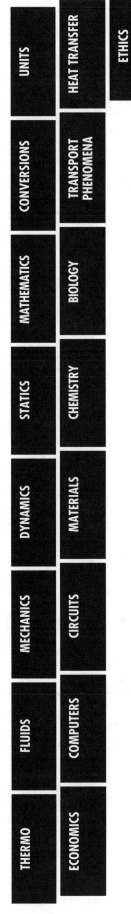

Table of Contents

Preface xiii

Acknowledgments xv

How to Use This Book xvii

Engineering Registration in the United States xxiii

Comments from Recent Examinees xxix

Topic I: Units and Fundamental Constants
Chapter 1   Units 1-1
Chapter 2   Fundamental Constants 2-1

Topic II: Conversion Factors
Chapter 3   Conversion Factors 3-1

Topic III: Mathematics
Diagnostic Examination for Mathematics . . . . DE III-1
Chapter 4   Analytic Geometry and Trigonometry 4-1
Chapter 5   Algebra and Linear Algebra 5-1
Chapter 6   Probability and Statistics 6-1
Chapter 7   Calculus 7-1
Chapter 8   Differential Equations and Transforms 8-1
Chapter 9   Numerical Analysis 9-1

Topic IV: Statics
Diagnostic Examination for Statics . . . . DE IV-1
Chapter 10   Systems of Forces 10-1
Chapter 11   Trusses 11-1
Chapter 12   Pulleys, Cables, and Friction 12-1
Chapter 13   Centroids and Moments of Inertia 13-1

Topic V: Dynamics
Diagnostic Examination for Dynamics . . . . DE V-1
Chapter 14   Kinematics 14-1
Chapter 15   Kinetics 15-1
Chapter 16   Kinetics of Rotational Motion 16-1
Chapter 17   Energy and Work 17-1

Topic VI: Mechanics of Materials
Diagnostic Examination for Mechanics of Materials . . . . DE VI-1
Chapter 18   Stress and Strain 18-1
Chapter 19   Thermal, Hoop, and Torsional Stress 19-1
Chapter 20   Beams 20-1
Chapter 21   Columns 21-1

Topic VII: Fluid Mechanics

Diagnostic Examination for Fluid Mechanics . . . . DE VII-1
Chapter 22   Fluid Properties 22-1
Chapter 23   Fluid Statics 23-1
Chapter 24   Fluid Dynamics 24-1
Chapter 25   Fluid Measurements and Similitude 25-1

Topic VIII: Thermodynamics

Diagnostic Examination for Thermodynamics and Heat Transfer . . . . . DE VIII-1
Chapter 26   Properties of Substances 26-1
Chapter 27   First Law of Thermodynamics 27-1
Chapter 28   Power Cycles and Entropy 28-1
Chapter 29   Mixtures of Gases, Vapors, and Liquids 29-1
Chapter 30   Combustion 30-1

Topic IX: Heat Transfer

Chapter 31   Heat Transfer 31-1

Topic X: Transport Phenomena

Chapter 32   Transport Phenomena 32-1

Topic XI: Biology

Diagnostic Examination for Biology . . . . DE XI-1
Chapter 33   Cellular Biology 33-1
Chapter 34   Toxicology 34-1
Chapter 35   Industrial Hygiene 35-1
Chapter 36   Bioprocessing 36-1

Topic XII: Chemistry

Diagnostic Examination for Chemistry . . . . DE XII-1
Chapter 37   Atoms, Elements, and Compounds 37-1
Chapter 38   Chemical Reactions 38-1
Chapter 39   Solutions 39-1

Topic XIII: Materials Science/Structure of Matter

Diagnostic Examination for Materials Science/Structure of Matter . . . DE XIII-1
Chapter 40   Crystallography and Atomic Bonding 40-1
Chapter 41   Material Testing 41-1
Chapter 42   Metallurgy 42-1

Topic XIV: Electric Circuits

Diagnostic Examination for Electric Circuits . . . . DE XIV-1
Chapter 43   Complex Numbers and Electrostatics 43-1
Chapter 44   Direct-Current Circuits 44-1
Chapter 45   Alternating-Current Circuits 45-1
Chapter 46   Rotating Machines 46-1

Topic XV: Computers, Measurement, and Controls

Diagnostic Examination for Computers, Measurement, and Controls . . DE XV-1
Chapter 47   Computer Hardware 47-1
Chapter 48   Computer Software 48-1
Chapter 49   Measurement 49-1
Chapter 50   Controls 50-1

Topic XVI: Engineering Economics

Diagnostic Examination for Engineering Economics . . . .   DE XVI-1

Chapter 51   Cash Flow and Equivalence.   51-1

Chapter 52   Depreciation and Special Topics   52-1

Chapter 53   Comparison of Alternatives.   53-1

Topic XVII: Ethics

Diagnostic Examination for Ethics   . . . .   DE XVII-1

Chapter 54   Ethics   54-1

Sample Examination with Solutions (Morning Section) SEAM-1

Sample Examination with Solutions (Afternoon Section) SEPM-1

Index I-1

Preface

This edition parallels significant changes to the Fundamentals of Engineering (FE) examination knowledge base. These changes were announced in October 2005 by the National Council of Examiners for Engineering and Surveying (NCEES), the publisher of the standardized national examination. The changes will affect all administrations of the FE exam after April 2006.

Although this book is an efficient review of the subjects you were exposed to during your college years, the book's primary purpose is to direct your preparation for the FE exam. So, quite simply, when the exam adds and drops subjects, this book must also.

The changes to the examination (and this book) are part of an evolving body of knowledge that defines a modern engineer. Certainly, the handheld calculator has made logarithm tables obsolete. Similarly, other activities that used to consume engineering time in the past no longer do so. And, new subjects continue become important to all engineers. Most noteworthy are those subjects that deal with public health, the environment, pollutants, hazardous waste disposal, and so on.

The major changes to the general FE examination that drove this new edition are:

1. Significant changes in the emphasis placed on many subjects, resulting in a sometimes dramatic change in the numbers of questions on the exam allocated to those subjects

2. The addition of cell biology, toxicology, industrial hygiene, wastewater processing, and other biological topics to the exam specifications

3. The elimination of electronics and three-phase electricity from the specifications for the general examination

4. The elimination of certain types of computer science subjects that might be solvable by handheld calculators from the exam specifications

You'll notice the qualifying phrase "... from (or to) the exam specifications" in some of the preceding descriptions. The intent of this wording is to let you know that these changes have been made to the exam specifications, but not necessarily to the exam. In the previous administrations of the FE exam, the NCEES has illustrated just how difficult it is to distill the entirety of engineering knowledge into a single booklet, the NCEES Fundamentals of Engineering Supplied-Reference Handbook, now in its seventh edition. The exam continues to require knowledge and formulas that are not specified in the exam specifications nor directly supported by the NCEES reference document. NCEES is pushing the limits and discovering what types of questions it can ask without providing direct support in its reference material. For these questions, NCEES expects you to draw on your own knowledge and to extrapolate from what is provided in the NCEES Handbook. Though the preface of NCEES' Handbook warns of this, it is a less-than-desirable situation for anyone wanting to know "exactly" what kinds of questions will appear.

Since the new exam specifications contain the caveat that you might need to show knowledge of unlisted topics, neither you nor I can ever be certain that topics previously part of the examination won't reappear on future exams. To address that possibility, I have collected approximately 50 questions that previously appeared in this book but have subsequently been removed for no other reason than they are no longer justified on the basis of the exam specifications. They're all great questions, and because it's hard for me to believe that NCEES doesn't want you to know what a diode is or how to convert base 10 numbers into hexadecimal, I'm offering them to you as a bonus. You may access these "questions that the author omitted" at **www.ppi2pass.com/extraFERM**. Consider them a reward for your having read this far in the preface.

This book is merely a tour guide, a roadmap to the exam. There is nothing magical about this book. It doesn't contain actual exam questions for you to memorize. And, it is only 872 pages long. Accordingly, the number of questions in each topic may be inadequate to bring you up to speed in that topic. For example, how many problems will you have to work in order to come up to speed in the subjects of divergence, curl, differential equations, and linear algebra? (Answer: Probably more than are in this book.) So, some out-of-book experience is inevitable if you want to enter the examination room adequately prepared.

The only effective option is for you to use the most targeted review materials available as a necessary-but-not-necessarily-sufficient guide to what's on the exam.

I continue to listen to suggestions from those who have taken the exam and have learned firsthand what knowledge is needed. So, when the next level of exam sophistication evolves (long after you've passed the exam!), this book will have evolved with it.

In the meantime, I invite you to visit PPI's website at **www.ppi2pass.com** to obtain the latest news, hints, advice, and scuttlebutt about the exam.

Michael R. Lindeburg, PE

p.s. Most Prefaces in books don't have postscripts. If having a "p.s." here is a serious error, PPI's editorial staff will tell me. What the editorial staff cannot tell me, however, is if this book contains a calculation or analysis error. After all, editors are not engineers. That's where you come in. If you don't tell me about an error you find, I won't know it's there, and that error will go on confusing other engineers for even longer.

PPI has established an easy way for you and I to communicate about an error you might find. To get there using pilotage, just go to **www.ppi2pass.com** and click on the "Support" tab. Or, if you prefer to navigate by dead reckoning, go to **www.ppi2pass.com/errata**. Look for the "How to Report Errata" link.

By the way, PPI's editors refer to an error as an "erratum." You and I can just continue on talking like regular people about the "error." It will be our little secret joke.

Speaking of secrets, the probability of finding an error in this book increases as you plow through the pages. Most of the errors live in the last third of the book's pages. I don't write an entire book in a single sitting, so fatigue isn't a factor in sprinkling more errors near the end. Over the years, I've learned that my propensity for inserting errors into my books is pretty constant throughout the writing process. I write at a certain level of proficiency and the space between that level and the top floor (to which I don't have an elevator key) generates errors at a predictable, albeit random and unpredictable, rate. Rather, the fact that a disproportionate number of errors lurk near the book's end is because many readers have already reported errors in the first chapters, but they didn't finish going through the entire book. If you are starting your review three or more months before your exam, you'll have time to find something with which to embarrass me. Otherwise, you'll probably have to read the book backwards in order to get to the good stuff.

Acknowledgments

The project of converting the first edition of this book into a second edition was spearheaded by Sarah Hubbard, PPI's Editorial Department Manager. She did more than manage her department — she was the visionary, coordinator, project editor. Along the way, she approached closer and closer to being an author, something that I have not failed to notice.

There were three people who contributed significantly to the content and revision of their respective sections. Ashok V. Naimpally, PhD, PE authored much of the biology material, as well as numerous original problems for the diagnostic and sample exams. Kirsten S. Rosselot, PE reviewed and edited the biology material, and she served as PPI's technical expert on the subject during the editing process.

Gregg Wagener, PE, contributed significantly to the electrical and computer chapters. Not only did he substantially edit and revise the existing chapters to correspond to the NCEES examination specifications, but he also wrote a new chapter, Rotating Machines, to replace the Three-Phase Systems and Electronics chapter that was no longer required per those specifications. Gregg has been a bulletproof contributor to many PPI books over the years. His contributions to the engineering profession are acknowledged.

The Editorial Department staff that turned out stubby pencils by the dozens include the following: Dennis Rowcliffe was the Project Editor, which at PPI means he shouldered all of the work of keeping this high-priority project on track; Heather Kinser edited the electrical and computer chapters; Jessica Holden came back from FERM retirement to edit and proof the preexisting material against changes in PPI's Style Guide, and to eliminate the spurious consequences of having revised the macros in our typesetting system; Scott Marley and Sarah Hubbard revised and updated the sample exam, including developing new problems; and, lastly-but-not-leastly, Jenny Lindeburg, MA, my daughter, who seems to be following in her father's publishing footsteps, did a thorough proofing of the mark-up round as well as editing some of the final round material. Finally, the contribution of newborn Chance Hubbard is duly noted, since he held off his arrival until after this project was 99.9% complete.

Under the direction of Cathy Schrott, Production Department Manager, who guided the whole production (typesetting and illustrating) process and kept the Editorial Department on its toes, are the following masters of their respective trades: Amy Schwertman used her sense of logic and proportion, line and color to develop a killer professional new cover; Tom Bergston drew a plethora of new and replacement illustrations according to PPI's revised Style Guide; Miriam Hanes did an extensive macro/style update early into the project so that these changes could be incorporated into all of the book, not just the new material, and then, later, provided excellent production keyboarding of the new material, proofreading and editing as she went along. I've labeled these people "masters" because bringing out technical books is nothing like bringing out novels. These PPI staffers use some pretty sophisticated tools in order to give you an engineering textbook that should exceed your expectations.

Among others, many engineering editorial contributors have helped this book to become a reality. I'd like to thank Dave Barksdale, MS; Wendy Holforty, PhD; Anthony Ponko, PhD; and Karen Vaughan, MS. I'd also like to thank Gretchen Rau, MS, PE, who was instrumental in organizing the original version of this book.

Over the years, there have been many technical reviewers for this book, including Mahesh Aggarwal, PhD; LeRoy Friel, PhD, PE; Victor Gerez, PhD, PE; Gordon Goff, PE; Richard H. Heist, PhD; Shahin A. Mansour, PhD, PE; N. S. Nandagopal, PhD; James S. Noble, PhD, PE; and A. H. Tabrizi, PhD, PE.

I have had a lot of engineers provide suggestions that greatly enhanced the book, including Fred Beaufait, PhD; Frank R. Cole, PhD; Jay Goldman, DSc; Irving Greene, PE; Ken Schneider, PE; Leighton Sissom; and Dan Turner, PhD, PE.

I'd also like to especially thank the members of the FE Examination Advisory Board for the first edition of this book: Robert H. Easton, PE (Rochester Institute of Technology); Leroy Friel, PhD, PE (Montana Tech); Jay Goldman, DSc, PE (University of Alabama at Birmingham); Irving Greene, PE (Rutgers University); Robert Helgeland, PE (University of Massachusetts, Darmouth); Thomas E. Hulbert, PE (Northeastern University); Neil J. Illenberg, PE (Course Director for the FE Review in Rochester, NY); John P. Klus, PhD (University of Wisconsin, Madison); and, Ronald E. Scott, DSc, PE (Northeastern University). Their reviews and comments gave me the confidence to know that we did the very best we could to bring you an accurate and easy-to-use review book.

Additionally, thousands of mini-authors have contributed to my bank of problems. I can't list all of the students and engineers who took the time to write problems for this book. Even if I could, the list would always be out of date since the process of collecting problems is ongoing as new problems are added based on each FE exam.

I'd also like to extend my appreciation to the PPI Production team who worked on previous editions and printings of this book. Mia Laurence acted as the original project editor, copyeditor, and proofreader. The original typesetters were Sylvia Osias, Cathy Schrott, and Jessica Whitney-Holden, and eventually on subsequent printings, Patricia Hoffman. The proofreaders were Lisa Rominger and Shelley Arenson. Additional editing was the contribution of Jessica Whitney-Holden, and additional proofreading was performed by Tracey Brown. Charles P. Oey did the original illustrations and the page and cover designs.

And finally back to real-time, as always but more so than ever, I thank my wife, Elizabeth, for permitting, accepting, and participating in a writer's life that is full-to-overflowing. Now that our children are on their own, we have less time than we had before. As a corollary to Aristotle's "Nature abhors a vacuum," I suggest: "Work expands to fill the void."

Thanks, one and all! Your efforts have made this book the best review any engineer could find for the FE exam.

Michael R. Lindeburg, PE

How to Use This Book

HOW EXAMINEES CAN USE THIS BOOK

This book is written for one purpose and one purpose only: to get you ready for the FE examination. I have no illusions about your using this as a reference book after the examination. I expect you to use this book, scribble and draw all over it, and spill coffee on it. Then, after the examination, you can throw this book away. Along the way, I hope you will have acquired a copy of my *Engineer-In-Training Reference Manual* to use throughout your career.

This book is not intended as a reference book, because you cannot use it in the examination. On the other hand, the book that may be used during the exam (the NCEES Handbook) is not suitable for use as a study aid. You certainly need to become familiar with its format, layout, and organization, however. This book, which follows the NCEES Handbook's subject sequence, satisfies your needs to become familiar with the NCEES Handbook and to study from something better.

You can start by contacting PPI to see what the current edition and printing of this book is, particularly if you are using a borrowed copy. This book is reprinted frequently to keep it up-to-date and relevant.

Because this book was designed in a certain way, there aren't too many options for how you can use it. When I designed it, here's what I had in mind.

- You should study every subject in this book. NCEES has already greatly reduced the scope of the FE exam and simplified the problems. There is no longer any reason for a mechanical engineer to skip the electricity problems, or for an electrical engineer to skip the thermodynamics problems. Difficulty is no longer an issue. The only complicating issue is how fast you can work problems outside of your favorite subjects. Given enough time to work all of the problems at your own pace, you should be able to score 100 percent on the FE exam.

- You need to decide on a study schedule. I separated the subjects and chose the quantity of problems in this book so that you can easily review a chapter a day in an hour or so—even if you are taking a review course with other homework.

There are 54 chapters and a sample exam in this book. So, you need at least 52 study days. I would plan to take every fourth or fifth study day off. Give yourself a week to take the realistic final exam and if you didn't get 100 percent, figure out why.

There is a blank study schedule at the end of this section. Using it, you will have to begin 72 days before the exam. This requires you to treat every day the same and work through weekends. Some weekdays you might have off, and some weekends you might have to study. If you'd rather take all the weekends off, retain the rest and review days, and still stick with the one-chapter-per-study-day concept, you will have to begin approximately 99 days before the exam.

You will have to fill in your own calendar dates, but the sample study schedule should help you in scheduling your work. Use the days off to rest, review, and study problems from other books. If you are pressed for time, you don't have to take the days off. That will be your choice.

- Get a copy of the NCEES Handbook.* You should use it to try to solve most of the problems in this book. By the end of your review, you should know the order of the chapters, what data is included, and the approximate locations of important figures and tables.

- Start from the beginning and systematically work your way through the *FE Review Manual*. The sequencing of subjects corresponds to the sequence of subjects in the NCEES Handbook. Studying in this manner will help you become familiar with where things are in the NCEES Handbook.

- Use the thirteen subject, or topic, diagnostic exams to determine how much you should study. You can take a diagnostic exam before studying the corresponding subject to determine if you can postpone or skip reviewing that subject. Alternatively, you can take each diagnostic exam after studying the subject to determine if you are ready

---

*If you do not receive a copy of this from your state engineering licensing board, you can obtain a copy from PPI.

to move on. In either case, a score of 50% or below during the allowed time should be considered an indication that you need additional review in that subject.

- Solve every problem in this book. Even though there are hundreds of problems, there are approximately only 25 to solve each day. And, each will only take a few minutes. Don't skip any of them. The problems in this book were chosen for a reason. Don't short-circuit your review by skipping problems that you know are important.

- The FE exam primarily uses SI units. Therefore, the need to work problems in both unit systems is greatly diminished. You need to learn the SI system if you are not already familiar with it.

- I tell my youth soccer players to "live with" their soccer balls during the summer. My advice to you: Live with a different chapter each day. You should study the chapter on the bus or train and during lunch at work. Review it after dinner.

- Don't turn directly to the problems without first reviewing the text. Unlike reference books that you skim or merely refer to when needed, you need to read each chapter. That's going to be your only review. There isn't much text to read in the first place, there aren't any derivations or proofs, and everything has a high probability of showing up on the exam. So, read the text part of the chapter.

- Review the sample problems in the chapter you are studying. There are four to seven of these in every chapter. You can either try to solve them (covering up the solutions) or just read through them and make sure you know what is going on. Personally, I think it is much harder to evaluate whether or not you know how to work a problem when you don't actually work the problem. However, I'll leave that choice to you.

- Then, when you have a few minutes and some space, get out your calculator and attempt the first set of the FE-style exam problems. The solutions to these problems have been placed at the end of the chapter to help you solve the problems under realistic conditions. Remember that you can refer back to the reference material in the chapter. While the explanatory text isn't present in the actual NCEES Handbook, the formulas, tables, and figures are all analogous.

- Finally, attempt the second set of FE-style exam problems using the NCEES Handbook as your only reference.

- After you have finished reviewing all 54 chapters, take the Sample Examination in this book. As Hari Seldon said in Isaac Asimov's *Foundation and Empire*, your course of action will be obvious from that point on.

HOW INSTRUCTORS CAN USE THIS BOOK

(FE course instructors are invited to view PPI's web page, **www.ppi2pass.com/instruct**, and utilize the Resources for Instructors. This page contains information on obtaining instructor-only suppor material based on this book including course outlines, lecture notes, and problem sets, as well as helpful information on course design and exam statistics.)

If you are teaching a review course for the FE examination without the benefit of recent, firsthand experience, you can use this book as a guide to preparing your lectures. You should spend most of your lecture time discussing the subjects in each chapter.

In solving problems in your lecture, everything you do should be tied back to the NCEES Handbook. You will be doing your students a great disservice if you get them accustomed to using your own handouts or notes to solve problems. They can't use your notes in the exam, so train them to use what references they are allowed to use.

That's why this book is written the way it is. It is organized in the same sequence as the NCEES Handbook, using the same terminology and nomenclature. The tables are analogous. You can feel confident that I had your students and the success of your course in mind when I designed this book. That is why I solicited comments from hundreds of review course instructors such as yourself prior to writing this book.

NCEES prominently displays the following warning in its Handbook:

> The *FE Supplied-Reference Handbook* is not designed to assist in all parts of the FE examination. For example, some of the basic theories, conversions, formulas, and definitions that examinees are expected to know have not been included ... In no event will NCEES be liable for not providing reference material to support all the questions in the FE examination.

While it is true that the exam draws upon a body of knowledge that has more breadth than the NCEES Handbook, very few problems will appear that require formulas not present in the Handbook. In its attempt to make the FE exam secure and pilferage-free, NCEES has been forced to limit the scope of the exam to what is in its Handbook. Therefore, instructors shouldn't deviate too much from the subject matter of each chapter.

I have taught more than 50 FE review courses, and I have supervised the offerings of hundreds more. It has always been my goal in FE exam review courses to over-prepare my students. That hasn't changed. What has changed is my definition of "overpreparation." In the past, overpreparation meant exposure to a wide variety of subjects, including those on the fringe. With the current limited-scope exam format, overpreparation now means repeated exposure and reinforcement. The basic concepts tested for in the FE exam can be reinforced by repeated exposure to exam-like problems. This will also improve the examinee's problem-solving and recall speeds.

I have ensured overpreparation by making sure my students work to their own individual capacities. If I assign ten hours of practice problems per week, and a student can only put in five hours of preparation, that student will have worked to capacity. Another student might be maxed out at three hours, and another might be able to put in the full ten. After the actual FE examination, your students will honestly say that they could not have prepared any more than they did in your course.

Students like to see and work lots of problems. They derive great comfort in exposure to exam-like problems. They experience great reassurance in finding out how easy the problems are and that they can solve these kinds of problems. Therefore, the repetition and reinforcement should come from working additional problems, not from more lecture.

However, it is unlikely that they will be working to capacity if their work is limited to what is in this book. You will have to provide more problems if the concept of capacity assignment appeals to you.*

I assign, but don't grade, individual homework assignments in my courses. Instead, the students are given the solutions to all practice problems in advance. However, I do answer individual questions. When each student turns in a completed set of problems for credit each week, I always address special needs or questions written on the assignment.

I have found that a 14-week format works well for a commercial or in-house industrial FE exam review course. Each week, I lecture for two or two-and-a-half hours, with an intermediate break. The following table outlines the basic course format that has worked well for me. I feel comfortable teaching the subjects in the order listed because my lectures build in that direction.

---

*PPI has a variety of old, new, and developing products that you can use to assign additional problems, including *1001 Solved Engineering Fundamentals Problems* and *FE/EIT Sample Examinations*. Contact PPI's Customer Care Department to discuss what would be best for you.

However, you may want to take the subjects in the order they appear in the NCEES Handbook.

Recommended 14-Week FE Exam Review Course Format (for Commercial Review Courses)

meeting	subjects covered	book chapters
1	Algebra; Trigonometry; Geometry; Properties of Areas and Solids	4, 5, 13
2	Probability; Statistics; Calculus; Differential Equations	6, 7, 8, 9
3	Engineering Economics; Ethics	51, 52, 53, 54
4	Biology, Inorganic and Organic Chemistry	33, 34, 35, 36, 37, 38, 39
5	Statics	10, 11, 12
6	Kinematics; Kinetics; Energy, Work, and Power	14, 15, 16, 17
7	Fluid Statics and Fluid Dynamics	22, 23, 24, 25
8	Thermodynamic Properties; Transport Phenomena	26, 27, 29, 32
9	Thermodynamics; Cycles; Combustion; Heat Transfer	28, 30, 31
10	Materials Science; Material Testing	40, 41, 42
11	Stress and Strain	18, 19
12	Beams and Columns	20, 21
13	Electrostatics; DC and AC Circuits; Electronics	43, 44, 45, 46
14	Computers; Measurement; Controls	47, 48, 49, 50

A 14-week course is too long for junior and senior engineering majors. Students and professors don't have that much time, and it is difficult for them to keep 14 lecture slots open. Also, students don't need as thorough a review as do working engineers who have forgotten the more basic elements of engineering. I would hope that engineering majors can get by with the most cursory of reviews in some of these subjects, such as mathematics, statics, and DC electricity.

I strongly believe in the need to expose my students to a realistic sample examination, but I no longer use an in-class sample exam. Since the review course usually ends only a few days before the real FE examination, I have hesitated to make students sit for four hours in the late evening to take a final exam. For commercial and industrial review courses, I distribute a sample examination at the first meeting of the review course and assign it as a take-home exam.

For engineering majors, I recommend a three-week, six-lecture review course, followed by a Saturday four-hour or eight-hour mock exam. The lectures are approximately two hours in length. The format consists of a forced march through all subjects except mathematics, with the major emphasis being on problem-solving. For engineering majors, the main goals are to keep the students focused and to wake up the latent memories, not to teach the subjects.

There are many other ways to organize an FE exam review course depending on the available time, budget, and intended audience. However, all good course formats have the same result: the students breeze through the examination. That's my wish for you and your students.

**Recommended 3-Week
FE Exam Review Course Format
(for Students)**

meeting	subjects covered	book chapters
1	Engineering Economics; Ethics; Biology;	51, 52, 53, 54
	Chemistry	33, 34, 35, 36, 37, 38, 39
2	Statics; Kinetics; Energy,	10, 11, 12
	Work, and Power	14, 15, 16, 17
3	Fluid Statics and	
	Dynamics	22, 23, 24, 25
4	Thermodynamics; Cycles;	
	Combustion;	26, 27, 28,
	Heat Transfer	29, 30, 31
5	Materials Science;	
	Material Testing;	40, 41, 42
	Stress and Strain;	18, 19
	Beams and Columns	20, 21
6	Electrostatics; DC and AC Circuits;	
	Rotating Machines;	43, 44, 45, 46,
	Electronics; Computers	47, 48, 49, 50
7	Mock Exam	

Sample Study Schedule
(for Individuals)

Time required to complete study schedule:

    52 days for a "crash course," going straight through, with no rest and review days, no weekends, and no final exam

    72 days going straight through, taking off rest and review days, but no weekends

    99 days using only the five-day workweek, taking off rest and review days, and weekends

Your examination date: _____

Number of days: _____

Your latest starting date: _____

day no.	date	weekday	chap. no.	subject
1	_____	_____	Intro, 1, 2, 3	Introduction; Units and Fundamental Constants; Conversion Factors
2	_____	_____	4	Geometry; Trigonometry
3	_____	_____	5	Algebra; Linear Algebra
4	_____	_____	6	Probability; Statistics
5	_____	_____	None	**Rest; Review**
6	_____	_____	7	Calculus
7	_____	_____	8	Differential Equations
8	_____	_____	9	Numerical Analysis
9	_____	_____	None	**Rest; Review**
10	_____	_____	10	Systems of Forces
11	_____	_____	11	Trusses
12	_____	_____	12	Pulleys, Cables, Friction
13	_____	_____	13	Centroids; Moments of Inertia
14	_____	_____	None	**Rest; Review**
15	_____	_____	14	Kinematics
16	_____	_____	15	Kinetics
17	_____	_____	16	Rotational Motion
18	_____	_____	17	Energy and Work
19	_____	_____	None	**Rest; Review**
20	_____	_____	18	Stress and Strain I
21	_____	_____	19	Stress and Strain II
22	_____	_____	20	Beams
23	_____	_____	21	Columns
24	_____	_____	None	**Rest; Review**
25	_____	_____	22	Fluid Properties
26	_____	_____	23	Fluid Statics
27	_____	_____	24	Fluid Dynamics
28	_____	_____	25	Fluid Measurement; Models
29	_____	_____	None	**Rest; Review**
30	_____	_____	26	Properties of Substances
31	_____	_____	27	First Law
32	_____	_____	28	Basic Cycles
33	_____	_____	29	Gas Mixtures
34	_____	_____	30	Combustion
35	_____	_____	None	**Rest; Review**

day no.	date	weekday	chap. no.	subject
36	————	————	31	Heat Transfer
37	————	————	32	Transport Phenomena
38	————	————	33	Cellular Biology
39	————	————	34	Toxicology
40	————	————	35	Industrial Hygiene
41	————	————	None	**Rest; Review**
42	————	————	36	Bioprocessing
43	————	————	37	Atoms, Elements, and Compounds
44	————	————	38	Chemical Reactions
45	————	————	39	Solutions
46	————	————	None	**Rest; Review**
47	————	————	40	Crystallography; Bonding
48	————	————	41	Material Testing
49	————	————	42	Metallurgy
50	————	————	None	**Rest; Review**
51	————	————	43	Electrostatics
52	————	————	44	DC Circuits
53	————	————	45	AC Circuits
54	————	————	46	Rotating Machines; Electronics
55	————	————	None	**Rest; Review**
56	————	————	47	Computer Hardware
57	————	————	48	Computer Software
58	————	————	49	Measurement
59	————	————	50	Controls
60	————	————	None	**Rest; Review**
61	————	————	51	Cash Flow; Equivalence
62	————	————	52	Costs and Depreciation
63	————	————	53	Alternative Comparison
64	————	————	54	Ethics
65	————	————	None	**Rest; Review**
66–72	————	————		Sample Examination
73	————	————	None	E-I-T/FE Examination

Engineering Registration in the United States

ENGINEERING REGISTRATION

Engineering registration (also known as *engineering licensing*) in the United States is an examination process by which a state's board of engineering licensing (i.e., registration board) determines and certifies that you have achieved a minimum level of competence. This process protects the public by preventing unqualified individuals from offering engineering services.

Most engineers do not need to be registered. In particular, most engineers who work for companies that design and manufacture products are exempt from the licensing requirement. This is known as the *industrial exemption*. Nevertheless, there are many good reasons for registering. For example, you cannot offer consulting engineering design services in any state unless you are registered in that state. Even within a product-oriented corporation, however, you may find that employment, advancement, or managerial positions are limited to registered engineers.

Once you have met the registration requirements, you will be allowed to use the titles Professional Engineer (PE), Registered Engineer (RE), and Consulting Engineer (CE).

Although the registration process is similar in all 50 states, each state has its own registration law. Unless you offer consulting engineering services in more than one state, however, you will not need to register in other states.

The U.S. Registration Procedure

The registration procedure is similar in most states. You will take two eight-hour written examinations. The first is the *Fundamentals of Engineering Examination*, also known as the *Engineer-In-Training Examination* and the *Intern Engineer Exam*. The initials FE, EIT, and IE are also used. This examination covers basic subjects from all of the mathematics, physics, chemistry, and engineering classes you took during your first four university years.

In rare cases, you may be allowed to skip this first examination. However, the actual details of registration qualifications, experience requirements, minimum education levels, fees, oral interviews, and examination schedules vary from state to state. Contact your state's registration board for more information.

The second eight-hour examination is the *Principles & Practice of Engineering Exam*. The initials PE are also used. This examination covers subjects only from your areas of specialty.

National Council of Examiners for Engineering and Surveying

The National Council of Examiners for Engineering and Surveying (NCEES) in Clemson, South Carolina, produces, distributes, and scores the national FE and PE examinations. The individual states purchase the examinations from NCEES and administer them themselves. NCEES does not distribute applications to take the examinations, administer the examinations or appeals, or notify you of the results. These tasks are all performed by the states.

Reciprocity Among States

All states use the NCEES examinations. If you take and pass the FE or PE examination in one state, your certificate will be honored by all of the other states. Although there may be other special requirements imposed by a state, it will not be necessary to retake the FE and PE examinations. The issuance of an engineering license based on another state's license is known as *reciprocity* or *comity*.

With minor exceptions, having a license from one state will not permit you to practice engineering in another state. You must have a professional engineering license from each state in which you work. For most engineers, this is not a problem, but for some, it is. Luckily, it is not too difficult to get a license from every state you work in once you have a license from one state.

The simultaneous administration of identical examinations in all states has led to the term *uniform examination*. However, each state is still free to choose its own minimum passing score and to add special questions and requirements to the examination process. Therefore, the use of a uniform examination has not, by itself, ensured reciprocity among states.

THE FE EXAMINATION

Applying for the Examination

Each state charges different fees, specifies different requirements, and uses different forms. Therefore, it will be necessary to request an application from the state in which you want to become registered. Generally, it is sufficient for you to phone for this application. At **www.ppi2pass.com/stateboards**, you'll find contact information (websites, telephone numbers, email addresses, etc.) for all U.S. state and territorial boards of registration.

Keep a copy of your examination application, and send the original application by certified mail, requesting a receipt of delivery. Keep your proof of mailing and delivery with your copy of the application.

Examination Dates

The national FE and PE examinations are administered twice a year (usually in mid-April and late October), on the same weekends in all states. Check for a current exam schedule at **www.ppi2pass.com/fefaq**.

FE Examination Format

The NCEES Fundamentals of Engineering examination has the following format and characteristics.

- There are two four-hour sessions separated by a one-hour lunch.

- Examination questions are distributed in a bound examination booklet. A different examination booklet is used for each of these two sessions.

- The morning session (also known as the *A.M. session*) has 120 multiple-choice questions, each with four possible answers lettered (A) to (D). Each problem in the morning session is worth one point. The total score possible in the morning is 120 points. Guessing is valid; no points are subtracted for incorrect answers.

There are questions on the morning session examination from most of the undergraduate engineering degree program subjects. Questions from the same subject are all grouped together, and the subjects are labeled. The percentages of questions on each subject in the morning session are given in the following table.

Morning FE Exam Subjects

subject	percentage of total questions (%)
mathematics	15
engineering probability and statistics	7
chemistry	9
computers	7
ethics and business practices	7
engineering economics	8
engineering mechanics (statics and dynamics)	10
strength of materials	7
material properties	7
fluid mechanics	7
electricity and magnetism	9
thermodynamics	7

- There are seven different versions of the afternoon session (also known as the *P.M. session*). One is a general exam and the other six correspond to a specific engineering discipline: chemical, civil, electrical and computer, environmental, industrial, and mechanical engineering. The seventh version of the afternoon examination is a general examination suitable for anyone, but in particular, for engineers whose specialties are not one of the other six disciplines. NCEES calls this seventh version the "Other/General" exam.

Each version of the afternoon session consists of 60 questions that count two points each, for a total of 120 points. All questions are mandatory. Each question consists of a problem statement followed by multiple-choice questions. Four answer choices lettered (A) through (D) are given, from which you must choose the best answer. Questions in each subject may be grouped into related problem sets containing between two and ten questions each.

Questions on the afternoon examination are intended to cover concepts learned in the last two years of a four-year degree program. Unlike morning questions, these questions may deal with more than one basic concept per question.

The percentages of questions on each subject in the general afternoon session examination are given in the following table. (This book assumes you will elect to take the general examination.)

Afternoon FE Exam Subjects (Other/General Exam)

subject	percentage of total questions (%)
advanced engineering mathematics	10
engineering probability and statistics	9
biology	5
engineering economics	10
application of engineering mechanics	13
engineering of materials	11
fluids	15
electricity and magnetism	12
thermodynamics and heat transfer	15

The percentages of questions on each subject in the discipline-specific afternoon session examination are listed at the end of this section. The discipline-specific afternoon examinations cover substantially different bodies of knowledge than the morning examination. Formulas and tables of data needed to solve questions in these examinations will be included in either the NCEES

Handbook or in the body of the question statement itself.

- The scores from the morning and afternoon sessions are added together to determine your total score. No points are subtracted for guessing or incorrect answers. Both sessions are given equal weight. It is not necessary to achieve any minimum score on either the morning or afternoon sessions.

- All grading is done by computer optical sensing. Responses must be recorded on special answer sheets with the mechanical pencils provided to examinees by NCEES.

Use of SI Units on the FE Exam

Metric, or SI, units are used in virtually all subjects, except some civil engineering and surveying subjects that typically use only customary U.S. (i.e., English) units. Some of the remaining questions may be presented on the exam in both metric and English units. These questions are actually stated twice, once in metric units and once in English units. Dual dimensioning is not used. However, these dual-statement questions are rapidly being phased out.

It is the goal of NCEES to use SI units that are consistent with ANSI/IEEE standard 268-1992 (the American Standard for Metric Practice). Non-SI metric units might still be used when common or where needed for consistency with tabulated data (e.g., use of bars in pressure measurement).

Grading and Scoring the FE Exam

The FE exam is not graded on a curve, and there is no guarantee that a certain percent of examinees will pass. Rather, NCEES uses a modification of the Angoff procedure to determine the suggested passing score (the cutoff point or cut score).

With this method, a group of engineering professors and other experts estimate the fraction of minimally qualified engineers that will be able to answer each question correctly. The summation of the estimated fractions for all test questions becomes the passing score. The passing score in recent years has been somewhat less than 50 percent (i.e., a raw score of approximately 110 points out of 240). Because the law in most states requires engineers to achieve a score of 70 percent to become licensed, you may be reported as having achieved a score of 70 percent if your raw score is greater than the passing score established by NCEES, regardless of the raw percentage. The actual score may be slightly more or slightly less than 110 as determined from the performance of all examinees on the equating subtest.

Approximately 20 percent of each FE exam consists of questions repeated from previous examinations—this is the *equating subtest*. Since the performance of previous examinees on the equating subtest is known, comparisons can be made between the two examinations and examinee populations. These comparisons are used to adjust the passing score.

The individual states are free to adopt their own passing score, but all adopt NCEES's suggested passing score because the states believe this cutoff score can be defended if challenged.

You will receive the results of your examination from your state board (not NCEES) by mail. Allow at least four months for notification. Candidates will receive a pass or fail notice only, and will no longer receive a numerical score. A diagnostic report is provided to those who fail.

See **www.ppi2pass.com/fepassrates** for recent and historic FE pass rates.

Permitted Reference Material

Since October 1993, the FE examination has been what NCEES calls a "limited-reference" exam. This means that no books or references other than those supplied by NCEES may be used. Therefore, the FE examination is really an "NCEES-publication only" exam. NCEES provides its own Handbook for use during the examination. No books from other publishers may be used.

What Does *Most Nearly* Really Mean?

One of the more disquieting aspects of these questions is that the available answer choices are seldom exact. Answer choices generally have only two or three significant digits. Exam questions instruct you to complete the sentence, "The value is most nearly...", or they ask "Which answer choice is closest to the correct value?" A lot of self-confidence is required to move on to the next question when you don't find an exact match for the answer you calculated, or if you have had to split the difference because no available answer choice is close.

NCEES describes it like this: Many of the questions on NCEES exams require calculations to arrive at a numerical answer. Depending on the method of calculation used, it is very possible that examinees working correctly will arrive at a range of answers. The phrase 'most nearly' is used to accommodate all these answers that have been derived correctly but which may be slightly different from the correct answer choice given on the exam. You should use good engineering judgment when selecting your choice of answer. For example, if the question asks you to calculate an electrical current or determine the load on a beam, you should literally select the answer option that is most nearly what you calculated, regardless of whether it is more or less than your calculated value. However, if the question asks you to select a fuse or circuit breaker to protect against a calculated current or to size a beam to carry a load, you should select an answer option that will safely carry the current or load. Typically, this requires selecting a value that is closest to but larger than the current or load.

The difference is significant. Suppose you were asked to calculate "most nearly" the volumetric pure water flow required to dilute a contaminated stream to an acceptable concentration. Suppose, also, that you calculated 823 gpm. If the answer choices were (A) 600 gpm, (B) 800 gpm, (C) 1000 gpm, and (D) 1200 gpm, you would go with answer choice (B), because it is most nearly what you calculated. If, however, you were asked to select a pump or pipe with the same rated capacities, you would have to go with choice (C). Got it? If not, stop reading until you understand the distinction.

CALCULATORS

To prevent unauthorized transcription and distribution of the examination questions, calculators with communicating and text editing capabilities have been banned by all states. Calculators permitted by the NCEES are listed at **www.ppi2pass.com/calculators**. You cannot share calculators with other examinees.

It is essential that a calculator used for engineering examinations have the following functions.

- trigonometric functions
- inverse trigonometric functions
- hyperbolic functions
- pi
- square root and x^2
- common and natural logarithms
- y^x and e^x

For maximum speed, your calculator should also have or be programmed for the following functions.

- extracting roots of quadratic and higher-order equations
- converting between polar (phasor) and rectangular vectors
- finding standard deviations and variances
- calculating determinants of 3×3 matrices
- linear regression
- economic analysis and other financial functions

STRATEGIES FOR PASSING THE FE EXAM

The most successful strategy to pass the FE exam is to prepare in all of the examination subjects. Do not limit the number of subjects you study in hopes of finding enough questions in your particular areas of knowledge to pass.

Fast recall and stamina are essential to doing well. You must be able to quickly recall solution procedures, formulas, and important data. You will not have time during the exam to derive solutions methods—you must know them instinctively. This ability must be maintained for eight hours. If you are using this book to prepare for the FE examination, the best way to develop fast recall and stamina is to work the practice problems at the end of the chapters. Be sure to gain familiarity with the NCEES Handbook by using it as your only reference for most of the problems you work.

In order to get exposure to all examination subjects, it is imperative that you develop and adhere to a review schedule. If you are not taking a classroom review course (where the order of your preparation is determined by the lectures), prepare your own review schedule. For example, plan on covering this book at the rate of one chapter per day in order to finish before the examination date.

There are also physical demands on your body during the examination. It is very difficult to remain tense, alert, and attentive for eight hours or more. Unfortunately, the more time you study, the less time you have to maintain your physical condition. Thus, most examinees arrive at the examination site in peak mental condition but in deteriorated physical condition. While preparing for the FE exam is not the only good reason for embarking on a physical conditioning program, it can serve as a good incentive to get in shape.

It will be helpful to make a few simple decisions prior to starting your review. You should be aware of the different options available to you. For example, you should decide early on to

- use SI units in your preparation
- perform electrical calculations with effective (rms) or maximum values
- take calculations out to a maximum of four significant digits
- prepare in all examination subjects, not just your specialty areas

At the beginning of your review program, you should locate a spare calculator. It is not necessary to buy a spare if you can arrange to borrow one from a friend or the office. However, if possible, your primary and spare calculators should be identical. If your spare calculator is not identical to the primary calculator, spend a few minutes familiarizing yourself with its functions.

A Few Days Before the Exam

There are a few things you should do a week or so before the examination date. For example, visit the exam site in order to find the building, parking areas, examination room, and rest rooms. You should also make arrangements for child care and transportation. Since the examination does not always start or end at the designated times, make sure that your child care and transportation arrangements can tolerate a later-than-usual completion.

Second in importance to your scholastic preparation is the preparation of your two examination kits. The first kit consists of a bag or box containing items to bring with you into the examination room. NCEES provides mechanical pencils for use in the exam. It is not necessary (nor is it permitted) for you to bring your own pencils or erasers.

[ ] letter admitting you to the examination
[ ] photographic identification
[ ] main calculator
[ ] spare calculator
[ ] extra calculator batteries
[ ] unobtrusive snacks
[ ] travel pack of tissues
[ ] headache remedy
[ ] $2.00 in change
[ ] light, comfortable sweater
[ ] loose shoes or slippers
[ ] cushion for your chair
[ ] small hand towel
[ ] earplugs
[ ] wristwatch
[ ] wire coat hanger
[ ] extra set of car keys

The second kit consists of the following items and should be left in a separate bag or box in your car in case they are needed.

[ ] copy of your application
[ ] proof of delivery
[ ] this book
[ ] other references
[ ] regular dictionary
[ ] scientific dictionary
[ ] course notes in three-ring binders
[ ] cardboard box (use as a bookcase)
[ ] instruction booklets for all your calculators
[ ] light lunch
[ ] beverages in thermos and cans
[ ] sunglasses
[ ] extra pair of prescription glasses

[ ] raincoat, boots, gloves, hat, and umbrella
[ ] street map of the examination site
[ ] note to the parking patrol for your windshield
[ ] battery powered desk lamp

The Day Before the Exam

Take the day before the examination off from work to relax. Do not cram the last night. A good prior night's sleep is the best way to start the examination. If you live far from the examination site, consider getting a hotel room in which to spend the night.

Make sure your exam kits are packed and ready to go.

The Day of the Exam

You should arrive at least 30 minutes before the examination starts. This will allow time for finding a convenient parking place, bringing your materials to the examination room, and making room and seating changes. Be prepared, though, to find that the examination room is not open or ready at the designated time.

Once the examination has started, observe the following suggestions.

- Set your wristwatch alarm for five minutes before the end of each four-hour session and use that remaining time to guess at all of the remaining unsolved problems. Do not work up until the very end. You will be successful with about 25 percent of your guesses, and these points will more than make up for the few points you might earn by working during the last five minutes.

- Do not spend more than two minutes per morning question. (The average time available per problem is two minutes.) If you have not finished a question in that time, make a note of it and continue on.

- Do not spend time trying to ask your proctors technical questions. Even if they are knowledgeable in engineering, they will not be permitted to answer your questions.

- Make a quick mental note about any problems for which you cannot find a correct response or for which you believe there are two correct answers. Errors in the exam are rare, but they do occur. Being able to point out an error later might give you the margin you need to pass. Since such problems are almost always discovered during the scoring process and discounted from the examination, it is not necessary to tell your proctor, but be sure to mark the one best answer before moving on.

- Make sure all of your responses on the answer sheet are dark and completely fill the bubbles.

Afternoon FE Exam Subjects (Discipline-Specific Exams)

The following tables list the FE afternoon discipline-specific exam subjects with their corresponding categories and percentage of each type of question as released by NCEES.

CHEMICAL ENGINEERING

subject	percentage of questions (%)
chemistry	10
material/energy balances	15
chemical engineering thermodynamics	10
fluid dynamics	10
heat transfer	10
mass transfer	10
chemical reaction engineering	10
process design and economic optimization	10
computer usage in chemical engineering	5
process control	5
safety, health, and environmental	5

CIVIL ENGINEERING

subject	percentage of questions (%)
surveying	11
hydraulics and hydrologic systems	12
soil mechanics and foundations	15
environmental engineering	12
transportation	12
structural analysis	10
structural design	10
construction management	10
materials	8

ELECTRICAL ENGINEERING

subject	percentage of questions (%)
circuits	16
power	13
electromagnetics	7
control systems	10
communications	9
signal processing	8
electronics	15
digital systems	12
computer systems	10

ENVIRONMENTAL ENGINEERING

subject	percentage of questions (%)
water resources	25
water and wastewater engineering	30
air quality engineering	15
solid and hazardous waste engineering	15
environmental science and management	15

INDUSTRIAL ENGINEERING

subject	percentage of questions (%)
engineering economics	15
probability and statistics	15
modeling and computation	12
industrial management	10
manufacturing and production systems	13
facilities and logistics	12
human factors, productivity, ergonomics, and work design	12
quality	11

MECHANICAL ENGINEERING

subject	percentage of questions (%)
mechanical design and analysis	15
kinematics, dynamics, and vibrations	15
materials and processing	10
measurements, instrumentation, and controls	10
thermodynamics and energy conversion processes	15
fluid mechanics and fluid machinery	15
heat transfer	10
refrigeration and HVAC	10

Comments from Recent Examinees

This is a compilation of comments from engineers who have had recent experience with the FE examination. Their comments and suggestions were collected from surveys submitted by examinees in all 50 states immediately after the examination.

In many cases, the comments duplicate and reinforce one another. However, the few contradictory statements serve to illustrate the diversity of engineering backgrounds represented by the examinees.

(Some of the comments have been edited for spelling, punctuation, grammar, clarity, and consistency.)

General Comments

"I have taken the FE exam six times. I... passed this time due to your excellent books."

"Don't put off studying for the exam."

"Just review. It is not as hard as everyone says if you just review."

"Pray you don't have to take it again."

When to Take the Exam

"Take the exam during your senior year."

"Do it as early as possible after completing the school work. It's tougher later."

"Take it as soon as possible after graduation."

"Take it as soon as possible."

"Take it as soon as possible after college."

"Don't wait so many years after graduation to take this test."

How Much to Study

"Study a lot more than you think you need to."

"Study at least two months before the exam."

"Start studying two to three months prior to the exam."

"Study three months in advance."

"Start studying at least four months before the exam."

"Study at least six months ahead."

"Make sure you have enough time to review the material."

"Study. Study. Study."

"Work as many practice problems as possible, especially in the areas that are not still fresh in your mind."

"Evening and part-time students should study twice as much."

"Give yourself enough preparation time to complete the entire book."

"Study more!"

"Study hard."

"Take the test as soon as you graduate. If you do this, no more than six weeks is all you should need to prepare for the test."

"Start reviewing early."

"Start preparing early."

"Start as early as possible and take it seriously."

Which Subjects to Study

"Do every kind of practice problem that is on this planet. You cannot do enough homework problems in preparation for this test. This test is geared toward the graduating senior, not for people who have been out of school for a while."

"Study everything."

"Getting used to all of the topics is really helpful."

"Attempt every problem in the *FE Review Manual*. There will be no surprises in the exam, which will help you maintain focus and confidence, even if you make errors."

"Make sure to do the sample problems (all) in the *FE Review Manual*."

"Study in areas where you are not proficient."

"Study hard so that new subjects don't pop up."

"Be sure to give yourself enough time to cover all FE-style problems."

"Study all chapters in depth..."

"Practice math."

"Do a lot of math problems."

"Study electrical and chemistry."

"Learn economics concepts. Learn thermo cycles."

"Make sure you know and study thermodynamics, chemistry, and dynamics first."

"Learn seven subjects. Learn them well."

"Study everything."

"Study the areas that you did not specialize in during college."

"Review all of the material—not just specific areas."

"Forget the subjects you don't use on a daily basis at work."

"Learn the basics. Don't get caught up in all the detail. It will only slow you down on the exam. Most of the questions are specific. So, study your strengths. (Hopefully, you do not have many weaknesses. The exam will find them.)"

"Review everything."

"Study the book thoroughly."

"Study even the parts that you are planning to skip during the exam. Many times, the solution is just an application of a simple formula, which is easier than guessing."

How to Study

"Study in 20 to 30 minute sessions every night. The [*FE Review Manual*] takes longer than one hour per chapter."

"Do practice problems. Do practice problems. Do practice problems."

"Do all of the problems in the *Review Manual* at least once. A week before the test, put your pencil down and review every problem that you have worked."

"I found that, after many years out of school, your speed at problem solving is not what it was. I would take a couple of simulated practice tests prior to the real exam."

"Do mock exams."

"Complete as many eight-hour exams as possible."

"Take a practice exam to prepare you for the long day ahead."

"Prepare with practice [exams] so you know how quickly you have to work. Then, relax and just do it."

"Practice timed testing to increase speed and accuracy."

"I cannot stress working practice problems enough."

"Do practice problems."

"Do as many sample problems as possible."

"Do as many different practice problems as possible."

"Do as many problems as you can find..."

"...just keep working practice problems."

"Practice plenty of problems."

"Work lots of problems."

"Work practice problems. Then, work more practice problems."

"Work as many practice problems as possible."

"Do as many problems as you have time for."

"Work as many sample problems from each section as possible."

"Work all the problems in the *Review Manual* and take the practice tests. I worked the problems, but did not realistically take the sample test."

"Study hard."

"Do as many sample problems as possible."

Speed of the Exam

"Pace yourself when you take the exam, and don't panic. Just keep moving."

"Time is essential."

Review Courses

"Take a review course."

"Take a review course, particularly if you have been out of school for a while."

"Study the *Review Manual* first, then take a review course."

NCEES Handbook

"Don't get bogged down with memorization exercises. All the equations you need to pass the test are in the NCEES Handbook. You just need to know how to apply them."

"Study the NCEES Handbook."

"Carefully review the NCEES Handbook so you know where you look quickly for certain equations."

"Become familiar with the NCEES Handbook."

"Be familiar with the NCEES reference book before the exam. Actually 'study' the book to be familiar with the location of topics. This way, when you see a problem, you will have an idea whether or not you can find information required for that particular problem."

Calculators

"Use a standard scientific calculator. Matrix operations, graphing, integration, etc., are all nice functions to have but are unnecessary for the FE exam."

"Take a calculator that is easy to use."

"Make sure that you know how to use your calculator before the exam."

"If you are going to buy a new calculator, make sure you have enough time to become familiar with it before the exam."

General Versus Discipline-Specific (DS) Exam

"Take the general exam in the afternoon."

"Don't take the afternoon discipline-specific test."

"Don't take the general second half if you are an industrial engineer."

"Take your major ... in the afternoon."

Miscellaneous Strategies

"Study from a book that uses the same format/nomenclature used in the exam."

"Stay in a hotel close to the place where the exam is given."

"Read the questions carefully. Sometimes, a quick read suggests the wrong approach or even an incorrect interpretation of what's being asked for."

"Just go through the *FE Review Manual* ... and stick with the suggested review schedule."

"Follow the guide in the book on what to bring to the test. If I hadn't followed the guide, I would have been freezing during the first half of the test, even though it was warm outside."

"Follow the program given in the *FE Review Manual*."

"Use the review manual with the study schedule."

"Study a lot. Pray a lot."

"You can skip some questions if you do not have time."

"Focus on qualitative problems."

"Don't stress."

"Study with one or two people when reviewing for the exam."

"Jump over lengthy problems or 'I don't know how to do ...' problems, then come back to them at the end. And, don't forget to guess. (But use your judgment when you guess.)"

"Do neck exercises to strengthen the muscles that allow the test taker to focus down on the test materials."

"Use review books. It is the only way to pass."

"Do not guess [all the answers] to a particular subject when you get the test. Each section has 'gifts' that may be definitions straight out of the reference handbook."

About PPI Books

"Use the *FE Review Manual*. It has the most logical format and best approach to review."

"The *FE Review Manual* has exactly what you need to know for the exam."

"Go buy the *FE Review Manual*; get confident."

"Get the *FE Review Manual*."

"I'd recommend both of the Lindeburg FE books. One for the more extensive background; one for focusing in on the areas likely to be on the exam. The *Reference Manual* makes a good general reference to keep after the exam, too."

"Buy one of Michael's books and study."

"Buy the PPI books."

"Purchase the *FE Review Manual* and read the whole book."

"Use Lindeburg's review books."

"Use PPI's review materials."

"Use PPI materials."

"Obtain PPI's books."

"I'd recommend the *FE Review Manual*."

Topic I: Units and Fundamental Constants

Chapter 1   Units
Chapter 2   Fundamental Constants

Hint: For the most current information about the exam, visit www.ppi2pass.com/fefaqs.html regularly. Use the Exam Forum to compare notes with other FE examinees.

1 Units

Subjects

INTRODUCTION 1-1
COMMON UNITS OF MASS 1-1
MASS AND WEIGHT 1-1
ACCELERATION OF GRAVITY 1-2
CONSISTENT SYSTEMS OF UNITS 1-2
THE ENGLISH ENGINEERING
    SYSTEM 1-2
OTHER FORMULAS AFFECTED BY
    INCONSISTENCY 1-3
WEIGHT AND SPECIFIC WEIGHT 1-3
THE ENGLISH GRAVITATIONAL
    SYSTEM 1-3
METRIC SYSTEMS OF UNITS 1-4
SI UNITS (THE mks SYSTEM) 1-4
RULES FOR USING THE SI SYSTEM 1-6

INTRODUCTION

The purpose of this chapter is to eliminate some of the confusion regarding the many units available for each engineering variable. In particular, an effort has been made to clarify the use of the so-called English systems, which for years have used the *pound* unit both for force and mass—a practice that has resulted in confusion for even those familiar with it.

It is expected that most engineering problems will be stated and solved in either English engineering or SI units. Therefore, a discussion of these two systems occupies the majority of this chapter.

COMMON UNITS OF MASS

The choice of a mass unit is the major factor in determining which system of units will be used in solving a problem. Obviously, you will not easily end up with a force in pounds if the rest of the problem is stated in meters and kilograms. Actually, the choice of a mass unit determines more than whether a conversion factor will be necessary to convert from one system to another (e.g., between the SI and English systems). An inappropriate choice of a mass unit may actually require a conversion factor *within* the system of units.

The common units of mass are the gram, pound, kilogram, and slug. There is nothing mysterious about these units. All represent different quantities of matter, as Fig. 1.1 illustrates. In particular, note that the pound and slug do not represent the same quantity of matter. One slug is equal to 32.1740 pounds-mass.

Figure 1.1 Common Units of Mass

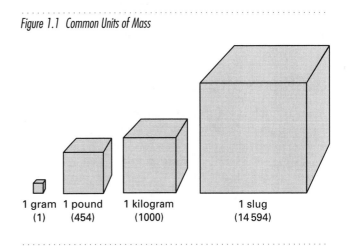

1 gram 1 pound 1 kilogram 1 slug
(1)   (454)   (1000)   (14 594)

MASS AND WEIGHT

The SI system uses *kilograms* for mass and *newtons* for weight (force). The units are different, and there is no confusion between the variables. However, for years, the term *pound* has been used for both mass and weight. This usage has obscured the distinction between the two: mass is a constant property of an object; weight varies with the gravitational field. Even the conventional use of the abbreviations *lbm* and *lbf* (to distinguish between pounds-mass and pounds-force) has not helped eliminate the confusion.

An object with a mass of one pound will have an earthly weight of one pound, but this is true only on the earth. The weight of the same object will be much less on the moon. Therefore, care must be taken when working with mass and force in the same problem.

The relationship that converts mass to weight is familiar to every engineering student:

$$W = mg \qquad 1.1$$

Equation 1.1 illustrates that an object's weight will depend on the local acceleration of gravity as well as the object's mass. The mass will be constant, but gravity will depend on location. Mass and weight are not the same.

ACCELERATION OF GRAVITY

Gravitational acceleration on the earth's surface is usually taken as 32.2 ft/sec^2 or 9.81 m/s^2. These values are rounded from the more exact standard values of 32.1740 ft/sec^2 and 9.8066 m/s^2. However, the need for greater accuracy must be evaluated on a problem-by-problem basis. Usually, three significant digits are adequate, since gravitational acceleration is not constant anyway, but is affected by location (primarily latitude and altitude) and major geographical features.

CONSISTENT SYSTEMS OF UNITS

A set of units used in a calculation is said to be *consistent* if no conversion factors are needed. (The terms *homogeneous* and *coherent* are also used to describe a consistent set of units.) For example, a moment is calculated as the product of a force and a lever arm length.

$$M = dF \qquad 1.2$$

A calculation using Eq. 1.2 would be consistent if M was in newton-meters, F was in newtons, and d was in meters. The calculation would be inconsistent if M was in ft-kips, F was in kips, and d was in inches (because a conversion factor of 1/12 would be required).

The concept of a consistent calculation can be extended to a system of units. A *consistent system of units* is one in which no conversion factors are needed for any calculation. For example, Newton's second law of motion can be written without conversion factors. Newton's second law simply states that the force required to accelerate an object is proportional to the acceleration of the object. The constant of proportionality is the object's mass.

$$F = ma \qquad 1.3$$

Notice that Eq. 1.3 is $F = ma$, not $F = Wa/g$ or $F = ma/g_c$. Equation 1.3 is consistent: It requires no conversion factors. This means that in a consistent system where conversion factors are not used, once the units of m and a have been selected, the units of F are fixed. This has the effect of establishing units of work and energy, power, fluid properties, and so on.

The decision to work with a consistent set of units is desirable but unnecessary, depending often on tradition and environment. Problems in fluid flow and thermodynamics are routinely solved in the United States with inconsistent units. This causes no more of a problem than working with inches and feet when calculating moment. It is necessary only to use the proper conversion factors.

THE ENGLISH ENGINEERING SYSTEM

Through common and widespread use, pounds-mass (lbm) and pounds-force (lbf) have become the standard units for mass and force in the *English Engineering System*.

There are subjects in the United States where the practice of using pounds for mass is firmly entrenched. For example, most thermodynamics, fluid flow, and heat transfer problems have traditionally been solved using the units of lbm/ft^3 for density, Btu/lbm for enthalpy, and Btu/lbm-°F for specific heat. Unfortunately, some equations contain both lbm-related and lbf-related variables, as does the steady flow conservation of energy equation, which combines enthalpy in Btu/lbm with pressure in lbf/ft^2.

The units of pounds-mass and pounds-force are as different as the units of gallons and feet, and they cannot be canceled. A mass conversion factor, g_c, is needed to make the equations containing lbf and lbm dimensionally consistent. This factor is known as the *gravitational constant* and has a value of 32.1740 lbm-ft/lbf-sec^2. The numerical value is the same as the standard acceleration of gravity, but g_c is not the local gravitational acceleration, g. (It is acceptable, and recommended, that g_c be rounded to the same number of significant digits as g. Therefore, a value of 32.2 for g_c would typically be used.) g_c is a conversion constant, just as 12.0 is the conversion factor between feet and inches.

The English Engineering System is an inconsistent system, as defined according to Newton's second law. $F = ma$ cannot be written if lbf, lbm, and ft/sec^2 are the units used. The g_c term must be included.

$$F \text{ in lbf} = \frac{(m \text{ in lbm})\left(a \text{ in } \dfrac{\text{ft}}{\text{sec}^2}\right)}{g_c \text{ in } \dfrac{\text{lbm-ft}}{\text{lbf-sec}^2}} \qquad 1.4$$

Note that in Eq. 1.4, g_c does more than "fix the units." Since g_c has a numerical value of 32.174, it actually changes the calculation numerically. A force of 1.0 pound will not accelerate a 1.0 pound-mass at the rate of 1.0 ft/sec^2.

In the English Engineering System, work and energy are typically measured in ft-lbf (mechanical systems) or in

British thermal units, Btu (thermal and fluid systems). One Btu is equal to 778.26 ft-lbf.

Example 1.1

Calculate the weight, in lbf, of a 1.0 lbm object in a gravitational field of 27.5 ft/sec^2.

Solution

From Eq. 1.4,

$$F = \frac{ma}{g_c} = \frac{(1 \text{ lbm}) \left(27.5 \frac{\text{ft}}{\text{sec}^2} \right)}{32.2 \frac{\text{lbm-ft}}{\text{lbf-sec}^2}}$$

$$= 0.854 \text{ lbf}$$

OTHER FORMULAS AFFECTED BY INCONSISTENCY

It is not a significant burden to include g_c in a calculation, but it may be difficult to remember when g_c should be used. Knowing when to include the gravitational constant can be learned through repeated exposure to the formulas in which it is needed, but it is safer to carry the units along in every calculation.

The following is a representative (but not exhaustive) list of formulas that require the g_c term. In all cases, it is assumed that the standard English Engineering System units will be used.

- kinetic energy

$$E = \frac{mv^2}{2g_c} \quad \text{[in ft-lbf]} \qquad 1.5$$

- potential energy

$$E = \frac{mgz}{g_c} \quad \text{[in ft-lbf]} \qquad 1.6$$

- pressure at a depth

$$p = \frac{\rho g h}{g_c} \quad \text{[in lbf/ft}^2\text{]} \qquad 1.7$$

- specific weight

$$\gamma = \frac{\rho g}{g_c} \qquad 1.8$$

- shear stress

$$\tau = \left(\frac{\mu}{g_c} \right) \left(\frac{dv}{dy} \right) \qquad 1.9$$

Example 1.2

A rocket with a mass of 4000 lbm travels at 27,000 ft/sec. What is its kinetic energy in ft-lbf?

Solution

From Eq. 1.5,

$$E_k = \frac{mv^2}{2g_c} = \frac{(4000 \text{ lbm}) \left(27{,}000 \frac{\text{ft}}{\text{sec}} \right)^2}{(2) \left(32.2 \frac{\text{lbm-ft}}{\text{lbf-sec}^2} \right)}$$

$$= 4.53 \times 10^{10} \text{ ft-lbf}$$

WEIGHT AND SPECIFIC WEIGHT

Weight is a force exerted on an object due to its placement in a gravitational field. If a consistent set of units is used, Eq. 1.1 can be used to calculate the weight of a mass. In the English Engineering System, however, Eq. 1.10 must be used.

$$W = \frac{mg}{g_c} \qquad 1.10$$

Both sides of Eq. 1.10 can be divided by the volume of an object to derive the *specific weight* (*unit weight, weight density*), γ, of the object. Equation 1.11 illustrates that the weight density (in lbf/ft^3) can also be calculated by multiplying the mass density (in lbm/ft^3) by g/g_c. Since g and g_c usually have the same numerical values, the only effect of Eq. 1.12 is to change the units of density.

$$\frac{W}{V} = \left(\frac{m}{V} \right) \left(\frac{g}{g_c} \right) \qquad 1.11$$

$$\gamma = \frac{W}{V} = \left(\frac{m}{V} \right) \left(\frac{g}{g_c} \right)$$

$$= \frac{\rho g}{g_c} \qquad 1.12$$

Weight does not occupy volume; only mass has volume. The concept of weight density has evolved to simplify certain calculations, particularly fluid calculations. For example, pressure at a depth is calculated from Eq. 1.13. (Compare this to Eq. 1.7.)

$$p = \gamma h \qquad 1.13$$

THE ENGLISH GRAVITATIONAL SYSTEM

Not all English systems are inconsistent. Pounds can still be used as the unit of force as long as pounds are not used as the unit of mass. Such is the case with the consistent *English Gravitational System*.

If acceleration is given in ft/sec², the units of mass for a consistent system of units can be determined from Newton's second law.

$$\text{units of } m = \frac{\text{units of } F}{\text{units of } a}$$

$$= \frac{\text{lbf}}{\dfrac{\text{ft}}{\text{sec}^2}} = \frac{\text{lbf-sec}^2}{\text{ft}} \qquad 1.14$$

The combination of units in Eq. 1.14 is known as a *slug*. g_c is not needed since this system is consistent. It would be needed only to convert slugs to another mass unit.

Slugs and pounds-mass are not the same, as Fig. 1.1 illustrates. However, both are units for the same quantity: mass. Equation 1.15 will convert between slugs and pounds-mass.

$$\text{no. of slugs} = \frac{\text{no. of lbm}}{g_c} \qquad 1.15$$

The number of slugs is not derived by dividing the number of pounds-mass by the local gravity. g_c is used regardless of the local gravity. The conversion between feet and inches is not dependent on local gravity; neither is the conversion between slugs and pounds-mass.

Since the English Gravitational System is consistent, Eq. 1.16 can be used to calculate weight. Notice that the local gravitational acceleration is used.

$$W \text{ in lbf} = (m \text{ in slugs}) \left(g \text{ in } \frac{\text{ft}}{\text{sec}^2} \right) \qquad 1.16$$

Figure 1.2 Common Force Units

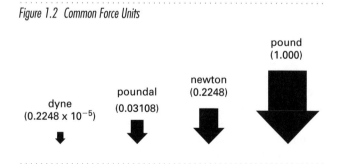

METRIC SYSTEMS OF UNITS

Strictly speaking, a *metric system* is any system of units that is based on meters or parts of meters. This broad definition includes *mks systems* (based on meters, kilograms, and seconds) as well as *cgs systems* (based on centimeters, grams, and seconds).

Metric systems avoid the pounds-mass versus pounds-force ambiguity in two ways. First, matter is not measured in units of force. All quantities of matter are specified as mass. Second, force and mass units do not share a common name.

The term *metric system* is not explicit enough to define which units are to be used for any given variable. For example, within the cgs system there is variation in how certain electrical and magnetic quantities are represented (resulting in the ESU and EMU systems). Also, within the mks system, it is common engineering practice today to use kilocalories as the unit of thermal energy, while the SI system requires the use of joules. Thus, there is a lack of uniformity even within the metricated engineering community.

The "metric" parts of this book are based on the SI system, which is the most developed and codified of the so-called metric systems. It is expected that there will be occasional variances with local engineering custom, but it is difficult to anticipate such variances within a book that must be consistent.

SI UNITS (THE mks SYSTEM)

SI units comprise an *mks system* (so named because it uses the meter, kilogram, and second as dimensional units). All other units are derived from the dimensional units, which are completely listed in Table 1.1. This system is fully consistent, and there is only one recognized unit for each physical quantity (variable).

Two types of units are used: base units and derived units. The *base units* (Table 1.1) are dependent only on accepted standards or reproducible phenomena. The previously unclassified *supplementary units*, radian and steradian, have been classified as derived units. The *derived units* (Tables 1.2 and 1.3) are made up of combinations of base and supplementary units.

Table 1.1 SI Base Units

quantity	name	symbol
length	meter	m
mass	kilogram	kg
time	second	s
electric current	ampere	A
temperature	kelvin	K
amount of substance	mole	mol
luminous intensity	candela	cd

Table 1.2 Some SI Derived Units with Special Names

quantity	name	symbol	expressed in terms of other units
frequency	hertz	Hz	
force	newton	N	
pressure, stress	pascal	Pa	N/m^2
energy, work, quantity of heat	joule	J	$N \cdot m$
power, radiant flux	watt	W	J/s
quantity of electricity, electric charge	coulomb	C	
electric potential, potential difference, electromotive force	volt	V	W/A
electric capacitance	farad	F	C/V
electric resistance	ohm	Ω	V/A
electric conductance	siemen	S	A/V
magnetic flux	weber	Wb	$V \cdot s$
magnetic flux density	tesla	T	Wb/m^2
inductance	henry	H	Wb/A
luminous flux	lumen	lm	
illuminance	lux	lx	lm/m^2
plane angle	radian	rad	
solid angle	steradian	sr	

In addition, there is a set of non-SI units that may be used. This concession is primarily due to the significance and widespread acceptance of these units. Use of the non-SI units listed in Table 1.4 will usually create an inconsistent expression requiring conversion factors.

The units of force can be derived from Newton's second law.

$$\text{units of force} = (m \text{ in kg}) \left(a \text{ in } \frac{m}{s^2} \right)$$

$$= \frac{kg \cdot m}{s^2} \qquad \qquad 1.17$$

This combination of units for force is known as a *newton*.

Energy variables in the SI system have units of N·m, or equivalently, $kg \cdot m^2/s^2$. Both of these combinations are known as a *joule*. The units of power are joules per second, equivalent to a *watt*.

Example 1.3

A 10 kg block hangs from a cable. What is the tension in the cable? (Standard gravity is 9.81 m/s^2.)

Solution

$$F = mg = (10 \text{ kg}) \left(9.81 \frac{m}{s^2} \right)$$

$$= 98.1 \frac{kg \cdot m}{s^2}$$

$$= 98.1 \text{ N}$$

Table 1.3 Some SI Derived Units

quantity	description	expressed in terms of other units
area	square meter	m^2
volume	cubic meter	m^3
speed—linear	meter per second	m/s
—angular	radian per second	rad/s
acceleration—linear	meter per second squared	m/s^2
—angular	radian per second squared	rad/s^2
density, mass density	kilogram per cubic meter	kg/m^3
concentration (of amount of substance)	mole per cubic meter	mol/m^3
specific volume	cubic meter per kilogram	m^3/kg
luminance	candela per square meter	cd/m^2
absolute viscosity	pascal second	$Pa \cdot s$
kinematic viscosity	square meters per second	m^2/s
moment of force	newton meter	$N \cdot m$
surface tension	newton per meter	N/m
heat flux density, irradiance	watt per square meter	W/m^2
heat capacity, entropy	joule per kelvin	J/K
specific heat capacity, specific entropy	joule per kilogram kelvin	$J/(kg \cdot K)$
specific energy	joule per kilogram	J/kg
thermal conductivity	watt per meter kelvin	$W/(m \cdot K)$
energy density	joule per cubic meter	J/m^3
electric field strength	volt per meter	V/m
electric charge density	coulomb per cubic meter	C/m^3
surface density of charge, flux density	coulomb per square meter	C/m^2
permittivity	farad per meter	F/m
current density	ampere per square meter	A/m^2
magnetic field strength	ampere per meter	A/m
permeability	henry per meter	H/m
molar energy	joule per mole	J/mol
molar entropy, molar heat capacity	joule per mole kelvin	$J/(mol \cdot K)$
radiant intensity	watt per steradian	W/sr

Example 1.4

A 10 kg block is raised vertically 3 m. What is the change in potential energy?

Solution

$$\Delta E_p = mg\Delta h = (10 \text{ kg}) \left(9.81 \frac{m}{s^2} \right) (3 \text{ m})$$

$$= 294 \frac{kg \cdot m^2}{s^2}$$

$$= 294 \text{ N} \cdot m$$

$$= 294 \text{ J}$$

Table 1.4 Acceptable Non-SI Units

quantity	unit name	symbol name	relationship to SI unit
area	hectare	ha	1 ha = 10 000 m^2
energy	kilowatt-hour	kWh	1 kWh = 3.6 MJ
mass	metric ton[a]	t	1 t = 1000 kg
plane angle	degree (of arc)	°	1° = 0.017453 rad
speed of rotation	revolution per minute	r/min	1 r/min=2π/60 rad/s
temperature interval	degree Celsius	°C	1°C = 1K
time	minute	min	1 min = 60 s
	hour	h	1 h = 3600 s
	day (mean solar)	d	1 d = 86 400 s
	year (calendar)	a	1 a = 31 536 000 s
velocity	kilometer per hour	km/h	1 km/h = 0.278 m/s
volume	liter[b]	L	1 L = 0.001 m^3

[a] The international name for metric ton is *tonne*. The metric ton is equal to the *megagram* (Mg).

[b] The international symbol for liter is the lowercase l, which can be easily confused with the numeral 1. Several English-speaking countries have adopted the script ℓ and uppercase L as a symbol for liter in order to avoid any misinterpretation.

RULES FOR USING THE SI SYSTEM

In addition to having standardized units, the SI system also has rigid syntax rules for writing the units and combinations of units. Each unit is abbreviated with a specific symbol. The following rules for writing and combining these symbols should be adhered to.

- The expressions for derived units in symbolic form are obtained by using the mathematical signs of multiplication and division; for example, units of velocity are m/s, and units of torque are N·m (not N-m or Nm).

- Scaling of most units is done in multiples of 1000.

- The symbols are always printed in roman type, regardless of the type used in the rest of the text. The only exception to this is in the use of the symbol for liter, where the use of the lowercase el (l) may be confused with the numeral one (1). In this case, "liter" should be written out in full, or the script ℓ or L should be used.

- Symbols are not pluralized: 1 kg, 45 kg (not 45 kgs).

- A period after a symbol is not used, except when the symbol occurs at the end of a sentence.

- When symbols consist of letters, there is always a full space between the quantity and the symbols: 45 kg (not 45kg). However, when the first character of a symbol is not a letter, no space

is left: 32°C (not 32° C or 32 °C); or 42°12′45″ (not 42° 12′ 45″).

- All symbols are written in lowercase, except when the unit is derived from a proper name: m for meter; s for second; A for ampere, Wb for weber, N for newton, W for watt.

- Prefixes are printed without spacing between the prefix and the unit symbol (e.g., km is the symbol for kilometer).

- In text, symbols should be used when associated with a number. However, when no number is involved, the unit should be spelled out: The area of the carpet is 16 m^2, not 16 square meters. Carpet is sold by the square meter, not by the m^2.

- A practice in some countries is to use a comma as a decimal marker, while the practice in North America, the United Kingdom, and some other countries is to use a period (or dot) as the decimal marker. Furthermore, in some countries that use the decimal comma, a dot is frequently used to divide long numbers into groups of three. Because of these differing practices, spaces must be used instead of commas to separate long lines of digits into easily readable blocks of three digits with respect to the decimal marker: 32 453.246 072 5. A space (half-space preferred) is optional with a four-digit number: 1 234 or 1 234.

Table 1.5 SI Prefixes[a]

prefix	symbol	value
exa	E	10^{18}
peta	P	10^{15}
tera	T	10^{12}
giga	G	10^9
mega	M	10^6
kilo	k	10^3
hecto	h	10^2
deka	da	10^1
deci	d	10^{-1}
centi	c	10^{-2}
milli	m	10^{-3}
micro	μ	10^{-6}
nano	n	10^{-9}
pico	p	10^{-12}
femto	f	10^{-15}
atto	a	10^{-18}

[a] There is no "B" (billion) prefix. In fact, the word billion means 10^9 in the United States but 10^{12} in most other countries. This unfortunate ambiguity is handled by avoiding the use of the term billion.

- Where a decimal fraction of a unit is used, a zero should always be placed before the decimal marker: 0.45 kg (not .45 kg). This practice draws attention to the decimal marker and helps avoid errors of scale.

- Some confusion may arise with the word "tonne" (1000 kg). When this word occurs in French text of Canadian origin, the meaning may be a ton of 2000 pounds.

SAMPLE PROBLEMS

Problem 1

What is a metric ton?

   (A) 200 kg
   (B) 1000 kg
   (C) 2000 kg
   (D) 2 kN

Solution

A metric ton, also known as a *tonne*, is 1000 kg.

Answer is B.

Problem 2

What is a kip?

   (A) 1000 in-lbf (torque)
   (B) 1000 lbm (mass)
   (C) 1000 lbf (force)
   (D) 1000 psi (pressure)

Solution

The abbreviation "kip" is used for "kilo-pound": 1000 pounds of force.

Answer is C.

Problem 3

What basic SI unit is equal to $kg \cdot m^2/s^2$?

   (A) joule
   (B) pascal
   (C) tesla
   (D) watt

Solution

Kinetic energy is calculated in the SI system as $(1/2)mv^2$, with units of joules (J).

Answer is A.

2 Fundamental Constants

quantity	symbol	English	SI
Charge			
electron	e		-1.6022×10^{-19} C
proton	p		$+1.6021 \times 10^{-19}$ C
Density			
air [STP, 32°F, (0°C)]		0.0805 lbm/ft^3	1.29 kg/m^3
air [70°F, (20°C), 1 atm]		0.0749 lbm/ft^3	1.20 kg/m^3
earth [mean]		345 lbm/ft^3	5520 kg/m^3
mercury		849 lbm/ft^3	1.360×10^4 kg/m^3
seawater		64.0 lbm/ft^3	1025 kg/m^3
water [mean]		62.4 lbm/ft^3	1000 kg/m^3
Distance [mean]			
earth radius		2.09×10^7 ft	6.370×10^6 m
earth-moon separation		1.26×10^9 ft	3.84×10^8 m
earth-sun separation		4.89×10^{11} ft	1.49×10^{11} m
moon radius		5.71×10^6 ft	1.74×10^6 m
sun radius		2.28×10^9 ft	6.96×10^8 m
first Bohr radius	a_0	1.736×10^{-10} ft	5.292×10^{-11} m
Gravitational Acceleration			
earth [mean]	g	32.174 (32.2) ft/sec^2	9.8067 (9.81) m/s^2
moon [mean]		5.47 ft/sec^2	1.67 m/s^2
Mass			
atomic mass unit	u	3.66×10^{-27} lbm	1.6606×10^{-27} kg
earth		1.32×10^{25} lbm	6.00×10^{24} kg
electron [rest]	m_e	2.008×10^{-30} lbm	9.109×10^{-31} kg
moon		1.623×10^{23} lbm	7.36×10^{22} kg
neutron [rest]	m_n	3.693×10^{-27} lbm	1.675×10^{-27} kg
proton [rest]	m_p	3.688×10^{-27} lbm	1.673×10^{-27} kg
sun		4.387×10^{30} lbm	1.99×10^{30} kg
Pressure, atmospheric		14.696 (14.7) lbf/in^2	1.0133×10^5 Pa
Temperature, standard		32°F (492°R)	0°C (273K)
Velocity			
earth escape		3.67×10^4 ft/sec	1.12×10^4 m/s
light [vacuum]	c	9.84×10^8 ft/sec	$2.99792 \ (3.00) \times 10^8$ m/s
sound [air, STP]	a	1090 ft/sec	331 m/s
[air, 70°F (20°C)]		1130 ft/sec	344 m/s
Volume, molal ideal gas [STP]	V_m	359 ft^3/lbmol	22.414 m^3/kmol
			22414 L/kmol
Fundamental Constants			
Avogadro's number	N_A		6.022×10^{23} mol^{-1}
Bohr magneton	μ_B		9.2732×10^{-24} J/T
Boltzmann constant	k	5.65×10^{-24} ft-lbf/°R	1.3807×10^{-23} J/K
Faraday constant	F		96 485 C/mol
gravitational constant	g_c	32.174 lbm-ft/lbf-sec^2	
gravitational constant	G	3.44×10^{-8} ft^4/lbf-sec^4	6.673×10^{-11} N·m^2/kg^2
nuclear magneton	μ_N		5.050×10^{-27} J/T
permeability of a vacuum	μ_0		1.2566×10^{-6} N/A^2 (H/m)
permittivity of a vacuum	ϵ_0		8.854×10^{-12} C^2/N·m^2 (F/m)
Planck's constant	h		6.6256×10^{-34} J·s
Rydberg constant	R_∞		1.097×10^7 m^{-1}
specific gas constant, air	R	53.3 ft-lbf/lbm-°R	287 J/kg·K
Stefan-Boltzmann constant		1.71×10^{-9} Btu/ft^2-hr-°R^4	5.670×10^{-8} W/m^2·K^4
triple point, water		32.02°F, 0.0888 psia	0.01109°C, 0.6123 kPa
universal gas constant	$\overline{R}$	1545 ft-lbf/lbmol-°R	8314 J/kmol·K
	$\overline{R}$	1.986 Btu/lbmol-°R	8.314 kPa·m^3/kmol·K

Topic II: Conversion Factors

Chapter 3   Conversion Factors

3 Conversion Factors

(Atmospheres are standard; calories are gram-calories; gallons are U.S. liquid; miles are statute; pounds-mass are avoirdupois.)

multiply	by	to obtain
ac	43,560	ft^2
angstrom	1×10^{-10}	m
atm	1.01325	bar
atm	76.0	cm Hg
atm	33.90	ft water
atm	29.92	in Hg
atm	14.696	lbf/in^2
atm	101.3	kPa
atm	1.013×10^5	Pa
bar	0.9869	atm
bar	10^5	Pa
Btu	778.17	ft-lbf
Btu	1055	J
Btu	2.928×10^{-4}	kW·h
Btu	10^{-5}	therm
Btu/hr	0.216	ft-lbf/sec
Btu/hr	3.929×10^{-4}	hp
Btu/hr	0.2931	W
Btu/lbm	2.326	kJ/kg
Btu/lbm-°R	4.1868	kJ/kg·K
cal (gm-cal)	3.968×10^{-3}	Btu
cal (gm-cal)	4.1868	J
cm	0.03281	ft
cm	0.3937	in
eV	1.602×10^{-19}	J
ft	0.3048	m
ft^2	2.2957×10^{-5}	ac
ft^3	7.481	gal
ft-lbf	1.285×10^{-3}	Btu
ft-lbf	1.35582	J
ft-lbf	3.766×10^{-7}	kW·h
ft-lbf	1.3558	N·m
gal	0.13368	ft^3
gal	3.785	L
gal	3.7854×10^{-3}	m^3
gal/min	0.002228	ft^3/sec
g/cm^3	1000	kg/m^3
g/cm^3	62.428	lbm/ft^3
hp	2545	Btu/hr
hp	33,000	ft-lbf/min
hp	550	ft-lbf/sec
hp	0.7457	kW
hp-hr	2545	Btu
in	2.54	cm
J	9.478×10^{-4}	Btu
J	6.2415×10^{18}	eV
J	0.73756	ft-lbf
J	1.0	N·m

multiply	by	to obtain
J/s	1.0	W
kg	2.20462	lbm
kip	1000	lbf
kip	4448	N
kJ	0.9478	Btu
kJ	737.56	ft-lbf
kJ/kg	0.42992	Btu/lbm
kJ/kg·K	0.23885	Btu/lbm-°R
km	3280.8	ft
km	0.6214	mi
km/h	0.6214	mi/hr
kPa	9.8693×10^{-3}	atm
kPa	0.14504	lbf/in^2
kW	3413	Btu/hr
kW	0.9481	Btu/sec
kW	737.6	ft-lbf/sec
kW	1.341	hp
kW·h	3413	Btu
kW·h	3.6×10^6	J
L	0.03531	ft^3
L	61.02	in^3
L	0.2642	gal
L	0.001	m^3
L/s	2.119	ft^3/min
L/s	15.85	gal/min
lbf	4.4482	N
lbf/in^2	0.06805	atm
lbf/in^2	2.307	ft water
lbf/in^2	2.036	in Hg
lbf/in^2	6894.8	Pa
lbm	0.4536	kg
lbm/ft^3	0.016018	g/cm^3
lbm/ft^3	16.018	kg/m^3
m	3.28083	ft
m/sec	196.8	ft/min
mi	5280	ft
mi	1.6093	km
micron	1×10^{-6}	m
N	0.22481	lbf
N·m	0.7376	ft-lbf
N·m	1.0	J
Pa	1.4504×10^{-4}	lbf/in^2
therm	10^5	Btu
W	3.413	Btu/hr
W	0.7376	ft-lbf/sec
W	1.341×10^{-3}	hp
W	1.0	J/sec

Commonly Used Equivalents

1 gal of water weighs	8.34 lbf
1 ft^3 of water weighs	62.4 lbf
1 in^3 of mercury weighs	0.491 lbf
The mass of 1 m^3 of water is	1000 kg

Temperature Conversions

$$°F = 1.8(°C) + 32$$

$$°C = \frac{°F - 32}{1.8}$$

$$°R = °F + 459.69$$

$$K = °C + 273.16$$

$$\Delta T_{°R} = \Delta T_{°F}$$

$$\Delta T_{K} = \Delta T_{°C}$$

Topic III: Mathematics

Diagnostic Examination for Mathematics

Chapter 4   Analytic Geometry and Trigonometry

Chapter 5   Algebra and Linear Algebra

Chapter 6   Probability and Statistics

Chapter 7   Calculus

Chapter 8   Differential Equations and Transforms

Chapter 9   Numerical Analysis

MATHEMATICS

Hint: For the most current information about the exam, visit www.ppi2pass.com/fefaqs.html regularly. Use the Exam Forum to compare notes with other FE examinees.

Diagnostic Examination

TOPIC III: MATHEMATICS

TIME LIMIT: 45 MINUTES

1. What is the general form of the equation for a line whose x-intercept is 4 and y-intercept is -6?

- (A) $2x - 3y - 18 = 0$
- (B) $2x + 3y + 18 = 0$
- (C) $3x - 2y - 12 = 0$
- (D) $3x + 2y + 12 = 0$

2. For some angle θ, $\csc \theta = -8/5$. What is $\cos 2\theta$?

- (A) $7/32$
- (B) $1/4$
- (C) $3/8$
- (D) $5/8$

3. What is the rectangular form of the following polar equation?

$$r^2 = 1 - \tan^2 \theta$$

- (A) $-x^2 + x^4 y^2 + y^2 = 0$
- (B) $x^2 + x^2 y^2 - y^2 + y^4 = 0$
- (C) $-x^4 + y^2 = 0$
- (D) $x^4 - x^2 + x^2 y^2 + y^2 = 0$

4. For three matrices $\mathbf{A}$, $\mathbf{B}$, and $\mathbf{C}$, which of the following statements is not necessarily true?

- (A) $\mathbf{A} + (\mathbf{B} + \mathbf{C}) = (\mathbf{A} + \mathbf{B}) + \mathbf{C}$
- (B) $\mathbf{A}(\mathbf{B} + \mathbf{C}) = \mathbf{AB} + \mathbf{AC}$
- (C) $(\mathbf{B} + \mathbf{C})\mathbf{A} = \mathbf{AB} + \mathbf{AC}$
- (D) $\mathbf{A} + (\mathbf{B} + \mathbf{C}) = \mathbf{C} + (\mathbf{A} + \mathbf{B})$

5. For the three vectors $\mathbf{A}$, $\mathbf{B}$, and $\mathbf{C}$, what is the product $\mathbf{A} \cdot (\mathbf{B} \times \mathbf{C})$?

$$\mathbf{A} = 6\mathbf{i} + 8\mathbf{j} + 10\mathbf{k}$$
$$\mathbf{B} = \mathbf{i} + 2\mathbf{j} + 3\mathbf{k}$$
$$\mathbf{C} = 3\mathbf{i} + 4\mathbf{j} + 5\mathbf{k}$$

- (A) 0
- (B) 64
- (C) 80
- (D) 216

6. The second and sixth terms of a geometric progression are $3/10$ and $243/160$, respectively. What is the first term of this sequence?

- (A) $1/10$
- (B) $1/5$
- (C) $3/5$
- (D) $3/2$

7. A marksman can hit a bull's-eye from 100 m with three out of every four shots. What is the probability that he will hit a bull's-eye with at least one of his next three shots?

- (A) $3/4$
- (B) $15/16$
- (C) $31/32$
- (D) $63/64$

8. The final scores of students in a graduate course are distributed normally with a mean of 72 and a standard deviation of 10. What is the probability that a student's score will be between 65 and 78?

- (A) 0.4196
- (B) 0.4837
- (C) 0.5161
- (D) 0.6455

9. Evaluate the following limit.

$$\lim_{x \to 0} \left(\frac{1 - e^{3x}}{4x} \right)$$

- (A) $-\infty$
- (B) $-3/4$
- (C) 0
- (D) $1/4$

10. The radius of a snowball rolling down a hill is increasing at a rate of 20 cm/min. How fast is its volume increasing when its diameter is 1 m?

- (A) $0.034 \text{ m}^3/\text{min}$
- (B) $0.52 \text{ m}^3/\text{min}$
- (C) $0.63 \text{ m}^3/\text{min}$
- (D) $0.84 \text{ m}^3/\text{min}$

11. Evaluate the following indefinite integral.

$$\int \cos^2 x \, \sin x \, dx$$

(A) $-\dfrac{2}{3} \sin^3 x + C$

(B) $-\dfrac{1}{3} \cos^3 x + C$

(C) $\dfrac{1}{3} \sin^3 x + C$

(D) $\dfrac{1}{2} \sin^2 x \cos^2 x + C$

12. What is the area of the region bounded by the curve $y = \sin x$ and the x-axis on the interval between $x = \pi/2$ and $x = 2\pi$?

(A) 1
(B) 2
(C) 3
(D) 4

13. What is the general solution to the following differential equation?

$$y'' - 8y' + 16y = 0$$

(A) $y = C_1 e^{4x}$
(B) $y = (C_1 + C_2 x)e^{4x}$
(C) $y = C_1 e^{-4x} + C_2 e^{4x}$
(D) $y = C_1 e^{2x} + C_2 e^{4x}$

14. Find the Laplace transform of the equation with the given boundary conditions.

$$f''(t) + f(t) = \sin \beta t$$
$$f'(0) = 0$$
$$f(0) = 0$$

(A) $F(s) = \left(\dfrac{1}{1+s^2}\right)\left(\dfrac{\beta}{s^2+\beta^2}\right)$

(B) $F(s) = \left(\dfrac{1}{1+s^2}\right)\left(\dfrac{\beta}{s^2-\beta^2}\right)$

(C) $F(s) = \left(\dfrac{1}{1-s^2}\right)\left(\dfrac{\beta}{s^2+\beta^2}\right)$

(D) $F(s) = \left(\dfrac{1}{1-s^2}\right)\left(\dfrac{s}{s^2+\beta^2}\right)$

15. Newton's method is being used to find the roots of the equation $f(x) = (x-2)^2 - 1$. What is the third approximation of the root if 9.33 is chosen as the first approximation?

(A) 1.0
(B) 2.0
(C) 3.0
(D) 4.0

SOLUTIONS TO DIAGNOSTIC EXAMINATION TOPIC III

1. Find the slope of the line.

$$m = \frac{y_2 - y_1}{x_2 - x_1}$$
$$= \frac{-6 - 0}{0 - 4}$$
$$= 3/2$$

Once the slope and y-intercept are known, the slope-intercept form is convenient to use.

$$y = mx + b$$
$$mx - y + b = 0$$
$$\frac{3}{2}x - y + (-6) = 0$$
$$3x - 2y - 12 = 0$$

Answer is C.

2. Use the cosine double angle formula.

$$\cos 2\theta = 1 - 2\sin^2 \theta$$
$$= 1 - 2\left(\frac{1}{\csc \theta}\right)^2$$
$$= 1 - (2)\left(\frac{1}{\dfrac{-8}{5}}\right)^2$$
$$= 7/32 \quad (0.21875)$$

Answer is A.

3. Use the identities relating r and θ to x and y.

$$r = \sqrt{x^2 + y^2}$$
$$\theta = \tan^{-1}\frac{y}{x}$$
$$r^2 = 1 - \tan^2\theta$$
$$\left(\sqrt{x^2+y^2}\right)^2 = 1 - \tan^2\left(\tan^{-1}\frac{y}{x}\right)$$
$$x^2 + y^2 = 1 - \frac{y^2}{x^2}$$
$$x^4 - x^2 + x^2 y^2 + y^2 = 0$$

Answer is D.

4. Matrix addition is both associative and commutative, so choices (A) and (D) are true. Multiplication is distributive (B), but not commutative (C).

Answer is C.

5. Find the cross product $\mathbf{B} \times \mathbf{C}$. The augmented matrix method is the easiest approach.

$$\mathbf{B} \times \mathbf{C} = \mathbf{i}(2)(5) + \mathbf{j}(3)(3) + \mathbf{k}(1)(4)$$
$$- \mathbf{i}(4)(3) - \mathbf{j}(5)(1) - \mathbf{k}(3)(2)$$
$$= -2\mathbf{i} + 4\mathbf{j} - 2\mathbf{k}$$

Now calculate the dot product.

$$\mathbf{A} \cdot (\mathbf{B} \times \mathbf{C}) = (6)(-2) + (8)(4) + (10)(-2)$$
$$= 0$$

Another way of looking at this problem is to use the graphical definitions of dot and cross products. The product of $\mathbf{B} \times \mathbf{C}$ is a vector that is perpendicular to both $\mathbf{B}$ and $\mathbf{C}$. Because $\mathbf{A}$ is a scalar multiple of $\mathbf{C}$, $\mathbf{B} \times \mathbf{C}$ is also perpendicular to $\mathbf{A}$, which means that $\mathbf{A} \cdot (\mathbf{B} \times \mathbf{C})$ is defined to be zero.

Answer is A.

6. Apply the formula for a geometric sequence. Solve this problem by dividing the expressions for the two given terms.

$$l_n = ar^{n-1}$$

$$\frac{l_6}{l_2} = \frac{ar^{6-1}}{ar^{2-1}} = r^4$$

$$r = \sqrt[4]{\frac{l_6}{l_2}} = \sqrt[4]{\frac{\frac{243}{160}}{\frac{3}{10}}}$$

$$= 3/2$$

$$l_2 = ar$$

$$a = \frac{l_2}{r} = \frac{\frac{3}{10}}{\frac{3}{2}}$$

$$= 1/5$$

Answer is B.

7. The easiest way to find the probability of making at least one bull's-eye is actually to solve for its complementary probability, that of making zero bull's-eyes.

$$P(\text{miss}) = 1 - P(\text{hit})$$
$$= 1 - \frac{3}{4}$$
$$= 1/4$$

$$P(\text{at least one}) = 1 - P(\text{none})$$
$$= 1 - (P(\text{miss}) \times P(\text{miss}) \times P(\text{miss}))$$
$$= 1 - \left(\frac{1}{4}\right)\left(\frac{1}{4}\right)\left(\frac{1}{4}\right)$$
$$= 63/64$$

Answer is D.

8. Calculate standard normal values for the points of interest, 65 and 78.

$$x = \frac{X_0 - \mu}{\sigma}$$

$$x_{65} = \frac{65 - 72}{10}$$
$$= -0.70$$

$$x_{78} = \frac{78 - 72}{10}$$
$$= 0.60$$

The probability of a score falling between 65 and 78 is equal to the area under the unit normal curve between these two standard normal values. Determine this area by subtracting $F(x_{65})$ from $F(x_{78})$. Although the $F(x)$ statistic is not tabulated for negative x values, the curve's symmetry allows the $R(x)$ statistic to be used instead.

$$F(x) = R(-x)$$

$$P(65 < X < 78) = F(x_{78}) - F(x_{65})$$
$$= F(x_{78}) - R(-x_{65})$$
$$= F(0.60) - R(0.70)$$
$$= 0.7257 - 0.2420$$
$$= 0.4837$$

Answer is B.

9. This limit has the indeterminate form $0/0$, so use L'Hôpital's rule.

$$\lim_{x \to a}\left(\frac{f(x)}{g(x)}\right) = \lim_{x \to a}\left(\frac{f'(x)}{g'(x)}\right)$$

$$\lim_{x \to 0} \left(\frac{1 - e^{3x}}{4x} \right) = \lim_{x \to 0} \left(\frac{-3e^{3x}}{4} \right)$$

$$= -3/4$$

Answer is B.

10. The goal of this problem is to find the rate of change of volume with respect to time, dV/dt. Derive this rate by differentiating the formula for the volume of a sphere with respect to time.

$$V = \frac{4}{3}\pi r^3$$

$$\frac{dV}{dt} = \frac{4}{3}\pi r^2 (3) \frac{dr}{dt}$$

$$= (4\pi)(0.5 \text{ m})^2 \left(0.2 \, \frac{\text{m}}{\text{min}} \right)$$

$$= 0.63 \text{ m}^3/\text{min}$$

Answer is C.

11. Simplify this problem by using the variable substitution $u = \cos x$.

$$u = \cos x$$

$$du = (-\sin x)dx$$

$$\int \cos^2 x \sin x \, dx = \int u^2 (-du)$$

$$= -\frac{1}{3}u^3 + C$$

$$= -\frac{1}{3}\cos^3 x + C$$

Answer is B.

12. Integrate $\sin x$ with respect to x to obtain the area between the curve and the x-axis. The calculation must be done in pieces, because over the region where the value of $\sin x$ is negative, the integral is negative. Take the absolute value of the negative area and add it to the result from the positive section to get the total area.

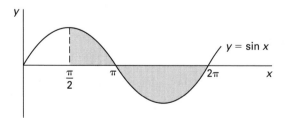

$$A = \int y \, dx$$

$$= \int_{\frac{\pi}{2}}^{\pi} \sin x \, dx + \left| \int_{\pi}^{2\pi} \sin x \, dx \right|$$

$$= -\cos x \Big|_{\frac{\pi}{2}}^{\pi} + \left| -\cos x \Big|_{\pi}^{2\pi} \right|$$

$$= \left(-\cos \pi - \left(-\cos \frac{\pi}{2} \right) \right) + \left| -\cos 2\pi - (-\cos \pi) \right|$$

$$= -(-1) + 0 + \left| -1 + (-1) \right|$$

$$= 1 + 2 = 3$$

Answer is C.

13. Find the roots of the characteristic equation.

$$r^2 + 2ar + b = 0 = r^2 - 8r + 16$$

$$2a = -8$$

$$a = -4$$

$$b = 16$$

$$r = -a \pm \sqrt{a^2 - b}$$

$$= -(-4) \pm \sqrt{(-4)^2 - 16}$$

$$= 4$$

Because $a^2 = b$, the characteristic equation has one double root, and the solution takes the form

$$y = (C_1 + C_2 x)e^{rx}$$

$$= (C_1 + C_2 x)e^{4x}$$

Answer is B.

14. Take the Laplace transform of both sides of the equation using the linearity theorem; then isolate $F(s)$.

$$\mathcal{L}\big(f''(t) + f(t)\big) = \mathcal{L}(\sin \beta t)$$

$$\mathcal{L}\big(f''(t)\big) + \mathcal{L}\big(f(t)\big) = \mathcal{L}(\sin \beta t)$$

$$\big(s^2 F(s) - sf(0) - f'(0)\big) + F(s) = \frac{\beta}{s^2 + \beta^2}$$

$$s^2 F(s) - 0 - 0 + F(s) = \frac{\beta}{s^2 + \beta^2}$$

$$F(s) = \left(\frac{1}{1 + s^2} \right) \left(\frac{\beta}{s^2 + \beta^2} \right)$$

Answer is A.

15. Perform two iterations of Newton's method with an initial guess of 9.33.

$$f(x) = (x - 2)^2 - 1$$

$$f'(x) = (2)(x - 2)$$

$$f(x_1) = (9.33 - 2)^2 - 1$$

$$= 52.73$$

$$f'(x_1) = (2)(9.33 - 2)$$

$$= 14.66$$

$$x_{n+1} = x_n - \frac{f(x_n)}{f'(x_n)}$$

$$x_2 = x_1 - \frac{f(x_1)}{f'(x_1)}$$

$$= 9.33 - \frac{f(9.33)}{f'(9.33)}$$

$$= 9.33 - \frac{52.73}{14.66}$$

$$= 5.73$$

$$f(x_2) = (5.73 - 2)^2 - 1$$

$$= 12.91$$

$$f'(x_2) = (2)(5.73 - 2)$$

$$= 7.46$$

$$x_3 = x_2 - \frac{f(x_2)}{f'(x_2)}$$

$$= 5.73 - \frac{f(5.73)}{f'(5.73)}$$

$$= 5.73 - \frac{12.91}{7.46}$$

$$= 4.0$$

Answer is D.

4 Analytic Geometry and Trigonometry

Subjects

STRAIGHT LINE 4-1
QUADRATIC EQUATION 4-1
CONIC SECTIONS 4-2
    Parabola 4-2
    Ellipse 4-3
    Hyperbola 4-3
    Circle 4-4
QUADRIC SURFACE (SPHERE) 4-4
DISTANCE BETWEEN POINTS 4-4
RIGHT TRIANGLES 4-4
TRIGONOMETRIC IDENTITIES 4-5
GENERAL TRIANGLES 4-6
MENSURATION OF AREAS 4-6
MENSURATION OF VOLUMES 4-7

STRAIGHT LINE

Figure 4.1 is a straight line in two-dimensional space. The *slope* of the line is m, the y-intercept is b, and the x-intercept is a. A known point on the line is represented as (x_1, y_1). The equation of the line can be represented in several forms, and the procedure for finding the equation depends on the form chosen to represent the line. In general, the procedure involves substituting one or more known points on the line into the equation in order to determine the constants.

Figure 4.1 Straight Line

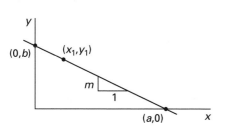

The *general form* of the equation of a line is
$$Ax + By + C = 0 \qquad 4.1$$

The *standard form*, also known as the *slope-intercept form*, is
$$y = mx + b \qquad 4.2$$

The *point-slope form* is
$$y - y_1 = m(x - x_1) \qquad 4.3$$

The equation for the slope of a straight line is
$$m = \frac{y_2 - y_1}{x_2 - x_1} \qquad 4.4$$

Two intersecting lines in two-dimensional space are shown in Fig. 4.2.

Figure 4.2 Two Lines Intersecting in Two-Dimensional Space

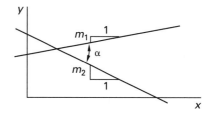

The slopes of the two lines are m_1 and m_2, and the angle, α, between the lines is
$$\alpha = \arctan\left(\frac{m_2 - m_1}{1 + m_2 m_1}\right) \qquad 4.5$$

The slopes of two lines that are perpendicular are related by
$$m_1 = \frac{-1}{m_2} \qquad 4.6$$

The smallest distance, d, between two points (x_1, y_1) and (x_2, y_2) is
$$d = \sqrt{(y_2 - y_1)^2 + (x_2 - x_1)^2} \qquad 4.7$$

QUADRATIC EQUATION

A *quadratic equation* is an equation of the general form $ax^2 + bx + c = 0$. The roots, x_1 and x_2, of the equation are the two values of x that satisfy it.
$$x_1, x_2 = \frac{-b \pm \sqrt{b^2 - 4ac}}{2a} \qquad 4.8$$

The types of roots of the equation can be determined from the *discriminant* (i.e., the quantity under the radical in Eq. 4.8).

- If $b^2 - 4ac > 0$, the roots are real and unequal.
- If $b^2 - 4ac = 0$, the roots are real and equal. This is known as a *double root*.
- If $b^2 - 4ac < 0$, the roots are complex and unequal.

CONIC SECTIONS

A *conic section* is any one of several curves produced by passing a plane through a cone as shown in Fig. 4.3. If θ is the angle between the vertical axis and the cutting plane and ϕ is the *cone-generating angle*, then the *eccentricity*, e, of the conic section is

$$e = \frac{\cos\theta}{\cos\phi} \qquad 4.9$$

All conic sections are described by second-degree polynomials (i.e., are quadratic equations) of the *general form*

$$Ax^2 + Bxy + Cy^2 + Dx + Ey + F = 0 \qquad 4.10$$

If $A = C$, then B is automatically 0 for a conic section. If $A = C = 0$, the conic section is a *line*, and if $A = C \neq 0$, the conic section is a *circle*. If $A \neq C$, then

- if $B^2 - 4AC < 0$, the conic section is an *ellipse*.
- if $B^2 - 4AC > 0$, the conic section is a *hyperbola*.
- if $B^2 - 4AC = 0$, the conic section is a *parabola*.

In the general form (Eq. 4.10), the figure axes can be at any angle relative to the coordinate axes. The *standard forms* presented in the following sections pertain to figures whose axes coincide with the coordinate axes, thereby eliminating the Bxy term, and, for some conic sections, other terms of the general equation.

Parabola

A *parabola* is the locus of points equidistant from the focus (point F in Fig. 4.4) and a line called the *directrix*. A parabola is symmetric with respect to its *parabolic axis*. The line normal to the parabolic axis and passing through the focus is known as the *latus rectum*. The eccentricity of a parabola is one.

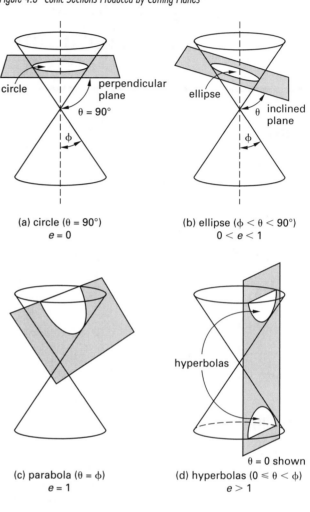

Figure 4.3 Conic Sections Produced by Cutting Planes

(a) circle ($\theta = 90°$)
$e = 0$

(b) ellipse ($\phi < \theta < 90°$)
$0 < e < 1$

(c) parabola ($\theta = \phi$)
$e = 1$

(d) hyperbolas ($0 \leq \theta < \phi$)
$e > 1$

The *general form* of the equation for a parabola is the same as for a general conic section, where $B^2 - 4AC = 0$.

$$Ax^2 + Bxy + Cy^2 + Dx + Ey + F = 0 \qquad 4.11$$

Equation 4.12 is the *standard form* of the equation of a parabola with *vertex* (also known as the parabola's *center*) at (h,k), *focus* at $(h + p/2, k)$, and *directrix* at $x = h - p/2$, and that opens horizontally.

$$(y - k)^2 = 2p(x - h) \qquad 4.12$$

The parabola opens to the right (points to the left) if $p > 0$, and it opens to the left (points to the right) if $p < 0$. If the vertex is at the origin ($h = k = 0$), then the focus is at $(p/2, 0)$ and the directrix is at $x = -p/2$.

Figure 4.4 Parabola

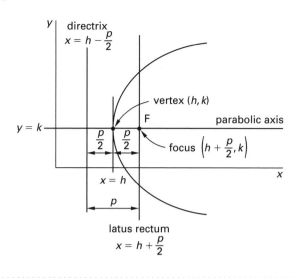

Equation 4.14 is the *standard form* of the equation of an ellipse with center at (h, k), *semimajor distance* a, and *semiminor distance* b.

$$\frac{(x-h)^2}{a^2} + \frac{(y-k)^2}{b^2} = 1 \qquad 4.14$$

When the center is at the origin $(h = k = 0)$, then the focus is located at $(ae, 0)$, the directrix is located at $x = a/e$, and the eccentricity and semiminor distance are given by Eqs. 4.15 and 4.16, respectively.

$$e = \sqrt{1 - \left(\frac{b}{a}\right)^2} = \frac{c}{a} \qquad 4.15$$

$$b = a\sqrt{1 - e^2} \qquad 4.16$$

Ellipse

An *ellipse* (Fig. 4.5) has two foci separated along the *major axis* by a distance $2c$. The line perpendicular to the major axis passing through the center of the ellipse is the *minor axis*. The lines perpendicular to the major axis passing through the foci are the *latus recta*. The distance between the two vertices is $2a$. The ellipse is the locus of points such that the sum of the distances from the two foci is $2a$. The eccentricity of the ellipse is always less than one. If the eccentricity is zero, the figure is a circle.

Hyperbola

As shown in Fig. 4.6, a *hyperbola* has two foci separated along the *transverse axis* by a distance $2c$. Lines perpendicular to the transverse axis passing through the foci are the *conjugate axes*. The distance between the two vertices is $2a$, and the distance along a conjugate axis between two points on the hyperbola is $2b$. The hyperbola is the locus of points such that the difference in distances from the two foci is $2a$.

Figure 4.6 Hyperbola

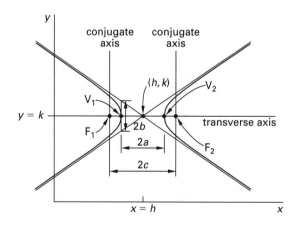

The *general form* of the equation for a hyperbola is

$$Ax^2 + Bxy + Cy^2 + Dx + Ey + F = 0 \qquad 4.17$$

Equation 4.18 is the *standard form* of the equation of a hyperbola with center at (h, k) and opening horizontally.

$$\frac{(x-h)^2}{a^2} - \frac{(y-k)^2}{b^2} = 1 \qquad 4.18$$

Figure 4.5 Ellipse

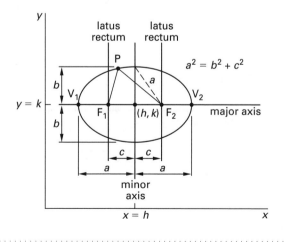

The *general form* of the equation for an ellipse is

$$Ax^2 + Bxy + Cy^2 + Dx + Ey + F = 0 \qquad 4.13$$

When the hyperbola is centered at the origin ($h = k = 0$), the focus is located at $(ae, 0)$, the directrix is located at $x = a/e$, and the eccentricity and half-length of the conjugate axis are given by Eqs. 4.19 and 4.20, respectively.

$$e = \sqrt{1 + \left(\frac{b}{a}\right)^2} \qquad 4.19$$

$$b = a\sqrt{e^2 - 1} \qquad 4.20$$

Circle

The *general form* of the equation for a circle is

$$Ax^2 + Ay^2 + Dx + Ey + F = 0 \qquad 4.21$$

Equation 4.22 is the *standard form* (also called the *center-radius form*) of the equation of a circle with center at (h,k) and radius $r = \sqrt{(x-h)^2 + (y-k)^2}$. Figure 4.7 shows such a circle.

$$(x - h)^2 + (y - k)^2 = r^2 \qquad 4.22$$

Figure 4.7 Circle

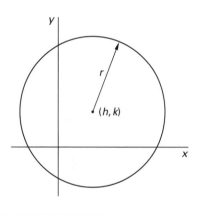

The two forms can be converted by use of Eqs. 4.23 through 4.25.

$$h = \frac{-D}{2A} \qquad 4.23$$

$$k = \frac{-E}{2A} \qquad 4.24$$

$$r^2 = \frac{D^2 + E^2 - 4AF}{4A^2} \qquad 4.25$$

If the right-hand side of Eq. 4.25 is positive, the figure is a circle. If it is zero, the circle shrinks to a point. If the right-hand side is negative, the figure is imaginary. A *degenerate circle* is one in which the right-hand side is less than or equal to zero.

The length, t, of a *tangent* to a circle from a point (x', y') in two-dimensional space is illustrated in Fig. 4.8 and is given by Eq. 4.26.

$$t^2 = (x' - h)^2 + (y' - k)^2 - r^2 \qquad 4.26$$

Figure 4.8 Tangent to a Circle from a Point

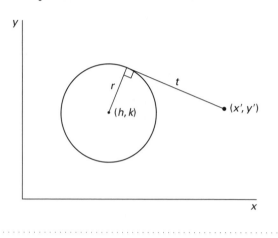

QUADRIC SURFACE (SPHERE)

The *general form* of the equation for a sphere is

$$Ax^2 + Ay^2 + Az^2 + Bx + Cy + Dz + E = 0 \quad 4.27$$

Equation 4.28 is the *standard form* of the equation of a sphere centered at (h, k, m) with radius r.

$$(x - h)^2 + (y - k)^2 + (z - m)^2 = r^2 \qquad 4.28$$

DISTANCE BETWEEN POINTS

The distance between two points (x_1, y_1, z_1) and (x_2, y_2, z_2) in three-dimensional space is

$$d = \sqrt{(x_2 - x_1)^2 + (y_2 - y_1)^2 + (z_2 - z_1)^2} \quad 4.29$$

RIGHT TRIANGLES

A *right triangle* is a triangle in which one of the angles is $90°$ ($\pi/2$ rad), as shown in Fig. 4.9. Choosing one of the acute angles as a reference, the sides of the triangle are called the *adjacent side*, x, the *opposite side*, y, and the *hypotenuse*, r.

MATHEMATICS
Geo/Trig

Figure 4.9 Right Triangle

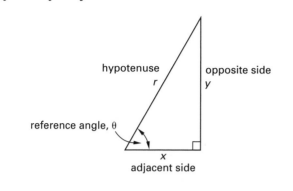

The trigonometric functions are calculated from the sides of the right triangle.

$$\sin \theta = \frac{y}{r} \qquad 4.30$$

$$\cos \theta = \frac{x}{r} \qquad 4.31$$

$$\tan \theta = \frac{y}{x} \qquad 4.32$$

$$\cot \theta = \frac{x}{y} \qquad 4.33$$

$$\csc \theta = \frac{r}{y} \qquad 4.34$$

$$\sec \theta = \frac{r}{x} \qquad 4.35$$

The trigonometric functions correspond to the lengths of various line segments in a right triangle in a unit circle. Figure 4.10 shows such a triangle inscribed in a unit circle.

Figure 4.10 Trigonometric Functions in a Unit Circle

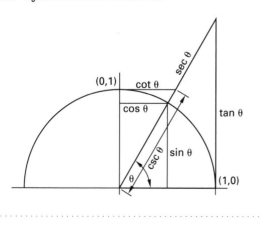

TRIGONOMETRIC IDENTITIES

Three of the trigonometric functions are reciprocals of the others. The prefix "co-" is not a good way to remember the reciprocal functions; while the tangent and cotangent functions are reciprocals of each other, two other pairs—the sine and cosine functions and the secant and cosecant functions—are not.

$$\csc \theta = \frac{1}{\sin \theta} \qquad 4.36$$

$$\sec \theta = \frac{1}{\cos \theta} \qquad 4.37$$

$$\cot \theta = \frac{1}{\tan \theta} \qquad 4.38$$

Equations 4.39 through 4.66 are commonly used identities.

• *General formulas*

$$\tan \theta = \frac{\sin \theta}{\cos \theta} \qquad 4.39$$

$$\cot \theta = \frac{\cos \theta}{\sin \theta} \qquad 4.40$$

$$\sin^2 \theta + \cos^2 \theta = 1 \qquad 4.41$$

$$\tan^2 \theta + 1 = \sec^2 \theta \qquad 4.42$$

$$\cot^2 \theta + 1 = \csc^2 \theta \qquad 4.43$$

• *Double-angle formulas*

$$\sin 2\alpha = 2 \sin \alpha \cos \alpha \qquad 4.44$$

$$\cos 2\alpha = \cos^2 \alpha - \sin^2 \alpha$$

$$= 1 - 2 \sin^2 \alpha$$

$$= 2 \cos^2 \alpha - 1 \qquad 4.45$$

$$\tan 2\alpha = \frac{2 \tan \alpha}{1 - \tan^2 \alpha} \qquad 4.46$$

$$\cot 2\alpha = \frac{\cot^2 \alpha - 1}{2 \cot \alpha} \qquad 4.47$$

• *Two-angle formulas*

$$\sin(\alpha + \beta) = \sin \alpha \cos \beta + \cos \alpha \sin \beta \qquad 4.48$$

$$\cos(\alpha + \beta) = \cos \alpha \cos \beta - \sin \alpha \sin \beta \qquad 4.49$$

$$\tan(\alpha + \beta) = \frac{\tan \alpha + \tan \beta}{1 - \tan \alpha \tan \beta} \qquad 4.50$$

$$\cot(\alpha + \beta) = \frac{\cot \alpha \cot \beta - 1}{\cot \alpha + \cot \beta} \qquad 4.51$$

$$\sin(\alpha - \beta) = \sin \alpha \cos \beta - \cos \alpha \sin \beta \qquad 4.52$$

$$\cos(\alpha - \beta) = \cos\alpha\cos\beta + \sin\alpha\sin\beta \qquad 4.53$$

$$\tan(\alpha - \beta) = \frac{\tan\alpha - \tan\beta}{1 + \tan\alpha\,\tan\beta} \qquad 4.54$$

$$\cot(\alpha - \beta) = \frac{\cot\alpha\,\cot\beta + 1}{\cot\beta - \cot\alpha} \qquad 4.55$$

- *Half-angle formulas*

$$\sin\frac{\alpha}{2} = \pm\sqrt{\frac{1 - \cos\alpha}{2}} \qquad 4.56$$

$$\cos\frac{\alpha}{2} = \pm\sqrt{\frac{1 + \cos\alpha}{2}} \qquad 4.57$$

$$\tan\frac{\alpha}{2} = \pm\sqrt{\frac{1 - \cos\alpha}{1 + \cos\alpha}} \qquad 4.58$$

$$\cot\frac{\alpha}{2} = \pm\sqrt{\frac{1 + \cos\alpha}{1 - \cos\alpha}} \qquad 4.59$$

- *Miscellaneous formulas*

$$\sin\alpha\sin\beta = \frac{1}{2}\left(\cos(\alpha - \beta) - \cos(\alpha + \beta)\right) \qquad 4.60$$

$$\cos\alpha\cos\beta = \frac{1}{2}\left(\cos(\alpha - \beta) + \cos(\alpha + \beta)\right) \qquad 4.61$$

$$\sin\alpha\cos\beta = \frac{1}{2}\left(\sin(\alpha + \beta) + \sin(\alpha - \beta)\right) \qquad 4.62$$

$$\sin\alpha + \sin\beta = 2\sin\left(\frac{\alpha + \beta}{2}\right)\cos\left(\frac{\alpha - \beta}{2}\right) \qquad 4.63$$

$$\sin\alpha - \sin\beta = 2\cos\left(\frac{\alpha + \beta}{2}\right)\sin\left(\frac{\alpha - \beta}{2}\right) \qquad 4.64$$

$$\cos\alpha + \cos\beta = 2\cos\left(\frac{\alpha + \beta}{2}\right)\cos\left(\frac{\alpha - \beta}{2}\right) \qquad 4.65$$

$$\cos\alpha - \cos\beta = -2\sin\left(\frac{(\alpha + \beta)}{2}\right)\sin\left(\frac{\alpha - \beta}{2}\right) \qquad 4.66$$

GENERAL TRIANGLES

A *general triangle*, as shown in Fig. 4.11, is any triangle that is not specifically a right triangle. For a general triangle, the *law of sines* relates the sines of the angles and their opposite sides.

$$\frac{a}{\sin A} = \frac{b}{\sin B} = \frac{c}{\sin C} \qquad 4.67$$

Figure 4.11 General Triangle

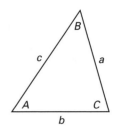

The *law of cosines* relates the cosine of an angle to an opposite side.

$$a^2 = b^2 + c^2 - 2bc\cos A \qquad 4.68$$

$$b^2 = a^2 + c^2 - 2ac\cos B \qquad 4.69$$

$$c^2 = a^2 + b^2 - 2ab\cos C \qquad 4.70$$

MENSURATION OF AREAS

The dimensions, perimeter, area, and other geometric properties constitute the *mensuration* (i.e., the measurements) of a geometric shape. For the following figures, A is the total surface area, p is the perimeter, and V is the volume.

Figure 4.12 Parabola

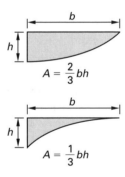

$$A = \frac{2}{3}bh$$

$$A = \frac{1}{3}bh$$

Figure 4.13 Ellipse

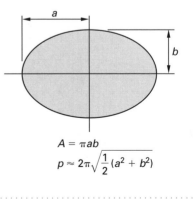

$$A = \pi ab$$

$$p \approx 2\pi\sqrt{\frac{1}{2}(a^2 + b^2)}$$

Figure 4.14 Circular Segment

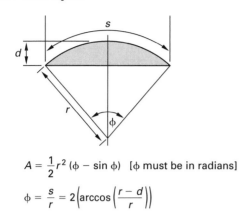

$$A = \frac{1}{2}r^2 (\phi - \sin \phi) \quad [\phi \text{ must be in radians}]$$

$$\phi = \frac{s}{r} = 2\left(\arccos\left(\frac{r-d}{r}\right)\right)$$

Figure 4.15 Circular Sector

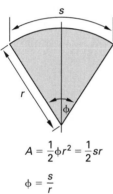

$$A = \frac{1}{2}\phi r^2 = \frac{1}{2}sr$$

$$\phi = \frac{s}{r}$$

Figure 4.16 Parallelogram

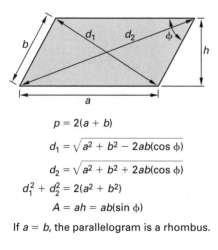

$$p = 2(a + b)$$

$$d_1 = \sqrt{a^2 + b^2 - 2ab(\cos \phi)}$$

$$d_2 = \sqrt{a^2 + b^2 + 2ab(\cos \phi)}$$

$$d_1^2 + d_2^2 = 2(a^2 + b^2)$$

$$A = ah = ab(\sin \phi)$$

If $a = b$, the parallelogram is a rhombus.

Figure 4.17 Regular Polygon (n equal sides)

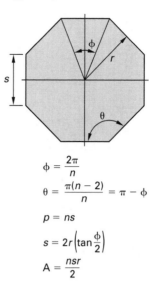

$$\phi = \frac{2\pi}{n}$$

$$\theta = \frac{\pi(n-2)}{n} = \pi - \phi$$

$$p = ns$$

$$s = 2r\left(\tan\frac{\phi}{2}\right)$$

$$A = \frac{nsr}{2}$$

MENSURATION OF VOLUMES

Figure 4.18 Sphere

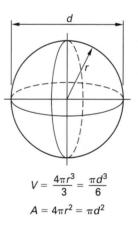

$$V = \frac{4\pi r^3}{3} = \frac{\pi d^3}{6}$$

$$A = 4\pi r^2 = \pi d^2$$

Figure 4.19 Right Circular Cone

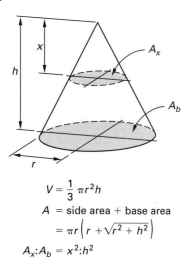

$$V = \frac{1}{3}\pi r^2 h$$

$$A = \text{side area} + \text{base area}$$

$$= \pi r \left(r + \sqrt{r^2 + h^2} \right)$$

$$A_x : A_b = x^2 : h^2$$

Figure 4.20 Right Circular Cylinder

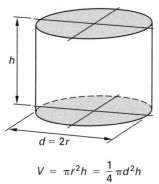

$$d = 2r$$

$$V = \pi r^2 h = \frac{1}{4}\pi d^2 h$$

$$A = \text{side area} + \text{end areas}$$

$$= 2\pi r (h + r)$$

Figure 4.21 Paraboloid of Revolution

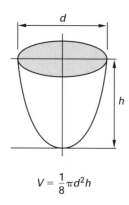
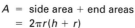

$$V = \frac{1}{8}\pi d^2 h$$

SAMPLE PROBLEMS

Problem 1

Which of the following lines is parallel to a line with the equation $y = \frac{1}{4}x + 6$?

    (A) $y = -\frac{1}{4}x - 6$

    (B) $y = -4x + 6$

    (C) $y = 2x - 3$

    (D) $x = 4y - 3$

Solution

For the two lines to be parallel, they must have the same slope. Rewrite each equation in the standard form to identify the slopes of the lines.

equation	standard form	slope
$y = -\frac{1}{4}x - 6$	same	$-\frac{1}{4}$
$y = -4x + 6$	same	-4
$y = 2x - 3$	same	2
$x = 4y - 3$	$y = \frac{1}{4}x + \frac{3}{4}$	$\frac{1}{4}$

Choice (D) has the same slope and, therefore, is parallel.

Answer is D.

Problem 2

What is the distance between the points $(3,2,-1)$ and $(4,-5,0)$?

    (A) $3\sqrt{3}$

    (B) $4\sqrt{3}$

    (C) 7

    (D) $\sqrt{51}$

Solution

$$d = \sqrt{(x_2 - x_1)^2 + (y_2 - y_1)^2 + (z_2 - z_1)^2}$$

$$= \sqrt{(4 - 3)^2 + (-5 - 2)^2 + (0 - (-1))^2}$$

$$= \sqrt{51}$$

Answer is D.

Problem 3

What is the value of θ (less than 2π) that will satisfy the following equation?

$$\sin^2 \theta + 4 \sin \theta + 3 = 0$$

(A) $\dfrac{\pi}{4}$

(B) $\dfrac{\pi}{2}$

(C) π

(D) $\dfrac{3\pi}{2}$

Solution

Factor the quadratic.

$$\sin^2 \theta + 4\sin \theta + 3 = 0$$
$$(\sin \theta + 3)(\sin \theta + 1) = 0$$
$$\sin \theta = -3 \text{ or } \sin \theta = -1$$

$\sin \theta = -3$ is not possible.

$$\theta = \sin^{-1}(-1) = \left[\cdots \frac{-5\pi}{2}, -\frac{\pi}{2}, \frac{3\pi}{2}, \frac{7\pi}{2}, \cdots \right]$$

Answer is D.

Problem 4

What are the coordinates of the center and the radius, respectively, of the following equation for a circle?

$$x^2 + y^2 + 12y - 2x + 12 = 0$$

(A) $(1,-6)$; 12
(B) $(-1,6)$; $\sqrt{12}$
(C) $(-1,6)$; 25
(D) $(1,-6)$; 5

Solution

The equation can be factored into the standard form with center (h, k) and radius r.

$$(x - h)^2 + (y - k)^2 = r^2$$
$$x^2 + y^2 + 12y - 2x + 12 = 0$$
$$(x^2 - 2x + 1) + (y^2 + 12y + 36) = -12 + 1 + 36$$
$$(x - 1)^2 + (y + 6)^2 = 25$$

The center is at

$$(h, k) = (1, -6)$$

The radius is
$$r = \sqrt{25} = 5$$

Answer is D.

Problem 5

What are the solutions to the following equation?

$$x^2 - x - 12 = 0$$

(A) $x_1 = 1$; $x_2 = 12$
(B) $x_1 = 4$; $x_2 = -3$
(C) $x_1 = -1$; $x_2 = 4$
(D) $x_1 = 6$; $x_2 = -2$

Solution

There are two ways to solve the equation. The first method is to factor the equation.

$$x^2 - x - 12 = (x + 3)(x - 4) = 0$$
$$x = -3 \text{ or } x = 4$$

The second method is to use the quadratic equation.

$$\begin{aligned} x_1, x_2 &= \frac{-b \pm \sqrt{b^2 - 4ac}}{2a} \\ &= \frac{-(-1) \pm \sqrt{(-1)^2 - (4)(1)(-12)}}{(2)(1)} \\ &= \frac{1 \pm 7}{2} \\ x_1 &= 4 \\ x_2 &= -3 \end{aligned}$$

Answer is B.

FE-STYLE EXAM PROBLEMS

1. What is the length of the line with slope $4/3$, from the point $(6, 4)$ to the y-axis?

(A) 10
(B) 25
(C) 50
(D) 75

2. A line goes through the point $(4, -6)$ and is perpendicular to the line $y = 4x + 10$. What is the equation of the line?

(A) $y = mx - 20$

(B) $y = -\dfrac{1}{4}x - 5$

(C) $y = \dfrac{1}{5}x + 5$

(D) $y = \dfrac{1}{4}x + 5$

MATHEMATICS Geo/Trig

3. The expression $\csc\theta\cos^3\theta\tan\theta$ is equivalent to which of the following expressions?

    (A) $\sin\theta$
    (B) $\cos\theta$
    (C) $1-\sin^2\theta$
    (D) $1+\sin^2\theta$

4. In the following illustration, angles 2 and 5 are $90°$, $AD = 15$, $DC = 20$, and $AC = 25$. What are the lengths BC and BD, respectively?

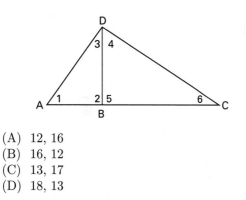

    (A) 12, 16
    (B) 16, 12
    (C) 13, 17
    (D) 18, 13

5. What are the x- and y-coordinates of the focus of the conic section described by the following equation? (Angle α corresponds to a right triangle with adjacent side x, opposite side y, and hypotenuse r.)

$$r\sin^2\alpha = \cos\alpha$$

    (A) $\left(-\dfrac{1}{2}, 0\right)$
    (B) $(0, 0)$
    (C) $\left(0, \dfrac{\pi}{2}\right)$
    (D) $\left(\dfrac{1}{4}, 0\right)$

For the following problems use the NCEES Handbook as your only reference.

6. What is the equation of the ellipse with center at $(0,0)$ that passes through the points $(2,0)$, $(0,3)$, and $(-2,0)$?

    (A) $\dfrac{x^2}{9} - \dfrac{y^2}{4} = 1$
    (B) $\dfrac{x^2}{4} - \dfrac{y^2}{9} = 1$
    (C) $\dfrac{x^2}{9} + \dfrac{y^2}{4} = 1$
    (D) $\dfrac{x^2}{4} + \dfrac{y^2}{9} = 1$

7. What is the area of the portion of the circle shown?

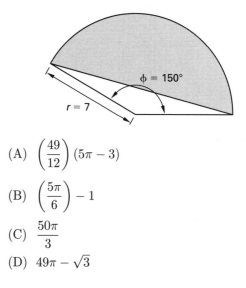

    (A) $\left(\dfrac{49}{12}\right)(5\pi - 3)$
    (B) $\left(\dfrac{5\pi}{6}\right) - 1$
    (C) $\dfrac{50\pi}{3}$
    (D) $49\pi - \sqrt{3}$

8. What is the equation of the circle passing through the (x,y) points $(0,0)$, $(0,4)$ and $(-4,0)$?

    (A) $(x-2)^2 + (y-2)^2 = \sqrt{8}$
    (B) $(x-2)^2 + (y-2)^2 = 8$
    (C) $(x+2)^2 + (y-2)^2 = 8$
    (D) $(x+2)^2 + (y+2)^2 = \sqrt{8}$

9. Which of the following equations describes a circle with center at $(2,3)$ and passing through the point $(-3,-4)$?

    (A) $(x+3)^2 + (y+4)^2 = 85$
    (B) $(x+3)^2 + (y+2)^2 = \sqrt{74}$
    (C) $(x-3)^2 + (y-2)^2 = 74$
    (D) $(x-2)^2 + (y-3)^2 = 74$

10. To find the width of a river, a surveyor sets up a transit at point C on one river bank and sights directly across to point B on the other bank. The surveyor then walks along the bank for a distance of 275 m to point A. The angle CAB is $57°28'$. What is the approximate width of the river?

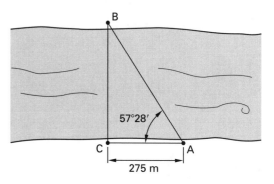

(A) 148 m
(B) 231 m
(C) 326 m
(D) 431 m

11. Which of the following expressions is equivalent to $\sin 2\theta$?

(A) $2\sin\theta\cos\theta$

(B) $\cos^2\theta - \sin^2\theta$

(C) $\sin\theta\cos\theta$

(D) $\dfrac{1-\cos 2\theta}{2}$

12. The x- and y-coordinates of a particle moving in the x-y plane are $x = 8\sin t$ and $y = 6\cos t$. Which of the following equations describes the path of the particle?

(A) $36x^2 + 64y^2 = 2304$
(B) $6x^2 + 8y^2 = 10$
(C) $64x^2 + 36y^2 = 2304$
(D) $64x^2 - 36y^2 = 2304$

13. All three sides of a triangle are initially 4 m in length. One of the triangle's sides is oriented horizontally. The triangle is scaled down in size without changing any of the angles. What is the new height of the triangle when the area is exactly half of the original triangle's area?

(A) 1.2 m
(B) 1.5 m
(C) 1.7 m
(D) 2.5 m

14. The vertical angle to the top of a flagpole from point A on the ground is observed to be $37°11'$. The observer walks 17 m directly away from point A and the flagpole to point B and finds the new angle to be $25°43'$. What is the approximate height of the flagpole?

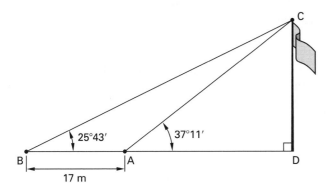

15. Two 20 m diameter circles are placed so that the circumference of each just touches the center of the other. What is the area common to each circle?

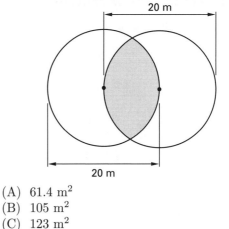

(A) 61.4 m^2
(B) 105 m^2
(C) 123 m^2
(D) 166 m^2

16. What conic section is described by the following equation?
$$4x^2 - y^2 + 8x + 4y = 15$$

(A) circle
(B) ellipse
(C) parabola
(D) hyperbola

17. What is the equation of a parabola with a vertex at (4,8) and a directrix at $y = 5$?

(A) $(x-8)^2 = 12(y-4)$
(B) $(x-4)^2 = 12(y-8)$
(C) $(x-4)^2 = 6(y-8)$
(D) $(y-8)^2 = 12(x-4)$

18. Which of the following statements is false for all non-circular ellipses?

(A) The eccentricity, e, is less than one.
(B) An ellipse has two foci.
(C) The sum of the two distances from the two foci to any point on the ellipse is $2a$ (i.e., twice the semimajor distance).
(D) The coefficients A and C preceding the x^2 and y^2 terms in the general form of the equation are equal.
$$Ax^2 + Bxy + Cy^2 + Dx + Ey + F = 0$$

(A) 10 m
(B) 22 m
(C) 82 m
(D) 300 m

19. What is the radius of a sphere with a center at the origin and that passes through the point (8,1,6)?

(A) 10
(B) 65
(C) $\sqrt{101}$
(D) 100

SOLUTIONS TO FE-STYLE EXAM PROBLEMS

1. The equation of the line is of the form

$$y = mx + b$$

$m = 4/3$, and a known point is $(x, y) = (6, 4)$.

$$4 = \left(\frac{4}{3}\right)(6) + b$$

$$b = 4 - \left(\frac{4}{3}\right)(6)$$

$$= -4$$

The complete equation is

$$y = \frac{4}{3}x - 4$$

The intersection with the y-axis is at point $(0, -4)$.

$$d = \sqrt{(y_2 - y_1)^2 + (x_2 - x_1)^2}$$

$$= \sqrt{(4 - (-4))^2 + (6 - 0)^2}$$

$$= 10$$

Answer is A.

2. The slopes of two lines that are perpendicular are related by

$$m_1 = \frac{-1}{m_2}$$

The slope of the line perpendicular to the line with slope $m_1 = 4$ is

$$m_2 = \frac{-1}{m_1} = -\frac{1}{4}$$

The equation of the line is given in the form

$$y = mx + b$$

$m = -1/4$, and a known point is $(x, y) = (4, -6)$.

$$-6 = \left(-\frac{1}{4}\right)(4) + b$$

$$b = -6 - \left(-\frac{1}{4}\right)(4)$$

$$= -5$$

The equation of the line is

$$y = -\frac{1}{4}x - 5$$

Answer is B.

3. Use trigonometric identities to simplify the expression.

$$\csc \theta \cos^3 \theta \tan \theta = \left(\frac{1}{\sin \theta}\right) \cos^3 \theta \left(\frac{\sin \theta}{\cos \theta}\right)$$

$$= \cos^2 \theta$$

$$= 1 - \sin^2 \theta$$

Answer is C.

4. For triangle ABD,

$$(BD)^2 + x^2 = (15)^2$$

For triangle DBC,

$$(BD)^2 + (25 - x)^2 = (20)^2$$

$$(15)^2 - x^2 = (20)^2 - (25 - x)^2$$

$$625 - 50x + x^2 - x^2 = 175$$

$$625 - 50x = 175$$

$$x = 9$$

$$= AB$$

$$BC = 25 - AB$$

$$= 16$$

$$(BD)^2 + (9)^2 = (15)^2$$

$$BD = 12$$

Alternatively, this problem can be solved using the law of cosines.

Answer is B.

5. Use the trigonometric functions of a right triangle to change the equation to x- and y-coordinates.

$$r \sin^2 \alpha = \cos \alpha$$

$$r \left(\frac{y}{r}\right)^2 = \frac{x}{r}$$

$$y^2 = x$$

The equation describes a parabola. The standard form for a parabola is

$$(y - k)^2 = 2p(x - h)$$
$$k = 0$$
$$h = 0$$
$$p = \frac{1}{2}$$

The focus of a parabola is at $(p/2, 0)$.

$$\left(\frac{p}{2}, 0\right) = \left(\frac{\frac{1}{2}}{2}, 0\right)$$
$$= \left(\frac{1}{4}, 0\right)$$

Answer is D.

6. An ellipse has the standard form

$$\frac{(x - h)^2}{a^2} + \frac{(y - k)^2}{b^2} = 1$$

The center is at $(h, k) = (0, 0)$.

$$\frac{(x - 0)^2}{a^2} + \frac{(y - 0)^2}{b^2} = 1$$

Substitute the known values of (x, y) to determine a and b.

For $(x, y) = (2, 0)$,

$$\frac{(2)^2}{a^2} + \frac{(0)^2}{b^2} = 1$$
$$a^2 = 4$$
$$a = 2$$

For $(x, y) = (0, 3)$,

$$\frac{(0)^2}{a^2} + \frac{(3)^2}{b^2} = 1$$
$$b^2 = 9$$
$$b = 3$$

For $(x, y) = (-2, 0)$,

$$\frac{(-2)^2}{a^2} + \frac{(0)^2}{b^2} = 1$$
$$a^2 = 4$$
$$a = 2 \quad \begin{bmatrix} \text{This step is not necessary} \\ \text{as } a \text{ and } b \text{ are determined} \\ \text{from the first point.} \end{bmatrix}$$

The equation of the ellipse is

$$\frac{x^2}{(2)^2} + \frac{y^2}{(3)^2} = 1$$
$$\frac{x^2}{4} + \frac{y^2}{9} = 1$$

Answer is D.

7.
$$A = \frac{r^2(\phi - \sin \phi)}{2}$$

$$\phi = (150°)\left(\frac{2\pi \text{ rad}}{360°}\right) = \frac{5\pi}{6} \text{ rad}$$

$$A = \frac{(7)^2\left(\frac{5\pi}{6} - \sin\left(\frac{5\pi}{6}\right)\right)}{2}$$
$$= \left(\frac{49}{2}\right)\left(\frac{5\pi}{6} - \frac{1}{2}\right)$$
$$= \left(\frac{49}{12}\right)(5\pi - 3)$$

(Be sure your calculator is set to "radians" when evaluating the sine.)

Answer is A.

8. The center-radius form of the equation of a circle is

$$(x - h)^2 + (y - k)^2 = r^2$$

Substitute the first two points, (0,0) and (0,4).

$$(0 - h)^2 + (0 - k)^2 = r^2$$
$$(0 - h)^2 + (4 - k)^2 = r^2$$

Since both are equal to the unknown r^2, set the left-hand sides equal. Simplify and solve for k.

$$h^2 + k^2 = h^2 + (4 - k)^2$$
$$k^2 = (4 - k)^2$$
$$k = 2$$

Substitute the third point, $(-4, 0)$, into the center-radius form.

$$(-4 - h)^2 + (0 - k)^2 = r^2$$

Set this third equation equal to the first equation. Simplify and solve for h.

$$(-4 - h)^2 + k^2 = h^2 + k^2$$
$$(-4 - h)^2 = h^2$$
$$h = -2$$

Now that h and k are known, substitute into the first equation to determine r^2.

$$h^2 + k^2 = r^2$$
$$(-2)^2 + (2)^2 = 8$$

Substitute the known values of h, k, and r^2 into the center-radius form.

$$(x + 2)^2 + (y - 2)^2 = 8$$

Answer is C.

9. Substitute the known points into the center-radius form of the equation of a circle.

$$(x - h)^2 + (y - k)^2 = r^2$$
$$(-3 - 2)^2 + (-4 - 3)^2 = 25 + 49$$
$$= 74$$

Therefore, the equation of the circle is

$$(x - 2)^2 + (y - 3)^2 = 74$$

Notice that $r^2 = 74$. The radius is $\sqrt{74}$.

Answer is D.

10.
$$57°28' = 57.467°$$
$$BC = CA \tan 57.467°$$
$$= (275 \text{ m}) \tan 57.467°$$
$$= 431.1 \text{ m} \quad (431 \text{ m})$$

Answer is D.

11. The double-angle identity is

$$\sin 2\theta = 2 \sin \theta \cos \theta$$

Answer is A.

12. Solve the two coordinate equations for $\sin t$ and $\cos t$.

$$\sin t = \frac{x}{8}$$
$$\cos t = \frac{y}{6}$$

Use the trigonometric identity.

$$\sin^2 \theta + \cos^2 \theta = 1$$
$$\left(\frac{x}{8}\right)^2 + \left(\frac{y}{6}\right)^2 = 1$$

Multiply both sides by $(8 \times 6)^2$ to clear the fractions.

$$36x^2 + 64y^2 = 2304$$

Answer is A.

13. The original area of the triangle is

$$A = \frac{1}{2}bh$$

The scaled down area of the triangle is

$$A' = \frac{1}{2}b'h'$$

Each dimension, side, and length has been scaled down by some factor, F. Since $A' = A/2$,

$$\frac{1}{2}b'h' = \left(\frac{1}{2}\right)\left(\frac{1}{2}bh\right)$$
$$2 = \left(\frac{b}{b'}\right)\left(\frac{h}{h'}\right) = F^2$$
$$F = \sqrt{2}$$

This is an equilateral triangle. Each angle is 60°. The original height of the triangle is

$$h = (4 \text{ m}) \sin 60°$$
$$= 3.464 \text{ m}$$

The new height of the triangle will be

$$h' = \frac{h}{F} = \frac{3.464 \text{ m}}{\sqrt{2}}$$
$$= 2.45 \text{ m}$$

Answer is D.

14.

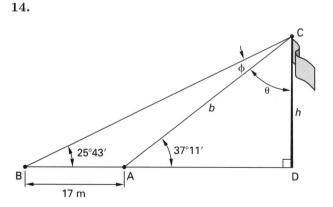

Work with triangle ADC.

$$37°11' + 90° + \theta = 180°$$
$$\theta = 52°49'$$

Work with triangle BDC.

$$25°43' + 90° + 52°49' + \phi = 180°$$
$$\phi = 11°28'$$

Work with triangle BAC. Use the law of sines to find side b.

$$\frac{\sin 11°28'}{17 \text{ m}} = \frac{\sin 25°43'}{b}$$

$$b = 37.11 \text{ m}$$

Work with triangle ADC. The flagpole height is

$$h = (37.11 \text{ m}) \sin 37°11'$$
$$= 22.43 \text{ m} \quad (22 \text{ m})$$

Answer is B.

15.

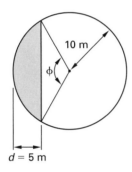

Refer to Fig. 4.14. The distance d is 5 m. The angle ϕ is given by

$$\phi = (2) \left(\arccos \left(\frac{r - d}{r} \right) \right)$$

$$= (2) \left(\arccos \left(\frac{10 \text{ m} - 5 \text{ m}}{10 \text{ m}} \right) \right)$$

$$= 120°$$

Convert ϕ to radians.

$$\phi = (120°) \left(\frac{2\pi}{360°} \right) = 2.094 \text{ rad}$$

Also from Fig. 4.14, the area of a circular segment is

$$A = \frac{1}{2} r^2 (\phi - \sin \phi)$$

$$= \left(\frac{1}{2} \right) (10 \text{ m})^2 (2.094 \text{ rad} - \sin(2.094 \text{ rad}))$$

$$= 61.4 \text{ m}^2$$

The common area is twice this amount.

$$A = (2)(61.4 \text{ m}^2) = 122.8 \text{ m}^2 \quad (123 \text{ m}^2)$$

(This answer can also be obtained by calculating the area of the sector and then subtracting the triangular area.)

Answer is C.

16. The general form of a conic section is

$$Ax^2 + Bxy + Cy^2 + Dx + Ey + F = 0$$

In this case, $A = 4$, $B = 0$, and $C = -1$.

Since $A \neq C$, the conic section is not a circle or line.

Calculate the discriminant.

$$B^2 - 4AC = (0)^2 - (4)(4)(-1) = 16$$

Since this is greater than zero, the section is a hyperbola.

Answer is D.

17. The directrix (the line described by $y = 5$) is parallel to the x-axis. Therefore, this is a vertical parabola. The parabola opens upward since the vertex (at $y = 8$) is above the directrix.

The distance from the vertex to the directrix is

$$\frac{p}{2} = 8 - 5 = 3$$
$$p = 6$$

The focus is located a distance $p/2$ from the vertex. Therefore, the focus is at $(4, 8 + 3)$ or $(4, 11)$.

The standard form equation for a parabola with vertex at (h, k) and opening upward is

$$(x - h)^2 = 2p(y - k)$$
$$(x - 4)^2 = (2)(6)(y - 8)$$
$$(x - 4)^2 = 12(y - 8)$$

Answer is B.

18. The coefficients preceding the squared terms in the general equation are equal only for a circle, not for an ellipse.

Answer is D.

19. Calculate the distance between the center and the point.

$$r^2 = (x - h)^2 + (y - k)^2 + (z - l)^2$$
$$r = \sqrt{(8 - 0)^2 + (1 - 0)^2 + (6 - 0)^2}$$
$$= \sqrt{101}$$

Answer is C.

5 Algebra and Linear Algebra

Subjects

LOGARITHMS 5-1
    Identities 5-1
COMPLEX NUMBERS 5-1
    Definition 5-1
    Polar Coordinates 5-2
    Roots 5-2
MATRICES 5-2
    Multiplication 5-3
    Addition 5-3
    Identity Matrix 5-3
    Transpose 5-3
    Determinant 5-3
    Cofactor and Classical Adjoint 5-4
    Inverse 5-4
    Simultaneous Linear Equations 5-4
VECTORS 5-5
    Vector Operations 5-5
    Vector Identities 5-6
PROGRESSIONS AND SERIES 5-6
    Arithmetic Progression 5-7
    Geometric Progression 5-7
    Properties of Series 5-7
    Taylor's Series 5-7

LOGARITHMS

Logarithms can be considered to be exponents. For example, the exponent c in the expression $b^c = x$ is the logarithm of x to the base b. Therefore, the two expressions $\log_b x = c$ and $b^c = x$ are equivalent.

$$\log_b x = c \Longrightarrow b^c = x \qquad 5.1$$

The base for *common logs* is 10. Usually, *log* will be written when common logs are desired, although log_{10} appears occasionally. The base for *natural logs* is 2.71828..., an irrational number that is given the symbol e. When natural logs are desired, usually *ln* will be written, although log_e is also used.

Identities

Logarithmic identities are useful in simplifying expressions containing exponentials and other logarithms.

$$\log_b b^n = n \qquad 5.2$$
$$\log x^c = c \log x \qquad 5.3$$
$$\operatorname{antilog}(\log x^c) = x^c = \operatorname{antilog}(c \log x) \qquad 5.4$$
$$\log xy = \log x + \log y \qquad 5.5$$
$$\log_b b = 1 \qquad 5.6$$
$$\log 1 = 0 \qquad 5.7$$
$$\log \frac{x}{y} = \log x - \log y \qquad 5.8$$
$$\log_b x = \frac{\log_a x}{\log_a b} \qquad 5.9$$

These identities can be used to derive the conversion between logarithms in one base to logarithms in another base (e.g., from common logs to natural logs).

$$\log_{10} x = \ln x \, \log_{10} e \qquad 5.10$$
$$\ln x = \frac{\log_{10} x}{\log_{10} e}$$
$$\approx 2.302585 \log_{10} x \qquad 5.11$$

COMPLEX NUMBERS

Definition

Complex numbers consist of combinations of *real* and *imaginary numbers*. Real numbers are *rational* and *irrational numbers*, and imaginary numbers are square roots of negative numbers. The symbols i and j are both used to represent the square root of -1.

$$i = \sqrt{-1} \qquad 5.12$$
$$j = \sqrt{-1} \qquad 5.13$$

When expressed as a sum (e.g., $a + ib$), the complex number is said to be in *rectangular* or *trigonometric form*. In Eq. 5.14, a is the real component and b is the imaginary component.

$$z \equiv a + ib \qquad \text{[rectangular form]} \qquad 5.14$$

Figure 5.1 Graphical Representation of a Complex Number

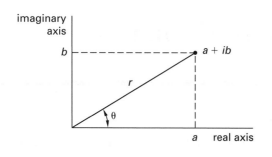

Most algebraic operations (addition, multiplication, exponentiation, etc.) work with complex numbers. When adding two complex numbers, real parts are added to real parts and imaginary parts are added to imaginary parts.

$$(a + ib) + (c + id) = (a + c) + i(b + d) \qquad 5.15$$
$$(a + ib) - (c + id) = (a - c) + i(b - d) \qquad 5.16$$

From these rules, it follows that

$$(a + ib) + (a - ib) = 2a \qquad 5.17$$
$$(a + ib) - (a - ib) = 2ib \qquad 5.18$$

Multiplication of two complex numbers in rectangular form is accomplished by use of the algebraic distributive law and the equivalency $i^2 = -1$.

$$(a + ib)(c + id) = (ac - bd) + i(ad + bc) \qquad 5.19$$

From Eq. 5.19, it follows that

$$(a + ib)(a - ib) = a^2 + b^2 \qquad 5.20$$

Division of complex numbers in rectangular form requires use of the *complex conjugate*. The complex conjugate of the complex number $(a + ib)$ is $(a - ib)$. By multiplying the numerator and the denominator by the complex conjugate of the denominator, the denominator will be converted to the real number $a^2 + b^2$. This technique is known as *rationalizing* the denominator.

$$\frac{a + ib}{c + id} = \frac{(a + ib)(c - id)}{(c + id)(c - id)}$$
$$= \frac{(ac + bd) + i(bc - ad)}{c^2 + d^2} \qquad 5.21$$

Polar Coordinates

The complex number $z = a + ib$ (see Fig. 5.1) can also be expressed in the *polar form*.

$$z \equiv r(\cos\theta + i\sin\theta) \qquad \text{[polar form]} \qquad 5.22$$

The rectangular form can be determined from r and θ.

$$x = r\cos\theta \qquad 5.23$$
$$y = r\sin\theta \qquad 5.24$$

Similarly, the polar form can be determined from x and y.

$$r = \sqrt{x^2 + y^2} \qquad 5.25$$
$$\theta = \tan^{-1}\frac{y}{x} \qquad 5.26$$

The multiplication and division rules defined for complex numbers expressed in rectangular form can be applied to complex numbers expressed in polar form. Using trigonometric identities, the rules reduce to

$$r_1(\cos\theta_1 + i\sin\theta_1)$$
$$\times r_2(\cos\theta_2 + i\sin\theta_2) = r_1 r_2(\cos(\theta_1 + \theta_2)$$
$$+ i\sin(\theta_1 + \theta_2)) \qquad 5.27$$

$$\frac{r_1(\cos\theta_1 + i\sin\theta_1)}{r_2(\cos\theta_2 + i\sin\theta_2)} = \left(\frac{r_1}{r_2}\right)(\cos(\theta_1 - \theta_2)$$
$$+ i\sin(\theta_1 - \theta_2)) \qquad 5.28$$

Equation 5.29 is called *de Moivre's formula*.

$$(x + iy)^n = (r(\cos\theta + i\sin\theta))^n$$
$$= r^n(\cos(n\theta) + i\sin(n\theta)) \qquad 5.29$$

Another notation for expressing the vector in polar coordinates is

$$z \equiv re^{i\theta} \qquad 5.30$$

In Eq. 5.30,

$$e^{i\theta} = \cos\theta + i\sin\theta \qquad 5.31$$
$$e^{-i\theta} = \cos\theta - i\sin\theta \qquad 5.32$$
$$\cos\theta = \frac{e^{i\theta} + e^{-i\theta}}{2} \qquad 5.33$$
$$\sin\theta = \frac{e^{i\theta} - e^{-i\theta}}{2i} \qquad 5.34$$

Roots

The *k*th root, w, of a complex number $z = r(\cos\theta + i\sin\theta)$ is found from Eq. 5.35.

$$w = \sqrt[k]{r}\left(\cos\left(\frac{\theta}{k} + n\frac{360°}{k}\right) + i\sin\left(\frac{\theta}{k} + n\left(\frac{360°}{k}\right)\right)\right) \qquad 5.35$$

MATRICES

A *matrix* is an ordered set of *entries* (*elements*) arranged rectangularly and set off by brackets. The entries can be variables or numbers. A matrix by itself has no particular value; it is merely a convenient method of representing a set of numbers.

The size of a matrix is given by the number of rows and columns, and the nomenclature $m \times n$ is used for a matrix with m rows and n columns. For a square matrix, the number of rows and columns are the same and are equal to the *order of the matrix*.

Matrix entries are represented by lowercase letters with subscripts, for example, a_{ij}. The term a_{23} would be the entry in the second row and third column of matrix $\mathbf{A}$.

Multiplication

A matrix can be multiplied by a scalar, an operation known as *scalar multiplication*, in which case all entries of the matrix are multiplied by that scalar. For example, for the 2×2 matrix $\mathbf{A}$,

$$k\mathbf{A} = \begin{bmatrix} ka_{11} & ka_{12} \\ ka_{21} & ka_{22} \end{bmatrix} \qquad 5.36$$

A matrix can be multiplied by another matrix, but only if the left-hand matrix has the same number of columns as the right-hand matrix has rows. *Matrix multiplication* occurs by multiplying the elements in each left-hand matrix row by the entries in each right-hand matrix column, adding the products, and placing the sum at the intersection point of the participating row and column. If $\mathbf{A}$ is an $m \times n$ matrix and $\mathbf{B}$ is an $n \times s$ matrix, then $\mathbf{AB}$ is an $m \times s$ matrix with the entries

$$\mathbf{C} \equiv c_{ij} \equiv \sum_{l=1}^{n} a_{il}b_{lj} \qquad 5.37$$

Addition

Addition and subtraction of two matrices are possible only if both matrices have the same shape and size. They are accomplished by adding or subtracting the corresponding entries of the two matrices. If $\mathbf{A}$ is an $m \times n$ matrix and $\mathbf{B}$ is also an $m \times n$ matrix, then $\mathbf{A} + \mathbf{B}$ is an $m \times n$ matrix with the entries

$$\mathbf{C} \equiv c_{ij} \equiv a_{ij} + b_{ij} \qquad 5.38$$

Identity Matrix

The *identity matrix* is a diagonal matrix, meaning that it is a square matrix with all zero entries except for the a_{ij} entries, for which $i = j$. The identity matrix is usually designated as $\mathbf{I}$. All nonzero entries are equal to one, and the matrix has the property that $\mathbf{AI} = \mathbf{IA} = \mathbf{A}$. For example, the 4×4 identity matrix is

$$\mathbf{I} = \begin{bmatrix} 1 & 0 & 0 & 0 \\ 0 & 1 & 0 & 0 \\ 0 & 0 & 1 & 0 \\ 0 & 0 & 0 & 1 \end{bmatrix}$$

Transpose

The *transpose*, $\mathbf{A}^T$, of an $m \times n$ matrix $\mathbf{A}$ is an $n \times m$ matrix constructed by taking the ith row and making it the ith column. The diagonal is unchanged. For example,

$$\mathbf{A} = \begin{bmatrix} 1 & 6 & 9 \\ 2 & 3 & 4 \\ 7 & 1 & 5 \end{bmatrix}$$

$$\mathbf{A}^T = \begin{bmatrix} 1 & 2 & 7 \\ 6 & 3 & 1 \\ 9 & 4 & 5 \end{bmatrix}$$

Determinant

A *determinant* is a scalar calculated from a square matrix. The determinant of matrix $\mathbf{A}$ can be represented as $\mathrm{D}\{\mathbf{A}\}$, $\mathrm{Det}(\mathbf{A})$, or $|\mathbf{A}|$. The following rules can be used to simplify the calculation of determinants.

- If $\mathbf{A}$ has a row or column of zeros, the determinant is zero.
- If $\mathbf{A}$ has two identical rows or columns, the determinant is zero.
- If $\mathbf{B}$ is obtained from $\mathbf{A}$ by adding a multiple of a row (column) to another row (column) in $\mathbf{A}$, then $|\mathbf{B}| = |\mathbf{A}|$.
- If $\mathbf{A}$ is *triangular* (a square matrix with zeros in all positions above or below the diagonal), the determinant is equal to the product of the diagonal entries.
- If $\mathbf{B}$ is obtained from $\mathbf{A}$ by multiplying one row or column in $\mathbf{A}$ by a scalar k, then $|\mathbf{B}| = k|\mathbf{A}|$.
- If $\mathbf{B}$ is obtained from the $n \times n$ matrix $\mathbf{A}$ by multiplying by the scalar matrix k, then $\mathbf{B} = |\mathbf{k} \times \mathbf{A}| = k^n|\mathbf{A}|$.
- If $\mathbf{B}$ is obtained from $\mathbf{A}$ by switching two rows or columns in $\mathbf{A}$, then $|\mathbf{B}| = -|\mathbf{A}|$.

Calculation of determinants is laborious for all but the smallest or simplest of matrices. For a 2×2 matrix, the formula used to calculate the determinant is easy to remember.

$$\mathbf{A} = \begin{bmatrix} a & b \\ c & d \end{bmatrix} \qquad 5.39$$

$$|\mathbf{A}| = \begin{vmatrix} a & b \\ c & d \end{vmatrix} = ad - bc \qquad 5.40$$

Two methods are commonly used for calculating the determinant of 3×3 matrices by hand. The first uses an augmented matrix constructed from the original matrix and the first two columns. The determinant is calculated as the sum of the products in the left-to-right downward diagonals less the sum of the products in the left-to-right upward diagonals.

$$\mathbf{A} = \begin{bmatrix} a & b & c \\ d & e & f \\ g & h & i \end{bmatrix}$$

$$\text{augmented } \mathbf{A} = \begin{bmatrix} \overset{+}{a} & \overset{+}{b} & \overset{+}{c} & \overset{-}{d} & \overset{-}{b} \\ d & e & f & d & e \\ g & h & i & g & h \end{bmatrix} \qquad 5.41$$

$$|\mathbf{A}| = aei + bfg + cdh - gec - hfa - idb \qquad 5.42$$

The second method of calculating the determinant is somewhat slower than the first for a 3×3 matrix but illustrates the method that must be used to calculate determinants of 4×4 and larger matrices. This method is known as *expansion by cofactors* (cofactors are explained in the following section). One row (column) is selected as the base row (column). The selection is arbitrary, but the number of calculations required to obtain the determinant can be minimized by choosing the row (column) with the most zeros. The determinant is equal to the sum of the products of the entries in the base row (column) and their corresponding cofactors.

$$\mathbf{A} = \begin{bmatrix} a & b & c \\ d & e & f \\ g & h & i \end{bmatrix} \qquad \begin{bmatrix} \text{first column chosen} \\ \text{as base column} \end{bmatrix}$$

$$|\mathbf{A}| = a \begin{vmatrix} e & f \\ h & i \end{vmatrix} - d \begin{vmatrix} b & c \\ h & i \end{vmatrix} + g \begin{vmatrix} b & c \\ e & f \end{vmatrix}$$

$$= a(ei - fh) - d(bi - ch) + g(bf - ce)$$

$$= aei - afh - dbi + dch + gbf - gce \qquad 5.43$$

Cofactor and Classical Adjoint

Cofactors are determinants of submatrices associated with particular entries in the original square matrix. The *minor* of entry a_{ij} is the determinant of a submatrix resulting from the elimination of the single row i and the single column j. For example, the minor corresponding to entry a_{12} in a 3×3 matrix $\mathbf{A}$ is the determinant of the matrix created by eliminating row 1 and column 2.

$$\text{minor of } a_{12} = \begin{vmatrix} a_{21} & a_{23} \\ a_{31} & a_{33} \end{vmatrix} \qquad 5.44$$

The cofactor of entry a_{ij} is the minor of a_{12} multiplied by either $+1$ or -1, depending on the position of the entry (i.e., the cofactor either exactly equals the minor or it differs only in sign). The sign of the cofactor of a_{ij} is positive if $(i + j)$ is even, and it is negative if $(i + j)$ is odd. For a 3×3 matrix, the multipliers in each position are

$$\begin{bmatrix} +1 & -1 & +1 \\ -1 & +1 & -1 \\ +1 & -1 & +1 \end{bmatrix}$$

For example, the cofactor of entry a_{12} in a 3×3 matrix $\mathbf{A}$ is

$$\text{cofactor of } a_{12} = - \begin{vmatrix} a_{21} & a_{23} \\ a_{31} & a_{33} \end{vmatrix} \qquad 5.45$$

The *classical adjoint* is the transpose of the cofactor matrix. The resulting matrix can be designated as $\mathbf{A}_{\text{adj}}$, adj$\{\mathbf{A}\}$ or $\mathbf{A}^{\text{adj}}$.

Inverse

The product of a matrix $\mathbf{A}$ and its *inverse*, $\mathbf{A}^{-1}$, is the identity matrix, $\mathbf{I}$. Only square matrices have inverses, but not all square matrices are invertible. A matrix has an inverse if and only if it is *nonsingular* (i.e., its determinant is nonzero).

$$\mathbf{A} \times \mathbf{A}^{-1} = \mathbf{A}^{-1} \times \mathbf{A} = \mathbf{I} \qquad 5.46$$

$$(\mathbf{A} \times \mathbf{B})^{-1} = \mathbf{B}^{-1} \times \mathbf{A}^{-1} \qquad 5.47$$

The inverse of a 2×2 matrix is easily determined by formula.

$$\mathbf{A} = \begin{bmatrix} a & b \\ c & d \end{bmatrix} \qquad 5.48$$

$$\mathbf{A}^{-1} = \frac{\begin{bmatrix} d & -b \\ -c & a \end{bmatrix}}{|\mathbf{A}|} \qquad 5.49$$

For a 3×3 or larger matrix, the inverse is determined by dividing every entry in the classical adjoint by the determinant of the original matrix.

$$\mathbf{A}^{-1} = \frac{\text{adj}(\mathbf{A})}{|\mathbf{A}|} \qquad 5.50$$

Simultaneous Linear Equations

Matrices are used to simplify the presentation and solution of sets of simultaneous linear equations. For example, the following three methods of presenting simultaneous linear equations are equivalent.

$$\left. \begin{aligned} a_{11}x_1 + a_{12}x_2 &= b_1 \\ a_{21}x_1 + a_{22}x_2 &= b_2 \end{aligned} \right\} \quad 5.51$$

$$\begin{bmatrix} a_{11} & a_{12} \\ a_{21} & a_{22} \end{bmatrix} \begin{bmatrix} x_1 \\ x_2 \end{bmatrix} = \begin{bmatrix} b_1 \\ b_2 \end{bmatrix} \qquad 5.52$$

$$\mathbf{A}\mathbf{X} = \mathbf{B} \qquad 5.53$$

In the second and third representations, $\mathbf{A}$ is known as the *coefficient matrix*, $\mathbf{X}$ as the *variable matrix*, and $\mathbf{B}$ as the *constant matrix*.

Determinants can be used to calculate the solution to linear simultaneous equations through a procedure known as *Cramer's rule*. The procedure calculates determinants of the original coefficient matrix $\mathbf{A}$ and of the n matrices resulting from the systematic replacement of a column in $\mathbf{A}$ by the constant matrix $\mathbf{B}$. For a system of three equations in three unknowns, there are three substitutional matrices, $\mathbf{A}_1$, $\mathbf{A}_2$, and $\mathbf{A}_3$, as well as the original coefficient matrix, for a total of four matrices whose determinants must be calculated.

The values of the unknowns that simultaneously satisfy all of the linear equations are

$$x_1 = \frac{|\mathbf{A}_1|}{|\mathbf{A}|} \qquad 5.54$$

$$x_2 = \frac{|\mathbf{A}_2|}{|\mathbf{A}|} \qquad 5.55$$

$$x_3 = \frac{|\mathbf{A}_3|}{|\mathbf{A}|} \qquad 5.56$$

VECTORS

A physical property or quantity can be a scalar, vector, or tensor. A *scalar* has only magnitude. Knowing its value is sufficient to define a scalar. Mass, enthalpy, density, and speed are examples of scalars.

Force, momentum, displacement, and velocity are examples of *vectors*. A vector is a directed straight line with a specific magnitude. Thus, a vector is specified completely by its direction (consisting of the vector's *angular orientation* and its *sense*) and magnitude. A vector's *point of application* (*terminal point*) is not needed to define the vector. Two vectors with the same direction and magnitude are said to be equal vectors even though their *lines of action* may be different.

Unit vectors are vectors with unit magnitudes (i.e., magnitudes of one). They are represented in the same notation as other vectors. Although they can have any direction, the standard unit vectors (i.e., the *Cartesian unit vectors* $\mathbf{i}$, $\mathbf{j}$, and $\mathbf{k}$) have the directions of the x-, y-, and z-coordinate axes, respectively, and constitute the *Cartesian triad*.

A vector $\mathbf{A}$ can be written in terms of unit vectors and its components.

$$\mathbf{A} = a_x\mathbf{i} + a_y\mathbf{j} + a_z\mathbf{k} \qquad 5.57$$

A *tensor* has magnitude in a specific direction, but the direction is not unique. A tensor in three-dimensional space is defined by nine components, compared with the three that are required to define vectors. These components are written in matrix form. Stress, dielectric constant, and magnetic susceptibility are examples of tensors.

Vector Operations

Addition of two vectors by the *polygon method* is accomplished by placing the tail of the second vector at the head (tip) of the first. The sum (i.e., the *resultant vector*) is a vector extending from the tail of the first vector to the head of the second (Fig. 5.2). Alternatively, the two vectors can be considered as two of the sides of a parallelogram, while the sum represents the diagonal. This is known as addition by the *parallelogram method*. The components of the resultant vector are the sums of the components of the added vectors.

$$\mathbf{A} + \mathbf{B} = (a_x + b_x)\mathbf{i} + (a_y + b_y)\mathbf{j} + (a_z + b_z)\mathbf{k} \quad 5.58$$

Figure 5.2   Addition of Two Vectors

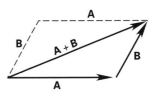

Similarly, for subtraction of two vectors,

$$\mathbf{A} - \mathbf{B} = (a_x - b_x)\mathbf{i} + (a_y - b_y)\mathbf{j} + (a_z - b_z)\mathbf{k} \quad 5.59$$

The *dot product* (*scalar product*) of two vectors is a scalar that is proportional to the length of the projection of the first vector onto the second vector. (See Fig. 5.3.)

Figure 5.3   Vector Dot Product

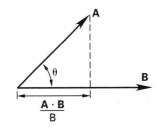

The dot product can be calculated in two ways, as Eq. 5.60 indicates. θ is limited to 180° and is the angle between the two vectors.

$$\mathbf{A} \cdot \mathbf{B} = a_xb_x + a_yb_y + a_zb_z$$
$$= |\mathbf{A}||\mathbf{B}|\cos\theta \qquad 5.60$$

The *cross product* (*vector product*), $\mathbf{A} \times \mathbf{B}$, of two vectors is a vector that is orthogonal (perpendicular) to the plane of the two vectors. (See Fig. 5.4.) The unit vector representation of the cross product can be calculated as a third-order determinant. $\mathbf{n}$ is the unit vector in the direction perpendicular to the plane containing $\mathbf{A}$ and $\mathbf{B}$.

$$\mathbf{A} \times \mathbf{B} = \begin{vmatrix} \mathbf{i} & \mathbf{j} & \mathbf{k} \\ a_x & a_y & a_z \\ b_x & b_y & b_z \end{vmatrix}$$

$$= |\mathbf{A}||\mathbf{B}|\mathbf{n} \sin\theta \qquad 5.61$$

Figure 5.4 Vector Cross Product

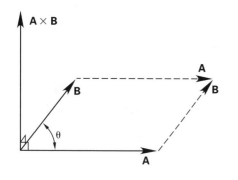

The direction of the cross-product vector corresponds to the direction a right-hand screw would progress if vectors $\mathbf{A}$ and $\mathbf{B}$ were placed tail to tail in the plane they define and $\mathbf{A}$ is rotated into $\mathbf{B}$. The direction can also be found from the *right-hand rule* (Fig. 5.5): Place the two vectors tail to tail. Close your right hand and position it over the pivot point. Rotate the first vector into the second vector, and position your hand such that your fingers curl in the same direction as the first vector rotates. Your extended thumb will coincide with the direction of the cross product. It is perpendicular to the plane of the two vectors.

Figure 5.5 Right-Hand Rule

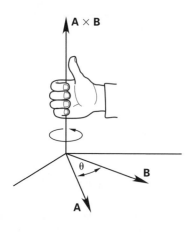

Vector Identities

The dot product for vectors is commutative and distributive.

$$\mathbf{A} \cdot \mathbf{B} = \mathbf{B} \cdot \mathbf{A} \qquad 5.62$$
$$\mathbf{A} \cdot (\mathbf{B} + \mathbf{C}) = \mathbf{A} \cdot \mathbf{B} + \mathbf{A} \cdot \mathbf{C} \qquad 5.63$$
$$\mathbf{A} \cdot \mathbf{A} = |\mathbf{A}|^2 \qquad 5.64$$

For unit vectors,

$$\mathbf{i} \cdot \mathbf{i} = \mathbf{j} \cdot \mathbf{j} = \mathbf{k} \cdot \mathbf{k} = 1 \qquad 5.65$$

The dot product can be used to determine whether a vector is a unit vector and to show that two vectors are orthogonal (perpendicular). For two non-null (nonzero) orthogonal vectors,

$$\mathbf{A} \cdot \mathbf{B} = 0 \qquad \text{[orthogonal]} \qquad 5.66$$
$$\mathbf{i} \cdot \mathbf{j} = \mathbf{i} \cdot \mathbf{k} = \mathbf{j} \cdot \mathbf{k} = 0 \qquad 5.67$$

Vector cross multiplication is distributive but not commutative.

$$\mathbf{A} \times \mathbf{B} = -\mathbf{B} \times \mathbf{A} \qquad 5.68$$
$$\mathbf{A} \times (\mathbf{B} + \mathbf{C}) = (\mathbf{A} \times \mathbf{B}) + (\mathbf{A} \times \mathbf{C}) \qquad 5.69$$
$$(\mathbf{B} + \mathbf{C}) \times \mathbf{A} = (\mathbf{B} \times \mathbf{A}) + (\mathbf{C} \times \mathbf{A}) \qquad 5.70$$

If two non-null vectors are parallel, their cross product will be zero.

$$\mathbf{A} \times \mathbf{B} = 0 \qquad \text{[parallel]} \qquad 5.71$$
$$\mathbf{i} \times \mathbf{i} = \mathbf{j} \times \mathbf{j} = \mathbf{k} \times \mathbf{k} = 0 \qquad 5.72$$

If two non-null vectors are normal (perpendicular), their vector cross product will be perpendicular to both vectors.

$$\mathbf{i} \times \mathbf{j} = -\mathbf{j} \times \mathbf{i} = \mathbf{k} \qquad 5.73$$
$$\mathbf{j} \times \mathbf{k} = -\mathbf{k} \times \mathbf{j} = \mathbf{i} \qquad 5.74$$
$$\mathbf{k} \times \mathbf{i} = -\mathbf{i} \times \mathbf{k} = \mathbf{j} \qquad 5.75$$

PROGRESSIONS AND SERIES

A *sequence*, $\{\mathbf{A}\}$, is an ordered progression of numbers a_i, such as 1, 4, 9, 16, 25, ... The *terms* in a sequence can be all positive, negative, or of alternating signs. ℓ is the last term and is also known as the *general term* of the sequence.

$$\{\mathbf{A}\} = a_1, a_2, a_3, \dots, \ell \qquad 5.76$$

A sequence is said to *diverge* (i.e., be *divergent*) if the terms approach infinity, and it is said to *converge* (i.e., be *convergent*) if the terms approach any finite value (including zero).

A *series* is the sum of terms in a sequence. There are two types of series: A *finite series* has a finite number of terms. An *infinite series* has an infinite number of terms, but this does not imply that the sum is infinite. The main tasks associated with series are determining the sum of the terms and determining whether the series converges. A series is said to converge if the sum, S_n, of its terms exists. A finite series is always convergent.

Arithmetic Progression

The *arithmetic sequence* is a standard sequence that diverges. It has the form

$$\ell = a + (n-1)d \qquad 5.77$$

In Eq. 5.77, a is the *first term*, d is a constant called the *common difference*, and n is the number of terms.

The difference of adjacent terms is constant in arithmetic progressions. The sum of terms in a finite arithmetic series is

$$S_n = \sum_{i=1}^{n}(a + (i-1)d) = \left(\frac{n}{2}\right)(a + \ell)$$

$$= \frac{n(2a + (n-1)d)}{2} \qquad 5.78$$

Geometric Progression

The *geometric sequence* is another standard sequence. The quotient of adjacent terms is constant in geometric progressions. It converges for $-1 < r < 1$ and diverges otherwise.

$$\ell = ar^{n-1} \qquad 5.79$$

In Eq. 5.79, a is the first term and r is known as the *common ratio*.

The sum of a finite geometric series is

$$S_n = \sum_{i=1}^{n} ar^{i-1} = \frac{a - r\ell}{1 - r}$$

$$= \frac{a(1 - r^n)}{1 - r} \qquad 5.80$$

The sum of an infinite geometric series is

$$S_n = \sum_{i=1}^{\infty} ar^{i-1}$$

$$= \frac{a}{1 - r} \qquad 5.81$$

Properties of Series

A *power series* is a series of the form

$$\sum_{i=1}^{n} a_i x^{i-1} = a_1 + a_2 x + a_3 x^2 + \cdots + a_n x^{n-1} \quad 5.82$$

The *interval of convergence* of a power series consists of the values of x for which the series is convergent. Due to the exponentiation of terms, an infinite power series can only be convergent in the interval $-1 < x < 1$.

A power series may be used to represent a function that is continuous over the interval of convergence of the series. The *power series representation* may be used to find the derivative or integral of that function.

The following rules are valid for power series.

$$\sum_{i=1}^{n} c = nc \qquad 5.83$$

$$\sum_{i=1}^{n} cx_i = c \sum_{i=1}^{n} x_i \qquad 5.84$$

$$\sum_{i=1}^{n}(x_i + y_i - z_i) = \sum_{i=1}^{n} x_i + \sum_{i=1}^{n} y_i - \sum_{i=1}^{n} z_i \quad 5.85$$

$$\sum_{x=1}^{n} x = \frac{n + n^2}{2} \qquad 5.86$$

Power series behave similarly to polynomials: They may be added together, subtracted from each other, multiplied together, or divided term by term within the interval of convergence. They may also be differentiated and integrated within their interval of convergence. If $f(x) = \sum_{i=1}^{n} a_i x^i$, then over the interval of convergence,

$$f'(x) = \sum_{i=1}^{n} \frac{d(a_i x^i)}{dx} \qquad 5.87$$

$$\int f(x)dx = \sum_{i=1}^{n} \int a_i x^i dx \qquad 5.88$$

Taylor's Series

Taylor's formula (*series*) can be used to expand a function around a point (i.e., to approximate the function at one point based on the function's value at another point). The approximation consists of a series, each term composed of a derivative of the original function and a polynomial. Using Taylor's formula requires that the original function be continuous in the interval $[a, b]$. To expand a function, $f(x)$, around a point, a, in order to obtain $f(b)$, Taylor's formula is

$$f(b) = f(a) + \left(\frac{f'(a)}{1!}\right)(b - a) + \left(\frac{f''(a)}{2!}\right)(b - a)^2$$

$$+ \cdots + \left(\frac{f^n(a)}{n!}\right)(b - a)^n \qquad 5.89$$

If $a = 0$, Eq. 5.89 is known as the *Maclaurin series*.

To be a useful approximation, point a must satisfy two requirements: It must be relatively close to point b, and the function and its derivatives must be known or be easy to calculate.

SAMPLE PROBLEMS

Problem 1

Which of the following numbers is equal to $\log_8 50$?

- (A) 0
- (B) 0.53
- (C) 0.79
- (D) 1.88

Solution

There are two methods to solve this problem. The first method is to use the definition of logarithms and trial and error.

$$\log_8 50 = c$$
$$8^c = 50$$

By trial and error, $c \approx 1.88$.

The second method uses an identity.

$$\log_a x = \log_b x \, \log_a b$$
$$\log_{10} 50 = \log_8 50 \, \log_{10} 8$$
$$\log_8 50 = \frac{\log 50}{\log 8}$$
$$= 1.88$$

Answer is D.

Problem 2

What is the polar form of the complex number $z = 3 + 4i$?

- (A) $(3)(\cos 36.87° + i \sin 36.87°)$
- (B) $(3)(\cos 53.15° + i \sin 36.87°)$
- (C) $(4)(\cos 53.15° + i \sin 53.15°)$
- (D) $(5)(\cos 53.13° + i \sin 53.13°)$

Solution

$$r = \sqrt{x^2 + y^2} = \sqrt{(3)^2 + (4)^2}$$
$$= 5$$
$$\theta = \tan^{-1}\left(\frac{y}{x}\right) = \tan^{-1}\left(\frac{4}{3}\right)$$
$$= 53.13°$$
$$z = (5)(\cos 53.13° + i \sin 53.13°)$$

Answer is D.

Problem 3

What is the determinant of the following matrix?

$$\mathbf{A} = \begin{bmatrix} 2 & 3 & 4 \\ 5 & 6 & 7 \\ 7 & 8 & 9 \end{bmatrix}$$

- (A) -8
- (B) -4
- (C) 0
- (D) 4

Solution

There are two methods for solving this problem. The first method is to use the augmented matrix.

$$\begin{bmatrix} 2 & 3 & 4 & 2 & 3 \\ 5 & 6 & 7 & 5 & 6 \\ 7 & 8 & 9 & 7 & 8 \end{bmatrix}$$

$$|\mathbf{A}| = (2)(6)(9) + (3)(7)(7) + (4)(5)(8)$$
$$- (7)(6)(4) - (8)(7)(2) - (9)(5)(3)$$
$$= 0$$

The second method is to use expansion by cofactors.

$$|\mathbf{A}| = 2\begin{vmatrix} 6 & 7 \\ 8 & 9 \end{vmatrix} - 5\begin{vmatrix} 3 & 4 \\ 8 & 9 \end{vmatrix} + 7\begin{vmatrix} 3 & 4 \\ 6 & 7 \end{vmatrix}$$
$$= (2)((6)(9) - (7)(8)) - (5)((3)(9) - (4)(8))$$
$$+ (7)((3)(7) - (4)(6))$$
$$= -4 + 25 - 21$$
$$= 0$$

Answer is C.

Problem 4

Given vectors $\mathbf{A}$, $\mathbf{B}$, and $\mathbf{C}$, what is the value of $(\mathbf{A} + \mathbf{B}) \cdot (\mathbf{B} + \mathbf{C})$?

$$\mathbf{A} = 8\mathbf{i} + 2\mathbf{j} + 2\mathbf{k}$$
$$\mathbf{B} = 4\mathbf{i} + 2\mathbf{j} + 4\mathbf{k}$$
$$\mathbf{C} = 6\mathbf{i} + 8\mathbf{j} + 10\mathbf{k}$$

- (A) 52
- (B) 104
- (C) 132
- (D) 244

Solution

Sum the like components of the vectors being added.

$$\mathbf{A} + \mathbf{B} = \begin{array}{r} 8\mathbf{i} + 2\mathbf{j} + 2\mathbf{k} \\ 4\mathbf{i} + 2\mathbf{j} + 4\mathbf{k} \\ \hline 12\mathbf{i} + 4\mathbf{j} + 6\mathbf{k} \end{array}$$

$$\mathbf{B} + \mathbf{C} = \begin{array}{r} 4\mathbf{i} + 2\mathbf{j} + 4\mathbf{k} \\ 6\mathbf{i} + 8\mathbf{j} + 10\mathbf{k} \\ \hline 10\mathbf{i} + 10\mathbf{j} + 14\mathbf{k} \end{array}$$

The dot product is the sum of the products of the like components.

$$(\mathbf{A} + \mathbf{B}) \cdot (\mathbf{B} + \mathbf{C}) = (12)(10) + (4)(10) + (6)(14)$$
$$= 244$$

Answer is D.

Problem 5

Solve the following set of simultaneous linear equations for A, B, and C.

$$2A + 3B - C = -10$$
$$-A + 4B + 2C = -4$$
$$2A - 2B + 5C = 35$$

- (A) $-2, -3, 5$
- (B) $2, -3, 5$
- (C) $2, 0, 5$
- (D) $2, 3, -5$

Solution

Use Cramer's rule to solve the simultaneous linear equations.

The coefficient matrix is

$$\mathbf{D} = \begin{bmatrix} 2 & 3 & -1 \\ -1 & 4 & 2 \\ 2 & -2 & 5 \end{bmatrix}$$

The determinant is

$$|\mathbf{D}| = 2 \begin{vmatrix} 4 & 2 \\ -2 & 5 \end{vmatrix} + 1 \begin{vmatrix} 3 & -1 \\ -2 & 5 \end{vmatrix} + 2 \begin{vmatrix} 3 & -1 \\ 4 & 2 \end{vmatrix}$$
$$= (2)(20 + 4) + (1)(15 - 2) + (2)(6 + 4)$$
$$= 81$$

The determinants of the substitutional matrices are

$$|\mathbf{A}_1| = \begin{vmatrix} -10 & 3 & -1 \\ -4 & 4 & 2 \\ 35 & -2 & 5 \end{vmatrix} = 162$$

$$|\mathbf{A}_2| = \begin{vmatrix} 2 & -10 & -1 \\ -1 & -4 & 2 \\ 2 & 35 & 5 \end{vmatrix} = -243$$

$$|\mathbf{A}_3| = \begin{vmatrix} 2 & 3 & -10 \\ -1 & 4 & -4 \\ 2 & -2 & 35 \end{vmatrix} = 405$$

$$A = \frac{162}{81} = 2$$

$$B = \frac{-243}{81} = -3$$

$$C = \frac{405}{81} = 5$$

Answer is B.

Problem 6

What is the angle between the two given vectors $\mathbf{A}$ and $\mathbf{B}$?

$$\mathbf{A} = 4\mathbf{i} + 12\mathbf{j} + 6\mathbf{k}$$
$$\mathbf{B} = 24\mathbf{i} - 8\mathbf{j} + 6\mathbf{k}$$

- (A) $-84.32°$
- (B) $84.32°$
- (C) $101.20°$
- (D) $122.36°$

Solution

There are two ways to calculate the dot product of vectors. Apply both, and set their results equal to one another.

$$\mathbf{A} \cdot \mathbf{B} = |\mathbf{A}||\mathbf{B}| \cos\theta$$
$$= \sqrt{(4)^2 + (12)^2 + (6)^2}$$
$$\times \left(\sqrt{(24)^2 + (-8)^2 + (6)^2} \right) \cos\theta$$
$$= (14)(26) \cos\theta$$

$$\mathbf{A} \cdot \mathbf{B} = a_x b_x + a_y b_y + a_z b_z$$
$$= (4)(24) + (12)(-8) + (6)(6)$$
$$= 36$$
$$(14)(26) \cos\theta = 36$$
$$\cos\theta = 0.0989$$
$$\theta = \cos^{-1} 0.0989$$
$$= 84.32°$$

Answer is B.

FE-STYLE EXAM PROBLEMS

1. What value of A satisfies the expression $A^{-6/8} = 0.001$?

- (A) 0
- (B) 100
- (C) 1000
- (D) 10,000

2. What expression is equivalent to $\log(x/(y+z))$?

- (A) $\log x - \log y - \log z$
- (B) $\log x - \log(y + z)$
- (C) $\dfrac{\log x}{\log y + \log z}$
- (D) $e^{x/(y+z)}$

Problems 3–5 refer to the following system of equations.

$$10x + 3y + 10z = 5$$
$$8x - 2y + 9z = 5$$
$$8x + y - 10z = 5$$

3. What is the cofactor matrix of the coefficient matrix?

(A) $\begin{bmatrix} 11 & 152 & 24 \\ 40 & -180 & 14 \\ 47 & -10 & -44 \end{bmatrix}$

(B) $\begin{bmatrix} 11 & -152 & 24 \\ -40 & 180 & 14 \\ 47 & 10 & 44 \end{bmatrix}$

(C) $\begin{bmatrix} 29 & -8 & -8 \\ -20 & -20 & 34 \\ 7 & 170 & 4 \end{bmatrix}$

(D) $\begin{bmatrix} 29 & 8 & -8 \\ 20 & -20 & -34 \\ 7 & -170 & 4 \end{bmatrix}$

4. What is the classical adjoint of the coefficient matrix?

(A) $\begin{bmatrix} 29 & -20 & 7 \\ -8 & -20 & 170 \\ -8 & 34 & 4 \end{bmatrix}$

(B) $\begin{bmatrix} 11 & 40 & 47 \\ 152 & -180 & -10 \\ 24 & 14 & -44 \end{bmatrix}$

(C) $\begin{bmatrix} 29 & 20 & 7 \\ 8 & -20 & -170 \\ -8 & -34 & 4 \end{bmatrix}$

(D) $\begin{bmatrix} 40 & 47 & 160 \\ 144 & -140 & 1800 \\ 32 & 658 & 18 \end{bmatrix}$

5. What is the inverse of the coefficient matrix?

(A) $\begin{bmatrix} 0.014 & -0.050 & 0.058 \\ -0.189 & 0.223 & 0.012 \\ 0.030 & 0.017 & 0.055 \end{bmatrix}$

(B) $\begin{bmatrix} 0.032 & 0.022 & 0.008 \\ 0.009 & -0.022 & -0.188 \\ -0.009 & -0.038 & 0.004 \end{bmatrix}$

(C) $\begin{bmatrix} \frac{29}{906} & \frac{-10}{453} & \frac{7}{906} \\ \frac{-4}{453} & \frac{-10}{453} & \frac{85}{453} \\ \frac{-4}{453} & \frac{17}{453} & \frac{2}{453} \end{bmatrix}$

(D) $\begin{bmatrix} \frac{11}{806} & \frac{20}{403} & \frac{47}{806} \\ \frac{76}{403} & \frac{-90}{403} & \frac{-5}{403} \\ \frac{12}{403} & \frac{7}{403} & \frac{-22}{403} \end{bmatrix}$

For the following problems use the NCEES Handbook as your only reference.

6. What is the solution to the following system of simultaneous linear equations?

$$10x + 3y + 10z = 5$$
$$8x - 2y + 9z = 3$$
$$8x + y - 10z = 7$$

(A) $x = 0.326$; $y = -0.192$; $z = 0.586$
(B) $x = 0.148$; $y = 1.203$; $z = 0.099$
(C) $x = 0.625$; $y = 0.186$; $z = -0.181$
(D) $x = 0.282$; $y = -1.337$; $z = -0.131$

7. Find the length of the resultant of the following vectors.

$$3i + 4j - 5k$$
$$7i + 2j + 3k$$
$$-16i - 14j + 2k$$

(A) 3
(B) 4
(C) 10
(D) 14

8. Find the unit vector (i.e., the direction vector) associated with the vector $18i + 3j + 29k$.

(A) $0.525i + 0.088j + 0.846k$
(B) $0.892i + 0.178j + 0.416k$
(C) $1.342i + 0.868j + 2.437k$
(D) $6i + j + \frac{29}{3}k$

9. What is the cross product, $A \times B$, of vectors A and B?

$$A = i + 4j + 6k$$
$$B = 2i + 3j + 5k$$

(A) $\mathbf{i} - \mathbf{j} - \mathbf{k}$
(B) $-\mathbf{i} + \mathbf{j} + \mathbf{k}$
(C) $2\mathbf{i} + 7\mathbf{j} - 5\mathbf{k}$
(D) $2\mathbf{i} + 7\mathbf{j} + 5\mathbf{k}$

10. What is the value of $\ln((7.3891)^{xy})$?

(A) $2/xy$
(B) $0.5xy$
(C) $0.8686xy$
(D) $2xy$

11. If the $\log_{10}$ of 4 is $0.703x$, what is the $\log_{10}$ of $1/4$?

(A) $-1.703x$
(B) $-0.703x$
(C) $0.297x$
(D) $0.703x$

12. What are the polar (r, θ) coordinates of the point that has rectangular (x, y) coordinates of $(4, 6)$?

(A) $(4, 6°)$
(B) $(4, 56.3°)$
(C) $(7.21, 33.7°)$
(D) $(7.21, 56.3°)$

13. The polar (r, θ) coordinates of a point are $(4, 120°)$. What are the rectangular (x, y) coordinates?

(A) $(2, 3.46)$
(B) $(3.46, 2)$
(C) $(-2, 3.46)$
(D) $(-2, -3.46)$

14. What is the matrix product $\mathbf{AB}$ of matrices $\mathbf{A}$ and $\mathbf{B}$?

$$\mathbf{A} = \begin{bmatrix} 2 & 1 \\ 1 & 0 \end{bmatrix} \qquad \mathbf{B} = \begin{bmatrix} 4 & 3 \\ 2 & 1 \end{bmatrix}$$

(A) $\begin{bmatrix} 10 & 4 \\ 7 & 3 \end{bmatrix}$

(B) $\begin{bmatrix} 11 & 4 \\ 5 & 2 \end{bmatrix}$

(C) $\begin{bmatrix} 8 & 3 \\ 2 & 0 \end{bmatrix}$

(D) $\begin{bmatrix} 10 & 7 \\ 4 & 3 \end{bmatrix}$

15. What is the matrix product $\mathbf{AB}$ of matrices $\mathbf{A}$ and $\mathbf{B}$?

$$\mathbf{A} = \begin{bmatrix} 1 & 2 & 3 \end{bmatrix} \qquad \mathbf{B} = \begin{bmatrix} 2 \\ -3 \\ 4 \end{bmatrix}$$

(A) $\begin{bmatrix} 2 \\ -6 \\ 12 \end{bmatrix}$

(B) $[8]$
(C) $[20]$
(D) $\begin{bmatrix} 2 & -6 & 12 \end{bmatrix}$

16. What is the matrix difference $\mathbf{A} - \mathbf{B}$ of matrices $\mathbf{A}$ and $\mathbf{B}$?

$$\mathbf{A} = \begin{bmatrix} 3 & 4 & 1 \\ 5 & 2 & -2 \end{bmatrix} \qquad \mathbf{B} = \begin{bmatrix} 5 & 2 & -6 \\ -5 & 4 & 2 \end{bmatrix}$$

(A) $\begin{bmatrix} -2 & 2 & 5 \\ 0 & -2 & -4 \end{bmatrix}$

(B) $\begin{bmatrix} 2 & -2 & -7 \\ -10 & 2 & 4 \end{bmatrix}$

(C) $\begin{bmatrix} 2 & 2 & -5 \\ 10 & -2 & -4 \end{bmatrix}$

(D) $\begin{bmatrix} -2 & 2 & 7 \\ 10 & -2 & -4 \end{bmatrix}$

17. The determinant of matrix $\mathbf{A}$ is -5. What is the missing value a?

$$\mathbf{A} = \begin{bmatrix} 1 & -3 & 2 \\ 0 & 1 & 2 \\ 3 & -1 & a \end{bmatrix}$$

(A) -9
(B) $5/22$
(C) 7
(D) 17

18. If the determinant of matrix $\mathbf{A}$ is -40, what is the determinant of matrix $\mathbf{B}$?

$$\mathbf{A} = \begin{bmatrix} 4 & 3 & 2 & 1 \\ 0 & 1 & 2 & -1 \\ 2 & 3 & -1 & 1 \\ 1 & 1 & 1 & 2 \end{bmatrix} \qquad \mathbf{B} = \begin{bmatrix} 2 & 1.5 & 1 & 0.5 \\ 0 & 1 & 2 & -1 \\ 2 & 3 & -1 & 1 \\ 1 & 1 & 1 & 2 \end{bmatrix}$$

(A) -80
(B) -40
(C) -20
(D) 0.5

19. What is the determinant of matrix $\mathbf{A}$?

$$\mathbf{A} = \begin{bmatrix} 3 & 6 \\ 2 & 4 \end{bmatrix}$$

(A) 0
(B) 15
(C) 14
(D) 26

20. What is the inverse of matrix $\mathbf{A}$?

$$\mathbf{A} = \begin{bmatrix} 2 & 3 \\ 1 & 1 \end{bmatrix}$$

(A) $\begin{bmatrix} 2 & 3 \\ 1 & 1 \end{bmatrix}$

(B) $\begin{bmatrix} 3 & 2 \\ 1 & 1 \end{bmatrix}$

(C) $\begin{bmatrix} 1 & -3 \\ -1 & 2 \end{bmatrix}$

(D) $\begin{bmatrix} -1 & 3 \\ 1 & -2 \end{bmatrix}$

21. What is the cube root of the complex number $(8, 60°)$?

(A) $(2)(\cos 60° + i \sin 60°)$
(B) $(2)(i \cos 20° + \sin 20°)$
(C) $(2.7)(\cos 20° + i \sin 20°)$
(D) $(2)(\cos(20° + 120°n) + i \sin(20° + 120°n))$

22. Which of the following statements is false for the identity matrix, $\mathbf{I}$?

(A) $\mathbf{I}$ is a square matrix.
(B) All entries in $\mathbf{I}$ are zero, except the entries whose row and column numbers are equal.
(C) All nonzero entries are equal to 1.
(D) $\mathbf{AI} = \mathbf{A}^{-1}$

23. What is the transpose of matrix $\mathbf{A}$?

$$\mathbf{A} = \begin{bmatrix} 5 & 8 & 5 & 8 \\ 8 & 7 & 6 & 2 \end{bmatrix}$$

(A) $\begin{bmatrix} 8 & 7 & 6 & 2 \\ 5 & 8 & 5 & 8 \end{bmatrix}$

(B) $\begin{bmatrix} 2 & 6 & 7 & 8 \\ 8 & 5 & 8 & 5 \end{bmatrix}$

(C) $\begin{bmatrix} 8 & 5 \\ 7 & 8 \\ 6 & 5 \\ 2 & 8 \end{bmatrix}$

(D) $\begin{bmatrix} 5 & 8 \\ 8 & 7 \\ 5 & 6 \\ 8 & 2 \end{bmatrix}$

24. What is the sum of the following finite sequence of terms?

$$18, 25, 32, 39, \ldots, 67$$

(A) 181
(B) 213
(C) 234
(D) 340

25. What is the sum of the following finite sequence of terms?

$$32, 80, 200, \ldots, 19531.25$$

(A) 21,131.25
(B) 24,718.25
(C) 31,250
(D) 32,530.75

26. Which of the following statements is true for a power series with the general term $a_i x^i$?

I.   An infinite power series converges for $x < 1$.
II. Power series can be added together or subtracted within their interval of convergence.
III. Power series can be integrated within their interval of convergence.

(A) I only
(B) II only
(C) I and III
(D) II and III

27. Expand the function $f(x)$ about $a = 0$ to obtain $f(b)$. What are the first two terms of the Taylor series?

$$f(x) = \frac{1}{3x^3 + 4x + 8}$$

(A) $\dfrac{1}{16} + \dfrac{b}{8}$

(B) $\dfrac{1}{8} - \dfrac{b}{16}$

(C) $\dfrac{1}{8} + \dfrac{b}{16}$

(D) $\dfrac{1}{4} - \dfrac{b}{16}$

28. Which of the following choices is closest to the rationalized form of the complex number

$$\frac{7 + j5.2}{3 + j4}$$

(A) $-0.030 + j1.8$
(B) $1.67 - j0.5$
(C) $2.33 + j1.2$
(D) $2.33 + j1.3$

Problems 29 and 30 refer to the vectors **A** and **B**.

$$\mathbf{A} = 2\mathbf{i} + 4\mathbf{j} + 8\mathbf{k}$$
$$\mathbf{B} = -2\mathbf{i} + \mathbf{j} - 4\mathbf{k}$$

29. What is the dot product, **A·B**, of the vectors?

(A) $-4\mathbf{i} + 4\mathbf{j} - 32\mathbf{k}$
(B) $-4\mathbf{i} - 4\mathbf{j} - 32\mathbf{k}$
(C) -40
(D) -32

30. What is the cross product, **A × B**, of the vectors?

(A) $-24\mathbf{i} - 8\mathbf{j} + 10\mathbf{k}$
(B) $-24\mathbf{i} + 8\mathbf{j} + 10\mathbf{k}$
(C) $-4\mathbf{i} - 4\mathbf{j} - 32\mathbf{k}$
(D) $-4\mathbf{i} + 4\mathbf{j} - 32\mathbf{k}$

31. What is the volume of a parallelepiped with sides represented by the zero-based vectors **A**, **B**, and **C**?

$$\mathbf{A} = 2\mathbf{i} - 2\mathbf{j} + \mathbf{k}$$
$$\mathbf{B} = 4\mathbf{i} + 2\mathbf{j} + 2\mathbf{k}$$
$$\mathbf{C} = \mathbf{i} + 5\mathbf{j} + 4\mathbf{k}$$

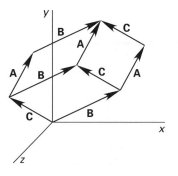

(A) 14
(B) 28
(C) 35
(D) 42

32. The cofactor matrix of matrix **A** is **C**. What is the inverse of matrix **A**?

$$\mathbf{A} = \begin{bmatrix} 4 & 2 & 3 \\ 3 & 2 & 2 \\ 2 & 1 & 4 \end{bmatrix} \qquad \mathbf{C} = \begin{bmatrix} 6 & -8 & -1 \\ -5 & 10 & 0 \\ -2 & 1 & 2 \end{bmatrix}$$

(A) $\begin{bmatrix} 0.25 & 0 & 0 \\ 0 & 0.50 & 0 \\ 0 & 0 & 0.25 \end{bmatrix}$

(B) $\begin{bmatrix} 0.25 & 0.50 & 0.33 \\ 0.33 & 0.50 & 0.50 \\ 0.50 & 1.0 & 0.25 \end{bmatrix}$

(C) $\begin{bmatrix} 1.2 & -1.0 & -0.40 \\ -1.6 & 2.0 & 0.20 \\ -0.20 & 0 & 0.40 \end{bmatrix}$

(D) $\begin{bmatrix} 0.80 & 0.40 & -0.60 \\ 0.20 & -0.40 & 0.40 \\ -0.40 & 0.60 & 0.80 \end{bmatrix}$

SOLUTIONS TO FE-STYLE EXAM PROBLEMS

1.
$$A^{-6/8} = 0.001$$
$$= 10^{-3}$$
$$(A^{-6/8})^{-8/6} = (10^{-3})^{-8/6}$$
$$A = 10^4$$
$$= 10{,}000$$

Answer is D.

2. This is a log of a quotient of two terms.

$$\log\left(\frac{x}{y + z}\right) = \log x - \log(y + z)$$

Answer is B.

3. The entries in the cofactor matrix are the determinants of submatrices resulting from elimination of row i and column j for entry a_{ij}. The entry is multiplied by $+1$ or -1, depending on its position. The coefficient matrix is

$$\begin{pmatrix} 10 & 3 & 10 \\ 8 & -2 & 9 \\ 8 & 1 & -10 \end{pmatrix}$$

The entries in the cofactor matrix are

$$a_{11} = (+1)((-2)(-10) - (9)(1)) = 11$$
$$a_{12} = (-1)((8)(-10) - (9)(8)) = 152$$
$$a_{13} = (+1)((8)(1) - (-2)(8)) = 24$$

$a_{21} = (-1)((3)(-10) - (10)(1)) = 40$

$a_{22} = (+1)((10)(-10) - (10)(8)) = -180$

$a_{23} = (-1)((10)(1) - (3)(8)) = 14$

$a_{31} = (+1)((3)(9) - (10)(-2)) = 47$

$a_{32} = (-1)((10)(9) - (10)(8)) = -10$

$a_{33} = (+1)((10)(-2) - (3)(8)) = -44$

The cofactor matrix is

$$\begin{bmatrix} 11 & 152 & 24 \\ 40 & -180 & 14 \\ 47 & -10 & -44 \end{bmatrix}$$

(Note: Solving this type of problem on the FE Exam would require finding only one or two entries in the cofactor matrix.)

Answer is A.

4. The classical adjoint is the transpose of the cofactor matrix.

$$\mathbf{A}_{\text{adj}} = \begin{bmatrix} 11 & 40 & 47 \\ 152 & -180 & -10 \\ 24 & 14 & -44 \end{bmatrix}$$

Answer is B.

5. The inverse of a 3×3 matrix is found by dividing every entry in the classical adjoint by the determinant of the original matrix.

$$|\mathbf{A}| = (10)(-2)(-10) + (3)(9)(8) + (10)(8)(1)$$
$$- (8)(-2)(10) - (1)(9)(10) - (-10)(8)(3)$$
$$= 806$$

$$\mathbf{A}^{-1} = \frac{1}{806} \begin{bmatrix} 11 & 40 & 47 \\ 152 & -180 & -10 \\ 24 & 14 & -44 \end{bmatrix}$$

$$= \begin{bmatrix} \dfrac{11}{806} & \dfrac{40}{806} & \dfrac{47}{806} \\ \dfrac{152}{806} & \dfrac{-180}{806} & \dfrac{-10}{806} \\ \dfrac{24}{806} & \dfrac{14}{806} & \dfrac{-44}{806} \end{bmatrix}$$

$$= \begin{bmatrix} \dfrac{11}{806} & \dfrac{20}{403} & \dfrac{47}{806} \\ \dfrac{76}{403} & \dfrac{-90}{403} & \dfrac{-5}{403} \\ \dfrac{12}{403} & \dfrac{7}{403} & \dfrac{-22}{403} \end{bmatrix}$$

(Note: Solving this type of problem on the FE Exam would require finding only one or two entries in the inverse matrix.)

Answer is D.

6. There are several ways of solving this problem.

$$\mathbf{AX} = \mathbf{B}$$

$$\begin{bmatrix} 10 & 3 & 10 \\ 8 & -2 & 9 \\ 8 & 1 & -10 \end{bmatrix} \begin{bmatrix} x \\ y \\ z \end{bmatrix} = \begin{bmatrix} 5 \\ 3 \\ 7 \end{bmatrix}$$

$$\mathbf{AA}^{-1}\mathbf{X} = \mathbf{A}^{-1}\mathbf{B}$$

$$\mathbf{IX} = \mathbf{A}^{-1}\mathbf{B}$$

$$\mathbf{X} = \mathbf{A}^{-1}\mathbf{B}$$

$$\mathbf{X} = \begin{bmatrix} \dfrac{11}{806} & \dfrac{20}{403} & \dfrac{47}{806} \\ \dfrac{76}{403} & \dfrac{-90}{403} & \dfrac{-5}{403} \\ \dfrac{12}{403} & \dfrac{7}{403} & \dfrac{-22}{403} \end{bmatrix} \begin{bmatrix} 5 \\ 3 \\ 7 \end{bmatrix}$$

$$= \begin{bmatrix} (5)\left(\dfrac{11}{806}\right) + (3)\left(\dfrac{20}{403}\right) + (7)\left(\dfrac{47}{806}\right) \\ (5)\left(\dfrac{76}{403}\right) + (3)\left(\dfrac{-90}{403}\right) + (7)\left(\dfrac{-5}{403}\right) \\ (5)\left(\dfrac{12}{403}\right) + (3)\left(\dfrac{7}{403}\right) + (7)\left(\dfrac{-22}{403}\right) \end{bmatrix}$$

$$= \begin{bmatrix} 0.625 \\ 0.186 \\ -0.181 \end{bmatrix}$$

(Note: Direct substitution of the four answer choices into the original equations is probably the fastest way of solving this type of problem on the FE exam.)

Answer is C.

7. The resultant is produced by adding the vectors.

$$3\mathbf{i} + 4\mathbf{j} - 5\mathbf{k}$$
$$7\mathbf{i} + 2\mathbf{j} + 3\mathbf{k}$$
$$\underline{-16\mathbf{i} - 14\mathbf{j} + 2\mathbf{k}}$$
$$-6\mathbf{i} - 8\mathbf{j} + 0\mathbf{k}$$

The length of the resultant vector is

$$L = \sqrt{(-6)^2 + (-8)^2 + (0)^2}$$
$$= \sqrt{100} = 10$$

Answer is C.

8. The unit vector of a particular vector is the vector itself divided by its length.

$$\text{unit vector} = \frac{18\mathbf{i} + 3\mathbf{j} + 29\mathbf{k}}{\sqrt{(18)^2 + (3)^2 + (29)^2}}$$
$$= 0.525\mathbf{i} + 0.088\mathbf{j} + 0.846\mathbf{k}$$

Answer is A.

9. The cross product of two vectors is the determinant of a third-order matrix, shown as follows.

$$\mathbf{A} \times \mathbf{B} = \begin{bmatrix} \mathbf{i} & \mathbf{j} & \mathbf{k} \\ a_x & a_y & a_z \\ b_x & b_y & b_z \end{bmatrix}$$
$$= \begin{bmatrix} \mathbf{i} & \mathbf{j} & \mathbf{k} \\ 1 & 4 & 6 \\ 2 & 3 & 5 \end{bmatrix}$$
$$= \mathbf{i}[(4)(5) - (6)(3)] - \mathbf{j}[(1)(5) - (6)(2)]$$
$$+ \mathbf{k}[(1)(3) - (4)(2)]$$
$$= 2\mathbf{i} + 7\mathbf{j} - 5\mathbf{k}$$

Answer is C.

10. Using logarithmic identities,

$$\ln((7.3891)^{xy}) = xy\ln(7.3891) = 2xy$$

Answer is D.

11. Apply the log identities, then use the value given for $\log(4)$.

$$\log\left(\frac{1}{4}\right) = \log(4)^{-1} = -\log(4)$$
$$= -0.703x$$

Answer is B.

12. The radius and angle of the polar form can be determined from the x- and y-coordinates.

$$r = \sqrt{x^2 + y^2} = \sqrt{(4)^2 + (6)^2}$$
$$= 7.211$$
$$\theta = \tan^{-1}\left(\frac{y}{x}\right) = \tan^{-1}\left(\frac{6}{4}\right)$$
$$= 56.3°$$

Answer is D.

13. This point is in the second quadrant.

$$\tan\theta = \frac{y}{x}$$
$$y = x\,\tan 120°$$
$$= -1.7321x$$
$$r^2 = x^2 + y^2$$
$$(4)^2 = x^2 + (-1.7321x)^2$$
$$= 4x^2$$
$$x = \pm 2$$

Since this point is in the second quadrant, $x = -2$.

$$y = (-1.7321)(-2) = 3.46$$

Answer is C.

14. Multiply the elements of each row in matrix $\mathbf{A}$ by the elements of the corresponding column in matrix $\mathbf{B}$.

$$\mathbf{AB} = \begin{bmatrix} 2 \times 4 + 1 \times 2 & 2 \times 3 + 1 \times 1 \\ 1 \times 4 + 0 \times 2 & 1 \times 3 + 0 \times 1 \end{bmatrix}$$
$$= \begin{bmatrix} 10 & 7 \\ 4 & 3 \end{bmatrix}$$

Answer is D.

15. $\mathbf{AB} = [1 \times 2 + 2 \times (-3) + 3 \times 4] = [8]$

Answer is B.

16. The entries in the difference matrix are the differences of the corresponding entries in the original two matrices.

$$\mathbf{A} - \mathbf{B} = \begin{bmatrix} 3 - 5 & 4 - 2 & 1 - (-6) \\ 5 - (-5) & 2 - 4 & -2 - 2 \end{bmatrix}$$
$$= \begin{bmatrix} -2 & 2 & 7 \\ 10 & -2 & -4 \end{bmatrix}$$

Answer is D.

17. Calculate the determinant by expanding the matrix along its first column. The second term drops out because its coefficient is zero.

$$|\mathbf{A}| = (1)(1 \times a + 1 \times 2) + 0 + (3)(-3 \times 2 - 2 \times 1)$$
$$-5 = (a + 2) + (3)(-8)$$
$$= a - 22$$
$$a = 17$$

Answer is D.

18. The first row of matrix **B** is half that of **A**. The determinant of **B** is half the determinant of **A**.

Answer is C.

19. For a square 2×2 matrix,

$$\begin{vmatrix} a & b \\ c & d \end{vmatrix} = ad - bc$$

$$\begin{vmatrix} 3 & 6 \\ 2 & 4 \end{vmatrix} = 3 \times 4 - 6 \times 2 = 0$$

Answer is A.

20. Find the determinant.

$$|\mathbf{A}| = 2 \times 1 - 1 \times 3 = -1$$

The inverse of a 2×2 matrix is

$$\mathbf{A}^{-1} = \frac{\begin{bmatrix} d & -b \\ -c & a \end{bmatrix}}{|\mathbf{A}|}$$

$$= \frac{\begin{bmatrix} 1 & -3 \\ -1 & 2 \end{bmatrix}}{-1}$$

$$= \begin{bmatrix} -1 & 3 \\ 1 & -2 \end{bmatrix}$$

Answer is D.

21. The kth root, w, of a complex number is

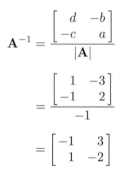

$$w = \sqrt[k]{r} \left(\begin{array}{c} \cos\left(\dfrac{\theta}{k} + n\left(\dfrac{360°}{k}\right)\right) \\ + i\sin\left(\dfrac{\theta}{k} + n\left(\dfrac{360°}{k}\right)\right) \end{array} \right)$$

$$= \sqrt[3]{8} \left(\begin{array}{c} \cos\left(\dfrac{60°}{3} + n\left(\dfrac{360°}{3}\right)\right) \\ + i\sin\left(\dfrac{60°}{3} + n\left(\dfrac{360°}{3}\right)\right) \end{array} \right)$$

$$[n = 0, 1, 2]$$

$$= (2)(\cos(20° + n(120°)) + i\sin(20° + n(120°)))$$

$$[n = 0, 1, 2]$$

Answer is D.

22. The main property of an identity matrix is that $\mathbf{AI} = \mathbf{A}$. Therefore, choice (D) is false.

Answer is D.

23. The transpose of a matrix is constructed by taking the ith row and making it the ith column.

Answer is D.

24. Each term is 7 more than the previous term. This is an arithmetic sequence. The general mathematical representation for an arithmetic sequence is

$$\ell = a + (n - 1)d$$

In this case, the difference term is $d = 7$. The first term is $a = 18$, and the last term is $\ell = 67$.

The sum of n terms is

$$S_n = \frac{n(2a + (n-1)d)}{2}$$

$$= \frac{(8)(2 \times 18 + (8-1)(7))}{2}$$

$$= 340$$

Answer is D.

25. Decimal numbers will not result from the addition of whole numbers. Therefore, multiplication is involved in the general term. Assume that this is a geometric series. The common ratio is

$$r = \frac{80}{32} = \frac{200}{80} = 2.5$$

Since the ratio and both the initial and final terms are known, the sum is

$$S_n = \frac{a - r\ell}{1 - r}$$

$$= \frac{32 - (2.5)(19531.25)}{1 - 2.5}$$

$$= 32,530.75$$

Answer is D.

26. Power series can be added together, subtracted from each other, differentiated, and integrated within their interval of convergence. The interval of convergence is $-1 < x < 1$.

Answer is D.

27.
$$f(0) = ((3)(0)^3 + (4)(0) + 8)^{-1}$$
$$= 1/8$$
$$f'(x) = (-1)(3x^3 + 4x + 8)^{-2}(9x^2 + 4)$$
$$f'(0) = -1/16$$

Use the first two terms of the Taylor's series.

$$f(b) = f(a) + \frac{f'(a)(b-a)}{1!}$$

$$= \frac{1}{8} + \frac{\left(\frac{-1}{16}\right)(b-0)}{1}$$

$$= \frac{1}{8} - \frac{b}{16}$$

Answer is B.

28. Multiply the numerator and denominator by the complex conjugate of the denominator. The denominator becomes a real number.

$$\frac{(7+j5.2)(3-j4)}{(3+j4)(3-j4)}$$

$$= \frac{(7)(3)+(7)(-j4)+(j5.2)(3)+(j5.2)(-j4)}{(3)(3)+(3)(-j4)+(j4)(3)+(j4)(-j4)}$$

$$= \frac{21-j28+j15.6+20.8}{9-j12+j12+16}$$

$$= \frac{41.8-j12.4}{25}$$

$$= 1.672 - j0.496 \quad (1.67 - j0.5)$$

Answer is B.

29. Multiply the corresponding components of the vectors and add the resulting products.

$$\mathbf{A}\cdot\mathbf{B} = a_x b_x + a_y b_y + a_z b_z$$
$$= (2)(-2)+(4)(1)+(8)(-4)$$
$$= -32$$

Answer is D.

30. Find the determinant of the 3×3 matrix whose rows are the unit vector and the coefficients of vectors A and B.

$$\mathbf{A}\times\mathbf{B} = \begin{vmatrix} \mathbf{i} & \mathbf{j} & \mathbf{k} \\ a_x & a_y & a_z \\ b_x & b_y & b_z \end{vmatrix}$$

$$= \mathbf{i}(a_y b_z - b_y a_z) - \mathbf{j}(a_x b_z - b_x a_z)$$
$$\quad + \mathbf{k}(a_x b_y - b_x a_y)$$

$$= \mathbf{i}((4)(-4)-(1)(8)) - \mathbf{j}((2)(-4)-(-2)(8))$$
$$\quad + \mathbf{k}((2)(1)-(-2)(4))$$

$$= -24\mathbf{i} - 8\mathbf{j} + 10\mathbf{k}$$

Answer is A.

31. The volume of the parallelepiped is given by the vector multiplication of $\mathbf{A}\cdot(\mathbf{B}\times\mathbf{C})$. While the cross product $\mathbf{B}\times\mathbf{C}$ can be found and then dotted with $\mathbf{A}$, it is more expedient to use the mixed triple product equation.

$$\mathbf{V}_1\cdot(\mathbf{V}_2\times\mathbf{V}_3) = \begin{vmatrix} V_{1x} & V_{1y} & V_{1z} \\ V_{2x} & V_{2y} & V_{2z} \\ V_{3x} & V_{3y} & V_{3z} \end{vmatrix}$$

$$= \begin{vmatrix} 2 & -2 & 1 \\ 4 & 2 & 2 \\ 1 & 5 & 4 \end{vmatrix}$$

Expand this determinant by the first column.

$$(2)((2)(4)-(5)(2)) - (4)((-2)(4)-(5)(1))$$
$$+ (1)((-2)(2)-(2)(1))$$

$$= (2)(-2)-(4)(-13)+(1)(-6)$$

$$= 42$$

Answer is D.

32. The classical adjoint is the transpose of the cofactor matrix.

$$\mathrm{adj}(\mathbf{A}) = \mathbf{C}^T = \begin{bmatrix} 6 & -5 & -2 \\ -8 & 10 & 1 \\ -1 & 0 & 2 \end{bmatrix}$$

Calculate the determinant of $\mathbf{A}$ by expanding along the top row.

$$|\mathbf{A}| = (4)(8-2)-(2)(12-4)+(3)(3-4)$$
$$= 24 - 16 - 3$$
$$= 5$$

Divide the classical adjoint by the determinant.

$$\mathbf{A}^{-1} = \frac{\mathrm{adj}(\mathbf{A})}{|\mathbf{A}|}$$

$$= \frac{\begin{bmatrix} 6 & -5 & -2 \\ -8 & 10 & 1 \\ -1 & 0 & 2 \end{bmatrix}}{5}$$

$$= \begin{bmatrix} 1.2 & -1.0 & -0.40 \\ -1.6 & 2.0 & 0.20 \\ -0.20 & 0 & 0.40 \end{bmatrix}$$

Answer is C.

6 Probability and Statistics

Subjects

COMBINATIONS AND PERMUTATIONS . 6-1
LAWS OF PROBABILITY 6-1
   General Character of Probability 6-1
   Law of Total Probability 6-1
   Law of Compound or Joint Probability . 6-1
NUMERICAL EVENTS 6-2
MEASURES OF CENTRAL TENDENCY . . 6-2
MEASURES OF DISPERSION 6-2
PROBABILITY DENSITY FUNCTIONS . . 6-3
PROBABILITY DISTRIBUTION
   FUNCTIONS 6-3
   Binomial Distribution 6-3
   Normal Distribution 6-3
   t-Distribution 6-4
CONFIDENCE LEVELS 6-4
CONFIDENCE INTERVALS 6-4
   Sampling Distribution Known 6-4
   Sampling Distribution Not Known . . . 6-5
HYPOTHESIS TESTING 6-5
SUM AND DIFFERENCE OF MEANS . . . 6-5

COMBINATIONS AND PERMUTATIONS

There are a finite number of ways in which n elements can be combined into distinctly different groups of r items. For example, suppose a farmer has a chicken, a rooster, a duck, and a cage that holds only two birds. The possible *combinations* of three birds taken two at a time are (chicken, rooster), (chicken, duck), and (rooster, duck). The birds in the cage will not remain stationary, so the combination (rooster, chicken) is not distinctly different from (chicken, rooster). That is, combinations are not *order conscious*.

The number of combinations of n items taken r at a time is written $C(n, r)$, C_r^n, $_nC_r$, or $\binom{n}{r}$ (pronounced "n choose r"). It is sometimes referred to as the *binomial coefficient* and is given by Eq. 6.1.

$$\binom{n}{r} = C(n, r) = \frac{n!}{r!(n-r)!} \quad [r \leq n] \qquad 6.1$$

An order-conscious subset of r items taken from a set of n items is the *permutation* $P(n, r)$, also written P_r^n and $_nP_r$. The permutation is order conscious because the arrangement of two items (e.g., a_i and b_i) as $a_i b_i$ is different from the arrangement $b_i a_i$. The number of permutations is

$$P(n, r) = \frac{n!}{(n-r)!} \quad [r \leq n] \qquad 6.2$$

LAWS OF PROBABILITY

Probability theory determines the relative likelihood that a particular event will occur. An *event*, E, is one of the possible outcomes of a *trial*. The *probability* of E occurring is denoted as $P(E)$.

Property 1 — General Character of Probability

Probabilities are real numbers in the range of zero to one. If an event A is certain to occur, then the probability $P(A)$ of the event is equal to one. If the event is certain *not* to occur, then the probability $P(A)$ of the event is equal to zero. The probability of any other event is between zero and one.

The probability of an event occurring is equal to one minus the probability of the event not occurring. This is known as a *complementary probability*.

$$P(E) = 1 - P(\text{not } E) \qquad 6.3$$

Property 2 — Law of Total Probability

Equation 6.4 gives the probability that either event A or B, or both, will occur. $P(A, B)$ is the probability that both A and B occur.

$$P(A + B) = P(A) + P(B) - P(A, B) \qquad 6.4$$

Property 3 — Law of Compound or Joint Probability

Equation 6.5 gives the probability that both events A and B will occur. $P(B/A)$ is the *conditional probability* that B occurs given that A has already occurred. Likewise, $P(A/B)$ is the conditional probability that A occurs given that B has already occurred. If the events are independent, then $P(B/A) = P(B)$ and $P(A/B) = P(A)$. Examples of dependent events for which the probability is conditional include drawing objects from a container or cards from a deck, without replacement.

$$P(A, B) = P(A)P(B/A)$$
$$= P(B)P(A/B) \qquad 6.5$$

NUMERICAL EVENTS

A *discrete numerical event* is an occurrence that can be described by an integer. For example, 27 cars passing through a bridge toll booth in an hour is a discrete numerical event. Most numerical events are *continuously distributed* and are not constrained to discrete or integer values. For example, the resistance of a 10 percent $1\ \Omega$ resistor may be any value between 0.9 and $1.1\ \Omega$.

MEASURES OF CENTRAL TENDENCY

It is often unnecessary to present experimental data in their entirety, either in tabular or graphic form. In such cases, the data and distribution can be represented by various parameters. One type of parameter is a measure of *central tendency*. The mode, median, and mean are measures of central tendency.

The *mode* is the observed value that occurs most frequently. The mode may vary greatly between series of observations. Therefore, its main use is as a quick measure of the central value since little or no computation is required to find it. Beyond this, the usefulness of the mode is limited.

The *median* is the point in the distribution that partitions the total set of observations into two parts containing equal numbers of observations. It is not influenced by the extremity of scores on either side of the distribution. The median is found by counting from either end through an ordered set of data until half of the observations have been accounted for. If the number of data points is odd, the median will be the exact middle value. If the number of data points is even, the median will be the average of the middle two values.

The *arithmetic mean* is the arithmetic average of the observations. The *sample mean*, $\overline{X}$, can be used as an unbiased estimator of the *population mean*, μ. The term *unbiased estimator* means that on the average, the sample mean is equal to the population mean (i.e., the mean of many determinations of the sample mean is equal to the population mean). The mean may be found without ordering the data (as was necessary to find the mode and median) from the following formula.

$$\overline{X} = \left(\frac{1}{n}\right)(X_1 + X_2 + \cdots + X_n)$$
$$= \frac{1}{n}\sum_{i=1}^{n} X_i \qquad 6.6$$

If some observations are considered to be more significant than others, a *weighted mean* can be calculated. Equation 6.7 defines a *weighted arithmetic average*, where w_i is the weight assigned to observation X_i.

$$\overline{X}_w = \frac{\displaystyle\sum_{i=1}^{n} w_i X_i}{\displaystyle\sum_{i=1}^{n} w_i} \qquad 6.7$$

The *geometric mean* is used occasionally when it is necessary to average ratios. The geometric mean is calculated as

$$\text{geometric mean} = \sqrt[n]{X_1 X_2 X_3 \cdots X_n} \quad [X_i > 0] \quad 6.8$$

The *root-mean-square* (rms) value of a series of observations is defined as

$$X_{\text{rms}} = \sqrt{\frac{\displaystyle\sum_{i=1}^{n} X_i^2}{n}} \qquad 6.9$$

MEASURES OF DISPERSION

Measures of dispersion describe the variability in observed data. One measure of dispersion is the *standard deviation*, defined in Eq. 6.10. N is the total population size, not the sample size, n.

$$\sigma = \sqrt{\frac{\displaystyle\sum_{i=1}^{N} (X_i - \mu)^2}{N}} \qquad 6.10$$

The *standard deviation of a sample* (particularly a small sample) of n items is a *biased estimator* of (i.e., on the average it is not equal to) the population standard deviation. A different measure of dispersion called the *sample standard deviation*, s (not the same as the standard deviation of a sample), is an unbiased estimator of the population standard deviation.

$$s = \sqrt{\frac{\displaystyle\sum_{i=1}^{n} (X_i - \overline{X})^2}{n - 1}} \qquad 6.11$$

The *variance* is the square of the standard deviation. Since there are two standard deviations, there are two variances. The *variance of the population* (i.e., the *population variance*) is σ^2, and the *sample variance* is s^2.

From Eqs. 6.10 and 6.11,

$$\sigma^2 = \frac{1}{N}((X_1 - \mu)^2 + (X_2 - \mu)^2 + \cdots + (X_N - \mu)^2)$$

$$= \frac{1}{N}\sum_{i=1}^{N}(X_i - \mu)^2 \qquad 6.12$$

$$s^2 = \frac{1}{n-1}\sum_{i=1}^{n}(X_i - \overline{X})^2 \qquad 6.13$$

The *relative dispersion* is defined as a measure of dispersion divided by a measure of central tendency. The *coefficient of variation*, CV, is a relative dispersion calculated from the sample standard deviation and the mean.

$$\text{CV} = \frac{s}{\overline{X}} \qquad 6.14$$

PROBABILITY DENSITY FUNCTIONS

If a random variable X is continuous over an interval, then a non-negative *probability density function* of that variable exists over the interval as defined by Eq. 6.15.

$$P(x_1 \le X \le x_2) = \int_{x_1}^{x_2} f(x)\,dx \qquad 6.15$$

Various mathematical models are used to describe probability density functions. The area under the probability density function is the probability that the variable will assume a value between the limits of evaluation. The total probability, or the probability that the variable will assume any value over the interval, is one. The probability of an exact numerical event is zero. That is, there is no chance that a numerical event will be exactly x. (Since the variable can take on any value and has an infinite number of significant digits, one can infinitely continue to increase the precision of the value.) It is possible to determine only the probability that a numerical event will be less than x, greater than x, or between the values of x_1 and x_2.

PROBABILITY DISTRIBUTION FUNCTIONS

A *probability distribution function* gives the cumulative probability that a numerical event will occur or the probability that the numerical event will be less than or equal to event X.

For a *discrete random variable*, the probability distribution function is the sum of the individual probabilities of all possible events up to the limits of evaluation.

$$F(X_n) = \sum_{k=1}^{n} P(X_k) = P(X \le X_n) \qquad 6.16$$

For a continuous random variable, the probability distribution function is the integral of the probability density function over an interval from negative infinity to the limit of evaluation.

$$F(X_1) = \int_{-\infty}^{X_1} f(x)\,dx = P(x \le X_1) \qquad 6.17$$

The *expected value* of a function $g(x)$ of a discrete random variable x is given by Eq. 6.18.

$$E\{g(x)\} = \sum_{i=1}^{N} g(x_i)P(x_i) \qquad 6.18$$

For a continuous random variable x, the expected value of $g(x)$ is given by Eq. 6.19, where $f(x)$ is the probability density function.

$$E\{g(x)\} = \int_{-\infty}^{+\infty} g(x)f(x)\,dx \qquad 6.19$$

Binomial Distribution

The *binomial probability function* is used when all outcomes can be categorized as either successes or failures. The probability of success in a single trial is designated as p, and the probability of failure is the complement, $q = 1 - p$. The population is assumed to be infinite in size so that sampling does not change the values of p and q. The binomial distribution can also be used with finite populations when sampling is with replacement.

Equation 6.20 gives the probability of x successes in n independent successive trials. The quantity $C(n, x)$ is the *binomial coefficient*, identical to the number of combinations of n items taken x at a time (Eq. 6.1).

$$F(x) = C(n, x)\,p^x q^{n-x} = \frac{n!}{x!(n-x)!}p^x q^{n-x} \qquad 6.20$$

Normal Distribution

The *normal distribution (Gaussian distribution)* is a symmetrical continuous distribution commonly referred to as the *bell-shaped curve*, which describes the distribution of outcomes of many real-world experiments, processes, and phenomena. The probability density function for the normal distribution with population mean μ and population variance σ^2 is illustrated in Fig. 6.1 and is represented by Eq. 6.21.

$$f(x) = \frac{1}{\sigma\sqrt{2\pi}}e^{-(x-\mu)^2/2\sigma^2} \qquad 6.21$$

Figure 6.1 Normal Distribution

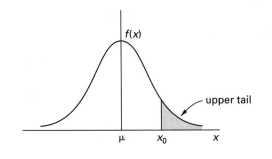

Since $f(x)$ is difficult to integrate (i.e., Eq. 6.17 is difficult to evaluate), Eq. 6.21 is seldom used directly and a *unit normal table* (see Table 6.2 at the end of this chapter) is used instead. The unit normal table (also called the *standard normal table*) is based on a normal distribution with a mean of zero and a standard deviation of one. The standard normal distribution is given by Eq. 6.22.

$$f(x) = \frac{1}{\sqrt{2\pi}} e^{-x^2/2} \qquad 6.22$$

Since the range of values from an experiment or phenomenon will not generally correspond to the standard normal table, a value, X_0, must be converted to a *standard normal value*, x. In Eq. 6.23, μ and σ are the population mean and standard deviation, respectively, of the distribution from which X_0 comes.

$$x = \frac{X_0 - \mu}{\sigma} \qquad 6.23$$

To use Eq. 6.23, the population parameters μ and σ must be known. The unbiased estimators for μ and σ are $\overline{X}$ and s, respectively, when a sample is used to estimate the population parameters. Both $\overline{X}$ and s approach the population values as the sample size (n) increases.

t-Distribution

The *t-test* is a method of comparing two variables, usually to test the significance of the difference between samples. For example, the *t*-test can be used to test whether the populations from which two samples are drawn have the same means. In Eq. 6.24, x is a unit normal variable and r is the root-mean-squared value of n other random variables (i.e., the sample size is n).

$$t = \frac{x}{r} \qquad 6.24$$

The probability distribution function for the *t-distribution* (commonly referred to as *Student's t-distribution*) with n *degrees of freedom* is

$$F(t) = \left(\frac{\Gamma\left(\frac{n+1}{2}\right)}{\Gamma\left(\frac{n}{2}\right)\sqrt{n\pi}} \right) \left(\frac{1}{\left(1 + \frac{t^2}{n}\right)^{(n+1)/2}} \right) \qquad 6.25$$

$$\Gamma(n) = \int_0^\infty t^{n-1} e^{-t} \, dt \qquad 6.26$$

The *t*-distribution is tabulated in Table 6.3, with t as a function of n and α. Note that since the *t*-distribution is symmetric about zero, $t_{1-\alpha,n} = -t_{\alpha,n}$. The *t*-distribution approaches the normal distribution as n increases.

$$\alpha = \int_{t_{\alpha,n}}^\infty f(t) \, dt \qquad 6.27$$

CONFIDENCE LEVELS

The results of experiments are seldom correct 100 percent of the time. Recognizing this, researchers accept a certain probability of being wrong. In order to minimize this probability, an experiment is repeated several times. The number of repetitions required depends on the level of confidence wanted in the results. For example, if the results have a 5 percent probability of being wrong, the *confidence level*, C, is 95 percent that the results are correct.

CONFIDENCE INTERVALS

Sampling Distribution Known

The properties of the underlying distribution (when known) can be used to calculate *confidence intervals* (*confidence limits*). With a particular confidence level, the true value will be found within these two limits.

As a consequence of the *central limit theorem*, means of samples of n items taken from a distribution that is normally distributed with mean μ and standard deviation σ will be normally distributed with mean μ and variance σ^2/n. Thus, the probability that any given average, $\overline{X}$, exceeds some value, L, is

$$p\{\overline{X} > L\} = p\left\{ x > \left| \frac{L - \mu}{\frac{\sigma}{\sqrt{n}}} \right| \right\} \qquad 6.28$$

L is the *confidence limit* for the confidence level $1 - p\{\overline{X} > L\}$ (expressed as a percent). Values of x are read directly from the standard normal table. As an example, $x = 1.645$ for a 95 percent confidence level since only 5 percent of the curve is above that x in the upper tail. Similar values are given in Table 6.1. This is known as a *one-tail confidence limit* because all of the probability is given to one side of the variation.

Table 6.1 Values of x for Various Confidence Levels

confidence level, C	one-tail limit x	two-tail limit x
90%	1.28	1.645
95%	1.645	1.96
97.5%	1.96	2.17
99%	2.33	2.575
99.5%	2.575	2.81
99.75%	2.81	3.00

With *two-tail confidence limits*, the probability is split between the two sides of variation. There will be upper and lower confidence limits: UCL and LCL, respectively. This is appropriate when it is not specifically known that the calculated parameter is too high or too low.

$$p\{\text{LCL} < \overline{X} < \text{UCL}\}$$

$$= p\left\{ \frac{\text{LCL} - \mu}{\frac{\sigma}{\sqrt{n}}} < x < \frac{\text{UCL} - \mu}{\frac{\sigma}{\sqrt{n}}} \right\} \qquad 6.29$$

Sampling Distribution Not Known

If the standard deviation, σ, of the underlying distribution is not known, it must be estimated from the sample standard deviation, s. Accordingly, the standard normal variable is replaced by the t-distribution parameter with $n - 1$ degrees of freedom, where n is the sample size. In that case, the upper and lower confidence limits are

$$\text{LCL} = \overline{x} - t_{C/2, n-1}\left(\frac{s}{\sqrt{n}}\right) \qquad 6.30$$

$$\text{UCL} = \overline{x} + t_{C/2, n-1}\left(\frac{s}{\sqrt{n}}\right) \qquad 6.31$$

HYPOTHESIS TESTING

A *hypothesis test* is a procedure that answers the question "Did these data come from [a particular type of] distribution?" There are many types of tests, depending on the distribution and parameter being evaluated. The most simple hypothesis test determines whether an average value obtained from n repetitions of an experiment could have come from a population with known mean μ and standard deviation σ. A practical application of this question is whether a manufacturing process has changed from what it used to be or should be. Of course, the answer (i.e., yes or no) cannot be given with absolute certainty—there will be a confidence level associated with the answer.

The following procedure is used to determine whether the average of n measurements can be assumed (with a given confidence level) to have come from a known population.

step 1: Assume random sampling from a normal population.

step 2: Choose the desired confidence level, C.

step 3: Decide on a one-tail or two-tail test. If the hypothesis being tested is that the average has or has not *increased* or *decreased*, use a one-tail test. If the hypothesis being tested is that the average has or has not *changed*, use a two-tail test.

step 4: Use Table 6.1 or the standard normal table to determine the x-value corresponding to the confidence level and number of tails.

step 5: Calculate the actual standard normal variable, x'.

$$x' = \frac{\overline{X} - \mu}{\frac{\sigma}{\sqrt{n}}} \qquad 6.32$$

step 6: If $x' \geq x$, the average can be assumed (with confidence level C) to have come from a different distribution.

SUM AND DIFFERENCE OF MEANS

When two variables are sampled from two different standard normal variables, their sums will be distributed with mean $\mu_{\text{new}} = \mu_1 + \mu_2$ and variance $\sigma^2_{\text{new}} = \sigma^2_1/n_1 + \sigma^2_2/n_2$. The sample sizes, n_1 and n_2, do not have to be the same. The relationships for confidence intervals and hypothesis testing can be used for a new variable $x_{\text{new}} = x_1 + x_2$ if μ is replaced by μ_{new} and σ is replaced by σ_{new}. (Notice that the standard deviation is the square root of the variance.)

The difference in two standard normal variables will be distributed with mean $\mu_{\text{new}} = \mu_1 - \mu_2$ and variance $\sigma^2_{\text{new}} = \sigma^2_1/n_1 + \sigma^2_2/n_2$. Thus, the mean is the difference in two population means, but the variance is the sum, as it was for the sum of two standard normal variables.

SAMPLE PROBLEMS

Problem 1

What is the probability that either two heads or three heads will be thrown if six fair coins are tossed at once?

    (A) 0.35
    (B) 0.55
    (C) 0.59
    (D) 0.63

Solution

$P(2 \text{ heads}) = $ the probability that 2 heads are thrown

$$= \frac{\text{total number of ways 2 heads can occur}}{\text{total number of possible outcomes}}$$

$$= \frac{C(6,2)}{(2)^6}$$

$$= \frac{\frac{6!}{2!(6-2)!}}{64}$$

$$= \frac{15}{64}$$

$$P(3 \text{ heads}) = \frac{C(6,3)}{(2)^6}$$

$$= \frac{\frac{6!}{3!(6-3)!}}{64}$$

$$= \frac{20}{64}$$

These two outcomes are mutually exclusive (i.e., both cannot occur). From the law of total probability, the probability that either of these outcomes will occur is the sum of the individual probabilities.

$$P(2 \text{ heads or 3 heads}) = P(2 \text{ heads}) + P(3 \text{ heads})$$

$$= \frac{15}{64} + \frac{20}{64}$$

$$= \frac{35}{64} = 0.547$$

Answer is B.

Problem 2

Three standard 52-card decks are used in a probability experiment. One card is drawn from each deck. What is the probability that a diamond is drawn from the first deck, an ace from the second, and the ace of hearts from the third?

(A) 0.000062
(B) 0.00015
(C) 0.00037
(D) 0.0062

Solution

The trials are independent of each other because different decks are used. (The same probability would be calculated if one deck was used with replacement of the card after each trial.) From the law of joint probability, the probability that all three events will occur is the product of the three individual probabilities.

$$\begin{aligned}\frac{P(\text{diamond and ace}}{\text{and ace of hearts})} &= \frac{P(\text{diamond}) \times P(\text{ace})}{\times P(\text{ace of hearts})} \\ &= \left(\frac{13}{52}\right)\left(\frac{4}{52}\right)\left(\frac{1}{52}\right) \\ &= 0.00037 \end{aligned}$$

Answer is C.

Problem 3

What are the mean and sample standard deviation of the following numbers?

$$\begin{array}{c} 71.3 \\ 74.0 \\ 74.25 \\ 78.54 \\ 80.6 \end{array}$$

(A) 74.3, 2.7
(B) 74.3, 3.7
(C) 75.0, 2.7
(D) 75.7, 3.8

Solution

The arithmetic mean is

$$\overline{X} = \frac{1}{n}\sum_{i=1}^{n} X_i$$

$$= \frac{1}{5}(71.3 + 74.0 + 74.25 + 78.54 + 80.6)$$

$$= 75.738$$

$$s = \sqrt{\frac{1}{n-1}\sum_{i=1}^{n}(X_i - \overline{X})^2}$$

$$= \sqrt{\frac{1}{5-1}\left(\begin{array}{c}(71.3 - 75.738)^2 + (74.0 - 75.738)^2 \\ + (74.25 - 75.738)^2 \\ + (78.54 - 75.738)^2 \\ + (80.6 - 75.738)^2\end{array}\right)}$$

$$= 3.756$$

Answer is D.

Problem 4

The water content of soil from a borrow site is normally distributed with a mean of 14.2% and a standard deviation of 2.3%. What is the probability that a sample taken from the site will have a water content above 16% or below 12%?

(A) 0.13
(B) 0.25
(C) 0.37
(D) 0.42

Solution

Find the standard normal values (*x*-values) for the two points of interest.

$$x_{16\%} = \frac{16\% - 14.2\%}{2.3\%} = 0.78 \quad [\text{use } 0.80]$$

$$x_{12\%} = \frac{12\% - 14.2\%}{2.3\%} = -0.96 \quad [\text{use } -1.00]$$

From the normal distribution table, Table 6.2, $R(0.80) = 0.2119$ and $R(1.00) = 0.1587$. The probability that the sample will fall outside these values is the sum of the two values.

$$P(x < 12\% \text{ or } x > 16\%) = 0.2119 + 0.1587$$
$$= 0.3706$$

Answer is C.

FE-STYLE EXAM PROBLEMS

1. Four fair coins are tossed at once. What is the probability of obtaining three heads and one tail?

(A) 1/4
(B) 3/8
(C) 1/2
(D) 3/4

2. Two students are working independently on a problem. Their respective probabilities of solving the problem are 1/3 and 3/4. What is the probability that at least one of them will solve the problem?

(A) 1/2
(B) 5/8
(C) 2/3
(D) 5/6

3. What is the sample variance of the following numbers?

$$2, 4, 6, 8, 10, 12, 14$$

(A) 4.32
(B) 5.29
(C) 8.00
(D) 18.7

4. What is the probability of picking an orange ball and a white ball out of a bag containing seven orange balls, eight green balls, and two white balls?

(A) 0.071
(B) 0.10
(C) 0.36
(D) 0.53

5. A cat has a litter of seven kittens. If the probability is 0.52 that a kitten will be female, what is the probability that exactly two of the seven will be male?

(A) 0.07
(B) 0.18
(C) 0.23
(D) 0.29

For the following problems use the NCEES Handbook as your only reference.

6. What is the approximate probability that no two people in a group of seven have the same birthday?

(A) 0.056
(B) 0.43
(C) 0.92
(D) 0.94

7. What is the approximate probability of exactly two people in a group of seven having a birthday on April 15?

(A) 1.2×10^{-18}
(B) 2.4×10^{-17}
(C) 7.4×10^{-6}
(D) 1.6×10^{-4}

8. What is the approximate probability of exactly two people in a group of seven having the same birthday?

(A) 0.00016
(B) 0.026
(C) 0.055
(D) 0.12

9. A bag contains 100 balls numbered 1 to 100. One ball is drawn from the bag. What is the probability that the number on the ball selected will be odd or greater than 80?

(A) 0.1
(B) 0.5
(C) 0.6
(D) 0.7

Problems 10–12 refer to the following situation.

100 random samples were taken from a large population. A particular numerical characteristic of sampled items was measured. The results of the measurements were as follows.

- 45 measurements were between 0.859 and 0.900.
- 0.901 was observed once.

- 0.902 was observed three times.
- 0.903 was observed twice.
- 0.904 was observed four times.
- 45 measurements were between 0.905 and 0.958.

The smallest value was 0.859, and the largest value was 0.958. The sum of all 100 measurements was 91.170. Except those noted, no measurements occurred more than twice.

10. What is the mean of the measurements?

(A) 0.9029
(B) 0.9050
(C) 0.9055
(D) 0.9117

11. What is the median of the measurements?

(A) 0.901
(B) 0.902
(C) 0.9025
(D) 0.903

12. What is the mode of the measurements?

(A) 0.902
(B) 0.903
(C) 0.904
(D) 0.909

Problems 13–15 refer to the following situation.

Samples of aluminum-alloy channels were tested for stiffness. The following frequency distribution was obtained. The distribution is assumed to be normal.

stiffness	frequency
2480	23
2440	35
2400	40
2360	33
2320	21

13. What is the approximate mean of the population from which the samples were taken?

(A) 2367
(B) 2398
(C) 2402
(D) 2419

14. What is the approximate standard deviation of the population from which the samples were taken?

(A) 19
(B) 23
(C) 37
(D) 51

15. What is the approximate probability that stiffness would be less than 2350 for any given channel section?

(A) 0.08
(B) 0.16
(C) 0.23
(D) 0.36

16. For the probability density function shown, what is the probability of the random variable x being less than 1/3?

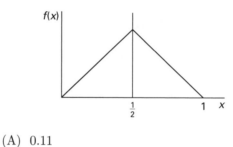

(A) 0.11
(B) 0.22
(C) 0.25
(D) 0.33

17. If seven fair coins are simultaneously tossed in the air, what is the probability that at least one will land heads up?

(A) 0.144
(B) 0.286
(C) 0.971
(D) 0.992

18. The required probability that a two-stage rocket will perform successfully is 0.97. The reliability for the first stage is 0.99. The reliabilities of the two stages are independent. What must be the reliability for the second stage?

(A) 0.90
(B) 0.95
(C) 0.97
(D) 0.98

19. What is the sample variance of the following data?

0.50, 0.80, 0.75, 0.52, 0.60

(A) 0.0146
(B) 0.0183
(C) 0.1128
(D) 0.1209

20. What is the t-test (based on Student's t-distribution) useful for?

(A) testing the distribution of outcomes to see if they come from a normal distribution
(B) determining if the function is symmetric about zero
(C) comparing values about the mean
(D) determining if the differences between samples means is significant

21. When it is operating properly, a chemical plant has a daily production rate that is normally distributed with a mean of 880 tons/day and a standard deviation of 21 tons/day. During an analysis period, the output is measured with random sampling on 50 consecutive days, and the mean output is found to be 871 tons/day. With a 95 percent confidence level, determine if the plant is operating properly.

(A) There is at least a 5 percent probablity that the plant is operating properly.
(B) There is at least a 95 percent probablity that the plant is operating properly.
(C) There is at least a 5 percent probablity that the plant is not operating properly.
(D) There is at least a 95 percent probablity that the plant is not operating properly.

SOLUTIONS TO FE-STYLE EXAM PROBLEMS

1. The probability of obtaining three heads and one tail is equivalent to the probability that exactly three heads (or exactly one tail) are thrown.

$$P(3 \text{ heads}) = \frac{C(4,3)}{(2)^4} = \frac{\dfrac{4!}{3!(4-3)!}}{16}$$

$$= \frac{4}{16}$$

$$= \frac{1}{4}$$

Alternatively, the binomial probability function can be used to determine the probability of three heads in four trials.

$$p = P(\text{heads}) = 0.5$$
$$q = P(\text{not heads}) = 0.5$$
$$n = \text{number of trials} = 4$$
$$x = \text{number of successes} = 3$$

$$F(x) = C(n,x)p^x q^{n-x}$$

$$= \left(\frac{n!}{x!(n-x)!} \right) p^x q^{n-x}$$

$$= \left(\frac{4!}{3!(4-3)!} \right)(0.5)^3(0.5)$$

$$= 0.25 \quad (1/4)$$

Answer is A.

2. The probability that either or both of the students solve the problem is given by the laws of total and joint probability.

$$P(A) = \frac{1}{3}$$

$$P(B) = \frac{3}{4}$$

$$P(A+B) = P(A) + P(B) - P(A,B)$$

$$= \frac{1}{3} + \frac{3}{4} - \left(\frac{1}{3}\right)\left(\frac{3}{4}\right)$$

$$= \frac{10}{12} = \frac{5}{6}$$

Answer is D.

3. Find the mean.

$$\sum_{i=1}^{n} X_i = 2 + 4 + 6 + 8 + 10 + 12 + 14 = 56$$

$$n = 7$$

$$\overline{X} = \frac{1}{n}\sum_{i=1}^{n} X_i = \left(\frac{1}{7}\right)(56)$$

$$= 8$$

The sample variance is the square of the sample standard deviation.

$$s^2 = \left(\frac{1}{n-1}\right)\sum_{i=1}^{n}(X_i - \overline{X})^2$$

$$= \left(\frac{1}{7-1}\right)$$

$$\times \left(\begin{array}{c} (2-8)^2 + (4-8)^2 + (6-8)^2 + (8-8)^2 \\ + (10-8)^2 + (12-8)^2 + (14-8)^2 \end{array} \right)$$

$$= 18.67 \quad (18.7)$$

Answer is D.

4. The possible successful outcomes are that either a white ball is picked and then an orange ball, or that an orange ball is picked and then a white ball.

$$P(\text{orange then white or white then orange}) = \left(\frac{7}{17}\right)\left(\frac{2}{16}\right) + \left(\frac{2}{17}\right)\left(\frac{7}{16}\right)$$
$$= 0.05147 + 0.05147$$
$$= 0.1029$$

Answer is B.

5. The trials follow a binomial distribution. The probability of a male kitten is defined as a success.

$$p = 1 - 0.52 = 0.48 = P(\text{male kitten})$$
$$q = 0.52 = P(\text{female kitten})$$
$$n = 7 \text{ trials}$$
$$x = 2 \text{ success}$$
$$P(x) = C(n, x)p^x q^{n-x}$$
$$= \left(\frac{n!}{x!(n-x)!}\right) p^x q^{n-x}$$
$$= \left(\frac{7!}{2!(7-2)!}\right)(0.48)^2(0.52)^{(7-2)}$$
$$= 0.184 \quad (0.18)$$

Answer is B.

6. This is the classic "birthday problem." The problem is to find the probability that all seven people have distinctly different birthdays. The solution can be found from simple counting.

The first person considered can be born on any day. The probability that the first person will be born on one of the 365 days in the year is

$$P(1) = \frac{365}{365}$$

The second person cannot have been born on the same day as the first person. However, the second person can be born on any other of the 364 days. The probability that the second person is born on any other day is

$$P(2) = \frac{365-1}{365} = \frac{364}{365}$$

The third person cannot have been born on either of the same days as the first and second people. The probability that the third person is born on any other day is

$$P(3) = \frac{365-2}{365} = \frac{363}{365}$$

This logic continues to the seventh person. The probability that all seven conditions are simultaneously satisfied is

$P(7 \text{ distinct birthdays})$
$$= P(1) \times P(2) \times P(3) \times P(4) \times P(5)$$
$$\times P(6) \times P(7)$$
$$= \left(\frac{365}{365}\right)\left(\frac{364}{365}\right)\left(\frac{363}{365}\right)\left(\frac{362}{365}\right)\left(\frac{361}{365}\right)$$
$$\times \left(\frac{360}{365}\right)\left(\frac{359}{365}\right)$$
$$= 0.9438 \quad (0.94)$$

Answer is D.

7. This can be considered a sampling problem: a sample of seven is taken from a large population. Thus, the sample size, n, is 7. The probability that a person will have been born on April 15 is $1/365$. Therefore, the probability of "success," p, is $1/365$, and the probability of "failure," $q = 1 - p$, is $364/365$.

The binomial probability function can be used to calculate the probability that two of the seven samples will have been born on April 15. Thus, $x = 2$.

$$F(x) = C(n, x)p^x q^{n-x}$$
$$= \left(\frac{n!}{x!(n-x)!}\right) p^x q^{n-x}$$
$$F(2) = \left(\frac{7!}{2!(7-2)!}\right)\left(\frac{1}{365}\right)^2\left(\frac{364}{365}\right)^{7-2}$$
$$= (21)\left(\frac{1}{365}\right)^2\left(\frac{364}{365}\right)^5$$
$$= 1.555 \times 10^{-4} \quad (1.6 \times 10^{-4})$$

Answer is D.

8. Problem 7 was concerned with any 2 people having an April 15 birthday. It did not matter if the remaining 5 people shared a birthday, as long as the birthday was not April 15.

Problem 8 is concerned with (exactly) 2 people having any birthday in common. This is more complex than a simple multiple of the Prob. 7 probability. In order to satisfy the condition that only 2 people share the same birthday, the other 5 must be prevented from sharing a birthday.

Take April 15 as an example. The probability that persons 1 and 2 are born on April 15 is $(1/365)^2$. The probability that person 3 is born on a day different from persons 1 and 2 is $364/365$. The probability that person 4 is born on a day different from persons 1, 2, and 3 is $363/365$. And, so on. The probability that persons 1 and 2 are born on April 15 and that persons 3 through 7 do not share a common birthday is

$$\left(\frac{1}{365}\right)^2 \left(\frac{364}{365}\right)\left(\frac{363}{365}\right)\left(\frac{362}{365}\right)\left(\frac{361}{365}\right)\left(\frac{360}{365}\right)$$

However, there are $C(7,2) = 21$ ("seven choose two") ways of combining 7 people into unique groups of 2, so the probability of any two people being born on April 15 is

$$(21)\left(\frac{1}{365}\right)^2 \left(\frac{364}{365}\right)\left(\frac{363}{365}\right)\left(\frac{362}{365}\right)\left(\frac{361}{365}\right)\left(\frac{360}{365}\right)$$

Finally, there are 365 distinct days (not just April 15) in the year, so the probability of (exactly) any 2 people sharing any birthday is

$$(21)(365)\left(\frac{1}{365}\right)^2 \left(\frac{364}{365}\right)\left(\frac{363}{365}\right)\left(\frac{362}{365}\right)$$
$$\times \left(\frac{361}{365}\right)\left(\frac{360}{365}\right)$$
$$= 0.0552 \quad (0.055)$$

Answer is C.

9. Find the probability of each of the outcomes being tested for. Including ball 100, there are 20 balls with numbers greater than 80.

$$P(\text{ball is odd}) = \frac{50}{100} = 0.5$$

$$P(\text{ball} > 80) = \frac{20}{100} = 0.2$$

It is possible for the number on the selected ball to be both odd and greater than 80. Use the law of total probability.

$$P(A+B) = P(A) + P(B) - P(A,B)$$
$$= P(A) + P(B) - P(A)P(B)$$
$$P(\text{odd or} > 80) = 0.5 + 0.2 - (0.5)(0.2)$$
$$= 0.6$$

Answer is C.

10. The mean is the sum divided by the number of measurements.

$$\overline{X} = \frac{\sum X_i}{n} = \frac{91.170}{100}$$
$$= 0.9117$$

Answer is D.

11. The median is the middle result. The middle result is the average of the 50th and 51st observed values. Both of these values are 0.903.

Answer is D.

12. The mode is the measurement observed the most frequently. 0.904 was observed four times, more often than any other value.

Answer is C.

13. The number of samples is

$$n = 23 + 35 + 40 + 33 + 21$$
$$= 152$$

Find the mean of the samples.

$$\overline{X} = \frac{\sum X_i}{n} = \frac{\begin{array}{c}(23)(2480) + (35)(2440) + (40)(2400)\\ + (33)(2360) + (21)(2320)\end{array}}{152}$$
$$= 2401.6 \quad (2402)$$

This is the mean of the sample, not the mean of the population. However, the mean of the sample is the unbiased estimator of the population mean.

Answer is C.

14. The sample standard deviation, s, is the unbiased estimator of the population standard deviation, σ.

$$s = \sqrt{\frac{\sum(X_i - \overline{X})^2}{n-1}}$$

$$= \sqrt{\frac{\begin{array}{c}(23)(2480 - 2402)^2 + (35)(2440 - 2402)^2\\ + (40)(2400 - 2402)^2 + (33)(2360 - 2402)^2\\ + (21)(2320 - 2402)^2\end{array}}{152 - 1}}$$

$$= 50.82 \quad (51)$$

Answer is D.

15. Find the standard normal variable corresponding to 2350.
$$x = \frac{X_0 - \mu}{\sigma} = \frac{2350 - 2402}{50.81}$$
$$= -1.02 \quad [\text{use } -1.0]$$

Since the normal distribution is symmetrical about $x = 0$, the probability of x being in the interval $[-\infty, -1]$

is the same as x being in the interval $[+1, +\infty]$. From Table 6.2, $R(1.0) = 0.1587$.

$$
\begin{aligned}
P\{X < 2350\} &= P\{x < -1.0\} \\
&= R(1.0) \\
&= 0.1587 \quad (0.16)
\end{aligned}
$$

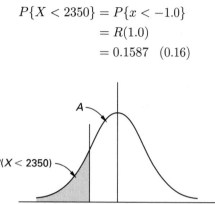

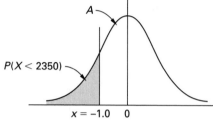

Answer is B.

16. The area under the curve represents the cumulative probability and is equal to 1. Therefore, the height of the curve at its peak is 2. The equation of the line up to $x = 1/2$ is

$$f(x) = 4x \quad [x < \tfrac{1}{2}]$$

The probability that $x < 1/3$ is equal to the area under the curve between 0 and 1/3.

$$
\begin{aligned}
P\left(x < \frac{1}{3}\right) &= \int_0^{\frac{1}{3}} f(x)\ dx \\
&= \int_0^{\frac{1}{3}} 4x\ dx \\
&= 2x^2 \Big|_0^{\frac{1}{3}} \\
&= 2\left(\frac{1}{3}\right)^2 - 0 \\
&= 0.222 \quad (0.22)
\end{aligned}
$$

Answer is B.

17. The probability of obtaining at least one head is equal to one minus the probability of obtaining no heads. The only way that there could be no heads would be for all coins to land tails up. The probability of this happening is

$$
\begin{aligned}
P(\text{all tails up}) &= P(\text{coin 1 tails up}) \\
&\quad \times P(\text{coin 2 tails up}) \\
&\quad \times \cdots \\
&\quad \times P(\text{coin 7 tails up}) \\
&= (0.5)^7 \\
&= 0.0078
\end{aligned}
$$

$$
\begin{aligned}
P(\text{at least one head}) &= 1 - P(\text{all tails}) = 1 - 0.0078 \\
&= 0.9922 \quad (0.992)
\end{aligned}
$$

Answer is D.

18. The probability that both stages will perform successfully is

$$
\begin{aligned}
P(\text{both successful}) &= P(\text{1st successful}) \\
&\quad \times P(\text{2nd successful}) \\
0.97 &= (0.99) \times P(\text{2nd successful}) \\
P(\text{2nd successful}) &= 0.98
\end{aligned}
$$

Answer is D.

19. The mean is

$$
\overline{X} = \frac{\sum X_i}{n} = \frac{0.50 + 0.80 + 0.75 + 0.52 + 0.60}{5}
$$
$$
= 0.634
$$

Use the mean to find the variance.

$$
\begin{aligned}
s^2 &= \frac{\sum (X_i - \overline{X})^2}{n - 1} \\[4pt]
&= \frac{\begin{aligned}(0.50 - 0.634)^2 + (0.80 - 0.634)^2 \\ + (0.75 - 0.634)^2 + (0.52 - 0.634)^2 \\ + (0.60 - 0.634)^2\end{aligned}}{5 - 1} \\[4pt]
&= 0.0183
\end{aligned}
$$

Answer is B.

20. The t-test is used to determine if the differences between samples are significant or merely the result of random variations.

Answer is D.

21.

step 1: The sampling is random and from a normal distribution.

step 2: $C = 0.95$ is given.

step 3: Since a specific direction in the variation is not given (i.e., the example does not ask if the average has decreased), use a two-tail hypothesis test.

step 4: From Table 6.1 or the fractile rows in Table 6.2, $x = 1.96$.

step 5: From Eq. 6.32,

$$x' = \left| \frac{\overline{x} - \mu}{\frac{\sigma}{\sqrt{n}}} \right| = \left| \frac{871 - 880}{\frac{21}{\sqrt{50}}} \right| = 3.03$$

Since $3.03 > 1.96$, the distributions are not the same. There is at least a 95 percent probability that the plant is not operating correctly.

Answer is D.

Table 6.2 Unit Normal Distribution

x	$f(x)$	$F(x)$	$R(x)$	$2R(x)$	$W(x)$
0.0	0.3989	0.5000	0.5000	1.0000	0.0000
0.1	0.3970	0.5398	0.4602	0.9203	0.0797
0.2	0.3910	0.5793	0.4207	0.8415	0.1585
0.3	0.3814	0.6179	0.3821	0.7642	0.2358
0.4	0.3683	0.6554	0.3446	0.6892	0.3108
0.5	0.3521	0.6915	0.3085	0.6171	0.3829
0.6	0.3332	0.7257	0.2743	0.5485	0.4515
0.7	0.3123	0.7580	0.2420	0.4839	0.5161
0.8	0.2897	0.7881	0.2119	0.4237	0.5763
0.9	0.2661	0.8159	0.1841	0.3681	0.6319
1.0	0.2420	0.8413	0.1587	0.3173	0.6827
1.1	0.2179	0.8643	0.1357	0.2713	0.7287
1.2	0.1942	0.8849	0.1151	0.2301	0.7699
1.3	0.1714	0.9032	0.0968	0.1936	0.8064
1.4	0.1497	0.9192	0.0808	0.1615	0.8385
1.5	0.1295	0.9332	0.0668	0.1336	0.8664
1.6	0.1109	0.9452	0.0548	0.1096	0.8904
1.7	0.0940	0.9554	0.0446	0.0891	0.9109
1.8	0.0790	0.9641	0.0359	0.0719	0.9281
1.9	0.0656	0.9713	0.0287	0.0574	0.9426
2.0	0.0540	0.9772	0.0228	0.0455	0.9545
2.1	0.0440	0.9821	0.0179	0.0357	0.9643
2.2	0.0355	0.9861	0.0139	0.0278	0.9722
2.3	0.0283	0.9893	0.0107	0.0214	0.9786
2.4	0.0224	0.9918	0.0082	0.0164	0.9836
2.5	0.0175	0.9938	0.0062	0.0124	0.9876
2.6	0.0136	0.9953	0.0047	0.0093	0.9907
2.7	0.0104	0.9965	0.0035	0.0069	0.9931
2.8	0.0079	0.9974	0.0026	0.0051	0.9949
2.9	0.0060	0.9981	0.0019	0.0037	0.9963
3.0	0.0044	0.9987	0.0013	0.0027	0.9973
Fractiles					
1.2816	0.1755	0.9000	0.1000	0.2000	0.8000
1.6449	0.1031	0.9500	0.0500	0.1000	0.9000
1.9600	0.0584	0.9750	0.0250	0.0500	0.9500
2.0537	0.0484	0.9800	0.0200	0.0400	0.9600
2.3263	0.0267	0.9900	0.0100	0.0200	0.9800
2.5758	0.0145	0.9950	0.0050	0.0100	0.9900

Table 6.3 *t-Distribution*
(values of t for n degrees of freedom; $1 - \alpha$ confidence level)

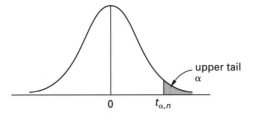

	area under the upper tail					
n	$\alpha = 0.10$	$\alpha = 0.05$	$\alpha = 0.025$	$\alpha = 0.01$	$\alpha = 0.005$	n
1	3.078	6.314	12.706	31.821	63.657	1
2	1.886	2.920	4.303	6.965	9.925	2
3	1.638	2.353	3.182	4.541	5.841	3
4	1.533	2.132	2.776	3.747	4.604	4
5	1.476	2.015	2.571	3.365	4.032	5
6	1.440	1.943	2.447	3.143	3.707	6
7	1.415	1.895	2.365	2.998	3.499	7
8	1.397	1.860	2.306	2.896	3.355	8
9	1.383	1.833	2.262	2.821	3.250	9
10	1.372	1.812	2.228	2.764	3.169	10
11	1.363	1.796	2.201	2.718	3.106	11
12	1.356	1.782	2.179	2.681	3.055	12
13	1.350	1.771	2.160	2.650	3.012	13
14	1.345	1.761	2.145	2.624	2.977	14
15	1.341	1.753	2.131	2.602	2.947	15
16	1.337	1.746	2.120	2.583	2.921	16
17	1.333	1.740	2.110	2.567	2.898	17
18	1.330	1.734	2.101	2.552	2.878	18
19	1.328	1.729	2.093	2.539	2.861	19
20	1.325	1.725	2.086	2.528	2.845	20
21	1.323	1.721	2.080	2.518	2.831	21
22	1.321	1.717	2.074	2.508	2.819	22
23	1.319	1.714	2.069	2.500	2.807	23
24	1.318	1.711	2.064	2.492	2.797	24
25	1.316	1.708	2.060	2.485	2.787	25
26	1.315	1.706	2.056	2.479	2.779	26
27	1.314	1.703	2.052	2.473	2.771	27
28	1.313	1.701	2.048	2.467	2.763	28
29	1.311	1.699	2.045	2.462	2.756	29
∞	1.282	1.645	1.960	2.326	2.576	∞

MATHEMATICS
Prob/Stat

Table 6.4   Critical Values of F

For a particular combination of numerator and denominator degrees of freedom, entry represents the critical values of F corresponding to a specified upper tail area (α).

$\alpha = 0.05$

$F(\alpha, df_1, df_2)$

numerator df_1

denomi-nator df_2	1	2	3	4	5	6	7	8	9	10	12	15	20	24	30	40	60	120	∞
1	161.4	199.5	215.7	224.6	230.2	234.0	236.8	238.9	240.5	241.9	243.9	245.9	248.0	249.1	250.1	251.1	252.2	253.3	254.3
2	18.51	19.00	19.16	19.25	19.30	19.33	19.35	19.37	19.38	19.40	19.41	19.43	19.45	19.45	19.46	19.47	19.48	19.49	19.50
3	10.13	9.55	9.28	9.12	9.01	8.94	8.89	8.85	8.81	8.79	8.74	8.70	8.66	8.64	8.62	8.59	8.57	8.55	8.53
4	7.71	6.94	6.59	6.39	6.26	6.16	6.09	6.04	6.00	5.96	5.91	5.86	5.80	5.77	5.75	5.72	5.69	5.66	5.63
5	6.61	5.79	5.41	5.19	5.05	4.95	4.88	4.82	4.77	4.74	4.68	4.62	4.56	4.53	4.50	4.46	4.43	4.40	4.36
6	5.99	5.14	4.76	4.53	4.39	4.28	4.21	4.15	4.10	4.06	4.00	3.94	3.87	3.84	3.81	3.77	3.74	3.70	3.67
7	5.59	4.74	4.35	4.12	3.97	3.87	3.79	3.73	3.68	3.64	3.57	3.51	3.44	3.41	3.38	3.34	3.30	3.27	3.23
8	5.32	4.46	4.07	3.84	3.69	3.58	3.50	3.44	3.39	3.35	3.28	3.22	3.15	3.12	3.08	3.04	3.01	2.97	2.93
9	5.12	4.26	3.86	3.63	3.48	3.37	3.29	3.23	3.18	3.14	3.07	3.01	2.94	2.90	2.86	2.83	2.79	2.75	2.71
10	4.96	4.10	3.71	3.48	3.33	3.22	3.14	3.07	3.02	2.98	2.91	2.85	2.77	2.74	2.70	2.66	2.62	2.58	2.54
11	4.84	3.98	3.59	3.36	3.20	3.09	3.01	2.95	2.90	2.85	2.79	2.72	2.65	2.61	2.57	2.53	2.49	2.45	2.40
12	4.75	3.89	3.49	3.26	3.11	3.00	2.91	2.85	2.80	2.75	2.69	2.62	2.54	2.51	2.47	2.43	2.38	2.34	2.30
13	4.67	3.81	3.41	3.18	3.03	2.92	2.83	2.77	2.71	2.67	2.60	2.53	2.46	2.42	2.38	2.34	2.30	2.25	2.21
14	4.60	3.74	3.34	3.11	2.96	2.85	2.76	2.70	2.65	2.60	2.53	2.46	2.39	2.35	2.31	2.27	2.22	2.18	2.13
15	4.54	3.68	3.29	3.06	2.90	2.79	2.71	2.64	2.59	2.54	2.48	2.40	2.33	2.29	2.25	2.20	2.16	2.11	2.07
16	4.49	3.63	3.24	3.01	2.85	2.74	2.66	2.59	2.54	2.49	2.42	2.35	2.28	2.24	2.19	2.15	2.11	2.06	2.01
17	4.45	3.59	3.20	2.96	2.81	2.70	2.61	2.55	2.49	2.45	2.38	2.31	2.23	2.19	2.15	2.10	2.06	2.01	1.96
18	4.41	3.55	3.16	2.93	2.77	2.66	2.58	2.51	2.46	2.41	2.34	2.27	2.19	2.15	2.11	2.06	2.02	1.97	1.92
19	4.38	3.52	3.13	2.90	2.74	2.63	2.54	2.48	2.42	2.38	2.31	2.23	2.16	2.11	2.07	2.03	1.98	1.93	1.88
20	4.35	3.49	3.10	2.87	2.71	2.60	2.51	2.45	2.39	2.35	2.28	2.20	2.12	2.08	2.04	1.99	1.95	1.90	1.84
21	4.32	3.47	3.07	2.84	2.68	2.57	2.49	2.42	2.37	2.32	2.25	2.18	2.10	2.05	2.01	1.96	1.92	1.87	1.81
22	4.30	3.44	3.05	2.82	2.66	2.55	2.46	2.40	2.34	2.30	2.23	2.15	2.07	2.03	1.98	1.94	1.89	1.84	1.78
23	4.28	3.42	3.03	2.80	2.64	2.53	2.44	2.37	2.32	2.27	2.20	2.13	2.05	2.01	1.96	1.91	1.86	1.81	1.76
24	4.26	3.40	3.01	2.78	2.62	2.51	2.42	2.36	2.30	2.25	2.18	2.11	2.03	1.98	1.94	1.89	1.84	1.79	1.73
25	4.24	3.39	2.99	2.76	2.60	2.49	2.40	2.34	2.28	2.24	2.16	2.09	2.01	1.96	1.92	1.87	1.82	1.77	1.71
26	4.23	3.37	2.98	2.74	2.59	2.47	2.39	2.32	2.27	2.22	2.15	2.07	1.99	1.95	1.90	1.85	1.80	1.75	1.69
27	4.21	3.35	2.96	2.73	2.57	2.46	2.37	2.31	2.25	2.20	2.13	2.06	1.97	1.93	1.88	1.84	1.79	1.73	1.67
28	4.20	3.34	2.95	2.71	2.56	2.45	2.36	2.29	2.24	2.19	2.12	2.04	1.96	1.91	1.87	1.82	1.77	1.71	1.65
29	4.18	3.33	2.93	2.70	2.55	2.43	2.35	2.28	2.22	2.18	2.10	2.03	1.94	1.90	1.85	1.81	1.75	1.70	1.64
30	4.17	3.32	2.92	2.69	2.53	2.42	2.33	2.27	2.21	2.16	2.09	2.01	1.93	1.89	1.84	1.79	1.74	1.68	1.62
40	4.08	3.23	2.84	2.61	2.45	2.34	2.25	2.18	2.12	2.08	2.00	1.92	1.84	1.79	1.74	1.69	1.64	1.58	1.51
60	4.00	3.15	2.76	2.53	2.37	2.25	2.17	2.10	2.04	1.99	1.92	1.84	1.75	1.70	1.65	1.59	1.53	1.47	1.39
120	3.92	3.07	2.68	2.45	2.29	2.17	2.09	2.02	1.96	1.91	1.83	1.75	1.66	1.61	1.55	1.50	1.43	1.35	1.25
∞	3.84	3.00	2.60	2.37	2.21	2.10	2.01	1.94	1.88	1.83	1.75	1.67	1.57	1.52	1.46	1.39	1.32	1.22	1.00

7 Calculus

Subjects

DIFFERENTIAL CALCULUS 7-1
   Derivatives 7-1
   Critical Points 7-1
   Partial Derivatives 7-2
   Curvature 7-2
   L'Hôpital's Rule 7-2
INTEGRAL CALCULUS 7-3
   Fundamental Theorem of Calculus 7-3
   Methods of Integration 7-3
CENTROIDS AND MOMENTS OF
  INERTIA 7-4
GRADIENT, DIVERGENCE, AND CURL . . 7-5
   Vector Del Operator 7-5
   Gradient of a Scalar Function 7-5
   Divergence of a Vector Field 7-5
   Curl of a Vector Field 7-5
   Laplacian of a Scalar Function 7-5

DIFFERENTIAL CALCULUS

Derivatives

In most cases, it is possible to transform a continuous function, $f(x_1, x_2, x_3, \ldots)$, of one or more independent variables into a derivative function. In simple cases, the *derivative* can be interpreted as the slope (tangent or rate of change) of the curve described by the original function.

Since the slope of a curve depends on x, the derivative function will also depend on x. The derivative, $f'(x)$, of a function $f(x)$ is defined mathematically by Eq. 7.1. However, limit theory is seldom needed to actually calculate derivatives.

$$f'(x) = \lim_{\Delta x \to 0} \left(\frac{f(x + \Delta x) - f(x)}{\Delta x} \right) \qquad 7.1$$

The derivative of a function $y = f(x)$, also known as the *first derivative*, is written in various ways, including the following.

$$f'(x), \frac{df(x)}{dx}, \mathbf{D}f(x), \mathbf{D}_x f(x), \mathbf{D}_x y, y', \frac{dy}{dx}$$

Formulas for the derivatives of several common functional forms are listed in Table 7.1 at the end of this chapter.

Critical Points

Derivatives are used to locate the local critical points, that is, *extreme points* (also known as *maximum* and *minimum points*) as well as the *inflection points* (*points of contraflexure*) of functions of one variable. The plurals *extrema*, *maxima*, and *minima* are used without the word "points." These points are illustrated in Fig. 7.1. There is usually an inflection point between two adjacent local extrema.

Figure 7.1 Critical Points

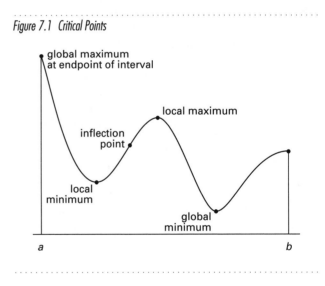

The first derivative, $f'(x)$, is calculated to determine where the critical points are. The second derivative, $f''(x)$, is calculated to determine whether a critical point is a local maximum, minimum, or inflection point. With this method, no distinction is made between local and global extrema. Therefore, the extrema should be compared to the function values at the endpoints of the interval.

Critical points are located where the first derivative is zero. That is, for a function $y = f(x)$, the point $x = a$ is a critical point if

$$f'(a) = 0 \qquad 7.2$$

- *Test for a maximum*: For a function $f(x)$ with an extreme point at $x = a$, if the point is a maximum, then

$$f'(a) = 0$$
$$f''(a) < 0 \qquad\qquad 7.3$$

- *Test for a minimum*: For a function $f(x)$ with a critical point at $x = a$, if the point is a minimum, then

$$f'(a) = 0$$
$$f''(a) > 0 \qquad\qquad 7.4$$

- *Test for a point of inflection*: For a function $f(x)$ with a critical point at $x = a$, if the point is a point of inflection, then

$$f''(a) = 0 \qquad\qquad 7.5$$

Partial Derivatives

Derivatives can be taken with respect to only one independent variable at a time. For example, $f'(x)$ is the derivative of $f(x)$ and is taken with respect to the independent variable x. If a function, $f(x_1, x_2, x_3, \ldots)$, has more than one independent variable, a *partial derivative* can be found, but only with respect to one of the independent variables. All other variables are treated as constants. Symbols for a partial derivative of $f(x, y)$ taken with respect to variable x are $\partial f / \partial x$ and $f_x(x, y)$.

The geometric interpretation of a partial derivative $\partial f / \partial x$ is the slope of a line tangent to the surface (a sphere, an ellipsoid, etc.) described by the function when all variables except x are held constant. In three-dimensional space with a function described by $z = f(x, y)$, the partial derivative $\partial f / \partial x$ (equivalent to $\partial z / \partial x$) is the slope of the line tangent to the surface in a plane of constant y. Similarly, the partial derivative $\partial f / \partial y$ (equivalent to $\partial z / \partial y$) is the slope of the line tangent to the surface in a plane of constant x.

Curvature

The sharpness of a curve between two points on the curve can be defined as the rate of change of the inclination of the curve with respect to the distance traveled along the curve. As shown in Fig. 7.2, the rate of change of the inclination of the curve is the change in the angle formed by the tangents to the curve at each point and the x-axis. The distance, s, traveled along the curve is the arc length of the curve between points 1 and 2. The sharpness of the curve at one point is known as the *curvature*, K, and is given by Eq. 7.6.

$$K = \lim_{\Delta s \to 0} \left(\frac{\Delta \alpha}{\Delta s} \right) = \frac{d\alpha}{ds} \qquad 7.6$$

Figure 7.2 Curvature

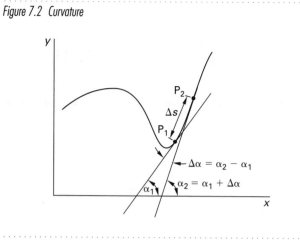

If the equation of a curve $f(x, y)$ is given in rectangular coordinates, the curvature is defined by Eq. 7.7.

$$K = \frac{y''}{[1 + (y')^2]^{\frac{3}{2}}} \qquad 7.7$$

If the function $f(x, y)$ is easier to differentiate with respect to y instead of x, then Eq. 7.8 may be used.

$$K = \frac{-x''}{[1 + (x')^2]^{\frac{3}{2}}} \qquad \left[x' = \frac{dx}{dy} \right] \qquad 7.8$$

The *radius of curvature*, R, of a curve describes the radius of a circle whose center lies on the concave side of the curve and whose tangent coincides with the tangent to the curve at that point. Radius of curvature is the absolute value of the reciprocal of the curvature.

$$R = \frac{1}{|K|} = \frac{[1 + (y')^2]^{\frac{3}{2}}}{|y''|} \qquad 7.9$$

L'Hôpital's Rule

A *limit* is the value a function approaches when an independent variable approaches a target value. For example, suppose the value of $y = x^2$ is desired as x approaches five. This could be written as

$$y(5) = \lim_{x \to 5} x^2$$

The power of limit theory is wasted on simple calculations such as this one, but is appreciated when the function is undefined at the target value. The object of limit theory is to determine the limit without having to evaluate the function at the target. The general case of a limit evaluated as x approaches the target value a is written as

$$\lim_{x \to a} f(x) \qquad 7.10$$

It is not necessary for the actual value $f(a)$ to exist for the limit to be calculated. The function $f(x)$ may be undefined at point a. However, it is necessary that $f(x)$ be defined on both sides of point a for the limit to exist. If $f(x)$ is undefined on one side, or if $f(x)$ is discontinuous at $x = a$, as in Fig. 7.3(c) and (d), the limit does not exist at $x = a$.

Figure 7.3 Existence of Limits

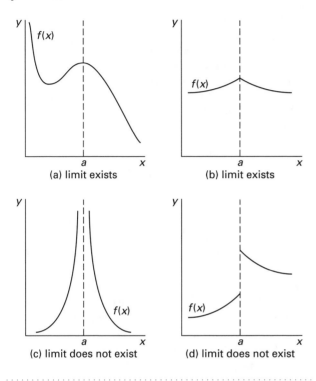

L'Hôpital's rule may be used only when the numerator and denominator of the expression both approach zero or both approach infinity at the limit point. $f^k(x)$ and $g^k(x)$ are the kth derivatives of the functions $f(x)$ and $g(x)$, respectively. L'Hôpital's rule can be applied repeatedly as required as long as the numerator and denominator are both indeterminate.

$$\lim_{x \to a}\left(\frac{f(x)}{g(x)}\right) = \lim_{x \to a}\left(\frac{f^k(x)}{g^k(x)}\right) \qquad 7.11$$

INTEGRAL CALCULUS

Fundamental Theorem of Calculus

Integration is the inverse operation of differentiation. There are two types of integrals: *definite integrals*, which are restricted to a specific range of the independent variable, and *indefinite integrals*, which are unrestricted. Indefinite integrals are sometimes referred to as *antiderivatives*.

The definition of a definite integral is given by the *fundamental theorem of integral calculus*.

$$\int_a^b f(x)dx = \lim_{n \to \infty} \sum_{i=1}^n f(x_i)\Delta x_i \qquad 7.12$$

Although expressions can be functions of several variables, integrals can only be taken with respect to one variable at a time. The *differential term* (dx in Eq. 7.12) indicates that variable. In Eq. 7.12, the function $f(x)$ is the *integrand* and x is the *variable of integration*.

When $f'(x) = h(x)$, the indefinite integral is defined as

$$\int h(x)dx = f(x) + C \qquad 7.13$$

While most of the function $f(x)$ can be recovered through integration of its derivative, $f'(x)$, any constant term will have been lost. This is because the derivative of a constant term vanishes (i.e., is zero), leaving nothing to recover from. A *constant of integration*, C, is added to the integral to recognize the possibility of such a term.

Elementary integration operations on common functional forms are usually memorized or listed in tables (see Table 7.1 at the end of this chapter). More complicated integrations may be performed by one of the following methods.

Methods of Integration

Integration by parts

If $f(x)$ and $g(x)$ are functions, then the integral of $f(x)$ with respect to $g(x)$ is found by the method of *integrations by parts*, given by Eq. 7.14.

$$\int f(x)dg(x) = f(x)g(x) - \int g(x)df(x) + C \quad 7.14$$

Integration by substitution

Integration by substitution means that an integrand (that is difficult to integrate) and the corresponding differential are replaced by equivalent expressions with known solutions. Substitutions that may be used include trigonometric substitutions, as illustrated in Fig. 7.4 and Eqs. 7.15 through 7.17.

Figure 7.4 Geometric Interpretation of Trigonometric Substitutions

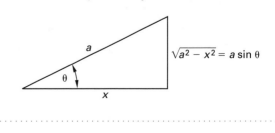

- $\sqrt{a^2 - x^2}$:   substitute $x = a\sin\theta$ *7.15*
- $\sqrt{a^2 + x^2}$:   substitute $x = a\tan\theta$ *7.16*
- $\sqrt{x^2 - a^2}$:   substitute $x = a\sec\theta$ *7.17*

Separation of rational fractions into partial fractions

The *method of partial fractions* is used to transform a proper polynomial fraction of two polynomials into a sum of simpler expressions, a procedure known as *resolution*. The technique can be considered to be the act of "unadding" a sum to obtain all of the addends.

Suppose $H(x)$ is a proper polynomial fraction of the form $P(x)/Q(x)$. The object of the resolution is to determine the partial fractions u_1/v_1, u_2/v_2, and so on, such that

$$H(x) = \frac{P(x)}{Q(x)} = \frac{u_1}{v_1} + \frac{u_2}{v_2} + \frac{u_3}{v_3} + \cdots \qquad 7.18$$

The form of the denominator polynomial $Q(x)$ will be the main factor in determining the form of the partial fractions. The task of finding the u_i and v_i is simplified by categorizing the possible forms of $Q(x)$.

case 1: $Q(x)$ factors into n different linear terms. That is,

$$Q(x) = (x - a_1)(x - a_2)\cdots(x - a_n)$$

Then,

$$H(x) = \sum_{i=1}^{n} \frac{A_i}{x - a_i} \qquad 7.19$$

case 2: $Q(x)$ factors into n identical linear terms. That is,

$$Q(x) = (x - a)(x - a)\cdots(x - a)$$

Then,

$$H(x) = \sum_{i=1}^{n} \frac{A_i}{(x - a)^i} \qquad 7.20$$

case 3: $Q(x)$ factors into n different quadratic terms $(x^2 + p_i x + q_i)$. Then,

$$H(x) = \sum_{i=1}^{n} \frac{A_i x + B_i}{x^2 + p_i x + q_i} \qquad 7.21$$

case 4: $Q(x)$ factors into n identical quadratic terms $(x^2 + px + q)$. Then,

$$H(x) = \sum_{i=1}^{n} \frac{A_i x + B_i}{(x^2 + px + q)^i} \qquad 7.22$$

Once the general forms of the partial fractions have been determined from inspection, the *method of undetermined coefficients* is used. The partial fractions are all cross multiplied to obtain $Q(x)$ as the denominator, and the coefficients are found by equating $P(x)$ and the cross-multiplied numerator.

CENTROIDS AND MOMENTS OF INERTIA

Applications of integration include the determination of the *centroid of an area* and the *area moment of inertia*. The centroid of an area is analogous to the *center of gravity* of a homogeneous body. The location of the centroid of the area bounded by the x- and y-axes and the mathematical function $y = f(x)$ can be found from Eqs. 7.23 through 7.26.

$$x_c = \int \frac{x\,dA}{A} \qquad 7.23$$

$$y_c = \int \frac{y\,dA}{A} \qquad 7.24$$

$$A = \int f(x)\,dx \qquad 7.25$$

$$dA = f(x)dx = g(y)dy \qquad 7.26$$

The quantity $\int x\,dA$ is known as the *first moment of the area* or *first area moment* with respect to the y-axis. Similarly, $\int y\,dA$ is known as the *first moment of the area* with respect to the x-axis. By rearranging Eqs. 7.23 and 7.24, it is obvious that the first moment of the area can be calculated from the area and centroidal distance.

$$M_y = \int x\,dA = x_c A \qquad 7.27$$

$$M_x = \int y\,dA = y_c A \qquad 7.28$$

The moment of inertia, I, of an area is needed in mechanics of materials problems. The symbol I_x is used to represent a moment of inertia with respect to the x-axis. Similarly, I_y is the moment of inertia with respect to the y-axis.

$$I_x = \int y^2\,dA \qquad 7.29$$

$$I_y = \int x^2\,dA \qquad 7.30$$

The moment of inertia taken with respect to an axis passing through the area's centroid is known as the *centroidal moment of inertia*, I_c. The centroidal moment of inertia is the smallest possible moment of inertia for the shape.

If the moment of inertia is known with respect to one axis, the moment of inertia with respect to another parallel axis can be calculated from the *parallel axis theorem* also known as the *transfer axis theorem*. This theorem is used to evaluate the moment of inertia of areas that are composed of two or more basic shapes. In Eq. 7.31, d is the distance between the centroidal axis and the second, parallel axis.

$$I_{\text{parallel axis}} = I_c + Ad^2 \qquad 7.31$$

The integration method for determining centroids and moments of inertia is not necessary for basic shapes. Formulas for basic shapes are tabulated into tables, as discussed in Ch. 11.

GRADIENT, DIVERGENCE, AND CURL

Vector Del Operator

The *vector del operator*, ∇, is defined as

$$\nabla = \frac{\partial}{\partial x}\mathbf{i} + \frac{\partial}{\partial y}\mathbf{j} + \frac{\partial}{\partial z}\mathbf{k} \qquad 7.32$$

Gradient of a Scalar Function

The slope of a function is the change in one variable with respect to a distance in a chosen direction. Usually, the direction is parallel to a coordinate axis. However, the maximum slope at a point on a surface may not be in a direction parallel to one of the coordinate axes.

The *gradient vector function* $\nabla f(x,y,z)$ (pronounced "del f") gives the maximum rate of change of the function $f(x,y,z)$.

$$\nabla f(x,y,z) = \left(\frac{\partial f(x,y,z)}{\partial x}\right)\mathbf{i} + \left(\frac{\partial f(x,y,z)}{\partial y}\right)\mathbf{j}$$
$$+ \left(\frac{\partial f(x,y,z)}{\partial z}\right)\mathbf{k} \qquad 7.33$$

Divergence of a Vector Field

The *divergence*, div $\mathbf{F}$, of a vector field $\mathbf{F}(x,y,z)$ is a scalar function defined by Eqs. 7.34 and 7.35. The divergence of $\mathbf{F}$ can be interpreted as the *accumulation* of flux (i.e., a flowing substance) in a small region (i.e., at a point). One of the uses of the divergence is to determine whether flow (represented in direction and magnitude by $\mathbf{F}$) is compressible. Flow is incompressible if div $\mathbf{F} = 0$, since the substance is not accumulating.

$$\mathbf{F} = P(x,y,z)\mathbf{i} + Q(x,y,z)\mathbf{j} + R(x,y,z)\mathbf{k} \quad 7.34$$
$$\text{div } \mathbf{F} = \frac{\partial P}{\partial x} + \frac{\partial Q}{\partial y} + \frac{\partial R}{\partial z} = \nabla \cdot \mathbf{F} \qquad 7.35$$

If there is no divergence, then the dot product calculated in Eq. 7.35 is zero.

Curl of a Vector Field

The *curl*, curl $\mathbf{F}$, of a vector field $\mathbf{F}(x,y,z)$ is a vector field defined by Eq. 7.37. The curl $\mathbf{F}$ can be interpreted as the *vorticity* per unit area of flux (i.e., a flowing substance) in a small region (i.e., at a point). One of the uses of the curl is to determine whether flow (represented in direction and magnitude by $\mathbf{F}$) is rotational. Flow is irrotational if curl $\mathbf{F} = 0$.

$$\mathbf{F} = P(x,y,z)\mathbf{i} + Q(x,y,z)\mathbf{j} + R(x,y,z)\mathbf{k} \quad 7.36$$

$$\text{curl } \mathbf{F} = \nabla \times \mathbf{F}$$
$$= \begin{vmatrix} \mathbf{i} & \mathbf{j} & \mathbf{k} \\ \dfrac{\partial}{\partial x} & \dfrac{\partial}{\partial y} & \dfrac{\partial}{\partial z} \\ P(x,y,z) & Q(x,y,z) & R(x,y,z) \end{vmatrix} \qquad 7.37$$

Laplacian of a Scalar Function

The *Laplacian* of a scalar function, $\phi = f(x,y,z)$, is interpreted as the divergence of the gradient function. (This is essentially the second derivative of a scalar function.) A function that satisfies Laplace's equation $\nabla^2 = 0$ is known as a *potential function*.

$$\nabla^2 \phi = \frac{\partial^2 \phi}{\partial x^2} + \frac{\partial^2 \phi}{\partial y^2} + \frac{\partial^2 \phi}{\partial z^2} \qquad 7.38$$

SAMPLE PROBLEMS

Problem 1

What is the maximum value of the following function on the interval $x \leq 0$?

$$y = 2x^3 + 12x^2 - 30x + 10$$

(A) -210
(B) -36
(C) -5
(D) 210

Solution

The critical points are located where $dy/dx = 0$.

$$\frac{dy}{dx} = 6x^2 + 24x - 30$$

$$6x^2 + 24x - 30 = 0$$
$$x^2 + 4x - 5 = 0$$
$$(x+5)(x-1) = 0$$
$$x = -5 \quad \text{or} \quad x = 1$$

A critical point is a maximum if $d^2y/dx^2 < 0$.

$$\frac{d^2y}{dx^2} = 12x + 24$$

At $x = -5$, the second derivative is

$$(12)(-5) + 24 = -36$$

At $x = 1$, the second derivative is

$$(12)(1) + 24 = 36$$

$x = -5$ is a maximum, and $x = 1$ is a minimum.

The maximum value is

$$(2)(-5)^3 + (12)(-5)^2 - (30)(-5) + 10 = 210$$

Answer is D.

Problem 2
Which of the following is *not* a correct derivative?

(A) $\dfrac{d}{dx}(\cos x) = -\sin x$

(B) $\dfrac{d}{dx}(1 - x)^3 = (-3)(1 - x)^2$

(C) $\dfrac{d}{dx}\left(\dfrac{1}{x}\right) = -\dfrac{1}{x^2}$

(D) $\dfrac{d}{dx}(\csc x) = -\cot x$

Solution
Determine each of the derivatives.

$$\frac{d}{dx}(\cos x) = -\sin x \qquad \text{[ok]}$$

$$\frac{d}{dx}(1 - x)^3 = (3)(1 - x)^2(-1) = (-3)(1 - x)^2 \quad \text{[ok]}$$

$$\frac{d}{dx}\left(\frac{1}{x}\right) = \frac{d}{dx}(x^{-1}) = (-1)(x^{-2}) = \frac{-1}{x^2} \qquad \text{[ok]}$$

$$\frac{d}{dx}(\csc x) = -\cot x \qquad \text{[incorrect]}$$

Answer is D.

Problem 3
What is dy/dx if $y = (2x)^x$?

(A) $(2x)^x(2 + \ln 2x)$
(B) $2x(1 + \ln 2x)^x$
(C) $(2x)^x(\ln 2x^2)$
(D) $(2x)^x(1 + \ln 2x)$

Solution
From the table of derivatives,

$$\mathbf{D}(f(x))^{g(x)} = g(x)(f(x))^{g(x)-1}\mathbf{D}f(x)$$
$$+ \ln(f(x))(f(x))^{g(x)}\mathbf{D}g(x)$$
$$f(x) = 2x$$
$$g(x) = x$$
$$\frac{d(2x)^x}{dx} = x(2x)^{x-1}(2) + (\ln 2x)(2x)^x(1)$$
$$= (2x)^x + (2x)^x \ln 2x$$
$$= (2x)^x(1 + \ln 2x)$$

(Recall $(2x)(2x)^n = (2x)^{n+1}$.)

Answer is D.

Problem 4
If $f(x, y) = x^2y^3 + xy^4 + \sin x + \cos^2 x + \sin^3 y$, what is $\partial f/\partial x$?

(A) $(2x + y)y^3 + 3\sin^2 y \cos y$
(B) $(4x - 3y^2)xy^2 + 3\sin^2 y \cos y$
(C) $(3x + 4y^2)xy + 3\sin^2 y \cos y$
(D) $(2x + y)y^3 + (1 - 2\sin x)\cos x$

Solution
The partial derivative with respect to x is found by treating all other variables as constants. Therefore, all terms that do not contain x have zero derivatives.

$$\frac{\partial f}{\partial x} = 2xy^3 + y^4 + \cos x + 2\cos x(-\sin x)$$
$$= (2x + y)y^3 + (1 - 2\sin x)\cos x$$

Answer is D.

Problem 5
Evaluate the following limit.

$$\lim_{x \to 2}\left(\frac{x^2 - 4}{x - 2}\right)$$

(A) 0
(B) 2
(C) 4
(D) ∞

Solution
Use L'Hôpital's rule because the expression approaches $0/0$ at the limit.

$$\frac{(2)^2 - 4}{2 - 2} = \frac{0}{0}$$

L'Hôpital's rule states that the limit of the expression is the same as the limit of the derivatives of the numerator and denominator.

$$\lim_{x \to 2}\left(\frac{x^2 - 4}{x - 2}\right) = \lim_{x \to 2}\left(\frac{2x}{1}\right) = \frac{(2)(2)}{1} = 4$$

This could also have been solved by factoring the numerator.

Answer is C.

FE-STYLE EXAM PROBLEMS

1. What are the minimum and maximum values, respectively, of the equation $f(x) = 5x^3 - 2x^2 + 1$ on the interval $(-2, 2)$?

  (A) -47, 33
  (B) -4, 4
  (C) 0.95, 1
  (D) 0, 0.27

2. What is the first derivative, dy/dx, of the following expression?

$$(xy)^x = e$$

  (A) 0

  (B) $\dfrac{-x}{y}\left(1 - x \ln x\right)$

  (C) $\dfrac{-y}{x}\left(1 + \ln xy\right)$

  (D) $\dfrac{y}{x}$

3. What is the standard form of the equation of the line tangent to a circle centered at the origin with a radius of 5 at the point (3,4)?

  (A) $x = \dfrac{-4}{3}y - \dfrac{25}{4}$

  (B) $y = \dfrac{3}{4}x + \dfrac{25}{4}$

  (C) $y = \dfrac{-3}{4}x + \dfrac{9}{4}$

  (D) $y = \dfrac{-3}{4}x + \dfrac{25}{4}$

Illustration for Problem 3

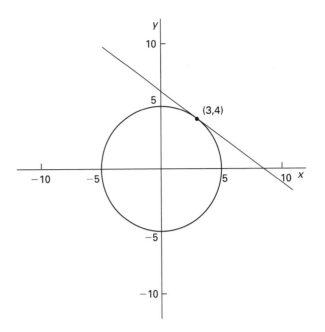

4. Evaluate dy/dx for the following expression.

$$y = e^{-x}\sin 2x$$

  (A) $e^{-x}\left(2\cos 2x - \sin 2x\right)$
  (B) $-e^{-x}\left(2\sin 2x + \cos 2x\right)$
  (C) $e^{-x}\left(2\sin 2x + \cos 2x\right)$
  (D) $-e^{-x}\left(2\cos 2x - \sin 2x\right)$

5. Evaluate the following limit. (x is in radians.)

$$\lim_{x \to \pi}\left(\frac{x^2 - \pi x + \sin x}{-\sin x}\right)$$

  (A) 0
  (B) 1
  (C) $\pi - 1$
  (D) $(2)(\pi - 1)$

For the following problems use the NCEES Handbook as your only reference.

6. What is the partial derivative $\partial v/\partial y$ of the following function?

$$v = 3x^2 + 9xy - \frac{y}{\ln z} + \cos(z^2 + x)$$

(A) $9x - \dfrac{1}{\ln z}$

(B) $6x + 9x - \dfrac{1}{\ln z} - \sin(z^2 + x)$

(C) $3x^2 y + \dfrac{9xy^2}{2} - \dfrac{y^2}{2 \ln z} + \dfrac{\sin(z^2 + x)}{z^2 + x}$

(D) $9x + \dfrac{1}{\ln z}$

7. Determine the following indefinite integral.

$$\int \frac{x^3 + x + 4}{x^2} dx$$

(A) $\dfrac{x}{4} + \ln |x| - \dfrac{4}{x} + C$

(B) $\dfrac{-x}{2} + \log x - 8x + C$

(C) $\dfrac{x^2}{2} + \ln |x| - \dfrac{2}{x^2} + C$

(D) $\dfrac{x^2}{2} + \ln |x| - \dfrac{4}{x} + C$

8. Find the shaded area between line 1, line 2, and the x-axis.

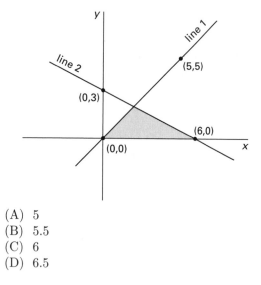

(A) 5
(B) 5.5
(C) 6
(D) 6.5

9. Evaluate the following integral.

$$\int \frac{4}{8 + 2x^2} dx$$

(A) $4 \ln |8 + 2x^2| + C$

(B) $\dfrac{1}{2} - 6x^{-3} + C$

(C) $\tan^{-1} \left(\dfrac{x}{2} \right) + C$

(D) $(-4)(8 + 2x^2)^{-2} + C$

10. What is the approximate area under the curve $y = 1/x$ between $y = 2$ and $y = 10$?

(A) 0.48
(B) 1.6
(C) 2.1
(D) 3.0

11. What is the approximate area bounded by the curves $y = 8 - x^2$ and $y = -2 + x^2$?

(A) 22.4
(B) 26.8
(C) 29.8
(D) 44.7

12. What is the approximate total area bounded by $y = \sin x$ and $y = 0$ over the interval $0 \leq x \leq 2\pi$? (x is in radians.)

(A) 0
(B) $\pi/2$
(C) 2
(D) 4

13. What is the approximate radius of curvature of the function $f(x)$ at the point $(x, y) = (8, 16)$?

$$f(x) = x^2 + 6x - 96$$

(A) 1.90×10^{-4}
(B) 9.80
(C) 96.0
(D) 5340

14. A peach grower estimates that if he picks his crop now, he will obtain 1000 lugs of peaches, which he can sell at $1.00 per lug. However, he estimates that his crop will increase by an additional 60 lugs of peaches for each week that he delays picking, but the price will drop at a rate of $0.025 per lug per week. In addition, he will experience a spoilage rate of approximately 10 lugs for each week he delays. In order to maximize his revenue, how many weeks should he wait before picking the peaches?

(A) 2 weeks
(B) 5 weeks
(C) 7 weeks
(D) 10 weeks

15. What is the derivative, dy/dt, of $y = \sin^2 \omega t$?

(A) $2\omega \sin \omega t$
(B) $\cos^2 \omega t$
(C) $2 \sin \omega t \cos \omega t$
(D) $2\omega \sin \omega t \cos \omega t$

16. What is the slope of the curve $y = 10x^2 - 3x - 1$ when it crosses the positive part of the x-axis?

(A) $3/20$
(B) $1/5$
(C) $1/3$
(D) 7

17. What is the volume of revolution from $x = 0$ to $x = 3/2$ when the function $f(x) = 2x^2$ is revolved around the y-axis?

(A) $3\pi/2$
(B) $27\pi/16$
(C) $35\pi/16$
(D) $81\pi/16$

18. What is the derivative, dy/dx, of the expression $x^2 y - e^{2x} = \sin y$?

(A) $\dfrac{2e^{2x}}{x^2 - \cos y}$

(B) $\dfrac{2e^{2x} - 2xy}{x^2 - \cos y}$

(C) $2e^{2x} - 2xy$
(D) $x^2 - \cos y$

Problems 19 through 21 are based on the following statement.

A two-dimensional function, $f(x, y)$, is defined as

$$f(x, y) = 2x^2 - y^2 + 3x - y$$

19. What is the gradient vector for this function?

(A) $\nabla f(x, y) = (2x^2 + 3x)\mathbf{i} + (-y^2 - y)\mathbf{j}$
(B) $\nabla f(x, y) = \left(x^3 + \frac{3}{2}x^2\right)\mathbf{i} + \left(-\frac{1}{3}y^3 - \frac{1}{2}y^2\right)\mathbf{j}$
(C) $\nabla f(x, y) = (4x + 3)\mathbf{i} + (-2y - 1)\mathbf{j}$
(D) $\nabla f(x, y) = (3x + 4)\mathbf{i} - (2y + 1)\mathbf{j}$

20. What is the direction of the line passing through the point $(1, -2)$ that has the maximum slope?

(A) $4\mathbf{i} + 2\mathbf{j}$
(B) $7\mathbf{i} + 3\mathbf{j}$
(C) $7\mathbf{i} + 4\mathbf{j}$
(D) $9\mathbf{i} - 7\mathbf{j}$

21. What is the magnitude of the slope at the point $(1, -2)$?

(A) 2.1
(B) 3.5
(C) 7.6
(D) 8.7

22. Determine the divergence of the vector function $\mathbf{F}(x, y, z)$.

$$\mathbf{F}(x, y, z) = xz\mathbf{i} + e^x y\mathbf{j} + 7x^3 y\mathbf{k}$$

(A) $z + e^x$
(B) $z + ye^x + 21x^2 y$
(C) $x + y$
(D) $x + ye^x$

23. Determine the curl of the vector function $\mathbf{F}(x, y, z)$.

$$\mathbf{F}(x, y, z) = 3x^2\mathbf{i} + 7e^x y\mathbf{j}$$

(A) $7e^x y$
(B) $7e^x y\mathbf{i}$
(C) $7e^x y\mathbf{j}$
(D) $7e^x y\mathbf{k}$

24. Determine the Laplacian of the scalar function $\frac{1}{3}x^3 - 9y + 5$ at the point $(3, 2, 7)$.

(A) 0
(B) 1
(C) 6
(D) 18

25. Find dy/dx for the parametric equations given.

$$x = 2t^2 - t$$
$$y = t^3 - 2t + 1$$

(A) $3t^2$
(B) $3t^2/2$
(C) $4t - 1$
(D) $(3t^2 - 2)/(4t - 1)$

SOLUTIONS TO FE-STYLE EXAM PROBLEMS

1. The critical points are located where the first derivative is zero.

$$f(x) = 5x^3 - 2x^2 + 1$$
$$f'(x) = 15x^2 - 4x$$
$$15x^2 - 4x = 0$$
$$x(15x - 4) = 0$$
$$x = 0 \quad \text{or} \quad x = 4/15$$

Test for a maximum, minimum, or inflection point.

$$f''(x) = 30x - 4$$
$$f''(0) = (30)(0) - 4$$
$$= -4$$
$$f''(a) < 0 \qquad \text{[maximum]}$$
$$f''\left(\frac{4}{15}\right) = (30)\left(\frac{4}{15}\right) - 4$$
$$= 4$$
$$f''(a) > 0 \qquad \text{[minimum]}$$

These could be a local maximum and minimum. Check the endpoints of the interval and compare with the function values at the critical points.

$$f(-2) = (5)(-2)^3 - (2)(-2)^2 + 1 = -47$$
$$f(2) = (5)(2)^3 - (2)(2)^2 + 1 = 33$$
$$f(0) = (5)(0)^3 - (2)(0)^2 + 1 = 1$$
$$f\left(\frac{4}{15}\right) = (5)\left(\frac{4}{15}\right)^3 - (2)\left(\frac{4}{15}\right)^2 + 1$$
$$= 0.95$$

The minimum and maximum values of the equation are at the endpoints, -47 and 33, respectively.

Answer is A.

2. The expression can be simplified by taking the natural logarithm of both sides. Then, use implicit differentiation.

$$(xy)^x = e$$
$$\ln(xy)^x = \ln(e)$$
$$x \ln(xy) = 1$$
$$\frac{d}{dx}(x \ln(xy)) = \frac{d}{dx}(1)$$

Use the product rule and chain rule to differentiate the left side.

$$x\left(\frac{1}{xy}\right)(xy' + y) + (1) \ln xy = 0$$
$$\left(\frac{1}{y}\right)(xy' + y) + \ln xy = 0$$
$$xy' + y = -y \ln xy$$
$$y' = \frac{-y - y \ln xy}{x}$$
$$= \left(\frac{-y}{x}\right)(1 + \ln xy)$$

Answer is C.

3. The slope of the radius line to point $(3,4)$ is $4/3$. Therefore, the slope of the tangent line is $-3/4$.

The point-slope form of a straight line with $(x_1, y_1) = (3, 4)$ is

$$y - 4 = \left(\frac{-3}{4}\right)(x - 3)$$

The standard form is

$$y = \frac{-3}{4}x + \left(\frac{9}{4} + 4\right)$$
$$= \frac{-3}{4}x + \frac{25}{4}$$

Answer is D.

4. Use the product rule.

$$\frac{d}{dx}\left(e^{-x} \sin 2x\right) = e^{-x} \frac{d}{dx}\left(\sin 2x\right)$$
$$+ (\sin 2x)\frac{d}{dx}\left(e^{-x}\right)$$
$$= e^{-x}\left(\cos 2x\right)(2)$$
$$+ (\sin 2x)(e^{-x})(-1)$$
$$= e^{-x}\left(2 \cos 2x - \sin 2x\right)$$

Answer is A.

5. The value of the function at the limit is

$$\frac{\pi^2 - \pi^2 + \sin \pi}{-\sin \pi} = \frac{0}{0}$$

L'Hôpital's rule can be used.

$$\lim_{x \to \pi}\left(\frac{x^2 - \pi x + \sin x}{-\sin x}\right) = \lim_{x \to \pi}\left(\frac{2x - \pi + \cos x}{-\cos x}\right)$$
$$= \frac{2\pi - \pi + \cos \pi}{-\cos \pi}$$
$$= \frac{\pi - 1}{-(-1)}$$
$$= \pi - 1$$

Answer is C.

6. To evaluate partial derivatives, all variables are taken as constants except the variable that the function's derivative is taken with respect to.

$$\frac{\partial}{\partial y}\left(\begin{array}{c}3x^2 + 9xy - \dfrac{y}{\ln z} \\ + \cos(z^2 + x)\end{array}\right) = 0 + 9x - \frac{1}{\ln z} + 0$$
$$= 9x - \frac{1}{\ln z}$$

Answer is A.

7. Break the fraction into parts and integrate each one separately.

$$\int \frac{x^3 + x + 4}{x^2}\,dx = \int \frac{x^3}{x^2}\,dx + \int \frac{x}{x^2}\,dx + \int \frac{4}{x^2}\,dx$$

$$= \int x\,dx + \int \frac{1}{x}\,dx + 4\int \frac{1}{x^2}\,dx$$

$$= \frac{x^2}{2} + \ln|x| + 4\left(\frac{x^{-1}}{-1}\right) + C$$

$$= \frac{x^2}{2} + \ln|x| - \frac{4}{x} + C$$

Answer is D.

8. For line 1,

$$\text{slope} = m = \frac{5-0}{5-0} = 1$$

$$(x_1, y_1) = (0,0)$$

The point-slope form is

$$y - 0 = (1)(x - 0)$$

The standard form is

$$y = x$$

For line 2,

$$m = \frac{y_1 - y_2}{x_1 - x_2} = \frac{3-0}{0-6}$$

$$= \frac{3}{-6} = -\frac{1}{2}$$

$$(x_1, y_1) = (0,3)$$

The point-slope form is

$$y - 3 = -\frac{1}{2}(x - 0)$$

The standard form is

$$y = -\frac{1}{2}x + 3$$

The intersection of lines 1 and 2 is where $y_1 = y_2$.

$$x = -\frac{1}{2}x + 3$$

$$\frac{3}{2}x = 3$$

$$x = 2$$

$$y = x = 2$$

The area of a triangle is

$$A = \tfrac{1}{2}bh = \left(\frac{1}{2}\right)(6)(2) = 6$$

Answer is C.

9. Simplify the expression.

$$\int \left(\frac{4}{8 + 2x^2}\right)dx = \frac{1}{2}\int \frac{dx}{1 + \frac{x^2}{4}}$$

Use a variable substitution for x.

$$u = \frac{x}{2}$$

$$du = \frac{dx}{2}$$

$$\frac{1}{2}\int \frac{dx}{1 + \frac{x^2}{4}} = \int \frac{du}{1 + u^2}$$

$$= \tan^{-1} u + C$$

$$= \tan^{-1} \frac{x}{2} + C$$

Answer is C.

10. Since the function's independent variable is x, convert the limits of integration to x. The function in terms of y is $x = 1/y$. The x limits are $1/2$ and $1/10$.

The integral of $f(x)$ represents the area under the curve $f(x)$ between the limits of integration.

$$A = \int_{x_1}^{x_2} f(x)\,dx$$

$$= \int_{\frac{1}{10}}^{\frac{1}{2}} \frac{1}{x}\,dx$$

$$= \ln x \Big|_{\frac{1}{10}}^{\frac{1}{2}}$$

$$= \ln\left(\frac{1}{2}\right) - \ln\left(\frac{1}{10}\right)$$

$$= 1.61 \quad (1.6)$$

Answer is B.

11. Find the intersection points by setting the two functions equal.

$$-2 + x^2 = 8 - x^2$$

$$2x^2 = 10$$

$$x = \pm\sqrt{5}$$

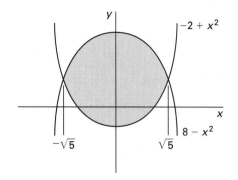

The integral of $f_1(x) - f_2(x)$ represents the area between the two curves between the limits of integration.

$$A = \int_{x_1}^{x_2} \left(f_1(x) - f_2(x) \right) \, dx$$

$$= \int_{-\sqrt{5}}^{\sqrt{5}} \left((8 - x^2) - (-2 + x^2) \right) \, dx$$

$$= \int_{-\sqrt{5}}^{\sqrt{5}} (10 - 2x^2) \, dx$$

$$= \left(10x - \left(\frac{2}{3} \right) x^3 \right) \Bigg|_{-\sqrt{5}}^{\sqrt{5}}$$

$$= 29.8$$

Answer is C.

12. The integral of $f(x)$ represents the area under the curve $f(x)$ between the limits of integration. However, since the value of $\sin x$ is negative in the range $\pi \leq x \leq 2\pi$, the total area would be calculated as zero if the integration was carried out in one step. The integral could be calculated over two ranges, but it is easier to exploit the symmetry of the sine curve.

$$A = \int_{x_1}^{x_2} f(x) \, dx$$

$$= \int_{0}^{2\pi} |\sin x| \, dx$$

$$= 2 \int_{0}^{\pi} \sin x \, dx$$

$$= -2 \, \cos x \Big|_{0}^{\pi}$$

$$= (-2)((-1) - 1) = 4$$

Answer is D.

13. The first and second derivatives are
$$f'(x) = 2x + 6$$
$$f'(8) = (2)(8) + 6 = 22$$
$$f''(x) = 2$$

The radius of curvature, R, is

$$R = \frac{[1 + f'(x)^2]^{\frac{3}{2}}}{|f''(x)|}$$

$$= \frac{[1 + (22)^2]^{\frac{3}{2}}}{2}$$

$$= 5340.5 \quad (5340)$$

Answer is D.

14. Let x represent the number of weeks.

The equation describing the price as a function of time is
$$\frac{\text{price}}{\text{lug}} = \$1 - \$0.025x$$

The equation describing the yield is
$$\text{lugs sold} = 1000 + (60 - 10)x$$
$$= 1000 + 50x$$

The revenue function is

$$R = \left(\frac{\text{price}}{\text{lug}} \right) (\text{lugs sold})$$

$$= (1 - 0.025x)(1000 + 50x)$$

$$= 1000 + 50x - 25x - 1.25x^2$$

$$= 1000 + 25x - 1.25x^2$$

To maximize the revenue function, set its derivative equal to zero.

$$\frac{dR}{dx} = 25 - 2.5x = 0$$

$$x = 10 \text{ weeks}$$

Answer is D.

15. The compound derivative identity is

$$Df(g(t)) = D_g f(g) D_t g(t)$$

In this case, $f = \sin^2$ and $g = \omega t$.

$$\frac{dy}{dt} = 2 \sin \omega t \left(\frac{d}{dt} \right) \sin \omega t$$

Apply the compound derivative identity once again.

$$\frac{dy}{dt} = (2 \sin \omega t)(\cos \omega t)(\omega)$$

$$= 2\omega \sin \omega t \cos \omega t$$

Answer is D.

16. The curve crosses the x-axis when $y = 0$.

$$10x^2 - 3x - 1 = 0$$

Use the quadratic equation or complete the square to determine the two values of x where the curve crosses the x-axis.

$$x^2 - 0.3x = 0.1$$
$$(x - 0.15)^2 = 0.1 + (0.15)^2$$
$$x = \pm 0.35 + 0.15$$
$$= -0.2, \ 0.5$$

The slope of the function is the first derivative.

$$\frac{dy}{dx} = 20x - 3$$

$$x = -0.2 : \ \frac{dy}{dx} = (20)(-0.2) - 3 = -7$$

$$x = 0.5 : \ \frac{dy}{dx} = (20)(0.5) - 3 = 7$$

Answer is D.

17. This can be solved by using calculus as

$$V = 2\pi \int_a^b x f(x) dx$$

$$= 2\pi \int_0^{\frac{3}{2}} x(2x^2) dx = 4\pi \int_0^{\frac{3}{2}} x^3 dx$$

$$= \pi x^4 \Big|_0^{\frac{3}{2}} = \frac{81\pi}{16}$$

However, it is just as expedient to solve as a paraboloid of revolution. From Fig. 4.21,

$$d = (2)\left(\frac{3}{2}\right) = 3$$

$$h = f\left(\frac{3}{2}\right) = (2)\left(\frac{3}{2}\right)^2 = 4.5$$

$$V = \frac{1}{8}\pi d^2 h = \frac{1}{8}\pi (3)^2 (4.5)$$

$$= 5.0625\pi \quad (81\pi/16)$$

Answer is D.

18.
$$x^2 y - e^{2x} = \sin y$$
$$f(x, y) = x^2 y - e^{2x} - \sin y = 0$$

Take the partial derivatives with respect to x and y.

$$\frac{\partial f(x, y)}{\partial x} = 2xy - 2e^{2x}$$
$$\frac{\partial f(x, y)}{\partial y} = x^2 - \cos y$$

Use implicit differentiation.

$$\frac{dy}{dx} = \frac{-\dfrac{\partial f(x, y)}{\partial x}}{\dfrac{\partial f(x, y)}{\partial y}} = \frac{2e^{2x} - 2xy}{x^2 - \cos y}$$

Answer is B.

19. It is necessary to calculate two partial derivatives in order to use Eq. 7.33.

$$\frac{\partial f(x, y)}{\partial x} = 4x + 3$$
$$\frac{\partial f(x, y)}{\partial y} = -2y - 1$$
$$\nabla f(x, y) = (4x + 3)\mathbf{i} + (-2y - 1)\mathbf{j}$$

Answer is C.

20. The direction of the line passing through $(1, -2)$ with maximum slope is found by inserting $x = 1$ and $y = -2$ into the gradient vector function.

$$\mathbf{V} = [(4)(1) + 3]\mathbf{i} + [(-2)(-2) - 1]\mathbf{j} = 7\mathbf{i} + 3\mathbf{j}$$

Answer is B.

21. The magnitude of the slope is

$$|\mathbf{V}| = \sqrt{(7)^2 + (3)^2} = \sqrt{58}$$
$$= 7.62 \quad (7.6)$$

Answer is C.

22. $\quad \text{div } \mathbf{F} = \dfrac{\partial}{\partial x}(xz) + \dfrac{\partial}{\partial y}(e^x y) + \dfrac{\partial}{\partial z}(7x^3 y)$

$$= z + e^x + 0$$
$$= z + e^x$$

Answer is A.

23. Using Eq. 7.37,

$$\text{curl } \mathbf{F} = \begin{vmatrix} \mathbf{i} & \mathbf{j} & \mathbf{k} \\ \dfrac{\partial}{\partial x} & \dfrac{\partial}{\partial y} & \dfrac{\partial}{\partial z} \\ 3x^2 & 7e^x y & 0 \end{vmatrix}$$

Expand the determinant across the top row.

$$\mathbf{i}\left(\frac{\partial}{\partial y}(0) - \frac{\partial}{\partial z}(7e^x y)\right) - \mathbf{j}\left(\frac{\partial}{\partial x}(0) - \frac{\partial}{\partial z}(3x^2)\right)$$

$$+ \mathbf{k}\left(\frac{\partial}{\partial x}(7e^x y) - \frac{\partial}{\partial y}(3x^2)\right)$$

$$= \mathbf{i}(0 - 0) - \mathbf{j}(0 - 0) + \mathbf{k}(7e^x y - 0)$$

$$= 7e^x y \mathbf{k}$$

Answer is D.

24. The Laplacian of the function is

$$\nabla^2 \left(\tfrac{1}{3}x^3 - 9y + 5\right) = \frac{\partial^2 \left(\tfrac{1}{3}x^3 - 9y + 5\right)}{\partial x^2}$$

$$+ \frac{\partial^2 \left(\tfrac{1}{3}x^3 - 9y + 5\right)}{\partial y^2}$$

$$+ \frac{\partial^2 \left(\tfrac{1}{3}x^3 - 9y + 5\right)}{\partial z^2}$$

$$= 2x + 0 + 0 = 2x$$

At $(3, 2, 7)$, $2x = (2)(3) = 6$.

Answer is C.

25.

$$\frac{dy}{dt} = 3t^2 - 2$$

$$\frac{dx}{dt} = 4t - 1$$

$$\frac{dy}{dx} = \frac{\dfrac{dy}{dt}}{\dfrac{dx}{dt}}$$

$$= \frac{3t^2 - 2}{4t - 1}$$

Answer is D.

Table 7.1 Derivatives and Indefinite Integrals

derivatives

$$\mathbf{D}k = 0$$

$$\mathbf{D}x^n = nx^{n-1}$$

$$\mathbf{D}\ln x = \frac{1}{x}$$

$$\mathbf{D}e^{ax} = ae^{ax}$$

$$\mathbf{D}\sin x = \cos x$$

$$\mathbf{D}\cos x = -\sin x$$

$$\mathbf{D}\tan x = \sec^2 x$$

$$\mathbf{D}\cot x = -\csc^2 x$$

$$\mathbf{D}\sec x = \sec x \tan x$$

$$\mathbf{D}\csc x = -\csc x \cot x$$

$$\mathbf{D}\arcsin x = \frac{1}{\sqrt{1-x^2}}$$

$$\mathbf{D}\arccos x = -\mathbf{D}\arcsin x$$

$$\mathbf{D}\arctan x = \frac{1}{1+x^2}$$

$$\mathbf{D}\operatorname{arccot} x = -\mathbf{D}\arctan x$$

$$\mathbf{D}\operatorname{arcsec} x = \frac{1}{x\sqrt{x^2-1}}$$

$$\mathbf{D}\operatorname{arccsc} x = -\mathbf{D}\operatorname{arcsec} x$$

$$\mathbf{D}kf(x) = k\mathbf{D}f(x)$$

$$\mathbf{D}(f(x) \pm g(x)) = \mathbf{D}f(x) \pm \mathbf{D}g(x)$$

$$\mathbf{D}(f(x) \cdot g(x)) = f(x)\mathbf{D}g(x) + g(x)\mathbf{D}f(x)$$

$$\mathbf{D}\left(\frac{f(x)}{g(x)}\right) = \frac{g(x)\mathbf{D}f(x) - f(x)\mathbf{D}g(x)}{(g(x))^2}$$

$$\mathbf{D}(f(x))^n = n(f(x))^{n-1}\mathbf{D}f(x)$$

$$\mathbf{D}f(g(x)) = \mathbf{D}_g f(g)\mathbf{D}_x g(x)$$

$$\mathbf{D}(f(x))^{g(x)} = g(x)(f(x))^{g(x)-1}\mathbf{D}f(x)$$
$$+ \ln(f(x))(f(x))^{g(x)}\mathbf{D}g(x)$$

indefinite integrals

$$\int k\,dx = kx + C$$

$$\int x^m\,dx = \frac{x^{m+1}}{m+1} + C \quad [m \neq -1]$$

$$\int \frac{1}{x}\,dx = \ln|x| + C$$

$$\int e^{kx}\,dx = \frac{e^{kx}}{k} + C$$

$$\int xe^{kx}\,dx = \frac{e^{kx}(kx-1)}{k^2} + C$$

$$\int k^{ax}\,dx = \frac{k^{ax}}{a\ln k} + C$$

$$\int \ln x\,dx = x\ln x - x + C$$

$$\int \sin x\,dx = -\cos x + C$$

$$\int \cos x\,dx = \sin x + C$$

$$\int \tan x\,dx = \ln\sec x + C$$

$$\int \cot x\,dx = \ln\sin x + C$$

$$\int \sec x\,dx = \ln(\sec x + \tan x) + C$$

$$\int \csc x\,dx = \ln(\csc x - \cot x) + C$$

$$\int \frac{dx}{k^2 + x^2} = \frac{1}{k}\arctan\frac{x}{k} + C$$

$$\int \sin^2 x\,dx = \frac{1}{2}x - \frac{1}{4}\sin 2x + C$$

$$\int \cos^2 x\,dx = \frac{1}{2}x + \frac{1}{4}\sin 2x + C$$

$$\int \tan^2 x\,dx = \tan x - x + C$$

$$\int kf(x)\,dx = k\int f(x)\,dx$$

$$\int (f(x) + g(x))\,dx = \int f(x)\,dx + \int g(x)\,dx$$

$$\int \frac{f'(x)}{f(x)}\,dx = \ln f(x) + C$$

$$\int f(x)\,dg(x) = f(x)\int dg(x) - \int g(x)\,df(x) + C$$
$$= f(x)g(x) - \int g(x)\,df(x) + C$$

MATHEMATICS
Calculus

8 Differential Equations and Transforms

Subjects

DIFFERENTIAL EQUATIONS 8-1
    Linear Homogeneous Differential
      Equations with Constant
      Coefficients 8-1
    Linear Nonhomogeneous Differential
      Equations with Constant
      Coefficients 8-2
FOURIER SERIES 8-2
PARSEVAL RELATION 8-3
LAPLACE TRANSFORMS 8-3
DIFFERENCE EQUATIONS 8-4
z-TRANSFORMS 8-4
EULER'S APPROXIMATION 8-5

DIFFERENTIAL EQUATIONS

A *differential equation* is a mathematical expression combining a function (e.g., $y = f(x)$) and one or more of its derivatives. The *order* of a differential equation is the highest derivative in it. *First-order differential equations* contain only first derivatives of the function, *second-order differential equations* contain second derivatives (and may contain first derivatives as well), and so on.

A *linear differential equation* can be written as a sum of products of multipliers of the function and its derivatives. If the multipliers are scalars, the differential equation is said to have *constant coefficients*. Equation 8.1 shows the general form of a linear differential equation with constant coefficients.

$$b_N \frac{d^N y(x)}{dx^N} + \cdots + b_1 \frac{dy(x)}{dx} + b_0 y(x) = f(x)$$
$$[b_i \text{ are constants}] \quad 8.1$$

If the function or one of its derivatives is raised to some power (other than one) or is embedded in another function (e.g., y embedded in $\sin y$ or e^y), the equation is said to be *nonlinear*.

Linear Homogeneous Differential Equations with Constant Coefficients

Each term of a *homogeneous differential equation* contains either the function or one of its derivatives. That is, the sum of the function and its derivative terms is equal to zero.

$$b_N \frac{d^N y(x)}{dx^N} + \cdots + b_1 \frac{dy(x)}{dx} + b_0 y(x) = 0 \quad 8.2$$

A *characteristic equation* can be written for a homogeneous linear differential equation with constant coefficients, regardless of order. This characteristic equation is simply the polynomial formed by replacing all derivatives with variables raised to the power of their respective derivatives.

$$P(r) = b_N r^N + b_{N-1} r^{N-1} + \cdots + b_1 r + b_0 \quad 8.3$$

Homogeneous linear differential equations are most easily solved by finding the n roots of the characteristic polynomial $P(r)$. There are two cases for real roots. If the roots of Eq. 8.3 are real and different, the solution is

$$y_h(x) = C_1 e^{r_1 x} + C_2 e^{r_2 x} + \cdots + C_N e^{r_N x} \quad 8.4$$

If the roots are real and the same, the solution is

$$y_h(x) = C_1 e^{rx} + C_2 x e^{rx} + \cdots + C_N x^{n-1} e^{rx} \quad 8.5$$

A homogeneous, first-order linear differential equation with constant coefficients has the general form of Eq. 8.6.

$$y' + ay = 0 \quad 8.6$$

The characteristic equation is $r + a = 0$ and has a root of $r = -a$. Equation 8.7 is the solution.

$$y(x) = Ce^{-ax} \quad 8.7$$

A second-order, homogeneous, linear differential equation has the general form

$$y'' + 2ay' + by = 0 \quad 8.8$$

The characteristic equation is

$$r^2 + 2ar + b = 0 \quad 8.9$$

The roots of the characteristic equation are

$$r_{1,2} = -a \pm \sqrt{a^2 - b} \quad 8.10$$

If $a^2 > b$, then the two roots are real and different, and the solution is

$$y = C_1 e^{r_1 x} + C_2 e^{r_2 x} \quad \text{[overdamped]} \qquad 8.11$$

If $a^2 = b$, then the two roots are real and the same, and the solution is

$$y = (C_1 + C_2 x)e^{rx} \quad \text{[critically damped]} \qquad 8.12$$

If $a^2 < b$, then the two roots are imaginary and of the form $(\alpha + i\beta)$ and $(\alpha - i\beta)$, and the solution is

$$y = e^{\alpha x}(C_1 \cos \beta x + C_2 \sin \beta x)$$

$$\text{[underdamped]} \quad 8.13$$

$$\alpha = -a \qquad 8.14$$

$$\beta = \sqrt{b - a^2} \qquad 8.15$$

Linear Nonhomogeneous Differential Equations with Constant Coefficients

In a nonhomogeneous differential equation, the sum of derivative terms is equal to a nonzero *forcing function* of the independent variable (i.e., $f(x)$ in Eq. 8.1 is nonzero). In order to solve a nonhomogeneous equation, it is often necessary to solve the homogeneous equation first. The homogeneous equation corresponding to a nonhomogeneous equation is known as a *reduced equation* or *complementary equation*. The complete solution to the nonhomogeneous differential equation is

$$y(x) = y_h(x) + y_p(x) \qquad 8.16$$

In Eq. 8.16, the term $y_h(x)$ is the *complementary solution*, which solves the complementary (i.e., homogeneous) case. The *particular solution*, $y_p(x)$, is any specific solution to the nonhomogeneous Eq. 8.1 that is known or can be found. Initial values are used to evaluate any unknown coefficients in the complementary solution after $y_h(x)$ and $y_p(x)$ have been combined. The particular solution will not have any unknown coefficients.

Two methods are available for finding a particular solution. The *method of undetermined coefficients*, as presented here, can be used only when $f(x)$ takes on one of the forms in Table 8.1.

The particular solution can be read from Table 8.1 if the forcing function is one of the forms given. Of course, the coefficients A_i and B_i are not known—these are the *undetermined coefficients*. The exponent s is the smallest non-negative number (and will be zero, one, or two, etc.), which ensures that no term in the particular solution is also a solution to the complementary equation. s must be determined prior to proceeding with the solution procedure.

Table 8.1 Particular Solutions

form of $f(x)$	form of $y_p(x)$
A	B
$Ae^{\alpha x}$	$Be^{\alpha x}$
$A_1 \sin \omega x + A_2 \cos \omega x$	$B_1 \sin \omega x + B_2 \cos \omega x$
$P_n(x) = a_0 x^n + a_1 x^{n-1} + \cdots + a_n$	$x^s(A_0 x^n + A_1 x^{n-1} + \cdots + A_n)$
$P_n(x)e^{\alpha x}$	$x^s(A_0 x^n + A_1 x^{n-1} + \cdots + A_n)e^{\alpha x}$
$P_n(x)e^{\alpha x} \begin{Bmatrix} \sin \omega x \\ \cos \omega x \end{Bmatrix}$	$x^s[(A_0 x^n + A_1 x^{n-1} + \cdots + A_n)e^{\alpha x} \cos \omega x + (B_0 x^n + B_1 x^{n-1} + \cdots + B_n)e^{\alpha x} \sin \omega x]$

$P_n(x)$ is a polynomial of degree n.

Once $y_p(x)$ (including s) is known, it is differentiated to obtain $dy_p(x)/dx$, $d^2 y_p(x)/dx^2$, and all subsequent derivatives. All of these derivatives are substituted into the original nonhomogeneous equation. The resulting equation is rearranged to match the forcing function, $f(x)$, and the unknown coefficients are determined, usually by solving simultaneous equations.

The purpose of solving a differential equation is to derive an expression for the function in terms of the independent variable. The expression does not need to be explicit in the function, but there can be no derivatives in the expression. Since, in the simplest cases, solving a differential equation is equivalent to finding an indefinite integral, it is not surprising that *constants of integration* must be evaluated from knowledge of how the system behaves. Additional data are known as *initial values*, and any problem that includes them is known as an *initial value problem*.

The presence of an exponential of the form e^{rx} in the solution indicates that *resonance* is present to some extent.

FOURIER SERIES

Any periodic waveform can be written as the sum of an infinite number of sinusoidal terms (i.e., an infinite series), known as *harmonic terms*. Such a sum of terms is known as a *Fourier series*, and the process of finding the terms is *Fourier analysis*. Since most series converge rapidly, it is possible to obtain a good approximation to the original waveform with a limited number of sinusoidal terms.

Fourier's theorem is Eq. 8.17. The object of a Fourier analysis is to determine the *Fourier coefficients* a_n and

b_n. The term $a_0/2$ can often be determined by inspection since it is the average value of the waveform.

$$f(t) = \frac{a_0}{2} + \sum_{n=1}^{\infty} [a_n \cos(n\omega t) + b_n \sin(n\omega t)] \quad \textit{8.17}$$

ω is the *natural (fundamental) frequency* of the waveform. It depends on the actual waveform *period*, τ. It is assumed that the Fourier analysis is performed over a full period, τ.

$$\omega = \frac{2\pi}{\tau} \quad \textit{8.18}$$

The coefficients a_n and b_n are found from the following relationships.

$$a_n = \frac{2}{\tau} \int_0^{\tau} f(t) \cos(n\omega t)\, dt \quad \textit{8.19}$$

$$b_n = \frac{2}{\tau} \int_0^{\tau} f(t) \sin(n\omega t)\, dt \quad \textit{8.20}$$

PARSEVAL RELATION

The *Parseval relation* (also known as *Parseval's equality*) can be used to calculate the root-mean-square (rms) value of a Fourier series that has been truncated after N terms. The rms value, f_N, is the square root of Eq. 8.21.

$$f_N^2 = \left(\frac{a_0}{2}\right)^2 + \left(\frac{1}{2}\right) \sum_{n=1}^{N} (a_n^2 + b_n^2) \quad \textit{8.21}$$

LAPLACE TRANSFORMS

Traditional methods of solving nonhomogeneous differential equations by hand are usually difficult and/or time consuming. *Laplace transforms* can be used to reduce many solution procedures to simple algebra.

Every mathematical function, $f(t)$, has a Laplace transform, written as $\mathcal{L}(f)$ or $F(s)$. The transform is written in the s-domain, regardless of the independent variable in the original function. The variable s is equivalent to a derivative operator, although it may be handled in the equations as a simple variable. Eq. 8.22 converts a function into a Laplace transform.

$$\mathcal{L}(f(t)) = F(s) = \int_0^{\infty} f(t)e^{-st}\, dt \quad \textit{8.22}$$

Generally, it is unnecessary to actually obtain a function's Laplace transform by use of Eq. 8.22. Tables of these transforms are readily available (see Table 8.2).

Extracting a function from its transform is the *inverse Laplace transform* operation. Although Eq. 8.23 could

be used and other methods exist, this operation is almost always done by finding the transform in a set of tables.

$$f(t) = \mathcal{L}^{-1}(F(s)) = \frac{1}{2\pi i} \int_{\sigma-i\infty}^{\sigma+i\infty} F(s)e^{st}\, dt \quad \textit{8.23}$$

The *initial value theorem* (IVT) is

$$\lim_{t \to 0} f(t) = \lim_{s \to \infty} sF(s) \quad \text{[if limits exist]} \quad \textit{8.24}$$

The *final value theorem* (FVT) is

$$\lim_{t \to \infty} f(t) = \lim_{s \to 0} sF(s) \quad \text{[if limits exist]} \quad \textit{8.25}$$

Working with Laplace transforms is simplified by the following two theorems.

- *linearity theorem:* If c is constant, then

$$\mathcal{L}(cf(t)) = c\mathcal{L}(f(t)) = cF(s) \quad \textit{8.26}$$

- *superposition theorem:* If $f(t)$ and $g(t)$ are different functions, then

$$\mathcal{L}(f(t) \pm g(t)) = \mathcal{L}(f(t)) \pm \mathcal{L}(g(t)) = F(s) \pm G(s) \quad \textit{8.27}$$

The Laplace transform method can be used with any linear differential equation with constant coefficients. Assuming the dependent variable is t, the basic procedure is as follows.

step 1: Put the differential equation in standard form (i.e., isolate the y'' term).

$$y'' + b_1 y' + b_2 y = f(t) \quad \textit{8.28}$$

step 2: Take the Laplace transform of both sides. Use the linearity and superposition theorems, Eqs. 8.26 and 8.27.

$$\mathcal{L}(y'') + b_1 \mathcal{L}(y') + b_2 \mathcal{L}(y) = \mathcal{L}(f(t)) \quad \textit{8.29}$$

step 3: Use the following relationships to expand the equation.

$$\mathcal{L}(y'') = s^2 \mathcal{L}(y) - sy(0) - y'(0) \quad \textit{8.30}$$
$$\mathcal{L}(y') = s\mathcal{L}(y) - y(0) \quad \textit{8.31}$$

step 4: Use algebra to solve for $\mathcal{L}(y)$.

step 5: If needed, use partial fractions to simplify the expression for $\mathcal{L}(y)$.

step 6: Take the inverse transform to find $y(t)$.

$$y(t) = \mathcal{L}^{-1}(\mathcal{L}(y)) \quad \textit{8.32}$$

Table 8.2 Laplace Transforms

$f(t)$	$\mathcal{L}(f(t))$
$\delta(t)$ (unit impulse at $t = 0$)	1
$\delta(t - c)$ (unit impulse at $t = c$)	e^{-cs}
$u(t)$ (unit step at $t = 0$)	$\dfrac{1}{s}$
u_c (unit step at $t = c$)	$\dfrac{e^{-cs}}{s}$
$tu(t)$ (unit ramp at $t = 0$)	$\dfrac{1}{s^2}$
t^n (n is a positive integer)	$\dfrac{n!}{s^{n+1}}$
$e^{-\alpha t}$	$\dfrac{1}{s + \alpha}$
$te^{-\alpha t}$	$\dfrac{1}{(s + \alpha)^2}$
$e^{-\alpha t} \sin \beta t$	$\dfrac{\beta}{(s + \alpha)^2 + \beta^2}$
$e^{-\alpha t} \cos \beta t$	$\dfrac{s + \alpha}{(s + \alpha)^2 + \beta^2}$
$\mathcal{L}(f^{(n)}(t))$ [n^{th} derivative]	$-f^{(n-1)}(0) - sf^{n-2}(0)$ $\cdots - s^{n-1}f(0)$ $+ s^n F(s)$
$\int_0^t f(u)du$	$\dfrac{1}{s}F(s)$
$tf(t)$	$-\dfrac{dF}{ds}$
$\dfrac{1}{t}f(t)$	$\int_s^\infty F(u)du$

DIFFERENCE EQUATIONS

Many processes can be accurately modeled by differential equations. However, in some cases, exact solutions to these models are difficult to obtain. In such cases, discrete versions of the original differential equations can be produced. These discrete equations are known as *finite difference equations* or just *difference equations*. Communication signal processing, heat transfer, and traffic flow are just a few of the applications of difference equations.

Difference equations are also ideal for modeling processes whose states or values are restricted to certain specified (equally spaced) points in time or space. Although simple difference equations can be solved by hand, in practice, they are solved by computer using numerical analysis techniques.

A difference equation is a relationship between a function and its differences over some interval of integers. (This is analogous to a differential equation that is a relationship of functions and their derivatives over some interval of real numbers.) The *order* of the difference equation is the number of differences that are in the equation. For example, a *first-order difference equation* is a relationship between the values of some function at two consecutive points in time or space.

$$y_k = f(y_{k-1}) \qquad 8.33$$

A loan balance can be modeled as a first-order difference equation. Given an effective compounding rate per period of i, the balance on the loan in the previous $(k-1)$ period, and a payment made in the previous period, the balance in the current (k) period is

$$y_k = y_{k-1}(1 + i) - A \qquad 8.34$$

The general form of a difference equation combines all of the difference terms into the left side. Thus, the general form of the loan balance model would be

$$y_k - y_{k-1}(1 + i) = -A \qquad 8.35$$

In a *homogeneous difference equation*, the right-hand side is zero. In a *linear difference equation*, all of the difference terms are multiplied by scalars.

A *second-order difference equation* is a relationship between the values of some function at three consecutive points in time or space.

$$y_k = f(f_{k-1},\ f_{k-2}) \qquad 8.36$$

Some forecasting techniques (used for sales and inventory modeling) make use of exponential smoothing. The forecasted quantity for the current period is some scalar combination of the quantity in two (or more) periods. The scalar coefficients, c_i, are determined experimentally or by analysis to complete the model. This results in a homogeneous, second-order difference equation.

$$Q_k = c_1 Q_{k-1} + c_2 Q_{k-2} \qquad 8.37$$

$$Q_k - c_1 Q_{k-1} - c_2 Q_{k-2} = 0 \qquad 8.38$$

z-TRANSFORMS

z-transforms are powerful tools used for solving difference equations. *z*-transforms are used extensively in discrete signal processing (DSP). In DSP, the *z*-transform changes discrete time domain signals into the complex-variable frequency domain. *z*-transforms do for discrete signals what the Laplace transform does for continuous time domain: They convert complex problems into simpler problems that can be solved algebraically.

The z-transform is derived from the Laplace transform of a train of pulses. For a function, $f(k)$, in the k-domain, the transform, $F(z)$, in the z-domain is found as the weighted summation of a series of values of the function. The value of the transform for various simple applications is given in Table 8.3.

$$F(z) = \sum f(n)z^{-n} \qquad 8.39$$

If the summation in Eq. 8.39 is from $n = -\infty$ to $n = +\infty$, a *two-sided z-transform* will be derived. This definition is for a noncausal signal (i.e., one for which $f(n)$ is known for $n < 0$). Most applications are causal (i.e., one for which $f(n)$ is not known for $n < 0$. If the summation is from $n = 0$ to $n = +\infty$, a *one-sided z-transform* will be derived.

A unit impulse will have a value of 1 at $n = 0$. (In the time domain, this would be a value of 1 at $t = 0$.) From Eq. 8.39, the z-transform is

$$F(\text{unit impulse}) = f(0)z^{-0} = (1)(1) = 1 \qquad 8.40$$

A unit step will have a value of 1 for $n \geq 0$. Since $f(n) = 1$ at all times, from Eq. 8.39,

$$F(\text{unit step}) = \sum z^{-n} = \sum (z^{-1})^n \qquad 8.41$$

Using the series expansion $\sum a^n = 1/(1-a)$,

$$F(\text{unit step}) = \frac{1}{1 - z^{-1}} \qquad 8.42$$

More z-transform pairs are given in Table 8.3.

Table 8.3 z-Transforms

$f(k)$	$F(z)$
$\delta(k)$, impulse at $k = 0$	1
$u(k)$, step at $k = 0$	$\dfrac{1}{1 - z^{-1}}$
β^k	$\dfrac{1}{1 - \beta z^{-1}}$
$y(k-1)$	$z^{-1}Y(z) + y(-1)$
$y(k-2)$	$z^{-2}Y(z) + y(-2) + y(-1)z^{-1}$
$y(k+1)$	$zY(z) - zy(0)$
$y(k+2)$	$z^2Y(z) - z^2y(0) - zy(1)$
$\displaystyle\sum_{m=0}^{\infty} X(k-m)h(m)$	$H(z)X(z)$

The *initial value theorem* (IVT) for z-transforms is

$$\lim_{k \to 0} f(k) = \lim_{z \to \infty} F(z) \qquad 8.43$$

The *final value theorem* (FVT) for z-transforms is

$$\lim_{k \to \infty} f(k) = \lim_{z \to 1}(1 - z^{-1})F(z) \qquad 8.44$$

EULER'S APPROXIMATION

Euler's approximation is a method for estimating the value of a function given the value and slope of the function at an adjacent location. The simplicity of the concept is illustrated by writing Euler's approximation in terms of the traditional two-dimensional x-y coordinate system.

$$y(x_2) = y(x_1) + (x_2 - x_1)y'(x_1) \qquad 8.45$$

As long as the derivative can be evaluated, Euler's method can be used to predict the value of any function whose values are limited to discrete, sequential points in time or space (e.g., a difference equation).

$$x_{i+1} = x_i + \frac{dx_i}{dt}\Delta t \qquad 8.46$$

The error associated with Euler's approximation is zero for linear systems. Euler's approximation can be used as a quick estimate for nonlinear systems as long as the presence of error is recognized. Figure 8.1 shows the geometric interpretation of Euler's approximation for a curvilinear function.

Figure 8.1 Geometric Interpretation of Euler's Approximation

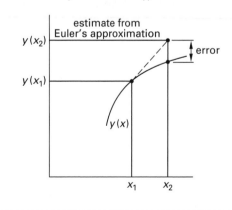

SAMPLE PROBLEMS

Problem 1

What is the general solution to the following second-order differential equation if A and B are constants?

$$y'' + 9y = 0$$

(A) $y = Ae^{3x} + Bxe^{-3x}$
(B) $y = Ae^{3x} + Be^{-3x}$
(C) $y = Ae^{x} - Bxe^{-x}$
(D) $y = A\cos 3x + B\sin 3x$

Solution

A second-order, homogeneous equation of the form $y'' + 2ay' + by = 0$ has a characteristic equation $r^2 + 2ar + b = 0$ and roots $-a \pm \sqrt{a^2 - b}$. The characteristic equation is

$$r^2 + 9 = 0$$

The roots are

$$r_{1,2} = \pm\sqrt{-9}$$
$$= \pm 3i \qquad [i = \sqrt{-1}]$$

Since $a^2 < b$, the solution to the equation is of the form

$$y = e^{\alpha x}(A\cos\beta x + B\sin\beta x)$$
$$\alpha = -a = 0$$
$$\beta = \sqrt{b - a^2} = \sqrt{9 - (0)^2}$$
$$= 3$$
$$y = e^{0x}(A\cos 3x + B\sin 3x)$$
$$= A\cos 3x + B\sin 3x$$

Answer is D.

Problem 2

Which of the following is the general solution to the differential equation and boundary conditions?

$$\frac{dy}{dt} - 5y = 0$$
$$y(0) = 3$$

(A) $-\dfrac{1}{3}e^{-5t}$

(B) $3e^{5t}$

(C) $5e^{-3t}$

(D) $\dfrac{1}{5}e^{-3t}$

Solution

This is a first-order, linear differential equation. The characteristic equation is

$$r - 5 = 0$$

The root is

$$r = 5$$

The solution is of the form

$$y = Ce^{5t}$$

The initial condition is used to find C.

$$y(0) = Ce^{(5)(0)} = 3$$
$$C = 3$$
$$y = 3e^{5t}$$

Answer is B.

Problem 3

What is the Laplace transform of a time-dependent function of magnitude e^{-at}?

(A) $\dfrac{1}{a}$

(B) $\dfrac{a}{s}$

(C) $\dfrac{1}{s + a}$

(D) $\dfrac{1}{s - a}$

Solution

By definition,

$$F(s) = \int_0^\infty f(t)e^{-st}\,dt$$
$$f(t) = e^{-at}$$
$$F(s) = \int_0^\infty e^{-at}e^{-st}\,dt = \int_0^\infty e^{-(s+a)t}\,dt$$
$$= \frac{-1}{s + a}\int_0^\infty e^{-(s+a)t}(-(s + a))\,dt$$
$$= \frac{-e^{-(s+a)t}}{s + a}\Bigg|_0^\infty$$
$$= 0 - \left(-\frac{1}{s + a}\right)$$
$$F(s) = \frac{1}{s + a}$$

This could also be determined directly from a table.

Answer is C.

FE-STYLE EXAM PROBLEMS

1. Solve the following differential equation.

$$y'' + 4y' + 4y = 0$$
$$y(0) = 1$$
$$y'(0) = 0$$

(A) $y = (1 - 2x)e^{2x}$
(B) $y = (2 - x)e^{-2x}$
(C) $y = (2 + x)e^{-2x}$
(D) $y = (1 + 2x)e^{-2x}$

2. What is the correct general solution for the following differential equation?

$$\frac{d^2y}{dx^2} + 2\frac{dy}{dx} + 2y = 0$$

(A) $y = C_1 \sin x - C_2 \cos x$
(B) $y = C_1 \cos x - C_2 \sin x$
(C) $y = C_1 \cos x + C_2 \sin x$
(D) $y = e^{-x}(C_1 \cos x + C_2 \sin x)$

Problems 3–5 refer to the following equation and initial conditions.

$$8y = e^{-2x} - 10y' - 2y''$$
$$y(0) = 1$$
$$y'(0) = -\frac{3}{2}$$

3. What type of differential equation is shown?

(A) nonlinear, second-order, nonhomogeneous
(B) linear, second-order, homogeneous
(C) linear, second-order, nonhomogeneous
(D) linear, third-order, nonhomogeneous

4. Which of the following statements is true for the equation?

(A) The system defined by the equation is unstable.
(B) The complete solution to the equation will be the sum of a complementary solution and a particular solution.
(C) The equation may be solved by successive integrations.
(D) An integrating factor must be used to solve the equation.

5. Which of the following is a complete solution to the equation?

(A) $y = \frac{9}{4}e^x - \ln(2x)$

(B) $y = \frac{9}{4}e^x - 2e^{4x}$

(C) $y = \frac{41}{108}e^{-x} - \frac{11}{108}e^{-4x} + \frac{1}{36}e^{-2x}$

(D) $y = e^{-x} + \frac{1}{4}e^{-4x} - \frac{1}{4}e^{-2x}$

For the following problems use the NCEES Handbook as your only reference.

6. What is the Laplace transform of $\sin t \cos t$?

(A) $\dfrac{1}{s^2 + 4}$

(B) $\dfrac{1}{s + 2}$

(C) $\dfrac{1}{2t + 2}$

(D) $\dfrac{1}{2s + 4}$

Problems 7 and 8 refer to the transform function $\mathcal{L}(s)$.

$$\mathcal{L}(s) = \frac{20}{s(s + 10)}$$

7. What is the partial fraction expansion of $\mathcal{L}(s)$?

(A) $\dfrac{1}{s} - \dfrac{1}{s + 10}$

(B) $\dfrac{1}{s} - \dfrac{2}{s + 10}$

(C) $\dfrac{2}{s} - \dfrac{1}{s + 10}$

(D) $\dfrac{2}{s} - \dfrac{2}{s + 10}$

8. What is the inverse Laplace transform of $\mathcal{L}(s)$?

(A) Ate^{-10t}
(B) $Ae^{-t} + Be^{-10t}$
(C) $A + Be^{-10t}$
(D) $Ae^{-5t} \sin 5t$

Problems 9 and 10 refer to the Laplace transform function $\mathcal{L}(s)$.

$$\mathcal{L}(s) = \frac{s(s + 10)}{(s + 5)(s + 15)}$$

9. What is the partial fraction expansion of $\mathcal{L}(s)$?

(A) $\dfrac{2.5}{s+5} - \dfrac{7.5}{s+15}$

(B) $\dfrac{15}{s+5} - \dfrac{5}{s+15}$

(C) $1 - \dfrac{75}{s+5} - \dfrac{25}{s+15}$

(D) $1 - \dfrac{2.5}{s+5} - \dfrac{7.5}{s+15}$

10. What is the inverse Laplace transform of $\mathcal{L}(s)$?

(A) $C_1 e^{-5t} + C_2 e^{-15t}$
(B) $C_1 e^{-3.75t} + C_2 e^{-2t}$
(C) $C_1 e^{-10t} + C_2 e^{-5t} + C_3 e^{-15t}$
(D) $C_1 + C_2 e^{-5t} + C_3 e^{-15t}$

11. What is the inverse transform of the Laplace transform function, $\mathcal{L}(s)$?

$$\mathcal{L}(s) = \frac{20s}{s^2 + 20s + 75}$$

(A) $Ae^{-15t} + Be^{-5t}$
(B) $Ate^{-7.5t}$
(C) $Ae^{-5t}\cos 15t$
(D) $Ae^{-15t}\cos 5t$

12. What are the first three sets of terms in the Fourier series of the repeating function illustrated?

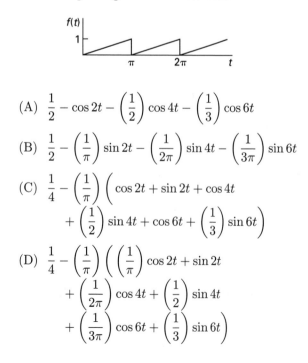

(A) $\dfrac{1}{2} - \cos 2t - \left(\dfrac{1}{2}\right)\cos 4t - \left(\dfrac{1}{3}\right)\cos 6t$

(B) $\dfrac{1}{2} - \left(\dfrac{1}{\pi}\right)\sin 2t - \left(\dfrac{1}{2\pi}\right)\sin 4t - \left(\dfrac{1}{3\pi}\right)\sin 6t$

(C) $\dfrac{1}{4} - \left(\dfrac{1}{\pi}\right)\left(\cos 2t + \sin 2t + \cos 4t \right.$
$\left. + \left(\dfrac{1}{2}\right)\sin 4t + \cos 6t + \left(\dfrac{1}{3}\right)\sin 6t\right)$

(D) $\dfrac{1}{4} - \left(\dfrac{1}{\pi}\right)\left(\left(\dfrac{1}{\pi}\right)\cos 2t + \sin 2t \right.$
$+ \left(\dfrac{1}{2\pi}\right)\cos 4t + \left(\dfrac{1}{2}\right)\sin 4t$
$\left. + \left(\dfrac{1}{3\pi}\right)\cos 6t + \left(\dfrac{1}{3}\right)\sin 6t\right)$

13. What is the complementary solution to the following differential equation?

$$y'' - 4y' + \left(\frac{25}{4}\right)y = 10\cos 8x$$

(A) $2C_1 x + C_2 x - C_3 x$
(B) $C_1 e^{2x} + C_2 e^{1.5x}$
(C) $C_1 e^{2x}\cos 1.5x + C_2 e^{2x}\sin 1.5x$
(D) $C_1 e^x \tan x + C_2 e^x \cot x$

14. What is the Laplace transform of the step function $f(t)$?

$$f(t) = u(t-1) + u(t-2)$$

(A) $e^{-s} + e^{-2s}$

(B) $\dfrac{e^{-s} + e^{-2s}}{s}$

(C) $1 + \dfrac{e^{-2s}}{s}$

(D) $\dfrac{e^s}{s} + \dfrac{e^{2s}}{s}$

15. A man borrows $10,000 now at a 15% effective interest per payment period. Each loan payment totals $3000. After the first payment is made, what is the outstanding balance on the loan?

(A) $7000
(B) $8050
(C) $8500
(D) $13,000

16. A woman bought a car and financed $15,500 of the purchase price. The effective interest rate per payment period was 1%. She made two payments of $350 and then decided to pay off the remaining balance in one lump sum. What was the approximate balance?

(A) $14,800
(B) $14,900
(C) $15,000
(D) $15,100

17. The values of an unknown function follow a Fibonacci number sequence. It is known that $f(1) = 4$ and $f(2) = 1.3$. What is $f(4)$?

(A) -4.1
(B) 0.33
(C) 2.7
(D) 6.6

18. What is the complete solution to the following differential equation?

$$y' + 8y = 16$$
$$y(0) = 5$$

(A) $-11e^{-8x} + 16$
(B) $e^{-2x} + 4$
(C) $3e^{-8x} + 2$
(D) $5e^{-8x} + 2$

19. What is the inverse-transform of the z-transform function $F(z)$?

$$F(z) = \frac{z}{z-5}$$

(A) e^{-5k}
(B) $2k$
(C) 2^k
(D) 5^k

20. Find the general solution of the following differential equation.

$$(x^2 + 9)\frac{dy}{dx} = xy$$

(A) $y = \sqrt{x^2+9} + C$
(B) $y = x^2 + 9$
(C) $y = C\sqrt{x^2+9}$
(D) $y = x^2 - 9$

21. Use Euler's approximation to determine $x(1.5)$, the value of a function at 1.5, given increments of time of 0.25, $x(1) = 1$, and $dx/dt = 2x$.

(A) 0
(B) 0.75
(C) 1.5
(D) 2.25

SOLUTIONS TO FE-STYLE EXAM PROBLEMS

1. The characteristic equation for the second-order, homogeneous linear equation is

$$r^2 + 4r + 4 = 0$$

The roots are

$$r_{1,2} = -a \pm \sqrt{a^2 - b}$$
$$= -2 \pm \sqrt{(2)^2 - 4}$$
$$= -2, -2$$

Since $a^2 = b$, the solution is

$$y = (C_1 + C_2 x)e^{rx}$$
$$= (C_1 + C_2 x)e^{-2x}$$

Evaluate the equation at the initial conditions.

$$y(0) = 1$$
$$1 = (C_1 + C_2(0))e^{(-2)(0)}$$
$$C_1 = 1$$

Differentiate the expression for y to obtain y'.

$$y' = (C_1 + C_2 x)(e^{-2x})(-2) + C_2 e^{-2x}$$
$$y'(0) = 0$$
$$0 = -2e^{(-2)(0)}(C_1 + C_2(0)) + C_2 e^{(-2)(0)}$$
$$= (-2)(1) + C_2$$
$$C_2 = 2$$
$$y = (1 + 2x)e^{-2x}$$

Answer is D.

2. The characteristic equation is

$$r^2 + 2r + 2 = 0$$

The roots are

$$r_{1,2} = -a \pm \sqrt{a^2 - b}$$
$$= -1 \pm \sqrt{(1)^2 - 2}$$
$$= (-1+i), (-1-i)$$

Since $a^2 < b$, the solution is

$$y = e^{\alpha x}(C_1 \cos \beta x + C_2 \sin \beta x)$$
$$\alpha = -a = -1$$
$$\beta = \sqrt{b - a^2} = \sqrt{2 - (1)^2}$$
$$= 1$$
$$y = e^{-x}(C_1 \cos x + C_2 \sin x)$$

Answer is D.

3. Rearrange the terms into the general form of a differential equation.

$$2y'' + 10y' + 8y = e^{-2x}$$

This is a linear, second-order, nonhomogeneous differential equation.

Answer is C.

4. Choice (B) is always true. The other choices are either false or are special cases.

Answer is B.

5. The complete solution is the sum of the complementary and particular solutions.

Complementary Solution

The homogeneous equation is

$$2y'' + 10y' + 8y = 0$$

The characteristic equation is

$$r^2 + \frac{10}{2}r + \frac{8}{2} = 0$$

$$r^2 + (2)\left(\frac{5}{2}\right)r + 4 = 0$$

The roots are

$$r_{1,2} = -\frac{5}{2} \pm \sqrt{\left(\frac{5}{2}\right)^2 - 4}$$

$$= -\frac{5}{2} \pm \sqrt{\frac{25 - 16}{4}}$$

$$= -1, -4$$

Therefore,

$$a^2 = \left(\frac{5}{2}\right)^2 = \frac{25}{4} > b = 4$$

$$y_h = C_1 e^{r_1 x} + C_2 e^{r_2 x}$$

$$= C_1 e^{-x} + C_2 e^{-4x}$$

Particular Solution

Assume the particular solution is of the form e^{-2x} since that is the form of the nonhomogeneous forcing function.

$$y_p = C_3 e^{-2x}$$

The first and second derivatives are

$$y_p' = -2C_3 e^{-2x}$$
$$y_p'' = 4C_3 e^{-2x}$$

$$2y'' + 10y' + 8y = e^{-2x}$$
$$(2)(4C_3 e^{-2x}) + (10)(-2C_3 e^{-2x})$$
$$+ (8)(C_3 e^{-2x}) = e^{-2x}$$
$$8C_3 - 20C_3 + 8C_3 = 1$$
$$C_3 = -\frac{1}{4}$$

Complete Solution

$$y = y_h + y_p$$

$$= C_1 e^{-x} + C_2 e^{-4x} - \frac{1}{4} e^{-2x}$$

Evaluate the unknown coefficients.

$$y(0) = 1 = C_1 e^{-(0)} + C_2 e^{-(4)(0)} - \frac{1}{4} e^{(-2)(0)}$$

$$1 = C_1 + C_2 - \frac{1}{4}$$

$$C_1 + C_2 = \frac{5}{4}$$

$$y'(0) = -\frac{3}{2}$$

$$= -C_1 e^{-(0)} - 4C_2 e^{(-4)(0)} + \frac{1}{2} e^{(-2)(0)}$$

$$-\frac{3}{2} = -C_1 - 4C_2 + \frac{1}{2}$$

$$-2 = -C_1 - 4C_2$$

$$-2 = -\left(\frac{5}{4} - C_2\right) - 4C_2$$

$$C_2 = \frac{1}{4}$$

$$C_1 = \frac{5}{4} - \frac{1}{4} = \frac{4}{4}$$
$$= 1$$

$$y = e^{-x} + \frac{1}{4} e^{-4x} - \frac{1}{4} e^{-2x}$$

Answer is D.

6. The double-angle formula for $\sin 2t$ is

$$2 \sin t \cos t = \sin 2t$$
$$\sin t \cos t = \frac{1}{2} \sin 2t$$
$$\mathcal{L}\{\sin t \cos t\} = \frac{1}{2}\mathcal{L}\{\sin 2t\}$$
$$= \frac{1}{2}\left(\frac{2}{s^2 + 4}\right) \quad \text{[using } \alpha = 0\text{]}$$
$$= \frac{1}{s^2 + 4}$$

Answer is A.

7. The general form of the partial expansion is

$$\frac{A}{s} + \frac{B}{s + 10}$$

A and B are unknown constants that are to be determined. Use the method of undetermined coefficients. Multiply the first term by $(s+10)/(s+10)$. Multiply the second term by s/s. Combine numerators over a common denominator.

$$\frac{A(s+10)+Bs}{s(s+10)}$$

Since the original function is $20/s(s+10)$, the numerators must be equal. (The denominators are already identical.)

$$A(s+10)+Bs=20$$
$$(A+B)s+10A=20$$

Since the constant terms are equal, $A=2$. Since there are no s terms in the original numerator, $A+B=0$; $B=-2$.

The partial fraction expansion is

$$\frac{2}{s}-\frac{2}{s+10}$$

Answer is D.

8. From Prob. 7, the Laplace transform is

$$\mathcal{L}(s)=\frac{2}{s}-\frac{2}{s+10}$$

Taking the inverse,

$$\mathcal{L}^{-1}(\mathcal{L}(s))=\mathcal{L}^{-1}\left(\frac{2}{s}\right)-\mathcal{L}^{-1}\left(\frac{2}{s+10}\right)$$

$$=2-2e^{-10t} \quad [\text{in the form of } A+Be^{-10t}]$$

Answer is C.

9. The denominator of $\mathcal{L}(s)$ is

$$(s+5)(s+15)=s^2+20s+75$$

The numerator of $\mathcal{L}(s)$ is

$$s(s+10)=s^2+10s$$

Since the order of the numerator and denominator are both second, use long division to obtain an equivalent expression with a lower-order numerator.

$$s^2+20s+75 \enclose{longdiv}{s^2+10s+0} \quad \overset{1}{}$$
$$-(s^2+20s+75)$$
$$\overline{0-10s-75}$$

Therefore,

$$\frac{s(s+10)}{(s+5)(s+15)}=1+\frac{-10s-75}{s^2+20s+75}$$

The general form of the partial expansion is assumed to be

$$1-\frac{A}{s+5}-\frac{B}{s+15}$$

A and B are unknown constants that are to be determined. Use the method of undetermined coefficients on the second and third terms to determine A and B. Multiply the second term by $(s+15)/(s+15)$. Multiply the third term by $(s+5)/(s+5)$. Combine numerators over a common denominator.

$$\frac{A(s+15)+B(s+5)}{(s+5)(s+15)}$$

Since the original function is $(10s+75)/(s+5)(s+15)$, the numerators must be equal. (The denominators are already identical.)

$$A(s+15)+B(s+5)=10s+75$$
$$(A+B)s+15A+5B=10s+75$$

Therefore, the simultaneous equations are

$$A+B=10$$
$$15A+5B=75$$

These have a solution of $A=2.5$ and $B=7.5$.

The partial fraction expansion is

$$\mathcal{L}(s)=1-\frac{2.5}{s+5}-\frac{7.5}{s+15}$$

Answer is D.

10. From Prob. 9, the partial fraction expansion of $\mathcal{L}(s)$ is

$$\mathcal{L}(s)=1-\frac{2.5}{s+5}-\frac{7.5}{s+15}$$

$$\mathcal{L}^{-1}(\mathcal{L}(s))=\mathcal{L}^{-1}(1)-\mathcal{L}^{-1}\left(\frac{2.5}{s+5}\right)$$
$$-\mathcal{L}^{-1}\left(\frac{7.5}{s+15}\right)$$

$$=\delta(t)-2.5e^{-5t}-7.5e^{-15t}$$
$$(C_1-C_2e^{-5t}-C_3e^{-15t})$$

Answer is D.

11. Factor the denominator into two linear terms.

$$s^2+20s+75=(s+5)(s+15)$$

Once this step is done, the solution proceeds normally.

The general form of the partial expansion is assumed to be

$$\frac{A}{s+5} + \frac{B}{s+15}$$

A and B are unknown constants that are to be determined. Use the method of undetermined coefficients on the first and second terms to determine A and B. Multiply the first term by $(s+15)/(s+15)$. Multiply the second term by $(s+5)/(s+5)$. Combine numerators over a common denominator.

$$\frac{A(s+15) + B(s+5)}{(s+5)(s+15)}$$

Since the original function is $20s/(s+5)(s+15)$, the numerators must be equal. (The denominators are already identical.)

$$A(s+15) + B(s+5) = 20s$$
$$(A+B)s + 15A + 5B = 20s$$

Therefore, the simultaneous equations are

$$A + B = 20$$
$$15A + 5B = 0$$

These have a solution of $A = -10$ and $B = 30$.

Insert the known values of A and B.

$$\mathcal{L}(s) = \frac{-10}{s+5} + \frac{30}{s+15}$$
$$\mathcal{L}^{-1}(\mathcal{L}(s)) = -10e^{-5t} + 30e^{-15t} \quad \left[\begin{array}{c}\text{in the form of}\\ Ae^{-15t} + Be^{-5t}\end{array}\right]$$

Answer is A.

12. A Fourier series has the form of

$$f(t) = \frac{a_0}{2}$$
$$+ a_1 \cos\omega t + b_1 \sin\omega t$$
$$+ a_2 \cos 2\omega t + b_2 \sin 2\omega t$$
$$+ a_3 \cos 3\omega t + b_3 \sin 3\omega t + \cdots$$

The constant term $a_0/2$ corresponds to the average of the function. The average is seen by observation to be $1/2$. Therefore, $a_0/2 = 1/2$.

The waveform has odd symmetry. (In other words, $f(-t) = -f(t)$. This is more obvious if the t-axis is raised so that it passes through $1/2$ and the $f(t)$ axis is shifted $\pi/2$. This changes the average value, but not the shape of the waveform.) Therefore, all a_n are zero, and only sine terms appear in the Fourier series.

Only choice (B) satisfies both of these requirements.

Alternatively, the values can be derived, though this would be a lengthy process.

Answer is B.

13. The complementary solution to a nonhomogeneous differential equation is the solution of the homogeneous differential equation.

The characteristic equation is

$$r^2 + 2ar + b = 0$$

So, $a = -2$ and $b = 25/4$.

The roots are

$$r_{1,2} = -a \pm \sqrt{a^2 - b}$$
$$= -(-2) \pm \sqrt{(-2)^2 - \frac{25}{4}}$$
$$= 2 \pm 1.5i$$

Since the roots are imaginary, the homogeneous solution has the form of

$$y = e^{\alpha x}(C_1 \cos \beta x + C_2 \sin \beta x)$$
$$\alpha = 2$$
$$\beta = \pm 1.5$$

The complementary solution is

$$y = e^{2x}(C_1 \cos(1.5x) + C_2 \sin(1.5x))$$
$$= C_1 e^{2x} \cos 1.5x + C_2 e^{2x} \sin 1.5x$$

Answer is C.

14. The Laplace transform of a unit step at $t = c$ is e^{-cs}/s. The notation used indicates that unit steps are applied at $t = 1$ and $t = 2$. Use superposition.

$$\mathcal{L}(f(t)) = \mathcal{L}(u(t-1)) + \mathcal{L}(u(t-2))$$
$$= \frac{e^{-s}}{s} + \frac{e^{-2s}}{s}$$
$$= \frac{e^{-s} + e^{-2s}}{s}$$

Answer is B.

15. Use the first-order difference equation. The balance at the end of period k after making a payment of A at the end of period k is

$$P_k = P_{k-1}(1+i) - A$$
$$P_1 = P_0(1+i) - A$$
$$= (\$10{,}000)(1+0.15) - \$3000$$
$$= \$8500$$

Answer is C.

16. Use the first-order difference equation. The balance at the end of period k after making a payment of A at the end of period k is

$$P_k = P_{k-1}(1+i) - A$$
$$P_1 = P_0(1+i) - A$$
$$= (\$15{,}500)(1+0.01) - \$350$$
$$= \$15{,}305$$
$$P_2 = (\$15{,}305)(1+0.01) - \$350$$
$$= \$15{,}108.05 \quad (\$15{,}100)$$

Answer is D.

17. The value of a number in a Fibonacci sequence is the sum of the previous two numbers in the sequence.

Use the second-order difference equation.

$$f(k) = f(k-1) + f(k-2)$$
$$f(3) = f(2) + f(1) = 1.3 + 4$$
$$= 5.3$$
$$f(4) = f(3) + f(2) = 5.3 + 1.3$$
$$= 6.6$$

Answer is D.

18. Find the homogeneous solution. The characteristic equation is

$$r + 8 = 0$$

The root is $r = -8$. Therefore, the homogeneous solution is

$$y_h = C_1 e^{-ax}$$
$$= C_1 e^{-8x}$$

Find a particular solution. The forcing function is a constant, so (from Table 8.1) the particular solution is a constant.

$$y_p = C_2$$
$$y_p' = 0$$

Substitute the particular solution into the original equation.

$$y' + 8y = 0 + (8)(C_2)$$
$$= 16$$
$$C_2 = 2$$

The solution is

$$y(x) = y_h + y_p = C_1 e^{-8x} + 2$$

Use the initial condition to determine the coefficient C_1.

$$y(0) = C_1 e^{(-8)(0)} + 2$$
$$= C_1 + 2$$
$$= 5$$
$$C_1 = 3$$

The complete solution is

$$y(x) = 3e^{-8x} + 2$$

Answer is C.

19.
$$F^{-1}(F(z)) = F^{-1}\left(\frac{z}{z-5}\right)$$
$$= 5^k$$

Answer is D.

20. Separate the variables.

$$(x^2 + 9)dy = (xy)dx$$
$$\frac{dy}{y} = \left(\frac{x}{x^2 + 9}\right)dx$$

Integrate.

$$\int \frac{dy}{y} = \int \left(\frac{x}{x^2 + 9}\right)dx$$
$$\ln|y| = \left(\frac{1}{2}\right)\ln(x^2 + 9) + C$$
$$= \ln\sqrt{x^2 + 9} + C$$
$$|y| = e^C \sqrt{x^2 + 9}$$
$$y = \pm e^C \sqrt{x^2 + 9}$$

$y = 0$ is also a solution. The general solution is

$$y = C\sqrt{x^2 + 9}$$

Answer is C.

21. Euler's approximation is

$$x_{i+1} = x_i + \Delta t \frac{dx_i}{dt}$$
$$x(1 + 0.25) = x(1) + \Delta t (2x(1))$$
$$x(1.25) = 1 + (0.25)(2)(1) = 1.5$$
$$x(1.25 + 0.25) = x(1.25) + \Delta t (2x(1.25))$$
$$x(1.5) = 1.5 + (0.25)(2)(1.5)$$
$$= 2.25$$

Answer is D.

9 Numerical Analysis

Subjects

NUMERICAL METHODS 9-1
FINDING ROOTS:
   BISECTION METHOD 9-1
FINDING ROOTS:
   NEWTON'S METHOD 9-2

NUMERICAL METHODS

Although the roots of second-degree polynomials are easily found by a variety of methods (by factoring, completing the square, or using the quadratic equation), easy methods of solving cubic and higher-order equations exist only for specialized cases. However, cubic and higher-order equations occur frequently in engineering, and they are difficult to factor. Trial and error solutions, including graphing, are usually satisfactory for finding only the general region in which the root occurs.

Numerical analysis is a general subject that covers, among other things, iterative methods for evaluating roots to equations. The most efficient numerical methods are too complex to present and, in any case, work by hand. However, some of the simpler methods are presented here. Except in critical problems that must be solved in real time, a few extra calculator or computer iterations will make no difference.[1]

FINDING ROOTS: BISECTION METHOD

The *bisection method* is an iterative method that "brackets" (also known as "straddles") an interval containing the *root* or *zero* of a particular equation.[2] The size of the interval is halved after each iteration. As the method's name suggests, the best estimate of the root after any iteration is the midpoint of the interval. The maximum error is half the interval length. The procedure continues until the size of the maximum error is "acceptable."[3]

---

[1]Most advanced hand-held calculators now have "root finder" functions that use numerical methods to iteratively solve equations.
[2]The equation does not have to be a pure polynomial. The bisection method requires only that the equation be defined and determinable at all points in the interval.
[3]The bisection method is not a closed method. Unless the root actually falls on the midpoint of one iteration's interval, the method

The disadvantages of the bisection method are (a) the slowness in converging to the root, (b) the need to know the interval containing the root before starting, and (c) the inability to determine the existence of or find other real roots in the starting interval.

The bisection method starts with two values of the independent variable, $x = L_0$ and $x = R_0$, which straddle a root. Since the function passes through zero at a root, $f(L_0)$ and $f(R_0)$ will have opposite signs. The following algorithm describes the remainder of the bisection method.

Let n be the iteration number. Then, for $n = 0, 1, 2, \ldots$, perform the following steps until sufficient accuracy is attained.

step 1: Set $m = \left(\dfrac{1}{2}\right)(L_n + R_n)$.   *9.1*

step 2: Calculate $f(m)$.

step 3: If $f(L_n)f(m) \leq 0$, set $L_{n+1} = L_n$ and $R_{n+1} = m$. Otherwise, set $L_{n+1} = m$ and $R_{n+1} = R_n$.

step 4: $f(x)$ has at least one root in the interval (L_{n+1}, R_{n+1}). The estimated value of that root, x^*, is

$$x^* \approx \left(\frac{1}{2}\right)(L_{n+1} + R_{n+1}) \qquad 9.2$$

The maximum error is $(1/2)(R_{n+1} - L_{n+1})$.

Example 9.1

Use two iterations of the bisection method to find a root of

$$f(x) = x^3 - 2x - 7$$

Solution

The first step is to find L_0 and R_0, which are the values of x that straddle a root and have opposite signs. A table can be made and values of $f(x)$ calculated for random values of x.

---

continues indefinitely. Eventually, the magnitude of the maximum error is small enough as to not matter.

x	-2	-1	0	$+1$	$+2$	$+3$
$f(x)$	-11	-6	-7	-8	-3	$+14$

Since $f(x)$ changes sign between $x = 2$ and $x = 3$, $L_0 = 2$ and $R_0 = 3$.

For the first iteration, $n = 0$.

$$m = \left(\frac{1}{2}\right)(2 + 3) = 2.5$$
$$f(2.5) = (2.5)^3 - (2)(2.5) - 7$$
$$= 3.625$$

Since $f(2.5)$ is positive, a root must exist in the interval $(2, 2.5)$. Therefore, $L_1 = 2$ and $R_1 = 2.5$. At this point, the best estimate of the root is

$$x^* \approx \left(\frac{1}{2}\right)(2 + 2.5) = 2.25$$

The maximum error is $(1/2)(2.5 - 2) = 0.25$.

For the second iteration, $n = 1$.

$$m = \left(\frac{1}{2}\right)(2 + 2.5) = 2.25$$
$$f(2.25) = (2.25)^3 - (2)(2.25) - 7$$
$$= -0.1094$$

Since $f(2.25)$ is negative, a root must exist in the interval $(2.25, 2.5)$. Therefore, $L_2 = 2.25$ and $R_2 = 2.5$. The best estimate of the root is

$$x^* \approx \left(\frac{1}{2}\right)(2.25 + 2.5) = 2.375$$

The maximum error is $(1/2)(2.5 - 2.25) = 0.125$.

FINDING ROOTS: NEWTON'S METHOD

Many other methods have been developed to overcome one or more of the disadvantages of the bisection method. These methods have their own disadvantages.[4]

Newton's method is a particular form of *fixed-point iteration*. In this sense, "fixed point" is often used as a synonym for "root" or "zero." However, fixed-point iterations get their name from functions with the characteristic property $x = g(x)$ such that the limit of $g(x)$ is the fixed point (i.e., is the root).

All fixed-point techniques require a starting point. Preferably, the starting point will be close to the actual

root.[5] And, while Newton's method converges quickly, it requires the function to be continuously differentiable.

Newton's method algorithm is simple. At each iteration ($n = 0$, 1, 2, etc.), Eq. 9.3 estimates the root. The maximum error is determined by looking at how much the estimate changes after each iteration. If the change between the previous and current estimates (representing the magnitude of error in the estimate) is too large, the current estimate is used as the independent variable for the subsequent iteration.[6]

$$x_{n+1} = g(x_n) = x_n - \frac{f(x_n)}{f'(x_n)} \qquad 9.3$$

Example 9.2

Solve Ex. 9.1 using two iterations of Newton's method. Use $x_0 = 2$.

Solution

The function and its first derivative are

$$f(x) = x^3 - 2x - 7$$
$$f'(x) = 3x^2 - 2$$

For the first iteration, $n = 0$.

$$x_0 = 2$$
$$f(x_0) = f(2) = (2)^3 - (2)(2) - 7$$
$$= -3$$
$$f'(x_0) = f'(2) = (3)(2)^2 - 2$$
$$= 10$$
$$x_1 = x_0 - \frac{f(x_0)}{f'(x_0)} = 2 - \frac{-3}{10}$$
$$= 2.3$$

For the second iteration, $n = 1$.

$$x_1 = 2.3$$
$$f(x_1) = (2.3)^3 - (2)(2.3) - 7$$
$$= 0.567$$
$$f'(x_1) = (3)(2.3)^2 - 2$$
$$= 13.87$$
$$x_2 = x_1 - \frac{f(x_1)}{f'(x_1)}$$
$$= 2.3 - \frac{0.567}{13.87}$$
$$= 2.259$$

---

[4]The *regula falsi (false position) method* converges faster than the bisection method but is unable to specify a small interval containing the root. The *secant method* is prone to round-off errors and gives no indication of the remaining distance to the root.

[5]Theoretically, the only penalty for choosing a starting point too far away from the root will be a slower convergence to the root.
[6]Actually, the theory defining the maximum error is more definite than this. For example, for a large enough value of n, the error decreases approximately linearly. Therefore, the consecutive values of x_n converge linearly to the root as well.

Topic IV: Statics

Diagnostic Examination for Statics

Chapter 10   Systems of Forces

Chapter 11   Trusses

Chapter 12   Pulleys, Cables, and Friction

Chapter 13   Centroids and Moments of Inertia

STATICS

Hint: For the most current information about the exam, visit www.ppi2pass.com/fefaqs.html regularly.
Use the Exam Forum to compare notes with other FE examinees.

Diagnostic Examination

TOPIC IV: STATICS

TIME LIMIT: 45 MINUTES

1. A 100 kg block rests on an incline. The coefficient of static friction between the block and the ramp is 0.2. The mass of the cable is negligible, and the pulley at point C is frictionless. What is the smallest mass of block B that will start the 100 kg block moving up the incline?

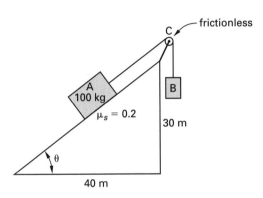

(A) 44 kg
(B) 65 kg
(C) 76 kg
(D) 92 kg

2. An inclined force, F, is applied to a block of mass m. What is the minimum coefficient of static friction between the block and the ramp surface such that no motion occurs?

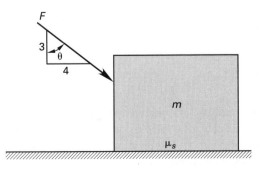

(A) $\dfrac{4F}{3F + 5mg}$

(B) $\dfrac{F \tan \theta}{mg}$

(C) $\dfrac{3F}{4F + 5mg}$

(D) $\dfrac{F}{mg}$

3. A ladder with a length of 10 m rests against a frictionless wall. The coefficient of static friction between the ladder and the floor is 0.4. The combined weight of the ladder and an individual can be idealized as an 800 N force applied at point B, as shown. What is the horizontal frictional force between the ladder and floor?

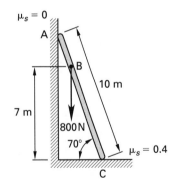

(A) 178 N
(B) 217 N
(C) 266 N
(D) 320 N

4. A sign has a mass of 150 kg. The sign is attached to the wall by a pin at point B and is supported by a cable between points A and C. Determine the tension in the cable.

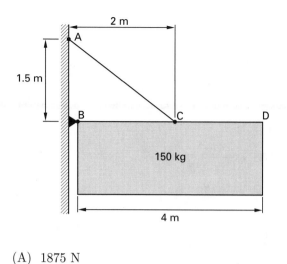

(A) 1875 N
(B) 2450 N
(C) 3750 N
(D) 5000 N

5. A 100 kg block rests on a frictionless incline. Forces are applied to the block as shown. What is the minimum force, P, such that no downward motion occurs?

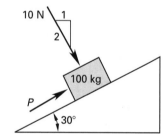

(A) 50 N
(B) 200 N
(C) 490 N
(D) 850 N

6. An experiment is performed to measure the frictional force between a block and an inclined plane. The block is initially at rest on the inclined plane. A constantly increasing force, F, is slowly applied to the block. After a period of time, the block begins to move. Which of the following graphs most likely depicts the relationship between the applied force, F, and the frictional force, F_f?

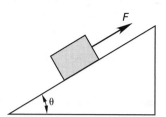

(A)

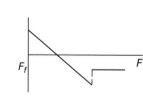

(B)

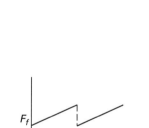

(C)

(D)

7. Two spheres, one with a mass of 7.5 kg and the other with a mass of 10.0 kg, are in equilibrium as shown. If all surfaces are frictionless, what is the magnitude of the reaction at point B?

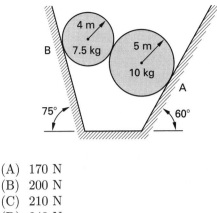

(A) 170 N
(B) 200 N
(C) 210 N
(D) 240 N

8. A uniform log of length L is inclined 30° from the horizontal when supported by a frictionless rock located $0.6L$ from its left end. The mass of the log is 200 kg. An engineer with a mass of 53.5 kg walks along the log from the left to the right until the log is balanced horizontally. How far from the left end of the log is the engineer when the log is horizontal?

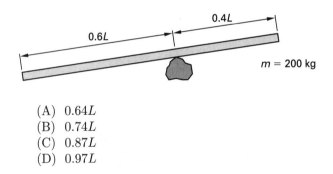

(A) $0.64L$
(B) $0.74L$
(C) $0.87L$
(D) $0.97L$

9. Three nonparallel, nonzero forces act on a rigid body. What must be true of all of these forces for the body to be in equilibrium?

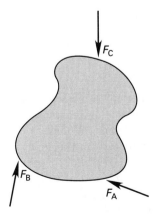

(A) The forces must be concurrent.
(B) The forces must be equal in magnitude but opposite in direction.
(C) The forces must have the same line of action.
(D) The forces must be at complementary angles to one another.

10. A 250 kg block is supported by the pulley system as shown. The system is in static equilibrium. Assuming the masses of the pulleys and cables are negligible, what is mass m_2?

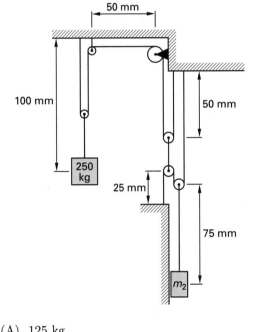

(A) 125 kg
(B) 250 kg
(C) 375 kg
(D) 500 kg

11. What are the x- and y-coordinates of the centroid of the area?

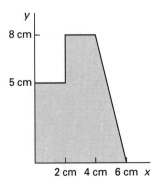

(A) (2.51 cm, 3.41 cm)
(B) (2.80 cm, 3.25 cm)
(C) (3.41 cm, 3.71 cm)
(D) (3.20 cm, 4.21 cm)

(A) 1.5 kN
(B) 2.9 kN
(C) 3.8 kN
(D) 5.1 kN

12. The cantilever truss shown supports a vertical force of 600 000 N applied at point G. What is the force in member CF?

14. A boat is moored to a dock with a rope tied to a pole as shown. If the tension in the rope is 1000 N, what is the torque exerted on the pole at point O, the origin?

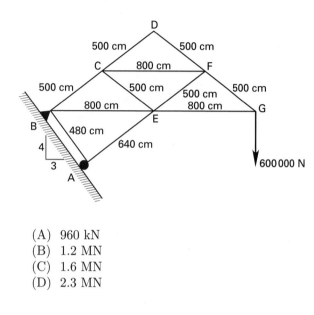

(A) 960 kN
(B) 1.2 MN
(C) 1.6 MN
(D) 2.3 MN

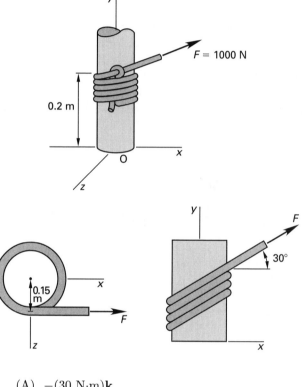

13. A 500 kg crate is supported by three cables as shown. If the force in cable AD is 5950 N, what is the force in cable BD?

(A) $-(30 \text{ N·m})\mathbf{k}$
(B) $(75 \text{ N·m})\mathbf{i} + (73 \text{ N·m})\mathbf{k}$
(C) $(100 \text{ N·m})\mathbf{j} - (130 \text{ N·m})\mathbf{k}$
(D) $-(75 \text{ N·m})\mathbf{i} + (130 \text{ N·m})\mathbf{j} - (173 \text{ N·m})\mathbf{k}$

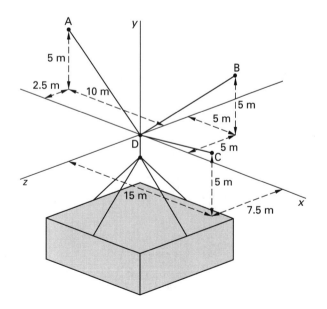

15. A rope supporting a mass passes around four simple frictionless pulleys at points A, B, C, and D, and around two fixed drums at points E and F. The mass exerts a force of 1000 N on the rope. The radius of all pulleys and drums is 75 mm. If the coefficient of friction between the rope and all the surfaces it contacts is 0.3, what is the minimum force, P, necessary to hold the weight in position?

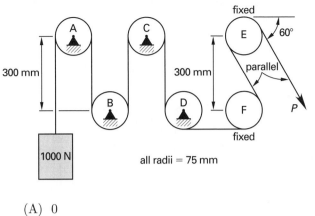

2.

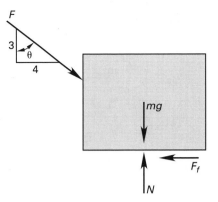

(A) 0
(B) 208 N
(C) 786 N
(D) 1154 N

$$\sum F_y = 0$$
$$= N - F \cos\theta - mg$$
$$N = F \cos\theta + mg$$

$$\sum F_x = 0$$
$$= F \sin\theta - F_f$$
$$= F \sin\theta - \mu_s N$$

$$\mu_s = \frac{F \sin\theta}{N}$$
$$= \frac{F \sin\theta}{F \cos\theta + mg}$$
$$= \frac{F\left(\dfrac{4}{5}\right)}{F\left(\dfrac{3}{5}\right) + mg}$$
$$= \frac{4F}{3F + 5mg}$$

Answer is A.

SOLUTIONS TO DIAGNOSTIC EXAMINATION TOPIC IV

1. Choose coordinate axes parallel and perpendicular to the incline.

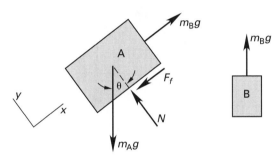

$$\sum F_y = 0$$
$$= N - m_A g \cos\theta$$
$$N = m_A g \cos\theta$$

$$\sum F_x = 0$$
$$= m_B g - m_A g \sin\theta - F_f$$
$$= m_B g - m_A g \sin\theta - \mu_s N$$
$$= m_B g - m_A g \sin\theta - \mu_s m_A g \cos\theta$$
$$= m_B g - m_A g(\sin\theta + \mu_s \cos\theta)$$

$$m_B = m_A(\sin\theta + \mu_s \cos\theta)$$
$$= (100 \text{ kg})\left(\frac{3}{5} + (0.2)\left(\frac{4}{5}\right)\right)$$
$$= 76 \text{ kg}$$

(Note that the frictional force serves to increase, not decrease, the force required to accelerate the block.)

Answer is C.

3.

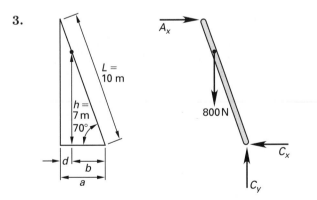

Since the wall is frictionless, the floor must support all of the vertical force.

$$\sum F_y = 0$$
$$= C_y - 800 \text{ N}$$
$$C_y = 800 \text{ N}$$

By trigonometry,

$$a = L \cos \theta$$
$$= (10 \text{ m}) \cos 70°$$
$$= 3.42 \text{ m}$$

$$b = \frac{h}{\tan \theta} = \frac{7 \text{ m}}{\tan 70°}$$
$$= 2.55 \text{ m}$$

$$d = a - b$$
$$= 3.42 \text{ m} - 2.55 \text{ m}$$
$$= 0.87 \text{ m}$$

Sum moments about point A.

$$\sum M_A = 0$$
$$= (800 \text{ N})d - C_y a + C_x L \sin 70°$$
$$= (800 \text{ N})(0.87 \text{ m}) - (800 \text{ N})(3.42 \text{ m})$$
$$+ C_x (10 \text{ m}) \sin 70°$$

$$C_x = \frac{(800 \text{ N})(3.42 \text{ m}) - (800 \text{ N})(0.87 \text{ m})}{(10 \text{ m}) \sin 70°}$$
$$= 217.1 \text{ N} \quad (217 \text{ N})$$

$(C_x < \mu_s N$ since motion is not impending.)

Answer is B.

4.

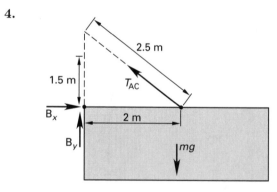

$$AC = \sqrt{(1.5 \text{ m})^2 + (2 \text{ m})^2}$$
$$= 2.5 \text{ m}$$

The vertical component of the tension in the cable is

$$T_{AC,y} = \left(\frac{1.5 \text{ m}}{2.5 \text{ m}} \right) T_{AC}$$
$$= 0.6 T_{AC}$$

Sum moments about point B.

$$\sum M_B = mg(2 \text{ m}) - T_{AC,y}(2 \text{ m})$$
$$= (150 \text{ kg}) \left(9.81 \frac{\text{m}}{\text{s}^2} \right) (2 \text{ m}) - (0.6) T_{AC}(2 \text{ m})$$

$$T_{AC} = \frac{(150 \text{ kg}) \left(9.81 \frac{\text{m}}{\text{s}^2} \right) (2 \text{ m})}{1.2 \text{ m}}$$
$$= 2452 \text{ N} \quad (2450 \text{ N})$$

Answer is B.

5. Choose coordinate axes that are parallel and perpendicular to the incline.

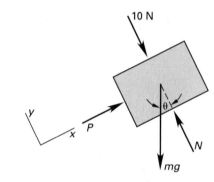

$$\sum F_x = 0$$
$$= P - mg \sin \theta$$
$$P = mg \sin \theta$$
$$= (100 \text{ kg}) \left(9.81 \frac{\text{m}}{\text{s}^2} \right) \sin 30°$$
$$= 490 \text{ N}$$

Answer is C.

6. Choose coordinate axes parallel and perpendicular to the incline.

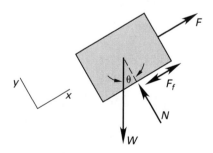

Let W be the weight of the block.

$$\sum F_x = 0$$
$$= F - W \sin\theta \pm F_f$$
$$\mp F_f = F - W \sin\theta$$

When $0 < F < W \sin\theta$, F_f must be positive to prevent the block from sliding down the incline. As F is increased, F_f will decrease at the same rate. When $F = W \sin\theta$, equilibrium requires that $F_f = 0$. At this point the applied forces and the forces due to gravity will be exactly balanced. When F becomes greater than $W \sin\theta$, F_f must change direction to satisfy equilibrium. For the range $W \sin\theta < F < W \sin\theta + \mu_s N$, the magnitude of F_f will increase at the same rate as F and equilibrium will be maintained. When $W \sin\theta + \mu_s N < F$, the frictional force will have reached its maximum value and the combined effects of friction and gravity can no longer prevent the block from moving up the incline. The block will begin to move, and the frictional force will change from static friction to kinetic friction.

Answer is D.

7.

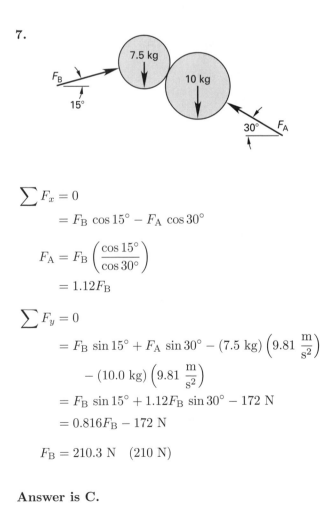

$$\sum F_x = 0$$
$$= F_B \cos 15° - F_A \cos 30°$$

$$F_A = F_B \left(\frac{\cos 15°}{\cos 30°} \right)$$
$$= 1.12 F_B$$

$$\sum F_y = 0$$
$$= F_B \sin 15° + F_A \sin 30° - (7.5 \text{ kg}) \left(9.81 \frac{\text{m}}{\text{s}^2} \right)$$
$$\quad - (10.0 \text{ kg}) \left(9.81 \frac{\text{m}}{\text{s}^2} \right)$$
$$= F_B \sin 15° + 1.12 F_B \sin 30° - 172 \text{ N}$$
$$= 0.816 F_B - 172 \text{ N}$$

$$F_B = 210.3 \text{ N} \quad (210 \text{ N})$$

Answer is C.

8.

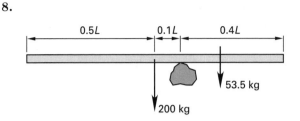

Sum moments about the pivot point.

$$\sum M = 0$$
$$= m_{\log} g(0.1L) - m_{\text{engineer}} g(x - 0.6L)$$
$$= (200 \text{ kg})(0.1L) - (53.5 \text{ kg})(x - 0.6L)$$

(The g terms cancel.)

$$x = \frac{(200 \text{ kg})(0.1L)}{53.5 \text{ kg}} + 0.6L$$
$$= 0.974L \quad (0.97L)$$

Answer is D.

9. Equilibrium requires $\sum M = 0$ about any point in the body. For a general three-force system, this is possible only if the forces all act through the same point (and balance statically).

Answer is A.

10. Draw the free-body diagrams.

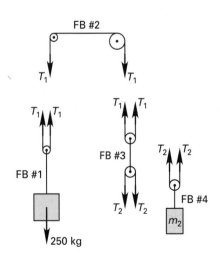

Since the system is in static equilibrium, the tension in the cables remains constant.

From the first free-body diagram,

$$\sum F_y = 0$$
$$= 2T_1 - (250 \text{ kg})g$$
$$T_1 = \left(\frac{1}{2}\right)(250 \text{ kg})g$$
$$= (125 \text{ kg})g$$

From the third free-body diagram,

$$\sum F_y = 0$$
$$= 2T_1 - 2T_2$$
$$T_2 = T_1$$

From the fourth free-body diagram,

$$\sum F_y = 0$$
$$= 2T_2 - m_2 g$$
$$= (2)(125 \text{ kg})g - m_2 g$$
$$m_2 = 250 \text{ kg}$$

Answer is B.

11.

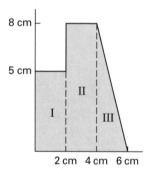

8 cm

5 cm

II

I

III

2 cm 4 cm 6 cm

Divide the shape into regions. Calculate the area and locate the centroid for each region.

For region I (rectangular),

$$A = (2 \text{ cm})(5 \text{ cm})$$
$$= 10 \text{ cm}^2$$
$$x_c = \frac{2 \text{ cm}}{2}$$
$$= 1 \text{ cm}$$
$$y_c = \frac{5 \text{ cm}}{2}$$
$$= 2.5 \text{ cm}$$

For region II (rectangular),

$$A = (2 \text{ cm})(8 \text{ cm}) = 16 \text{ cm}^2$$
$$x_c = \frac{2 \text{ cm} + 4 \text{ cm}}{2} = 3 \text{ cm}$$
$$y_c = \frac{8 \text{ cm}}{2} = 4 \text{ cm}$$

For region III (triangular),

$$A = \left(\frac{1}{2}\right)(2 \text{ cm})(8 \text{ cm})$$
$$= 8 \text{ cm}^2$$
$$x_c = 4 \text{ cm} + \left(\frac{1}{3}\right)(2 \text{ cm})$$
$$= 4.67 \text{ cm}$$
$$y_c = \frac{8 \text{ cm}}{3}$$
$$= 2.67 \text{ cm}$$

Calculate x_c and y_c.

$$x_c = \frac{\sum x_{c,i} A_i}{\sum A_i}$$
$$= \frac{(1 \text{ cm})(10 \text{ cm}^2) + (3 \text{ cm})(16 \text{ cm}^2) + (4.67 \text{ cm})(8 \text{ cm}^2)}{10 \text{ cm}^2 + 16 \text{ cm}^2 + 8 \text{ cm}^2}$$
$$= \frac{95.36 \text{ cm}^3}{34 \text{ cm}^2}$$
$$= 2.80 \text{ cm}$$
$$y_c = \frac{\sum y_{c,i} A_i}{\sum A_i}$$
$$= \frac{(2.5 \text{ cm})(10 \text{ cm}^2) + (4 \text{ cm})(16 \text{ cm}^2) + (2.67 \text{ cm})(8 \text{ cm}^2)}{10 \text{ cm}^2 + 16 \text{ cm}^2 + 8 \text{ cm}^2}$$
$$= \frac{110.36 \text{ cm}^3}{34 \text{ cm}^2}$$
$$= 3.25 \text{ cm}$$

Answer is B.

12.

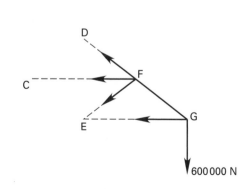

D

C

F

E

G

600 000 N

Equilibrium of pin D requires that the forces in members CD and DF are zero.

Use the method of sections. Sum moments about joint E.

$$\sum M_E = 0$$
$$= (600\,000 \text{ N})(800 \text{ cm}) - CF(300 \text{ cm})$$
$$CF = \frac{(600\,000 \text{ N})(800 \text{ cm})}{300 \text{ cm}}$$
$$= 1\,600\,000 \text{ N} \quad (1.6 \text{ MN})$$

Answer is C.

13. Determine the directions in which the forces act. The unit vectors are

$$\mathbf{u_{AD}} = \frac{x_{AD}\mathbf{i} + y_{AD}\mathbf{j} + z_{AD}\mathbf{k}}{\sqrt{x_{AD}^2 + y_{AD}^2 + z_{AD}^2}}$$
$$= \frac{-\big((10 \text{ m})\mathbf{i} + (5 \text{ m})\mathbf{j} - (2.5 \text{ m})\mathbf{k}\big)}{\sqrt{(10 \text{ m})^2 + (5 \text{ m})^2 + (2.5 \text{ m})^2}}$$
$$= \frac{-\big((10 \text{ m})\mathbf{i} + (5 \text{ m})\mathbf{j} - (2.5 \text{ m})\mathbf{k}\big)}{\sqrt{131.25 \text{ m}^2}}$$
$$= -0.873\mathbf{i} + 0.436\mathbf{j} - 0.218\mathbf{k}$$

Since $AD = 5950$,

$$A_x = (-0.873)(5950 \text{ N})$$
$$= -5194 \text{ N}$$
$$A_y = (0.436)(5950 \text{ N})$$
$$= 2594 \text{ N}$$
$$A_z = (-0.218)(5950 \text{ N})$$
$$= -1297 \text{ N}$$

Write the equilibrium conditions for cable AD.

$$\sum F_x = 0: \ -5194 \text{ N} + B_x + C_x = 0$$
$$\sum F_z = 0: \ -1297 \text{ N} + B_z + C_z = 0$$

By observation, based on the coordinates of the cable end-points,

$$B_z = -B_x$$
$$C_x = \left(\frac{15 \text{ m}}{5 \text{ m}}\right) C_y$$
$$= 3C_y$$
$$C_z = \left(\frac{7.5 \text{ m}}{5 \text{ m}}\right) C_y$$
$$= 1.5C_y$$
$$B_x + 3C_y = 5194 \text{ N}$$
$$-B_x + 1.5C_y = 1297 \text{ N}$$

Solving simultaneously,

$$C_y = 1442 \text{ N}$$
$$B_x = 868 \text{ N}$$

Since $B_x = B_y = -B_z$,

$$B = \sqrt{B_x^2 + B_y^2 + (-B_z)^2}$$
$$= \sqrt{(868 \text{ N})^2 + (868 \text{ N})^2 + (-868 \text{ N})^2}$$
$$= 1503 \text{ N} \quad (1.5 \text{ kN})$$

Answer is A.

14. Determine the force vector.
$$\mathbf{F} = F \cos\theta_x \mathbf{i} + F \cos\theta_y \mathbf{j} + F \cos\theta_z \mathbf{k}$$
$$= F \cos(30°)\mathbf{i} + F \cos(60°)\mathbf{j} + F \cos(90°)\mathbf{k}$$
$$= (1000 \text{ N})(0.866)\mathbf{i} + (1000 \text{ N})(0.5)\mathbf{j} + (1000 \text{ N})(0)\mathbf{k}$$
$$= (866 \text{ N})\mathbf{i} + (500 \text{ N})\mathbf{j}$$

Determine the position vector for the point at which the force acts.

$$\mathbf{r} = x_p\mathbf{i} + y_p\mathbf{j} + z_p\mathbf{k}$$
$$= (0.2 \text{ m})\mathbf{j} + (0.15 \text{ m})\mathbf{k}$$

The torque (moment) is the cross product of the moment arm and the force.

$$M_O = \mathbf{r} \times \mathbf{F}$$

Use an augmented matrix to calculate the cross product.

$$M_O = \begin{vmatrix} \mathbf{i} & \mathbf{j} & \mathbf{k} \\ 0 & 0.2 & 0.15 \\ 866 & 500 & 0 \end{vmatrix} \begin{matrix} \mathbf{i} & \mathbf{j} \\ 0 & 0.2 \\ 866 & 500 \end{matrix}$$

$$= -(866 \text{ N})(0.2 \text{ m})\mathbf{k} - (500 \text{ N})(0.15 \text{ m})\mathbf{i} - (0)(0)\mathbf{j}$$
$$\quad + (500 \text{ N})(0)\mathbf{k} + (866 \text{ N})(0.15 \text{ m})\mathbf{j}$$
$$\quad + (0)(0.2 \text{ m})\mathbf{i}$$
$$= -(75 \text{ N·m})\mathbf{i} + (130 \text{ N·m})\mathbf{j} - (173 \text{ N·m})\mathbf{k}$$

Answer is D.

15.

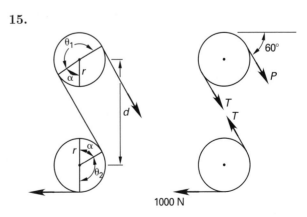

$$\cos\alpha = \frac{r}{\frac{1}{2}d}$$

$$\alpha = \cos^{-1}\left(\frac{2r}{d}\right)$$

$$= \cos^{-1}\left(\frac{(2)(75 \text{ mm})}{300 \text{ mm}}\right)$$

$$= \cos^{-1}(0.5)$$

$$= 60° \quad (\pi/3 \text{ rad})$$

$$\theta_1 = 180° + 60° - \alpha$$

$$= 180° + 60° - 60°$$

$$= 180°$$

$$\theta_2 = 180° - \alpha$$

$$= 180° - 60°$$

$$= 120°$$

$$T = Pe^{\mu_s\theta_1}$$

$$W = Te^{\mu_s\theta_2}$$

$$= Pe^{\mu_s\theta_1}e^{\mu_s\theta_2}$$

$$= Pe^{\mu_s(\theta_1+\theta_2)}$$

The angles must be expressed in radians.

$$\theta_1 + \theta_2 = \frac{(180° + 120°)2\pi}{360°}$$

$$= 5.24 \text{ rad}$$

$$P = \frac{W}{e^{\mu_s(\theta_1+\theta_2)}}$$

$$= \frac{1000 \text{ N}}{e^{(0.3)(5.24)}}$$

$$= 208 \text{ N}$$

Answer is B.

10 Systems of Forces

Subjects

FORGES 10-1
   Resultant 10-1
   Resolution of a Force 10-1
   Moments 10-1
   Couples 10-2
SYSTEMS OF FORCES 10-3
   Equilibrium Requirements 10-3
   Concurrent Forces 10-3
   Two- and Three-Force Members . . . 10-3
PROBLEM-SOLVING APPROACHES . . . 10-4
   Determinacy 10-4
   Free-Body Diagrams 10-4
   Reactions 10-4

Nomenclature

A	area	ft^2	m^2
d	distance	ft	m
F	force	lbf	N
M	moment	ft-lbf	N·m
r	radius	ft	m
R	resultant	lbf	N
T	tension	lbf	N
w	load per unit length	lbf/ft	N/m

Symbols

θ	angle	deg	rad

FORCES

Statics is the study of rigid bodies that are stationary. To be stationary, a rigid body must be in static equilibrium. In the language of statics, a stationary rigid body has no *unbalanced forces* acting on it.

Force is a push or a pull that one body exerts on another, including gravitational, electrostatic, magnetic, and contact influences. Force is a vector quantity, having a magnitude, direction, and point of application.

Strictly speaking, actions of other bodies on a rigid body are known as *external forces*. If unbalanced, an external force will cause motion of the body. *Internal forces* are the forces that hold together parts of a rigid body. Although internal forces can cause deformation of a body, motion is never caused by internal forces.

Forces are frequently represented in terms of unit vectors and force components. A *unit vector* is a vector of unit length directed along a coordinate axis. Unit vectors are used in vector equations to indicate direction without affecting magnitude. In the rectangular coordinate system, there are three unit vectors, **i**, **j**, and **k**. In two dimensions,

$$\mathbf{F} = F_x\mathbf{i} + F_y\mathbf{j} \qquad \left[{\text{two} \atop \text{dimensional}}\right] \qquad 10.1$$

Resultant

The *resultant*, or sum, of n two-dimensional forces is equal to the sum of the components.

$$\mathbf{F} = \mathbf{i}\sum_{i=1}^{n} F_{x,i} + \mathbf{j}\sum_{i=1}^{n} F_{y,i} \qquad \left[{\text{two} \atop \text{dimensional}}\right] \quad 10.2$$

$$R = \sqrt{\left(\sum_{i=1}^{n} F_{x,i}\right)^2 + \left(\sum_{i=1}^{n} F_{y,i}\right)^2} \qquad 10.3$$

$$\theta = \tan^{-1}\left(\frac{\displaystyle\sum_{i=1}^{n} F_{y,i}}{\displaystyle\sum_{i=1}^{n} F_{x,i}}\right) \qquad 10.4$$

Resolution of a Force

The components of a two- or three-dimensional force can be found from its *direction cosines*, the cosines of the true angles made by the force vector with the x-, y-, and z-axes.

$$F_x = F\cos\theta_x \qquad 10.5$$
$$F_y = F\cos\theta_y \qquad 10.6$$
$$F_z = F\cos\theta_z \qquad 10.7$$

Moments

Moment is the name given to the tendency of a force to rotate, turn, or twist a rigid body about an actual or assumed pivot point. (Another name for moment is *torque*, although torque is used mainly with shafts and other power-transmitting machines.) When acted upon by a moment, unrestrained bodies rotate. However, rotation is not required for the moment to exist. When a

Figure 10.1 Components and Direction Angles of a Force

Figure 10.1 Components and Direction Angles of a Force

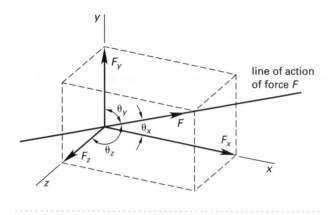

Figure 10.2 Right-Hand Rule

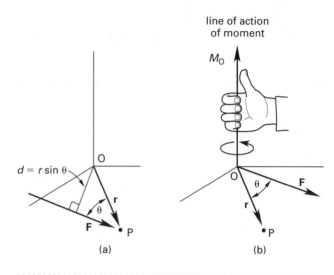

(a)   (b)

restrained body is acted upon by a moment, there is no rotation.

An object experiences a moment whenever a force is applied to it. Only when the line of action of the force passes through the center of rotation (i.e., the actual or assumed pivot point) will the moment be zero. (The moment may be zero, as when the moment arm length is zero, but there is a trivial moment nevertheless.)

Moments have primary dimensions of length × force. Typical units are foot-pounds, inch-pounds, and newton-meters. To avoid confusion with energy units, moments may be expressed as pound-feet, pound-inches, and meter-newtons.

Moments are vectors. The moment vector, $\mathbf{M}_O$, for a force about point O is the *cross product* of the force, $\mathbf{F}$, and the vector from point O to the point of application of the force, known as the *position vector*, $\mathbf{r}$. The scalar product $|\mathbf{r}|\ \sin\theta$ is known as the *moment arm, d*.

$$\mathbf{M}_O = \mathbf{r} \times \mathbf{F} \qquad\qquad 10.8$$

$$M_O = |\mathbf{M}_O| = |\mathbf{r}||\mathbf{F}|\ \sin\theta = d|\mathbf{F}| \quad [\theta \le 180°] \quad 10.9$$

The line of action of the moment vector is normal to the plane containing the force vector and the position vector. The sense (i.e., the direction) of the moment is determined from the *right-hand rule*.

> *Right-hand rule*: Place the position and force vectors tail to tail. Close your right hand and position it over the pivot point. Rotate the position vector into the force vector and position your hand such that your fingers curl in the same direction as the position vector rotates. Your extended thumb will coincide with the direction of the moment.

The direction cosines of a force can be used to determine the components of the moment about the coordinate axes.

$$M_x = M\ \cos\theta_x \qquad\qquad 10.10$$
$$M_y = M\ \cos\theta_y \qquad\qquad 10.11$$
$$M_z = M\ \cos\theta_z \qquad\qquad 10.12$$

Alternatively, the following three equations can be used to determine the components of the moment from the component of a force applied at point (x, y, z) referenced to an origin at $(0, 0, 0)$.

$$M_x = yF_z - zF_y \qquad\qquad 10.13$$
$$M_y = zF_x - xF_z \qquad\qquad 10.14$$
$$M_z = xF_y - yF_x \qquad\qquad 10.15$$

The resultant moment magnitude can be reconstituted from its components.

$$M = \sqrt{M_x^2 + M_y^2 + M_z^2} \qquad\qquad 10.16$$

Couples

Any pair of equal, opposite, and parallel forces constitute a *couple*. A couple is equivalent to a single moment vector. Since the two forces are opposite in sign, the x-, y-, and z-components of the forces cancel out. Therefore, a body is induced to rotate without translation. A couple can be counteracted only by another couple. A couple can be moved to any location without affecting the equilibrium requirements. (Such a moment is known as a *free moment, moment of a couple,* or *coupling moment.*)

In Fig. 10.3, the equal but opposite forces produce a moment vector $\mathbf{M}_O$ of magnitude Fd. The two forces can

Figure 10.3 Couple

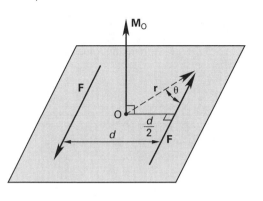

be replaced by this moment vector that can be moved to any location on a body.

$$M_O = 2rF \sin\theta = Fd \qquad 10.17$$

If a force, F, is moved a distance, d, from the original point of application, a couple, M, equal to Fd must be added to counteract the induced couple. The combination of the moved force and the couple is known as a *force-couple system*. Alternatively, a force-couple system can be replaced by a single force located a distance $d = M/F$ away.

SYSTEMS OF FORCES

Any collection of forces and moments in three-dimensional space is statically equivalent to a single resultant force vector plus a single resultant moment vector. (Either or both of these resultants can be zero.)

The x-, y-, and z-components of the resultant force are the sums of the x-, y-, and z-components of the individual forces, respectively.

$$\mathbf{R} = \sum \mathbf{F}_n$$
$$= \mathbf{i} \sum_{i=1}^{n} F_{x,i} + \mathbf{j} \sum_{i=1}^{n} F_{y,i} + \mathbf{k} \sum_{i=1}^{n} F_{z,i} \quad \begin{bmatrix} \text{three} \\ \text{dimensional} \end{bmatrix}$$
$$10.18$$

The resultant moment vector is more complex. It includes the moments of all system forces around the reference axes plus the components of all system moments.

$$\mathbf{M} = \sum \mathbf{M}_n \qquad 10.19$$

$$M_x = \sum_i (yF_z - zF_y)_i + \sum_i (M \cos\theta_x)_i \quad 10.20$$

$$M_y = \sum_i (zF_x - xF_z)_i + \sum_i (M \cos\theta_y)_i \quad 10.21$$

$$M_z = \sum_i (xF_y - yF_x)_i + \sum_i (M \cos\theta_z)_i \quad 10.22$$

Equilibrium Requirements

An object is static when it is stationary. To be stationary, all of the forces on the object must be in equilibrium. For an object to be in equilibrium, the resultant force and moment vectors must both be zero.

$$\mathbf{R} = 0 \qquad 10.23$$
$$R = \sqrt{R_x^2 + R_y^2 + R_z^2} = 0 \qquad 10.24$$
$$\mathbf{M} = 0 \qquad 10.25$$
$$M = \sqrt{M_x^2 + M_y^2 + M_z^2} = 0 \qquad 10.26$$

Since the square of any nonzero quantity is positive, Eqs. 10.27 through 10.32 follow directly from Eqs. 10.23 through 10.26.

$$R_x = 0 \qquad 10.27$$
$$R_y = 0 \qquad 10.28$$
$$R_z = 0 \qquad 10.29$$
$$M_x = 0 \qquad 10.30$$
$$M_y = 0 \qquad 10.31$$
$$M_z = 0 \qquad 10.32$$

Equations 10.27 through 10.32 seem to imply that six simultaneous equations must be solved in order to determine whether a system is in equilibrium. While this is true for general three-dimensional systems, fewer equations are necessary with most problems.

Concurrent Forces

A *concurrent force system* is a category of force systems wherein all of the forces act at the same point.

If the forces on a body are all concurrent forces, then only force equilibrium is necessary to ensure complete equilibrium. In two dimensions,

$$\sum F_x = 0 \qquad 10.33$$
$$\sum F_y = 0 \qquad 10.34$$

In three dimensions,

$$\sum F_x = 0 \qquad 10.35$$
$$\sum F_y = 0 \qquad 10.36$$
$$\sum F_z = 0 \qquad 10.37$$

Two- and Three-Force Members

Members limited to loading by two or three forces are special cases of equilibrium. A *two-force member* can be in equilibrium only if the two forces have the same line of action (i.e., are collinear) and are equal but opposite.

In most cases, two-force members are loaded axially, and the line of action coincides with the member's longitudinal axis. By choosing the coordinate system so that one axis coincides with the line of action, only one equilibrium equation is needed.

A *three-force member* can be in equilibrium only if the three forces are concurrent or parallel. Stated another way, the force polygon of a three-force member in equilibrium must close on itself.

PROBLEM-SOLVING APPROACHES

Determinacy

When the equations of equilibrium are independent, a rigid body force system is said to be *statically determinate*. A statically determinate system can be solved for all unknowns, which are usually reactions supporting the body. Examples of determinate beam types are illustrated in Fig. 10.4.

Figure 10.4 Types of Determinate Systems

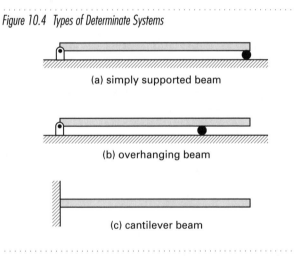

(a) simply supported beam

(b) overhanging beam

(c) cantilever beam

When the body has more supports than are necessary for equilibrium, the force system is said to be *statically indeterminate*. In a statically indeterminate system, one or more of the supports or members can be removed or reduced in restraint without affecting the equilibrium position. Those supports and members are known as *redundant supports* and *redundant members*. The number of redundant members is known as the *degree of indeterminacy*. Figure 10.5 illustrates several common indeterminate structures.

A body that is statically indeterminate requires additional equations to supplement the equilibrium equations. The additional equations typically involve deflections and depend on mechanical properties of the body.

Figure 10.5 Examples of Indeterminate Systems

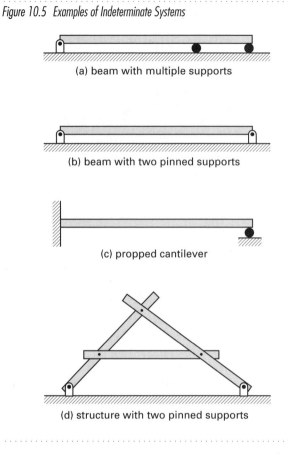

(a) beam with multiple supports

(b) beam with two pinned supports

(c) propped cantilever

(d) structure with two pinned supports

Free-Body Diagrams

A *free-body diagram* is a representation of a body in equilibrium, showing all applied forces, moments, and reactions. Free-body diagrams do not consider the internal structure or construction of the body, as Fig. 10.6 illustrates.

Since the body is in equilibrium, the resultants of all forces and moments on the free body are zero. In order to maintain equilibrium, any portions of the body that are conceptually removed must be replaced by the forces and moments those portions impart to the body. Typically, the body is isolated from its physical supports in order to help evaluate the reaction forces. In other cases, the body may be sectioned (i.e., cut) in order to determine the forces at the section.

Reactions

The first step in solving most statics problems, after drawing the free-body diagram, is to determine the reaction forces (i.e., the *reactions*) supporting the body. The manner in which a body is supported determines the type, location, and direction of the reactions. Conventional symbols are often used to define the type of support (i.e., pinned, roller, etc.). Examples of the symbols are shown in Table 10.1.

Figure 10.6 Bodies and Free Bodies

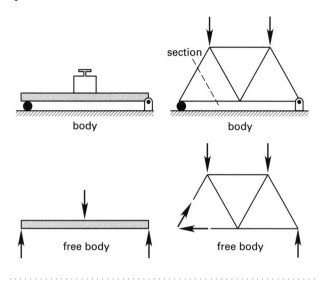

body   body

free body   free body

For beams, the two most common types of supports are the roller support and the pinned support. The *roller support*, shown as a cylinder supporting the beam, supports vertical forces only. Rather than support a horizontal force, a roller support simply rolls into a new equilibrium position. Only one equilibrium equation (i.e., the sum of vertical forces) is needed at a roller support. Generally, the terms *simple support* and *simply supported* refer to a roller support.

The *pinned support*, shown as a pin and clevis, supports both vertical and horizontal forces. Two equilibrium equations are needed.

Generally, there will be vertical and horizontal components of a reaction when one body touches another. However, when a body is in contact with a *frictionless surface*, there is no frictional force component parallel to the surface. Therefore, the reaction is normal to the contact surfaces. The assumption of frictionless contact is particularly useful when dealing with systems of spheres and cylinders in contact with rigid supports. Frictionless contact is also assumed for roller and rocker supports.

The procedure for finding determinate reactions in two-dimensional problems is straightforward. Determinate structures will have either a roller support and pinned support or two roller supports.

step 1: Establish a convenient set of coordinate axes. (To simplify the analysis, one of the coordinate directions should coincide with the direction of the forces and reactions.)

Table 10.1 Types of Two-Dimensional Supports

type of support	reactions and moments	number of unknowns[a]
simple, roller, rocker, ball, or frictionless surface	reaction normal to surface, no moment	1
cable in tension, or link	reaction in line with cable or link, no moment	1
frictionless guide or collar	reaction normal to rail, no moment	1
built-in, fixed support	two reaction components, one moment	3
frictionless hinge, pin connection, or rough surface	reaction in any direction, no moment	2

[a] The number of unkowns is valid for two-dimensional problems only.

step 2: Draw the free-body diagram.

step 3: Resolve the reaction at the pinned support (if any) into components normal and parallel to the coordinate axes.

step 4: Establish a positive direction of rotation (e.g., clockwise) for purposes of taking moments.

step 5: Write the equilibrium equation for moments about the pinned connection. (By choosing the pinned connection as the point about which to take moments, the pinned connection reactions do not enter into the equation.) This will usually determine the vertical reaction at the roller support.

step 6: Write the equilibrium equation for the forces in the vertical direction. Usually, this equation will have two unknown vertical reactions.

step 7: Substitute the known vertical reaction from step 5 into the equilibrium equation from step 6. This will determine the second vertical reaction.

step 8: Write the equilibrium equation for the forces at the horizontal direction. Since there is a minimum of one unknown reaction component in the horizontal direction, this step will determine that component.

step 9: If necessary, combine the vertical and horizontal force components at the pinned connection into a resultant reaction.

SAMPLE PROBLEMS

Problem 1

The five forces shown act at point A. What is the magnitude of the resultant force?

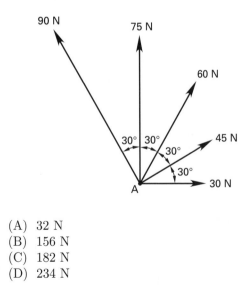

(A) 32 N
(B) 156 N
(C) 182 N
(D) 234 N

Solution

$$\sum F_x = 30 \text{ N} + (45 \text{ N})\cos 30° + (60 \text{ N})\cos 60°$$
$$+ (75 \text{ N})\cos 90° + (90 \text{ N})\cos 120°$$
$$= 54 \text{ N}$$

$$\sum F_y = (30 \text{ N})\sin 0° + (45 \text{ N})\sin 30°$$
$$+ (60 \text{ N})\sin 60° + 75 \text{ N}$$
$$+ (90 \text{ N})\sin 120°$$
$$= 227.4 \text{ N}$$

$$R = \sqrt{(54 \text{ N})^2 + (227.4 \text{ N})^2}$$
$$= 233.7 \text{ N} \quad (234 \text{ N})$$

Answer is D.

Problem 2

An angle bracket is subjected to the forces and couple shown. Determine the equivalent force-couple system at point A.

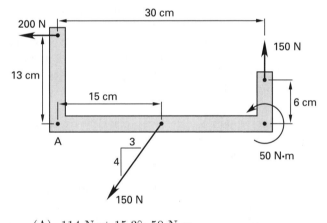

(A) 114 N at 15.3°; 50 N·m
(B) 292 N at 174.1°; 103 N·m
(C) 292 N at 185.9°; −103 N·m
(D) 333 N at 42.9°; 53 N·m

Solution

Use clockwise from the horizontal as the positive direction for angles and moments.

The inclined force's orientation is described by a 3-4-5 triangle.

$$R_x = -200 \text{ N} - \left(\frac{3}{5}\right)(150 \text{ N})$$
$$= -290 \text{ N}$$
$$R_y = \left(\frac{4}{5}\right)(-150 \text{ N}) + 150 \text{ N}$$
$$= 30 \text{ N}$$

$$R = \sqrt{(-290 \text{ N})^2 + (30 \text{ N})^2}$$
$$= 291.5 \text{ N}$$
$$\theta = \tan^{-1}\left(\frac{30 \text{ N}}{-290 \text{ N}}\right)$$
$$= 185.9° \quad [(180° + 5.9°) \text{ clockwise from the horizontal}]$$
$$M_A = \sum M_n \quad [\text{clockwise positive}]$$
$$= (-200 \text{ N})(0.13 \text{ m}) - (150 \text{ N})(0.30 \text{ m})$$
$$+ \left(\frac{4}{5}\right)(150 \text{ N})(0.15 \text{ m}) - 50 \text{ N·m}$$
$$= -103 \text{ N·m}$$

The force-couple system at A is

$$F = 291.5 \text{ N at } 185.9°$$
$$M = -103 \text{ N·m}$$

Answer is C.

Problem 3

Block D sits freely on the homogeneous bar and experiences a gravitation force of 50 N. Homogeneous bar AB experiences a gravitational force of 25 N. What is the force between the bar and block D?

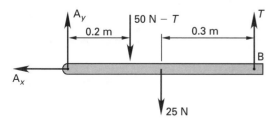

(A) 15 N
(B) 19 N
(C) 21 N
(D) 28 N

Solution
The cable tension is the same everywhere.

The free-body diagram of the bar is

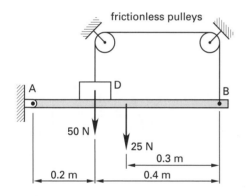

$$\sum M_A = 0$$
$$= (50 \text{ N} - T)(0.2 \text{ m}) + (25 \text{ N})(0.3 \text{ m})$$
$$- T(0.6 \text{ m})$$
$$T = 21.88 \text{ N}$$

The free-body diagram of block D is,

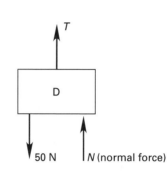

$$\sum F_y = 0$$
$$= T - 50 \text{ N} + N$$
$$N = 50 \text{ N} - 21.88 \text{ N}$$
$$= 28.12 \text{ N} \quad (28 \text{ N})$$

Answer is D.

Problem 4

What is the reaction at point A for the simply supported beam shown?

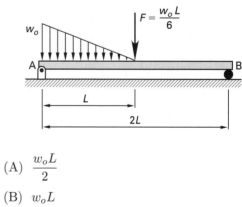

(A) $\dfrac{w_o L}{2}$

(B) $w_o L$

(C) $\dfrac{2w_o L}{3}$

(D) $\dfrac{w_o L}{3}$

Solution

The triangular load is equivalent to a concentrated load of $(1/2)w_o L$ acting at the centroid of the triangular distribution.

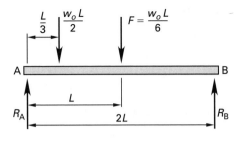

$$\sum M_B = 0$$

$$= R_A(2L) - \left(\frac{w_oL}{6}\right)L - \left(\frac{w_oL}{2}\right)\left(L + \frac{2}{3}L\right)$$

$$R_A = \left(\frac{1}{2L}\right)\left(\frac{w_oL^2}{6}\right)(1 + 5)$$

$$= w_oL/2$$

Answer is A.

Problem 5

Determine the reaction at point C.

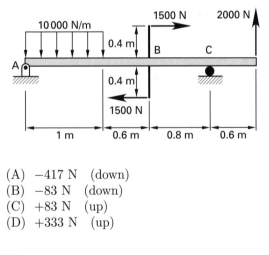

(A) −417 N   (down)
(B) −83 N   (down)
(C) +83 N   (up)
(D) +333 N   (up)

Solution

The free body of the beam is

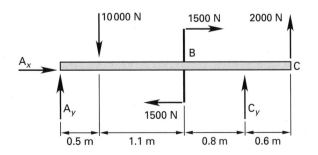

$$\sum M_A = 0$$

$$= \left(10\,000\ \frac{N}{m}\right)(1\ m)(0.5\ m)$$

$$+ (2)(1500\ N)(0.4\ m)$$

$$- C_y(2.4\ m) - (2000\ N)(3\ m)$$

$$C_y = 83.33\ N \qquad [\text{up}]$$

Answer is C.

FE-STYLE EXAM PROBLEMS

1. What is the resultant R of the system of forces shown?

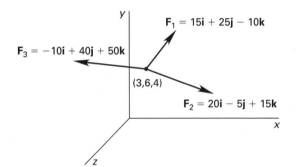

(A)

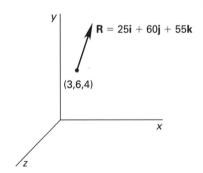

(B)

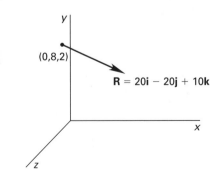

(C)

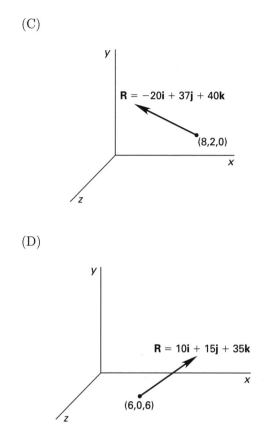

R = −20**i** + 37**j** + 40**k**

(8,2,0)

(D)

R = 10**i** + 15**j** + 35**k**

(6,0,6)

2. Resolve the 300 N force into two components, one along line P and the other along line Q. (*F*, P, and Q are coplanar.)

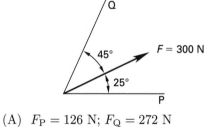

Q

45°

F = 300 N

25°

P

(A) F_P = 126 N; F_Q = 272 N
(B) F_P = 186 N; F_Q = 232 N
(C) F_P = 226 N; F_Q = 135 N
(D) F_P = 226 N; F_Q = 212 N

3. Which type of load is NOT resisted by a pinned joint?

(A) moment
(B) shear
(C) axial
(D) compression

4. Three forces act on a hook. Determine the magnitude of the resultant of the forces. Neglect hook bending.

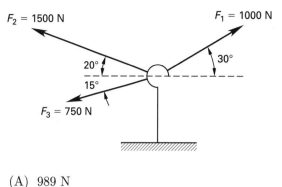

F_2 = 1500 N F_1 = 1000 N

20°   30°

15°

F_3 = 750 N

(A) 989 N
(B) 1140 N
(C) 1250 N
(D) 1510 N

5. The loading shown requires a resisting moment of 20 N·m at the support. Calculate the value of force *F*.

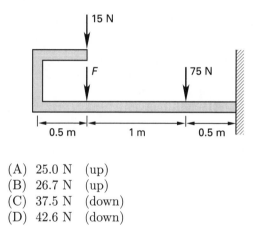

15 N

F   75 N

0.5 m   1 m   0.5 m

(A) 25.0 N   (up)
(B) 26.7 N   (up)
(C) 37.5 N   (down)
(D) 42.6 N   (down)

For the following problems use the NCEES Handbook as your only reference.

6. A bent beam is acted upon by a moment and several concentrated forces, as shown. Find the missing force *F* and distance *x* that will maintain equilibrium on the member shown.

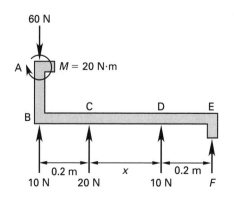

60 N

A   *M* = 20 N·m

C   D   E

B

0.2 m   *x*   0.2 m

10 N   20 N   10 N   F

STATICS
Forces

(A) $F = 5$ N; $x = 0.8$ m
(B) $F = 10$ N; $x = 0.6$ m
(C) $F = 20$ N; $x = 0.2$ m
(D) $F = 20$ N; $x = 0.4$ m

7. Four bolts (not shown) connect support A to the ground. Determine the design load for each of the four bolts.

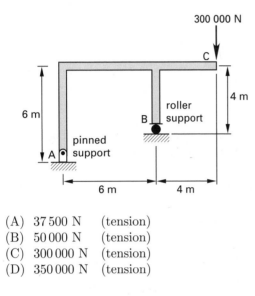

(A) 37 500 N   (tension)
(B) 50 000 N   (tension)
(C) 300 000 N   (tension)
(D) 350 000 N   (tension)

Problems 8–10 refer to the frame shown. The 700 N·m moment is applied at point B.

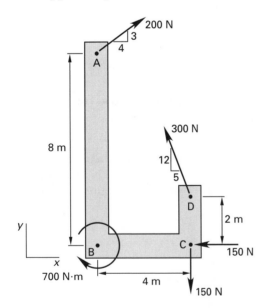

8. What is the x-component of the 300 N force at point D?

(A) 115 N
(B) 125 N
(C) 180 N
(D) 240 N

9. A single force (not shown) is applied at point B in the y-direction, in line with points A and B. What should this force be in order for the frame to be in equilibrium in that direction?

(A) −120 N   (down)
(B) −180 N   (down)
(C) −250 N   (down)
(D) −280 N   (down)

10. If the frame is pinned so that it rotates around point B, what counteracting moment must be applied at point A to put the frame in equilibrium?

(A) 650 N·m
(B) 890 N·m
(C) 1150 N·m
(D) 1240 N·m

11. A force is defined by the vector $\mathbf{A} = 3.5\mathbf{i} - 1.5\mathbf{j} + 2.0\mathbf{k}$. $\mathbf{i}$, $\mathbf{j}$, and $\mathbf{k}$ are unit vectors in the x-, y-, and z-directions, respectively. What is the angle that the force makes with the positive y-axis?

(A) 20.4°
(B) 66.4°
(C) 69.6°
(D) 110°

12. In the structure shown, the beam is pinned at point B. Point E is a roller support. The beam is loaded with a distributed load from point A to point B of 400 N/m, a 500 N·m couple at point C, and a vertical 900 N force at point D. If the distributed load and the vertical load are removed and replaced with a vertically upward force of 1700 N at point F, what moment at point F would be necessary to keep the reaction at point E the same?

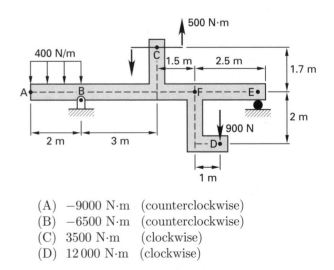

(A) −9000 N·m   (counterclockwise)
(B) −6500 N·m   (counterclockwise)
(C) 3500 N·m   (clockwise)
(D) 12 000 N·m (clockwise)

Problems 13 and 14 refer to the following information and beam shown.

The overhanging beam shown is supported by a roller and a pinned support as shown. The moment is removed and replaced by a couple consisting of forces applied at points A and C.

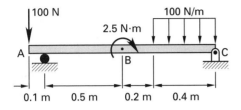

13. What is the magnitude of the couple that exactly replaces the moment that is removed?

   (A) 0.08 N·m
   (B) 0.16 N·m
   (C) 2.5 N·m
   (D) 15 N·m

14. What is the magnitude of the forces that constitute the moment?

   (A) 2.1 N
   (B) 4.2 N
   (C) 6.3 N
   (D) 8.3 N

15. A disk-shaped body with a 4 cm radius has a 320 N force acting through the center at an unknown angle θ, and two 40 N loads acting as a couple, as shown. All of these forces are removed and replaced by a single 320 N force at point B, parallel to the original 320 N force. What is the angle θ?

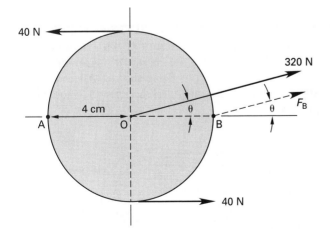

   (A) 0°
   (B) 7.6°
   (C) 15°
   (D) 29°

16. A rigid body is subjected to three concurrent, coplanar forces. What is the minimum number of independent equations that are necessary to establish the equilibrium conditions?

   (A) 0
   (B) 1
   (C) 2
   (D) 3

17. Three concurrent forces act as shown. If the forces are in equilibrium and F_2 is 11 N, what is the magnitude of F_1?

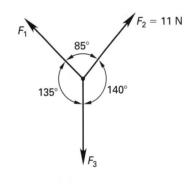

   (A) 8 N
   (B) 10 N
   (C) 11 N
   (D) 12 N

18. Where can a couple be moved on a rigid body to have an equivalent effect?

   (A) along the line of action
   (B) in a parallel plane
   (C) along the perpendicular bisector joining the two original forces
   (D) anywhere on the rigid body

19. Which of the structures shown is statically determinant and stable with the loadings shown?

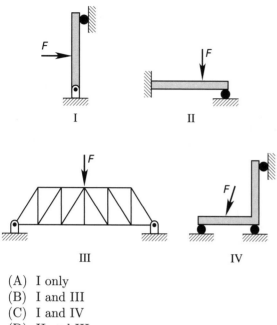

I

II

III

IV

(A) I only
(B) I and III
(C) I and IV
(D) II and III

20. Three coplanar forces are in equilibrium on the surface of a steel plate, as shown. Two of the forces are known to be 10 N. What is the angle, θ, of the third force?

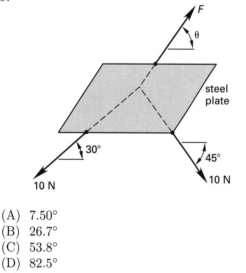

(A) 7.50°
(B) 26.7°
(C) 53.8°
(D) 82.5°

21. A signal arm carries two traffic signals and a sign, as shown. The signals and sign are rigidly attached to the arm. Each traffic signal is 0.2 m² in frontal area and weighs 210 N. The sign weighs 60 N/m². The design wind pressure is 575 N/m². The maximum moment that the connection between the arm and pole can withstand due to wind is 6000 N·m, and the maximum permitted moment due to the loads is 4000 N·m. As limited by moment on the connection, what is the maximum area of the sign?

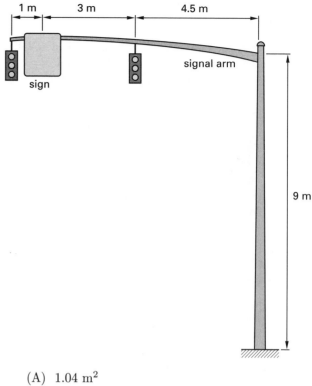

(A) 1.04 m²
(B) 1.15 m²
(C) 5.65 m²
(D) 8.03 m²

22. Two meshing spur gears are arranged such that neither gear is turning and both are in equilibrium. Gear 1 has a radius of 4 cm. Gear 1's shaft carries a torsional moment of 65 N·m from an external motor. Gear 2 has a radius of 6 cm. Assuming a 100% transmission efficiency, what torque is transmitted by the shaft of gear 2?

(A) 65 N·m
(B) 97.5 N·m
(C) 101 N·m
(D) 107 N·m

23. A hinged arch is composed of two pin-connected curved members supported on two pinned supports, as shown. Both members are rigid. A horizontal force of 1000 N is applied to pin B, as shown. All coordinates are in meters. What are the reactions and moments at joint A?

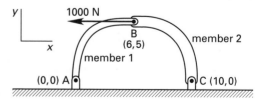

(A) $A_x = 500$ N; $A_y = 600$ N; $M_A = 5000$ N·m
(B) $A_x = 600$ N; $A_y = 500$ N; $M_A = 0$
(C) $A_x = 680$ N; $A_y = 400$ N; $M_A = 5000$ N·m
(D) $A_x = 616$ N; $A_y = 480$ N; $M_A = 0$

SOLUTIONS TO FE-STYLE EXAM PROBLEMS

1. The resultant force has the same point of application since the forces are concurrent, and choice (A) is the only one that originates at (3,6,4). The resultant force is calculated by summing the magnitudes in each coordinate direction.

$$\mathbf{R} = (15\mathbf{i} + 20\mathbf{i} - 10\mathbf{i}) + (25\mathbf{j} - 5\mathbf{j} + 40\mathbf{j})$$
$$+ (-10\mathbf{k} + 15\mathbf{k} + 50\mathbf{k})$$
$$= 25\mathbf{i} + 60\mathbf{j} + 55\mathbf{k}$$

Answer is A.

2. $\sum F_y = (300 \text{ N}) \sin 25°$
$$= F_Q \sin 70°$$
$$F_Q = (300 \text{ N}) \left(\frac{\sin 25°}{\sin 70°} \right)$$
$$= 134.9 \text{ N} (135 \text{ N})$$
$$\sum F_x = (300 \text{ N}) \cos 25°$$
$$= F_P + F_Q \cos 70°$$
$$F_P = (300 \text{ N}) \cos 25° - (134.9 \text{ N}) \cos 70°$$
$$= 225.8 \text{ N} (226 \text{ N})$$

Answer is C.

3. A pinned support will transmit or support all forces but not moments.

Answer is A.

4. $\sum F_x = (1000 \text{ N}) \cos 30° - (750 \text{ N}) \cos 15°$
$$- (1500 \text{ N}) \cos 20°$$
$$= -1268 \text{ N}$$
$$\sum F_y = (1000 \text{ N}) \sin 30° - (750 \text{ N}) \sin 15°$$
$$+ (1500 \text{ N}) \sin 20°$$
$$= 818.9 \text{ N}$$
$$R = \sqrt{(-1268 \text{ N})^2 + (818.9 \text{ N})^2}$$
$$= 1509 \text{ N} (1510 \text{ N})$$

Answer is D.

5. $\sum M_A = 0$
$$= 20 \text{ N·m} - (75 \text{ N})(0.5 \text{ m}) - F(1.5 \text{ m})$$
$$- (15 \text{ N})(1.5 \text{ m})$$
$$F = -26.7 \text{ N} (26.7 \text{ N}) [\text{up}]$$

Answer is B.

6.

$$\sum F_y = 0$$
$$= -60 \text{ N} + 10 \text{ N} + 20 \text{ N} + 10 \text{ N} + F$$
$$F = 20 \text{ N}$$
$$\sum M_A = 0$$
$$= 20 \text{ N·m} - (20 \text{ N})(0.2 \text{ m})$$
$$- (10 \text{ N})(0.2 \text{ m} + x) - (20 \text{ N})(0.4 \text{ m} + x)$$

$$4 + 2 + 10x + 8 + 20x = 20$$
$$30x = 6$$
$$x = 0.2 \text{ m}$$

Answer is C.

7.

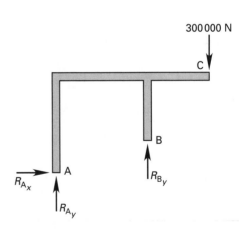

$$R_{A_x} = 0 [\text{all loads vertical}]$$
$$\sum M_B = 0$$
$$= R_{A_y}(6 \text{ m}) + (300\,000 \text{ N})(4 \text{ m})$$
$$R_{A_y} = -200\,000 \text{ N}$$
$$= 200\,000 \text{ N} [\text{down}]$$
$$\frac{200\,000 \text{ N}}{4 \text{ bolts}} = 50\,000 \text{ N/bolt} \left[\begin{array}{c} \text{member in} \\ \text{tension} \end{array} \right]$$

Answer is B.

8. Use the Pythagorean theorem to calculate the hypotenuse of the inclined force triangle. (Alternatively, recognize that this is a 5-12-13 triangle.)

$$\sqrt{(12)^2 + (5)^2} = 13$$

The x-component of the force is

$$F_x = F \cos\theta = (300 \text{ N})\left(\frac{5}{13}\right)$$
$$= 115.4 \text{ N} \quad (115 \text{ N})$$

Answer is A.

9. The direction triangle for the 200 N force is a 3-4-5 triangle. The vertical component of this force is

$$F_{y,1} = F \cos\theta_1 = (200 \text{ N})\left(\frac{3}{5}\right)$$
$$= 120 \text{ N}$$

The direction triangle for the 300 N force is a 5-12-13 triangle. The vertical component of this force is

$$F_{y,2} = F \sin\theta_2 = (300 \text{ N})\left(\frac{12}{13}\right)$$
$$= 277 \text{ N}$$

There is a -150 N vertical force acting downward. The moment is applied at point B and does not affect a force applied through that point.

The total force in the y-direction is

$$F_y + F_{y,1} + F_{y,2} + F_{y,3} = 0$$
$$F_y + 120 \text{ N} + 277 \text{ N} - 150 \text{ N} = 0$$
$$F_y = -247 \text{ N} \quad (-250 \text{ N})$$

Answer is C.

10. Let clockwise moments be positive. Take moments about point B.

$$\sum M_B = 700 \text{ N·m} + (150 \text{ N})(4 \text{ m})$$
$$- (300 \text{ N})\left(\frac{5}{13}\right)(2 \text{ m})$$
$$- (300 \text{ N})\left(\frac{12}{13}\right)(4 \text{ m})$$
$$+ (200 \text{ N})\left(\frac{4}{5}\right)(8 \text{ m})$$
$$= 1242 \text{ N·m} \quad (1240 \text{ N·m})$$

The application point of the moment is irrelevant.

Answer is D.

11. The magnitude of the force **A** is

$$\mathbf{A} = \sqrt{A_x^2 + A_y^2 + A_z^2}$$
$$= \sqrt{(3.5)^2 + (-1.5)^2 + (2.0)^2}$$
$$= 4.3$$
$$\theta = \cos^{-1}\left(\frac{A_y}{A}\right) = \cos^{-1}\left(\frac{-1.5}{4.3}\right)$$
$$= 110.4° \quad (110°)$$

Answer is D.

12. The reaction at point E is unknown, but it is irrelevant. Since the reaction is to be unchanged, it is necessary only to calculate the change in the loading.

Assume clockwise moments are positive. Take moments about point B for the forces that are removed and added.

$$\sum M_B = -\left(400 \, \frac{\text{N}}{\text{m}}\right)(2 \text{ m})\left(\frac{2 \text{ m}}{2}\right)$$
$$+ (900 \text{ N})(1 \text{ m} + 1.5 \text{ m} + 3 \text{ m})$$
$$- (-1700 \text{ N})(1.5 \text{ m} + 3 \text{ m})$$
$$= 11\,800 \text{ N·m} \quad (12\,000 \text{ N·m}) \quad [\text{clockwise}]$$

This is the moment that is applied by the forces that are removed, reduced by the moment of the new force that is applied. An 11 800 N·m clockwise moment must be applied to counteract this change. Round answer to 12 000 N·m. The location of the new moment is not relevant.

Answer is D.

13. A couple is a moment. When a moment of 2.5 N·m is removed, it must be replaced by the same moment.

Answer is C.

14. The distance between the forces is

$$d = 0.4 \text{ m} + 0.2 \text{ m} + 0.5 \text{ m} + 0.1 \text{ m}$$
$$= 1.2 \text{ m}$$

The magnitude of a couple is

$$M = Fd$$
$$F = \frac{M}{d} = \frac{2.5 \text{ N·m}}{1.2 \text{ m}}$$
$$= 2.08 \text{ N} \quad (2.1 \text{ N})$$

Answer is A.

15. Assume clockwise moments are positive. Take moments about the center for the original forces. The 320 N force has no moment arm, so it does not contribute to the moment. The couple is

$$M = Fd$$
$$= -(40 \text{ N})(8 \text{ cm})$$
$$= -320 \text{ N·cm}$$

The replacement force must produce a moment of -320 N·cm. The horizontal component of the replacement force acts through the center. Therefore, only the vertical component of the force contributes to the moment.

$$M = -320 \text{ N·cm} = Fr = -(320 \text{ N}) \sin \theta \, (4 \text{ cm})$$
$$\theta = 14.47° \quad (15°)$$

Answer is C.

16. Only force equilibrium equations are needed to analyze systems of concurrent forces. For planar (two-dimensional) forces, only two equations are necessary to establish equilibrium.

Answer is C.

17. Establish the line of action of force F_3 as the y-axis. The x-components of forces F_1 and F_2 must be equal and opposite. The sum of x-components of the forces is

$$\sum F_x: F_{2,x} + F_{1,x} = 0$$
$$(11 \text{ N}) \cos(140° - 90°) - F_1 \cos(135° - 90°) = 0$$
$$F_1 = 10 \text{ N}$$

Answer is B.

18. Since a couple is composed of two equal but opposite forces, the x- and y-components will always cancel, no matter what the orientation. Therefore, only the moment produced by the couple remains.

Answer is D.

19. Structure I is simply supported and determinant. Structure II is a propped cantilever beam, always indeterminant by one degree. Structure III is a truss that is pinned at both ends, also indeterminant by one degree. Structure IV is a beam with three rollers, two in the vertical direction and one in the horizontal direction. It is determinant but not stable.

Answer is A.

20. Let the horizontal direction correspond to the x-axis. The condition for equilibrium in the y-direction (in the plane of the plate) is

$$\sum F_y: (-10 \text{ N}) \sin 30° + (-10 \text{ N}) \sin 45° + F \sin \theta = 0$$
$$F = \frac{12.07 \text{ N}}{\sin \theta}$$

The condition for equilibrium in the x-direction is

$$\sum F_x: (-10 \text{ N}) \cos 30° + (10 \text{ N}) \cos 45° + F \cos \theta = 0$$
$$F = \frac{1.59 \text{ N}}{\cos \theta}$$

Set these two expressions for F equal.

$$\frac{12.07 \text{ N}}{\sin \theta} = \frac{1.59 \text{ N}}{\cos \theta}$$
$$\frac{\sin \theta}{\cos \theta} = \tan \theta = \frac{12.07 \text{ N}}{1.59 \text{ N}}$$
$$= 7.59$$
$$\theta = \tan^{-1}(7.59) = 82.5°$$

Answer is D.

21. The length of the signal arm is

$$1 \text{ m} + 3 \text{ m} + 4.5 \text{ m} = 8.5 \text{ m}$$

Set the moment on the arm due to the wind equal to the maximum allowed.

$$M_{\text{wind}}: (0.2 \text{ m}^2) \left(575 \, \frac{\text{N}}{\text{m}^2}\right) (8.5 \text{ m} + 4.5 \text{ m})$$
$$+ A_{\text{sign}} \left(575 \, \frac{\text{N}}{\text{m}^2}\right) (7.5 \text{ m}) = 6000 \text{ N·m}$$
$$A_{\text{sign}} = 1.04 \text{ m}^2$$

Set the moment on the arm due to vertical loading equal to the maximum allowed.

$$M_{\text{loads}}: (210 \text{ N})(8.5 \text{ m} + 4.5 \text{ m})$$
$$+ \left(60 \, \frac{\text{N}}{\text{m}^2}\right) A_s(7.5 \text{ m}) = 4000 \text{ N·m}$$
$$A_{\text{sign}} = 2.82 \text{ m}^2$$

The maximum area of the sign is the smaller of these two values, 1.04 m^2.

Answer is A.

22. The force on each gear tooth can be found from the torque transmitted through it. For convenience, assume only one pair of teeth is in contact.

$$T = Fr$$

$$F_1 = \frac{T_1}{r_1}$$

$$F_2 = \frac{T_2}{r_2}$$

The forces are the same for all pairs of meshing gear teeth.

$$F_1 = F_2$$

$$\frac{T_1}{r_1} = \frac{T_2}{r_2}$$

$$T_2 = \frac{T_1 r_2}{r_1}$$

$$= \frac{(65 \text{ N·m})(6 \text{ cm})}{4 \text{ cm}}$$

$$= 97.5 \text{ N·m}$$

Answer is B.

23. Sum moments about point C. The x-component of the reaction at point A has a zero moment arm.

$$\sum M_{\mathrm{C}} = A_y 10 - (1000 \text{ N})(5) = 0$$

$$A_y = 500 \text{ N}$$

Point B is a pin, which transmits no moment. Sum moments to the left of point B.

$$\sum M_{\mathrm{B}} = A_y 6 - A_x 5$$

$$= 0$$

$$= (500 \text{ N})(6) - A_x 5$$

$$= 0$$

$$A_x = 600 \text{ N}$$

Point A is a pin, which transmits no moment.

Answer is B.

11 Trusses

Subjects

STATICALLY DETERMINATE
TRUSSES 11-1
    Method of Joints 11-2
    Method of Sections 11-2

Nomenclature

F	force	lbf	N
M	moment	ft-lbf	N·m
R	reaction	lbf	N

STATICALLY DETERMINATE TRUSSES

A *truss* or *frame* is a set of pin-connected axial *members* (i.e., *two-force members*). The connection points are known as *joints*. Member weights are disregarded, and truss loads are applied only at joints. A *structural cell* consists of all members in a closed loop of members. For the truss to be stable (i.e., to be a *rigid truss*), all of the structural cells must be triangles. Figure 11.1 identifies *chords*, *end posts*, *panels*, and other elements of a typical *bridge truss*.

Figure 11.1 Parts of a Bridge Truss

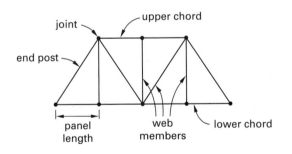

A *trestle* is a braced structure spanning a ravine, gorge, or other land depression in order to support a road or rail line. Trestles are usually indeterminate, have multiple earth contact points, and are more difficult to evaluate than simple trusses.

Several types of trusses have been given specific names. Some of the more common named trusses are shown in Fig. 11.2.

Figure 11.2 Special Types of Trusses

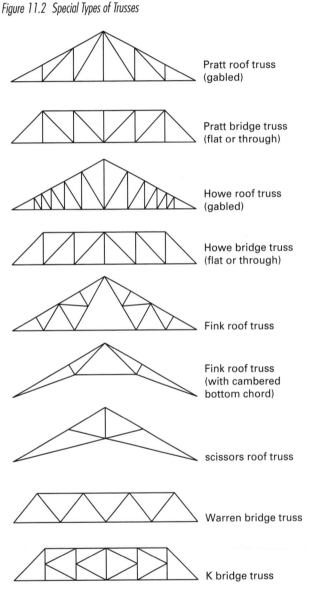

Pratt roof truss (gabled)

Pratt bridge truss (flat or through)

Howe roof truss (gabled)

Howe bridge truss (flat or through)

Fink roof truss

Fink roof truss (with cambered bottom chord)

scissors roof truss

Warren bridge truss

K bridge truss

Truss loads are considered to act only in the plane of a truss. Therefore, trusses are analyzed as two-dimensional structures. Forces in truss members hold the various truss parts together and are known as *internal forces*. The internal forces are found by drawing free-body diagrams.

Although free-body diagrams of truss members can be drawn, this is not usually done. Instead, free-body

STATICS
Trusses

diagrams of the pins (i.e., the joints) are drawn. A pin in compression will be shown with force arrows pointing toward the pin, away from the member. Similarly, a pin in tension will be shown with force arrows pointing away from the pin, toward the member.

With typical bridge trusses supported at the ends and loaded downward at the joints, the upper chords are almost always in compression, and the end panels and lower chords are almost always in tension.

Since truss members are axial members, the forces on the truss joints are concurrent forces. Therefore, only force equilibrium needs to be enforced at each pin; the sum of the forces in each of the coordinate directions equals zero.

Figure 11.3 Zero-Force Members

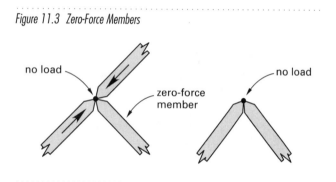

Forces in truss members can sometimes be determined by inspection. One of these cases is *zero-force members*. A third member framing into a joint already connecting two collinear members carries no internal force unless there is a load applied at that joint. Similarly, both members forming an apex of the truss are zero-force members unless there is a load applied at the apex.

A truss will be *statically determinate* if Eq. 11.1 holds true.

$$\text{no. of members} = (2)(\text{no. of joints}) - 3 \qquad 11.1$$

If the left-hand side of Eq. 11.1 is greater than the right-hand side (i.e., there are *redundant members*), the truss is statically indeterminate. If the left-hand side is less than the right-hand side, the truss is unstable and will collapse under certain types of loading.

Method of Joints

The *method of joints* is one of the methods that can be used to find the internal forces in each truss member. This method is useful when most or all of the truss member forces are to be calculated. Because this method advances from joint to adjacent joint, it is inconvenient when a single isolated member force is to be calculated.

The method of joints is a direct application of the equations of equilibrium in the x- and y-directions. Traditionally, the method begins by finding the reactions supporting the truss. Next the joint at one of the reactions is evaluated, which determines all the member forces framing into the joint. Then, knowing one or more of the member forces from the previous step, an adjacent joint is analyzed. The process is repeated until all the unknown quantities are determined.

At a joint, there may be up to two unknown member forces, each of which can have dependent x- and y-components. Since there are two equilibrium equations, the two unknown forces can be determined. Even though determinate, however, the sense of a force will often be unknown. If the sense cannot be determined by logic, an arbitrary decision can be made. If the incorrect direction is chosen, the force will be negative.

Occasionally, there will be three unknown member forces. In that case, an additional equation must be derived from an adjacent joint.

Method of Sections

The *method of sections* is a direct approach to finding forces in any truss member. This method is convenient when only a few truss member forces are unknown.

As with the previous method, the first step is to find the support reactions. Then a cut is made through the truss, passing through the unknown member. (Knowing where to cut the truss is the key part of this method. Such knowledge is developed only by repeated practice.) Finally, all three conditions of equilibrium are applied as needed to the remaining truss portion. Since there are three equilibrium equations, the cut cannot pass through more than three members in which the forces are unknown.

SAMPLE PROBLEMS

Problem 1

Identify the zero-force members in the truss shown.

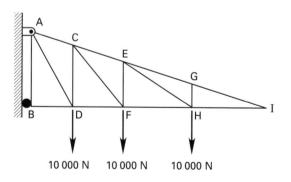

(A) GI, HI
(B) AB, GH
(C) AB, HI, GI
(D) AB, GH, GI, HI, EG

Solution

At joint B, the reaction and member BD are collinear; member AB is a zero-force member.

At joint G, members EG and GI are collinear; member GH is a zero-force member.

At joint I, there is no member or external load to offset the vertical component of force in member GI, so GI is a zero-force member.

If GI is a zero-force member, so is HI.

If GI and GH are zero-force members, so is EG.

Answer is D.

Problem 2

Determine the force in member AG for the pin-connected truss shown.

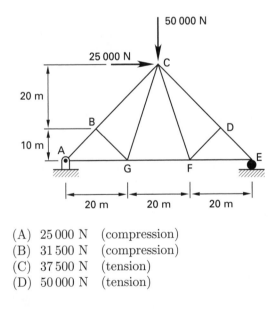

(A) 25 000 N   (compression)
(B) 31 500 N   (compression)
(C) 37 500 N   (tension)
(D) 50 000 N   (tension)

Solution

Choose the positive directions as upward and to the right. Choose positive moments as clockwise. Find reactions at A.

$$\sum M_E = 0$$
$$= R_{A_y}(60 \text{ m}) + (25\,000 \text{ N})(10 \text{ m} + 20 \text{ m})$$
$$- (50\,000 \text{ N})\left(\frac{60 \text{ m}}{2}\right)$$

$$R_{A_y} = 12\,500 \text{ N} \quad [\text{upward}]$$
$$\sum F_x = 0 = R_{A_x} + 25\,000 \text{ N}$$
$$R_{A_x} = -25\,000 \text{ N} \quad [\text{to the left}]$$

Use the method of joints.

For pin A,

$$\sum F_y = 0 = 12\,500 \text{ N} + AB \sin 45°$$
$$AB = -17\,678 \text{ N} \quad [\text{compression}]$$
$$\sum F_x = 0 = -25\,000 \text{ N} - (17\,678 \text{ N}) \cos 45° + AG$$
$$AG = 37\,500 \text{ N} \quad [\text{tension}]$$

Answer is C.

Problem 3

Find the force in member BC.

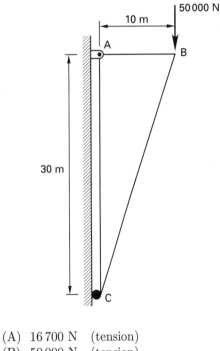

(A) 16 700 N   (tension)
(B) 50 000 N   (tension)
(C) 50 000 N   (compression)
(D) 52 700 N   (compression)

Solution

Find the reaction at C.

$$\sum M_A = 0 = -R_{C_x}(30 \text{ m}) + (50\,000 \text{ N})(10 \text{ m})$$
$$R_{C_x} = 16\,667 \text{ N} \quad [\text{to the right}]$$
$$\sum F_x = 0 = 16\,667 \text{ N} + R_{A_x}$$
$$R_{A_x} = -16\,667 \text{ N} \quad [\text{to the left}]$$

Use the method of joints. The length of member BC is

$$\sqrt{(10 \text{ m})^2 + (30 \text{ m})^2} = 31.62 \text{ m}$$

The x-component of force BC is 10/31.62 of the total.

For pin C,

$$\sum F_x = 0 = 16\,667 \text{ N} - \text{BC}\left(\frac{10 \text{ m}}{31.62 \text{ m}}\right)$$
$$\text{BC} = 52\,701 \text{ N} \quad (52\,700 \text{ N}) \quad [\text{compression}]$$

Answer is D.

Problem 4

Determine the force in member FH for the pin-connected truss shown.

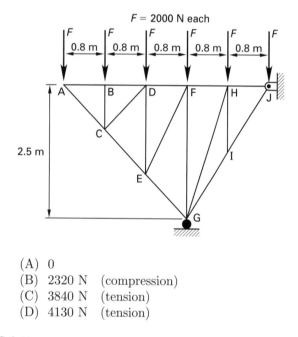

F = 2000 N each

(A) 0
(B) 2320 N (compression)
(C) 3840 N (tension)
(D) 4130 N (tension)

Solution

Use the method of sections. Cut the truss as shown.

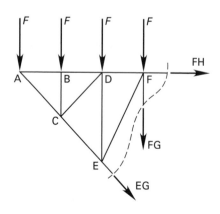

Sum moments about point A.

$$\sum M_A = 0$$
$$= (2000 \text{ N})(0.8 \text{ m} + 1.6 \text{ m} + 2.4 \text{ m})$$
$$+ \text{FG}(2.4 \text{ m})$$

$$\text{FG} = -4000 \text{ N} \quad [\text{compression}]$$

$$\sum M_E = 0$$
$$= (2000 \text{ N})(-1.6 \text{ m} - 0.8 \text{ m} + 0.8 \text{ m})$$
$$- (4000 \text{ N})(0.8 \text{ m}) + \text{FH}(1.67 \text{ m})$$

$$\text{FH} = 3839 \text{ N} \quad (3840 \text{ N}) \quad [\text{tension}]$$

Alternatively, sum moments about point G.

$$\sum M_G = 0$$
$$= (2000 \text{ N})(0.8 \text{ m} + 1.6 \text{ m} + 2.4 \text{ m})$$
$$- \text{FH}(2.5 \text{ m})$$

$$\text{FH} = 3840 \text{ N} \quad [\text{tension}]$$

Answer is C.

FE-STYLE EXAM PROBLEMS

1. In the pin-jointed truss shown, what is the force in member DE?

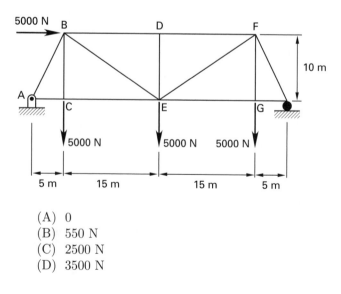

(A) 0
(B) 550 N
(C) 2500 N
(D) 3500 N

2. What is the reaction at point A?

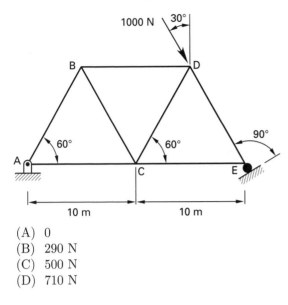

(A) 0
(B) 290 N
(C) 500 N
(D) 710 N

3. Determine the force in member BC.

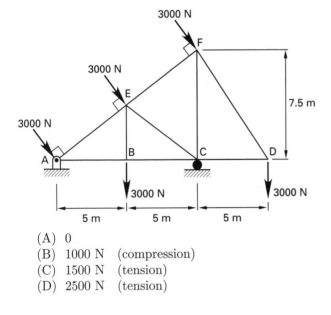

(A) 0
(B) 1000 N   (compression)
(C) 1500 N   (tension)
(D) 2500 N   (tension)

4. Find the force in member DE.

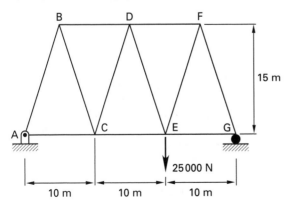

(A) 0
(B) 6300 N   (tension)
(C) 8800 N   (tension)
(D) 10 000 N   (compression)

Problems 5 and 6 refer to the truss shown.

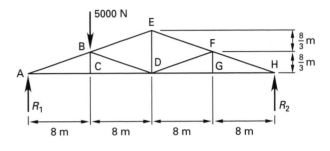

5. What are R_1 and R_2?

(A) $R_1 = 1000$ N; $R_2 = 4000$ N
(B) $R_1 = 1250$ N; $R_2 = 3750$ N
(C) $R_1 = 2500$ N; $R_2 = 2500$ N
(D) $R_1 = 3750$ N; $R_2 = 1250$ N

6. What are the forces in members AC and BD?

(A) AC = 11 000 N   (tension);
     BD = −7900 N   (compression)
(B) AC = 0; BD = −2000 N   (compression)
(C) AC = 1100 N   (tension);
     BD = 2500 N   (tension)
(D) AC = 0; BD = −7900 N   (compression)

For the following problems use the NCEES Handbook as your only reference.

7. The braced frame shown is constructed with pin-connected members and supports. All applied forces are horizontal. What is the force in the diagonal member AB?

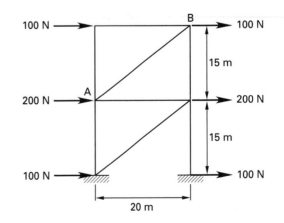

(A) 0
(B) 160 N
(C) 200 N
(D) 250 N

8. The pedestrian bridge truss shown has 10 000 N applied loads at points I, J, and K. What is the force in member IJ?

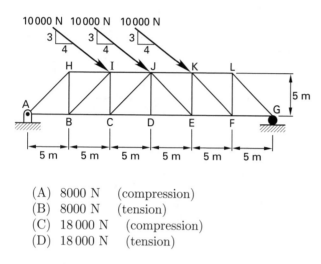

(A) 8000 N   (compression)
(B) 8000 N   (tension)
(C) 18 000 N   (compression)
(D) 18 000 N   (tension)

Problems 9 and 10 refer to the cantilever truss shown.

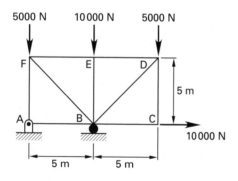

9. What is the reaction at point B?

(A) 5000 N
(B) 10 000 N
(C) 15 000 N
(D) 20 000 N

10. What is the force in member AF?

(A) 0
(B) 5000 N
(C) 10 000 N
(D) 15 000 N

SOLUTIONS TO FE-STYLE EXAM PROBLEMS

1. Member DE is a zero-force member. Members BD and DF are collinear, and there is no force applied at the joint.

Answer is A.

2. The applied load is normal to the support at E and collinear with member DE. The full load is resisted by the support at E, so there is no reaction at A.

Answer is A.

3. $\sum M_A = 0$

$$
\begin{aligned}
= &(3000 \text{ N})(6.25 \text{ m}) + (3000 \text{ N})(12.5 \text{ m}) \\
&+ (3000 \text{ N})(5 \text{ m}) - R_{C_y}(10 \text{ m}) \\
&+ (3000 \text{ N})(15 \text{ m})
\end{aligned}
$$

$$R_{C_y} = 11\,625 \text{ N} \quad [\text{upward}]$$

From trigonometry, the applied forces are inclined from the horizontal at arctan $(7.5 \text{ m}/(5 \text{ m} + 5 \text{ m})) = 36.87°$.

$$
\begin{aligned}
\sum F_y &= 0 \\
&= R_{A_y} - (3)(3000 \text{ N})\cos 36.87° \\
&\quad - (2)(3000 \text{ N}) + 11\,625 \text{ N}
\end{aligned}
$$

$$R_{A_y} = 1575 \text{ N} \quad [\text{upward}]$$

$$
\begin{aligned}
\sum F_x &= 0 \\
&= R_{A_x} + (3)(3000 \text{ N})\sin 36.87°
\end{aligned}
$$

$$R_{A_x} = -5400 \text{ N} \quad [\text{to the left}]$$

Use the method of sections.

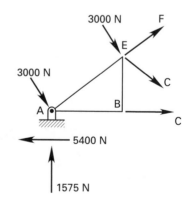

The vertical downward force at point B passes through point E, and it does not generate a moment.

$$\sum M_E = 0$$
$$= (5400 \text{ N})(3.75 \text{ m}) + (1575 \text{ N})(5 \text{ m})$$
$$- (3000 \text{ N})(6.25 \text{ m}) - BC(3.75 \text{ m})$$

$$BC = 2500 \text{ N}$$

Answer is D.

4. Take moments about point G.

$$\sum M_G = 0$$
$$= (-25\,000 \text{ N})(10 \text{ m}) + R_{A_y}(30 \text{ m})$$

$$R_{A_y} = 8333 \text{ N} \quad [\text{upward}]$$

Use the method of sections.

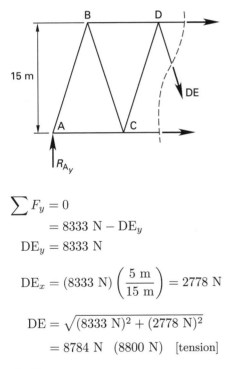

$$\sum F_y = 0$$
$$= 8333 \text{ N} - DE_y$$
$$DE_y = 8333 \text{ N}$$

$$DE_x = (8333 \text{ N})\left(\frac{5 \text{ m}}{15 \text{ m}}\right) = 2778 \text{ N}$$

$$DE = \sqrt{(8333 \text{ N})^2 + (2778 \text{ N})^2}$$
$$= 8784 \text{ N} \quad (8800 \text{ N}) \quad [\text{tension}]$$

Answer is C.

5. $$\sum M_A = 0 = (5000 \text{ N})(8 \text{ m}) - R_2(32 \text{ m})$$

$$R_2 = 1250 \text{ N} \quad [\text{upward}]$$

$$\sum F_y = 0 = R_1 + 1250 \text{ N} - 5000 \text{ N}$$

$$R_1 = 3750 \text{ N} \quad [\text{upward}]$$

Answer is D.

6. The angle made by the inclined members with the horizontal is

$$\tan^{-1}\left(\frac{\frac{8}{3} \text{ m}}{8 \text{ m}}\right) = 18.435°$$

(Alternatively, the force components could be determined from geometry.)

Use the method of joints.

For pin A,

$$\sum F_y = 0$$
$$= R_1 + AB \sin 18.435°$$
$$= 3750 \text{ N} + AB \sin 18.435°$$

$$AB = -11\,859 \text{ N} \quad [\text{compression}]$$

$$\sum F_x = 0$$
$$= (-11\,859 \text{ N}) \cos 18.435° + AC$$

$$AC = 11\,250 \text{ N} \quad (11\,000 \text{ N}) \quad [\text{tension}]$$

For pin C,

$$\sum F_y = 0$$

$$BC = 0 \quad [\text{zero-force member}]$$

For pin B,

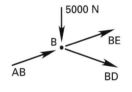

$$\sum F_x = 0$$
$$= AB \cos 18.435° + BE \cos 18.435°$$
$$+ BD \cos 18.435°$$

$$0 = AB + BE + BD$$

$$\sum F_y = 0$$
$$= AB \sin 18.435° + BE \sin 18.435°$$
$$- BD \sin 18.435° - 5000 \text{ N}$$

$$BD \sin 18.435° = AB \sin 18.435° + (-AB - BD)$$
$$\times (\sin 18.435°) - 5000 \text{ N}$$
$$= \sin 18.435° (AB - AB - BD)$$
$$- 5000 \text{ N}$$

$$2BD \sin 18.435° = -5000 \text{ N}$$

$$BD = -7906 \text{ N} \quad (-7900 \text{ N})$$
$$\begin{bmatrix}\text{compression—in opposite} \\ \text{direction shown}\end{bmatrix}$$

Answer is A.

7. Determine the length of member AB by recognizing this configuration to be a 3-4-5 triangle.

$$L_{AB} = 25 \text{ m}$$

Use the method of sections. Cut the frame horizontally through member AB.

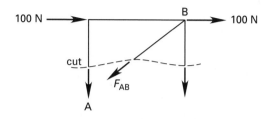

By inspection, the horizontal component of F_{AB} balances the two applied horizontal loads.

$$F_{AB,x} = 100 \text{ N} + 100 \text{ N} = 200 \text{ N}$$

By similar triangles,

$$F = \left(\frac{5}{4}\right)(200 \text{ N})$$

$$= 250 \text{ N}$$

Answer is D.

8. Recognize that the directions of the applied inclined loads correspond to a 3-4-5 triangle. By similar triangles, the x- and y-components of the applied loads are

$$F_x = (10\,000 \text{ N})\left(\frac{4}{5}\right) = 8000 \text{ N}$$

$$F_y = (10\,000 \text{ N})\left(\frac{3}{5}\right) = 6000 \text{ N}$$

Find the vertical reaction at point A by taking moments about point G.

$$\sum M_G: (6000 \text{ N})(10 \text{ m}) + (6000 \text{ N})(15 \text{ m})$$
$$+ (6000 \text{ N})(20 \text{ m}) - (30 \text{ m})R_{A,y}$$
$$- (3)(8000 \text{ N})(5 \text{ m}) = 0$$

$$R_{A,y} = 5000 \text{ N}$$

Use the method of sections. Cut the truss vertically through IJ.

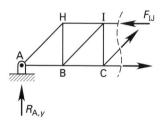

Take moments about point C. Assume a direction for force F_{IJ}. The forces in members CJ and CD and the horizontal component of the reaction at point A have zero moment arms.

$$M_C: F_{IJ}(5 \text{ m}) - (8000 \text{ N})(5 \text{ m}) - (5000 \text{ N})(10 \text{ m}) = 0$$

$$F_{IJ} = 18\,000 \text{ N}$$

The sign of F_{IJ} is positive, so the direction (as shown) was chosen correctly. The force is compressing the member.

Answer is C.

9. Take moments about point A. (The 10 000 N horizontal force has no moment arm.)

$$\sum M_A: R_B(5 \text{ m}) - (10\,000 \text{ N})(5 \text{ m})$$
$$- (5000 \text{ N})(10 \text{ m}) = 0$$

$$R_B = 20\,000 \text{ N}$$

Answer is D.

10. By inspection, the force in member AF is equal to the vertical component of the reaction at point A.

Answer is A.

12 Pulleys, Cables, and Friction

Subjects

PULLEYS 12-1
CABLES 12-1
FRICTION 12-2
    Belt Friction 12-2
SQUARE SCREW THREAD 12-2

Nomenclature

F	force	lbf	N
D	diameter	ft	m
g	acceleration due to gravity	ft/sec^2	m/s^2
g_c	gravitational constant	lbm-ft/lbf-sec^2	–
m	mass	lbm	kg
n	number of sheaves	–	–
N	normal force	lbf	N
P	power	ft-lbf/sec	W
r	radius	ft	m
R	reaction force	lbf	N
T	tension	lbf	N
T	torque	ft-lbf	N·m
v	velocity	ft/sec	m/s
W	weight	lbf	N

Symbols

θ	wrap angle	rad	rad
μ	coefficient of friction	–	–
ϕ	inclination angle	deg	deg

Subscripts

s	static
f	friction
t	tangential

PULLEYS

A *pulley* (also known as a *sheave*) is used to change the direction of an applied tensile force. A series of pulleys working together (known as a *block and tackle*) can also provide *pulley advantage* (i.e., *mechanical advantage*).

Figure 12.1 Mechanical Advantage of Rope-Operated Machines

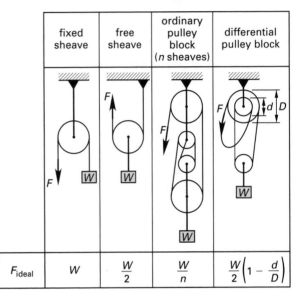

	fixed sheave	free sheave	ordinary pulley block (*n* sheaves)	differential pulley block
F_{ideal}	W	$\dfrac{W}{2}$	$\dfrac{W}{n}$	$\dfrac{W}{2}\left(1 - \dfrac{d}{D}\right)$

If the pulley is attached by a bracket to a fixed location, it is said to be a *fixed pulley*. If the pulley is attached to a load, or if the pulley is free to move, it is known as a *free pulley*.

Most simple problems disregard friction and assume that all ropes (fiber ropes, wire ropes, chains, belts, etc.) are parallel. In such cases, the pulley advantage is equal to the number of ropes coming to and going from the load-carrying pulley. The diameters of the pulleys are not factors in calculating the pulley advantage.

CABLES

An *ideal cable* is assumed to be completely flexible, massless, and incapable of elongation; therefore, it acts as an axial tension member between points of concentrated loading. In fact, the term *tension* or *tensile force* is commonly used in place of member force when dealing with cables.

The methods of joints and sections used in truss analysis can be used to determine the tensions in cables carrying concentrated loads. After separating the reactions into x- and y-components, it is particularly useful to sum moments about one of the reaction points. All cables will be found to be in tension, and (with vertical loads only) the horizontal tension component will be the same in all cable segments. Unlike the case of a rope passing over a series of pulleys, however, the total tension in the cable will not be the same in every cable segment.

Figure 12.2 Cable with Concentrated Load

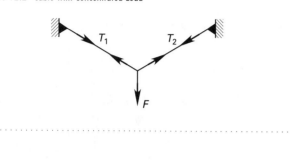

FRICTION

Friction is a force that always resists motion or impending motion. It always acts parallel to the contacting surfaces. The frictional force, F, exerted on a stationary body is known as *static friction*, *Coulomb friction*, and *fluid friction*. If the body is moving, the friction is known as *dynamic friction* and is less than the static friction.

The actual magnitude of the frictional force depends on the *normal force*, N, and the *coefficient of friction*, μ, between the body and the surface.

$$F = \mu N \qquad 12.1$$

For a body resting on a horizontal surface, the normal force is the weight of the body.

$$N = mg \qquad \text{[SI]} \qquad 12.2a$$

$$N = \frac{mg}{g_c} \qquad \text{[U.S.]} \qquad 12.2b$$

If a body rests on a plane inclined at an angle ϕ from the horizontal, the normal force is

$$N = mg \cos \phi \qquad \text{[SI]} \qquad 12.3a$$

$$N = \frac{mg \cos \phi}{g_c} \qquad \text{[U.S.]} \qquad 12.3b$$

Belt Friction

Friction between a belt, rope, or band wrapped around a pulley or sheave is responsible for the transfer of torque. Except when stationary, one side of the belt (the tight side) will have a higher tension than the other

(the slack side). The basic relationship between the belt tensions and the coefficient of friction neglects centrifugal effects and is given by Eq. 12.4. F_1 is the tension on the tight side (direction of movement); F_2 is the tension on the other side. The *angle of wrap*, θ, must be expressed in radians.

$$F_1 = F_2 e^{\mu\theta} \qquad 12.4$$

Figure 12.3 Belt Friction

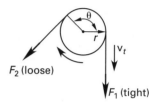

The net transmitted torque is

$$T = (F_1 - F_2)r \qquad 12.5$$

The power transmitted to a belt running at tangential velocity v_t is

$$P = (F_1 - F_2)\text{v}_t \qquad 12.6$$

SQUARE SCREW THREADS

A *power screw* changes angular position into linear position (i.e., changes rotary motion into traversing motion). The linear positioning can be horizontal (as in vices and lathes) or vertical (as in a jacks). Square, Acme, and 10-degree modified screw threads are commonly used in power screws. A square screw thread is shown in Fig. 12.4.

Figure 12.4 Square Screw Thread

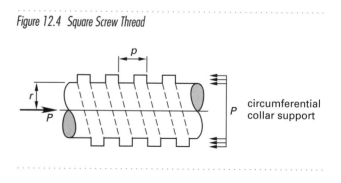

A square screw thread is designated by a mean radius, r, pitch, p, and *pitch angle*, α. The *pitch*, p, is the distance between corresponding points on a thread. The *lead* is the distance the screw advances each revolution. Often, double- and triple-threaded screws are used. The lead is one, two, or three times the pitch for single-, double-, and triple-threaded screws, respectively.

$$p = 2\pi r \tan \alpha \qquad 12.7$$

The coefficient of friction, μ, between the threads can be designated directly or by way of a *thread friction angle*, ϕ.

$$\mu = \tan \phi \qquad 12.8$$

The torque or external moment, M, required to turn a square screw in motion against an axial force, P (i.e., "raise" the load), is

$$M = Pr \tan(\alpha + \phi) \qquad 12.9$$

The torque required to turn the screw in motion in the direction of the applied axial force (i.e., "lower" the load) is given by Eq. 12.10. If the torque is zero or negative (as it would be if the lead is large or friction is low), then the screw is not self-locking and the load will lower by itself, causing the screw to spin (i.e, it will "overhaul"). The screw will be self-locking when $\tan \alpha \leq \mu$.

$$M = Pr \tan(\alpha - \phi) \qquad 12.10$$

The torque calculated in Eqs. 12.9 and 12.10 is required to overcome thread friction and to raise the load (i.e., axially compress the screw). Typically, only 10–15 percent of the torque goes into axial compression of the screw. The remainder is used to overcome friction. The mechanical efficiency of the screw is the ratio of torque without friction to the torque with friction. The torque without friction can be calculated from Eq. 12.9 or 12.10 (depending on the travel direction) using $\phi = 0$.

$$\eta_m = \frac{T_{f=0}}{T} \qquad 12.11$$

In the absence of an antifriction ring, an additional torque will be required to overcome friction in the collar. Since the collar is generally flat, the normal force is the jack load, P, for the purpose of calculating the frictional force.

$$T_{\text{collar}} = P\mu_{\text{collar}}r_{\text{collar}} \qquad 12.12$$

SAMPLE PROBLEMS

Problem 1

Find the tension, T, that must be applied to pulley A to lift the 1200 N weight.

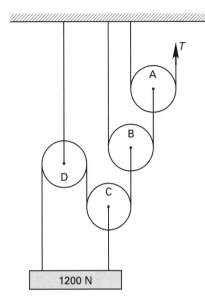

(A) 100 N
(B) 300 N
(C) 400 N
(D) 600 N

Solution

The free bodies of the system are shown.

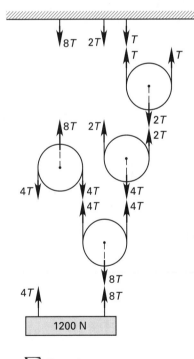

$$\sum F_y = 0$$
$$= -1200 \text{ N} + 4T + 8T$$

$$12T = 1200 \text{ N}$$

$$T = 100 \text{ N}$$

Answer is A.

Problems 2 and 3 refer to the following illustration.

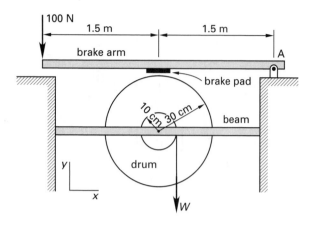

Problem 2

The coefficient of friction between the brake pad and drum is 0.3. Assuming that the beam supporting the cable drum is rigid and more than adequate for the loads involved, what load, W, can be held stationary?

(A) 33 N
(B) 90 N
(C) 100 N
(D) 180 N

Solution

The free bodies of the brake arm and drum are

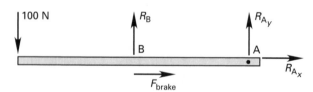

(a) free body of the brake arm

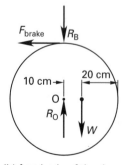

(b) free body of the drum

The brake reaction (normal force), R_B, is found by summing moments about the lever pivot point.

$$\sum M_A = 0$$

$$= -(100 \text{ N})(3 \text{ m}) + R_B(1.5 \text{ m})$$

$$R_B = 200 \text{ N}$$

$$F_{\text{brake}} = \mu R_B = (0.3)(200 \text{ N})$$

$$= 60 \text{ N}$$

From the free body of the drum,

$$\sum M_O = 0$$

$$= W(0.1 \text{ m}) - (60 \text{ N})(0.3 \text{ m})$$

$$W = 180 \text{ N} \quad [\text{maximum without slipping}]$$

Answer is D.

Problem 3

If $W = 80$ N, what are the reactions at point A?

(A) $0\,\mathbf{i} + 100\,\mathbf{j}$ N
(B) $0\,\mathbf{i} + 180\,\mathbf{j}$ N
(C) $27\,\mathbf{i}$ N $+ 100\,\mathbf{j}$ N
(D) $-27\,\mathbf{i}$ N $- 100\,\mathbf{j}$ N

Solution

Refer to the free-body diagram of the brake arm in Prob. 2.

$$F_{\text{brake}}(0.3 \text{ m}) = W(0.1 \text{ m})$$

$$F_{\text{brake}} = \frac{W}{3} = \frac{80 \text{ N}}{3}$$

$$= 26.67 \text{ N} \quad [\text{to the right}]$$

$$\sum F_x = 0$$

$$= F_{\text{brake}} + R_{A_x}$$

$$R_{A_x} = -F_{\text{brake}}$$

$$= -26.67 \text{ N} \quad [\text{to the left}]$$

R_B is a function only of the applied 100 N force and geometry.

$$\sum F_y = 0$$

$$= -100 \text{ N} + R_B + R_{A_y}$$

$$R_{A_y} = 100 \text{ N} - 200 \text{ N}$$

$$= -100 \text{ N}$$

$$R = -27\,\mathbf{i} \text{ N} - 100\,\mathbf{j} \text{ N}$$

Answer is D.

Problem 4

The 285 kg plate shown is suspended horizontally by four wires of equal length, and the tension of each wire is equal. If wire D snaps, the tension in the three remaining wires is redistributed. Determine the tension in each wire after wire D snaps.

Problem 5

The cabinet shown weighs 110 N and is supported on wheels. Which of the following cases yields the largest value of F that will move the cabinet to the right?

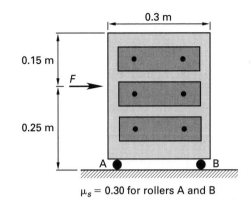

$\mu_s = 0.30$ for rollers A and B

I.   rollers A and B are locked
II.   roller B is locked, and roller A is free to rotate
III. roller A is locked, and roller B is free to rotate

(A)   case I
(B)   case II
(C)   cases I and II
(D)   cases II and III

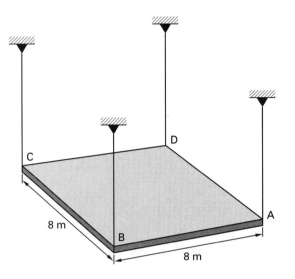

(A) $T_A = 699$ N; $T_B = 1398$ N; $T_C = 699$ N
(B) $T_A = 1398$ N; $T_B = 0$ N; $T_C = 1398$ N
(C) $T_A = 699$ N; $T_B = 699$ N; $T_C = 1398$ N
(D) $T_A = 1398$ N; $T_B = 1398$ N; $T_C = 0$ N

Solution

Sum moments about edge AB. Assume all of the weight acts at the plate's centroid.

$$T_D = 0$$

$$\sum M_{AB} = 0$$

$$= (285 \text{ kg})\left(9.81 \ \frac{\text{m}}{\text{s}^2}\right)(4 \text{ m}) - T_C(8 \text{ m})$$

$$T_C = 1398 \text{ N}$$

Sum moments about edge CB.

$$\sum M_{CB} = 0$$

$$= (285 \text{ kg})\left(9.81 \ \frac{\text{m}}{\text{s}^2}\right)(4 \text{ m}) - T_A(8 \text{ m})$$

$$T_A = 1398 \text{ N}$$

$$\sum F_y = 0$$

$$= T_A + T_B + T_C - mg$$

$$T_B = -1398 \text{ N} - 1398 \text{ N} + (285 \text{ kg})\left(9.81 \ \frac{\text{m}}{\text{s}^2}\right)$$

$$= 0$$

Answer is B.

Solution

For case I,

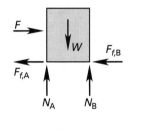

$$\sum F_y = 0 = N_A + N_B - 110 \text{ N}$$

$$\sum F_x = 0 = F - F_{f,A} - F_{f,B}$$

$$= F - \mu_s(N_A + N_B)$$

$$F = (0.30)(110 \text{ N})$$

$$= 33 \text{ N}$$

For case II,

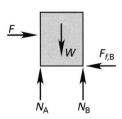

$$\sum F_y = 0$$
$$= N_A + N_B - 110 \text{ N}$$

$$\sum F_x = 0 = F - F_{f,B}$$

$$F = F_{f,B} = \mu_s N_B$$

$$\sum M_A = 0$$
$$= (-0.3 \text{ m})N_B + (110 \text{ N})(0.15 \text{ m}) + F(0.25 \text{ m})$$

$$N_B = \frac{16.5 + 0.25F}{0.3}$$

$$F_{f,B} = \mu_s N_B$$
$$= (0.30)\left(\frac{16.5 + 0.25F}{0.3}\right)$$

Since $F_{f,B} = F$,

$$0.75F = 16.5$$

$$F = 22 \text{ N}$$

For case III,

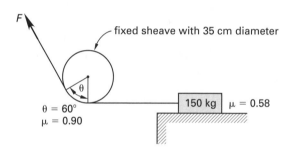

$$\sum F_x = 0 = F - F_{f,A}$$

$$F = F_{f,A} = \mu_s N_A$$

$$\sum M_B = 0$$
$$= (0.3 \text{ m})N_A - (110 \text{ N})(0.15 \text{ m})$$
$$+ F(0.25 \text{ m})$$

$$N_A = \frac{16.5 - 0.25F}{0.3}$$

$$F = (0.30)\left(\frac{16.5 - 0.25F}{0.3}\right)$$

$$1.25F = 16.5$$

$$F = 13.2 \text{ N}$$

Case I yields the largest value of F.

Answer is A.

FE-STYLE EXAM PROBLEMS

1. The system shown is in static equilibrium. Find W.

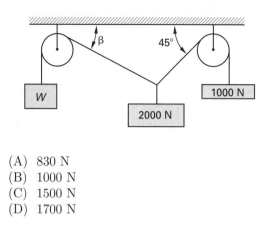

(A) 830 N
(B) 1000 N
(C) 1500 N
(D) 1700 N

Problems 2 and 3 refer to the following illustration.

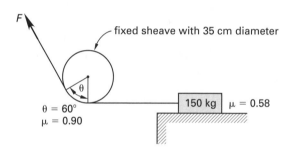

2. A block with a mass of 150 kg is pulled over a horizontal surface by a cable guided by a fixed sheave as shown. The coefficients of friction are 0.58 between the surface and the block and 0.90 between the cable and the pulley. What force, F, must be applied to the cable for the block to move?

(A) 900 N
(B) 1700 N
(C) 2200 N
(D) 2500 N

3. What total torque is applied to the pulley?

(A) 0
(B) 230 N·m
(C) 280 N·m
(D) 300 N·m

4. What is the tension in cable AB?

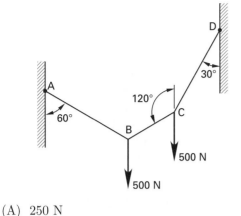

(A) 250 N
(B) 430 N
(C) 500 N
(D) 870 N

For the following problems use the NCEES Handbook as your only reference.

5. A 2 kg block rests on a 34° incline. If the coefficient of static friction is 0.2, how much additional force, F, must be applied to keep the block from sliding down the incline?

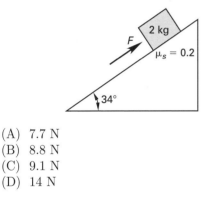

(A) 7.7 N
(B) 8.8 N
(C) 9.1 N
(D) 14 N

6. The cylinder shown is acted on by couple M. Wall A is frictionless ($\mu_s = 0$), but the coefficient of static friction between the cylinder and wall B is $\mu_s = 0.3$. The cylinder has a weight of 200 N. What is the largest value of the couple M for which the cylinder will not turn?

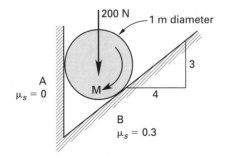

(A) 31 N·m
(B) 48 N·m
(C) 72 N·m
(D) 96 N·m

7. A rope passes over a fixed sheave as shown. The two rope ends are parallel. A fixed load on one end of the rope is supported by a constant force on the other end. The coefficient of friction between the rope and the sheave is 0.30. What is the maximum ratio of tensile forces in the two rope ends?

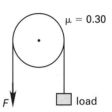

(A) 1.1
(B) 1.2
(C) 1.6
(D) 2.6

8. The nuts on a collar are each tightened to 18 N·m torque. 17% of this torque is used to overcome screw thread friction. The bolts have a nominal diameter of 10 mm. The threads are a simple square cut with a pitch angle of 15°. The coefficient of friction in the threads is 0.10. What is the approximate tensile force in each bolt?

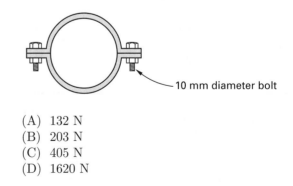

(A) 132 N
(B) 203 N
(C) 405 N
(D) 1620 N

9. A box has uniform density and a total weight of 600 N. It is suspended by three equal-length cables, AE, BE, and CE, as shown. Point E is 0.5 m directly above the center of the box's top surface. What is the tension in cable CE?

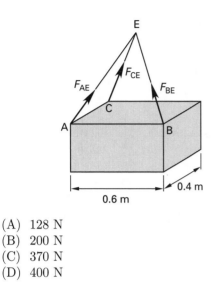

(A) 128 N
(B) 200 N
(C) 370 N
(D) 400 N

10. A rope is wrapped over a 6 cm diameter pipe to support a bucket of tools being lowered. The coefficient of friction between the rope and the pipe is 0.20. The combined mass of bucket and tools is 100 kg. What is the range of force that can be applied to the free end of the rope such that the bucket remains stationary?

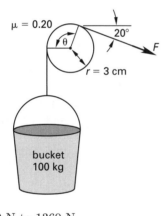

(A) 560 N to 1360 N
(B) 670 N to 1440 N
(C) 720 N to 1360 N
(D) 720 N to 1510 N

11. The two cables shown carry a 100 N vertical load. What is the tension in cable AB?

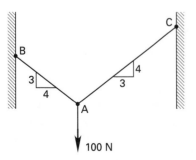

(A) 40 N
(B) 50 N
(C) 60 N
(D) 80 N

12. A cable passes over a stationary sheave and supports a 60 kg bucket, as shown. The coefficient of friction between the cable and the sheave is 0.10. The cable has a uniform mass per unit length of 0.4 kg/m. The cable is in the shape of a catenary due to its own weight. The tension of the cable at the pulley is given by $T = wy$, where w is the weight per unit length and the constant y (for this configuration) is known to be 151 m. How much more mass can be added to the bucket before the cable slips over the pulley?

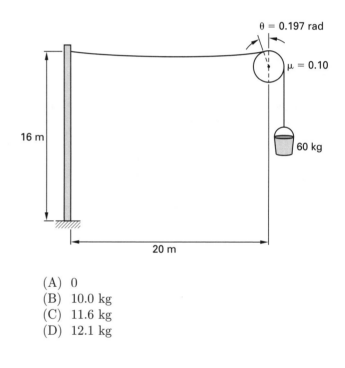

(A) 0
(B) 10.0 kg
(C) 11.6 kg
(D) 12.1 kg

SOLUTIONS TO FE-STYLE EXAM PROBLEMS

1. The free-body diagrams are

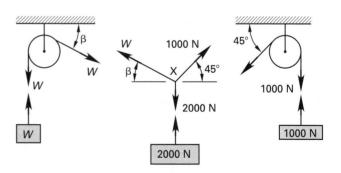

For the free body of point X,

$$\sum F_x = 0$$
$$= (1000 \text{ N}) \cos 45° - W \cos \beta$$

$$\sum F_y = 0$$
$$= -2000 \text{ N} + (1000 \text{ N}) \sin 45° + W \sin \beta$$

$$\frac{W \sin \beta}{W \cos \beta} = \tan \beta$$

$$= \frac{2000 \text{ N} - (1000 \text{ N}) \sin 45°}{(1000 \text{ N}) \cos 45°}$$

$$= 1.828$$

$$\beta = \tan^{-1} 1.828$$
$$= 61.3°$$

$$W = \frac{(1000 \text{ N}) \cos 45°}{\cos 61.3°}$$
$$= 1472 \text{ N} \quad (1500 \text{ N})$$

Answer is C.

2.

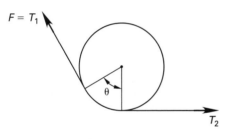

$$F = T_1$$
$$T_2 = \mu_1 N$$
$$= \mu_1 mg$$
$$= (0.58)(150 \text{ kg}) \left(9.81 \frac{\text{m}}{\text{s}^2}\right)$$
$$= 853.5 \text{ N}$$

$$\theta = (60°) \left(\frac{2\pi \text{ rad}}{360°}\right)$$
$$= 1.0472 \text{ rad}$$

$$T_1 = T_2 e^{\mu\theta}$$
$$= (853.5 \text{ N}) e^{(0.9)(1.0472)}$$
$$= 2190 \text{ N} \quad (2200 \text{ N})$$

Answer is C.

3. The torque is

$$T = (F_{\max} - F_{\min}) r$$
$$= (T_1 - T_2) \left(\frac{D}{2}\right)$$
$$= (2190 \text{ N} - 853.5 \text{ N}) \left(\frac{0.35 \text{ m}}{2}\right)$$
$$= 233.9 \text{ N·m} \quad (230 \text{ N·m})$$

Answer is B.

4.

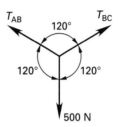

This is a three-force member. The law of sines can be used to solve the force triangle.

$$\frac{T_{AB}}{\sin 120°} = \frac{500 \text{ N}}{\sin 120°}$$
$$T_{AB} = 500 \text{ N}$$

Alternatively, moments could be taken about point D.

Answer is C.

5.

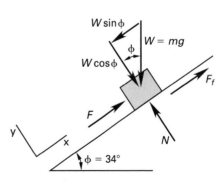

Choose coordinate axes parallel and perpendicular to the incline.

$$\sum F_x = 0$$
$$= F + F_f - W \sin \phi$$

$$F = W \sin\phi - F_f$$
$$= mg \sin\phi - \mu_s N$$
$$= mg \sin\phi - \mu_s mg \cos\phi$$
$$= mg(\sin\phi - \mu \cos\phi)$$
$$= (2 \text{ kg})\left(9.81 \frac{\text{m}}{\text{s}^2}\right)(\sin 34° - (0.2)\cos 34°)$$
$$= 7.7 \text{ N}$$

Answer is A.

6.

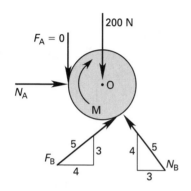

Frictional force F_B and N_B both have vertical components to balance the weight: $F_B = \mu N_B$.

$$\sum F_y = 0: \frac{4}{5}N_B + \frac{3}{5}F_B - W = 0$$

$$\frac{4}{5}N_B + \left(\frac{3}{5}\right)(0.30)N_B - 200 \text{ N} = 0$$

$$N_B = 204.1 \text{ N}$$

$$F_{f,A} = 0$$

$$F_{f,B} = \mu_s N_B = (0.30)(204.1 \text{ N})$$
$$= 61.2 \text{ N}$$

$$\sum M_O = 0$$
$$= M - rF_{f,B}$$
$$= M - (0.5 \text{ m})(61.2 \text{ N})$$

$$M = 30.6 \text{ N·m} \quad (31 \text{ N·m})$$

Answer is A.

7. The angle of wrap, θ, is 180°, but it must be expressed in radians.

$$F_1 = F_2 e^{\mu\theta}$$
$$\frac{F_1}{F_2} = e^{(0.3)(\pi \text{ rad})}$$
$$= 2.57 \quad (2.6)$$

Either side could be the tight side. Thus, the restraining force could be 2.6 times smaller or larger than the load tension.

Answer is D.

8. The equation for power screw threads is

$$M = Pr \tan(\alpha \pm \phi)$$

r is the mean thread radius, M is the torque on the screw, and P is the tensile or compressive force in the screw (i.e., is the load being raised or lowered). The angles are added for tightening operations; they are subtracted for loosening. This equation assumes that all of the torque is used to raise or lower the load.

There are two sources of friction when bolts are used in clamping situations. As with a power screw, there is friction between the contacting threads. There is also friction between the annular area of contact of the bolt head and the collar. (The nut is assumed not to twist. There are no washers in this application.) Only the screw thread friction (17% of the total torque in this application) contributes to the tensile force in the bolt.

The friction angle, ϕ, is

$$\phi = \tan^{-1}\mu = \tan^{-1}(0.10)$$
$$= 5.71°$$

$$P = \frac{M}{r \tan(\alpha + \phi)}$$
$$= \frac{(0.17)(18 \text{ N·m})}{\left(\dfrac{0.01 \text{ m}}{2}\right)\tan(15° + 5.71°)}$$
$$= 1619 \text{ N} \quad (1620 \text{ N})$$

Answer is D.

9. The length of the diagonal is

$$BC = \sqrt{(0.4 \text{ m})^2 + (0.6 \text{ m})^2}$$
$$= 0.721 \text{ m}$$

The cable length is

$$BE = \sqrt{\left(\frac{0.721 \text{ m}}{2}\right)^2 + (0.5 \text{ m})^2}$$
$$= 0.616 \text{ m}$$

There is nothing to balance the component force in the direction from point A to the opposite corner. Therefore, the force in cable AE is zero. Cables BE and CE each carry half of the box weight. Therefore, the vertical component of force in each cable is 300 N.

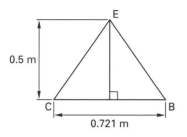

By similar triangles, the tensile force in each cable is

$$T = \frac{(300 \text{ N})(0.616 \text{ m})}{0.5 \text{ m}} = 370 \text{ N}$$

Answer is C.

10. The angle of wrap is

$$\theta = (90° + 20°)\left(\frac{2\pi \text{ rad}}{360°}\right) = 1.92 \text{ rad}$$

(This must be expressed in radians.)

The tensile force in the rope due to the bucket's mass is

$$F = mg = (100 \text{ kg})\left(9.81 \ \frac{\text{m}}{\text{s}^2}\right)$$
$$= 981 \text{ N}$$

The free end of the rope can either be on the tight or loose side. These two options define the range of force that will keep the bucket stationary.

The ratio of tight-side to loose-side forces is

$$\frac{F_1}{F_2} = e^{\mu\theta} = e^{(0.20)(1.92 \text{ rad})}$$
$$= 1.468$$

Multiply and divide the bucket-end tension by this ratio.

$$\text{minimum tension: } \frac{981 \text{ N}}{1.468} = 668 \text{ N} \quad (670 \text{ N})$$

$$\text{maximum tension: } (1.468)(981 \text{ N}) = 1440 \text{ N}$$

Answer is B.

11. Recognize that the orientations of both cables are defined by 3-4-5 triangles.

The equilibrium condition for horizontal forces at point A is

$$F_x: T_{\text{AC},x} - T_{\text{AB},x} = 0$$

$$\frac{3}{5}T_{\text{AC}} - \frac{4}{5}T_{\text{AB}} = 0$$

$$T_{\text{AC}} = \frac{4}{3}T_{\text{AB}}$$

The equilibrium condition for vertical forces at point A is

$$F_y: T_{\text{AB},y} + T_{\text{AC},y} - 100 \text{ N} = 0$$

$$\frac{3}{5}T_{\text{AB}} + \frac{4}{5}T_{\text{AC}} - 100 \text{ N} = 0$$

Substitute $(4/3)T_{\text{AB}}$ for T_{AC}.

$$\frac{3}{5}T_{\text{AB}} + \left(\frac{4}{5}\right)\left(\frac{4}{3}\right)T_{\text{AB}} = 100 \text{ N}$$

$$\left(\frac{3}{5} + \frac{16}{15}\right)T_{\text{AB}} = 100 \text{ N}$$

$$T_{\text{AB}} = 60 \text{ N}$$

Answer is C.

12. The angle of wrap is

$$\theta = 0.197 \text{ rad} + \frac{\pi}{2} \text{ rad} = 1.768 \text{ rad}$$

The tension in the fixed end of the cable is

$$wy = mgy$$
$$= \left(0.4 \ \frac{\text{kg}}{\text{m}}\right)\left(9.81 \ \frac{\text{m}}{\text{s}^2}\right)(151 \text{ m})$$
$$= 592.5 \text{ N}$$

The ratio of tight-side to loose-side tensions in the cable is

$$\frac{F_1}{F_2} = e^{\mu\theta} = e^{(0.10)(1.768 \text{ rad})}$$
$$= 1.193$$

The maximum tension in the bucket-end of the cable would be

$$(1.193)(592.5 \text{ N}) = 706.9 \text{ N}$$

The current tension in the bucket-end of the cable is

$$(60 \text{ kg})\left(9.81 \ \frac{\text{m}}{\text{s}^2}\right) = 588.6 \text{ N}$$

The mass that can be added to the bucket is

$$m = \frac{\Delta F}{g} = \frac{706.9 \text{ N} - 588.6 \text{ N}}{9.81 \ \frac{\text{m}}{\text{s}^2}}$$

$$= 12.06 \text{ kg} \quad (12.1 \text{ kg})$$

Answer is D.

13 Centroids and Moments of Inertia

Subjects

CENTROID 13-1
    First Moment 13-1
    Centroid of a Line 13-1
    Centroid of an Area 13-1
    Centroid of a Volume 13-1
MOMENT OF INERTIA 13-2
    Polar Moment of Inertia 13-2
    Parallel Axis Theorem 13-2
    Radius of Gyration 13-3
    Product of Inertia 13-3

Nomenclature

A	area	in^2	m^2
d	distance	ft	m
I	moment of inertia	in^4	m^4
J	polar moment of inertia	in^4	m^4
L	length	in	m
M	statical moment	in^3	m^3
r	radius of gyration	in	m
V	volume	in^3	m^3

Subscripts

c	centroidal
p	polar

CENTROID

Centroids of continuous functions can be found by the methods of integral calculus. For most engineering applications, though, the functions to be integrated are regular shapes such as the rectangular, circular, or composite rectangular shapes of beams. For these shapes, simple formulas are readily available and should be used. Formulas for basic shapes are compiled in Table 13.1 at the end of this chapter.

First Moment

The quantity $\int x\,dA$ is known as the *first moment of the area* or *first area moment* with respect to the y-axis. Similarly, $\int y\,dA$ is known as the first moment of the area with respect to the x-axis. For regular shapes with areas A_n,

$$M_y = \int x\,dA = \sum x_{c,n}A_n \qquad 13.1$$

$$M_x = \int y\,dA = \sum y_{c,n}A_n \qquad 13.2$$

The two primary applications of the first moment are determining centroidal locations and shear stress distributions. In the latter application, the first moment of the area is known as the *statical moment*.

Centroid of a Line

The location of the centroid of a line is defined by Eqs. 13.3 through 13.5. For a composite line of total length $L = \sum L_n$,

$$x_c = \frac{\int x\,dL}{L} = \frac{\sum x_n L_n}{L} \qquad 13.3$$

$$y_c = \frac{\int y\,dL}{L} = \frac{\sum y_n L_n}{L} \qquad 13.4$$

$$z_c = \frac{\int z\,dL}{L} = \frac{\sum z_n L_n}{L} \qquad 13.5$$

Centroid of an Area

The *centroid* of an area is often described as the point at which a thin homogeneous plate would balance. This definition, however, combines the definitions of centroid and center of gravity, and implies gravity is required to identify the centroid, which is not true. Nonetheless, this definition provides some intuitive understanding of the centroid.

The location of the centroid of an area depends only on the geometry of the area, and it is identified by the coordinates (x_c, y_c). For a composite area with total area given by $A = \sum A_n$,

$$x_c = \frac{\int x\,dA}{A} = \frac{\sum x_{c,n}A_n}{A} = \frac{M_y}{A} \qquad 13.6$$

$$y_c = \frac{\int y\,dA}{A} = \frac{\sum y_{c,n}A_n}{A} = \frac{M_x}{A} \qquad 13.7$$

$$z_c = \frac{\int z\,dA}{A} = \frac{\sum z_{c,n}A_n}{A} = \frac{M_z}{A} \qquad 13.8$$

Centroid of a Volume

The location of the centroid of a volume is defined by Eqs. 13.9 through 13.11, which are analogous to the equations used for centroids of areas and lines. For a composite volume of total volume $V = \sum V_n$,

$$x_c = \frac{\int x\, dV}{V} = \frac{\sum x_n V_n}{V} \qquad 13.9$$

$$y_c = \frac{\int y\, dV}{V} = \frac{\sum y_n V_n}{V} \qquad 13.10$$

$$z_c = \frac{\int z\, dV}{V} = \frac{\sum z_n V_n}{V} \qquad 13.11$$

A solid body will have both a center of gravity and a centroid, but the locations of these two points will not necessarily coincide. The earth's attractive force, which is called *weight*, can be assumed to act through the *center of gravity* (also known as the *center of mass*). Only when the body is homogeneous will the *centroid of the volume* coincide with the center of gravity.

MOMENT OF INERTIA

The *moment of inertia*, I, of an area is needed in mechanics of materials problems. It is convenient to think of the moment of inertia of a beam's cross-sectional area as a measure of the beam's ability to resist bending. Thus, given equal loads, a beam with a small moment of inertia will bend more than a beam with a large moment of inertia.

Since the moment of inertia represents a resistance to bending, it is always positive. Since a beam can be unsymmetrical (e.g., a rectangular beam) and be stronger in one direction than another, the moment of inertia depends on orientation. Therefore, a reference axis or direction must be specified.

The symbol I_x is used to represent a moment of inertia with respect to the x-axis. Similarly, I_y is the moment of inertia with respect to the y-axis. I_x and I_y do not combine and are not components of some resultant moment of inertia.

Any axis can be chosen as the reference axis, and the value of the moment of inertia will depend on the reference selected. The moment of inertia taken with respect to an axis passing through the area's centroid is known as the *centroidal moment of inertia*, $I_{c,x}$ or $I_{c,y}$. The centroidal moment of inertia is the smallest possible moment of inertia for the area.

Integration can be used to calculate the moment of inertia of a function that is bounded by the x- and y-axes and a curve $y = f(x)$. From Eqs. 13.12 and 13.13, it is apparent why the moment of inertia is also known as the *second moment of the area* or *second area moment*.

$$I_y = \int x^2\, dA \qquad 13.12$$

$$I_x = \int y^2\, dA \qquad 13.13$$

Moments of inertia of the basic shapes are listed in Table 13.1.

Polar Moment of Inertia

The *polar moment of inertia*, J or I_z, is required in torsional shear stress calculations. It can be thought of as a measure of an area's resistance to torsion (twisting). The definition of a polar moment of inertia of a two-dimensional area requires three dimensions because the reference axis for a polar moment of inertia of a plane area is perpendicular to the plane area.

The polar moment of inertia can be derived from Eq. 13.14.

$$J = I_z = \int (x^2 + y^2)\, dA \qquad 13.14$$

It is often easier to use the *perpendicular axis theorem* to quickly calculate the polar moment of inertia.

> *Perpendicular axis theorem:* The moment of inertia of a plane area about an axis normal to the plane is equal to the sum of the moments of inertia about any two mutually perpendicular axes lying in the plane and passing through the given axis.

$$J = I_x + I_y \qquad 13.15$$

Since the two perpendicular axes can be chosen arbitrarily, it is most convenient to use the centroidal moments of inertia.

$$J_c = I_{c,x} + I_{c,y} \qquad 13.16$$

Parallel Axis Theorem

If the moment of inertia is known with respect to one axis, the moment of inertia with respect to another, parallel axis can be calculated from the *parallel axis theorem*, also known as the *transfer axis theorem*. This theorem is used to evaluate the moment of inertia of areas that are composed of two or more basic shapes. In Eq. 13.17, d is the distance between the centroidal axis and the second, parallel axis.

$$I_{\text{parallel}} = I_c + Ad^2 \qquad 13.17$$

The second term in Eq. 13.17 is often much larger than the first term. Areas close to the centroidal axis do not affect the moment of inertia considerably. This principle is exploited in the design of structural steel shapes that derive bending resistance from *flanges* located far from the centroidal axis. The *web* does not contribute significantly to the moment of inertia.

**STATICS
Centroids**

Figure 13.1 Structural Steel Shape

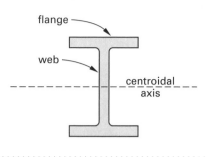

Radius of Gyration

Every nontrivial area has a centroidal moment of inertia. Usually, some portions of the area are close to the centroidal axis, and other portions are farther away. The *radius of gyration*, r, is an imaginary distance from the centroidal axis at which the entire area can be assumed to exist without changing the moment of inertia. Despite the name "radius," the radius of gyration is not limited to circular shapes or polar axes. This concept is illustrated in Fig. 13.2.

Figure 13.2 Radius of Gyration of Two Equivalent Areas

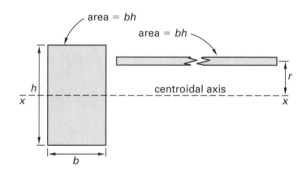

The radius of gyration, r, is given by

$$I = r^2 A \qquad 13.18$$

$$r_x = \sqrt{\frac{I_x}{A}} \qquad 13.19$$

$$r_y = \sqrt{\frac{I_y}{A}} \qquad 13.20$$

The analogous quantity in the polar system is

$$r_p = r_z = \sqrt{\frac{J}{A}} \qquad 13.21$$

Just as the polar moment of inertia, J, can be calculated from the two rectangular moments of inertia, the polar radius of gyration can be calculated from the two rectangular radii of gyration.

$$r_p^2 = r_x^2 + r_y^2 \qquad 13.22$$

Product of Inertia

The *product of inertia*, I_{xy}, of a two-dimensional area is found by multiplying each differential element of area by its x- and y-coordinate and then summing over the entire area.

$$I_{xy} = \int xy\, dA \qquad 13.23$$

$$I_{xz} = \int xz\, dA \qquad 13.24$$

$$I_{yz} = \int yz\, dA \qquad 13.25$$

The product of inertia is zero when either axis is an axis of symmetry. Since the axes can be chosen arbitrarily, the area may be in one of the negative quadrants, and the product of inertia may be negative.

The transfer theorem for products of inertia is given by Eq. 13.26. (Both axes are allowed to move to new positions.) d_x and d_y are the distances to the centroid in the new coordinate system, and $I_{x_c y_c}$ is the centroidal product of inertia in the old system.

$$I_{x'y'} = I_{x_c y_c} + d_x d_y A \qquad 13.26$$

SAMPLE PROBLEMS

Problems 1–6 refer to the area shown.

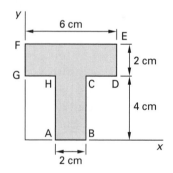

Problem 1

What are the x- and y-coordinates of the centroid of the area?

(A) 2.4 cm; 3.4 cm
(B) 3.0 cm; 3.6 cm
(C) 3.0 cm; 3.8 cm
(D) 3.0 cm; 4.0 cm

Solution

The centroids of the areas are located by inspection.

$$A = (2\text{ cm})(4\text{ cm}) + (6\text{ cm})(2\text{ cm})$$
$$= 20\text{ cm}^2$$

$$x_c = \frac{\sum x_{c,n} A_n}{A}$$
$$= \frac{(3 \text{ cm})((2 \text{ cm})(4 \text{ cm})) + (3 \text{ cm})((6 \text{ cm})(2 \text{ cm}))}{20 \text{ cm}^2}$$
$$= 3.0 \text{ cm}$$

Since the figure is symmetrical about a line parallel to the y-axis, this coordinate could be obtained by inspection.

$$y_c = \frac{\sum y_{c,n} A_n}{A}$$
$$= \frac{(2 \text{ cm})((2 \text{ cm})(4 \text{ cm})) + (5 \text{ cm})((6 \text{ cm})(2 \text{ cm}))}{20 \text{ cm}^2}$$
$$= 3.8 \text{ cm}$$

Answer is C.

Problem 2

What are the x- and y-coordinates of the centroid of the perimeter line?

    (A) 1.0 cm; 3.8 cm
    (B) 1.0 cm; 4.0 cm
    (C) 3.0 cm; 3.7 cm
    (D) 3.0 cm; 3.8 cm

Solution

$$\begin{aligned} L = {} & 2 \text{ cm} + 4 \text{ cm} + 2 \text{ cm} + 2 \text{ cm} + 6 \text{ cm} \\ & + 2 \text{ cm} + 2 \text{ cm} + 4 \text{ cm} \\ = {} & 24 \text{ cm} \end{aligned}$$

$$x_c = \frac{\sum x_n L_n}{L}$$

$$= \frac{\begin{array}{c} (0 \text{ cm})(2 \text{ cm}) + (1 \text{ cm})(2 \text{ cm}) + (2 \text{ cm})(4 \text{ cm}) \\ + (3 \text{ cm})(2 \text{ cm} + 6 \text{ cm}) + (4 \text{ cm})(4 \text{ cm}) \\ + (5 \text{ cm})(2 \text{ cm}) + (6 \text{ cm})(2 \text{ cm}) \end{array}}{24 \text{ cm}}$$

$$= 3.0 \text{ cm}$$

$$y_c = \frac{\sum y_n L_n}{L}$$

$$= \frac{\begin{array}{c} (0 \text{ cm})(2 \text{ cm}) + (2)(2 \text{ cm})(4 \text{ cm}) \\ + (2)(4 \text{ cm})(2 \text{ cm}) + (2)(5 \text{ cm})(2 \text{ cm}) \\ + (6 \text{ cm})(6 \text{ cm}) \end{array}}{24 \text{ cm}}$$

$$= 3.67 \text{ cm} \quad (3.7 \text{ cm})$$

Note that while, due to symmetry, the locations of the centroids of the area and perimeter are equal about a vertical axis, they are not equal about a horizontal axis.

Answer is C.

Problem 3

What is the area moment of inertia about the x-axis?

    (A) 47 cm^4
    (B) 59 cm^4
    (C) 170 cm^4
    (D) 350 cm^4

Solution

The moment of inertia about an edge for rectangular shapes is given in Table 13.1.

For rectangle ABCH,

$$\begin{aligned} I_{x,1} &= \frac{bh^3}{3} \\ &= \frac{(2 \text{ cm})(4 \text{ cm})^3}{3} \\ &= 42.67 \text{ cm}^4 \end{aligned}$$

Use the parallel axis theorem to calculate the moment of inertia of rectangle CDEFGH. $d = 5$ cm is the distance from the centroid of CDEFGH to the x-axis.

$$\begin{aligned} I_{x,2} &= \frac{bh^3}{12} + Ad^2 \\ &= \frac{(6 \text{ cm})(2 \text{ cm})^3}{12} + (6 \text{ cm})(2 \text{ cm})(5 \text{ cm})^2 \\ &= 304 \text{ cm}^4 \\ I_x &= I_{x,1} + I_{x,2} \\ &= 42.67 \text{ cm}^4 + 304 \text{ cm}^4 \\ &= 346.7 \text{ cm}^4 \quad (350 \text{ cm}^4) \end{aligned}$$

Answer is D.

Problem 4

What is the centroidal moment of inertia with respect to the x-axis?

    (A) 39 cm^4
    (B) 58 cm^4
    (C) 82 cm^4
    (D) 200 cm^4

Solution

From Prob. 1, the centroidal y-coordinate is 3.8 cm. Use the parallel axis theorem to move the centroidal moment of inertia of both rectangular areas.

$$\text{ABCH} = \text{area 1}$$
$$\text{CDEFGH} = \text{area 2}$$

$$I_{c,x} = (I_{c,1} + A_1 d_1^2) + (I_{c,2} + A_2 d_2^2)$$

$$= \left(\begin{array}{c} \dfrac{(2\text{ cm})(4\text{ cm})^3}{12} + (8\text{ cm}^2) \\ \times (3.8\text{ cm} - 2.0\text{ cm})^2 \end{array} \right)$$

$$+ \left(\begin{array}{c} \dfrac{(6\text{ cm})(2\text{ cm})^3}{12} + (12\text{ cm}^2) \\ \times (5.0\text{ cm} - 3.8\text{ cm})^2 \end{array} \right)$$

$$= 57.9\text{ cm}^4 \quad (58\text{ cm}^4)$$

Answer is B.

Problem 5

What is the centroidal polar moment of inertia?

- (A) 79 cm^4
- (B) 82 cm^4
- (C) 97 cm^4
- (D) 140 cm^4

Solution

The centroidal polar moment of inertia is given by Eq. 13.16.

$$J_c = I_{c,x} + I_{c,y}$$

$I_{c,x}$ was calculated in Prob. 4 as 57.9 cm^4. Proceeding similarly,

$$I_{c,y} = \frac{(2\text{ cm})^3(4\text{ cm})}{12} + \frac{(6\text{ cm})^3(2\text{ cm})}{12}$$

$$= 38.7\text{ cm}^4$$

Note that since the centroids of the individual rectangles coincide with the centroid of the composite area about the y-axis, $I_{c,y}$ is simply the sum of the moments of inertia of the individual areas.

$$J_c = I_{c,x} + I_{c,y}$$
$$= 57.9\text{ cm}^4 + 38.7\text{ cm}^4$$
$$= 96.6\text{ cm}^4 \quad (97\text{ cm}^4)$$

Answer is C.

Problem 6

What is the radius of gyration about a horizontal axis passing through the centroid?

- (A) 0.86 cm
- (B) 1.7 cm
- (C) 2.3 cm
- (D) 3.7 cm

Solution

By definition, the radius of gyration is calculated with respect to the centroidal axis.

$$r_x = \sqrt{\frac{I_{c,x}}{A}} = \sqrt{\frac{57.9\text{ cm}^4}{20\text{ cm}^2}}$$

$$= 1.70\text{ cm}$$

Answer is B.

FE-STYLE EXAM PROBLEMS

Problems 1 and 2 refer to the composite area shown.

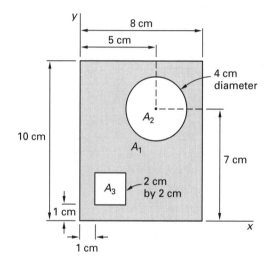

1. What are the x- and y-coordinates of the centroid of the area?

- (A) 3.40 cm; 5.60 cm
- (B) 3.50 cm; 5.50 cm
- (C) 3.93 cm; 4.79 cm
- (D) 4.00 cm; 5.00 cm

2. Assume that the centroidal moment of inertia of area A_2 with respect to the composite centroidal x-axis is 73.94 cm^4. The moment of inertia of area A_3 with respect to the composite centroidal horizontal axis is 32.47 cm^4. What is the moment of inertia of the composite area with respect to its centroidal x-axis?

- (A) 350 cm^4
- (B) 460 cm^4
- (C) 480 cm^4
- (D) 560 cm^4

3. Find the x- and y-coordinates of the centroid of wire ABC.

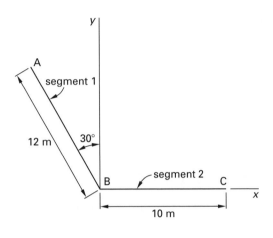

(A) 0.43 m; 1.29 m
(B) 0.64 m; 2.83 m
(C) 2.71 m; 1.49 m
(D) 3.33 m; 2.67 m

4. The moment of inertia about the x'-axis of the cross section shown is $245\,833$ cm^4. If the cross-sectional area is 250 cm^2 and the thicknesses of the web and the flanges are the same, what is the moment of inertia about the centroidal axis?

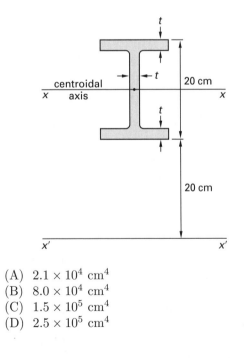

(A) 2.1×10^4 cm^4
(B) 8.0×10^4 cm^4
(C) 1.5×10^5 cm^4
(D) 2.5×10^5 cm^4

5. The structure shown is formed of three separate solid aluminum cylindrical rods, each with a 1 cm diameter. What is the x-coordinate of the centroid of volume for the structure?

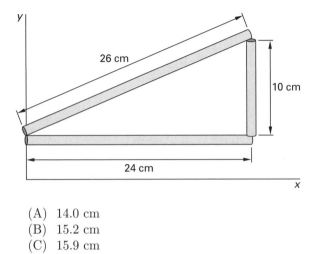

(A) 14.0 cm
(B) 15.2 cm
(C) 15.9 cm
(D) 16.0 cm

For the following problems use the NCEES Handbook as your only reference.

Problems 6 and 7 refer to the composite plane areas shown.

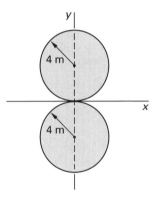

6. What is the polar moment of inertia about the composite centroid?

(A) 1020 m^4
(B) 1260 m^4
(C) 1530 m^4
(D) 2410 m^4

7. What is the polar radius of gyration?

(A) 3.6 m
(B) 4.0 m
(C) 4.2 m
(D) 4.9 m

8. A 28 mm diameter circular area is reduced by a 21 mm diameter circular area that is cut out. Both circles are tangent to the y-axis. What is the moment of inertia about the y-axis of the remaining (shaded) area?

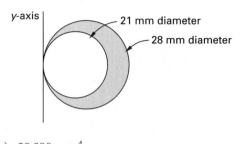

(A) 20 600 mm^4
(B) 103 000 mm^4
(C) 330 000 mm^4
(D) 1 340 000 mm^4

9. What is the x-coordinate of the centroid of the area under the curve $y = \cos x$ between $x = 0$ and $x = \pi/2$?

(A) $1 - \dfrac{2}{\pi}$

(B) $\dfrac{\pi}{2} - 1$

(C) $\dfrac{\pi}{6}$

(D) $\dfrac{\pi}{4}$

Problems 10–13 refer to the complex shape shown.

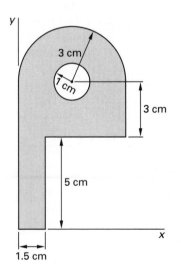

10. What is the x-coordinate of the centroid?

(A) 2.45 cm
(B) 2.54 cm
(C) 2.76 cm
(D) 3.23 cm

11. What is the y-coordinate of the centroid?

(A) 4.55 cm
(B) 4.74 cm
(C) 6.09 cm
(D) 6.62 cm

12. What is the moment of inertia about the x-axis?

(A) 1665 cm^4
(B) 1859 cm^4
(C) 1990 cm^4
(D) 2002 cm^4

13. If the moment of inertia about the y-axis is 352 cm^4, what is the radius of gyration with respect to the y-axis?

(A) 1.89 cm
(B) 3.11 cm
(C) 3.27 cm
(D) 4.01 cm

14. An area is a composite of a semicircle and a triangle, as shown. What is the distance between the x-axis and the centroid?

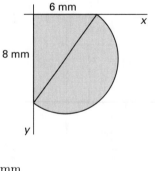

(A) 3.46 mm
(B) 3.68 mm
(C) 4.28 mm
(D) 5.35 mm

Problems 15 and 16 refer to the cross section of the angle shown. The area moments of inertia of the section about the respective edges are $I_{xx} = 4.7$ cm^4 and $I_{yy} = 20.9$ cm^4.

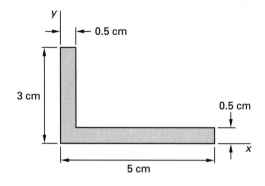

15. What is the approximate polar moment of inertia of the area taken about the intersection of the x- and y-axes?

    (A) 16.2 cm^4
    (B) 21.4 cm^4
    (C) 25.6 cm^4
    (D) 27.3 cm^4

16. What is the x-coordinate of the centroid of the perimeter line?

    (A) 1.56 cm
    (B) 1.66 cm
    (C) 1.75 cm
    (D) 1.80 cm

17. A model T-beam is constructed from five balsa boards. Refer to the illustration for the as-built dimensions. What is the approximate centroidal moment of inertia about an axis parallel to the x-axis?

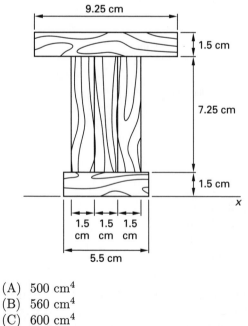

    (A) 500 cm^4
    (B) 560 cm^4
    (C) 600 cm^4
    (D) 660 cm^4

18. What is the y-coordinate of the centroid of the two-dimensional shape shown?

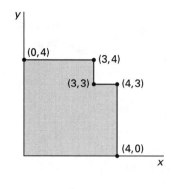

    (A) 1.7
    (B) 1.8
    (C) 1.9
    (D) 2.1

SOLUTIONS TO FE-STYLE EXAM PROBLEMS

1. $A = \sum A_n$

$$= (8 \text{ cm})(10 \text{ cm}) - \frac{\pi(4 \text{ cm})^2}{4} - (2 \text{ cm})(2 \text{ cm})$$

$$= 63.43 \text{ cm}^2$$

$$y_c = \frac{\sum y_{c,n} A_n}{A}$$

$$= \frac{(5 \text{ cm})(80 \text{ cm}^2) - \left(\frac{\pi}{4}\right)(4 \text{ cm})^2(7 \text{ cm})}{63.43 \text{ cm}^2}$$
$$\qquad \frac{- (2 \text{ cm})(4 \text{ cm}^2)}{63.43 \text{ cm}^2}$$

$$= 4.79 \text{ cm}$$

$$x_c = \frac{\sum x_{c,n} a_n}{A}$$

$$= \frac{(4 \text{ cm})(80 \text{ cm}^2) - (5 \text{ cm})\left(\frac{\pi}{4}\right)(4 \text{ cm})^2}{63.43 \text{ cm}^2}$$
$$\qquad \frac{- (2 \text{ cm})(4 \text{ cm}^2)}{63.43 \text{ cm}^2}$$

$$= 3.93 \text{ cm}$$

Answer is C.

2. $I_{c,x} = (I_{c,x,1} + A_1 d_1^2) - I_{c,x',2} - I_{c,x',3}$

$$= \left(\begin{array}{c} \dfrac{(8 \text{ cm})(10 \text{ cm})^3}{12} \\ + (8 \text{ cm})(10 \text{ cm})(5 \text{ cm} - 4.79 \text{ cm})^2 \end{array} \right)$$
$$\qquad - 73.94 \text{ cm}^4 - 32.47 \text{ cm}^4$$

$$= 563.8 \text{ cm}^4 \quad (560 \text{ cm}^4)$$

Answer is D.

3. The centroid of a line is given by

$$y_c = \frac{\sum y_{c,n} L_n}{L}$$

$$x_c = \frac{\sum x_{c,n} L_n}{L}$$

$$L = 12 \text{ m} + 10 \text{ m} = 22 \text{ m}$$

$$x_c = \frac{\left(\dfrac{(-12 \text{ m})\sin 30°}{2}\right)(12 \text{ m}) + \left(\dfrac{10 \text{ m}}{2}\right)(10 \text{ m})}{22 \text{ m}}$$

$$= 0.64 \text{ m}$$

$$y_c = \frac{\left(\dfrac{(12 \text{ m})\cos 30°}{2}\right)(12 \text{ m}) + (0 \text{ m})(10 \text{ m})}{22 \text{ m}}$$

$$= 2.83 \text{ m}$$

Answer is B.

4.
$$I'_x = I_{c,x} + Ad^2$$
$$I_{c,x} = I_{x'} - Ad^2$$
$$= 245\,833 \text{ cm}^4 - (250 \text{ cm}^2)(30 \text{ cm})^2$$
$$= 20\,833 \text{ cm}^4 \quad (2.1 \times 10^4 \text{ cm}^4)$$

Answer is A.

5. $x_c = \dfrac{\sum x_{c,n} V_n}{V}$

$$V_1 = \left(\frac{\pi}{4}\right)(1 \text{ cm})^2(24 \text{ cm})$$
$$= 18.85 \text{ cm}^3$$
$$V_2 = \left(\frac{\pi}{4}\right)(1 \text{ cm})^2(10 \text{ cm})$$
$$= 7.85 \text{ cm}^3$$
$$V_3 = \left(\frac{\pi}{4}\right)(1 \text{ cm})^2(26 \text{ cm})$$
$$= 20.42 \text{ cm}^3$$
$$V = 18.85 \text{ cm}^3 + 7.85 \text{ cm}^3 + 20.42 \text{ cm}^3$$
$$= 47.12 \text{ cm}^3$$
$$x_c = \frac{\begin{array}{c}(12 \text{ cm})(18.85 \text{ cm}^3) + (24 \text{ cm})(7.85 \text{ cm}^3) \\ + (12 \text{ cm})(20.42 \text{ cm}^3)\end{array}}{47.12 \text{ cm}^3}$$
$$= 14.0 \text{ cm}$$

(Note that the $\pi/4$ and area terms all cancel and could have been omitted.)

Answer is A.

6. For a circle,

$$I_{c,x} = I_{c,y} = \frac{\pi r^4}{4}$$

Use the parallel axis theorem for a composite area.

$$I_{c,x} = (2)\left(\frac{\pi r^4}{4} + (\pi r^2)(r)^2\right) = \frac{5\pi r^4}{2}$$

$$I_{c,y} = (2)\left(\frac{\pi r^4}{4}\right) = \frac{\pi r^4}{2}$$

$$J_c = I_{c,x} + I_{c,y} = \frac{5\pi r^4}{2} + \frac{\pi r^4}{2}$$
$$= 3\pi r^4$$
$$= 3\pi(4 \text{ m})^4$$
$$= 2413 \text{ m}^4 \quad (2410 \text{ m}^4)$$

Answer is D.

7.
$$r_p = \sqrt{\frac{J}{A}} = \sqrt{\frac{3\pi r^4}{2\pi r^2}}$$
$$= r\sqrt{\frac{3}{2}}$$
$$= (4 \text{ m})\sqrt{\frac{3}{2}}$$
$$= 4.9 \text{ m}$$

Answer is D.

8. The moment of inertia of a circle about a centroidal axis parallel to the y-axis is

$$I_c = \frac{\pi r^4}{4}$$

Use the parallel axis theorem. The moment of inertia about the y-axis located a distance $d = r$ away is

$$I_y = I_c + Ar^2 = \frac{\pi r^4}{4} + (\pi r^2)r^2$$
$$= \frac{5}{4}\pi r^4$$

The moment of inertia of the shaded area is

$$I_y = \frac{5}{4}\pi r^4_{\text{larger}} - \frac{5}{4}\pi r^4_{\text{smaller}}$$
$$= \frac{5}{4}\pi\left(\left(\frac{28 \text{ mm}}{2}\right)^4 - \left(\frac{21 \text{ mm}}{2}\right)^4\right)$$
$$= 103\,126 \text{ mm}^4 \quad (103\,000 \text{ mm}^4)$$

Answer is B.

9.
$$dA = y\, dx$$

$$A = \int dA = \int_0^{\frac{\pi}{2}} y\, dx$$

$$= \int_0^{\frac{\pi}{2}} \cos x\, dx$$

$$= \sin x \Big|_0^{\frac{\pi}{2}}$$

$$= 1 - 0 = 1$$

$$\int_0^{\frac{\pi}{2}} x\, dA = \int_0^{\frac{\pi}{2}} xy\, dx$$

$$= \int_0^{\frac{\pi}{2}} x \cos x\, dx$$

$$= (\cos x + x \sin x) \Big|_0^{\frac{\pi}{2}}$$

$$= \left(0 + \frac{\pi}{2}\right) - (1 + 0) = \frac{\pi}{2} - 1$$

$$x_c = \frac{\int x\, dA}{A} = \frac{\frac{\pi}{2} - 1}{1}$$

$$= \frac{\pi}{2} - 1$$

Answer is B.

10. Separate the shape into regions. Calculate the area and locate the centroid for each region.

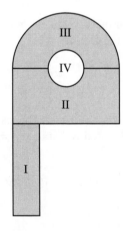

For region I (rectangular),
$$A = (1.5 \text{ cm})(5 \text{ cm})$$
$$= 7.5 \text{ cm}^2$$

$$x_c = \frac{1.5 \text{ cm}}{2}$$
$$= 0.75 \text{ cm}$$

For region II (rectangular),
$$A = (2)(3 \text{ cm})(3 \text{ cm})$$
$$= 18 \text{ cm}^2$$

$$x_c = \frac{(2)(3 \text{ cm})}{2}$$
$$= 3 \text{ cm}$$

For region III (semicircular),
$$A = \frac{\pi r^2}{2} = \frac{\pi (3 \text{ cm})^2}{2}$$
$$= \frac{9\pi}{2}$$

$$x_c = 3 \text{ cm} \quad \text{[by inspection]}$$

For region IV (circular),
$$A = \pi r^2 = \pi (1 \text{ cm})^2$$
$$= \pi$$

$$x_c = 3 \text{ cm} \quad \text{[by inspection]}$$

$$x_c = \frac{\sum x_{c,i} A_i}{\sum A_i}$$

$$= \frac{(0.75 \text{ cm})(7.5 \text{ cm}^2) + (3 \text{ cm})(18 \text{ cm}^3) + (3 \text{ cm})\left(\dfrac{9\pi}{2} \text{ cm}^2\right) - (3 \text{ cm})(\pi \text{ cm}^2)}{7.5 \text{ cm}^2 + 18 \text{ cm}^2 + \dfrac{9\pi}{2} \text{ cm}^2 - \pi \text{ cm}^2}$$

$$= \frac{92.61 \text{ cm}^3}{36.50 \text{ cm}^2}$$

$$= 2.54 \text{ cm}$$

Answer is B.

11. Refer to Prob. 10.

For region I,
$$y_c = \frac{5 \text{ cm}}{2}$$
$$= 2.5 \text{ cm}$$

For region II,
$$y_c = 5 \text{ cm} + \frac{3 \text{ cm}}{2}$$
$$= 6.5 \text{ cm}$$

For region III,
$$y_c = 5 \text{ cm} + 3 \text{ cm} + \frac{4r}{3\pi}$$
$$= 8 \text{ cm} + \frac{(4)(3 \text{ cm})}{3\pi}$$
$$= 8 \text{ cm} + \frac{4}{\pi} \text{ cm}$$

For region IV,
$$y_c = 5 \text{ cm} + 3 \text{ cm} = 8 \text{ cm}$$

Use the areas calculated in Prob. 10.

$$y_c = \frac{\sum y_{c,i} A_i}{\sum A_i}$$

$$= \frac{\left(\begin{array}{c}(2.5 \text{ cm})(7.5 \text{ cm}^2) + (6.5 \text{ cm})(18 \text{ cm}^2) \\ + \left(8 \text{ cm} + \dfrac{4}{\pi} \text{ cm}\right)\left(\dfrac{9\pi}{2} \text{ cm}^2\right) \\ - (8 \text{ cm})(\pi \text{ cm}^2)\end{array}\right)}{7.5 \text{ cm}^2 + 18 \text{ cm}^2 + \dfrac{9\pi}{2} \text{ cm}^2 - \pi \text{ cm}^2}$$

$$= \frac{241.7 \text{ cm}^3}{36.50 \text{ cm}^2}$$

$$= 6.62 \text{ cm}$$

Answer is D.

12. Calculate the centroidal moment of inertia and the distance from that centroid to the x-axis for each region. Use the areas calculated in Prob. 10.

For region I,

$$I_c = \frac{bh^3}{12} = \frac{(1.5 \text{ cm})(5 \text{ cm})^3}{12}$$

$$= 15.625 \text{ cm}^4$$

$$d = \frac{5 \text{ cm}}{2} = 2.5 \text{ cm}$$

$$A = 7.5 \text{ cm}^2$$

$$d^2 A = (2.5 \text{ cm})^2 (7.5 \text{ cm}^2)$$

$$= 46.875 \text{ cm}^4$$

For region II,

$$I_c = \frac{bh^3}{12} = \frac{(6 \text{ cm})(3 \text{ cm})^3}{12}$$

$$= 13.5 \text{ cm}^4$$

$$d = 5 \text{ cm} + \frac{3 \text{ cm}}{2}$$

$$= 6.5 \text{ cm}$$

$$A = 18 \text{ cm}^2$$

$$d^2 A = (6.5 \text{ cm})^2 (18 \text{ cm})$$

$$= 760.5 \text{ cm}^4$$

For region III,

$$I_c = 0.1098 r^4$$

$$= (0.1098)(3 \text{ cm})^4$$

$$= 8.8938 \text{ cm}^4$$

$$d = 5 \text{ cm} + 3 \text{ cm} + \frac{4r}{3\pi}$$

$$= 8 \text{ cm} + \frac{(4)(3 \text{ cm})}{3\pi}$$

$$= 8 \text{ cm} + \frac{4}{\pi} \text{ cm}$$

$$A = \frac{9\pi}{2}$$

$$d^2 A = \left(8 \text{ cm} + \frac{4}{\pi} \text{ cm}\right)^2 \left(\frac{9\pi}{2}\right)$$

$$= 1215.7 \text{ cm}^4$$

For region IV,

$$I_c = \frac{\pi r^4}{4} = \frac{\pi (1 \text{ cm})^4}{4}$$

$$= \frac{\pi}{4} \text{ cm}^4$$

$$d = 5 \text{ cm} + 3 \text{ cm}$$

$$= 8 \text{ cm}$$

$$A = \pi$$

$$d^2 A = (8 \text{ cm})^2 \pi$$

$$= 64\pi \text{ cm}^4$$

Use the parallel axis theorem for each shape.

$$I_x = \sum I_{c,x} + \sum d^2 A$$

$$= 15.625 \text{ cm}^4 + 46.875 \text{ cm}^4$$

$$+ 13.5 \text{ cm}^4 + 760.5 \text{ cm}^4$$

$$+ 8.8938 \text{ cm}^4 + 1215.7 \text{ cm}^4$$

$$- \frac{\pi}{4} \text{ cm}^4 - 64\pi \text{ cm}^4$$

$$= 1859 \text{ cm}^4$$

Answer is B.

13. $$r_y = \sqrt{\frac{I_y}{A}} = \sqrt{\frac{352 \text{ cm}^4}{36.5 \text{ cm}^2}}$$

$$= 3.11 \text{ cm}$$

Answer is B.

14. Triangle:

$$A_{\text{triangle}} = \frac{1}{2} bh$$

$$= (0.5)(8 \text{ mm})(6 \text{ mm})$$

$$= 24 \text{ mm}^2$$

$$y_{c,\text{triangle}} = \left(\frac{1}{3}\right)(8 \text{ mm})$$

$$= 2.667 \text{ mm}$$

Semicircle:

This is a 3-4-5 triangle. The diameter of the circle is 10 mm. The radius is 5 mm.

$$A_{\text{semicircle}} = \left(\frac{1}{2}\right)\left(\frac{\pi}{4}\right)d^2$$
$$= \frac{\pi(10 \text{ mm})^2}{8}$$
$$= 39.27 \text{ mm}^2$$

Refer to the following illustration. Distance y_1 is

$$y_1 = \left(\frac{1}{2}\right)(8 \text{ mm}) = 4 \text{ mm}$$

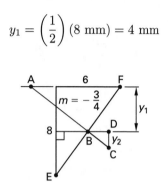

From Table 13.1, the distance between the centroid of a semicircle and its chord (diameter), distance BC, is

$$\text{BC} = \frac{4r}{3\pi} = \frac{(4)(5 \text{ mm})}{3\pi}$$
$$= 2.122 \text{ mm}$$

Since this is a 3-4-5 triangle, distance y_2 is found from similar triangles.

$$y_2 = \left(\frac{3}{5}\right)(2.122 \text{ mm})$$
$$= 1.273 \text{ mm}$$

The y-coordinate of the semicircular area is

$$y_{c,\text{semicircle}} = 4.00 \text{ mm} + 1.273 \text{ mm}$$
$$= 5.273 \text{ mm}$$

The y-coordinate of the composite area is

$$y_c = \frac{\sum y_i A_i}{\sum A_i}$$
$$= \frac{\left(\begin{array}{c}(2.667 \text{ mm})(24 \text{ mm}^2) \\ + (5.273 \text{ mm})(39.27 \text{ mm}^2)\end{array}\right)}{24 \text{ mm}^2 + 39.27 \text{ mm}^2}$$
$$= 4.28 \text{ mm}$$

Answer is C.

15. I_{xx} and I_{yy} are given.

Use the perpendicular axis theorem.

$$J = I_{xx} + I_{yy}$$
$$= 4.7 \text{ cm}^4 + 20.9 \text{ cm}^4$$
$$= 25.6 \text{ cm}^4$$

Answer is C.

16. The length of the perimeter is

$$L = 5 \text{ cm} + 0.5 \text{ cm} + 4.5 \text{ cm}$$
$$+ 2.5 \text{ cm} + 0.5 \text{ cm} + 3 \text{ cm}$$
$$= 16 \text{ cm}$$

The lengths and locations of the centroids of each line segment are found from observation.

$$x_c = \frac{\sum x_i L_i}{\sum L_i}$$
$$= \left(\frac{1}{16}\right)((2.5 \text{ cm})(5 \text{ cm}) + (5 \text{ cm})(0.5 \text{ cm})$$
$$+ (2.75 \text{ cm})(5 \text{ cm} - 0.5 \text{ cm})$$
$$+ (0.5 \text{ cm})(3 \text{ cm} - 0.5 \text{ cm})$$
$$+ (0.25 \text{ cm})(0.5 \text{ cm}) + (0)(3 \text{ cm}))$$
$$= 1.80 \text{ cm}$$

Answer is D.

17. First, find the y-coordinate of the centroid. Centroidal locations of each section are found by observation. (All dimensions and lengths are in centimeters.)

$$y_c = \frac{\sum y_i A_i}{\sum A_i}$$
$$= \frac{\left(\begin{array}{c}(0.75)(1.5)(5.5) + (3)\left(1.5 + \left(\frac{1}{2}\right)(7.25)\right) \\ \times (1.5)(7.25) + (1.5 + 7.25 + 0.75) \\ \times (1.5)(9.25)\end{array}\right)}{(1.5)(5.5) + (3)(7.25)(1.5) + (1.5)(9.25)}$$
$$= \frac{305.2 \text{ cm}^3}{54.75 \text{ cm}^2}$$
$$= 5.57 \text{ cm}$$

From Table 13.1, the centroidal moment of inertia of a rectangular section is $bh^3/12$.

Use the parallel axis theorem. (All distances are in centimeters.)

$$I_x = I_c + Ad^2 = \frac{bh^3}{12} + Ad^2$$

$$= \left(\frac{1}{12}\right)(5.5)(1.5)^3 + (5.5)(1.5)(5.57 - 0.75)^2$$

$$+ \left(\frac{1}{12}\right)(3)(1.5)(7.25)^3$$

$$+ (3)(1.5)(7.25)\left(5.57 - \left(1.5 + \left(\frac{1}{2}\right)(7.25)\right)\right)^2$$

$$+ \left(\frac{1}{12}\right)(9.25)(1.5)^3$$

$$+ (9.25)(1.5)(5.57 - (1.5 + 7.25 + 0.75))^2$$

$$= 193.21 \text{ cm}^4 + 149.36 \text{ cm}^4 + 216.90 \text{ cm}^4$$

$$= 559.5 \text{ cm}^4 \quad (560 \text{ cm}^4)$$

Answer is B.

18. The shape is a composite of a 4 by 4 square and a 1 by 1 (missing) square. By quick mental calculation, the areas of these shapes are 16 and 1, respectively. By inspection, the centroids of these two shapes are located at $y = 2$ and $y = 3.5$, respectively. The y-coordinate of the composite centroid is

$$y_c = \frac{\sum A_i y_i}{\sum A_i} = \frac{(16)(2) + (-1)(3.5)}{16 - 1}$$

$$= 1.9$$

Answer is C.

Figure 13.1 Centroids and Area Moments of Inertia for Basic Shapes

shape		x_c	y_c	A	I, J	r
rectangle	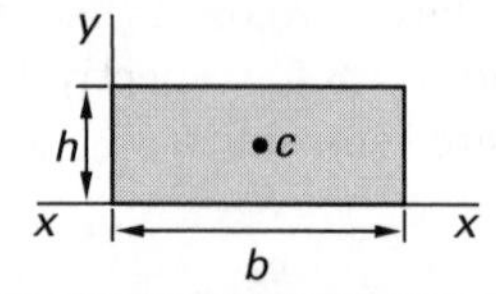	$\dfrac{b}{2}$	$\dfrac{h}{2}$	bh	$I_x = \dfrac{bh^3}{3}$ $I_{c,x} = \dfrac{bh^3}{12}$ $J_c = \dfrac{1}{12}bh(b^2 + h^2)$	$r_x = \dfrac{h}{\sqrt{3}}$ $r_{c,x} = \dfrac{h}{2\sqrt{3}}$
triangular area	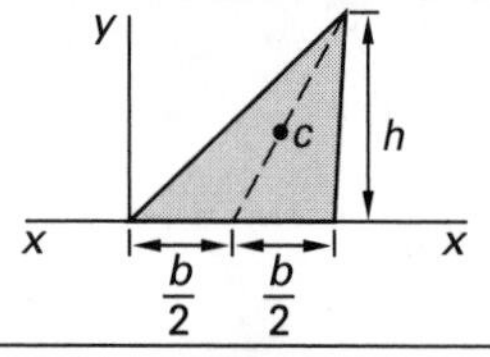		$\dfrac{h}{3}$	$\dfrac{bh}{2}$	$I_x = \dfrac{bh^3}{12}$ $I_{c,x} = \dfrac{bh^3}{36}$	$r_x = \dfrac{h}{\sqrt{6}}$ $r_{c,x} = \dfrac{h}{3\sqrt{2}}$
trapezoid	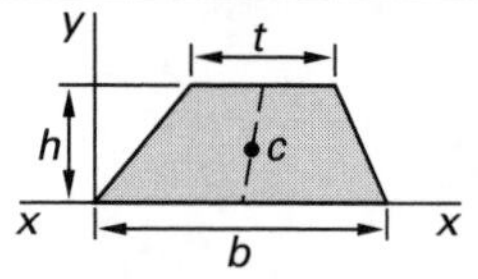		$h\left(\dfrac{b+2t}{3b+3t}\right)$	$\dfrac{(b+t)h}{2}$	$I_x = \dfrac{(b+3t)h^3}{12}$ $I_{c,x} = \dfrac{(b^2 + 4bt + t^2)h^3}{(36)(b+t)}$	$r_x = \dfrac{h}{\sqrt{6}}\sqrt{\dfrac{b+3t}{b+t}}$ $r_{c,x} = \dfrac{h\sqrt{2(b^2 + 4bt + t^2)}}{(6)(b+t)}$
circle	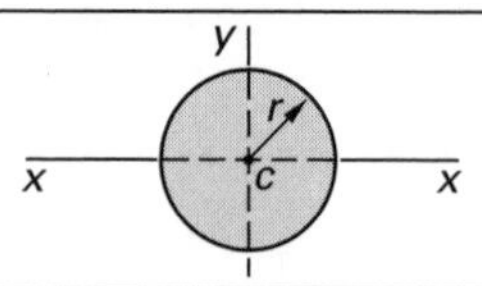	0	0	πr^2	$I_x = I_y = \dfrac{\pi r^4}{4}$ $J_c = \dfrac{\pi r^4}{2}$	$r_x = \dfrac{r}{2}$
quarter-circular area	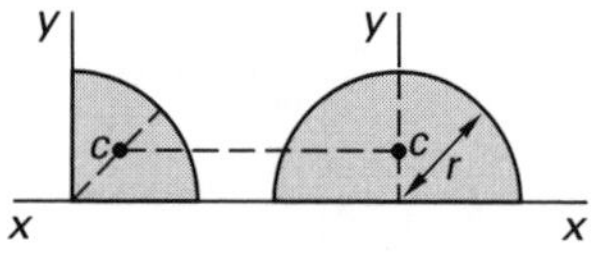	$\dfrac{4r}{3\pi}$	$\dfrac{4r}{3\pi}$	$\dfrac{\pi r^2}{4}$	$I_x = I_y = \dfrac{\pi r^4}{16}$ $J_o = \dfrac{\pi r^4}{8}$	
semicircular area		0	$\dfrac{4r}{3\pi}$	$\dfrac{\pi r^2}{2}$	$I_x = I_y = \dfrac{\pi r^4}{8}$ $I_{c,x} = 0.1098r^4$ $J_o = \dfrac{\pi r^4}{4}$ $J_c = 0.5025r^4$	$r_x = \dfrac{r}{2}$ $r_{c,x} = 0.264r$
quarter-elliptical area	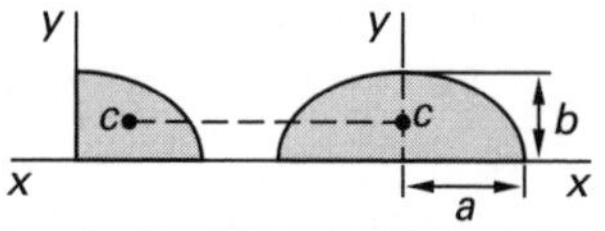	$\dfrac{4a}{3\pi}$	$\dfrac{4b}{3\pi}$	$\dfrac{\pi ab}{4}$	$I_x = \dfrac{\pi ab^3}{8}$ $I_y = \dfrac{\pi a^3 b}{8}$ $J_o = \dfrac{\pi ab(a^2 + b^2)}{8}$	
semielliptical area		0	$\dfrac{4b}{3\pi}$	$\dfrac{\pi ab}{2}$		
semiparabolic area	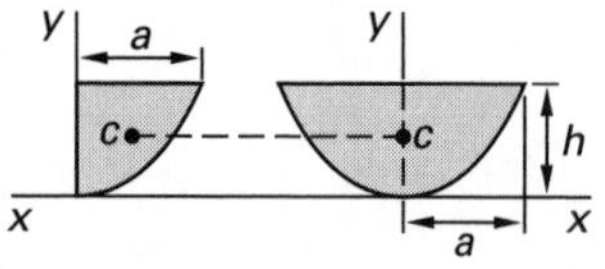	$\dfrac{3a}{8}$	$\dfrac{3h}{5}$	$\dfrac{2ah}{3}$		
parabolic area		0	$\dfrac{3h}{5}$	$\dfrac{4ah}{3}$	$I_x = \dfrac{4ah^3}{7}$ $I_y = \dfrac{4ha^3}{15}$	$r_x = h\sqrt{\dfrac{3}{7}}$ $r_y = \dfrac{a}{\sqrt{5}}$
parabolic spandrel	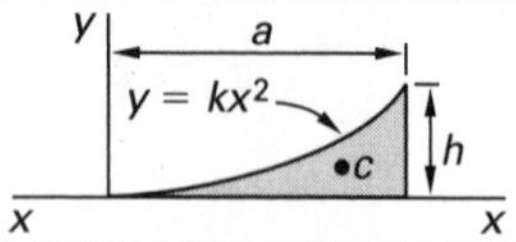	$\dfrac{3a}{4}$	$\dfrac{3h}{10}$	$\dfrac{ah}{3}$	$I_x = \dfrac{ah^3}{21}$ $I_y = \dfrac{3ha^3}{15}$	
general spandrel	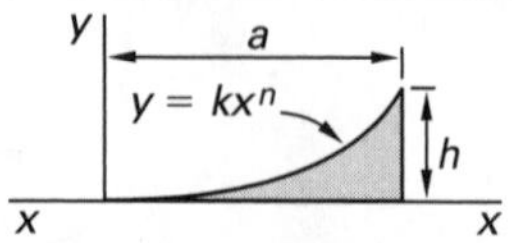	$\left(\dfrac{n+1}{n+2}\right)a$	$\left(\dfrac{n+1}{4n+2}\right)h$	$\dfrac{ah}{n+1}$		
circular sector	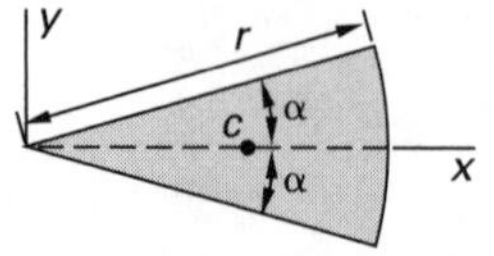	$\dfrac{2r\sin\alpha}{3\alpha}$	0	αr^2		

Topic V: Dynamics

Diagnostic Examination for Dynamics

Chapter 14   Kinematics

Chapter 15   Kinetics

Chapter 16   Kinetics of Rotational Motion

Chapter 17   Energy and Work

DYNAMICS

Hint: For the most current information about the exam, visit www.ppi2pass.com/fefaqs.html regularly. Use the Exam Forum to compare notes with other FE examinees.

Diagnostic Examination

TOPIC V: DYNAMICS

TIME LIMIT: 45 MINUTES

1. The velocity (in m/s) of a falling ball is described by the equation $v = 32 + t + 6t^2$. What is the acceleration at time $t = 2$ s?

(A) 9.8 m/s^2
(B) 25 m/s^2
(C) 32 m/s^2
(D) 58 m/s^2

2. A particle starting from rest experienced an acceleration of 3 m/s^2 for 2 s. The particle then returned to rest in an additional distance of 8 m. Assuming all accelerations were uniform, what was the total time elapsed for the particle's motion?

(A) 2.67 s
(B) 4.00 s
(C) 4.67 s
(D) 5.33 s

3. The location of a particle moving in the x-y plane is given by the parametric equations $x = t^2 + 4t$ and $y = (1/4)t^4 - 60t$, where x and y are in meters and t is in seconds. What is the particle's velocity at $t = 4$ s?

(A) 8.95 m/s
(B) 11.3 m/s
(C) 12.6 m/s
(D) 16.0 m/s

4. A projectile whose mass is 10 g is fired directly upward from ground level with an initial velocity of 1000 m/s. Neglecting the effects of air resistance, what will be the speed of the projectile when it impacts the ground?

(A) 707 m/s
(B) 981 m/s
(C) 1000 m/s
(D) 1414 m/s

5. A fisherman cuts his boat's engine as it is entering a harbor. The boat comes to a dead stop with its front end touching the dock. The fisherman's mass is 80 kg. He moves 5 m from his seat in the back to the front of the boat in 5 s, expecting to be able to reach the dock.

If the empty boat has a mass of 300 kg, how far will the fisherman have to jump to reach the dock?

(A) 1.1 m
(B) 1.3 m
(C) 1.9 m
(D) 5.0 m

6. The elevator in a 12-story building has a mass of 1000 kg. Its maximum velocity and maximum acceleration are 2 m/s and 1 m/s^2, respectively. A passenger with a mass of 75 kg stands on a bathroom scale in the elevator as the elevator ascends at its maximum acceleration. What is the scale reading just as the elevator reaches its maximum acceleration?

(A) 75 N
(B) 150 N
(C) 811 N
(D) 886 N

7. A 100 kg block is pulled along a smooth, flat surface by an external 500 N force. If the coefficient of friction between the block and the surface is 0.15, what acceleration is experienced by the block due to the external force?

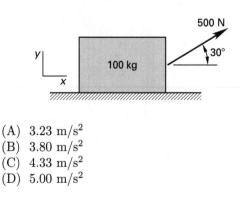

(A) 3.23 m/s^2
(B) 3.80 m/s^2
(C) 4.33 m/s^2
(D) 5.00 m/s^2

8. A 2000 kg car pulls a 500 kg trailer. The car and trailer accelerates from 50 km/h to 75 km/h at a rate of 1 m/s^2. What linear impulse does the car impart on the trailer?

(A) 3470 N·s
(B) 8680 N·s
(C) 12 500 N·s
(D) 17 400 N·s

9. The block-spring system shown oscillates once every 3 s. There is no friction between the block and the surface. If the spring constant is 6 N/m, what is the approximate mass of the block?

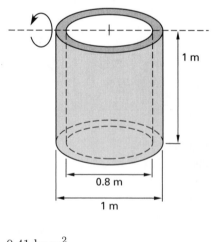

(A) 1.37 kg
(B) 5.47 kg
(C) 26.3 kg
(D) 72.0 kg

10. A hollow cylinder has a mass of 2 kg, a height of 1 m, an outer diameter of 1 m, and an inner diameter of 0.8 m. What is the cylinder's mass moment of inertia about an axis perpendicular to the cylinder's longitudinal axis and located at the cylinder's end?

(A) 0.41 kg·m^2
(B) 0.79 kg·m^2
(C) 0.87 kg·m^2
(D) 1.49 kg·m^2

11. A wheel with a radius of 80 cm rolls along a flat surface at 3 m/s. If arc AB on the wheel's perimeter measures 90°, what is the velocity of point A when point B contacts the ground?

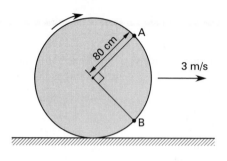

(A) 3.00 m/s
(B) 3.39 m/s
(C) 3.75 m/s
(D) 4.24 m/s

12. A 10 kg block is resting on a horizontal circular disk (e.g., turntable) at a radius of 0.5 m from the center. The coefficient of friction between the block and disk is 0.2. The disk rotates with a uniform angular velocity. What is the minimum angular velocity of the disk that will cause the block to slip?

(A) 1.40 rad/s
(B) 1.98 rad/s
(C) 3.92 rad/s
(D) 4.43 rad/s

13. A perfect sphere moves up a frictionless incline. Which of the following quantities increases?

(A) angular velocity
(B) total energy
(C) potential energy
(D) linear momentum

14. A ball is dropped from rest at a point 12 m above the ground into a smooth, frictionless chute. The ball exits the chute 2 m above the ground and at an angle 45° from the horizontal. Air resistance is negligible. Approximately how far will the ball travel in the horizontal direction before hitting the ground?

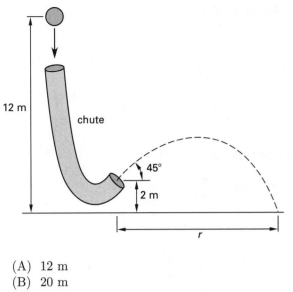

12 m

chute

45°

2 m

r

(A) 12 m
(B) 20 m
(C) 22 m
(D) 24 m

15. A 6 kg sphere moving at 3 m/s collides with a 10 kg sphere traveling 2.5 m/s in the same direction. The 6 kg ball comes to a complete stop after the collision. What is the new velocity of the 10 kg ball immediately after the collision?

(A) 0.5 m/s
(B) 2.8 m/s
(C) 4.3 m/s
(D) 5.5 m/s

SOLUTIONS TO DIAGNOSTIC EXAMINATION TOPIC V

1.
$$a = \frac{dv}{dt} = \frac{d(32 + t + 6t^2)}{dt}$$
$$= 1 + 12t$$

At $t = 2$,
$$a = 1 + (12)(2 \text{ s})$$
$$= 25 \text{ m/s}^2$$

Answer is B.

2. Separate the acceleration and deceleration phases. Apply the equations for straight line motion with constant acceleration to each phase.

For the acceleration phase,
$$v_{f1} = v_{0,1} + a_1 t_1$$
$$= 0 \frac{\text{m}}{\text{s}} + \left(3 \frac{\text{m}}{\text{s}^2}\right)(2 \text{ s})$$
$$= 6 \text{ m/s}$$

For the deceleration phase,
$$v_{f2} = v_{0,2} + a_2 t_2 = v_{f1} + a_2 t_2$$

$$a_2 = \frac{v_{f2} - v_{f1}}{t_2}$$
$$= \frac{0 \frac{\text{m}}{\text{s}} - 6 \frac{\text{m}}{\text{s}}}{t_2}$$
$$= \frac{-\left(6 \frac{\text{m}}{\text{s}}\right)}{t_2}$$

$$s_2 = v_{0,2} t_2 + \frac{1}{2} a_2 t_2^2 = v_{f1} t_2 + \frac{1}{2} a_2 t_2^2$$

$$8 \text{ m} = \left(6 \frac{\text{m}}{\text{s}}\right) t_2 + \left(\frac{1}{2}\right)\left(\frac{-\left(6 \frac{\text{m}}{\text{s}}\right)}{t_2}\right) t_2^2$$

$$t_2 = \frac{8}{3} \text{ s} = 2.67 \text{ s}$$

$$t_{\text{total}} = t_1 + t_2 = 2 \text{ s} + 2.67 \text{ s}$$
$$= 4.67 \text{ s}$$

Answer is C.

3.
$$v_x = \frac{dx}{dt} = \frac{d(t^2 + 4t)}{dt} = 2t + 4$$
$$v_y = \frac{dy}{dt} = d\left(\frac{t^4}{4} - 60t\right) = t^3 - 60$$

At $t = 4$ s,

$$v_x = (2)(4 \text{ s}) + 4 = 12 \text{ m/s}$$

$$v_y = (4 \text{ s})^3 - 60 = 4 \text{ m/s}$$

$$v = \sqrt{v_x^2 + v_y^2}$$
$$= \sqrt{\left(12 \frac{\text{m}}{\text{s}}\right)^2 + \left(4 \frac{\text{m}}{\text{s}}\right)^2}$$
$$= 12.6 \text{ m/s}$$

Answer is C.

4. Use conservation of energy.

$$PE_1 + KE_1 = PE_2 + KE_2$$

$$mgh_1 + \frac{1}{2}mv_1^2 = mgh_2 + \frac{1}{2}mv_2^2$$

$$(mgh_1 - mgh_2) + \frac{1}{2}(mv_1^2 - mv_2^2) = 0$$

$$0 + \left(\frac{m}{2}\right)(v_1^2 - v_2^2) = 0$$

$$v_2^2 = v_1^2$$

$$v_2 = \pm v_1$$

$$= \pm 1000 \text{ m/s}$$

If air resistance is neglected, the impact velocity will be the same as the initial velocity.

Answer is C.

5. The velocity of the fisherman relative to the boat is

$$v = \frac{s}{t} = \frac{5 \text{ m}}{5 \text{ s}}$$

$$= 1 \text{ m/s}$$

The velocity of the fisherman relative to the dock is

$$v'_{\text{fisherman}} = 1\frac{\text{m}}{\text{s}} + v'_{\text{boat}}$$

Use the conservation of momentum.

$$\sum m_i v_i = \sum m_i v'_i$$

$$\begin{array}{c} m_{\text{fisherman}} v_{\text{fisherman}} \\ + m_{\text{boat}} v_{\text{boat}} \end{array} = \begin{array}{c} m_{\text{fisherman}} v'_{\text{fisherman}} \\ + m_{\text{boat}} v'_{\text{boat}} \end{array}$$

However, $v_{\text{fisherman}} = v_{\text{boat}} = 0$ initially, so

$$0 = m_{\text{fisherman}}\left(1\frac{\text{m}}{\text{s}} + v'_{\text{boat}}\right) + m_{\text{boat}} v'_{\text{boat}}$$

$$v'_{\text{boat}} = \frac{-m_{\text{fisherman}}\left(1\frac{\text{m}}{\text{s}}\right)}{m_{\text{fisherman}} + m_{\text{boat}}}$$

$$= \frac{(-80 \text{ kg})\left(1\frac{\text{m}}{\text{s}}\right)}{80 \text{ kg} + 300 \text{ kg}}$$

$$= -0.211\frac{\text{m}}{\text{s}}$$

$$s = v'_{\text{boat}} t = \left(-0.211\frac{\text{m}}{\text{s}}\right)(5 \text{ s})$$

$$= -1.05 \text{ m} \quad (-1.1 \text{ m}) \quad [\text{backward}]$$

Answer is A.

6. This is a direct application of Newton's second law. The acceleration of the elevator adds to the gravitational acceleration.

$$F = ma = m(a_1 + a_2)$$

$$= (75 \text{ kg})\left(9.81\frac{\text{m}}{\text{s}^2} + 1\frac{\text{m}}{\text{s}^2}\right)$$

$$= 811 \text{ N}$$

Answer is C.

7. The gravitational normal force is reduced by the vertical component of the applied force. The frictional force is

$$F_f = \mu N = \mu(mg - F_y)$$

$$= (0.15)\left((100 \text{ kg})\left(9.81\frac{\text{m}}{\text{s}^2}\right) - (500 \text{ N})\sin 30°\right)$$

$$= 110 \text{ N}$$

Use Newton's second law.

$$ma_x = \sum F_x = F_x - F_f$$

$$a_x = \frac{F_x - F_f}{m}$$

$$\frac{(500 \text{ N})\cos 30° - 110 \text{ N}}{100 \text{ kg}} = 3.23 \text{ m/s}^2$$

Answer is A.

8. The impulse delivered to a system is equal to the change in its momentum (the impulse-momentum principle).

$$\text{Imp} = \Delta p = \Delta mv = m(\Delta v)$$

$$= m(v_2 - v_1)$$

$$= (500 \text{ kg})\left(75\frac{\text{km}}{\text{h}} - 50\frac{\text{km}}{\text{h}}\right)\left(1000\frac{\text{m}}{\text{km}}\right)$$

$$\times \left(\frac{1 \text{ h}}{3600 \text{ s}}\right)$$

$$= 3472 \text{ N·s} \quad (3470 \text{ N·s})$$

Answer is A.

9. The angular frequency is

$$\omega = \sqrt{\frac{k}{m}} = \frac{2\pi}{T}$$

$$\frac{k}{m} = \frac{4\pi^2}{T^2}$$

$$m = \frac{kT^2}{4\pi^2} = \frac{\left(6\frac{\text{N}}{\text{m}}\right)(3 \text{ s})^2}{4\pi^2}$$

$$= 1.37 \text{ kg}$$

Answer is A.

10. Use the tabulated formulas.

$$R_1 = \frac{1 \text{ m}}{2} = 0.5 \text{ m}$$

$$R_2 = \frac{0.8 \text{ m}}{2} = 0.4 \text{ m}$$

$$I = \frac{m(3R_1^2 + 3R_2^2 + 4h^2)}{12}$$

$$= \frac{(2 \text{ kg})((3)(0.5 \text{ m})^2 + (3)(0.4 \text{ m})^2 + (4)(1.0 \text{ m})^2)}{12}$$

$$= 0.87 \text{ kg·m}^2$$

Answer is C.

11. The wheel's radius is 0.8 m. Point B becomes the instantaneous center of rotation when it is in contact with the ground.

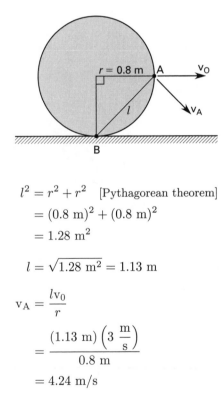

$$l^2 = r^2 + r^2 \quad \text{[Pythagorean theorem]}$$
$$= (0.8 \text{ m})^2 + (0.8 \text{ m})^2$$
$$= 1.28 \text{ m}^2$$

$$l = \sqrt{1.28 \text{ m}^2} = 1.13 \text{ m}$$

$$v_A = \frac{l v_0}{r}$$

$$= \frac{(1.13 \text{ m}) \left(3 \frac{\text{m}}{\text{s}}\right)}{0.8 \text{ m}}$$

$$= 4.24 \text{ m/s}$$

Answer is D.

12. For the block to begin to slip, the centrifugal force must equal the frictional force.

$$F_c = F_f$$

$$mr\omega^2 = \mu N = \mu mg$$

$$\omega^2 = \frac{\mu g}{r}$$

$$\omega = \sqrt{\frac{\mu g}{r}}$$

$$= \sqrt{\frac{(0.2) \left(9.81 \frac{\text{m}}{\text{s}^2}\right)}{0.5 \text{ m}}}$$

$$= 1.98 \text{ rad/s}$$

Answer is B.

13. Since the system is frictionless, there is no moment causing the sphere to rotate or stop rotating. Therefore, angular velocity is constant. Since the system is frictionless, total energy is constant. Kinetic energy is converted to potential energy. As the linear velocity decreases, so does the linear momentum.

Answer is C.

14. The change in elevation between points A and B represents a decrease in the ball's potential energy. The kinetic energy increases correspondingly.

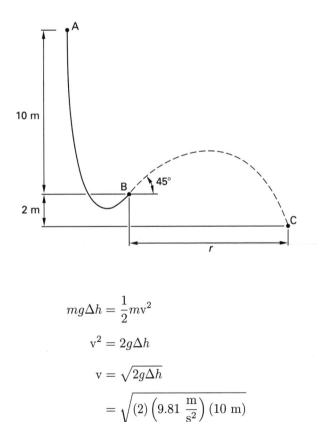

$$mg\Delta h = \frac{1}{2}mv^2$$

$$v^2 = 2g\Delta h$$

$$v = \sqrt{2g\Delta h}$$

$$= \sqrt{(2) \left(9.81 \frac{\text{m}}{\text{s}^2}\right)(10 \text{ m})}$$

$$= 14.0 \text{ m/s}$$

The ball follows the path of a projectile between points B and C.

$$y = v_0 t \sin\theta - \frac{1}{2}gt^2$$

$$\left(\frac{g}{2}\right)t^2 - (v_0 \sin\theta)t + y = 0$$

Because the landing point is below the chute exit, $y = -2$ m.

$$\left(\frac{9.81 \frac{m}{s^2}}{2}\right)t^2 - \left(14.0 \frac{m}{s}\right)(\sin 45°)t - 2 \text{ m} = 0$$

$$\left(4.91 \frac{m}{s^2}\right)t^2 - \left(9.9 \frac{m}{s}\right)t - 2 \text{ m} = 0$$

Solve for t using the quadratic formula.

$$t = \frac{-b \pm \sqrt{b^2 - 4ac}}{2a}$$

$$= \frac{9.9 \pm \sqrt{(9.9)^2 - (4)(4.91)(-2.0)}}{(2)(4.91)}$$

$$= 2.2 \text{ s}, \; -0.2 \text{ s}$$

Calculate the distance traveled from the x-component of the velocity.

$$x = v_0 t \cos\theta$$

$$= \left(14.0 \frac{m}{s}\right)(2.2 \text{ s})\cos 45°$$

$$= 21.8 \text{ m} \quad (22 \text{ m})$$

Answer is C.

15. Use conservation of momentum.

$$m_1 v_1 + m_2 v_2 = m_1 v_1' + m_2 v_2'$$

$$(6 \text{ kg})\left(3 \frac{m}{s}\right) + (10 \text{ kg})\left(2.5 \frac{m}{s}\right)$$

$$= (6 \text{ kg})\left(0 \frac{m}{s}\right) + (10 \text{ kg})v_2'$$

$$v_2' = \frac{43 \frac{kg \cdot m}{s}}{10 \text{ kg}}$$

$$= 4.3 \text{ m/s}$$

Answer is C.

14 Kinematics

Subjects

INTRODUCTION TO KINEMATICS 14-1
RECTILINEAR MOTION 14-1
    Rectangular Coordinates 14-1
    Constant Acceleration 14-2
CURVILINEAR MOTION 14-2
    Transverse and Radial Components . . 14-2
    Tangential and Normal Components . . 14-3
    Plane Circular Motion 14-3
RELATIONSHIPS BETWEEN LINEAR
    AND ROTATIONAL VARIABLES 14-3
PROJECTILE MOTION 14-3

Nomenclature

a	acceleration	ft/sec^2	m/s^2
g	gravitational acceleration	ft/sec^2	m/s^2
r	position	ft	m
r	radius	ft	m
s	distance	ft	m
t	time	sec	s
v	velocity	ft/sec	m/s

Symbols

α	angular acceleration	rad/sec^2	rad/s^2
θ	angular position	rad	rad
ω	angular velocity	rad/sec	rad/s

Subscripts

0	initial
f	final
n	normal
r	radial
t	tangential
θ	transverse

INTRODUCTION TO KINEMATICS

Dynamics is the study of moving objects. The subject is divided into kinematics and kinetics. *Kinematics* is the study of a body's motion independent of the forces on the body. It is a study of the geometry of motion without consideration of the causes of motion. Kinematics deals only with relationships among position, velocity, acceleration, and time.

A body in motion can be considered a particle if rotation of the body is absent or insignificant. A particle does not possess rotational kinetic energy. All parts of a particle have the same instantaneous displacement, velocity, and acceleration.

A *rigid body* does not deform when loaded and can be considered a combination of two or more particles that remain at a fixed, finite distance from each other. At any given instant, the parts (particles) of a rigid body can have different displacements, velocities, and accelerations if the body has rotational as well as translational motion.

If $\mathbf{r}$ is the position vector of a particle, the instantaneous velocity and acceleration are

$$\mathbf{r} \qquad \text{[position]} \qquad 14.1$$

$$\mathbf{v} = \frac{d\mathbf{r}}{dt} \qquad \text{[velocity]} \qquad 14.2$$

$$\mathbf{a} = \frac{d\mathbf{v}}{dt} = \frac{d^2\mathbf{r}}{dt^2} \qquad \text{[acceleration]} \qquad 14.3$$

RECTILINEAR MOTION

A *rectilinear system* is one in which particles move only in straight lines. (Another name is *linear system*.) The relationships among position, velocity, and acceleration for a linear system are given by Eqs. 14.4 through 14.6.

$$s(t) = \int v(t)dt = \int \int a(t)dt^2 \qquad 14.4$$

$$v(t) = \frac{ds(t)}{dt} = \int a(t)dt \qquad 14.5$$

$$a(t) = \frac{dv(t)}{dt} = \frac{d^2s(t)}{dt^2} \qquad 14.6$$

Rectangular Coordinates

The position of a particle is specified with reference to a coordinate system. Three coordinates are necessary to identify the position in three-dimensional space; in two dimensions, two coordinates are necessary. A coordinate can represent a linear position, as in the rectangular coordinate system, or it can represent an angular position, as in the polar system.

Consider the particle shown in Fig. 14.1. Its position, as well as its velocity and acceleration, can be specified in three primary forms: vector form, rectangular coordinate form, and unit vector form.

Figure 14.1 Rectangular Coordinates

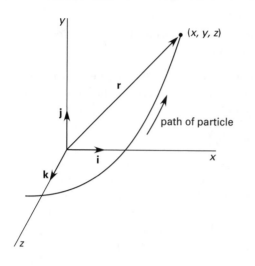

The *vector form* of the particle's position is $\mathbf{r}$, where the vector $\mathbf{r}$ has both magnitude and direction. The *rectangular coordinate form* is (x, y, z). The *unit vector form* is

$$\mathbf{r} = x\mathbf{i} + y\mathbf{j} + z\mathbf{k} \qquad \textit{14.7}$$

The velocity and acceleration are the first two derivatives of the position vector, as shown in Eqs. 14.8 and 14.9.

$$\begin{aligned} \mathbf{v} &= \frac{d\mathbf{r}}{dt} \\ &= \dot{x}\mathbf{i} + \dot{y}\mathbf{j} + \dot{z}\mathbf{k} \qquad \textit{14.8} \\ \mathbf{a} &= \frac{d\mathbf{v}}{dt} = \frac{d^2\mathbf{r}}{dt^2} \\ &= \ddot{x}\mathbf{i} + \ddot{y}\mathbf{j} + \ddot{z}\mathbf{k} \qquad \textit{14.9} \end{aligned}$$

Constant Acceleration

Acceleration is a constant in many cases, such as a free-falling body with constant acceleration g. If the acceleration is constant, the acceleration term can be taken out of the integrals in Eqs. 14.4 and 14.5. The initial distance from the origin is s_0; the initial velocity is a constant, v_0; and a constant acceleration is denoted a_0.

$$a(t) = a_0 \qquad \textit{14.10}$$

$$v(t) = a_0 \int dt = v_0 + a_0 t \qquad \textit{14.11}$$

$$s(t) = a_0 \iint dt^2$$

$$= s_0 + v_0 t + \frac{a_0 t^2}{2} \qquad \textit{14.12}$$

$$v^2(t) = v_0^2 + 2a_0(s - s_0) \qquad \textit{14.13}$$

CURVILINEAR MOTION

Curvilinear motion describes the motion of a particle along a path that is not a straight line. Special examples of curvilinear motion include plane circular motion and projectile motion. For particles traveling along curvilinear paths, the position, velocity, and acceleration may be specified in rectangular coordinates as they were for rectilinear motion, or it may be more convenient to express the kinematic variables in terms of other coordinate systems (e.g., polar coordinates).

Transverse and Radial Components

In polar coordinates, the position of a particle is described by a radius, r, and an angle, θ. The position may also be expressed as a vector of magnitude r and direction specified by unit vector $\mathbf{e}_r$. Since the velocity of a particle is not usually directed radially out from the center of the coordinate system, it can be divided into two components, called *radial* and *transverse*, which are parallel and perpendicular, respectively, to the unit radial vector. Figure 14.2 illustrates the radial and transverse components of velocity in a polar coordinate system, and the unit radial and unit transverse vectors, $\mathbf{e}_r$ and $\mathbf{e}_\theta$, used in the vector forms of the motion equations.

Figure 14.2 Radial and Transverse Coordinates

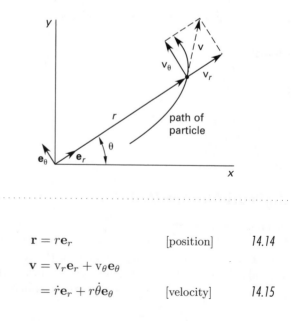

$$\mathbf{r} = r\mathbf{e}_r \qquad \text{[position]} \qquad \textit{14.14}$$

$$\begin{aligned} \mathbf{v} &= v_r\mathbf{e}_r + v_\theta\mathbf{e}_\theta \\ &= \dot{r}\mathbf{e}_r + r\dot{\theta}\mathbf{e}_\theta \qquad \text{[velocity]} \qquad \textit{14.15} \end{aligned}$$

$$\mathbf{a} = a_r \mathbf{e}_r + a_\theta \mathbf{e}_\theta$$

$$= (\ddot{r} - r\dot\theta^2)\mathbf{e}_r$$

$$+ (r\ddot\theta + 2\dot{r}\dot\theta)\mathbf{e}_\theta \quad \text{[acceleration]} \quad \textit{14.16}$$

Tangential and Normal Components

A particle moving in a curvilinear path will have instantaneous linear velocity and linear acceleration. These linear variables will be directed tangentially to the path, and, therefore, are known as *tangential velocity*, v_t, and *tangential acceleration*, a_t, respectively. The force that constrains the particle to the curved path will generally be directed toward the center of rotation, and the particle will experience an inward acceleration perpendicular to the tangential velocity and acceleration, known as the *normal acceleration*, a_n. The resultant acceleration, $\mathbf{a}$, is the vector sum of the tangential and normal accelerations. Normal and tangential components of acceleration are illustrated in Fig. 14.3. The unit vectors $\mathbf{e}_n$ and $\mathbf{e}_t$ are normal and tangential to the path, respectively. ρ is the principal *radius of curvature*.

$$\mathbf{v} = v_t \mathbf{e}_t \quad\quad \textit{14.17}$$

$$\mathbf{a} = \left(\frac{dv_t}{dt}\right)\mathbf{e}_t + \left(\frac{v_t^2}{\rho}\right)\mathbf{e}_n \quad \textit{14.18}$$

Figure 14.3 Tangential and Normal Coordinates

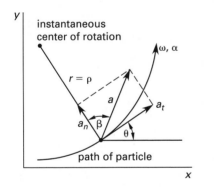

Plane Circular Motion

Plane circular motion (also known as *rotational particle motion*, *angular motion*, or *circular motion*) is motion of a particle around a fixed circular path. The behavior of a rotating particle is defined by its angular position, θ, angular velocity, ω, and angular acceleration, α. These variables are analogous to the s, v, and a variables for linear systems. Angular variables can be substituted one-for-one in place of linear variables in most equations.

$$\theta \quad\quad \text{[angular position]} \quad \textit{14.19}$$

$$\omega = \frac{d\theta}{dt} \quad \text{[angular velocity]} \quad \textit{14.20}$$

$$\alpha = \frac{d\omega}{dt}$$

$$= \frac{d^2\theta}{dt^2} \quad \text{[angular acceleration]} \quad \textit{14.21}$$

RELATIONSHIPS BETWEEN LINEAR AND ROTATIONAL VARIABLES

$$s = r\theta \quad\quad \textit{14.22}$$

$$v_t = r\omega \quad\quad \textit{14.23}$$

$$a_t = r\alpha = \frac{dv_t}{dt} \quad\quad \textit{14.24}$$

$$a_n = \frac{v_t^2}{r} = r\omega^2 \quad\quad \textit{14.25}$$

PROJECTILE MOTION

A projectile is placed into motion by an initial impulse. (Kinematics deals only with dynamics during the flight. The force acting on the projectile during the launch phase is covered in kinetics.) Neglecting air drag, once the projectile is in motion, it is acted upon only by the downward gravitational acceleration (i.e., its own weight). Thus, projectile motion is a special case of motion under constant acceleration.

Consider a general projectile set into motion at an angle of θ from the horizontal plane, and initial velocity v_0, as shown in Fig. 14.4. In the absence of air drag, the following rules apply to the case of travel over a horizontal plane.

Figure 14.4 Projectile Motion

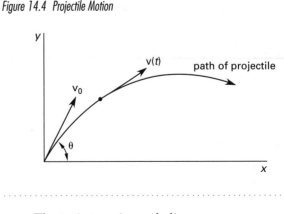

- The trajectory is parabolic.
- The impact velocity is equal to initial velocity, v_0.
- The range is maximum when $\theta = 45°$.

- The time for the projectile to travel from the launch point to the apex is equal to the time to travel from apex to impact point.
- The time for the projectile to travel from the apex of its flight path to impact is the same time an initially stationary object would take to fall straight down from that height.

The following solutions to most common projectile problems are derived from the laws of uniform acceleration and conservation of energy.

$$a_x = 0 \qquad\qquad 14.26$$

$$a_y = -g \qquad\qquad 14.27$$

$$v_x = v_{x0} = v_0 \cos\theta \qquad\qquad 14.28$$

$$v_y = v_{y0} - gt = v_0 \sin\theta - gt \qquad\qquad 14.29$$

$$x = v_{x0}t = v_0 t \cos\theta \qquad\qquad 14.30$$

$$y = v_{y0}t - \frac{1}{2}gt^2 = v_0 t \sin\theta - \frac{1}{2}gt^2 \quad 14.31$$

SAMPLE PROBLEMS

Problems 1–3 refer to a particle whose curvilinear motion is represented by the equation $s = 20t + 4t^2 - 3t^3$.

Problem 1

What is the particle's initial velocity?

- (A) 20 m/s
- (B) 25 m/s
- (C) 30 m/s
- (D) 32 m/s

Solution

$$v = \frac{ds}{dt} = 20 + 8t - 9t^2$$

At $t = 0$,

$$v = 20 + (8)(0) - (9)(0)^2$$
$$= 20 \text{ m/s}$$

Answer is A.

Problem 2

What is the acceleration of the particle at time $t = 0$?

- (A) 2 m/s^2
- (B) 3 m/s^2
- (C) 5 m/s^2
- (D) 8 m/s^2

Solution

$$a = \frac{d^2s}{dt^2} = 8 - 18t$$

At $t = 0$,

$$a = 8 \text{ m/s}^2$$

Answer is D.

Problem 3

What is the maximum speed reached by the particle?

- (A) 21.8 m/s
- (B) 27.9 m/s
- (C) 34.6 m/s
- (D) 48.0 m/s

Solution

The maximum of the velocity function is found by equating the derivative of the velocity function to zero and solving for t.

$$v = 20 + 8t - 9t^2$$

$$\frac{dv}{dt} = 8 - 18t = 0$$

$$t = \frac{8}{18} \text{ s} = 0.444 \text{ s}$$

$$v_{max} = 20 + (8)(0.444 \text{ s}) - (9)(0.444 \text{ s})^2$$
$$= 21.8 \text{ m/s}$$

Answer is A.

Problem 4

Choose the equation that best represents a rigid body or particle under constant acceleration.

- (A) $a = 9.81 \text{ m/s}^2 + v_0/t$
- (B) $v = v_0 + a_0 t$
- (C) $v = v_0 + \int_0^t a(t)dt$
- (D) $a = v_t^2/r$

Solution

Choice (B) is the expression for the velocity of a linear system under constant acceleration.

$$v(t) = a_0 \int dt = v_0 + a_0 t$$

The other answer choices can be eliminated. Choice (A) is an expression for acceleration that varies with time; choice (C) is an expression for velocity with a generalized time-varying acceleration. The expression in choice (D) relates tangential and normal accelerations, respectively, along a curved path, to the tangential velocity. For a generalized curved path, these accelerations are not constant.

Answer is B.

Problem 5

A roller coaster train climbs a hill with a constant gradient. Over a 10 s period, the acceleration is constant at 0.4 m/s², and the average velocity of the train is 40 km/h. Find the final velocity.

(A) 9.1 m/s
(B) 11.1 m/s
(C) 13.1 m/s
(D) 15.1 m/s

Solution

Use the constant acceleration equations for straight line motion.

$$v_{ave} = \left(40 \ \frac{km}{h}\right)\left(1000 \ \frac{m}{km}\right)\left(\frac{1}{60 \ \frac{min}{h}}\right)\left(\frac{1}{60 \ \frac{s}{min}}\right)$$

$$= 11.11 \ m/s$$

The distance traveled in 10 s is

$$\left(11.11 \ \frac{m}{s}\right)(10 \ s) = 111.1 \ m$$

$$s(t) = s_0 + v_0 t + \frac{1}{2}a_0 t^2$$

$$v_0 = \frac{s(t) - s_0 - \frac{1}{2}a_0 t^2}{t}$$

$$= \frac{111.1 \ m - 0 - \left(\frac{1}{2}\right)\left(0.4 \ \frac{m}{s^2}\right)(10 \ s)^2}{10 \ s}$$

$$= 9.11 \ m/s$$

$$v_f = v_0 + a_0 t$$

$$= 9.11 \ \frac{m}{s} + \left(0.4 \ \frac{m}{s^2}\right)(10 \ s)$$

$$= 13.11 \ m/s$$

Alternate Solution

$$v = v_0 + a_0 t$$

$$v_{ave} = \frac{\Delta s}{\Delta t} = \frac{1}{t_f - t_0}\int_{t_0}^{t_f}(v_0 + a_0 t)dt$$

$$= \left(\frac{1}{t_f - t_0}\right)\left(v_0 t + \frac{1}{2}a_0 t^2\right)\Big|_{t_0}^{t_f}$$

$$= \left(\frac{1}{t_f - t_0}\right)\left(v_0 t_f + \frac{1}{2}a_0 t_f^2 - v_0 t_0 - \frac{1}{2}a_0 t_0^2\right)$$

Arbitrarily, let $t_0 = 0$ and $t_f = 10$.

$$v_{ave} = \left(\frac{1}{t_f}\right)\left(v_0 t_f + \left(\frac{a_0}{2}\right)t_f^2\right)$$

$$= v_0 + \frac{a_0 t_f}{2}$$

$$v_0 = v_{ave} - \frac{a_0 t_f}{2}$$

$$= 11.11 \ \frac{m}{s} - \left(\frac{0.4 \ \frac{m}{s^2}}{2}\right)(10 \ s)$$

$$= 9.11 \ m/s$$

$$v_{ave} = \frac{v_0 + v_f}{2}$$

$$v_f = 2v_{ave} - v_0$$

$$= (2)\left(11.11 \ \frac{m}{s}\right) - 9.11 \ \frac{m}{s}$$

$$= 13.11 \ m/s \quad (13.1 \ m/s)$$

Answer is C.

Problem 6

A projectile is fired from a cannon with an initial velocity of 1000 m/s and at an angle of 30° from the horizontal. What distance from the cannon will the projectile strike the ground if the point of impact is 1500 m below the point of release?

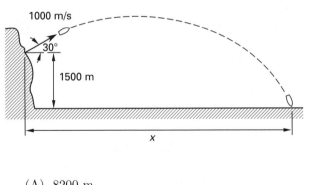

(A) 8200 m
(B) 67 300 m
(C) 78 200 m
(D) 90 800 m

Solution

$$y = v_0 t \sin\theta - \frac{gt^2}{2}$$

$$\frac{g}{2}t^2 - v_0 t \sin\theta + y = 0$$

$y = -1500$ m since it is below the launch plane.

$$\left(\frac{9.81 \ \frac{m}{s^2}}{2}\right)t^2 - \left(1000 \ \frac{m}{s}\right)t \sin 30° - 1500 \ m = 0$$

$$\left(4.905 \ \frac{m}{s^2}\right)t^2 - \left(500 \ \frac{m}{s}\right)t - 1500 \ m = 0$$

$$t = \frac{-b \pm \sqrt{b^2 - 4ac}}{2a} \quad \text{[quadratic formula]}$$

$$= \frac{500 \pm \sqrt{(-500)^2 - (4)(4.905)(-1500)}}{(2)(4.905)}$$

$$= +104.85 \text{ s}, \ -2.9166 \text{ s}$$

$$x = v_0 t \cos\theta$$

$$= \left(1000 \ \frac{\text{m}}{\text{s}}\right)(104.85 \text{ s}) \cos 30°$$

$$= 90\,803 \text{ m} \quad (90\,800 \text{ m})$$

Answer is D.

FE-STYLE EXAM PROBLEMS

Problems 1 and 2 refer to a particle for which the position is defined by

$$s(t) = 2 \sin t\mathbf{i} + 4 \cos t\mathbf{j} \quad [t \text{ in radians}]$$

1. What is the magnitude of the particle's velocity at $t = 4$ rad?

    (A) 2.61
    (B) 2.75
    (C) 3.30
    (D) 4.12

2. What is the magnitude of the particle's acceleration at $t = \pi$?

    (A) 2.00
    (B) 2.56
    (C) 3.14
    (D) 4.00

3. For the reciprocating pump shown, the radius of the crank is $r = 0.3$ m, and the rotational speed is $n = 350$ rpm. What is the tangential velocity of point A on the crank corresponding to an angle of $\theta = 35°$ from the horizontal?

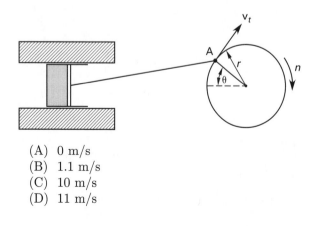

    (A) 0 m/s
    (B) 1.1 m/s
    (C) 10 m/s
    (D) 11 m/s

4. A golfer on level ground attempts to drive a golf ball across a 50 m wide pond, hitting the ball so that it travels initially at 25 m/s. The ball travels at an initial angle of 45° to the horizontal plane. How far will the golf ball travel, and does it clear the pond?

    (A) 32 m; the ball does not clear the pond
    (B) 45 m; the ball does not clear the pond
    (C) 58 m; the ball clears the pond
    (D) 64 m; the ball clears the pond

5. Rigid link AB is 12 m long. It rotates counterclockwise about point A at 12 rev/min. A thin disk with radius 1.75 m is pinned at its center to the link at point B. The disk rotates counterclockwise at 60 rev/min with respect to point B. What is the maximum tangential velocity seen by any point on the disk?

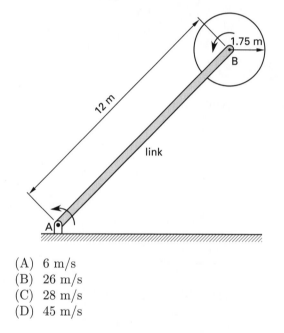

    (A) 6 m/s
    (B) 26 m/s
    (C) 28 m/s
    (D) 45 m/s

For the following problems use the NCEES Handbook as your only reference.

6. A particle has a tangential acceleration of a_t (represented by the equation given) when it moves around a point in a curve with instantaneous radius of 1 m. What is the instantaneous angular velocity (in rad/s) of the particle?

$$a_t = 2t - \sin t + 3 \ \cot t \quad [\text{in m/s}^2]$$

    (A) $t^2 + \cos t + 3 \ \ln \ |\csc t|$
    (B) $t^2 - \cos t + 3 \ \ln \ |\csc t|$
    (C) $t^2 - \cos t + 3 \ \ln \ |\sin t|$
    (D) $t^2 + \cos t + 3 \ \ln \ |\sin t|$

7. A stone is dropped down a well. 2.47 s after the stone is released, a splash is heard. If the velocity of sound in air is 342 m/s, find the distance to the surface of the water in the well.

    (A) 2.4 m
    (B) 7.2 m
    (C) 28 m
    (D) 30 m

Problems 8 and 9 refer to the following situation.

A motorist is traveling at 70 km/h when he sees a traffic light in an intersection 250 m ahead turn red. The light's red cycle is 15 s. The motorist wants to enter the intersection without stopping his vehicle, just as the light turns green.

8. What uniform deceleration of the vehicle will just put the motorist in the intersection when the light turns green?

    (A) 0.18 m/s^2
    (B) 0.25 m/s^2
    (C) 0.37 m/s^2
    (D) 1.3 m/s^2

9. If the vehicle decelerates at a constant rate of 0.5 m/s^2, what will be its speed when the light turns green?

    (A) 43 km/h
    (B) 52 km/h
    (C) 59 km/h
    (D) 63 km/h

Problems 10 and 11 refer to the following information.

The position (in radians) of a car traveling around a curve is described by the following function of time (in seconds).

$$\theta(t) = t^3 - 2t^2 - 4t + 10$$

10. What is the angular velocity at $t = 3$ s?

    (A) -16 rad/s
    (B) -4 rad/s
    (C) 11 rad/s
    (D) 15 rad/s

11. What is the angular acceleration at $t = 5$ s?

    (A) 4 rad/s^2
    (B) 6 rad/s^2
    (C) 26 rad/s^2
    (D) 30 rad/s^2

12. The rotor of a steam turbine is rotating at 7200 rev/min when the steam supply is suddenly cut off. The rotor decelerates at a constant rate and comes to rest after 5 min. What was the angular deceleration of the rotor?

    (A) 0.40 rad/s^2
    (B) 2.5 rad/s^2
    (C) 5.8 rad/s^2
    (D) 16 rad/s^2

13. A flywheel rotates at 7200 rev/min when the power is suddenly cut off. The flywheel decelerates at a constant rate of 2.1 rad/s^2 and comes to rest 6 min later. How many revolutions does the flywheel make before coming to rest?

    (A) 18 000 rev
    (B) 22 000 rev
    (C) 72 000 rev
    (D) 390 000 rev

Problems 14–16 refer to the following situation.

A projectile has an initial velocity of 110 m/s and a launch angle of 20° from the horizontal. The surrounding terrain is level, and air friction is to be disregarded.

14. What is the flight time of the projectile?

    (A) 3.8 s
    (B) 7.7 s
    (C) 8.9 s
    (D) 12 s

15. What is the horizontal distance traveled by the projectile?

    (A) 80 m
    (B) 400 m
    (C) 800 m
    (D) 1200 m

16. What is the maximum elevation achieved by the projectile?

    (A) 72 m
    (B) 140 m
    (C) 350 m
    (D) 620 m

SOLUTIONS TO FE-STYLE EXAM PROBLEMS

1.
$$\mathbf{v}(t) = \frac{ds(t)}{dt} = 2\cos t\mathbf{i} - 4\sin t\mathbf{j}$$

**DYNAMICS
Kinematics**

At $t = 4$ rad,

$$v(4) = 2 \cos(4 \text{ rad})\mathbf{i} - 4 \sin(4 \text{ rad})\mathbf{j}$$
$$= -1.31\mathbf{i} - (-3.03)\mathbf{j}$$
$$|v(4)| = \sqrt{(-1.31)^2 + (3.03)^2}$$
$$= 3.30$$

Answer is C.

2. From Prob. 1,

$$v(t) = 2 \cos t \mathbf{i} - 4 \sin t \mathbf{j}$$
$$a(t) = \frac{dv(t)}{dt} = -2 \sin t \mathbf{i} - 4 \cos t \mathbf{j}$$
$$a(\pi) = -2 \sin \pi \mathbf{i} - 4 \cos \pi \mathbf{j}$$
$$= 0\mathbf{i} + 4.0\mathbf{j}$$
$$|a(\pi)| = \sqrt{(0)^2 + (4.0)^2}$$
$$= 4.0$$

Answer is D.

3. Use the relationship between the tangential and rotational variables.

$$v_t = r\omega$$
$$\omega = \text{angular velocity of the crank}$$
$$= \left(350 \, \frac{\text{rev}}{\text{min}}\right)\left(2\pi \, \frac{\text{rad}}{\text{rev}}\right)\left(\frac{1}{60 \, \frac{\text{s}}{\text{min}}}\right)$$
$$= 36.65 \, \text{rad/s}$$
$$v_t = (0.3 \, \text{m})\left(36.65 \, \frac{\text{rad}}{\text{s}}\right)$$
$$= 11.0 \, \text{m/s}$$

This value is the same for any point on the crank at $r = 0.3$ m.

Answer is D.

4. The elevation of the ball above the ground is

$$y = v_{y0}t - \frac{gt^2}{2} = v_0 t \sin\theta - \frac{gt^2}{2}$$

When the ball hits the ground, $y = 0$, and

$$v_0 t \sin\theta = \frac{gt^2}{2}$$

Solving for t, the time to impact is

$$t = \frac{2v_0 \sin\theta}{g}$$

Substitute the time of impact into the expression for x to obtain an expression for the range.

$$x = v_0 t \cos\theta = v_0 \left(\frac{2v_0 \sin\theta}{g}\right)\cos\theta$$
$$= \left(\frac{2v_0^2}{g}\right)\sin\theta \cos\theta$$
$$= \left(\frac{(2)\left(25 \, \frac{\text{m}}{\text{s}}\right)^2}{9.81 \, \frac{\text{m}}{\text{s}^2}}\right)\sin 45° \cos 45°$$
$$= 63.7 \, \text{m} \quad (64 \, \text{m})$$

Answer is D.

5. The maximum tangential velocity of an extension of line AB to an extreme point on the disk, due to the rotation of the link only, with respect to point A is

$$v_{t,\text{B}|\text{A}} = r\omega = r(2\pi f)$$
$$= \frac{(12 \, \text{m} + 1.75 \, \text{m})\left(2\pi \, \frac{\text{rad}}{\text{rev}}\right)\left(12 \, \frac{\text{rev}}{\text{min}}\right)}{60 \, \frac{\text{s}}{\text{min}}}$$
$$= 17.28 \, \text{m/s}$$

The maximum tangential velocity of the periphery of the disk with respect to point B is

$$v_{t,\text{disk}|\text{B}} = r\omega = r(2\pi f)$$
$$= \frac{(1.75 \, \text{m})\left(2\pi \, \frac{\text{rad}}{\text{rev}}\right)\left(60 \, \frac{\text{rev}}{\text{min}}\right)}{60 \, \frac{\text{s}}{\text{min}}}$$
$$= 11.00 \, \text{m/s}$$

The velocities combine when the two velocity vectors coincide in direction. The maximum velocity of the periphery of the disk with respect to point A is the sum of the magnitudes of the two velocities.

$$v_{t,\text{disk}|\text{B}} = v_{t,\text{B}|\text{A}} + v_{t,\text{disk}|\text{B}}$$
$$= 17.28 \, \frac{\text{m}}{\text{s}} + 11.00 \, \frac{\text{m}}{\text{s}}$$
$$= 28.28 \, \text{m/s} \quad (28 \, \text{m/s})$$

Answer is C.

6.
$$a_t = \frac{dv_t}{dt}$$

$$v_t = \int a_t \, dt$$

$$= \int (2t - \sin t + 3 \, \cot t) \, dt$$

$$= t^2 + \cos t + 3 \, \ln \, |\sin t| \quad [\text{in m/s}]$$

$$\omega = \frac{v_t}{r} = \frac{t^2 + \cos t + 3 \, \ln \, |\sin t|}{1 \text{ m}}$$

$$= t^2 + \cos t + 3 \, \ln \, |\sin t| \quad [\text{in rad/s}]$$

Answer is D.

7. The elapsed time is the sum of the time for the stone to fall and the time for the sound to return to the listener.

The distance, x, traveled by the stone under the influence of a constant gravitational acceleration is

$$x = x_0 + v_0 t + \frac{1}{2} g t^2$$

x_0 and v_0 are both zero.

$$x = \frac{1}{2} g t^2$$

Solving for t, the time for the stone to drop is

$$t_1 = \sqrt{\frac{2x}{g}}$$

The time for the sound (traveling at velocity c) to return to the listener is
$$t_2 = \frac{x}{c}$$

The total time taken is

$$t_1 + t_2 = 2.47 \text{ s}$$

$$\sqrt{\frac{2x}{g}} + \frac{x}{c} = 2.47 \text{ s}$$

Substitute values for g and c.

$$\sqrt{\frac{2x}{9.81 \, \frac{\text{m}}{\text{s}^2}}} + \frac{x}{342 \, \frac{\text{m}}{\text{s}}} = 2.47 \text{ s}$$

By trial and error with the answer choices given (or by solving the quadratic equation),

$$x = 28 \text{ m}$$

Answer is C.

8. The initial speed of the vehicle is

$$v_0 = \frac{\left(70 \, \frac{\text{km}}{\text{h}}\right)\left(1000 \, \frac{\text{m}}{\text{km}}\right)}{3600 \, \frac{\text{s}}{\text{h}}}$$

$$= 19.44 \text{ m/s}$$

The distance traveled under a constant deceleration is

$$x = x_0 + v_0 t - \frac{1}{2} a t^2$$

Letting $x_0 = 0$, the acceleration required to travel a distance x in time t starting with velocity v_0 is

$$a = \frac{(2)(v_0 t - x)}{t^2}$$

$$= \frac{(2)\left(\left(19.44 \, \frac{\text{m}}{\text{s}}\right)(15 \text{ s}) - 250 \text{ m}\right)}{(15 \text{ s})^2}$$

$$= 0.37 \text{ m/s}^2$$

Answer is C.

9.
$$v = v_0 + at$$

$$= 70 \, \frac{\text{km}}{\text{h}} + \frac{\left(-0.5 \, \frac{\text{m}}{\text{s}^2}\right)(15 \text{ s})\left(3600 \, \frac{\text{s}}{\text{h}}\right)}{1000 \, \frac{\text{m}}{\text{km}}}$$

$$= 43 \text{ km/h}$$

Answer is A.

10.
$$\omega(t) = \frac{d\theta}{dt} = 3t^2 - 4t - 4$$

$$\omega(3) = (3)(3)^2 - (4)(3) - 4$$

$$= 11 \text{ rad/s}$$

Answer is C.

11.
$$\alpha(t) = \frac{d\omega(t)}{dt}$$
$$= 6t - 4$$

$$\alpha(5) = (6)(5) - 4$$
$$= 26 \text{ rad/s}^2$$

Answer is C.

12.
$$\omega = \omega_0 - \alpha t$$

$$0 = \frac{\left(7200 \ \frac{\text{rev}}{\text{min}}\right)\left(2\pi \ \frac{\text{rad}}{\text{rev}}\right)}{60 \ \frac{\text{s}}{\text{min}}}$$

$$- \alpha(5 \ \text{min})\left(60 \ \frac{\text{s}}{\text{min}}\right)$$

$$\alpha = 2.51 \ \text{rad/s}^2 \quad (2.5 \ \text{rad/s}^2)$$

Answer is B.

13.
$$\theta = \theta_0 + \omega_0 t - \frac{1}{2}\alpha t^2$$

$$= 0 + \left(7200 \ \frac{\text{rev}}{\text{min}}\right)(2\pi)(6 \ \text{min})$$

$$- \left(\frac{1}{2}\right)\left(2.1 \ \frac{\text{rad}}{\text{s}^2}\right)$$

$$\times \left((6 \ \text{min})\left(60 \ \frac{\text{s}}{\text{min}}\right)\right)^2$$

$$= 135.4 \times 10^3 \ \text{rad}$$

$$\theta = \frac{135.4 \times 10^3 \ \text{rad}}{2\pi}$$

$$= 21.5 \times 10^3 \ \text{rev}$$

Alternate solution: The average rotational speed during deceleration is

$$\overline{\omega} = \frac{\omega_1 - \omega_2}{2} = \frac{7200 \ \frac{\text{rev}}{\text{min}} - 0}{2}$$

$$= 3600 \ \text{rev/min}$$

$$\theta = \overline{\omega}t = \left(3600 \ \frac{\text{rev}}{\text{min}}\right)(6 \ \text{min})$$

$$= 21\,600 \ \text{rev} \quad (22\,000 \ \text{rev})$$

Answer is B.

14. The vertical component of velocity is zero at the apex.

$$v_y = v_0 \sin\theta - gt$$

$$0 = \left(110 \ \frac{\text{m}}{\text{s}}\right)\sin 20° - \left(9.81 \ \frac{\text{m}}{\text{s}^2}\right)t$$

$$t = 3.84 \ \text{s}$$

The projectile takes an equal amount of time to return to the ground from the apex. The total flight time is

$$T = (2)(3.84 \ \text{s}) = 7.68 \ \text{s} \quad (7.7 \ \text{s})$$

Answer is B.

15. Calculate the range from the horizontal component of velocity.

$$x = v_x t = v_0 \cos\theta t$$

$$= \left(110 \ \frac{\text{m}}{\text{s}}\right)\cos 20°(7.68 \ \text{s})$$

$$= 794 \ \text{m} \quad (800 \ \text{m})$$

Answer is C.

16. The elevation at time t is

$$y = v_0 t \sin\theta - \frac{1}{2}gt^2$$

$$= \left(110 \ \frac{\text{m}}{\text{s}}\right)(3.84 \ \text{s})\sin 20°$$

$$- \left(\frac{1}{2}\right)\left(9.81 \ \frac{\text{m}}{\text{s}^2}\right)(3.84 \ \text{s})^2$$

$$= 72 \ \text{m}$$

Answer is A.

15 Kinetics

Subjects

INTRODUCTION TO KINETICS 15-1
MOMENTUM 15-1
NEWTON'S FIRST AND SECOND LAWS
   OF MOTION 15-1
WEIGHT 15-2
FRICTION 15-2
KINETICS OF A PARTICLE 15-3
   Rectangular Coordinates 15-3
   Tangential and Normal Components . . 15-3
   Radial and Transverse Components . . 15-3
FREE VIBRATION 15-3

Nomenclature

a	acceleration	ft/sec^2	m/s^2
f	linear frequency	Hz	Hz
F	force	lbf	N
g	gravitational acceleration	ft/sec^2	m/s^2
g_c	gravitational constant (32.2)	lbm-ft/lbf-sec^2	–
k	spring constant	lbf/ft	N/m
m	mass	lbm	kg
N	normal force	lbf	N
P	linear momentum	lbf-sec	N·s
r	position	ft	m
r	radius	ft	m
R	resultant force	lbf	N
s	distance	ft	m
t	time	sec	s
T	period	sec	s
v	velocity	ft/sec	m/s
W	weight	lbf	N

Symbols

α	angular acceleration	rad/sec^2	rad/s^2
δ	deflection	ft	m
θ	angular position	rad	rad
μ	coefficient of friction	–	–
ϕ	angle	deg	deg
ω	natural frequency	rad/sec	rad/s

Subscripts

0	initial
f	friction
k	dynamic
n	normal
r	radial
R	resultant
s	static
t	tangential
θ	transverse

INTRODUCTION TO KINETICS

Kinetics is the study of motion and the forces that cause motion. Kinetics includes an analysis of the relationship between the force and mass for translational motion and between torque and moment of inertia for rotational motion. Newton's laws form the basis of the governing theory in the subject of kinetics.

MOMENTUM

The vector *linear momentum* (*momentum*) is defined by Eq. 15.1. It has the same direction as the velocity vector. Momentum has units of force × time (e.g., lbf-sec or N·s).

$$\mathbf{p} = m\mathbf{v} \qquad \text{[SI]} \qquad 15.1a$$

$$\mathbf{p} = \frac{m\mathbf{v}}{g_c} \qquad \text{[U.S.]} \qquad 15.1b$$

Momentum is conserved when no external forces act on a particle. If no forces act on the particle, the velocity and direction of the particle are unchanged. The *law of conservation of momentum* states that the linear momentum is unchanged if no unbalanced forces act on the particle. This does not prohibit the mass and velocity from changing, however. Only the product of mass and velocity is constant.

NEWTON'S FIRST AND SECOND LAWS OF MOTION

Newton's first law of motion states that a particle will remain in a state of rest or will continue to move with constant velocity unless an unbalanced external force acts on it.

DYNAMICS Kinetics

This law can also be stated in terms of conservation of momentum: If the resultant external force acting on a particle is zero, then the linear momentum of the particle is constant.

Newton's second law of motion states that the acceleration of a particle is directly proportional to the force acting on it and is inversely proportional to the particle mass. The direction of acceleration is the same as the direction of force.

This law can be stated in terms of the force vector required to cause a change in momentum: The resultant force is equal to the rate of change of linear momentum.

$$\mathbf{F} = \frac{d\mathbf{p}}{dt} \qquad 15.2$$

For a fixed mass,

$$\mathbf{F} = \frac{d\mathbf{p}}{dt} = \frac{d(m\mathbf{v})}{dt}$$

$$= m\frac{d\mathbf{v}}{dt}$$

$$= m\mathbf{a} \qquad \text{[SI]} \qquad 15.3a$$

$$\mathbf{F} = \frac{m\mathbf{a}}{g_c} \qquad \text{[U.S.]} \qquad 15.3b$$

WEIGHT

The *weight*, W, of an object is the force the object exerts due to its position in a gravitational field, g.

$$W = mg \qquad \text{[SI]} \qquad 15.4a$$

$$W = \frac{mg}{g_c} \qquad \text{[U.S.]} \qquad 15.4b$$

g_c is the gravitational constant, approximately 32.2 lbm-ft/lbf-sec^2.

FRICTION

Friction is a force that always resists motion or impending motion. It always acts parallel to the contacting surfaces. If the body is moving, the friction is known as *dynamic friction*. If the body is stationary, friction is known as *static friction*.

The magnitude of the frictional force depends on the normal force, N, and the *coefficient of friction*, μ, between the body and the contacting surface.

$$F_f = \mu N \qquad 15.5$$

The static coefficient of friction is usually denoted with the subscript s, while the dynamic coefficient of friction is denoted with the subscript k. μ_k is often assumed

to be 75 percent of the value of μ_s. These coefficients are complex functions of surface properties. Experimentally determined values for various contacting conditions can be found in handbooks.

For a body resting on a horizontal surface, the *normal force* is the weight of the body. If the body rests on an inclined surface, the normal force is calculated as the component of weight normal to that surface, as illustrated in Fig. 15.1.

$$N = mg \cos\phi \qquad \text{[SI]} \qquad 15.6a$$

$$N = \frac{mg \cos\phi}{g_c} \qquad \text{[U.S.]} \qquad 15.6b$$

Figure 15.1 Frictional and Normal Forces

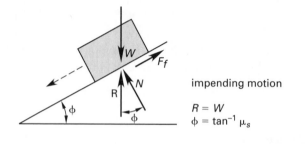

The frictional force acts only in response to a disturbing force, and it increases as the disturbing force increases. The motion of a stationary body is impending when the disturbing force reaches the maximum frictional force, $\mu_s N$. Figure 15.1 shows the condition of impending motion for a block on a plane. Just before motion starts, the resultant, R, of the frictional force and normal force equals the weight of the block. The angle at which motion is just impending can be calculated from the coefficient of static friction.

$$\phi = \tan^{-1}\mu_s \qquad 15.7$$

Once motion begins, the coefficient of friction drops slightly, and a lower frictional force opposes movement. This is illustrated in Fig. 15.2.

Figure 15.2 Frictional Force versus Disturbing Force

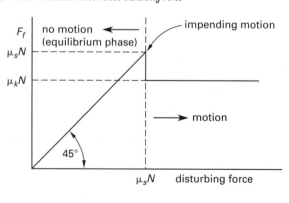

KINETICS OF A PARTICLE

Newton's second law can be applied separately to any direction in which forces are resolved into components. The law can be expressed in rectangular coordinate form (i.e., in terms of x- and y-component forces), in polar coordinate form (i.e., in tangential and normal components), or in radial and transverse component form.

Rectangular Coordinates

Equation 15.8 is Newton's second law in rectangular coordinate form and refers to motion in the x-direction. Similar equations can be written for the y-direction or any other coordinate direction.

$$F_x = ma_x \qquad \text{[SI]} \qquad 15.8$$

In general, F_x may be a function of time, displacement, and/or velocity. If F_x is a function of time only, then the motion equations are

$$v_x(t) = v_{x0} + \int \left(\frac{F_x(t)}{m} \right) dt \qquad \text{[SI]} \qquad 15.9$$

$$x(t) = x_0 + v_{x0}t + \int v_x(t)dt \qquad 15.10$$

If F_x is constant (i.e., is independent of time, displacement, or velocity), then the motion equations become

$$F_x = ma_x \qquad \text{[SI]} \qquad 15.11$$

$$v_x(t) = v_{x0} + \left(\frac{F_x}{m} \right) t$$
$$= v_{x0} + a_x t \qquad 15.12$$

$$x(t) = x_0 + v_{x0}t + \frac{F_x t^2}{2m}$$
$$= x_0 + v_{x0}t + \frac{a_x t^2}{2} \qquad 15.13$$

Tangential and Normal Components

For a particle moving along a circular path, the tangential and normal components of force, acceleration, and velocity are related.

$$\sum F_n = ma_n = m \left(\frac{v_t^2}{r} \right) \qquad \text{[SI]} \qquad 15.14$$

$$\sum F_t = ma_t = m \left(\frac{dv_t}{dt} \right) \qquad \text{[SI]} \qquad 15.15$$

Radial and Transverse Components

For a particle moving along a circular path, the radial and transverse components of force are

$$\sum F_r = ma_r \qquad \text{[SI]} \qquad 15.16$$

$$\sum F_\theta = ma_\theta \qquad \text{[SI]} \qquad 15.17$$

FREE VIBRATION

Vibration is an oscillatory motion about an equilibrium point. If the motion is the result of a disturbing force that is applied once and then removed, the motion is known as *natural* (or *free*) *vibration*. If a continuous force or single impulse is applied repeatedly to a system, the motion is known as *forced vibration*.

A simple application of free vibration is a mass suspended from a vertical spring, as shown in Fig. 15.3. After the mass is displaced and released, it will oscillate up and down. If there is no friction (i.e., the vibration is undamped), the oscillations will continue forever.

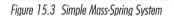

Figure 15.3 Simple Mass-Spring System

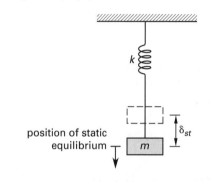

position of static equilibrium

The system shown in Fig. 15.3 is initially at rest. The mass is hanging on the spring, and the equilibrium position is the static deflection, δ_{st}. This is the deflection due to the gravitational force alone.

$$mg = k\delta_{st} \qquad \text{[SI]} \qquad 15.18a$$

$$\frac{mg}{g_c} = k\delta_{st} \qquad \text{[U.S.]} \qquad 15.18b$$

The system is then disturbed by a downward force (i.e., the mass is pulled downward from its static deflection and released). After the initial disturbing force is removed, the mass will be acted upon by the restoring force $(-kx)$ and the inertial force $(-mg)$. Both of these forces are proportional to the displacement from the equilibrium point, and they are opposite in sign from the displacement. The equation of motion is

$$F = ma \qquad \text{[SI]} \qquad 15.19$$
$$mg - k(x + \delta_{st}) = m\ddot{x} \qquad \text{[SI]} \qquad 15.20$$
$$k\delta_{st} - k(x + \delta_{st}) = m\ddot{x} \qquad \text{[SI]} \qquad 15.21$$
$$m\ddot{x} + kx = 0 \qquad \text{[SI]} \qquad 15.22$$

The solution to this second-order differential equation is

$$x(t) = C_1 \cos \omega t + C_2 \sin \omega t \qquad 15.23$$

C_1 and C_2 are constants of integration that depend on the initial displacement and velocity of the mass. ω is known as the *natural frequency of vibration* or *angular frequency*. It has units of radians per second. It is not the same as the linear frequency, f, which has units of hertz. The *period of oscillation*, T, is the reciprocal of the linear frequency.

$$\omega = \sqrt{\frac{k}{m}} \qquad \text{[SI]} \qquad 15.24a$$

$$\omega = \sqrt{\frac{kg_c}{m}} \qquad \text{[U.S.]} \qquad 15.24b$$

$$f = \frac{\omega}{2\pi} = \frac{1}{T} \qquad 15.25$$

$$T = \frac{1}{f} = \frac{2\pi}{\omega} \qquad 15.26$$

For the general case where the initial displacement is x_0 and the initial velocity is v_0, the solution of the equation of motion is

$$x(t) = x_0 \cos \omega t + \left(\frac{v_0}{\omega} \right) \sin \omega t \qquad 15.27$$

For the special case where the initial displacement is x_0 and the initial velocity is zero, the solution of the equation of motion is

$$x(t) = x_0 \cos \omega t \qquad 15.28$$

SAMPLE PROBLEMS

Problem 1

For which of the following situations is the net force acting on a particle necessarily equal to zero?

(A) The particle is traveling at constant velocity around a circle.
(B) The particle has constant linear momentum.
(C) The particle has constant kinetic energy.
(D) The particle has constant angular momentum.

Solution

This is a restatement of Newton's first law of motion, which says that if the resultant external force acting on a particle is zero, then the linear momentum of the particle is constant.

Answer is B.

Problem 2

One newton is the force required to

(A) give a 1 g mass an acceleration of 1 m/s².
(B) accelerate a 10 kg mass at a rate of 0.10 m/s².
(C) accelerate a 1 kg mass at a rate of 1.00 cm/s².
(D) accelerate a 1 kg mass at a rate of 9.81 m/s².

Solution

Newton's second law can be expressed in the form of $F = ma$. The unit of force in SI units is the newton, which has fundamental units of kg·m/s². A newton is the force required to accelerate a 1 kg mass at a rate of 1 m/s² or a 10 kg mass at a rate of 0.10 m/s².

Answer is B.

Problem 3

A 550 kg mass initially at rest is acted upon by a force of $50e^t$ N. What are the acceleration, speed, and displacement of the mass at $t = 4$ s?

(A) 4.96 m/s²; 4.87 m/s; 4.51 m
(B) 4.96 m/s²; 4.87 m/s; 19.5 m
(C) 4.96 m/s²; 135.5 m/s; 2466 m
(D) 4.96 m/s²; 271 m/s; 3900 m

Solution

$$a = \frac{F}{m} = \frac{50e^4 \text{ N}}{550 \text{ kg}} = 4.96 \text{ m/s}^2$$

$$v = \int a \, dt = \int \left(\frac{F}{m} \right) dt$$

$$= \int \left(\frac{50e^t}{550} \right) dt$$

$$= \frac{1}{11} \int e^t \, dt$$

$$= \left(\frac{1}{11} \right) \left(e^t + C_1 \right)$$

Since $v = 0$ at $t = 0$,

$$0 = \left(\frac{1}{11} \right) \left(e^0 + C_1 \right)$$

$$C_1 = -1$$

$$v(4) = \left(\frac{1}{11} \right) \left(e^4 - 1 \right)$$

$$= 4.87 \text{ m/s}$$

Similarly,

$$s = \int v \, dt = \int \frac{1}{11} \left(e^t - 1 \right) dt$$

$$= \frac{1}{11} \int \left(e^t - 1 \right) dt$$

$$= \left(\frac{1}{11} \right) \left(e^t - t + C_2 \right)$$

Since $s = 0$ at $t = 0$,

$$0 = \left(\frac{1}{11}\right)\left(e^0 - 0 + C_2\right)$$

$$C_2 = -1$$

$$s(4) = \left(\frac{1}{11}\right)\left(e^4 - 4 - 1\right)$$

$$= 4.51$$

Answer is A.

Problems 4 and 5 refer to the following situation.

- A 5 kg block begins from rest and slides down an inclined plane.
- After 4 s, the block has a velocity of 6 m/s.

Problem 4

If the angle of inclination is 45°, how far has the block traveled after 4 s?

(A) 1.5 m
(B) 3 m
(C) 6 m
(D) 12 m

Solution

$$v(t) = v_0 + a_0 t$$

$$a_0 = \frac{v(t) - v_0}{t} = \frac{6 \,\frac{m}{s} - 0}{4 \text{ s}}$$

$$= 1.5 \text{ m/s}^2$$

$$s(t) = s_0 + v_0 t + \frac{1}{2} a_0 t^2$$

$$= 0 + 0 + \left(\frac{1}{2}\right)\left(1.5 \,\frac{m}{s^2}\right)(4 \text{ s})^2$$

$$= 12 \text{ m}$$

Answer is D.

Problem 5

What is the coefficient of friction between the plane and the block?

(A) 0.15
(B) 0.22
(C) 0.78
(D) 0.85

Solution

Choose a coordinate system so that the x-direction is parallel to the inclined plane.

$$\sum F_x = ma_x = mg_x - F_f$$

$$ma_x = mg \sin 45° - \mu mg \cos 45°$$

$$\mu = \frac{mg \sin 45° - ma_x}{mg \cos 45°}$$

$$= \frac{g \sin 45° - a_x}{g \cos 45°}$$

$$= \frac{\left(9.81 \,\frac{m}{s^2}\right) \sin 45° - 1.5 \,\frac{m}{s^2}}{\left(9.81 \,\frac{m}{s^2}\right) \cos 45°}$$

$$= 0.78$$

Answer is C.

Problem 6

A constant force of 750 N is applied through a pulley system to lift a mass of 50 kg as shown. Neglecting the mass and friction of the pulley system, what is the acceleration of the 50 kg mass?

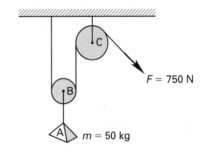

(A) 5.20 m/s²
(B) 8.72 m/s²
(C) 16.2 m/s²
(D) 20.2 m/s²

Solution

Apply Newton's second law to the mass and to the two frictionless, massless pulleys. Refer to the following free-body diagrams.

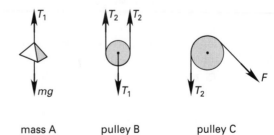

mass A   pulley B   pulley C

mass A: $T_1 - mg = ma$

pulley B: $2T_2 - T_1 = 0$

pulley C: $T_2 = F = 750 \text{ N}$

$$a = \frac{T_1 - mg}{m} = \frac{2T_2 - mg}{m} = \frac{2F - mg}{m}$$

$$= \frac{(2)(750 \text{ N}) - (50 \text{ kg})\left(9.81 \frac{m}{s^2}\right)}{50 \text{ kg}}$$

$$= 20.2 \text{ m/s}^2$$

Answer is D.

Problem 7

A mass of 10 kg is suspended from a vertical spring with a spring constant of 10 N/m. What is the period of vibration?

(A) 0.30 s
(B) 0.60 s
(C) 0.90 s
(D) 6.3 s

Solution

$$T = 2\pi\sqrt{\frac{m}{k}} = 2\pi\sqrt{\frac{10 \text{ kg}}{10 \frac{N}{m}}}$$

$$= 6.3 \text{ s}$$

Answer is D.

FE-STYLE EXAM PROBLEMS

1. If the sum of the forces on a particle is not equal to zero, the particle is

(A) moving with constant velocity in the direction of the resultant force.
(B) accelerating in a direction opposite to the resultant force.
(C) accelerating in the same direction as the resultant force.
(D) moving with a constant velocity opposite to the direction of the resultant force.

2. A varying force acts on a 40 kg mass as shown in the following force versus time diagram. What is the object's velocity at $t = 4$ s if the object starts from rest?

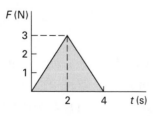

(A) 0 m/s
(B) 0.075 m/s
(C) 0.15 m/s
(D) 0.30 m/s

Problems 3 and 4 refer to the following situation.

- The 52 kg block shown starts from rest at position A and slides down the inclined plane to position B.
- When the block reaches position B, a 383 N horizontal force is applied.
- The block comes to a complete stop at position C.
- The coefficient of friction between the block and the plane is $\mu = 0.15$.

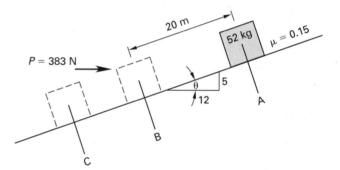

3. Find the velocity at position B.

(A) 2.41 m/s
(B) 4.12 m/s
(C) 6.95 m/s
(D) 9.83 m/s

4. Find the distance between positions B and C.

(A) 3.23 m
(B) 4.78 m
(C) 7.78 m
(D) 10.1 m

Problems 5 and 6 refer to the following pulley system. In standard gravity, block A exerts a force of 10 000 N and block B exerts a force of 7500 N. Both blocks are initially held stationary. There is no friction and the pulleys have no mass.

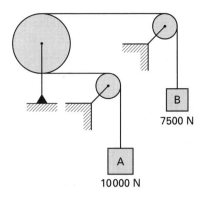

7500 N

B

A

10000 N

5. Find the acceleration of block A after the blocks are released.

(A) 0 m/s²
(B) 1.4 m/s²
(C) 2.5 m/s²
(D) 5.6 m/s²

6. Find the velocity of block A 2.5 s after the blocks are released.

(A) 0 m/s
(B) 3.5 m/s
(C) 4.4 m/s
(D) 4.9 m/s

For the following problems use the NCEES Handbook as your only reference.

7. What is the period of a pendulum that passes the center point 20 times a minute?

(A) 0.2 s
(B) 0.3 s
(C) 3 s
(D) 6 s

8. A variable force of $(40\text{ N})\cos\theta$ is attached to the end of a spring whose spring constant is 50 N/m. There is no deflection when $\theta = 90°$. At what angle, θ, will the spring deflect 20 cm from its equilibrium position?

(A) −14°
(B) 25°
(C) 64°
(D) 76°

9. A spring has a constant of 50 N/m. The spring is hung vertically, and a mass is attached to its end. The spring end displaces 30 cm from its equilibrium position. The same mass is removed from the first spring and attached to the end of a second (different) spring, and

the displacement is 25 cm. What is the spring constant of the second spring?

(A) 46 N/m
(B) 56 N/m
(C) 60 N/m
(D) 63 N/m

10. A cannonball of mass 10 kg is fired from a cannon of mass 250 kg. The initial velocity of the cannonball is 1000 km/h. All of the cannon's recoil is absorbed by a spring with a spring constant of 520 N/cm. What is the maximum recoil distance of the cannon?

(A) 0.35 m
(B) 0.59 m
(C) 0.77 m
(D) 0.92 m

11. A child keeps a 1 kg toy airplane flying horizontally in a circle by holding onto a 1.5 m long string attached to its wing tip. The string is always in the plane of the circular flight path. If the plane flies at 10 m/s, what is the tension in the string?

(A) 7 N
(B) 15 N
(C) 28 N
(D) 67 N

12. A car with a mass of 1530 kg tows a trailer (mass of 200 kg) at 100 km/h. What is the total momentum of the car-trailer combination?

(A) 4600 N·s
(B) 22000 N·s
(C) 37000 N·s
(D) 48000 N·s

13. A car is pulling a trailer at 100 km/h. A 5 kg cat riding on the roof of the car jumps from the car to the trailer. What is the change in the cat's momentum?

(A) −25 N·s (loss)
(B) 0 N·s
(C) 25 N·s (gain)
(D) 1300 N·s (gain)

14. A 3500 kg car accelerates from rest. The constant forward tractive force of the car is 1000 N, and the constant drag force is 150 N. What distance will the car travel in 3 s?

(A) 0.19 m
(B) 1.1 m
(C) 1.3 m
(D) 15 m

SOLUTIONS TO FE-STYLE EXAM PROBLEMS

1. Newton's second law, $F = ma$, can be applied separately to any direction in which forces are resolved into components, including the resultant direction.

$$F_R = ma_R$$

Since force and acceleration are both vectors, and mass is a scalar, the direction of acceleration is the same as the resultant force.

$$\mathbf{a}_R = \frac{\mathbf{F}_R}{m}$$

Answer is C.

2. Use the impulse-momentum principle. The impulse is the area under the $\mathbf{F}$-t curve. There are two right triangles.

$$\mathbf{F} = \frac{d\mathbf{p}}{dt}$$

$$\mathbf{F}\Delta t = m\Delta \mathbf{v}$$

$$\Delta \mathbf{v} = \frac{\mathbf{F}\Delta t}{m} = \frac{(2)\left(\left(\frac{1}{2}\right)(3\,\text{N})(2\,\text{s})\right)}{40\,\text{kg}}$$

$$= 0.15 \text{ m/s}$$

Answer is C.

3. Choose a coordinate system parallel and perpendicular to the plane, as shown.

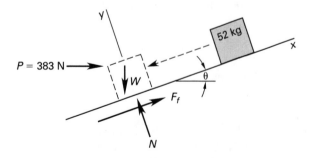

$$\sum F_x = ma_x$$

$$W_x - \mu N = ma_x$$

$$mg\sin\theta - \mu mg\cos\theta = ma_x$$

$$a_x = g\sin\theta - \mu g\cos\theta$$

$$= \left(9.81\,\frac{\text{m}}{\text{s}^2}\right)\left(\frac{5}{13} - (0.15)\left(\frac{12}{13}\right)\right)$$

$$= 2.415 \text{ m/s}^2$$

$$v^2 = v_0^2 + 2a_0(s - s_0)$$

$$v_0 = s_0 = 0$$

$$v^2 = 2a_0 s$$

$$= (2)\left(2.415\,\frac{\text{m}}{\text{s}^2}\right)(20\,\text{m})$$

$$= 96.6 \text{ m}^2/\text{s}^2$$

$$v = \sqrt{96.6\,\frac{\text{m}^2}{\text{s}^2}} = 9.83 \text{ m/s}$$

Answer is D.

4.
$$\sum F_x = ma$$

$$W_x - P_x - \mu N = ma$$

$$mg\sin\theta - P\cos\theta - \mu(mg\cos\theta + P\sin\theta) = ma$$

$$a = \left(\frac{1}{m}\right)(mg\sin\theta - P\cos\theta - \mu(mg\cos\theta + P\sin\theta))$$

$$= g\sin\theta - \mu g\cos\theta - \frac{P}{m}(\cos\theta + \mu\sin\theta)$$

$$= \left(9.81\,\frac{\text{m}}{\text{s}^2}\right)\left(\frac{5}{13} - (0.15)\left(\frac{12}{13}\right)\right)$$

$$- \left(\frac{383\,\text{N}}{52\,\text{kg}}\right)\left(\frac{12}{13} + (0.15)\left(\frac{5}{13}\right)\right)$$

$$= -4.809 \text{ m/s}^2$$

$$v^2 = v_0^2 + 2a_0(s - s_0)$$

$$v_0 = 9.83 \text{ m/s} \quad \text{[from Prob. 3]}$$

$$v = s_0 = 0$$

$$s = \frac{-v_0^2}{2a_0} = \frac{-\left(9.83\,\frac{\text{m}}{\text{s}}\right)^2}{(2)\left(-4.809\,\frac{\text{m}}{\text{s}^2}\right)}$$

$$= 10.05 \text{ m} \quad (10.1 \text{ m})$$

Answer is D.

5. Refer to the following free-body diagrams.

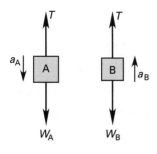

Apply Newton's second law to the free body of mass A.

$$\sum F_y = m_A a_A = W_A - T_A \quad \text{[down]}$$

But $m = W/g$, so

$$\left(\frac{10\,000 \text{ N}}{9.81 \, \frac{\text{m}}{\text{s}^2}}\right) a_A = 10\,000 \text{ N} - T_A$$

Apply Newton's second law to the free body of mass B.

$$\sum F_y = m_B a_B = W_B - T_B \quad \text{[up]}$$

$$\left(\frac{7500 \text{ N}}{9.81 \, \frac{\text{m}}{\text{s}^2}}\right) a_B = 7500 \text{ N} - T_B$$

Combine the equations and solve for a_A, setting $T_A = T_B$ and $a_A = -a_B$.

$$W_A - m_A a_A = W_B - m_B a_B = W_B + m_B a_A$$

$$a_A = \frac{W_A - W_B}{m_B + m_A} = \frac{g(W_A - W_B)}{W_B + W_A}$$

$$= \left(9.81 \, \frac{\text{m}}{\text{s}^2}\right)\left(\frac{10\,000 \text{ N} - 7500 \text{ N}}{7500 \text{ N} + 10\,000 \text{ N}}\right)$$

$$= 1.4 \text{ m/s}^2$$

Alternate Solution

$$|a_A| = |a_B|$$

But $F = ma$, so

$$\frac{F_A}{m_A} = \frac{F_B}{m_B}$$

Since the tension in the rope is the same everywhere,

$$\frac{W_A - T}{m_A} = \frac{T - W_B}{m_B}$$

$$\frac{10\,000 \text{ N} - T}{10\,000 \text{ N}} = \frac{T - 7500 \text{ N}}{7500 \text{ N}}$$

$$T = 8571 \text{ N}$$

From block A,

$$a = \frac{F_A}{m_A} = \frac{W_A - T}{\dfrac{W}{g}}$$

$$= \frac{g(W_A - T)}{W_A}$$

$$= \frac{\left(9.81 \, \frac{\text{m}}{\text{s}^2}\right)(10\,000 \text{ N} - 8571 \text{ N})}{10\,000 \text{ N}}$$

$$= 1.4 \text{ m/s}^2$$

Answer is B.

6.
$$v_A = v_0 + a_A t = 0 + \left(1.4 \, \frac{\text{m}}{\text{s}^2}\right)(2.5 \text{ s})$$

$$= 3.5 \text{ m/s}$$

Answer is B.

7. A pendulum will pass the center point two times during each complete cycle. Therefore, 10 cycles are completed in 60 s.

$$T = \frac{\text{elapsed time}}{\text{no. of cycles}} = \frac{60 \text{ s}}{10}$$

$$= 6 \text{ s}$$

Answer is D.

8. From Hooke's law, the relationship between force, F, and deflection, x, for a linear spring is

$$F = kx$$

$$(40 \text{ N})(\cos\theta) = \frac{\left(50 \, \frac{\text{N}}{\text{m}}\right)(20 \text{ cm})}{100 \, \frac{\text{cm}}{\text{m}}}$$

$$\cos\theta = 0.25$$

$$\theta = \cos^{-1}(0.25)$$

$$= 75.5° \quad (76°)$$

Answer is D.

9. The gravitational force on the mass is the same for both springs. From Hooke's law,

$$F = k_1 x_1 = k_2 x_2$$

$$k_2 = \frac{k_1 x_1}{x_2} = \frac{\left(50 \, \frac{\text{N}}{\text{m}}\right)(30 \text{ cm})}{25 \text{ cm}}$$

$$= 60 \text{ N/m}$$

Answer is C.

10. Use the conservation of momentum equation to determine the velocity of the cannon after the ball is fired. Initially, the cannon and cannonball are both at rest. Since the cannon recoils, its velocity direction will be opposite (i.e., negative) to the direction of the cannonball.

$$\sum m_i v_i = \sum m_i v_{i'}$$

$$(10 \text{ kg})(0) + (250 \text{ kg})(0)$$
$$= (10 \text{ kg})\left(1000 \ \frac{\text{km}}{\text{h}}\right) + (250 \text{ kg})v_c$$

$$v_c = -40 \text{ km/h}$$

Use the conservation of energy principle to determine the compression of the spring. The kinetic energy of the cannon will be equal to the elastic potential energy stored in the spring.

$$KE = PE$$

$$\frac{1}{2}m_c v_c^2 = \frac{1}{2}kx^2$$

$$(0.5)(250 \text{ kg})\left(\frac{\left(40 \ \frac{\text{km}}{\text{h}}\right)\left(1000 \ \frac{\text{m}}{\text{km}}\right)}{3600 \ \frac{\text{s}}{\text{h}}}\right)^2$$

$$= (0.5)\left(520 \ \frac{\text{N}}{\text{cm}}\right)\left(100 \ \frac{\text{cm}}{\text{m}}\right)x^2$$

$$x = 0.77 \text{ m}$$

Answer is C.

11. The normal acceleration (perpendicular to the path of the airplane) is

$$a_n = \frac{v_t^2}{r} = \frac{\left(10 \ \frac{\text{m}}{\text{s}}\right)^2}{1.5 \text{ m}}$$

$$= 66.7 \text{ m/s}^2$$

The tension in the string is equal to the centripetal force.

$$F = ma_n$$
$$= (1 \text{ kg})\left(66.7 \ \frac{\text{m}}{\text{s}^2}\right)$$
$$= 66.7 \text{ N} \quad (67 \text{ N})$$

Answer is D.

12.
$$v = \frac{\left(100 \ \frac{\text{km}}{\text{h}}\right)\left(1000 \ \frac{\text{m}}{\text{km}}\right)}{3600 \ \frac{\text{s}}{\text{h}}}$$

$$= 27.78 \text{ m/s}$$

$$P = mv$$
$$= (1530 \text{ kg} + 200 \text{ kg})\left(27.78 \ \frac{\text{m}}{\text{s}}\right)$$
$$= 48\,060 \text{ N·s} \quad (48\,000 \text{ N·s})$$

Answer is D.

13. The law of conservation of momentum states that the linear momentum is unchanged if no unbalanced forces act on an object. This does not prohibit the mass and velocity from changing; only the product of mass and velocity is constant. In this case, both the total mass and the velocity are constant. Thus, there is no change.

Answer is B.

14.
$$F = 1000 \text{ N} - 150 \text{ N}$$
$$= 850 \text{ N}$$
$$a = \frac{F}{m}$$
$$= \frac{850 \text{ N}}{3500 \text{ kg}}$$
$$= 0.243 \text{ m/s}^2$$
$$s = v_0 t + \frac{1}{2}at^2$$
$$= 0 + \left(\frac{1}{2}\right)\left(0.243 \ \frac{\text{m}}{\text{s}^2}\right)(3 \text{ s})^2$$
$$= 1.09 \text{ m} \quad (1.1 \text{ m})$$

Answer is B.

16 Kinetics of Rotational Motion

Subjects

MASS MOMENT OF INERTIA 16-1
PLANE MOTION OF A RIGID BODY . . . 16-2
    Rotation About a Fixed Axis 16-2
    Instantaneous Center of Rotation . . . 16-2
CENTRIPETAL AND CENTRIFUGAL
    FORCES 16-3
BANKING OF CURVES 16-3
TORSIONAL FREE VIBRATION 16-3

Nomenclature

a	acceleration	ft/sec^2	m/s^2
A	area	ft^2	m^2
d	distance	ft	m
F	force	lbf	N
g	gravitational acceleration	ft/sec^2	m/s^2
g_c	gravitational constant (32.2)	lbm-ft/lbf-sec^2	–
G	shear modulus	lbf/ft^2	Pa
h	angular momentum	ft-lbf-sec	N·m·s
I	mass moment of inertia	lbm-ft^2	kg·m^2
J	area polar moment of inertia	ft^4	m^4
k_t	torsional spring constant	ft-lbf/rad	N·m/rad
l	length	ft	m
m	mass	lbm	kg
M	moment	ft-lbf	N·m
r	radius	ft	m
r	radius of gyration	ft	m
R	moment arm	ft	m
t	time	sec	s
v	velocity	ft/sec	m/s
W	weight	lbf	N

Symbols

α	angular acceleration	rad/sec^2	rad/s^2
θ	angular position	rad	rad
θ	superelevation angle	deg	deg
μ	coefficient of friction	–	–
ρ	density	lbm/ft^3	kg/m^3
ω	angular velocity	rad/sec	rad/s
ω	natural frequency	rad/sec	rad/s

Subscripts

0	initial
c	centroidal
f	friction
n	normal or natural
O	origin or center
s	static
t	tangential or torsional

MASS MOMENT OF INERTIA

The *mass moment of inertia* measures a solid object's resistance to changes in rotational speed about a specific axis. I_x, I_y, and I_z are the mass moments of inertia with respect to the x-, y-, and z-axes, respectively. They are not components of a resultant value.

$$I_x = \int (y^2 + z^2)dm \qquad 16.1$$

$$I_y = \int (x^2 + z^2)dm \qquad 16.2$$

$$I_z = \int (x^2 + y^2)dm \qquad 16.3$$

The *centroidal mass moment of inertia*, I_c, is obtained when the origin of the axes coincides with the object's center of gravity. Once the centroidal mass moment of inertia is known, the *parallel axis theorem* is used to find the mass moment of inertia about any parallel axis. In Eq. 16.4, d is the distance from the center of mass to the parallel axis.

$$I_{\text{any parallel axis}} = I_c + md^2 \qquad 16.4$$

For a composite object, the parallel axis theorem must be applied for each of the constituent objects.

$$I = I_{c,1} + m_1d_1^2 + I_{c,2} + m_2d_2^2 + \cdots \qquad 16.5$$

The *radius of gyration*, r, of a solid object represents the distance from the rotational axis at which the object's entire mass could be located without changing the mass moment of inertia.

$$r = \sqrt{\frac{I}{m}} \qquad 16.6$$

$$I = r^2m \qquad 16.7$$

Table 16.1 (at the end of this chapter) lists the mass moments of inertia and radii of gyration for some standard shapes.

PLANE MOTION OF A RIGID BODY

General rigid body plane motion, such as rolling wheels, gear sets, and linkages, can be represented in two dimensions (i.e., the plane of motion). Plane motion can be considered as the sum of a translational component and a rotation about a fixed axis, as illustrated in Fig. 16.1.

Figure 16.1 Components of Plane Motion

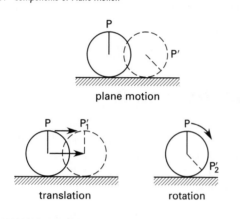

Rotation About a Fixed Axis

Rotation about a fixed axis describes a motion in which all particles within the body move in concentric circles about the axis of rotation.

The *angular momentum* taken about a point O is the moment of the linear momentum vector. Angular momentum has units of distance × force × time (e.g., ft-lbf-sec or N·m·s). It has the same direction as the rotation vector and can be determined from the vectors by use of the right-hand rule (cross product).

$$\mathbf{h}_O = \mathbf{r} \times m\mathbf{v} \qquad \text{[SI]} \qquad 16.8a$$

$$\mathbf{h}_O = \mathbf{r} \times \frac{m\mathbf{v}}{g_c} \qquad \text{[U.S.]} \qquad 16.8b$$

For a rigid body rotating about an axis passing through its center of gravity located at point O, the scalar value of angular momentum is given by Eq. 16.9.

$$h_O = I\omega \qquad \text{[SI]} \qquad 16.9a$$

$$h_O = \frac{I\omega}{g_c} \qquad \text{[U.S.]} \qquad 16.9b$$

Although Newton's laws do not specifically deal with rotation, there is an analogous relationship between applied moment and change in angular momentum. For a rotating body, the moment (torque), M, required to change the angular momentum is

$$\mathbf{M} = \frac{d\mathbf{h}_O}{dt} \qquad 16.10$$

The rotation of a rigid body will be about the center of gravity unless the body is constrained otherwise. If the moment of inertia is constant, the scalar form of Eq. 16.10 is

$$M = I\frac{d\omega}{dt} = I\alpha \qquad \text{[SI]} \qquad 16.11a$$

$$M = \left(\frac{I}{g_c}\right)\frac{d\omega}{dt} = \frac{I\alpha}{g_c} \qquad \text{[U.S.]} \qquad 16.11b$$

Velocity and position in terms of rotational variables can be determined by integrating the expression for acceleration.

$$\alpha = \frac{M}{I} \qquad 16.12$$

$$\omega = \int \alpha\,dt = \omega_0 + \left(\frac{M}{I}\right)t \qquad 16.13$$

$$\theta = \int\int \alpha\,dt^2 = \theta_0 + \omega_0 t + \left(\frac{M}{2I}\right)t^2 \qquad 16.14$$

Instantaneous Center of Rotation

Analysis of the rotational component of a rigid body's plane motion can sometimes be simplified if the location of the body's *instantaneous center* is known. Using the instantaneous center reduces many relative motion problems to simple geometry. The instantaneous center (also known as the *instant center* and IC) is a point at which the body could be fixed (pinned) without changing the instantaneous angular velocities of any point on the body. Thus, for angular velocities, the body seems to rotate about a fixed instantaneous center.

The instantaneous center is located by finding two points for which the absolute velocity directions are known. Lines drawn perpendicular to these two velocities will intersect at the instantaneous center. (This graphic procedure is slightly different if the two velocities are parallel, as Fig. 16.2 shows.) For a rolling wheel, the instantaneous center is the point of contact with the supporting surface.

Figure 16.2 Graphic Method of Finding the Instantaneous Center

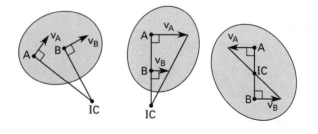

The absolute velocity of any point, P, on a wheel rolling (Fig. 16.3) with translational velocity, v_O, can be found by geometry. Assume that the wheel is pinned at point C and rotates with its actual angular velocity, $\dot{\theta} = \omega = v_O/r$. The direction of the point's velocity will be perpendicular to the line of length l between the instantaneous center and the point.

$$v = l\omega = \frac{lv_O}{r} \qquad 16.15$$

Figure 16.3 Instantaneous Center of a Rolling Wheel

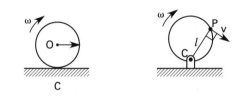

CENTRIPETAL AND CENTRIFUGAL FORCES

Newton's second law states that there is a force for every acceleration that a body experiences. For a body moving around a curved path, the total acceleration can be separated into tangential and normal components. By Newton's second law, there are corresponding forces in the tangential and normal directions. The force associated with the normal acceleration is known as the *centripetal force*. The centripetal force is a real force on the body toward the center of rotation. The so-called *centrifugal force* is an apparent force on the body directed away from the center of rotation. The centripetal and centrifugal forces are equal in magnitude but opposite in sign.

Equation 16.16 gives the centrifugal force on a body of mass m with distance r from the center of rotation to the center of mass.

$$F_c = ma_n = \frac{mv_t^2}{r} = mr\omega^2 \quad \text{[SI]} \qquad 16.16a$$

$$F_c = \frac{ma_n}{g_c} = \frac{mv_t^2}{g_c r} = \frac{mr\omega^2}{g_c} \quad \text{[U.S.]} \qquad 16.16b$$

BANKING OF CURVES

If a vehicle travels in a circular path on a flat plane with instantaneous radius r and tangential velocity v_t, it will experience an apparent centrifugal force. The centrifugal force is resisted by a combination of roadway banking (superelevation) and sideways friction. The vehicle weight, W, corresponds to the normal force. For small banking angles, the maximum frictional force is

$$F_f = \mu_s N = \mu_s W \qquad 16.17$$

For large banking angles, the centrifugal force contributes to the normal force. If the roadway is banked so that friction is not required to resist the centrifugal force, the superelevation angle, θ, can be calculated from Eq. 16.18.

$$\tan\theta = \frac{v_t^2}{gr} \qquad 16.18$$

TORSIONAL FREE VIBRATION

The *torsional pendulum* in Fig. 16.4 can be analyzed in a manner similar to the spring-mass combination. Disregarding the mass and moment of inertia of the shaft, the differential equation is

$$\ddot{\theta} + \omega_n^2\theta = 0 \qquad 16.19$$

Figure 16.4 Torsional Pendulum

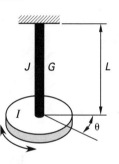

For the torsional pendulum, the torsional spring constant k_t can be written

$$k_t = \frac{GJ}{L} = \omega^2 I \qquad 16.20$$

The solution to Eq. 16.20 is directly analogous to the solution for the spring-mass system.

$$\theta(t) = \theta_0 \cos\omega_n t + \left(\frac{\omega_0}{\omega_n}\right)\sin\omega_n t \qquad 16.21$$

SAMPLE PROBLEMS

Problem 1

Why does a spinning ice skater's angular velocity increase as she brings her arms in toward her body?

    (A) Her mass moment of inertia is reduced.
    (B) Her angular momentum is constant.
    (C) Her radius of gyration is reduced.
    (D) all of the above

DYNAMICS
Rotation

Solution

As the skater brings her arms in, her radius of gyration and mass moment of inertia decrease. However, in the absence of friction, her angular momentum, h, is constant. From Eq. 16.9,

$$\omega = \frac{h}{I}$$

Since angular velocity, ω, is inversely proportional to the mass moment of inertia, the angular velocity increases when the mass moment of inertia decreases.

Answer is D.

Problem 2

Link AB of the linkage mechanism shown in the illustration rotates with an instantaneous counterclockwise angular velocity of 10 rad/s. What is the instantaneous angular velocity of link BC when link AB is horizontal and link CD is vertical?

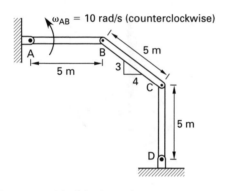

(A) 2.25 rad/s (clockwise)
(B) 3.25 rad/s (counterclockwise)
(C) 5.50 rad/s (clockwise)
(D) 12.5 rad/s (clockwise)

Solution

Find the instantaneous center of rotation. The absolute velocity directions at points B and C are known. The instantaneous center is located by drawing perpendiculars to these velocities as shown. The angular velocity of any point on rigid body link BC is the same at this instant.

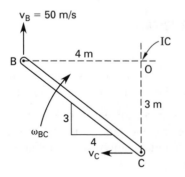

$$v_B = AB\omega_{AB}$$

$$= (5 \text{ m}) \left(10 \ \frac{\text{rad}}{\text{s}} \right)$$

$$= 50 \text{ m/s}$$

$$\omega_{BC} = \frac{v_B}{OB} = \frac{50 \ \dfrac{\text{m}}{\text{s}}}{4 \text{ m}}$$

$$= 12.5 \text{ rad/s} \qquad [\text{clockwise}]$$

Answer is D.

Problem 3

Two 2 kg blocks are linked as shown. Assuming that the surfaces are frictionless, what is the velocity of block B if block A is moving at a speed of 3 m/s?

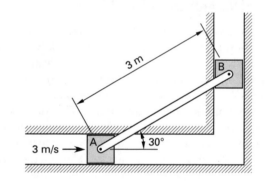

(A) 0 m/s
(B) 1.30 m/s
(C) 1.73 m/s
(D) 5.20 m/s

Solution

The instantaneous center of rotation for the slider rod assembly can be found by extending perpendiculars from the velocity vectors, as shown. Both blocks can be assumed to rotate about point C with angular velocity ω.

$$\omega = \frac{v_A}{CA} = \frac{v_B}{BC}$$

$$v_B = \frac{v_A BC}{CA}$$

$$= \frac{\left(3 \ \dfrac{\text{m}}{\text{s}} \right) (3 \text{ m}) \cos 30°}{(3 \text{ m}) \sin 30°}$$

$$= 5.20 \text{ m/s}$$

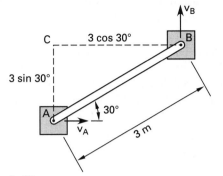

Answer is D.

Problem 4

An automobile travels on a perfectly horizontal, un-banked circular track of radius r. The coefficient of friction between the tires and the track is 0.3. If the car's velocity is 10 m/s, what is the smallest radius it may travel without skidding?

(A) 10 m
(B) 34 m
(C) 50 m
(D) 68 m

Solution

The automobile uses friction to resist the centrifugal force and stay on the curved track.

The centrifugal force is

$$F_c = \frac{mv^2}{r}$$

The frictional force is

$$F_f = \mu N = \mu mg$$
$$\frac{mv^2}{r} = \mu mg$$

$$r = \frac{v^2}{\mu g} = \frac{\left(10 \, \frac{m}{s}\right)^2}{(0.3)\left(9.81 \, \frac{m}{s^2}\right)}$$

$$= 34 \text{ m}$$

Answer is B.

Problem 5

If the car described in Prob. 4 moves along a track that is banked 5°, what is the smallest radius it can travel without skidding?

(A) 6 m
(B) 18 m
(C) 26 m
(D) 47 m

Solution

The car uses a combination of friction and supereleva-tion to resist the centrifugal force and stay on the curved track. Assume the banking angle is small.

$$F_c = F_f + F_\theta$$
$$F_c = \frac{mv^2}{r}$$
$$F_f = \mu N = \mu mg$$
$$F_\theta = mg \, \tan\theta$$
$$\frac{mv^2}{r} = \mu mg + mg \, \tan\theta$$

$$r = \frac{v^2}{g(\mu + \tan\theta)}$$

$$= \frac{\left(10 \, \frac{m}{s}\right)^2}{\left(9.81 \, \frac{m}{s^2}\right)(0.3 + \tan 5°)}$$

$$= 26 \text{ m}$$

(If the banking angle is assumed to be large enough to increase the normal force, the radius will be 25.6 m.)

Answer is C.

FE-STYLE EXAM PROBLEMS

1. A 2 kg mass swings in a vertical plane at the end of a 2 m cord. When $\theta = 30°$, the magnitude of the tangential velocity of the mass is 1 m/s. What is the tension in the cord at this position?

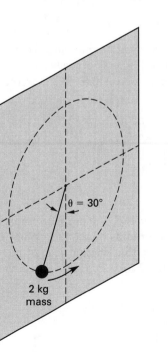

(A) 18.0 N
(B) 19.6 N
(C) 24.5 N
(D) 29.4 N

2. A 2 kg mass swings in the horizontal plane of a circle of radius 1.5 m and is held by a taut cord. The tension in the cord is 100 N. What is the angular momentum of the mass?

(A) 5.77 N·m·s
(B) 26.0 N·m·s
(C) 113 N·m·s
(D) 150 N·m·s

3. A disk rolls along a flat surface at a constant speed of 10 m/s. Its diameter is 0.5 m. At a particular instant, point P on the edge of the disk is 45° from the horizontal. What is the velocity of point P at that instant?

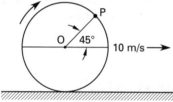

(A) 10.0 m/s
(B) 15.0 m/s
(C) 16.2 m/s
(D) 18.5 m/s

4. A car travels around an unbanked 50 m radius curve without skidding. The coefficient of friction between the tires and road is 0.3. What is the car's maximum speed?

(A) 14 km/h
(B) 25 km/h
(C) 44 km/h
(D) 54 km/h

5. Traffic travels at 100 km/h around a banked highway curve with a radius of 1000 m. What banking angle is necessary such that friction will not be required to resist the centrifugal force?

(A) 1.4°
(B) 2.8°
(C) 4.5°
(D) 46°

For the following problems use the NCEES Handbook as your only reference.

6. The center of gravity of a roller coaster car is 0.5 m above the rails. The rails are 1 m apart. What is the maximum speed that the car can travel around an unbanked curve of radius 15 m without the inner wheel losing contact with the top of the rail?

(A) 8.58 m/s
(B) 12.1 m/s
(C) 17.2 m/s
(D) 24.2 m/s

7. A 50 kg cylinder has a height of 3 m and a radius of 50 cm. The cylinder sits on the x-axis and is oriented with its major axis parallel to the y-axis. What is the mass moment of inertia about the x-axis?

(A) 4.1 kg·m^2
(B) 16 kg·m^2
(C) 41 kg·m^2
(D) 150 kg·m^2

8. A uniform thin disk has a radius of 30 cm and a mass of 2 kg. A constant force of 10 N is applied tangentially at a varying, but unknown, distance from the center of the disk. The disk accelerates about its axis at $3t$ rad/s^2. What is the distance from the center of the disk at which the force is applied at $t = 12$ s?

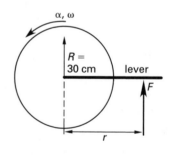

(A) 32.4 cm
(B) 36.0 cm
(C) 54.0 cm
(D) 108 cm

9. A torsional pendulum consists of a 5 kg uniform disk with a diameter of 50 cm attached at its center to a rod 1.5 m in length. The torsional spring constant is 0.625 N·m/rad. Disregarding the mass of the rod, what is the natural frequency of the torsional pendulum?

(A) 1.0 rad/s
(B) 1.2 rad/s
(C) 1.4 rad/s
(D) 2.0 rad/s

10. A 3 kg disk with a diameter of 0.6 m is rigidly attached at point B to a 1 kg rod 1 m in length. The rod-disk combination rotates around point A. What is the mass moment of inertia about point A for the combination?

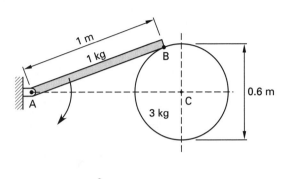

(A) 0.47 kg·m^2
(B) 0.56 kg·m^2
(C) 0.87 kg·m^2
(D) 3.7 kg·m^2

11. A 1 kg uniform rod 1 m long is suspended from the ceiling by a frictionless hinge. The rod is free to pivot. What is the product of inertia of the rod about the pivot point?

(A) 0 kg·m^2
(B) 0.045 kg·m^2
(C) 0.13 kg·m^2
(D) 0.33 kg·m^2

12. A wheel with a radius of 0.75 m starts from rest and accelerates clockwise. The angular acceleration (in rad/s^2) of the wheel is defined by $\alpha = 6t - 4$. What is the resultant linear acceleration of a point on the wheel rim at $t = 2$ s?

(A) 6 m/s^2
(B) 12 m/s^2
(C) 13 m/s^2
(D) 18 m/s^2

13. A uniform rod (AB) of length L and weight W is pinned at point C and restrained by cable OA. The cable is suddenly cut. The rod starts to rotate about point C, with point A moving down and point B moving up. What is the instantaneous linear acceleration of point B?

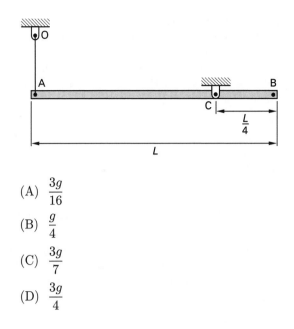

(A) $\dfrac{3g}{16}$

(B) $\dfrac{g}{4}$

(C) $\dfrac{3g}{7}$

(D) $\dfrac{3g}{4}$

14. A uniform rod (AB) of length L and weight W is pinned at point C. The rod starts from rest and accelerates with an angular acceleration (in rad/s^2) of $\alpha = 12g/7L$. What is the instantaneous reaction at point C at the moment rotation begins?

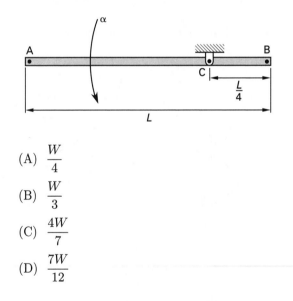

(A) $\dfrac{W}{4}$

(B) $\dfrac{W}{3}$

(C) $\dfrac{4W}{7}$

(D) $\dfrac{7W}{12}$

15. A 1530 kg car is towing a 300 kg trailer. The coefficient of friction between all tires and the road is 0.80. How fast can the car and trailer travel around an unbanked curve of radius 200 m without either the car or trailer skidding?

(A) 40.0 km/h
(B) 75.2 km/h
(C) 108.1 km/h
(D) 143 km/h

16. A 1530 kg car is towing a 300 kg trailer. The coefficient of friction between all tires and the road is 0.80. The car and trailer are traveling at 100 km/h around a banked curve of radius 200 m. What is the necessary banking angle such that tire friction will not be necessary to prevent skidding?

    (A) 8°
    (B) 21°
    (C) 36°
    (D) 78°

17. A wheel with a 0.75 m radius has a mass of 200 kg. The wheel is pinned at its center and has a radius of gyration of 0.25 m. A rope is wrapped around the wheel and supports a hanging 100 kg block. When the wheel is released, the rope begins to unwind. What is the angular acceleration of the wheel?

    (A) 5.9 rad/s²
    (B) 6.5 rad/s²
    (C) 11 rad/s²
    (D) 14 rad/s²

SOLUTIONS TO FE-STYLE EXAM PROBLEMS

1. Use tangential and normal components.

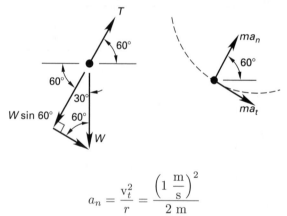

$$a_n = \frac{v_t^2}{r} = \frac{\left(1 \, \frac{\text{m}}{\text{s}}\right)^2}{2 \text{ m}}$$

$$= 0.5 \text{ m/s}^2$$

Sum forces in the normal direction.

$$\sum F_n = ma_n = T - W \sin 60°$$

$$T = ma_n + mg \sin 60°$$

$$= (2 \text{ kg})\left(0.5 \, \frac{\text{m}}{\text{s}^2}\right) + (2 \text{ kg})\left(9.81 \, \frac{\text{m}}{\text{s}^2}\right)\sin 60°$$

$$= 18.0 \text{ N}$$

Answer is A.

2.
$$\text{tension} = \text{centripetal force}$$

$$T = \frac{mv_t^2}{r} = \frac{m(r\omega)^2}{r}$$

$$= mr\omega^2$$

$$\omega = \sqrt{\frac{T}{mr}} = \sqrt{\frac{100 \text{ N}}{(2 \text{ kg})(1.5 \text{ m})}}$$

$$= 5.77 \text{ rad/s}$$

$$h_O = rmv = r^2 m\omega$$

$$= (1.5 \text{ m})^2 (2 \text{ kg})\left(5.77 \, \frac{\text{rad}}{\text{s}}\right)$$

$$= 26.0 \text{ N·m·s}$$

Answer is B.

3. Use the instantaneous center of rotation to solve this problem. Assume the wheel is pinned at point A.

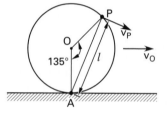

$$l^2 = (2)(0.25 \text{ m})^2 - (2)(0.25 \text{ m})^2 \cos 135°$$

                                    [law of cosines]

$$= 0.2134 \text{ m}^2$$

$$l = \sqrt{0.2134 \text{ m}^2} = 0.462 \text{ m}$$

$$v_P = \frac{lv_O}{r} = \frac{(0.462 \text{ m})\left(10 \, \frac{\text{m}}{\text{s}}\right)}{0.25 \text{ m}}$$

$$= 18.5 \text{ m/s}$$

The velocity of point P is perpendicular to the line AP.

Answer is D.

4. The car uses friction to resist the centrifugal force.

$$F_c = \frac{mv^2}{r}$$

$$F_f = \mu N = \mu mg$$

$$\frac{mv^2}{r} = \mu mg$$

$$v = \sqrt{\mu g r}$$

$$= \sqrt{(0.3)\left(9.81 \; \frac{m}{s^2}\right)(50 \; m)}$$

$$= 12.13 \; m/s$$

$$v = \frac{\left(12.13 \; \frac{m}{s}\right)\left(3600 \; \frac{s}{h}\right)}{1000 \; \frac{m}{km}}$$

$$= 43.67 \; km/h \quad (44 \; km/h)$$

Answer is C.

5. Since there is no friction force, the superelevation angle, θ, can be determined directly.

$$\tan\theta = \frac{v_t^2}{gr}$$

$$\theta = \tan^{-1}\left(\frac{v_t^2}{gr}\right)$$

$$= \tan^{-1} \frac{\left(\frac{\left(100 \; \frac{km}{h}\right)\left(1000 \; \frac{m}{km}\right)}{3600 \; \frac{s}{h}}\right)^2}{\left(9.81 \; \frac{m}{s^2}\right)(1000 \; m)}$$

$$= 4.50°$$

Answer is C.

6. The wheel will lose contact with the top of the rail when the reaction on the rail is zero. Refer to the following illustration. Wheel A is the inner wheel.

The forces acting on the car are the centrifugal force, F_c, its weight, and the reaction at the outer wheel. (The reaction at the inner wheel is zero.) Take moments about rail B.

$$\sum M_B = 0 = Wx_{CG} - F_c y_{CG}$$

$$= mgx_{CG} - \frac{mv_t^2 y_{CG}}{r}$$

$$v_t = \sqrt{\frac{gx_{CG}r}{y_{CG}}}$$

$$= \sqrt{\frac{\left(9.81 \; \frac{m}{s^2}\right)\left(\frac{1 \; m}{2}\right)(15 \; m)}{0.5 \; m}}$$

$$= 12.1 \; m/s$$

Answer is B.

7. Find the formula for I_x in the Mass Moments of Inertia table.

$$I_x = \frac{m(3R^2 + 4h^2)}{12}$$

$$= \frac{(50 \; kg)((3)(0.5 \; m)^2 + (4)(3 \; m)^2)}{12}$$

$$= 153.1 \; kg{\cdot}m^2 \quad (150 \; kg{\cdot}m^2)$$

Answer is D.

8. The centroidal mass moment of inertia is

$$I = \frac{1}{2}mR^2$$

$$= (0.5)(2 \; kg)(0.3 \; m)^2$$

$$= 0.09 \; kg{\cdot}m^2$$

The acceleration varies with time. The acceleration at $t = 12$ s is

$$\alpha = 3t = \left(3 \; \frac{rad}{s}\right)(12 \; s)$$

$$= 36 \; rad/s^2$$

$$M_0 = Fr = I\alpha$$

$$r = \frac{I\alpha}{F}$$

$$= \frac{(0.09 \; kg{\cdot}m^2)\left(36 \; \frac{rad}{s^2}\right)}{10 \; N}$$

$$= 0.324 \; m \quad (32.4 \; cm)$$

Answer is A.

9. The radius of the disk is

$$R = \frac{1}{2}D$$

$$= \frac{50 \; cm}{(2)\left(100 \; \frac{cm}{m}\right)}$$

$$= 0.25 \; m$$

Using a hollow cylinder with $R_{\text{inner}} = 0$, from Table 16.1, the mass moment of inertia of the disk is

$$I = \frac{1}{2}MR^2 = \left(\frac{1}{2}\right)(5 \text{ kg})(0.25 \text{ m})^2$$

$$= 0.15625 \text{ kg·m}^2$$

Natural frequency can be determined from the equation for the torsional spring constant.

$$k_t = \omega^2 I$$

$$\omega = \sqrt{\frac{k_t}{I}} = \sqrt{\frac{0.625 \dfrac{\text{N·m}}{\text{rad}}}{0.15625 \text{ kg·m}^2}}$$

$$= 2 \text{ rad/s}$$

Answer is D.

10. The mass moment of inertia of the rod about its end is

$$I_{\text{rod,A}} = \frac{ML^2}{3} = \frac{(1 \text{ kg})(1 \text{ m})^2}{3}$$

$$= 0.33 \text{ kg·m}^2$$

The mass moment of inertia of the disk about its own center is

$$I_{\text{disk,C}} = \frac{MR^2}{2} = \frac{(3 \text{ kg})\left(\dfrac{0.6 \text{ m}}{2}\right)^2}{2}$$

$$= 0.135 \text{ kg·m}^2$$

The distance AC is

$$\text{AC} = \sqrt{\text{AB}^2 + \text{BC}^2}$$

$$= \sqrt{(1 \text{ m})^2 + \left(\frac{0.6 \text{ m}}{2}\right)^2}$$

$$= 1.04 \text{ m}$$

Using the parallel axis theorem, the mass moment of inertia of the disk about point A is

$$I_{\text{disk,A}} = I_{\text{disk,C}} + m_{\text{disk}}\text{AC}^2$$

$$= 0.135 \text{ kg·m}^2 + (3 \text{ kg})(1.04 \text{ m})^2$$

$$= 3.38 \text{ kg·m}^2$$

The total moment of inertia of the rod and disk is

$$I_A = I_{\text{rod,A}} + I_{\text{disk,A}}$$

$$= 0.33 \text{ kg·m}^2 + 3.38 \text{ kg·m}^2$$

$$= 3.71 \text{ kg·m}^2 \quad (3.7 \text{ kg·m}^2)$$

Answer is D.

11. The product of inertia for the rod is zero because the pivot point lies on an axis of symmetry.

Answer is A.

12. The angular acceleration at $t = 2$ s is

$$\alpha(2 \text{ s}) = (6)(2 \text{ s}) - 4$$

$$= 8 \text{ rad/s}^2$$

The equation for the angular velocity is

$$\omega = \int \alpha(t)dt = \int (6t - 4)dt$$

$$= 3t^2 - 4t + \omega_0$$

However, the wheel starts from rest, so $\omega_0 = 0$.

At $t = 2$ s, the angular velocity is

$$\omega(2 \text{ s}) = (3)(2 \text{ s})^2 - (4)(2 \text{ s})$$

$$= 4 \text{ rad/s}$$

The tangential acceleration of the point is

$$a_t = r\alpha = (0.75 \text{ m})\left(8 \frac{\text{rad}}{\text{s}^2}\right)$$

$$= 6 \text{ m/s}^2$$

The normal acceleration (directed toward the center of the wheel) is

$$a_n = r\omega^2$$

$$= (0.75 \text{ m})\left(4 \frac{\text{rad}}{\text{s}}\right)^2$$

$$= 12 \text{ m/s}^2$$

The resultant acceleration is

$$a = \sqrt{a_t^2 + a_n^2} = \sqrt{\left(6 \frac{\text{m}}{\text{s}^2}\right)^2 + \left(12 \frac{\text{m}}{\text{s}^2}\right)^2}$$

$$= 13.4 \text{ m/s}^2 \quad (13 \text{ m/s}^2)$$

Answer is C.

13. Point C is $L/4$ from the center of gravity of the rod. The mass moment of inertia about point C is

$$I_C = I_{\text{CG}} + md^2 = \left(\frac{1}{12}\right)mL^2 + m\left(\frac{L}{4}\right)^2$$

$$= mL^2\left(\frac{1}{12} + \frac{1}{16}\right)$$

$$= \left(\frac{7}{48}\right)mL^2$$

The sum of moments on the rod is

$$\sum M_C = \sum Fr$$

$$= \left(\frac{3W}{4}\right)\left(\frac{\frac{3L}{4}}{2}\right) - \left(\frac{W}{4}\right)\left(\frac{\frac{L}{4}}{2}\right)$$

$$= \frac{WL}{4}$$

$$= \frac{mgL}{4}$$

The angular acceleration is

$$\alpha = \frac{\sum M_C}{I_C} = \frac{\dfrac{mgL}{4}}{\left(\dfrac{7}{48}\right)mL^2}$$

$$= \frac{12g}{7L}$$

The tangential acceleration of point B is

$$a_{t,B} = r\alpha = \left(\frac{L}{4}\right)\left(\frac{12g}{7L}\right)$$

$$= \frac{3g}{7}$$

Answer is C.

14. The mass moment of inertia of the rod about its center of gravity is

$$I_{CG} = \left(\frac{1}{12}\right)mL^2 = \left(\frac{1}{12}\right)\left(\frac{W}{g}\right)L^2$$

Take moments about the center of gravity of the rod. All moments due to gravitational forces will cancel. The only unbalanced force acting on the rod will be the vertical reaction force, R_C, at point C.

$$\sum M_{CG} = R_C\left(\frac{L}{4}\right)$$

$$\sum M_{CG} = I_{CG}\alpha_{CG}$$

$$R_C\left(\frac{L}{4}\right) = \left(\frac{1}{12}\right)\left(\frac{W}{g}\right)L^2\left(\frac{12g}{7L}\right)$$

$$R_C = \frac{4W}{7}$$

The angular velocity is zero, so the center of the mass does not have a component of acceleration in the horizontal direction. There is no horizontal force component at point C.

Answer is C.

15. To keep the vehicle and trailer from skidding, the centripetal force must be less than or equal to the frictional force. At the limit,

$$F_c = F_f$$

$$ma_n = \mu N$$

$$a_n = \frac{\mu N}{m} = \frac{\mu mg}{m}$$

$$= \mu g$$

$$= (0.8)\left(9.81\ \frac{\text{m}}{\text{s}^2}\right)$$

$$= 7.848\ \text{m/s}^2$$

The normal acceleration can be calculated from the tangential velocity.

$$a_n = \frac{v_t^2}{r}$$

$$v_t = \sqrt{a_n r} = \sqrt{\left(7.848\ \frac{\text{m}}{\text{s}^2}\right)(200\ \text{m})}$$

$$= 39.6\ \text{m/s}$$

$$= \frac{\left(39.6\ \dfrac{\text{m}}{\text{s}}\right)\left(3600\ \dfrac{\text{s}}{\text{h}}\right)}{1000\ \dfrac{\text{m}}{\text{km}}}$$

$$= 143\ \text{km/h}$$

Answer is D.

16. The velocity is

$$v = \frac{\left(100\ \dfrac{\text{km}}{\text{h}}\right)\left(1000\ \dfrac{\text{m}}{\text{km}}\right)}{3600\ \dfrac{\text{s}}{\text{h}}}$$

$$= 27.78\ \text{m/s}$$

The necessary superelevation angle without relying on friction (equivalent to setting $\mu = 0$) is

$$\theta = \tan^{-1}\left(\frac{v^2}{gr}\right)$$

$$= \tan^{-1}\left(\frac{\left(27.78\ \dfrac{\text{m}}{\text{s}}\right)^2}{\left(9.81\ \dfrac{\text{m}}{\text{s}^2}\right)(200\ \text{m})}\right)$$

$$= 21.47°\quad(21°)$$

Answer is B.

17. The mass moment of inertia of the wheel is

$$I = m_{\text{wheel}}r^2$$

The unbalanced torque (moment) on the wheel is

$$M = FR = (mg - ma)R = mR(g - a)$$
$$= m_{block}R(g - R\alpha)$$

The acceleration is given by

$$M = I\alpha$$
$$m_{block}R(g - R\alpha) = m_{wheel}r^2\alpha$$

$$\alpha = \frac{m_{block}Rg}{m_{wheel}r^2 + m_{block}R^2}$$

$$= \frac{(100\ kg)(0.75\ m)\left(9.81\ \dfrac{m}{s^2}\right)}{(200\ kg)(0.25\ m)^2 + (100\ kg)(0.75\ m)^2}$$

$$= 10.7\ rad/s^2 \quad (11\ rad/s^2)$$

Answer is C.

Table 16.1 *Mass Moments of Inertia*

figure	mass and centroid	mass moment of inertia	(radius of gyration)^2	product of inertia
slender rod	$M = \rho LA$ $x_c = L/2$ $y_c = 0$ $z_c = 0$	$I_x = I_{x_c} = 0$ $I_{y_c} = I_{z_c} = ML^2/12$ $I_y = I_z = ML^2/3$	$r_x^2 = r_{x_c}^2 = 0$ $r_{y_c}^2 = r_{z_c}^2 = L^2/12$ $r_y^2 = r_z^2 = L^2/3$	$I_{x_c y_c}$, etc. $= 0$ I_{xy}, etc. $= 0$
slender ring	$M = 2\pi\rho RA$ $x_c = R$ $y_c = R$ $z_c = 0$	$I_{x_c} = I_{y_c} = MR^2/2$ $I_{z_c} = MR^2$ $I_x = I_y = 3MR^2/2$ $I_z = 3MR^2$	$r_{x_c}^2 = r_{y_c}^2 = R^2/2$ $r_{z_c}^2 = R^2$ $r_x^2 = r_y^2 = 3R^2/2$ $r_z^2 = 3R^2$	$I_{x_c y_c}$, etc. $= 0$ $I_{xy} = MR^2$ $I_{xz} = I_{yz} = 0$
cylinder	$M = \pi\rho R^2 h$ $x_c = 0$ $y_c = h/2$ $z_c = 0$	$I_{x_c} = I_{z_c} = M(3R^2 + h^2)/12$ $I_{y_c} = I_y = MR^2/2$ $I_x = I_z = M(3R^2 + 4h^2)/12$	$r_{x_c}^2 = r_{z_c}^2 = (3R^2 + h^2)/12$ $r_{y_c}^2 = r_y^2 = R^2/2$ $r_x^2 = r_z^2 = (3R^2 + 4h^2)/12$	$I_{x_c y_c}$, etc. $= 0$ I_{xy}, etc. $= 0$
hollow cylinder	$M = \pi\rho h\,(R_1^2 - R_2^2)$ $x_c = 0$ $y_c = h/2$ $z_c = 0$	$I_{x_c} = I_{z_c}$ $\quad = M(3R_1^2 + 3R_2^2 + h^2)/12$ $I_{y_c} = I_y = M(R_1^2 + R_2^2)/2$ $I_x = I_z$ $\quad = M(3R_1^2 + 3R_2^2 + 4h^2)/12$	$r_{x_c}^2 = r_{z_c}^2$ $\quad = (3R_1^2 + 3R_2^2 + h^2)/12$ $r_{y_c}^2 = r_y^2 = (R_1^2 + R_2^2)/2$ $r_x^2 = r_z^2$ $\quad = (3R_1^2 + 3R_2^2 + 4h^2)/12$	$I_{x_c y_c}$, etc. $= 0$ I_{xy}, etc. $= 0$
sphere	$M = 4\pi\rho R^3/3$ $x_c = 0$ $y_c = 0$ $z_c = 0$	$I_{x_c} = I_x = 2MR^2/5$ $I_{y_c} = I_y = 2MR^2/5$ $I_{z_c} = I_z = 2MR^2/5$	$r_{x_c}^2 = r_x^2 = 2R^2/5$ $r_{y_c}^2 = r_y^2 = 2R^2/5$ $r_{z_c}^2 = r_z^2 = 2R^2/5$	$I_{x_c y_c}$, etc. $= 0$

17 Energy and Work

Subjects

ENERGY AND WORK 17-1
    Kinetic Energy 17-1
    Potential Energy 17-2
    Elastic Potential Energy 17-2
ENERGY CONSERVATION PRINCIPLE . 17-2
LINEAR IMPULSE 17-2
IMPACTS 17-2

Nomenclature

a	acceleration	ft/sec^2	m/s^2
e	coefficient of restitution	–	–
E	energy	ft-lbf	J
F	force	lbf	N
g	gravitational acceleration	ft/sec^2	m/s^2
g_c	gravitational constant (32.2)	lbm-ft/lbf-sec^2	–
h	height above datum	ft	m
I	mass moment of inertia	lbm-ft^2	kg·m^2
Imp	impulse	lbf-sec	N·s
k	spring constant	lbf/ft	N/m
m	mass	lbm	kg
p	linear momentum	lbm-ft/sec	kg·m/s
r	distance	ft	m
t	time	sec	s
v	velocity	ft/sec	m/s
W	work	ft-lbf	J
x	displacement	ft	m

Symbols

ω	angular velocity	rad/sec	rad/s

ENERGY AND WORK

The *energy* of a mass represents the capacity of the mass to do work. Such energy can be stored and released. There are many forms that the stored energy can take, including mechanical, thermal, electrical, and magnetic energies. Energy is a positive, scalar quantity, although the change in energy can be either positive or negative.

Work, W, is the act of changing the energy of a mass. Work is a signed, scalar quantity. Work is positive when a force acts in the direction of motion and moves a mass from one location to another. Work is negative when a force acts to oppose motion. (Friction, for example, always opposes the direction of motion and can only do negative work.) The net work done on a mass by more than one force can be found by superposition.

The work performed by a force is calculated as a dot product of the force acting through a displacement.

$$W = \int \mathbf{F} \cdot d\mathbf{r} \qquad 17.1$$

Kinetic Energy

Kinetic energy is a form of mechanical energy associated with a moving or rotating body. The *linear kinetic energy* of a body moving with instantaneous linear velocity v is

$$\mathrm{KE} = \frac{1}{2}m\mathrm{v}^2 \qquad \text{[SI]} \qquad 17.2a$$

$$\mathrm{KE} = \frac{m\mathrm{v}^2}{2g_c} \qquad \text{[U.S.]} \qquad 17.2b$$

The *rotational kinetic energy* of a body moving with instantaneous angular velocity ω is

$$\mathrm{KE} = \frac{1}{2}I\omega^2 \qquad \text{[SI]} \qquad 17.3a$$

$$\mathrm{KE} = \frac{I\omega^2}{2g_c} \qquad \text{[U.S.]} \qquad 17.3b$$

For general plane motion in which there are translational and rotational components, the kinetic energy is the sum of the translational and rotational forms.

The change in kinetic energy is calculated from the difference of squares of velocity, not from the square of the velocity difference.

$$\Delta\mathrm{KE} = \frac{1}{2}m(\mathrm{v}_2^2 - \mathrm{v}_1^2) \neq \frac{1}{2}m(\mathrm{v}_2 - \mathrm{v}_1)^2 \qquad 17.4a$$

$$\Delta\mathrm{KE} = \frac{m(\mathrm{v}_2^2 - \mathrm{v}_1^2)}{2g_c} \neq \frac{m(\mathrm{v}_2 - \mathrm{v}_1)^2}{2g_c} \qquad 17.4b$$

DYNAMICS Energy

Potential Energy

Potential energy (also known as *gravitational potential energy*) is a form of mechanical energy possessed by a mass due to its relative position in a gravitational field. Potential energy is lost when the elevation of a mass decreases. The lost potential energy usually is converted to kinetic energy or heat.

$$\text{PE} = mgh \qquad \text{[SI]} \qquad 17.5a$$

$$\text{PE} = \frac{mgh}{g_c} \qquad \text{[U.S.]} \qquad 17.5b$$

Elastic Potential Energy

A spring is an energy storage device because a compressed spring has the ability to perform work. In a perfect spring, the amount of energy stored is equal to the work required to compress the spring initially. The stored spring energy does not depend on the mass of the spring. Given a spring with *spring constant (stiffness)* k, the spring's *elastic potential energy* is

$$\text{PE} = \frac{1}{2}kx^2 \qquad 17.6$$

ENERGY CONSERVATION PRINCIPLE

According to the *energy conservation principle*, energy cannot be created or destroyed. However, energy can be transformed into different forms. Therefore, the sum of all energy forms of a system is constant.

$$\Sigma E = \text{constant} \qquad 17.7$$

For many problems, the total energy of the mass is equal to the sum of the potential (gravitational and elastic) and kinetic energies.

Because energy can neither be created nor destroyed, external work performed on a conservative system must go into changing the system's total energy. This is known as the *work-energy principle*.

$$W = E_2 - E_1 \qquad 17.8$$

Generally, the principle of conservation of energy is applied to mechanical energy problems (i.e., conversion of work into kinetic or potential energy).

Conversion of one form of energy into another does not violate the conservation of energy law. Most problems involving conversion of energy are really special cases. For example, consider a falling body that is acted upon by a gravitational force. The conversion of potential energy into kinetic energy can be interpreted as equating the work done by the constant gravitational force to the change in kinetic energy.

LINEAR IMPULSE

Impulse is a vector quantity equal to the change in momentum. Units of linear impulse are the same as those for linear momentum: lbf-sec or N·s. Figure 17.1 illustrates that impulse is represented by the area under the force-time curve.

$$\mathbf{Imp} = \int_{t_1}^{t_2} \mathbf{F}\,dt \qquad 17.9$$

Figure 17.1 Impulse

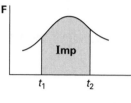

If the applied force is constant, impulse is easily calculated.

$$\mathbf{Imp} = \mathbf{F}(t_2 - t_1) \qquad 17.10$$

The change in momentum is equal to the impulse. This is known as the *impulse-momentum principle*. For a linear system with constant force and mass,

$$\mathbf{Imp} = \Delta\mathbf{p} \qquad 17.11$$

Rewriting this equation for a constant force and mass moving in any direction demonstrates that the impulse-momentum principle follows directly from Newton's second law.

$$\mathbf{F}(t_2 - t_1) = \Delta(m\mathbf{v}) \qquad \text{[SI]} \qquad 17.12$$

$$F\,dt = m\,dv \qquad \text{[SI]} \qquad 17.13$$

$$F = \frac{m\,dv}{dt} = ma \qquad \text{[SI]} \qquad 17.14$$

IMPACTS

According to Newton's second law, momentum is conserved unless a body is acted upon by an external force such as gravity or friction. In an impact or collision contact is very brief, and the effect of external forces is insignificant. Therefore, momentum is conserved, even though energy may be lost through heat generation and deforming the bodies.

Consider two particles, initially moving with velocities v_1 and v_2 on a collision path, as shown in Fig. 17.2. The conservation of momentum equation can be used to find the velocities after impact, v'_1 and v'_2.

$$m_1 v_1 + m_2 v_2 = m_1 v'_1 + m_2 v'_2 \qquad \text{[always true]} \quad 17.15$$

Figure 17.2 Direct Central Impact

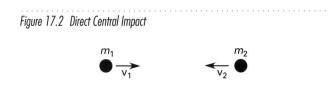

The impact is said to be an *inelastic impact* if kinetic energy is lost. The impact is said to be *perfectly inelastic* or *perfectly plastic* if the two particles stick together and move on with the same final velocity. The impact is said to be an *elastic impact* only if kinetic energy is conserved.

$$m_1 v_1^2 + m_2 v_2^2 = m_1 v_1'^2 + m_2 v_2'^2 \quad \text{[elastic only]} \quad 17.16$$

A simple way of determining whether the impact is elastic or inelastic is by calculating the *coefficient of restitution, e.* The coefficient of restitution is the ratio of relative velocity differences along a mutual straight line. The collision is inelastic if $e < 1.0$, perfectly inelastic if $e = 0$, and elastic if $e = 1.0$. (When both impact velocities are not directed along the same straight line, the coefficient of restitution should be calculated separately for each velocity component.)

$$e = \frac{\text{relative separation velocity}}{\text{relative approach velocity}}$$
$$= \frac{v_1' - v_2'}{v_2 - v_1} \qquad 17.17$$

SAMPLE PROBLEMS

Problem 1
The first derivative of kinetic energy with respect to time is

    (A) force.
    (B) momentum.
    (C) work.
    (D) power.

Solution

$$\text{KE} = \frac{1}{2} m v^2$$
$$\frac{d(\text{KE})}{dt} = (2)\left(\frac{1}{2}\right)(mv)\frac{dv}{dt}$$
$$= mva$$

This combination of variables ($\text{kg·m}^2/\text{s}^3$) corresponds to a watt (i.e., power).

Answer is D.

Problems 2 and 3 refer to the following illustration. A 50 kg block is released down a curved, frictionless surface. The radius of the curve is 5 m.

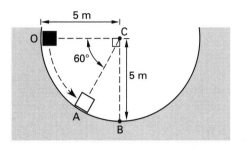

Problem 2
What is the tangential velocity of the block at point A?

    (A) 6.52 m/s
    (B) 9.22 m/s
    (C) 42.5 m/s
    (D) 85.0 m/s

Solution
The vertical distance between points O and A is

$$h = (5 \text{ m}) \sin 60°$$
$$= 4.33 \text{ m}$$

This drop in elevation corresponds to a decrease in potential energy and an increase in kinetic energy.

$$\Delta\text{PE} = \Delta\text{KE}$$
$$mg\Delta h = \frac{mv^2}{2}$$
$$v = \sqrt{2g\Delta h}$$
$$= \sqrt{(2)\left(9.81 \frac{\text{m}}{\text{s}^2}\right)(4.33 \text{ m})}$$
$$= 9.22 \text{ m/s}$$

Answer is B.

Problem 3
What is the instantaneous acceleration in the direction of travel of the block at point B?

    (A) 0 m/s^2
    (B) 2.45 m/s^2
    (C) 4.91 m/s^2
    (D) 19.6 m/s^2

Solution
At point B, all of the energy of the mass is kinetic and the velocity is maximum. Acceleration is the rate of

change of velocity. Since the velocity is maximum at point B, the acceleration is zero. There is centripetal acceleration directed away from point C, but this is not in the direction of travel.

Answer is A.

Problem 4

A 1 kg disk with a diameter of 10 cm and a width of 4 cm is placed on edge at the top of an inclined ramp 1 m high. The ramp is inclined at 10°. At the bottom of the ramp is a spring whose spring constant is 2000 N/m. The disk rolls down the ramp and compresses the spring while coming to a complete stop. What is the maximum compression of the spring?

    (A) 9.9 cm
    (B) 11.4 cm
    (C) 11.7 cm
    (D) 14.1 cm

Solution

At the top of the ramp, all of the energy is gravitational potential energy; at the bottom, the energy is spring potential energy. Neglecting the small toss of gravitational potential energy in deflecting the spring, the energy balance is

$$mgh = \frac{1}{2}kx^2$$

$$x = \sqrt{\frac{2mgh}{k}} = \sqrt{\frac{(2)(1 \text{ kg})\left(9.81 \frac{\text{m}}{\text{s}^2}\right)(1 \text{ m})}{2000 \frac{\text{N}}{\text{m}}}}$$

$$= 0.099 \text{ m} \quad (9.9 \text{ cm})$$

Answer is A.

Problems 5 and 6 refer to the following situation.

Two balls, both of mass 2 kg, collide head on. The velocity of each ball at the time of the collision is 2 m/s. The coefficient of restitution is 0.5.

Problem 5

What are the final velocities of the balls?

    (A) 1 m/s and −1 m/s
    (B) 2 m/s and −2 m/s
    (C) 3 m/s and −3 m/s
    (D) 4 m/s and −4 m/s

Solution

From the definition of coefficient of restitution,

$$e = \frac{v_1' - v_2'}{v_2 - v_1}$$

$$v_1' - v_2' = e(v_2 - v_1)$$

$$= (0.5)\left(-2 \frac{\text{m}}{\text{s}} - 2 \frac{\text{m}}{\text{s}}\right)$$

$$= -2 \text{ m/s} \qquad [\text{I}]$$

From the conservation of momentum,

$$m_1 v_1 + m_2 v_2 = m_1 v_1' + m_2 v_2'$$

But, $m_1 = m_2$.

$$v_1 + v_2 = v_1' + v_2'$$

Since $v_1 = 2$ m/s and $v_2 = -2$ m/s,

$$v_1 + v_2 = 2 \frac{\text{m}}{\text{s}} + \left(-2 \frac{\text{m}}{\text{s}}\right) = 0$$

So,

$$v_1' + v_2' = 0 \qquad [\text{II}]$$

Solve Eqs. I and II simultaneously by adding them.

$$v_1' = -1 \text{ m/s}$$

$$v_2' = 1 \text{ m/s}$$

Answer is A.

Problem 6

What is the loss of energy in the collision?

    (A) 1.4 N·m
    (B) 2.3 N·m
    (C) 6.0 N·m
    (D) 8.6 N·m

Solution

Each ball possesses kinetic energy before and after the collision. The velocity of each ball is reduced from $|2$ m/s$|$ to $|1$ m/s$|$.

$$\Delta KE = (KE)_i - (KE)_f$$

$$= (2)\left(\frac{mv_i^2}{2} - \frac{mv_f^2}{2}\right)$$

$$= (2)\left(\frac{(2 \text{ kg})\left(2 \frac{\text{m}}{\text{s}}\right)^2}{2} - \frac{(2 \text{ kg})\left(1 \frac{\text{m}}{\text{s}}\right)^2}{2}\right)$$

$$= 6 \text{ N·m}$$

Answer is C.

Problem 7

A 2 kg ball of clay moving at 40 m/s collides with a 5 kg ball of clay moving at 10 m/s directly toward the first ball. What is the final velocity if both balls stick together after the collision?

(A) 4.29 m/s
(B) 23.0 m/s
(C) 30.0 m/s
(D) 42.9 m/s

Solution

Since the balls stick together, $v_1' = v_2'$ and $e = 0$. Thus, the collision is perfectly inelastic. Only momentum is conserved.

$$m_1 v_1 + m_2 v_2 = mv'$$

$$(2 \text{ kg})\left(40 \ \frac{\text{m}}{\text{s}}\right) + (5 \text{ kg})\left(-10 \ \frac{\text{m}}{\text{s}}\right) = (2 \text{ kg} + 5 \text{ kg})v'$$

$$v' = 4.29 \text{ m/s}$$

Answer is A.

FE-STYLE EXAM PROBLEMS

Problems 1–4 refer to the following situation.

- The mass m in the following illustration is guided by the frictionless rail and has a mass of 40 kg.
- The spring constant, k, is 3000 N/m.
- The spring is compressed sufficiently and released, such that the mass barely reaches point B.

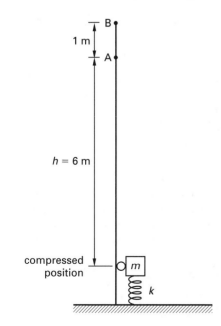

1. What is the initial spring compression?

(A) 0.96 m
(B) 1.3 m
(C) 1.4 m
(D) 1.8 m

2. What is the kinetic energy of the mass at point A?

(A) 19.8 J
(B) 219 J
(C) 392 J
(D) 2350 J

3. What is the velocity of the mass at point A?

(A) 3.13 m/s
(B) 4.43 m/s
(C) 9.80 m/s
(D) 19.6 m/s

4. What is the energy stored in the spring if the spring is compressed 0.5 m?

(A) 375 J
(B) 750 J
(C) 1500 J
(D) 2100 J

5. A hockey puck traveling at 30 km/h hits a massive wall at an angle of 30° from the wall. What are its final velocity and deflection angle if the coefficient of restitution is 0.63?

(A) 9.5 km/h at 30°
(B) 19 km/h at 30°
(C) 28 km/h at 20°
(D) 30 km/h at 20°

For the following problems use the NCEES Handbook as your only reference.

6. The *impulse and momentum principle* is mostly useful for solving problems involving

(A) force, velocity, and time.
(B) force, acceleration, and time.
(C) velocity, acceleration, and time.
(D) force, velocity, and acceleration.

7. A 12 kg aluminum box is dropped from rest onto a large wooden beam. The box travels 20 cm before contacting the beam. After impact, the box bounces 5 cm above the beam's surface. What impulse does the beam impart on the box?

(A) 8.6 N·s
(B) 12 N·s
(C) 36 N·s
(D) 42 N·s

Problems 8–10 refer to the following diagram and information.

A uniform rod is 1 m long and has a mass of 10 kg. It is pinned at point A, a frictionless pivot. The rod is released from a rest position 45° from the horizontal. The top of an undeflected ideal spring is located at point B to contact the tip of the rod when the rod is horizontal. The spring has a constant of 98 kN/m.

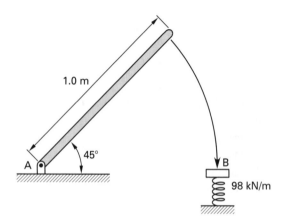

8. What is the velocity of the rod's tip when the rod just becomes horizontal?

(A) 2.6 m/s
(B) 4.6 m/s
(C) 6.9 m/s
(D) 8.9 m/s

9. How far up will the tip of the rod bounce after striking the spring at B?

(A) 35 cm
(B) 50 cm
(C) 71 cm
(D) 100 cm

10. Assume the rod is dropped from an unknown angle, and the angular velocity of the rod is 10 rad/s when the rod hits the spring. What will be the maximum deflection of the spring?

(A) 0.2 cm
(B) 1.8 cm
(C) 3.1 cm
(D) 5.9 cm

11. A pickup truck is traveling forward at 25 m/s. The bed is loaded with boxes whose coefficient of friction with the bed is 0.40. What is the shortest time that the truck can be brought to a stop such that the boxes do not shift?

(A) 2.3 s
(B) 4.7 s
(C) 5.9 s
(D) 6.4 s

Problems 12 and 13 refer to the following situation.

A 1500 kg car traveling at 100 km/h is towing a 250 kg trailer. The coefficient of friction between the tires and the road is 0.8 for both the car and trailer.

12. What energy is dissipated by the brakes if the car and trailer are braked to a complete stop?

(A) 96 kJ
(B) 385 kJ
(C) 579 kJ
(D) 675 kJ

13. What is the horizontal force on the trailer hitch when the car accelerates from 60 km/h to 100 km/h at a rate of 2.5 m/s^2?

(A) 156 N
(B) 625 N
(C) 3380 N
(D) 4380 N

14. A 60 000 kg railcar moving at 1 km/h is instantaneously coupled to a stationary 40 000 kg railcar. What is the speed of the coupled cars?

(A) 0.40 km/h
(B) 0.60 km/h
(C) 0.88 km/h
(D) 1.0 km/h

15. A 60 000 kg railcar moving at 1 km/h is coupled to a second, stationary railcar. If the velocity of the two cars after coupling is 0.2 m/s (in the original direction of motion) and the coupling is completed in 0.5 s, what is the average impulsive force on the 60 000 kg railcar?

(A) 520 N
(B) 990 N
(C) 3100 N
(D) 9300 N

Problems 16 and 17 refer to the following situation.

A 3500 kg car traveling at 65 km/h skids and hits a wall 3 s later. The coefficient of friction between the tires and the road is 0.60.

16. What is the speed of the car when it hits the wall?

- (A) 0.14 m/s
- (B) 0.40 m/s
- (C) 5.1 m/s
- (D) 6.2 m/s

17. Assuming that the speed of the car when it hits the wall is 0.20 m/s, what energy must the bumper absorb in order to prevent damage to the car?

- (A) 70 J
- (B) 140 J
- (C) 220 J
- (D) 360 kJ

SOLUTIONS TO FE-STYLE EXAM PROBLEMS

1. At the point just before the spring is released, all of the energy in the system is elastic potential energy; while at point B, all of the energy is potential energy due to gravity.

$$\frac{1}{2}kx^2 = mgh$$

$$x = \sqrt{\frac{2mgh}{k}}$$

$$= \sqrt{\frac{(2)(40 \text{ kg})\left(9.81 \frac{m}{s^2}\right)(6 \text{ m} + 1 \text{ m})}{3000 \frac{N}{m}}}$$

$$= 1.35 \text{ m}\quad(1.4 \text{ m})$$

Answer is C.

2. At point A, the energy of the mass is a combination of kinetic and gravitational potential energies. The total energy of the system is constant, and the kinetic energy at B is 0.

$$E_A = E_B$$

$$(PE)_A + (KE)_A = (PE)_B$$

$$mgh + \frac{mv^2}{2} = mg(h + 1 \text{ m})$$

$$(KE)_A = mg(h + 1 \text{ m}) - mgh$$

$$= mg(1 \text{ m})$$

$$= (40 \text{ kg})\left(9.81 \frac{m}{s^2}\right)(1 \text{ m})$$

$$= 392 \text{ J}$$

Answer is C.

3. From Prob. 2,

$$(KE)_A = \frac{mv^2}{2} = 392 \text{ J}$$

$$v = \sqrt{\frac{(2)(KE)_A}{m}}$$

$$= \sqrt{\frac{(2)(392 \text{ J})}{40 \text{ kg}}}$$

$$= 4.43 \text{ m/s}$$

Answer is B.

4.

$$PE = \frac{1}{2}kx^2$$

$$= \left(\frac{1}{2}\right)\left(3000 \frac{N}{m}\right)(0.5 \text{ m})^2$$

$$= 375 \text{ J}$$

Answer is A.

5. For the case of an object rebounding from a massive, stationary plane, only the object's velocity component normal to the plane is changed. This is an impact where $m_2 = \infty$ and $v_2 = 0$. For the component normal to the plane,

$$e = \frac{v'_{1y} - v'_{2y}}{v_{2y} - v_{1y}} = \frac{v'_{1y}}{-v_{1y}}$$

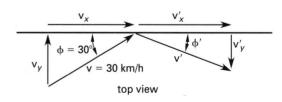

top view

The new normal component of velocity is

$$v'_y = ev_y = (0.63)\left(30 \frac{km}{h}\right)\sin 30°$$

$$= 9.45 \text{ km/h}$$

The parallel component of velocity is unchanged.

$$v'_x = v_x = \left(30 \ \frac{\text{km}}{\text{h}}\right) \cos 30°$$

$$= 25.98 \text{ km/h}$$

The resultant velocity is

$$v' = \sqrt{v'^2_y + v'^2_x}$$

$$= \sqrt{\left(9.45 \ \frac{\text{km}}{\text{h}}\right)^2 + \left(25.98 \ \frac{\text{km}}{\text{h}}\right)^2}$$

$$= 27.65 \text{ km/h}   (28 \text{ km})$$

The deflection angle is

$$\phi' = \arctan\left(\frac{v'_y}{v'_x}\right)$$

$$= \arctan\left(\frac{9.45 \ \frac{\text{km}}{\text{h}}}{25.98 \ \frac{\text{km}}{\text{h}}}\right)$$

$$= 19.99°   (20°)$$

Answer is C.

6. Impulse is calculated from force and time. Momentum is calculated from mass and velocity. Therefore, the *impulse and momentum principle* is useful in solving problems involving force, time, velocity, and mass.

Answer is A.

7. Initially, the box has potential energy only. (This takes the beam's upper surface as the reference plane.) When the box reaches the beam, all of the potential energy will have been converted to kinetic energy.

$$mgh_1 = \frac{1}{2}mv_1^2$$

$$v_1 = \sqrt{2gh_1}$$

$$= \sqrt{(2)\left(9.81 \ \frac{\text{m}}{\text{s}^2}\right)(0.2 \text{ m})}$$

$$= 1.98 \text{ m/s}   [\text{downward}]$$

When the box rebounds to its highest point, all of its remaining energy will be potential energy once again.

$$mgh_2 = \frac{1}{2}mv_2^2$$

$$v_2 = \sqrt{2gh_2}$$

$$= \sqrt{(2)\left(9.81 \ \frac{\text{m}}{\text{s}^2}\right)(0.05 \text{ m})}$$

$$= 0.99 \text{ m/s}   [\text{upward}]$$

Use the impulse-momentum principle. (Downward is taken as the positive velocity direction.)

$$\textbf{Imp} = \Delta p = m(v_1 - v_2)$$

$$= (12 \text{ kg})\left(1.98 \ \frac{\text{m}}{\text{s}} - \left(-0.99 \ \frac{\text{m}}{\text{s}}\right)\right)$$

$$= 35.7 \text{ N·s}   (36 \text{ N·s})$$

Answer is C.

8. Consider all of the rod's mass to be located at its centroid. When the rod is inclined 45°, that centroid is located at elevation (above the reference plane)

$$h = \frac{1}{2}L \sin 45°$$

The rod's initial potential energy is mgh.

The rod's mass moment of inertia taken about one end is

$$I_{\text{rod}} = \frac{mL^2}{3}$$

When the rod is moving with angular velocity ω, the rod's kinetic energy will be

$$\frac{I\omega^2}{2}$$

Initially, all of the rod's energy is potential energy. When the rod is horizontal, the potential energy will have been converted to rotational kinetic energy. Use the conservation of energy principle.

$$mgh = \frac{I\omega^2}{2}$$

$$mg\left(\frac{1}{2}L \sin 45°\right) = \left(\frac{mL^2}{3}\right)\left(\frac{\omega^2}{2}\right)$$

$$\omega = \sqrt{\frac{3g \sin 45°}{L}}$$

$$= \sqrt{\frac{(3)\left(9.81 \ \frac{\text{m}}{\text{s}^2}\right)\sin 45°}{1.0 \text{ m}}}$$

$$= 4.56 \text{ rad/s}$$

The linear velocity of the rod's tip is

$$v = r\omega = (1.0 \text{ m})\left(4.56 \ \frac{\text{rad}}{\text{s}}\right)$$

$$= 4.56 \text{ m/s}   (4.6 \text{ m/s})$$

Answer is B.

9. Since the pivot is frictionless and the spring is ideal, energy is conserved. The tip of the rod will return to its original height.

$$h = r \sin 45°$$
$$= (1.0 \text{ m}) \sin 45° = 0.71 \text{ m}$$
$$= 71 \text{ cm}$$

Answer is C.

10. The mass moment of inertia of the rod taken about one end is

$$I_{\text{rod}} = \frac{mL^2}{3} = \frac{(10 \text{ kg})(1 \text{ m})^2}{3}$$
$$= 3.33 \text{ kg·m}^2$$

When the rod contacts the spring, the rod will continue to drop below the horizontal position, giving up additional potential energy. Thus, the rod has potential energy and kinetic energy coming into contact with the spring. Once the spring has compressed completely, all energy will be stored as elastic potential energy in the spring. The additional drop of the rod tip equals the spring compression.

$$(PE)_{\text{rod}} + (KE)_{\text{rod}} = (PE)_{\text{spring}}$$

Let the spring compression be x. The center of mass (located halfway along the rod's length) drops $x/2$.

$$mg\left(\frac{x}{2}\right) + \frac{I\omega^2}{2} = \frac{1}{2}kx^2$$

$$\frac{(10 \text{ kg})\left(9.81 \frac{\text{m}}{\text{s}^2}\right)x}{2}$$
$$+ \frac{(3.33 \text{ kg·m}^2)\left(10 \frac{\text{rad}}{\text{s}}\right)^2}{2} = (0.5)\left(98\,000 \frac{\text{N}}{\text{m}}\right)x^2$$

$$49.05x + 167 = 49\,000x^2$$

Solving using the quadratic formula, completing the square, or using an equation solver,

$$x = 0.059 \text{ m}   (5.9 \text{ cm})$$

Answer is D.

11. The frictional force is the only force preventing the boxes from shifting. The forces on each box are its weight, the normal force, and the frictional force. The normal force on each box is equal to the box weight.

$$N = W = mg$$

The frictional force is

$$F_f = \mu N = \mu mg$$

Use the impulse-momentum principle. $v_2 = 0$. The frictional force is opposite of the direction of motion, hence, negative.

$$\textbf{Imp} = \Delta p$$
$$F_f \Delta t = m\Delta v$$
$$\Delta t = \frac{m(v_2 - v_1)}{F_f} = \frac{-mv_1}{-\mu mg}$$
$$= \frac{v_1}{\mu g}$$
$$= \frac{25 \dfrac{\text{m}}{\text{s}}}{(0.4)\left(9.81 \dfrac{\text{m}}{\text{s}^2}\right)}$$
$$= 6.37 \text{ s}   (6.4 \text{ s})$$

Answer is D.

12. The original velocity of the car and trailer is

$$v = \frac{\left(100 \dfrac{\text{km}}{\text{h}}\right)\left(1000 \dfrac{\text{m}}{\text{km}}\right)}{3600 \dfrac{\text{s}}{\text{h}}}$$
$$= 27.78 \text{ m/s}$$

Since the final velocity is zero, the energy dissipated is the original kinetic energy.

$$\Delta KE = \frac{mv^2}{2}$$
$$= \frac{(1500 \text{ kg} + 250 \text{ kg})\left(27.78 \dfrac{\text{m}}{\text{s}}\right)^2}{2}$$
$$= 675\,262 \text{ J}   (675 \text{ kJ})$$

Answer is D.

13. Use Newton's second law of motion to determine the force.
$$F = ma$$
$$= (250 \text{ kg})\left(2.5 \frac{\text{m}}{\text{s}^2}\right)$$
$$= 625 \text{ N}$$

Answer is B.

14. Use the conservation of momentum principle.

$$m_1 v_1 + m_2 v_2 = (m_1 + m_2)v'$$

$$(60\,000 \text{ kg})\left(1 \frac{\text{km}}{\text{h}}\right) + (40\,000 \text{ kg})\left(0 \frac{\text{km}}{\text{h}}\right)$$

$$= (60\,000 \text{ kg} + 40\,000 \text{ kg})v'$$

$$v' = 0.6 \text{ km/h}$$

Answer is B.

15. The original velocity of the 60 000 kg railcar is

$$v = \frac{\left(1 \frac{\text{km}}{\text{h}}\right)\left(1000 \frac{\text{m}}{\text{km}}\right)}{3600 \frac{\text{s}}{\text{h}}}$$

$$= 0.2777 \text{ m/s}$$

Use the impulse-momentum principle.

$$F\Delta t = m\Delta v$$

$$F = \frac{m\Delta v}{\Delta t}$$

$$= \frac{(60\,000 \text{ kg})\left(0.2777 \frac{\text{m}}{\text{s}} - 0.2 \frac{\text{m}}{\text{s}}\right)}{0.5 \text{ s}}$$

$$= 9320 \text{ N} \quad (9300 \text{ N})$$

Answer is D.

16. The initial velocity of the car is

$$v = \frac{65\,000 \frac{\text{m}}{\text{h}}}{3600 \frac{\text{s}}{\text{h}}}$$

$$= 18.06 \text{ m/s}$$

The frictional force decelerating the car is

$$F_f = \mu N = \mu m g$$

$$= (0.60)(3500 \text{ kg})\left(9.81 \frac{\text{m}}{\text{s}^2}\right)$$

$$= 20\,600 \text{ N}$$

Use the impulse-momentum principle.

$$F_f \Delta t = m\Delta v$$

$$(20\,600 \text{ N})(3 \text{ s}) = (3500 \text{ kg})\left(18.06 \frac{\text{m}}{\text{s}} - v_2\right)$$

$$v_2 = 0.403 \text{ m/s} \quad (0.40 \text{ m/s})$$

Answer is B.

17. The kinetic energy of the car is

$$KE = \frac{1}{2}mv^2$$

$$= \left(\frac{1}{2}\right)(3500 \text{ kg})\left(0.2 \frac{\text{m}}{\text{s}}\right)^2$$

$$= 70 \text{ J}$$

Answer is A.

Topic VI: Mechanics of Materials

Diagnostic Examination for Mechanics of Materials

Chapter 18   Stress and Strain

Chapter 19   Thermal, Hoop, and Torsional Stress

Chapter 20   Beams

Chapter 21   Columns

MECHANICS

Hint: For the most current information about the exam, visit www.ppi2pass.com/fefaqs.html regularly.
Use the Exam Forum to compare notes with other FE examinees.

Diagnostic Examination

TOPIC VI: MECHANICS OF MATERIALS

TIME LIMIT: 45 MINUTES

Problems 1–3 refer to the following situation.

A 25 mm diameter aluminum rod is loaded axially in tension as shown. Aluminum has a modulus of elasticity of 69 GPa and a Poisson's ratio of 0.35.

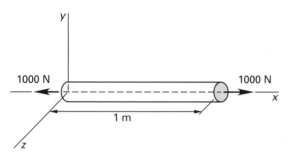

1. What is the decrease in diameter of the rod due to the applied load?

   (A) −260 nm
   (B) −165 nm
   (C) −73 nm
   (D) −30 nm

2. If the rod decreases in diameter by 258 nm while the length increases, what is the percent change in volume of the rod?

   (A) −0.09%   (decrease)
   (B) −0.0009%   (decrease)
   (C)   0.0009%   (increase)
   (D)   0.09%   (increase)

3. What is the total stored strain energy in the rod as a result of the loading?

   (A) 0.007 J
   (B) 0.015 J
   (C) 0.030 J
   (D) 0.045 J

4. A 10 m × 5 m rectangular steel plate is loaded in compression by two opposing triangular distributed loads as shown. Steel has a modulus of elasticity of 2.1×10^{11} Pa and a Poisson's ratio of 0.3. Buckling is to be disregarded. What is the shear stress, τ_{yx}, along line line A-A?

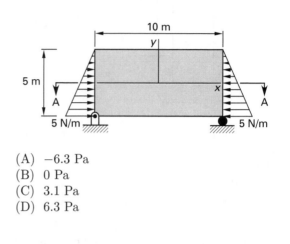

   (A) −6.3 Pa
   (B) 0 Pa
   (C) 3.1 Pa
   (D) 6.3 Pa

5. A thin-walled pressure vessel is constructed by rolling a 6 mm thick steel sheet into a cylindrical shape, welding the seam along line A-B, and capping the ends. The vessel is subjected to an internal pressure of 1.25 MPa. What is the normal stress normal to line A-B?

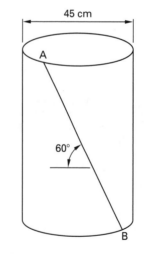

   (A) 10 MPa
   (B) 29 MPa
   (C) 41 MPa
   (D) 52 MPa

6. A steel pipe fixed at one end is subjected to a torque of 100 000 N·m. Steel has a modulus of elasticity of 2.1×10^{11} Pa and a Poisson's ratio of 0.3. What is the resulting angle of twist, α, of the pipe?

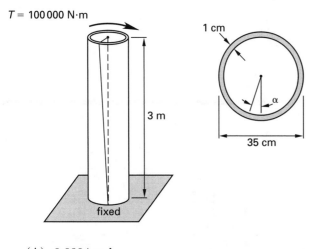

$T = 100\,000$ N·m

3 m

fixed

1 cm

35 cm

α

(A) 0.0004 rad
(B) 0.0008 rad
(C) 0.012 rad
(D) 0.024 rad

I

II

III

IV

(A) I only
(B) III only
(C) I and IV
(D) II, III, and IV

7. A 5 m long steel bar with a cross-sectional area of 0.01 m^2 is connected to a spring as shown. The spring has a stiffness of 2×10^8 N/m and is initially undeformed. The bar is fixed at its base. The temperature of the bar is increased by 70°C. Steel has a modulus of elasticity of 210 GPa and a coefficient of thermal expansion of 11.7×10^{-6} 1/°C. What is the resulting force in the spring?

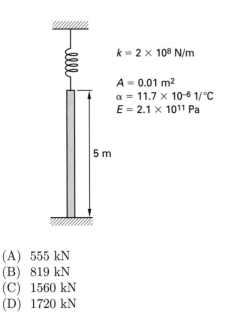

$k = 2 \times 10^8$ N/m

$A = 0.01$ m^2
$\alpha = 11.7 \times 10^{-6}$ 1/°C
$E = 2.1 \times 10^{11}$ Pa

5 m

(A) 555 kN
(B) 819 kN
(C) 1560 kN
(D) 1720 kN

8. In which of the steel structures shown would a temperature change produce internal stresses? Disregard thermal roller strain.

9. For the beam and loading shown, which of the following diagrams correctly represents the shape of the shear diagram?

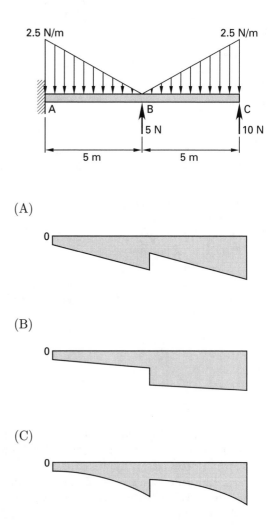

2.5 N/m   2.5 N/m

A   B   C
5 N   10 N

5 m   5 m

(A)

0

(B)

0

(C)

0

(D)

10. The shear diagram for a simply supported beam is as shown. What is the maximum moment in the beam?

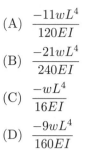

100.8 N

44.8 N

| 4.8 m | 2.8 m | 2.8 m | 4.8 m |

A   B   C

44.8 N

100.8 N

(A) 390 N·m
(B) 425 N·m
(C) 490 N·m
(D) 740 N·m

11. A beam has a triangular cross section as shown. What is the maximum compressive stress in the beam?

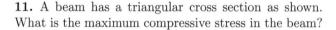

$w = 80$ N/m

10 cm   10 cm

12 cm   10 m

(A) 7.8 MPa
(B) 15.6 MPa
(C) 23.4 MPa
(D) 31.3 MPa

12. A cantilever beam is loaded with a triangular distributed load as shown. What is the deflection at the free end?

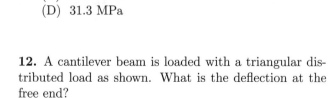

w

A   B

L

(A) $\dfrac{-11wL^4}{120EI}$

(B) $\dfrac{-21wL^4}{240EI}$

(C) $\dfrac{-wL^4}{16EI}$

(D) $\dfrac{-9wL^4}{160EI}$

13. A beam is simultaneously loaded by a pair of eccentric 1000 N forces and a 50 N·m moment. Neglect buckling effects. What is the maximum compressive stress?

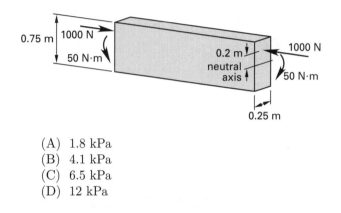

0.75 m   1000 N

50 N·m

0.2 m   1000 N

neutral
axis   50 N·m

0.25 m

(A) 1.8 kPa
(B) 4.1 kPa
(C) 6.5 kPa
(D) 12 kPa

14. A symmetrical pin-connected truss is loaded at its midpoint by a vertical force, Q. All members have the same cross section and are constructed of the same material. As the force is increased, which member of the truss will buckle first?

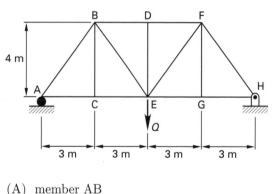

B   D   F

4 m

A

H

C   E   G

Q

| 3 m | 3 m | 3 m | 3 m |

(A) member AB
(B) member AC
(C) member BD
(D) member BE

15. A rigid beam, BD, is supported by a hinge at end B and a pin-connected structural link, AC, as shown.

Point A is directly below point B. Member AC is a solid rod with a radius of 0.025 m. The modulus of elasticity of member AC is 210 GPa. What vertical downward force, F, applied at point D will cause member AC to buckle?

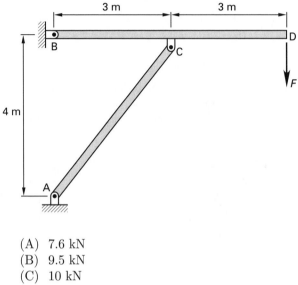

(A) 7.6 kN
(B) 9.5 kN
(C) 10 kN
(D) 25 kN

SOLUTIONS TO DIAGNOSTIC EXAMINATION TOPIC VI

1.
$$\sigma_x = \frac{P}{A} = \frac{P}{\left(\frac{\pi}{4}\right)d^2} = \frac{4P}{\pi d^2}$$

$$= \frac{(4)(1000 \text{ N})}{\pi(0.025 \text{ m})^2}$$

$$= 2.037 \times 10^6 \text{ Pa}$$

$$\epsilon_y = \epsilon_z = \frac{-\nu\sigma_x}{E}$$

$$= \frac{-(0.35)(2.037 \times 10^6 \text{ Pa})}{6.9 \times 10^{10} \text{ Pa}}$$

$$= -1.033 \times 10^{-5} \text{ m/m}$$

$$\delta_d = \epsilon_y d$$
$$= -\left(1.033 \times 10^{-5} \frac{\text{m}}{\text{m}}\right)(0.025 \text{ m})$$

$$= -2.58 \times 10^{-7} \text{ m} \quad (-260 \text{ nm})$$

Answer is A.

2. The change in length of the rod is

$$\delta_L = \epsilon_x L = \left(\frac{\sigma_x}{E}\right)L = \left(\frac{\frac{P}{A}}{E}\right)L$$

$$= \frac{PL}{AE}$$

$$= \frac{4PL}{\pi d^2 E}$$

$$= \frac{(4)(1000 \text{ N})(1 \text{ m})}{\pi(0.025 \text{ m})^2(6.9 \times 10^{10} \text{ Pa})}$$

$$= 2.95 \times 10^{-5} \text{ m}$$

The change in diameter of the rod was given.

$$\delta_d = -258 \text{ nm}$$

The percent change in volume of the rod is

$$\%\left(\frac{\delta V}{V_o}\right) = \frac{V - V_o}{V_o} \times 100\%$$

$$= \left(\frac{\left(\frac{\pi}{4}\right)(d+\delta_d)^2(L+\delta_L)}{\left(\frac{\pi}{4}\right)d^2 L}\right) \times 100\%$$

$$= \left(\frac{(d+\delta_d)^2(L+\delta_L) - d^2 L}{d^2 L}\right) \times 100\%$$

$$= \left(\frac{\begin{array}{c}(0.025 \text{ m} - 2.58 \times 10^{-7} \text{ m})^2 \\ \times (1 \text{ m} + 2.95 \times 10^{-5} \text{ m}) \\ - (0.025 \text{ m})^2(1 \text{ m})\end{array}}{(0.025 \text{ m})^2(1 \text{ m})}\right) \times 100\%$$

$$= 0.0009\%$$

Answer is C.

3.
$$U = \left(\frac{1}{2}\right)P\delta_L = \left(\frac{1}{2}\right)P\left(\frac{PL}{AE}\right)$$

$$= \left(\frac{1}{2}\right)\left(\frac{P^2 L}{AE}\right)$$

$$= \frac{\left(\frac{1}{2}\right)P^2 L}{\left(\frac{\pi}{4}\right)d^2 E} = \frac{2P^2 L}{\pi d^2 E}$$

$$= \frac{(2)(1000 \text{ N})^2(1 \text{ m})}{\pi(0.025 \text{ m})^2(6.9 \times 10^{10} \text{ Pa})}$$

$$= 0.015 \text{ J}$$

Answer is B.

4. The plate is loaded in pure compression. Loading is one-dimensional, so the compressive stress is the principal stress. Shear stress is zero in the direction of the principal stress.

Answer is B.

5.

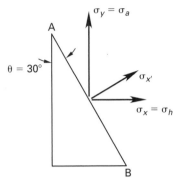

The tensile hoop stress is

$$\sigma_h = \frac{pD}{2t} = \frac{(1.25 \times 10^6 \text{ Pa})(0.45 \text{ m})}{(2)(0.006 \text{ m})}$$

$$= 4.688 \times 10^7 \text{ Pa} \quad (46.88 \text{ MPa})$$

The tensile axial (long) stress is

$$\sigma_a = \frac{pD}{4t} = \frac{\sigma_h}{2} = \frac{46.88 \text{ MPa}}{2}$$

$$= 23.44 \text{ MPa}$$

The normal stress on the weld can be computed from the hoop and axial stresses. There is no torsional shear stress.

$$\sigma_{x'} = \frac{\sigma_x + \sigma_y}{2} + \left(\frac{\sigma_x - \sigma_y}{2}\right)\cos 2\theta + \tau_{xy} \sin 2\theta$$

$$= \frac{46.88 \text{ MPa} + 23.44 \text{ MPa}}{2}$$

$$+ \left(\frac{46.88 \text{ MPa} - 23.44 \text{ MPa}}{2}\right)\cos\left((2)(30°)\right)$$

$$= 41.02 \text{ MPa} \quad (41 \text{ MPa})$$

Answer is C.

6. Calculate the polar moment of inertia of the column.

$$J = \left(\frac{\pi}{32}\right)(D_o^4 - D_i^4)$$

$$= \left(\frac{\pi}{32}\right)\left((0.35 \text{ m})^4 - (0.33 \text{ m})^4\right)$$

$$= 3.0896 \times 10^{-4} \text{ m}^4$$

The shear modulus of steel can be calculated from the modulus of elasticity and Poisson's ratio.

$$G = \frac{E}{(2)(1 + \nu)}$$

$$= \frac{2.1 \times 10^{11} \text{ Pa}}{(2)(1 + 0.3)}$$

$$= 8.0769 \times 10^{10} \text{ Pa}$$

The angle of twist is

$$\phi = \frac{TL}{GJ}$$

$$= \frac{(100\,000 \text{ N·m})(3 \text{ m})}{(8.0769 \times 10^{10} \text{ Pa})(3.0896 \times 10^{-4} \text{ m}^4)}$$

$$= 0.012 \text{ rad}$$

Answer is C.

7. If the spring were not present, the bar would elongate by an amount equal to

$$\delta_L = \alpha L \Delta T$$

Under the action of the spring force alone, the bar would contract by an amount equal to

$$\delta_p = \frac{PL}{AE}$$

The deformation of the bar must equal the deformation in the spring, P/k.

By superposition,

$$\alpha L \Delta T - \frac{PL}{AE} = \frac{P}{k}$$

$$P\left(\frac{1}{k} + \frac{L}{AE}\right) = \alpha L \Delta T$$

$$P = \frac{\alpha L \Delta T}{\dfrac{1}{k} + \dfrac{L}{AE}}$$

$$= \frac{\left(11.7 \times 10^{-6} \dfrac{1}{°\text{C}}\right)(5 \text{ m})(70°\text{C})}{\dfrac{1}{2 \times 10^8 \dfrac{\text{N}}{\text{m}}} + \dfrac{5 \text{ m}}{(0.01 \text{ m}^2)(2.1 \times 10^{11} \text{ Pa})}}$$

$$= 554\,806 \text{ N} \quad (555 \text{ kN})$$

Answer is A.

8. Thermal changes can only produce stresses in structures that are constrained against movement. Structure I cannot expand axially. Structure IV cannot flex upward (i.e., bend).

Answer is C.

9.

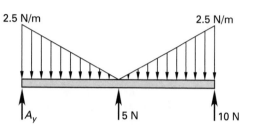

Solve for the vertical reaction at point A.

$$\sum F_y = 0$$

$$A_y - (2)\left(\frac{1}{2}\right)\left(2.5\ \frac{N}{m}\right)(5\ m) + 5\ N + 10\ N = 0$$

$$A_y = -2.5\ N$$

The shear force just to the right of the support is equal and opposite to the vertical reaction at point A.

Just to the left of point B, the shear force is

$$\sum F_y = 0$$

$$A_y - \left(\frac{1}{2}\right)\left(2.5\ \frac{N}{m}\right)(5\ m) - V = 0$$

$$V = A_y - \left(\frac{1}{2}\right)\left(2.5\ \frac{N}{m}\right)(5\ m)$$

$$= -2.5\ N - \left(\frac{1}{2}\right)\left(2.5\ \frac{N}{m}\right)(5\ m)$$

$$= -8.75\ N$$

Because the load varies linearly between points A and B, the shear force will vary quadratically between these points. And since the load is decreasing, the shear diagram, when drawn according to the sign convention, will be concave between these points.

The shear diagram is discontinuous at the location of a concentrated load.

Just to the right of point B, the shear force is

$$\sum F_y = 0$$

$$A_y - \left(\frac{1}{2}\right)\left(2.5\ \frac{N}{m}\right)(5\ m) + P - V = 0$$

$$V = A_y - \left(\frac{1}{2}\right)\left(2.5\ \frac{N}{m}\right)(5\ m) + P$$

$$= -2.5\ N - \left(\frac{1}{2}\right)\left(2.5\ \frac{N}{m}\right)(5\ m) + 5\ N$$

$$= -3.75\ N$$

Just to the left of point C, the shear force is equal to

$$\sum F_y = 0$$

$$A_y - \left(2.5\ \frac{N}{m}\right)(5\ m) + 5\ N - V = 0$$

$$V = A_y - \left(2.5\ \frac{N}{m}\right)(5\ m) + 5\ N$$

$$= -2.5\ N - \left(2.5\ \frac{N}{m}\right)(5\ m) + 5\ N$$

$$= -10\ N$$

Because the load varies linearly between points B and C, the shear force will vary quadratically between these points. And since the load is increasing, the shear diagram, when drawn according to the sign convention, will be convex between these points.

As a final check, note that the shear just to the left of point C is equal to the concentrated load at point C.

Answer is D.

10. The maximum moment occurs at point B where the shear is zero.

The relationship between shear and moment is

$$V(x) = \frac{dM(x)}{dx}$$

Integrating,

$$M_B - M_A = \int_{x_A}^{x_B} V(x)\,dx$$

$M_A = 0$ since the beam is simply supported.

The integral is equal to the area under the shear force diagram between the two points.

$$M_B = (44.8\ N)(4.8\ m)$$

$$+ \left(\frac{1}{2}\right)(100.8\ N - 44.8\ N)(4.8\ m)$$

$$+ \left(\frac{1}{2}\right)(100.8\ N)(2.8\ m)$$

$$= 491\ N{\cdot}m \quad (490\ N{\cdot}m)$$

Answer is C.

11.

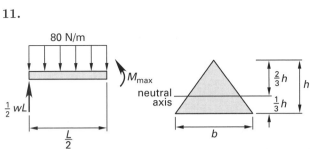

12.

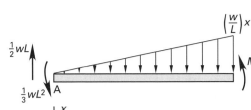

Due to symmetry of the applied load, the two vertical reactions are equal. Each vertical reaction is half of the total vertical load.

$$R = \left(\frac{1}{2}\right) wL$$

The maximum moment on a simply supported beam with a uniformly distributed load occurs at the center of the span.

$$M_{\max} = \left(\frac{1}{2}\right) wL \left(\frac{L}{2}\right) - w\left(\frac{L}{2}\right)\left(\frac{L}{4}\right)$$

$$= \left(\frac{1}{8}\right) wL^2$$

The centroidal moment of inertia of the triangular cross section is given by

$$I = \frac{bh^3}{36}$$

The neutral axis for this cross section is at $h/3$ from the bottom of the beam. The maximum compressive stress will occur at the top of the beam, a distance of $2h/3$ from the neutral axis.

The maximum stress in the beam can be computed as

$$\sigma = \frac{Mc}{I}$$

$$= \frac{\left(\frac{1}{8}\right) wL^2 \left(\frac{2h}{3}\right)}{\frac{bh^3}{36}}$$

$$= \frac{3wL^2}{bh^2}$$

$$= \frac{(3)\left(80\ \frac{\text{N}}{\text{m}}\right)(10\ \text{m})^2}{(0.12\ \text{m})(0.08\ \text{m})^2}$$

$$= 3.125 \times 10^7\ \text{Pa} \quad (31.3\ \text{MPa})$$

Answer is D.

Determine the reaction at the built-in end (point A). This is equal to the total vertical load on the beam.

$$\sum F_y = 0$$

$$R_A - \left(\frac{1}{2}\right) wL = 0$$

$$R_A = \left(\frac{1}{2}\right) wL$$

Determine the moment at the built-in end.

$$\sum M_A = 0$$

$$M_A - \left(\frac{1}{2}\right) wL \left(\frac{2}{3}\right) L = 0$$

$$M_A = \left(\frac{1}{3}\right) wL^2$$

Considering the equilibrium of the free body, the moment at any point in the beam is

$$M + \left(\frac{1}{3}\right) wL^2 - \left(\frac{1}{2}\right) wLx$$

$$+ \left(\frac{w}{L}\right) x \left(\frac{1}{2}\right) x \left(\frac{1}{3}\right) x$$

$$= 0$$

$$M = \left(\frac{1}{2}\right) wLx - \left(\frac{1}{3}\right) wL^2 - \left(\frac{1}{6}\right)\left(\frac{w}{L}\right) x^3$$

The deflection of the beam can be found by the double-integration method.

$$\frac{d^2y}{dx^2} = \frac{M}{EI}$$

$$= \left(\frac{w}{6EI}\right)\left(3Lx - 2L^2 - \frac{x^3}{L}\right)$$

Integrating the curvature yields the slope.

$$\frac{dy}{dx} = \left(\frac{w}{6EI}\right)\left(\left(\frac{3}{2}\right) Lx^2 - 2L^2x - \left(\frac{1}{4}\right)\left(\frac{x^4}{L}\right)\right) + C_1$$

At the support, where $x = 0$, the slope is zero. From this boundary condition, $C_1 = 0$.

MECHANICS

Integrating the slope yields the deflection.

$$y = \left(\frac{w}{6EI}\right)\left(\left(\frac{1}{2}\right)Lx^3 - L^2x^2 - \left(\frac{1}{20}\right)\left(\frac{x^5}{L}\right)\right) + C_2$$

At the support, the deflection is zero. From this boundary condition, $C_2 = 0$.

The deflection at any point in the beam is given by

$$y = \left(\frac{w}{6EI}\right)\left(\left(\frac{1}{2}\right)Lx^3 - L^2x^2 - \left(\frac{1}{20}\right)\left(\frac{x^5}{L}\right)\right)$$

At $x = L$, the deflection is

$$y = \frac{-11wL^4}{120EI}$$

Answer is A.

13. The eccentric load and applied moment are opposite in direction.

The maximum compressive stress is

$$\sigma = \frac{F}{A} \pm \frac{Mc}{I} = \frac{F}{A} \pm \frac{(Fe - M)c}{I}$$

$$= \frac{F}{A} \pm \frac{(Fe - M)c}{\left(\frac{1}{12}\right)bh^3}$$

$$= \frac{F}{A} \pm \frac{(12)(Fe - M)c}{bh^3}$$

$$= \frac{1000 \text{ N}}{(0.25 \text{ m})(0.75 \text{ m})}$$

$$\pm \frac{(12)\big((1000 \text{ N})(0.2 \text{ m}) - 50 \text{ N·m}\big)(0.375 \text{ m})}{(0.25 \text{ m})(0.75 \text{ m})^3}$$

$$= 5333 \text{ Pa} \pm 6400 \text{ Pa}$$

$$= 11\,733 \text{ Pa} \quad (12 \text{ kPa})$$

Answer is D.

14.

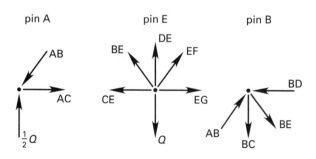

Since the truss is loaded at its midpoint,

$$R_A = R_H = \frac{Q}{2}$$

The length of all of the diagonal members is 5 (3-4-5 triangle).

Determine the compressive members and the forces in the members.

From the equilibrium of pin A,

$$\sum F_y = 0 = R_A + AB_y$$

$$= \frac{Q}{2} + \left(\frac{4}{5}\right)AB$$

$$AB = \left(\frac{5}{8}\right)Q \quad \text{[compression]}$$

Considering the equilibrium of pin D, DE = 0.

From the equilibrium of pin E (by symmetry),

$$\sum F_y = 0 = -Q + BE_y + EF_y = -Q + 2BE_y$$

$$= -Q + (2)\left(\frac{4}{5}\right)BE$$

$$BE = \left(\frac{5}{8}\right)Q \quad \text{[tension]}$$

BC is also a zero-force member. From the equilibrium of pin B,

$$\sum F_x = AB_x + BD + BE_x = 0$$

$$= \left(\frac{5}{8}\right)Q\left(\frac{3}{5}\right) + BD + \left(\frac{5}{8}\right)Q\left(\frac{3}{5}\right)$$

$$BD = \left(\frac{3}{4}\right)Q \quad \text{[compression]}$$

A quick check of the other pins will reveal that all the other members, with the exception of the symmetric counterparts to AB and BD, must be in tension.

The member with the largest compressive force will not necessarily buckle first. Short compressive members can support large loads. The length must be considered. Since the buckling load is inversely proportional to the length of the member, $P_{\text{cr}_{AB}}$ will be smaller than $P_{\text{cr}_{BD}}$.

$$P_{\text{cr}_{AB}} = \frac{\pi^2 EI}{(kl)^2_{AB}} = \frac{\pi^2 EI}{(5 \text{ m})^2}$$

$$= \frac{\pi^2 EI}{25 \text{ m}^2}$$

$$= C/25 \quad [C \text{ is a constant equal to } \pi^2 EI]$$

$$P_{\text{cr}_{BD}} = \frac{\pi^2 EI}{(kl)^2_{BD}} = \frac{\pi^2 EI}{(3 \text{ m})^2}$$

$$= \frac{\pi^2 EI}{9 \text{ m}^2}$$

$$= C/9$$

Compute the ratio of the force in each member to its buckling load.

$$\frac{P_{\text{cr}_{AB}}}{AB} = \frac{\dfrac{C}{25}}{\left(\dfrac{5}{8}\right) Q} = \frac{\left(\dfrac{8}{125}\right) C}{Q}$$

$$AB = \left(\frac{125}{8}\right)\left(\frac{Q}{C}\right) P_{\text{cr}_{AB}}$$

$$\frac{P_{\text{cr}_{BD}}}{BD} = \frac{\dfrac{C}{9}}{\left(\dfrac{3}{4}\right) Q} = \frac{\left(\dfrac{4}{27}\right) C}{Q}$$

$$BD = \left(\frac{27}{4}\right)\left(\frac{Q}{C}\right) P_{\text{cr}_{BD}}$$

The force in AB is a higher percentage of its critical load than the force in BD. Member AB will buckle before member BD.

Answer is A.

15. The length of member AC is 5 (3-4-5 triangle).

The buckling load for member AC is

$$P_{\text{cr}} = \frac{\pi^2 EI}{(kL)^2} = \frac{\pi^2 E \left(\dfrac{\pi r^4}{4}\right)}{(kl)^2}$$

$$= \frac{\pi^3 (2.1 \times 10^{11} \text{ Pa})(0.025 \text{ m})^4}{(4)(5 \text{ m})^2}$$

$$= 25\,435 \text{ N}$$

Compute the force in AC as a function of F. Sum moments about point B.

$$\sum M_{\text{B}}: \text{AC}\left(\frac{4}{5}\right)(3 \text{ m}) - F(6 \text{ m}) = 0$$

$$\text{AC} = 2.5F$$

Buckling will occur when the force in AC equals the critical load.

$$P_{\text{cr}} = 2.5F$$

$$25\,435 \text{ N} = 2.5F$$

$$F = 10\,174 \text{ N} \quad (10 \text{ kN})$$

Answer is C.

MECHANICS

18 Stress and Strain

Subjects

DEFINITIONS 18-1
UNIAXIAL LOADING AND
    DEFORMATION 18-2
ELASTIC STRAIN ENERGY IN
    UNIAXIAL LOADING 18-2
BIAXIAL AND TRIAXIAL LOADING . . . 18-2
TRANSFORMATION OF AXES 18-3
PRINCIPAL STRESSES 18-3
MOHR'S CIRCLE 18-4
GENERAL STRAIN 18-4
FAILURE THEORIES 18-5
    Maximum Normal Stress Theory 18-5
    Maximum Shear Stress Theory 18-5
    Distortion Energy Theory 18-5

Nomenclature

A	area	in^2	m^2
E	modulus of elasticity	lbf/in^2	MPa
FS	factor of safety	–	–
G	shear modulus	lbf/in^2	MPa
L	length	in	m
P	force	lbf	N
S	strengh	lbf/in^2	MPa
u	strain energy per unit volume	lbf/in^2	MPa
U	energy	in-lbf	N·m
W	work	in-lbf	N·m

Symbols

γ	shear strain	–	–
δ	deformation	in	m
ϵ	linear strain	–	–
θ	angle	rad	rad
ν	Poisson's ratio	–	–
σ	normal stress	lbf/in^2	MPa
τ	shear stress	lbf/in^2	MPa

Subscripts

a	allowable
f	final
s	shear
t	tension
u	ultimate
y	yield

DEFINITIONS

Mechanics of materials deals with the elastic behavior of materials and the stability of members. Mechanics of materials concepts are used to determine the stress and deformation of axially loaded members, connections, torsional members, thin-walled pressure vessels, beams, eccentrically loaded members, and columns.

Stress is force per unit area. Typical units of stress are lbf/in^2, ksi, and MPa. There are two primary types of stress: *normal stress* and *shear stress*. With normal stress, σ, the force is normal to the surface area. With shear stress, τ, the force is parallel to the surface area.

$$\sigma = \frac{P_{\text{normal to area}}}{A} \qquad 18.1$$

$$\tau = \frac{P_{\text{parallel to area}}}{A} \qquad 18.2$$

Linear strain (normal strain, longitudinal strain, axial strain), ϵ, is a change of length per unit of length. Linear strain may be listed as having units of in/in, mm/mm, percent, or no units at all. *Shear strain*, γ, is an angular deformation resulting from shear stress. Shear strain may be presented in units of radians, percent, or no units at all.

$$\epsilon = \frac{\delta}{L} \qquad 18.3$$

$$\gamma = \frac{\delta_{\text{parallel to area}}}{\text{height}} = \tan \theta \approx \theta$$
$$[\theta \text{ in radians}] \qquad 18.4$$

Hooke's law is a simple mathematical statement of the relationship between elastic stress and strain: stress is proportional to strain. For normal stress, the constant of proportionality is the *modulus of elasticity (Young's Modulus)*, E.

$$\sigma = E\epsilon \qquad 18.5$$

Poisson's ratio, ν, is a constant that relates the lateral strain to the axial strain for axially loaded members.

$$\nu = -\frac{\epsilon_{\text{lateral}}}{\epsilon_{\text{axial}}} \qquad 18.6$$

Theoretically, Poisson's ratio could vary from zero to 0.5, but *typical values* are 0.35 for aluminum and 0.3 for steel.

Hooke's law may also be applied to a plane element in pure shear. For such an element, the shear stress is linearly related to the shear strain, γ, by the *shear modulus* (also known as the *modulus of rigidity*), G.

$$\tau = G\gamma \qquad 18.7$$

For an elastic, isotropic material, the modulus of elasticity, shear modulus, and Poisson's ratio are related by Eq. 18.8 and Eq. 18.9.

$$G = \frac{E}{2(1+\nu)} \qquad 18.8$$

$$E = 2G(1+\nu) \qquad 18.9$$

Table 18.1 Material Properties

material	units[a]	steel	aluminum	cast iron	wood fir
modulus of elasticity, E	Mpsi	29.0	10.0	14.5	1.6
	GPa	200.0	69.0	100.0	11.0
modulus of rigidity, G	Mpsi	11.5	3.8	6.0	0.6
	GPa	80	26	41.4	4.1
Poisson's ratio, ν		0.30	0.33	0.21	0.33

[a]Mpsi = millions of pounds per square inch

UNIAXIAL LOADING AND DEFORMATION

The deformation, δ, of an axially loaded member of original length L can be derived from Hooke's law. Tension loading is considered to be positive; compressive loading is negative. The sign of the deformation will be the same as the sign of the loading.

$$\delta = L\epsilon = L\left(\frac{\sigma}{E}\right) = \frac{PL}{AE} \qquad 18.10$$

This expression for axial deformation assumes that the linear strain is proportional to the normal stress ($\epsilon = \sigma/E$) and that the cross-sectional area is constant.

When an axial member has distinct sections differing in cross-sectional area or composition, superposition is used to calculate the total deformation as the sum of individual deformations.

$$\delta = \sum \frac{PL}{AE} = P \sum \frac{L}{AE} \qquad 18.11$$

When one of the variables (e.g., A), varies continuously along the length,

$$\delta = \int \frac{P\,dL}{AE} = P\int \frac{dL}{AE} \qquad 18.12$$

The new length of the member including the deformation is given by Eq. 18.13. The algebraic sign of the deformation must be observed.

$$L_f = L + \delta \qquad 18.13$$

ELASTIC STRAIN ENERGY IN UNIAXIAL LOADING

Strain energy, also known as *internal work*, is the energy per unit volume stored in a deformed material. The strain energy is equivalent to the work done by the applied force. Simple work is calculated as the product of a force moving through a distance.

$$\text{work} = \text{force} \times \text{distance}$$

$$= \int F\,dL \qquad 18.14$$

$$\text{work per volume} = \int \frac{F\,dL}{AL}$$

$$= \int \sigma\,d\epsilon \qquad 18.15$$

Work per unit volume corresponds to the area under the stress-strain curve. Units are in-lbf/in^3, usually shortened to lbf/in^2 (MPa).

For an axially loaded member below the proportionality limit, the total strain energy is given by Eq. 18.16.

$$U = \frac{1}{2}P\delta = \frac{P^2 L}{2AE} \qquad 18.16$$

The strain energy per unit volume is

$$u = \frac{U}{AL} = \frac{\sigma^2}{2E} \qquad 18.17$$

BIAXIAL AND TRIAXIAL LOADING

Triaxial loading on an infinitesimal solid element is illustrated in Fig. 18.1.

Figure 18.1 Stress Components with Triaxial Loading

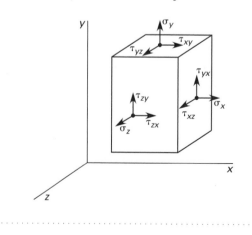

Normal and shear stresses exist on each face of the element. The stresses on the face normal to the x-axis are defined by Eqs. 18.18 through 18.20. Stresses on the other faces are similarly defined.

$$\sigma_x = \lim_{\Delta A_x \to 0} \left(\frac{\Delta F_x}{\Delta A_x} \right) \qquad 18.18$$

$$\tau_{xy} = \lim_{\Delta A_x \to 0} \left(\frac{\Delta F_y}{\Delta A_x} \right) \qquad 18.19$$

$$\tau_{xz} = \lim_{\Delta A_x \to 0} \left(\frac{\Delta F_z}{\Delta A_x} \right) \qquad 18.20$$

Loading is rarely confined to a single direction. All real structural members are three dimensional and most experience *triaxial loading*, but most problems can be analyzed with two dimensions because the normal stresses in one direction are either zero or negligible. This two-dimensional loading of the member is called *plane stress* or *biaxial loading*.

Biaxial loading on an infinitesimal element is illustrated in Fig. 18.2.

Figure 18.2 *Sign Conventions for Positive Stresses in Two Dimensions*

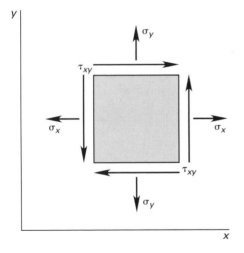

TRANSFORMATION OF AXES

If the normal and shear stresses are known for one set of orthogonal planes in an infinitesimal element, then the stresses on any other plane within the element can also be found. Equations 18.21 through 18.23 give the transformation of stress on the plane that is inclined at an angle from the vertical plane. The sign conventions for positive normal and shear stresses are shown in Fig. 18.2. Tensile normal stresses are positive; compressive normal stresses are negative. Shear stresses acting on opposite and parallel faces of an element are designated as positive when they form a clockwise couple.

$$\sigma_{x'} = \frac{\sigma_x + \sigma_y}{2} + \left(\frac{\sigma_x - \sigma_y}{2} \right) \cos 2\theta$$
$$+ \tau_{xy} \sin 2\theta \qquad 18.21$$

$$\sigma_{y'} = \frac{\sigma_x + \sigma_y}{2} - \left(\frac{\sigma_x - \sigma_y}{2} \right) \cos 2\theta$$
$$- \tau_{xy} \sin 2\theta \qquad 18.22$$

$$\tau_{x'y'} = - \left(\frac{\sigma_x - \sigma_y}{2} \right) \sin 2\theta + \tau_{xy} \cos 2\theta \qquad 18.23$$

Figure 18.3 *Transformation of Axes*

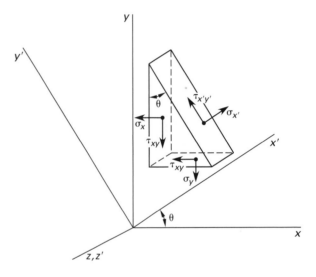

PRINCIPAL STRESSES

For any point in a loaded specimen, a plane can be found where the shear stress is zero. The normal stresses associated with this plane are known as the *principal stresses*, which are the maximum and minimum normal stresses acting at that point.

The maximum and minimum normal stresses may be found by differentiating Eq. 18.21 (or 18.22) with respect to θ, setting the derivative equal to zero, and substituting back into Eq. 18.21 (or 18.22). A similar procedure is used to derive the minimum and maximum shear stresses from Eq. 18.23.

$$\sigma_1, \sigma_2 = \frac{1}{2}(\sigma_x + \sigma_y) \pm \tau_1$$
$$= \frac{1}{2}(\sigma_x + \sigma_y) \pm \sqrt{\left(\frac{\sigma_x - \sigma_y}{2} \right)^2 + \tau_{xy}^2} \qquad 18.24$$

$$\tau_1, \tau_2 = \pm \frac{1}{2} \sqrt{(\sigma_x - \sigma_y)^2 + (2\tau_{xy})^2}$$

$$= \pm \frac{\sigma_1 - \sigma_2}{2} \qquad 18.25$$

The angles of the planes on which the principal stresses act are given by Eq. 18.26. θ is measured from the x-axis, clockwise if positive. Equation 18.26 will yield two angles, 90 degrees apart. These angles can be substituted back into Eq. 18.21 to determine which angle corresponds to the minimum normal stress and which angle corresponds to the maximum normal stress.

Alternatively, Eq. 18.26 can be used to determine the direction of the principal planes. Let σ_x be the algebraically larger of the two given normal stresses. The angle between the direction of σ_x and the direction of σ_1, the algebraically larger principal stress, will always be less than 45 degrees.

$$\theta_{\sigma_1, \sigma_2} = \frac{1}{2} \tan^{-1} \left(\frac{2\tau_{xy}}{\sigma_x - \sigma_y} \right) \qquad 18.26$$

The angles of the planes on which the shear stress is minimum and maximum are given by Eq. 18.27. These planes will be 90 degrees apart and will be rotated 45 degrees from the planes of principal normal stresses. As with Eq. 18.26, θ is measured clockwise if positive and counterclockwise if negative.

$$\theta_{\tau_1, \tau_2} = \frac{1}{2} \tan^{-1} \left(\frac{\sigma_x - \sigma_y}{-2\tau_{xy}} \right) \qquad 18.27$$

MOHR'S CIRCLE

Mohr's circle can be constructed to graphically determine the principal normal and shear stresses. In some cases, this procedure may be faster than using the preceding equations, but a solely graphical procedure is less accurate. By convention, tensile stresses are positive; compressive stresses are negative. Clockwise shear stresses are positive; counterclockwise shear stresses are negative.

step 1:   Determine the applied stresses: σ_x, σ_y, and τ_{xy}. Observe the correct sign conventions.

step 2:   Draw a set of σ-τ axes.

step 3:   Locate the center of the circle, point c, by calculating $\sigma_c = \frac{1}{2}(\sigma_x + \sigma_y)$.

step 4:   Locate the point $p_1 = (\sigma_x, -\tau_{xy})$. (Alternatively, locate p_1' at $(\sigma_y, +\tau_{xy})$.)

step 5:   Draw a line from point p_1 through the center, c, and extend it an equal distance above the σ axis to p_1'. This is the diameter of the circle.

step 6:   Using the center, c, and point p_1, draw the circle. An alternative method is to draw a circle of radius r about point c.

$$r = \sqrt{\frac{1}{4}(\sigma_x - \sigma_y)^2 + \tau_{xy}^2} \qquad 18.28$$

step 7:   Point p_2 defines the smaller principal stress, σ_2. Point p_3 defines the larger principal stress, σ_1.

step 8:   Determine the angle θ as half of the angle 2θ on the circle. This angle corresponds to the larger principal stress, σ_1. On Mohr's circle, angle 2θ is measured from the p_1-p_1' line to the horizontal axis. θ (measured from the x-axis to the plane of principal stress) is in the same direction as 2θ (measured from line p_1-p_1' to the σ-axis).

step 9:   The top and bottom of the circle define the largest and smallest shear stresses.

Figure 18.4 Mohr's Circle for Stress

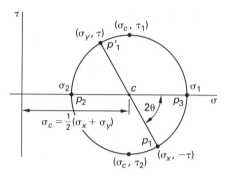

Mohr's circle can also be used to determine stresses when the axes are rotated (i.e., transformed) through an angle θ, as in Fig. 18.3. The p_1-p_1' diameter is rotated in the same direction, but through an angle of 2θ. The endpoints and coordinates of the rotated diameter will define $\sigma_{x'}$, $\sigma_{y'}$, and $\tau_{x'y'}$. The principal stresses will, of course, be unchanged.

GENERAL STRAIN

Hooke's law, previously defined for axial loads and for pure shear, can be derived for three-dimensional stress-strain relationships and written in terms of the three elastic constants, E, G, and ν. The following equations can be used to find the stresses and strains on the differential element in Fig. 18.1.

$$\epsilon_x = \frac{1}{E} \big(\sigma_x - \nu(\sigma_y + \sigma_z) \big) \qquad 18.29$$

$$\epsilon_y = \frac{1}{E}\left(\sigma_y - \nu(\sigma_z + \sigma_x)\right) \qquad 18.30$$

$$\epsilon_z = \frac{1}{E}\left(\sigma_z - \nu(\sigma_x + \sigma_y)\right) \qquad 18.31$$

$$\gamma_{xy} = \frac{\tau_{xy}}{G} \qquad 18.32$$

$$\gamma_{yz} = \frac{\tau_{yz}}{G} \qquad 18.33$$

$$\gamma_{zx} = \frac{\tau_{zx}}{G} \qquad 18.34$$

FAILURE THEORIES

Maximum Normal Stress Theory

The *maximum normal stress theory* predicts the failure stress reasonably well for brittle materials under static biaxial loading. Failure is assumed to occur if the largest tensile principal stress, σ_1, is greater than the ultimate tensile strength, or if the largest compressive principal stress, σ_2, is greater than the ultimate compressive strength. Brittle materials generally have much higher compressive than tensile strengths, so both tensile and compressive stresses must be checked.

Stress concentration factors are applicable to brittle materials under static loading. The factor of safety, FS, is the ultimate strength, S_u, divided by the actual stress, σ. Where a factor of safety is known in advance, the *allowable stress*, S_a, can be calculated by dividing the ultimate strength by it.

$$\text{FS} = \frac{S_u}{\sigma} \qquad 18.35$$

$$S_a = \frac{S_u}{\text{FS}} \qquad 18.36$$

The failure criterion is

$$\sigma > \frac{S_u}{\text{FS}} \qquad 18.37$$

Maximum Shear Stress Theory

For ductile materials (e.g., steel) under static loading (the conservative *maximum shear stress theory*), shear stress can be used to predict yielding (i.e., failure). Despite the theory's name, however, loading is not limited to shear and torsion. Loading can include normal stresses as well as shear stresses. According to the maximum shear stress theory, yielding occurs when the maximum shear stress exceeds the yield strength in shear. It is implicit in this theory that the yield strength in shear is half of the tensile yield strength.

$$S_{ys} = \frac{S_{yt}}{2} \qquad 18.38$$

From combined-stress theory, the maximum shear stress, $\tau_{\max}$, is the maximum of the three combined shear stresses. (For biaxial loading, only Eq. 18.39 is used.)

$$\tau_{12} = \frac{\sigma_1 - \sigma_2}{2} \qquad 18.39$$

$$\tau_{23} = \frac{\sigma_2 - \sigma_3}{2} \qquad 18.40$$

$$\tau_{31} = \frac{\sigma_3 - \sigma_1}{2} \qquad 18.41$$

$$\tau_{\max} = \max(\tau_{12},\ \tau_{23},\ \tau_{31}) \qquad 18.42$$

The failure criterion is

$$\tau_{\max} > S_{ys} \qquad 18.43$$

The factor of safety with the maximum shear stress theory is

$$\text{FS} = \frac{S_{yt}}{2\tau_{\max}} \qquad 18.44$$

Distortion Energy Theory

The *distortion energy theory* is similar in development to the strain energy method but is more strict. It is commonly used to predict tensile and shear failure in steel and other ductile parts subjected to static loading. The *von Mises stress* (also known as the *effective stress*), σ', is calculated from the principal stresses.

$$\sigma' = \sqrt{\sigma_1^2 + \sigma_2^2 - \sigma_1\sigma_2} \qquad 18.45$$

For triaxial loading, the von Mises stress is

$$\sigma' = \sqrt{\frac{1}{2}\left((\sigma_1 - \sigma_2)^2 + (\sigma_2 - \sigma_3)^2 + (\sigma_3 - \sigma_1)^2\right)} \qquad 18.46$$

The failure criterion is

$$\sigma' > S_{yt} \qquad 18.47$$

The factor of safety is

$$\text{FS} = \frac{S_{yt}}{\sigma'} \qquad 18.48$$

If the loading is pure torsion at failure, then $\sigma_1 = \sigma_2 = \pm\tau_{\max}$, and $\sigma_3 = 0$. If $\tau_{\max}$ is substituted for σ in Eq. 18.46 (with $\sigma_3 = 0$), an expression for the yield strength in shear is derived. Equation 18.49 predicts a larger yield strength in shear than did the maximum shear stress theory ($0.5S_{yt}$).

$$S_{ys} = \tau_{\max,\text{failure}} = \frac{S_{yt}}{\sqrt{3}} = 0.577 S_{yt} \qquad 18.49$$

SAMPLE PROBLEMS

Problem 1

A steel bar with the dimensions shown is subjected to an axial compressive load of 265 kN. The modulus of elasticity of the steel is 210 GPa, and Poisson's ratio is 0.3. What is the final thickness of the bar? Neglect buckling.

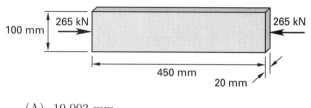

(A) 19.003 mm
(B) 19.006 mm
(C) 20.00 mm
(D) 20.004 mm

Solution

Compressive normal stresses are negative.

$$\delta_{\text{axial}} = \frac{PL}{AE}$$
$$= \frac{(-265 \times 10^3 \text{ N})(0.45 \text{ m})}{(0.1 \text{ m})(0.02 \text{ m})(210 \times 10^9 \text{ Pa})}$$
$$= -0.000284 \text{ m}$$

$$\epsilon_{\text{axial}} = \frac{\delta}{L} = \frac{-0.000284 \text{ m}}{0.45 \text{ m}}$$
$$= -0.00063 \text{ m/m}$$

$$\nu = -\frac{\epsilon_{\text{lateral}}}{\epsilon_{\text{axial}}}$$

$$\epsilon_{\text{lateral}} = -\nu(\epsilon_{\text{axial}})$$
$$= -(0.3)\left(-0.00063 \frac{\text{m}}{\text{m}}\right)$$
$$= 0.000189 \text{ m/m}$$

$$\Delta t = \epsilon_{\text{lateral}} t$$
$$= \left(0.000189 \frac{\text{m}}{\text{m}}\right)(0.02 \text{ m})$$
$$= 0.00000378 \text{ m} \quad (0.00378 \text{ mm})$$

$$t = 20 \text{ mm} + 0.00378 \text{ mm}$$
$$= 20.00378 \text{ mm} \quad (20.004 \text{ mm})$$

Answer is D.

Problem 2

A steel bar with the dimensions and cross section shown is suspended vertically. Three concentric downward loads are applied to the bar: 22 250 N at the lower end,

13 500 N at 0.35 m above the lower end, and 9000 N at 1 m above the lower end. The modulus of elasticity of the steel is 210 GPa. What is the total change in length of the bar?

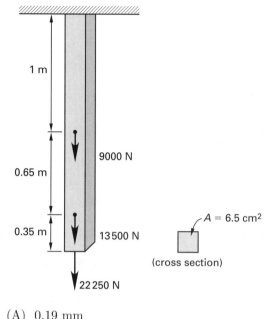

(A) 0.19 mm
(B) 0.56 mm
(C) 0.66 mm
(D) 0.98 mm

Solution

$$\delta_1 = \frac{PL}{AE} = \frac{(22\,250 \text{ N})(0.35 \text{ m})}{(6.5 \text{ cm}^2)\left(\frac{1 \text{ m}}{100 \text{ cm}}\right)^2 (210 \times 10^9 \text{ Pa})}$$
$$= 5.71 \times 10^{-5} \text{ m}$$

$$\delta_2 = \frac{(35\,750 \text{ N})(0.65 \text{ m})}{(6.5 \text{ cm}^2)\left(\frac{1 \text{ m}}{100 \text{ cm}}\right)^2 (210 \times 10^9 \text{ Pa})}$$
$$= 1.7 \times 10^{-4} \text{ m}$$

$$\delta_3 = \frac{(44\,750 \text{ N})(1 \text{ m})}{(6.5 \text{ cm}^2)\left(\frac{1 \text{ m}}{100 \text{ cm}}\right)^2 (210 \times 10^9 \text{ Pa})}$$
$$= 3.28 \times 10^{-4} \text{ m}$$

The total change in length is

$$\delta_1 + \delta_2 + \delta_3 = 0.057 \text{ mm} + 0.17 \text{ mm} + 0.328 \text{ mm}$$
$$= 0.555 \text{ mm} \quad (0.56 \text{ mm})$$

Answer is B.

Problems 3–5 refer to the following illustration. The element is subjected to the plane stress condition shown.

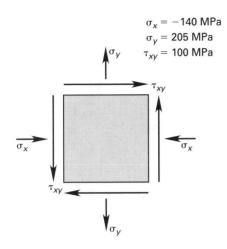

$$\sigma_x = -140 \text{ MPa}$$
$$\sigma_y = 205 \text{ MPa}$$
$$\tau_{xy} = 100 \text{ MPa}$$

Problem 3

What is the maximum shear stress?

(A) 100 MPa
(B) 160 MPa
(C) 200 MPa
(D) 210 MPa

Solution

There are two methods for solving the problem. The first method is to use Eq. 18.25; the second method is to draw Mohr's circle.

Solving by Eq. 18.25,

$$\tau_{max} = \pm\frac{1}{2}\sqrt{(\sigma_x - \sigma_y)^2 + (2\tau)^2}$$

$$= \frac{1}{2}\sqrt{(-140 \text{ MPa} - 205 \text{ MPa})^2 + \left((2)(100 \text{ MPa})\right)^2}$$

$$= 199.4 \text{ MPa} \quad (200 \text{ MPa})$$

Solving by Mohr's circle,

step 1:

$$\sigma_x = -140 \text{ MPa}$$
$$\sigma_y = 205 \text{ MPa}$$
$$\tau_{xy} = 100 \text{ MPa}$$

step 2: Draw σ-τ axes.

step 3: The circle center is

$$\sigma_c = \frac{1}{2}(\sigma_x + \sigma_y)$$

$$= \left(\frac{1}{2}\right)(-140 \text{ MPa} + 205 \text{ MPa})$$

$$= 32.5 \text{ MPa}$$

step 4: Plot points (−140 MPa, −100 MPa) and (205 MPa, 100 MPa).

step 5: Draw the diameter of the circle.

step 6: Draw the circle.

step 7: Find the radius of the circle.

step 8: Maximum shear stress is at the top of the circle, $\tau_{max} = 199.4$ MPa. (200 MPa)

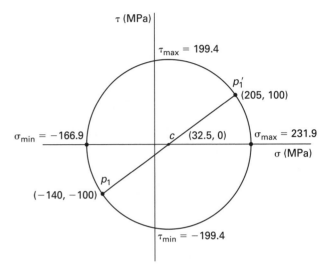

Answer is C.

Problem 4

What are the principal stresses?

(A) 140 MPa; −210 MPa
(B) 200 MPa; −140 MPa
(C) 230 MPa; −200 MPa
(D) 230 MPa; −170 MPa

Solution

$$\sigma_{max}, \sigma_{min} = \frac{1}{2}(\sigma_x + \sigma_y) \pm \tau_{max}$$

$$= \left(\frac{1}{2}\right)(-140 \text{ MPa} + 205 \text{ MPa}) \pm 199.4 \text{ MPa}$$

$$= 32.5 \text{ MPa} \pm 199.4 \text{ MPa}$$

$$\sigma_{max} = 231.9 \text{ MPa} \quad (230 \text{ MPa})$$

$$\sigma_{min} = -166.9 \text{ MPa} \quad (-170 \text{ MPa})$$

Alternatively, the principal stresses may be found from the Mohr's circle. (See illustration in next problem.)

Answer is D.

Problem 5

What are the orientations of the principal stress planes (relative to the x-axis)?

(A) $-75°$; $15°$
(B) $-35°$; $73°$
(C) $-27°$; $86°$
(D) $-15°$; $75°$

Solution

$$\theta = \frac{1}{2} \tan^{-1} \left(\frac{2\tau_{xy}}{\sigma_x - \sigma_y} \right) \quad \text{[Eq. 18.26]}$$

$$\theta = \frac{1}{2} \tan^{-1} \left(\frac{(2)(100 \text{ MPa})}{-140 \text{ MPa} - 205 \text{ MPa}} \right)$$

$$= -15.05° \text{ and } 74.95° \quad (-15° \text{ and } 75°)$$

Alternatively, the orientations can be found through the Mohr's circle construction as shown.

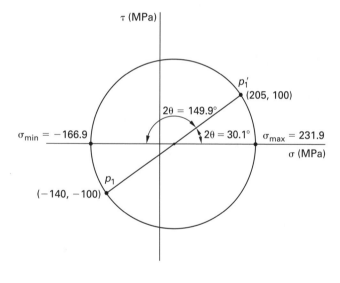

Answer is D.

FE-STYLE EXAM PROBLEMS

Problems 1–3 refer to the following situation.

A plane element in a body is subjected to a normal tensile stress in the x-direction of 84 000 kPa, as well as shear stresses of 28 000 kPa, as shown.

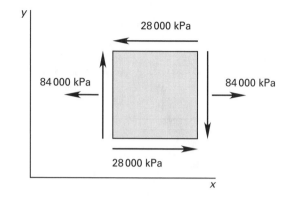

1. What are the principal stresses?

(A) 70 000 kPa and 14 000 kPa
(B) 84 000 kPa and 28 000 kPa
(C) 92 000 kPa and -8500 kPa
(D) 112 000 kPa and $-28 000$ kPa

2. What is the maximum shear stress?

(A) 28 000 kPa
(B) 33 500 kPa
(C) 42 000 kPa
(D) 50 500 kPa

3. What is the angle between the plane of the maximum (positive) shear stress and original plane of stress?

(A) $0°$
(B) $22°$
(C) $37°$
(D) $62°$

4. A 10 kg axial load is uniformly carried by an aluminum alloy pipe with an outside diameter of 10 cm and an inside diameter of 9.6 cm. The pipe is 1.2 m long. Young's modulus for the aluminum alloy is 7.5×10^4 MPa. How much is the pipe compressed? Neglect buckling.

(A) 0.00026 mm
(B) 0.0026 mm
(C) 0.11 mm
(D) 25 mm

5. A straight bar of uniform cross section is tested in tension. The bar's cross-sectional area is 6.5 cm^2, and the length is 4 m. When the tensile load reaches 85 kN, the total elongation is 2.5 mm. What is the modulus of elasticity?

(A) 210 GPa
(B) 240 GPa
(C) 270 GPa
(D) 300 GPa

For the following problems use the NCEES Hand-book as your only reference.

6. A solid round steel rod 6.25 mm in diameter and 375 mm long is rigidly connected to the end of a solid square brass rod 25 mm on a side and 300 mm long. The geometric axes of the bars are along the same line. An axial tensile force of 5.4 kN is applied at the extreme ends of the assemby. For steel, $E = 200$ GPa and for brass, $E = 90$ GPa. Determine the total elongation for the assembly.

    (A) 0.14 mm
    (B) 0.36 mm
    (C) 0.79 mm
    (D) 1.30 mm

7. A steel bar with a cross-sectional area of 8.5 cm^2 is subjected to axial tensile forces of 65 kN applied at each end of the bar. Determine the normal stress and the magnitude of the shearing stress on a plane inclined 30° from the direction of loading.

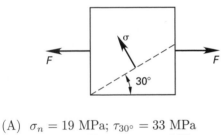

    (A) $\sigma_n = 19$ MPa; $\tau_{30°} = 33$ MPa
    (B) $\sigma_n = 25$ MPa; $\tau_{30°} = 18$ MPa
    (C) $\sigma_n = 35$ MPa; $\tau_{30°} = 24$ MPa
    (D) $\sigma_n = 57$ MPa; $\tau_{30°} = 33$ MPa

Problems 8 and 9 refer to the plane element shown. The element is subjected to shear stress and bidirectional tensile stresses.

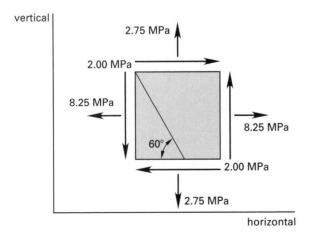

8. What is the approximate normal stress on a plane inclined at 60° from the horizontal as shown?

    (A) 5.1 MPa
    (B) 5.9 MPa
    (C) 7.0 MPa
    (D) 8.6 MPa

9. What is the shear stress parallel to a plane inclined 60° from the horizontal?

    (A) −1.4 MPa
    (B) −1.0 MPa
    (C) 1.0 MPa
    (D) 1.4 MPa

Problems 10–12 refer to the plane element shown. The element is acted upon by combined stresses. The material has a modulus of elasticity of 200 GPa and a Poisson's ratio of 0.27.

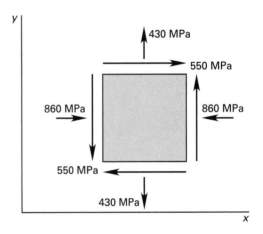

10. What is the approximate strain in the y-direction?

    (A) -4.9×10^{-3}
    (B) 0.58×10^{-3}
    (C) 0.99×10^{-3}
    (D) 3.3×10^{-3}

11. What is the shear modulus for this material?

    (A) 54 GPa
    (B) 80 GPa
    (C) 100 GPa
    (D) 110 GPa

12. What is the approximate shear strain on the x-y plane?

(A) -4.0×10^{-3}
(B) 2.5×10^{-3}
(C) 4.0×10^{-3}
(D) 7.0×10^{-3}

13. A horizontal beam carries a triangular distributed load over section AB. The horizontal beam has a mass of 148 kg/m. The beam is simply supported at point B. The cantilever end BC is restrained by a thin aluminum rod, CD. The aluminum rod has a cross-sectional area of 3.25 cm^2 and a modulus of elasticity of 6.9×10^{10} Pa. What is the change in length of the aluminum rod CD?

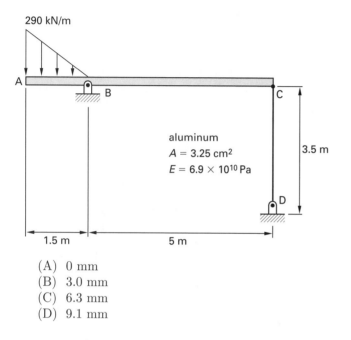

(A) 0 mm
(B) 3.0 mm
(C) 6.3 mm
(D) 9.1 mm

Problems 14–16 refer to the beam ABD illustrated. The beam supports an 80 kg vertical mass applied at D. The beam is pinned to the wall at point A and is pinned to vertical bar BC at point B. Vertical bar BC has a length of 5 m, a square cross section of 15 cm × 15 cm, a modulus of elasticity of 70 GPa, and a Poisson's ratio of 0.35.

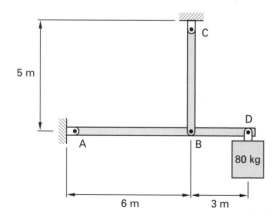

14. What is the normal stress in bar BC?

(A) 1.2 kPa
(B) 2.4 kPa
(C) 5.3 kPa
(D) 52 kPa

15. If the force in bar CB is 1200 N, what is the elongation of bar BC?

(A) 3.8 μm
(B) 8.7 μm
(C) 0.88 mm
(D) 0.38 cm

16. If the elongation of bar BC is 5.0 μm, what is the change in thickness of bar BC?

(A) −53 nm
(B) −3.9 nm
(C) −2.9 μm
(D) −3.9 μm

Problems 17 and 18 refer to the plane element shown. The element is acted upon by combined stresses.

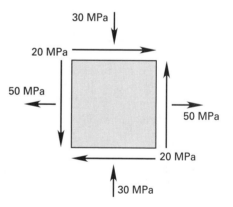

17. As determined from Mohr's circle, what is the maximum shear stress on the element?

(A) 20 MPa
(B) 42 MPa
(C) 45 MPa
(D) 50 MPa

18. As determined from Mohr's circle, what is the maximum normal stress on the element?

(A) 45 MPa
(B) 55 MPa
(C) 64 MPa
(D) 80 MPa

SOLUTIONS TO FE-STYLE EXAM PROBLEMS

1. τ_{xy} is negative according to the sign convention in Fig. 18.2.

$$\sigma_{\text{max,min}} = \frac{1}{2}(\sigma_x + \sigma_y) \pm \tau_{\text{max}}$$

$$= \frac{1}{2}(\sigma_x + \sigma_y) \pm \frac{1}{2}\sqrt{(\sigma_x - \sigma_y)^2 + (2\tau_{xy})^2}$$

$$= \frac{1}{2}(84\,000 \text{ kPa} + 0 \text{ kPa})$$

$$\pm \frac{1}{2}\sqrt{\begin{array}{c}(84\,000 \text{ kPa} - 0 \text{ kPa})^2 \\ + \big((2)(-28\,000 \text{ kPa})\big)^2\end{array}}$$

$$= 42\,000 \text{ kPa} \pm 50\,478 \text{ kPa}$$

$$= 92\,478 \text{ kPa}; -8478 \text{ kPa}$$

$$\quad (92\,000 \text{ kPa}; -8500 \text{ kPa})$$

Answer is C.

2.

$$\tau_{\text{max}} = \frac{1}{2}\sqrt{(\sigma_x - \sigma_y)^2 + (2\tau_{xy})^2}$$

$$= \frac{1}{2}\sqrt{(84\,000 \text{ kPa} - 0 \text{ kPa})^2 + \big((2)(-28\,000 \text{ kPa})\big)^2}$$

$$= 50\,478 \text{ kPa} \quad (50\,500 \text{ kPa})$$

Answer is D.

3.

$$\theta = \frac{1}{2}\tan^{-1}\left(\frac{\sigma_x - \sigma_y}{-2\tau_{xy}}\right)$$

$$= \frac{1}{2}\tan^{-1}\left(\frac{84\,000 \text{ kPa} - 0 \text{ kPa}}{(-2)(-28\,000 \text{ kPa})}\right)$$

$$= 28.15°; -61.85° \quad (28°; -62°)$$

Mohr's circle is constructed as follows. The circle center is

$$\frac{1}{2}(84\,000 \text{ kPa}) = 42\,000 \text{ kPa}$$

Solving Mohr's circle graphically shows that the orientation of the maximum shear stress is 28.15° (28°) and the orientation of the minimum shear stress is −61.85° (62°).

Answer is D.

4.

$$\delta = \frac{PL}{AE}$$

$$= \frac{(10 \text{ kg})\left(9.81 \frac{\text{m}}{\text{s}^2}\right)(1.2 \text{ m})}{\frac{\pi}{4}\big((0.10 \text{ m})^2 - (0.096 \text{ m})^2\big)(75 \times 10^9 \text{ Pa})}$$

$$= 2.55 \times 10^{-6} \text{ m} \quad (0.0026 \text{ mm})$$

Answer is B.

5.

$$\delta = \frac{PL}{AE}$$

$$E = \frac{PL}{A\delta}$$

$$= \frac{(85\,000 \text{ N})(4 \text{ m})\left(100 \frac{\text{cm}}{\text{m}}\right)^2}{(6.5 \text{ cm}^2)(0.0025 \text{ m})}$$

$$= 2.09 \times 10^{11} \text{ Pa} \quad (210 \text{ GPa})$$

Answer is A.

6. For steel,

$$\delta_{\text{steel}} = \frac{PL}{AE}$$

$$= \frac{(5.4 \times 10^3 \text{ N})(375 \times 10^{-3} \text{ m})}{\frac{\pi}{4}(625 \times 10^{-5} \text{ m})^2\left(200 \times 10^9 \frac{\text{N}}{\text{m}^2}\right)}$$

$$= 3.3002 \times 10^{-4} \text{ m}$$

For brass,

$$\delta_{\text{brass}} = \frac{PL}{AE} = \frac{(5.4 \times 10^3 \text{ N})(300 \times 10^{-3} \text{ m})}{(25 \times 10^{-3} \text{ m})^2\left(90 \times 10^9 \frac{\text{N}}{\text{m}^2}\right)}$$

$$= 2.8800 \times 10^{-5} \text{ m}$$

$$\delta_{\text{total}} = \delta_{\text{steel}} + \delta_{\text{brass}}$$

$$= 3.3002 \times 10^{-4} \text{ m} + 2.8800 \times 10^{-5} \text{ m}$$

$$= 3.5882 \times 10^{-4} \text{ m} \quad (0.36 \text{ mm})$$

Answer is B.

7.

$$\sigma_{\text{axial}} = \frac{P}{A}$$

$$= \frac{(65 \times 10^3 \text{ N})\left(100 \frac{\text{cm}}{\text{m}}\right)^2}{8.5 \text{ cm}^2}$$

$$= 7.647 \times 10^7 \text{ Pa} \quad (76.47 \text{ MPa})$$

The normal stress on the 30° plane is the stress oriented 120° (30° + 90°) to the axial plane.

$$\sigma_{\text{normal}} = \frac{\sigma_x + \sigma_y}{2} + \left(\frac{\sigma_x - \sigma_y}{2}\right)$$
$$\times \cos 2\theta + \tau_{xy}\sin 2\theta$$
$$= \frac{76.47\text{ MPa} + 0\text{ MPa}}{2}$$
$$+ \left(\frac{76.47\text{ MPa} - 0\text{ MPa}}{2}\right)$$
$$\times \cos\left((2)(120°)\right) + 0\text{ MPa}$$
$$= 19.12\text{ MPa}\quad(19\text{ MPa})$$

$$\tau_{30°} = -\left(\frac{\sigma_x - \sigma_y}{2}\right)$$
$$\times \sin 2\theta + \tau_{xy}\cos 2\theta$$
$$= -\left(\frac{76.47\text{ MPa} - 0\text{ MPa}}{2}\right)$$
$$\times \sin\left((2)(120°)\right) + 0\text{ MPa}$$
$$= 33.11\text{ MPa}\quad(33\text{ MPa})$$

Answer is A.

8.
$$\sigma_{x'} = \frac{1}{2}(\sigma_x + \sigma_y) + \frac{1}{2}(\sigma_x - \sigma_y)$$
$$\times \cos 2\theta + \tau_{xy}\sin 2\theta$$

As illustrated in Fig. 18.3, the θ is the angle inclined from the vertical. A plane inclined 60° from the horizontal is inclined 30° from the vertical. Therefore, $\theta = 30°$.

Both normal stresses are tensile, therefore, σ_x and σ_y are both positive. The shear stresses are in the same directions as shown in Fig. 18.2. Therefore, τ_{xy} is positive.

$$\sigma_{30°} = \left(\frac{1}{2}\right)(8.25\text{ MPa} + 2.75\text{ MPa})$$
$$+ \left(\frac{1}{2}\right)(8.25\text{ MPa} - 2.75\text{ MPa})\cos\left((2)(30°)\right)$$
$$+ (2.00\text{ MPa})\sin\left((2)(30°)\right)$$
$$= 8.61\text{ MPa}\quad(8.6\text{ MPa})$$

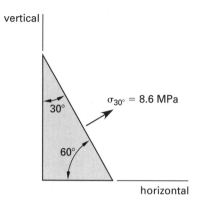

Answer is D.

9.
$$\tau_{x'y'} = -\frac{1}{2}(\sigma_x - \sigma_y)\sin 2\theta + \tau_{xy}\cos 2\theta$$

As in Prob. 8, $\theta = 30°$.

$$\tau_{30°} = \left(-\frac{1}{2}\right)(8.25\text{ MPa} - 2.75\text{ MPa})\sin\left((2)(30°)\right)$$
$$+ (2.00\text{ MPa})\cos\left((2)(30°)\right)$$
$$= -1.38\text{ MPa}\quad(-1.4\text{ MPa})$$

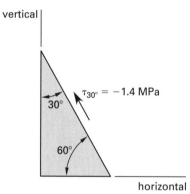

Answer is A.

10. The normal stress in the y-direction is tensile. Therefore, σ_y is positive. The normal stress in the x-direction is compressive. Therefore, σ_x is negative.

The modulus of elasticity is

$$E = (200\text{ GPa})\left(1000\,\frac{\text{MPa}}{\text{GPa}}\right) = 2 \times 10^5\text{ MPa}$$

The axial strain is

$$\epsilon_y = \frac{1}{E}\left(\sigma_y - \nu(\sigma_z + \sigma_x)\right)$$
$$= \left(\frac{1}{2 \times 10^5\text{ MPa}}\right)\left(\begin{array}{c}430\text{ MPa} - (0.27)\\\times(0\text{ MPa} - 860\text{ MPa})\end{array}\right)$$
$$= 3.31 \times 10^{-3}\quad(3.3 \times 10^{-3})$$

Answer is D.

11. The shear modulus is

$$G = \frac{E}{2(1+\nu)} = \frac{200 \text{ GPa}}{(2)(1+0.27)}$$

$$= 78.74 \text{ GPa} \quad (80 \text{ GPa})$$

Answer is B.

12. The shear stresses are in the directions shown in Fig. 18.2. Therefore, τ_{xy} is positive.

From Prob. 11, the shear modulus is

$$G = (78.74 \text{ GPa})\left(1000 \, \frac{\text{MPa}}{\text{GPa}}\right)$$

$$= 7.874 \times 10^4 \text{ MPa}$$

$$\gamma_{xy} = \frac{\tau_{xy}}{G} = \frac{550 \text{ MPa}}{7.874 \times 10^4 \text{ MPa}}$$

$$= 6.99 \times 10^{-3} \quad (7.0 \times 10^{-3})$$

Answer is D.

13. The triangular distribution has a total magnitude of

$$\frac{1}{2}bh = \left(\frac{1}{2}\right)(1.5 \text{ m})\left(290 \times 10^3 \, \frac{\text{N}}{\text{m}}\right)$$

$$= 217\,500 \text{ N}$$

This force can be assumed to act through the centroid of the triangular distribution located 2/3 of the length AB from support B. This is at

$$\left(\frac{2}{3}\right)(1.5 \text{ m}) = 1 \text{ m}$$

The beam self-loading for 1.5 m on either side of support B is equal. Only the last 5 m − 1.5 m = 3.5 m of the right side of the beam are unbalanced. The unbalanced beam load is

$$\left(148 \, \frac{\text{kg}}{\text{m}}\right)\left(9.81 \, \frac{\text{m}}{\text{s}^2}\right)(3.5 \text{ m}) = 5082 \text{ N}$$

Measuring distance from point B to the right, this unbalanced beam load acts at

$$1.5 \text{ m} + \left(\frac{1}{2}\right)(3.5 \text{ m}) = 3.25 \text{ m}$$

Take moments about point B to determine the tensile force in the aluminum rod CD.

$$M_{\text{B}}: (217\,500 \text{ N})(1 \text{ m}) - (5082 \text{ N})(3.25 \text{ m})$$

$$+ R_{\text{C}}(5 \text{ m}) = 0$$

$$R_{\text{C}} = -40\,197 \text{ N} \quad \text{[downward]}$$

The elongation is

$$\delta = \frac{PL}{AE}$$

$$= \frac{(40\,197 \text{ N})(3.5 \text{ m})\left(100 \, \frac{\text{cm}}{\text{m}}\right)^2}{(3.25 \text{ cm}^2)(6.9 \times 10^{10} \text{ Pa})}$$

$$= 0.00627 \text{ m} \quad (6.3 \text{ mm})$$

Answer is C.

14. Find the tensile stress in bar BC. Take moments about point A.

$$M_{\text{A}}: (-80 \text{ kg})\left(9.81 \, \frac{\text{m}}{\text{s}^2}\right)(6 \text{ m} + 3 \text{ m})$$

$$+ F_{\text{CB}}(6 \text{ m}) = 0$$

$$F_{\text{CB}} = 1177 \text{ N}$$

The cross-sectional area of bar BC is

$$A_{\text{BC}} = \frac{(15 \text{ cm})^2}{10^4 \, \frac{\text{cm}^2}{\text{m}^2}}$$

$$= 0.0225 \text{ m}^2$$

$$\sigma = \frac{F}{A} = \frac{1177 \text{ N}}{0.0225 \text{ m}^2}$$

$$= 52\,311 \text{ Pa} \quad (52 \text{ kPa})$$

Answer is D.

15. $\delta = \dfrac{PL}{AE} = \dfrac{(1200 \text{ N})(5 \text{ m})}{(0.0225 \text{ m}^2)(70 \times 10^9 \text{ Pa})}$

$$= 3.81 \times 10^{-6} \text{ m} \quad (3.8 \, \mu\text{m})$$

Answer is A.

16. The axial strain is

$$\epsilon_{\text{axial}} = \frac{\delta_{\text{axial}}}{L}$$

$$= \frac{(5.0 \, \mu\text{m})\left(1 \times 10^{-6} \, \frac{\text{m}}{\mu}\right)}{5 \text{ m}}$$

$$= 1 \times 10^{-6}$$

The lateral strain is

$$\epsilon_{\text{lateral}} = -\nu\epsilon_{\text{axial}}$$

$$= -(0.35)(1 \times 10^{-6})$$

$$= -3.5 \times 10^{-7}$$

The thickness of bar BC is 15 cm (0.15 m). The change in thickness is

$$\delta_{\text{lateral}} = \epsilon_{\text{lateral}} \times \text{thickness}$$
$$= (-3.5 \times 10^{-7})(0.15 \text{ m})$$
$$= -5.25 \times 10^{-8} \text{ m} \quad (-53 \text{ nm})$$

Answer is A.

17. Use Mohr's circle to solve for the maximum shear stress.

step 1: The stress in the x-direction is tensile, therefore, $\sigma_x = +50$ MPa. The stress in the y-direction is compressive, therefore, $\sigma_y = -30$ MPa. The shear stresses are in the same direction as shown in Fig. 18.2, so $\tau_{xy} = +20$ MPa.

step 2: Draw σ-τ axes.

step 3: Find the center of Mohr's circle. The center is located at

$$\sigma_{\text{center}} = \frac{1}{2}(\sigma_x + \sigma_y)$$
$$= \left(\frac{1}{2}\right)(50 \text{ MPa} + (-30 \text{ MPa}))$$
$$= 10 \text{ MPa}$$

step 4: Plot the point p_1 and p_1'.

$$p_1: (\sigma_{x'}, -\tau_{xy}) = (50 \text{ MPa}, -20 \text{ MPa})$$

step 5: Draw the circle with radius from the center at (10, 0 MPa) to p_1 at (50 MPa, −20 MPa).

step 6: Draw the complete circle.

step 7: The maximum shear stress corresponds to the top of the circle. At that point,

$$\tau_{\text{max}} \approx 45 \text{ MPa}$$

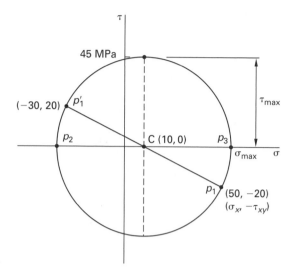

Alternatively, the actual radius can be calculated. (However, this is equivalent to solving Eq. 18.25 and is not strictly a graphical solution.)

$$r = \sqrt{\frac{1}{4}(\sigma_x - \sigma_y)^2 + \tau_{xy}^2}$$
$$= \sqrt{\left(\frac{1}{4}\right)(50 \text{ MPa} + 30 \text{ MPa})^2 + (20 \text{ MPa})^2}$$
$$= 44.7 \text{ MPa} \quad (45 \text{ MPa})$$

Answer is C.

18. The maximum (i.e., principal) stress on the element corresponds to point p_3. By observation, this is approximately 55 MPa. Alternatively, the σ value for the center could be added to the radius.

$$\sigma_{\text{max}} = 10 \text{ MPa} + 45 \text{ MPa}$$
$$= 55 \text{ MPa}$$

Answer is B.

19 Thermal, Hoop, and Torsional Stress

Subjects

THERMAL STRESS 19-1
THIN-WALLED TANKS 19-2
    Hoop Stress 19-2
    Axial Stress 19-2
    Principal Stresses in Tanks 19-2
    Thin-Walled Spherical Tanks 19-2
THICK-WALLED PRESSURE VESSELS . . 19-3
TORSIONAL STRESS 19-3
    Shafts 19-3
    Hollow, Thin-Walled Shells 19-3

Nomenclature

A	area	in^2	m^2
D	diameter	in	m
E	modulus of elasticity	lbf/in^2	MPa
G	shear modulus	lbf/in^2	MPa
J	polar moment of inertia	in^4	m^4
L	length	in	m
p	pressure	lbf/in^2	MPa
P	force	lbf	N
q	shear flow	lbf/in	N/m
r	radius	in	m
s	side length	in	m
t	temperature	°F	°C
t	thickness	in	m
T	torque	in-lbf	N·m

Symbols

α	coefficient of linear thermal expansion	1/°F	1/°C
γ	shear strain	–	–
δ	deformation	in	m
ϵ	axial strain	–	–
σ	normal stress	lbf/in^2	MPa
τ	shear stress	lbf/in^2	MPa
ϕ	angle of twist	rad	rad

Subscripts

0	initial
a	axial
h	hoop
i	inner
m	mean
o	outer
th	thermal

THERMAL STRESS

If the temperature of an object is changed, the object will experience length, area, and volume changes. The magnitude of these changes will depend on the *coefficient of linear thermal expansion*, α. The deformation is given by

$$\delta_{th} = \alpha L(t - t_0) \qquad 19.1$$

Changes in temperature affect all dimensions the same way. An increase in temperature will cause an increase in the dimensions, and likewise, a decrease in temperature will cause a decrease in the dimensions. It is a common misconception that a hole in a plate will decrease in size when the plate is heated (because the surrounding material "squeezes in" on the hole). In this case, the circumference of the hole is a linear dimension that follows Eq. 19.1. As the circumference increases, the hole area also increases.

Figure 19.1 Thermal Expansion of an Area

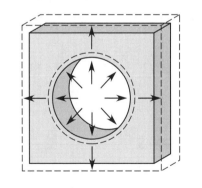

If Eq. 19.1 is rearranged, an expression for the *thermal strain* is obtained.

$$\epsilon_{th} = \frac{\delta_{th}}{L} = \alpha(t - t_0) \qquad 19.2$$

Thermal strain is handled in the same manner as strain due to an applied load. For example, if a bar is heated but is not allowed to expand, the thermal stress can be calculated from the thermal strain and Hooke's law.

$$\sigma_{th} = E\epsilon_{th} \qquad 19.3$$

MECHANICS
Stresses

Low values of the coefficient of expansion, such as with Pyrex™ glassware, result in low thermally induced stresses and insensitivity to temperature extremes. Intentional differences in the coefficients of expansion of two materials are used in *bimetallic elements*, such as thermostatic springs and strips.

Table 19.1 Average Coefficients of Linear Thermal Expansion

(Multiply all values by 10^{-6}.)

substance	1/°F	1/°C
aluminum alloy	12.8	23.0
brass	10.0	18.0
cast iron	5.6	10.1
chromium	3.8	6.8
concrete	6.7	12.0
copper	8.9	16.0
glass (plate)	4.9	8.8
glass (Pyrex™)	1.8	3.2
invar	0.39	0.7
lead	15.6	28.0
magnesium alloy	14.5	26.1
marble	6.5	11.7
platinum	5.0	9.0
quartz, fused	0.2	0.4
steel	6.5	11.7
tin	14.9	26.8
titanium alloy	4.9	8.8
tungsten	2.4	4.3
zinc	14.6	26.3

Multiply 1/°F by 9/5 to obtain 1/°C.
Multiply 1/°C by 5/9 to obtain 1/°F.

THIN-WALLED TANKS

Tanks under internal pressure experience circumferential, longitudinal, and radial stresses. If the wall thickness is small, the radial stress component is negligible and can be disregarded. A cylindrical tank can be assumed to be a *thin-walled tank* if the ratio of thickness-to-internal radius is less than approximately 0.1.

$$\frac{t}{R_i} < 0.1 \quad \text{[thin walled]} \qquad 19.4$$

A cylindrical tank with a wall thickness-to-radius ratio greater than 0.1 should be considered a *thick-walled pressure vessel*. In thick-walled tanks, radial stress is significant and cannot be disregarded, and for this reason, the radial and circumferential stresses vary with location through the tank wall.

Tanks under external pressure usually fail by buckling, not by yielding. For this reason, thin-wall equations cannot be used for tanks under external pressure.

Hoop Stress

The *hoop stress*, σ_h, also known as *circumferential stress* and *tangential stress*, for a cylindrical thin-walled tank under internal pressure, p, is derived from the free-body diagram of a cylinder. If the cyclinder tank is truly thin walled, it is not important which diameter, D (e.g., inner, mean, or outer), is used in Eq. 19.5. Although the inner diameter is used by common convention, the mean diameter will provide more accurate values as the wall thickness increases. The hoop stress is given by Eq. 19.5.

$$\sigma_h = \frac{pD}{2t} \qquad 19.5$$

Figure 19.2 Stresses in a Thin-Walled Tank

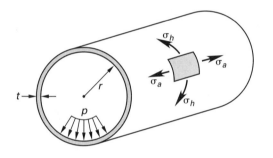

Axial Stress

When the cylindrical tank is closed at the ends like a soft drink can, the axial force on the ends produces a stress directed along the longitudinal axis known as the *longitudinal, long,* or *axial stress, σ_a*.

$$\sigma_a = \frac{pD}{4t} = \frac{\sigma_h}{2} \qquad 19.6$$

Principal Stresses in Tanks

The hoop and axial stresses are the principal stresses for pressure vessels when internal pressure is the only loading. If a three-dimensional portion of the shell is considered, the stress on the outside surface is zero. For this reason, the largest shear stress in three dimensions is $\sigma_h/2$ and is oriented at 45° to the surface.

Thin-Walled Spherical Tanks

Because of symmetry, the surface (tangential) stress of a spherical tank is the same in all directions.

$$\sigma = \frac{pD}{4t} \qquad 19.7$$

THICK-WALLED PRESSURE VESSELS

A thick-walled cylinder has a wall thickness-to-radius ratio greater than 0.1. In thick-walled tanks, radial stress is significant and cannot be disregarded. In *Lame's solution*, a thick-walled cylinder is assumed to be made up of thin laminar rings. This method shows that the radial and tangential (circumferential or hoop) stresses vary with location within the tank wall.

At every point in the cylinder, the tangential, radial, and long stresses are the principal stresses. Unless an external torsional shear stress is added, it is not necessary to use the combined stress equations.

The maximum radial, tangential, and shear stresses occur at the inner surface for both internal and external pressurization. (The terms *tangential stress* or *circumferential stress* are preferred over *hoop stress* when dealing with thick-walled cylinders.) Compressive stresses are negative.

For internal pressurization alone,

$$\sigma_{t,\max} = \frac{p_i(r_o^2 + r_i^2)}{r_o^2 - r_i^2} \qquad 19.8$$

$$\sigma_{r,\max} = -p_i \qquad 19.9$$

Cylinders under internal pressurization will also experience an axial stress in the direction of the end caps. This axial stress is calculated as the axial force divided by the annular area of the wall material.

$$\sigma_{\text{axial}} = \frac{F}{A} = \frac{p_i \pi r_i^2}{\pi(r_o^2 - r_i^2)} = \frac{p_i r_i^2}{r_o^2 - r_i^2} \qquad 19.10$$

For external pressurization alone,

$$\sigma_{t,\max} = \frac{-p_o(r_o^2 + r_i^2)}{r_o^2 - r_i^2} \qquad 19.11$$

$$\sigma_{r,\max} = -p_o \qquad 19.12$$

TORSIONAL STRESS

Shafts

Shear stress occurs when a shaft is placed in *torsion*. The shear stress at the outer surface of a bar of radius r, which is torsionally loaded by a torque, T, is

$$\tau = \frac{Tr}{J} \qquad 19.13$$

The *polar moment of inertia*, J, of a solid round shaft is

$$J = \frac{\pi r^4}{2} = \frac{\pi D^4}{32} \qquad 19.14$$

For a hollow round shaft,

$$J = \frac{\pi}{2}(r_o^4 - r_i^4) = \frac{\pi}{32}(D_o^4 - D_i^4) \qquad 19.15$$

If a shaft of length L carries a torque T, the angle of twist (in radians) will be

$$\phi = \frac{TL}{GJ} \quad \text{[radians]} \qquad 19.16$$

The *torsional stiffness* (*torsional spring constant* or *twisting moment per radian of twist*), denoted by the symbol k or c, is given by Eq. 19.17.

$$k = \frac{T}{\phi} = \frac{GJ}{L} \qquad 19.17$$

Hollow, Thin-Walled Shells

Shear stress due to torsion in a thin-walled, noncircular shell (also known as a *closed box*) acts around the perimeter of the tube, as shown in Fig. 19.3. The shear stress, τ, is given by Eq. 19.18. A_m is the area enclosed by the centerline of the shell.

$$\tau = \frac{T}{2A_m t} \qquad 19.18$$

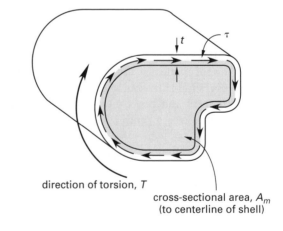

Figure 19.3 Torsion in Thin-Walled Shells

direction of torsion, T

cross-sectional area, A_m (to centerline of shell)

The shear stress at any point is not proportional to the distance from the centroid of the cross section. Rather, the *shear flow*, q, around the shell is constant, regardless of whether the wall thickness is constant or variable. The shear flow is the shear per unit length of the centerline path. At any point where the shell thickness is t,

$$q = \tau t = \frac{T}{2A_m} \quad \text{[constant]} \qquad 19.19$$

SAMPLE PROBLEMS

Problem 1

A steel pipe with a 50 mm outside diameter and a 43.75 mm inside diameter surrounds a solid brass rod 37.5 mm in diameter as shown. Both materials are joined to a rigid cover plate at each end. The assembly is free to expand longitudinally. The assembly is stress free at a temperature of 27°C. For steel, Young's modulus is 200 GN/m², and the coefficient of linear thermal expansion is 11.7×10^{-6} 1/°C. For brass, Young's modulus is 93.33 GN/m², and the coefficient of linear thermal expansion is 1.872×10^{-5} 1/°C. What is the stress in the steel tube when the temperature is raised to 121°C?

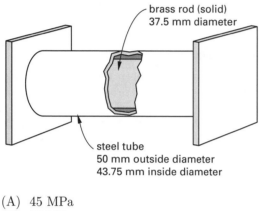

brass rod (solid)
37.5 mm diameter

steel tube
50 mm outside diameter
43.75 mm inside diameter

- (A) 45 MPa
- (B) 70 MPa
- (C) 85 MPa
- (D) 120 MPa

Solution

The brass has a higher coefficient of thermal expansion than the steel. The steel will not "let" the brass expand as much as it would if unconstrained. So, the brass is in compression. By similar reasoning, the steel is in tension. Both the steel and brass develop axial stresses. The steel's elongation is a combination of the thermal strain and the brass' tendency for greater thermal expansion.

For the steel,

$$\delta_{\text{steel}} = \alpha_{\text{steel}} L_{\text{steel}} \Delta t + \frac{P_{\text{steel}} L_{\text{steel}}}{A_{\text{steel}} E_{\text{steel}}}$$

For the brass, the thermal strain is resisted by the compressive strain.

$$\delta_{\text{brass}} = \alpha_{\text{brass}} L_{\text{brass}} \Delta t - \frac{P_{\text{brass}} L_{\text{brass}}}{A_{\text{brass}} E_{\text{brass}}}$$

Since both brass and steel are fixed to the same plates, their elongations are the same.

$$\alpha_{\text{steel}} L_{\text{steel}} \Delta t + \frac{P_{\text{steel}} L_{\text{steel}}}{A_{\text{steel}} E_{\text{steel}}}$$

$$= \alpha_{\text{brass}} L_{\text{brass}} \Delta t - \frac{P_{\text{brass}} L_{\text{brass}}}{A_{\text{brass}} E_{\text{brass}}}$$

$$L_{\text{steel}} = L_{\text{brass}} = L$$

$$\alpha_{\text{steel}} \Delta t + \frac{P_{\text{steel}}}{A_{\text{steel}} E_{\text{steel}}}$$

$$= \alpha_{\text{brass}} \Delta t - \frac{P_{\text{brass}}}{A_{\text{brass}} E_{\text{brass}}}$$

$$P_{\text{steel}} = P_{\text{brass}} = P$$

$$\Delta t (\alpha_{\text{brass}} - \alpha_{\text{steel}})$$
$$= P \left(\frac{1}{A_{\text{steel}} E_{\text{steel}}} + \frac{1}{A_{\text{brass}} E_{\text{brass}}} \right)$$

$$P = \frac{\Delta t (\alpha_{\text{brass}} - \alpha_{\text{steel}})}{\dfrac{1}{A_{\text{steel}} E_{\text{steel}}} + \dfrac{1}{A_{\text{brass}} E_{\text{brass}}}}$$

$$= \frac{(121°\text{C} - 27°\text{C}) \left(\begin{array}{c} 1.872 \times 10^{-5} \, \frac{1}{°\text{C}} \\ - 11.7 \times 10^{-6} \, \frac{1}{°\text{C}} \end{array} \right)}{ \begin{array}{c} \dfrac{1}{\left(\frac{\pi}{4}\right) \left((0.05 \text{ m})^2 - (0.04375 \text{ m})^2\right) \left(200 \times 10^9 \, \frac{\text{N}}{\text{m}^2}\right)} \\ + \dfrac{1}{\left(\frac{\pi}{4}\right) (0.0375 \text{ m})^2 \left(93.33 \times 10^9 \, \frac{\text{N}}{\text{m}^2}\right)} \end{array} }$$

$$= 32\,085 \text{ N}$$

$$\sigma_{\text{steel}} = \frac{P}{A_{\text{steel}}} = \frac{32\,085 \text{ N}}{\left(\frac{\pi}{4}\right) \left((0.05 \text{ m})^2 - (0.04375 \text{ m})^2\right)}$$

$$= 6.97 \times 10^7 \text{ N/m}^2 \quad (70 \text{ MPa})$$

Answer is B.

Problem 2

A compressed gas cylinder for use in a laboratory has an internal gage pressure of 8 MPa at the time of delivery. The outside diameter of the cylinder is 25 cm. If the steel has an allowable stress of 90 MPa, what is the required thickness of the wall?

- (A) 0.69 cm
- (B) 0.95 cm
- (C) 1.0 cm
- (D) 1.9 cm

Solution

Assume a thin-walled tank.

$$\sigma_h = \frac{pD}{2t}$$

$$t = \frac{pD}{2\sigma_h}$$

$$= \frac{p(\text{outside diameter} - 2t)}{2\sigma_h}$$

$$= \frac{(8 \text{ MPa})(25 \text{ cm} - 2t)}{(2)(90 \text{ MPa})}$$

$$180t = 200 - 16t$$

$$t = \frac{200}{196} = 1.01 \text{ cm} \quad (1.0 \text{ cm})$$

(Note: $t = 1.11$ cm if the distinction between inside and outside diameters is not made. $t = 1.06$ cm if the mean diameter is used.)

Check the thin-wall assumption.

$$\frac{t}{R_i} = \frac{t}{\dfrac{D_o - 2t}{2}} = \frac{1.01 \text{ cm}}{\dfrac{25 \text{ cm} - (2)(1.01 \text{ cm})}{2}}$$

$$= 0.088 < 0.1 \quad [\text{thin wall}]$$

Answer is C.

Problem 3

The maximum torque on a 15 cm diameter solid shaft is 13 500 N·m. What is the maximum shear stress in the shaft?

(A) 20.4 MPa
(B) 22.6 MPa
(C) 27.7 MPa
(D) 33.5 MPa

Solution

$$J = \frac{\pi r^4}{2} = \left(\frac{\pi}{2}\right)\left(\frac{0.15 \text{ m}}{2}\right)^4$$

$$= 4.97 \times 10^{-5} \text{ m}^4$$

$$\tau = \frac{Tr}{J}$$

$$= \frac{(13\,500 \text{ N·m})\left(\dfrac{0.15 \text{ m}}{2}\right)}{4.97 \times 10^{-5} \text{ m}^4}$$

$$= 20.37 \text{ MPa} \quad (20.4 \text{ MPa})$$

Answer is A.

Problem 4

One end of the hollow aluminum shaft is fixed, and the other end is connected to a gear with an outside diameter of 40 cm as shown. The gear is subjected to a tangential gear force of 45 kN. The shear modulus of the aluminum is 2.8×10^{10} Pa. What are the maximum angle of twist and the shear stress in the shaft?

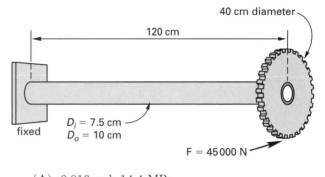

(A) 0.016 rad; 14.4 MPa
(B) 0.025 rad; 216 MPa
(C) 0.057 rad; 67.1 MPa
(D) 0.250 rad; 195 MPa

Solution

$$T = rF = \left(\frac{0.40 \text{ m}}{2}\right)(45\,000 \text{ N})$$

$$= 9000 \text{ N·m}$$

$$J = \frac{\pi}{2}(r_o^4 - r_i^4)$$

$$= \left(\frac{\pi}{2}\right)\left((0.05 \text{ m})^4 - (0.0375 \text{ m})^4\right)$$

$$= 6.71 \times 10^{-6} \text{ m}^4$$

$$\theta = \frac{TL}{GJ} = \frac{(9000 \text{ N·m})\left(\dfrac{120 \text{ cm}}{100 \dfrac{\text{cm}}{\text{m}}}\right)}{(2.8 \times 10^{10} \text{ Pa})(6.71 \times 10^{-6} \text{ m}^4)}$$

$$= 0.057 \text{ rad}$$

$$\tau = \frac{Tr}{J} = \frac{(9000 \text{ N·m})\left(\dfrac{0.10 \text{ m}}{2}\right)}{6.71 \times 10^{-6} \text{ m}^4}$$

$$= 67.06 \text{ MPa} \quad (67.1 \text{ MPa})$$

Answer is C.

FE-STYLE EXAM PROBLEMS

1. The glass window shown is subjected to a temperature change from 0°C to 50°C. The coefficient of thermal expansion for the glass is 8.8×10^{-6} 1/°C. What is the change in area of the glass?

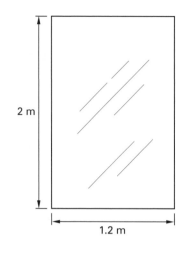

(A) 0.0004 m²
(B) 0.0013 m²
(C) 0.0021 m²
(D) 0.0028 m²

2. A rectangular steel beam is held between two rigid, unyielding walls 2.25 m apart. The modulus of elasticity of the steel is 210 GPa, the coefficient of thermal expansion is 11.7×10^{-6} 1/°C, and the cross-sectional area of the beam is 6.5 cm². If the beam temperature is increased by 30°C, what is the change in stress in the beam?

(A) 67.5 MPa   (compression)
(B) 73.7 MPa   (compression)
(C) 99.5 MPa   (tension)
(D) 166 MPa   (tension)

3. Which of the following statements is true for a pressurized cylindrical tank?

(A) Tangential stresses are independent of the radius of the vessel.
(B) Both longitudinal and tangential stresses are dependent on the radius of the vessel.
(C) Longitudinal stresses are greater than both the radial and tangential stresses.
(D) Longitudinal stresses are independent of the radius of the vessel.

4. The cylindrical steel tank shown is 3.5 m in diameter, 5 m high, and filled with a brine solution. Brine has a density of 1198 kg/m³. The thickness of the steel shell is 12.5 mm. What is the hoop stress in the steel 0.65 m above the rigid concrete pad? Neglect the weight of the tank.

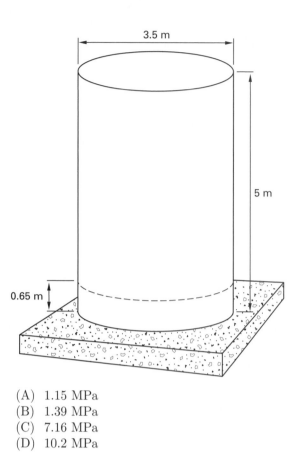

(A) 1.15 MPa
(B) 1.39 MPa
(C) 7.16 MPa
(D) 10.2 MPa

5. A spherical tank for storing gas under pressure is 25 m in diameter and made of steel 15 mm thick. The yield point of the material is 240 MPa. A factor of safety of 2.5 is desired. What is the maximum permissible internal pressure?

(A) 90 kPa
(B) 230 kPa
(C) 430 kPa
(D) 570 kPa

For the following problems use the NCEES Handbook as your only reference.

6. An aluminum (shear modulus $= 2.8 \times 10^{10}$ Pa) rod is 25 mm in diameter and 50 cm long. One end is rigidly fixed to a support. What torque must be applied to twist the rod 4.5° about its longitudinal axis?

(A) 26 N·m
(B) 84 N·m
(C) 110 N·m
(D) 170 N·m

7. A steel shaft of 200 mm diameter is twisted by a torque of 135.6 kN·m. What is the maximum shear stress in the shaft?

(A) 86 MPa
(B) 110 MPa
(C) 160 MPa
(D) 190 MPa

8. A circular shaft subjected to pure torsion will display which of the following?

(A) constant shear stress throughout the shaft
(B) maximum shear stress at the center of the shaft
(C) no shear stress throughout the shaft
(D) maximum shear stress at the outer fibers

9. A small cylindrical pressure tank has an internal gage pressure of 1600 Pa. The outside diameter is 75 mm and the wall thickness is 3 mm. What is the axial stress of the tank?

(A) 7700 Pa
(B) 9200 Pa
(C) 11 000 Pa
(D) 18 000 Pa

10. A point is located on the inside wall of a thin-walled tank under internal pressure. The tank does not experience any torsion. What is the maximum shear stress at that point, considering all three orthogonal directions?

(A) zero
(B) 25% of the hoop stress
(C) 50% of the hoop stress
(D) 75% of the hoop stress

11. A hollow, thin-walled shell has a wall thickness of 12.5 mm. The shell is acted upon by a 280 N·m torque. Find the approximate torsional shear stress in the shell's wall.

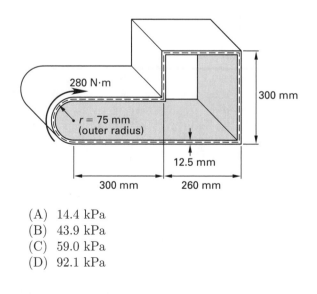

(A) 14.4 kPa
(B) 43.9 kPa
(C) 59.0 kPa
(D) 92.1 kPa

12. A 12.5 mm diameter steel rod is pinned between two rigid walls. The rod is initially unstressed. The rod's temperature subsequently increases 50°C. The rod is adequately stiffened and supported such that buckling does not occur. The coefficient of linear thermal expansion for steel is 11.7×10^{-6} 1/°C. The modulus of elasticity for steel is 210 GPa. What is the approximate axial force in the rod?

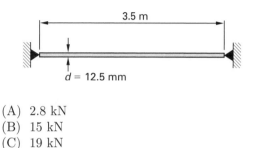

(A) 2.8 kN
(B) 15 kN
(C) 19 kN
(D) 58 kN

13. 10 km of steel railroad track are placed when the temperature is 20°C. The linear coefficient of thermal expansion for the rails is 11×10^{-6} 1/°C. The track is free to slide forward. How far apart will the ends of the track be when the temperature reaches 50°C?

(A) 10.0009 km
(B) 10.0027 km
(C) 10.0033 km
(D) 10.0118 km

14. The pressure gauge in an air cylinder reads 850 kPa. The cylinder is constructed of 6 mm rolled plate steel with an internal diameter of 350 mm. What is the tangential stress in the tank?

(A) 2.06 MPa
(B) 12.4 MPa
(C) 16.5 MPa
(D) 24.8 MPa

15. The pressure gauge in an air cylinder reads 850 kPa. The cylinder is constructed of 6 mm rolled plate steel with an internal diameter of 350 mm. What is the circumferential stress in the tank?

(A) 2.06 MPa
(B) 12.4 MPa
(C) 16.5 MPa
(D) 24.8 MPa

16. A deep-submersible diving bell has a cylindrical pressure hull with an outside diameter of 2.5 m and a wall thickness of 15 cm. The hull is expected to experience an external pressure of 50 MPa. How should the design proceed?

(A) The logarithmic-mean area should be used as the design area in stress calculations.

(B) A factor of safety of at least 4 should be used if the hull material is ductile, and of at least 8 if the hull material is notch-brittle.

(C) The hull can be designed as a thin-walled pressure vessel because the ratio of hull thickness to hull diameter is less than 0.1.

(D) The hull should be designed as a thick-walled pressure vessel because vessels under external pressure fail by buckling, not by yielding.

17. A structural steel tube with a 203 mm × 203 mm square cross section has an average wall thickness of 6.35 mm. The tube resists a torque of 8 N·m. What is the average shear flow?

(A) 100 N/m
(B) 200 N/m
(C) 400 N/m
(D) 800 N/m

18. A cantilever horizontal tube is acted upon by a vertical force and a torque at its free end. Where is the maximum stress in the cylinder?

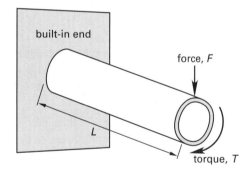

(A) at the upper surface at midlength ($L/2$)
(B) at the lower surface at the built-in end
(C) at the upper surface at the built-in end
(D) at both the upper and lower surfaces at the built-in end

SOLUTIONS TO FE-STYLE EXAM PROBLEMS

1. Changes in temperature affect each linear dimension.

$$\delta_{\text{width}} = \alpha L(t - t_o)$$
$$= \left(8.8 \times 10^{-6} \ \frac{1}{\degree\text{C}}\right)(1.2 \text{ m})(50\degree\text{C})$$
$$= 0.000528 \text{ m}$$

$$\delta_{\text{height}} = \left(8.8 \times 10^{-6} \ \frac{1}{\degree\text{C}}\right)(2 \text{ m})(50\degree\text{C})$$
$$= 0.00088 \text{ m}$$

$$A_{\text{initial}} = (2 \text{ m})(1.2 \text{ m}) = 2.4 \text{ m}^2$$

$$A_{\text{final}} = (2 \text{ m} + 0.00088 \text{ m})$$
$$\times (1.2 \text{ m} + 0.000528 \text{ m})$$
$$= 2.40211 \text{ m}^2$$

$$\text{change in area} = A_{\text{final}} - A_{\text{initial}}$$
$$= 2.40211 \text{ m}^2 - 2.4 \text{ m}^2$$
$$= 0.00211 \text{ m}^2 \quad (0.0021 \text{ m}^2)$$

Alternate solution: The area coefficient of thermal expansion is, for all practical purposes, equal to 2α.

The change in area is

$$\Delta A = 2\alpha A_o \Delta t$$
$$= (2)\left(8.8 \times 10^{-6} \ \frac{1}{\degree\text{C}}\right)(2.4 \text{ m}^2)(50\degree\text{C})$$
$$= 0.00211 \text{ m}^2 \quad (0.0021 \text{ m}^2)$$

Answer is C.

2.
$$\delta = \alpha L(t - t_o) - \frac{PL}{AE}$$
$$0 = \alpha L(t - t_o) - \frac{\sigma L}{E}$$
$$\sigma = \alpha(t - t_o)E$$
$$= \left(11.7 \times 10^{-6} \ \frac{1}{\degree\text{C}}\right)(30\degree\text{C})(2.1 \times 10^{11} \text{ Pa})$$
$$= 7.37 \times 10^7 \text{ Pa} \quad (73.7 \text{ MPa})$$

Answer is B.

3. In general, a tank under pressure will experience longitudinal, circumferential (also known as hoop or tangential), and radial stresses. All of these stresses are dependent on the radius of the tank, except that the radial stress is negligible if the tank is thin walled. All stresses are dependent on wall thickness.

Answer is B.

4.
$$\frac{t}{R} = \frac{0.0125 \text{ m}}{\dfrac{3.5 \text{ m}}{2}} = 0.008 < 0.1$$

Use formulas for thin-walled cylindrical tanks. The pressure is

$$p = \rho g h$$
$$= \left(1198 \ \frac{\text{kg}}{\text{m}^3}\right)\left(9.81 \ \frac{\text{m}}{\text{s}^2}\right)(5 \text{ m} - 0.65 \text{ m})$$
$$= 51\,123 \text{ Pa}$$

$$\sigma_h = \frac{pD}{2t} = \frac{(51\,123 \text{ Pa})(3.5 \text{ m})}{(2)(0.0125 \text{ m})}$$
$$= 7.157 \times 10^6 \text{ Pa} \quad (7.16 \text{ MPa})$$

Answer is C.

5.
$$\frac{t}{R} = \frac{0.015 \text{ m}}{\dfrac{25 \text{ m}}{2}}$$
$$= 1.2 \times 10^{-3} < 0.10 \qquad \text{[thin wall]}$$

$$\text{allowable stress} = \frac{240 \text{ MPa}}{2.5}$$
$$= 96 \text{ MPa}$$

For a thin-walled, spherical tank,

$$\sigma = \frac{pD}{4t}$$

$$p = \frac{4t\sigma}{D} = \frac{(4)(0.015 \text{ m})(96 \times 10^6 \text{ Pa})}{25 \text{ m}}$$
$$= 230\,400 \text{ Pa} \quad (230 \text{ kPa})$$

Answer is B.

6.

$$\phi = (4.5°)\left(\frac{2\pi \text{ rad}}{360°}\right) = 7.854 \times 10^{-2} \text{ rad}$$

$$J = \frac{\pi}{2}r^4 = \left(\frac{\pi}{2}\right)(0.0125 \text{ m})^4$$
$$= 3.83 \times 10^{-8} \text{ m}^4$$

$$\phi = \frac{TL}{GJ}$$

$$T = \frac{\phi GJ}{L}$$
$$= \frac{(7.854 \times 10^{-2} \text{ rad})(2.8 \times 10^{10} \text{ Pa})(3.83 \times 10^{-8} \text{ m}^4)}{0.5 \text{ m}}$$
$$= 168.5 \text{ N·m} \quad (170 \text{ N·m})$$

Answer is D.

7.
$$\tau = \frac{Tr}{J} = \frac{Tr}{\dfrac{\pi}{2}r^4}$$
$$= \frac{(135.6 \text{ kN·m})\left(\dfrac{0.2 \text{ m}}{2}\right)}{\left(\dfrac{\pi}{2}\right)\left(\dfrac{0.2 \text{ m}}{2}\right)^4}$$
$$= 86\,326 \text{ kN/m}^2 \quad (86 \text{ MPa})$$

Answer is A.

8. The shear stress increases from the center of a circular shaft to the outermost fiber, where the maximum shear stress is experienced.

Answer is D.

9. Check the ratio of wall thickness to radius.

$$\frac{t}{R} \approx \frac{3 \text{ mm}}{\dfrac{75 \text{ mm}}{2}} = 0.08$$

Since $t/R < 0.1$, this can be evaluated as a thin-walled tank.

It is not particularly important which diameter (i.e., inside, outside, or midpoint) is used. However, the axial force on each end of the tank depends on the inside diameter, so use it. The inside diameter is

$$D_i = D_o - 2t$$
$$= 75 \text{ mm} - (2)(3 \text{ mm})$$
$$= 69 \text{ mm}$$

The axial tensile stress is

$$\sigma_a = \frac{pD}{4t}$$
$$= \frac{(1600 \text{ Pa})(0.069 \text{ m})}{(4)(0.003 \text{ m})}$$
$$= 9200 \text{ Pa}$$

Answer is B.

10. The x-axis corresponds to the direction of the hoop (tangential, circumferential) stress. The y-axis will correspond to the direction of the radial stress.

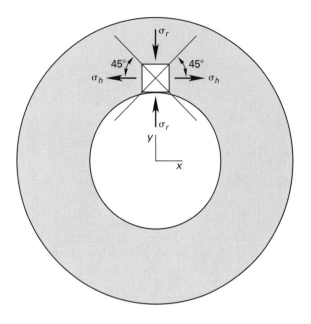

Calculate the maximum shear stress. Since there is no torsion, $\tau = 0$.

$$\tau_{1,2} = \pm\frac{1}{2}\sqrt{(\sigma_x - \sigma_y)^2 + (2\tau)^2}$$

$$= \pm\frac{1}{2}\sqrt{(\sigma_h - \sigma_r)^2 + ((2)(0))^2}$$

$$= \pm\frac{1}{2}(\sigma_h - \sigma_r)$$

The radial stress is nearly zero in a thin-walled tank.

$$\tau_{1,2} = \pm\frac{1}{2}\sigma_h \quad \text{(50\% of hoop stress)}$$

Answer is C.

---

11. The enclosed area to the centerline of the shell is

$$A = (0.3 \text{ m} - 0.0125 \text{ m})(0.26 \text{ m} - 0.00625 \text{ m})$$

$$+ \big((2)(0.075 \text{ m}) - 0.0125 \text{ m}\big)(0.3 \text{ m})$$

$$+ \left(\frac{\pi}{2}\right)(0.075 \text{ m} - 0.00625 \text{ m})^2$$

$$= 0.1216 \text{ m}^2$$

The torsional shear stress is

$$\tau = \frac{T}{2At} = \frac{280 \text{ N·m}}{(2)(0.1216 \text{ m}^2)(0.0125 \text{ m})}$$

$$= 92\,105 \text{ Pa} \quad \text{(92.1 kPa)}$$

Answer is D.

---

12. The thermal strain is

$$\epsilon_{\text{th}} = \alpha\Delta T$$

$$= \left(11.7 \times 10^{-6} \, \frac{1}{^\circ\text{C}}\right)(50^\circ\text{C})$$

$$= 0.000585 \text{ m/m}$$

The thermal stress is

$$\sigma_{\text{th}} = E\epsilon_{\text{th}}$$

$$= (210 \times 10^9 \text{ Pa})\left(0.000585 \, \frac{\text{m}}{\text{m}}\right)$$

$$= 1.2285 \times 10^8 \text{ Pa} \quad \text{(123 MPa)}$$

(This is less than the yield strength of steel.)

The force in the rod is

$$F = \sigma A$$

$$= (1.2285 \times 10^8 \text{ Pa})\left(\frac{\pi}{4}\right)(0.0125 \text{ m})^2$$

$$= 15\,076 \text{ N} \quad \text{(15 kN)}$$

Answer is B.

13.
$$\delta_{\text{th}} = \alpha L_0(T - T_0)$$

$$= \left(11 \times 10^{-6} \, \frac{1}{^\circ\text{C}}\right)(10\,000 \text{ m})$$

$$\times (50^\circ\text{C} - 20^\circ\text{C})$$

$$= 3.3 \text{ m}$$

$$L = L_0 + \delta_{\text{th}}$$

$$= 10 \text{ km} + 0.0033 \text{ km}$$

$$= 10.0033 \text{ km}$$

Answer is C.

---

14. Tangential stress is the same as hoop stress.

$$\sigma_h = \frac{pD}{2t}$$

$$= \frac{(850 \times 10^3 \text{ Pa})(0.35 \text{ m})}{(2)(0.006 \text{ m})}$$

$$= 2.479 \times 10^7 \text{ Pa} \quad \text{(24.8 MPa)}$$

Answer is D.

15. Circumferential stress is the same as hoop stress. This is the same as Prob. 14.

Answer is D.

16. Tanks under external pressure fail by buckling (i.e., collapse), not by yielding. They should not be designed using the simplistic formulas commonly used for thin-walled tanks under internal pressure.

Answer is D.

17. The mean area is

$$A_m = s^2 = (0.203 \text{ m} - 0.00635 \text{ m})^2$$
$$= 0.03867 \text{ m}^2$$

The shear flow is

$$q = \frac{T}{2A_m} = \frac{8 \text{ N·m}}{(2)(0.03867 \text{ m}^2)}$$
$$= 103 \text{ N/m} \quad (100 \text{ N/m})$$

Answer is A.

18. The torsional shear stress is the same everywhere in the tube. The maximum moment occurs at the built-in end, tensile at the upper surface and compressive at the lower surface. The absolute value of the combined stress at the upper and lower surfaces at the built-in end will be the same.

Answer is D.

**MECHANICS
Stresses**

20 Beams

Subjects

SHEARING FORCE AND
    BENDING MOMENT 20-1
        Sign Conventions 20-1
        Shear and Moment
            Relationships 20-2
        Shear and Moment Diagrams 20-2
STRESSES IN BEAMS 20-2
        Bending Stress 20-2
        Shear Stress 20-3
DEFLECTION OF BEAMS 20-4
COMPOSITE BEAMS 20-4

Nomenclature

A	area	in^2	m^2
b	width	in	m
c	distance to extreme fiber	in	m
C	couple	in-lbf	N·m
d	distance	in	m
E	modulus of elasticity	lbf/in^2	MPa
F	force	lbf	N
I	moment of inertia	in^4	m^4
M	moment	in-lbf	N·m
n	modular ratio	–	–
Q	statical moment	in^3	m^3
r	radius	in	m
R	reaction force	lbf	N
V	shear	lbf	N
w	load per unit length	lbf/in	N/m
y	distance from neutral axis	in	m

Symbols

ϵ	axial strain	–	–
ρ	radius of curvature	in	m
σ	normal stress	lbf/in^2	MPa
τ	shear stress	lbf/in^2	MPa

Subscripts

b	bending
c	centroidal
l	left
o	original
r	right
t	transformed
x	in x-direction
y	in y-direction

SHEARING FORCE AND BENDING MOMENT

Sign Conventions

The internal *shear* at a section is the sum of all vertical forces acting on an object up to that section. It has units of pounds, kips, newtons, and so on. Shear is not the same as shear stress, since the area of the object is not considered.

The most typical application is shear, V, at a section on a beam defined as the sum of all vertical forces between the section and one of the ends. The direction (i.e., to the left or right of the section) in which the summation proceeds is not important. Since the values of shear will differ only in sign for summation to the left and right ends, the direction that results in the fewest calculations should be selected.

$$V = \sum_{\substack{\text{[section to} \\ \text{one end]}}} F_i \qquad 20.1$$

Shear is positive when there is a net upward force to the left of a section, and it is negative when there is a net downward force to the left of the section.

Figure 20.1 Shear Sign Conventions

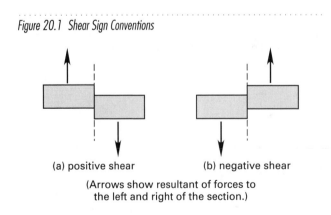

(a) positive shear   (b) negative shear

(Arrows show resultant of forces to
the left and right of the section.)

The *moment*, M, will be the algebraic sum of all moments and couples located between the section and one of the ends.

$$M = \sum_{\substack{\text{[section to} \\ \text{one end]}}} F_i d_i \; + \sum_{\substack{\text{[section to} \\ \text{one end]}}} C_i \qquad 20.2$$

MECHANICS
Beams

Moments in a beam are positive when the upper surface of the beam is in compression and the lower surface is in tension. Positive moments cause lengthening of the lower surface and shortening of the upper surface. A useful image with which to remember this convention is to imagine the beam "smiling" when the moment is positive.

Figure 20.2 Bending Moment Sign Conventions

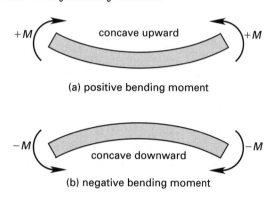

(a) positive bending moment

(b) negative bending moment

Shear and Moment Relationships

The change in magnitude of the shear at any point is equal to the integral of the load function, $w(x)$, or the area under the load diagram up to that point.

$$V_2 - V_1 = \int_{x_1}^{x_2} w(x)dx \qquad 20.3$$

$$w(x) = \frac{dV(x)}{dx} \qquad 20.4$$

The change in magnitude of the moment at any point is equal to the integral of the shear function, or the area under the shear diagram up to that point.

$$M_2 - M_1 = \int_{x_1}^{x_2} V(x)dx \qquad 20.5$$

$$V(x) = \frac{dM(x)}{dx} \qquad 20.6$$

Shear and Moment Diagrams

Both shear and moment can be described mathematically for simple loadings by the preceding equations, but the formulas become discontinuous as the loadings become more complex. It is more convenient to describe complex shear and moment functions graphically. Graphs of shear and moment as functions of position along the beam are known as *shear and moment diagrams*.

The following guidelines and conventions should be observed when constructing a *shear diagram*.

- The shear at any section is equal to the sum of the loads and reactions from the section to the left end.
- The magnitude of the shear at any section is equal to the slope of the moment function at that section.
- Loads and reactions acting upward are positive.
- The shear diagram is straight and sloping for uniformly distributed loads.
- The shear diagram is straight and horizontal between concentrated loads.
- The shear is undefined at points of concentrated loads.

The following guidelines and conventions should be observed when constructing a bending *moment diagram*. By convention, the moment diagram is drawn on the compression side of the beam.

- The moment at any section is equal to the sum of the moments and couples from the section to the left end.

- The change in magnitude of the moment at any section is the integral of the shear diagram, or the area under the shear diagram. A concentrated moment will produce a jump or discontinuity in the moment diagram.

- The maximum or minimum moment occurs where the shear is either zero or passes through zero.

- The moment diagram is parabolic and is curved downward for downward uniformly distributed loads.

STRESSES IN BEAMS

Bending Stress

Normal stress occurs in a bending beam, as shown in Fig. 20.3. Although it is a normal stress, the term *bending stress* or *flexural stress* is used to indicate the source of the stress. For positive bending moment, the lower surface of the beam experiences tensile stress while the upper surface of the beam experiences compressive stress. The bending stress distribution passes through zero at the centroid, or *neutral axis*, of the cross section. The distance from the neutral axis is y, and the distance from the neutral axis to the *extreme fiber* (i.e., the top or bottom surface most distant from the neutral axis) is c.

Bending stress varies with location (depth) within the beam. It is zero at the neutral axis, and increases linearly with distance from the neutral axis, as predicted by Eq. 20.7.

$$\sigma_b = -\frac{My}{I} \qquad \text{20.7}$$

Figure 20.3 Bending Stress Distribution at a Section in a Beam

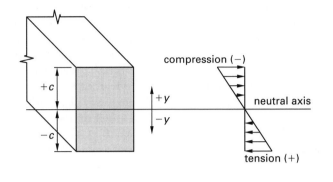

In Eq. 20.7, I is the centroidal area moment of inertia of the beam. The negative sign in Eq. 20.7, required by the convention that compression is negative, is commonly omitted.

Since the maximum stress will govern the design, y can be set equal to c to obtain the extreme fiber stress.

$$\sigma_{b,\max} = \frac{Mc}{I} \qquad \text{20.8}$$

Equation 20.8 shows that the maximum bending stress will occur at the section where the moment is maximum.

For standard structural shapes, I and c are fixed. Therefore, for design, the *elastic section modulus*, S, is often used.

$$S = \frac{I}{c} \qquad \text{20.9}$$

$$\sigma_b = \frac{M}{S} \qquad \text{20.10}$$

For a rectangular $b \times h$ section, the centroidal moment of inertia and section modulus are

$$I = \frac{bh^3}{12} \qquad \text{20.11}$$

$$S_{\text{rectangular}} = \frac{bh^2}{6} \qquad \text{20.12}$$

Shear Stress

The shear stresses in a vertical section of a beam consist of both horizontal and transverse (vertical) shear stresses.

The exact value of shear stress is dependent on the location, y, within the depth of the beam. The shear stress distribution is given by Eq. 20.13. The shear stress is zero at the top and bottom surfaces of the beam. For a regular shaped beam, the shear stress is maximum at the neutral axis.

$$\tau_{xy} = \frac{QV}{Ib} \qquad \text{20.13}$$

Figure 20.4 Dimensions for Shear Stress Calculations

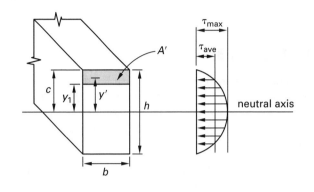

In Eq. 20.13, I is the area moment of inertia and b is the width or thickness of the beam at the depth y within the beam where the shear stress is to be found. The *first* (or *statical*) *moment of the area* of the beam with respect to the neutral axis, Q, is defined by Eq. 20.14.

$$Q = \int_{y_1}^{c} y\, dA \qquad \text{20.14}$$

For rectangular beams, $dA = b\,dy$. Then, the moment of the area A' above layer y is equal to the product of the area and the distance from the centroidal axis to the centroid of the area.

$$Q = y'A' \qquad \text{20.15}$$

For a rectangular beam, Eq. 20.13 can be simplified. The maximum shear stress is 50% higher than the average shear stress.

$$\tau_{\max,\text{rectangular}} = \frac{3V}{2A} = \frac{3V}{2bh} = 1.5\tau_{\text{ave}} \qquad \text{20.16}$$

For a beam with a circular cross section, the maximum shear stress is

$$\tau_{\max,\text{circular}} = \frac{4V}{3A} = \frac{4V}{3\pi r^2} \qquad \text{20.17}$$

For a steel beam with web thickness t_{web} and depth d, the web shear stress is approximated by

$$\tau_{\text{ave}} = \frac{V}{A_{\text{web}}} = \frac{V}{dt_{\text{web}}} \qquad \text{20.18}$$

Figure 20.5 Dimensions of a Steel Beam

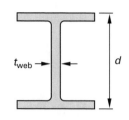

DEFLECTION OF BEAMS

The curvature of a beam caused by a bending moment is given by Eq. 20.19, where ρ is the *radius of curvature*, c is the largest distance from the neutral axis of the beam, and $\epsilon_{\max}$ is the maximum longitudinal normal strain in the beam.

$$\frac{1}{\rho} = \frac{\epsilon_{\max}}{c} = \frac{M}{EI} = \frac{d^2y}{dx^2} = \frac{d\theta}{dx} \qquad 20.19$$

$$\epsilon_{\max} = \frac{c}{\rho} \qquad 20.20$$

Using the preceding relationships, the deflection and slope of a loaded beam are related to the moment $M(x)$, shear $V(x)$, and load $w(x)$ by Eqs. 20.21 through 20.25.

$$y = \text{deflection} \qquad 20.21$$

$$y' = \frac{dy}{dx} = \text{slope} \qquad 20.22$$

$$y'' = \frac{d^2y}{dx^2} = \frac{M(x)}{EI} \qquad 20.23$$

$$y''' = \frac{d^3y}{dx^3} = \frac{V(x)}{EI} \qquad 20.24$$

$$y'''' = \frac{d^4y}{dx^4} = \frac{w(x)}{EI} \qquad 20.25$$

If the moment function, $M(x)$, is known for a section of the beam, the deflection at any point on that section can be found from Eq. 20.26. The constants of integration are determined from the beam boundary conditions in Table 20.1.

$$EIy = \int \int M(x)dx \qquad 20.26$$

Table 20.1 Beam Boundary Conditions

end condition	y	y'	y''	V	M
simple support	0				0
built-in support	0	0			
free end			0	0	0
hinge					0

Commonly used beam deflection formulas are compiled into Table 20.2 at the end of this chapter. These formulas should never need to be derived and should be used whenever possible.

When multiple loads act simultaneously on a beam, all of the loads contribute to deflection. The principle of *superposition* permits the deflections at a point to be calculated as the sum of the deflections from each individual load acting singly. Superposition can also be used to calculate the shear and moment at a point and to draw the shear and moment diagrams. This principle is valid as long as the normal stress and strain are related by the modulus of elasticity, E. Generally this is true when the deflections are not excessive and all stresses are kept less than the yield point of the beam material.

COMPOSITE BEAMS

A *composite structure* is one in which two or more different materials are used. Each material carries part of the applied load. Examples of composite structures include steel-reinforced concrete and timber beams with bolted-on steel plates.

Most simple composite structures can be analyzed using the *method of consistent deformations*, also known as the *transformation method*. This method assumes that the strains are the same in both materials at the interface between them. Although the strains are the same, the stresses in the two adjacent materials are not equal, since stresses are proportional to the moduli of elasticity.

The transformation method starts by determining the modulus of elasticity for each (usually two in number) of the materials in the composite beam and then calculating the *modular ratio*, n. E_{weaker} is the smaller modulus of elasticity.

$$n = \frac{E}{E_{\text{weaker}}} \qquad 20.27$$

The area of the stronger material is increased by a factor of n. The transformed area is used to calculate the transformed composite area, $A_{c,t}$, or transformed moment of inertia, $I_{c,t}$. For compression and tension members, the stresses in the weaker and stronger materials are

$$\sigma_{\text{weaker}} = \frac{F}{A_{c,t}} \qquad 20.28$$

$$\sigma_{\text{stronger}} = \frac{nF}{A_{c,t}} \qquad 20.29$$

For beams in bending, the bending stresses in the weaker and stronger materials are

$$\sigma_{\text{weaker}} = \frac{Mc_{\text{weaker}}}{I_{c,t}} \qquad 20.30$$

$$\sigma_{\text{stronger}} = \frac{nMc_{\text{stronger}}}{I_{c,t}} \qquad 20.31$$

SAMPLE PROBLEMS

Problem 1

For the beam loaded as shown, which of the following diagrams correctly represents the shape of the shear diagram? (Diagrams not to scale.)

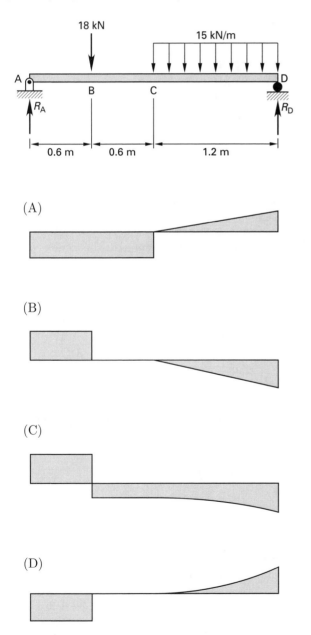

(A)

(B)

(C)

(D)

Problem 2

For the beam shown, where does the maximum moment occur?

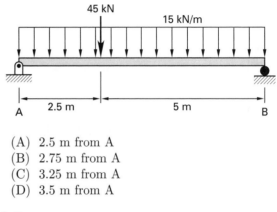

(A) 2.5 m from A
(B) 2.75 m from A
(C) 3.25 m from A
(D) 3.5 m from A

Solution

Due to symmetry of the moments caused by the applied loads, the reactions at A and B are equal. The shear on the left end of the beam is equal to +18 kN. The shear is constant from A to B, decreases by 18 kN at B, and is constant from B to C. From C to D the shear decreases linearly from zero to −18 kN.

Answer is B.

Solution

Draw the shear and bending moment diagrams.

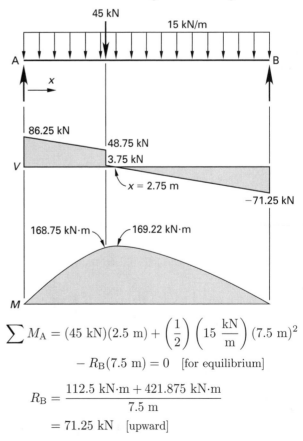

$$\sum M_{\text{A}} = (45 \text{ kN})(2.5 \text{ m}) + \left(\frac{1}{2}\right)\left(15 \ \frac{\text{kN}}{\text{m}}\right)(7.5 \text{ m})^2$$

$$- R_{\text{B}}(7.5 \text{ m}) = 0 \quad \text{[for equilibrium]}$$

$$R_{\text{B}} = \frac{112.5 \text{ kN·m} + 421.875 \text{ kN·m}}{7.5 \text{ m}}$$

$$= 71.25 \text{ kN} \quad \text{[upward]}$$

$$\sum F_y = R_A - 45 \text{ kN} - \left(15 \ \frac{\text{kN}}{\text{m}}\right)(7.5 \text{ m})$$
$$+ 71.25 \text{ kN}$$
$$= 0$$

$$R_A = 86.25 \text{ kN} \quad [\text{upward}]$$

Point of zero shear is where

$$86.25 \text{ kN} - 45 \text{ kN} - \left(15 \ \frac{\text{kN}}{\text{m}}\right)x = 0$$

$$x = 2.75 \text{ m} \quad [\text{from left end}]$$

Answer is B.

Problem 3

For the beam shown, find the vertical shear at point B.

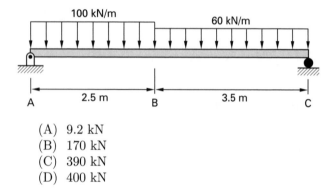

100 kN/m   60 kN/m

A   2.5 m   B   3.5 m   C

(A) 9.2 kN
(B) 170 kN
(C) 390 kN
(D) 400 kN

Solution

First, find the vertical reaction at A.

$$\sum M_C = R_A(6 \text{ m})$$
$$- \left(100 \ \frac{\text{kN}}{\text{m}}\right)(2.5 \text{ m})\left(3.5 \text{ m} + \frac{2.5 \text{ m}}{2}\right)$$
$$- \left(60 \ \frac{\text{kN}}{\text{m}}\right)(3.5 \text{ m})\left(\frac{3.5 \text{ m}}{2}\right)$$
$$= 0$$

$$R_A = 259.2 \text{ kN}$$

Summing forces from the left end to point B,

$$V_B = 259.2 \text{ kN} - \left(100 \ \frac{\text{kN}}{\text{m}}\right)(2.5 \text{ m})$$
$$= 9.2 \text{ kN}$$

Answer is A.

Problem 4

Find the maximum compressive stress in the beam shown.

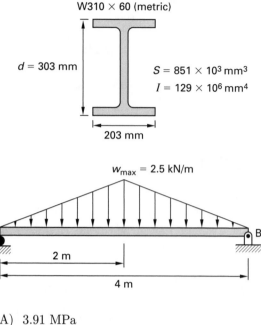

W310 × 60 (metric)

$d = 303$ mm

$S = 851 \times 10^3 \text{ mm}^3$
$I = 129 \times 10^6 \text{ mm}^4$

203 mm

$w_{\text{max}} = 2.5$ kN/m

A   2 m   B

4 m

(A) 3.91 MPa
(B) 4.23 MPa
(C) 5.00 MPa
(D) 5.05 MPa

Solution

Due to symmetry of the applied load, $R_A = R_B$.

$$R_A = R_B = \left(\frac{1}{2}\right)\left(2.5 \ \frac{\text{kN}}{\text{m}}\right)(2 \text{ m}) = 2.5 \text{ kN}$$

The maximum moment occurs at the center of the beam, where the shear is zero.

$$M_{\text{max}} = R_x - W_{\bar{x}}$$
$$= (2.5 \text{ kN})(2 \text{ m})$$
$$- \left(\frac{1}{2}\right)(2 \text{ m})\left(2.5 \ \frac{\text{kN}}{\text{m}}\right)\left(\frac{2 \text{ m}}{3}\right)$$
$$= 3.33 \text{ kN·m}$$
$$\sigma_{\text{max}} = \frac{Mc}{I}$$
$$= \frac{(3.33 \times 10^3 \text{ n·m})\left(\frac{0.303 \text{ m}}{2}\right)\left(1000 \ \frac{\text{mm}}{\text{m}}\right)^4}{129 \times 10^6 \text{ mm}^4}$$
$$= 3.91 \times 10^6 \text{ Pa} \quad (3.91 \text{ MPa})$$

Answer is A.

Problem 5

A 25 mm × 25 mm beam is loaded at its tip by a pair of 9 kN forces as shown. The modulus of elasticity of the beam is 210 GPa. What is the vertical deflection of the structure at A?

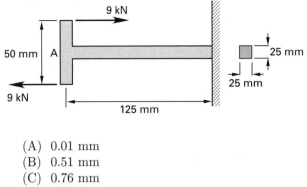

(A) 0.01 mm
(B) 0.51 mm
(C) 0.76 mm
(D) 1.02 mm

Solution

This is an example of a cantilever with an end moment (Case 3).

$$M = Fx = (9000 \text{ N})(0.05 \text{ m})$$

$$= 450 \text{ N·m}$$

$$\delta_{\max} = \frac{M_o L^2}{2EI}$$

$$= \frac{(450 \text{ N·m})(0.125 \text{ m})^2}{(2)(210 \times 10^9 \text{ Pa})\left(\dfrac{(0.025 \text{ m})(0.025 \text{ m})^3}{12}\right)}$$

$$= 5.14 \times 10^{-4} \text{ m} \quad (0.514 \text{ mm}) \quad [\text{at } x = 0]$$

Answer is B.

FE-STYLE EXAM PROBLEMS

1. What is the bending stress at a section of a loaded beam at its neutral axis?

(A) a combination of both shear and moment at that section
(B) equal to the shear at that section
(C) the maximum stress at that section
(D) zero

2. For a simply supported beam, where does the maximum shear stress occur?

(A) at the section of maximum moment
(B) at the top fibers
(C) at the bottom fibers
(D) at the supports

Problems 3 and 4 refer to the following simply supported beam.

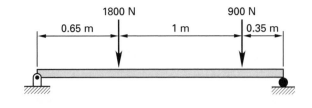

3. What is the maximum bending moment?

(A) 340 N·m
(B) 460 N·m
(C) 660 N·m
(D) 890 N·m

4. What is the maximum shear?

(A) 430 N
(B) 900 N
(C) 1330 N
(D) 1370 N

5. For the fixed steel rod shown, what is the force, F, necessary to deflect the rod a vertical distance of 7.5 mm?

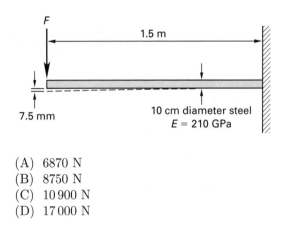

(A) 6870 N
(B) 8750 N
(C) 10 900 N
(D) 17 000 N

For the following problems use the NCEES Handbook as your only reference.

6. For the beam shown, what is the maximum compressive stress at section D-D, 1.5 m from the left end?

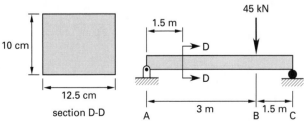

section D-D

(A) 63 MPa
(B) 108 MPa
(C) 225 MPa
(D) 334 MPa

7. If the beam in Prob. 6 has a tee-shaped cross section (instead of rectangular) with a moment of inertia of 1192.8 cm^4, what is the maximum tensile stress at section D-D?

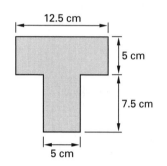

(A) 75 MPa
(B) 100 MPa
(C) 120 MPa
(D) 140 MPa

8. A rectangular beam has a cross section of 5 cm wide by 10 cm deep and a maximum shear of 2250 N. What is the maximum shear stress in the beam?

(A) 416 kPa
(B) 567 kPa
(C) 675 kPa
(D) 790 kPa

9. A cantilever beam is 2.5 m long and has a 5 cm × 15 cm cross section. An unknown tensile force is applied 2.5 cm from the upper surface of the beam, directed downward at 20° from the horizontal. The maximum allowable shear stress is 205 MPa. Neglect the beam's own mass and buckling. What is the approximate maximum tensile force the beam can support?

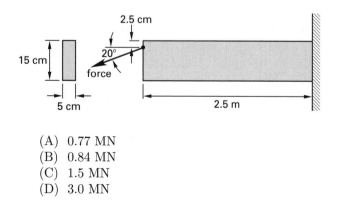

(A) 0.77 MN
(B) 0.84 MN
(C) 1.5 MN
(D) 3.0 MN

Problems 10–12 refer to the cantilevered structural section shown. The beam is manufactured from steel with a modulus of elasticity of 210 GPa. The beam's cross-sectional area is 37.9 cm^2; its moment of inertia is 2880 cm^4. The beam has a mass of 45.9 kg/m. A 6000 N force is applied at the top of the beam, at an angle of 30° from the horizontal. Neglect buckling.

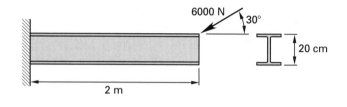

10. What is the maximum shear force in the beam?

(A) 3000 N
(B) 3900 N
(C) 5200 N
(D) 6100 N

11. What is the approximate maximum bending moment in the beam?

(A) 6070 N·m
(B) 6380 N·m
(C) 6850 N·m
(D) 7470 N·m

12. What is the deflection at the tip of the beam due to the external force alone (i.e., neglecting the beam's own mass)?

(A) 3.15 mm   (downward)
(B) 2.31 mm   (downward)
(C) 1.32 mm   (downward)
(D) 1.15 mm   (downward)

13. A simply supported beam supports a triangular distributed load as shown. The peak load at the right end of the beam is 5 N/m. What is the bending moment at a point 7 m from the left end of the beam?

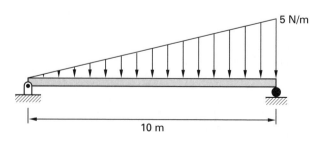

(A) 15 N·m
(B) 17 N·m
(C) 28 N·m
(D) 30 N·m

Problems 14 and 15 refer to the propped cantilever shown. The beam shown is fixed at one end and simply supported at the other end. The beam has a mass of 30.6 kg/m. The modulus of elasticity of the beam is 210 GPa; the moment of inertia is 2880 cm^4.

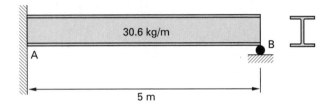

14. What is the reaction at the simply supported end?

(A) 72 N
(B) 510 N
(C) 560 N
(D) 770 N

15. The support at B is removed and replaced by a variable force. The force is adjusted until it is 900 N. What is the net deflection of the beam at a point 3.5 m from the fixed end?

(A) 0.29 mm (downward)
(B) 0.32 mm (downward)
(C) −1.2 mm (upward)
(D) −1.9 mm (upward)

SOLUTIONS TO FE-STYLE EXAM PROBLEMS

1. The neutral axis is the plane in the beam where the normal stress is zero.

Answer is D.

2. The shear stress is maximum where the shear is maximum. For a simply supported beam, the maximum shear is usually at the supports. Within the cross section of a beam, the maximum shear stress is at the neutral axis, not at the extreme fibers.

Answer is D.

3. Draw the shear and bending moment diagrams.

$$\sum M_B = -R_A(0.65 \text{ m} + 1 \text{ m} + 0.35 \text{ m})$$
$$+ (1800 \text{ N})(1 \text{ m} + 0.35 \text{ m})$$
$$+ (900 \text{ N})(0.35 \text{ m})$$
$$= 0$$
$$R_A = 1372.5 \text{ N}$$
$$\sum F_y = R_B + 1372.5 \text{ N} - 1800 \text{ N} - 900 \text{ N}$$
$$= 0$$
$$R_B = 1327.5 \text{ N}$$

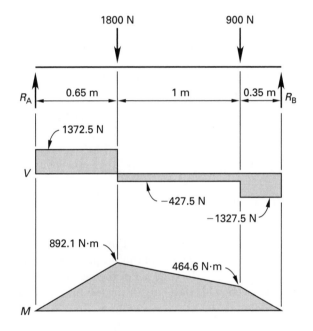

The maximum moment occurs 0.65 m from the left end of the beam and is equal to 892 N·m.

Answer is D.

4. From the shear and bending moment diagram in Prob. 3, the maximum shear is 1372.5 N.

Answer is D.

5. From Table 20.2 for a cantilever beam loaded at its tip (Case 1), with $a = L$,

$$\delta_{\max} = \left(\frac{Pa^2}{6EI}\right)(3L - a) = \frac{PL^3}{3EI}$$

$$P = \frac{3EI\delta_{max}}{L^3}$$

$$= \frac{(3)(210 \times 10^9 \text{ Pa})\left(\frac{\pi}{4}\right)(0.05 \text{ m})^4(0.0075 \text{ m})}{(1.5 \text{ m})^3}$$

$$= 6872 \text{ N} \quad (6870 \text{ N}) \quad \text{[downward]}$$

Answer is A.

6. Find the reaction at A.

$$\sum M_C = R_A(4.5 \text{ m}) - (45 \text{ kN})(1.5 \text{ m})$$
$$= 0$$
$$R_A = 15 \text{ kN}$$

The bending moment at section D-D is

$$(15 \text{ kN})(1.5 \text{ m}) = 22.5 \text{ kN·m}$$

The maximum compressive stress is at the top fiber of the beam section.

$$\sigma_{max} = \frac{Mc}{I} = \frac{(22.5 \times 10^3 \text{ N·m})(0.05 \text{ m})}{\dfrac{(0.125 \text{ m})(0.10 \text{ m})^3}{12}}$$

$$= 108 \text{ MPa}$$

Answer is B.

7. The moment is positive, so the maximum tensile stress occurs at the bottom of the beam. The distance, c, is measured from the centroid of the beam. The location of the centroid, measured from the bottom of the beam, is

$$y_c = \frac{\sum y_i A_i}{\sum A_i}$$

$$= \frac{(3.75 \text{ cm})(5 \text{ cm})(7.5 \text{ cm}) + (10 \text{ cm})(12.5 \text{ cm})(5 \text{ cm})}{(5 \text{ cm})(7.5 \text{ cm}) + (12.5 \text{ cm})(5 \text{ cm})}$$

$$= 7.65625 \text{ cm} \quad \begin{bmatrix} \text{distance from centroid to} \\ \text{bottom of beam section} \end{bmatrix}$$

From Prob. 6, $M = 22.5$ kN·m.

$$\sigma_{max,tension} = \frac{Mc}{I}$$

$$= \frac{(22.5 \times 10^3 \text{ N·m})(0.0765625 \text{ m}) \times \left(100 \dfrac{\text{cm}}{\text{m}}\right)^4}{1192.8 \text{ cm}^4}$$

$$= 144.4 \text{ MPa} \quad (140 \text{ MPa})$$

Answer is D.

8.
$$\tau_{max} = \frac{3V}{2A} = \frac{(3)(2250 \text{ N})}{(2)(0.05 \text{ m})(0.1 \text{ m})}$$
$$= 675 \text{ kPa}$$

Answer is C.

9. Determine the maximum vertical shear that the beam can sustain.

$$\tau_{max} = \frac{3V}{2A}$$

$$V = \frac{2A\tau_{max}}{3} = \frac{(2)(0.05 \text{ m})(0.15 \text{ m})(205 \times 10^6 \text{ Pa})}{3}$$

$$= 1.025 \times 10^6 \text{ N} \quad (1.025 \text{ MN})$$

The unknown force is

$$P = \frac{V}{\sin 20°}$$

$$= \frac{1.025 \text{ MN}}{\sin 20°}$$

$$= 2.997 \text{ MN} \quad (3 \text{ MN})$$

The 2.5 cm eccentricity will affect the bending stress, but it does not affect the shear stress.

Answer is D.

10. The maximum vertical shear in the beam will occur at the fixed end.

$$V = mg + F_y$$

$$= \left(45.9 \frac{\text{kg}}{\text{m}}\right)\left(9.81 \frac{\text{m}}{\text{s}^2}\right)(2 \text{ m}) + (6000 \text{ N})\sin 30°$$

$$= 3900 \text{ N}$$

Answer is B.

11. The maximum bending moment will occur at the fixed end of the beam. The moment will be affected by the distributed load and the external force. Since the force does not act through the centroid of the beam (i.e., the force is eccentric), both the vertical and the horizontal components of the external force must be included.

The moment due to the beam's own mass is

$$M_1 = \frac{1}{2}wL^2 = \left(\frac{1}{2}\right)\left(45.9 \frac{\text{kg}}{\text{m}}\right)\left(9.81 \frac{\text{m}}{\text{s}^2}\right)(2 \text{ m})^2$$

$$= 900 \text{ N·m}$$

The moment due to the vertical component of the external force is

$$M_2 = F_y L = (6000 \text{ N}) \sin 30°(2 \text{ m})$$
$$= 6000 \text{ N·m}$$

The force is not applied through the beam's centroid. The horizontal component of the force causes the beam to bend upward, while the other forces bend the beam downward. The moment due to the eccentricity is

$$M_3 = -F_x e = -(6000 \text{ N}) \cos 30° \left(\frac{0.20 \text{ m}}{2}\right)$$
$$= -519.6 \text{ N·m}$$

The total moment is

$$M = M_1 + M_2 + M_3$$
$$= 900 \text{ N·m} + 6000 \text{ N·m} - 519.6 \text{ N·m}$$
$$= 6380.4 \text{ N·m} \quad (6380 \text{ N·m})$$

Answer is B.

12. From Table 20.2, Case 1, with $a = L$ the deflection due to the vertical component of the force is

$$\delta_1 = \left(\frac{Pa^2}{6EI}\right)(3L - a) = \frac{PL^3}{3EI}$$
$$= \frac{(6000 \text{ N}) \sin 30°(2 \text{ m})^3 \left(100 \frac{\text{cm}}{\text{m}}\right)^4}{(3)(210 \times 10^9 \text{ Pa})(2880 \text{ cm}^4)}$$
$$= 0.001323 \text{ m} \quad (1.32 \text{ mm}) \quad \text{[downward]}$$

The eccentric application of the force causes an upward deflection. From Table 20.2, Case 3, the deflection due to the end moment is

$$\delta_2 = \frac{-ML^2}{2EI}$$
$$= \frac{-(519.6 \text{ N·m})(2 \text{ m})^2 \left(100 \frac{\text{cm}}{\text{m}}\right)^4}{(2)(210 \times 10^9 \text{ Pa})(2880 \text{ cm}^4)}$$
$$= -0.000172 \text{ m} \quad (-0.172 \text{ mm}) \quad \text{[upward]}$$

The total deflection due to the external force alone is

$$\delta = \delta_1 + \delta_2$$
$$= 1.32 \text{ mm} - 0.172 \text{ mm}$$
$$= 1.148 \text{ mm} \quad (1.15 \text{ mm})$$

Answer is D.

13. The total force from the distributed load is

$$\left(\frac{1}{2}\right)(10 \text{ m})\left(5 \frac{\text{N}}{\text{m}}\right) = 25 \text{ N}$$

This force can be assumed to act at two-thirds of the beam length from the left end, or one-third of the beam length from the right end.

Sum the moments around the right end to find the left reaction.

$$\sum M_{\text{right end}}: (25 \text{ N})\left(\frac{10 \text{ m}}{3}\right) - R_{\text{left}}(10 \text{ m}) = 0$$
$$R_{\text{left}} = 8.33 \text{ N}$$

The load increases linearly to 5 N/m at 10 m. At 7 m, the loading is $(0.7)(5 \text{ N/m})$. The total distributed force over the first 7 m of the beam is

$$\left(\frac{1}{2}\right)(7 \text{ m})\left((0.7)\left(5 \frac{\text{N}}{\text{m}}\right)\right) = 12.25 \text{ N}$$

Sum moments from the point of interest (7 m from the left end) to either end. The calculation is easier from the left end.

$$\sum M = (12.25 \text{ N})\left(\frac{7 \text{ m}}{3}\right) - (8.33 \text{ N})(7 \text{ m})$$
$$= -29.73 \text{ N·m} \quad (30 \text{ N·m})$$

Answer is D.

14. Propped cantilevers are statically indeterminate and must be solved using criteria (usually equal deflections at some known point) other than equilibrium.

The deflection at the supported end is known to be zero. Therefore, the deflection due to the distributed load combined with the deflection due to the concentrated reaction load must sum to zero.

The deflection for a distributed load is found from Table 20.2, Case 2.

$$\delta_1 = \frac{wL^4}{8EI} \quad \text{[downward]}$$

The deflection due to a concentrated load is found from Table 20.2, Case 1, with $a = L$.

$$\delta_2 = \left(\frac{-Pa^2}{6EI}\right)(3L - a) = \frac{-PL^3}{3EI} \quad \text{[upward]}$$

Since $\delta_1 + \delta_2 = 0$,

$$\frac{wL^4}{8EI} = \frac{PL^3}{3EI}$$
$$P = \frac{3wL}{8} = \frac{(3)\left(30.6 \frac{\text{kg}}{\text{m}}\right)\left(9.81 \frac{\text{m}}{\text{s}^2}\right)(5 \text{ m})}{8}$$
$$= 562.8 \text{ N} \quad (560 \text{ N})$$

Answer is C.

15. Use the principle of superposition to determine the deflection. The total deflection is the upward deflection due to the concentrated force less the downward deflection due to the weight of the beam.

The upward deflection due to the concentrated force is found from Table 20.2, Case 1, with $a = L = 5$ m. Distance x is measured from the fixed end.

$$\delta_{x,1} = \left(\frac{Px^2}{6EI}\right)(-x + 3a) \quad [x \le a]$$

$$= \frac{-(900\,\text{N})(3.5\,\text{m})^2\left(100\frac{\text{cm}}{\text{m}}\right)^4}{(6)(210 \times 10^9\,\text{Pa})(2880\,\text{cm}^4)}$$

$$\times\left(-3.5\,\text{m} + (3)(5\,\text{m})\right)$$

$$= -0.00349\,\text{m} \quad (-3.5\,\text{mm}) \quad [\text{upward}]$$

The downward deflection due to the beam's own mass is found from Table 20.2, Case 2. Distance x is measured from the fixed end. The load per unit length is

$$w = mg = \left(30.6\,\frac{\text{kg}}{\text{m}}\right)\left(9.81\,\frac{\text{m}}{\text{s}^2}\right)$$

$$= 300\,\text{N/m}$$

$$\delta_{x,2} = \left(\frac{wx^2}{24EI}\right)(x^2 + 6L^2 - 4Lx)$$

$$= \left(\frac{\left(300\,\frac{\text{N}}{\text{m}}\right)(3.5\,\text{m})^2\left(100\frac{\text{cm}}{\text{m}}\right)^4}{(24)(210 \times 10^9\,\text{Pa})(2880\,\text{cm}^4)}\right)$$

$$\times\left((3.5\,\text{m})^2 + (6)(5\,\text{m})^2 - (4)(5\,\text{m})(3.5\,\text{m})\right)$$

$$= 0.00232\,\text{m} \quad (2.3\,\text{mm}) \quad [\text{downward}]$$

$$\delta = \delta_{x,1} + \delta_{x,2} = -3.5\,\text{mm} + 2.3\,\text{mm}$$

$$= -1.2\,\text{mm} \quad [\text{upward}]$$

Answer is C.

Table 20.2 Beam Deflection Formulas (δ is positive downward)

case		δ	δ_{max}	ϕ_{max}
1		$\delta = \left(\dfrac{Pa^2}{6EI}\right)(3x - a)$, for $x > a$ $\delta = \left(\dfrac{Px^2}{6EI}\right)(-x + 3a)$, for $x \le a$	$\delta_{max} = \left(\dfrac{Pa^2}{6EI}\right)(3L - a)$	$\phi_{max} = \dfrac{Pa^2}{2EI}$
2		$\delta = \left(\dfrac{w_o x^2}{24EI}\right)(x^2 + 6L^2 - 4Lx)$	$\delta_{max} = \dfrac{w_o L^4}{8EI}$	$\phi_{max} = \dfrac{w_o L^3}{6EI}$
3		$\delta = \dfrac{M_o x^2}{2EI}$	$\delta_{max} = \dfrac{M_o L^2}{2EI}$	$\phi_{max} = \dfrac{M_o L}{EI}$
4		$\delta = \left(\dfrac{Pb}{6LEI}\right)\left(\dfrac{L}{b}(x - a)^3 - x^3 + (L^2 - b^2)x\right)$, for $x > a$ $\delta = \left(\dfrac{Pb}{6LEI}\right)\left(-x^3 + (L^2 - b^2)x\right)$, for $x \le a$	$\delta_{max} = \dfrac{Pb(L^2 - b^2)^{3/2}}{9\sqrt{3}\,LEI}$ at $x = \sqrt{\dfrac{L^2 - b^2}{3}}$	$\phi_1 = \dfrac{Pa(2L - a)}{6LEI}$ $\phi_2 = \dfrac{Pab(2L - b)}{6LEI}$
5		$\delta = \left(\dfrac{w_o x}{24EI}\right)(L^3 - 2Lx^2 + x^3)$	$\delta_{max} = \dfrac{5w_o L^4}{384EI}$	$\phi_1 = \phi_2 = \dfrac{w_o L^3}{24EI}$
6		$\delta = \left(\dfrac{M_o Lx}{6EI}\right)\left(1 - \dfrac{x^2}{L^2}\right)$	$\delta_{max} = \dfrac{M_o L^2}{9\sqrt{3}\,EI}$ at $x = \dfrac{L}{\sqrt{3}}$	$\phi_1 = \dfrac{M_o L}{6EI}$ $\phi_2 = \dfrac{M_o L}{3EI}$

MECHANICS
Beams

21 Columns

Subjects

BEAM-COLUMNS 21-1
LONG COLUMNS 21-1

Nomenclature

A	area	in^2	m^2
c	distance to extreme fiber	in	m
D	diameter	in	m
e	eccentricity	in	m
E	modulus of elasticity	lbf/in^2	MPa
F	force	lbf	N
I	moment of inertia	in^4	m^4
k	end-restraint constant	–	–
l	unbraced length	in	m
M	moment	in-lbf	N·m
P	force	lbf	N
r	radius of gyration	in	m
S	strength	lbf/in^2	MPa

Symbols

σ	normal stress	lbf/in^2	MPa

Subscripts

cr	critical
y	yield

BEAM-COLUMNS

If a load is applied through the centroid of a tension or compression member's cross section, the loading is said to be *axial loading* or *concentric loading*. *Eccentric loading* occurs when the load is not applied through the centroid. In Fig. 21.1, distance e is known as the *eccentricity*.

If an axial member is loaded eccentrically, it will bend and experience bending stress in the same manner as a beam. Since the member experiences both axial stress and bending stress, it is known as a *beam-column*.

Both the axial stress and bending stress are normal stresses oriented in the same direction; therefore, simple addition can be used to combine them.

Figure 21.1 Eccentric Loading of a Beam-Column

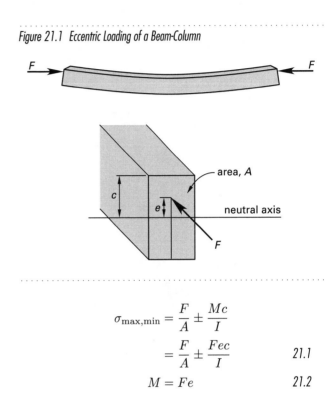

$$\sigma_{\mathrm{max,min}} = \frac{F}{A} \pm \frac{Mc}{I}$$
$$= \frac{F}{A} \pm \frac{Fec}{I} \qquad 21.1$$
$$M = Fe \qquad 21.2$$

If a pier or column (primarily designed as a compression member) is loaded with an eccentric compressive load, part of the section can still be in tension. Tension will exist when the Mc/I term in Eq. 21.1 is larger than the F/A term. It is particularly important to eliminate or severely limit tensile stresses in concrete and masonry piers, since these materials cannot support much tension.

Regardless of the size of the load, there will be no tension as long as the eccentricity is low enough. In a rectangular member, the load must be kept within a rhombus-shaped area formed from the middle thirds of the centroidal axes. This area is known as the *core*, *kern*, or *kernel*. Figure 21.2 illustrates the kernel for other cross sections.

LONG COLUMNS

Short columns, called *piers* or *pedestals*, will fail by compression of the material. *Long columns* will *buckle* in the transverse direction that has the smallest radius

Figure 21.2 Kernel for Various Column Shapes

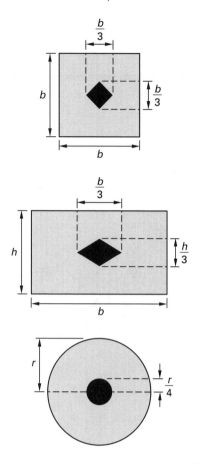

of gyration. Buckling failure is sudden, often without significant warning. If the material is wood or concrete, the material will usually fracture (because the yield stress is low); however, if the column is made of steel, the column will usually fail by local buckling, followed later by twisting and general yielding failure. Intermediate length columns will usually fail by a combination of crushing and buckling.

The load at which a long column fails is known as the *critical load* or *Euler load*. The Euler load is the theoretical maximum load that an initially straight column can support without transverse buckling. For columns with frictionless or pinned ends, this load is given by Eq. 21.3, known as *Euler's formula*.

$$P_{cr} = \frac{\pi^2 EI}{l^2} \qquad 21.3$$

The corresponding column stress is given by Eq. 21.4. This stress cannot exceed the yield strength of the column material.

$$\sigma_{cr} = \frac{P_{cr}}{A} = \frac{\pi^2 E}{\left(\frac{l}{r}\right)^2} \qquad [\sigma_{cr} \leq S_y] \qquad 21.4$$

l is the longest unbraced column length. If a column is braced against buckling at some point between its two ends, the column is known as a *braced column*, and l will be less than the full column height.

The quantity l/r is known as the *slenderness ratio*. Long columns have high slenderness ratios. The smallest slenderness ratio for which Eq. 21.4 is valid is the *critical slenderness ratio*, which can be calculated from the material's yield strength and modulus of elasticity. Typical slenderness ratios range from 80 to 120. The critical slenderness ratio becomes smaller as the compressive yield strength increases.

Most columns have two radii of gyration, r_x and r_y, and therefore, have two slenderness ratios. The largest slenderness ratio will govern the design.

Columns do not usually have frictionless or pinned ends. Often, a column will be fixed at its top and base. In such cases, the *effective length*, kl, which is the distance between inflection points on the column, must be used in place of l in Eqs. 21.3 and 21.4.

$$P_{cr} = \frac{\pi^2 EI}{(kl)^2} \qquad 21.5$$

k is the *end-restraint coefficient*, which theoretically varies from 0.5 to 2.0 according to Table 21.1. For design, values of k should be modified using engineering judgment based on realistic assumptions regarding end fixity.

SAMPLE PROBLEMS

Problem 1

A rectangular steel bar 37.5 mm wide and 50 mm thick is pinned at each end and subjected to axial compression. The bar has a length of 1.75 m. The modulus of elasticity is 200 GN/m². What is the critical buckling load?

(A) 60 kN
(B) 93 kN
(C) 110 kN
(D) 140 kN

Solution

Use Euler's formula. $k = 1$ since both ends are pinned. Calculate I for the weak direction.

$$P_{cr} = \frac{\pi^2 EI}{l^2}$$

$$= \frac{\pi^2 \left(200 \ \frac{GN}{m^2}\right)\left(10^9 \ \frac{N}{GN}\right)\left(\frac{(0.05 \ m)(0.0375 \ m)^3}{12}\right)}{(1.75 \ m)^2}$$

$$= 141\,624 \ N \quad (140 \ kN)$$

Answer is D.

Table 21.1 End-Restraint Conditions

illus.	end conditions	k theoretical	k design
(a)	both ends pinned	1	1.00
(b)	both ends built in	0.5	0.65
(c)	one end pinned, one end built in	0.7	0.8
(d)	one end built in, one end free	2	2.10
(e)	one end built in, one end fixed against rotation but free	1	1.20
(f)	one end pinned, one end fixed against rotation but free	2	2.0

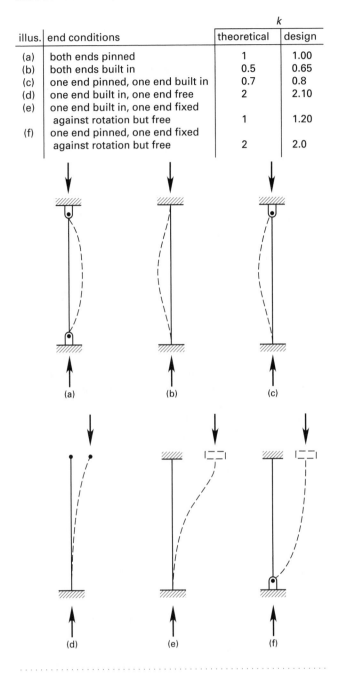

(a)   (b)   (c)

(d)   (e)   (f)

Problem 2

Determine the maximum resultant normal stress at A for the cantilever beam shown.

   (A) 7240 kPa
   (B) 9430 kPa
   (C) 9770 kPa
   (D) 9900 kPa

Solution

The beam experiences both axial tension and bending stresses, so it should be analyzed as a beam-column.

Illustration for Problem 2

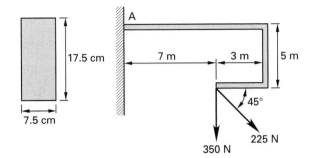

$$\sum M_A = (350 \text{ N})(7 \text{ m}) + (225 \text{ N})\sin 45°(7 \text{ m})$$
$$- (225 \text{ N})\cos 45°(5 \text{ m})$$
$$= 2768 \text{ N·m}$$
$$\sigma = \frac{P}{A} + \frac{Mc}{I}$$

$$\sigma_{max} = \frac{(225 \text{ N})\cos 45°}{(0.075 \text{ m})(0.175 \text{ m})}$$

$$+ \frac{(2768 \text{ N·m})\left(\dfrac{0.175 \text{ m}}{2}\right)}{\dfrac{(0.075 \text{ m})(0.175 \text{ m})^3}{12}}$$

$$= 7.24 \times 10^6 \text{ Pa} (7240 \text{ kPa})$$

Answer is A.

FE-STYLE EXAM PROBLEMS

1. The length, l, of a column divided by r is one of the terms in the equation for the buckling of a column subjected to compression loads. What does r stand for in the l/r ratio?

   (A) radius of the column
   (B) radius of gyration
   (C) least radius of gyration
   (D) slenderness

Problems 2 and 3 refer to the cantilever rod shown.

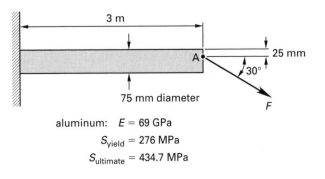

aluminum: $E = 69$ GPa
         $S_{yield} = 276$ MPa
       $S_{ultimate} = 434.7$ MPa

2. What force, F, will cause plastic deformation?

(A) 1900 N
(B) 3770 N
(C) 3900 N
(D) 7530 N

3. What is the maximum elastic deflection of the rod at point A due to the concentrated force, F, found in Prob. 2?

(A) 25.7 mm
(B) 86.6 mm
(C) 109 mm
(D) 320 mm

4. A real rectangular steel bar supports a concentric load of 58.5 kN. Both ends are fixed (i.e., built in). If the modulus of elasticity is 210 GPa, what is the maximum length the rod can be without experiencing buckling failure?

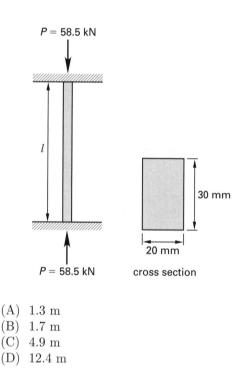

(A) 1.3 m
(B) 1.7 m
(C) 4.9 m
(D) 12.4 m

5. A pin-connected tripod frame consists of the three round steel bars AB, AC, and AD, as shown. Each bar is 75 cm long and has a cross-sectional area of 3 cm^2. The modulus of elasticity for the steel is 210 GPa. A vertical 9000 N load is supported at common point A. If column buckling based on Euler's formula is the only criterion, what is the approximate maximum force allowed in each rod?

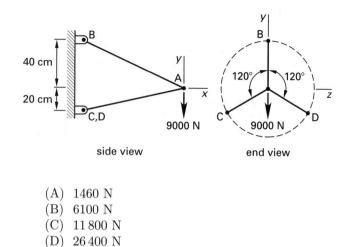

side view   end view

(A) 1460 N
(B) 6100 N
(C) 11 800 N
(D) 26 400 N

For the following problems use the NCEES Handbook as your only reference.

6. A square steel column with a solid cross section is pinned at both its base and top. The column is 5 m in height and supports a load of 3.5 MN. The modulus of elasticity of the steel is 210 GPa. What is the minimum cross-sectional size to avoid buckling?

(A) 2.5 cm × 2.5 cm
(B) 7.5 cm × 7.5 cm
(C) 15 cm × 15 cm
(D) 25 cm × 25 cm

7. A steel column with a cross section of 12 cm × 16 cm is 4 m in height and fixed at its base. The column is pinned at its top. The column's modulus of elasticity is 2.1×10^5 MPa. What is the maximum theoretical vertical load the column can support without buckling?

(A) 1 MN
(B) 5 MN
(C) 6 MN
(D) 11 MN

8. A steel column with a cross section of 12 cm × 16 cm is 4 m in height and fixed at its base. The column is braced against bucking in its weak direction only by a pin connection at the top but is unbraced in its strong direction. The column's modulus of elasticity is 2.1×10^5 MPa. What is the maximum theoretical vertical load the column can support without buckling?

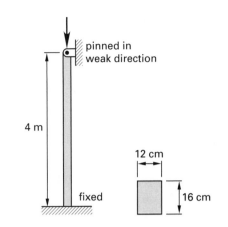

pinned in weak direction

4 m

12 cm

fixed

16 cm

(A) 1.3 MN
(B) 5.2 MN
(C) 6.1 MN
(D) 11 MN

9. A 10 cm × 10 cm square column supports a compressive force of 9000 N. The load is applied with an eccentricity of 2.5 cm along one of the lines of symmetry. What is the maximum tensile stress in the column?

(A) 450 kPa
(B) 900 kPa
(C) 1350 kPa
(D) 2250 kPa

10. A square column with a solid cross section is placed in a building to support a load of 5 MN. The maximum allowable stress in the column is 350 MPa. The column reacts linearly to all loads. If the contractor is permitted to load the column anywhere in the central one-fifth of the column's cross section, what are the smallest possible dimensions of the column?

(A) 12 cm × 12 cm
(B) 14 cm × 14 cm
(C) 16 cm × 16 cm
(D) 18 cm × 18 cm

SOLUTIONS TO FE-STYLE EXAM PROBLEMS

1. The length, l, of the column divided by r is the slenderness ratio. r is the radius of gyration of the column. For most columns, there are two radii of gyration, and the smallest one is used for the slenderness ratio in design.

Answer is C.

2. The moment is

$$M = \sum M_{\text{wall}}$$
$$= F \cos 30°(0.0125 \text{ m}) + F \sin 30°(3 \text{ m})$$
$$= 1.511F$$

The axial load is

$$F_{\text{axial}} = F \cos 30°$$
$$= 0.866F$$

To prevent plastic deformation,

$$\sigma_{\text{max}} \leq 276 \text{ MPa}$$

For an eccentrically loaded beam-column,

$$\sigma_{\text{max}} = \frac{F_{\text{axial}}}{A} + \frac{Mc}{I}$$
$$= \frac{0.866F}{\pi(0.0375 \text{ m})^2} + \frac{1.511F(0.0375 \text{ m})}{\left(\frac{\pi}{4}\right)(0.0375 \text{ m})^4}$$
$$= 36\,678F$$

$$36\,678F_{\text{max}} = 276 \times 10^6 \text{ Pa}$$

$$F_{\text{max}} = 7525 \text{ N} \quad (7530 \text{ N})$$

Answer is D.

3. The concentrated force is the sum of the horizontal and vertical components. The horizontal axial component causes a moment on the beam.

From Table 20.2, for a cantilever with an end moment,

$$\delta_{\text{max}} = \frac{M_o L^2}{2EI} = \frac{F \cos \theta e L^2}{2EI}$$
$$= \frac{(7525 \text{ N}) \cos 30°(0.0125 \text{ m})(3 \text{ m})^2}{(2)(69 \times 10^9 \text{ Pa})\left(\frac{\pi}{4}\right)(0.0375 \text{ m})^4}$$
$$= 0.00342 \text{ m} \quad (3.42 \text{ mm}) \quad [\text{down}]$$

The deflection caused by the vertical component is given by Table 20.2 for a cantilever with an end load.

$$\delta_{\text{max}} = \frac{-PL^3}{3EI}$$
$$= \frac{-(7525 \text{ N}) \sin 30°(3 \text{ m})^3}{(3)(69 \times 10^9 \text{ Pa})\left(\frac{\pi}{4}\right)(0.0375 \text{ m})^4}$$
$$= 0.3160 \text{ m} \quad (316.0 \text{ mm}) \quad [\text{down}]$$

The total vertical deflection is

$$y_{\text{max}} = 3.42 \text{ mm} + 316.0 \text{ mm}$$
$$= 319.42 \text{ mm} \quad (320 \text{ mm})$$

Answer is D.

4.
$$P_{cr} = \frac{\pi^2 EI}{(kl)^2}$$

For a real column fixed at both ends, use the design value of $k = 0.65$.

$$I = \frac{bh^3}{12} = \frac{(0.03 \text{ m})(0.02 \text{ m})^3}{12}$$
$$= 2 \times 10^{-8} \text{ m}^4$$

$$(0.65l)^2 = \frac{\pi^2 EI}{P_{cr}}$$
$$= \frac{\pi^2 (210 \times 10^9 \text{ Pa})(2 \times 10^{-8} \text{ m}^4)}{58.5 \times 10^3 \text{ N}}$$
$$= 0.709 \text{ m}^2$$

$$l = 1.30 \text{ m}$$

Answer is A.

5. The moment of inertia of a round cross section is

$$I = \frac{\pi r^4}{4}$$

The buckling force for a long column is given by Eq. 21.3. Since each rod is pinned at both ends, from Table 21.1, the end-restraint coefficient, k, is 1. The effective length is l.

$$P_{cr} = \frac{\pi^2 EI}{l^2} = \frac{\pi^2 E \pi r^4}{4l^2} = \frac{\pi E A^2}{4l^2}$$
$$= \frac{\pi (210 \times 10^9 \text{ Pa})(3 \text{ cm}^2)^2 \left(\frac{1 \text{ m}}{100 \text{ cm}}\right)^4}{(4)(0.75 \text{ m})^2}$$
$$= 26\,389 \text{ N} \quad (26\,400 \text{ N})$$

This solution of this problem did not require finding the forces in each leg.

Answer is D.

6. Since the column is pinned at both ends, the end-restraint coefficient, k, is 1 (Table 21.1). The effective length is l.

Solve Eq. 21.3 for the moment of inertia.

$$P_{cr} = \frac{\pi^2 EI}{l^2}$$

$$I = \frac{P_{cr} l^2}{\pi^2 E}$$
$$= \frac{(3.5 \times 10^6 \text{ N})(5 \text{ m})^2}{\pi^2 (210 \times 10^9 \text{ Pa})}$$
$$= 0.0000422 \text{ m}^4$$

The centroidal moment of inertia for a square section is

$$I = \frac{b^4}{12}$$

$$b = \sqrt[4]{12I} = \sqrt[4]{(12)(0.0000422 \text{ m}^4)}$$
$$= 0.15 \text{ m} \quad (15 \text{ cm})$$

Answer is C.

7. Since the column is fixed at one end and pinned at the other, the end-restraint coefficient, k, is 0.7 (Table 21.1). The theoretical effective length for buckling in the weak direction is

$$l' = kl$$
$$= (0.7)(4 \text{ m})$$
$$= 2.8 \text{ m}$$

There are two moments of inertia, corresponding to buckling in two different directions. The smaller moment of inertia is

$$I = \frac{bh^3}{12} = \frac{(16 \text{ cm})(12 \text{ cm})^3}{(12)\left(100 \frac{\text{cm}}{\text{m}}\right)^4}$$
$$= 2.3 \times 10^{-5} \text{ m}^4$$

Calculate the critical buckling force from Euler's formula, Eq. 21.3.

$$P_{cr} = \frac{\pi^2 EI}{l^2}$$

$$= \frac{\pi^2 (2.1 \times 10^5 \text{ MPa})\left(10^6 \frac{\text{Pa}}{\text{MPa}}\right)(2.3 \times 10^{-5} \text{ m}^4)}{(2.8 \text{ m})^2}$$
$$= 6.08 \times 10^6 \text{ N} \quad (6 \text{ MN})$$

Answer is C.

8. Since the column is fixed at one end and pinned at the other, the theoretical end-restraint coefficient, k, is 0.7 (Table 21.1). The effective length for buckling in the weak direction is

$$l' = kl$$
$$= (0.7)(4 \text{ m})$$
$$= 2.8 \text{ m}$$

The moment of inertia for buckling in the weak direction is

$$I = \frac{bh^3}{12} = \frac{(16 \text{ cm})(12 \text{ cm})^3}{(12)\left(100 \frac{\text{cm}}{\text{m}}\right)^4}$$
$$= 2.3 \times 10^{-5} \text{ m}^4$$

Calculate the critical buckling force from Euler's formula, Eq. 21.3.

$$P_{cr} = \frac{\pi^2 EI}{l^2}$$

$$= \frac{\pi^2 (2.1 \times 10^5 \text{ MPa}) \left(10^6 \frac{\text{Pa}}{\text{MPa}}\right)(2.3 \times 10^{-5} \text{ m}^4)}{(2.8 \text{ m})^2}$$

$$= 6.08 \times 10^6 \text{ N} \quad (6.1 \text{ MN})$$

Check the buckling force in the strong direction. The column is not braced in that direction, so from Table 21.1 for a column fixed at one end and free at the other, $k = 2$.

$$l' = 2l$$
$$= (2)(4 \text{ m})$$
$$= 8 \text{ m}$$

The moment of inertia for buckling in the strong direction is

$$I = \frac{bh^3}{12} = \frac{(12 \text{ cm})(16 \text{ cm})^3}{(12)\left(100 \frac{\text{cm}}{\text{m}}\right)^4}$$

$$= 4.1 \times 10^{-5} \text{ m}^4$$

Calculate the critical buckling force from Euler's formula, Eq. 21.3.

$$P_{cr} = \frac{\pi^2 EI}{l^2}$$

$$= \frac{\pi^2 (2.1 \times 10^5 \text{ MPa}) \left(10^6 \frac{\text{Pa}}{\text{MPa}}\right)(4.1 \times 10^{-5} \text{ m}^4)}{(8 \text{ m})^2}$$

$$= 1.3 \times 10^6 \text{ N} \quad (1.3 \text{ MN})$$

This is less than for buckling in the weak direction. This force controls.

Answer is A.

9. The cross-sectional area is

$$A = b^2 = (0.1 \text{ m})^2$$
$$= 0.01 \text{ m}^2$$

The moment of inertia of the square cross section is

$$I = \frac{b^4}{12} = \frac{(0.1 \text{ m})^4}{12}$$

$$= 8.33 \times 10^{-6} \text{ m}^4$$

The distance from the neutral axis to the extreme fibers is

$$c = \frac{b}{2} = \frac{0.1 \text{ m}}{2}$$
$$= 0.05 \text{ m}$$

$$\sigma = \frac{F}{A} \pm \frac{Fec}{I}$$

$$= \frac{-9000 \text{ N}}{0.01 \text{ m}^2} \pm \frac{(-9000 \text{ N})(0.025 \text{ m})(0.05 \text{ m})}{8.33 \times 10^{-6} \text{ m}^4}$$

$$= -9 \times 10^5 \text{ Pa} \pm 1.35 \times 10^6 \text{ Pa}$$
$$(-900 \text{ kPa} \pm 1350 \text{ kPa})$$

The first term is due to the compressive column load and is compressive. (Compressive forces and stresses are usually given a negative sign.) The second term is due to the eccentricity. The second term increases the compressive stress at the inner face. It counteracts the compressive stress at the outer face.

The maximum tensile stress is

$$\sigma_{t,\text{max}} = -900 \text{ kPa} + 1350 \text{ kPa}$$
$$= 450 \text{ kPa}$$

Answer is A.

10. The middle one-fifth of the column is a square with dimensions of $b/5 \times b/5$ ($0.2b \times 0.2b$).

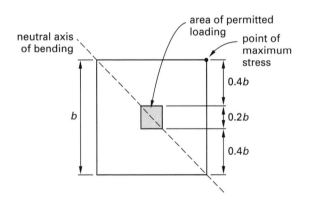

The maximum stress will be induced when the middle one-fifth square is loaded at one of its corners.

The cross-sectional area is

$$A = b^2$$

The moment of inertia of the square cross section is

$$I = \frac{b^4}{12}$$

The distance from the neutral axis to the extreme fibers is

$$c = \frac{b}{2}$$

The maximum eccentricity is

$$e = 0.1b$$

The stress at the extreme corner is

$$\sigma = \frac{F}{A} \pm \frac{F e_x c_x}{I_x} + \frac{F e_y c_y}{I_y}$$

$$= F \left(\frac{1}{b^2} \pm \frac{(2)(0.1b)\left(\dfrac{b}{2}\right)}{\dfrac{b^4}{12}} \right)$$

$$= F \left(\frac{1}{b^2} \pm \frac{1.2}{b^2} \right) = \frac{2.2F}{b^2}$$

$$b = \sqrt{\frac{2.2F}{\sigma}}$$

$$= \sqrt{\frac{(2.2)(5 \text{ MN})\left(10^6 \dfrac{\text{N}}{\text{MN}}\right)}{(350 \text{ MPa})\left(10^6 \dfrac{\text{Pa}}{\text{MPa}}\right)}}$$

$$= 0.177 \text{ m} \quad (18 \text{ cm})$$

Answer is D.

Topic VII: Fluid Mechanics

Diagnostic Examination for Fluid Mechanics

Chapter 22 Fluid Properties

Chapter 23 Fluid Statics

Chapter 24 Fluid Dynamics

Chapter 25 Fluid Measurements and Similitude

FLUIDS

Hint: For the most current information about the exam, visit www.ppi2pass.com/fefaqs.html regularly. Use the Exam Forum to compare notes with other FE examinees.

Diagnostic Examination

TOPIC VII: FLUID MECHANICS

TIME LIMIT: 45 MINUTES

1. 10.0 L of an incompressible liquid exert a force of 20 N at the earth's surface. What force would 2.3 L of this liquid exert on the surface of the moon? The gravitational acceleration on the surface of the moon is 1.67 m/s^2.

    (A) 0.39 N
    (B) 0.78 N
    (C) 3.4 N
    (D) 4.6 N

2. A sliding-plate viscometer is used to measure the viscosity of a Newtonian fluid. A force of 25 N is required to keep the top plate moving at a constant velocity of 5 m/s. What is the viscosity of the fluid?

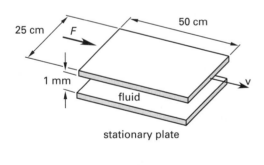

    (A) 0.005 N·s/m^2
    (B) 0.04 N·s/m^2
    (C) 0.2 N·s/m^2
    (D) 5.0 N·s/m^2

3. A 2 mm (inside diameter) glass tube is placed in a container of mercury. An angle of 40° is measured as illustrated. The density and surface tension of mercury are $13\,550 \text{ kg/m}^3$ and $37.5 \times 10^{-2} \text{ N/m}$, respectively. How high will the mercury rise or be depressed in the tube as a result of capillary action?

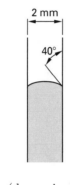

    (A) −4.3 mm (depression)
    (B) −1.6 mm (depression)
    (C) 4.2 mm (rise)
    (D) 6.4 mm (rise)

4. An open water manometer is used to measure the pressure in a tank. The tank is half-filled with 50 000 kg of a liquid chemical that is not miscible in water. The manometer tube is filled with liquid chemical. What is the pressure in the tank relative to the atmospheric pressure?

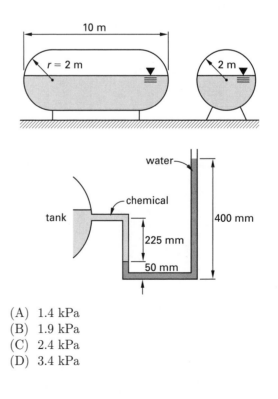

    (A) 1.4 kPa
    (B) 1.9 kPa
    (C) 2.4 kPa
    (D) 3.4 kPa

5. A gravity dam has the cross section shown. What is the magnitude of the resultant water force (per meter of width) acting on the face of the dam?

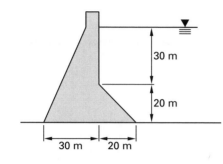

(A) 7.85 MN/m
(B) 12.3 MN/m
(C) 14.6 MN/m
(D) 20.2 MN/m

6. A 6 m × 6 m × 6 m vented cubical tank is half-filled with water; the remaining space is filled with oil (SG = 0.8). What is the total force on one side of the tank?

(A) 690 kN
(B) 900 kN
(C) 950 kN
(D) 1.0 MN

7. A 35 cm diameter solid sphere ($\rho = 4500$ kg/m^3) is suspended by a cable as shown. Half of the sphere is in one fluid ($\rho = 1200$ kg/m^3) and the other half of the sphere is in another ($\rho = 1500$ kg/m^3). What is the tension in the cable?

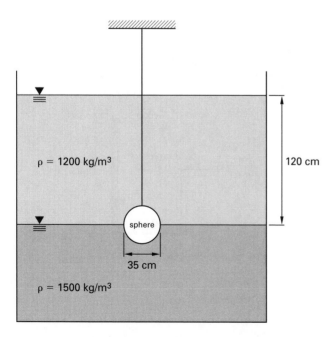

(A) 297 N
(B) 593 N
(C) 694 N
(D) 826 N

8. The diameter of a water pipe gradually changes from 5 cm at point A to 15 cm at point B. Point A is 5 m lower than point B. The pressure is 700 kPa at point A and 664 kPa at point B. Friction between the water and the pipe walls is negligible. What is the rate of discharge at point B?

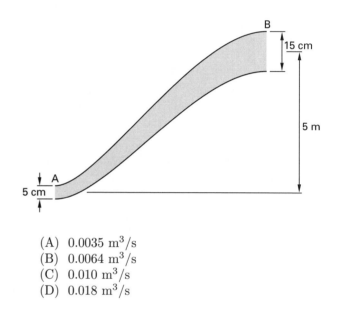

(A) 0.0035 m^3/s
(B) 0.0064 m^3/s
(C) 0.010 m^3/s
(D) 0.018 m^3/s

9. What is the hydraulic radius of the trapezoidal irrigation canal shown?

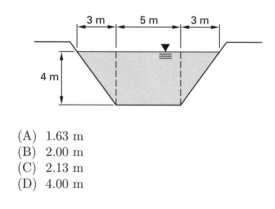

(A) 1.63 m
(B) 2.00 m
(C) 2.13 m
(D) 4.00 m

10. A liquid with a specific gravity of 0.9 is stored in a pressurized, closed storage tank. The tank is cylindrical with a 10 m diameter. The absolute pressure in the tank above the liquid is 200 kPa. What is the initial velocity of a fluid jet when a 5 cm diameter orifice is opened at point A?

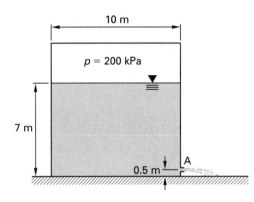

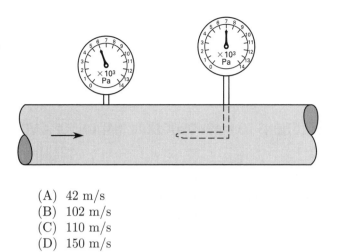

(A) 42 m/s
(B) 102 m/s
(C) 110 m/s
(D) 150 m/s

(A) 11.3 m/s
(B) 18.0 m/s
(C) 18.6 m/s
(D) 23.9 m/s

13. Water flows out of a tank at 12.5 m/s from an orifice located 9 m below the surface. The cross-sectional area of the orifice is 0.002 m², and the coefficient of discharge is 0.85. What is the diameter, D, at the vena contracta?

11. Water is flowing at 50 m/s through a 15 cm diameter pipe. The pipe makes a 90° turn, as shown. What is the reaction on the water in the z-direction at the bend?

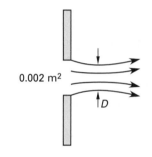

(A) 4.2 cm
(B) 4.5 cm
(C) 4.7 cm
(D) 4.8 cm

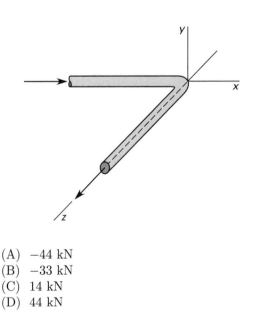

(A) −44 kN
(B) −33 kN
(C) 14 kN
(D) 44 kN

14. A nuclear submarine is capable of a top underwater speed of 65 km/h. How fast would a 1/20 scale model of the submarine have to be moved through a testing pool filled with seawater for the forces on the submarine and model to be dimensionally similar?

(A) 0.90 m/s
(B) 18 m/s
(C) 180 m/s
(D) 360 m/s

FLUIDS

12. The density of air flowing in a duct is 1.15 kg/m³. A pitot tube is placed in the duct as shown. The static pressure in the duct is measured with a wall tap and pressure gage. Use the gage readings to determine the velocity of the air.

15. A jet aircraft is flying at a speed of 1700 km/h. The air temperature is 20°C. The molecular weight of air is 29 g/mol. What is the Mach number of the aircraft?

(A) 0.979
(B) 1.38
(C) 1.92
(D) 5.28

SOLUTIONS TO DIAGNOSTIC EXAMINATION TOPIC VII

1. The density of an incompressible fluid is constant. It is not a function of pressure or temperature. The mass is calculated from Newton's second law.

$$m = \frac{F}{a} = \frac{20 \text{ N}}{9.81 \frac{\text{m}}{\text{s}^2}}$$

$$= 2.04 \text{ kg}$$

The density is

$$\rho = \frac{m}{V} = \frac{2.04 \text{ kg}}{10 \text{ L}}$$

$$= 0.204 \text{ kg/L}$$

The force of 2.3 L of this liquid on the moon is

$$F = ma = \rho V a$$

$$= \left(0.204 \frac{\text{kg}}{\text{L}}\right)(2.3 \text{ L})\left(1.67 \frac{\text{m}}{\text{s}^2}\right)$$

$$= 0.784 \text{ N} \quad (0.78 \text{ N})$$

Answer is B.

2. The force required to maintain the velocity is given by

$$F = \frac{\mu \text{v} A}{\delta}$$

Rearranging to solve for viscosity directly,

$$\mu = \frac{F\delta}{\text{v}A}$$

$$= \frac{(25 \text{ N})(0.001 \text{ m})}{\left(5 \frac{\text{m}}{\text{s}}\right)(0.5 \text{ m})(0.25 \text{ m})}$$

$$= 0.04 \text{ N·s/m}^2$$

Answer is B.

3.

$\beta = 180° - 40° = 140°$

The contact angle is the angle made by the liquid with the wetted tube wall. If this angle is greater than $90°$, the fluid level in the tube will be depressed. The equation for capillary action is

$$h = \frac{4\sigma \cos \beta}{\rho d_{\text{tube}} g}$$

$$= \frac{(4)\left(37.5 \times 10^{-2} \frac{\text{N}}{\text{m}}\right) \cos 140°}{\left(13\,550 \frac{\text{kg}}{\text{m}^3}\right)(0.002 \text{ m})\left(9.81 \frac{\text{m}}{\text{s}^2}\right)}$$

$$= -4.32 \times 10^{-3} \text{ m} \quad (-4.3 \text{ mm}) \quad [\text{depression}]$$

Answer is A.

4. Calculate the density of the chemical from the volume and mass. The total volume of the tank is

$$V = \left(\frac{4}{3}\right)\pi r^3 + \pi r^2 (L - 2r)$$

$$= \left(\frac{4}{3}\right)\pi (2 \text{ m})^3 + \pi (2 \text{ m})^2 \big(10 \text{ m} - (2)(2 \text{ m})\big)$$

$$= 108.9 \text{ m}^3$$

Since the contents have a mass of $50\,000$ kg and the tank is half-full, the density of the chemical is

$$\rho_{\text{chem}} = \frac{50\,000 \text{ kg}}{\left(\frac{1}{2}\right)(108.9 \text{ m}^3)}$$

$$= 918.3 \text{ kg/m}^3$$

The relative pressure is

$$p_o - p_2 = \rho_{\text{water}} g h_2 - \rho_{\text{chem}} g h_1$$

$$= \left(1000 \frac{\text{kg}}{\text{m}^3}\right)\left(9.81 \frac{\text{m}}{\text{s}^2}\right)(0.40 \text{ m} - 0.05 \text{ m})$$

$$- \left(918.3 \frac{\text{kg}}{\text{m}^3}\right)\left(9.81 \frac{\text{m}}{\text{s}^2}\right)(0.225 \text{ m})$$

$$= 1407 \text{ Pa} \quad (1.4 \text{ kPa})$$

Answer is A.

5. The maximum water pressure at the base of the dam is

$$p_{\text{max}} = \rho g h$$

$$= \left(1000 \ \frac{\text{kg}}{\text{m}^3}\right)\left(9.81 \ \frac{\text{m}}{\text{s}^2}\right)(50 \ \text{m})$$

$$= 490\,500 \ \text{Pa}$$

The total horizontal force on the dam face is equal to the area of the triangular pressure distribution.

$$F_x = \frac{1}{2}p_{\text{max}}A$$

$$= \frac{1}{2}p_{\text{max}}Lw$$

Per unit width, w,

$$\frac{F_x}{w} = \left(\frac{1}{2}\right)(490\,500 \ \text{Pa})(20 \ \text{m} + 30 \ \text{m})$$

$$= 12.3 \times 10^6 \ \text{N/m} \quad (12.3 \ \text{MN/m})$$

The total vertical force acting on the dam face is equal to the weight of the water column above the face.

$$F_y = \rho g V = \rho g A w$$

$$\frac{F_y}{w} = \left(1000 \ \frac{\text{kg}}{\text{m}^3}\right)\left(9.81 \ \frac{\text{m}}{\text{s}^2}\right)$$

$$\times \left((30 \ \text{m})(20 \ \text{m}) + \left(\frac{1}{2}\right)(20 \ \text{m})(20 \ \text{m})\right)$$

$$= 7.85 \times 10^6 \ \text{N/m} \quad (7.85 \ \text{MN/m})$$

The resultant force is

$$F = \sqrt{F_x^2 + F_y^2}$$

$$= \sqrt{\left(12.3 \ \frac{\text{MN}}{\text{m}}\right)^2 + \left(7.85 \ \frac{\text{MN}}{\text{m}}\right)^2}$$

$$= 14.6 \ \text{MN/m}$$

Answer is C.

6.

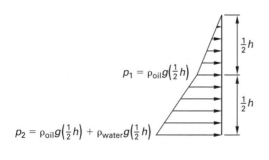

$$p_1 = \rho_{\text{oil}}g\left(\tfrac{1}{2}h\right)$$

$$p_2 = \rho_{\text{oil}}g\left(\tfrac{1}{2}h\right) + \rho_{\text{water}}g\left(\tfrac{1}{2}h\right)$$

$$\rho_{\text{oil}} = (\text{SG})\rho_{\text{water}}$$

$$= (0.8)\left(1000 \ \frac{\text{kg}}{\text{m}^3}\right)$$

$$= 800 \ \text{kg/m}^3$$

Since the oil has a lower specific gravity than the water, it will float on the water. The pressure at the oil/water interface is due to the weight of the oil above the interface.

$$p_1 = \rho_{\text{oil}}g\left(\frac{h}{2}\right)$$

$$= \left(800 \ \frac{\text{kg}}{\text{m}^3}\right)\left(9.81 \ \frac{\text{m}}{\text{s}^2}\right)\left(\frac{6 \ \text{m}}{2}\right)$$

$$= 23\,544 \ \text{Pa}$$

The pressure at the bottom of the box is

$$p_2 = p_1 + \rho_{\text{water}}g\left(\frac{h}{2}\right)$$

$$= 23\,544 \ \text{Pa} + \left(1000 \ \frac{\text{kg}}{\text{m}^3}\right)\left(9.81 \ \frac{\text{m}}{\text{s}^2}\right)\left(\frac{6 \ \text{m}}{2}\right)$$

$$= 52\,974 \ \text{Pa}$$

The total force acting on the face of the box is

$$F = \left(\frac{1}{2}\right)(23\,544 \ \text{Pa})(3 \ \text{m})(6 \ \text{m})$$

$$+ (23\,544 \ \text{Pa})(3 \ \text{m})(6 \ \text{m})$$

$$+ \left(\frac{1}{2}\right)(52\,974 \ \text{Pa} - 23\,544 \ \text{Pa})(3 \ \text{m})(6 \ \text{m})$$

$$= 9.01 \times 10^5 \ \text{N} \quad (900 \ \text{kN})$$

Answer is B.

7. By Archimedes' principle, the buoyant force exerted on a submerged body is equal to the weight of the fluid displaced by the body.

The weight of the fluid displaced by the body is

$$W_{\text{fluid}} = \rho_1 g\left(\frac{1}{2}\right)V + \rho_2 g\left(\frac{1}{2}\right)V$$

$$= g\left(\frac{1}{2}\right)V(\rho_1 + \rho_2)$$

$$= g\left(\frac{1}{2}\right)\left(\frac{4}{3}\right)\pi r^3(\rho_1 + \rho_2)$$

$$= \left(9.81 \ \frac{\text{m}}{\text{s}^2}\right)\left(\frac{2}{3}\right)\pi(0.175 \ \text{m})^3$$

$$\times \left(1200 \ \frac{\text{kg}}{\text{m}^3} + 1500 \ \frac{\text{kg}}{\text{m}^3}\right)$$

$$= 297.3 \ \text{N}$$

FLUIDS

The tension in the cable can be calculated by statics.

$$\sum F_y = 0$$

$$T - W_{\text{sphere}} + F_{\text{buoyant}} = 0$$

$$T - \rho_{\text{sphere}}gV + F_{\text{buoyant}} = 0$$

$$T - \rho_{\text{sphere}}g\left(\frac{4}{3}\right)\pi r^3 + F_{\text{buoyant}} = 0$$

$$T - \left(4500 \ \frac{\text{kg}}{\text{m}^3}\right)\left(9.81 \ \frac{\text{m}}{\text{s}^2}\right)\left(\frac{4}{3}\right)$$

$$\times \pi \left(\frac{0.35 \ \text{m}}{2}\right)^3 + 297.3 \ \text{N} = 0$$

$$T - 991.0 \ \text{N} + 297.3 \ \text{N} = 0$$

$$T = 693.7 \ \text{N} \quad (697 \ \text{N})$$

Answer is C.

8. The velocity at point B can be found from the Bernoulli equation.

$$\frac{p_A}{\rho g} + \frac{v_A^2}{2g} + z_A = \frac{p_B}{\rho g} + \frac{v_B^2}{2g} + z_B$$

$$v_B^2 = v_A^2 + \frac{2(p_A - p_B)}{\rho} + 2g(z_A - z_B)$$

The velocities are related.

$$Q_A = Q_B$$

$$v_A A_A = v_B A_B$$

$$v_A \left(\frac{\pi}{4}\right)d_A^2 = v_B \left(\frac{\pi}{4}\right)d_B^2$$

$$v_A = v_B \left(\frac{d_B}{d_A}\right)^2$$

$$v_B^2 \left(1 - \left(\frac{d_B}{d_A}\right)^4\right) = \frac{2(p_A - p_B)}{\rho} + 2g(z_A - z_B)$$

$$v_B^2 \left(1 - \left(\frac{15 \ \text{cm}}{5 \ \text{cm}}\right)^4\right) = \frac{(2)(7 \times 10^5 \ \text{Pa} - 6.64 \times 10^5 \ \text{Pa})}{1000 \ \frac{\text{kg}}{\text{m}^3}}$$

$$+ (2)\left(9.81 \ \frac{\text{m}}{\text{s}^2}\right)(0 \ \text{m} - 5 \ \text{m})$$

$$(-80)v_B^2 = -26.1 \ \frac{\text{m}^2}{\text{s}^2}$$

$$v_B = \sqrt{\frac{-26.1 \ \frac{\text{m}^2}{\text{s}^2}}{-80}} = 0.571 \ \text{m/s}$$

$$Q = A_B v_B = \frac{\pi}{4}d_B^2 v_B$$

$$= \left(\frac{\pi}{4}\right)(0.15 \ \text{m})^2 \left(0.571 \ \frac{\text{m}}{\text{s}}\right)$$

$$= 0.01 \ \text{m}^3/\text{s}$$

Answer is C.

9. The two triangular sections, when put together, have the same cross-sectional area as a 3 m × 4 m rectangle. Since the triangular portions have sides of 3 m and 4 m, the diagonal length is 5 m (3-4-5 triangle).

$$R_H = \frac{\text{cross-sectional area}}{\text{wetted perimeter}}$$

$$= \frac{(3 \ \text{m} + 5 \ \text{m})(4 \ \text{m})}{5 \ \text{m} + 5 \ \text{m} + 5 \ \text{m}}$$

$$= \frac{32 \ \text{m}^2}{15 \ \text{m}}$$

$$= 2.13 \ \text{m}$$

Answer is C.

10. The density of the liquid is

$$\rho = (\text{SG})\rho_{\text{water}}$$

$$= (0.9)\left(1000 \ \frac{\text{kg}}{\text{m}^3}\right)$$

$$= 900 \ \text{kg/m}^3$$

Use Bernoulli's equation with point B corresponding to the top of the liquid surface.

$$\frac{p_A}{\rho g} + \frac{v_A^2}{2g} + z_A = \frac{p_B}{\rho g} + \frac{v_B^2}{2g} + z_B$$

The velocity, v_B, at the top of the liquid surface is zero. The gage pressure, p_A, at point A is zero. The pressure in the tank is an absolute pressure and must be converted to gage pressure.

$$\frac{p_B}{\rho g} + z_B = \frac{v_A^2}{2g} + z_A$$

$$v_A^2 = \frac{2p_B}{\rho} + 2g(z_B - z_A)$$

$$= \frac{(2)(2 \times 10^5 \ \text{Pa} - 1.013 \times 10^5 \ \text{Pa})}{(0.9)\left(1000 \ \frac{\text{kg}}{\text{m}^3}\right)}$$

$$+ (2)\left(9.81 \ \frac{\text{m}}{\text{s}^2}\right)(7 \ \text{m} - 0.5 \ \text{m})$$

$$= 346.86 \ \text{m}^2/\text{s}^2$$

$$v_A = \sqrt{346.86 \, \frac{m^2}{s^2}}$$
$$= 18.6 \text{ m/s}$$

Answer is C.

11. The force exerted on the fluid is found from the impulse-momentum principle.

$$\sum F_z = Q_2 \rho_2 v_{z2} - Q_1 \rho_1 v_{z1}$$
$$= Q\rho(v_{z2} - v_{z1})$$
$$= vA\rho(v_{z2} - v_{z1})$$
$$= v\pi r^2 \rho(v_{z2} - v_{z1})$$
$$= \left(50 \, \frac{m}{s}\right) \pi \left(\frac{0.15 \text{ m}}{2}\right)^2 \left(1000 \, \frac{kg}{m^3}\right)\left(50 - 0 \, \frac{m}{s}\right)$$
$$= 44\,179 \text{ N} \quad (44 \text{ kN})$$

Answer is D.

12. The static pressure is read from the static pressure gage as 6000 Pa. The impact pressure is 7000 Pa.

$$v = \sqrt{\frac{2(p_o - p_s)}{\rho}}$$
$$= \sqrt{\frac{(2)(7000 \text{ Pa} - 6000 \text{ Pa})}{1.15 \, \frac{kg}{m^3}}}$$
$$= 41.7 \text{ m/s} \quad (42 \text{ m/s})$$

Answer is A.

13. Determine the actual coefficient of velocity.

$$C_v = \frac{v}{v_{ideal}} = \frac{v}{\sqrt{2gh}}$$
$$= \frac{12.5 \, \frac{m}{s}}{\sqrt{(2)\left(9.81 \, \frac{m}{s^2}\right)(9 \text{ m})}}$$
$$= 0.941$$

Calculate the coefficient of contraction.

$$C_c = \frac{C}{C_v} = \frac{0.85}{0.941}$$
$$= 0.903$$

$$A_{\text{venacontracta}} = C_c A_{\text{opening}}$$
$$= (0.903)(0.002 \text{ m}^2)$$
$$= 0.00181 \text{ m}^2$$

$$A = \left(\frac{1}{4}\right)\pi d^2$$

$$d = \sqrt{\frac{4A}{\pi}} = \sqrt{\frac{(4)(0.00181 \text{ m}^2)}{\pi}}$$
$$= 0.048 \text{ m} \quad (4.8 \text{ cm})$$

Answer is D.

14. For completely submerged objects, viscous and inertial forces are significant. The Reynolds number is the ratio of these forces. The Reynolds numbers of the model (m for model) and real submarine (p for prototype) must be equal.

$$Re_p = Re_m$$
$$\left(\frac{vl\rho}{\mu}\right)_p = \left(\frac{vl\rho}{\mu}\right)_m$$
$$v_m = \frac{v_p}{\frac{l_m}{l_p}} = \frac{\left(65 \, \frac{km}{h}\right)\left(1000 \, \frac{m}{km}\right)}{\left(\frac{1}{20}\right)\left(3600 \, \frac{s}{h}\right)}$$
$$= 361.1 \text{ m/s} \quad (360 \text{ m/s})$$

Answer is D.

15. The speed of sound in the air is

$$c = \sqrt{kRT}$$

The molecular weight has the same value regardless of whether the units are g/mol, kg/kmol, or lbm/lbmol. The ratio of specific heats, k, of air is 1.4. The specific gas constant is

$$R = \frac{\overline{R}}{MW}$$
$$c = \sqrt{\frac{k\overline{R}T}{(MW)_{air}}}$$
$$= \sqrt{\frac{(1.4)\left(8314 \, \frac{J}{kmol \cdot K}\right)(20°C + 273.15)}{29 \, \frac{kg}{kmol}}}$$
$$= 343 \text{ m/s}$$

FLUIDS

The Mach number is

$$M = \frac{v}{c}$$

$$= \frac{\left(1700 \ \frac{km}{h}\right)\left(1000 \ \frac{m}{km}\right)}{\left(343 \ \frac{m}{s}\right)\left(3600 \ \frac{s}{h}\right)}$$

$$= 1.38$$

Answer is B.

22 Fluid Properties

Subjects

FLUIDS 22-1
DENSITY 22-1
SPECIFIC VOLUME 22-1
SPECIFIC WEIGHT 22-1
SPECIFIC GRAVITY 22-2
PRESSURE 22-2
STRESS 22-2
VISCOSITY 22-3
SURFACE TENSION 22-3
CAPILLARITY 22-4

Nomenclature

A	area	ft²	m²
F	force	lbf	N
g	gravitational acceleration	ft/sec²	m/s²
g_c	gravitational constant (32.2)	lbm-ft/lbf-sec²	–
L	length	ft	m
m	mass	lbm	kg
SG	specific gravity	–	–
v	velocity	ft/sec	m/s
W	weight	lbf	N
y	distance	ft	m

Symbols

β	angle of contact	deg	deg
γ	specific weight	lbf/ft³	N/m³
δ	thickness	ft	m
μ	absolute viscosity	lbf-sec/ft²	Pa·s
ν	kinematic viscosity	ft²/sec	m²/s
ρ	density	lbm/ft³	kg/m³
σ	surface tension	lbf/ft	N/m
τ	stress	lbf/ft²	Pa
v	specific volume	ft³/lbm	m³/kg
$\forall$	volume	ft³	m³

Subscripts

n	normal
t	tangential

FLUIDS

Fluids are substances in either the liquid or gas phase. Fluids cannot support shear, and they deform continuously to minimize applied shear forces.

In fluid mechanics, the fluid is modeled as a *continuum*—that is, a substance that can be divided into infinitesimally small volumes, with properties that are continuous functions over the entire volume. For the infinitesimally small volume $\Delta\forall$, Δm is the infinitesimal mass, and ΔW is the infinitesimal weight.

DENSITY

The *density*, ρ, also called *mass density*, of a fluid is its mass per unit volume. The density of a fluid in a liquid form is usually given, known in advance, or easily obtained from tables. If $\Delta\forall$ is the volume of an infinitesimally small element,

$$\rho = \lim_{\Delta\forall \to 0}\left(\frac{\Delta m}{\Delta\forall}\right) \qquad 22.1$$

In SI units, density is measured in kg/m³. In a consistent English system, density is measured in slugs/ft³, even though fluid density has traditionally been reported in lbm/ft³. Most English fluid data are reported on a per pound basis.

SPECIFIC VOLUME

Specific volume, v, is the volume occupied by a unit mass of fluid.

$$v = \frac{1}{\rho} \qquad 22.2$$

Since specific volume is the reciprocal of density, typical units will be ft³/lbm or m³/kg.

SPECIFIC WEIGHT

Specific weight, γ, is the weight of fluid per unit volume.

$$\gamma = \lim_{\Delta\forall \to 0}\left(\frac{\Delta W}{\Delta\forall}\right) \qquad 22.3$$

$$\gamma = \lim_{\Delta\forall \to 0}\left(\frac{g\,\Delta m}{\Delta\forall}\right) = \rho g \qquad \text{[SI]} \qquad 22.3a$$

FLUIDS Properties

$$\gamma = \lim_{\Delta \forall \rightarrow 0} \left(\frac{g \Delta m}{g_c \Delta \forall} \right) = \frac{\rho g}{g_c} \quad \text{[U.S.]} \qquad 22.3b$$

The use of specific weight is most often encountered in civil engineering work in the United States, where it is commonly called *density*. Mechanical and chemical engineers seldom encounter the term. The usual units of specific weight are lbf/ft^3. Specific weight is not an absolute property of a fluid since it depends on the local gravitational field.

If the gravitational acceleration is $32.2 \ ft/sec^2$, as it is almost everywhere on the earth, the specific weight in lbf/ft^3 will be numerically equal to the density in lbm/ft^3. For example, if the density of water is $62.4 \ lbm/ft^3$, the specific weight of water will be

$$\gamma = \rho \frac{g}{g_c}$$

$$= \left(62.4 \ \frac{lbm}{ft^3} \right) \left(\frac{32.2 \ \frac{ft}{sec^2}}{32.2 \ \frac{lbm\text{-}ft}{sec^2\text{-}lbf}} \right)$$

$$= 62.4 \ lbf/ft^3 \qquad 22.4$$

SPECIFIC GRAVITY

Specific gravity, SG, is a dimensionless ratio of a fluid's density to a standard reference density. For liquids and solids, the reference is the density of pure water. However, there is some variation in this reference density, since the temperature at which the water density is evaluated is not standardized. Fortunately, the density of water is the same to three significant digits over the normal ambient temperature range: $62.4 \ lbm/ft^3$, $1.94 \ slugs/ft^3$, or $1000 \ kg/m^3$.

$$SG = \frac{\rho}{\rho_{water}} = \frac{\gamma}{\gamma_{water}} \qquad 22.5$$

Since the SI density of water is very nearly $1.000 \ g/cm^3$ ($1000 \ kg/m^3$), the numerical values of density in g/cm^3 and specific gravity are the same.

PRESSURE

Fluid pressures are measured with respect to two pressure references: zero pressure and atmospheric pressure. Pressures measured with respect to a true zero pressure reference are known as *absolute pressures*. Pressures measured with respect to atmospheric pressure are known as *gage pressures*. Most pressure gauges read the excess of the test pressure over atmospheric pressure

(i.e., the gage pressure). To distinguish between these two pressure measurements, the letters "a" and "g" are traditionally added to the unit symbols in the English unit system (e.g., 14.7 psia and 4015 psig). In SI, the actual words "gauge" and "absolute" can be added to the measurement (e.g., 25.1 kPa absolute). Alternatively, the pressure is assumed to be absolute unless the "g" is used (e.g., 15 kPag).

Absolute and gage pressures are related by Eq. 22.6. Note that $p_{atmospheric}$ in Eq. 22.6 is the actual atmospheric pressure existing when the gage measurement is taken. It is not standard atmospheric pressure unless that pressure is implicitly or explicitly applicable. (*Standard atmospheric pressure* is equal to 14.696 psia, 29.921 inches of mercury, or 101.3 kPa.) Also, since a barometer measures atmospheric pressure, *barometric pressure* is synonymous with atmospheric pressure.

$$p_{absolute} = p_{gage} + p_{atmospheric} \qquad 22.6$$

A *vacuum* measurement is implicitly a pressure below atmospheric pressure (i.e., a negative gage pressure). It must be assumed that any measured quantity given as a vacuum is a quantity to be subtracted from the atmospheric pressure. Thus, when a condenser is operating with a vacuum of 4.0 inches of mercury, the absolute pressure is approximately $29.92 - 4.0 = 25.92$ inches of mercury (25.92 in Hg). Vacuums are always stated as positive numbers.

$$p_{absolute} = p_{atmospheric} - p_{vacuum} \qquad 22.7$$

STRESS

Stress, τ, is force per unit area. For a point **P** with infinitesimal area ΔA and subjected to a force $\Delta \mathbf{F}$, the stress is defined as

$$\tau(\mathbf{P}) = \lim_{\Delta A \rightarrow 0} \left(\frac{\Delta \mathbf{F}}{\Delta A} \right) \qquad 22.8$$

There are two primary types of stress, differing in the orientation of the loaded area. With *normal stress*, τ_n, the area is normal to the force carried. Normal stress is equal to the pressure of the fluid. With *tangential* (or *shear*) *stress*, τ_t, the area is parallel to the force.

$$\tau_n = p \qquad 22.9$$

Ideal fluids that are inviscid and incompressible respond to normal stresses, but they cannot support shear, and they deform continuously to minimize applied shear forces.

VISCOSITY

The *viscosity* of a fluid is a measure of that fluid's resistance to flow when acted upon by an external force, such as a pressure gradient or gravity.

Viscosity of fluids can be determined with a *sliding plate viscometer* test. Consider two plates of area A separated by a fluid with thickness δ. The bottom plate is fixed, and the top plate is kept in motion at a constant velocity, v, by a force, F.

Figure 22.1 Sliding Plate Viscometer

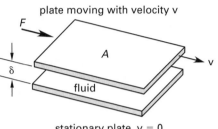

Experiments with many fluids have shown that the force, F, required to maintain the velocity, v, is proportional to the velocity and the area but is inversely proportional to the separation of the plates.

$$F \propto \frac{vA}{\delta} \qquad 22.10$$

The constant of proportionality is the *absolute viscosity*, μ, also known as the *absolute dynamic viscosity*. For a linear velocity profile, $v/\delta = dv/dy$. F/A is the *fluid shear stress*, τ_t.

$$\tau_t = \mu \frac{dv}{dy} \qquad 22.11$$

The quantity dv/dy is known by various names, including *rate of strain, shear rate, velocity gradient*, and *rate of shear formation*. Equation 22.11 is known as *Newton's law of viscosity*, from which Newtonian fluids get their name. (Not all fluids are Newtonian, although most are.) For a Newtonian fluid, strains are proportional to the applied shear stress (i.e., the stress versus strain curve is a straight line with slope μ). The straight line will be closer to the τ axis if the fluid is highly viscous. For low-viscosity fluids, the straight line will be closer to the dv/dy axis. Typical units for absolute viscosity are lbf-sec/ft^2 (lbm/ft-sec) (N·s/m^2), and poise (dyne·s/cm^2). One centipoise is equal to 0.01 poise and 0.001 Pa·s. 479 poise are equal to 1 lbf-sec/ft^2.

Equation 22.11 is applicable only to Newtonian fluids, for which the relationship is linear. The fluid shear stress for most non-Newtonian fluids can be predicted by the *power law*, Eq. 22.12. In Eq. 22.12, the constant K is known as the *consistency index*. For *pseudoplastic non-Newtonian fluids*, $n < 1$; for *dilatant non-Newtonian fluids*, $n > 1$. For Newtonian fluids, $n = 1$.

$$\tau_t = K \left(\frac{dv}{dy} \right)^n \qquad 22.12$$

Another quantity with the name *viscosity* is the ratio of absolute viscosity to mass density. This combination of variables, known as *kinematic viscosity*, ν, appears often in fluids and other problems and warrants its own symbol and name. Thus, kinematic viscosity is merely the name given to a frequently occuring combination of variables. The primary dimensions of kinematic viscosity are L^2/θ. Typical units are ft^2/sec and m^2/s.

$$\nu = \frac{\mu}{\rho} \qquad \text{[SI]} \qquad 22.13a$$

$$\nu = \frac{\mu g_c}{\rho} \qquad \text{[U.S.]} \qquad 22.13b$$

SURFACE TENSION

The membrane or "skin" that seems to form on the free surface of a fluid is caused by intermolecular cohesive forces and is known as *surface tension*, σ. Surface tension is the reason that insects are able to sit on a pond and a needle is able to float on the surface of a glass of water. Surface tension also causes bubbles and droplets to form in spheres, since any other shape would have more surface area per unit volume.

Surface tension can be interpreted as the tensile force between two points a unit distance apart on the surface, or as the amount of work required to form a new unit of surface area in an apparatus similar to that shown in Fig. 22.2. Typical units of surface tension are lbf/ft and N/m.

$$\sigma = \frac{F}{L} \qquad 22.14$$

Figure 22.2 Wire Frame for Stretching a Film

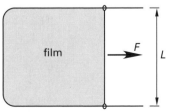

The apparatus shown in Fig. 22.2 consists of a wire frame with a sliding side that has been dipped in a liquid to form a film. Surface tension is determined by measuring the force necessary to keep the sliding side stationary against the surface tension pull of the film. Since the film has two surfaces (i.e., two surface tensions), the surface tension is

$$\sigma = \frac{F}{2L} \qquad \left[\begin{matrix}\text{wire frame}\\\text{apparatus}\end{matrix}\right] \qquad 22.15$$

Alternatively, surface tension can also be measured by measuring the force required to pull a Du Nouy wire ring out of the liquid, as shown in Fig. 22.3. Because the ring's inner and outer sides are in contact with the liquid, the wetted perimeter is twice the circumference. The surface tension is

$$\sigma = \frac{F}{4\pi r} \qquad \left[\begin{matrix}\text{Du Nouy ring}\\\text{apparatus}\end{matrix}\right] \qquad 22.16$$

CAPILLARITY

Capillary action is the name given to the behavior of a liquid in a thin-bore tube. Capillary action is caused by surface tension between the liquid and a vertical solid surface. In water, the adhesive forces between the liquid molecules and the surface are greater than (i.e., dominate) the cohesive forces between the water molecules themselves. The adhesive forces cause the water to attach itself to and climb a solid vertical surface; thus the

Figure 22.3 Du Nouy Ring Surface Tension Apparatus

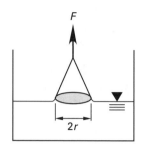

water rises above the general water surface level. The curved surface of the liquid within the tube is known as a *meniscus.*

For a few liquids, such as mercury, the molecules have a strong affinity for each other (i.e., the cohesive forces dominate). These liquids avoid contact with the tube surface. In such liquids, the meniscus will be below the general surface level.

The *angle of contact*, β, is an indication of whether adhesive or cohesive forces dominate. For contact angles less than 90°, adhesive forces dominate. For contact angles greater than 90°, cohesive forces dominate. For water in a glass tube, the contact angle is zero.

Equation 22.17 can be used to predict the capillary rise in a small-bore tube. Surface tension is a material property of a fluid, and contact angles are specific to a particular fluid-solid interface. Either may be obtained from tables.

Table 22.1 Properties of Water

temperature °C	specific weight γ, kN/m^3	density ρ, kg/m^3	viscosity $\mu \times 10^3$, Pa·s	kinematic viscosity $\nu \times 10^6$, m^2/s	vapor pressure p_a, kPa
0	9.805	999.8	1.781	1.785	0.61
5	9.807	1000.0	1.518	1.518	0.87
10	9.804	999.7	1.307	1.306	1.23
15	9.798	999.1	1.139	1.139	1.70
20	9.789	998.2	1.002	1.003	2.34
25	9.777	997.0	0.890	0.893	3.17
30	9.764	995.7	0.798	0.800	4.24
40	9.730	992.2	0.653	0.658	7.38
50	9.689	988.0	0.547	0.553	12.33
60	9.642	983.2	0.466	0.474	19.92
70	9.589	977.8	0.404	0.413	31.16
80	9.530	971.8	0.354	0.364	47.34
90	9.466	965.3	0.315	0.326	70.10
100	9.399	958.4	0.282	0.294	101.33

$$h = \frac{4\sigma \cos \beta}{\rho d_{\text{tube}} g} \qquad \text{[SI]} \qquad 22.17a$$

$$h = \frac{4\sigma g_c \cos \beta}{\rho g d_{\text{tube}}} = \frac{4\sigma \cos \beta}{\gamma d_{\text{tube}}} \qquad \text{[U.S.]} \qquad 22.17b$$

SAMPLE PROBLEMS

Problem 1

A vessel is initially connected to a reservoir open to the atmosphere. The connecting valve is then closed, and a vacuum of 65.5 kPa is applied to the vessel. What is the absolute pressure in the vessel? Assume standard atmospheric pressure.

(A) 36 kPa
(B) 66 kPa
(C) 86 kPa
(D) 110 kPa

Solution

For vacuum pressures,

$$p_{\text{absolute}} = p_{\text{atmospheric}} - p_{\text{vacuum}}$$
$$= 1.013 \times 10^5 \text{ Pa} - 65.5 \times 10^3 \text{ Pa}$$
$$= 35\,800 \text{ Pa} \quad (36 \text{ kPa})$$

Answer is A.

Problem 2

At a particular temperature, the surface tension of water is 0.073 N/m. Under ideal conditions, the contact angle between glass and water is zero. A student in a laboratory observes water in a glass capillary tube with a diameter of 0.1 mm. What is the theoretical height of the capillary rise?

(A) 0.00020 m
(B) 0.013 m
(C) 0.045 m
(D) 0.30 m

Solution

$$h = \frac{4\sigma \cos \beta}{\rho g d}$$

$$= \frac{(4)\left(0.073 \; \dfrac{\text{N}}{\text{m}}\right)\cos 0^\circ}{\left(1000 \; \dfrac{\text{kg}}{\text{m}^3}\right)\left(9.81 \; \dfrac{\text{m}}{\text{s}^2}\right)(0.0001 \text{ m})}$$

$$= 0.2977 \text{ m} \quad (0.30 \text{ m})$$

Answer is D.

FE-STYLE EXAM PROBLEMS

1. What is the atmospheric pressure on a planet if the absolute pressure is 100 kPa and the gage pressure is 10 kPa?

(A) 10 kPa
(B) 80 kPa
(C) 90 kPa
(D) 100 kPa

2. 100 g of water are mixed with 150 g of another fluid ($\rho = 790$ kg/m^3). What is the specific volume of the resulting mixture, assuming that the volumes are additive and the mixture is homogeneous?

(A) 0.63 cm^3/g
(B) 0.82 cm^3/g
(C) 0.88 cm^3/g
(D) 1.20 cm^3/g

3. 100 g of water are mixed with 150 g of another fluid ($\rho = 790$ kg/m^3). What is the specific gravity of the resulting mixture, assuming that the volumes are additive and the mixture is homogeneous?

(A) 0.63
(B) 0.82
(C) 0.86
(D) 0.95

4. Kinematic viscosity can be expressed in which of the following units?

(A) m^2/s
(B) s^2/m
(C) kg·s^2/m
(D) kg/s

5. Which of the following does not affect the rise or fall of liquid in a small-diameter capillary tube?

(A) adhesive forces
(B) cohesive forces
(C) surface tension
(D) viscosity of the fluid

For the following problems use the NCEES Handbook as your only reference.

6. The film width in a surface tension experiment is 10 cm. If mercury is the fluid (surface tension = 0.52 N/m), what is the maximum force that can be applied without breaking the membrane? Neglect gravitational force.

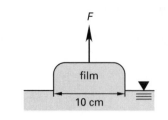

(A) 0.1 N
(B) 1.0 N
(C) 2.0 N
(D) 3.4 N

Problems 7 and 8 refer to the following situation.

The cross section of a concrete dam is shown. The density of concrete is 2400 kg/m^3. The dam has a constant slope along surface BC and is parabolic according to $y = 0.24x^2$ along surface AD.

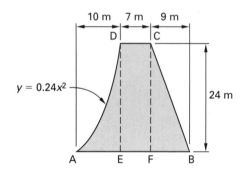

7. What is the approximate mass of concrete contained in a cross section having a width of 1 m and bounded by points F, B, and C?

(A) 102 Mg
(B) 195 Mg
(C) 226 Mg
(D) 259 Mg

8. What is the mass of concrete contained in a cross section having a width of 1 m and bounded by points A, D, and E?

(A) 192 Mg
(B) 240 Mg
(C) 288 Mg
(D) 384 Mg

9. Oil with a specific gravity of 0.72 is used as the indicating fluid in a manometer. If the differential pressure across the ends of the manometer is 7 kPa, what will be the difference in oil levels in the manometer?

(A) 0.23 m
(B) 0.44 m
(C) 0.53 m
(D) 1.0 m

10. A pressure of 35 kPa is measured 4 m below the surface of an unknown liquid. What is the specific gravity of the liquid?

(A) 0.09
(B) 0.89
(C) 0.93
(D) 1.85

11. A dimensional analysis of a ship model is being performed using a total of four different dimensional quantities: mass, length, time, and temperature. Six different variables (density, surface tension, viscosity, gravity, power, and mass) have been identified as potentially affecting the velocity of the boat. Use the Buckingham π-theorem to predict how many π-groups will be needed?

(A) 1
(B) 2
(C) 10
(D) 24

SOLUTIONS TO FE-STYLE EXAM PROBLEMS

1.
$$p_{\text{atmospheric}} = p_{\text{absolute}} - p_{\text{gage}}$$
$$= 100 \text{ kPa} - 10 \text{ kPa}$$
$$= 90 \text{ kPa}$$

Answer is C.

2.
$$\rho_{\text{water}} = 1 \text{ g/cm}^3$$
$$\rho_{\text{fluid}} = 790 \text{ kg/m}^3 = 0.79 \text{ g/cm}^3$$
$$V = \frac{\text{mass}}{\text{density}}$$
$$V_{\text{water}} + V_{\text{fluid}} = \frac{100 \text{ g}}{1.0 \, \dfrac{\text{g}}{\text{cm}^3}} + \frac{150 \text{ g}}{0.79 \, \dfrac{\text{g}}{\text{cm}^3}}$$
$$= 289.87 \text{ cm}^3$$
$$\rho_{\text{mixture}} = \frac{m_{\text{water}} + m_{\text{fluid}}}{V_{\text{water}} + V_{\text{fluid}}} = \frac{100 \text{ g} + 150 \text{ g}}{289.87 \text{ cm}^3}$$
$$= 0.862 \text{ g/cm}^3$$

$$v = \frac{1}{\rho} = \frac{1}{0.862 \, \frac{g}{cm^3}}$$

$$= 1.16 \text{ cm}^3/\text{g} (1.20 \text{ cm}^3/\text{g})$$

Answer is D.

3. $\rho_{\text{mixture}} = 0.862 \text{ g/cm}^3$ [from Prob. 2]

$$SG = \frac{\rho_{\text{mixture}}}{\rho_{\text{water}}} = \frac{0.862 \, \frac{g}{cm^3}}{1.0 \, \frac{g}{cm^3}}$$

$$= 0.862 (0.86)$$

Answer is C.

4. The units of kinematic viscosity are m^2/s (SI) or ft^2/sec (U.S.).

Answer is A.

5. The height of capillary rise is determined from

$$h = \frac{4\sigma \cos \beta}{\gamma d}$$

σ is the surface tension of the fluid, d is the diameter of the tube, β is the contact angle, and γ is the specific weight of the liquid.

The viscosity of the fluid is not directly relevant to the height of capillary rise.

Answer is D.

6. $\sigma = \dfrac{F}{2L}$

$$F = 2\sigma L = (2) \left(0.52 \, \frac{N}{m} \right) \left(\frac{10 \text{ cm}}{100 \, \frac{cm}{m}} \right)$$

$$= 0.104 \text{ N} (0.1 \text{ N})$$

Answer is A.

7. The area of the cross section FBC is

$$A_{\text{FBC}} = \frac{1}{2}bh = \frac{1}{2}(9 \text{ m})(24 \text{ m})$$

$$= 108 \text{ m}^2$$

The volume of concrete is

$$V = Aw$$

$$= (108 \text{ m}^2)(1 \text{ m})$$

$$= 108 \text{ m}^3$$

The mass is

$$m = \rho V = \left(2400 \, \frac{kg}{m^3} \right) (108 \text{ m}^3)$$

$$= 2.59 \times 10^5 \text{ kg} (259 \text{ Mg})$$

Answer is D.

8. The area of cross section AED is found by integration.

$$A_{\text{AED}} = \int_0^{10} 0.24x^2 \, dx = 0.08x^3 \Big|_0^{10}$$

$$= 80 \text{ m}^2$$

$$V = Aw = (80 \text{ m}^2)(1 \text{ m})$$

$$= 80 \text{ m}^3$$

$$m = \rho V = \left(2400 \, \frac{kg}{m^3} \right) (80 \text{ m}^3)$$

$$= 1.92 \times 10^5 \text{ kg} (192 \text{ Mg})$$

Answer is A.

9. The density of the oil is

$$\rho_{\text{oil}} = (SG)\rho_{\text{water}}$$

$$= (0.72) \left(1000 \, \frac{kg}{m^3} \right)$$

$$= 720 \text{ kg/m}^3$$

$$h = \frac{\Delta p}{\rho g} = \frac{7000 \, \frac{N}{m^2}}{\left(720 \, \frac{kg}{m^3} \right) \left(9.81 \, \frac{m}{s^2} \right)}$$

$$= 0.99 \text{ m} (1 \text{ m})$$

Answer is D.

10. $p = \rho g h = (SG)\rho_{\text{water}} g h$

$$SG = \frac{p}{\rho_{\text{water}} g h}$$

$$= \frac{\left(35\,000 \, \frac{N}{m^2} \right)}{\left(1000 \, \frac{kg}{m^3} \right) \left(9.81 \, \frac{m}{s^2} \right) (4 \text{ m})}$$

$$= 0.89$$

Answer is B.

11. The Buckingham π-theorem predicts the number of π-groups as $k = m - n$, where m is the number of variables and n is the number of dimensional quantities. In this case, $k = 6 - 4 = 2$.

Answer is B.

FLUIDS Properties

23 Fluid Statics

Subjects

HYDROSTATIC PRESSURE 23-1
MANOMETRY 23-2
BAROMETERS 23-2
FORCES ON SUBMERGED PLANE
   SURFACES 23-3
CENTER OF PRESSURE 23-4
BUOYANCY 23-4

Nomenclature

A	area	ft^2	m^2
F	force	lbf	N
g	gravitational acceleration	ft/sec^2	m/s^2
g_c	gravitational constant (32.2)	lbm-ft/lbf-sec^2	–
h	vertical depth	ft	m
I	moment of inertia	ft^4	m^4
I	product of inertia	ft^4	m^4
M	moment	ft-lbf	N·m
p	pressure	lbf/ft^2	N/m^2
R	resultant force	lbf	N
W	weight	lbf	–
z	inclined distance	ft	m

Symbols

α	angle	deg	deg
γ	specific weight	lbf/ft^3	N/m^3
ρ	density	lbm/ft^3	kg/m^3

Subscripts

a	atmospheric
b	buoyant
c	centroidal
f	fluid
m	manometer
o	static (enclosed vessel)
R	resultant
v	vapor

HYDROSTATIC PRESSURE

Hydrostatic pressure is the pressure a fluid exerts on an immersed object or on container walls. The term *hydrostatic* is used with all fluids, not only with water.

Pressure is equal to the force per unit area of surface.

$$p = \frac{F}{A} \qquad\qquad 23.1$$

Hydrostatic pressure in a stationary, incompressible fluid behaves according to the following characteristics.

- Pressure is a function of vertical depth (and density) only. The pressure will be the same at two points with identical depths.

- Pressure varies linearly with (vertical) depth. The relationship between pressure and depth for an incompressible fluid is given by Eq. 23.2.

$$p = \rho g h \qquad\qquad \text{[SI]} \qquad 23.2a$$

$$p = \frac{\rho g h}{g_c} = \gamma h \qquad \text{[U.S.]} \qquad 23.2b$$

Since ρ and g are constants, Eq. 23.2 shows that p and h are linearly related. One determines the other.

- Pressure is independent of an object's area and size, and of the weight (mass) of water above the object. Figure 23.1 illustrates the *hydrostatic paradox*. The pressures at depth h are the same in all four columns because pressure depends only on depth, not on volume.

Figure 23.1 Hydrostatic Paradox

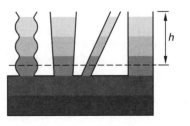

- Pressure at a point has the same magnitude in all directions (*Pascal's law*). Thus, pressure is a scalar quantity.

- Pressure is always normal to a surface, regardless of the surface's shape or orientation. (This is a result of the fluid's inability to support shear stress.)

MANOMETRY

Manometers can be used to indicate small pressure differences, and for this purpose, they provide good accuracy. A difference in manometer fluid surface heights indicates a pressure difference. When both ends of the manometer are connected to pressure sources, the name *differential manometer* is used. If one end of the manometer is open to the atmosphere, the name *open manometer* is used. An open manometer indicates gage pressure. It is theoretically possible, but impractical, to have a manometer indicate absolute pressure, since one end of the manometer would have to be exposed to a perfect vacuum.

Consider the simple manometer in Fig. 23.2. The pressure difference $p_2 - p_1$ causes the difference h_m in manometer fluid surface heights. Fluid column h_2 exerts a hydrostatic pressure on the manometer fluid, forcing the manometer fluid to the left. This increase must be subtracted out. Similarly, the column h_1 restricts the movement of the manometer fluid. The observed measurement must be increased to correct for this restriction. The typical way to solve for pressure differences in a manometer is to start with the pressure on one side, and then add or subtract changes in hydrostatic pressure at known points along the column until the pressure on the other side is reached.

Figure 23.2 Manometer Requiring Corrections

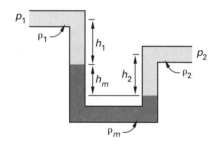

$$p_2 = p_1 + \rho_1 g h_1 + \rho_m g h_m - \rho_2 g h_2 \quad \text{[SI]} \quad 23.3a$$

$$p_2 = p_1 + \gamma_1 h_1 + \gamma_m h_m - \gamma_2 h_2 \quad \text{[U.S.]} \quad 23.3b$$

Figure 23.3 illustrates an open manometer. Neglecting the air in the open end, the pressure difference is given by Eq. 23.4. Notice that $p_o - p_a$ is the gage pressure in the vessel.

$$p_o - p_a = \rho_2 g h_2 - \rho_1 g h_1 \quad \text{[SI]} \quad 23.4a$$

$$p_o - p_a = \gamma_2 h_2 - \gamma_1 h_1 \quad \text{[U.S.]} \quad 23.4b$$

If fluid 1 is absent or has a low density (i.e., is a gas), or if distance h_1 is so small as to be insignificant, then the pressure difference will be

Figure 23.3 Open Manometer

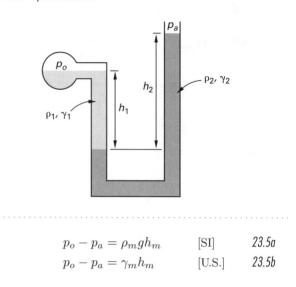

$$p_o - p_a = \rho_m g h_m \quad \text{[SI]} \quad 23.5a$$

$$p_o - p_a = \gamma_m h_m \quad \text{[U.S.]} \quad 23.5b$$

BAROMETERS

The *barometer* is a common device for measuring the absolute pressure of the atmosphere. It is constructed by filling a long tube open at one end with mercury (or alcohol or some other liquid) and inverting the tube such that the open end is below the level of a mercury-filled container. If the vapor pressure of the mercury in the tube is neglected, the fluid column will be supported only by the atmospheric pressure transmitted through the container fluid at the lower, open end. The atmospheric pressure is given by Eq. 23.6.

$$p_a = \rho g h \quad \text{[SI]} \quad 23.6a$$

$$p_a = \frac{\rho g h}{g_c} = \gamma h \quad \text{[U.S.]} \quad 23.6b$$

Figure 23.4 Barometer

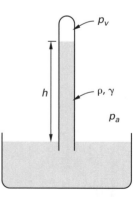

If the vapor pressure of the barometer liquid is significant (as it would be with alcohol or water), the vapor pressure effectively reduces the height of the fluid column, as Eq. 23.7 illustrates.

$$p_a - p_v = \rho g h \qquad \text{[SI]} \qquad 23.7a$$

$$p_a - p_v = \frac{\rho g h}{g_c} = \gamma h \qquad \text{[U.S.]} \qquad 23.7b$$

FORCES ON SUBMERGED PLANE SURFACES

The pressure on a horizontal plane surface is uniform over the surface because the depth of the fluid above is uniform. The resultant of the pressure distribution acts through the center of pressure of the surface, which corresponds to the centroid of the surface.

Figure 23.5 Hydrostatic Pressure on a Horizontal Plane Surface

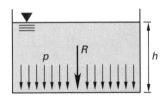

The total vertical force on the horizontal plane of area A is given by Eq. 23.8.

$$R = pA \qquad 23.8$$

It is not always correct to calculate the vertical force on a submerged surface as the weight of the fluid above it. Such an approach works only when there is no change in the cross-sectional area of the fluid above the surface. This is a direct result of the hydrostatic paradox. (See Fig. 23.1.) Figure 23.6 illustrates two containers with the same pressure distribution (force) on their bottom surfaces.

Figure 23.6 Two Containers with the Same Pressure Distribution

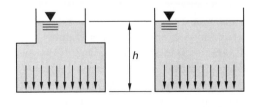

The pressure on a vertical rectangular plane surface increases linearly with depth. The pressure distribution will be triangular, as in Fig. 23.7(a), if the plane surface extends to the surface; otherwise the distribution will be trapezoidal, as in Fig. 23.7(b).

Figure 23.7 Hydrostatic Pressure on a Vertical Plane Surface

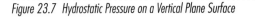

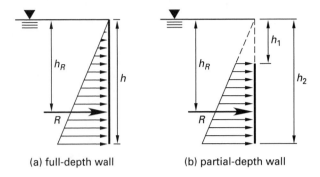

(a) full-depth wall   (b) partial-depth wall

The resultant force is calculated from the *average pressure*, which is also the pressure at the location of the centroid of the plane area.

$$\bar{p} = \tfrac{1}{2}(p_1 + p_2) \qquad 23.9$$

$$\bar{p} = \tfrac{1}{2}\rho g(h_1 + h_2) \qquad \text{[SI]} \qquad 23.10a$$

$$\bar{p} = \frac{\tfrac{1}{2}\rho g(h_1 + h_2)}{g_c}$$
$$= \tfrac{1}{2}\gamma(h_1 + h_2) \qquad \text{[U.S.]} \qquad 23.10b$$

$$R = \bar{p}A \qquad 23.11$$

Although the resultant is calculated from the average depth, the resultant does not act at the average depth. The resultant of the pressure distribution passes through the centroid of the pressure distribution. For the triangular distribution of Fig. 23.7(a), the resultant is located at a depth of $h_R = \tfrac{2}{3}h$. For the more general case, the center of pressure can be calculated by the method described in the next section.

The average pressure and resultant force on an inclined rectangular plane surface are calculated in much the same fashion as for the vertical plane surface. The pressure varies linearly with depth. The resultant is calculated from the average pressure, which, in turn, depends on the average depth.

Figure 23.8 Hydrostatic Pressure on an Inclined Rectangular Plane Surface

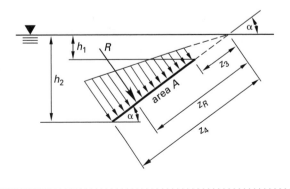

The average pressure and resultant on an inclined plane surface are given by Eqs. 23.12 through 23.15. As with the vertical plane surface, the resultant acts at the centroid of the pressure distribution, not at the average depth.

$$\bar{p} = \tfrac{1}{2}(p_1 + p_2) \qquad\qquad 23.12$$

$$\bar{p} = \tfrac{1}{2}\rho g(h_1 + h_2) \qquad [\text{SI}] \qquad 23.13a$$

$$\bar{p} = \tfrac{1}{2}\left(\frac{\rho g(h_1 + h_2)}{g_c}\right)$$

$$\quad = \tfrac{1}{2}\gamma(h_1 + h_2) \qquad [\text{U.S.}] \qquad 23.13b$$

$$\bar{p} = \tfrac{1}{2}\rho g(z_3 + z_4)\sin\alpha \qquad [\text{SI}] \qquad 23.14a$$

$$\bar{p} = \tfrac{1}{2}\gamma(z_3 + z_4)\sin\alpha \qquad [\text{U.S.}] \qquad 23.14b$$

$$R = \bar{p}A \qquad\qquad 23.15$$

CENTER OF PRESSURE

For the case of pressure on a general plane surface, the resultant force depends on the average pressure and acts through the *center of pressure* (CP). Figure 23.9 illustrates a nonrectangular plane surface of area A that may or may not extend to the liquid surface and that may or may not be inclined. The average pressure is calculated from the location of the plane surface's centroid (C), where z_c is measured parallel to the plane surface. That is, if the plane surface is inclined, z_c is an inclined distance. p_o is the external pressure at the liquid surface and can be disregarded if the surface is exposed to the atmosphere and all pressures are gage pressures.

Figure 23.9 Hydrostatic Pressure on a General Plane Surface

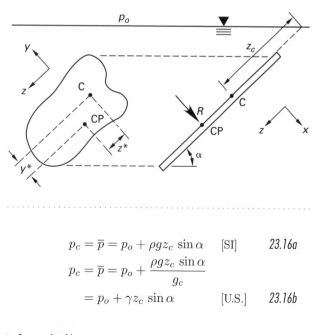

$$p_c = \bar{p} = p_o + \rho g z_c \sin\alpha \qquad [\text{SI}] \qquad 23.16a$$

$$p_c = \bar{p} = p_o + \frac{\rho g z_c \sin\alpha}{g_c}$$

$$\quad = p_o + \gamma z_c \sin\alpha \qquad [\text{U.S.}] \qquad 23.16b$$

The resultant force acts normal to the plane surface at the center of pressure, which has coordinates y^* and z^* as shown in Fig. 23.9, relative to the centroid. I_{yy} and I_{yz} are the centroidal area moment of inertia and product of inertia, respectively, both with dimensions of L^4 (length^4), about an axis parallel to the surface.

$$y^* = \frac{\rho g I_{yz}\sin\alpha}{p_c A} \qquad [\text{SI}] \qquad 23.17a$$

$$y^* = \frac{\rho g I_{yz}\sin\alpha}{g_c p_c A}$$

$$\quad = \frac{\gamma I_{yz}\sin\alpha}{p_c A} \qquad [\text{U.S.}] \qquad 23.17b$$

$$z^* = \frac{\rho g I_{yy}\sin\alpha}{p_c A} \qquad [\text{SI}] \qquad 23.18a$$

$$z^* = \frac{\rho g I_{yy}\sin\alpha}{g_c p_c A}$$

$$\quad = \frac{\gamma I_{yy}\sin\alpha}{p_c A} \qquad [\text{U.S.}] \qquad 23.18b$$

If the surface is open to the atmosphere, then $p_o = 0$, and

$$p_c = \bar{p} = \rho g z_c \sin\alpha \qquad [\text{SI}] \qquad 23.19a$$

$$p_c = \bar{p} = \frac{\rho g z_c \sin\alpha}{g_c}$$

$$\quad = \gamma z_c \sin\alpha \qquad [\text{U.S.}] \qquad 23.19b$$

$$y_{cp} - y_c = y^* = \frac{I_{yz}}{z_c A} \qquad\qquad 23.20$$

$$z_{cp} - z_c = z^* = \frac{I_{yy}}{z_c A} \qquad\qquad 23.21$$

The center of pressure is always at least as deep as the area's centroid. In most cases, it is deeper.

BUOYANCY

Buoyant force is an upward force that acts on all objects that are partially or completely submerged in a fluid. The fluid can be a liquid or a gas. There is a buoyant force on all submerged objects, not only on those that are stationary or ascending. A buoyant force caused by displaced air also exists, although it may be insignificant. Examples include the buoyant force on a rock sitting at the bottom of a pond, the buoyant force on a rock sitting exposed on the ground (since the rock is "submerged" in air), and the buoyant force on partially exposed floating objects, such as icebergs.

Buoyant force always acts to cancel the object's weight (i.e., buoyancy acts against gravity). The magnitude of the buoyant force is predicted from *Archimedes' principle* (the *buoyancy theorem*), which states that the buoyant force on a submerged or floating object is equal to the weight of the displaced fluid. An equivalent statement of Archimedes' principle is that a floating object displaces liquid equal in weight to its own weight. In the situation of an object floating at the interface between two immiscible liquids of different densities, the buoyant force equals the sum of the weights of the two displaced fluids.

In the case of stationary (i.e., not moving vertically) floating or submerged objects, the buoyant force and object weight are in equilibrium. If the forces are not in equilibrium, the object will rise or fall until equilibrium is reached—that is, the object will sink until its remaining weight is supported by the bottom, or it will rise until the weight of liquid is reduced by breaking the surface.

The two forces acting on a stationary floating object are the *buoyant force* and the *object's weight*. The buoyant force acts upward through the centroid of the displaced volume (not the object's volume). This centroid is known as the *center of buoyancy*. The gravitational force on the object (i.e., the object's weight) acts downward through the entire object's center of gravity.

SAMPLE PROBLEMS

Problem 1

The average specific gravity of seawater is 1.15. What is the absolute pressure at the bottom of 3000 m of sea?

(A) 2.1 kPa
(B) 2.5 MPa
(C) 28 MPa
(D) 34 MPa

Solution

$$p = p_o + \gamma h$$
$$= p_o + \rho g h$$

The density of cold water is approximately 1000 kg/m^3. Standard atmospheric pressure is 1.013×10^5 Pa.

$$p = 1.013 \times 10^5 \text{ Pa}$$
$$+ (1.15)\left(1000 \ \frac{\text{kg}}{\text{m}^3}\right)\left(9.81 \ \frac{\text{m}}{\text{s}^2}\right)(3000 \text{ m})$$
$$= 3.39 \times 10^7 \text{ Pa} \quad (34 \text{ MPa})$$

Answer is D.

Problem 2

The specific gravity of mercury is 13.6, and the specific gravity of glycerine is 1.26. For the manometer shown, calculate the difference in pressure between points A and B.

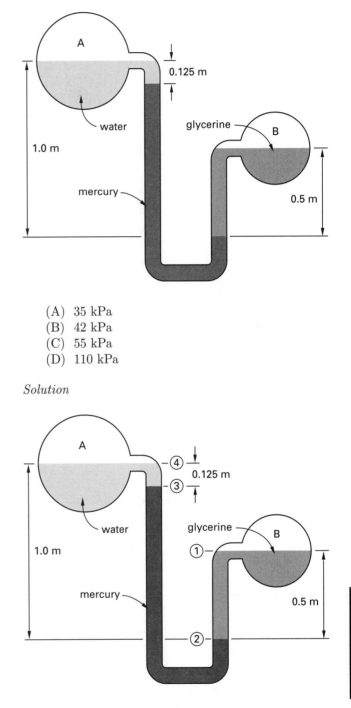

(A) 35 kPa
(B) 42 kPa
(C) 55 kPa
(D) 110 kPa

Solution

The pressure at level 2 is the same in both (left and right) legs of the manometer.

Working up the right leg from level 2,

$$p_2 = p_\text{B} + \rho_\text{glycerine} g h_{1-2}$$

Working up the left leg from level 2,

$$p_2 = p_A + \rho_{water}gh_{3-4} + \rho_{Hg}gh_{2-3}$$

Equating these two equations for p_2 and solving for the pressure difference $p_A - p_B$,

$$
\begin{aligned}
p_A - p_B &= g(\rho_{glycerine}h_{1-2} - \rho_{water}h_{3-4} - \rho_{Hg}h_{2-3}) \\
&= g\rho_{water}(SG_{glycerine}h_{1-2} - SG_{water}h_{3-4} \\
&\quad - SG_{Hg}h_{2-3}) \\
&= \left(9.81\ \frac{m}{s^2}\right)\left(1000\ \frac{kg}{m^3}\right)\left(0.001\ \frac{kN}{N}\right) \\
&\quad \times \big((1.26)(0.5\ m) - (1.00)(0.125\ m) \\
&\quad - (13.6)(1.0\ m - 0.125\ m)\big) \\
&= -111.8\ kPa \quad (110\ kPa)
\end{aligned}
$$

Answer is D.

Problems 3–5 refer to the following diagram of a water-filled tank.

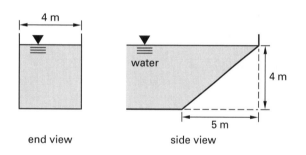

end view   side view

Problem 3

What is the resultant force on the inclined wall?

(A) 222 kN
(B) 395 kN
(C) 503 kN
(D) 526 kN

Solution

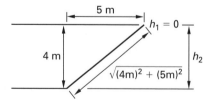

$$\bar{p} = \tfrac{1}{2}\rho g(h_1 + h_2)$$

$$= \left(\frac{1}{2}\right)\left(1000\ \frac{kg}{m^3}\right)\left(9.81\ \frac{m}{s^2}\right)(0\ m + 4\ m)$$

$$= 19\,620\ Pa$$

$$R = \bar{p}A$$

$$= (19\,620\ Pa)(4\ m)\left(\sqrt{(4\ m)^2 + (5\ m)^2}\right)$$

$$= 502\,517\ N \quad (503\ kN)$$

Answer is C.

Problem 4

What is the vertical force on the inclined wall?

(A) 197 kN
(B) 392 kN
(C) 486 kN
(D) 544 kN

Solution

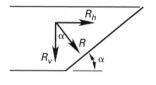

$$R_v = R\cos\alpha$$

$$= (502\,517\ N)\left(\frac{5\ m}{\sqrt{(4\ m)^2 + (5\ m)^2}}\right)$$

$$= 392\,400\ N \quad (392\ kN)$$

Answer is B.

Problem 5

What is the horizontal force on the inclined wall?

(A) 197 kN
(B) 314 kN
(C) 421 kN
(D) 540 kN

Solution

$$R_h = R\sin\alpha$$

$$= (502\,517\ N)\left(\frac{4\ m}{\sqrt{(4\ m)^2 + (5\ m)^2}}\right)$$

$$= 313\,920\ N \quad (314\ kN)$$

Answer is B.

Problem 6

Archimedes established his principle while investigating a suspected fraud in the construction of a crown. The crown was made from an alloy of gold and silver instead of from pure gold. Assume that the volume of the alloy was the combined volumes of the components (the density of gold is 19.3 g/cm^3, the density of silver is

10.5 g/cm^3). If the crown had a weight of 1000 g in air and 940 g in pure water, what percentage of gold (by weight) was it?

(A) 53.1%
(B) 67.4%
(C) 81.2%
(D) 91.3%

Solution

Archimedes' principle states that the buoyant force on a submerged object is equal to the weight of the displaced fluid. If V is the volume of the crown and ρ_{crown} its average density in air, then

$$F_b = \rho_{water}gV$$

The weight of the crown in air is

$$W = \rho_{crown}gV$$

The ratio of W to F_b gives the specific gravity of the crown, SG_{crown}.

$$\frac{W}{F_b} = \frac{\rho_{crown}}{\rho_{water}} = SG_{crown}$$

The buoyant force is also the difference between the weight in air and the weight in water (assuming the buoyant force in air to be negligible). If W' is the weight in water, then

$$F_b = W - W'$$

But,

$$\frac{W}{F_b} = \frac{W}{W - W'} = SG_{crown}$$

$$\frac{W}{W - W'} = \frac{mg}{mg - m'g} = \frac{m}{m - m'}$$

$$= \frac{1000 \text{ g}}{1000 \text{ g} - 940 \text{ g}}$$

$$SG_{crown} = 16.67$$

The volume of the alloy is the combined volumes of the components. Let x be the volume of silver in 1 cm^3 of alloy. Then,

$$\left(16.67 \ \frac{\text{g}}{\text{cm}^3}\right)(1 \text{ cm}^3) = (1 - x)\left(19.3 \ \frac{\text{g}}{\text{cm}^3}\right)$$

$$+ x\left(10.5 \ \frac{\text{g}}{\text{cm}^3}\right)$$

$$16.67 \text{ g} = 19.3 \text{ g} - \left(8.8 \ \frac{\text{g}}{\text{cm}^3}\right)x$$

$$x = 0.2989 \text{ cm}^3$$

The volume of gold in 1 cm^3 is

$$1 \text{ cm}^3 - 0.2989 \text{ cm}^3 = 0.7011 \text{ cm}^3$$

The mass of gold in 1 cm^3 is

$$\left(19.3 \ \frac{\text{g}}{\text{cm}^3}\right)(0.7011 \text{ cm}^3) = 13.53 \text{ g}$$

The mass of alloy in 1 cm^3 is 16.67 g.

The percentage of gold in the crown is

$$\frac{13.53 \text{ g}}{16.67 \text{ g}} = 0.812 \quad (81.2\%)$$

Answer is C.

Problem 7

What is the depth of the center of pressure on the vertical plate if the upper edge is 1.5 m below the water surface?

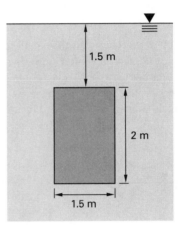

(A) 2.12 m
(B) 2.32 m
(C) 2.50 m
(D) 2.63 m

Solution

The centroidal moment of inertia about an axis parallel to the surface is

$$I = \frac{bh^3}{12} = \frac{(1.5 \text{ m})(2 \text{ m})^3}{12}$$

$$= 1 \text{ m}^4$$

From Eq. 23.20,

$$y_{cp} = y_c + y^* = y_c + \frac{I}{y_c A}$$

$$= 2.5 \text{ m} + \frac{1 \text{ m}^4}{(1.5 \text{ m})(2 \text{ m})(2.5 \text{ m})}$$

$$= 2.63 \text{ m}$$

Answer is D.

FE-STYLE EXAM PROBLEMS

1. What height of mercury column is equivalent to a pressure of 700 kPa? The density of mercury is $13\,500$ kg/m^3.

    (A) 0.75 m
    (B) 1.5 m
    (C) 3.4 m
    (D) 5.3 m

2. A fluid with a vapor pressure of 0.2 Pa and a specific gravity of 12 is used in a barometer. If the fluid's column height is 1 m, what is the atmospheric pressure?

    (A) 9.80 kPa
    (B) 11.8 kPa
    (C) 101 kPa
    (D) 118 kPa

3. One leg of a mercury U-tube manometer is connected to a pipe containing water under a gage pressure of 100 kPa. The mercury in this leg stands 750 mm below the water. What is the height of mercury in the other leg, which is open to the air? The specific gravity of mercury is 13.5.

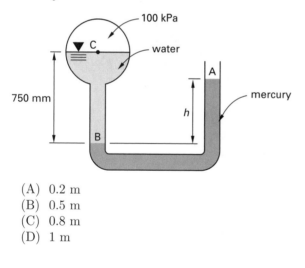

    (A) 0.2 m
    (B) 0.5 m
    (C) 0.8 m
    (D) 1 m

4. What is the resultant hydrostatic force on one side of a 25 cm diameter vertical circular plate standing at the bottom of a 3 m pool of water?

    (A) 1.38 kN
    (B) 1.63 kN
    (C) 1.91 kN
    (D) 2.72 kN

5. A closed tank with the dimensions shown contains water. If the pressure of the air in the tank is 700 kPa, what is the pressure at point P, which is located halfway up the inclined wall?

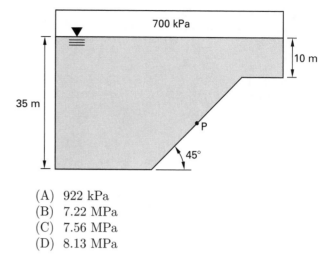

    (A) 922 kPa
    (B) 7.22 MPa
    (C) 7.56 MPa
    (D) 8.13 MPa

For the following problems use the NCEES Handbook as your only reference.

6. A triangular gate with a horizontal base 1.5 m long and an altitude of 2 m is inclined 45° from the vertical with the vertex pointing upward. The hinged horizontal base of the gate is 3 m below the water surface. What normal force must be applied at the vertex of the gate to keep it closed?

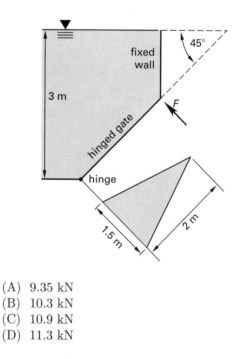

    (A) 9.35 kN
    (B) 10.3 kN
    (C) 10.9 kN
    (D) 11.3 kN

7. Which of the following statements concerning buoyancy are false?

I.   Buoyancy is the tendency of a fluid to exert a supporting force on a body placed in that fluid.
II. Buoyancy is the ability of a body to return to its original position after being tilted on its horizontal axis.
III. The buoyant force is measured by multiplying the specific weight of the object by the displaced volume of the fluid.
IV. Buoyant forces occur both when an object floats in a fluid and when an object sinks in a fluid.
V.   The buoyant force acts vertically upward through the centroid of the displaced volume.

(A) I, II, and III
(B) III, IV, and V
(C) I and V
(D) II and III

Problems 8–10 refer to the following situation.

A 300 mm long rigid metal cylindrical container with a diameter of 200 mm is closed at one end. The container is held vertically, barely submerged, and closed-end-up in water, as shown. The atmospheric pressure is 101.3 kPa. The water rises 75 mm inside the container under these conditions.

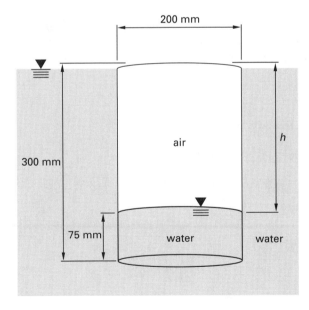

8. What is the approximate total pressure of the air inside the container?

(A) 101.3 kPa
(B) 102.1 kPa
(C) 103.5 kPa
(D) 110.4 kPa

9. The container is slowly moved vertically downward until the pressure in the container is 105 kPa. What will be the depth of the water surface measured from the free water surface (i.e., what is the vertical distance between the free water surface and the water surface in the container)?

(A) 160 mm
(B) 170 mm
(C) 300 mm
(D) 380 mm

10. The container is slowly moved vertically downward until the pressure in the container is 105 kPa. What will be the height of the air space in the container (i.e., the vertical distance between the closed upper end and the water surface in the container)?

(A) 217 mm
(B) 222 mm
(C) 227 mm
(D) 230 mm

SOLUTIONS TO FE-STYLE EXAM PROBLEMS

1. Pressure increases linearly with depth.

$$p = \rho g h$$

$$h = \frac{p}{\rho g}$$

$$= \frac{700 \times 10^3 \ \dfrac{N}{m^2}}{\left(13\,500 \ \dfrac{kg}{m^3}\right)\left(9.81 \ \dfrac{m}{s^2}\right)}$$

$$= 5.29 \ m \quad (5.3 \ m)$$

Answer is D.

2.
$$p_a = p_v + \rho g h$$

$$= 0.2 \ Pa + (12)\left(1000 \ \frac{kg}{m^3}\right)$$

$$\times \left(9.81 \ \frac{m}{s^2}\right)(1 \ m)$$

$$= 117\,720 \ Pa \quad (118 \ kPa)$$

Answer is D.

3. $p_C = p_A + g\rho_{Hg}h - g\rho_{water}(0.75 \text{ m})$

$$h = \frac{p_C - p_A + \rho_{water}g(0.75 \text{ m})}{\text{SG}_{Hg}\rho_{water}g}$$

$$= \frac{\begin{array}{c}1 \times 10^5 \text{ Pa} - 0 \text{ Pa} \\ + \left(1000 \dfrac{\text{kg}}{\text{m}^3}\right)\left(9.81 \dfrac{\text{m}}{\text{s}^2}\right)(0.75 \text{ m})\end{array}}{(13.5)\left(1000 \dfrac{\text{kg}}{\text{m}^3}\right)\left(9.81 \dfrac{\text{m}}{\text{s}^2}\right)}$$

$$= 0.81 \text{ m} \quad (0.8 \text{ m})$$

Answer is C.

4. The resultant force is calculated from the average pressure on the plate, which is the pressure at the plate's centroid.

$$h_c = 3 \text{ m} - 0.125 \text{ m}$$
$$= 2.875 \text{ m}$$

$$p_c = \bar{p} = \rho g h_c$$
$$= \left(1000 \frac{\text{kg}}{\text{m}^3}\right)\left(9.81 \frac{\text{m}}{\text{s}^2}\right)(2.875 \text{ m})$$
$$= 28\,204 \text{ Pa}$$

$$R = \bar{p}A$$
$$= (28\,204 \text{ Pa})\pi(0.125 \text{ m})^2$$
$$= 1384 \text{ N} \quad (1.38 \text{ kN})$$

Although the resultant force is calculated from the depth of the centroid, the resultant force does not act at the centroid.

Answer is A.

5.

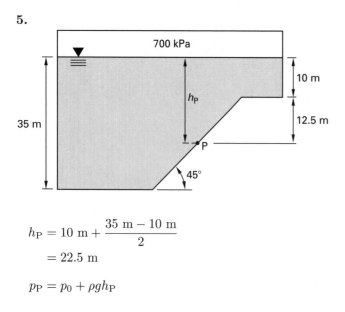

$$h_P = 10 \text{ m} + \frac{35 \text{ m} - 10 \text{ m}}{2}$$
$$= 22.5 \text{ m}$$

$$p_P = p_0 + \rho g h_P$$

$$p_P = 7 \times 10^5 \text{ Pa} + \left(1000 \frac{\text{kg}}{\text{m}^3}\right)\left(9.81 \frac{\text{m}}{\text{s}^2}\right)(22.5 \text{ m})$$
$$= 9.2 \times 10^5 \text{ Pa} \quad (922 \text{ kPa})$$

Answer is A.

6. The gate and its geometry are shown.

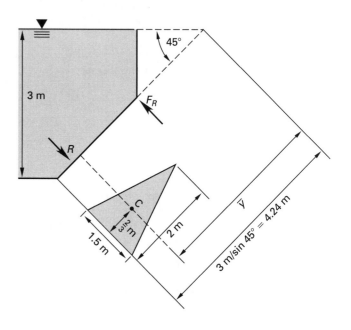

The resultant force is calculated from the average pressure using the depth to the centroid (measured parallel to the plane surface).

$$h_c = \bar{y} = \frac{3 \text{ m}}{\sin 45°} - \frac{2}{3} \text{ m}$$
$$= 3.58 \text{ m}$$

$$p_c = p_0 + \rho g h_c \sin \alpha$$
$$= 0 \text{ Pa} + \left(1000 \frac{\text{kg}}{\text{m}^3}\right)\left(9.81 \frac{\text{m}}{\text{s}^2}\right)(3.58 \text{ m})\sin 45°$$
$$= 24\,833 \text{ Pa}$$

$$R = p_c A$$
$$= (24\,833 \text{ Pa})\left(\frac{1}{2}\right)(1.5 \text{ m})(2 \text{ m})$$
$$= 37\,250 \text{ N}$$

The resultant force acts at the center of pressure. Since the water surface is at atmospheric pressure,

$$I_{yy} = \frac{bh^3}{36} = \frac{(1.5 \text{ m})(2 \text{ m})^3}{36}$$
$$= 0.333 \text{ m}^4 \quad \begin{bmatrix}\text{centroidal axis} \\ \text{parallel to base}\end{bmatrix}$$

$$z^* = \frac{I_{yy}}{Az_c}$$

$$= \frac{0.333 \text{ m}^4}{\left(\dfrac{1}{2}\right)(1.5 \text{ m})(2 \text{ m})(3.58 \text{ m})}$$

$$= 0.062 \text{ m}$$

Take the sum of moments about the base of the gate to find the force, F_R, needed to keep the gate closed.

$$\sum M_{\text{base}} = (37\,250 \text{ N})\left(\frac{2}{3} \text{ m} - 0.062 \text{ m}\right) - F_R(2 \text{ m})$$

$$F_R = \frac{(37\,250 \text{ N})(0.605 \text{ m})}{2 \text{ m}}$$

$$= 11\,268 \text{ N} \quad (11.3 \text{ kN})$$

Answer is D.

7. Statements II and III are false. Statement II describes stability, not buoyancy. The buoyant force is determined by multiplying the specific weight of the fluid (not of the object) by the displaced volume of the fluid.

Answer is D.

8. $p = p_a + \rho g h$

$$= 1.013 \times 10^5 \text{ Pa} + \left(1000 \ \frac{\text{kg}}{\text{m}^3}\right)\left(9.81 \ \frac{\text{m}}{\text{s}^2}\right)$$

$$\times (300 \text{ mm} - 75 \text{ mm})\left(\frac{1}{1000 \ \dfrac{\text{mm}}{\text{m}}}\right)$$

$$= 1.035 \times 10^5 \text{ Pa} \quad (103.5 \text{ kPa})$$

Answer is C.

9. $p = p_a + \rho g h$

$$105 \times 10^3 \text{ Pa} = 1.013 \times 10^5 \text{ Pa}$$

$$+ \left(1000 \ \frac{\text{kg}}{\text{m}^3}\right)\left(9.81 \ \frac{\text{m}}{\text{s}^2}\right) h$$

$$h = 0.377 \text{ m} \quad (380 \text{ mm})$$

Answer is D.

10. Since the container is moved slowly, the compression of the gas is assumed to be an isothermal process. From Prob. 8, the pressure of the air in the container was 103.5 kPa when the height of the void space was 225 mm. When the pressure increases, the void volume decreases.

$$p_1 V_1 = p_2 V_2$$

$$p_1 A h_1 = p_2 A h_2$$

$$h_2 = \frac{p_1 h_1}{p_2} \quad [A \text{ is constant}]$$

$$= \frac{(103.5 \text{ kPa})(225 \text{ mm})}{105 \text{ kPa}}$$

$$= 221.8 \text{ mm} \quad (222 \text{ mm})$$

Answer is B.

24 Fluid Dynamics

Subjects

CONSERVATION LAWS	24-1
Conservation of Mass	24-1
Fluid Energy	24-2
Hydraulic Grade Line	24-2
Energy Line	24-2
Conservation of Energy	24-2
FLOW OF A REAL FLUID	24-3
Reynolds Number	24-3
FLOW DISTRIBUTION	24-4
STEADY INCOMPRESSIBLE FLOW IN PIPES AND CONDUITS	24-4
Friction Loss: Darcy Equation	24-4
Friction Loss: Hagen-Poiseuille Equation	24-6
Flow in Noncircular Conduits	24-6
Minor Losses in Pipe Fittings, Contractions, and Expansions	24-6
IMPULSE-MOMENTUM PRINCIPLE	24-6
Pipe Bends, Enlargements, and Contractions	24-7
Jet Propulsion	24-8
Deflectors and Blades	24-8
MULTIPATH PIPELINES	24-9
DRAG	24-10
OPEN CHANNEL AND PARTIAL-AREA PIPE FLOW	24-11

Nomenclature

A	area	ft^2	m^2
C	coefficient	–	–
C	Hazen-Williams roughness coefficient	–	–
D	diameter	ft	m
E	specific energy	ft-lbf/lbm	J/kg
f	Darcy friction factor	–	–
F	force	lbf	N
g	gravitational acceleration	ft/sec^2	m/s^2
g_c	gravitational constant (32.2)	lbm-ft/lbf-sec^2	–
h	height or head	ft	m
I	impulse	lbf-sec	N·s
L	length	ft	m
$\dot{m}$	mass flow rate	lbm/sec	kg/s
n	Manning's roughness coefficient	–	–

p	pressure	lbf/ft^2	N/m^2
P	momentum	lbm-ft/sec	kg·m/s
P	power	ft-lbf/sec	W
Q	flow rate	ft^3/sec	m^3/s
R	radius	ft	m
Re	Reynolds number	–	–
S	hydraulic grade	ft/ft	m/m
t	time	sec	s
v	velocity	ft/sec	m/s
W	weight	lbf	N
z	elevation	ft	m

Symbols

α	angle	deg	deg
γ	specific weight	lbf/ft^3	N/m^3
ϵ	specific roughness	ft	m
μ	absolute viscosity	lbf-sec/ft^2	Pa·s
ν	kinematic viscosity	ft^2/sec	m^2/s
ρ	density	lbm/ft^3	kg/m^3

Subscripts

D	drag
f	friction
H	hydraulic
L	minor losses
p	pressure
v	velocity
z	elevation

CONSERVATION LAWS

Conservation of Mass

Fluid mass is always conserved in fluid systems, regardless of the pipeline complexity, orientation of the flow, or type of fluid flowing. This single concept is often sufficient to solve simple fluid problems.

$$\dot{m}_1 = \dot{m}_2 \qquad 24.1$$

When applied to fluid flow, the conservation of mass law is known as the *continuity equation*. The continuity equation states that the flow passing any two points in a system is the same, as illustrated in Fig. 24.1.

$$\dot{m} = \rho A \mathrm{v} = \rho Q \qquad 24.2$$

$$\rho_1 A_1 \mathrm{v}_1 = \rho_2 A_2 \mathrm{v}_2 \qquad 24.3$$

If the fluid is incompressible, then $\rho_1 = \rho_2$.

$$Q = A_1 v_1 = A_2 v_2 \qquad \text{24.4}$$

Figure 24.1 Generalized Flow Conservation

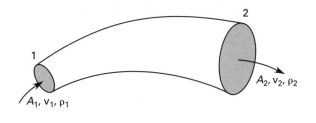

Fluid Energy

Work is performed and energy is expended when a substance is compressed. Thus, a mass of fluid at high pressure will have more energy than an identical mass of fluid at a lower pressure. The energy is the *pressure energy* of the fluid, E_p. Equation 24.5 gives the pressure energy per unit mass of fluid (or *specific pressure energy*) at pressure p.

$$E_p = \frac{p}{\rho} \qquad \text{24.5}$$

The quantity known as *pressure head* can be calculated from the pressure energy.

$$h_p = \frac{E_p}{g} = \frac{p}{\rho g} \qquad \text{[SI]} \qquad \text{24.6a}$$

$$h_p = \frac{E_p g_c}{g} = \frac{p g_c}{\rho g} = \frac{p}{\gamma} \qquad \text{[U.S.]} \qquad \text{24.6b}$$

Energy is required to accelerate a stationary body. Thus, a moving mass of fluid possesses more energy than an identical, stationary mass. The energy is the *kinetic energy* of the fluid. If the kinetic energy is evaluated per unit mass, the term *specific kinetic energy* is used. Equation 24.7 gives the specific kinetic energy for a turbulent flow with uniform velocity, v. For laminar flow with an average velocity, v, the kinetic energy is two times that calculated from Eq. 24.7.

$$E_v = \frac{v^2}{2} \qquad \text{[SI]} \qquad \text{24.7a}$$

$$E_v = \frac{v^2}{2g_c} \qquad \text{[U.S.]} \qquad \text{24.7b}$$

The specific kinetic energy is used to calculate the quantity known as the *velocity head*.

$$h_v = \frac{E_v}{g} = \frac{v^2}{2g} \qquad \text{[SI]} \qquad \text{24.8a}$$

$$h_v = E_v \left(\frac{g_c}{g}\right) = \frac{v^2}{2g} \qquad \text{[U.S.]} \qquad \text{24.8b}$$

Work is performed in elevating a body. Thus, a mass of fluid at high elevation will have more energy than an identical mass of fluid at a lower elevation. The energy is the *potential energy* of the fluid. Equation 24.9 gives the potential energy per unit mass of fluid (or *specific potential energy*) at an elevation, z.

$$E_z = zg \qquad \text{[SI]} \qquad \text{24.9a}$$

$$E_z = \frac{zg}{g_c} \qquad \text{[U.S.]} \qquad \text{24.9b}$$

The quantity known as the *gravity, gravitational, potential,* or *elevation head* can be calculated from the potential energy.

$$h_z = \frac{E_z}{g} = z \qquad \text{[SI]} \qquad \text{24.10a}$$

$$h_z = E_z \left(\frac{g_c}{g}\right) = z \qquad \text{[U.S.]} \qquad \text{24.10b}$$

Hydraulic Grade Line

The *hydraulic grade line*, HGL, is the graph of the pressure head, plotted as a position along the pipeline. The hydraulic grade line represents the height of the water column at any point along the pipe, if a piezometer tap were installed. Since the pressure head can increase or decrease depending on changes in velocity head, the HGL can also change in elevation if the flow area changes.

Figure 24.2 Hydraulic Grade Line in a Horizontal Pipe

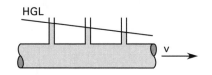

Energy Line

The *energy line*, EL, or *energy grade line*, EGL, is a graph of the total energy along a length of pipe. In a frictionless pipe without pumps or turbines, the total specific energy is constant, and the EL will be horizontal. (This is a restatement of the Bernoulli equation.) The EL is equal to the HGL plus the velocity head $(v^2/2g)$.

Conservation of Energy

The *Bernoulli equation*, also known as the *field equation*, is an energy conservation equation that is valid for incompressible, frictionless flow. The Bernoulli equation states that the total energy of a fluid flowing without friction losses in a pipe is constant. The total energy possessed by the fluid is the sum of its pressure, kinetic, and potential energies. In other words, the Bernoulli equation states that the total head at any two points is the same.

$$\frac{p_1}{\rho_1 g} + \frac{v_1^2}{2g} + z_1 = \frac{p_2}{\rho_2 g} + \frac{v_2^2}{2g} + z_2 \qquad \text{[SI]} \qquad 24.11a$$

$$\frac{p_1}{\gamma_1} + \frac{v_1^2}{2g} + z_1 = \frac{p_2}{\gamma_2} + \frac{v_2^2}{2g} + z_2 \qquad \text{[U.S.]} \qquad 24.11b$$

FLOW OF A REAL FLUID

Consider the steady flow of fluid through the pipe in Fig. 24.3. If the fluid is incompressible, $\rho_1 = \rho_2 = \rho$.

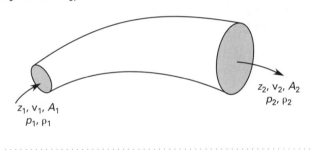

Figure 24.3 Energy Elements in Generalized Flow

z_1, v_1, A_1
p_1, ρ_1

z_2, v_2, A_2
p_2, ρ_2

The original Bernoulli equation assumes frictionless flow and does not consider the effects of pumps and turbines. In actual practice, friction occurs during fluid flow. This friction acts as an energy sink, so that the fluid at the end of a pipe section has less energy than it does at the beginning.

The *head loss due to friction* is denoted by the symbol h_f. This loss is added into the original Bernoulli equation to restore the equality. The *extended Bernoulli equation* accounting for friction is

$$\frac{p_1}{\rho g} + \frac{v_1^2}{2g} + z_1 = \frac{p_2}{\rho g} + \frac{v_2^2}{2g} + z_2 + h_f \qquad \text{[SI]} \qquad 24.12a$$

$$\frac{p_1}{\gamma} + \frac{v_1^2}{2g} + z_1 = \frac{p_2}{\gamma} + \frac{v_2^2}{2g} + z_2 + h_f \qquad \text{[U.S.]} \qquad 24.12b$$

The pipe in Fig. 24.4 is constant-diameter and horizontal. An incompressible fluid flows through it at a steady rate. Since the elevation of the pipe does not change, the potential energy is constant. Since the pipe has a constant area, the kinetic energy (velocity) is constant. Therefore, the friction energy loss must show up as a decrease in pressure energy. Since the fluid is incompressible, this can only occur if the pressure decreases in the direction of flow.

$$h_f = \frac{p_1 - p_2}{\rho g} \qquad \text{[SI]} \qquad 24.13a$$

$$h_f = \frac{p_1 - p_2}{\gamma} \qquad \text{[U.S.]} \qquad 24.13b$$

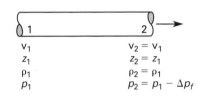

Figure 24.4 Pressure Drop in a Pipe

v_1 $v_2 = v_1$
z_1 $z_2 = z_1$
ρ_1 $\rho_2 = \rho_1$
p_1 $p_2 = p_1 - \Delta p_f$

Reynolds Number

The *Reynolds number*, Re, is a dimensionless number interpreted as the ratio of inertial forces to viscous forces in the fluid.

The inertial forces are proportional to the flow diameter, velocity, and fluid density. (Increasing these variables will increase the momentum of the fluid in flow.) The viscous force is represented by the fluid's absolute viscosity, μ.

$$\text{Re} = \frac{vD\rho}{\mu} \qquad \text{[SI]} \qquad 24.14a$$

$$\text{Re} = \frac{vD\rho}{g_c \mu} \qquad \text{[U.S.]} \qquad 24.14b$$

Since μ/ρ is the *kinematic viscosity*, ν, Eq. 24.14 can be simplified.

$$\text{Re} = \frac{vD}{\nu} \qquad 24.15$$

If all of the fluid particles move in paths parallel to the overall flow direction (i.e., in layers), the flow is said to be *laminar*. This occurs when the Reynolds number is less than approximately 2100. *Laminar flow* is typical when the flow channel is small, the velocity is low, and the fluid is viscous. Viscous forces are dominant in laminar flow.

Turbulent flow is characterized by a three-dimensional movement of the fluid particles superimposed on the overall direction of motion. A fluid is said to be in turbulent flow if the Reynolds number is greater than approximately 4000. (This is the most common situation.)

The flow is said to be in the *critical zone* or *transition region* when the Reynolds number is between 2100 and 4000. These numbers are known as the lower and upper *critical Reynolds numbers*, respectively.

Equations 24.14 and 24.15 are specifically for Newtonian fluids. For non-Newtonian fluids, *power law* parameters must be used when calculating the Reynolds number, Re′. In Eq. 24.16, the constant K is known as the *consistency index*. For *pseudoplastic non-Newtonian fluids*, $n < 1$; for *dilatant non-Newtonian fluids*, $n > 1$. For Newtonian fluids, $n = 1$, and Eq. 24.16 reduces to Eq. 24.14.

$$\text{Re}' = \frac{\text{v}^{2-n} D^n \rho}{K \left(\dfrac{3n+1}{4n}\right)^n 8^{n-1}} \qquad 24.16$$

FLOW DISTRIBUTION

With laminar flow in a circular pipe or between two parallel plates, viscosity makes some fluid particles adhere to the wall. The closer to the wall, the greater the tendency will be for the fluid to adhere. In general, the fluid velocity will be zero at the wall and will follow a parabolic distribution away from the wall. The *average flow velocity* (also known as the *bulk velocity*) is found from the flow rate and cross-sectional area.

$$\text{v} = \frac{Q}{A} \quad \text{[average]} \qquad 24.17$$

Because of the parabolic distribution, velocity will be maximum at the centerline, midway between the two walls (i.e., at the center of a pipe). The maximum velocity for laminar flow is

$$\text{v}_{\text{max}} = 2\text{v} \quad \text{[flow in circular pipe]} \qquad 24.18$$
$$\text{v}_{\text{max}} = 1.5\text{v} \quad \text{[flow between plates]} \qquad 24.19$$

For flow through a pipe with diameter $2R$ or between parallel plates with separation distance $2R$, the velocity at any point a distance r from the centerline is

$$\text{v}_r = \text{v}_{\text{max}} \left(1 - \left(\frac{r}{R}\right)^2\right) \qquad 24.20$$

The shear stress also varies with location. The shear stress at any point a distance r from the centerline can be found from the shear stress at the wall.

$$\frac{\tau}{\tau_{\text{wall}}} = \frac{r}{R} \qquad 24.21$$

With turbulent flow, a distinction between velocities of particles near the pipe wall or centerline is usually not

made. All of the fluid particles are assumed to flow at the bulk velocity. In reality, no flow is completely turbulent, and there is a slight difference between the centerline velocity and the average velocity. For fully turbulent flow (Re > 10 000), a good approximation of the average velocity is approximately 85% of the maximum velocity.

Figure 24.5 Laminar and Turbulent Velocity Distributions

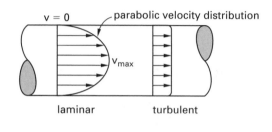

STEADY INCOMPRESSIBLE FLOW IN PIPES AND CONDUITS

The extended *field* (or *energy*) *equation* for steady incompressible flow is identical to Eq. 24.12.

$$\frac{p_1}{\rho g} + \frac{\text{v}_1^2}{2g} + z_1 = \frac{p_2}{\rho g} + \frac{\text{v}_2^2}{2g} + z_2 + h_f \quad \text{[SI]} \qquad 24.22a$$

$$\frac{p_1}{\gamma} + \frac{\text{v}_1^2}{2g} + z_1 = \frac{p_2}{\gamma} + \frac{\text{v}_2^2}{2g} + z_2 + h_f \quad \text{[U.S.]} \qquad 24.22b$$

For a pipe of constant cross-sectional area and constant elevation, the pressure change from one point to another is given by

$$p_1 - p_2 = \rho g h_f \quad \text{[SI]} \qquad 24.23a$$
$$p_1 - p_2 = \gamma h_f \quad \text{[U.S.]} \qquad 24.23b$$

Friction Loss: Darcy Equation

The *Darcy equation* is one method for calculating the frictional energy loss for fluids. It can be used for both laminar and turbulent flow.

$$h_f = \frac{fL\text{v}^2}{2Dg} \qquad 24.24$$

The *Darcy friction factor*, f, is one of the parameters that is used to calculate the friction loss. One of the advantages to using the Darcy equation is that the assumption of laminar or turbulent flow does not need to be confirmed if f is known. The friction factor is not constant, but decreases as the Reynolds number (fluid velocity) increases, up to a certain point, known as *fully turbulent flow*. Once the flow is fully turbulent, the friction factor remains constant and depends only on the relative roughness of the pipe surface and not on the Reynolds number. For very smooth pipes, fully turbulent flow is achieved only at very high Reynolds numbers.

The friction factor is not dependent on the material of the pipe, but is affected by its roughness. For example, for a given Reynolds number, the friction factor will be the same for any smooth pipe material (glass, plastic, smooth brass, copper, etc.).

The friction factor is determined from the *relative roughness*, ϵ/D, and the Reynolds number, Re. The relative roughness is calculated from the *specific roughness* of the material, ϵ, given in tables, and the diameter of the pipe. The *Moody friction factor chart* (also known as the *Stanton diagram*), Fig. 24.6, presents the friction factor graphically. There are different lines

for selected discrete values of relative roughness. Because of the complexity of this graph, it is easy to incorrectly locate the Reynolds number or use the wrong curve. Nevertheless, the Moody chart remains the most common method of obtaining the friction factor.

Table 24.1 Specific Roughness of Typical Materials

material	ϵ	
	ft	mm
riveted steel	0.003–0.03	0.9–9.0
concrete	0.001–0.01	0.3–3.0
galvanized iron	0.00085	0.15
commercial steel		
or wrought iron	0.00015	0.046
drawn tubing	0.000005	0.0015

Figure 24.6 Moody Friction Factor Chart

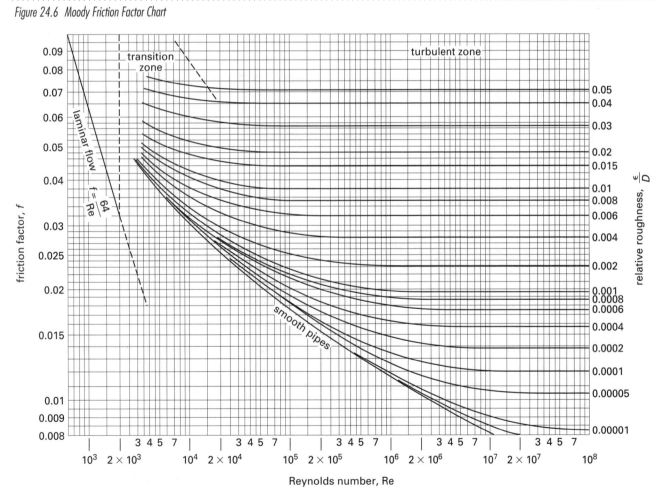

Reproduced from *Principles of Engineering Heat Transfer*, Giedt, published by D. Van Nostrand Company, Inc., 1957, with permission from Wadsworth Publishing Company, Inc., Belmont, CA.

Professional Publications, Inc.

Friction Loss: Hagen-Poiseuille Equation

If the flow is laminar and the fluid is flowing in a circular pipe, then the *Hagen-Poiseuille equation* can be used to calculate the flow rate. In Eq. 24.25, the Hagen-Poiseuille equation is presented in the form of a pressure drop, Δp_f.

$$Q = \frac{\pi R^4 \Delta p_f}{8\mu L} = \frac{\pi D^4 \Delta p_f}{128\mu L} \qquad 24.25$$

Flow in Noncircular Conduits

The *hydraulic radius* is defined as the area in flow divided by the *wetted perimeter*. The area in flow is the cross-sectional area of the fluid flowing. When a fluid is flowing under pressure in a pipe (i.e., *pressure flow*), the area in flow will be the internal area of the pipe. However, the fluid may not completely fill the pipe and may flow simply because of a sloped surface (i.e., *gravity flow* or *open channel flow*).

The wetted perimeter is the length of the line representing the interface between the fluid and the pipe or channel. It does not include the *free surface* length (i.e., the interface between fluid and atmosphere).

$$R_H = \frac{\text{area in flow}}{\text{wetted perimeter}} \qquad 24.26$$

For a circular pipe flowing completely full, the area in flow is πR^2. The wetted perimeter is the entire circumference, $2\pi R$. The hydraulic radius is

$$R_H = \frac{\pi R^2}{2\pi R} = \frac{R}{2} = \frac{D}{4} \qquad 24.27$$

The hydraulic radius of a pipe flowing half full is also $R/2$, since the flow area and wetted perimeter are both halved.

Many fluid, thermodynamic, and heat transfer processes are dependent on the physical length of an object. The general name for this controlling variable is *characteristic dimension*. The characteristic dimension in evaluating fluid flow is the *equivalent diameter* (also known as the *hydraulic diameter*). The equivalent diameter for a full-flowing circular pipe is simply its inside diameter. If the hydraulic radius of a noncircular duct is known, it can be used to calculate the equivalent diameter.

$$R_H = \frac{D_H}{4} \qquad 24.28$$

$$D_H = 4R_H$$

$$= 4 \times \frac{\text{area in flow}}{\text{wetted perimeter}} \qquad 24.29$$

The frictional energy loss by a fluid flowing in a rectangular, annular, or other noncircular duct can be calculated from the Darcy equation by using the equivalent diameter (*hydraulic diameter*), D_H, in place of the diameter, D. The friction factor, f, is determined in any of the conventional manners.

Minor Losses in Pipe Fittings, Contractions, and Expansions

In addition to the frictional energy lost due to viscous effects, friction losses also result from fittings in the line, changes in direction, and changes in flow area. These losses are known as *minor losses*, since they are usually much smaller in magnitude than the pipe wall frictional loss.

The energy conservation equation accounting for minor losses is

$$\frac{p_1}{\rho g} + \frac{v_1^2}{2g} + z_1 = \frac{p_2}{\rho g} + \frac{v_2^2}{2g} + z_2 + h_f + h_{L,\text{fitting}}$$
$$\text{[SI]} \qquad 24.30a$$

$$\frac{p_1}{\gamma} + \frac{v_1^2}{2g} + z_1 = \frac{p_2}{\gamma} + \frac{v_2^2}{2g} + z_2 + h_f + h_{L,\text{fitting}}$$
$$\text{[U.S.]} \qquad 24.30b$$

The minor losses can be calculated using the *method of loss coefficients*. Each fitting has a *loss coefficient*, C, associated with it, which, when multiplied by the kinetic energy, gives the head loss. Thus, a loss coefficient is the minor head loss expressed in fractions (or multiples) of the velocity head.

$$h_{L,\text{fitting}} = C\left(\frac{v^2}{2g}\right) \qquad 24.31$$

Loss coefficients for specific fittings and valves must be known in order to be used. They cannot be derived theoretically.

Losses at pipe exits and entrances in tanks also fall under the category of minor losses. The following values of C account for minor losses in various exit and entrance conditions.

exit/entrance condition	C value
exit, sharp	1.0
exit, protruding	0.8
entrance, sharp	0.5
entrance, rounded	0.1
entrance, gradual, smooth	0.04

IMPULSE-MOMENTUM PRINCIPLE

The *impulse-momentum principle* states that the impulse applied to a body is equal to the change in momentum.

$$\mathbf{I} = \Delta \mathbf{P} \qquad 24.32$$

The *impulse*, **I**, of a constant force is calculated as the product of the force's magnitude and the length of time the force is applied.

$$\mathbf{I} = \mathbf{F}\Delta t \qquad 24.33$$

The *momentum*, **P**, of a moving object is a vector quantity defined as the product of the object's mass and velocity.

$$\mathbf{P} = m\mathbf{v} \qquad \text{[SI]} \qquad 24.34a$$

$$\mathbf{P} = \frac{m\mathbf{v}}{g_c} \qquad \text{[U.S.]} \qquad 24.34b$$

$$\begin{aligned} F\Delta t &= m\Delta \mathrm{v} \\ &= m(\mathrm{v}_2 - \mathrm{v}_1) \qquad \text{[SI]} \qquad 24.35a \end{aligned}$$

$$\begin{aligned} F\Delta t &= m\Delta \mathrm{v} \\ &= \frac{m(\mathrm{v}_2 - \mathrm{v}_1)}{g_c} \qquad \text{[U.S.]} \qquad 24.35b \end{aligned}$$

For fluid flow, there is a mass flow rate, $\dot{m}$, but no mass per se. Since $\dot{m} = m/\Delta t$, the impulse-momentum equation can be rewritten as

$$F = \dot{m}\Delta \mathrm{v} \qquad \text{[SI]} \qquad 24.36a$$

$$F = \frac{\dot{m}\Delta \mathrm{v}}{g_c} \qquad \text{[U.S.]} \qquad 24.36b$$

Substituting for the mass flow rate, $\dot{m} = \rho A\mathrm{v}$. The quantity $Q\rho \mathrm{v}$ is the *rate of momentum*.

$$\begin{aligned} F &= \rho A\mathrm{v}\Delta \mathrm{v} \\ &= Q\rho\Delta \mathrm{v} \qquad \text{[SI]} \qquad 24.37a \end{aligned}$$

$$\begin{aligned} F &= \rho A\mathrm{v}\Delta \mathrm{v} \\ &= \frac{Q\rho\Delta \mathrm{v}}{g_c} \qquad \text{[U.S.]} \qquad 24.37b \end{aligned}$$

The impulse-momentum principle applied to a control volume is

$$\sum \mathbf{F} = Q_2\rho_2\mathbf{v}_2 - Q_1\rho_1\mathbf{v}_1 \qquad \text{[SI]} \qquad 24.38a$$

$$\sum \mathbf{F} = \frac{Q_2\rho_2\mathbf{v}_2 - Q_1\rho_1\mathbf{v}_1}{g_c} \qquad \text{[U.S.]} \qquad 24.38b$$

Pipe Bends, Enlargements, and Contractions

The impulse-momentum principle illustrates that fluid momentum is not always conserved when the fluid is acted upon by an external force. Examples of external forces are gravity (considered zero for horizontal pipes), gage pressure, friction, and turning forces from walls and vanes. Only if these external forces are absent is fluid momentum conserved.

When a fluid enters a pipe fitting or bend, as illustrated in Fig. 24.7, momentum is changed. Since the fluid is confined, the forces due to static pressure must be included in the analysis. The effects of gravity and friction are neglected.

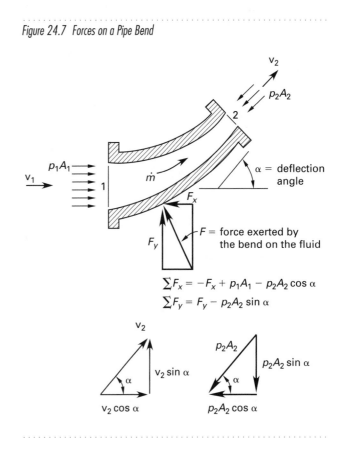

Figure 24.7 Forces on a Pipe Bend

$$\sum F_x = -F_x + p_1 A_1 - p_2 A_2 \cos\alpha$$
$$\sum F_y = F_y - p_2 A_2 \sin\alpha$$

Applying Eq. 24.38 to the fluid in the pipe bend in Fig. 24.7, the following equations for the force of the bend on the fluid are obtained. m_{fluid} and W_{fluid} are the mass and weight, respectively, of the fluid in the bend (often neglected).

$$\begin{aligned} -F_x &= p_2 A_2 \cos\alpha - p_1 A_1 \\ &+ Q\rho(\mathrm{v}_2 \cos\alpha - \mathrm{v}_1) \qquad \text{[SI]} \qquad 24.39a \end{aligned}$$

$$\begin{aligned} -F_x &= p_2 A_2 \cos\alpha - p_1 A_1 \\ &+ \frac{Q\rho(\mathrm{v}_2 \cos\alpha - \mathrm{v}_1)}{g_c} \qquad \text{[U.S.]} \qquad 24.39b \end{aligned}$$

$$F_y = (p_2 A_2 + Q\rho \mathrm{v}_2)\sin\alpha + m_{\text{fluid}}g \qquad \text{[SI]} \qquad 24.40a$$

$$F_y = \left(p_2 A_2 + \frac{Q\rho \mathrm{v}_2}{g_c}\right)\sin\alpha + W_{\text{fluid}} \qquad \text{[U.S.]} \qquad 24.40b$$

Jet Propulsion

A basic application of the impulse-momentum principle is *jet propulsion*. The velocity of a fluid jet issuing from an orifice in a tank can be determined by comparing the total energies at the free fluid surface and at the jet itself. At the fluid surface, $p_1 = 0$ (atmospheric) and $v_1 = 0$. The only energy the fluid has is potential energy. At the jet, $p_2 = 0$ and $z_2 = 0$. All of the potential energy difference has been converted to kinetic energy. The change in momentum of the fluid produces a force.

For Bernoulli's equation (Eq. 24.11), it is easy to calculate the initial jet velocity (known as *Torricelli's speed of efflux*).

$$v = \sqrt{2gh} \qquad 24.41$$

The governing equation for jet propulsion is Eq. 24.42.

$$
\begin{aligned}
F &= \dot{m}(v_2 - v_1) \\
&= \dot{m}(v_2 - 0) \\
&= Q\rho v_2 \\
&= v_2 A_2 \rho v_2 \\
&= A_2 \rho v_2^2 \\
&= A_2 \rho \left(\sqrt{2gh}\right)^2 \\
&= 2g\rho h A_2 \\
&= 2\gamma h A_2 \qquad 24.42
\end{aligned}
$$

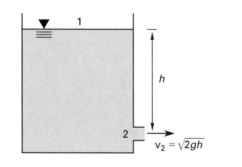

Figure 24.8 Fluid Jet Issuing from a Tank Orifice

Deflectors and Blades

Fixed Blade

Figure 24.9 illustrates a fluid jet being turned through an angle, α, by a *fixed blade* (also called a *fixed* or *stationary vane*). It is common to assume that $|v_2| = |v_1|$, although this will not be strictly true if friction between the blade and fluid is considered. Since the fluid is both retarded (in the x-direction) and accelerated (in the y-direction), there will be two components of blade force on the fluid.

$$-F_x = Q\rho(v_2 \cos\alpha - v_1) \qquad \text{[SI]} \qquad 24.43a$$

$$-F_x = \frac{Q\rho(v_2 \cos\alpha - v_1)}{g_c} \qquad \text{[U.S.]} \qquad 24.43b$$

$$F_y = Q\rho v_2 \sin\alpha \qquad \text{[SI]} \qquad 24.44a$$

$$F_y = \frac{Q\rho v_2 \sin\alpha}{g_c} \qquad \text{[U.S.]} \qquad 24.44b$$

Figure 24.9 Open Jet on a Stationary Blade

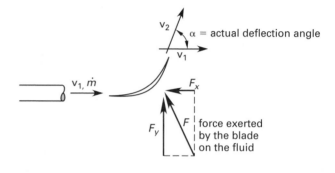

Moving Blade

If a blade is moving away at velocity v from the source of the fluid jet, only the *relative velocity difference* between the jet and blade produces a momentum change. Furthermore, not all of the fluid jet overtakes the moving blade.

$$-F_x = -Q\rho(v_1 - v)(1 - \cos\alpha) \qquad \text{[SI]} \qquad 24.45a$$

$$-F_x = \frac{-Q\rho(v_1 - v)(1 - \cos\alpha)}{g_c} \qquad \text{[U.S.]} \qquad 24.45b$$

$$F_y = Q\rho(v_1 - v)\sin\alpha \qquad \text{[SI]} \qquad 24.46a$$

$$F_y = \frac{Q\rho(v_1 - v)\sin\alpha}{g_c} \qquad \text{[U.S.]} \qquad 24.46b$$

Figure 24.10 Open Jet on a Moving Blade

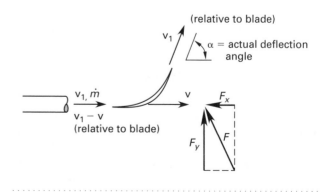

Impulse Turbine

An *impulse turbine* consists of a series of blades (buckets or vanes) mounted around a wheel. The power transferred from a fluid jet to the blades of a turbine is calculated from the x-component of force on the blades. The y-component of force does no work. v is the tangential blade velocity.

$$P = Q\rho(\mathrm{v}_1 - \mathrm{v})(1 - \cos\alpha)\mathrm{v} \quad \text{[SI]} \qquad 24.47a$$

$$P = \frac{Q\rho(\mathrm{v}_1 - \mathrm{v})(1 - \cos\alpha)\mathrm{v}}{g_c} \quad \text{[U.S.]} \qquad 24.47b$$

Figure 24.11 Impulse Turbine

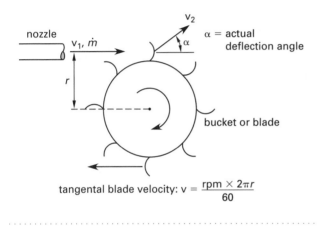

$$\text{tangental blade velocity: } \mathrm{v} = \frac{\text{rpm} \times 2\pi r}{60}$$

The maximum theoretical blade velocity is the velocity of the jet: $\mathrm{v} = \mathrm{v}_1$. This is known as the *runaway speed* and can only occur when the turbine is unloaded. If Eq. 24.47 is maximized with respect to v, however, the maximum power will be found to occur when the blade is traveling at half of the jet velocity: $\mathrm{v} = \mathrm{v}_1/2$. The power (force) is also affected by the deflection angle of the blade. Power is maximized when $\alpha = 180°$. Figure 24.12 illustrates the relationship between power and the variables α and v.

$$P_{\max} = Q\rho\left(\frac{\mathrm{v}_1^2}{4}\right)(1 - \cos\alpha) \quad \text{[SI]} \qquad 24.48a$$

$$P_{\max} = \frac{Q\rho\left(\dfrac{\mathrm{v}_1^2}{4}\right)(1 - \cos\alpha)}{g_c} \quad \text{[U.S.]} \qquad 24.48b$$

Substituting $\alpha = 180°$ and $\mathrm{v} = \mathrm{v}_1/2$ into Eq. 24.47,

$$P_{\max} = \frac{Q\rho\mathrm{v}_1^2}{2} \quad \text{[SI]} \qquad 24.49a$$

$$P_{\max} = \frac{Q\gamma\mathrm{v}_1^2}{2g} \quad \text{[U.S.]} \qquad 24.49b$$

Figure 24.12 Turbine Power

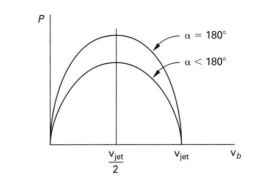

MULTIPATH PIPELINES

A *pipe loop* is a set of two pipes placed in parallel, both originating and terminating at the same junction. Adding a second pipe in parallel with a first is a standard method of increasing the capacity of a line.

Figure 24.13 Parallel Pipe Loop System

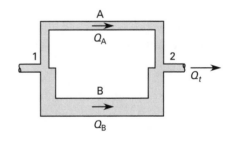

The following three principles govern the distribution of flow between the two branches.

- The flow divides in such a manner as to make the head loss in each branch the same.

$$h_{f,\mathrm{A}} = h_{f,\mathrm{B}} \qquad 24.50$$

$$\frac{f_\mathrm{A} L_\mathrm{A} \mathrm{v}_\mathrm{A}^2}{2 D_\mathrm{A} g} = \frac{f_\mathrm{B} L_\mathrm{B} \mathrm{v}_\mathrm{B}^2}{2 D_\mathrm{B} g} \qquad 24.51$$

- The head loss between the two junctions is the same as the head loss in each branch.

$$h_{f,1-2} = h_{f,\mathrm{A}} = h_{f,\mathrm{B}} \qquad 24.52$$

- The total flow rate is the sum of the flow rates in the two branches.

$$Q_t = Q_A + Q_B \qquad 24.53$$

$$\frac{\pi}{4} D_1^2 v_1 = \frac{\pi}{4} D_A^2 v_A + \frac{\pi}{4} D_B^2 v_B$$

$$= \frac{\pi}{4} D_2^2 v_2 \qquad 24.54$$

If the pipe diameters are known, Eqs. 24.50 through 24.54 can be solved simultaneously for the branch velocities. In such problems, it is common to neglect minor losses, the velocity head, and the variation in the friction factor, f, with velocity.

DRAG

Drag is a frictional force that acts parallel but opposite to the direction of motion. Drag is made up of several components (e.g., skin friction and pressure drag), but the total drag force can be calculated from a dimensionless *drag coefficient*, C_D. Dimensional analysis shows that the drag coefficient depends only on the Reynolds number.

$$F_D = \frac{C_D A \rho v^2}{2} \qquad \text{[SI]} \qquad 24.55a$$

$$F_D = \frac{C_D A \rho v^2}{2g_c} \qquad \text{[U.S.]} \qquad 24.55b$$

In most cases, the area, A, to be used is the projected area (i.e., the frontal area) normal to the stream. This is appropriate for spheres, disks, and vehicles. It is also appropriate for cylinders and ellipsoids that are oriented such that their longitudinal axes are perpendicular to the flow. In a few cases (e.g., airfoils and flat plates parallel to the flow), the area is the projection of the object onto a plane parallel to the stream.

Drag coefficients vary considerably with Reynolds numbers, often showing regions of distinctly different behavior. For that reason, the drag coefficient is often plotted. (Figure 24.14 illustrates the drag coefficient for spheres and circular flat disks oriented perpendicular to the flow.) Semiempirical equations can be used to calculate drag coefficients as long as the applicable ranges of Reynolds numbers are stated. For example, for flat plates placed parallel to the flow, Eqs. 24.56 and 24.57 can be used.

$$C_D = \frac{1.33}{\text{Re}^{0.5}} \qquad [10^4 < \text{Re} < 5 \times 10^5] \qquad 24.56$$

$$C_D = \frac{0.031}{\text{Re}^{1/7}} \qquad [10^6 < \text{Re} < 10^9] \qquad 24.57$$

Figure 24.14 Drag Coefficients for Spheres and Circular Flat Disks

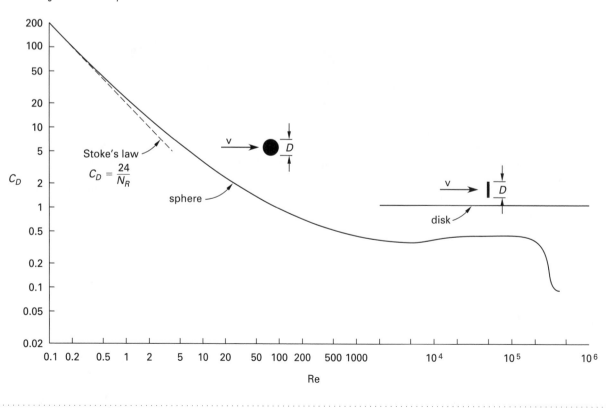

OPEN CHANNEL AND PARTIAL-AREA PIPE FLOW

Manning's equation has typically been used to estimate the velocity of flow in any open channel. It depends on hydraulic radius, R_H, the slope of the energy grade line, S, and a dimensionless *Manning's roughness coeffient*, n. The slope of the energy grade line is the terrain grade (slope) for uniform flow. The Manning roughness coefficient is typically taken as 0.013 for concrete.

$$v = \left(\frac{1}{n}\right) R_H^{2/3} S^{1/2} \qquad \text{[SI]} \qquad 24.58a$$

$$v = \left(\frac{1.486}{n}\right) R_H^{2/3} S^{1/2} \qquad \text{[U.S.]} \qquad 24.58b$$

Although Manning's equation could be used for circular pipes flowing less than full, the *Hazen-Williams equation* is used more often. The *Hazen-Williams roughness coefficient*, C, has a typical range of 100 to 130 as shown in Table 24.1, although very smooth materials can have higher values.

$$v = 0.849 C R_H^{0.63} S^{0.54} \qquad \text{[SI]} \qquad 24.59a$$

$$v = 1.318 C R_H^{0.63} S^{0.54} \qquad \text{[U.S.]} \qquad 24.59b$$

Table 24.1 Typical Values of Hazen-Williams Roughness Coefficient

material	Hazen-Williams roughness coefficient, C
asbestos-cement	140
brick (sewers)	100
cast iron, new	130
cast iron, 5 years old	120
cast iron, 10 years old	100
clay, vitrified	110
concrete (any age)	130
plastic	150
steel, riveted, new	110
steel, welded, new	120
wood stave (any age)	120

SAMPLE PROBLEMS

Problem 1

Consider water flowing through a converging channel as shown and discharging freely to the atmosphere at the exit. What is the gage pressure at the inlet? Assume the flow to be incompressible, and neglect any frictional effects.

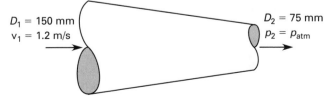

D_1 = 150 mm
v_1 = 1.2 m/s

D_2 = 75 mm
$p_2 = p_{atm}$

(A) 10.2 kPa
(B) 10.8 kPa
(C) 11.3 kPa
(D) 12.7 kPa

Solution

From the continuity equation,

$$A_1 v_1 = A_2 v_2$$

$$v_2 = \frac{A_1 v_1}{A_2} = \left(\frac{D_1}{D_2}\right)^2 v_1$$

$$= \left(\frac{150 \text{ mm}}{75 \text{ mm}}\right)^2 \left(1.2 \, \frac{\text{m}}{\text{s}}\right)$$

$$= 4.8 \text{ m/s}$$

From Bernoulli's equation,

$$\frac{p_2}{\rho g} + \frac{v_2^2}{2g} + z_2 = \frac{p_1}{\rho g} + \frac{v_1^2}{2g} + z_1$$

$$z_1 = z_2$$

$$p_2 = 0 \quad \text{[gage]}$$

$$p_1 = \left(\frac{\rho}{2}\right)(v_2^2 - v_1^2)$$

$$= \left(\frac{1000 \, \frac{\text{kg}}{\text{m}^3}}{2}\right)\left(\left(4.8 \, \frac{\text{m}}{\text{s}}\right)^2 - \left(1.2 \, \frac{\text{m}}{\text{s}}\right)^2\right)$$

$$= 10\,800 \text{ Pa} \quad (10.8 \text{ kPa})$$

Answer is B.

Problem 2

A steel pipe with an inside diameter of 25 mm is 20 m long and carries water at a rate of 4.5 m³/h. Assuming the specific roughness of the pipe is 0.00005 m, the water has an absolute viscosity of 1.00×10^{-3} Pa·s and a density of 1000 kg/m³, what is the friction factor?

(A) 0.023
(B) 0.024
(C) 0.026
(D) 0.028

Solution

$$v = \frac{Q}{A} = \frac{\left(4.5 \, \frac{\text{m}^3}{\text{h}}\right)\left(\frac{1}{3600 \, \frac{\text{s}}{\text{h}}}\right)}{\left(\frac{\pi}{4}\right)(0.025 \text{ m})^2}$$

$$= 2.55 \text{ m/s}$$

$$\text{Re} = \frac{\rho v D}{\mu}$$

$$= \frac{\left(1000 \ \frac{\text{kg}}{\text{m}^3}\right)\left(2.55 \ \frac{\text{m}}{\text{s}}\right)(0.025 \ \text{m})}{1 \times 10^{-3} \ \text{Pa·s}}$$

$$= 6.4 \times 10^4$$

$$\frac{\epsilon}{D} = \frac{0.00005 \ \text{m}}{0.025 \ \text{m}}$$

$$= 0.002$$

From the Moody chart,

$$f = 0.026$$

Answer is C.

Problem 3

A steel pipe has an inside diameter of 25 mm, is 20 m long, and carries 10°C water at a rate of 4.5 m³/h. At this rate, the friction factor of the pipe is 0.0259. If the static pressure at the inlet is 70 kPa, what is the static pressure of the water at the outlet?

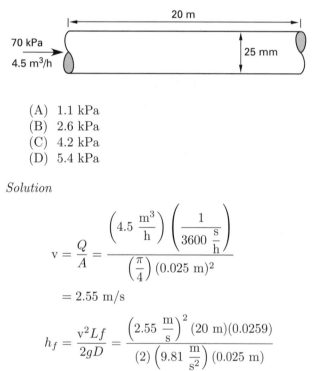

(A) 1.1 kPa
(B) 2.6 kPa
(C) 4.2 kPa
(D) 5.4 kPa

Solution

$$v = \frac{Q}{A} = \frac{\left(4.5 \ \frac{\text{m}^3}{\text{h}}\right)\left(\dfrac{1}{3600 \ \frac{\text{s}}{\text{h}}}\right)}{\left(\dfrac{\pi}{4}\right)(0.025 \ \text{m})^2}$$

$$= 2.55 \ \text{m/s}$$

$$h_f = \frac{v^2 L f}{2gD} = \frac{\left(2.55 \ \frac{\text{m}}{\text{s}}\right)^2 (20 \ \text{m})(0.0259)}{(2)\left(9.81 \ \frac{\text{m}}{\text{s}^2}\right)(0.025 \ \text{m})}$$

$$= 6.87 \ \text{m}$$

Use the energy equation.

$$\frac{p_1}{\rho g} + \frac{v_1^2}{2g} + z_1 = \frac{p_2}{\rho g} + \frac{v_2^2}{2g} + z_2 + h_f$$

$$z_1 = z_2$$

$$v_1 = v_2$$

Therefore, the field equation becomes

$$p_2 = p_1 - \rho g h_f$$

$$= 70 \times 10^3 \ \text{Pa} - \left(1000 \ \frac{\text{kg}}{\text{m}^3}\right)\left(9.81 \ \frac{\text{m}}{\text{s}^2}\right)(6.87 \ \text{m})$$

$$= 2605 \ \text{Pa} \quad (2.6 \ \text{kPa})$$

Answer is B.

Problem 4

A 90° reducing elbow is in the vertical plane, and water flows through it. What is the horizontal force required to hold the reducer elbow in a stationary position?

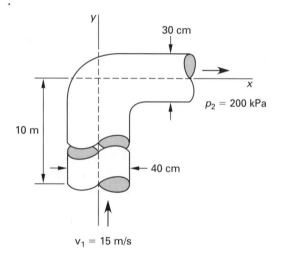

(A) 24.20 kN to the right
(B) 57.45 kN to the left
(C) 64.43 kN to the right
(D) 71.17 kN to the left

Solution
The free-body diagram of the fluid control volume in the reducer is

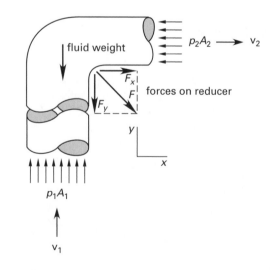

$$A_1 = \frac{\pi D_1^2}{4} = \frac{\pi(0.4 \text{ m})^2}{4}$$
$$= 0.1257 \text{ m}^2$$

$$A_2 = \frac{\pi D_2^2}{4} = \frac{\pi(0.3 \text{ m})^2}{4}$$
$$= 0.0707 \text{ m}^2$$

From the continuity equation,

$$Q = A_1 v_1 = A_2 v_2$$

$$v_2 = \frac{A_1 v_1}{A_2} = \frac{(0.1257 \text{ m}^2)\left(15 \dfrac{\text{m}}{\text{s}}\right)}{0.0707 \text{ m}^2}$$
$$= 26.67 \text{ m/s}$$

Comparing the problem statement with Fig. 24.7, the x- and y-axes are reversed. Therefore, F_y in Fig. 24.7 corresponds to F_x in this problem, and downward in Fig. 24.7 corresponds to the left in this problem.

From Eq. 24.40, recognizing that $\sin \alpha = \sin 90° = 1$,

$$F_{x,\text{this problem}} = p_2 A_2 + Q\rho v_2$$

$$= p_2 A_2 + A_2 \rho v_2^2$$

$$= (200 \text{ kPa})(0.0707 \text{ m}^2)$$
$$+ \frac{(0.0707 \text{ m}^2)\left(1000 \dfrac{\text{kg}}{\text{m}^3}\right)\left(26.67 \dfrac{\text{m}}{\text{s}}\right)^2}{1000 \dfrac{\text{N}}{\text{kN}}}$$

$$= 64.43 \text{ kN} \quad \text{[to the right]}$$

The force exerted by the fluid on the reducer is equal and opposite to this force. Therefore, the x-component of the resultant force on the reducer is $F_x = -64.43$ kN to the left. The horizontal force, F_x, required to hold the reducer in a stationary position is $F_x = 64.43$ kN to the right.

Answer is C.

Problem 5

The Darcy friction factor for both of the pipes shown is 0.024. The total flow rate is 300 m³/h. What is the flow rate through the 250 mm pipe?

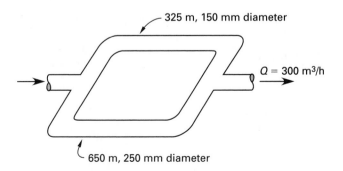

325 m, 150 mm diameter

$Q = 300$ m³/h

650 m, 250 mm diameter

(A) 0.04 m³/s
(B) 0.05 m³/s
(C) 0.06 m³/s
(D) 0.07 m³/s

Solution

Neglect minor losses through the pipe bends.

$$A_1 = \frac{\pi}{4}(0.15 \text{ m})^2$$
$$= 0.0177 \text{ m}^2$$

$$A_2 = \frac{\pi}{4}(0.25 \text{ m})^2$$
$$= 0.0491 \text{ m}^2$$

$$h_{f,1} = h_{f,2}$$

$$f\left(\frac{L_1}{D_1}\right)\left(\frac{v_1^2}{2g}\right) = f\left(\frac{L_2}{D_2}\right)\left(\frac{v_2^2}{2g}\right)$$

$$f_1 = f_2 = 0.024$$

Therefore,

$$\left(\frac{L_1}{D_1}\right)v_1^2 = \left(\frac{L_2}{D_2}\right)v_2^2$$

$$v_1^2 = \left(\frac{650 \text{ m}}{0.25 \text{ m}}\right)\left(\frac{0.15 \text{ m}}{325 \text{ m}}\right)v_2^2$$

$$v_1 = 1.095 v_2$$

$$Q_t = Q_1 + Q_2 = A_1 v_1 + A_2 v_2$$
$$= 300 \text{ m}^3/\text{h}$$

$$(0.0177 \text{ m}^2)(1.095 v_2) + (0.0491 \text{ m}^2)v_2 = 300 \text{ m}^3/\text{h}$$

$$v_2 = 4381 \text{ m/h} \quad (1.22 \text{ m/s})$$

$$Q_2 = v_2 A_2 = \left(1.22 \frac{\text{m}}{\text{s}}\right)(0.0491 \text{ m}^2)$$

$$= 0.06 \text{ m}^3/\text{s}$$

Answer is C.

FE-STYLE EXAM PROBLEMS

1. Water flows through a multisectional pipe placed horizontally on the ground. The velocity is 3.0 m/s at the entrance and 2.1 m/s at the exit. What is the pressure difference between these two points? Neglect friction.

FLUIDS Dynamics

(A) 0.2 kPa
(B) 2.3 kPa
(C) 28 kPa
(D) 110 kPa

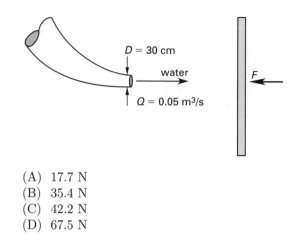

2. What is the mass flow rate of a liquid ($\rho = 0.690$ g/cm³) flowing through a 5 cm (inside diameter) pipe at 8.3 m/s?

(A) 11 kg/s
(B) 69 kg/s
(C) 140 kg/s
(D) 340 kg/s

(A) 17.7 N
(B) 35.4 N
(C) 42.2 N
(D) 67.5 N

7. Water flows with a velocity of 6 m/s through 6 m of cast-iron pipe (specific roughness = 0.0003 m). The pipe has an inside diameter of 43 mm. The kinematic viscosity of the water is 1.00×10^{-6} m²/s. The loss coefficient for the standard elbow is 0.9. What percentage of the total head loss is caused by the elbow?

3. The mean velocity of 40°C water in a 44.7 mm (inside diameter) tube is 1.5 m/s. The kinematic viscosity is $\nu = 6.58 \times 10^{-7}$ m²/s. What is the Reynold's number?

(A) 8.13×10^3
(B) 8.54×10^3
(C) 9.06×10^4
(D) 1.02×10^5

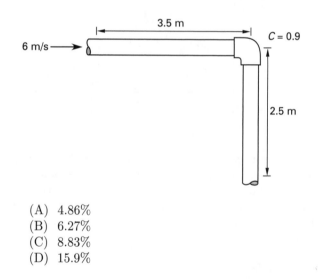

4. What is the head loss for water flowing through a horizontal pipe if the gage pressure at point 1 is 1.03 kPa, the gage pressure at point 2 downstream is 1.00 kPa, and the velocity is constant?

(A) 3.1×10^{-3} m
(B) 3.1×10^{-2} m
(C) 2.3×10^{-2} m
(D) 2.3 m

(A) 4.86%
(B) 6.27%
(C) 8.83%
(D) 15.9%

5. The *hydraulic radius* is

(A) the mean radius of the pipe.
(B) the radius of the pipe bend on the line.
(C) the wetted perimeter of a conduit divided by the area of flow.
(D) the cross-sectional fluid area divided by the wetted perimeter.

8. The pipe manifold shown is at a steady-state condition. What is the fluid velocity v_3 (in m/s) in the 50 mm diameter outlet?

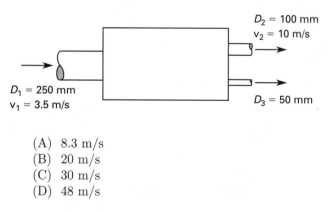

For the following problems use the NCEES Handbook as your only reference.

6. What horizontal force is required to hold the plate stationary against the water jet? (All of the water leaves parallel to the plate.)

(A) 8.3 m/s
(B) 20 m/s
(C) 30 m/s
(D) 48 m/s

Problems 9–12 refer to the following situation.

A rectangular open channel has a base of length $2b$. Water is flowing through the channel at a depth of b.

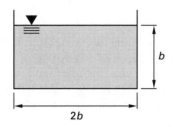

9. What is the wetted perimeter of the channel?

(A) $\dfrac{b}{8}$

(B) $\dfrac{b}{4}$

(C) $2b$

(D) $4b$

10. What is the area in flow?

(A) $\dfrac{2}{3}b^2$

(B) $1.5b^2$

(C) $2b^2$

(D) $3b^2$

11. What is the hydraulic radius?

(A) $\dfrac{b}{4}$

(B) $\dfrac{b}{3}$

(C) $\dfrac{b}{2}$

(D) $\dfrac{2b}{3}$

12. If the flow rate in the channel is 35 m³/s, what is the critical depth (i.e., the depth of flow that minimizes the total energy of flow)?

(A) $2.12b^{-2/3}$ m

(B) $3.15b^{-2/3}$ m

(C) $3.36b^{-2/3}$ m

(D) $5.00b^{-2/3}$ m

13. What are minor losses?

(A) decreases in pressure due to friction in fully developed turbulent flow through pipes of constant area

(B) decreases in pressure due to friction in valves, tees, and elbows, and other frictional effects

(C) decreases in pressure due to friction that can usually be ignored

(D) decreases in pressure due to friction in fully developed turbulent flow in nonconstant area pipes

14. A nozzle directs a jet of water vertically upward with a velocity v and a flow rate Q. A horizontal plate is located directly above the nozzle at a height h. If the density of water is ρ, what reaction force is required to keep the plate stationary against the force of the water jet?

(A) $Q\rho \text{v}$

(B) $Q\rho\sqrt{2gh}$

(C) $\dfrac{Q\rho gh}{\text{v}}$

(D) $Q\rho\sqrt{\text{v}^2 - 2gh}$

Problems 15–17 refer to the following situation.

A 1 m penstock is anchored by a thrust block at a point where the flow makes a 20° change in direction. The water flow rate is 5.25 m³/s. The water pressure is 140 kPa everywhere in the penstock.

15. Assuming the initial flow direction is parallel to the x-direction, what is the magnitude of the force on the thrust block in the x-direction?

(A) 6.8 kN

(B) 8.3 kN

(C) 8.7 kN

(D) 9.2 kN

16. Assuming the initial flow direction is perpendicular to the y-direction, what is the force on the thrust block in the y-direction?

(A) 40.4 kN

(B) 44.7 kN

(C) 47.2 kN

(D) 49.6 kN

17. What is the resultant force on the thrust block?

(A) 10.8 kN
(B) 43.5 kN
(C) 50.4 kN
(D) 146 kN

18. A pipe with a radius of 1.2 m flows partially full as shown. What is the approximate hydraulic radius?

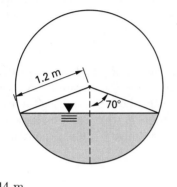

(A) 0.44 m
(B) 0.88 m
(C) 1.30 m
(D) 1.80 m

19. Water jets horizontally from a nozzle installed near the base of a tank. The water level is 10 m above the level of the nozzle. The nozzle necks down from a 75 mm diameter to a 25 mm diameter. The coefficient of velocity, C_v, for the nozzle is 0.962. What is the maximum power that can be extracted from the water jet?

(A) 550 W
(B) 600 W
(C) 650 W
(D) 1200 W

20. A lawn sprinkler consists of a rotating runner with two nozzles. The nozzles are oriented at right angles to the runner. The diameter of the runner is 20 mm; the diameter of the sprinkler nozzles is 10 mm. Water is supplied by the attached hose (not shown) at a rate of 14 m³/h. What single force must be placed on one side of the runner at a distance of 100 mm from the center of rotation in order to stop the sprinkler from rotating?

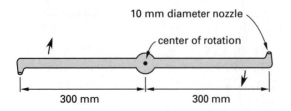

(A) 12 N
(B) 50 N
(C) 75 N
(D) 290 N

21. An incompressible fluid is flowing through two 75 mm diameter pipes at a velocity of 1 m/s in each pipe. The pipes join together and the fluid flows through a 150 mm diameter pipe. What is the velocity of the fluid in the 150 mm diameter pipe?

(A) 0.5 m/s
(B) 1 m/s
(C) 2 m/s
(D) 3 m/s

22. What is the correct definition of the "hydraulic radius" of a fluid conduit?

(A) the mean radius from the center of flow to the wetted side of the conduit
(B) the cross-sectional area of the conduit divided by the wetted perimeter
(C) the wetted perimeter of the conduit divided by the area of flow
(D) the cross-sectional area in flow divided by the wetted perimeter

23. A rectangular flume is 13 cm wide and 7 cm high. The freeboard is 25 percent of the flume height. What is the hydraulic radius?

(A) 2.3 cm
(B) 2.9 cm
(C) 3.2 cm
(D) 3.4 cm

24. A 2 m wide, 3000 m long rectangular channel carries 2 m³/s of water. The depth of flow is 1 m. The channel is constructed from rough-formed concrete with a roughness coefficient of $n = 0.017$. What is the difference in elevation between the two ends of the channel?

(A) 1.2 m
(B) 1.6 m
(C) 2.2 m
(D) 2.7 m

Problems 25–27 refer to the open channel shown in the following figure. The channel is constructed of smooth concrete.

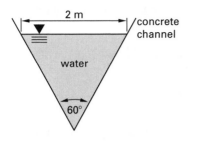

2 m

concrete channel

water

60°

25. The width of flow is 2 m at the surface. What is the hydraulic radius of the channel?

(A) $\dfrac{1}{6}$ m

(B) $\dfrac{\sqrt{3}}{4}$ m

(C) $\dfrac{1}{2}$ m

(D) $\dfrac{\sqrt{3}}{2}$ m

26. [Note: For this problem, use a reasonable value of the Manning's roughness coefficient.] Water flows at 3 m³/s in uniform flow. What is the minimum slope?

(A) 0.0002
(B) 0.001
(C) 0.002
(D) 0.01

27. [Note: For this problem, use a slope of 0.005 and a reasonable value of the Hazen-Williams roughness coefficient.] The channel's water level is adjusted so that the fluid depth is reduced to 50% of its original value. What is the flow rate?

(A) 0.80 m³/s
(B) 1.00 m³/s
(C) 1.45 m³/s
(D) 2.20 m³/s

SOLUTIONS TO FE-STYLE EXAM PROBLEMS

1. From the Bernoulli equation,

$$\frac{p_2}{\rho g} + \frac{v_2^2}{2g} + z_2 = \frac{p_1}{\rho g} + \frac{v_1^2}{2g} + z_1$$

$z_2 = z_1$ [since the pipe is on the ground]

$$\Delta p = p_2 - p_1$$

$$= \rho g \left(\frac{v_1^2 - v_2^2}{2g} \right)$$

$$= \left(\frac{\rho}{2} \right) (v_1^2 - v_2^2)$$

$$= \left(\frac{1000 \ \frac{\text{kg}}{\text{m}^3}}{2} \right) \left(\left(3.0 \ \frac{\text{m}}{\text{s}} \right)^2 - \left(2.1 \ \frac{\text{m}}{\text{s}} \right)^2 \right)$$

$$= 2295 \text{ Pa} \quad (2.3 \text{ kPa})$$

Answer is B.

2. $\dot{m} = \rho A v$

$$= \left(0.690 \ \frac{\text{g}}{\text{cm}^3} \right) \left(\left(\frac{\pi}{4} \right) (5 \text{ cm})^2 \right)$$

$$\times \left(8.3 \ \frac{\text{m}}{\text{s}} \right) \left(100 \ \frac{\text{cm}}{\text{m}} \right)$$

$$= 11\,245 \text{ g/s} \quad (11 \text{ kg/s})$$

Answer is A.

3. $\text{Re} = \dfrac{\rho v D}{\mu}$

$$= \frac{v D}{\nu}$$

$$= \frac{\left(1.5 \ \frac{\text{m}}{\text{s}} \right) (0.0447 \text{ m})}{6.58 \times 10^{-7} \ \frac{\text{m}^2}{\text{s}}}$$

$$= 1.02 \times 10^5$$

Answer is D.

4. From the Bernoulli equation,

$$\frac{p_1}{\rho g} + \frac{v_1^2}{2g} + z_1 = \frac{p_2}{\rho g} + \frac{v_2^2}{2g} + z_2 + h_f$$

$$z_1 = z_2$$

$$v_1 = v_2$$

$$\frac{p_1}{\rho g} = \frac{p_2}{\rho g} + h_f$$

$$h_f = \frac{p_1 - p_2}{\rho g} = \frac{p_1 - p_2}{\rho g}$$

$$= \frac{(1.03 \text{ kPa} - 1.0 \text{ kPa}) \left(1000 \ \frac{\text{Pa}}{\text{kPa}} \right)}{\left(1000 \ \frac{\text{kg}}{\text{m}^3} \right) \left(9.81 \ \frac{\text{m}}{\text{s}^2} \right)}$$

$$= 3.1 \times 10^{-3} \text{ m}$$

Answer is A.

5.
$$R_H = \frac{\text{area in flow}}{\text{wetted perimeter}}$$

Answer is D.

6. The force exerted by the flat plate on the fluid is

$$v_1 = \frac{Q}{A} = \frac{0.05 \ \frac{m^3}{s}}{\left(\frac{\pi}{4}\right)(0.3 \ m)^2}$$

$$= 0.707 \ m/s$$

$$F_x = Q\rho(v_2 - v_1)$$

$$= \left(0.05 \ \frac{m^3}{s}\right)\left(1000 \ \frac{kg}{m^3}\right)\left(0 - 0.707 \ \frac{m}{s}\right)$$

$$= -35.4 \ N \quad \text{[to the left]}$$

The force exerted by the water on the plate is opposite to this.

$$F = 35.4 \ N \quad \text{[to the right]}$$

Answer is B.

7. The pressure drop between the entrance and exit is caused by a combination of pipe friction and minor losses through the elbow.

Solve for the friction.

$$\frac{\epsilon}{D} = \frac{0.0003 \ m}{0.043 \ m} = 0.007$$

$$Re = \frac{vD}{\nu} = \frac{\left(6 \ \frac{m}{s}\right)(0.043 \ m)}{1.0 \times 10^{-6} \ \frac{m^2}{s}}$$

$$= 2.58 \times 10^5$$

From the Moody diagram,

$$f = 0.034$$

$$h_f = \frac{fLv^2}{2Dg}$$

$$= \frac{(0.034)(3.5 \ m + 2.5 \ m)\left(6 \ \frac{m}{s}\right)^2}{(2)(0.043 \ m)\left(9.81 \ \frac{m}{s}\right)}$$

$$= 8.7 \ m$$

Solve for minor losses through the elbow.

$$h_L = C\left(\frac{v^2}{2g}\right) = \frac{(0.9)\left(6 \ \frac{m}{s}\right)^2}{(2)\left(9.81 \ \frac{m}{s}\right)}$$

$$= 1.65 \ m$$

The percentage of pressure drop caused by the standard elbow is

$$\frac{1.65 \ m}{8.7 \ m + 1.65 \ m} = 0.159 \quad (15.9\%)$$

Answer is D.

8. From the continuity equation, since $\rho_1 = \rho_2 = \rho_3$,

$$Q_1 = Q_2 + Q_3$$

$$A_1v_1 = A_2v_2 + A_3v_3$$

$$\frac{\pi}{4}v_1D_1^2 = \frac{\pi}{4}\left(v_2D_2^2 + v_3D_3^2\right)$$

$$v_3 = \frac{v_1D_1^2 - v_2D_2^2}{D_3^2}$$

$$= \frac{\left(3.5 \ \frac{m}{s}\right)(0.25 \ m)^2 - \left(10 \ \frac{m}{s}\right)(0.1 \ m)^2}{(0.05 \ m)^2}$$

$$= 47.5 \ m/s \quad (48 \ m/s)$$

Answer is D.

9. The wetted perimeter is the length of the line tracing the interface between the fluid and the channel.

$$\text{wetted perimeter} = b + b + 2b = 4b$$

Answer is D.

10.
$$\text{area in flow} = (2b)(b) = 2b^2$$

Answer is C.

11.
$$R_H = \frac{\text{area in flow}}{\text{wetted perimeter}} = \frac{2b^2}{4b} = \frac{b}{2}$$

Answer is C.

12. The critical depth is the depth of flow, z, that minimizes the energy of flow. The total energy of flow is

$$E_t = E_p + E_v + E_z$$

$$= \frac{p}{\rho} + \frac{v^2}{2} + zg$$

Since the area of flow for a rectangular channel is the depth times the width ($A = zw = 2zb$),

$$E_t = \frac{p}{\rho} + \frac{Q^2}{2A^2} + zg$$

$$= \frac{p}{\rho} + \frac{Q^2}{8b^2z^2} + zg$$

Differentiate E_t with respect to the depth, z, and set the derivative equal to zero to obtain the minimum.

$$\frac{dE_t}{dz} = \frac{-2Q^2}{8b^2z^3} + g = 0$$

Solve for the critical depth.

$$z_c^3 = \frac{Q^2}{4gb^2}$$

$$z_c = \sqrt[3]{\frac{\left(35 \ \frac{\text{m}^3}{\text{s}}\right)^2}{(4)\left(9.81 \ \frac{\text{m}}{\text{s}^2}\right)b^2}}$$

$$= 3.15b^{-2/3} \ \text{m}$$

Answer is B.

13. Minor losses are friction losses caused by fittings, changes in direction, and changes in flow area. The flow regime (laminar or turbulent) and the pipe cross section are irrelevant.

Answer is B.

14. Use the conservation of energy principle to find the velocity of the water as it hits the plate. As the water leaves the nozzle, it possesses kinetic energy. As it hits the plate, it will have both kinetic and potential energy.

$$\frac{1}{2}mv^2 = \frac{1}{2}mv_f^2 + mgh$$

The velocity of the water as it hits the plate is

$$v_f = \sqrt{v^2 - 2gh}$$

Find the force of the jet on the plate.

$$F = Q\rho v_f = Q\rho\sqrt{v^2 - 2gh}$$

Answer is D.

15. The pipe area is

$$A = \frac{\pi}{4}d^2 = \left(\frac{\pi}{4}\right)(1 \ \text{m})^2$$

$$= 0.7854 \ \text{m}^2$$

$$v = \frac{Q}{A} = \frac{5.25 \ \frac{\text{m}^3}{\text{s}}}{0.7854 \ \text{m}^2}$$

$$= 6.68 \ \text{m/s}$$

Use the thrust equation for a pipe bend.

$$-F_x = p_2 A_2 \cos\alpha - p_1 A_1 + Q\rho(v_2 \cos\alpha - v_1)$$

$$= \left(1.4 \times 10^5 \ \frac{\text{N}}{\text{m}^2}\right)(0.7854 \ \text{m}^2)(\cos 20° - 1)$$

$$+ \left(5.25 \ \frac{\text{m}^3}{\text{s}}\right)\left(1000 \ \frac{\text{kg}}{\text{m}^3}\right)\left(6.68 \ \frac{\text{m}}{\text{s}}\right)$$

$$\times (\cos 20° - 1)$$

$$= -8746 \ \text{N} \quad (8.7 \ \text{kN})$$

Answer is C.

16.
$$F_y = (p_2 A_2 + Q\rho v_2)\sin\alpha$$

$$= \left(\left(1.4 \times 10^5 \ \frac{\text{N}}{\text{m}^2}\right)(0.7854 \ \text{m}^2)\right.$$

$$+ \left(5.25 \ \frac{\text{m}^3}{\text{s}}\right)$$

$$\left.\times \left(1000 \ \frac{\text{kg}}{\text{m}^3}\right)\left(6.68 \ \frac{\text{m}}{\text{s}}\right)\right)\sin 20°$$

$$= 49\,602 \ \text{N} \quad (49.6 \ \text{kN})$$

Answer is D.

17.
$$F = \sqrt{F_x^2 + F_y^2}$$

$$= \sqrt{(-8746 \ \text{N})^2 + (49\,602 \ \text{N})^2}$$

$$= 50\,367 \ \text{N} \quad (50.4 \ \text{kN})$$

Answer is C.

18. The total interior angle is twice 70°, or 140°. Convert this angle to radians.

$$\phi = (140°)\left(\frac{2\pi}{360°}\right) = 2.44 \ \text{rad}$$

Determine the area in flow.

$$A = \left(\frac{1}{2}\right)r^2(\phi - \sin\phi)$$

$$= \frac{1}{2}(1.2 \ \text{m})^2(2.44 \ \text{rad} - \sin(2.44 \ \text{rad}))$$

$$= 1.29 \ \text{m}^2$$

The wetted perimeter is

$$P = r\phi = (1.2 \ \text{m})(2.44 \ \text{rad})$$

$$= 2.93 \ \text{m}$$

$$R_H = \frac{\text{area in flow}}{\text{wetted perimeter}} = \frac{1.29 \ \text{m}^2}{2.93 \ \text{m}}$$

$$= 0.44 \ \text{m}$$

Answer is A.

19. The diameter of the nozzle entrance is irrelevant. The area of the jet is

$$A = \frac{\pi}{4}d^2 = \left(\frac{\pi}{4}\right)(0.025 \text{ m})^2$$
$$= 4.91 \times 10^{-4} \text{ m}^2$$

The jet velocity is

$$v_j = C_v\sqrt{2gh}$$
$$= (0.962)\sqrt{(2)\left(9.81 \frac{\text{m}}{\text{s}^2}\right)(10 \text{ m})}$$
$$= 13.47 \text{ m/s}$$
$$Q = Av_j$$
$$= (4.91 \times 10^{-4} \text{ m}^2)(13.47 \frac{\text{m}}{\text{s}})$$
$$= 0.0066 \text{ m}^3/\text{s}$$

The power is

$$P = \dot{m}E_v = C_v^2 Q\rho gh = Q\rho\left(\frac{v_j^2}{2}\right)$$

$$= \left(0.0066 \frac{\text{m}^3}{\text{s}}\right)\left(1000 \frac{\text{kg}}{\text{m}^3}\right)\frac{\left(13.47 \frac{\text{m}}{\text{s}}\right)^2}{2}$$

$$= 599 \text{ W} \quad (600 \text{ W})$$

Answer is B.

20. The velocity of each of the two jets is

$$v = \frac{Q}{A} = \frac{\left(14 \frac{\text{m}^3}{\text{h}}\right)\left(\dfrac{1}{3600 \frac{\text{s}}{\text{h}}}\right)}{(2)\left(\frac{\pi}{4}\right)(0.01 \text{ m})^2}$$

$$= 24.8 \text{ m/s}$$

Each jet force is

$$F = Q\rho v$$

$$= \frac{\left(14 \frac{\text{m}^3}{\text{h}}\right)\left(\dfrac{1}{3600 \frac{\text{s}}{\text{h}}}\right)\left(1000 \frac{\text{kg}}{\text{m}^3}\right)\left(24.8 \frac{\text{m}}{\text{s}}\right)}{2}$$

$$= 48.2 \text{ N}$$

Take moments about the center of rotation.

$$\sum M_O = (2)(-48.2 \text{ N})(0.3 \text{ m}) + F(0.1 \text{ m})$$
$$= 0$$
$$F = 289.2 \text{ N} \quad (290 \text{ N})$$

Answer is D.

21. Use the continuity equation.

$$Q_{\text{in}} = Q_{\text{out}}$$
$$v_1 A_1 + v_2 A_2 = v_{\text{out}} A_{\text{out}}$$

For circular pipes, the area is $(\pi/4)d^2$. The common $\pi/4$ term can be omitted.

$$(2)\left(1 \frac{\text{m}}{\text{s}}\right)(0.075 \text{ m})^2 = v_{\text{out}}(0.15 \text{ m})^2$$

$$v_{\text{out}} = 0.5 \text{ m/s}$$

Answer is A.

22. The hydraulic radius is defined as the area in flow divided by the wetted perimeter.

$$R_H = \frac{\text{area in flow}}{\text{wetted perimeter}}$$

Answer is D.

23. The depth of flow is

$$d = (1 - 0.25)(7 \text{ cm})$$

The area in flow is

$$A = wd$$
$$= (13 \text{ cm})(1 - 0.25)(7 \text{ cm})$$
$$= 68.25 \text{ cm}^2$$

The wetted perimeter is

$$P = w + 2d$$
$$= 13 \text{ cm} + (2)(1 - 0.25)(7 \text{ cm})$$
$$= 23.5 \text{ cm}$$

The hydraulic radius is

$$R_H = \frac{A}{P} = \frac{68.25 \text{ cm}^2}{23.5 \text{ cm}}$$
$$= 2.90 \text{ cm}$$

Answer is B.

24. The area in flow is

$$A = wd = (2 \text{ m})(1 \text{ m})$$
$$= 2 \text{ m}^2$$

The wetted perimeter is

$$P = w + 2d = 2 \text{ m} + (2)(1 \text{ m})$$
$$= 4 \text{ m}$$

The hydraulic radius is

$$R_H = \frac{A}{P} = \frac{2 \text{ m}^2}{4 \text{ m}}$$
$$= 0.5 \text{ m}$$

The velocity of flow is

$$\text{v} = \frac{Q}{A} = \frac{2 \dfrac{\text{m}^3}{\text{s}}}{2 \text{ m}^2}$$
$$= 1 \text{ m/s}$$

Solve Manning's equation for the slope.

$$\text{v} = \frac{1}{n} R_H^{2/3} S^{1/2}$$

$$S = \left(\frac{\text{v}n}{R_H^{2/3}} \right)^2 = \left(\frac{\left(1 \dfrac{\text{m}}{\text{s}} \right)(0.017)}{(0.5)^{2/3}} \right)^2$$

$$= 0.0007282$$

The slope is small, so the horizontal distance is essentially the same as the total length. For uniform flow, the energy gradient and geometric gradient are the same. The change in elevation is

$$y = Sx$$
$$= (0.0007282)(3000 \text{ m})$$
$$= 2.18 \text{ m} \quad (2.2 \text{ m})$$

Answer is C.

25. The two unknown angles are the same.

$$\left(\frac{1}{2} \right)(180° - 60°) = 60°$$

This is an equilatural triangle, and the wetted side length is the same as the surface width, 2 m.

From the Pythagorean theorem, the maximum depth of the triangular channel is

$$d = \sqrt{(2 \text{ m})^2 - (1 \text{ m})^2}$$
$$= \sqrt{3} \text{ m}$$

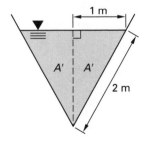

The area in flow is

$$A = 2A' = (2) \left(\frac{1}{2} \right) bd$$
$$= bd$$
$$= (1 \text{ m})(\sqrt{3} \text{ m})$$
$$= \sqrt{3} \text{ m}^2$$

The wetted perimeter is

$$P = 2 \text{ m} + 2 \text{ m}$$
$$= 4 \text{ m}$$

The hydraulic radius is

$$R_H = \frac{A}{P} = \frac{\sqrt{3} \text{ m}^2}{4 \text{ m}}$$
$$= \sqrt{3}/4 \text{ m}$$

Answer is B.

26. Use Manning's equation for the velocity. Since the concrete is smooth, use $n = 0.013$.

$$\text{v} = \frac{Q}{A} = \frac{1}{n} R_H^{2/3} S^{1/2}$$

$$S = \left(\frac{Qn}{A R_H^{2/3}} \right)^2$$

$$= \left(\frac{\left(3 \dfrac{\text{m}^3}{\text{s}} \right)(0.013)}{(\sqrt{3} \text{ m}^2) \left(\dfrac{\sqrt{3}}{4} \text{ m} \right)^{2/3}} \right)^2$$

$$= 0.001547 \quad (0.002)$$

Answer is C.

27. From Prob. 25, the original depth of flow was $\sqrt{3}$ m. The new depth of flow will be

$$d = \frac{\sqrt{3} \text{ m}}{2} \quad \text{or} \quad \frac{1}{2}\sqrt{3} \text{ m}$$

The width of flow at the surface is also halved. The area in flow is

$$A = 2A' = (2) \left(\frac{1}{2} \right) bd$$
$$= bd$$
$$= \left(\frac{1}{2} \text{ m} \right) \left(\frac{\sqrt{3}}{2} \text{ m} \right)$$
$$= \frac{1}{4}\sqrt{3} \text{ m}$$

The wetted perimeter is

$$P = 1 \text{ m} + 1 \text{ m}$$
$$= 2 \text{ m}$$

The hydraulic radius is

$$R_H = \frac{A}{P} = \frac{\dfrac{\sqrt{3}}{4} \text{ m}^2}{2 \text{ m}}$$
$$= \frac{1}{8}\sqrt{3} \text{ m}$$

Since the channel is smooth concrete, use a value of $C = 130$ for the Hazen-Williams roughness coefficient.

The flow rate is

$$Q = Av$$
$$= A(0.849)CR_H^{0.63}S^{0.54}$$
$$= \left(\frac{\sqrt{3}}{4} \text{ m}^2\right)(0.849)(130)$$
$$\times \left(\frac{\sqrt{3}}{8} \text{ m}\right)^{0.63}(0.005)^{0.54}$$
$$= 1.04 \text{ m}^3/\text{s} \quad (1.00 \text{ m}^3/\text{s})$$

Answer is B.

25 Fluid Measurements and Similitude

Subjects

PUMP POWER 25-1
IDEAL GAS 25-2
   Equation of State 25-2
   Speed of Sound 25-2
   Air 25-2
FLUID MEASUREMENTS 25-2
   Pitot Tube 25-2
   Venturi Meter 25-3
   Orifice Meter 25-3
   Submerged Orifice 25-4
   Orifice Discharging Freely 25-4
DIMENSIONAL ANALYSIS 25-4
SIMILITUDE 25-5

Nomenclature

A	area	ft^2	m^2
c	speed of sound	ft/sec	m/s
c	specific heat	Btu/lbm-°F	kJ/kg·K
C	flow coefficient	–	–
Ca	Cauchy number	–	–
E	modulus of elasticity	lbf/in^2	MPa
F	force	lbf	N
Fr	Froude number	–	–
g	gravitational acceleration	ft/sec^2	m/s^2
g_c	gravitational constant (32.2)	lbm-ft/lbf-sec^2	–
h	head or height	ft	m
k	ratio of specific heats	–	–
k	number of pi-groups	–	–
ℓ	length	ft	m
m	number of independent dimensionless quantities	–	–
$\dot{m}$	mass flow rate	lbm/sec	kg/s
M	Mach number	–	–
MW	molecular weight	lbm/lbmol	kg/kmol
n	number of independent variables	–	–
p	pressure	lbf/ft^2	N/m^2
P	power	ft-lbf/sec	W
Q	flow rate	ft^3/sec	m^3/s
R	specific gas constant	ft-lbf/lbm-°R	kJ/kg·K
$\overline{R}$	universal gas constant	ft-lbf/lbmol-°R	J/kmol·K
Re	Reynolds number	–	–
T	temperature	°R	K
v	velocity	ft/sec	m/s
V	volume	ft^3	m^3
W	work	ft-lbf	J
We	Weber number	–	–
z	elevation	ft	m

Symbols

γ	specific weight	lbf/ft^3	–
η	efficiency	–	–
μ	absolute viscosity	lbf-sec/ft^2	Pa·s
ν	kinematic viscosity	ft^2/sec	m^2/s
π	dimensionless group of variables	–	–
ρ	density	lbm/ft^3	kg/m^3
σ	surface tension	lbf/ft	N/m
υ	specific volume	ft^3/lbm	m^3/kg

Subscripts

0	stagnation (zero velocity)
c	contraction
E	elastic
G	gravity
I	inertia
m	manometer fluid, or model
p	constant pressure
p	prototype
P	pressure
s	static
T	surface tension force
v	constant volume
v	velocity
V	viscous

PUMP POWER

Pumps convert mechanical energy into fluid energy, thus increasing the energy of the fluid. Pump output power is known as *hydraulic power* or *water power*. Hydraulic power is the net power transferred to the fluid by the pump.

Horsepower is the unit of power used in the United States and other non-SI countries, which gives rise to the terms *hydraulic horsepower* and *water horsepower*. The unit of power in SI units is the watt.

FLUIDS Measurements

Hydraulic power is the net energy transferred to the fluid per unit time. The input power delivered by the motor to the pump is known as the *brake pump power*, P. Due to frictional losses between the fluid and the pump and mechanical losses in the pump itself, the brake pump power will be greater than the hydraulic power. The difference between the brake and hydraulic powers is accounted for by the *pump efficiency*, η.

In Eq. 25.1, h is the head that is added by the pump to the fluid.

$$
\begin{aligned}
P = \dot{W} &= \frac{Q\gamma h}{\eta} \\
&= \frac{Q\rho g h}{\eta} \\
&= \frac{\dot{m}gh}{\eta}
\end{aligned} \qquad 25.1
$$

IDEAL GAS

Equation of State

The *ideal gas law* is an *equation of state* for ideal gases. An equation of state is a relationship that predicts the state (i.e., a property, such as pressure, temperature, volume, etc.) from a set of two other independent properties.

$$
pv = \frac{p}{\rho} = RT \qquad 25.2
$$

R, the *specific gas constant*, can be determined from the *molecular weight* of the substance, MW, and the *universal gas constant*, $\overline{R}$.

$$
R = \frac{\overline{R}}{\text{MW}} \qquad 25.3
$$

The universal gas constant, $\overline{R}$, is "universal" (within a system of units) because the same value can be used for any gas. Its value depends on the units used for pressure, temperature, and volume, as well as on the units of mass.

The equation of state leads to another general relationship.

$$
\frac{p_1 v_1}{T_1} = \frac{p_2 v_2}{T_2} \qquad 25.4
$$

When temperature is held constant, this reduces to *Boyle's law*.

$$
pv = \text{constant} \qquad 25.5
$$

There is no heat loss in an *adiabatic process*. An *isentropic process* is an adiabatic process in which there is no change in system *entropy* (i.e., the process is reversible). For such a process, Eq. 25.6 is valid. For gases, the *ratio of specific heats*, k, is defined by Eq. 25.7, in which c_p is the *specific heat at constant pressure* and c_v is the *specific heat at constant volume*.

$$
pv^k = \text{constant} \qquad 25.6
$$

$$
k = \frac{c_p}{c_v} \qquad 25.7
$$

Speed of Sound

The *speed of sound*, c, in a fluid is a function of its bulk modulus, or equivalently, of its compressibility. Equation 25.8 gives the speed of sound in an ideal gas. The temperature, T, must be in degrees absolute (i.e., °R or K).

$$
c = \sqrt{kRT} \qquad \text{[SI]} \qquad 25.8a
$$

$$
c = \sqrt{kg_c RT} \qquad \text{[U.S.]} \qquad 25.8b
$$

The *Mach number* of an object is the ratio of the object's speed to the speed of sound in the medium through which the object is traveling.

$$
\text{M} = \frac{\text{v}}{c} \qquad 25.9
$$

Air

For air, the ratio of specific heats is $k = 1.40$, and the molecular weight is 29.0. The universal gas constant is $\overline{R} = 1545.3$ ft-lbf/lbmol-°R (8314 J/kmol·K).

FLUID MEASUREMENTS

Pitot Tube

A *pitot tube* is simply a hollow tube that is placed longitudinally in the direction of fluid flow, allowing the flow to enter one end at the fluid's *velocity of approach*. A pitot tube is used to measure velocity of flow.

Figure 25.1   Pitot Tube

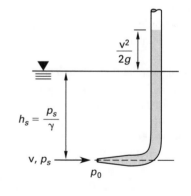

When the fluid enters the pitot tube, it is forced to come to a stop (at the *stagnation point*), and its kinetic energy is transformed into static pressure energy.

Bernoulli's equation can be used to predict the static pressure at the stagnation point. Since the velocity of the fluid within the pitot tube is zero, the upstream velocity can be calculated if the static (p_s) and *stagnation* (p_0) *pressures* are known.

$$\frac{p_s}{\rho} + \frac{v^2}{2} = \frac{p_0}{\rho} \qquad \text{[SI]} \qquad 25.10a$$

$$\frac{p_s}{\gamma} + \frac{v^2}{2g} = \frac{p_0}{\gamma} \qquad \text{[U.S.]} \qquad 25.10b$$

$$v = \sqrt{\frac{2(p_0 - p_s)}{\rho}} \qquad \text{[SI]} \qquad 25.11a$$

$$v = \sqrt{\frac{2g(p_0 - p_s)}{\gamma}} \qquad \text{[U.S.]} \qquad 25.11b$$

In reality, the fluid may be compressible. If the Mach number is less than approximately 0.3, Eq. 25.11 for incompressible fluids may be used.

Venturi Meter

Figure 25.2 illustrates a simple *venturi meter*. This flow-measuring device can be inserted directly into a pipeline. Since the diameter changes are gradual, there is very little friction loss. Static pressure measurements are taken at the throat and upstream of the diameter change. The difference in these pressures is directly indicated by a differential manometer.

Figure 25.2 Venturi Meter with Differential Manometer

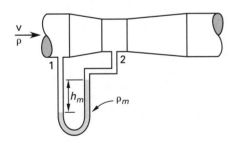

The pressure differential across the venturi meter shown in Fig. 25.2 can be calculated from Eq. 25.12 or 25.13.

$$p_1 - p_2 = (\rho_m - \rho)gh_m = (\gamma_m - \gamma)h_m \qquad 25.12$$

$$\frac{p_1 - p_2}{\rho} = \left(\frac{\rho_m}{\rho} - 1\right)gh_m \qquad \text{[SI]} \qquad 25.13a$$

$$\frac{p_1 - p_2}{\gamma} = \left(\frac{\gamma_m}{\gamma} - 1\right)h_m \qquad \text{[U.S.]} \qquad 25.13b$$

The flow rate, Q, can be calculated from venturi measurements as follows. For a horizontal venturi meter, $z_1 = z_2$.

$$Q = \left(\frac{C_v A_2}{\sqrt{1 - \left(\frac{A_2}{A_1}\right)^2}}\right) \sqrt{2\left(\frac{p_1}{\rho} + gz_1 - \frac{p_2}{\rho} - gz_2\right)}$$
$$\text{[SI]} \qquad 25.14a$$

$$Q = \left(\frac{C_v A_2}{\sqrt{1 - \left(\frac{A_2}{A_1}\right)^2}}\right) \sqrt{2g\left(\frac{p_1}{\gamma} + z_1 - \frac{p_2}{\gamma} - z_2\right)}$$
$$\text{[U.S.]} \qquad 25.14b$$

The *coefficient of velocity*, C_v, accounts for the small effect of friction and is very close to 1.0, usually 0.98 or 0.99.

Orifice Meter

The *orifice meter* (or *orifice plate*) is used more frequently than the venturi meter to measure flow rates in small pipes. It consists of a thin or sharp-edged plate with a central, round hole through which the fluid flows.

As with the venturi meter, pressure taps are used to obtain the static pressure upstream of the orifice plate and at the *vena contracta* (i.e., at the point of minimum pressure). A *differential manometer* connected to the two taps conveniently indicates the difference in static pressures. Equations 25.12 and 25.13, derived for the manometer in Fig. 25.2, are also valid for the manometer configuration of the orifice shown in Fig. 25.3.

Figure 25.3 Orifice Meter with Differential Manometer

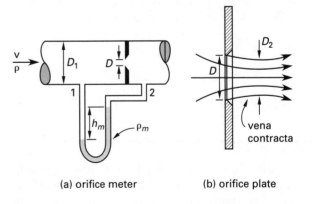

(a) orifice meter   (b) orifice plate

The area of the orifice is A, and the area of the pipeline is A_1. The area at the vena contracta, A_2, can be calculated from the orifice area and the *coefficient of contraction*, C_c.

$$A_2 = C_c A \qquad 25.15$$

The *flow coefficient of the meter*, C, combines the coefficients of velocity and contraction in a way that corrects the theoretical discharge of the meter for frictional flow and for contraction at the vena contracta. Approximate orifice coefficients are listed in Table 25.1.

$$C = \frac{C_v C_c}{\sqrt{1 - C_c^2 \left(\dfrac{A}{A_1}\right)^2}} \qquad 25.16$$

Table 25.1 Approximate Orifice Coefficients for Turbulent Water

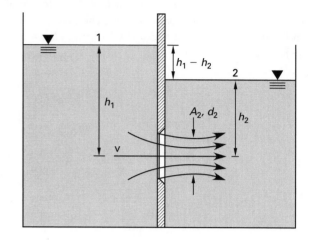

illustration	description	C	C_c	C_v
A	sharp-edged	0.61	0.62	0.98
B	round-edged	0.98	1.00	0.98
C	short tube (fluid separates from walls)	0.61	1.00	0.61
D	sharp tube (no separation)	0.80	1.00	0.80
E	sharp tube with rounded entrance	0.97	0.99	0.98
F	reentrant tube, length less than one-half of pipe diameter	0.54	0.55	0.99
G	reentrant tube, length 2–3 pipe diameters	0.72	1.00	0.72
H	Borda	0.51	0.52	0.98
(none)	smooth, well-tapered nozzle	0.98	0.99	0.99

The flow rate through the orifice meter is given by Eq. 25.17. Generally, z_1 and z_2 are equal.

$$Q = CA\sqrt{2\left(\frac{p_1}{\rho} + gz_1 - \frac{p_2}{\rho} - gz_2\right)} \quad \text{[SI]} \quad 25.17a$$

$$Q = CA\sqrt{2g\left(\frac{p_1}{\gamma} + z_1 - \frac{p_2}{\gamma} - z_2\right)} \quad \text{[U.S.]} \; 25.17b$$

Submerged Orifice

The flow rate of a jet issuing from a submerged orifice in a tank can be determined by modifying Eq. 25.17 in terms of the potential energy difference, or head difference, on either side of the orifice.

$$Q = A_2 v_2 = C_c C_v A\sqrt{2g(h_1 - h_2)} \qquad 25.18$$

The coefficients of velocity and contraction can be combined into the *coefficient of discharge*, C.

$$C = C_c C_v \qquad 25.19$$
$$Q = CA\sqrt{2g(h_1 - h_2)} \qquad 25.20$$

Figure 25.4 Submerged Orifice

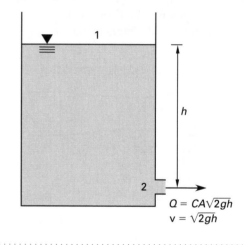

Orifice Discharging Freely

If the orifice discharges from a tank into the atmosphere, Eqs. 25.19 and 25.20 can be further simplified.

$$Q = CA\sqrt{2gh} \qquad 25.21$$

Figure 25.5 Orifice Discharging Freely into the Atmosphere

$$Q = CA\sqrt{2gh}$$
$$v = \sqrt{2gh}$$

DIMENSIONAL ANALYSIS

Dimensional analysis is a means of obtaining an equation that describes some phenomenon without understanding the mechanism of the phenomenon. The most serious limitation is the necessity of knowing beforehand which variables influence the phenomenon. Once the variables are known or assumed, dimensional analysis can be applied using a routine procedure.

Dimensional analysis is performed with a system of *primary dimensions*, usually the $ML\theta T$ system (mass, length, time, and temperature). A dimensionally homogeneous equation is one in which each term in the equation has the same dimensions. An equation must be dimensionally homogeneous to be valid.

A simplification of dimensional analysis is to combine the variables into dimensionless groups, called *pi-groups*. If these dimensionless groups are represented by π_1, π_2, π_3, ... π_k, the equation expressing the relationship between the variables is given by the *Buckingham π-theorem*, Eq. 25.23, in which m is the number of different variables and n is the number of different independent dimensional quantities. k dimensionless pi-groups are needed to describe a phenomenon. The pi-groups are usually found from the m variables according to an intuitive process.

$$f(\pi_1,\ \pi_2,\ \pi_3,\ \dots\ \pi_k) = 0 \qquad 25.22$$
$$k = m - n \qquad 25.23$$

SIMILITUDE

Similarity considerations between a *model* (subscript m) and a full-size object (subscript p, for *prototype*) imply that the model can be used to predict the performance of the prototype. Such a model is said to be *mechanically similar* to the prototype.

Complete mechanical similarity requires geometric, kinematic, and dynamic similarity. *Geometric similarity* means that the model is true to scale in length, area, and volume. *Kinematic similarity* requires that the flow regimes of the model and prototype be the same. *Dynamic similarity* means that the ratios of all types of forces are equal for the model and the prototype. These forces result from inertia, gravity, viscosity, elasticity (i.e., fluid compressibility), surface tension, and pressure.

For dynamic similarity, the number of possible ratios of forces is large. For example, the ratios of viscosity/inertia, inertia/gravity, and inertia/surface tension are only three of the ratios of forces that must match for every corresponding point on the model and prototype. Fortunately, some force ratios can be neglected because the forces are negligible or are self-canceling.

If the following five simultaneous equations are satisfied for two flow pictures, dynamic similarity will be achieved.

$$\left[\frac{F_I}{F_P}\right]_p = \left[\frac{F_I}{F_P}\right]_m = \left[\frac{\rho \mathrm{v}^2}{p}\right]_p = \left[\frac{\rho \mathrm{v}^2}{p}\right]_m \qquad 25.24$$

$$\left[\frac{F_I}{F_V}\right]_p = \left[\frac{F_I}{F_V}\right]_m = \left[\frac{\mathrm{v}\ell\rho}{\mu}\right]_p = \left[\frac{\mathrm{v}\ell\rho}{\mu}\right]_m$$
$$= \mathrm{Re}_p = \mathrm{Re}_m \qquad 25.25$$

$$\left[\frac{F_I}{F_G}\right]_p = \left[\frac{F_I}{F_G}\right]_m = \left[\frac{\mathrm{v}^2}{\ell g}\right]_p = \left[\frac{\mathrm{v}^2}{\ell g}\right]_m$$
$$= \mathrm{Fr}_p = \mathrm{Fr}_m \qquad 25.26$$

$$\left[\frac{F_I}{F_E}\right]_p = \left[\frac{F_I}{F_E}\right]_m = \left[\frac{\rho \mathrm{v}^2}{E}\right]_p = \left[\frac{\rho \mathrm{v}^2}{E}\right]_m$$
$$= \mathrm{Ca}_p = \mathrm{Ca}_m \qquad 25.27$$

$$\left[\frac{F_I}{F_T}\right]_p = \left[\frac{F_I}{F_T}\right]_m = \left[\frac{\rho\ell \mathrm{v}^2}{\sigma}\right]_p = \left[\frac{\rho\ell \mathrm{v}^2}{\sigma}\right]_m$$
$$= \mathrm{We}_p = \mathrm{We}_m \qquad 25.28$$

F_E = elastic force	E = modulus of elasticity
F_G = gravity force	Fr = Froude number
F_I = inertia force	ℓ = characteristic length
F_P = pressure force	(e.g., diameter)
F_T = surface tension force	Ca = Cauchy number
	Re = Reynolds number
F_V = viscous force	We = Weber number

SAMPLE PROBLEMS

Problem 1

A pump requires 75 kW to move water with a specific gravity of 1.0 at a certain flow rate to a given elevation. What power does the pump require if the flow rate and elevation conditions are the same, but the fluid pumped has a specific gravity of 0.8?

  (A) 45 kW
  (B) 60 kW
  (C) 75 kW
  (D) 90 kW

Solution

$$P_{\mathrm{water}} = 75 \times 10^3\ \mathrm{W} = \frac{Q\rho_{\mathrm{water}}gh}{\eta}$$

If SG = 0.8,

$$\rho_{\mathrm{fluid}} = 0.8\rho_{\mathrm{water}}$$

$$P_{\mathrm{fluid}} = 0.8\rho_{\mathrm{water}}g\left(\frac{Qh}{\eta}\right)$$
$$= (0.8)(75 \times 10^3\ \mathrm{W})$$
$$= 60 \times 10^3\ \mathrm{W}\quad (60\ \mathrm{kW})$$

Answer is B.

Problem 2

A pitot tube is used to measure the flow of an incompressible fluid ($\rho = 926\ \mathrm{kg/m^3}$). If the velocity is measured as 2 m/s and the stagnation pressure is 14.1 kPa, what is the static pressure of the fluid where the measurement is taken?

(A) 10.4 kPa
(B) 11.7 kPa
(C) 12.2 kPa
(D) 13.5 kPa

Solution

The pitot tube equation for velocity is

$$v = \sqrt{\frac{(2)(p_0 - p_s)}{\rho}}$$

$$p_s = p_0 - \frac{\rho v^2}{2}$$

$$= 14.1 \text{ kPa} - \frac{\left(926 \, \frac{\text{kg}}{\text{m}^3}\right)\left(2 \, \frac{\text{m}}{\text{s}}\right)^2}{(2)\left(1000 \, \frac{\text{Pa}}{\text{kPa}}\right)}$$

$$= 12.2 \text{ kPa}$$

Answer is C.

Problem 3

A sharp-edged orifice with a 50 mm diameter opening in the vertical side of a large tank discharges under a head of 5 m. If the coefficient of contraction is 0.62 and the coefficient of velocity is 0.98, what is the discharge?

(A) 0.0003 m³/s
(B) 0.004 m³/s
(C) 0.010 m³/s
(D) 0.012 m³/s

Solution

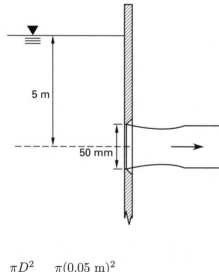

$$A = \frac{\pi D^2}{4} = \frac{\pi (0.05 \text{ m})^2}{4}$$

$$= 0.00196 \text{ m}^2$$

$$C = C_c C_v = (0.62)(0.98)$$

$$= 0.6076$$

$$Q = CA\sqrt{2gh}$$

$$= (0.6076)(0.00196 \text{ m}^2)\sqrt{(2)\left(9.81 \, \frac{\text{m}}{\text{s}^2}\right)(5 \text{ m})}$$

$$= 0.012 \text{ m}^3/\text{s}$$

Answer is D.

Problem 4

Which dimensionless number represents the ratio of inertial forces to gravitational forces?

(A) Reynolds number
(B) Froude number
(C) Grashof number
(D) Weber number

Solution

$$Fr = \frac{F_I}{F_G} = \frac{v^2}{gL}$$

The Froude number is the dimensionless ratio of inertial forces to gravitational forces.

Answer is B.

FE-STYLE EXAM PROBLEMS

1. A 70% efficient pump pumps water from ground level to a height of 5 m. How much power is used if the flow rate is 10 m³/s?

(A) 80 kW
(B) 220 kW
(C) 700 kW
(D) 950 kW

2. The acoustic velocity in a specific gas depends only on which of the following variables?

(A) c_p, specific heat at constant pressure
(B) k, ratio of specific heats
(C) c_v, specific heat at constant temperature
(D) T, absolute temperature

3. Which of the following cannot be directly determined with the use of a pitot tube?

(A) velocity of a flowing fluid
(B) stagnation pressure
(C) discharge rate
(D) total pressure

4. The velocity of the water in the stream is 1.2 m/s. What is the height of water in the pitot tube?

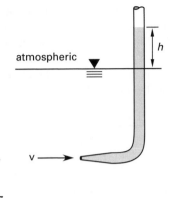

(A) 3.7 cm
(B) 4.6 cm
(C) 7.3 cm
(D) 9.2 cm

5. A venturi meter with a diameter of 15 cm at the throat is installed in an 45 cm water main. A differential manometer gauge is partly filled with mercury (the remainder of the tube is filled with water) and connected to the meter at the throat and inlet. The mercury column stands 37.5 cm higher in one leg than in the other. Neglecting friction, what is the flow through the meter? The specific gravity of mercury is 13.6.

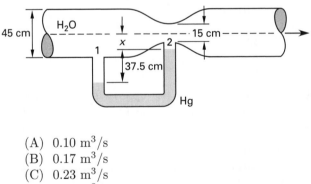

(A) 0.10 m³/s
(B) 0.17 m³/s
(C) 0.23 m³/s
(D) 0.28 m³/s

For the following problems use the NCEES Handbook as your only reference.

6. What is the velocity of water under an 18 m head discharging through a 25 mm diameter round-edged orifice?

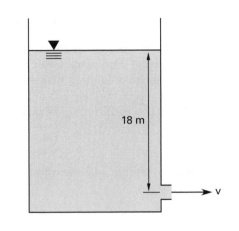

(A) 1.18 m/s
(B) 3.22 m/s
(C) 8.21 m/s
(D) 18.4 m/s

7. A 1:1 model of a torpedo is tested in a wind tunnel according to the Reynolds number criterion. At the testing temperature, $\nu_{\text{air}} = 1.41 \times 10^{-5}$ m²/s and $\nu_{\text{water}} = 1.31 \times 10^{-6}$ m²/s. If the velocity of the torpedo in water is 7 m/s, what should be the air velocity in the wind tunnel?

(A) 0.6 m/s
(B) 7.0 m/s
(C) 18 m/s
(D) 75 m/s

Problems 8 and 9 refer to the following situation.

A sharp-edged orifice with a 50 mm diameter opening is located in the vertical side of a large tank. The coefficient of contraction for the orifice is 0.62, and the coefficient of velocity is 0.98. The orifice discharges under a hydraulic head of 5 m.

8. What is the minimum diameter of the jet?

(A) 31.1 mm
(B) 39.4 mm
(C) 50 mm
(D) 63.7 mm

9. What is the velocity at the vena contracta?

(A) 1.71 m/s
(B) 3.33 m/s
(C) 5.36 m/s
(D) 9.71 m/s

Problems 10–13 refer to the following situation.

The bottom of a tall tank sits on level ground. The tank is kept filled to a depth of 5 m, while water discharges at a constant rate through a 300 mm diameter hole in the tank side. The center of the hole is 3.5 m from the water surface above. The coefficient of velocity for the hole is essentially 1.0.

10. What horizontal distance will the water jet travel before hitting the ground?

    (A) 2.1 m
    (B) 2.3 m
    (C) 2.5 m
    (D) 4.6 m

11. What is the velocity of the water jet?

    (A) 7.2 m/s
    (B) 8.3 m/s
    (C) 8.8 m/s
    (D) 10.0 m/s

12. If the hole is represented by a sharp-edged orifice with a coefficient of discharge of 0.61, what will be the rate of discharge?

    (A) 0.32 m^3/s
    (B) 0.36 m^3/s
    (C) 0.39 m^3/s
    (D) 0.44 m^3/s

13. Assume the orifice can be moved to any point on the side of the tank. What distance below the water surface should the orifice be located such that the horizontal distance traveled by the jet (before hitting the ground) is the greatest?

    (A) 2.5 m
    (B) 2.9 m
    (C) 3.3 m
    (D) 3.7 m

14. A 2 m tall, 0.5 m inside diameter tank is filled with water. A 10 cm hole is opened 0.75 m from the bottom of the tank. What is the velocity of the exiting water? Ignore all orifice losses.

    (A) 4.75 m/s
    (B) 4.80 m/s
    (C) 4.85 m/s
    (D) 4.95 m/s

Problems 15–17 refer to the following situation.

Air flowing through a cylindrical duct encounters a constricted flow area, as shown. The initial flow area is 2.5 m^2. The air entering the constriction has a temperature of 300K and a pressure of 97 kPa. The constricted area is 0.1 m^2. The air in the constriction has a density of 1.416 kg/m^3. The differences in pressures across the constriction is relatively low, and the air velocity through the constriction is 10 m/s.

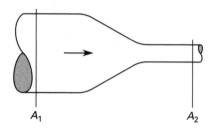

A_1 A_2

15. What is the density of the air before it encounters the constriction?

    (A) 0.0389 kg/m^3
    (B) 1.00 kg/m^3
    (C) 1.13 kg/m^3
    (D) 1.18 kg/m^3

16. What is the velocity of the air approaching the constriction?

    (A) 0.40 m/s
    (B) 0.50 m/s
    (C) 2.1 m/s
    (D) 3.3 m/s

17. Assume the air velocity approaching the constriction is 1.0 m/s. What is the Mach number at that point?

    (A) 0.0029
    (B) 0.014
    (C) 0.29
    (D) 1.0

Problems 18–20 refer to the following situation.

A large pump for a metropolitan water system transfers water from a lower reservoir to an upper reservoir 350 m above. The flow rate is 8.5 m^3/min. The water is discharged freely to the upper reservoir. A gas turbine drives the pump, and effective hydraulic power is 625 kW after losses.

18. What is the approximate theoretical hydraulic head developed across the pump?

(A) 20 m
(B) 90 m
(C) 180 m
(D) 450 m

19. What is the friction loss in the pipe between the two reservoirs?

(A) 6 m
(B) 14 m
(C) 52 m
(D) 100 m

20. If the pipe between the two reservoirs is perfectly insulated, what would be the increase in water temperature due to the frictional loss of the water flowing between the two reservoirs?

(A) 0.23°C
(B) 0.52°C
(C) 0.70°C
(D) 1.0°C

Problems 21 and 22 refer to the following situation.

2750 kg of 15°C water are pumped each minute through 55 m of 150 mm diameter pipe. The pipe has a Darcy friction factor of 0.02. The water is moved vertically 13 m. The pump efficiency is 76%.

21. What is the friction loss over the entire length of the pipe?

(A) 0.96 m
(B) 1.7 m
(C) 2.1 m
(D) 2.5 m

22. What power must be delivered to the pump?

(A) 7.6 kW
(B) 9.2 kW
(C) 28 kW
(D) 42 kW

SOLUTIONS TO FE-STYLE EXAM PROBLEMS

1. $\dot{W} = \dfrac{Q\rho g h}{\eta}$

$$= \frac{\left(10 \ \frac{m^3}{s}\right)\left(1000 \ \frac{kg}{m^3}\right)\left(9.81 \ \frac{m}{s^2}\right)(5 \ m)}{0.70}$$

$$= 700\,714 \ W \quad (700 \ kW)$$

Answer is C.

2.
$$c = \sqrt{kRT}$$
$$k = \frac{c_p}{c_v}$$

R, k, c_p, and c_v are properties of the gas. Only T can vary independently.

Answer is D.

3. The pitot tube is used to measure the velocity of flow. The stagnation pressure (also known as *total pressure*) is measured by a pitot tube. The discharge rate could be determined from the velocity but would require additional knowledge about the pipe size and velocity distribution within the pipe. Therefore, discharge rate cannot be directly determined.

Answer is C.

4. The difference in height between the pitot tube and the free-water surface is a measure of the difference in static and stagnation pressures.

$$v = \sqrt{\frac{2(p_0 - p_s)}{\rho}} = \sqrt{\frac{2\rho g h}{\rho}}$$
$$= \sqrt{2gh}$$

$$h = \frac{v^2}{2g} = \frac{\left(1.2 \ \frac{m}{s}\right)^2}{(2)\left(9.81 \ \frac{m}{s^2}\right)}$$

$$= 0.073 \ m \quad (7.3 \ cm)$$

Answer is C.

5. For the venturi meter,

$$Q = \left(\frac{C_v A_2}{\sqrt{1 - \left(\frac{A_2}{A_1}\right)^2}}\right)\sqrt{2\left(\frac{p_1}{\rho} + gz_1 - \frac{p_2}{\rho} - gz_2\right)}$$

This can be written in terms of a manometer fluid reading. Assuming horizontal flow so that $z_1 = z_2$,

$$Q = \left(\frac{C_v A_2}{\sqrt{1 - \left(\frac{A_2}{A_1}\right)^2}}\right)\sqrt{\frac{2g(\rho_m - \rho)h}{\rho}}$$

Since friction is to be neglected, $C_v = 1$. (For venturi meters, C_v is usually very close to one because the diameter changes are gradual and there is little friction loss.)

$$A_2 = \left(\frac{\pi}{4}\right)(0.15\ \text{m})^2 = 0.0177\ \text{m}^2$$

$$A_1 = \left(\frac{\pi}{4}\right)(0.45\ \text{m})^2 = 0.159\ \text{m}^2$$

$$Q = \left(\frac{(1)(0.0177\ \text{m}^2)}{\sqrt{1 - \left(\frac{0.0177\ \text{m}^2}{0.159\ \text{m}^2}\right)^2}}\right)$$

$$\times \sqrt{\frac{(2)\left(9.81\ \frac{\text{m}}{\text{s}^2}\right)\left(1000\ \frac{\text{kg}}{\text{m}^3}\right)}{\times (13.6-1)(0.375\ \text{m})}{1000\ \frac{\text{kg}}{\text{m}^3}}}$$

$$= 0.171\ \text{m}^3/\text{s} \quad (0.17\ \text{m}^3/\text{s})$$

Answer is B.

6. For an orifice discharging freely into the atmosphere,

$$Q = CA\sqrt{2gh}$$

$$v = \frac{Q}{A} = C\sqrt{2gh}$$

For a round-edged orifice, the coefficient of discharge is approximately 0.98.

$$v = (0.98)\sqrt{(2)\left(9.81\ \frac{\text{m}}{\text{s}^2}\right)(18\ \text{m})}$$

$$= 18.4\ \text{m/s}$$

Answer is D.

7.
$$p = \text{prototype (water)}$$
$$m = \text{model (air)}$$

$$\frac{\ell_p}{\ell_m} = 1$$

Use the Reynolds criterion.

$$\text{Re}_m = \text{Re}_p$$

$$\left[\frac{v\ell\rho}{\mu}\right]_m = \left[\frac{v\ell\rho}{\mu}\right]_p$$

$$\frac{\mu}{\rho} = \nu$$

$$\left[\frac{v\ell}{\nu}\right]_m = \left[\frac{v\ell}{\nu}\right]_p$$

$$v_m = v_p\left(\frac{\ell_p}{\ell_m}\right)\left(\frac{\nu_m}{\nu_p}\right)$$

$$= \left(7\ \frac{\text{m}}{\text{s}}\right)(1)\left(\frac{1.41 \times 10^{-5}\ \frac{\text{m}^2}{\text{s}}}{1.31 \times 10^{-6}\ \frac{\text{m}^2}{\text{s}}}\right)$$

$$= 75.3\ \text{m/s} \quad (75\ \text{m/s})$$

Answer is D.

8.
$$A_2 = C_c A$$

$$\frac{\pi}{4}d_2^2 = C_c\left(\frac{\pi}{4}\right)d^2$$

$$d_2 = d\sqrt{C_c} = (50\ \text{mm})\left(\sqrt{0.62}\right)$$

$$= 39.4\ \text{mm}$$

Answer is B.

9.
$$v_2 = C_v\sqrt{2gh}$$

$$= (0.98)\sqrt{(2)\left(9.81\ \frac{\text{m}}{\text{s}^2}\right)(5\ \text{m})}$$

$$= 9.71\ \text{m/s}$$

Answer is D.

10. The vertical distance that the water falls is

$$y = 5\ \text{m} - 3.5\ \text{m} = 1.5\ \text{m}$$

The distance that an object falls in time t under constant acceleration and starting from rest is

$$y = \frac{1}{2}gt^2$$

The time taken by the water to fall is

$$t = \sqrt{\frac{2y}{g}}$$

The horizontal distance traveled by the water with constant velocity v is

$$x = vt = \sqrt{2gh}\sqrt{\frac{2y}{g}}$$

$$= 2\sqrt{hy}$$

$$= 2\sqrt{(3.5\ \text{m})(1.5\ \text{m})}$$

$$= 4.58\ \text{m} \quad (4.6\ \text{m})$$

Answer is D.

11.
$$v = C_v \sqrt{2gh}$$
$$= (1.0)\sqrt{(2)\left(9.81 \ \frac{m}{s^2}\right)(3.5 \ m)}$$
$$= 8.29 \ m/s \quad (8.3 \ m/s)$$

Answer is B.

12. $A = \frac{\pi}{4}d^2 = \left(\frac{\pi}{4}\right)(0.3 \ m)^2$
$$= 0.071 \ m^2$$
$$Q = CA\sqrt{2gh}$$
$$= (0.61)(0.071 \ m^2)\sqrt{(2)\left(9.81 \ \frac{m}{s^2}\right)(3.5 \ m)}$$
$$= 0.36 \ m^3/s$$

Answer is B.

13. From Prob. 10, the horizontal distance traveled is
$$x = vt = \sqrt{2gh}\sqrt{\frac{2y}{g}}$$
$$= 2\sqrt{hy}$$

Since $h + y = 5$ m,
$$x = 2\sqrt{h(5-h)} = 2\sqrt{5h - h^2}$$

Differentiate x with respect to h.
$$\frac{dx}{dh} = \frac{(2)\left(\frac{1}{2}\right)(5-2h)}{\sqrt{5h - h^2}}$$

Setting $dx/dh = 0$ requires the numerator to be zero.
$$5 \ m - 2h = 0$$
$$h = \frac{5 \ m}{2}$$
$$= 2.5 \ m$$

Answer is A.

14. The hydraulic head at the hole is
$$h = 2 \ m - 0.75 \ m = 1.25 \ m$$
$$v = \sqrt{2gh} = \sqrt{(2)\left(9.81 \ \frac{m}{s^2}\right)(1.25 \ m)}$$
$$= 4.95 \ m/s$$

Answer is D.

15. The specific gas constant for air is
$$R = \frac{\overline{R}}{MW} = \frac{8314 \ \frac{J}{kmol \cdot K}}{29 \ \frac{kg}{kmol}}$$
$$= 287 \ J/kg \cdot K$$

Use the equation of state for an ideal gas.
$$\rho = \frac{p}{RT} = \frac{(97 \ kPa)\left(1000 \ \frac{Pa}{kPa}\right)}{\left(287 \ \frac{J}{kg \cdot K}\right)(300K)}$$
$$= 1.1266 \ kg/m^3 \quad (1.13 \ kg/m^3)$$

Answer is C.

16. Use the conservation of mass law.
$$\dot{m}_1 = \dot{m}_2$$
$$\rho_1 v_1 A_1 = \rho_2 v_2 A_2$$

The density of the entering air was found in Prob. 15 to be $1.1266 \ kg/m^3$.
$$v_1 = \frac{\rho_2 v_2 A_2}{\rho_1 A_1}$$
$$= \frac{\left(1.416 \ \frac{kg}{m^3}\right)\left(10 \ \frac{m}{s}\right)(0.1 \ m^2)}{\left(1.1266 \ \frac{kg}{m^3}\right)(2.5 \ m^2)}$$
$$= 0.503 \ m/s \quad (0.50 \ m/s)$$

(Note that if the density of the air in the constriction had not been specifically given, a reasonable interpretation could have been drawn from the flow velocity. Since the flow velocity is much less than the sonic velocity, the flow probably could be assumed to be incompressible. In that case, the densities would have been the same.)

Answer is B.

17. The ratio of specific heats for air is $k = 1.4$. From Prob. 15, the specific gas constant for air is $287 \ J/kg \cdot K$.

Calculate the speed of sound.
$$c = \sqrt{kRT} = \sqrt{(1.4)\left(287 \ \frac{J}{kg \cdot k}\right)(300K)}$$
$$= 347 \ m/s$$

The Mach number is

$$M = \frac{\text{v}}{c} = \frac{1 \frac{\text{m}}{\text{s}}}{347 \frac{\text{m}}{\text{s}}}$$

$$= 0.00288 \quad (0.0029)$$

Answer is A.

18. The head developed by the pump is

$$h = \frac{\eta P_{\text{ideal}}}{Q\rho g} = \frac{P_{\text{hydraulic}}}{Q\rho g}$$

$$= \frac{625 \times 10^3 \text{ W}}{\left(8.5 \frac{\text{m}^3}{\text{min}}\right)\left(\frac{1}{60 \frac{\text{s}}{\text{min}}}\right)\left(1000 \frac{\text{kg}}{\text{m}^3}\right)\left(9.81 \frac{\text{m}}{\text{s}^2}\right)}$$

$$= 449.7 \text{ m} \quad (450 \text{ m})$$

Answer is D.

19. Since entrance and exit velocities and pressures are the same, all of the pump power goes into raising the elevation and friction.

$$h = \Delta z + h_f$$
$$h_f = h - \Delta z = 449.7 \text{ m} - 350 \text{ m}$$
$$= 99.7 \text{ m} \quad (100 \text{ m})$$

Answer is D.

20. Convert the frictional head loss to specific energy. (m^2/s^2 is the same as J/kg.)

$$E_f = h_f g$$

$$= (99.7 \text{ m})\left(9.81 \frac{\text{m}}{\text{s}^2}\right)$$

$$= 978 \text{ m}^2/\text{s}^2 \quad (980 \text{ J/kg})$$

Water has a specific heat of 4.18 kJ/kg·k. The temperature increase is

$$\Delta T = \frac{E_f}{c_p} = \frac{978 \frac{\text{J}}{\text{kg}}}{4180 \frac{\text{J}}{\text{kg·K}}}$$

$$= 0.234\text{K} \quad (0.23°\text{C})$$

Answer is A.

21. The flow area is

$$A = \frac{\pi}{4}d^2 = \left(\frac{\pi}{4}\right)(0.15 \text{ m})^2$$

$$= 0.0177 \text{ m}^2$$

The density of water at 15°C is approximately 1000 kg/m^3. The flow rate is

$$Q = \frac{\dot{m}}{\rho} = \frac{2750 \frac{\text{kg}}{\text{min}}}{\left(60 \frac{\text{s}}{\text{min}}\right)\left(1000 \frac{\text{kg}}{\text{m}^3}\right)}$$

$$= 0.0458 \text{ m}^3/\text{s}$$

The flow velocity is

$$\text{v} = \frac{Q}{A} = \frac{0.0458 \frac{\text{m}^3}{\text{s}}}{0.0177 \text{ m}^2}$$

$$= 2.59 \text{ m/s}$$

The friction loss is

$$h_f = \frac{fLv^2}{2Dg}$$

$$= \frac{(0.02)(55 \text{ m})\left(2.59 \frac{\text{m}}{\text{s}}\right)^2}{(2)(0.15 \text{ m})\left(9.81 \frac{\text{m}}{\text{s}^2}\right)}$$

$$= 2.51 \text{ m} \quad (2.5 \text{ m})$$

Answer is D.

22. The total head developed by the pump is

$$h = h_z + h_f = 13 \text{ m} + 2.5 \text{ m}$$
$$= 15.5 \text{ m}$$

The pumping power is

$$P = \frac{\dot{m}gh}{\eta}$$

$$= \frac{\left(2750 \frac{\text{kg}}{\text{min}}\right)\left(9.81 \frac{\text{m}}{\text{s}^2}\right)(15.5 \text{ m})}{\left(60 \frac{\text{s}}{\text{min}}\right)(0.76)}$$

$$= 9170 \text{ W} \quad (9.2 \text{ kW})$$

Answer is B.

Topic VIII: Thermodynamics

Diagnostic Examination for Thermodynamics
and Heat Transfer

Chapter 26   Properties of Substances
Chapter 27   First Law of Thermodynamics
Chapter 28   Power Cycles and Entropy
Chapter 29   Mixtures of Gases, Vapors, and Liquids
Chapter 30   Combustion

THERMO

Hint: For the most current information about the exam, visit www.ppi2pass.com/fefaqs.html regularly.
Use the Exam Forum to compare notes with other FE examinees.

Diagnostic Examination

TOPICS VIII and IX: THERMODYNAMICS AND HEAT TRANSFER

TIME LIMIT: 45 MINUTES

---

1. A closed rigid vessel contains 40% liquid water and 60% water vapor by volume. The liquid-vapor mixture is in equilibrium at 150°C. What is the quality of the mixture?

(A) 4.15×10^{-3}
(B) 0.40
(C) 0.60
(D) 0.996

2. A 35 m × 15 m swimming pool is filled with water to a depth of 3 m. How much heat is required to raise the temperature of the water in the pool from 10°C to 25°C?

(A) 99 GJ
(B) 230 GJ
(C) 970 GJ
(D) 1900 GJ

3. A balloon is filled with pure oxygen at 10°C. The balloon is released from the bottom of a water-filled testing pool and floats 30 m to the water surface (where the atmospheric pressure is standard). The initial (submerged) volume of the balloon is 3000 cm³. The temperature is 25°C at the surface of the water, and the water temperature at the bottom of the pool is 10°C. What is the volume of the balloon at the surface after it has come into thermal equilibrium with the surroundings?

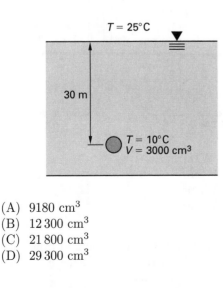

(A) 9180 cm³
(B) 12 300 cm³
(C) 21 800 cm³
(D) 29 300 cm³

4. 2 m³ of an ideal gas are compressed from 100 kPa to 200 kPa. As a result of the process, the internal energy of the gas increases by 10 kJ, and 140 kJ of heat is transferred to the surroundings. How much work was done by the gas during the process?

(A) −150 kJ
(B) −130 kJ
(C) −85 kJ
(D) −45 kJ

5. A cylinder fitted with a frictionless piston contains an ideal gas at temperature T and pressure p. The gas expands isothermally and reversibly until the pressure is $p/3$. Which statement is true regarding the work done by the gas during expansion?

(A) It is equal to the change in enthalpy of the gas.
(B) It is equal to the change in internal energy of the gas.
(C) It is equal to the heat absorbed by the gas.
(D) It is greater than the heat absorbed by the gas.

6. During steady-state operation, the condenser of a steam-power plant receives 7 kg/s of 0.1 MPa steam. Superheated steam enters the condenser at a temperature of 900°C. Saturated cooling water exits the condenser at 100°C. At what rate is heat transferred in the condenser?

(A) −27.8 MW
(B) −13.2 MW
(C) −1.89 MW
(D) −270 kW

7. An ideal heat engine is constructed by cycling 1 kg of oxygen gas (O_2), initially at 1 atm and 25°C, through a sequence of four processes. The sequence consists of isothermal expansion to 2.35 m³, followed by adiabatic expansion to 8.21 m³, followed by isothermal compression to 2.67 m³, and finally, adiabatic compression back to the initial state. All steps are reversible. For oxygen, the specific heat at constant pressure (c_p) is

0.918 kJ/kg·K, and the specific heat at constant volume (c_v) is 0.658 kJ/kg·K. What is the thermal efficiency of this heat engine?

(A) 20%
(B) 40%
(C) 53%
(D) 61%

8. Argon gas (molecular weight of 40) is compressed in a closed system from 100 kPa and 30°C to 500 kPa and 170°C. For argon, the specific heat at constant pressure (c_p) is 0.520 kJ/kg·K, and the specific heat at constant volume (c_v) is 0.312 kJ/kg·K. What is the specific entropy change?

(A) −10.0 kJ/kg·K
(B) −0.14 kJ/kg·K
(C) 0.37 kJ/kg·K
(D) 10.5 kJ/kg·K

9. A gas sample undergoes the closed cycle shown. The difference in internal energy between state B and state A is 600 J. The heat absorption from state B to state C is −1200 J. What is the heat absorption from state A to state B?

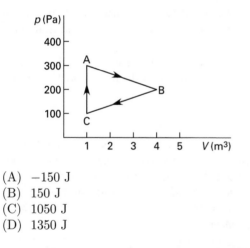

(A) −150 J
(B) 150 J
(C) 1050 J
(D) 1350 J

10. A container is partially filled with a mixture of two volatile liquids. The mixture consists of 2 mol of liquid A and 3 mol of liquid B. The vapor pressures of the pure liquids A and B are 6.0 kPa and 8.5 kPa, respectively. The mixture behaves ideally. After equilibrium, what is the mole fraction of component B in the vapor?

(A) 0.32
(B) 0.59
(C) 0.68
(D) 0.74

11. Moist atmospheric air with a dry-bulb temperature of 30°C and a relative humidity of 40% is cooled under

constant pressure until condensation of the water vapor is observed. What is the final temperature of the air?

(A) 10°C
(B) 16°C
(C) 20°C
(D) 30°C

12. How much energy per unit mass of dry air is released during the cooling process described in Prob. 11?

(A) 0 kJ/kg
(B) 16 kJ/kg
(C) 63 kJ/kg
(D) 100 kJ/kg

13. A small gas turbine burns liquid octane (C_8H_{18}) with 300% excess air. How many moles of nitrogen gas (N_2) in the combustion products are produced per mole of fuel consumed?

(A) 47 N_2
(B) 94 N_2
(C) 141 N_2
(D) 188 N_2

14. A wall is constructed of 25 mm thick white pine boards and insulated with 75 mm thick fiberglass insulation. The thermal conductivities of the two materials are 0.11 W/m·K and 0.048 W/m·K, respectively. On a given day, the internal and external temperatures are 22°C and −10°C, respectively. What is the rate of heat loss per unit area of wall?

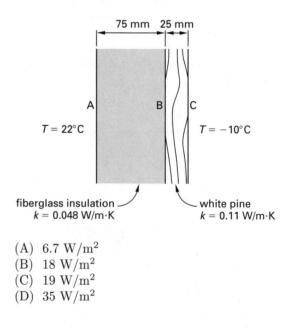

(A) 6.7 W/m²
(B) 18 W/m²
(C) 19 W/m²
(D) 35 W/m²

15. Two coaxial cylinders are produced from a hollow cylinder with a 25 cm radius supported concentrically within a long hollow cylinder of 50 cm radius. The cylinders are at temperatures of 1200°C and 750°C, respectively. Both cylinders display the characteristics of black bodies. Neglecting conduction, convection, and end effects, what is the net radiation exchange per unit length between the two cylinders?

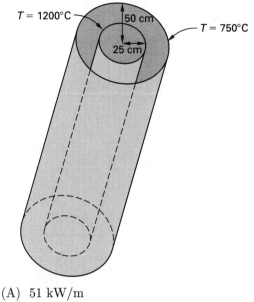

T = 1200°C
50 cm
T = 750°C
25 cm

(A) 51 kW/m
(B) 160 kW/m
(C) 320 kW/m
(D) 640 kW/m

SOLUTIONS TO DIAGNOSTIC EXAMINATION TOPIC VIII

1. The quality, x, is the fraction of the total mass that is vapor.

$$x = \frac{m_g}{m_t} = \frac{m_g}{m_g + m_f}$$

$$= \frac{\dfrac{V_g}{v_g}}{\dfrac{V_g}{v_g} + \dfrac{V_f}{v_f}}$$

$$= \frac{V_g}{V_g + \left(\dfrac{v_g}{v_f}\right) V_f}$$

The specific volumes of saturated liquid water and saturated liquid water vapor at 150°C can be read from the steam table.

$$x = \frac{0.6V}{0.6V + \left(\dfrac{0.3928 \,\dfrac{m^3}{kg}}{0.001091 \,\dfrac{m^3}{kg}}\right)(0.4V)}$$

$$= 4.15 \times 10^{-3}$$

Answer is A.

2. $V = \text{area} \times \text{depth}$
$$= (35 \text{ m})(15 \text{ m})(3 \text{ m}) = 1575 \text{ m}^3$$

$$m = \rho V = \left(1000 \,\frac{kg}{m^3}\right)(1575 \text{ m}^3)$$
$$= 1.575 \times 10^6 \text{ kg}$$

$$Q = mc_p\Delta T = (1.575 \times 10^6 \text{ kg})\left(4.18 \,\frac{kJ}{kg\cdot K}\right)$$
$$\times (25°C - 10°C)$$
$$= 98.8 \times 10^6 \text{ kJ} \quad (99 \text{ GJ})$$

Answer is A.

3. $$p_2 = 1.013 \times 10^5 \text{ Pa}$$

At the bottom of the pool, the absolute pressure is the combination of atmospheric and hydrostatic pressures.

$$p_1 = p_2 + p_{\text{water}}$$
$$= p_2 + \rho g h$$
$$= 1.013 \times 10^5 \text{ Pa} + \left(1000 \,\frac{kg}{m^3}\right)\left(9.81 \,\frac{m}{s^2}\right)(30 \text{ m})$$
$$= 3.956 \times 10^5 \text{ Pa}$$

The volume at the surface can be found from the ideal gas law.

$$V_2 = \frac{p_1 V_1 T_2}{p_2 T_1}$$

$$= \frac{(3.956 \times 10^5 \text{ Pa})(3000 \text{ cm}^3)(25°C + 273)}{(1.013 \times 10^5 \text{ Pa})(10°C + 273)}$$

$$= 12\,337 \text{ cm}^3 \quad (12\,300 \text{ cm}^3)$$

Answer is B.

4. Apply the first law of thermodynamics.

$$W = Q - \Delta U$$

Heat transferred to the surroundings is negative.

$$W = -140 \text{ kJ} - 10 \text{ kJ}$$
$$= -150 \text{ kJ}$$

The minus sign indicates that work was done on the gas.

Answer is A.

5. Use the first law of thermodynamics.

$$Q = \Delta U + W$$

THERMO

Since the internal energy remains constant during an isothermal process, $\Delta U = 0$.

$$Q = W$$

The work done by the gas is equal to the heat absorbed by the gas.

Answer is C.

6. The conservation of energy equation for open systems is

$$\sum \dot{m}_{in}\left(h_{in} + \frac{1}{2}v_{in}^2 + gz_{in}\right) + \dot{Q}$$
$$= \sum \dot{m}_{out}\left(h_{out} + \frac{1}{2}v_{out}^2 + gz_{out}\right) + \dot{W}$$

The mass flow rates are equal.

Potential and kinetic energy terms are negligible.

$$\dot{Q} = \dot{m}(h_{out} - h_{in})$$
$$= \left(7\,\frac{kg}{s}\right)\left(419.04\,\frac{kJ}{kg} - 4396.1\,\frac{kJ}{kg}\right)$$
$$= -27\,839\text{ kW}\quad(-27.8\text{ MW})$$

Answer is A.

7. Reversible adiabatic processes are isentropic processes. The sequence described is a Carnot Cycle. The efficiency depends only on the temperatures.

The initial volume is calculated from the ideal gas law.

$$p_A V_A = n\overline{R}T_A$$

$$V_A = \frac{n\overline{R}T_A}{p_A}$$

$$= \frac{\left(\dfrac{1\text{ kg}}{32\,\frac{kg}{kmol}}\right)\left(8.314\,\frac{kJ}{kmol\cdot K}\right)(25°C + 273)}{101.35\text{ kPa}}$$

$$= 0.764\text{ m}^3$$

For the constant entropy step between point D and point A,

$$T_A V_A^{k-1} = T_D V_D^{k-1}$$

$$\frac{T_D}{T_A} = \left(\frac{V_A}{V_D}\right)^{k-1} = \left(\frac{V_A}{V_D}\right)^{\frac{c_p}{c_v}-1}$$

$$= \left(\frac{0.764\text{ m}^3}{2.67\text{ m}^3}\right)^{\frac{0.918\,\frac{kJ}{kg\cdot K}}{0.658\,\frac{kJ}{kg\cdot K}}-1}$$

$$= 0.610$$

The efficiency of a heat engine operating on the Carnot Cycle is

$$\eta_{th,\text{Carnot}} = \frac{T_H - T_L}{T_H} = 1 - \frac{T_D}{T_A}$$
$$= 1 - 0.610$$
$$= 0.39\quad(40\%)$$

Answer is B.

8.

$$\Delta s = c_v \ln\left(\frac{T_2}{T_1}\right) + R\ln\left(\frac{V_2}{V_1}\right)$$

$$= c_v \ln\left(\frac{T_2}{T_1}\right) + \left(\frac{\overline{R}}{MW}\right)\ln\left(\frac{p_1 T_2}{p_2 T_1}\right)$$

$$= \left(\left(0.312\,\frac{kJ}{kg\cdot K}\right)\ln\left(\frac{170°C + 273}{30°C + 273}\right)\right)$$

$$+ \left(\left(\frac{8.314\,\frac{kJ}{kmol\cdot K}}{40\,\frac{kg}{kmol}}\right)\right.$$

$$\left. \times \ln\left(\frac{(100\text{ kPa})(170°C + 273)}{(500\text{ kPa})(30°C + 273)}\right)\right)$$

$$= -0.137\text{ kJ/kg·K}\quad(-0.14\text{ kJ/kg·K})$$

Answer is B.

9. The work done by the gas between state A and state B is equal to the area under the p-V diagram between those two points.

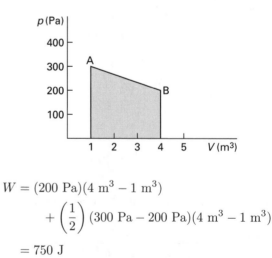

$$W = (200\text{ Pa})(4\text{ m}^3 - 1\text{ m}^3)$$
$$+ \left(\frac{1}{2}\right)(300\text{ Pa} - 200\text{ Pa})(4\text{ m}^3 - 1\text{ m}^3)$$
$$= 750\text{ J}$$

The heat absorbed by the gas can be calculated from the first law.

$$Q = \Delta U + W = 600\text{ J} + 750\text{ J}$$
$$= 1350\text{ J}$$

Answer is D.

10. The mole fractions are

$$x_A = \frac{N_A}{N} = \frac{2}{5}$$
$$= 0.4$$

$$x_B = 1 - x_A = 1 - 0.4$$
$$= 0.6$$

Use Raoult's law.

$$p_A = p_A^* x_A = (6\text{ kPa})(0.4)$$
$$= 2.4\text{ kPa}$$

$$p_B = p_B^* x_B = (8.5\text{ kPa})(0.6)$$
$$= 5.1\text{ kPa}$$

$$p = p_A + p_B = 2.4\text{ kPa} + 5.1\text{ kPa}$$
$$= 7.5\text{ kPa}$$

The mole fraction of component B in the vapor can be determined from Henry's law.

$$y_B = \frac{p_B}{p} = \frac{5.1\text{ kPa}}{7.5\text{ kPa}}$$
$$= 0.68$$

Answer is C.

11. Cooling is represented by a horizontal line (constant humidity ratio) on the psychrometric chart. From the intersection of the curve corresponding to 40% relative humidity and the line corresponding to 30°C, follow the psychrometric chart horizontally to the saturation line. At that point, the dry-bulb, wet-bulb, and dew-point temperatures are all the same.

$$T_{dp} \approx 16°C$$

Answer is B.

12. The energy released during cooling is equal to the change in enthalpies between the two states. The enthalpy values can be read from the psychrometric chart. The minus sign indicates that heat is transferred to the surroundings.

$$Q = h_2 - h_1 = 42\ \frac{\text{kJ}}{\text{kg}} - 58\ \frac{\text{kJ}}{\text{kg}}$$
$$= -16\text{ kJ/kg}\quad (16\text{ kJ/kg})$$

Answer is B.

13. There are approximately 3.76 volumes of nitrogen for every volume of oxygen. Write the unbalanced combustion reaction.

$$C_8H_{18} + (O_2 + 3.76N_2) \rightarrow CO_2 + H_2O + N_2$$

Balance the combustion reaction.

$$C_8H_{18} + (12.5)(O_2 + 3.76N_2) \rightarrow$$
$$8CO_2 + 9H_2O + (12.5)(3.76N_2)$$

Add excess air to the equation. With 300% excess air, the total air is four times the stoichiometric amount.

$$C_8H_{18} + (4)(12.5)(O_2 + 3.76N_2) \rightarrow$$
$$8CO_2 + 9H_2O + (12.5)(3.76N_2)$$
$$+ (3)(12.5)(O_2 + 3.76N_2)$$

Simplifying,

$$C_8H_{18} + (50)(O_2 + 3.76N_2) \rightarrow$$
$$8CO_2 + 9H_2O + 37.5O_2 + 188N_2$$

Answer is D.

14. For a composite wall,

$$\dot{Q} = \frac{-(T_C - T_A)}{R_1 + R_2}$$
$$= \frac{-(T_C - T_A)}{\dfrac{L_1}{k_1 A} + \dfrac{L_2}{k_2 A}}$$

$$\frac{\dot{Q}}{A} = \frac{-(T_C - T_A)}{\dfrac{L_1}{k_1} + \dfrac{L_2}{k_2}}$$
$$= \frac{-(-10°C - 22°C)}{\dfrac{0.075\text{ m}}{0.048\ \frac{\text{W}}{\text{m·K}}} + \dfrac{0.025\text{ m}}{0.11\ \frac{\text{W}}{\text{m·K}}}}$$
$$= 17.88\text{ W/m}^2\quad (18\text{ W/m}^2)$$

Answer is B.

15.
$$\dot{Q}_{12} = \sigma A_1 F_{12}(T_1^4 - T_2^4)$$
$$= \sigma(2\pi r_1)L F_{12}(T_1^4 - T_2^4)$$

$$\frac{\dot{Q}_{12}}{L} = \sigma(2\pi r_1)F_{12}(T_1^4 - T_2^4)$$

The shape factor, F_{12}, for a black body enclosed completely within another black body is equal to 1.

$$\frac{\dot{Q}_{12}}{L} = \left(5.670 \times 10^{-8}\ \frac{\text{W}}{\text{m}^2\text{·K}^4}\right) 2\pi(0.25\text{ m})(1)$$
$$\times ((1200°C + 273)^4 - (750°C + 273)^4)$$
$$= 321\,744\text{ W/m}\quad (320\text{ kW/m})$$

Answer is C.

26 Properties of Substances

Subjects

PHASES OF A PURE SUBSTANCE 26-1
STATE FUNCTIONS (PROPERTIES) ... 26-3
    Mass 26-4
    Pressure 26-4
    Temperature 26-4
    Specific Volume 26-4
    Internal Energy 26-4
    Enthalpy 26-5
    Entropy 26-5
    Gibbs' Function 26-5
    Helmholtz Function 26-5
    Heat Capacity (Specific Heat) 26-5
LIQUID-VAPOR MIXTURES 26-6
IDEAL GASES 26-7
SPEED OF SOUND AND
    MACH NUMBER 26-8
    Air 26-8

Nomenclature

a	Helmholtz function	Btu/lbm	kJ/kg
A	molar Helmholtz function	Btu/lbmol	kJ/kmol
c	specific heat	Btu/lbm-°F	kJ/kg·K
C	molar specific heat	Btu/lbmol-°F	kJ/kmol·K
g	Gibbs' function	Btu/lbm	kJ/kg
g_c	gravitational constant	ft-lbm/lbm-sec^2	–
G	molar Gibbs' function	Btu/lbmol	kJ/kmol
h	enthalpy	Btu/lbm	kJ/kg
H	molar enthalpy	Btu/lbmol	kJ/kmol
J	Joule's constant (778)	ft-lbf/Btu	n.a.
k	ratio of specific heats	–	–
m	mass	lbm	kg
M	Mach number	–	–
MW	molecular weight	lbm/lbmol	kg/kmol
n	number of moles	–	–
N_A	Avogadro's number (6.022×10^{23})	–	1/mol
p	pressure	lbf/in^2	Pa
q	heat	Btu/lbm	kJ/kg
R	specific gas constant	ft-lbf/lbm-°R	kJ/kg·K
$\overline{R}$	universal gas constant	ft-lbf/lbmol-°R	kJ/kmol·K
s	entropy	Btu/lbm-°R	kJ/kg·K
S	molar entropy	Btu/lbmol-°R	kJ/kmol·K
T	absolute temperature	°R	K
u	internal energy	Btu/lbm	kJ/kg
U	molar internal energy	Btu/lbmol	kJ/kmol
v	velocity	ft/sec	m/s
V	volume	ft^3	m^3
V	molar specific volume	ft^3/lbmol	m^3/kmol
x	quality	–	–

Symbols

ρ	density	lbm/ft^3	kg/m^3
υ	specific volume	ft^3/lbm	m^3/kg

Subscripts

f	fluid (liquid)
fg	liquid-to-gas (vaporization)
g	gas (vapor)
l	liquid
p	constant pressure
v	constant volume

PHASES OF A PURE SUBSTANCE

Thermodynamics is the study of a substance's energy-related properties. The properties of a substance and the procedures used to determine those properties depend on the state and the phase of the substance. The thermodynamic *state* of a substance is defined by two or more independent thermodynamic properties. For example, the temperature and pressure of a substance are two properties commonly used to define the state of a superheated vapor.

The common *phases* of a substance are solid, liquid, and gas. However, because substances behave according to different rules, it is convenient to categorize them into more than only these three phases.

THERMO
Properties

Solid: A solid does not take on the shape or volume of its container.

Saturated liquid: A saturated liquid has absorbed as much heat energy as it can without vaporizing. Liquid water at standard atmospheric pressure and 212°F (100°C) is an example of a saturated liquid.

Subcooled liquid: If a liquid is not saturated (i.e., the liquid is not at its boiling point), it is said to be subcooled. Water at 1 atm and room temperature is subcooled, as it can absorb additional energy without vaporizing.

Liquid-vapor mixture: A liquid and vapor of the same substance can coexist at the same temperature and pressure. This is called a two-phase, liquid-vapor mixture.

Perfect gas: A perfect gas is an ideal gas whose specific heats (and hence ratio of specific heats) are constant.

Saturated vapor: A vapor (e.g., steam at standard atmospheric pressure and 212°F (100°C)) that is on the verge of condensing is said to be saturated.

Superheated vapor: A superheated vapor is one that has absorbed more energy than is needed merely to vaporize it. A superheated vapor will not condense when small amounts of energy are removed.

Ideal gas: A gas is a highly superheated vapor. If the gas behaves according to the ideal gas law, $pV = nRT$, it is called an ideal gas.

Real gas: A real gas does not behave according to the ideal gas laws.

Gas mixtures: Most gases mix together freely. Two or more pure gases together constitute a gas mixture.

Vapor-gas mixtures: Atmospheric air is an example of a mixture of several gases and water vapor.

It is theoretically possible to develop a three-dimensional surface that predicts a substance's phase based on the properties of pressure, temperature, and specific volume. Such a three-dimensional p-v-T diagram is illustrated in Fig. 26.1.

If one property is held constant during a process, a two-dimensional projection of the p-v-T diagram can be used. Figure 26.2 is an example of this projection, which is known as an *equilibrium diagram* or a *phase diagram*.

The most important part of a phase diagram is limited to the liquid-vapor region. A general phase diagram showing this region and the bell-shaped dividing line (known as the *vapor dome*) is shown in Fig. 26.3.

The vapor dome region can be drawn with many variables for the axes. For example, either temperature

Figure 26.1 Three-Dimensional p-v-T Phase Diagram

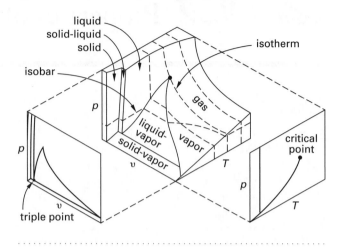

Figure 26.2 Pressure-Volume Phase Diagram

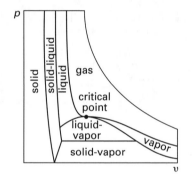

Figure 26.3 Vapor Dome with Isobars

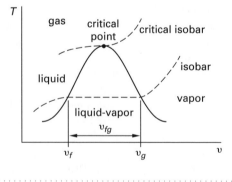

or pressure can be used for the vertical axis. Internal energy, enthalpy, specific volume, or entropy can be chosen for the horizontal axis. However, the principles presented here apply to all combinations.

The left-hand part of the vapor dome curve separates the liquid phase from the liquid-vapor phase. This part of the line is known as the *saturated liquid line*. Similarly, the right-hand part of the line separates the liquid-vapor phase from the vapor phase. This line is called the *saturated vapor line*.

Lines of constant pressure (*isobars*) can be superimposed on the vapor dome. Each isobar is horizontal as it passes through the two-phase region, verifying that both temperature and pressure remain unchanged as a liquid vaporizes.

Notice that there is no dividing line between liquid and vapor at the top of the vapor dome. Above the vapor dome, the phase is a gas.

The implicit dividing line between liquid and gas is the isobar that intersects the top-most part of the vapor dome. This is known as the *critical isobar*, and the highest point of the vapor dome is known as the *critical point*. This critical isobar also provides a way to distinguish between a vapor and a gas. A substance below the critical isobar (but to the right of the vapor dome) is a vapor. Above the critical isobar, it is a gas.

The *triple point* of a substance is a unique state at which solid, liquid, and gaseous phases can coexist. For instance, the triple point of water occurs at a pressure of 0.00592 atm and a temperature of 491.71°R (273.16K).

Figure 26.4 illustrates a vapor dome for which pressure has been chosen as the vertical axis and enthalpy has been chosen as the horizontal axis. The shape of the dome is essentially the same, but the lines of constant temperature (*isotherms*) have slopes of different signs than the isobars.

Figure 26.4 also illustrates the subscripting convention used to identify points on the saturation line. The subscript f (fluid) is used to indicate a saturated liquid. The subscript g (gas) is used to indicate a saturated vapor. The subscript fg is used to indicate the difference in saturation properties.

Figure 26.4 Vapor Dome with Isotherms

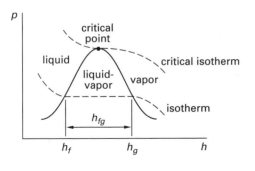

The vapor dome is a good tool for illustration, but it cannot be used to determine a substance's phase. Such a determination must be made based on the substance's pressure and temperature according to the following rules.

rule 1: A substance is a subcooled liquid if its temperature is less than the saturation temperature corresponding to its pressure.

rule 2: A substance is in the liquid-vapor region if its temperature is equal to the saturation temperature corresponding to its pressure.

rule 3: A substance is a superheated vapor if its temperature is greater than the saturation temperature corresponding to its pressure.

rule 4: A substance is a subcooled liquid if its pressure is greater than the saturation pressure corresponding to its temperature.

rule 5: A substance is in the liquid-vapor region if its pressure is equal to the saturation pressure corresponding to its temperature.

rule 6: A substance is a superheated vapor if its pressure is less than the saturation pressure corresponding to its temperature.

STATE FUNCTIONS (PROPERTIES)

The thermodynamic state or condition of a substance is determined by its properties. *Intensive properties* are independent of the amount of substance present. Temperature, pressure, and stress are examples of intensive properties. *Extensive properties* are dependent on (i.e., are proportional to) the amount of substance present. Examples are volume, strain, charge, and mass.

In most books on thermodynamics, both lowercase and uppercase forms of the same characters are used to represent property variables. The two forms are used to distinguish between the units of mass. For example, lowercase h represents specific enthalpy (usually called "enthalpy") in units of Btu/lbm or kJ/kg. Uppercase H is used to represent the molar enthalpy in units of Btu/lbmol or kJ/kmol.

Properties of gases in tabulated form are useful or necessary for solving many thermodynamic problems. The properties of saturated and superheated steam are tabulated in Tables 26.5 and 26.6 at the end of this chapter. h-p diagrams for refrigerant HFC-134 in customary U.S. and SI units are presented in Figs. 26.6 and 26.7 at the end of this chapter.

Mass

The mass, m, of a substance is a measure of its quantity. Mass is independent of location and gravitational field strength. In thermodynamics, the customary U.S. and SI units of mass are pound-mass (lbm) and kilogram (kg), respectively.

Pressure

Customary U.S. pressure units are pounds per square inch (lbf/in^2). Standard SI pressure units are kPa or MPa, although bars are also used in tabulations of thermodynamic data. (1 bar = 10^5 Pa.)

Most pressure gauges read atmospheric pressures, but in general, thermodynamic calculations will be performed using absolute pressures. The values of a standard atmosphere in various units are given in Table 26.1.

Table 26.1 Standard Atmospheric Pressure

1.000 atm	(atmosphere)
14.696 psia	(pounds per square inch absolute)
2116.2 psfa	(pounds per square foot absolute)
407.1 in w.g.	(inches of water; inches water gage)
33.93 ft w.g.	(feet of water; feet water gage)
29.921 in Hg	(inches of mercury)
760.0 mm Hg	(millimeters of mercury)
760.0 torr	
1.013 bars	
1013 millibars	
1.013×10^5 Pa	(pascals)
101.3 kPa	(kilopascals)

Temperature

Temperature is a thermodynamic property of a substance that depends on energy content. Heat energy entering a substance will increase the temperature of that substance. Normally, heat energy will flow only from a hot object to a cold object. If two objects are in *thermal equilibrium* (are at the same temperature), no heat energy will flow between them.

If two systems are in thermal equilibrium, they must be at the same temperature. If both systems are in equilibrium with a third system, then all three systems are at the same temperature. This concept is known as the *Zeroth Law of Thermodynamics*.

The *absolute temperature* scale defines temperature independently of the properties of any particular substance. This is unlike the Celsius and Fahrenheit scales, which are based on the freezing point of water. The absolute temperature scale should be used for all thermodynamic calculations.

In the customary U.S. system, the absolute temperature scale is the *Rankine scale*.

$$T_{\circ R} = T_{\circ F} + 459.67 \qquad 26.1$$

$$\Delta T_{\circ R} = \Delta T_{\circ F} \qquad 26.2$$

The absolute temperature scale in SI is the *Kelvin scale*.

$$T_K = T_{\circ C} + 273.15 \qquad 26.3$$

$$\Delta T_K = \Delta T_{\circ C} \qquad 26.4$$

The relationships between the four temperature scales are illustrated in Fig. 26.5, which also defines the approximate *boiling point, triple point, ice point,* and *absolute zero* temperatures.

Figure 26.5 Temperature Scales

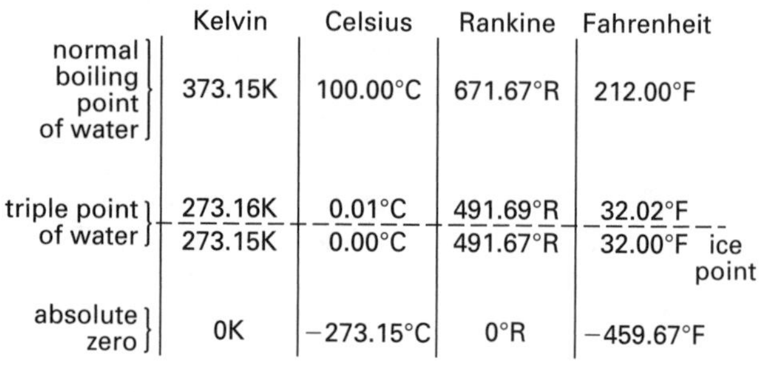

	Kelvin	Celsius	Rankine	Fahrenheit
normal boiling point of water	373.15K	100.00°C	671.67°R	212.00°F
triple point of water	273.16K	0.01°C	491.69°R	32.02°F
	273.15K	0.00°C	491.67°R	32.00°F — ice point
absolute zero	0K	−273.15°C	0°R	−459.67°F

Specific Volume

Specific volume, v, is the volume occupied by one unit mass of a substance. Customary U.S. units in tabulations of thermodynamic data are cubic feet per pound (ft^3/lbm). Standard SI specific volume units are cubic meters per kilogram (m^3/kg). Molar specific volume has units of ft^3/lbmole (m^3/kmol) and is seldom encountered. Specific volume is the reciprocal of density.

$$v = \frac{1}{\rho} \qquad 26.5$$

$$V = (MW) \times v \qquad 26.6$$

Internal Energy

Internal energy accounts for all of the energy of the substance excluding pressure, potential, and kinetic energy. The internal energy is a function of the state of a system. Examples of internal energy are the translational, rotational, and vibrational energies of the molecules and atoms in the substance. Since the movement of atoms and molecules increases with temperature, internal energy is a function of temperature. It does not depend on the process or path taken to reach a particular temperature.

In the United States, the *British thermal unit* (Btu) is used in thermodynamics to represent the various forms of energy. (One Btu is approximately the energy given off by burning one wooden match.) Standard units of specific internal energy (u) are Btu/lbm and kJ/kg. The units of molar internal energy (U) are Btu/lbmol and kJ/kmol. Equation 26.7 gives the relationship between the specific and molar quantities.

$$U = (MW) \times u \qquad 26.7$$

Enthalpy

Enthalpy represents the total useful energy of a substance. Useful energy consists of two parts: the internal energy, u, and the *flow energy* (also known as *flow work* and *p-V work*), pV. Therefore, enthalpy has the same units as internal energy.

$$h = u + pv \qquad 26.8$$

$$H = U + pV \qquad 26.9$$

$$H = (MW) \times h \qquad 26.10$$

Enthalpy is defined as useful energy because, ideally, all of it can be used to perform useful tasks. It takes energy to increase the temperature of a substance. If that internal energy is recovered, it can be used to heat something else (e.g., to vaporize water in a boiler). Also, it takes energy to increase pressure and volume (as in blowing up a balloon). If pressure and volume are decreased, useful energy is given up.

Strictly speaking, the customary U.S. units of Eq. 26.8 and 26.9 are not consistent, since flow work (as written) has units of ft-lbf/lbm, not Btu/lbm. (There is also a consistency problem if pressure is defined in lbf/ft^2 and given in lbf/in^2.) Therefore, Eq. 26.8 should be written as

$$h = u + \frac{pv}{J} \qquad 26.11$$

The conversion factor, J, in Eq. 26.11 is known as *Joule's constant*. It has a value close to 778.17 ft-lbf/Btu (which is often shortened to 778 to maintain three significant digits in calculations). (In SI units, Joule's constant has a value of 1.0 N·m/J and is unnecessary.) As in Eqs. 26.8 and 26.9, Joule's constant is often omitted from the statement of generic thermodynamic equations, but it is always needed with customary U.S. units for dimensional consistency.

Entropy

Entropy is a measure of the energy that is no longer available to perform useful work within the current environment. Other definitions (the "disorder of the system," the "randomness of the system," etc.) are frequently quoted. Although these alternate definitions

cannot be used in calculations, they are consistent with the third law of thermodynamics (also known as the *Nernst theorem*). This law states that the absolute entropy of a perfect crystalline solid in thermodynamic equilibrium is (approaches) zero when the temperature is (approaches) absolute zero. Equation 26.12 expresses the third law mathematically.

$$\lim_{T \to 0} s = 0 \qquad 26.12$$

An increase in entropy is known as *entropy production*. The total entropy in a system is equal to the summation of all entropy productions that have occurred over the life of the system.

$$s = \sum \Delta s \qquad 26.13$$

$$\Delta s = \int ds = \int \frac{dq_{\text{reversible}}}{T} \qquad 26.14$$

The units of specific entropy are Btu/lbm-°R and kJ/kg·K. For molar entropy, the units are Btu/lbmol-°R and kJ/kmol·K.

In practice, entropy is not referenced to absolute zero conditions, but is measured with respect to some other convenient thermodynamic state. For water, the reference condition is the liquid phase at the triple point.

Gibbs' Function

Gibbs' function for a pure substance is defined by Eqs. 26.15 through 26.17.

$$g = h - Ts = u + pv - Ts \qquad 26.15$$

$$G = H - TS = U + pV - TS \qquad 26.16$$

$$G = (MW) \times g \qquad 26.17$$

Gibbs' function is used in investigating latent heat changes and chemical reactions.

Helmholtz Function

The Helmholtz function is defined for a pure substance by Eqs. 26.18 through 26.20.

$$a = u - Ts = h - pv - Ts \qquad 26.18$$

$$A = U - TS = H - pV - TS \qquad 26.19$$

$$A = (MW) \times a \qquad 26.20$$

Like the Gibbs function, the Helmholtz function is used in investigating chemical reactions.

Heat Capacity (Specific Heat)

An increase in internal energy is needed to cause a rise in temperature. Different substances differ in the quantity of heat needed to produce a given temperature increase. The ratio of heat energy, Q, required to change the temperature of a mass, m, by an amount ΔT is called the *specific heat (heat capacity) of the substance*, c. Because specific heats of solids and liquids are slightly temperature dependent, the mean specific heats are used for processes covering large temperature ranges.

$$Q = mc\Delta T \qquad 26.21$$

$$c = \frac{Q}{m\Delta T} \qquad 26.22$$

The lowercase c implies that the units are Btu/lbm-°R or J/kg·K. The molar specific heat, designated by the symbol C, has units of Btu/lbmol-°F or J/kmol·K.

$$C = (MW) \times c \qquad 26.23$$

For gases, the specific heat depends on the type of process during which the heat exchange occurs. Specific heats for constant-volume and constant-pressure processes are designated by c_v and c_p, respectively.

$$Q = mc_v\Delta T \quad \text{[constant-volume process]} \qquad 26.24$$

$$Q = mc_p\Delta T \quad \text{[constant-pressure process]} \qquad 26.25$$

c_p and c_v for solids and liquids are essentially the same and are given in Table 26.2. Approximate values of c_p and c_v for common gases are given in Table 26.3.

LIQUID-VAPOR MIXTURES

Water is at its saturation pressure and temperature within the vapor dome. There are an infinite number of thermodynamic states in which the water can simultaneously exist in liquid and vapor phases. The *quality* is the fraction by weight of the total mass that is vapor.

$$x = \frac{m_{\text{vapor}}}{m_{\text{vapor}} + m_{\text{liquid}}}$$
$$= \frac{m_g}{m_g + m_f} \qquad 26.26$$

When the thermodynamic state of a substance is within the vapor dome, there is a one-to-one correspondence between the saturation temperature and saturation pressure. Knowing one determines the other. The thermodynamic state is uniquely defined by any two independent properties (temperature and quality, pressure and enthalpy, entropy and quality, etc.).

If the quality of a liquid-vapor mixture is known, it can be used to calculate all of the primary thermodynamic properties. If a thermodynamic property has a value

Table 26.2 Approximate Specific Heats of Selected Liquids and Solids

liquids and solids	c_p Btu/lbm-°R	c_p kJ/kg·K
aluminum, pure	0.23	0.960
aluminum, 2024-T4	0.2	0.840
ammonia	1.16	4.860
asbestos	0.20	0.840
benzene	0.41	1.720
brass, red	0.093	0.390
bronze	0.082	0.340
concrete	0.21	0.880
copper, pure	0.094	0.390
Freon-12	0.24	1.000
gasoline	0.53	2.200
glass	0.18	0.750
gold, pure	0.031	0.130
ice	0.49	2.050
iron, pure	0.11	0.460
iron, cast (4% C)	0.10	0.420
lead, pure	0.031	0.130
magnesium, pure	0.24	1.000
mercury	0.033	0.140
oil, light hydrocarbon	0.5	2.090
silver, pure	0.06	0.250
steel, 1010	0.10	0.420
steel, stainless 301	0.11	0.460
tin, pure	0.055	0.230
titanium, pure	0.13	0.540
tungsten, pure	0.032	0.130
water	1.0	4.190
wood (typical)	0.6	2.500
zinc, pure	0.088	0.370

(Multiply Btu/lbm-°R by 4.1868 to obtain kJ/kg·K. Values in cal/g-°C are the same as Btu/lbm-°R.)

between the saturated liquid and saturated vapor values (i.e., h is between h_f and h_g), any of the Eqs. 26.27 through 26.31 can be solved for the quality.

$$h = h_f + xh_{fg} \qquad 26.27$$

$$h_{fg} = h_g - h_f \qquad 26.28$$

$$s = s_f + xs_{fg} \qquad 26.29$$

$$u = u_f + xu_{fg} \qquad 26.30$$

$$v = v_f + xv_{fg} \qquad 26.31$$

IDEAL GASES

A gas can be considered to behave ideally if its pressure is very low or the temperature is much higher than

Table 26.3 Approximate Specific Heats of Selected Gases

gas	symbol	c_p Btu/lbm-°R	c_p J/kg·K	c_v Btu/lbm-°R	c_v J/kg·K
acetylene	C_2H_2	0.350	1465	0.274	1146
air		0.240	1005	0.171	718
ammonia	NH_3	0.523	2190	0.406	1702
argon	Ar	0.124	519	0.074	311
butane (n)	C_4H_{10}	0.395	1654	0.361	1511
carbon dioxide	CO_2	0.207	867	0.162	678
carbon monoxide	CO	0.249	1043	0.178	746
chlorine	Cl_2	0.115	481	0.087	364
ethane	C_2H_6	0.386	1616	0.320	1340
ethylene	C_2H_4	0.400	1675	0.329	1378
Freon-12[a]	CCl_2F_2	0.159	666	0.143	597
helium	He	1.240	5192	0.744	3115
hydrogen	H_2	3.420	14319	2.435	10195
hydrogen sulfide	H_2S	0.243	1017	0.185	773
krypton	Kr	0.059	247	0.035	148
methane	CH_4	0.593	2483	0.469	1965
neon	Ne	0.248	1038	0.150	626
nitrogen	N_2	0.249	1043	0.178	746
nitric oxide	NO	0.231	967	0.165	690
nitrous oxide	NO_2	0.221	925	0.176	736
octane	C_8H_{18}	0.407	1704	0.390	1631
oxygen	O_2	0.220	921	0.158	661
propane	C_3H_8	0.393	1645	0.348	1457
sulfur dioxide	SO_2	0.149	624	0.118	494
water vapor[a]	H_2O	0.445	1863	0.335	1402
xenon	Xe	0.038	159	0.023	96

[a] Values for steam and Freon are approximate and should be used only for low pressures and high temperatures.

its critical temperature. (Otherwise, the substance is in vapor form.) Under these conditions, the molecule size is insignificant compared with the distance between molecules, and molecules do not interact. By definition, an ideal gas behaves according to the various ideal gas laws.

An *equation of state* is a relationship that predicts the state (i.e., a property, such as pressure, temperature, volume, etc.) from a set of two other independent properties.

Avogadro's law states that equal volumes of different gases at the same temperature and pressure contain equal numbers of molecules. Avogadro's number, $N_A = 6.022 \times 10^{23}$, is the number of molecules of an ideal gas in one gram-mole (mol or gmol). For one mole of any gas, Avogadro's law can be stated as the *equation of state* for ideal gases. Temperature, T, in Eq. 26.32 must be in degrees absolute.

$$pV = \overline{R}T \qquad 26.32$$

In Eq. 26.32, $\overline{R}$ is known as the *universal gas constant*. It is "universal" (within a system of units) because the same value can be used with any gas. Its value depends on the units used for pressure, temperature, and volume, as well as on the units of mass. Values of the universal gas constant in various units are given in Table 26.4.

Table 26.4 Values of the Universal Gas Constant, $\overline{R}$

<u>units in SI and metric systems</u>

8.3143 kJ/kmol·K
8314.3 J/kmol·K
0.08206 atm·l/mol·K
1.986 cal/mol·K
8.314 J/mol·K
82.06 atm·cm^3/mol·K
0.08206 atm·m^3/kmol·K
8314.3 kg·m^2/s^2·kmol·K
8314.3 m^3·Pa/kmol·K
8.314×10^7 erg/mol·K

<u>units in English systems</u>

1545.33 ft-lbf/lbmol-°R
1.986 Btu/lbmol-°R
0.7302 atm-ft^3/lbmol-°R
10.73 ft^3-lbf/in^2-lbmol-°R

Since $\overline{R}$ is a constant, it follows that the quantity pV/T is constant for an ideal gas undergoing any process.

$$\frac{p_1 V_1}{T_1} = \frac{p_2 V_2}{T_2} \qquad 26.33$$

The ideal gas equation of state can be modified for more than one mole of gas. If there are n moles, then

$$pV = n\overline{R}T \qquad 26.34$$

The number of moles can be calculated from the substance's mass and molecular weight.

$$n = \frac{m}{\mathrm{MW}} \qquad 26.35$$

Equations 26.34 and 26.35 can be combined. R is the *specific gas constant*. It is specific because it is valid only for a gas with a molecular weight of MW.

$$pV = \frac{m\overline{R}T}{\mathrm{MW}} = m\left(\frac{\overline{R}}{\mathrm{MW}}\right)T$$

$$= mRT \qquad 26.36$$

$$R = \frac{\overline{R}}{\mathrm{MW}} \qquad 26.37$$

In terms of specific volume,

$$pv = RT \qquad\qquad 26.38$$

The specific heats of an ideal gas can be calculated from its specific gas constant.

$$c_p - c_v = R \qquad\qquad 26.39$$

Some relations for determining property changes in an ideal gas do not depend on the type of process. Changes in enthalpy, internal energy, and entropy are independent of the process.

$$\Delta h = c_p \Delta T \qquad\qquad 26.40$$

$$\Delta u = c_v \Delta T \qquad\qquad 26.41$$

$$\Delta s = c_p \ln\left(\frac{T_2}{T_1}\right) - R \ln\left(\frac{p_2}{p_1}\right) \qquad 26.42$$

$$\Delta s = c_v \ln\left(\frac{T_2}{T_1}\right) + R \ln\left(\frac{v_2}{v_1}\right) \qquad 26.43$$

The following relationships are valid for ideal gases undergoing *isentropic processes* (i.e., entropy is constant). The *ratio of specific heats*, k, is given by Eq. 26.47.

$$p_1 v_1^k = p_2 v_2^k \qquad\qquad 26.44$$

$$T_1 v_1^{k-1} = T_2 v_2^{k-1} \qquad\qquad 26.45$$

$$T_1 p_1^{(1-k)/k} = T_2 p_2^{(1-k)/k} \qquad\qquad 26.46$$

$$k = \frac{c_p}{c_v} \qquad\qquad 26.47$$

SPEED OF SOUND AND MACH NUMBER

The *speed of sound*, c, in a fluid is a function of its bulk modulus, or equivalently, of its compressibility. Equation 26.48 gives the speed of sound in an ideal gas. The temperature, T, must be in degrees absolute (i.e., °R or K).

$$c = \sqrt{kRT} \qquad \text{[SI]} \quad 26.48a$$

$$c = \sqrt{kg_cRT} \qquad \text{[U.S.]} \quad 26.48b$$

The *Mach number* of an object is the ratio of the object's speed to the speed of sound in the medium through which the object is traveling.

$$M = \frac{v}{c} \qquad\qquad 26.49$$

Air

For air, the ratio of specific heats is $k = 1.40$ and the molecular weight is 29.0. The specific gas constant is $R = 53.3$ ft-lbf/lbm-°R (287 J/kg·K).

SAMPLE PROBLEMS

Problem 1

What is the enthalpy of HFC-134a at 0.4 MPa and 85% quality?

   (A) 241 kJ/kg
   (B) 261 kJ/kg
   (C) 333 kJ/kg
   (D) 375 kJ/kg

Solution

Read approximate values of h_f and h_g from the HFC-134a p-h diagram (at the end of this chapter) at 0.4 MPa.

$$\begin{aligned}
h &= h_f + x h_{fg} \\
&= h_f + x(h_g - h_f) \\
&= (1-x)h_f + x h_g \\
&= (1 - 0.85)\left(212 \ \frac{\text{kJ}}{\text{kg}}\right) + (0.85)\left(404 \ \frac{\text{kJ}}{\text{kg}}\right) \\
&= 375.2 \ \text{kJ/kg} \quad (375 \ \text{kJ/kg})
\end{aligned}$$

The diagram also has quality isobars from which the enthalpy can be read directly.

Answer is D.

Problem 2

A fluid has a mass of 5 kg and occupies a volume of 1 m³ at a pressure of 150 kPa. If the internal energy is 2500 kJ/kg, what is the total enthalpy?

   (A) 2.5 MJ
   (B) 9.8 MJ
   (C) 12.5 MJ
   (D) 12.7 MJ

Solution

$$v = \frac{V}{m} = \frac{1 \ \text{m}^3}{5 \ \text{kg}} = 0.2 \ \text{m}^3/\text{kg}$$

$$\begin{aligned}
h &= u + pv \\
&= 2500 \ \frac{\text{kJ}}{\text{kg}} + \left(150 \ \frac{\text{kN}}{\text{m}^2}\right)\left(0.2 \ \frac{\text{m}^3}{\text{kg}}\right) \\
&= 2530 \ \text{kJ/kg}
\end{aligned}$$

$$\begin{aligned}
H &= mh = (5 \ \text{kg})\left(2530 \ \frac{\text{kJ}}{\text{kg}}\right) \\
&= 12\,650 \ \text{kJ} \quad (12.7 \ \text{MJ})
\end{aligned}$$

Answer is D.

Problem 3

An ideal gas at a gage pressure of 0.3 MPa and 25°C is heated in a closed container to 75°C. What is the final pressure?

(A) 0.35 MPa
(B) 0.47 MPa
(C) 0.90 MPa
(D) 1.20 MPa

Solution

For an ideal gas, the ratio pV/T is constant.

$$\frac{p_1 V_1}{T_1} = \frac{p_2 V_2}{T_2}$$

$$V_1 = V_2$$

$$p_2 = \frac{p_1 T_2}{T_1}$$

$$= \frac{(0.3 \text{ MPa} + 0.101 \text{ MPa})(75°C + 273)}{25°C + 273}$$

$$= 0.468 \text{ MPa} \quad (0.47 \text{ MPa})$$

Answer is B.

Problem 4

Steam exists at a pressure of 120 Pa and a temperature of 250K. How many molecules are present in 2 cm³ at these conditions?

(A) 5×10^{12}
(B) 5×10^{16}
(C) 7×10^{16}
(D) 7×10^{18}

Solution

$$pV = n\overline{R}T$$

$$n = \frac{pV}{\overline{R}T}$$

$$= \frac{(120 \text{ Pa})(2 \times 10^{-6} \text{ m}^3)}{\left(8314.3 \frac{\text{m}^3 \cdot \text{Pa}}{\text{kmol} \cdot \text{K}}\right)(250\text{K})}$$

$$= 1.155 \times 10^{-10} \text{ kmol}$$

$$\begin{array}{l}\text{no. of} \\ \text{molecules}\end{array} = nN_A$$

$$= \left(1.155 \times 10^{-10} \text{ kmol}\right)\left(1000 \frac{\text{mol}}{\text{kmol}}\right)$$

$$\times \left(6.022 \times 10^{23} \frac{\text{molecules}}{\text{mol}}\right)$$

$$= 6.96 \times 10^{16} \text{ molecules}$$

$$(7 \times 10^{16} \text{ molecules})$$

Answer is C.

Problem 5

The specific gas constant of oxygen is $R = 0.25983$ kJ/kg·K. If a 2 m³ tank contains 40 kg of oxygen at 40°C, what is the gage pressure in the tank?

(A) 61 kPa
(B) 110 kPa
(C) 160 kPa
(D) 1.53 MPa

Solution

$$pV = mRT$$

$$p = \frac{mRT}{V}$$

$$= \frac{(40 \text{ kg})\left(0.25983 \frac{\text{kJ}}{\text{kg}\cdot\text{K}}\right)(40\text{K} + 273\text{K})}{2 \text{ m}^3}$$

$$= 1627 \text{ kPa}$$

$$p_{\text{gage}} = p_{\text{absolute}} - p_{\text{atm}}$$

$$= 1627 \text{ kPa} - 101.3 \text{ kPa}$$

$$= 1526 \text{ kPa} \quad (1.53 \text{ MPa})$$

Answer is D.

FE-STYLE EXAM PROBLEMS

1. Steam at 2.0 kPa is saturated at 17.5°C. In what state will the steam be at 40°C if the pressure is 2.0 kPa?

(A) superheated
(B) subcooled
(C) saturated
(D) supersaturated

2. All real gases deviate to some extent from the ideal behavior described by the equation $pV = mRT$. For which of the following conditions are the deviations smallest?

(A) high temperatures and low volumes
(B) high temperatures and low pressures
(C) high pressures and low volumes
(D) high pressures and low temperatures

Problems 3–5 refer to the following conditions.

A 0.5 m³ rigid tank contains equal volumes of Freon-12 vapor and Freon-12 liquid at 312K. Additional Freon-12 is then slowly added to the tank until the total mass of Freon-12 (liquid and vapor) is 400 kg. Some vapor is bled off to maintain the original temperature and pressure. Freon-12 is saturated at 312K and 0.9334 MPa.

Under these saturated conditions, $v_f = 0.000795 \text{ m}^3/\text{kg}$ and $v_g = 0.01872 \text{ m}^3/\text{kg}$.

3. What is the final mass of Freon-12 vapor?

 (A) 10 kg
 (B) 39 kg
 (C) 100 kg
 (D) 300 kg

4. What is the final volume of Freon-12 liquid?

 (A) 0.16 m^3
 (B) 0.18 m^3
 (C) 0.31 m^3
 (D) 0.39 m^3

5. What was the mass of Freon-12 added to the tank?

 (A) 14 kg
 (B) 22 kg
 (C) 46 kg
 (D) 72 kg

For the following problems use the NCEES Handbook as your only reference.

6. What is the change in internal energy of air (assumed to be an ideal gas) cooled from 550°C to 100°C?

 (A) 320 kJ/kg
 (B) 390 kJ/kg
 (C) 450 kJ/kg
 (D) 550 kJ/kg

7. When the volume of an ideal gas is doubled while the temperature is halved, what happens to the pressure?

 (A) Pressure is doubled.
 (B) Pressure is halved.
 (C) Pressure is quartered.
 (D) Pressure is quadrupled.

8. Assuming air to be an ideal gas with a molecular weight of 28.967, what is the density of air at 1 atm and 600°C?

 (A) 0.12 kg/m^3
 (B) 0.40 kg/m^3
 (C) 0.59 kg/m^3
 (D) 0.68 kg/m^3

9. A boy on the beach holds a spherical balloon filled with air. At 10:00 a.m., the temperature on the beach is 20°C and the balloon has a diameter of 30 cm. Two hours later, the balloon diameter is 30.5 cm. Assuming that the air is an ideal gas and that no air was lost or added, what is the temperature on the beach at noon?

 (A) 21°C
 (B) 25°C
 (C) 32°C
 (D) 35°C

10. Which thermodynamic property is the best measure of the molecular activity of a substance?

 (A) enthalpy
 (B) internal energy
 (C) entropy
 (D) external energy

11. A liquid boils when its vapor pressure is equal to

 (A) one atmosphere of pressure.
 (B) the gage pressure.
 (C) the absolute pressure.
 (D) the ambient pressure.

12. A substance whose properties are uniform throughout a sample is referred to as

 (A) a solid.
 (B) an ideal substance.
 (C) a pure substance.
 (D) a standard substance.

13. For every gas there is a particular temperature above which the properties of the gas cannot be distinquished from the properties of a liquid no matter how great the pressure. This temperature is known as the

 (A) absolute temperature.
 (B) saturation temperature.
 (C) standard temperature.
 (D) critical temperature.

14. Steam initially at 1 MPa and 200°C expands in a turbine to 40°C and 83% quality. What is the change in entropy?

 (A) -0.35 kJ/kg·K
 (B) 0.00 kJ/kg·K
 (C) 0.26 kJ/kg·K
 (D) 0.73 kJ/kg·K

15. A 3 kg mixture of water and water vapor at 70°C is held at constant pressure while heat is added. The enthalpy of the water increases by 50 kJ/kg. What is the change in entropy?

(A) 0.111 kJ/kg·K
(B) 0.146 kJ/kg·K
(C) 0.158 kJ/kg·K
(D) 0.177 kJ/kg·K

16. Through an isentropic process, a piston compresses 2 kg of an ideal gas at 150 kPa and 35°C in a cylinder to a pressure of 300 kPa. The specific heat of the gas for constant pressure processes is 5 kJ/kg·K; for constant volume processes, the specific heat is 3 kJ/kg·K. What is the final temperature of the gas?

(A) 134°C
(B) 190°C
(C) 212°C
(D) 258°C

17. 3 kg of steam with quality of 30% has a pressure of 12.056 bar. At that pressure, the specific volume of a saturated fluid is $v_f = 1.5289$ cm^3/g. The specific volume of a saturated vapor is $v_g = 14.1889$ cm^3/g. What is the specific volume of the steam?

(A) 5.25 cm^3/g
(B) 5.33 cm^3/g
(C) 5.40 cm^3/g
(D) 5.48 cm^3/g

18. An ideal gas has a ratio of specific heats of 1.4 and a specific heat at constant pressure of 1100 kJ/kg·K. The gas flows through a partially closed valve and experiences a pressure drop from 700 kPa to 150 kPa. If the temperature before the valve is 100°C, what is the temperature after the valve?

(A) −193°C
(B) −33°C
(C) 64°C
(D) 100°C

19. What is the specific volume of refrigerant HFC-134a with an enthalpy of 570 kJ/kg and a pressure of 0.6 MPa?

(A) 0.006 m^3/kg
(B) 0.06 m^3/kg
(C) 0.6 m^3/kg
(D) 180 m^3/kg

20. Refrigerant HFC-134a is ideally throttled from an enthalpy of 480 kJ/kg at the evaporator pressure of 0.2 MPa to a pressure of 0.02 MPa. What is the final enthalpy?

(A) 370 kJ/kg
(B) 410 kJ/kg
(C) 450 kJ/kg
(D) 480 kJ/kg

21. The enthalpy of refrigerant HFC-134a is reduced from 440 kJ/kg at 0.8 MPa to 300 kJ/kg in a water-cooled condenser. What is the approximate final quality of the refrigerant?

(A) 32%
(B) 37%
(C) 63%
(D) 71%

22. Which property of state is not an extensive state?

(A) temperature
(B) volume
(C) number of molecules
(D) mass

SOLUTIONS TO FE-STYLE EXAM PROBLEMS

1. The temperature of the steam is higher than the saturation temperature for 2.0 kPa pressure. Therefore, the steam is superheated.

Answer is A.

2. A gas is considered to behave ideally when its pressure is very low and its temperature is much higher than the critical temperature. Under these conditions, the distance between molecules is great and the interactions between molecules are minimal. The higher the temperature and the lower the pressure, the more the gas behaves ideally. At low enough pressures, vapors behave nearly ideally, as well.

Answer is B.

3. The average specific volume after adding the Freon is

$$v = \frac{V}{m} = \frac{0.5 \text{ m}^3}{400 \text{ kg}}$$

$$= 0.00125 \text{ m}^3/\text{kg}$$

$$v = v_f + x(v_g - v_f)$$

$$x = \frac{v - v_f}{v_g - v_f}$$

$$= \frac{0.00125 \, \frac{m^3}{kg} - 0.000795 \, \frac{m^3}{kg}}{0.01872 \, \frac{m^3}{kg} - 0.000795 \, \frac{m^3}{kg}}$$

$$= 0.02538$$

From the definition of quality,

$$m_g = x(m_g + m_f) = xm$$
$$= (0.02538)(400 \text{ kg})$$
$$= 10.2 \text{ kg} \quad (10 \text{ kg})$$

Answer is A.

4. $$1 - x = \frac{m_f}{m_g + m_f}$$

$$m_f = (1 - x)m$$
$$= (1 - 0.02538)(400 \text{ kg})$$
$$= 389.8 \text{ kg}$$

$$V_f = m_f v_f$$
$$= (389.8 \text{ kg}) \left(0.000795 \, \frac{m^3}{kg} \right)$$
$$= 0.31 \text{ m}^3$$

Answer is C.

5. The liquid and vapor are coexisting, so they are at the saturation conditions. Since they both originally occupied half of the tank, the masses of vapor and liquid were

$$m_g = \frac{V}{v_g} = \frac{\frac{0.5 \text{ m}^3}{2}}{0.01872 \, \frac{m^3}{kg}}$$
$$= 13.35 \text{ kg}$$

$$m_l = \frac{\frac{0.5 \text{ m}^3}{2}}{0.000795 \, \frac{m^3}{kg}}$$
$$= 314.5 \text{ kg}$$

The mass added is

$$400 \text{ kg} - 13.35 \text{ kg} - 314.5 \text{ kg} = 72.2 \text{ kg} \quad (72 \text{ kg})$$

Answer is D.

6. Because of the ideal gas assumption, the internal energy change depends on temperature.

$$\Delta u = c_v \Delta T \quad \text{[for any process]}$$

$$c_v = 0.718 \text{ kJ/kg·K}$$

$$\Delta u = \left(0.718 \, \frac{kJ}{kg·K} \right) (550°C - 100°C)$$
$$= 323 \text{ kJ/kg} \quad (320 \text{ kJ/kg})$$

Note that $\Delta T_{°C} = \Delta T_K$ even though $T_{°C} \neq T_K$.

Answer is A.

7. Using the equation of state for an ideal gas,

$$\frac{p_1 V_1}{T_1} = \frac{p_2 V_2}{T_2}$$

$$V_2 = 2V_1$$

$$T_2 = \frac{T_1}{2}$$

$$p_2 = \frac{p_1 V_1 T_2}{T_1 V_2} = \frac{p_1 V_1 \left(\frac{T_1}{2} \right)}{T_1 (2V_1)}$$
$$= \frac{p_1}{4}$$

The pressure is quartered.

Answer is C.

8. $$R = \frac{\overline{R}}{\text{MW}} = \frac{8.3143 \, \frac{kJ}{kmol·K}}{28.967 \, \frac{kg}{kmol}}$$

$$= 0.2870 \text{ kJ/kg·K}$$

$$\rho = \frac{m}{V}$$

$$pV = mRT$$

$$\rho = \frac{p}{RT}$$

$$= \frac{(1 \text{ atm}) \left(101.3 \, \frac{kPa}{atm} \right)}{\left(0.2870 \, \frac{kJ}{kg·K} \right) (600°C + 273)}$$

$$= 0.404 \text{ kg/m}^3 \quad (0.40 \text{ kg/m}^3)$$

Answer is B.

9. From the equation of state for an ideal gas,

$$\frac{p_1 V_1}{T_1} = \frac{p_2 V_2}{T_2}$$

$$p_1 = p_2$$

$$\frac{V_1}{T_1} = \frac{V_2}{T_2}$$

$$T_2 = \frac{V_2 T_1}{V_1}$$

$$V_1 = \frac{4}{3}\pi r^3 = \frac{4}{3}\pi \left(\frac{0.3 \text{ m}}{2}\right)^3$$

$$= 0.01414 \text{ m}^3$$

$$V_2 = \frac{4}{3}\pi r^3 = \frac{4}{3}\pi \left(\frac{0.305 \text{ m}}{2}\right)^3$$

$$= 0.01486 \text{ m}^3$$

$$T_2 = \frac{(0.01486 \text{ m}^3)(20°C + 273)}{0.01414 \text{ m}^3}$$

$$= 308\text{K}$$

$$T_2 = 308\text{K} - 273$$

$$= 35°C$$

Answer is D.

10. Internal energy is a measure of the translational, rotational, and vibrational kinetic energies of the molecules of a substance.

Answer is B.

11. A liquid boils when its vapor pressure is equal to the surrounding pressure.

Answer is D.

12. Answer is C.

13. The critical temperature is the highest point on the vapor dome. The substance cannot exist in liquid form at a temperature above the critical temperature.

Answer is D.

14. Without a Mollier diagram, saturation tables must be used. Determine the entropy for 1 MPa and 200°C superheated steam.

$$s_1 = 6.6940 \text{ kJ/kg·K}$$

Using the saturated water table, determine s_f and s_g for 40°C saturated steam.

$$s_f = 0.5725 \text{ kJ/kg·K}$$

$$s_g = 8.2570 \text{ kJ/kg·K}$$

$$s_{fg} = s_g - s_f$$

$$= 8.2570 \frac{\text{kJ}}{\text{kg·K}} - 0.5725 \frac{\text{kJ}}{\text{kg·K}}$$

$$= 7.6845 \text{ kJ/kg·K}$$

Using the quality, determine the entropy for the saturated mixture.

$$s_2 = s_f + x s_{fg}$$

$$= 0.5725 \frac{\text{kJ}}{\text{kg·K}} + (0.83)\left(7.6845 \frac{\text{kJ}}{\text{kg·K}}\right)$$

$$= 6.9506 \text{ kJ/kg·K}$$

The change in entropy is

$$s_2 - s_1 = 6.9506 \frac{\text{kJ}}{\text{kg·K}} - 6.6940 \frac{\text{kJ}}{\text{kg·K}}$$

$$= 0.2566 \text{ kJ/kg·K} \quad (0.26 \text{ kJ/kg·K})$$

Answer is C.

15. Constant-pressure processes are also constant-temperature processes for liquid-vapor mixtures. This corresponds to a change of phase.

The heat flow, Q, is equal to the change in enthalpy.

$$\Delta s = \int \frac{dq}{T} = \frac{Q}{T_0}$$

$$= \frac{\Delta h}{T_0}$$

$$= \frac{50 \frac{\text{kJ}}{\text{kg}}}{70°C + 273}$$

$$= 0.1458 \text{ kJ/kg·K} \quad (0.146 \text{ kJ/kg·K})$$

Answer is B.

16. Find the ratio of specific heats.

$$k = \frac{c_p}{c_v} = \frac{5 \frac{\text{kJ}}{\text{kg·K}}}{3 \frac{\text{kJ}}{\text{kg·K}}}$$

$$= 1.67$$

$$T_1 p_1^{(1-k)/k} = T_2 p_2^{(1-k)/k}$$

$$T_2 = T_1 \left(\frac{p_1}{p_2}\right)^{(1-k)/k}$$

$$= (35°\text{C} + 273)\left(\frac{150 \text{ kPa}}{300 \text{ kPa}}\right)^{(1-1.67)/1.67}$$

$$= 406.7\text{K}$$

$$T_2 = 406.7\text{K} - 273$$

$$= 133.7°\text{C} (134°\text{C})$$

Answer is A.

17. Use the quality to find the specific volume for the liquid-vapor mixture.

$$v = v_f + x v_{fg}$$

$$= v_f + x(v_g - v_f)$$

$$= 1.5289 \; \frac{\text{cm}^3}{\text{g}} + (0.3)\left(14.1889 \; \frac{\text{cm}^3}{\text{g}} - 1.5289 \; \frac{\text{cm}^3}{\text{g}}\right)$$

$$= 5.327 \text{ cm}^3/\text{g} (5.33 \text{ cm}^3/\text{g})$$

Answer is B.

18. The general form of the equation for heat transfer is

$$Q = mc\Delta T$$

However, throttling is an adiabatic process. Therefore, $Q = 0$.

$$\Delta T = 0$$

$$T_1 = T_2 = 100°\text{C}$$

Answer is D.

19. The specific volume is read directly from the refrigerant chart at the intersection of the 570 kJ/kg line and the 0.6 MPa (or 6 bars) pressure axis.

Answer is B.

20. Enthalpy remains constant when an ideal gas is throttled. The HFC-134a is processed entirely in the superheated region and can be expected to approximate ideal gas behavior.

Answer is D.

21. The enthalpy reduction occurs in a condenser and, therefore, is a constant-pressure process. This is represented on the pressure-enthalpy chart by a horizontal line moving from the superheated region into the vapor dome along the 0.8 MPa isobar. The starting enthalpy is irrelevant. The final condition is determined by the intersection of the 300 kJ/kg enthalpy line and the 0.8 MPa pressure line. The quality is approximately 32%.

Answer is A.

22. An extensive state depends on the amount of substance present.

Answer is A.

Table 26.5 Properties of Saturated Steam by Temperature (SI units)

temp. (°C)	absolute pressure (bars)	specific volume (cm³/g)		internal energy (kJ/kg)		enthalpy (kJ/kg)			entropy (kJ/kg·K)		temp. (°C)
		sat. liquid v_f	sat. vapor v_g	sat. liquid u_f	sat. vapor u_g	sat. liquid h_f	evap. h_{fg}	sat. vapor h_g	sat. liquid s_f	sat. vapor s_g	
0.01	0.00611	1.0002	206136	0.00	2375.3	0.01	2501.3	2501.4	0.0000	9.1562	0.01
4	0.00813	1.0001	157232	16.77	2380.9	16.78	2491.9	2508.7	0.0610	9.0514	4
5	0.00872	1.0001	147120	20.97	2382.3	20.98	2489.6	2510.6	0.0761	9.0257	5
6	0.00935	1.0001	137734	25.19	2383.6	25.20	2487.2	2512.4	0.0912	9.0003	6
8	0.01072	1.0002	120917	33.59	2386.4	33.60	2482.5	2516.1	0.1212	8.9501	8
10	0.01228	1.0004	106379	42.00	2389.2	42.01	2477.7	2519.8	0.1510	8.9008	10
11	0.01312	1.0004	99857	46.20	2390.5	46.20	2475.4	2521.6	0.1658	8.8765	11
12	0.01402	1.0005	93784	50.41	2391.9	50.41	2473.0	2523.4	0.1806	8.8524	12
13	0.01497	1.0007	88124	54.60	2393.3	54.60	2470.7	2525.3	0.1953	8.8285	13
14	0.01598	1.0008	82848	58.79	2394.7	58.80	2468.3	2527.1	0.2099	8.8048	14
15	0.01705	1.0009	77926	62.99	2396.1	62.99	2465.9	2528.9	0.2245	8.7814	15
16	0.01818	1.0011	73333	67.18	2397.4	67.19	2463.6	2530.8	0.2390	8.7582	16
17	0.01938	1.0012	69044	71.38	2398.8	71.38	2461.2	2532.6	0.2535	8.7351	17
18	0.02064	1.0014	65038	75.57	2400.2	75.58	2458.8	2534.4	0.2679	8.7123	18
19	0.02198	1.0016	61293	79.76	2401.6	79.77	2456.5	2536.2	0.2823	8.6897	19
20	0.02339	1.0018	57791	83.95	2402.9	83.96	2454.1	2538.1	0.2966	8.6672	20
21	0.02487	1.0020	54514	88.14	2404.3	88.14	2451.8	2539.9	0.3109	8.6450	21
22	0.02645	1.0022	51447	92.32	2405.7	92.33	2449.4	2541.7	0.3251	8.6229	22
23	0.02810	1.0024	48574	96.51	2407.0	96.52	2447.0	2543.5	0.3393	8.6011	23
24	0.02985	1.0027	45883	100.70	2408.4	100.70	2444.7	2545.4	0.3534	8.5794	24
25	0.03169	1.0029	43360	104.88	2409.8	104.89	2442.3	2547.2	0.3674	8.5580	25
26	0.03363	1.0032	40994	109.06	2411.1	109.07	2439.9	2549.0	0.3814	8.5367	26
27	0.03567	1.0035	38774	113.25	2412.5	113.25	2437.6	2550.8	0.3954	8.5156	27
28	0.03782	1.0037	36690	117.42	2413.9	117.43	2435.2	2552.6	0.4093	8.4946	28
29	0.04008	1.0040	34733	121.60	2415.2	121.61	2432.8	2554.5	0.4231	8.4739	29
30	0.04246	1.0043	32894	125.78	2416.6	125.79	2430.5	2556.3	0.4369	8.4533	30
31	0.04496	1.0046	31165	129.96	2418.0	129.97	2428.1	2558.1	0.4507	8.4329	31
32	0.04759	1.0050	29540	134.14	2419.3	134.15	2425.7	2559.9	0.4644	8.4127	32
33	0.05034	1.0053	28011	138.32	2420.7	138.33	2423.4	2561.7	0.4781	8.3927	33
34	0.05324	1.0056	26571	142.50	2422.0	142.50	2421.0	2563.5	0.4917	8.3728	34
35	0.05628	1.0060	25216	146.67	2423.4	146.68	2418.6	2565.3	0.5053	8.3531	35
36	0.05947	1.0063	23940	150.85	2424.7	150.86	2416.2	2567.1	0.5188	8.3336	36
38	0.06632	1.0071	21602	159.20	2427.4	159.21	2411.5	2570.7	0.5458	8.2950	38
40	0.07384	1.0078	19523	167.56	2430.1	167.57	2406.7	2574.3	0.5725	8.2570	40
45	0.09593	1.0099	15258	188.44	2436.8	188.45	2394.8	2583.2	0.6387	8.1648	45
50	0.1235	1.0121	12032	209.32	2443.5	209.33	2382.7	2592.1	0.7038	8.0763	50
55	0.1576	1.0146	9568	230.21	2450.1	230.23	2370.7	2600.9	0.7679	7.9913	55
60	0.1994	1.0172	7671	251.11	2456.6	251.13	2358.5	2609.6	0.8312	7.9096	60
65	0.2503	1.0199	6197	272.02	2463.1	272.06	2346.2	2618.3	0.8935	7.8310	65
70	0.3119	1.0228	5042	292.95	2469.6	292.98	2333.8	2626.8	0.9549	7.7553	70
75	0.3858	1.0259	4131	313.90	2475.9	313.93	2321.4	2635.3	1.0155	7.6824	75
80	0.4739	1.0291	3407	334.86	2482.2	334.91	2308.8	2643.7	1.0753	7.6122	80
85	0.5783	1.0325	2828	355.84	2488.4	355.90	2296.0	2651.9	1.1343	7.5445	85
90	0.7014	1.0360	2361	376.85	2494.5	376.92	2283.2	2660.1	1.1925	7.4791	90
95	0.8455	1.0397	1982	397.88	2500.6	397.96	2270.2	2668.1	1.2500	7.4159	95
100	1.014	1.0435	1673	418.94	2506.5	419.04	2257.0	2676.1	1.3069	7.3549	100
110	1.433	1.0516	1210	461.14	2518.1	461.30	2230.2	2691.5	1.4185	7.2387	110
120	1.985	1.0603	891.9	503.50	2529.3	503.71	2202.6	2706.3	1.5276	7.1296	120
130	2.701	1.0697	668.5	546.02	2539.9	546.31	2174.2	2720.5	1.6344	7.0269	130
140	3.613	1.0797	508.9	588.74	2550.0	589.13	2144.7	2733.9	1.7391	6.9299	140

(continued)

Table 26.5 Properties of Saturated Steam by Temperature (SI units) (continued)

temp. (°C)	absolute pressure (bars)	specific volume (cm³/g)		internal energy (kJ/kg)		enthalpy (kJ/kg)			entropy (kJ/kg·K)		temp. (°C)
		sat. liquid v_f	sat. vapor v_g	sat. liquid u_f	sat. vapor u_g	sat. liquid h_f	evap. h_{fg}	sat. vapor h_g	sat. liquid s_f	sat. vapor s_g	
150	4.758	1.0905	392.8	631.68	2559.5	632.20	2114.3	2746.5	1.8418	6.8379	150
160	6.178	1.1020	307.1	674.86	2568.4	675.55	2082.6	2758.1	1.9427	6.7502	160
170	7.917	1.1143	242.8	718.33	2576.5	719.21	2049.5	2768.7	2.0419	6.6663	170
180	10.02	1.1274	194.1	762.09	2583.7	763.22	2015.0	2778.2	2.1396	6.5857	180
190	12.54	1.1414	156.5	806.19	2590.0	807.62	1978.8	2786.4	2.2359	6.5079	190
200	15.54	1.1565	127.4	850.65	2595.3	852.45	1940.7	2793.2	2.3309	6.4323	200
210	19.06	1.1726	104.4	895.53	2599.5	897.76	1900.7	2798.5	2.4248	6.3585	210
220	23.18	1.1900	86.19	940.87	2602.4	943.62	1858.5	2802.1	2.5178	6.2861	220
230	27.95	1.2088	71.58	986.74	2603.9	990.12	1813.8	2804.0	2.6099	6.2146	230
240	33.44	1.2291	59.76	1033.2	2604.0	1037.3	1766.5	2803.8	2.7015	6.1437	240
250	39.73	1.2512	50.13	1080.4	2602.4	1085.4	1716.2	2801.5	2.7927	6.0730	250
260	46.88	1.2755	42.21	1128.4	2599.0	1134.4	1662.5	2796.6	2.8838	6.0019	260
270	54.99	1.3023	35.64	1177.4	2593.7	1184.5	1605.2	2789.7	2.9751	5.9301	270
280	64.12	1.3321	30.17	1227.5	2586.1	1236.0	1543.6	2779.6	3.0668	5.8571	280
290	74.36	1.3656	25.57	1278.9	2576.0	1289.1	1477.1	2766.2	3.1594	5.7821	290
300	85.81	1.4036	21.67	1332.0	2563.0	1344.0	1404.9	2749.0	3.2534	5.7045	300
320	112.7	1.4988	15.49	1444.6	2525.5	1461.5	1238.6	2700.1	3.4480	5.5362	320
340	145.9	1.6379	10.80	1570.3	2464.6	1594.2	1027.9	2622.0	3.6594	5.3357	340
360	186.5	1.8925	6.945	1725.2	2351.5	1760.5	720.5	2481.0	3.9147	5.0526	360
374.14	220.9	3.155	3.155	2029.6	2029.6	2099.3	0	2099.3	4.4298	4.4298	374.14

Adapted from *Steam Tables—Metric Units*, by J.H. Keenan, F.G. Keyes, P.G. Hill, and J.G. Moore, © 1978 by J.H. Keenan and F.G. Keyes.

Table 26.6 Properties of Superheated Steam (SI units)

specific volume (v) in m^3/kg; enthalpy (h) in kJ/kg; entropy (s) in kJ/kg·K

absolute pressure (kPa) (sat. temp. °C)		temperature (°C)							
		100	150	200	250	300	360	420	500
10 (45.81)	v	17.196	19.512	21.825	24.136	26.445	29.216	31.986	35.679
	h	2687.5	2783.0	2879.5	2977.3	3076.5	3197.6	3320.9	3489.1
	s	8.4479	8.6882	8.9038	9.1002	9.2813	9.4821	9.6682	9.8978
50 (81.33)	v	3.418	3.889	4.356	4.820	5.284	5.839	6.394	7.134
	h	2682.5	2780.1	2877.7	2976.0	3075.5	3196.8	3320.4	3488.7
	s	7.6947	7.9401	8.1580	8.3556	8.5373	8.7385	8.9249	9.1546
75 (91.78)	v	2.270	2.587	2.900	3.211	3.520	3.891	4.262	4.755
	h	2679.4	2778.2	2876.5	2975.2	3074.9	3196.4	3320.0	3488.4
	s	7.5009	7.7496	7.9690	8.1673	8.3493	8.5508	8.7374	8.9672
100 (99.63)	v	1.6958	1.9364	2.172	2.406	2.639	2.917	3.195	3.565
	h	2672.2	2776.4	2875.3	2974.3	3074.3	3195.9	3319.6	3488.1
	s	7.3614	7.6134	7.8343	8.0333	8.2158	8.4175	8.6042	8.8342
150 (111.37)	v	...	1.2853	1.4443	1.6012	1.7570	1.9432	2.129	2.376
	h	...	2772.6	2872.9	2972.7	3073.1	3195.0	3318.9	3487.6
	s	...	7.4193	7.6433	7.8438	8.0720	8.2293	8.4163	8.6466
400 (143.63)	v	...	0.4708	0.5342	0.5951	0.6548	0.7257	0.7960	0.8893
	h	...	2752.8	2860.5	2964.2	3066.8	3190.3	3315.3	3484.9
	s	...	6.9299	7.1706	7.3789	7.5662	7.7712	7.9598	8.1913
700 (164.97)	v	...	...	0.2999	0.3363	0.3714	0.4126	0.4533	0.5070
	h	...	...	2884.8	2953.6	3059.1	3184.7	3310.9	3481.7
	s	...	...	6.8865	7.1053	7.2979	7.5063	7.6968	7.9299
1000 (179.91)	v	...	...	0.2060	0.2327	0.2579	0.2873	0.3162	0.3541
	h	...	...	2827.9	2942.6	3051.2	3178.9	3306.5	3478.5
	s	...	...	6.6940	6.9247	7.1229	7.3349	7.5275	7.7622
1500 (198.32)	v	...	...	0.13248	0.15195	0.16966	0.18988	0.2095	0.2352
	h	...	...	2796.8	2923.3	3037.6	3169.2	3299.1	3473.1
	s	...	...	6.4546	6.7090	6.9179	7.1363	7.3323	7.5698
2000 (212.42)	v	...	...	...	0.11144	0.12547	0.14113	0.15616	0.17568
	h	...	...	...	2902.5	3023.5	3159.3	3291.6	3467.6
	s	...	...	...	6.5453	6.7664	6.9917	7.1915	7.4317
2500 (223.99)	v	...	...	...	0.08700	0.09890	0.11186	0.12414	0.13998
	h	...	...	...	2880.1	3008.8	3149.1	3284.0	3462.1
	s	...	...	...	6.4085	6.6438	6.8767	7.0803	7.3234
3000 (233.90)	v	...	...	...	0.07058	0.08114	0.09233	0.10279	0.11619
	h	...	...	...	2855.8	2993.5	3138.7	3276.3	3456.5
	s	...	...	...	6.2872	6.5390	6.7801	6.9878	7.2338

Adapted from *Steam Tables—Metric Units*, by J.H. Keenan, F.G. Keyes, P.G. Hill, and J.G. Moore, © 1992 by J.H. Keenan and F.G. Keyes.

THERMO
Properties

Figure 26.6 *p-h* Diagram for Refrigerant HFC-134a (SI Units)

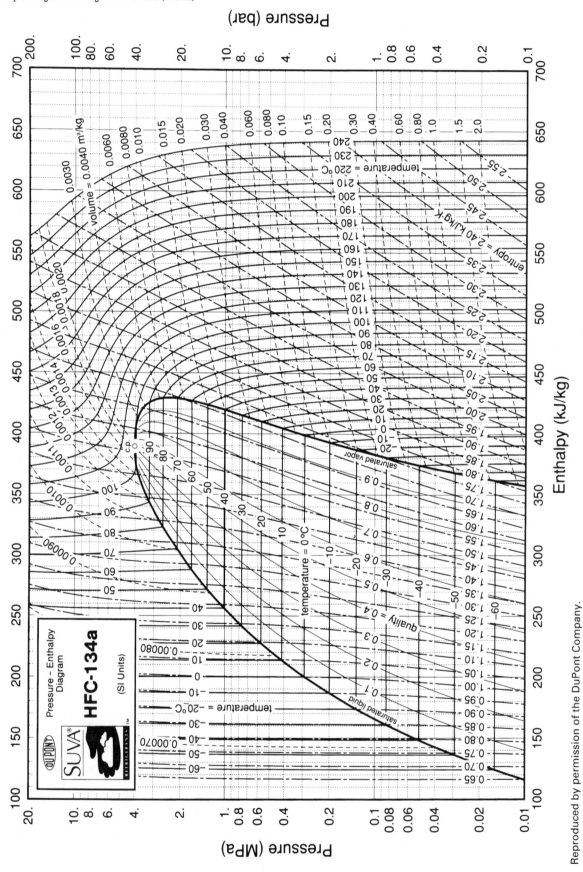

27 First Law of Thermodynamics

Subjects

TYPES OF PROCESSES 27-1
FIRST LAW OF THERMODYNAMICS . . 27-2
    Closed Systems 27-2
    Special Cases of Closed Systems
      (for Ideal Gases) 27-2
    Open Systems 27-3
    Special Cases of Open Systems
      (for Ideal Gases) 27-3
    Steady-State Systems 27-3
    Special Cases of Steady-Flow
      Energy Equation 27-4

Nomenclature

g	acceleration of gravity	ft/sec^2	m/s^2
h	enthalpy	Btu/lbm	kJ/kg
H	total enthalpy	Btu	kJ
J	Joule's constant (778)	ft-lbf/Btu	–
m	mass	lbm	kg
$\dot{m}$	mass flow rate	lbm/sec	kg/s
n	polytropic exponent	–	–
p	absolute pressure	lbf/ft^2	Pa
q	heat energy	Btu/lbm	kJ/kg
Q	total heat energy	Btu	kJ
$\dot{Q}$	rate of heat transfer	Btu/sec	kW
R	specific gas constant	ft-lbf/lbm-°R	kJ/kg·K
s	entropy	Btu/lbm-°R	kJ/kg·K
S	total entropy	Btu/°R	kJ/K
T	absolute temperature	°R	K
u	internal energy	Btu/lbm	kJ/kg
U	total internal energy	Btu	kJ
v	velocity	ft/sec	m/s
V	volume	ft^3	m^3
W	work	ft-lbf	kJ
$\dot{W}$	rate of work (power)	ft-lbf/sec	kW
z	elevation	ft	m

Symbols

η	efficiency	–	–
v	specific volume	ft^3/lbm	m^3/kg

Subscripts

e	exit
es	ideal (isentropic) exit state
f	fluid (liquid)
fg	liquid-to-gas (vaporization)
g	gas (vapor)
i	in (entrance)
rev	reversible
s	isentropic

TYPES OF PROCESSES

Changes in thermodynamic properties of a system often depend on the type of process experienced. This is particularly true for gaseous systems. The following list describes several common types of processes.

- *adiabatic process*—a process in which no energy crosses the system boundary. Adiabatic processes include isentropic and throttling processes.

- *isentropic process*—an adiabatic process in which there is no entropy production (i.e., it is reversible). Also known as a *constant entropy process*.

- *throttling process*—an adiabatic process in which there is no change in enthalpy, but for which there is a significant pressure drop.

- *constant pressure process*—also known as an *isobaric process*.

- *constant temperature process*—also known as an *isothermal process*.

- *constant volume process*—also known as an *isochoric* or *isometric process*.

- *polytropic process*—a process that obeys the polytropic equation of state, Eq. 27.1. Gases always constitute the system in polytropic processes. n is the *polytropic exponent*, a property of the equipment, not of the gas.

$$pv^n = \text{constant} \qquad 27.1$$

A system that is in equilibrium at the start and finish of a process may or may not be in equilibrium during the process. A *quasistatic process* (*quasiequilibrium process*) is one that can be divided into a series of infinitesimal deviations (steps) from equilibrium. During each

THERMO First Law

step, the property changes are small, and all intensive properties are uniform throughout the system. The interim equilibrium at each step is often called *quasiequilibrium*.

A *reversible process* is one that is performed in such a way that, at the conclusion of the process, both the system and the local surroundings can be restored to their initial states. Quasiequilibrium processes are assumed to be reversible processes.

FIRST LAW OF THERMODYNAMICS

There is a basic principle that underlies all property changes as a system undergoes a process: All energy must be accounted for. Energy that enters a system must either leave the system or be stored in some manner, and energy cannot be created or destroyed. These statements are the primary manifestations of the *first law of thermodynamics*: The net energy crossing the system boundary is the change in energy inside the system.

The first law applies whether or not a process is reversible. Hence, the first law can also be stated as: The work done in an adiabatic process depends only on the system's endpoint conditions, not on the nature of the process.

A *thermodynamic system* or *control volume* is defined as the matter enclosed within an arbitrary but precisely defined control volume. Everything external to the system is defined as the *surroundings*, *environment*, or *universe*. The environment and system are separated by the *system boundaries*. The surface of the control volume is known as the *control surface*. The control surface can be real (e.g., piston and cylinder walls) or imaginary.

If no mass crosses the system boundaries, the system is said to be a *closed system*. The matter in a closed system may be referred to as a *control mass*. Closed systems can have variable volumes. The gas compressed by a piston in a cylinder is an example of a closed system with a variable volume.

If mass flows through the system across system boundaries, the system is an *open system*. Examples are pumps, heat exchangers, and jet engines. An important type of open system is the *steady-flow open system*, in which matter enters and leaves at the same rate. Pumps, turbines, heat exchangers, and boilers are all steady-flow open systems.

In most cases, energy transferred by heat, by work, or as electrical energy can enter or leave any open or closed system. Systems closed to both matter and energy transfer are known as *isolated systems*.

A standard sign convention is used in calculating work, heat, and property changes in systems. This sign convention takes the system (not the environment) as the reference. For example, a net heat gain would mean the system gained energy and the environment lost energy.

- Heat, Q, transferred due to a temperature difference is positive if heat flows into the system.
- Work, W, is positive if the system does work on the surroundings.
- Changes in enthalpy, entropy, and internal energy (ΔH, ΔS, and ΔU) are positive if these properties increase within the system.

In accordance with the standard sign convention, Q will be negative if the net heat exchange is a loss of heat to the surroundings. ΔU will be negative if the internal energy of the system decreases. W will be negative if the surroundings do work on the system (e.g., a piston compressing gas in a cylinder).

Closed Systems

The first law of thermodynamics for closed systems can be written in most cases in finite terms.

$$Q = \Delta U + W \qquad 27.2$$

Equation 27.2 states that the heat energy, Q, entering a closed system can either increase the temperature (increase U) or be used to perform work (increase W) on the surroundings. The Q term is understood to be the net heat entering the system, which is the heat energy entering the system less the heat energy lost to the surroundings.

The work done by or on the system during a reversible process is calculated by the area under the curve in the p-V plane, and is called *reversible work*, *p-V work*, or *flow work*.

$$W_{\text{rev}} = \int p \, dV \qquad 27.3$$

Special Cases of Closed Systems (for Ideal Gases)
Constant Pressure

Equation 27.4 is known as *Charles' law* for constant pressure processes.

$$\frac{T}{v} = \text{constant} \qquad 27.4$$

$$W = p\Delta v \qquad 27.5$$

Constant Volume

$$\frac{T}{p} = \text{constant} \qquad 27.6$$

$$W = 0 \qquad 27.7$$

Constant Temperature

Equation 27.8 is known as *Boyle's law* for constant temperature processes.

$$pv = \text{constant} \qquad 27.8$$

$$W = RT \ln\left(\frac{v_2}{v_1}\right) = RT \ln\left(\frac{p_1}{p_2}\right) \qquad 27.9$$

Isentropic

$$pv^k = \text{constant} \qquad 27.10$$

$$W = \frac{p_2 v_2 - p_1 v_1}{1 - k}$$

$$= \frac{R(T_2 - T_1)}{1 - k}$$

$$= \left(\frac{RT_1}{k-1}\right)\left(1 - \left(\frac{p_2}{p_1}\right)^{\frac{k-1}{k}}\right) \qquad 27.11$$

Polytropic

$$pv^n = \text{constant} \qquad 27.12$$

$$W = \frac{p_2 v_2 - p_1 v_1}{1 - n} \qquad 27.13$$

Open Systems

The first law of thermodynamics can also be written for open systems, but more terms are required to account for the many energy forms. The first law formulation is essentially the *Bernoulli energy conservation equation* extended to non-adiabatic processes.

$$Q = \Delta U + \Delta E_p + \Delta E_k + W_{\text{rev}} + W_{\text{shaft}} \qquad 27.14$$

Q is the heat flow into the system, inclusive of any losses. It can be supplied from furnace flame, electrical heating, nuclear reaction, or other sources. If the system is adiabatic, Q is zero.

If the kinetic and potential energy terms are neglected and shaft work is zero, the reversible work (p-V work, flow work) performed by or on the system is given by Eq. 27.15.

$$W_{\text{rev}} = -\int v\, dp \qquad 27.15$$

The reversible (flow) work is the work required to cause flow into the system against the exit pressure.

Special Cases of Open Systems (for Ideal Gases)

Constant Volume

$$W = -v(p_2 - p_1) \qquad 27.16$$

Constant Pressure

$$W = 0 \qquad 27.17$$

Constant Temperature

$$pv = \text{constant} \qquad 27.18$$

$$W = RT \ln\left(\frac{v_2}{v_1}\right) = RT \ln\left(\frac{p_1}{p_2}\right) \qquad 27.19$$

Isentropic

$$pv^k = \text{constant} \qquad 27.20$$

$$W = \frac{k(p_2 v_2 - p_1 v_1)}{1 - k}$$

$$= \frac{kR(T_2 - T_1)}{1 - k}$$

$$= \left(\frac{k}{k-1}\right) RT_1 \left(1 - \left(\frac{p_2}{p_1}\right)^{\frac{k-1}{k}}\right) \qquad 27.21$$

Polytropic

$$pv^n = \text{constant} \qquad 27.22$$

$$W = \frac{n(p_2 v_2 - p_1 v_1)}{1 - n} \qquad 27.23$$

Steady-State Systems

If the mass flow rate is constant, the system is a *steady-flow system*, and the first law is known as the *steady-flow energy equation*, SFEE, Eq. 27.24. The subscripts i and e denote conditions at the in-point and exit of the control volume, respectively.

$$\sum \dot{m}_i\left(h_i + \frac{v_i^2}{2} + gz_i\right)$$
$$- \sum \dot{m}_e\left(h_e + \frac{v_e^2}{2} + gz_e\right)$$
$$+ \dot{Q} - \dot{W} = 0 \qquad \text{[SI]} \qquad 27.24a$$

$$\sum \dot{m}_i\left(h_i + \frac{v_i^2}{2g_c J} + \frac{gz_i}{g_c J}\right)$$
$$- \sum \dot{m}_e\left(h_e + \frac{v_e^2}{2g_c J} + \frac{gz_e}{g_c J}\right)$$
$$+ \dot{Q} - \frac{\dot{W}}{J} = 0 \qquad \text{[U.S.]} \qquad 27.24b$$

$\frac{1}{2}v^2 + gz$ represents the sum of the fluid's kinetic and potential energies. Generally, these terms are insignificant compared with the thermal energy.

$\dot{W}$ is the rate of *shaft work* (i.e., *shaft power*)—work that the steady-flow device does on the surroundings. Its name is derived from the output shaft that serves to transmit energy out of the system. For example, turbines and internal combustion engines have output shafts. $\dot{W}$ can be negative, as in the case of a pump or compressor.

The enthalpy, h, represents a combination of internal energy and reversible (flow) work.

Special Cases of Steady-Flow Energy Equation

Nozzles and Diffusers

Since a flowing fluid is in contact with nozzle, orifice, and valve walls for only a very short period of time, flow through them is essentially adiabatic. No work is done on the fluid as it passes through. If the potential energy changes are neglected, the SFEE reduces to

$$h_i + \frac{v_i^2}{2} = h_e + \frac{v_e^2}{2} \qquad \text{[SI]} \qquad 27.25a$$

$$h_i + \frac{v_i^2}{2g_c J} = h_e + \frac{v_e^2}{2g_c J} \qquad \text{[U.S.]} \qquad 27.25b$$

The *nozzle efficiency* is defined as

$$
\begin{aligned}
\eta &= \frac{\Delta h_{\text{actual}}}{\Delta h_{\text{ideal}}} \\
&= \frac{v_e^2 - v_i^2}{2(h_i - h_{es})} \qquad\qquad 27.26
\end{aligned}
$$

The subscript *es* refers to the exit condition for an isentropic (ideal) expansion.

Turbines, Pumps, and Compressors

A *pump* or *compressor* converts mechanical energy into fluid energy, increasing the total energy content of the fluid flowing through it. *Turbines* can generally be thought of as pumps operating in reverse. A turbine extracts energy from the fluid, converting fluid energy into mechanical energy.

These devices can be considered to be adiabatic because the fluid gains (or loses) very little heat during the short time it passes through them. The kinetic and potential energy terms can be neglected. Then, the SFEE reduces to Eq. 27.27. The rate of work, $\dot{W}$, is the same as the power.

$$\dot{W} = \dot{m}(h_i - h_e) \qquad \text{[SI]} \qquad 27.27a$$

$$\frac{\dot{W}}{J} = \dot{m}(h_i - h_e) \qquad \text{[U.S.]} \qquad 27.27b$$

On a per unit mass basis (i.e., $\dot{m} = 1$),

$$W_{\text{per kilogram}} = h_i - h_e \qquad \text{[SI]} \qquad 27.28a$$

$$\frac{W_{\text{per pound}}}{J} = h_i - h_e \qquad \text{[U.S.]} \qquad 27.28b$$

Equation 27.28 assumes that the pump or turbine is capable of isentropic compression. However, due to inefficiencies, the actual exit enthalpy will deviate from the ideal isentropic enthalpy, h_{es}. The actual efficiencies are given by Eqs. 27.29 and 27.30.

$$\eta_{\text{turbine}} = \frac{h_i - h_e}{h_i - h_{es}} \qquad 27.29$$

$$\eta_{\text{pump}} = \frac{h_{es} - h_i}{h_e - h_i} \qquad 27.30$$

Throttling Valves and Throttling Processes

In a *throttling process* there is no change in system enthalpy, but there is a significant pressure drop. The process is adiabatic and the SFEE reduces to

$$h_i = h_e \qquad 27.31$$

Boilers, Condensers, and Evaporators

A *boiler* is part of a steam generator that transfers combustion heat energy from a furnace to feedwater. Modern boilers are water-tube boilers (i.e., water passes through tubes surrounded by combustion gases).

Condensers are special-purpose heat exchangers that remove the heat of vaporization from fluids. This heat energy is transferred through the heat exchanger walls to cooling water or air and then to the environment.

Evaporators vaporize low-pressure liquid by heat absorption.

These devices are nonadiabatic. The SFEE reduces to

$$h_i + q = h_e \qquad 27.32$$

Heat Exchangers

A *heat exchanger* transfers heat energy from one fluid to another through a wall separating them. If the heat exchanger is considered as the control volume and the heat transfer takes place entirely within the control volume, then the process may be assumed to be adiabatic. For this condition, the changes in each fluid stream's energy are equal but opposite. No work is done within a heat exchanger, and the potential and kinetic energies of the fluids can be ignored. Therefore, the SFEE reduces to

$$\frac{\text{energy increase}}{\text{of fluid 1}} = \frac{\text{energy decrease}}{\text{of fluid 2}} \qquad 27.33$$

$$\dot{m}_1(h_{1i} - h_{1e}) = -\dot{m}_2(h_{2i} - h_{2e}) \qquad 27.34$$

Feedwater Heaters

A *feedwater heater* uses steam to increase the temperature of water entering the steam generator. The steam can come from any waste steam source but is usually bled off from a turbine. In this latter case, the heater is known as an *extraction heater*. The water that is heated usually comes from the condenser.

Open heaters (also known as *direct contact heaters* and *mixing heaters*) physically mix the steam and water. A *closed feedwater heater* is a traditional closed heat exchanger that can operate at either high or low pressures. There is no mixing of the water and steam in the feedwater heater. The cooled stream leaves the feedwater heater as a liquid. For adiabatic operation, the SFEE reduces to

$$\sum \dot{m}_i h_i = \sum \dot{m}_e h_e \qquad 27.35$$

$$\sum \dot{m}_i = \sum \dot{m}_e \qquad 27.36$$

SAMPLE PROBLEMS

Problem 1

A closed thermodynamic system consists of a stone with a mass of 10 kg and a bucket that contains 100 kg of water. Initially the stone is at rest 40.8 m above the water, and the stone, bucket, water, and environment are at the same temperature. The stone then falls into the water. No water is lost from the bucket.

What is the change in internal energy (ΔU), change in kinetic energy (ΔKE), change in potential energy (ΔPE), heat flow (Q), and work done (W), after the stone has been dropped into the water and the system has reached equilibrium at its original temperature?

(A) $\Delta U = 0$ kJ; $\Delta KE = 0$ kJ;
$\Delta PE = -4$ kJ; $Q = -4$ kJ; $W = 0$ kW
(B) $\Delta U = 4$ kJ; $\Delta KE = 0$ kJ;
$\Delta PE = -4$ kJ; $Q = 0$ kJ; $W = 0$ kW
(C) $\Delta U = 0$ kJ; $\Delta KE = 4$ kJ;
$\Delta PE = -4$ kJ; $Q = 0$ kJ; $W = 0$ kW
(D) $\Delta U = -4$ kJ; $\Delta KE = 0$ kJ;
$\Delta PE = 0$ kJ; $Q = -4$ kJ; $W = 0$ kW

Solution
The change in potential energy of the stone is

$$\Delta PE_{stone} = mg\Delta h$$

$$= (10 \text{ kg}) \left(9.81 \ \frac{\text{m}}{\text{s}^2}\right) (0 \text{ m} - 40.8 \text{ m})$$

$$= -4002 \text{ J} \quad (-4 \text{ kJ})$$

The system transfers 4 kJ to the environment, so $Q = -4$ kJ. Everything starts at rest and ends at rest, so $\Delta KE = 0$ kJ. Everything starts at the temperature of the environment and ends at the same temperature, so $\Delta U = 0$ kJ. The falling stone performs no work on the surroundings; therefore, $W = 0$ kW.

Answer is A.

Problem 2

During a process, 30 J of work are done by a closed stationary system on its surroundings. The internal energy of the system decreases by 40 J. What is the heat transfer?

(A) 10 J released into the surroundings
(B) 10 J absorbed by the system
(C) 70 J released into the surroundings
(D) 70 J absorbed by the system

Solution
From the first law,

$$Q = \Delta U + W$$
$$= -40 \text{ J} + 30 \text{ J}$$
$$= -10 \text{ J} \quad (10 \text{ J})$$

Answer is A.

Problem 3

A steam coil operating at steady state receives 30 kg/min of steam with an enthalpy of 2900 kJ/kg. If the steam leaves with an enthalpy of 1600 kJ/kg, what is the rate of heat transfer from the coil?

(A) 140 kJ/min
(B) 650 kJ/min
(C) 2300 kJ/min
(D) 39 000 kJ/min

Solution
The steady-flow energy equation is

$$\dot{m}\left(h_i + \frac{\text{v}_i^2}{2} + gz_i\right) - \dot{m}\left(h_e + \frac{\text{v}_e^2}{2} + gz_e\right) + \dot{Q} - \dot{W} = 0$$

$$\text{v}_i = \text{v}_e$$
$$z_i = z_e$$
$$\dot{W} = 0$$
$$\dot{m}(h_i - h_e) + \dot{Q} = 0$$
$$\dot{Q} = -\dot{m}(h_i - h_e)$$
$$= -\left(30 \ \frac{\text{kg}}{\text{min}}\right)\left(2900 \ \frac{\text{kJ}}{\text{kg}} - 1600 \ \frac{\text{kJ}}{\text{kg}}\right)$$
$$= -39\,000 \text{ kJ/min}$$

Answer is D.

Problem 4

A gas with a molecular weight of 55 initially at 2 MPa, 200°C, and 0.5 m³ expands in accordance with the relation $pV^{1.35} = C$ to 0.2 MPa. Determine the work for the expansion process.

(A) 1.3 MJ
(B) 2.3 MJ
(C) 4.1 MJ
(D) 5.9 MJ

Solution

$$p_1 V_1^n = p_2 V_2^n$$

$$V_2^n = \frac{p_1 V_1^n}{p_2}$$

$$V_2^{1.35} = \left(\frac{2 \text{ MPa}}{0.2 \text{ MPa}} \right) (0.5 \text{ m}^3)^{1.35}$$

$$= 3.923 \text{ m}^3$$

$$(1.35)\left(\ln V_2 \right) = \ln 3.923 \text{ m}^3$$

$$V_2 = 2.752 \text{ m}^3$$

For a polytropic closed system,

$$W = \frac{p_2 V_2 - p_1 V_1}{1 - n}$$

$$= \frac{(0.2 \text{ MPa})(2.752 \text{ m}^3) - (2 \text{ MPa})(0.5 \text{ m}^3)}{1 - 1.35}$$

$$= 1.285 \text{ MJ} \quad (1.3 \text{ MJ})$$

Answer is A.

Problem 5

2 m^3/min of a light oil are to be heated from 20°C to 100°C (with zero vaporization) in an exchanger using 250 kPa steam of 90% quality. The heat losses to the surrounding air have been estimated to be 5% of the heat transferred from the condensing steam to the oil. If the steam condensate leaves at its saturation point, what mass of steam per hour will be used in the exchanger?

For light oil,

$$\text{specific gravity} = 0.88$$
$$\text{specific heat} = 2.00 \text{ kJ/kg·K}$$

For steam at 250 kPa,

$$\text{saturated liquid} = h_f = 535.49 \text{ kJ/kg}$$
$$v_f = 0.001067 \text{ m}^3/\text{kg}$$
$$\text{saturated vapor} = h_g = 2716.8 \text{ kJ/kg}$$
$$v_g = 0.7188 \text{ m}^3/\text{kg}$$

(A) 1100 kg/h
(B) 4600 kg/h
(C) 8600 kg/h
(D) 9000 kg/h

Solution

As the steam enters the heat exchanger, it is at its saturation pressure and temperature.

$$h_i = h_f + x_i h_{fg}$$

$$= 535.49 \frac{\text{kJ}}{\text{kg}}$$

$$+ (0.90) \left(2716.8 \frac{\text{kJ}}{\text{kg}} - 535.49 \frac{\text{kJ}}{\text{kg}} \right)$$

$$= 2498.7 \text{ kJ/kg}$$

The steam condensate is at its saturation point with a quality of zero. The energy transferred inside the heat exchanger is the change in enthalpies.

$$q = h_e - h_i$$

$$= 535.49 \frac{\text{kJ}}{\text{kg}} - 2498.7 \frac{\text{kJ}}{\text{kg}}$$

$$= -1963.2 \text{ kJ/kg}$$

The energy required to heat the oil is

$$\dot{Q} = \dot{m} \Delta h = \dot{m} c \Delta T$$

$$= \left(2 \frac{\text{m}^3}{\text{min}} \right) \left(60 \frac{\text{min}}{\text{h}} \right) (0.88)$$

$$\times \left(1000 \frac{\text{kg}}{\text{m}^3} \right) \left(2.00 \frac{\text{kJ}}{\text{kg·K}} \right) (100°\text{C} - 20°\text{C})$$

$$= 1.69 \times 10^7 \text{ kJ/h}$$

The total energy required must account for heat losses.

$$\dot{Q}_{\text{total}} = (1 + 0.05) \left(1.69 \times 10^7 \frac{\text{kJ}}{\text{h}} \right)$$

$$= 1.775 \times 10^7 \text{ kJ/h}$$

The amount of steam required is

$$\dot{m}_{\text{steam}} = \frac{\dot{Q}_{\text{total}}}{q} = \frac{1.775 \times 10^7 \frac{\text{kJ}}{\text{h}}}{1963.2 \frac{\text{kJ}}{\text{kg}}}$$

$$= 9041 \text{ kg/h} \quad (9000 \text{ kg/h})$$

Answer is D.

FE-STYLE EXAM PROBLEMS

1. Air is compressed isentropically such that its pressure is increased by 50%. The initial temperature is 70°C. What is the final temperature?

(A) 80°C
(B) 110°C
(C) 140°C
(D) 240°C

2. Air is compressed in a piston-cylinder arrangement to 1/10 of its initial volume. If the initial temperature is 35°C and the process is frictionless and adiabatic, what is the final temperature?

(A) 350K
(B) 360K
(C) 620K
(D) 770K

3. Gas initially at 1 MPa and 150°C receives 7.2 MJ of work while 1.5 kW of heat are removed from the system. Calculate the internal energy change for the system over a period of an hour.

(A) −5.7 MJ
(B) 1.8 MJ
(C) 8.7 MJ
(D) 13 MJ

4. One kilogram of air is compressed from a volume of 1.0 m³ and a pressure of 100 kPa to a volume of 0.147 m³ and a pressure of 1000 kPa. Assuming that the compression follows the law $pv^n = $ constant, find the work done during the compression process.

(A) 70 kJ
(B) 100 kJ
(C) 118 kJ
(D) 235 kJ

5. Steam enters an adiabatic nozzle at 1 MPa, 250°C, and 30 m/s. At one point in the nozzle the enthalpy has dropped 40 kJ/kg from its inlet value. Determine the velocity at that point.

(A) 31 m/s
(B) 110 m/s
(C) 250 m/s
(D) 280 m/s

For the following problems use the NCEES Handbook as your only reference.

6. A boiler feedwater pump receives saturated liquid water at 50°C and compresses it isentropically to 1 MPa. For a water flow rate of 100 Mg/h, estimate the pump power.

(A) −39 kW
(B) −35 kW
(C) −28 kW
(D) −20 kW

7. The pump work required to compress water at 1.5 MPa and 50°C to 15 MPa is 15 kJ/kg. What is the efficiency of the pump?

(A) 88%
(B) 90%
(C) 91%
(D) 94%

8. Calculate the power required to compress 10 kg/s of air from 1 atm and 37°C to 2 atm and 707°C.

For low pressure air,

$$T = 310\text{K}; \ h = 290.4 \text{ kJ/kg}$$
$$T = 980\text{K}; \ h = 1023 \text{ kJ/kg}$$

(A) 5260 kW
(B) 7020 kW
(C) 7260 kW
(D) 7330 kW

9. Which of the following statements is true for an isentropic steady-flow process? (p is pressure, T is temperature, and V is volume.)

(A) $\dfrac{p_2}{p_1} = \left(\dfrac{V_1}{V_2}\right)^k$

(B) work $= -V(p_2 - p_1)$

(C) $\dfrac{p_2}{p_1} = \dfrac{V_1}{V_2}$

(D) $\dfrac{p_1}{T_1} = \dfrac{p_2}{T_2}$

10. Air is compressed from 100 kPa and 40°C to 1500 kPa and 130°C in a steady-flow operation. During the compression process, each kilogram of air loses 90 kJ as heat to the environment. Air is discharged at the rate of 10 m³/min. What power is required to drive the compressor?

(A) 126 kW
(B) 180 kW
(C) 195 kW
(D) 391 kW

11. Which of the following principles is the closest interpretation of the first law of thermodynamics for a closed system?

(A) The mass within a closed control volume does not change.

(B) The net energy crossing the system boundary is the change in energy inside the system.

(C) The change of total energy is equal to the rate of work performed.

(D) All real processes tend toward increased entropy.

12. The state of an ideal gas is changed in a steady-state open process from 400 kPa and 1.2 m^3 to 300 kPa and 1.5 m^3. The relationship between pressure and volume during the process is $pV^{1.3} = $ constant. What work is performed?

(A) 130 kJ
(B) 930 kJ
(C) 1210 kJ
(D) 9030 kJ

Problems 13–16 refer to the following situation.

0.5 m^3 of superheated steam at 400 kPa and 300°C is expanded behind a piston until the temperature is 210°C. The steam expands polytropically with a polytropic exponent of 1.3.

13. What is the mass of steam contained behind the piston?

(A) 0.040 kg
(B) 0.76 kg
(C) 42 kg
(D) 55 kg

14. What is the final pressure of the steam?

(A) 120 kPa
(B) 140 kPa
(C) 190 kPa
(D) 230 kPa

15. What is the final volume of the steam?

(A) 0.65 m^3
(B) 0.72 m^3
(C) 0.78 m^3
(D) 0.88 m^3

16. What is the work done during the expansion process?

(A) 24.2 kJ
(B) 105 kJ
(C) 326 kJ
(D) 416 kJ

Problems 17–19 refer to the following situation.

A gas goes through the following thermodynamic processes.

A to B: constant-temperature compression
B to C: constant-volume cooling
C to A: constant-pressure expansion

The pressure and volume at state C are 1.4 bars and 0.028 m^3, respectively. The net work during the C-to-A process is 10.5 kJ.

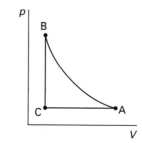

17. What is the volume at state A?

(A) 0.07 m^3
(B) 0.10 m^3
(C) 0.19 m^3
(D) 0.24 m^3

18. What work is performed in the A-to-B process?

(A) −19 kJ
(B) −13 kJ
(C) 0 kJ
(D) 5.3 kJ

19. What net work is derived from one complete A-B-C cycle?

(A) −21 kJ
(B) −8.3 kJ
(C) −6.5 kJ
(D) 4.8 kJ

SOLUTIONS TO FE-STYLE EXAM PROBLEMS

1. For a closed isentropic (constant entropy) process,

$$T_1 = 70°C + 273 = 343K$$

$$\frac{T_1}{T_2} = \left(\frac{p_2}{p_1}\right)^{(1-k)/k}$$

$$T_2 = T_1 \left(\frac{p_1}{p_2}\right)^{(1-k)/k}$$

$$= (343K)\left(\frac{1}{1.5}\right)^{(1-1.4)/1.4}$$

$$= 385.1K$$

$$T_2 = 385.1K - 273$$

$$= 112°C \quad (110°C)$$

Answer is B.

2. A frictionless adiabatic process is also isentropic. For an isentropic (constant entropy) process,

$$\frac{T_2}{T_1} = \left(\frac{v_1}{v_2}\right)^{k-1}$$

$$T_2 = T_1 \left(\frac{v_1}{v_2}\right)^{k-1}$$

$$= (35°C + 273)\left(\frac{10}{1}\right)^{1.4-1}$$

$$= 773.7K \quad (770K)$$

Answer is D.

3. For a closed system,

$$\Delta U = Q - W$$

$$= (-1500 \text{ W})\left(3600 \frac{\text{s}}{\text{h}}\right)(1 \text{ h}) + (7.2 \times 10^6 \text{ J})$$

$$= 1.8 \times 10^6 \text{ J} \quad (1.8 \text{ MJ})$$

Answer is B.

4. Since pv^n is constant,

$$p_1 v_1^n = p_2 v_2^n$$

$$\frac{p_2}{p_1} = \left(\frac{v_1}{v_2}\right)^n$$

$$n = \frac{\log\left(\frac{p_2}{p_1}\right)}{\log\left(\frac{v_1}{v_2}\right)} = \frac{\log\left(\frac{1000 \text{ kPa}}{100 \text{ kPa}}\right)}{\log\left(\frac{1.0 \text{ m}^3}{0.147 \text{ m}^3}\right)}$$

$$= 1.2$$

Use the equation for work for a closed polytropic process.

$$W = \frac{p_2 v_2 - p_1 v_1}{1 - n}$$

$$= \frac{(1000 \text{ kPa})(0.147 \text{ m}^3) - (100 \text{ kPa})(1 \text{ m}^3)}{1 - 1.2}$$

$$= -235 \text{ kJ} \quad (235 \text{ kJ})$$

(By convention, work performed by the system is positive. In this case, work is performed on the system and is negative.)

Answer is D.

5. For the nozzle,

$$h_i + \frac{v_i^2}{2} = h_e + \frac{v_e^2}{2}$$

$$v_e = \sqrt{2(h_i - h_e) + v_i^2}$$

$$= \sqrt{(2)\left(40 \frac{\text{kJ}}{\text{kg}}\right)\left(1000 \frac{\text{J}}{\text{kJ}}\right) + \left(30 \frac{\text{m}}{\text{s}}\right)^2}$$

$$= 284 \text{ m/s} \quad (280 \text{ m/s})$$

Answer is D.

6. From the saturated water table, at $50°C$, $p = 12.349$ kPa and $v = 0.001012$ m^3/kg. For steady state,

$$W = -\int v \, dp$$

$$= -v\Delta p$$

$$\dot{W} = \dot{m}(-v\Delta p)$$

$$= \left(\frac{100\,000 \frac{\text{kg}}{\text{h}}}{3600 \frac{\text{s}}{\text{h}}}\right)\left(-0.001012 \frac{\text{m}^3}{\text{kg}}\right)$$

$$\times (1000 \text{ kPa} - 12.349 \text{ kPa})$$

$$= -27.8 \text{ kW} \quad (-28 \text{ kW})$$

(The answer is negative since the surroundings do work on the water.)

Answer is C.

7. If water is considered to be incompressible, this is a constant volume process.

$$W_{\text{actual}} = 15 \text{ kJ/kg}$$

$$W_{\text{ideal}} = -\int v \, dp$$

$$= -v\Delta p$$

$$\eta = \text{pump efficiency} = \frac{W_{\text{ideal}}}{W_{\text{actual}}}$$

$$= \frac{\left(0.001012 \, \frac{\text{m}^3}{\text{kg}}\right)(15\,000 \text{ kPa} - 1500 \text{ kPa})}{15 \, \frac{\text{kJ}}{\text{kg}}}$$

$$= 0.91 \quad (91\%)$$

Answer is C.

8.
$$\dot{W} = \dot{m}(h_i - h_e)$$

$$= \left(10 \, \frac{\text{kg}}{\text{s}}\right)\left(1023 \, \frac{\text{kJ}}{\text{kg}} - 290.4 \, \frac{\text{kJ}}{\text{kg}}\right)$$

$$= 7326 \text{ kW} \quad (7330 \text{ kW})$$

Answer is D.

9. Choice (A) is valid for an isentropic process. Choice (B) is valid for an incompressible fluid in a constant-volume process. Choice (C) is valid for an ideal gas in a constant-temperature process. Choice (D) is valid for a constant-volume process.

Answer is A.

10. Use the first law of thermodynamics for steady-flow, open systems.

$$\dot{m}_1\left(h_1 + \frac{1}{2}v_1^2 + gz_1\right)$$

$$- \dot{m}_2\left(h_2 + \frac{1}{2}v_2^2 + gz_2\right)$$

$$+ \dot{Q} - \dot{W} = 0$$

The $(1/2)(v^2) + gz$ terms are the kinetic and potential energies. These terms are insignificant compared with the thermal energy and can generally be ignored. Since the flow is steady, the mass-flow terms are equal. Dividing all terms by $\dot{m}$ places Q and W on a per-unit-mass basis.

$$W = h_1 - h_2 + Q \quad [\text{per unit mass}]$$

Choose the air as the system. W will be positive if the system does work on the environment. (In this case, W

is negative because the environment does work on the air during the compression.) Q will be positive if heat energy is transferred into the air. (In this case, Q is negative because the air loses heat energy.)

Consider air to be an ideal gas. The change in the enthalpy is linearly dependent on the temperature change. The difference in temperatures is the same in °C as in K.

$$c_p = 1.005 \text{ kJ/kg·K}$$

$$h_2 - h_1 = c_p(T_2 - T_1)$$

$$= \left(1.005 \, \frac{\text{kJ}}{\text{kg·K}}\right)(130°\text{C} - 40°\text{C})$$

$$= 90.45 \text{ kJ/kg}$$

Reversing the order,

$$h_1 - h_2 = -90.45 \text{ kJ/kg}$$

The heat loss is 90 kJ/kg.

$$W_{\text{per kilogram}} = h_1 - h_2 + Q$$

$$= -90.45 \, \frac{\text{kJ}}{\text{kg}} - 90 \, \frac{\text{kJ}}{\text{kg}}$$

$$= -180.45 \text{ kJ/kg}$$

Use the ideal gas law to determine the mass flow rate. The molecular weight of air is approximately 29 kg/kmol.

$$pV = \frac{m\overline{R}T}{\text{MW}}$$

$$\dot{m} = \frac{p\dot{V}(\text{MW})}{\overline{R}T}$$

$$= \frac{(1500 \text{ kPa})\left(10 \, \frac{\text{m}^3}{\text{min}}\right)\left(29 \, \frac{\text{kg}}{\text{kmol}}\right)}{\left(60 \, \frac{\text{s}}{\text{min}}\right)\left(8.3143 \, \frac{\text{kJ}}{\text{kmol·K}}\right)(130°\text{C} + 273)}$$

$$= 2.164 \text{ kg/s}$$

The rate of work input (i.e., the power input) is

$$\dot{W} = W_{\text{per kilogram}}\dot{m}$$

$$= \left(180.45 \, \frac{\text{kJ}}{\text{kg}}\right)\left(2.164 \, \frac{\text{kg}}{\text{s}}\right)$$

$$= 390.5 \text{ kW} \quad (391 \text{ kW})$$

Answer is D.

11. A mathematical expression of the first law is

$$Q - W = \Delta U$$

Q is the heat transfer into the system across the system boundary. W is the energy transferred across the system boundary to the surroundings in the form of work. ΔU is the change in energy stored within the system in the form of internal energy. This expression shows that the energy transferred across the system boundary comes from a change in stored energy.

Answer is B.

12. The work for an open, polytropic process is

$$
\begin{aligned}
W &= \frac{n(p_2 V_2 - p_1 V_1)}{1 - n} \\
&= \frac{(1.3)((300 \text{ kPa})(1.5 \text{ m}^3) - (400 \text{ kPa})(1.2 \text{ m}^3))}{1 - 1.3} \\
&= 130 \text{ kJ}
\end{aligned}
$$

Notice that the work for a polytropic open system is not the same as that for a polytropic closed system.

Answer is A.

13. Since the steam is superheated, consider it to be an ideal gas. The molecular weight of water is 18 kg/kmol. The specific gas constant is

$$
\begin{aligned}
R = \frac{\overline{R}}{MW} &= \frac{8.314 \dfrac{\text{kJ}}{\text{kmol·K}}}{18 \dfrac{\text{kg}}{\text{kmol}}} \\
&= 0.4619 \text{ kJ/kg·K}
\end{aligned}
$$

Use the ideal gas law.

$$
\begin{aligned}
pV &= mRT \\
m &= \frac{pV}{RT} \\
&= \frac{(400 \text{ kPa})\left(1000 \dfrac{\text{Pa}}{\text{kPa}}\right)(0.5 \text{ m}^3)}{\left(0.4619 \dfrac{\text{kJ}}{\text{kg·K}}\right)\left(1000 \dfrac{\text{J}}{\text{kJ}}\right)(300^\circ\text{C} + 273)} \\
&= 0.7557 \text{ kg} \quad (0.76 \text{ kg})
\end{aligned}
$$

Answer is B.

14. Use the polytropic expression.

$$
\begin{aligned}
p_1 V_1^n &= p_2 V_2^n \\
p_2 &= p_1 \left(\frac{V_1}{V_2}\right)^n
\end{aligned}
$$

However, V_2 is not known. Use the ideal gas law.

$$
\begin{aligned}
p_2 V_2 &= mRT_2 \\
V_2 &= \frac{mRT_2}{p_2}
\end{aligned}
$$

Substitute V_2 into the polytropic expression.

$$
p_1 V_1^n = p_2 \left(\frac{mRT_2}{p_2}\right)^n = p_2^{1-n}(mRT_2)^n
$$

$$
p_2^{1-n} = p_1 \left(\frac{V_1}{mRT_2}\right)^n
$$

$$
\begin{aligned}
&= (400 \text{ kPa})\left(\frac{0.5 \text{ m}^3}{(0.7557 \text{ kg})\left(0.4618 \dfrac{\text{kJ}}{\text{kg·K}}\right) \times (210^\circ\text{C} + 273)}\right)^{1.3} \\
&= 0.207
\end{aligned}
$$

$$
p_2 = (0.207)^{1/(1-1.3)} = 190.7 \text{ kPa} \quad (190 \text{ kPa})
$$

Answer is C.

15. From Prob. 14,

$$
\begin{aligned}
V_2 &= \frac{mRT_2}{p_2} \\
&= \frac{(0.7557 \text{ kg})\left(0.4618 \dfrac{\text{kJ}}{\text{kg·K}}\right)\left(1000 \dfrac{\text{J}}{\text{kJ}}\right) \times (210^\circ\text{C} + 273)}{(190.7 \text{ kPa})\left(1000 \dfrac{\text{Pa}}{\text{kPa}}\right)} \\
&= 0.884 \text{ m}^3 \quad (0.88 \text{ m}^3)
\end{aligned}
$$

Answer is D.

16. This is a closed polytropic process.

$$
\begin{aligned}
W &= \frac{p_2 V_2 - p_1 V_1}{1 - n} \\
&= \frac{(190.7 \text{ kPa})(0.884 \text{ m}^3) - (400 \text{ kPa})(0.5 \text{ m}^3)}{1 - 1.3} \\
&= 104.7 \text{ kJ} \quad (105 \text{ kJ})
\end{aligned}
$$

Notice that the work for a polytropic open system is not the same as that for a polytropic closed system.

Answer is B.

17. C to A: The work done during this process is given. Use the formula for work of a constant-pressure process.

$$W_{\text{C-A}} = p(V_A - V_C)$$

$$(10.5 \text{ kJ}) \left(1000 \ \frac{\text{J}}{\text{kJ}}\right) = (1.4 \text{ bar}) \left(10^5 \ \frac{\text{Pa}}{\text{bar}}\right)$$
$$\times (V_A - 0.028 \text{ m}^3)$$
$$V_A = 0.103 \text{ m}^3 \quad (0.10 \text{ m}^3)$$

Answer is B.

18. The pressure at state A is the same as the pressure at state C. The work in constant-temperature processes per unit volume is

$$W_{\text{A-B}} = RT_A \ \ln \frac{p_A}{p_B}$$

Substitute $p_A V_A$ for RT_A (ideal gas law). Substitute V_B/V_A for p_A/p_B (Boyle's law).

$$W_{\text{A-B}} = p_A V_A \ \ln \frac{V_B}{V_A}$$

$$= (1.4 \text{ bar}) \left(10^5 \ \frac{\text{Pa}}{\text{bar}}\right)(0.103 \text{ m}^3) \ \ln \left(\frac{0.028 \text{ m}^3}{0.103 \text{ m}^3}\right)$$

$$= -18782 \text{ J} \quad (-19 \text{ kJ})$$

Answer is A.

19. The work for the A-to-B process was calculated in Prob. 18. The work in the C-to-A process was given. The work in a constant-volume process (i.e., B to C) is zero.

$$W_{\text{net}} = W_{\text{A-B}} + W_{\text{B-C}} + W_{\text{C-A}}$$
$$= -18.8 \text{ kJ} + 0 + 10.5 \text{ kJ}$$
$$= -8.3 \text{ kJ}$$

Answer is B.

28 Power Cycles and Entropy

Subjects

BASIC CYCLES 28-1
  Carnot Cycle 28-2
  Rankine Cycle 28-2
  Otto Cycle 28-4
  Refrigeration Cycles 28-4
ENTROPY CHANGES 28-5
  Inequality of Clausius 28-6
SECOND LAW OF THERMODYNAMICS . 28-6
  Kelvin-Planck Statement of
    Second Law (Power Cycles) 28-6
  Clausius Statement of Second Law
    (Refrigeration Cycles) 28-7
FINDING WORK AND HEAT
    GRAPHICALLY 28-7

Nomenclature

c	specific heat	Btu/lbm-°R	kJ/kg·K
COP	coefficient of performance	–	–
g	acceleration of gravity	ft/sec^2	m/s^2
h	enthalpy	Btu/lbm	kJ/kg
I	process irreversibility	ft-lbf	kJ
k	ratio of specific heats	–	–
m	mass	lbm	kg
MW	molecular weight	lbm/lbmole	kg/kmol
p	absolute pressure	lbf/ft^2	Pa
P	power	Btu/sec	kW
Q	total heat energy	Btu	kJ
$\dot{Q}$	rate of heat transfer	Btu/sec	kW
r_v	volumetric compression ratio	–	–
R	specific gas constant	ft-lbf/lbm-°R	J/kg·K
s	entropy	Btu/lbm-°R	kJ/kg·K
S	total entropy	Btu/°R	kJ/K
T	absolute temperature	°R	K
V	volume	ft^3	m^3
W	work	ft-lbf	kJ
z	elevation	ft	m

Symbols

η	efficiency	–	–
υ	specific volume	ft^3/lbm	m^3/kg
ϕ	closed-system availability	Btu/lbm	kJ/kg
Ψ	open-system availability	Btu/lbm	kJ/kg

Subscripts

C	Carnot
f	final or fluid (liquid)
fg	liquid-to-gas (vaporization)
g	gas (vapor)
H	high temperature
L	low temperature
th	thermal

BASIC CYCLES

It is convenient to show a source of energy as an infinite constant-temperature reservoir. Figure 28.1 illustrates a source of energy known as a *high-temperature reservoir* or *source* reservoir. By convention, the reservoir temperature is designated T_H, and the heat transfer from it is Q_H. The energy derived from such a theoretical source might actually be supplied by combustion, electrical heating, or nuclear reaction.

Similarly, energy is released (i.e., is "rejected") to a low-temperature reservoir known as a *sink reservoir* or *energy sink*. The most common practical sink is the local environment. T_L and Q_L are used to represent the reservoir temperature and energy absorbed. It is common to refer to Q_L as the "rejected energy" or "energy rejected to the environment."

Although heat can be extracted and work can be performed in a single process, a *cycle* is necessary to obtain work in a useful quantity and duration. A *cycle* is a series of processes that eventually brings the system back to its original condition. Most cycles are continually repeated.

A cycle is completely defined by the working substance, the high- and low-temperature reservoirs, the means of doing work on the system, and the means of removing energy from the system. (The Carnot cycle depends only on the source and sink temperatures, not on the working fluid. However, most practical cycles depend on the working fluid.)

A cycle will appear as a closed curve when plotted on p-V and T-s diagrams. The area within the p-V and T-s curves represents both the net work and net heat.

THERMO
Power Cycles

Figure 28.1   Energy Flow in Basic Cycles

Figure 28.1   Energy Flow in Basic Cycles

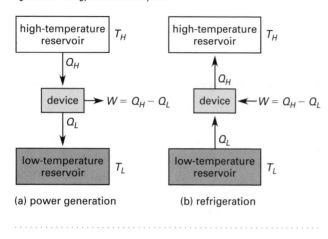

(a) power generation   (b) refrigeration

Figure 28.2   Carnot Cycle

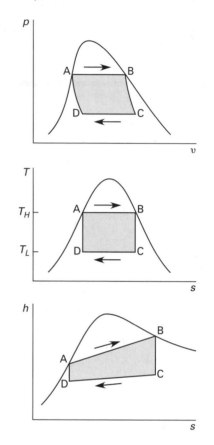

A *power cycle* is a cycle that takes heat and uses it to do work on the surroundings. The *heat engine* is the equipment needed to perform the cycle. The *thermal efficiency* of a power cycle is defined as the ratio of useful work output to the supplied input energy. (The effectiveness of refrigeration and compression cycles is measured by other parameters.) W_{net} in Eq. 28.1 is the net work, since some of the gross output work may be used to run certain parts of the cycle. For example, a small amount of turbine output power may run boiler feed pumps.

$$
\begin{aligned}
\eta_{\mathrm{th}} &= \frac{W_{\mathrm{out}} - W_{\mathrm{in}}}{Q_{\mathrm{in}}} = \frac{W_{\mathrm{net}}}{Q_{\mathrm{in}}} \\
&= \frac{Q_{\mathrm{in}} - Q_{\mathrm{out}}}{Q_{\mathrm{in}}} = \frac{Q_{\mathrm{net}}}{Q_{\mathrm{in}}}
\end{aligned}
\qquad 28.1
$$

Equation 28.1 shows that obtaining the maximum efficiency requires minimizing the Q_{out} term. Equation 28.1 also shows that $W_{\mathrm{net}} = Q_{\mathrm{net}}$. This follows directly from the first law.

Carnot Cycle

The *Carnot cycle* is an ideal power cycle that is impractical to implement. However, its theoretical work output sets the maximum attainable from any heat engine, as evidenced by the isentropic (reversible) processes between states (D and A) and (B and C) in Fig. 28.2. The working fluid in a Carnot cycle is irrelevant.

The processes involved are as follows.

A to B: isothermal expansion of saturated liquid
         to saturated vapor
B to C: isentropic expansion of vapor $(Q = 0, \Delta s = 0)$
C to D: isothermal compression of vapor
D to A: isentropic compression $(Q = 0, \Delta s = 0)$

The most efficient power cycle possible is the Carnot cycle. The thermal efficiency of the entire cycle is given by Eq. 28.2. Temperature must be expressed in the absolute scale.

$$
\eta_{\mathrm{th,Carnot}} = \frac{T_H - T_L}{T_H} = 1 - \frac{T_L}{T_H}
\qquad 28.2
$$

Rankine Cycle

The basic *Rankine cycle* is similar to the Carnot cycle except that the compression process occurs in the liquid region. The Rankine cycle is closely approximated in steam turbine plants. The efficiency of the Rankine cycle is lower than that of a Carnot cycle operating between the same temperature limits because the mean temperature at which heat is added to the system is lower than T_H.

The processes used in the basic Rankine cycle are as follows.

A to B: vaporization in the boiler
B to C: isentropic expansion in the turbine
C to D: condensation
D to E: isentropic compression to boiler pressure
E to A: heating liquid to saturation temperature

Figure 28.3 Basic Rankine Heat Engine

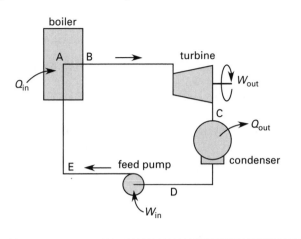

Figure 28.4 Basic Rankine Cycle

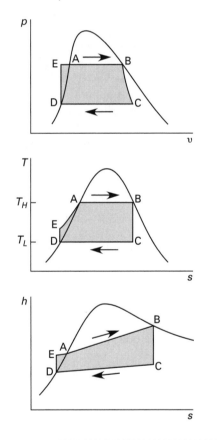

The thermal efficiency of the entire cycle is

$$\eta_{\text{th}} = \frac{W_{\text{out}} - W_{\text{in}}}{Q_{\text{in}}} = \frac{(h_B - h_C) - (h_E - h_D)}{h_B - h_E} \quad 28.3$$

Superheating occurs when heat in excess of that required to produce saturated vapor is added to the water. Superheat is used to raise the vapor above the critical temperature, to raise the mean effective temperature at which heat is added, and to keep the expansion primarily in the vapor region to reduce wear on the turbine blades.

The processes in the *Rankine cycle with superheat*, shown in Fig. 28.5, are similar to the basic Rankine cycle.

Figure 28.5 Rankine Cycle with Superheat

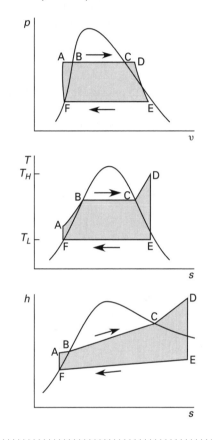

Otto Cycle

Combustion power cycles differ from vapor power cycles in that the combustion products cannot be returned to their initial conditions for reuse. Due to the computational difficulties of working with mixtures of fuel vapor and air, combustion power cycles are often analyzed as air-standard cycles.

An *air-standard cycle* is a hypothetical closed system using a fixed amount of ideal air as the working fluid. In contrast to a combustion process, the heat of combustion is included in the calculations without consideration of the heat source or delivery mechanism (i.e., the combustion process is replaced by a process of instantaneous heat transfer from high-temperature surroundings). Similarly, the cycle ends with an instantaneous transfer of waste heat to the surroundings. All processes are considered to be internally reversible. Because the air is assumed to be ideal, it has a constant specific heat.

Actual engine efficiencies for internal combustion engine cycles may be as much as 50% lower than the efficiencies calculated from air-standard analyses. Empirical corrections must be applied to theoretical calculations based on the characteristics of the engine. However, the large amount of excess air used in turbine combustion cycles results in better agreement (in comparison to reciprocating cycles) between actual and ideal performance.

The *air-standard Otto cycle* consists of the following processes and is illustrated in Fig. 28.6.

A to B: isentropic compression $(Q = 0, \Delta s = 0)$
B to C: constant volume heat addition
C to D: isentropic expansion $(Q = 0, \Delta s = 0)$
D to A: constant volume heat rejection

Figure 28.6 Air-Standard Otto Cycle

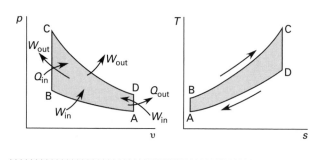

The Otto cycle is a four-stroke cycle because four separate piston movements (strokes) are required to accomplish all of the processes: the intake, compression, power, and exhaust strokes. Two complete crank revolutions are required for these four strokes. Therefore, each cylinder contributes one power stroke every other revolution.

The ideal thermal efficiency for the Otto cycle, Eq. 28.4, can be calculated from the *compression ratio*, Eq. 28.5.

$$\eta_{\text{th}} = 1 - r_v^{1-k} \qquad 28.4$$

$$r_v = \frac{V_A}{V_B} = \frac{V_D}{V_C} \qquad 28.5$$

Refrigeration Cycles

Opposite from heat engines, in refrigeration cycles, heat is transferred from a low-temperature area to a high-temperature area. Since heat flows spontaneously only from high- to low-temperature areas, refrigeration needs an external energy source to force the heat transfer to

occur. This energy source is a pump or compressor that does work in compressing the refrigerant. It is necessary to perform this work on the refrigerant in order to get it to discharge energy to the high-temperature area.

In a power (heat engine) cycle, heat from combustion is the input and work is the desired effect. Refrigeration cycles, though, are power cycles in reverse, and work is the input, with cooling the desired effect. (For every power cycle, there is a corresponding refrigeration cycle.) In a refrigerator, the heat is absorbed from a low-temperature area and is rejected to a high-temperature area. The pump work is also rejected to the high-temperature area.

General refrigeration devices consist of a coil (the evaporator) that absorbs heat, a condenser that rejects heat, a compressor, and a pressure-reduction device (the expansion valve or throttling valve).

In operation, liquid refrigerant passes through the evaporator where it picks up heat from the low-temperature area and vaporizes, becoming slightly superheated. The vaporized refrigerant is compressed by the compressor and in so doing, increases even more in temperature. The high-pressure, high-temperature refrigerant passes through the condenser coils, and because it is hotter than the high-temperature environment, it loses energy. Finally, the pressure is reduced in the expansion valve, where some of the liquid refrigerant also flashes into a vapor.

If the low-temperature area from which the heat is being removed is occupied space (i.e., air is being cooled), the device is known as an *air conditioner*; if the heat is being removed from water, the device is known as a *chiller*. An air conditioner produces cold air; a chiller produces cold water.

Rate of refrigeration (i.e., the rate at which heat is removed) is measured in *tons*. A ton of refrigeration corresponds to 200 Btu/min (12,000 Btu/hr). The ton is derived from the heat flow required to melt a ton of ice in 24 hours.

Heat pumps also operate on refrigeration cycles. Like standard refrigerators, they transfer heat from low-temperature areas to high-temperature areas. The device shown in Fig. 28.1(b) could represent either a heat pump or a refrigerator. There is no significant difference in the mechanisms or construction of heat pumps and refrigerators; the only difference is the purpose of each.

The main function of a refrigerator is to cool the low-temperature area. The useful energy transfer of a refrigerator is the heat removed from the cold area. A heat pump's main function is to warm the high-temperature area. The useful energy transfer is the heat rejected to the high-temperature area.

The concept of thermal efficiency is not used with devices operating on refrigeration cycles. Rather, the *coefficient of performance* (COP) is defined as the ratio of useful energy transfer to the work input. The higher the coefficient of performance, the greater the effect for a given work input will be. Since the useful energy transfer is different for refrigerators and heat pumps, the coefficients of performance will also be different.

$$(COP)_{\text{refrigerator}} = \frac{Q_L}{W} \qquad 28.6$$

$$(COP)_{\text{heat pump}} = \frac{Q_H}{W}$$

$$= (COP)_{\text{refrigerator}} + 1 \quad 28.7$$

The *Carnot refrigeration cycle* is a Carnot power cycle running in reverse. Because it is reversible, the Carnot refrigeration cycle has the highest coefficient of performance for any given temperature limits of all the refrigeration cycles. As shown in Fig. 28.7, all processes occur within the vapor dome.

A to B: isentropic expansion
B to C: isothermal heating (vaporization)
C to D: isentropic compression
D to A: isothermal cooling (condensation)

The coefficients of performance for a Carnot refrigeration cycle establish the upper limit of the COP.

$$(COP)_{\text{Carnot}} = \frac{T_L}{T_H - T_L} \qquad \begin{bmatrix} \text{Carnot} \\ \text{refrigerator} \end{bmatrix} \quad 28.8$$

$$(COP)_{\text{Carnot}} = \frac{T_H}{T_H - T_L} \qquad \begin{bmatrix} \text{Carnot} \\ \text{heat pump} \end{bmatrix} \quad 28.9$$

ENTROPY CHANGES

Entropy is a measure of the energy that is no longer available to perform useful work within the current environment. An increase in entropy is known as *entropy production*. The total entropy in a system is equal to the integral of all entropy productions that have occurred over the life of the system.

$$S_2 - S_1 = \int dS = \int \frac{dQ_{\text{reversible}}}{T} \qquad 28.10$$

$$dS = \frac{dQ_{\text{reversible}}}{T} \qquad 28.11$$

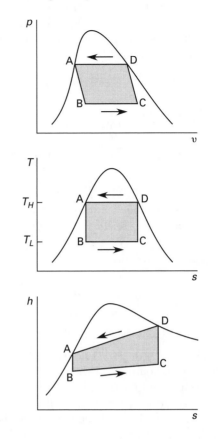

Figure 28.7 Carnot Refrigeration Cycle

Inequality of Clausius

The *inequality of Clausius* is a statement of the second law of thermodynamics.

$$\oint \frac{dQ}{T} \leq 0 \qquad 28.12$$

For a process taking place at a constant temperature, T_0 (i.e., discharging energy to a constant-temperature reservoir at T_0), the entropy production in the reservoir depends on the amount of energy transfer.

$$\Delta S = S_2 - S_1 = \frac{Q}{T_0} \qquad 28.13$$

For an isentropic process, there is no change in system entropy (i.e., the process is reversible).

$$\Delta S = S_2 - S_1 = 0 \qquad 28.14$$

For an adiabatic process,

$$\Delta S \geq 0 \qquad 28.15$$

For incompressible solids and liquids with constant (or mean) specific heats, the change in entropy per unit mass can be calculated from Eq. 28.16.

$$s_2 - s_1 = c \ln\left(\frac{T_2}{T_1}\right) \qquad 28.16$$

The maximum possible work that can be obtained from a cycle is known as the *availability*. Availability is independent of the device but is dependent on the temperature of the local environment. Both the first and second law must be applied to determine availability.

For a closed system, the availability is defined by Eq. 28.17.

$$\phi = u - u_0 - T_0(s - s_0) + p_0(v - v_0) \quad \text{[SI]} \qquad 28.17a$$

$$\phi = u - u_0 - T_0(s - s_0) + \frac{p_0(v - v_0)}{J} \quad \text{[U.S.]} \qquad 28.17b$$

$$W_{\text{reversible}} = \phi_1 - \phi_2 \qquad 28.18$$

For an open system, the steady-state availability function, Ψ, is given by Eq. 28.19.

$$\Psi = h - h_0 - T_0(s - s_0) + \frac{v^2}{2} + gz \quad \text{[SI]} \qquad 28.19a$$

$$\Psi = h - h_0 - T_0(s - s_0) + \frac{v^2}{2g_cJ} + \frac{gz}{g_cJ} \quad \text{[U.S.]} \qquad 28.19b$$

$$W_{\text{reversible}} = \Psi_1 - \Psi_2 \qquad 28.20$$

If the equality 28.19 (or 28.17) holds, both the process within the control volume and the energy transfers between the system and environment must be reversible. Maximum work output, therefore, will be obtained in a reversible process. The difference between the maximum and the actual work output is known as the *process irreversibility*, I.

$$I = W_{\text{reversible}} - W_{\text{actual}} \qquad 28.21$$

SECOND LAW OF THERMODYNAMICS

The *second law of thermodynamics* can be stated in several ways. Equation 28.22 is the mathematical relation defining the second law. The equality holds for reversible processes; the inequality holds for irreversible processes.

$$\Delta S \geq \int_{T_1}^{T_2} \frac{dQ}{T} \qquad 28.22$$

Equation 28.22 effectively states that net entropy must always increase in practical (irreversible) cyclical processes:

A natural process that starts in one equilibrium state and ends in another will go in the direction that causes the entropy of the system and the environment to increase.

Kelvin-Planck Statement of Second Law (Power Cycles)

The *Kelvin-Planck statement of the second law* effectively says that it is impossible to build a cyclical engine that will have a thermal efficiency of 100%:

It is impossible to operate an engine operating in a cycle that will have no other effect than to extract heat from a reservoir and turn it into an equivalent amount of work.

A corollary to this formulation of the second law is that the maximum possible efficiency of a heat engine is the Carnot cycle efficiency.

This formulation is not a contradiction of the first law of thermodynamics. The first law does not preclude the possibility of converting heat entirely into work—it only denies the possibility of creating or destroying energy. The second law says that if some heat is converted entirely into work, some other energy must be rejected to a low-temperature sink (i.e., lost to the surroundings). Figure 28.8(a) illustrates a violation of the second law.

Figure 28.8 Second Law Violations

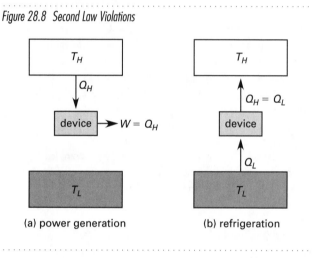

(a) power generation   (b) refrigeration

Clausius Statement of Second Law (Refrigeration Cycles)

The *Clausius statement* of the second law says that it is impossible to devise a cycle that produces, as its only effect, the transfer of heat from a low-temperature body to a high-temperature body. An input of work is always required for refrigeration cycles. A violation of this law is shown in Fig. 28.8(b).

A corollary to the Clausius statement is that the Carnot cycle COP is the highest possible COP for a refrigerator or heat pump.

FINDING WORK AND HEAT GRAPHICALLY

It is sometimes convenient to see what happens to the pressure and volume of a system by plotting the path on a p-V diagram. In addition, the work done by or on the system can be determined from the graph. This is possible because the integral calculating p-V *work* represents the area under the curve in the p-V plane.

$$W = \int_{V_1}^{V_2} p\, dV \qquad \text{28.23}$$

Similarly, the amount of heat absorbed or released from a system can be determined as the area under the path on the T-s diagram.

$$Q = \int_{s_1}^{s_2} T\, ds \qquad \text{28.24}$$

....

Figure 28.9 Process Work and Heat

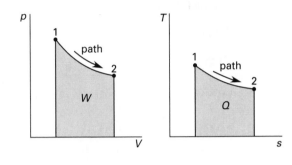

....

The variables p, V, T, and s are *point functions* because their values are independent of the path taken to arrive at the thermodynamic state. Work and heat (W and Q), however, are *path functions* because they depend on the path taken.

SAMPLE PROBLEMS

Problem 1

Two Carnot engines operate in series such that the heat rejected from one is the heat input to the other. The heat transfer from the high-temperature reservoir is 500 kJ. If the overall temperature limits are 1000K and 400K and both engines produce equal work, determine the intermediate temperature between the two engines.

    (A) 400K
    (B) 500K
    (C) 700K
    (D) 1000K

Solution

Let T_H, T_I, and T_L represent the high, intermediate, and low temperatures, respectively.

For the two cycles,

$$W_1 = W_2$$
$$\eta_{\text{th},1} Q_{\text{in},1} = \eta_{\text{th},2} Q_{\text{in},2}$$

However,

$$Q_{\text{in},2} = Q_{\text{out},1} = (1 - \eta_{\text{th},1}) Q_{\text{in},1}$$

$$\left(\frac{T_H - T_I}{T_H}\right) Q_{\text{in},1} = \left(\frac{T_I - T_L}{T_I}\right)\left(1 - \frac{T_H - T_I}{T_H}\right) Q_{\text{in},1}$$

Dividing through by $Q_{\text{in},1}$ and multiplying through by T_H,

$$T_H - T_I = \left(\frac{T_I - T_L}{T_I}\right)(T_H - T_H + T_I)$$

$$T_H - T_I = T_I - T_L$$

$$T_I = \frac{T_H + T_L}{2} = \frac{1000\text{K} + 400\text{K}}{2}$$

$$= 700\text{K}$$

Answer is C.

Problem 2

At the start of compression in an air-standard Otto cycle with a compression ratio of 10, air is at 100 kPa and 40°C. A heat addition of 2800 kJ/kg is made. Calculate the thermal efficiency.

    (A) 52%
    (B) 60%
    (C) 64%
    (D) 67%

Solution

$$\eta_{\text{th}} = 1 - r_v^{1-k} = 1 - 10^{1-1.4}$$
$$= 0.602 \quad (60\%)$$

Answer is B.

Problem 3

A Carnot refrigerating system receives heat from a cold reservoir at 0°C. The power input is 1750 W per ton of refrigeration. Find the system's coefficient of performance.

    (A) 1.4
    (B) 1.6
    (C) 1.8
    (D) 2.0

Solution

The coefficient of performance is the ratio of heat transfer to work input. In the SI system, one ton of refrigeration corresponds to 3516 W.

$$\text{COP} = \frac{Q_L}{W} = \frac{\dot{Q}_L}{P}$$

$$\text{COP} = \frac{3516\, \frac{\text{W}}{\text{ton}}}{1750\, \frac{\text{W}}{\text{ton}}} = 2.01 \quad (2.0)$$

Answer is D.

Problem 4

Which of the following is a proper statement of the second law of thermodynamics?

(A) It is impossible for a heat engine to produce net work in a complete cycle if it exchanges heat only with bodies at a lower temperature.

(B) It is impossible for a system working in a complete cycle to accomplish, as its sole effect, the transfer of heat from a body at a given temperature to a body at a higher temperature.

(C) It is impossible for a system working in a complete cycle to accomplish, as its sole effect, the transfer of heat from a body at a given temperature to a body at a lower temperature.

(D) It is impossible for a heat engine to produce net work in a complete cycle if it exchanges heat only with bodies exhibiting a temperature differential.

Solution

Choice (B) is the Clausius statement of the second law.

Answer is B.

Problem 5

In a heat treating process a 2 kg metal part (specific heat = 0.5 kJ/kg·K) initially at 800°C is quenched in a tank containing 200 kg of water initially at 20°C. Calculate the total entropy change of the process immediately after quenching.

(A) 2.3 kJ/K   (decrease)
(B) 0.65 kJ/K   (decrease)
(C) 0.90 kJ/K   (increase)
(D) 1.6 kJ/K   (increase)

Solution

$$(mc\Delta T)_{\text{metal}} + (mc\Delta T)_{\text{water}} = 0$$

$$(2 \text{ kg})\left(0.5 \; \frac{\text{kJ}}{\text{kg·K}}\right)(800°\text{C} - T_f)$$

$$+ (200 \text{ kg})\left(4.19 \; \frac{\text{kJ}}{\text{kg·K}}\right)(20°\text{C} - T_f)$$

$$= 0$$

$$T_f = 20.93°\text{C}$$

For a solid or liquid,

$$\Delta S = mc \ln\left(\frac{T_2}{T_1}\right)$$

Consider a system consisting of the metal part and the water in the quenching tank.

$$\Delta S_{\text{metal}} = mc \ln\left(\frac{T_2}{T_1}\right)$$

$$= (2 \text{ kg})\left(0.5 \; \frac{\text{kJ}}{\text{kg·K}}\right) \ln\left(\frac{21°\text{C} + 273}{800°\text{C} + 273}\right)$$

$$= -1.295 \text{ kJ/K}$$

$$\Delta S_{\text{water}} = mc \ln\left(\frac{T_2}{T_1}\right)$$

$$= (200 \text{ kg})\left(4.19 \; \frac{\text{kJ}}{\text{kg·K}}\right) \ln\left(\frac{21°\text{C} + 273}{20°\text{C} + 273}\right)$$

$$= 2.855 \text{ kJ/K}$$

$$\Delta S_{\text{total}} = \Delta S_{\text{metal}} + \Delta S_{\text{water}}$$

$$= -1.295 \; \frac{\text{kJ}}{\text{K}} + 2.855 \; \frac{\text{kJ}}{\text{K}}$$

$$= 1.56 \text{ kJ/K}   (1.6 \text{ kJ/K})$$

Answer is D.

FE-STYLE EXAM PROBLEMS

1. A Carnot engine receives 100 kJ of heat from a hot reservoir at 370°C and rejects 37 kJ of heat. Determine the temperature of the cold reservoir.

(A) −35°C
(B) 100°C
(C) 130°C
(D) 230°C

2. What is the maximum thermal efficiency possible for a power cycle operating between 600°C and 110°C?

(A) 47%
(B) 56%
(C) 63%
(D) 74%

3. An ideal Rankine cycle consists of which of the following?

(A) two constant volume and two isentropic processes

(B) two constant pressure and two isentropic processes

(C) two constant volume and two constant temperature processes

(D) two constant pressure and two constant temperature processes

4. What cycle or process does the temperature-entropy diagram represent?

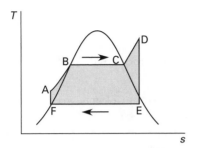

(A) Rankine cycle with superheated steam
(B) Carnot cycle
(C) diesel cycle
(D) refrigeration cycle

5. A Rankine cycle operates between the pressure limits of 600 kPa and 10 kPa. For saturated liquid water leaving the condenser and a turbine inlet temperature of 300°C, determine the thermal efficiency of the cycle.

(A) 13%
(B) 25%
(C) 37%
(D) 52%

For the following problems use the NCEES Handbook as your only reference.

6. A refrigeration cycle has a coefficient of performance 80% of the value of a Carnot refrigerator operating between the temperature limits of 50°C and −5°C. For 3 kW of cooling, what power input is required?

(A) 0.53 kW
(B) 0.62 kW
(C) 0.77 kW
(D) 0.89 kW

7. A heat pump takes heat from groundwater at 7°C and maintains a room at 21°C. What is the maximum coefficient of performance possible for this heat pump?

(A) 1.4
(B) 2.8
(C) 5.6
(D) 21

8. An inventor claims that an engine produces 130 kW with a fuel consumption of 20 kg/h. The fuel has a chemical energy of 40 000 kJ/kg. The energy is received at a mean temperature of 500°C and is rejected at a mean temperature of 50°C. Which laws of thermodynamics are violated?

(A) first law only
(B) second law only
(C) both first and second laws
(D) neither first nor second laws

9. A Carnot cycle operates between the temperature limits of 800K and 300K. If the entropy of the low-temperature reservoir increases 2.34 kJ/K, what is the work of the cycle?

(A) 230 kJ
(B) 440 kJ
(C) 670 kJ
(D) 1200 kJ

10. A ball is dropped onto a smooth floor. It deforms elastically and then returns to its original shape as it rebounds to its original height. Air friction is negligible. Which of the following statements best describes the thermodynamic change in the ball?

(A) The entropy of the ball increases.
(B) The entropy of the ball is unchanged.
(C) The temperature of the ball decreases.
(D) The enthalpy of the ball increases.

11. A 10 m^3 uninsulated tank contains nitrogen at 2 MPa and 250°C. The temperature of the environment surrounding the tank is 35°C. Disregard the mass of the tank. What entropy change do the surroundings experience after a long period of time?

(A) −600 kJ/K
(B) −76 kJ/K
(C) 67 kJ/K
(D) 120 kJ/K

12. A Carnot cycle refrigerator operates between −11°C and 22°C. What is the coefficient of performance?

(A) 0.25
(B) 1.1
(C) 4.3
(D) 7.9

13. A Rankine steam cycle operates between the pressure limits of 600 kPa and 10 kPa. The turbine inlet temperature is 300°C. The liquid water leaving the condenser is saturated. (The enthalpies of the steam at the various points on the cycle are shown in the illustration.) What is the thermal efficiency of the cycle?

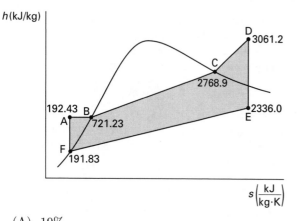

(A) 19%
(B) 25%
(C) 32%
(D) 48%

Problems 14–16 refer to a power plant operating on an ideal Rankine steam cycle. The plant operates between the pressure limits of 1 MPa and 5.628 kPa. The temperature of the steam entering the turbines is 500°C. The total gross power generated is 300 MW.

14. What is the quality of the steam after it has expanded in the turbines?

(A) 78%
(B) 92%
(C) 97%
(D) 100%

15. What is the enthalpy of the steam as it enters the condenser?

(A) 2380 kJ/kg
(B) 2420 kJ/kg
(C) 2560 kJ/kg
(D) 2600 kJ/kg

16. What is the total steam flow rate through all the turbines?

(A) 10 kg/s
(B) 110 kg/s
(C) 200 kg/s
(D) 270 kg/s

17. A household window air conditioner has an energy efficiency rating of 7.5 kJ/W·h. What is the coefficient of performance for the unit?

(A) 2.1
(B) 5.2
(C) 7.0
(D) 9.4

18. What are the processes in an ideal Otto combustion cycle?

(A) two constant volume processes and two isentropic processes
(B) two constant pressure processes and two isentropic processes
(C) two constant volume processes and two constant temperature processes
(D) two constant pressure processes and two constant temperature processes

19. A reversible thermodynamic system is made to follow the Carnot cycle between the temperature limits of 300°C and 75°C. 300 kW of heat are supplied per cycle to the system. What is the change in entropy during the heat addition?

(A) 0.52 kW/K
(B) 0.86 kW/K
(C) 1.0 kW/K
(D) 4.0 kW/K

20. Which of the following statements holds true for an ideal Otto cycle during the combustion portion of the cycle?

(A) The volume increases.
(B) The volume is constant.
(C) The pressure is constant.
(D) The temperature is constant.

21. Which cycle do most steam turbine power-generating plants operate on?

(A) Brayton
(B) Otto
(C) Rankine
(D) Ericsson

22. 1 kg of steam is initially at 400°C and 800 kPa. The steam expands adiabatically to 200°C and 400 kPa in a closed process, performing 450 kJ of work. The enthalpies, internal energies, and entropies of these two states are as follows.

At 400°C and 800 kPa,

$$u = 2959.7 \text{ kJ/kg}$$
$$h = 3267.1 \text{ kJ/kg}$$
$$s = 7.5716 \text{ kJ/kg·K}$$

At 200°C and 400 kPa,

$$u = 2646.8 \text{ kJ/kg}$$
$$h = 2860.5 \text{ kJ/kg}$$
$$s = 7.1706 \text{ kJ/kg·K}$$

Which law(s) of thermodynamics does this process violate?

(A) Zeroth law
(B) first law
(C) second law
(D) first and second law

Problems 23–25 refer to the following situation.

A system operating on the cycle shown uses water as the working fluid. The A-to-B and C-to-D processes are isothermal. The B-to-C and D-to-A processes are isentropic. The pressure of the process at points A and B is 800 kPa, and the pressure at points C and D is 2 kPa. All processes are within the liquid-vapor region for water.

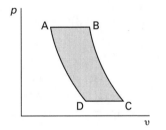

At 800 kPa,

$$h_f = 721.1 \text{ kJ/kg}$$
$$h_g = 2769.1 \text{ kJ/kg}$$
$$s_f = 2.0466 \text{ kJ/kg·K}$$
$$s_g = 6.6636 \text{ kJ/kg·K}$$

At 2 kPa,

$$h_f = 73.5 \text{ kJ/kg}$$
$$h_g = 2533.5 \text{ kJ/kg}$$
$$s_f = 0.2606 \text{ kJ/kg·K}$$
$$s_g = 8.7245 \text{ kJ/kg·K}$$

23. What is the approximate change in entropy from state A to state B?

(A) 0 kJ/kg·K
(B) 2.1 kJ/kg·K
(C) 4.6 kJ/kg·K
(D) 8.5 kJ/kg·K

24. What is the approximate change in entropy from state B to state C?

(A) 0 kJ/kg·K
(B) 2.1 kJ/kg·K
(C) 4.6 kJ/kg·K
(D) 6.7 kJ/kg·K

25. What is the approximate enthalpy at state C?

(A) 74 kJ/kg
(B) 720 kJ/kg
(C) 1800 kJ/kg
(D) 1900 kJ/kg

26. What does the term "tonnage" refer to in regard to refrigeration cycles?

(A) the rate at which heat is removed from a refrigeration cycle
(B) the coefficient of performance of a refrigeration cycle
(C) the energy efficiency rating of a refrigeration cycle
(D) the enthalpy change of a refrigeration cycle

27. In a particular power cycle, 350 MJ of heat are transferred into the system each cycle. The heat transferred out the system is 297.5 MJ per cycle. What is the thermal efficiency of this cycle?

(A) 1.0%
(B) 5.0%
(C) 7.5%
(D) 15%

SOLUTIONS TO FE-STYLE EXAM PROBLEMS

1. The thermal efficiency of the cycle is

$$\eta_{th} = \frac{Q_{net}}{Q_{in}} = \frac{100 \text{ kJ} - 37 \text{ kJ}}{100 \text{ kJ}}$$
$$= 0.63 \quad (63\%)$$

For a Carnot cycle, the thermal efficiency is

$$\eta_{th} = \frac{T_H - T_L}{T_H}$$
$$0.63 = \frac{(370°C + 273) - (T_L + 273)}{370°C + 273}$$
$$T_L = -35.09°C \quad (-35°C)$$

Answer is A.

2. The maximum possible efficiency is equal to the Carnot cycle efficiency.

$$\eta_{th} = 1 - \frac{T_L}{T_H}$$
$$= 1 - \frac{110°C + 273}{600°C + 273}$$
$$= 0.561 \quad (56\%)$$

Answer is B.

3. The ideal Rankine cycle consists of the processes of isentropic compression, constant pressure heat addition, isentropic expansion, and constant pressure heat rejection.

Answer is B.

4. The Rankine cycle is similar to the Carnot cycle except that compression occurs in the liquid region. In addition, the saturated vapor is usually heated above the critical temperature in the superheated region.

Answer is A.

5. Refer to Fig. 28.5.

$$h_D = 3061.6 \text{ kJ/kg}$$
$$[300°C; \ 0.6 \text{ MPa superheated steam}]$$
$$x_E = \frac{s_E - s_F}{s_{fg}}$$
$$= \frac{7.3724 \ \frac{\text{kJ}}{\text{kg·K}} - 0.6491 \ \frac{\text{kJ}}{\text{kg·K}}}{7.5019 \ \frac{\text{kJ}}{\text{kg·K}}}$$
$$= 0.8962$$
$$h_E = h_F + x_E h_{fg}$$
$$= 191.8 \ \frac{\text{kJ}}{\text{kg}} + (0.8962)\left(2393.0 \ \frac{\text{kJ}}{\text{kg}}\right)$$
$$= 2336.4 \text{ kJ/kg}$$

$$h_F = 191.8 \text{ kJ/kg}$$
$$h_A = h_F + v_F(p_A - p_F)$$
$$= 191.8 \ \frac{\text{kJ}}{\text{kg}} + \left(0.00101 \ \frac{\text{m}^3}{\text{kg}}\right)(600 \text{ kPa} - 10 \text{ kPa})$$
$$= 192.40 \text{ kJ/kg}$$

$$\eta_{th} = \frac{W}{Q_H} = \frac{Q_{in} - Q_{out}}{Q_{in}}$$
$$= \frac{(h_D - h_A) - (h_E - h_F)}{h_D - h_A}$$
$$= \frac{\left(3061.6 \ \frac{\text{kJ}}{\text{kg}} - 192.40 \ \frac{\text{kJ}}{\text{kg}}\right) - \left(2336.4 \ \frac{\text{kJ}}{\text{kg}} - 191.8 \ \frac{\text{kJ}}{\text{kg}}\right)}{3061.6 \ \frac{\text{kJ}}{\text{kg}} - 192.40 \ \frac{\text{kJ}}{\text{kg}}}$$
$$= 0.253 \quad (25\%)$$

Answer is B.

6. Find the COP of a Carnot cycle operating with the given temperatures, then use 80% of that value to calculate the necessary power input.

$$(\text{COP})_{ideal} = \frac{T_L}{T_H - T_L}$$
$$= \frac{-5°C + 273}{(50°C + 273) - (-5°C + 273)}$$
$$= 4.87$$
$$(\text{COP})_{actual} = (0.8)(4.87) = 3.90$$
$$(\text{COP})_{actual} = \frac{Q_L}{W}$$
$$W = \frac{Q_L}{\text{COP}_{actual}} = \frac{3 \text{ kW}}{3.90}$$
$$= 0.769 \text{ kW} \quad (0.77 \text{ kW})$$

Answer is C.

7. The upper limit for the COP of a heat pump is set by the COP of a Carnot heat pump.

$$\text{COP} = \frac{T_H}{T_H - T_L}$$
$$= \frac{21°C + 273}{(21°C + 273) - (7°C + 273)}$$
$$= 21$$

Answer is D.

8. The rate of energy entering the system is

$$\left(40\,000 \ \frac{\text{kJ}}{\text{kg}}\right)\left(20 \ \frac{\text{kg}}{\text{h}}\right) = 800\,000 \text{ kJ/h}$$

The output power is

$$P_{out} = (130 \text{ kW})\left(3600 \ \frac{\text{s}}{\text{h}}\right) = 468\,000 \text{ kJ/h}$$

The power produced is not greater than the energy put into the system, so the first law is not violated.

$$\eta_{\text{th}} = \frac{P_{\text{out}}}{\dot{Q}_{\text{in}}} = \frac{468\,000 \ \dfrac{\text{kJ}}{\text{h}}}{800\,000 \ \dfrac{\text{kJ}}{\text{h}}}$$

$$= 0.585 \quad (58.5\%)$$

$$\eta_{\text{th,ideal}} = \frac{T_H - T_L}{T_H}$$

$$= \frac{(500°\text{C} + 273) - (50°\text{C} + 273)}{500°\text{C} + 273}$$

$$= 0.582 \quad (58.2\%)$$

The efficiency of the engine is greater than the ideal efficiency. This violates the second law.

Answer is B.

9. Refer to the T-s diagram of Fig. 28.2. Since the area in the cycle is the net heat and the area is rectangular,

$$Q_{\text{net}} = (T_H - T_L)\Delta S$$

$$= (800\text{K} - 300\text{K})\left(2.34 \ \frac{\text{kJ}}{\text{K}}\right)$$

$$= 1170 \ \text{kJ} \quad (1200 \ \text{kJ})$$

However, in an adiabatic closed system,

$$W_{\text{net}} = Q_{\text{net}} = 1170 \ \text{kJ} \quad (1200 \ \text{kJ})$$

Answer is D.

10. Since the ball returns to its original height, the impact is perfectly elastic and energy is conserved. The temperature, enthalpy, and entropy are unchanged.

Answer is B.

11. After a long period of time, the nitrogen will have cooled to the temperature of the environment.

From the table of specific heats, the specific heats for nitrogen are

$$c_p = 1.04 \ \text{kJ/kg·K}$$
$$c_v = 0.743 \ \text{kJ/kg·K}$$

The specific gas constant for nitrogen is

$$R = c_p - c_v$$

$$= 1.04 \ \frac{\text{kJ}}{\text{kg·K}} - 0.743 \ \frac{\text{kJ}}{\text{kg·K}}$$

$$= 0.297 \ \text{kJ/kg·K}$$

The mass of nitrogen gas is

$$m = \frac{pV}{RT}$$

$$= \frac{(2 \ \text{MPa})\left(10^6 \ \dfrac{\text{Pa}}{\text{MPa}}\right)(10 \ \text{m}^3)}{\left(0.297 \ \dfrac{\text{kJ}}{\text{kg·K}}\right)\left(1000 \ \dfrac{\text{J}}{\text{kJ}}\right)(250°\text{C} + 273)}$$

$$= 128.8 \ \text{kg}$$

Cooling occurs at constant volume. Therefore, the heat loss is

$$Q = mc_v(T_2 - T_1)$$

$$= (128.8 \ \text{kg})\left(0.743 \ \frac{\text{kJ}}{\text{kg·K}}\right)(35°\text{C} - 250°\text{C})$$

$$= -2.058 \times 10^4 \ \text{kJ}$$

The negative sign means that heat is leaving the nitrogen. The sign is positive if the surroundings are taken as the system.

The entropy change of a constant-temperature reservoir is given by

$$S_2 - S_1 = \frac{Q}{T_0}$$

$$= \frac{2.058 \times 10^7 \ \text{J}}{35°\text{C} + 273}$$

$$= 66\,818 \ \text{J/K} \quad (67 \ \text{kJ/K}) \quad \text{[increase]}$$

Answer is C.

12. The coefficient of performance for a Carnot refrigeration cycle is

$$\text{COP} = \frac{T_L}{T_H - T_L}$$

$$= \frac{-11°\text{C} + 273}{(22°\text{C} + 273) - (-11°\text{C} + 273)}$$

$$= 7.9$$

Answer is D.

13. The thermal efficiency of the Rankine cycle is

$$\eta_{\text{th}} = \frac{W_{\text{out}} - W_{\text{in}}}{Q_{\text{in}}} = \frac{(h_D - h_E) - (h_A - h_F)}{h_D - h_A}$$

$$= \frac{\left(3061.2 \ \dfrac{\text{kJ}}{\text{kg}} - 2336.0 \ \dfrac{\text{kJ}}{\text{kg}}\right) - \left(192.43 \ \dfrac{\text{kJ}}{\text{kg}} - 191.83 \ \dfrac{\text{kJ}}{\text{kg}}\right)}{3061.2 \ \dfrac{\text{kJ}}{\text{kg}} - 192.43 \ \dfrac{\text{kJ}}{\text{kg}}}$$

$$= 0.2526 \quad (25\%)$$

Answer is B.

14. Use the superheated steam table. For 1 MPa, the saturation temperature is 179.91°C. Since the actual temperature is 500°C, the steam is superheated. Refer to the following illustrations.

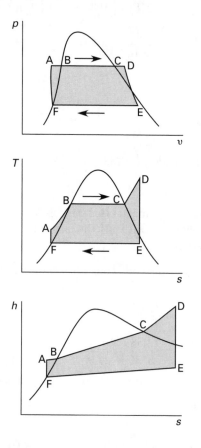

A to B: heating water to saturation temperature in the boiler
B to C: vaporization of water in the boiler
C to D: superheating steam in the superheater region of the boiler
D to E: adiabatic expansion in the turbine
E to F: condensation
F to A: adiabatic compression to the boiler pressure

From the superheated steam table,

$$p_{\mathrm{D}} = 1 \text{ MPa}$$
$$T_{\mathrm{D}} = 500°\mathrm{C}$$
$$h_{\mathrm{D}} = 3478.5 \text{ kJ/kg}$$
$$s_{\mathrm{D}} = 7.7622 \text{ kJ/kg·K}$$

From the saturated steam table,

$$p_{\mathrm{E}} = 5.628 \text{ kPa}$$
$$h_f = 146.68 \text{ kJ/kg}$$
$$h_{fg} = 2418.6 \text{ kJ/kg}$$
$$s_f = 0.5053 \text{ kJ/kg·K}$$
$$s_{fg} = 7.8478 \text{ kJ/kg·K}$$

Since the cycle is ideal, the expansion is isentropic. The entropy at point E is the same as the entropy at point D.

$$s_{\mathrm{E}} = s_f + x_{\mathrm{E}} s_{fg}$$

$$x_{\mathrm{E}} = \frac{s_{\mathrm{E}} - s_f}{s_{fg}}$$

$$= \frac{7.7622 \ \dfrac{\text{kJ}}{\text{kg·K}} - 0.5053 \ \dfrac{\text{kJ}}{\text{kg·K}}}{7.8478 \ \dfrac{\text{kJ}}{\text{kg·K}}}$$

$$= 0.925 \quad (92\%)$$

Answer is B.

15. The enthalpy at point E is

$$h_{\mathrm{E}} = h_f + x_{\mathrm{E}} h_{fg}$$

$$= 146.68 \ \frac{\text{kJ}}{\text{kg}} + (0.925)\left(2418.6 \ \frac{\text{kJ}}{\text{kg}}\right)$$

$$= 2383.9 \text{ kJ/kg} \quad (2380 \text{ kJ/kg})$$

Answer is A.

16. The work extracted by the turbine per kilogram of steam is

$$W_{\text{turbine}} = h_{\mathrm{D}} - h_{\mathrm{E}}$$

$$= 3478.5 \ \frac{\text{kJ}}{\text{kg}} - 2383.9 \ \frac{\text{kJ}}{\text{kg}}$$

$$= 1094.6 \text{ kJ/kg}$$

The total power generated is

$$P = \dot{m} W_{\text{turbine}}$$

$$\dot{m} = \frac{P}{W_{\text{turbine}}} = \frac{300\,000 \text{ kW}}{1094.6 \ \dfrac{\text{kJ}}{\text{kg}}}$$

$$= 274.1 \text{ kg/s} \quad (270 \text{ kg/s})$$

Answer is D.

17. The coefficient of performance for an air conditioner is the energy experienced at the cooling effect divided by the work required to obtain the cooling. This air conditioner has a cooling effect of 7.5 kJ/h for each watt of energy.

$$\mathrm{COP} = \frac{Q_{\text{out}}}{W_{\text{in}}} = \frac{\left(7.5 \ \dfrac{\text{kJ}}{\text{h}}\right)\left(1000 \ \dfrac{\text{J}}{\text{kJ}}\right)}{\left(3600 \ \dfrac{\text{s}}{\text{h}}\right)(1 \text{ W})}$$

$$= 2.08 \quad (2.1)$$

Answer is A.

18. An Otto combustion cycle consists of the following processes.

> A to B: isentropic compression
> B to C: constant volume heat addition
> C to D: isentropic expansion
> D to A: constant volume heat rejection

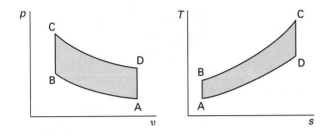

Answer is A.

19. The heat addition occurs during the isothermal process at the high temperature.

$$\Delta S = \frac{\dot{Q}}{T_0} = \frac{300 \text{ kW}}{300°\text{C} + 273}$$

$$= 0.524 \text{ kW/K} \quad (0.52 \text{ kW/K})$$

Answer is A.

20. In an Otto cycle, heat is added during the combustion process (i.e., the B-to-C process). The volume remains constant.

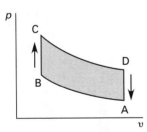

Answer is B.

21. Steam turbine cycles in commercial power-generating plants closely approximate the Rankine cycle.

Answer is C.

22. The internal energy change is

$$u_2 - u_1 = 2646.8 \frac{\text{kJ}}{\text{kg}} - 2959.7 \frac{\text{kJ}}{\text{kg}}$$

$$= -312.9 \text{ kJ/kg}$$

The Zeroth law, which concerns thermal equilibrium conditions, is not applicable in this case.

Based on the the first law, the heat transfer would have to be

$$Q = \Delta U + W$$

$$= -312.9 \frac{\text{kJ}}{\text{kg}} + 450 \frac{\text{kJ}}{\text{kg}}$$

$$= 137.1 \text{ kJ/kg}$$

However, this process is known to be adiabatic, for which $Q = 0$. Therefore, the first law is violated.

The entropy change is

$$s_2 - s_1 = 7.1706 \frac{\text{kJ}}{\text{kg·K}} - 7.5716 \frac{\text{kJ}}{\text{kg·K}}$$

$$= -0.4010 \text{ kJ/kg·K}$$

The entropy decreases. However, for an adiabatic process the entropy must either remain the same or increase. Therefore, this process violates the second law.

Answer is D.

23. The processes describe a Carnot cycle. The process from A to B is isothermal expansion of a saturated liquid to a saturated vapor.

$$s_\text{B} - s_\text{A} = 6.6636 \frac{\text{kJ}}{\text{kg·K}} - 2.0466 \frac{\text{kJ}}{\text{kg·K}}$$

$$= 4.617 \text{ kJ/kg·K} \quad (4.6 \text{ kJ/kg·K})$$

Answer is C.

24. The process from B to C is isentropic. Therefore, there is no change in entropy.

Answer is A.

25. The process from B to C is isentropic, so the entropy at C is the same as at B.

Find the quality, x, at C, knowing the entropy.

$$x_\text{C} = \frac{s_\text{C} - s_f}{s_{fg}} = \frac{s_\text{C} - s_f}{s_g - s_f}$$

$$= \frac{6.6636 \frac{\text{kJ}}{\text{kg·K}} - 0.2606 \frac{\text{kJ}}{\text{kg·K}}}{8.7245 \frac{\text{kJ}}{\text{kg·K}} - 0.2606 \frac{\text{kJ}}{\text{kg·K}}}$$

$$= 0.757$$

Calculate the enthalpy at state C.

$$h_C = h_f + x h_{fg}$$
$$= h_f + x(h_g - h_f)$$
$$= 73.5 \frac{\text{kJ}}{\text{kg}} + (0.757)$$
$$\times \left(2533.5 \frac{\text{kJ}}{\text{kg}} - 73.5 \frac{\text{kJ}}{\text{kg}} \right)$$
$$= 1935.7 \text{ kJ/kg} \quad (1900 \text{ kJ/kg})$$

Answer is D.

26. Tonnage is the rate at which heat is removed, expressed in tons of refrigeration.

Answer is A.

27.
$$\eta_{\text{th}} = \frac{Q_{\text{in}} - Q_{\text{out}}}{Q_{\text{in}}}$$
$$= \frac{350 \text{ MJ} - 297.5 \text{ MJ}}{350 \text{ MJ}}$$
$$= 0.15 \quad (15\%)$$

Answer is D.

29 Mixtures of Gases, Vapors, and Liquids

Subjects

IDEAL GAS MIXTURES 29-1
    Mass and Mole Fractions 29-1
    Partial Pressures 29-2
    Partial Volumes 29-2
    Other Properties 29-2
VAPOR-LIQUID MIXTURES 29-2
    Henry's Law 29-2
    Raoult's Law 29-2
PSYCHROMETRICS 29-3
    Psychrometric Chart 29-3
PHASE RELATIONS 29-5
    Gibbs' Phase Rule 29-5
    Free Energy 29-5

Symbols

v	specific volume	ft³/lbm	m³/kg
ϕ	relative humidity	–	–
ω	humidity ratio	–	–

Subscripts and Superscripts

0	standard conditions
a	dry air
db	dry bulb
dp	dew point
fg	liquid-to-gas (vaporization)
g	saturation
p	constant pressure
v	water vapor
wb	wet bulb
*	pure component

Nomenclature

a	Helmholtz function	Btu/lbm	kJ/kg
A	molar Helmholtz function	Btu/lbmol	kJ/kmol
c_p	specific heat at constant pressure	Btu/lbm-°R	kJ/kg·K
C	number of components	–	–
F	degrees of freedom	–	–
g	Gibbs' function	Btu/lbm	kJ/kg
G	molar Gibbs' function	Btu/lbmol	kJ/kmol
h	Henry's law constant	atm	atm
h	enthalpy	Btu/lbm	kJ/kg
m	mass	lbm	kg
MW	molecular weight	lbm/lbmol	kg/kmol
N	number of moles	–	–
p	absolute pressure	lbf/ft²	Pa
P	number of phases	–	–
R	specific gas constant	ft-lbf/lbm-°R	kJ/kg·K
s	entropy	Btu/lbm-°R	kJ/kg·K
T	absolute temperature	°R	K
u	internal energy	Btu/lbm	kJ/kg
V	volume	ft³	m³
w	mass fraction	–	–
x	mole fraction in condensed phase	–	–
y	mole fraction in gaseous phase	–	–

IDEAL GAS MIXTURES

Mass and Mole Fractions

An ideal gas mixture consists of a mixture of ideal gases, each behaving as if it alone occupied the space.

The *mass fraction*, w (also known as the *gravimetric fraction*), of a component i in a mixture of components $i = 1, 2, ..., n$ is the ratio of the component's mass to the total mixture mass.

$$w_i = \frac{m_i}{m} \qquad 29.1$$

$$m = \sum m_i \qquad 29.2$$

$$\sum w_i = 1 \qquad 29.3$$

The *mole fraction*, x_i, of a liquid component i is the ratio of the number of moles of a substance i to the total number of moles of all substances in the mixture. If the component is in a gaseous phase, the mole fraction is given the symbol y_i.

$$x_i = \frac{N_i}{N} \qquad 29.4$$

$$N = \sum N_i \qquad 29.5$$

$$\sum x_i = 1 \qquad 29.6$$

It is possible to convert from mole fraction to mass fraction through the molecular weight of the component, $(MW)_i$.

$$w_i = \frac{x_i(\text{MW})_i}{\sum x_i(\text{MW})_i} \qquad 29.7$$

$$(\text{MW})_{\text{mixture}} = \frac{m}{N}$$

$$= \sum x_i(\text{MW})_i \qquad 29.8$$

Similarly, it is possible to convert from mass fraction to mole fraction.

$$x_i = \frac{\dfrac{w_i}{(\text{MW})_i}}{\sum \dfrac{w_i}{(\text{MW})_i}} \qquad 29.9$$

Partial Pressures

The *partial pressure*, p_i, of gas component i in a mixture of nonreacting gases $i = 1, 2, ..., n$ is the pressure gas i alone would exert in the total volume at the temperature of the mixture.

$$p_i = \frac{m_i R_i T}{V} = x_i p \qquad 29.10$$

According to *Dalton's law of partial pressures*, the total pressure of a gas mixture is the sum of the partial pressures.

$$p = \sum p_i \qquad 29.11$$

Partial Volumes

The *partial volume*, V_i, of gas i in a mixture of nonreacting gases is the volume that gas i alone would occupy at the temperature and pressure of the mixture.

$$V_i = \frac{m_i R_i T}{p} \qquad 29.12$$

Amagat's law (also known as *Amagat-Leduc's rule*) states that the total volume of a mixture of nonreacting gases is equal to the sum of the partial volumes.

$$V = \sum V_i \qquad 29.13$$

For ideal gases, the mole fraction, partial pressure ratio, and volumetric fraction are the same.

$$x_i = \frac{p_i}{p} = \frac{V_i}{V} \qquad 29.14$$

Other Properties

A mixture's specific internal energy, enthalpy, specific heats, and specific gas constant are equal to the sum of the values of its individual components (i.e., the mixture average is gravimetrically weighted).

$$u = \sum w_i u_i \qquad 29.15$$

$$h = \sum w_i h_i \qquad 29.16$$

If the mixing is reversible and adiabatic, the entropy will be equal to the sum of the individual entropies. However, each individual entropy, s_i, must be evaluated at the temperature and volume of the mixture and at the individual partial pressure, p_i.

$$s = \sum w_i s_i \qquad 29.17$$

Equations 29.15 through 29.17 are mathematical formulations of *Gibbs' theorem* (also known as *Gibbs' rule*). This theorem states that the total property (e.g., u, h, or s) of a mixture of ideal gases is the sum of the properties that the individual gases would have if each occupied the total mixture volume alone at the same temperature.

While the *specific* mixture properties (i.e., u, h, s, c_p, c_v, and R) are all gravimetrically weighted, the *molar* properties are not. Molar U, H, S, C_p, and C_v, as well as the molecular weight and mixture density, are all volumetrically weighted.

VAPOR-LIQUID MIXTURES

Henry's Law

Henry's law states that the partial pressure of a slightly soluble gas above a liquid is proportional to the amount (i.e., mole fraction) of the gas dissolved in the liquid. This law applies separately to each gas to which the liquid is exposed, as if each gas were present alone. The algebraic form of Henry's law is given by Eq. 29.18, in which h is the *Henry's law constant*.

$$p_i = x_i h = y_i p \qquad 29.18$$

Vapor pressure is the pressure exerted by the solvent's vapor molecules when they are in equilibrium with the liquid. The symbol for the vapor pressure of a pure vapor over a pure solvent at a particular temperature is p_i^*. This equilibrium vapor pressure increases with increasing temperature.

Raoult's Law

Raoult's law (Eq. 29.19) states that the vapor pressure, p_i, of a solvent is proportional to the mole fraction of that substance in the solution.

$$p_i = x_i p_i^* \qquad 29.19$$

According to Raoult's law, the vapor pressure of a solution component will increase with increasing temperature (as p_i^* increases) and with increasing mole fraction of that component in the solution. Raoult's law applies to each of the substances in the solution. By Dalton's law, the total vapor pressure above the liquid is equal to the sum of the vapor pressures of each component.

PSYCHROMETRICS

The study of the properties and behavior of atmospheric air is known as *psychrometrics*. Properties of air are seldom evaluated from theoretical thermodynamic principles, however. Specialized techniques and charts have been developed for that purpose.

Air in the atmosphere contains small amounts of moisture and can be considered to be a mixture of two ideal gases—dry air and water vapor. All of the thermodynamic rules relating to the behavior of nonreacting gas mixtures apply to atmospheric air. From Dalton's law, for example, the total atmospheric pressure is the sum of the dry air partial pressure and the water vapor pressure.

$$p = p_a + p_v \qquad\qquad 29.20$$

At first, psychrometrics seems complicated by three different definitions of temperature. These three terms are not interchangeable.

- *dry-bulb temperature*, T_{db}: This is the temperature that a regular thermometer measures if exposed to air.
- *wet-bulb temperature*, T_{wb}: This is the temperature of air that has gone through an adiabatic saturation process. It is measured with a thermometer that is covered with a water-saturated cotton wick.
- *dew-point temperature*, T_{dp}: This is the dry-bulb temperature at which water starts to condense when moist air is cooled in a constant pressure process. The dew-point temperature is equal to the saturation temperature (read from steam tables) for the partial pressure of the vapor.

For every temperature, there is a unique equilibrium vapor pressure of water, p_g, called the *saturation pressure*. If the vapor pressure equals the saturation pressure, the air is said to be saturated. *Saturated air* is a mixture of dry air and water vapor at the saturation pressure. When the air is saturated, all three temperatures are equal.

$$T_{db} = T_{wb} = T_{dp} \qquad [\text{saturated}] \qquad 29.21$$

Unsaturated air is a mixture of dry air and superheated water vapor. When the air is unsaturated, the dew-point temperature will be less than the wet-bulb temperature.

$$T_{dp} < T_{wb} < T_{db} \qquad [\text{unsaturated}] \qquad 29.22$$

The amount of water in atmospheric air is specified by the *humidity ratio* (also known as the *specific humidity*), ω. The humidity ratio is the mass ratio of water vapor to dry air. If both masses are expressed in pounds (kilograms), the units of ω are lbm/lbm (kg/kg). However, since there is so little water vapor, the water vapor mass is often reported in *grains* or grams. (There are 7000 grains per pound.) Accordingly, the humidity ratio may have the units of grains per pound or grams per kg. The humidity ratio is expressed as Eq. 29.23 or 29.24. Notice that the humidity ratio is expressed per pound of dry air, not per pound of the total mixture. Equation 29.23 is derived from the ideal gas law, and 0.622 is the ratio of the specific gas constants for air and water vapor.

$$\omega = \frac{m_v}{m_a} \qquad\qquad 29.23$$

$$\omega = 0.622\left(\frac{p_v}{p_a}\right) = 0.622\left(\frac{p_v}{p - p_v}\right) \qquad 29.24$$

The *relative humidity*, ϕ, is another index of moisture content. The relative humidity is the partial pressure of the water vapor divided by the saturation pressure.

$$\phi = \frac{p_v}{p_g} = \frac{m_v}{m_g} \qquad\qquad 29.25$$

Psychrometric Chart

It is possible to develop mathematical relationships for enthalpy and specific volume (the two most useful thermodynamic properties) for atmospheric air. However, these relationships are almost never used. Rather, psychrometric properties are read directly from psychrometric charts, as illustrated in Fig. 29.1.

A psychrometric chart is easy to use, despite the multiplicity of scales. The thermodynamic state (i.e., the position on the chart) is defined by specifying the values of any two parameters on intersecting scales (e.g., dry-bulb and wet-bulb temperature, or dry-bulb temperature and relative humidity). Once the state has been located on the chart, all other properties can be read directly.

There are different psychrometric charts for low, medium, and high temperature ranges, as well as charts for different atmospheric pressures (i.e., elevations). The usage of several scales varies somewhat from chart to chart. In particular, the use of the enthalpy scale depends on the chart used. Furthermore, not all psychrometric charts contain all scales.

Enthalpy of the air-vapor mixture is the sum of the enthalpies of the air and water vapor. Enthalpy of the mixture per pound of dry air can be read from the psychrometric chart.

$$h = h_a + \omega h_v \qquad \begin{bmatrix} \text{per pound} \\ \text{of dry air} \end{bmatrix} \qquad 29.26$$

$$h_{\text{total}} = m_a h \qquad \begin{bmatrix} \text{for any mass} \\ \text{of dry air} \end{bmatrix} \qquad 29.27$$

THERMO Gases/Vapors

Figure 29.1 Psychrometric Chart, SI Units

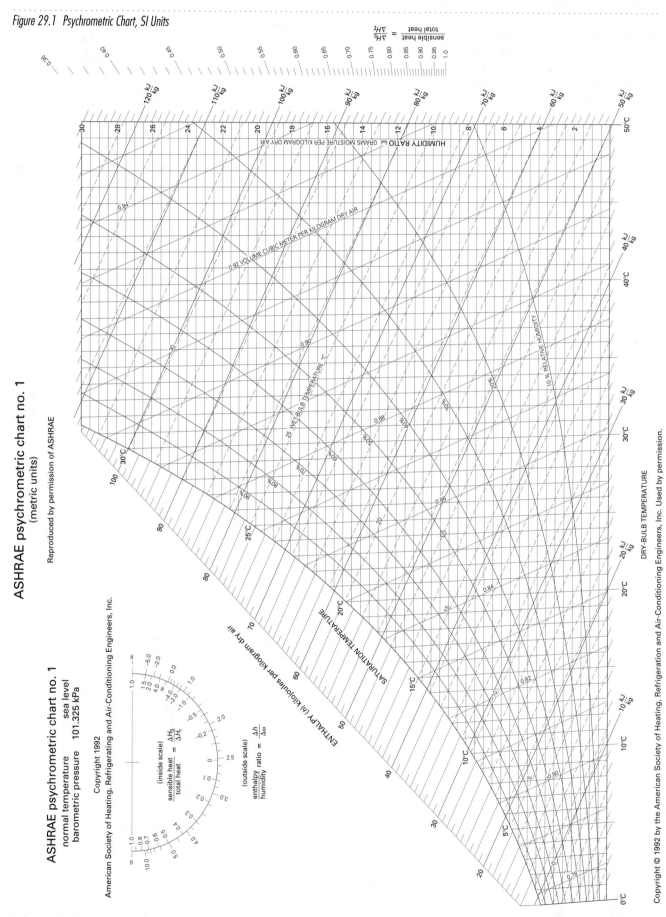

ASHRAE psychrometric chart no. 1
(metric units)

Reproduced by permission of ASHRAE

ASHRAE psychrometric chart no. 1

normal temperature   sea level
barometric pressure   101.325 kPa

Copyright 1992
American Society of Heating, Refrigerating and Air-Conditioning Engineers, Inc.

$$\frac{\text{sensible heat}}{\text{total heat}} = \frac{\Delta H_s}{\Delta H_t}$$

PHASE RELATIONS

The change in enthalpy during a phase transition, although it cannot be measured directly, can be determined from the pressure, temperature, and specific volume changes through the *Clapeyron equation*. $(dp/dT)_{\text{sat}}$ is the slope of the vapor-liquid saturation line.

$$\left(\frac{dp}{dT}\right)_{\text{sat}} = \frac{h_{fg}}{Tv_{fg}} = \frac{s_{fg}}{v_{fg}} \qquad 29.28$$

Gibbs' Phase Rule

Gibbs' phase rule defines the relationship between the number of phases and components in a mixture at equilibrium.

$$P + F = C + 2 \qquad 29.29$$

P is the number of phases existing simultaneously; F is the number of independent variables, known as *degrees of freedom*; and C is the number of components in the system. Composition, temperature, and pressure are examples of degrees of freedom that can be varied.

For example, if water is to be stored such that three phases (solid, liquid, gas) are present simultaneously, then $P = 3$, $C = 1$, and $F = 0$. That is, neither pressure nor temperature can be varied independently. This state is exemplified by water at its triple point.

Free Energy

The *Gibbs function* is defined for a pure substance by Eqs. 29.30 and 29.31.

$$g = h - Ts = u + pv - Ts \qquad 29.30$$
$$G = H - TS = U + pV - TS \qquad 29.31$$

The Gibbs function is used in investigating latent changes and chemical reactions. For a constant-temperature, constant-pressure, nonflow process approaching equilibrium, the Gibbs function approaches a minimum value.

$$(dG)_{T,p} < 0 \qquad \text{[nonequilibrium]} \qquad 29.32$$

Once the minimum value is obtained, equilibrium is attained and the Gibbs function will be constant.

$$(dG)_{T,p} = 0 \qquad \text{[equilibrium]} \qquad 29.33$$

The *Gibbs function of formation*, G^0, has been tabulated at the standard reference conditions of 25°C (77°F) and 1 atm. A chemical reaction can occur spontaneously only if the change in Gibbs' function is negative (i.e., the Gibbs function for the products is less than the Gibbs function for the reactants).

$$\sum_{\text{products}} nG^0 < \sum_{\text{reactants}} nG^0 \qquad 29.34$$

The *Helmholtz function* is defined for a pure substance by Eqs. 29.35 and 29.36.

$$a = u - Ts = h - pv - Ts \qquad 29.35$$
$$A = U - TS = H - pV - TS \qquad 29.36$$

Like the Gibbs function, the Helmholtz function is used in investigating equilibrium conditions. For a constant-temperature, constant-volume, nonflow process approaching equilibrium, the Helmholtz function approaches its minimum value.

$$(dA)_{T,V} < 0 \qquad \text{[nonequilibrium]} \qquad 29.37$$

Once the minimum value is obtained, equilibrium is attained, and the Helmholtz function will be constant.

$$(dA)_{T,V} = 0 \qquad \text{[equilibrium]} \qquad 29.38$$

The Helmholtz function is sometimes known as the *free energy of the system* because its change in a reversible isothermal process equals the maximum energy that can be "freed" and converted to mechanical work. The same term has also been used for the Gibbs function under analogous conditions. For example, the difference in standard Gibbs functions of reactants and products has often been called the "free energy difference."

Since there is a possibility for confusion, it is better to refer to the Gibbs and Helmholtz functions by their actual names.

SAMPLE PROBLEMS

Problem 1

A gas mixture with volumetric proportions of 30% carbon dioxide (specific heat, c_p, of 0.867 kJ/kg·K) and 70% nitrogen (specific heat, c_p, of 1.043 kJ/kg·K) is cooled at constant pressure from 150°C to 50°C. Determine the heat released.

    (A) -210 kJ/kg
    (B) -160 kJ/kg
    (C) -97 kJ/kg
    (D) -46 kJ/kg

Solution

For ideal gases, the mole and volumetric fractions are the same. The volumetric fractions are given. However, the gravimetric fractions are needed to calculate the mixture's specific heat. Find the mass fraction, and use that to apply Gibbs' rule.

$$w_i = \frac{x_i(\text{MW})_i}{\sum x_i(\text{MW})_i}$$

THERMO
Gases/Vapors

$$w_{CO_2} = \frac{(0.30)(44)}{(0.30)(44) + (0.70)(28)}$$
$$= 0.402$$

$$w_{N_2} = 1 - w_{CO_2} = 1 - 0.402$$
$$= 0.598$$

$$c_p = \sum w_i c_{pi}$$
$$= (0.402)\left(0.867 \ \frac{kJ}{kg \cdot K}\right)$$
$$+ (0.598)\left(1.043 \ \frac{kJ}{kg \cdot K}\right)$$
$$= 0.972 \ kJ/kg \cdot K$$

$$Q = mc_p \Delta T$$

$$\frac{Q}{m} = \left(0.972 \ \frac{kJ}{kg \cdot K}\right)(50°C - 150°C)$$
$$= -97.2 \ kJ/kg$$

Answer is C.

Problem 2

By weight, atmospheric air is approximately 23.15% oxygen and 76.85% nitrogen. What is the partial pressure of oxygen in the air at standard temperature and pressure?

(A) 21 kPa
(B) 23 kPa
(C) 26 kPa
(D) 30 kPa

Solution

Partial pressure is volumetrically (not gravimetrically) weighted. Calculate oxygen's mole fraction.

$$x_{O_2} = \frac{\dfrac{w_{O_2}}{(MW)_{O_2}}}{\dfrac{w_{O_2}}{(MW)_{O_2}} + \dfrac{w_{N_2}}{(MW)_{N_2}}} = \frac{\dfrac{0.2315}{32}}{\dfrac{0.2315}{32} + \dfrac{0.7685}{28}}$$
$$= 0.209$$

$$p_{O_2} = x_i p = (0.209)p_{atm}$$
$$= (0.209)(101 \ kPa)$$
$$= 21.1 \ kPa \quad (21 \ kPa)$$

Answer is A.

Problems 3–5 refer to the following air-water vapor mixture.

In an air-water vapor mixture at 30°C, the partial pressure of the water vapor is 1.5 kPa and the partial pressure of the dry air is 100.5 kPa. The saturation pressure

for water vapor at 30°C is 4.246 kPa. The gas constant for water vapor is 0.46152 kJ/kg·K, and the gas constant for air is 0.28700 kJ/kg·K.

Problem 3

What is the relative humidity of the air-water vapor mixture?

(A) 0.0093
(B) 0.042
(C) 0.35
(D) 0.65

Solution

$$\phi = \frac{p_v}{p_g} = \frac{1.5 \ kPa}{4.246 \ kPa}$$
$$= 0.3533 \quad (0.35)$$

Answer is C.

Problem 4

What is the humidity ratio of the air-water vapor mixture?

(A) 0.0093
(B) 0.26
(C) 0.35
(D) 0.38

Solution

$$w = (0.622)\left(\frac{p_v}{p_a}\right) = (0.622)\left(\frac{1.5 \ kPa}{100.5 \ kPa}\right)$$
$$= 0.00928 \quad (0.0093)$$

Answer is A.

Problem 5

If the mass of water vapor in the mixture is 5 kg, what is the mass of air?

(A) 13 kg
(B) 14 kg
(C) 190 kg
(D) 540 kg

Solution

$$w = \frac{m_{water}}{m_a}$$

$$m_a = \frac{m_{water}}{w} = \frac{5 \ kg}{0.00928}$$
$$= 538.8 \ kg \quad (540 \ kg)$$

Answer is D.

Problem 6

Water at 25°C has a vapor pressure of 3.1504 kPa. Compound B (MW = 52.135) at 25°C has a vapor pressure of 7.2601 kPa. If 75 g of liquid water are mixed with 45 g of liquid compound B, what is the resulting vapor pressure of the solution?

(A) 3.4 kPa
(B) 3.9 kPa
(C) 4.7 kPa
(D) 5.2 kPa

Solution

For water,

$$MW = 18$$

$$N = \frac{75 \text{ g}}{18 \ \frac{\text{g}}{\text{mol}}} = 4.17 \text{ mol}$$

For compound B,

$$N = \frac{m}{MW} = \frac{45 \text{ g}}{52.135 \ \frac{\text{g}}{\text{mol}}}$$

$$= 0.863 \text{ mol}$$

$$x_{\text{water}} = \frac{4.17}{4.17 + 0.863} = 0.829$$

$$x_{\text{B}} = \frac{0.863}{4.17 + 0.863} = 0.171$$

$$p_i = x_i p_i^*$$

$$p_{\text{water}} = (0.829)(3.1504 \text{ kPa})$$

$$= 2.6117 \text{ kPa}$$

$$p_{\text{B}} = (0.171)(7.2601 \text{ kPa})$$

$$= 1.2415 \text{ kPa}$$

$$p = \sum p_i = 2.6117 \text{ kPa} + 1.2415 \text{ kPa}$$

$$= 3.853 \text{ kPa} \quad (3.9 \text{ kPa})$$

Answer is B.

FE-STYLE EXAM PROBLEMS

1. What does Dalton's law of partial pressures state about gases?

(A) The total pressure of a gas mixture is the sum of the individual gases' partial pressures.
(B) The total volume of a nonreactive gas mixture is the sum of the individual gases' volumes.
(C) Each gas of a mixture has the same partial pressure as that of the mixture.
(D) The gas pressure of a mixture is the weighted average of the individual gas pressures.

2. Atmospheric air at 21°C has a relative humidity of 50%. What is the dew-point temperature?

(A) 7°C
(B) 10°C
(C) 17°C
(D) 24°C

3. How can the relative humidity of an air sample be determined?

(A) Measure the wet-bulb and dry-bulb temperatures and divide the wet-bulb temperature by the dry-bulb temperature.
(B) Measure the barometric pressure and look up the corresponding relative humidity on a psychrometric chart.
(C) Measure the wet-bulb and dry-bulb temperatures and look up the corresponding relative humidity on a psychrometric chart.
(D) Measure the wet-bulb and dry-bulb temperatures and look up the corresponding relative humidity on a Mollier chart.

4. One method of removing moisture from air is to cool the air so that the moisture condenses or freezes out. To what temperature must air at 100 atm be cooled at constant pressure in order to obtain a humidity ratio of 0.0001?

(A) −6°C
(B) 2°C
(C) 8°C
(D) 14°C

5. Which of the following statements is true if atmospheric air with relative humidity of 50% is heated from 10°C to 25°C at constant volume?

(A) The dew point of the air is lowered.
(B) The dew point of the air is not changed.
(C) The relative humidity of the air is increased.
(D) The heat absorbed is equal to the increase in heat content of the mixture.

For the following problems use the NCEES Handbook as your only reference.

6. Atmospheric air at 101.3 kPa, 27°C, and 50% relative humidity is heated at constant pressure to 43°C. What heat transfer is required?

(A) 17 kJ/kg
(B) 24 kJ/kg
(C) 39 kJ/kg
(D) 61 kJ/kg

7. How many phases may exist in equilibrium for a fixed proportion water-alcohol mixture held at constant pressure?

    (A) 0
    (B) 1
    (C) 2
    (D) 3

8. A mixture of oxygen and nitrogen exists at the following conditions.

$$m = 5 \text{ kg}$$
$$p = 700 \text{ kPa}$$
$$T = 300 \text{K}$$
$$V = 0.6 \text{ m}^3$$

What mass of oxygen must be added to change the mixture's volumetric percentage to 75% oxygen at the same temperature and volume?

    (A) 2.3 kg
    (B) 7.0 kg
    (C) 9.3 kg
    (D) 12 kg

9. A mixture of two gases, A and B, exists at pressure p_1, volume V, and temperature T_1. Gas A is subsequently removed from the mixture in a constant-volume process. The remaining gas B is found to have a pressure p_2, volume V, and temperature T_2. Express the ratio of the number of moles of gas B to the number of moles of gas A in the original sample in terms of the given pressures, volume, and temperatures.

    (A) $\dfrac{p_2 T_1}{p_1 T_2 - p_2 T_1}$

    (B) $\dfrac{p_2 R_B T_2}{p_1 R_A T_1}$

    (C) $\dfrac{p_1 R_B T_2 - p_2 R_B T_2}{p_2 R_A T_1}$

    (D) $\dfrac{p_2 R_A T_1 V}{(p_1 - p_2) R_B T_2}$

10. A 0.75 m³ tank contains nitrogen at 150 kPa and 40°C. 2.0 kg of oxygen are added to the tank. What is the final volumetric percentage of the nitrogen?

    (A) 14%
    (B) 17%
    (C) 38%
    (D) 41%

Problems 11–14 refer to the following situation.

Air at 27°C and 50% relative humidity is cooled in a sensible cooling process to 18°C. The air is then heated to 45°C in a sensible heating process. Finally, the air experiences an adiabatic saturation process that increases the relative humidity back to 50%.

11. Find the specific energy that is removed when the air is cooled to 18°C.

    (A) 6 kJ/kg
    (B) 10 kJ/kg
    (C) 19 kJ/kg
    (D) 34 kJ/kg

12. Find the relative humidity of the air after it has been heated.

    (A) 11%
    (B) 18%
    (C) 25%
    (D) 33%

13. What is the humidity ratio of the air during the sensible heating process?

    (A) 5 g/kg
    (B) 9 g/kg
    (C) 11 g/kg
    (D) 25 g/kg

14. What is the moisture content after the adiabatic saturation process?

    (A) 11 g/kg
    (B) 17 g/kg
    (C) 20 g/kg
    (D) 23 g/kg

SOLUTIONS TO FE-STYLE EXAM PROBLEMS

1. Dalton's law states that the total pressure of a gas mixture is the sum of the partial pressures.

Answer is A.

2. The dew-point temperature is the saturation temperature for the current vapor pressure conditions.

$$\phi = \frac{p_v}{p_g} = 0.5$$
$$p_g = 2.487 \text{ kPa} \qquad [\text{at } 21°\text{C}]$$
$$p_v = \phi p_g = (0.5)(2.487 \text{ kPa})$$
$$= 1.2435 \text{ kPa}$$

From the steam table at 1.2435 kPa,

$$T_{\text{sat}} = T_{\text{dp}} \approx 10°\text{C}$$

Alternatively, the psychrometric chart can be used. From the intersection of 21°C dry-bulb temperature and the curved 50% humidity line, follow the horizontal line to the left to where it intersects with $T_{\text{dp}} \approx 10°\text{C}$.

Answer is B.

3. To find the relative humidity, use the psychrometric chart. (The Mollier diagram is an enthalpy-entropy diagram for steam.) The relative humidity is located on the chart at the intersection of the wet-bulb and dry-bulb temperatures.

Answer is C.

4.
$$\omega = (0.622)\left(\frac{p_v}{p - p_v}\right)$$

$$p_v = \left(\frac{\omega}{0.622}\right)(p - p_v)$$

$$= \frac{\dfrac{\omega p}{0.622}}{1 + \dfrac{w}{0.622}}$$

$$= \frac{(0.0001)(100 \text{ atm})\left(101.35 \dfrac{\text{kPa}}{\text{atm}}\right)}{0.622}{1 + \dfrac{0.0001}{0.622}}$$

$$= 1.629 \text{ kPa}$$

From the steam table corresponding to 1.629 kPa,

$$T_{\text{sat}} \approx 14°\text{C}$$

Answer is D.

5. At constant volume, the air pressure increases with temperature, so the dew point (saturation temperature for vapor pressure conditions) also increases. The relative humidity decreases. The heat absorbed by the mixture increases the heat content of both the water and the air.

Answer is D.

6. Use the psychrometric chart. As long as no moisture is added or removed, the humidity ratio (ω) is constant. This is a horizontal line on the psychrometric chart. From the intersection of a dry-bulb temperature of 27°C and relative humidity of 50%, follow the horizontal line

to the right until it intersects at $T_{\text{db}} = 43°\text{C}$. Then, follow the diagonal enthalpy line upward and to the left to approximately $h_2 = 73 \text{ kJ/kg}$.

The initial enthalpy for 27°C and $\phi = 50\%$ is $h_1 = 56 \text{ kJ/kg}$.

$$Q = h_2 - h_1 = 73 \frac{\text{kJ}}{\text{kg}} - 56 \frac{\text{kJ}}{\text{kg}}$$

$$= 17 \text{ kJ/kg}$$

Answer is A.

7. Use Gibbs' phase rule.

$$P + F = C + 2$$
$$P = C + 2 - F$$

P = number of phases
C = number of components = 2   [water and alcohol]
F = degrees of freedom = 1   [temperature]
$P = 2 + 2 - 1 = 3$

Answer is D.

8.
$$pV = mRT$$

$$R_{\text{mixture}} = \frac{pV}{mT} = \frac{(700 \text{ kPa})(0.6 \text{ m}^3)}{(5 \text{ kg})(300\text{K})}$$

$$= 0.28 \text{ kJ/kg·K}$$

$$R_{\text{mixture}} = \sum y_i R_i = \overline{R} \sum \frac{y_i}{(\text{MW})_i}$$

$$0.28 \frac{\text{kJ}}{\text{kg·K}} = \left(8.3143 \frac{\text{kJ}}{\text{kmol·K}}\right)\left(\frac{y_{O_2}}{32} + \frac{y_{N_2}}{28}\right)$$

$$= \left(8.3143 \frac{\text{kJ}}{\text{kmol·K}}\right)\left(\frac{y_{O_2}}{32} + \frac{1 - y_{O_2}}{28}\right)$$

$$y_{O_2} = 0.46$$

$$y_{N_2} = 0.54$$

$$m_{O_2} = y_{O_2}m = (0.46)(5 \text{ kg})$$

$$= 2.3 \text{ kg}$$

$$N_{O_2} = \frac{m_{O_2}}{(\text{MW})_{O_2}} = \frac{2.3 \text{ kg}}{32 \dfrac{\text{kg}}{\text{kmol}}}$$

$$= 0.07188 \text{ kmol}$$

Similarly,

$$m_{N_2} = 5 \text{ kg} - 2.3 \text{ kg}$$

$$= 2.7 \text{ kg}$$

$$N_{N_2} = \frac{2.7 \text{ kg}}{28 \dfrac{\text{kg}}{\text{kmol}}}$$

$$= 0.09643 \text{ kmol}$$

THERMO
Gases/Vapors

Solve for the final condition.

$$\frac{V_i}{V} = 0.75 = \frac{N'_{O_2}}{N'_{O_2} + 0.09643 \text{ kmol}}$$

$$N'_{O_2} = 0.2893$$

$$m_{O_2} = N'_{O_2}(\text{MW})_{O_2} = (0.2893)(32)$$

$$= 9.258 \text{ kg}$$

$$\Delta m = 9.258 \text{ kg} - 2.3 \text{ kg}$$

$$= 6.958 \text{ kg} \quad (7.0 \text{ kg})$$

Answer is B.

9. Use the ideal gas law.

$$pV = N\overline{R}T$$

$$N = \frac{pV}{\overline{R}T}$$

$$\frac{N_B}{N_A} = \frac{\dfrac{p_B V_B}{\overline{R}T_B}}{\dfrac{p_A V_A}{\overline{R}T_A}}$$

The $\overline{R}$ terms cancel, and the volumes are the same.

$$\frac{N_B}{N_A} = \frac{p_B T_A}{T_B p_A}$$

Use Dalton's law of partial pressures.

$$p_1 = p_A + p_B$$

$$p_B = p_2\left(\frac{T_1}{T_2}\right)$$

$$p_A = p_1 - p_B = p_1 - p_2\left(\frac{T_1}{T_2}\right)$$

Also, $T_A = T_1$ and $T_B = T_2$.

$$\frac{N_B}{N_A} = \frac{p_2 T_1}{T_2\left(p_1 - p_2\left(\dfrac{T_1}{T_2}\right)\right)} = \frac{p_2 T_1}{p_1 T_2 - p_2 T_1}$$

Answer is A.

10. For ideal gases, the mole and volumetric fractions are the same. Calculate the number of moles of each gas.

For nitrogen,

$$pV = N\overline{R}T$$

$$N = \frac{pV}{\overline{R}T}$$

$$= \frac{(150 \text{ kPa})(0.75 \text{ m}^3)}{\left(8.3143 \dfrac{\text{kJ}}{\text{kmol·K}}\right)(40°\text{C} + 273)}$$

$$= 0.0432 \text{ kmol}$$

For oxygen,

$$N = \frac{m}{\text{MW}} = \frac{2 \text{ kg}}{32 \dfrac{\text{kg}}{\text{kmol}}}$$

$$= 0.0625 \text{ kmol}$$

The mole fraction (the same as the volumetric fraction) is

$$x = \frac{N_{N_2}}{N_{O_2} + N_{N_2}}$$

$$= \frac{0.0432 \text{ kmol}}{0.0625 \text{ kmol} + 0.0432 \text{ kmol}}$$

$$= 0.409 \quad (41\%)$$

Answer is D.

11. The specific energy (energy per unit mass) removed is the change in enthalpy. Use the psychrometric chart to determine the enthalpies.

Find the intersection of the dry-bulb temperature of 27°C and the relative humidity of 50%. Follow the enthalpy lines up and to the left, and read the value for enthalpy. (For more accuracy, align the enthalpy values with the corresponding scale along the bottom and right edges.)

$$h_1 = 56 \text{ kJ/kg}$$

In a sensible cooling process, the air and water vapor are both cooled without a change in moisture content. (That is, there is no condensation.) Sensible cooling and heating processes are represented by horizontal lines on the psychrometric chart. Follow a horizontal line from the initial point to the left to the vertical line corresponding to a dry-bulb temperature of 18°C. Read the corresponding enthalpy by following the lines up and to the left.

$$h_2 = 46 \text{ kJ/kg}$$

The change in enthalpy is

$$h = 56 \frac{\text{kJ}}{\text{kg}} - 46 \frac{\text{kJ}}{\text{kg}}$$

$$= 10 \text{ kJ/kg}$$

Answer is B.

12. Since a sensible heating process does not remove moisture, the path of the air and moisture is again represented by a horizontal line on the psychrometric chart. Follow the same horizontal line to the right to the vertical line corresponding to a dry-bulb temperature of 45°C. The relative humidity at that point is approximately 18% or 19%.

Answer is B.

13. The humidity ratio is read directly from the scale at the right as approximately 11.3 g (11 g) of moisture per kilogram of dry air.

Answer is C.

14. Since the process is adiabatic, the enthalpy of the air-moisture mixture is unchanged. The process follows a line of constant enthalpy. Since the process is one of saturation, the line of constant enthalpy is followed upward to the left to the 50% relative humidity line. From that point (at approximately 33.5°C dry bulb), follow a horizontal line to the right and read the humidity ratio of approximately 16.5 g/kg (17 g/kg). (Notice that the air is not saturated, even though the process is an "adiabatic saturation process.")

Answer is B.

30 Combustion

Subjects

COMBUSTION 30-1

Nomenclature

m mass   lbm   kg

COMBUSTION

Combustion reactions involving organic compounds and oxygen take place according to standard stoichiometric principles. *Stoichiometric air (ideal air)* is the exact quantity of air necessary to provide the oxygen required for complete combustion of the fuel. Stoichiometric oxygen volumes can be determined from the balanced chemical reaction equation. Table 30.1 contains some of the more common chemical reactions.

Table 30.1 Stoichiometric Combustion Reactions

fuel	formula	reaction equation (excluding nitrogen)[a]
carbon (to CO)	C	$2C + O_2 \longrightarrow 2CO$
carbon (to CO_2)	C	$2C + 2O_2 \longrightarrow 2CO_2$
sulfur (to SO_2)	S	$S + O_2 \longrightarrow SO_2$
sulfur (to SO_3)	S	$2S + 3O_2 \longrightarrow 2SO_3$
carbon monoxide	CO	$2CO + O_2 \longrightarrow 2CO_2$
methane	CH_4	$CH_4 + 2O_2 \longrightarrow CO_2 + 2H_2O$
acetylene	C_2H_2	$2C_2H_2 + 5O_2 \longrightarrow 4CO_2 + 2H_2O$
ethylene	C_2H_4	$C_2H_4 + 3O_2 \longrightarrow 2CO_2 + 2H_2O$
ethane	C_2H_6	$2C_2H_6 + 7O_2 \longrightarrow 4CO_2 + 6H_2O$
hydrogen	H_2	$2H_2 + O_2 \longrightarrow 2H_2O$
hydrogen sulfide	H_2S	$2H_2S + 3O_2 \longrightarrow 2H_2O + 2SO_2$
propane	C_3H_8	$C_3H_8 + 5O_2 \longrightarrow 3CO_2 + 4H_2O$
n-butane	C_4H_{10}	$2C_4H_{10} + 13O_2 \longrightarrow 8CO_2 + 10H_2O$
octane	C_8H_{18}	$2C_8H_{18} + 25O_2 \longrightarrow 16CO_2 + 18H_2O$
olefin series	C_nH_{2n}	$2C_nH_{2n} + 3nO_2 \longrightarrow 2nCO_2 + 2nH_2O$
paraffin series	C_nH_{2n+2}	$2C_nH_{2n+2} + (3n+1)O_2 \longrightarrow 2nCO_2 + (2n+2)H_2O$

[a]Multiply oxygen volume by 3.78 to get nitrogen volume.

Stoichiometric air requirements are usually stated in units of mass (pounds or kilograms) of air for solid and liquid fuels, and in units of volume (cubic feet or cubic meters) of air for gaseous fuels. When stated in terms of mass, the ratio of air to fuel masses is known as the *air-fuel ratio*, A/F.

$$\frac{A}{F} = \frac{m_{\text{air}}}{m_{\text{fuel}}} \qquad 30.1$$

Atmospheric air is a mixture of oxygen, nitrogen, and small amounts of carbon dioxide, water vapor, argon, and other inert gases. If all constituents except oxygen are grouped with the nitrogen, the air composition is as given in Table 30.2.

Table 30.2 Composition of Air

	% by weight	% by volume
oxygen (O_2)	23.15	20.9
nitrogen (N_2)[a]	76.85	79.1
ratio of nitrogen to oxygen	3.32	3.78 (3.76)

[a]Inert gases are included as N_2.

Stoichiometric air includes atmospheric nitrogen. For each volume (or mole) of oxygen, 3.78 volumes (or moles) of nitrogen pass unchanged through the reaction. (This value is often quoted as 3.76.) For example, the combustion of methane in air would be written as

$$CH_4 + 2O_2 + 2(3.76)N_2 \longrightarrow CO_2 + 2H_2O + 7.52N_2$$

Complete combustion occurs when all of the fuel is burned. If there is inadequate oxygen, there will be *incomplete combustion*, and some carbon will appear as carbon monoxide in the products of combustion. The *percent theoretical air* is the actual air-fuel ratio as a percentage of the theoretical air-fuel ratio calculated from the stoichiometric combustion equation.

$$\begin{array}{c} \text{percent} \\ \text{theoretical} \\ \text{air} \end{array} = \frac{\left(\dfrac{A}{F}\right)_{\text{actual}}}{\left(\dfrac{A}{F}\right)_{\text{stoichiometric}}} \times 100\% \qquad 30.2$$

Usually 10–50% excess air is required for complete combustion to occur. *Excess* air is expressed as a percentage of the stoichiometric air requirements. Excess oxygen appears as pure oxygen along with the products of combustion.

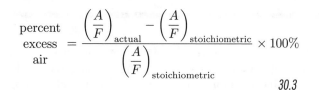

$$\begin{array}{c} \text{percent} \\ \text{excess} \\ \text{air} \end{array} = \frac{\left(\dfrac{A}{F}\right)_{\text{actual}} - \left(\dfrac{A}{F}\right)_{\text{stoichiometric}}}{\left(\dfrac{A}{F}\right)_{\text{stoichiometric}}} \times 100\%$$

30.3

SAMPLE PROBLEMS

Problem 1

Eleven grams of propane are burned with just enough pure oxygen for complete combustion. How many grams of combustion products are produced?

  (A) 31 g
  (B) 39 g
  (C) 41 g
  (D) 51 g

Solution

Follow the steps for stoichiometric problem-solving. First, balance the combustion reaction equation.

$$C_3H_8 + O_2 \longrightarrow CO_2 + H_2O \qquad \text{[unbalanced]}$$

Since there are three carbons on the left, there must be three (or a multiple of three) carbon dioxides on the right.

$$C_3H_8 + O_2 \longrightarrow 3CO_2 + H_2O \qquad \text{[unbalanced]}$$

Since there are eight hydrogens on the left, multiply H_2O by 4.

$$C_3H_8 + O_2 \longrightarrow 3CO_2 + 4H_2O \qquad \text{[unbalanced]}$$

Multiply the O_2 by 5 to balance the equation.

$$C_3H_8 + 5O_2 \longrightarrow 3CO_2 + 4H_2O \qquad \text{[balanced]}$$
molecular
 weights   44.1   (5)(32)   (3)(44) (4)(18)

The mass of combustion products produced is found by forming the ratio of product to reactant masses.

$$\frac{\text{product}}{\text{reactant}} = \frac{(3)(44) + (4)(18)}{44.1} = \frac{x}{11 \text{ g}}$$
$$x = 50.9 \text{ g} \quad (51 \text{ g})$$

Answer is D.

Problem 2

The combustion of a hydrocarbon fuel (C_xH_y) in an automotive engine results in the following dry exhaust gas analysis (% by volume).

$$\begin{array}{l} 11\% \ CO_2 \\ 0.5\% \ CO \\ 2\% \ CH_4 \\ 1.5\% \ H_2 \\ 6\% \ O_2 \\ 79\% \ N_2 \end{array}$$

Find the actual air-fuel ratio.

  (A) 8.7
  (B) 10
  (C) 14
  (D) 15

Solution

Balance the chemical combustion equation.

$$C_xH_y + (a+6)O_2 + bN_2 \longrightarrow$$
$$11CO_2 + 0.5CO + 2CH_4 + 1.5H_2$$
$$+ 79N_2 + cH_2O + 6O_2$$

$$b = 79$$

$$x = 11 + 0.5 + 2 = 13.5$$

$$a = \frac{b}{3.78} - 6 = \frac{79}{3.78} - 6$$
$$= 14.9$$

$$(2)(a+6) = (2)(14.9 + 6)$$
$$= (2)(11) + 0.5 + c + (2)(6)$$

$$c = 7.3$$

$$y = (2)(4) + (2)(1.5) + (7.3)(2)$$
$$= 25.6$$

The complete reaction equation is

$$C_{13.5}H_{25.6} + 20.9O_2 + 79N_2 \longrightarrow$$
$$11CO_2 + 0.5CO + 2CH_4 + 1.5H_2$$
$$+ 79N_2 + 7.3H_2O + 6O_2$$

The air-fuel ratio is

$$\frac{A}{F} = \frac{m_{\text{air}}}{m_{\text{fuel}}} = \frac{(20.9)(32) + (79)(28)}{(13.5)(12) + (25.6)(1)}$$
$$= 15.4 \quad (15)$$

Answer is D.

FE-STYLE EXAM PROBLEMS

1. Theoretically, how many kilograms of air are needed to completely burn 5 kg of ethane (C_2H_6) gas?

    (A) 0.8 kg
    (B) 19 kg
    (C) 81 kg
    (D) 330 kg

2. What are the products of complete combustion of a gaseous hydrocarbon?

    (A) carbon monoxide only
    (B) water, carbon monoxide, and carbon dioxide
    (C) carbon dioxide and water
    (D) carbon monoxide, water, and ammonia

3. Why is excess air required in combustion?

    (A) It allows the reaction to occur stoichiometrically.
    (B) It reduces air pollution.
    (C) It reduces the heat requirements.
    (D) It allows complete combustion.

SOLUTIONS TO FE-STYLE EXAM PROBLEMS

1. The balanced reaction equation is

$$2C_2H_6 + 7O_2 \longrightarrow 4CO_2 + 6H_2O \quad \text{[balanced]}$$

molecular
weights   (2)(30)   (7)(32)   (4)(44)   (6)(18)

By ratio,

$$\frac{O_2}{C_2H_6} = \frac{(7)(32)}{(2)(30)} = \frac{x}{5}$$

$$x = 18.67 \text{ kg}$$

Air is 23.15% O_2 by weight.

The mass of air required is

$$\frac{18.67 \text{ kg } O_2}{0.2315} = 80.6 \text{ kg} \quad (81 \text{ kg})$$

Answer is C.

2. A gaseous hydrocarbon reacts with oxygen to form carbon dioxide and water. Carbon monoxide only forms with incomplete combustion.

Answer is C.

3. Excess air is required for complete combustion to occur.

Answer is D.

Topic IX: Heat Transfer

Chapter 31   Heat Transfer

Hint: For the most current information about the exam, visit www.ppi2pass.com/fefaqs.html regularly.
Use the Exam Forum to compare notes with other FE examinees.

31

Heat Transfer

Subjects

HEAT TRANSFER 31-1
    Conduction 31-1
    Convection 31-2
    Radiation 31-2
FINS 31-3

Nomenclature

A	area	ft^2	m^2
h	enthalpy	Btu/lbm	kJ/kg
h	coefficient of heat transfer	Btu/hr-ft^2-°F	W/m^2·K
k	thermal conductivity	Btu/hr-ft-°F	W/m·K
L	thickness	ft	m
m	factor	1/ft	1/m
m	mass	lbm	kg
P	perimeter	ft	m
Q	heat energy	Btu	kJ
$\dot{Q}$	rate of heat transfer	Btu/hr	W
r	radius	ft	m
R	thermal resistance	ft^2-hr-°F/Btu	m^2·K/W
t	time	sec	s
T	absolute temperature	°R	K
U	overall coefficient of heat transfer	Btu/hr-ft^2-°F	W/m^2·K
V	volume	gal	L
$\dot{V}$	flow rate	gal/min	L/s

Symbols

α	absorptivity	–	–
ϵ	emissivity	–	–
ρ	reflectivity	–	–
σ	Stefan-Boltzmann constant	Btu/hr-ft^2-°R^4	W/m^2·K^4
τ	transmissivity	–	–

Subscripts

b	base
c	cross section, or corrected
fg	liquid-to-gas (vaporization)
i	inner
m	mean
o	outer
th	thermal
w	work (wall)
∞	bulk fluid

HEAT TRANSFER

Heat is thermal energy in motion. There are three distinct mechanisms by which thermal energy can move from one location to another. These mechanisms are distinguished by the media through which the energy moves.

If no medium (air, water, solid concrete, etc.) is required, the heat transfer occurs by *radiation*. If energy is transferred through a solid material by molecular vibration, the heat transfer mechanism is known as *conduction*. If energy is transferred from one point to another by a moving fluid, the mechanism is known as *convection*. *Natural convection* transfers heat by relying on density changes to cause fluid motion. *Forced convection* requires a pump, fan, or relative motion to move the fluid. (Change of phase—evaporation and condensation—is categorized as convection.)

In almost all problems, the energy transfer rate, q, will initially vary with time. This initial period is known as the *transient period*. Eventually, the rate of energy transfer becomes constant, and this is known as the *steady-state* or *equilibrium rate*.

Conduction

If energy is transferred through a solid material by molecular vibration, the heat transfer mechanism is known as *conduction*.

The steady-state heat transfer by conduction through a flat slab is specified by *Fourier's law*. Fourier's law is written with a minus sign to indicate that the heat flow is opposite the direction of the thermal gradient.

$$\dot{Q} = -kA\frac{dT}{dx} \qquad 31.1$$

$$\dot{Q} = -\frac{kA(T_2 - T_1)}{L} \qquad 31.2$$

The quantity L/kA is referred to as the *thermal resistance*, R_{th}, of the material. In a composite slab material, the thermal resistances are in series. The total thermal resistance of a composite slab is the sum of the individual thermal resistances. For example, in Fig. 31.1 the total thermal resistance is given by the sum of the two individual resistances.

$$R_1 = \frac{L_1}{k_1 A} \qquad 31.3$$

$$R_2 = \frac{L_2}{k_2 A} \qquad 31.4$$

Figure 31.1 Composite Slab Wall

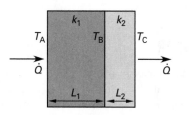

The temperature at any point within a single-layer or composite wall can be found if the heat transfer rate, $\dot{Q}$, is known. The thermal resistance is determined to the point or layer where the temperature is of interest. Then, Eq. 31.2 is solved for ΔT and the unknown temperature. For the composite wall in Fig. 31.1,

$$T_B = T_A - \dot{Q}R_1 \qquad 31.5$$

$$T_C = T_B - \dot{Q}R_2 \qquad 31.6$$

The Fourier equation is based on a uniform path length and a constant cross-sectional area. If the heat flow is through an area that is not constant, the *logarithmic mean area*, A_m, given by Eq. 31.7, should be used in place of the regular area. The log mean area should be used with heat transfer through thick pipe and cylindrical tank walls.

$$A_m = \frac{A_o - A_i}{\ln\left(\dfrac{A_o}{A_i}\right)} \qquad 31.7$$

$$\dot{Q} = \frac{A_m k (T_1 - T_2)}{L_{\text{radial}}}$$

$$= \frac{2\pi k L_{\text{longitudinal}} (T_1 - T_2)}{\ln\left(\dfrac{r_o}{r_i}\right)} \qquad 31.8$$

Convection

Unlike conduction, convective heat transfer depends on the movement of a heated fluid to transfer heat energy. Forced convection results when a fan, pump, or relative vehicle motion moves the fluid.

Figure 31.2 Cylindrical Wall

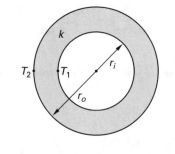

Newton's law of convection, Eq. 31.9, calculates the forced convection heat transfer. If a single film dominates the thermal resistance, as is often the case in heat exchangers and other thin-wall applications, the overall coefficient of heat transfer is h.

$$\dot{Q} = hA(T_w - T_\infty) \qquad 31.9$$

Radiation

Thermal radiation is electromagnetic radiation with a wavelength between 700 nm and 10^5 nm (7×10^{-7} and 1×10^{-4} m). If all of the radiation has the same wavelength, it is *monochromatic radiation*. The adjective "spectral" is used to denote a variation in some property with wavelength.

Radiation directed at a body can be absorbed, reflected, or transmitted through the body, with the total of the three resultant energy streams equaling the incident energy. If α is the fraction of energy being absorbed (i.e., the *absorptivity*), ρ is the fraction being reflected (i.e., the *reflectivity*), and τ is the fraction being transmitted through (i.e., the *transmissivity*), then the radiation conservation law is

$$\alpha + \rho + \tau = 1 \qquad 31.10$$

For opaque solids and some liquids, $\tau = 0$. Gases reflect very little radiant energy, so $\rho \approx 0$.

A *black body* (*ideal radiator*) is a body that absorbs all of the radiant energy that impinges on it (i.e., absorptivity, α, is equal to unity). A black body also emits the maximum possible energy when acting as a source.

Black bodies, like ideal gases, are never achieved in practice. Thus, all real bodies are "gray" bodies. The *emissivity*, ϵ, of a gray body is the ratio of the actual radiation emitted to that emitted by a black body.

$$\epsilon = \frac{\dot{Q}_{\text{gray}}}{\dot{Q}_{\text{black}}} \qquad 31.11$$

Notice that emissivity, ϵ, does not appear in the radiation conservation law, Eq. 31.10. However, for a black body, $\epsilon = \alpha = 1.0$. And, for any body in thermal equilibrium (i.e., radiating all energy that is being absorbed), $\epsilon = \alpha$.

In summary,

$$\alpha + \rho = 1 \quad \text{[opaque body; } \tau = 0] \qquad \textit{31.12}$$

$$\alpha + \rho = \epsilon + \rho = 1 \quad \text{[gray body; } \tau = 0; \alpha = \epsilon] \quad \textit{31.13}$$

$$\alpha = \epsilon = 1 \quad \begin{bmatrix} \text{black body; } \tau = 0; \\ \rho = 0; \alpha = \epsilon \end{bmatrix} \qquad \textit{31.14}$$

Radiant heat transfer is the name given to heat transfer by way of thermal radiation. The energy radiated by a hot body at absolute temperature T is given by the *Stefan-Boltzmann law*, also known as the *fourth-power law*. In Eq. 31.15, σ is the *Stefan-Boltzmann constant*. For a body with surface area A,

$$\dot{Q}_{\text{black}} = \epsilon \sigma A T^4 \qquad \textit{31.15}$$

$$\sigma = 5.670 \times 10^{-8} \text{ W/m}^2 \cdot \text{K}^4 \qquad \text{[SI]}$$

$$\sigma = 0.1713 \times 10^{-8} \text{ Btu/hr-ft}^2\text{-}^\circ\text{R}^4 \qquad \text{[U.S.]}$$

When two bodies can "see each other," each will radiate energy to and absorb energy from the other. The net radiant heat transfer between the two bodies is given by Eq. 31.16.

$$\dot{Q}_{12} = \sigma A_1 F_{1-2} (T_1^4 - T_2^4) \qquad \textit{31.16}$$

F_{1-2} is the *configuration factor*, or *shape factor*, which depends on the shapes, emissivities, and orientations of the two bodies. If body 1 is small and completely enclosed by body 2, then $F_{1-2} = \epsilon_1$.

FINS

Fins (also known as *extended surfaces*) are objects that receive and move thermal energy by conduction along their length and width, prior to convective and radiative heat removal. Some simple configurations can be considered and evaluated as fins even though heat removal is not their intended function.

Figure 31.3 Finite Straight Fin

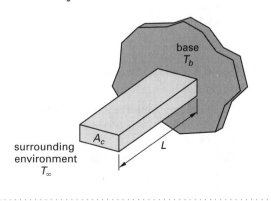

External fins are attached at their base to a source of thermal energy at temperature T_b. The temperature across the face of the fin at any point along its length is assumed to be constant. The far-field temperature of the surrounding environment is T_∞. For *rectangular fins* (also known as *straight fins* or *longitudinal fins*), the cross-sectional area, A_c, is uniform and is equal to the base area. The heat transfer rate from a straight (rectangular) fin is

$$\dot{Q} = \sqrt{hPkA_c} \times (T_b - T_\infty) \tan h(mL)$$
$$\text{[rectangular fin]} \qquad \textit{31.17}$$

$$m = \sqrt{\frac{hP}{kA_c}} \qquad \textit{31.18}$$

Most equations for heat transfer from a fin disregard the smaller amount of heat transfer from the exposed end. For that reason, the fin is assumed to possess an *adiabatic tip* or *insulated tip*. A simple approximation to the exact solution of a nonadiabatic tip can be obtained by replacing the actual fin length with a corrected length.

$$\dot{Q} = \sqrt{hPkA_c} \times (T_b - T_\infty) \tan h(mL_c) \quad \textit{31.19}$$

$$L_c = L + \frac{A_c}{P} \qquad \textit{31.20}$$

The *radiator efficiency (fin efficiency)*, η, is the ratio of the actual to ideal heat transfers assuming the entire fin is at the base temperature, T_b.

$$\eta_f = \frac{\dot{Q}_{\text{actual}}}{\dot{Q}_{\text{ideal}}}$$

$$= \frac{\dot{Q}_{\text{actual}}}{hA_f(T_b - T_\infty)} \qquad \textit{31.21}$$

SAMPLE PROBLEMS

Problem 1

A small distiller evaporates 1 L of water per half hour. Copper tubing exposed to the air serves as a condenser to recover the steam. The outside surface of the tubing is at 99.9°C. The ambient temperature is 20°C. The inside diameter of the tube is 0.75 cm, and the outside diameter is 1.2 cm. The thermal conductivity of copper is 388 W/m·K. h_{fg} for saturated steam at 100°C is 2257 kJ/kg. How long must the tube be to condense all of the steam?

    (A) 0.3 m
    (B) 1.2 m
    (C) 2.4 m
    (D) 8.3 m

Solution

The heat transfer is

$$\dot{Q} = \frac{mh_{fg}}{t} = \frac{(1\ L)\left(1\ \dfrac{kg}{L}\right)\left(2257\ \dfrac{kJ}{kg}\right)\left(1000\ \dfrac{\frac{W}{kJ}}{s}\right)}{(0.5\ h)\left(3600\ \dfrac{s}{h}\right)}$$

$$= 1254\ W$$

The difference in temperature is the same in °C and K. From the equation for heat transfer through curved surfaces, and recognizing that $r_o/r_i = d_o/d_i$,

$$L = \frac{\dot{Q}\ \ln\left(\dfrac{r_o}{r_i}\right)}{2\pi k(T_i - T_o)} = \frac{(1254\ W)\ \ln\left(\dfrac{0.012}{0.0075}\right)}{2\pi\left(388\ \dfrac{W}{m\cdot K}\right)(100°C - 99.9°C)}$$

$$= 2.4\ m$$

Answer is C.

Problem 2

What is the thermal resistance if the composite wall shown has an exposed surface area of 120 m²?

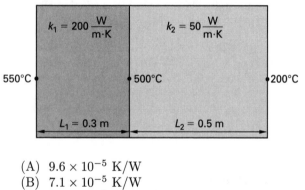

- (A) 9.6×10^{-5} K/W
- (B) 7.1×10^{-5} K/W
- (C) 6.3×10^{-5} K/W
- (D) 2.7×10^{-5} K/W

Solution

$$R = R_1 + R_2 = \frac{L_1}{k_1 A} + \frac{L_2}{k_2 A}$$

$$= \left(\frac{1}{A}\right)\left(\frac{L_1}{k_1} + \frac{L_2}{k_2}\right)$$

$$= \left(\frac{1}{120\ m^2}\right)\left(\frac{0.3\ m}{200\ \dfrac{W}{m\cdot K}} + \frac{0.5\ m}{50\ \dfrac{W}{m\cdot K}}\right)$$

$$= 9.58 \times 10^{-5}\ K/W \quad (9.6 \times 10^{-5}\ K/W)$$

Answer is A.

FE-STYLE EXAM PROBLEMS

1. The thermal resistance for one-dimensional steady conduction heat transfer through a cylindrical wall in the radial direction is which of the following?

- (A) linear
- (B) logarithmic
- (C) exponential
- (D) polynomial

2. A well-insulated copper rod (thermal conductivity of 388 W/m·K) is 50 cm long and has a cross-sectional area of 10 cm². One end of the rod is in contact with a block of ice at 0°C, and the other end is in boiling water at 100°C. Heat of vaporization for saturated steam at 100°C is 2257 kJ/kg. The heat of fusion for ice at 0°C is 334 kJ/kg. How much ice is melted in 1 min?

- (A) 0.2 g
- (B) 0.5 g
- (C) 1.1 g
- (D) 13.9 g

3. A well-insulated solid copper wire 1 m long with a cross-sectional area of 0.0839 cm² connects two reservoirs, one with boiling water at 100°C and the other with ice at 0°C. How long does it take for 100 cal of energy to move through the wire?

- (A) 3.5 min
- (B) 21 min
- (C) 86 min
- (D) 240 min

4. In 1 hr, how much black-body radiation escapes a 1 cm by 2 cm rectangular opening in a kiln whose internal temperature is 980°C?

- (A) 20 kJ
- (B) 100 kJ
- (C) 130 kJ
- (D) 150 kJ

5. Which of the following statements concerning radiation heat transfer is false?

- (A) The radiation emitted by a body is proportional to the fourth power of its absolute temperature.
- (B) For a body of a particular size and temperature, the maximum energy is emitted by a black body.

(C) For an opaque body, the sum of absorptivity and reflectivity is always equal to unity.

(D) Radiation energy cannot travel through a vacuum.

For the following problems use the NCEES Handbook as your only reference.

6. Heat transfer occurs by conduction through a composite wall as shown. The temperature on one side of the wall is $-5°C$; on the other, the temperature is $25°C$. (The thermal conductivities and film coefficients are shown in the illustration.) What is the heat transfer through the wall?

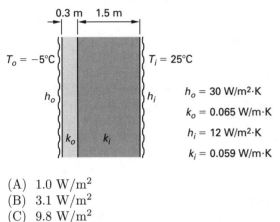

(A) 1.0 W/m^2
(B) 3.1 W/m^2
(C) 9.8 W/m^2
(D) 17 W/m^2

7. A composite wall constructed of 2.5 cm of steel ($k = 60.5$ W/m·K) and 5.0 cm of aluminum ($k = 177$ W/m·K) separates two liquids. The liquid on the steel side has a film coefficient of 15 W/m^2·K and a temperature of 400°C. The liquid on the aluminum side has a film coefficient of 30 W/m^2·K and a temperature of 100°C. Assuming that steady-state conditions have been reached, what is the approximate temperature at the steel-aluminum interface?

(A) 150°C
(B) 200°C
(C) 250°C
(D) 300°C

8. The ideal rate of heat transfer through a section of a uniform wall is

(A) directly proportional to the overall coefficient of heat transfer and directly proportional to the thickness of the wall.

(B) inversely proportional to the overall coefficient of heat transfer and directly proportional to the thickness of the wall.

(C) directly proportional to the overall coefficient of heat transfer and inversely proportional to the thickness of the wall.

(D) inversely proportional to the overall coefficient of heat transfer and to the thickness of the wall.

9. If T is the absolute temperature, the intensity of radiation from an ideal radiator will be proportional to

(A) T
(B) T^2
(C) T^3
(D) T^4

10. A 3 m × 3 m plate at 500°C is suspended vertically in a very large room. The plate has an emissivity of 0.13. The room is at 25°C. What is the net heat transfer from the plate?

(A) 8.3 kW
(B) 24 kW
(C) 46 kW
(D) 47 kW

11. At 3:00 p.m. in July, a stone wall reaches a maximum temperature of 50°C. The stone has an emissivity of 0.95. What is the radiant heat energy transfer from the wall per square meter of this wall at that temperature?

(A) 190 W/m^2
(B) 590 W/m^2
(C) 1100 W/m^2
(D) 3800 W/m^2

12. Hot air at an average temperature of 100°C flows through a 3 m long tube with an inside diameter of 60 mm. The temperature of the tube is 20°C along its entire length. The convective film coefficient is 20.1 W/m^2·K. Determine the rate of convective heat transfer from the air to the tube.

(A) 520 W
(B) 850 W
(C) 910 W
(D) 1070 W

13. An engine develops a brake power of 30 kW + 40 kW are absorbed by cooling water that is pumped through the water jacket and a radiator. The water enters the top of the engine's radiator at 95°C. The enthalpy of the water at that temperature is 397.96 kJ/kg. The water leaves the bottom of the radiator at 90°C and with an enthalpy of 376.92 kJ/kg. What is the water flow rate for steady-state operation?

(A) $4.2 \times 10^{-4} \text{ m}^3/\text{s}$
(B) $1.4 \times 10^{-3} \text{ m}^3/\text{s}$
(C) $1.9 \times 10^{-3} \text{ m}^3/\text{s}$
(D) $3.3 \times 10^{-3} \text{ m}^3/\text{s}$

SOLUTIONS TO FE-STYLE EXAM PROBLEMS

1. For heat flow radially through cylindrical pipe walls, the logarithmic mean area is used.

$$A_m = \frac{A_o - A_i}{\ln\left(\dfrac{A_o}{A_i}\right)}$$

Since thermal resistance is L/kA, the thermal resistance is logarithmic.

Answer is B.

2. Because the rod's heat loss to the environment is negligible, the heat transfer through the rod must exactly equal the heat absorbed by the melting ice.

$$\dot{Q} = \frac{kA(T_1 - T_2)}{L} = \frac{m h_{\text{fusion}}}{t}$$

$$m = \frac{kAt(T_1 - T_2)}{h_{\text{fusion}} L}$$

$$= \frac{\left(388 \dfrac{\text{W}}{\text{m·K}}\right)(10 \text{ cm}^2)\left(\dfrac{1}{100 \dfrac{\text{cm}}{\text{m}}}\right)^2}{\left(334 \dfrac{\text{kJ}}{\text{kg}}\right)(0.5 \text{ m})\left(1000 \dfrac{\text{J}}{\text{kJ}}\right)}$$
$$\times (60 \text{ s})(100°\text{C} - 0°\text{C})$$

$$= 1.39 \times 10^{-2} \text{ kg} \quad (13.9 \text{ g})$$

Answer is D.

3. Apply Fourier's law to determine the rate of heat conduction.

$$Q = \dot{Q}t = \frac{kAt(T_1 - T_2)}{L}$$

$$t = \frac{QL}{kA(T_1 - T_2)}$$

$$= \frac{(100 \text{ cal})\left(4.184 \dfrac{\text{J}}{\text{cal}}\right)(1 \text{ m})}{\left(388 \dfrac{\text{W}}{\text{m·K}}\right)(0.0839 \text{ cm}^2)}$$
$$\times \left(\dfrac{1}{100 \dfrac{\text{cm}}{\text{m}}}\right)^2 (100°\text{C} - 0°\text{C})\left(60 \dfrac{\text{s}}{\text{min}}\right)$$

$$= 21.4 \text{ min} \quad (21 \text{ min})$$

Answer is B.

4.
$$A = (1 \text{ cm})(2 \text{ cm})$$
$$= 2 \text{ cm}^2$$
$$\dot{Q}_{\text{black}} = \epsilon \sigma A T^4$$

$$= (1)\left(5.67 \times 10^{-8} \frac{\text{W}}{\text{m}^2\text{·K}^4}\right)(2 \text{ cm}^2)$$
$$\times \left(\frac{1}{100 \dfrac{\text{cm}}{\text{m}}}\right)^2 (980°\text{C} + 273)^4$$

$$= 28 \text{ W}$$

$$Q = \dot{Q}t$$
$$= (28 \text{ W})(1 \text{ h})\left(3600 \frac{\text{s}}{\text{h}}\right)$$
$$= 100\,800 \text{ J} \quad (100 \text{ kJ})$$

Answer is B.

5. All of the statements are true except (D). Radiation travels through a vacuum very well.

Answer is D.

6. The conductive thermal resistance of a substance is given by the quantity L/k. The thermal resistance through convective films is $1/h$. The thermal resistance of the films is added to the conductive resistance. The total thermal resistance per unit area is

$$R_{\text{th}} = \frac{1}{h_i} + \frac{L_i}{k_i} + \frac{L_o}{k_o} + \frac{1}{h_o}$$

$$= \frac{1}{12 \dfrac{\text{W}}{\text{m}^2\text{·K}}} + \frac{1.5 \text{ m}}{0.059 \dfrac{\text{W}}{\text{m·K}}}$$
$$+ \frac{0.3 \text{ m}}{0.065 \dfrac{\text{W}}{\text{m·K}}} + \frac{1}{30 \dfrac{\text{W}}{\text{m}^2\text{·K}}}$$

$$= 30.16 \text{ m}^2\text{·K/W}$$

The conductive heat transfer per unit area is

$$\frac{\dot{Q}}{A} = \frac{T_i - T_o}{R_{\text{th}}}$$

$$= \frac{25°\text{C} - (-5°\text{C})}{30.16 \dfrac{\text{m}^2\text{·K}}{\text{W}}}$$

$$= 0.995 \text{ W/m}^2 \quad (1.0 \text{ W/m}^2)$$

Answer is A.

7. First, determine the heat transfer.

$$R_{\text{th}} = \frac{1}{h_{\text{aluminum}}} + \frac{L_{\text{aluminum}}}{k_{\text{aluminum}}} + \frac{L_{\text{steel}}}{k_{\text{steel}}} + \frac{1}{h_{\text{steel}}}$$

$$= \frac{1}{30 \dfrac{\text{W}}{\text{m}^2 \cdot \text{K}}} + \frac{0.05 \text{ m}}{177 \dfrac{\text{W}}{\text{m} \cdot \text{K}}}$$

$$\quad + \frac{0.025 \text{ m}}{60.5 \dfrac{\text{W}}{\text{m} \cdot \text{K}}} + \frac{1}{15 \dfrac{\text{W}}{\text{m}^2 \cdot \text{K}}}$$

$$= 0.1007 \text{ m}^2 \cdot \text{K/W}$$

Notice that the aluminum and steel contribute little to the overall thermal resistance.

The conductive heat transfer per unit area is

$$\frac{\dot{Q}}{A} = \frac{T_{\text{hot}} - T_{\text{cold}}}{R_{\text{th}}}$$

$$= \frac{400°\text{C} - 100°\text{C}}{0.1007 \dfrac{\text{m}^2 \cdot \text{K}}{\text{W}}}$$

$$= 2979 \text{ W/m}^2$$

The thermal resistance from the steel-side fluid to the interface is

$$R_{\text{th}} = \frac{L_{\text{steel}}}{k_{\text{steel}}} + \frac{1}{h_{\text{steel}}}$$

$$= \frac{0.025 \text{ m}}{60.5 \dfrac{\text{W}}{\text{m} \cdot \text{K}}} + \frac{1}{15 \dfrac{\text{W}}{\text{m}^2 \cdot \text{K}}}$$

$$= 0.06708 \text{ m}^2 \cdot \text{K/W}$$

The heat transfer is known. Solve the conduction equation for the unknown temperature.

$$\frac{\dot{Q}}{A} = \frac{T_{\text{hot}} - T_{\text{interface}}}{R_{\text{th}}}$$

$$2979 \frac{\text{W}}{\text{m}^2} = \frac{400°\text{C} - T_{\text{interface}}}{0.06708 \dfrac{\text{m}^2 \cdot \text{K}}{\text{W}}}$$

$$T_{\text{interface}} = 200°\text{C}$$

Answer is B.

8. Three formulations for the heat transfer through a plane surface are

$$\dot{Q} = UA(T_{\text{hot}} - T_{\text{cold}})$$

$$\dot{Q} = \frac{A(T_{\text{hot}} - T_{\text{cold}})}{R_{\text{th}}}$$

$$\dot{Q} = \frac{kA(T_{\text{hot}} - T_{\text{cold}})}{L} \quad \text{[conduction only]}$$

The rate of heat transfer, $\dot{Q}$, is directly proportional to the overall coefficient of heat transfer, U, and inversely proportional to the thickness of the wall, L.

Answer is C.

9. The intensity of an ideal (black-body) radiator at absolute temperature T is given by the Stefan-Boltzmann law, also known as the "fourth power law."

$$\dot{Q}_{\text{black}} = \epsilon \sigma A T^4$$

Answer is D.

10. The plate has two surfaces. Disregarding the edges, the radiating area is

$$A = (2)(3 \text{ m})(3 \text{ m}) = 18 \text{ m}^2$$

The configuration factor for a fully enclosed body is the emissivity of the body. In this case,

$$F_{1-2} = \epsilon_1 = 0.13$$

$$\dot{Q}_{1-2} = \sigma A_1 F_{1-2}(T_1^4 - T_2^4)$$

$$= \left(5.670 \times 10^{-8} \frac{\text{W}}{\text{m}^2 \cdot \text{K}^4}\right)(18 \text{ m}^2)(0.13)$$

$$\quad \times ((500°\text{C} + 273)^4 - (25°\text{C} + 273)^4)$$

$$= 46\,325 \text{ W} \quad (46 \text{ kW})$$

Answer is C.

11. The energy radiated by a "gray" body is

$$\dot{Q} = \epsilon \sigma A T^4$$

$$\frac{\dot{Q}}{A} = \epsilon \sigma T^4$$

$$= (0.95)\left(5.670 \times 10^{-8} \frac{\text{W}}{\text{m}^2 \cdot \text{K}^4}\right)(50°\text{C} + 273)^4$$

$$= 586 \text{ W/m}^2 \quad (590 \text{ W/m}^2)$$

Answer is B.

12. The heat transfer area is

$$A = \pi d \times \text{length} = \frac{\pi (60 \text{ mm})(3 \text{ m})}{1000 \dfrac{\text{mm}}{\text{m}}}$$

$$= 0.565 \text{ m}^2$$

Use Newton's law of convection. The temperature difference is the same for temperatures expressed in °C or K.

$$\dot{Q} = hA(T_{\text{air}} - T_{\text{wall}})$$

$$= \left(20.1 \; \frac{\text{W}}{\text{m}^2 \cdot \text{K}}\right) (0.565 \; \text{m}^2)(100°\text{C} - 20°\text{C})$$

$$= 908.5 \; \text{W} \quad (910 \; \text{W})$$

Answer is C.

13. Although the rate of combustion energy generated is 40 kW + 30 kW, only the 40 kW contributes to water heating. The remaining 30 kW are dissipated as useful work by the mechanisms (tires, etc.) attached to the output shaft. If this were not true, then the 30 kW would be causing a heating effect in two places.

The enthalpy change is calculated from

$$\dot{Q} = \dot{m}(h_{\text{in}} - h_{\text{out}})$$

$$40 \; \text{kW} = \dot{m}\left(397.96 \; \frac{\text{kJ}}{\text{kg}} - 376.92 \; \frac{\text{kJ}}{\text{kg}}\right)$$

$$\dot{m} = 1.901 \; \text{kg/s}$$

The density of water is approximately 1000 kg/m³. (A more accurate value could be found as the reciprocal of the specific volume found from the steam table.)

$$\dot{V} = \frac{1.901 \; \dfrac{\text{kg}}{\text{s}}}{1000 \; \dfrac{\text{kg}}{\text{m}^3}}$$

$$= 0.001901 \; \text{m}^3/\text{s} \quad (1.9 \times 10^{-3} \; \text{m}^3/\text{s})$$

Answer is C.

Topic X: Transport Phenomena

Chapter 32   Transport Phenomena

TRANSPORT
PHENOMENA

Hint: For the most current information about the exam, visit www.ppi2pass.com/fefaqs.html regularly.
Use the Exam Forum to compare notes with other FE examinees.

32 Transport Phenomena

Subjects

INTRODUCTION 32-1
UNIT OPERATIONS 32-1
MOMENTUM TRANSFER 32-2
ENERGY TRANSFER 32-2
MASS TRANSFER 32-3
TRANSPORT PHENOMENA
    ANALOGIES 32-3

Subscripts

f	friction
H	heat
m	mass
M	mass
w	wall

Nomenclature

a	acceleration	ft/sec^2	m/s^2
A	area	ft^2	m^2
c_m	concentration	mol/ft^3	mol/m^3
c_p	specific heat	Btu/lbm-°R	kJ/kg·K
D	diameter	ft	m
D_m	diffusion coefficient	ft^2/sec	m^2/s
f	Moody friction factor	–	–
F	force	lbf	N
G	mass velocity	lbm/ft^2-sec	kg/m^2·s
h	film coefficient	Btu/hr-ft^2-°R	W/m^2·K
h	head loss	ft	m
j_H	heat j-number	–	–
j_M	mass j-number	–	–
k	thermal conductivity	Btu/hr-ft^2-°R	W/m^2·K
L	length	ft	m
m	mass	lbm	kg
p	pressure	lbf/ft^2	Pa
Pr	Prandtl number	–	–
$\dot{Q}$	heat transfer	Btu/hr	W
Re	Reynolds number	–	–
Sc	Schmidt number	–	–
Sh	Sherwood number	–	–
St	Stanton number	–	–
t	time	sec	s
T	temperature	°R	K
v	velocity	ft/sec	m/s
y	distance	ft	m
y	radial distance from inner wall to tube centerline	ft	m

Symbols

ρ	density	lbm/ft^3	kg/m^3
μ	absolute viscosity	lbf-sec/ft^2	N·s/m^2
τ	shear stress	lbf/ft^2	N/m^2

INTRODUCTION

Transport phenomena traditionally include the processes of energy transfer, momentum transfer, and mass transfer. Not all engineering students have studied all of these subjects; for instance, mass transfer is typically taught only to chemical engineers. Some students have studied a subject without recognizing it was a subset of transport phenomena; for example, heat transfer is an energy transport phenomenon.

Energy transfer is concerned with the transfer of thermal energy from one location to another. *Mass transfer* is concerned with the transfer of mass from one phase to another. *Momentum transfer* is concerned with the transfer of momentum from one location to another.

All of the transport phenomena have basic concepts in common. The forms of the equations and the methods of solutions are similar. Only the names of the constants and coefficients vary. An analogy can be drawn between the phenomena.

UNIT OPERATIONS

There are many applications of transport phenomena. Regardless of the industry or product, transport phenomena can be divided into separate and distinct steps called *unit operations*. The most important unit operations are:

- fluid flow
- heat transfer
- evaporation
- drying
- distillation
- absorption
- membrane separation

- liquid-liquid extraction
- liquid-solid leaching
- crystallization
- mechanical separation processes

MOMENTUM TRANSFER

Momentum transfer is generally studied and referred to as "fluid mechanics." The concepts of flow regime (laminar or turbulent) and viscosity are basic. The flow regime is determined by calculating the Reynolds number, Re, which is the ratio of inertia force to viscous force.

$$\text{Re} = \frac{Dv\rho}{\mu} \qquad 32.1$$

When a turbulent fluid flows in a circular tube, the fluid shear stress at the wall is a function of the velocity gradient, dv/dy. The constant of proportionality, μ, is the absolute viscosity.

$$\frac{F}{A} = \tau = -\mu\frac{dv}{dy} \qquad 32.2$$

Equation 32.2 is taught in every fluid dynamics course. The linear relationship is intrinsic to the definition of a Newtonian fluid. However, the connection with momentum transfer may not be obvious. For a constant mass flow rate, Newton's second law can be written as

$$F = ma = m\frac{dv}{dt} = \frac{d(mv)}{dt}$$
$$= \frac{d(\text{momentum})}{dt} \qquad 32.3$$

Therefore, the force, F, can be interpreted as the flux of fluid momentum in the radial direction (i.e., away from the wall). The shear stress, F/A, can be interpreted as the flux of fluid momentum in the radial direction (i.e., away from the wall) per unit area. The negative sign in Eq. 32.2 indicates that momentum transfer is high when the velocity is low (i.e., at the wall), and vice versa.

The *Fanning friction factor* is defined as the shear stress divided by the kinetic energy per unit volume. The *Moody friction factor*, f, is four times the Fanning friction factor, hence the factor "4" in Eq. 32.4.

$$f = \frac{4\tau}{\frac{\frac{1}{2}mv^2}{V}} = \frac{8\tau}{\rho v^2} \qquad 32.4$$

The Moody friction factor is used to calculate friction loss. The traditional Darcy equation for friction head loss along a length, L, of pipe is

$$h_f = \frac{\Delta p}{\rho g} = \frac{fLv^2}{2Dg} \qquad 32.5$$

Since $h_f = \Delta p/\rho g$, the Moody friction factor can also be written as

$$f = \frac{2D\Delta p}{\rho L v^2} \qquad 32.6$$

Combining Eqs. 32.2, 32.4, and 32.6, the momentum transfer in the radial direction per unit area is

$$\frac{F}{A} = \tau = -\mu\frac{dv}{dy}$$
$$= \frac{-f\rho v^2}{8} = \frac{-D\Delta p}{4L} \qquad 32.7$$

ENERGY TRANSFER

Energy transfer is generally referred to as "heat transfer." The concepts of transfer regime (laminar or turbulent) and thermal conductivity are basic. An important dimensionless number is the *Prandtl number*, Pr. The Prandtl number is the ratio of the shear component of momentum diffusivity to the heat diffusivity (i.e., the ratio of diffusion of momentum to the diffusion of heat).

$$\text{Pr} = \frac{c_p\mu}{k} \qquad 32.8$$

When a thermal gradient exists in a material, the heat flux is a function of the temperature gradient, dT/dy. The constant of proportionality, k, is the thermal conductivity of the substance.

$$\frac{\dot{Q}}{A} = -k\frac{dT}{dy} \qquad 32.9$$

While Eq. 32.9 is valid specifically for conduction in any substance, the relationship is somewhat different for heat transfer to or from a fluid moving past a solid surface (i.e., a turbulent fluid moving through a tube). The *heat transfer film coefficient*, h, usually just referred to as the *film coefficient*, is defined as the rate of heat transfer per degree temperature of difference. Unlike with Eq. 32.9, it is not necessary to know the thickness of the film in order to use the film coefficient. In Eqs. 32.10 and 32.11, ΔT is the difference in tube wall and bulk fluid temperatures.

$$h = \frac{\frac{\dot{Q}}{A}}{\Delta T} \qquad 32.10$$

$$\frac{\dot{Q}}{A} = h\Delta T \qquad 32.11$$

MASS TRANSFER

Mass transfer is generally studied by chemical engineers as background for designing and analyzing such processes as distillation, absorption, drying, and extraction. Mass transfer deals with the migration of a single substance from one location or phase to another.

Quantities of mass transfer are usually calculated per unit area. Although the units of molecules per unit volume (per unit time) could be used, it is more common to work with units of moles per unit volume (per unit time). For very large quantities, units of mass per unit area (per unit time) can be used. The symbol G and the names *mass velocity* or *mass flow rate per unit area* are widely used to represent mass transfer in units of lbm/ft^2-sec or kg/m$^2 \cdot$s.

The flow regime for the diffusion of a small amount of substance 1 through substance 2 (i.e., a dilute "mixture" or *dilute solution*) is determined by calculating the *Schmidt number*, Sc. This is the dimensionless ratio of the molecular momentum diffusivity (μ/ρ) to the molecular mass diffusivity, quantified by the diffusion coefficient for substance 1 moving through substance 2, D_m. Typical values for gases range from 0.5 to 2.0. For liquids, the range is about 100 to more than 10,000 for viscous liquids. The Schmidt number is essentially independent of temperature and pressure within "normal" operating conditions.

$$ \text{Sc} = \frac{\mu}{\rho D_m} \qquad 32.12 $$

Fick's law describes molecular diffusion of mass for dilute solutions. The number of molecules of substance 1 moving through substance 2 (i.e, the *molecular diffusion*) per unit area in the y-direction is given by Eq. 32.13. c_m is the concentration of substance 1 at any particular point. dc_m/dy is the *concentration gradient* of substance 1 in the y-direction. When moles are the units used, N/A is referred to as the *molar flux*.

$$ N/A = -D_m \frac{dc_m}{dy} \qquad 32.13 $$

For a fluid flowing past a solid surface (as a turbulent fluid flowing in a pipe), the relationship for the inward mass transfer flux, N/A, at the wall is very similar to Eq. 32.11. k_m is the *mass transfer coefficient* for the process, and Δc_m is the difference in the concentrations between the wall and the bulk fluid.

$$ N/A = k_m \Delta c_m \qquad 32.14 $$

For convenience, two other dimensionless numbers are used with mass transfer. The dimensionless *Sherwood number*, Sh, is the ratio of mass diffusivity to molecular diffusivity and is given by Eq. 32.15. For turbulent flow through a pipe, D is the pipe inside diameter.

$$ \text{Sh} = \frac{k_m D}{D_m} \qquad 32.15 $$

The *Stanton number*, St, is the ratio of heat transfer at the wall to the energy transported by the mass flow.

$$ \text{St} = \frac{h}{c_p G} \qquad 32.16 $$

TRANSPORT PHENOMENA ANALOGIES

It is clear that the structures of the equations that describe the three transport processes are similar. In that sense, momentum, energy, and mass transport processes are *analogous* processes. However, the phrase "transport phenomena analogy" actually has a slightly different meaning.

In some fluid processes, two or three transport processes occur simultaneously. For example, a turbulent fluid flowing through a cooled pipe will experience both frictional and thermal energy losses. It is convenient to use known data from one of the processes to predict the performance of the other process(es). For example, since fluid friction factors (a momentum transport property) are easily calculated, it is desirable to use them to predict the (more difficult) heat and mass transfer rates. Such correlations between the performance data are known as "analogies."

In the popular Chilton and Colburn analogy, several dimensionless numbers are correlated with dimensionless "*j*-factors" and friction factors.

$$ j_H = (\text{St})(\text{Pr})^{2/3} = \frac{f}{8} \qquad 32.17 $$

$$ j_M = (\text{Sh})(\text{Sc})^{2/3} = \frac{f}{8} \qquad 32.18 $$

Equations 32.17 and 32.18 are supported by theoretical derivations for laminar flow over flat plates as well as experimental data. For turbulent flow, the analogies are supported by experimental data for liquids and gases.

Topic XI: Biology

Diagnostic Examination for Biology

Chapter 33 Cellular Biology
Chapter 34 Toxicology
Chapter 35 Industrial Hygiene
Chapter 36 Bioprocessing

BIOLOGY

Hint: For the most current information about the exam, visit www.ppi2pass.com/fefaqs.html regularly. Use the Exam Forum to compare notes with other FE examinees.

Diagnostic Examination

TOPIC XI: BIOLOGY

TIME LIMIT: 45 MINUTES

Problems 1–3 refer to the following information.

An environmental engineer has been assigned to manage a laboratory. Table 1 shows the concentrations of possible toxicants the engineer found in the breathing zones at three workstations.

Table 1

		concentration for indicated duration (ppm)			
station	toxicant	2 h	4 h	2 h	peak < 15 min
1	butane	110	100	120	250
2	pentane	700	410	250	800
3	hexane	300	600	320	900

The engineer reviewed the American Conference of Governmental Industrial Hygienists (ACGIH)-TLVs for the three toxicants identified. Table 2 shows the TLV-TWAs (threshold limit values-time weighted averages) and STELs (short-term exposure limits) for the same toxicants.

Table 2

toxicant	8 h TLV-TWA (ppm)	15 min STEL (ppm)
butane	800	–
pentane	600	750
hexane	500	1000

1. The TWA for butane at station 1 is most nearly

- (A) 54 ppm
- (B) 74 ppm
- (C) 108 ppm
- (D) 140 ppm

2. The concentration of pentane at station 2

- (A) exceeds the 8 h TWA-TLV and the 15 min STEL
- (B) exceeds the 8 h TWA-TLV but not the 15 min STEL
- (C) exceeds the 15 min STEL but not the 8 h TWA-TLV
- (D) does not exceed either the 8 h TWA-TLV or the 15 min STEL

3. The concentration of hexane at station 3

- (A) exceeds the 8 h TWA-TLV and the 15 min STEL
- (B) exceeds the 8 h TWA-TLV but not the 15 min STEL
- (C) exceeds the 15 min STEL but not the 8 h TWA-TLV
- (D) does not exceed either the 8 h TWA-TLV or the 15 min STEL

4. A concentration of 0.8 ppm of carbon monoxide (CO) is measured in a landfill at a temperature of 25°C and a pressure of 1.06 atm. The concentration of CO in mg/m^3 is most nearly

- (A) $0.97 \ mg/m^3$
- (B) $1.1 \ mg/m^3$
- (C) $25 \ mg/m^3$
- (D) $35 \ mg/m^3$

Problems 5 and 6 refer to the following information.

During each day shift, workers in a sewage treatment plant are exposed to noise from a sludge pump, a blower, and a ventilation fan as indicated. All noise is at 1000 Hz and referenced to the dBA scale.

source	sound pressure level (dB)	period of operation (h)
sludge pump	95	8 am – 12 noon 4 pm – 5 pm
blower	90	8 am – 10 am 4 pm – 5 pm
ventilation fan	100	8 am – 10 am 11 am – 4 pm

5. The sound pressure level the workers are exposed to during the period 4 pm to 5 pm is most nearly

- (A) 90 dB
- (B) 95 dB
- (C) 96 dB
- (D) 100 dB

6. If the reaction rate constant, K, is 0.10/d, the time-weighted sound pressure level exposure for one 8 am to 5 pm shift is most nearly

(A) 90.5 dB
(B) 95.0 dB
(C) 99.5 dB
(D) 100.0 dB

Problems 7 and 8 refer to the following information.

A 15 mL unseeded domestic wastewater sample is mixed with 985 mL of distilled water. The initial dissolved oxygen level is 8.2 mg/L, and the level after 2 d is 6.2 mg/L. The rate constant for biodegradation of the wastewater, K, is 0.10/d.

7. The ultimate BOD of the original sample is most nearly

(A) 6.2 mg/L
(B) 8.2 mg/L
(C) 11 mg/L
(D) 740 mg/L

8. Given an ultimate BOD of the diluted sample of 11 mg/L, the 5 day BOD of the original sample is most nearly

(A) 63 mg/L
(B) 83 mg/L
(C) 290 mg/L
(D) 700 mg/L

9. An organic material with average composition, $C_3H_{14}O_2N$, is completely stabilized in an anaerobic system. The mass of ammonia gas released per gram of organic material is most nearly

(A) 0.18 g
(B) 0.25 g
(C) 0.33 g
(D) 1.00 g

10. Aerobic biodegradation of glucose with ammonia is carried out in an experimental reactor. If the yield is 90%, the mass of cells produced per kg of glucose is most nearly

(A) 0.10 g
(B) 0.12 g
(C) 0.16 g
(D) 0.48 g

11. During the exponential log growth phase of a batch culture, the instantaneous value of the number of cells is 2000, and the cell growth rate is 300/d. The specific growth rate is most nearly

(A) 0.15/d
(B) 0.20/d
(C) 0.30/d
(D) 0.40/d

12. A faculative pond has an inflow of 10 ML/d. The incoming concentration is 356 mg/L (BOD). Therefore, the BOD loading of the faculative pond is most nearly

(A) 1.2 kg/d
(B) 360 kg/d
(C) 3600 kg/d
(D) 36 000 kg/d

13. An aerobic digester has an influent average flow rate of 1100 m³/d. The influent suspended solids concentration is 2000 mg/L. The fraction of influent BOD_5 consisting of raw primary sludge is 55%. The influent BOD_5 is 200 mg/L. The reaction rate constant is 0.24/d. The suspended solids concentration is 1500 mg/L, the volatile fraction of suspended solids is 0.34, and the sludge age is 2 d. The volume of the aerobic digester is most nearly

(A) 2.00 m³
(B) 260 m³
(C) 1100 m³
(D) 2700 m³

14. For a logistic growth-batch growth model, the logistic growth constant is 0.8/h. If the initial concentration is 1.5 g/L and the carrying capacity is 2.5 g/L, the initial growth rate is most nearly

(A) 0.48 g/L·h
(B) 0.80 g/L·h
(C) 1.5 g/L·h
(D) 2.5 g/L·h

15. If the specific growth rate for a microorganism during the exponential (log) phase is 2.5/d, the growth rate of the microorganisms when 2500 cells are present is most nearly

(A) 2500 cells/d
(B) 5000 cells/d
(C) 6300 cells/d
(D) 8000 cells/d

SOLUTIONS TO DIAGNOSTIC EXAMINATION TOPIC XI

Solutions 1–3 refer to the following information.

The TWA is calculated by multiplying each concentration by the exposure time interval and dividing the sum by the total time period.

$$\text{TWA} = \frac{\sum C_i t_i}{\sum t_i}$$

1.

$$\text{TWA}_{\text{butane}} = \frac{\left(\begin{array}{c}(110\ \text{ppm})(2\ \text{h}) \\ + (100\ \text{ppm})(4\ \text{h}) \\ + (120\ \text{ppm})(2\ \text{h})\end{array}\right)}{2\ \text{h} + 4\ \text{h} + 2\ \text{h}}$$
$$= 107.5\ \text{ppm} \quad (108\ \text{ppm})$$

Answer is C.

2.

$$\text{TWA}_{\text{pentane}} = \frac{\left(\begin{array}{c}(700\ \text{ppm})(2\ \text{h}) \\ + (410\ \text{ppm})(4\ \text{h}) \\ + (250\ \text{ppm})(2\ \text{h})\end{array}\right)}{2\ \text{h} + 4\ \text{h} + 2\ \text{h}}$$
$$= 442.5\ \text{ppm}$$

This is less than the 8 h TWA-TLV and the 15 min STEL.

Answer is D.

3.

$$\text{TWA}_{\text{hexane}} = \frac{\left(\begin{array}{c}(300\ \text{ppm})(2\ \text{h}) \\ + (600\ \text{ppm})(4\ \text{h}) \\ + (320\ \text{ppm})(2\ \text{h})\end{array}\right)}{2\ \text{h} + 4\ \text{h} + 2\ \text{h}}$$
$$= 455\ \text{ppm}$$

This is less than the 8 h TWA-TLV, and below the 15 min STEL.

Answer is D.

4. The CO concentration can be converted from ppm to mg/m^3 by using the following formula.

$$C_{\text{CO,mg/m}^3} = \left(\frac{(C_{\text{CO,ppm}})(\text{MW})}{22.414\ \text{L}}\right)\left(\frac{273\text{K}}{T^\circ\text{K}}\right)p$$
$$\text{MW}_{\text{CO}} = \left(12\ \frac{\text{g}}{\text{mol}} + 16\ \frac{\text{g}}{\text{mol}}\right)$$
$$= 28\ \text{g/mol}$$

$$T = 25^\circ\text{C} + 273$$
$$= 298\text{K}$$
$$p = 1.06\ \text{atm}$$
$$C_{\text{CO,mg/m}^3} = (0.8\ \text{ppm})\left(28\ \frac{\text{g}}{\text{mol}}\right)\left(\frac{1\ \text{mol}}{22.414\ \text{L}}\right)$$
$$\times \left(\frac{273\text{K}}{298\text{K}}\right)(1.06\ \text{atm})$$
$$= 0.97\ \text{mg/m}^3$$

Answer is A.

5.
$$L_{\text{p,total}} = 10\log\sum_{i=1}^{3} 10^{L_i/10}$$
$$= 10\log(10^{95/10} + 10^{90/10} + 10^{0/10})$$
$$= 10\log(3.16\times10^9 + 1\times10^9 + 1)$$
$$= 10\log(4.16\times10^9)$$
$$= 96\ \text{dB}$$

Answer is C.

6. The sound pressure level for each time period, L_{pt}, can be computed. The results in turn can be used to compute the time-weighted sound pressure level.

8 am to 10 am

$$L_{\text{pt}} = 10\log\sum 10^{L_i/10}$$
$$= 10\log(10^{95/10} + 10^{90/10} + 10^{100/10})$$
$$= 10\log(10^{9.5} + 10^9 + 10^{10})$$
$$= 101.5\ \text{dB}$$

10 am to 11 am (Only the sludge pump is running.)

$$L_{\text{pt}} = 95\ \text{dB}$$

11 am to 12 noon

$$L_{\text{pt}} = 10\log(10^{95/10} + 10^0 + 10^{100/10})$$
$$= 10\log(10^{9.5} + 1 + 10^{10})$$
$$= 101.20\ \text{dB}$$

12 noon to 4 pm (Only the ventilation fan is running.)

$$L_{\text{pt}} = 100\ \text{dB}$$

4 pm to 5 pm

$$L_{\text{pt}} = 10\log(10^{95/10} + 10^{90/10})$$
$$= 10\log(10^{9.5} + 10^9)$$
$$= 96.2\ \text{dB}$$

The time-averaged value of the sound pressure for one shift is

$$
\frac{\left(\begin{array}{c} (101.5 \text{ dB})(2 \text{ h}) \\ + (95 \text{ dB})(1 \text{ h}) \\ + (101.20 \text{ dB})(1 \text{ h}) \\ + (100 \text{ dB})(4 \text{ h}) \\ + (96.2 \text{ dB})(1 \text{ h}) \end{array} \right)}{2 \text{ h} + 1 \text{ h} + 1 \text{ h} + 4 \text{ h} + 1 \text{ h}} = 99.5 \text{ dB}
$$

This is above the permissible average sound pressure for an 8 h exposure, so attenuation would be required.

Answer is C.

7.
$$
K = 0.1 \text{ day}^{-1}
$$
$$
Y_{t=2 \text{ d}} = \left(8.2 \, \frac{\text{mg}}{\text{L}} - 6.2 \, \frac{\text{mg}}{\text{L}} \right)
$$
$$
= 2.0 \text{ mg/L}
$$

The ultimate BOD of the diluted sample, L_d, is given by

$$
Y = L(1 - e^{-Kt}) = \text{DO}_0 - \text{DO}_t
$$
$$
L_d = \frac{Y}{1 - e^{Kt}}
$$
$$
= \frac{2 \, \frac{\text{mg}}{\text{L}}}{1 - e^{-(0.1 \; 1/\text{d})(2 \text{ d})}}
$$
$$
= 11 \text{ mg/L}
$$

The ultimate BOD, L, of the original sample (before dilution) is

$$
L = \frac{L_d(1000 \text{ mL})}{(15 \text{ mL})}
$$
$$
= \frac{\left(11 \, \frac{\text{mg}}{\text{L}} \right) (1000 \text{ mL})}{15 \text{ mL}}
$$
$$
= 740 \text{ mg/L}
$$

Answer is D.

8. The ultimate BOD, L, of the original sample before dilution is

$$
L = \frac{L_d(1000 \text{ mL})}{(15 \text{ mL})}
$$
$$
= \frac{\left(11 \, \frac{\text{mg}}{\text{L}} \right) (1000 \text{ mL})}{15 \text{ mL}}
$$
$$
= 740 \text{ mg/L}
$$

The 5 day BOD of the original sample is

$$
Y_5 = L_d(1 - e^{-Kt})
$$
$$
= \left(740 \, \frac{\text{mg}}{\text{L}} \right) (1 - e^{-0.5})
$$
$$
= 290 \text{ mg/L}
$$

Answer is C.

9. $C_3H_{14}O_2N + rH_2O \rightarrow mCH_4 + sCO_2 + NH_3$
   (1 mole)   (1 mole)

$$
\text{MW}_{\text{NH}_3} = (1)\left(14 \, \frac{\text{g}}{\text{mol}} \right) + (3)\left(1 \, \frac{\text{g}}{\text{mol}} \right)
$$
$$
= 17 \text{ g/mol}
$$
$$
\text{MW}_{\text{C}_3\text{H}_{14}\text{O}_2\text{N}} = (3)\left(12 \, \frac{\text{g}}{\text{mol}} \right) + (14)\left(1 \, \frac{\text{g}}{\text{mol}} \right)
$$
$$
+ (2)\left(16 \, \frac{\text{g}}{\text{mol}} \right) + (1)\left(14 \, \frac{\text{g}}{\text{mol}} \right)
$$
$$
= 96 \text{ g/mol}
$$

96 kg of organic material releases 1 mole, (17 kg), of ammonia.

So, 1 g of organic material releases

$$
(1 \text{ g}) \left(\frac{17 \, \frac{\text{kg}}{\text{mol}}}{96 \, \frac{\text{kg}}{\text{mol}}} \right) = 0.178 \text{ g} \quad (0.18 \text{ g})
$$

Answer is A.

10. The equation for biodegradation with ammonia is

$$
C_6H_{12}O_6 + 1.92O_2 + 0.77NH_3
$$
$$
\text{[cells]}
$$
$$
\rightarrow cCH_{1.8}O_{0.5}N_{0.2} + dCO_2 + eH_2O
$$

From a nitrogen balance,

$$
0.77 = 0.2c
$$
$$
c = 3.88
$$
$$
\text{MW}_{\text{cells}} = (1)\left(12 \, \frac{\text{g}}{\text{mol}} \right) + (1.8)\left(1 \, \frac{\text{g}}{\text{mol}} \right)
$$
$$
+ (0.5)\left(16 \, \frac{\text{g}}{\text{mol}} \right) + (0.2)\left(14 \, \frac{\text{g}}{\text{mol}} \right)
$$
$$
= 24.6 \text{ g/mol}
$$
$$
\text{MW}_{\text{glucose}} = (6)\left(12 \, \frac{\text{g}}{\text{mol}} \right) + (12)\left(1 \, \frac{\text{g}}{\text{mol}} \right)
$$
$$
+ (6)\left(16 \, \frac{\text{g}}{\text{mol}} \right)
$$
$$
= 180 \text{ g/mol}
$$

180 g of glucose produces

$$\left(24.6 \ \frac{g}{mol}\right)(0.9)(3.88 \ mol) = 85.9 \ g \ of \ cells$$

$$\frac{85.9 \ g}{180 \ g} = 0.477 \ g$$

1 g of glucose produces 0.48 g of cells.

Answer is D.

11. x = number of cells present

$$= 2000$$

$$\frac{dx}{dt} = \text{instantaneous growth rate}$$

$$= 300/d$$

μ = specific growth rate

$$= \frac{1}{x}\frac{dx}{dt}$$

$$= \left(\frac{1}{2000}\right)\left(\frac{300}{d}\right)$$

$$= 0.15/d$$

Answer is A.

12. BOD loading is

$$L_{BOD} = QC$$

$$Q = 10 \ ML/d$$

$$L_{BOD} = \left(10 \ \frac{ML}{d}\right)\left(356 \ \frac{mg}{L}\right)\left(10^6 \ \frac{L}{ML}\right)\left(\frac{1 \ kg}{10^6 \ mg}\right)$$

$$= 3600 \ kg/d$$

Answer is C.

13. The volume of the aerobic digester is given by

$$V = \frac{Q_i(X_i + FS_i)}{X_d\left(K_dP_v + \frac{1}{\theta_c}\right)}$$

Q_i = influent average flow rate to digester
$\quad$ = 1100 m^3/d
X_i = influent suspended solids
$\quad$ = 2000 mg/L
F = fraction of influent BOD$_5$ consisting of raw primary sewage
$\quad$ = 0.55
S_i = influent BOD$_5$
$\quad$ = 200 mg/L
X_d = digester suspended solids
$\quad$ = 1500 mg/L

K_d = reaction rate constant in d^{-1}
$\quad$ = 0.24 d^{-1}
P_v = volatile fraction of digester suspended solids
$\quad$ = 0.34
θ_c = sludge age
$\quad$ = 2 d

$$V = \frac{\left(1100 \ \frac{m^3}{d}\right)\left(2000 \ \frac{mg}{L} + (0.55)\left(200 \ \frac{mg}{L}\right)\right)}{1500 \ \frac{mg}{L}\left((0.24 \ d^{-1})(0.34) + \frac{1}{2 \ d}\right)}$$

$$= 2660 \ m \quad (2700 \ m^3)$$

Answer is D.

14. The growth rate dx/dt is given by

$$\frac{dx}{dt} = Kx\left(1 - \frac{x}{x_\infty}\right)$$

K = logistic growth constant
$\quad$ = 0.80/h
x_∞ = carrying capacity
$\quad$ = 2.500 g/L
$x = x_0$ = initial concentration
$\quad$ = 1.50 g/L

At $t = 0$,

$$\left(\frac{dx}{dt}\right)_0 = \left(\frac{0.8}{h}\right)\left(1.5 \ \frac{g}{L}\right)\left(1 - \frac{1.5 \ \frac{g}{L}}{2.5 \ \frac{g}{L}}\right)$$

$$= 0.48 \ g/L \cdot h$$

Answer is A.

15. The formula for specific growth rate is

$$\mu = \left(\frac{1}{x}\right)\frac{dx}{dt}$$

Solve for the growth rate.

$$\frac{dx}{dt} = x\mu$$

$$\mu = (2500 \ cells)\left(\frac{2.5}{d}\right)$$

$$= 6250 \ cells/d \quad (6300 \ cells/d)$$

Answer is C.

33 Cellular Biology

Subjects

CELL STRUCTURE 33-1
CELL TRANSPORT 33-2
ORGANISMAL GROWTH IN
   A BATCH CULTURE 33-2
MICROORGANISMS 33-2
   Pathogens 33-3
   Microbe Categorization 33-3
   Systemic Effects 33-3
   Viruses 33-3
   Bacteria 33-3
   Fungi 33-3
   Algae 33-4
   Protozoa 33-4
   Worms and Rotifers 33-4
   Mollusks 33-4
   Indicator Organisms 33-4
   Metabolism/Metabolic Processes 33-4
   Decomposition of Waste 33-4
   Aerobic Decomposition 33-4
   Anoxic Decomposition 33-5
   Anaerobic Decomposition 33-5
   Factors Affecting Disease
      Transmission 33-7
   AIDS 33-8
STOICHIOMETRY OF SELECTED
   BIOLOGICAL SYSTEMS 33-15

Nomenclature

BOD	biochemical oxygen demand	mg/L
DO	dissolved oxygen	mg/L
F	fraction of influent	–
k	logistic growth rate constant	d^{-1}
K	biodegradation rate constant	d^{-1}
K	reaction rate constant	d^{-1}
Ka	partition coefficient	–
K_{eq}	equilibrium constant	–
L	ultimate BOD	mg/L
MW	molecular weight	g/mol
pKa	ionization constant	–
P	fraction of digester suspended solids	–
Q	volumetric flow rate	m^3/s or m^3/d
t	time	h
TWA	time-weighted average	ppm
x	cell or organism number or concentration	–
X	suspended solids	–
Y	yield coefficient	–

Symbols

μ	specific growth rate	–
k	logistic growth constant	h
x_0	initial concentration	mg/L
x_∞	carrying capacity	mg/L

Subscripts

5	5 days
b	biomass
d	digester
d	diluted
i	influent
p	product
s	substrate
v	volatile

CELL STRUCTURE

A cell is the fundamental unit of living organisms. Organisms are classified as either prokaryotes or eukaryotes. A *prokaryote* is a cellular organism that does not have a distinct nucleus. Examples of prokaryotes are bacteria and blue-green algae. Figure 33.1 shows the features of a typical prokaryotic cell.

A *eukaryote* is an organism composed of one or more cells containing visibly evident nuclei and *organelles* (structures with specialized functions). Eukaryotic cells are found in protozoa, fungi, plants, and animals. For a long time, it was thought that eukaryotic cells were composed of an outer membrane, an inner nucleus, and a large mass of cytoplasm within the cell. Better experimental techniques revealed that eukaryotic cells also contain many organelles. Figure 33.2 shows the features of a typical animal and a typical plant cell. Note the size scale for the cells in Figures 33.1 and 33.2.

A cell membrane consists of a double phospholipid layer in which the polar ends of the molecules point to the outer and inner surfaces of the membrane, and the nonpolar ends point to the center. A cell membrane also

Figure 33.1 Prokaryotic Cell Features

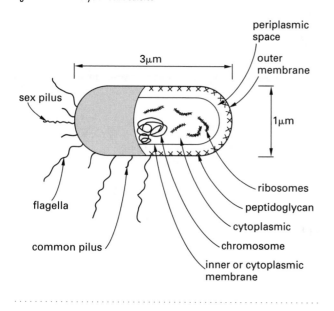

Figure 33.2 Eukaryotic Cell Features

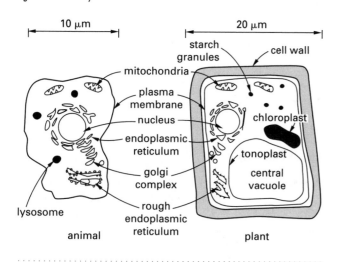

contains proteins, cholesterol, and glycoproteins. Some organisms, such as plant cells, have a cell wall made of carbohydrates that surrounds the cell membrane. Many single-celled organisms have specialized structures called *flagella* that extend outside the cell and help them to move.

Cell size is limited by the transport of materials through the cell membrane. The volume of a cell is proportional to the cubic power of its average linear dimension; the surface area of a cell is proportional to the square of the average linear dimension. The material (nutrients, etc.) transported through a cell's membrane is proportional to its surface area, while the material composing the cell is proportional to the volume of the cell. Therefore,

the potential per unit of cell mass for transport through the membrane is inversely proportional to cell size.

The *cytoplasm* includes all the contents of the cell other than the nucleus. *Cytosol* is the watery solution of the cytoplasm. It contains dissolved glucose and other nutrients, salts, enzymes, carbon dioxide, and oxygen. In eukaryotic cells, the cytoplasm also contains the organelles.

The *endoplasmic reticulum* is one example of an organelle that is found in all eukaryotic cells. Rough endoplasmic reticulum has sub-microscopic organelles called ribosomes attached to it. Amino acids are bound together at the surface of ribosomes to form proteins.

The *Golgi apparatus*, or Golgi complex, is an organelle found in most eukaryotic cells. It looks like a stack of flattened sacks. The Golgi apparatus is responsible for accepting materials (mostly proteins), making modifications, and packaging the materials for transport to specific areas of the cell.

The *mitochondria* are specialized organelles that are the major site of energy production via aerobic respiration in eukaryotic cells. They convert organic materials into energy.

Lysosomes are other organelles located within eukaryotic cells that serve the specialized function of digestion within the cell. They break lipids, carbohydrates, and proteins into smaller particles that can be used by the rest of the cell.

Chloroplasts are organelles located within plant cells and algae that conduct photosynthesis. Chlorophyll is the green pigment that absorbs energy from sunlight during photosynthesis. During photosynthesis, the energy from light is captured and eventually stored as sugar.

Vacuoles are found in some eukaryotic cells. They serve many functions including storage, separation of harmful materials from the remainder of the cell, and maintaining fluid balance or cell size. Vacuoles are separated from the cytoplasm by a single membrane called the tonoplast. Most mature plant cells contain a large central vacuole that occupies the largest volume of any single structure within the cell.

Two main types of nucleic acids are found in cells—*deoxyribonucleic acid* (DNA) and *ribonucleic acid*, (RNA). DNA is found within the nucleus of eukaryotic cells and within the cytoplasm in prokaryotic cells. DNA contains the genetic sequence that is passed on during reproduction. This sequence governs the functions of cells by determining the sequence of amino acids that are combined to form proteins. Each species manufactures its own unique proteins.

RNA is found both within the nucleus and in the cytoplasm of prokaryotic and eukaryotic cells. Most RNA molecules are involved in protein synthesis. Messenger RNA (mRNA) serves the function of carrying the DNA sequence from the nucleus to the rest of the cell. Ribosomal RNA (rRNA) makes up part of the ribosomes, where amino acids are bound together to form proteins. Transfer RNA (tRNA) carries amino acids to the ribosomes for protein synthesis.

CELL TRANSPORT

The transfer of materials across membrane barriers occurs by means of several mechanisms.

- Passive diffusion in cells is similar to the transfer that occurs in non-living systems. Material moves spontaneously from a region of high concentration to a region of low concentration. The rate of transfer obeys Fick's Law, the principle governing passive diffusion in dilute solutions. The rate is proportional to the concentration gradient across the membrane. In the case of living beings, passive diffusion is affected by lipid solubility (high solubility increases the rate of transport), the size of the molecules (the rate of transport increases with decreasing size of molecules), and the degree of ionization (the rate of transport increases with decreasing ionization). A *partition coefficient*, Ka, or the relative solubility of the solute in lipid to its solubility in water, can be used to describe the effect of lipid solubility on transport.

$$\text{Ka} = \frac{\text{concentration in lipid}}{\text{concentration in water}} \quad \textit{33.1}$$

It is more common to report log Ka than Ka because the relationship between pKa and permeability is fairly linear.

$$\text{pKa} = \log_{10}\left(\frac{\text{concentration in lipid}}{\text{concentration in water}}\right) \quad \textit{33.2}$$

The permeability of a molecule across a membrane of thickness x is

$$\text{permeability}_{\text{cm/s}} = \frac{\text{pKa}\left(\begin{array}{c}\text{diffusion}\\\text{coefficient}\\\text{in water}\end{array}\right)_{\text{cm}^2/\text{s}}}{x_{\text{cm}}} \quad \textit{33.3}$$

The pH of a solution will have an effect on the partition coefficient. Although the relationship is complex, for weakly acidic and basic solutions pKa and pH are approximately related by the *Henderson-Hasselbach equation*.

$$\text{pKa} - \text{pH} = \log_{10}\left(\frac{\text{nonionized form}}{\text{ionized form}}\right)$$
$$= \log_{10}\frac{\text{HA}}{\text{A}} \quad \text{[weakly acidic]} \quad \textit{33.4}$$

$$\text{pKa} - \text{pH} = \log_{10}\left(\frac{\text{ionized form}}{\text{nonionized form}}\right)$$
$$= \log_{10}\frac{\text{HB}^+}{\text{B}} \quad \text{[weakly basic]} \quad \textit{33.5}$$

- *Facilitated diffusion* is transfer during which a permease or membrane enzyme carries the substance across the membrane.
- Active diffusion, of which there are three categories, is transfer forced by a pressure gradient or "piggy-back" function.
  - *Membrane pumping*, where cell membrane proteins called permeases transport the substance in a direction opposite the direction of passive diffusion
  - *Endocytosis*, a process where cells absorb a material by surrounding it with their membrane
  - *Exocytosis*, during which a secretory vesicle expels material that was within the cell to an area outside the membrane
- Other specialized mechanisms for specific organs.

In the case of many bacteria, including *Escherichia coli* (*E. coli*), sugars and amino acids are transported across the inner plasma membrane by water-soluble proteins located in the periplasmic space between the two membranes.

ORGANISMAL GROWTH IN A BATCH CULTURE

Bacterial organisms can be grown (cultured) in a nutritive medium. The rate of growth follows the phases depicted in Fig. 33.3.

As Fig. 33.3 shows, the *lag phase* begins immediately after inoculation of the microbes into the nutrient medium. In this period, the microbial cells adapt to their new environment. The microbes might have to produce new enzymes to take advantage of new nutrients they are being exposed to, or they may need to adapt to different concentrations of solutes or to different temperatures than they are accustomed to. The length of the lag phase depends on the differences between the conditions the microbes experience before and after inoculation. The cells will start to divide when the requirements for growth are satisfied.

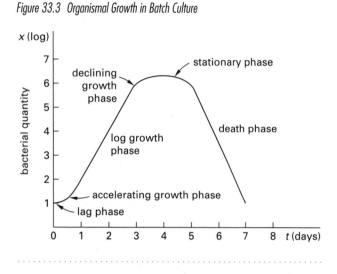

Figure 33.3 Organismal Growth in Batch Culture

The *exponential*, or *logarithmic, growth phase* follows the lag phase. During this phase the number of cells increases according to the following equation.

$$\mu = \frac{1}{x}\frac{dx}{dt}$$ *33.6*

In this equation, x is the cell or organism number or concentration, t is time, and μ is the specific growth rate during the exponential phase.

The declining growth phase follows the log phase. During this phase, one or more essential nutrients are depleted or a waste product is accumulated at levels that slow cell growth.

The stationary phase begins after the decelerating growth phase. During the stationary phase, the net growth rate of the cells is zero. In other words, new cells are created at the same rate at which cells are dying.

The death phase follows the stationary phase. During this phase, the death rate exceeds the growth rate.

The *logistic equation* is used to represent the population quantity up to (but excluding) the death phase because the curve takes on a *sigmoidal shape* also known as an *S-curve* or *bounded exponential growth curve*. In the logistic formulation, the specific growth rate is related to the *carrying capacity*, x_∞, which is the maximum population the environment can support. Carrying capacity depends on the specific culture, medium, and conditions. The equation for growth rate is

$$\mu = k\left(1 - \frac{x}{x_\infty}\right)$$ *33.7*

In this equation, k is the *logistic growth rate constant*, and x is the number of organisms at time t. Therefore,

$$\frac{dx}{dt} = kx\left(1 - \frac{x}{x_\infty}\right)$$ *33.8*

Integration of Eq. 33.8 gives the equation for the number of cells or organisms as a function of time.

$$x = \frac{x_0 e^{kt}}{1 - \dfrac{x_0}{x_\infty}(1 - e^{kt})}$$ *33.9*

MICROORGANISMS

In respect to public water supplies, microorganisms occur in untreated water and represent a potential human risk. However, other microorganisms are put to beneficial use in wastewater treatment. Microorganisms include viruses, bacteria, fungi, algae, protozoa, worms, rotifers, and crustaceans.

Microorganisms are organized into three broad groups based on their structural and functional differences. The groups are called *kingdoms*. The three kingdoms are animals (rotifers and crustaceans), plants (mosses and ferns), and *Protista* (bacteria, algae, fungi, and protozoa). Bacteria and protozoa of the kingdom Protista make up the major groups of microorganisms in the biological system that is used in secondary treatment of wastewater.

Pathogens

Many infectious diseases in humans or animals are caused by organisms categorized as *pathogens*. Pathogens are found in fecal wastes that are transmitted and transferred through the handling of wastewater. Pathogens will proliferate in areas where sanitary disposal of feces is not adequately practiced and where contamination of water supply from infected individuals is not properly controlled. The wastes may also be improperly discharged into surface waters, making the water *nonpotable* (unfit for drinking). Certain shellfish can become toxic when they concentrate pathogenic organisms in their tissues, increasing the toxic levels much higher than the levels in the surrounding waters.

Organisms that are considered to be pathogens include bacteria, protozoa, viruses, and helminths (worms). Table 33.1 lists potential waterborne diseases, the causative organisms, and the typical infection sources.

Not all microorganisms are considered pathogens. Some microorganisms are exploited for their usefulness in wastewater processing. Most wastewater engineering (and an increasing portion of environmental engineering) involves designing processes and operating facilities that use microorganisms to destroy organic and inorganic substances.

Table 33.1 Potential Pathogens

name of organism	major disease	source
Bacteria		
Salmonella typhi	typhoid fever	human feces
Salmonella paratyphi	paratyphoid fever	human feces
other *Salmonella*	salmonellosis	human/animal feces
Shigella	bacillary dysentery	human feces
Vibrio cholerae	cholera	human feces
Enteropathogenic coli	gastroenteritis	human feces
Yersinia enterocolitica	gastroenteritis	human/animal feces
Campylobacter jejuni	gastroenteritis	human/animal feces
Legionella pneumophila	acute respiratory illness	thermally enriched waters
Mycobacterium	tuberculosis	human respiratory exudates
other *Mycobacteria*	pulmonary illness	soil and water
Opportunistic bacteria	variable	natural waters
Enteric Viruses/Enteroviruses		
Polioviruses	poliomyelitis	human feces
Coxsackieviruses A	aseptic meningitis	human feces
Coxsackieviruses B	aseptic meningitis	human feces
Echoviruses	aseptic meningitis	human feces
other *Enteroviruses*	encephalitis	human feces
Reoviruses	upper respiratory and gastrointestinal illness	human/animal feces
Rotaviruses	gastroenteritis	human feces
Adenoviruses	upper respiratory and gastrointestinal illness	human feces
Hepatitis A virus	infectious hepatitis	human feces
Norwalk and related gastrointestinal viruses	gastroenteritis	human feces
Fungi		
Aspergillus	ear, sinus, lung, and skin infections	airborne spores
Candida	yeast infections	various
Protozoa		
Acanthamoeba castellani	amoebic meningoencephalitis	soil and water
Balantidium coli	balantidosis (dysentery)	human feces
Cryptosporidium[a]	cryptosporidiosis	human/animal feces
Entamoeba histolytica	amoebic dysentery	human feces
Giardia lamblia	giardiasis (gastroenteritis)	human/animal feces
Naegleria fowleri	amoebic meningoencephalitis	soil and water
Algae (blue-green)		
Anabaena flos-aquae	gastroenteritis (possible)	natural waters
Microcystis aeruginosa	gastroenteritis (possible)	natural waters
Alphanizomenon flos-aquae	gastroenteritis (possible)	natural waters
Schizothrix calciola	gastroenteritis (possible)	natural waters
Helminths (intestinal parasites/worms)		
Ascaris lumbricoides (roundworm)	digestive disturbances	ingested worm eggs
E. vericularis (pinworm)	any part of the body	ingested worm eggs
Hookworm	pneumonia, anemia	ingested worm eggs
Threadworm	abdominal pain, nausea, weight loss	ingested worm eggs
T. trichiuro (whipworm)	trichinosis	ingested worm eggs
Tapeworm	digestive disturbances	ingested worm eggs

[a]Disinfectants have little effect on *Cryptosporidia*. Most large systems now use filtration, the most effective treatment to date against *Cryptosporidia*.

Microbe Categorization

Carbon is the basic building block for cell synthesis, and it is prevalent in large quantities in wastewater. Wastewater treatment converts carbon into microorganisms that are subsequently removed from the water by settling. Therefore, the growth of organisms that use organic material as energy is encouraged.

If a microorganism uses organic material as a carbon supply, it is *heterotrophic*. *Autotrophs* require only carbon dioxide to supply their carbon needs. Organisms that rely only on the sun for energy are called *phototrophs*. *Chemotrophs* extract energy from organic or inorganic oxidation/reduction (redox) reactions. *Organotrophs* use organic materials, while *lithotrophs* oxidize inorganic compounds.

Most microorganisms in wastewater treatment processes are bacteria. Conditions in the treatment plant are readjusted so that chemoheterotrophs predominate.

Each species of bacteria reproduces most efficiently within a limited range of temperatures. Table 33.2 shows these types and their most viable temperature ranges.

Table 33.2   Best Temperatures for Bacterial Growth

bacteria type	best temperature range for growth
psychrophiles	below 68°F (20°C)
mesophiles	68°F (20°C) to 113°F (45°C)
thermophiles	113°F (45°C) to 140°F (60°C)
stenothermophiles	above 140°F (60°C)

Because most reactions proceed slowly at these temperatures, cells use *enzymes* to speed up the reactions and control the rate of growth. Enzymes are proteins, ranging from simple structures to complex conjugates, and are specialized for the reactions they catalyze.

The temperature ranges in Table 33.2 are qualitative and somewhat subjective. The growth range of facultative thermophiles extends from the thermophilic range into the mesophilic range. Bacteria will grow in a range of temperatures and will survive at a very large range of temperatures. *E. coli*, for example, is classified as a mesophile. It grows best at temperatures between 68°F (20°C) and 122°F (50°C) but can continue to reproduce at temperatures down to 32°F (0°C).

Nonphotosynthetic bacteria are classified into two groups, heterotrophic and autotrophic, by their sources of nutrients and energy. *Heterotrophs* use organic matter as both an energy source and a carbon source for synthesis. Heterotrophs are further subdivided into groups depending on their behavior toward free oxygen: aerobes, anaerobes, and facultative bacteria. *Obligate aerobes* require free dissolved oxygen while they decompose organic matter to gain energy for growth and reproduction. *Obligate anaerobes* oxidize organics in the complete absence of dissolved oxygen by using the oxygen bound in other compounds, such as nitrate and sulfate. *Facultative bacteria* comprise a group that uses free dissolved oxygen when available but that can also behave anaerobically in the absence of free dissolved oxygen also known as *anoxic conditions*. Under anoxic conditions, a group of facultative anaerobes, called *denitrifiers*, uses nitrites and nitrates instead of oxygen. Nitrate nitrogen is converted to nitrogen gas in the absence of oxygen. This process is called *anoxic denitrification*.

Autotrophic bacteria (*autotrophs*) oxidize inorganic compounds for energy, use free oxygen, and use carbon dioxide as a carbon source. Significant members of this group are the *Leptothrix* and *Crenothrix* families of *iron bacteria*. These have the ability to oxidize soluble ferrous iron into insoluble ferric iron. Because soluble iron is often found in well waters and iron pipe, these bacteria deserve some attention. They thrive in water pipes where dissolved iron is available as an energy source and bicarbonates are available as a carbon source. As the colonies die and decompose, they release foul tastes and odors and have the potential to cause staining of porcelain or fabrics.

Table 33.3 lists selected microbial cells and describes some of their characteristics.

Viruses

Viruses are parasitic organisms that pass through filters that retain bacteria, can only be seen with an electron microscope, and grow and reproduce only inside living cells, although they can survive outside the host. They are not cells, but particles composed of a protein sheath surrounding a nucleic-acid core. Most viruses of interest in supply water range in size from 10 to 25 nm.

The viral particles invade living cells and the viral genetic material redirects cell activities toward production of new viral particles. A large number of viruses are released to infect other cells when the infected cell dies. Viruses are host-specific, attacking only one type of organism.

There are more than 100 types of human enteric viruses. Those of interest in drinking water are *Hepatitis A*, *Norwalk*-type viruses, *Rotaviruses*, *Adenoviruses*, *Enteroviruses*, and *Reoviruses*.

Table 33.3 Characteristics of Selected Microbial Cells

organism genus or type	type	metabolism[1]	gram reaction[2]	morphological characteristics[3]
escherichia	bacteria	chemoorganotroph-facultative	negative	rod-may or may not be motile, variable extracellular material
enterobacter	bacteria	chemoorganotroph-facultative	negative	rod-motile; significant extracellular material
bacillus	bacteria	chemoorganotroph-aerobic	positive	rod-usually motile; spore; can be significant extracellular material
lactobacillus	bacteria	chemoogranotroph-facultative	variable	rod-chains-usually nonmotile; little extracellular material
staphylococcus	bacteria	chemoogranotroph-facultative	positive	cocci-nonmotile; moderate extracellular material
nitrobacter	bacteria	chemoautotroph-aerobic; can use nitrite as electron donor	negative	short rod-usually nonmotile; little extracellular material
rhizobium	bacteria	chemoorganotroph-aerobic; nitrogen fixing	negative	rods-motile; copius extracellular slime
pseudomonas	bacteria	chemoorganotroph-aerobic and some chemolithotroph facultative (using NO_3 as electron acceptor)	negative	rods-motile; little extracellular slime
thiobacillus	bacteria	chemoautotroph-facultative	negative	rods-motile; little extracellular slime
clostridium	bacteria	chemoorganotroph-anaerobic	positive	rods-usually motile; spore; some extracellular slime
methanobacterium	bacteria	chemoautotroph-anaerobic	unknown	rods or cocci-motility unknown; some extracellular slime
chromatium	bacteria	photoautotroph-anaerobic	n/a	rods-motile; some extracellular material
spirogyra	alga	photoautotroph-aerobic	n/a	rod/filaments; little extracellular material
aspergillus	mold	chemoorganotroph-aerobic and facultative	–	filamentous fan-like or cylindrical conidia and various spores
candida	yeast	chemoorganotroph-aerobic and facultative	–	usually oval, but can form elongated cells, mycelia and various spores
saccharomyces	yeast	chemoorganotroph-facultative	–	spherical or ellipsoidal; reproduced by budding; can form various spores

[1] aerobic – requires or can use oxygen as an electron receptor.

facultative – can vary the electron receptor from oxygen to organic materials.

anaerobic – organic or inorganics other than oxygen serve as electron acceptor.

chemoorganotrophs – derive energy and carbon from organic materials.

chemoautotrophs – derive energy from organic carbons and carbon from carbon dioxide. Some species can also derive energy from inorganic sources.

photolithotrophs – derive energy from light and carbon from CO_2. May be aerobic or anaerobic.

[2] Gram negative indicates a complex cell wall with a lipopolychaccharide outer layer; Gram positive indicates a less complicated cell wall with a peptide-based outer layer.

[3] Extracellular material production usually increases with reduced oxygen levels (e.g., facultative). Carbon source also affects production; extracellular material may be polysaccharides and/or proteins; statements above are to be understood as general in nature.

Pelczar, M.J., R.D. Reid, and E.C.S. Chan, Microbiology. McGraw-Hill. 1977.

Bacteria

Bacteria are microscopic plants having round, rodlike, spiral or filamentous single-celled or noncellular bodies. They are often aggregated into colonies. Bacteria use soluble food and reproduce through binary fission. Most bacteria are not pathogenic to humans, but they do play a significant role in the decomposition of organic material and can have an impact on the aesthetic quality of water.

Fungi

Fungi are aerobic, multicellular, nonphotosynthetic, heterotrophic, eukaryotic protists. Most fungi are saprophytes that degrade dead organic matter. Fungi grow in low-moisture areas, and they are tolerant of low-pH environments. Fungi release carbon dioxide and nitrogen during the breakdown of organic material.

Fungi are obligate aerobes that reproduce by a variety of methods including fission, budding, and spore formation. They form normal cell material with one-half the nitrogen required by bacteria. In nitrogen-deficient wastewater, they may replace bacteria as the dominant species.

Algae

Algae are autotrophic, photosynthetic organisms (*photoautotrophs*) and may be either unicellular or multicellular. They take on the color of the pigment that is the catalyst for photosynthesis. In addition to chlorophyll (green), different algae have different pigments, such as carotenes (orange), phycocyanin (blue), phycoerythrin (red), fucoxanthin (brown), and xanthophylls (yellow).

Algae derive carbon from carbon dioxide and bicarbonates in water. The energy required for cell synthesis is obtained through photosynthesis. Algae use oxygen for respiration in the absence of light. Algae and bacteria have a symbiotic relationship in aquatic systems, with the algae producing oxygen used by the bacterial population.

In the presence of sunlight, the photosynthetic production of oxygen is greater than the amount used in respiration. At night algae use up oxygen in respiration. If the daylight hours exceed the night hours by a reasonable amount, there is a net production of oxygen.

Excessive algal growth (*algal blooms*) can result in supersaturated oxygen conditions in the daytime and anaerobic conditions at night.

Some algae cause tastes and odors in natural water. While they are not generally considered pathogenic to humans, algae do cause turbidity, and turbidity provides a residence for microorganisms that are pathogenic.

Protozoa

Protozoa are single-celled animals that reproduce by *binary fission* (dividing in two). Most are aerobic chemoheterotrophs (*facultative heterotrophs*). Protozoa have complex digestive systems and use solid organic matter, including algae and bacteria, as food. Therefore, they are desirable in wastewater effluent because they act as polishers in consuming the bacteria.

Flagellated protozoa are the smallest protozoans. The *flagella* (long hairlike strands) provide mobility by a whiplike action. Amoeba move and take in food through the action of a mobile protoplasm. Free-swimming protozoa have *cilia* (small hairlike features) used for propulsion and gathering in organic matter.

Worms and Rotifers

A number of worms and rotifers are of importance to water quality. *Rotifers* are aerobic, multicellular chemoheterotrophs. The rotifer derives its name from the apparent rotating motion of two sets of cilia on its head. The cilia provide mobility and a mechanism for catching food. Rotifers consume bacteria and small particles of organic matter.

Many *worms* are aquatic parasites. *Flatworms* of the class *Trematoda* are known as *flukes*, and the *Cestoda* are tapeworms. *Nematodes* of public health concern are *Trichinella*, which causes trichinosis; *Necator*, which causes pneumonia; *Ascaris*, which is the common roundworm; and *Filaria*, which causes filariasis.

Mollusks

Mollusks, such as mussels and clams, are characterized by a shell structure. They are aerobic chemoheterotrophs that feed on bacteria and algae. They are a source of food for fish and are not found in wastewater treatment systems to any extent, except in underloaded lagoons. Their presence is indicative of a high level of dissolved oxygen and a very low level of organic matter.

Macrofouling is a term referring to infestation of water inlets and outlets by clams and mussels. For example, *Zebra mussels* were accidentally introduced into the United States in 1986 and are particularly troublesome for several reasons. First, young Zebra mussels are microscopic, and can easily pass through intake screens. Second, they attach to anything, even other mussels, which produces thick mussel colonies. Third, adult Zebra mussels quickly sense other biocides, most notably those that are halogen-based, like chlorine. They quickly close and remain closed for days or weeks.

The use of biocides to control the growth of zebra mussels is controversial. Chlorination treatment is recommended with some caution since it results in increased toxicity, affecting other species and THM (trihalomethane) production. An ongoing biocide program aimed at pre-adult mussels, combined with slippery polymer-based surface coatings, is most likely the best approach to prevention as once a pipe is colonized, mechanical removal by scraping or water blasting is the only practical option.

Indicator Organisms

The techniques for comprehensive bacteriological examination for pathogens are complex and time consuming. Isolating and identifying specific pathogenic microorganisms is a difficult and lengthy task. Many of these organisms require sophisticated tests that take several days to produce results. Because of these difficulties, and also because the number of pathogens relative to other microorganisms in water can be very small, *indicator organisms* are used as a measure of the quality of the water. The primary function of an indicator organism is to provide evidence of recent fecal contamination from warm-blooded animals.

Characteristics of a good indicator organism are:

(a) The indicator is always present when the pathogenic organism of concern is present. It is absent in clean, uncontaminated water.

(b) The indicator is present in fecal material in large numbers.

(c) The indicator responds to natural environmental conditions and to treatment processes in a manner similar to the pathogens of interest.

(d) The indicator is easy to isolate, identify, and enumerate.

(e) The ratio of indicator to pathogen should be high.

(f) The indicator and pathogen should come from the same source, such as gastrointestinal tract.

While there are several microorganisms that meet these criteria, *total coliform* and *fecal coliform* are the indicators generally used. *Total coliform* refers to the group of aerobic and facultatively anaerobic, gram-negative, nonspore-forming, rod-shaped bacteria that ferment lactose with gas formation within 48 hr at 95°F (35°C). This encompasses a variety of organisms, mostly of intestinal origin, including *E. coli*, which is the most numerous facultative bacterium in the feces of warm-blooded animals. Unfortunately, this group also includes *Enterobacter, Klebsiella,* and *Citrobacter*, which are present in wastewater but can be derived from other environmental sources such as soil and plant materials.

Fecal coliforms are a subgroup of the total coliforms that come from the intestines of warm-blooded animals. They are measured by running the standard total coliform fermentation test at an elevated temperature of 112°F (44.5°C), providing a means to distinguish false positives in the total coliform test.

Results of fermentation tests are reported as a *most probable number index* (MPN). This is an index of the number of coliform bacteria that, more than any other number, would give the results shown by the laboratory examination. MPN is not an actual enumeration.

Metabolism/Metabolic Processes

Metabolism is a term given to describe all chemical activities performed by a cell. The cell uses *adenosine triphosphate* (ATP) as the principal energy currency in all processes. Those processes that allow the bacterium to synthesize new cells from the energy stored within its body are called *anabolic*. All biochemical processes in which cells convert substrate into useful energy and waste products are called *catabolic*.

Decomposition of Waste

Decomposition of waste involves oxidation/reduction reactions and is classified as aerobic or anaerobic. The type of electron acceptor available for catabolism determines the type of decomposition used by a mixed culture of microorganisms. Each type of decomposition has peculiar characteristics that affect its use in waste treatment.

Aerobic Decomposition

Molecular oxygen, O_2, must be present as the terminal electron acceptor in order for decomposition to proceed by aerobic oxidation. As in natural water bodies, the dissolved oxygen content is measured. When oxygen is present, it is the only terminal electron acceptor used. Hence, the chemical end products of decomposition are primarily carbon dioxide, water, and new cell material as shown in Table 33.4. Odoriferous, gaseous end products are kept to a minimum. In healthy natural water systems, aerobic decomposition is the principal means of self-purification.

A wide spectrum of organic material can be oxidized by aerobic decomposition. Aerobic oxidation releases large amounts of energy, meaning most aerobic organisms are capable of high growth rates. Consequently, there is a relatively large production of new cells in comparison with the other oxidation systems. This means that more biological sludge is generated in aerobic oxidation than in the other oxidation systems.

A laboratory analysis of organic matter in water often includes a biochemical oxygen demand (BOD) test.

Table 33.4 Waste Decomposition End Products

	representative end products		
substrates	aerobic decomposition	anoxic decomposition	anaerobic decomposition
proteins and other organic nitrogen compounds	amino acids ammonia → nitrites → nitrates alcohols organic acids $\}$ → CO_2 + H_2O	amino acids nitrates → nitrites → N_2 alcohols organic acids $\}$ → CO_2 + H_2O	amino acids ammonia hydrogen sulfide methane carbon dioxide alcohols organic acids
carbohydrates	alcohols fatty acids $\}$ → CO_2 + H_2O	alcohols fatty acids $\}$ → CO_2 + H_2O	carbon dioxide alcohols fatty acids
fats and related substances	fatty acids + glycerol alcohols lower fatty acids $\}$ → CO_2 + H_2O	fatty acids + glycerol alcohols lower fatty acids $\}$ → CO_2 + H_2O	fatty acids + glycerol carbon dioxide alcohols lower fatty acids

Most water quality laboratories have a ready supply of water that is saturated with oxygen, obtained by sparging air overnight through the water. To measure BOD, a sample of wastewater is diluted with oxygen-saturated water, and a small amount of bacteria is added to the sample. Oxygen concentrations are measured at the beginning of the test and also on a daily basis. The difference in the oxygen concentration at the initial time and at time t, gives a measure of the concentration of organic compounds in the water and is called the *BOD exerted* at time t. In this way, BOD provides a measure of the concentration of organic compounds in water without the complexity of analyzing the different compounds.

Aerobic decomposition is the preferred method for large quantities of dilute (BOD_5 < 500 mg/L) wastewater because decomposition is rapid and efficient and has a low odor potential. For concentrated wastewater (BOD_5 > 1000 mg/L), aerobic decomposition is not suitable because of the difficulty in supplying enough oxygen and because of the large amount of biological sludge that is produced.

Anoxic Decomposition

Some microorganisms will use nitrates in the absence of oxygen to oxidize carbon. This is known as *dentrification*. The end products from such denitrification are nitrogen gas, carbon dioxide, water, and new cell material. The amount of energy made available to the cell during denitrification is about the same as that made during aerobic decomposition. The production of cells, though not as high as in aerobic decomposition, is relatively high.

Denitrification is especially important in wastewater treatment when nitrogen must be removed. In such cases, a special treatment step is added to the conventional process for removal of carbonaceous material. One other important aspect of denitrification is in final clarification of the treated wastewater. If the final clarifier becomes anoxic, the formation of nitrogen gas will cause large masses of sludge to float to the surface and escape from the treatment plant into the receiving water. Thus, it is necessary to ensure that anoxic conditions do not develop in the final clarifier.

Anaerobic Decomposition

In order to achieve anaerobic decomposition, molecular oxygen and nitrate must not be present as terminal electron acceptors. Sulfate, carbon dioxide, and organic compounds that can be reduced serve as terminal electron acceptors. The reduction of sulfate results in the production of hydrogen sulfide, H_2S, and a group of equally odoriferous organic sulfur compounds called *mercaptans*.

The anaerobic decomposition of organic matter, also known as *fermentation*, is generally considered to be a two-step process. In the first step, complex organic compounds are fermented to low molecular weight *fatty acids* (*volatile acids*). In the second step, the organic acids are converted to methane. Carbon dioxide serves as the electron acceptor.

Anaerobic decomposition yields carbon dioxide, methane, and water as the major end products. Additional end products include ammonia, hydrogen sulfide, and mercaptans. As a consequence of these last three compounds, anaerobic decomposition is characterized by a malodorous stench.

Because only small amounts of energy are released during anaerobic oxidation, the amount of cell production is low. Thus, sludge production is correspondingly low. Wastewater treatment based on anaerobic decomposition is used to stabilize sludge produced during aerobic and anoxic decomposition.

Direct anaerobic decomposition of wastewater generally is not feasible for dilute waste. The optimum growth temperature for the anaerobic bacteria is at the upper end of the mesophilic range. Therefore, to get reasonable biodegradation, the temperature of the culture must first be elevated. For dilute wastewater, this is not practical. For concentrated wastes ($BOD_5 > 1000$ mg/L), anaerobic digestion is quite appropriate.

FACTORS AFFECTING DISEASE TRANSMISSION

Waterborne disease transmission is influenced by the latency, persistence, and quantity (dose) of the pathogens. *Latency* is the period of time between excretion of a pathogen and its becoming infectious to a new host. *Persistence* is the length of time that a pathogen remains viable in the environment outside a human host. The *infective dose* is the number of organisms that must be ingested to result in disease.

AIDS

Acquired immunodeficiency syndrome (AIDS) is caused by the *human immunodeficiency virus* (HIV). HIV is present in virtually all body excretions of infected persons, and therefore is present in wastewater. The risk of contracting AIDS through contact with wastewater or working at a wastewater plant is small based on the following facts.

- HIV is relatively weak and does not remain viable for long periods of time in harsh environments such as wastewater.
- HIV is quickly inactivated by alcohol, chlorine, and exposure to air. The chlorine concentration present in water used to flush the toilet would probably deactivate HIV.
- HIV that survives disinfection would be too dilute to be infectious.
- HIV replicates in white blood cells, not in the human intestinal tract, and therefore, would not reproduce in wastewater.
- There is no evidence that HIV can be transmitted through water, air, food, or casual contact. HIV must enter the bloodstream directly, through a wound. It cannot enter through unbroken skin or through respiration.
- HIV is less infectious than the hepatitis virus.
- There are no reported AIDS cases linked to occupational exposure in wastewater collection and treatment.

STOICHIOMETRY OF SELECTED BIOLOGICAL SYSTEMS

This section shows four classes of biological reactions involving microorganisms, each with simplified stoichiometric equations.

Stoichiometric problems are known as *weight and proportion problems* because their solutions use simple ratios to determine the masses of reactants required to produce given masses of products, or vice versa. The procedure for solving these problems is essentially the same regardless of the reaction.

step 1: Write and balance the chemical equation.

step 2: Determine the atomic (molecular) weight of each element (compound) in the equation.

step 3: Multiply the atomic (molecular) weights by their respective coefficients and write the products under the formulas.

step 4: Write the given mass data under the weights determined in step 3.

step 5: Fill in the missing information by calculating simple ratios.

The first biological reaction is the production of biomass with a single extracellular product. Water and carbon dioxide are also produced as shown in the reaction equation.

$$CH_mO_n + aO_2 + bNH_3$$
[substrate]
$$\rightarrow cCH_\alpha O_\beta N_\delta + dCH_x O_y N_z + eH_2O + fCO_2$$
[biomass] [product] *33.10*

Equations 33.11–33.13 are used to calculate degrees of reduction (available electrons per unit of carbon) for substrate (s), biomass (b), and product (p).

$$\gamma_s = 4 + m - 2n \qquad 33.11$$
$$\gamma_b = 4 + \alpha - 2\beta - 3\delta \qquad 33.12$$
$$\gamma_p = 4 + x - 2y - 3z \qquad 33.13$$

Table 33.5 shows typical degrees of reduction, γ. A high *degree of reduction* denotes a low degree of oxidation. Solving for the coefficients in Eq. 33.10 requires satisfying the carbon, nitrogen and electron balances, plus

Table 33.5 Composition Data for Biomass and Selected Organic Compounds

compound	molecular formula	degree of reduction, γ	molecular weight (MW)
biomass	$CH_{1.64}N_{0.16}O_{0.52}$ $P_{0.0054}S_{0.005}$*	4.17 (NH_3) 4.65(N_3) 5.45(HNO_3)	24.5
methane	CH_4	8	16.0
n-alkane	C_4H_{32}	6.13	14.1
methanol	CH_4O	6.0	32.0
ethanol	C_2H_6O	6.0	23.0
glycerol	$C_2H_6O_3$	4.67	30.7
mannitol	$C_6H_{14}O_6$	4.33	30.3
acetic acid	$C_2H_4O_2$	4.0	30.0
lactic acid	$C_3H_6O_3$	4.0	30.0
glucose	$C_6H_{12}O_6$	4.0	30.0
formaldehyde	CH_2O	4.0	30.0
gluconic acid	$C_6H_{12}O_7$	3.67	32.7
succinic acid	$C_4H_6O_4$	3.50	29.5
citric acid	$C_6H_8O_7$	3.0	32.0
malic acid	$C_4H_6O_5$	3.0	33.5
formic acid	CH_2O_2	2.0	46.0
oxalic acid	$C_2H_2O_4$	1.0	45.0

*The stoichiometric coefficients represent theoretical balance conditions. Product yields in practice are frequently lower and may be adjusted by the application of yield coefficients.

B. Atkinson and F. Mavitona, *Biochemical Engineering and Biotechnology Handbook*, Macmillan, Inc., 1983. Used with permission of Nature Publishing Group (www.nature.com).

knowing the respiratory coefficient and a yield coefficient. The key biomass production and reduction factors involved in determining carbon, nitrogen, electron, and energy balances are shown in Eqs. 33.14–33.17.

$$c + d + f = 1 \quad \text{[carbon]} \qquad 33.14$$

$$c\delta + dz = b \quad \text{[nitrogen]} \qquad 33.15$$

$$c\gamma_b + d\gamma_p = \gamma_s - 4a \quad \text{[electron]} \qquad 33.16$$

$$Q_o C\gamma_b + Q_o d\gamma_p$$
$$= Q_o\gamma_s - Q_o 4a \quad \text{[energy]} \qquad 33.17$$

Q_o = heat evolved per equivalent of available electrons
  ≈ 26.95 kcal/gm of electrons

The *respiratory quotient* (RQ) is the CO_2 produced per unit of O_2.

$$RQ = \frac{f}{a} \qquad 33.18$$

The coefficients c and d in Eq. 33.10 are referred to as maximum theoretical *yield coefficients* when expressed per gram of substrate. The yield coefficient can be given either as grams of cells or grams of product per gram of substrate.

$$Y_{ideal,b} = \frac{m_b}{m_s} \qquad 33.19$$

$$Y_{ideal,p} = \frac{m_p}{m_s} \qquad 33.20$$

The ideal yield coefficients are related to the actual yield coefficients by the *yield factor*.

$$\text{yield factor} = \frac{Y_{actual}}{Y_{ideal}} \qquad 33.21$$

The second reaction is the aerobic biodegradation of glucose in the presence of oxygen and ammonia. In this reaction, cells are formed and carbon dioxide and water are the only products. The stoichiometric equation is

$$C_6H_{12}O_6 + aO_2 + bNH_3$$
[substrate]
$$\rightarrow cCH_{1.8}O_{0.5}N_{0.2} + dCO_2 + eH_2O$$
[cells] $\qquad 33.22$

For Eq. 33.19,

$$a = 1.94$$
$$b = 0.77$$
$$c = 3.88$$
$$d = 2.13$$
$$e = 3.68$$

The c coefficient is the theoretical maximum yield coefficient, which may be reduced by a yield factor.

The third reaction is the anaerobic (no oxygen) biodegradation of organic wastes with incomplete stabilization. Methane, carbon dioxide, ammonia, and water as well as smaller organic waste molecules are the products. The stoichiometric equation is

$$C_aH_bO_cN_d$$
$$\rightarrow nC_wH_xO_yN_z + mCH_4$$
$$+ sCO_2 + rH_2O + (d - nx)NH_3 \quad 33.23$$
$$s = a - nw - m \qquad 33.24$$
$$r = c - ny - 2s \qquad 33.25$$

Knowledge of product composition, yield coefficient, and a methane CO_2 ratio is needed.

The fourth reaction is the anaerobic biodegradation of organic wastes with complete stabilization. Besides organic waste, water is consumed in this reaction and the products are methane, carbon dioxide, and ammonia. The stoichiometric equation is

$$C_aH_bO_cN_d + rH_2O$$
$$\rightarrow mCH_4 + sCO_2 + dNH_3 \qquad 33.26$$
$$r = \frac{4a - b - 2c + 3d}{4} \qquad 33.27$$
$$s = \frac{4a - b + 2c + 3d}{8} \qquad 33.28$$
$$m = \frac{4a + b - 2c - 3d}{8} \qquad 33.29$$

Composition data for biomass and selected organic compounds is given in Table 33.5.

SAMPLE PROBLEMS

Problems 1 and 2 relate to a waste stabilization pond that will be used in a municipal wastewater treatment system.

Problem 1

Explain the function of bacteria and algae in the stabilization pond's application.

Solution

A waste stabilization pond's operation is dependent on the reaction of bacteria and algae. Organic matter is metabolized by bacteria to produce the principal products of carbon dioxide, water, and a small amount of ammonia nitrogen. Algae convert sunlight into energy through photosynthesis. They use the end products of cell synthesis and other nutrients to synthesize new cells and produce oxygen. The most important role of the algae is in the production of oxygen in the pond for use by aerobic bacteria. In the absence of sunlight, the algae will consume oxygen in the same manner as bacteria. Algae removal is important in producing a high quality effluent from the pond.

Problem 2

Explain why algae would be a problem if it were present in the discharge from the pond.

Solution

The discharge of algae increases suspended solids in the discharge and may present a problem in meeting water quality criteria. The algae exert an oxygen demand when they settle to the bottom of the stream and undergo respiration.

FE-STYLE EXAM PROBLEMS

Problems 1–9 refer to the following information

A fresh wastewater sample containing nitrate ions, sulfate ions, and dissolved oxygen is taken and placed into a sealed jar absent of air.

1. What is the sequence of oxidation of the compounds?

- (A) nitrate, dissolved oxygen, and then sulfate
- (B) sulfate, nitrate, and then dissolved oxygen
- (C) dissolved oxygen, nitrate, and then sulfate
- (D) none of the above

2. Obnoxious odors will

- (A) appear in the sample when the dissolved oxygen is exhausted.
- (B) appear in the sample when the dissolved oxygen and nitrates are exhausted.
- (C) appear in the sample when the dissolved oxygen, nitrate, and sulfate are exhausted.
- (D) not appear.

3. Bacteria will convert ammonia to nitrate if the bacteria are

- (A) phototropic.
- (B) autotropic.
- (C) thermophilic.
- (D) obligate anaerobes.

4. Bacteria will generally reduce nitrate to nitrogen gas only if the bacteria are

- (A) photosynthetic.
- (B) obligate aerobic.
- (C) facultative heterotropic.
- (D) aerobic phototropic.

5. A bacteriophage is a

- (A) bacterial enzyme.
- (B) virus that infects bacteria.
- (C) mesophilic organism.
- (D) virus that stimulates bacterial growth.

6. Algal growth in the wastewater

- (A) will be inhibited when the dissolved oxygen is exhausted.
- (B) will be inhibited when toxins are also found in the solution.
- (C) will be inhibited when toxins are found in the solution or when the dissolved oxygen is exhausted.
- (D) cannot be inhibited by chemical means.

7. In the presence of nitrifying bacteria, nontoxic inorganic compounds, and sunlight, algal growth in the wastewater sample will be

- (A) prevented.
- (B) inhibited.
- (C) unaffected.
- (D) enhanced.

8. The addition of protozoa to the wastewater will

(A) not change the wastewater's biochemical composition.

(B) increase the growth of algae in wastewater.

(C) increase the growth of bacteria in wastewater.

(D) decrease the growth of algae and bacteria in the wastewater.

9. Coliform bacteria in wastewater from human, animal, or soil sources

(A) can be distinguished in the multiple-tube fermentation test.

(B) can be categorized into only two groups: human/animal and soil.

(C) can be distinguished if multiple-tube fermentation and Eschericheiae coli (EC) tests are both used.

(D) cannot be distinguished.

SOLUTIONS TO FE-STYLE EXAM PROBLEMS

1. The sequence of oxygen usage reduced by bacteria is dissolved oxygen, nitrate, and then sulfate.

Answer is C.

2. Following the sequence of oxidation, obnoxious odors will occur when dissolved oxygen and nitrate are exhausted.

Answer is B.

3. Nitrification is performed by autotrophic bacteria to obtain energy for growth by synthesis of carbon dioxide in an aerobic environment.

Answer is B.

4. Facultative heterotropic bacteria can decompose organic matter to gain energy under anaerobic conditions by removing the oxygen from nitrate, releasing nitrogen gas.

Answer is C.

5. A bacteriophage is a virus that infects bacteria.

Answer is B.

6. Algae are photosynthetic (gaining energy from light), releasing oxygen during metabolism. They thrive in aerobic environments. The presence of nitrifying bacteria and toxins and the depletion of dissolved oxygen would inhibit the growth process.

Answer is C.

7. Nitrifying bacteria are nonphotosynthetic, obtaining energy by taking in oxygen to oxidize reduced inorganic nitrogen. The presence of inorganic nutrients would maintain the nitrification process, thereby limiting algal growth.

Answer is B.

8. Protozoa consume bacteria and algae in wastewater treatment and in the aquatic food chain.

Answer is D.

9. Fecal coliforms from humans and other warm-blooded animals are the same bacterial species. Coliforms originating from the soil can be separated by a confirmatory procedure using EC medium broth incubated at the elevated temperature of 44.5°C (112°F).

Answer is B.

34 Toxicology

Subjects

INTRODUCTION 34-1
EXPOSURE PATHWAYS 34-2
   Dermal Absorption 34-2
   Inhalation 34-2
   Ingestion 34-3
   The Eye 34-3
   Systemic Effects 34-3
EFFECTS OF EXPOSURE TO
   TOXICANTS 34-3
   Pulmonary Toxicity 34-3
   Cardiotoxicity 34-3
   Hematoxicity 34-3
   Hepatoxicity 34-4
   Nephrotoxicity 34-4
   Neurotoxicity 34-4
   Immunotoxicity 34-4
   Reproductive Toxicity 34-4
   Toxic Effects on the Eye 34-4
DOSE-RESPONSE RELATIONSHIPS . . . 34-4
   Dose-Responsive Curves 34-4
SAFE HUMAN DOSE 34-5
   EPA Methods 34-5
   ACGIH Methods 34-7
   NIOSH Methods 34-8
LEGAL (OSHA) STANDARDS FOR
   WORKER PROTECTION 34-18

Nomenclature

AT	averaging time	yr
BCF	bioconcentration factor, (mg/kg) in tissue/ (mg/L) in water	L/kg
BW	body mass (weight)	kg
C	concentration	ppm, mg/m^3
CDI	chronic daily intake	mg/kg·d
CPF	carcinogen potency factor (same as slope factor)	(mg/kg·d)$^{-1}$
CR	contact rate	day^{-1}
E	time weighted average exposure	ppm, mg/m^3
ED	exposure duration	yr
EED	estimated exposure dose	mg/kg·d
EF	exposure factor	–
E_m	equivalent exposure of mixture	–
EP	exposed population	–
k	decay constant	d^{-1}
L	exposure limit of particular contaminants	mg/m^3
LC$_{50}$	lethal concentration	mg/m^3
LD$_{50}$	lethal dose	mg/kg
LOAEL	lowest observed adverse effect level	mg/kg·d
LOEL	lowest observed effect level	mg/kg·d
MF	modifying factor	–
NOAEL	no observed adverse effect level	mg/kg·d
NOEL	no observed effect level	mg/kg·d
R	risk; probability of excess cancer	–
SF	slope factor (same as carcinogen potency factor)	(mg/kg·d)$^{-1}$
T	time of exposure	s
UF	uncertainty factor	–

Subscripts

m	mixture
o	initial condition
org	organism
t	time
w	water (or other medium)
x	toxicant

INTRODUCTION

Toxicology is defined as the study of adverse effects of chemicals on living organisms.

This chapter examines the pathways of human exposure available to chemicals and other toxicants, the effects on workers of exposure to toxicants, dose-response relationships, methods determining safe human doses, and standards for worker protection.

EXPOSURE PATHWAYS

The human body has three primary exposure pathways: dermal absorption, inhalation, and ingestion (see Fig. 34.1). The eyes are an additional exposure pathway because they are particularly vulnerable to damage in the workplace.

Dermal Absorption

The skin is composed of the epidermis, the dermis, and the subcutaneous layer. The *epidermis* is the upper layer, which is composed of several layers of flattened and scale-like cells. These cells do not contain blood vessels; they obtain their nutrients from the underlying dermis. The cells of the epidermis migrate to the surface, die, and leave behind a protein called *keratin*. Keratin is the most insoluble of all proteins and together with the scale-like cells, provides extreme resistance to substances and environmental conditions. Beneath the epidermis lies the *dermis*, which contains blood vessels, connective tissue, hair follicles, sweat glands, and other glands. The dermis supplies the nutrients for itself and for the epidermis. The innermost layer is called the *subcutaneous fatty tissue*, which provides a cushion for the skin and connection to the underlying tissue.

The condition of the skin, and the chemical nature of any toxic substance it contacts, affect whether and at what rate the skin absorbs that substance. The epidermis is impermeable to many gases, water, and chemicals. However, if the epidermis is damaged by cuts and abrasions, or is broken down by repeated exposure to soaps, detergents, or organic solvents, toxic substances can readily penetrate and enter the bloodstream.

Chemical burns, such as those from acids, can also destroy the protection afforded by the epidermis, allowing toxicants to enter the bloodstream. Inorganic chemicals (and organic chemicals that are dissolved in water) are not readily absorbed through healthy skin. However, many organic solvents are lipid- (fat-) soluble and can easily penetrate skin cells and enter the body. After a toxicant has penetrated the skin and entered the bloodstream, the blood can transport it to target organs in the body.

Inhalation

The respiratory tract consists of the nasal cavity, pharynx, larynx, trachea, primary bronchi, bronchioles, and alveoli. Molecular transfer of oxygen and carbon dioxide between the bloodstream and the lungs occurs in the millions of air sacs, known as *alveoli*. Toxicants that reach the alveoli can be transferred to the blood through the respiratory system.

Toxicants that reach the alveoli will not be transferred to the blood at the same rate. Various factors can increase or decrease the transfer rate of one toxicant

Figure 34.1 Exposure Routes for Chemical Agents

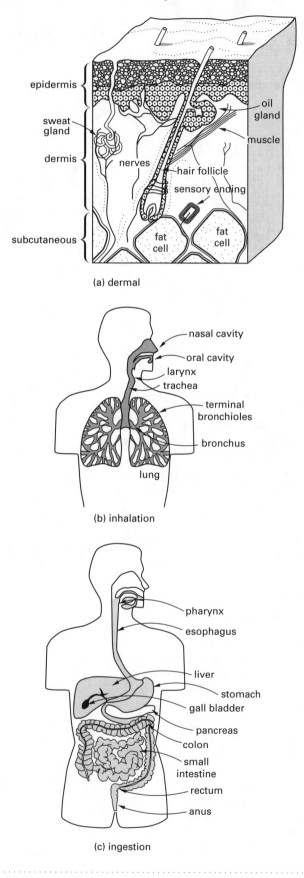

(a) dermal

(b) inhalation

(c) ingestion

relative to another. Both the respiration rate and the duration of exposure will affect the mass of toxicant transferred to the bloodstream over a given period.

Ingestion

The small intestine is the principal site in the digestive tract where the beneficial nutrients from food, and toxics from contaminated substances, are absorbed. Within the small intestine, millions of *villi* (projections) provide a huge surface area to absorb substances into the bloodstream.

Toxic substances will be absorbed in the intestines at various rates depending on the specific toxicant, its molecular size, and its degree of lipid (fat) solubility. Small molecular size and high lipid solubility facilitate diffusion of toxicants in the digestive tract.

The Eye

Transparent tissue in the front of the eye is known as the *cornea* and is the most likely eye tissue to come in contact with toxic substances.

Systemic Effects

For systemic effects to occur, the rate of accumulation of toxicants must exceed the body's ability to excrete (eliminate) it or to biotransform it (transform it to less harmful substances). A toxicant can be eliminated from the body through the *kidneys*, which are the primary organs for eliminating toxicants from the body. The kidneys biotransform a toxicant into a water-soluble form and then eliminate it though the urine. The *liver* is also an important organ for eliminating toxicants from the body, first by biotransformation, then by excretion into the bile where it is eliminated through the small intestine as feces.

A toxicant may also be stored in tissues for long periods before an effect occurs. Toxic substances stored in tissue (primarily in fat but also in bones, the liver, and the kidneys) may exert no effect for many years, or at all within the affected person's life. DDT, a pesticide, for instance, can be stored in body fat for many years and not exert any adverse effect on the body.

When toxic substances are not eliminated fast enough to keep up with the exposure, the liver, kidneys, and central nervous system are the target organs and are commonly affected systemically.

EFFECTS OF EXPOSURE TO TOXICANTS

After toxicants are absorbed into the body through one or more of the pathways, a wide variety of effects on the human body are possible. When the toxic agents concentrate in target tissue or organs, the agents may interfere with the normal functioning of enzymes and cells or may cause genetic mutations.

Pulmonary Toxicity

Pulmonary toxicity refers to adverse effects on the respiratory system from toxic agents. Examples are

- Damage to the nasal passages and nerve cells
- *Nasal cancer*
- *Bronchitis*, excessive mucus secretion
- *Pulmonary edema*, the excessive accumulation of fluid in the alveoli of the lungs
- *Fibrosis*, an increased amount of connective tissue
- *Silicosis*, the deposition of connective tissue around alveoli
- *Emphysema*, the inability of lungs to expand and contract

Cardiotoxicity

Cardiotoxicity refers to the effects of toxic agents on the heart.

- The heart rate may be changed, and the strength of contractions may be diminished.
- Certain metals can affect the contractions of the heart and can interfere with cell metabolism.
- Carbon monoxide can result in a decrease in the oxygen supply, causing improper functioning of the nervous system controlling the heart rate.

Hematoxicity

Hematoxicity refers to damage to the body's blood supply, which includes red blood cells, white blood cells, platelets, and plasma. The *red blood cells* transport oxygen to the body's cells and carbon dioxide to the lungs. *White blood cells* perform a variety of functions associated with the immune system.

- *Platelets* are important in blood clotting.
- *Plasma* is the noncellular portion of blood and contains proteins, nutrients, gases, and waste products.
- Benzene, lead, methylene chloride, nitrobenzene, naphthalene, and insecticides are capable of red blood cell destruction and can cause a decrease in the oxygen-carrying capacity of the blood. The resulting anemia can affect normal nerve cell functioning and control of the heart rate, and can cause shortness of breath, pale skin, and fatigue.
- Carbon tetrachloride, pesticides, benzene, and ionizing radiation can affect the ability of the bone marrow to produce red blood cells.
- Mercury, cadmium, and other toxicants can affect the ability of the kidneys to stimulate the bone

marrow to produce more red blood cells when needed to counteract low oxygen levels in the blood.

- Some chemicals, including carbon monoxide, can interfere with the blood's capacity to carry oxygen, resulting in lowered blood pressure, dizziness, fainting, increased heart rate, muscular weakness, nausea, and after prolonged exposure, death.
- Hydrogen cyanide and hydrogen sulfide can stimulate cells in the aorta, causing increased heart and respiratory rate. At high concentrations, death can result from respiratory failure.
- Benzene, carbon tetrachloride, and trinitrotoluene can suppress stem cell production and the production of white blood cells. This can affect the clotting mechanism and the immune system.
- Benzene can cause high levels of white blood cells, a condition known as *leukemia*.

Hepatoxicity

Hepatoxicity refers to adverse effects on the liver that impede its ability to function properly. The liver converts carbohydrates, fats, and proteins to maintain the proper levels of glucose in the blood and converts excess protein and carbohydrates to fat. It also converts excess amino acids to ammonia and urea, which are removed in the kidneys. The liver also provides storage of vitamins and beneficial metals, as well as carbohydrates, fats, and proteins. Red blood cells that have degenerated are removed by the liver. Substances needed for other metabolic processes are provided by the liver. Finally, the liver detoxifies metabolically produced substances and toxicants that enter the body.

- Hexavalent chromium and arsenic cause cell damage in the liver.
- Carbon tetrachloride and alcohol can cause damage and death of liver cells, a condition known as *cirrhosis of the liver*.
- Chemicals or viruses can cause inflammation of the liver, known as *hepatitis*. Cell death and enlargement of the liver can occur.

Nephrotoxicity

Nephrotoxicity refers to adverse effects on the kidneys. The kidneys excrete ammonia as urea to rid the body of metabolic wastes. They maintain blood pH by exchanging hydrogen ions for sodium ions, and maintain the ion and water balance by excreting excess ions or water as needed. They also secrete hormones needed to regulate blood pressure. Like the liver, the kidneys function to detoxify substances.

- Heavy metals—primarily lead, mercury, and cadmium—cause impaired cell function and cell death. These metals can be stored in the kidneys,

interfering with the functioning of enzymes in the kidneys.

- Chloroform and other organic substances can cause cell dysfunction, cell death, and cancer.
- Ethylene glycol can cause renal failure from obstruction of the normal flow of liquid through the kidneys.

Neurotoxicity

Neurotoxicity refers to toxic effects on the nervous system, which consists of the *central nervous system* (CNS) and the *peripheral nervous system* (PNS). The central nervous system includes the brain and the spinal cord, while the peripheral nervous system includes the remaining nerves, which are distinguished as sensory and motor nerves.

Neurotoxic effects fall into two basic types: *destruction* of nerve cells and *interference* with neurotransmission.

Immunotoxicity

Immunotoxicity refers to toxic effects on the immune system, which includes the lymph system, blood cells, and antibodies in the blood.

Reproductive Toxicity

The effect of toxicants on the male or female reproductive system is known as *reproductive toxicity*. For the male reproductive system, toxicants primarily affect the division of sperm cells and the development of healthy sperm. For the female reproductive system, toxicants can affect the endocrine system, the brain, and the reproductive tract.

Toxic Effects on the Eye

There are a wide variety of toxic substances that can cause damage to the eye through contact with the cornea.

DOSE-RESPONSE RELATIONSHIPS

Dose-response relationships can be used to relate the response of an organism to increasing dose levels of toxicants.

Dose-Response Curves

An objective of toxicity tests is to establish the dose-response curve, as illustrated in Fig. 34.2.

Several important features of a dose-response curve are illustrated and described for toxicant A of Fig. 34.2 as follows.

- *Response*: The ordinate.
- *Dose*: The abscissa.

- *No Observed Effect*: The range of the curve below which no effect is observed is the range of no observed effect. The upper end of the range is known as the *threshold*, the *no effect level* (NEL), or the *no observed effect level* (NOEL).

- *Lowest Observed Effect*: The dose where minor effects first can be measured, but the effects are not directly related to the response being measured, is known as the *lowest observed effect level* (LOEL).

- *No Observed Adverse Effect*: The dose where effects related to the response being measured first can be measured is known as the *no observed adverse effect level* (NOAEL). However, at this level, the effects observed at the higher doses are not observed.

- *Lowest Observed Adverse Effect*: The dose where effects related to the response being measured first can be measured, and are the same effects as the effects observed at the higher doses, is known as the *lowest observed adverse effect level* (LOAEL).

- *Frank Effect*: The *frank effect level* (FEL) dose marks the point where maximum effects are observed with little increase in effect for increasing dose.

For some toxicants, primarily those believed to be carcinogens, there is no apparent threshold. Any dose is considered to have an effect even though such effect may be unmeasurable at low doses. Such toxicants have no safe exposure level. This is illustrated as toxicant B in Fig. 34.2. Lead is an example of a toxicant with no threshold dose.

The *lethal dose* or *lethal concentration* is the concentration of toxicant at which a specified percentage of test animals die. The lethal dose is expressed as the mass of toxicant per unit mass of test animal. Thus, LD_{50} means the dose in milligrams of toxicant per kilogram of body mass at which 50% of the test animals died.

For acute tests involving inhalation as the exposure pathway, the concentration, in parts per million, of the toxicant in air is used. If the toxicant is in particulate form, the concentration in milligrams of toxic particles per cubic meter of air is used. Thus, LC_{50} means the concentration of the toxicant in air at which 50% of the test animals died.

SAFE HUMAN DOSE

EPA Methods

Several approaches exist for selecting a safe human dose from the data obtained from toxicological and epidemiological studies. The approach most likely to be encountered by the environmental engineer is the one

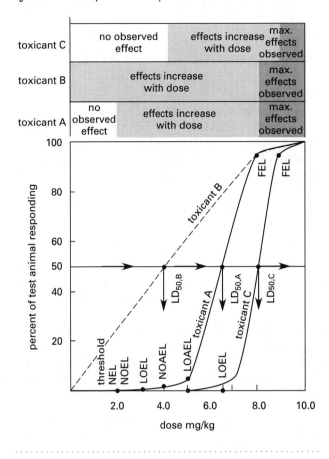

Figure 34.2 Dose-Response Relationships

recommended by the U.S. Environmental Protection Agency (EPA).

Carcinogens generally do not exhibit thresholds of response at low doses of exposure and any exposure is assumed to have an associated risk. The EPA model used to evaluate carcinogenic risk assumes no threshold and a linear response to any amount of exposure. *Noncarcinogens* (systemic toxicants) are chemicals that do not produce tumors (or gene mutations) but instead adversely interfere with the functions of enzymes in the body, thereby causing abnormal metabolic responses. Noncarcinogens have a dose threshold below which no adverse health response can be measured.

Noncarcinogens

The threshold below which adverse health effects in humans are not measurable or observable is defined by the EPA as the *reference dose* (RfD). The reference dose is the safe daily intake that is believed not to cause adverse health effects. The reference dose relates to the ingestion and dermal contact pathways and is route specific. For gases and vapors (exposure by the inhalation pathway), the threshold may be defined as the *reference concentration* (RfC). Sometimes the term "reference dose" is also used for exposure by the inhalation pathway.

After an RfD has been identified for one or more toxicants, the *hazard ratio* (HR) can be determined to assess whether exposures indicate an unacceptable hazard. The hazard ratio is the *estimated exposure dose* (EED) divided by the reference dose for each of the toxicants from all routes of exposure. If the sum of the ratios exceeds 1.0, the risk is unacceptable. Calculating the hazard ratio should be considered a preliminary assessment.

$$\mathrm{HR} = \frac{\mathrm{EED}}{\mathrm{RfD}} \qquad 34.1$$

Carcinogens

The distinguishing feature of cancer is the uncontrolled growth of cells into masses of tissue called *tumors*. Tumors may be *benign*, in which the mass of cells remains localized, or *malignant*, in which the tumors spread through the bloodstream to other sites within the body. This latter process is known as *metastasis* and determines whether the disease is characterized as cancer. The term *neoplasm* (new and abnormal tissue) is also used to describe tumors.

Cancer occurs in three stages: initiation, promotion, and progression. During the *initiation* stage, a cell mutates and the DNA is not repaired by the body's normal DNA repair mechanisms. During the *promotion* stage, the mutated cells increase in number and undergo differentiation to create new genes. During *progression*, the cancer cells invade adjacent tissue and move through the bloodstream to other sites in the body. It is believed that continued exposure to the agent that initiated genetic mutation is necessary for progression to continue. Many mutations are believed to be required for the progression of cancer cells to occur at remote sites in the body.

Direct Human Exposure

The EPA's classification system for carcinogenicity is based on a consensus of expert opinion called *weight of evidence*.

The EPA maintains a database of toxicological information known as the Integrated Risk Information System (IRIS). The IRIS data include chemical names, chemical abstract service registry numbers (CASRN), reference doses for systemic toxicants, carcinogen potency factors (CPF) for carcinogens, and the carcinogenicity group classification, which is shown in Table 34.1.

The dose-response for carcinogens differs substantially from that of noncarcinogens. For carcinogens it is believed that any dose can cause a response (mutation

Table 34.1 EPA Carcinogenicity Classification System

group	description
A	human carcinogen
B1 or B2	probable human carcinogen B1 indicates that human data are available. B2 indicates sufficient evidence in animals and inadequate or no evidence in humans.
C	possible human carcinogen
D	not classifiable as to human carcinogenicity
E	evidence of noncarcinogenicity for humans

of DNA). Since there are no levels (no thresholds) of carcinogens that could be considered safe for continued human exposure, a judgment must be made as to the acceptable level of exposure, which is typically chosen to be an excess lifetime cancer risk of 1×10^{-6} (0.0001%). *Excess lifetime cancer risk* refers to the incidence of cancers developed in the exposed animals minus the incidence in the unexposed control animals. For whole populations exposed to carcinogens, the number of total excess cancers, EC, is the product of the probability of excess cancer, R, and the total exposed population, EP.

$$\mathrm{EC} = (\mathrm{EP})(R) \qquad 34.2$$

Under the EPA approach, the *carcinogen potency factor* (CPF) is the slope of the dose-response curve at very low exposures. The CPF is also called the *potency factor* or *slope factor* and has dimensions of $(\mathrm{mg/kg \cdot d})^{-1}$. The CPF is pathway (route) specific. The CPF is obtained by extrapolation from the high doses typically used in toxicological studies.

The CPF is the probability of risk produced by lifetime exposure to 1.0 mg/kg·d of the known or potential human carcinogen. Thus, the slope factor can be multiplied by the long-term daily intake (*chronic daily intake*, CDI) to obtain the lifetime probability of risk, R, for daily doses other than 1.0 mg/kg·d. The CDI can be calculated from Eq. 34.3.

$$\mathrm{CDI}_{a/w} = \frac{\mathrm{C(CR)(EF)(ED)}}{\mathrm{(BW)(AT)}} \qquad 34.3$$

For less than lifetime exposure, the *exposure duration* must be used to calculate the total intake, which must

Table 34.2 EPA Standard Values for Intake Calculations

parameter	standard value
average body weight, adult	70 kg
average body weight, child	10 kg
daily water ingestion, adult	2 L
daily water ingestion, child	1 L
daily air breathed, adult	20 m^3
daily air breathed, child	5 m^3
daily fish consumed, adult	6.5 g
lifetime exposure period	70 yr

be divided by the *averaging duration* of 70 years for carcinogens. For noncarcinogens, the averaging duration is the same as the exposure duration.

Once the CDI is known, the probable risk of additional cancers for adults and children can be found from Eq. 34.4.

$$R = (SF)(CDI) \qquad 34.4$$

Bioconcentration Factors

Besides setting factors for direct human exposure to toxicants through water ingestion, inhalation, and skin contact, the EPA also has developed *bioconcentration factors* (BCF) so that the human intake from consumption of fish and other foods can be determined. Bioconcentration factors have been developed for many toxicants and provide a relationship between the toxicant concentration in the tissue of the organism and the concentration in the medium (e.g., water). The concentration in the organism equals the product of the BCF and the concentration in the medium. Not all chemicals or other substances will bioaccumulate, and the BCF pertains to a specific organism, such as fish.

$$C_{org} = (BCF)(C_w) \qquad 34.5$$

Selected bioconcentration factors (BCF) for selected chemicals in fish are given in Table 34.3. The substances are arranged in descending order of BCFs to illustrate the substances that have a high potential to bioaccumulate in fish. These substances are of great importance when the oral pathway is present in a particular situation.

The BCF factors can be applied to determine the total dose to humans who ingest fish from water contaminated with toxicants that bioaccumulate. This dose would be added to the dose received from drinking the contaminated water.

Table 34.3 Selected Bioconcentration Factors for Fish

substance	BCF (L/kg)
polychlorinated biphenyls	100 000
4,4' DDT	54 000
DDE	51 000
heptachlor	15 700
chlordane	14 000
toxaphene	13 100
mercury	5500
2,3,7,8 tetrachloro dibenzo-p-dioxin (TCDD)	5000
dieldrin	4760
copper	200
cadmium	81
lead	49
zinc	47
arsenic	44
tetrachloroethylene	31
aldrin	28
carbon tetrachloride	19
chromium	16
chlorobenzene	10
benzene	5.2
chloroform	3.75
vinyl chloride	1.17
antimony	1

ACGIH Methods

The American Conference of Governmental Industrial Hygienists (ACGIH) uses methods for determining the safe human dose that are somewhat different from the EPA methods previously described.

Threshold Limit Values

The ACGIH method uses predetermined *threshold limit values* (TLV) for both noncarcinogens and carcinogens. The TLVs are the concentrations in air that workers could be repeatedly exposed to on a daily basis without adverse health effects. The term TLV-TWA means the maximum time-weighted average concentration that all workers may be exposed to during an 8 hour day and 40 hour week. The TLV-TWA is for the inhalation route of exposure.

ACGIH also determines *short-term exposure limits* (TLV-STEL) for airborne toxicants, which are the recommended concentrations workers may be exposed to for short periods during the workday without suffering certain adverse health effects (irritation, chronic tissue damage, and narcosis). The TLV-STEL is the TWA concentration in air that should not be exceeded for more than 15 minutes of the workday. The TLV-STEL should not occur more than four times daily, and there should be at least 60 minutes between successive STEL exposures. In such cases, the excursions may exceed three times the TLV-TWA for no more than a total of 30 minutes during the workday, but shall not exceed five times the TLV-TWA under any circumstances. In all cases, the TLV-TWA may not be exceeded. Short-term exposure limits have not been established by ACGIH for some toxicants.

ACGIH also publishes *ceiling threshold limit values* (TLV-C) that should not be exceeded at any time during the workday. If instantaneous sampling is infeasible, the sampling period for the TLV-C can be up to 15 minutes in duration. Also, the TLV-TWA should not be exceeded.

For mixtures of substances, the *equivalent exposure* over 8 hours is the sum of the individual exposures.

$$E = \frac{1}{8}\sum_{i=1}^{n} C_i T_i \qquad 34.6$$

The *hazard ratio* is the concentration of the contaminant divided by the exposure limit of the contaminant. For mixtures of substances, the total hazard ratio is the sum of the individual hazard ratios and must not exceed unity. This is known as the *law of additive effects*. For this law to apply, the effects from the individual substances in the mixture must act on the same organ. If the effects do not act on the same organ, then each of the individual hazard ratios must not exceed unity.

The equivalent exposure of a mixture of gases is

$$E_m = \sum_{i=1}^{n} \frac{C_i}{L_i} \qquad 34.7$$

NIOSH Methods

The National Institute for Occupational Safety and Health (NIOSH) was established by the Occupational Safety and Health Act of 1970. NIOSH is part of the Centers for Disease Control and Prevention (CDC) and is the only federal institute responsible for conducting research and making recommendations for the prevention of work-related illnesses and injuries. The Institute's responsibilities include

- investigating hazardous working conditions as requested by employers or workers
- evaluating hazards ranging from chemicals to machinery
- creating and disseminating methods for preventing disease, injury, and disability
- conducting research and providing recommendations for protecting workers

- providing education and training to persons preparing for or actively working in the field of occupational safety and health

The NIOSH recommended exposure limits (RELs) are time-weighted average (TWA) concentrations for up to a 10 hour workday during a 40 hour workweek. A short-term exposure limit (STEL) is a 15 minute TWA exposure that should not be exceeded at any time during a workday. A ceiling REL should not be exceeded at any time. The "skin" designation means there is a potential for dermal absorption, so skin exposure should be prevented as necessary through the use of good work practices and gloves, coveralls, goggles, and other appropriate equipment.

LEGAL STANDARDS FOR WORKER PROTECTION

While the EPA provides exposure limitations and risk factors for environmental cleanup projects, ACGIH provides recommendations to industrial hygienists about workplace exposure, and NIOSH provides research and recommendations for workplace exposure limits (recommended exposure limits), the *Occupational Safety and Health Administration* (OSHA) sets the legally enforceable workplace exposure limits. OSHA standards are given in the 29 Consolidated Federal Regulations (CFR).

The calculation procedure for use of the *permissible exposure limits* (PELs) is the same as for ACGIH–TLVs. The procedure for mixtures is also the same.

FE-STYLE EXAM PROBLEMS

Problems 1 and 2 refer to the following information.

An environmental engineer has been assigned to manage a laboratory that analyzes hazardous waste samples. She decides to check the workplace air for compliance with the ACGIH values because the ventilation system has been working poorly. She obtained data on possible toxicants in the breathing zone at four workstations.

station	toxicant	concentration for indicated duration (ppm)			
		2 h	4 h	2 h	peak <15 min
1	butane	100	80	120	200
2	pentane	500	70	900	1000
3	hexane	100	200	200	1050
4	mixture of all of above				

The environmental engineer reviewed the ACGIH–TLVs for the toxicants identified.

toxicant	TLV-TWA (ppm)	STEL (ppm)	ceiling (ppm)
butane	800	–	–
pentane	600	–	–
hexane	500	1000	–

1. The TLV-TWA for butane at workstation 1 is most nearly

    (A) 80 ppm
    (B) 85 ppm
    (C) 90 ppm
    (D) 95 ppm

2. Assuming the effects of the toxicants are additive, the equivalent exposure hazard ratio of the mixture is most nearly

    (A) 0.77
    (B) 0.94
    (C) 1.1
    (D) 1.4

3. The major pathways of exposure to toxic agents are

    I.   inhalation
    II.   ingestion
    III. whole body
    IV. skin absorption
    V.   eyes and hands

    (A) I, II, V
    (B) I, III, IV
    (C) III, IV, V
    (D) I, II, IV

4. LD_{50} means the dose in

    (A) mg/d at which 50 of the test animals died
    (B) mg/d at which 50% of the test animals died
    (C) mg/kg body mass at which 50% of the test animals died
    (D) mg/kg body mass at which an adverse effect was observed in 50% of the test animals

5. Chemical A has an LD_{50} of 3 mg/kg. Chemical B has an LD_{50} of 9 mg/kg. Chemical A is

    (A) one-third as toxic as chemical B
    (B) three times as toxic as chemical B
    (C) not comparable to chemical B because the pathways are not specified
    (D) not comparable to chemical B because the target organs are not specified

6. The law of additive effects

    (A) applies to all mixtures of gases and vapors
    (B) applies only when the effects from the component gases and vapors occur within the same organs
    (C) applies to noncarcinogens less than the TLV
    (D) applies to carcinogens with a high slope factor

SOLUTIONS TO FE-STYLE EXAM PROBLEMS

1. From Eq. 34.6,

$$E = \frac{1}{8}\sum_{i=1}^{n} C_i T_i$$

$$E_{\text{butane}} = \frac{1}{8\ \text{h}}\left(\begin{array}{c}(100\ \text{ppm})(2\ \text{h}) + (80\ \text{ppm})(4\ \text{h}) \\ + (120\ \text{ppm})(2\ \text{h})\end{array}\right)$$

$$= 95\ \text{ppm}$$

Answer is D.

2. From Eq. 34.7,

$$E_m = \sum_{i=1}^{n} \frac{C_i}{L_i}$$

$$E_m = \frac{C_{\text{butane}}}{L_{\text{butane}}} + \frac{C_{\text{pentane}}}{L_{\text{pentane}}} + \frac{C_{\text{hexane}}}{L_{\text{hexane}}}$$

$$= \frac{95\ \text{ppm}}{800\ \text{ppm}} + \frac{385\ \text{ppm}}{600\ \text{ppm}} + \frac{175\ \text{ppm}}{500\ \text{ppm}}$$

$$= 0.119 + 0.642 + 0.35$$

$$= 1.11$$

Answer is C.

3. The major pathways are inhalation, ingestion, and skin absorption.

Answer is D.

4. Only C is true.

Answer is C.

5. Only B is true.

Answer is B.

6. Only B is true.

Answer is B.

35

Industrial Hygiene

Subjects

INDUSTRIAL HYGIENE 35-2
HAZARD IDENTIFICATION 35-2
    Overview of Hazards 35-2
    Hazard Communication 35-3
GASES, VAPORS, AND SOLVENTS 35-3
Exposure Factors for Gases
    and Vapors 35-3
    Solvents 35-3
Gases and Flammable or
    Combustible Liquids 35-3
    Evaluation and Control of Hazards . . 35-5
PARTICULATES 35-6
    Silica 35-6
    Asbestos 35-6
    Lead 35-6
    Beryllium 35-7
    Coal Dust 35-7
    Welding Fumes 35-7
    Radioactive Dust 35-7
    Biological Particulates 35-7
    Control of Particulates 35-7
NOISE 35-7
    Sound and Noise 35-7
    Loudness 35-8
    Hearing Loss 35-9
    Classes of Noise Exposure 35-9
    Noise Control 35-9
    Audiometry 35-9
RADIATION 35-10
    Nuclear Radiation 35-10
    Radioactive Decay 35-11
    Radiation Effects on Humans 35-11
    Safety Factors 35-11
HEAT AND COLD STRESS 35-11
    Thermal Stress 35-11
    Heat Stress 35-12
    Cold Stress 35-12
ERGONOMICS 35-12
    Work-Rest Cycles 35-12
    Manual Handling of Loads 35-12
    Cumulative Trauma Disorders 35-12
RECOGNITION OF CTDS 35-13
    Carpal Tunnel Syndrome 35-13
    Cubital Tunnel Syndrome 35-13
    Epicondylitis 35-13
    Ganglionitis 35-13

    Neck Tension Syndrome 35-13
    Pronator Syndrome 35-13
    Tendonitis 35-13
    Tenosynovitis 35-13
    Thoracic Outlet Syndrome 35-14
    Ulnar Artery Aneurysm 35-14
    Ulnar Nerve Entrapment 35-14
    White Finger 35-14
BIOLOGICAL HAZARDS 35-14
    Biological Agents 35-14
    Infection 35-14
Biohazardous Workplaces
    and Activities 35-15
    Blood-Borne Pathogens 35-16
    Bacteria- and Virus-Derived Toxins . . 35-16

Nomenclature

a	speed of sound	m/s
A	activity metabolism	W
A	radioactivity	Bq
AM	asymmetry multiplier for lifting	–
B	basal metabolism	W
C	concentration	ppm, mg/m^3
C	constant for calculating sound intensity	–
C	time of noise exposure at specified level	s
CL	ceiling heat limit	°C
CM	coupling multiplier for lifting	–
DM	distance multiplier for lifting	–
E	noise exposure	–
ECT	equivalent chill temperature	°C
f	frequency of sound	Hz
FM	frequency multiplier for lifting	–
HM	horizontal multiplier for lifting	–
I	sound intensity	W/m^2
IL	insertion loss	dB
k	ratio of specific heats	–
L	sound pressure or sound power level	dB
LC	load constant for lifting	kg
m	mass	kg
MW	molecular weight	g/mol
n	number of moles	–
p	total number of observations	–

p	pressure or partial pressure	Pa
P	posture metabolism	W
PEL	permissible exposure limit	mg/m^3
Q	heat flow	W
r	distance from sound source	m
R	specific gas constant	kJ/kg·K
R^*	universal gas constant	kJ/kmol·K
RAL	recommended heat alert limit	°C
REL	recommended heat exposure limit	°C
RWL	recommended weight limit	kg
t	time	s
t	rest time, percent of period	–
T	temperature	°C, K
TLV	threshhold limit value	–
V	velocity metabolism	W
V	volume	L, m^3
VM	vertical multiplier for lifting	–
W	sound power	W
WBGT	wet-bulb globe temperature	°C
x	mole fraction	–
x_{rms}	root mean square value of n observations	–

Symbols

ρ	density	kg/m^3
λ	decay constant for radionuclides	s^{-1}
λ	wavelength	m

Subscripts

0	initial condition, or reference
C	convection
db	dry bulb
E	evaporation
g	globe
in	indoor
m	mixed
max	maximum
M	metabolic
M/V	mass per unit volume
nwb	natural wet bulb
p	sound power
R	radiation
rest	resting
rms	root-mean-square
rms-ref	reference rms
S	storage
t	time, time period
W	sound power

INDUSTRIAL HYGIENE

Industrial hygiene is the art and science of identifying, evaluating, and controlling environmental factors (including stress) in the workplace that may cause sickness, health impairment, or discomfort among workers or citizens of the community. Industrial hygiene involves the recognition of health hazards associated with work operations, processes, evaluations, measurements of the magnitude of hazards, and determining applicable control methods. Occupational health hazards specifically involve illness or impairment for which a worker may be compensated under a worker protection program.

The fundamental law governing worker protection is the 1970 federal *Occupational Safety and Health Act*. It requires employers to provide a workplace that is free from hazards and to comply with specified safety and health standards. Employees must also comply with the safety and health standards of the 1970 Act that apply to their own conduct. The federal regulatory agency responsible for administering the Occupational Safety and Health Act is OSHA, the Occupational Safety and Health Administration. OSHA sets standards, investigates violations of the standards, performs inspections of plants and other facilities, investigates complaints, and takes enforcement action against violators. OSHA also funds state programs, which can be established by states if the state programs are at least as stringent as the federal program.

The 1970 Act also established the *National Institute for Occupational Safety and Health* (NIOSH). NIOSH is responsible for safety and health research and makes recommendations for regulations. The recommendations are known as *Recommended Exposure Limits* (RELs). Among other activities, NIOSH also publishes health and safety criteria and notifications of health hazard alerts, and is responsible for testing and certifying respiratory protective equipment.

HAZARD IDENTIFICATION

Overview of Hazards

There are four basic types of hazards with which industrial hygiene is concerned: chemical hazards, physical hazards, ergonomic hazards, and biological hazards. *Chemical hazards* result from airborne chemicals such as gases, vapors, or particulates in harmful concentrations. Besides inhalation, chemical hazards may affect workers by absorption through the skin. *Physical hazards* include radiation, noise, vibration, and excessive heat or cold. *Ergonomic hazards* include work procedures that involve motions that result in biomechanical stress and injury. *Biological hazards* result from exposure to biological organisms that may lead to illness.

In respect to chemical hazards, the terms "toxicity" and "hazard" are not synonymous. *Toxicity* is the capacity of the chemical to produce harm when it has reached a sufficient concentration at a particular site in the body. *Hazard* refers to the probability that this concentration will occur.

Hazard Communication

Two important preventative measures that are required by OSHA are *Material Safety Data Sheets* (MSDS) and labeling of containers of hazardous materials. A third important OSHA requirement is that all covered employers must provide the necessary information and training to affected workers. The OSHA Hazard Communication Standard is given in Title 29 of the *Code of Federal Regulations* (CFR) Part 1910.1200. Other OSHA requirements are also given in Title 29.

MSDS

MSDS sheets provide key information about a chemical or substance so that users or emergency responders can determine safe use procedures and necessary emergency response actions. An MSDS provides information on the identification of the material and its manufacturer, identification of hazardous components and their characteristics, physical and chemical characteristics of the ingredients, fire and explosion hazard data, reactivity data, health hazard data, precautions for safe handling and use, and recommended control measures for use of the material. The information on MSDS sheets is essential for dealing with hazardous chemicals and should be complete.

Container Labeling

Labels are required on hazardous materials containers. A label should provide essential information for the safe use and storage of hazardous materials. Failure to provide adequate labeling of hazardous material containers is a common violation of OSHA standards.

Worker Information and Training

The OSHA standard requires that employers provide workers with information about the potential health hazards from exposure to hazardous chemicals that they use in the workplace. It also requires employers to provide adequate training to workers on how to safely handle and use hazardous materials.

GASES, VAPORS, AND SOLVENTS

Exposure Factors for Gases and Vapors

The most frequently encountered hazard in the workplace is exposure to gases and vapors from solvents and chemicals. Several factors define the exposure potential for gases and vapors. The most important factors are how the material is used and what engineering or personal protective controls exist. If the inhalation route of entry is controlled, dermal contact may be the major route of exposure.

Vapor pressure of a chemical is related to temperature. Vapor pressure affects the concentration of the chemical in vapor form above the liquid and is dependent upon the temperature and the properties of the chemical. Processes that operate at lower temperatures are inherently less hazardous than processes that operate at higher temperatures because fewer chemicals have boiling points and vapor pressures in the operating temperature range.

Reactivity affects the hazard potential because the reactions determine whether highly volatile products are generated. The products may be volatile or nonvolatile depending on the properties of the combining substances.

Two terms are used to quantify the concentration of a gas in air that a worker can be safely exposed to. These are *threshold limit value* (TLV) and *permissible exposure limit* (PEL). The TLV is a concentration in air that nearly all workers can be exposed to daily wihtout any adverse effects. The PEL is a regulatory exposure limit for workers; OSHA publishes PELs as standards. A representative listing of toxic materials and their TLVs is shown in Table 35.1.

Solvents

Solvents are widely used throughout industry for many purposes, and their safe use is an important industrial hygiene concern. It is essential that accurate information be provided to employees on the physical properties and toxicological effects of exposure to solvents.

The MSDS provides information about the hazard content of an organic solvent.

Gases and Flammable or Combustible Liquids

Hazardous gases fall into four main types: cryogenic liquids, simple asphyxiants, chemical asphyxiants, and all other gases whose hazards depend on their properties.

Cryogenic liquids can vaporize rapidly, producing a gas that is more dense than air and displacing oxygen in confined spaces.

Simple asphyxiants, which include helium, neon, nitrogen, hydrogen, and methane, can dilute or displace oxygen. *Chemical asphyxiants*, which include carbon monoxide, hydrogen cyanide, and hydrogen sulfide, can pass into blood cells and tissue and interfere with blood-carrying oxygen.

Table 35.1 Representative Hazardous Concentrations in Air[a]

	Combustibles						Toxics				
Material	LEL (%/Vol)	UEL (%/Vol)	TLV/ TWA (ppm)	IDLH (ppm)	Density (Air =1.0)	Material	TLV/ TWA (ppm)	IDLH	LEL (ppm)	LEL (%/Vol)	Density (Air =1.0)
Acetone	2.5	12.8	750	2500	2.0	Acetone	750	2500	25 000	2.5	2.0
Acetylene	2.5	100.0	-A-	-A-	.9	Ammonia	25	300	160 000	16.0	0.6
Ammonia	15.0	28.0	25	300	0.6	Benzene	1.0	-C-	12 000	1.2	2.6
Benzene	1.2	7.8	1.0	500	2.6	Butane	800	-U-	16 000	1.6	2.0
Butane	1.6	8.4	800	-U-	2.0	n-Butyl					
n-Butyl						Acetate	150	1700	17 000	1.7	4.0
Acetate	1.7	7.6	150	1700	4.0	Carbon					
Diborane	0.8	88.0	0.1	15	1.0	Dioxide	5000	40 000	N/C	N/C	1.5
Ethane	3.0	12.5	-A-	-A-	1.0	Carbon					
Ethanol	3.3	19.0	1000	-U-	1.6	Monoxide	25	1200	125 000	12.5	1.0
Ethyl						Chlorine	0.5	10	N/C	N/C	2.5
Acetate	2.0	11.5	400	2000	3.0	Ethylene					
Ethyl						Oxide	1	-C-	30 000	3.0	1.5
Ether	1.9	36.0	400	1900	2.6	Ethyl					
Ethylene						Ether	400	19 000	19 000	1.9	2.6
Oxide	3.0	100.0	1	-C-	1.5	Gasoline	300	-U-	14 000	1.4	3-4.0
Gasoline	1.4	7.6	300	-U-	3-4.0	Heptane	400	750	10 500	1.05	3.5
Heptane	1.05	6.7	400	750	3.5	Hexane	50	1100	11 000	1.0	3.0
Hexane	1.1	7.5	50	1100	3.0	Hydrogen					
Hydrogen	4.0	75.0	-A-	-A-	0.1	Cyanide	10	50	56 000	5.6	0.9
Isopropyl						Hydrogen					
Alcohol	2.0	12.0	400	2000	2.1	Sulfide	10	100	40 000	4.0	1.2
Methane	5.0	15.0	-A-	-A-	0.6	Isopropyl					
Methanol	6.0	36.0	200	6000	1.1	Alcohol	400	2000	20 000	2.0	2.1
Methyl Ethyl						Methyl					
Ketone	1.4	11.4	200	3000	2.5	Acetate	200	3100	31 000	3.1	2.6
Pentane	1.5	7.8	600	15 000	2.5	Methanol	200	6000	60 000	6.0	1.1
Propane	2.1	9.5	1000	2100	1.6	Methyl					
Propylene						Chloride	50	2000	81 000	8.1	1.8
Oxide	2.3	36.0	20	400	2.0	Methyl Ethyl					
Styrene	0.9	6.8	50	700	3.6	Ketone	200	3000	14 000	1.4	2.5
Toluene	1.1	7.1	50	500	3.1	Methyl					
Turpentine	0.8	-U-	100	800	4.7	Methacrylate	100	1000	17 000	1.7	3.5
Vinyl						Nitric					
Acetate	2.6	13.4	10	-U-	3.0	Oxide	25	100	N/C	N/C	1.0
Vinyl						Nitrogen					
Chloride	3.6	33.0	1.0	-C-	2.2	Dioxide	3	20	N/C	N/C	1.6
Xylene	0.9	6.7	100	900	3.7	Pentane	600	15 000	15 000	1.5	2.5
						n-Propyl					
						Acetate	200	1700	17 000	1.7	3.5
						Styrene	50	700	9000	0.9	3.6
						Sulfur					
						Dioxide	2	100	N/C	N/C	2.2
						1,1,1-Trichloro-					
						ethane	350	700	75 000	7.5	4.6
						Toluene	50	500	11 000	1.1	3.2
						Trichloro-					
						ethylene	50	1000	80 000	8.0	4.5
						Turpentine	100	800	8000	0.8	4.7
						Vinyl					
						Chloride	1.0	-C-	36 000	3.6	2.2
						Xylene	100	900	9000	0.9	3.7

Key: A, asphyxiant; C, carcinogen; U, data not available; N/C, noncombustible
[a]Subject to change without notice

The term *flammable* refers to the ability of an ignition source to propagate a flame throughout the vapor-air mixture and have a closed-cup flash point below 37.8°C (100°F) and a vapor pressure not exceeding 272 atm at 37.8°C (100°F). The phrase *closed-cup flash point* refers to a method of testing for flash points of liquids. The term *combustible* refers to liquids with flash points above 37.8°C (100°F).

For each airborne flammable substance there are minimum and maximum concentrations in air between which propagation will occur. The lower concentration in air is known as the *lower explosive limit* (LEL) or *lower flammable limit* (LFL). The upper limit is known as the *upper explosive limit* (UEL) or *upper flammable limit* (UFL). Above the UEL, there is not enough air to propagate a flame. The lower the LEL, the greater the hazard from a flammable liquid. See Table 35.1 for a representative listing of combustible materials and their LELs and UELs. For many common liquids and gases, the LEL is a few percent and the UEL is 6% to 12%. Note that if a concentration in air is less than the PEL or the TLV, the concentration will be less than the LEL. The occupational safety requirements for handling and using flammable and combustible liquids are given in Subpart H of 29 CFR 1910.106.

Evaluation and Control of Hazards

The toxicological effects from aqueous solutions include dermatitis, throat irritation, and bronchitis.

Vapor-Hazard Ratio

One indicator of hazards from vapors and gases from solvents is the *vapor-hazard ratio number*, which is the equilibrium vapor pressure at 25°C in ppm divided by the TLV in ppm. The higher the ratio, the greater the hazard. To assess the overall hazard, the vapor-hazard ratio should be used with the TLV, ignition temperature, flash point, toxicological information, and degree of exposure. The vapor-hazard ratio accounts for the volatility of a solvent as well as its toxicity.

The best control method is not to use a solvent that is hazardous. Sometimes a process can be redesigned to eliminate the use of a solvent or hazardous chemical, but if use of a solvent cannot be eliminated altogether, then water should be used as the solvent. The following evaluation steps are recommended.

- Use an aqueous solution when possible.
- Use a *safety solvent* if it is not possible to use water. Safety solvents have inhibitors and high flash points.
- Use a different process when possible to avoid use of a hazardous solvent.
- Provide a properly designed ventilation system if toxic solvents must be used.

- Never use highly toxic or highly flammable solvents (benzene, carbon tetrachloride, gasoline).

Ventilation

The most effective way to prevent inhalation of vapors from solvents is to provide closed systems or adequate local exhaust ventilation. If limitations exist on the use of closed systems or local exhaust ventilation, then workers should be provided with personal protective equipment.

Personal Protective Equipment

Respirators provide emergency and backup protection but are unreliable as a primary source of protection from hazardous vapors because they leak around the edges of the face mask, can become contaminated around the edges, reduce the efficiency of the worker, and increase the lack of oxygen in oxygen-deficient areas. Other drawbacks are the need to have the respirator properly fitted to the worker and the need for the worker to be trained in its proper use. Additionally, the worker may feel a false sense of security while wearing a respirator.

Besides inhalation, dermal contact is an important concern when working with hazardous solvents. Mechanical equipment should be provided to keep the worker isolated from contact with the solvent. However, since some contact may occur even with mechanical equipment in use, protective clothing should be provided. Protective clothing includes aprons, face shields, goggles, and gloves. The manufacturer's recommendations should be followed for use of all protective clothing and equipment.

One common problem with protective clothing is incorrect selection or misuse of gloves. The time for particular solvents to penetrate gloves that are commonly thought of as "protective" is surprisingly short. Both the permeability and the abrasion resistance of gloves must be considered in their selection and use. For example, methyl chloride will permeate a neoprene glove in less than 15 minutes. The manufacturer should provide the *breakthrough time* and the *permeation rates* for the glove being evaluated. The breakthrough time and permeation rate are dependent on the specific chemical and the composition and thickness of the glove.

Protective eyewear should be provided where the risk of splashing of chemicals is present. Of course, mechanical equipment, barriers, guards, and other engineering measures should be provided as the first line of defense. For chemical splash protection, unvented chemical goggles, indirect-vented chemical goggles, or indirect-vented eyecup goggles should be used. A face shield may also be needed. Direct-vented goggles and normal eyeglasses should not be used, and contact lenses should not be worn.

PARTICULATES

Particulates include dusts, fumes, fibers, and mists. *Dusts* have a wide range of sizes and usually result from a mechanical process such as grinding. *Fumes* are extremely small particles, less than 1 μm in diameter, and result from combustion and other processes. *Fibers* are thin and long particulates, with asbestos being a prime example. *Mists* are suspended liquids that float in air, such as from the atomization of cutting oil. All of these types of particulate can pose an inhalation hazard if they reach the lungs.

With one known exception, particles larger than approximately 5 μm cannot reach the alveoli or inner recesses of the lungs before being trapped and expelled from the body through the digestive system or from the mouth and nose. That's due to the presence of mucus and cilia in the nasal passages, throat, larynx, trachea, and bronchi. The exception is asbestos fibers, which can reach the alveoli even though fibers may be larger.

There are four factors that affect the health risk from exposure to particulates: the types of particulate, the length of exposure, the concentration of particulates in the breathing zone, and the size of particulates in the breathing zone.

The type of particulate can determine the type of health effect that may result from the exposure. Both organic and inorganic dusts can produce allergic effects, dermatitis, and systemic toxic effects. Particulates that contain free silica can produce pneumoconiosis from chronic exposure. *Pneumoconiosis* is lung disease caused by fibrosis from exposure to both organic and inorganic particulates. Other particulates can cause systemic toxicity to the kidneys, blood, and central nervous system. Asbestos fibers can cause lung scarring and cancer.

The critical duration of exposure varies with the type of particulate.

The concentration of particulates in the breathing zone is the primary factor in determining the health risk from particulates. The American Conference of Governmental Industrial Hygienists (ACGIH) has established TLVs that should not be exceeded. OSHA establishes PELs for safe exposure levels in the workplace.

The fourth exposure factor is the size of the particulates. Particles larger than 5 μm will normally be filtered out through the upper respiratory system before reaching the alveoli of the lungs. Particles smaller than 5 μm are considered respirable dusts and pose an exposure hazard when present in the breathing zone.

Silica

Silica (SiO_2) is a common material used in industry that has several associated health hazards. The crystalline form of free silica (quartz) deposited in the lungs causes the growth of fibrous tissue around the deposit. The fibrous tissue reduces the amount of normal lung tissue, thereby reducing the ability of the lungs to transfer oxygen. When the heart tries to pump more blood to compensate, heart strain and permanent damage or death may result. This condition is known as *silicosis*. Mycobacterial infection occurs in about 25% of silicosis cases. Smokers exposed to silica dust have a significantly increased chance of developing lung cancer.

Asbestos

Asbestos can take several forms and is generically described as naturally occurring, fibrous, hydrated mineral silicates. Inhalation of short asbestos fibers can cause *asbestosis*, a kind of pneumoconiosis, as a nonmalignant scarring of the lungs. *Bronchogenic carcinoma* is a malignancy (cancer) of the lining of the lung's air passages. *Mesothelioma* is a diffuse malignancy of the lining of the chest cavity or the lining of the abdomen.

Asbestos mining, construction activities, and working in shipyards are possible exposure activities that may call for an environmental engineer.

The onset of illness seems to be correlated with length and diameter of inhaled asbestos fibers. Fibers 2 μm in length cause asbestosis. Mesothelioma is associated with fibers 5 μm long. Fibers longer than 10 μm produce lung cancer. Fiber diameters greater than 3 μm are more likely to cause asbestosis or lung cancer, while fibers 3 μm or less are associated with mesothelioma.

The OSHA regulations for protection from exposure to asbestos are extensive. They require an employer to perform a negative exposure assessment in many cases. Monitoring must be performed by a competent person who is capable of identifying asbestos hazards and selecting control strategies and who has the authority to make corrective changes. The regulations also specify when medical surveillance is required, when personal protection must be provided, and the engineering controls and work practices that must be implemented.

Lead

The body does not use lead for any metabolic purpose, so any exposure to lead is undesirable. Lead dust and fumes can pose a severe hazard. Acute large doses of lead can cause systemic poisoning or seizures. Chronic exposure can damage the blood-forming bone marrow and the urinary, reproductive, and nervous systems. Lead is a suspected human carcinogen.

Beryllium

Inhalation of metallic beryllium, beryllium oxide, or soluble beryllium compounds can lead to *chronic beryllium disease* (berylliosis). Ingestion and dermal contact do not pose a documented hazard, so maintaining beryllium dusts and fumes below the TLV in the breathing zone is a critical protection measure.

Chronic beryllium disease is characterized by granulomas on the lungs, skin, and other organs. The disease can result in lung and heart dysfunction and enlargement of certain organs. Beryllium has been classified as a suspected human carcinogen.

Coal Dust

Coal dust can cause chronic bronchitis, silicosis, and *coal worker's pneumoconiosis*, which may also be called *black lung disease.*

Welding Fumes

Exposure to welding fumes can cause a disease known as *metal fume fever.* This disease results from inhalation of extremely fine particles that have been freshly formed as fume. Zinc oxide fume is the most common source, but magnesium oxide, copper oxide, and other metallic oxides can also cause metal fume fever. Metal fume fever is of short duration, with symptoms including fever and shaking chills, appearing 4 to 12 hours after exposure.

Radioactive Dusts

Radioactive dusts can cause toxicity in addition to the effects from ionizing radiation. Inhalation of radioactive dust can result in deposition of the radionuclide in the body, which may enter the bloodstream and affect individual organs.

Control measures should be instituted to prevent workers from inhaling radioactive dust, either by restricting access or by providing appropriate personal protection such as respirators. Engineering controls to capture radioactive dust are an absolute necessity to minimize worker exposure.

Biological Particulates

A wide variety of biological organisms can be inhaled as particulates. Examples include dust that contains anthrax spores from the wool or bones of infected animals. Similarly, fungi from grain and other agricultural produce can be inhaled by workers, causing respiratory diseases and allergies.

Control of Particulates

Ventilation

Ventilation is the most effective method of control of particulates. Closed processes should be used wherever possible to eliminate the particulates from the work area. Equipment can be enclosed so that only the feed and discharge openings are open. With adequate pressure, enclosed equipment can be nearly as effective as closed processes. Large automated equipment can sometimes be placed in separate enclosures so that workers would have to wear personal protection to enter the enclosures, but not in general work areas. Local exhaust ventilation with hooded enclosures can be very effective at controlling particulate emissions into general work areas. Where complete enclosure and local exhaust methods are not sufficient, *dilution ventilation* will be necessary to control particulates in the work area. In some instances, the work process can be changed from a dry to a wet process to reduce particulate generation.

Personal Protection

Respirators are an effective means of controlling worker exposure to particulates that remain in the work area after engineering controls have been applied or when access to dusty areas is intermittent. Respirators may also be used to provide additional protection or comfort to workers in areas where local or general ventilation is effective. The NIOSH guidelines for selection of respirators should be followed to ensure that the respirator will be effective at removing the specific particulate to which the workers are exposed.

NOISE

Noise is vibration conducted through solids, liquids, or gases. Noise can cause psychological and physiological damage and can interfere with workers' communication, thereby affecting safety. Exposure to excessive noise for a sufficient time can result in hearing loss. The permissible noise exposures are given in Table 35.2.

A *hearing conservation program* should include noise measurements, noise control measures, hearing protection, audiometric testing of workers, and information and training programs. Employees are required to properly use the protective equipment provided by employers.

Sound and Noise

Characteristics of Sound

Sound is pressure variation in air, water, or some other medium that the human ear can detect. *Noise* is unwanted, unpleasant, or painful sound. The *frequency* of sound is the number of pressure variations per second, measured in cycles per second, or hertz (Hz). The frequency range of human audible sound is approximately 20 Hz to 20 000 Hz.

Table 35.2 Permissible Noise Exposures

duration per day (h)	sound level (dBA slow response[a])
8	90
6	92
4	95
3	97
2	100
1.5	102
1	105
0.5	110
0.25 or less	115

[a] "Slow response" refers to the metering circuit being set on slow mode to reduce rapid, hard-to-read needle excursions. Compliance with OSHA regulations requires measurement set on slow response mode.

Note: When the daily noise exposure is composed of two or more periods of noise exposure at different levels, their combined effect should be considered, rather than the individual effect of each. If the sum of the fractions $C_1/t_1 + C_2/t_2 \ldots C_n/t_n$ exceeds unity, then the mixed exposure should be considered to exceed the limit value. C_n indicates the total time of exposure at a specified noise level, and t_n indicates the total time of exposure permitted at that level.

Source: 29 CFR 1910.95, Table G–16

Velocity is the rate at which analogous successive sound pressure points pass a point. Velocity is called the *speed of sound* and is equal to the product of the wavelength and the frequency.

$$a = f\lambda \qquad 35.1$$

The speed of sound is dependent upon the medium, as illustrated in Table 35.3.

Table 35.3 Speed of Sound in Various Media

medium	speed of sound (m/s)	condition
air	330	1 atm, 0°C
water	1490	1 atm, 20°C
aluminum	4990	1 atm
steel	5150	1 atm

Sound Pressure

Sound pressure measures the intensity of sound and is the variation in atmospheric pressure caused by the disturbance of the air by a vibrating object. The *root-mean-square (rms) pressure* is used.

Figure 35.1 Characteristics of a Sound Wave

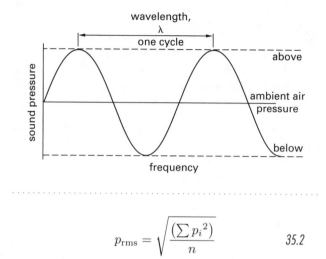

$$p_{\text{rms}} = \sqrt{\frac{\left(\sum p_i{}^2\right)}{n}} \qquad 35.2$$

Sound pressure level (L_p) is measured in decibels relative to a reference level of 20 μPa, the threshold of hearing at a reference frequency of 1000 Hz, as follows.

$$L_p = 10 \log \left(\frac{p}{p_0}\right)^2$$
$$= 20 \log \left(\frac{p}{p_0}\right) \qquad 35.3$$

Sound Power

Sound power, W, is the energy per unit time (in watts) due to an acoustic wave from a source. The sound power level, L_W, is measured in decibels relative to a reference level, W_0, of 10^{-12} W, as follows.

$$L_W = 10 \log \left(\frac{W}{W_0}\right) \qquad 35.4$$

Sound Intensity

Sound intensity is a function of the sound power of a source. The relationship of sound intensity to sound power is

$$I = \frac{W}{4\pi r^2} \qquad 35.5$$

Combining Sound Sources

The combined sound pressure level or sound power level from two or more sound sources can be determined by

$$L = 10 \log \sum 10^{L_i/10} \qquad 35.6$$

Loudness

Loudness is primarily determined by sound pressure but is also affected by frequency because the human ear is more sensitive to high-frequency sounds than low-frequency sounds.

Hearing Loss

Hearing loss can be caused by sudden intense noise over only a few exposures. This type of loss is known as *acoustic trauma*. Hearing loss can also be caused by exposure over a long duration (months or years) to hazardous noise levels. This type is known as *noise-induced hearing loss*. The permanence and nature of the injury depends on the type of hearing loss.

The main risk factors associated with hearing loss are the intensity of the noise (sound pressure level), the type of noise (frequency), daily exposure time (hours per day), and the total work duration (years of exposure). These are known as *noise exposure factors*. Generally, exposure to sound levels above 115 dBA is considered hazardous, and exposure to levels below 70 dBA is considered safe from risk of permanent hearing loss. Also, noise with predominant frequencies above 500 Hz is considered to have a greater potential to cause hearing loss than lower-frequency sounds.

Classes of Noise Exposure

Noise exposure can be classified as continuous noise, intermittent noise, and impact noise. *Continuous noise* is broadband noise of a nearly constant sound pressure level and frequency to which a worker is exposed 8 hours daily and 40 hours weekly. *Intermittent noise* involves exposure to a specific broadband sound pressure level several times a day. *Impact noise* is a sharp burst of short duration sound.

OSHA has established permissible noise exposures for a specific duration each workday, known as *permissible exposure levels* (PELs). The PELs are based on continuous 8 hour exposure at a sound pressure level of 90 dBA, which is established as 100%. For other exposure durations, OSHA has established relationships between the sound level and the exposure time. Every 5 dBA increase in noise cuts the allowable exposure time in half. Sound pressure levels below 90 dBA are not considered hazardous and do not have to be determined.

When workers are exposed to different noise levels during the day, the mixed exposure must be calculated as follows.

$$E_m = \sum_{i=1}^{n} \frac{C_i}{t_i} \qquad 35.7$$

If E_m equals or exceeds 1, the mixed exposure exceeds the OSHA standard.

For intermittent noise, the time characteristics of the noise must also be determined. Both short-term and long-term exposure must be measured. A dosimeter is typically used for intermittent noises.

For impact noises, workers should not be exposed to peaks of more than 140 dBA under any circumstances. The threshold limit values for impulse noise should not exceed the values provided in Table 35.4.

Table 35.4 Typical Threshold Limit Values for Impact Noise

peak sound level (dB)	maximum number of daily impacts
140	100
130	1000
120	10 000

Noise Control

After the noise exposure is compared with acceptable noise levels, the degree of noise reduction needed can be determined. Noise reduction measures can comprise the following three basic methods applied in order.

1. changing the engineering design of the process or equipment

2. limiting the exposure

3. using hearing protection

Administrative controls include changing the exposure of workers to high noise levels by modifying work schedules or locations so as to reduce workers' exposure times. Administrative controls include any administrative decision that limits a worker's exposure to noise.

Personal hearing protection is the final noise control measure that should be undertaken after engineering controls are implemented. Protective devices do not reduce the noise hazard and may not be totally effective. Therefore, engineering controls are preferred over hearing protection. Protective devices include helmets, earplugs, canal caps, and earmuffs. Earplugs may be used with helmets to increase the level of noise reduction.

An important aspect of selecting personal hearing protection is to determine the *noise reduction rating* (NRR) for a device. The NRR is established by the Environmental Protection Agency (EPA) and must be printed on the package of a device. The NRR can be used to determine whether a device provides sufficient hearing protection.

Audiometry

Audiometry is the measurement of hearing acuity. It is used to assess a worker's hearing ability by measuring the individual's threshold sound pressure level at various frequencies (250 Hz to 6000 Hz). The threshold audiogram can be used to create a baseline of hearing ability and to determine changes over time and identify changes resulting from noise control measures. Baseline and annual hearing tests are required where workers are exposed to more than 85 dBA (TWA). The average change from the baseline is used to measure the degree of hearing impairment.

RADIATION

Radiation can be either nonionizing or ionizing. *Nonionizing radiation* includes electric fields, magnetic fields, electromagnetic radiation, radio frequency and microwave radiation, and optical radiation and lasers.

Dealing with *ionizing radiation* requires special skills and knowledge. A specialist in health physics should be consulted whenever ionizing radiation is encountered.

Nuclear Radiation

Nuclear radiation is a term that applies to all forms of radiation energy that originate in the nucleus of a radioactive atom and subatomic particles. Nuclear radiation includes alpha particles, beta particles, neutrons, X-rays, and gamma rays. The common property of all nuclear radiation is its ability to be absorbed by and transfer energy to the absorbing body.

The unit of ionizing radiation given in NCRP's *Recommended Limitations for Exposure to Ionizing Radiation*, is the mSv. Sv is the symbol for sievert, which is the SI unit of absorbed dose times the quality factor of the radiation as compared to gamma radiation. The absorbed dose is measured in grays (Gy). (A sievert is equal to a gray times the *quality factor*.) The gray is a unit of *absorbed dose* equal to 1 J of absorbed energy per kilogram of matter.

A summary of ionizing radiation units is given in Table 35.5.

Alpha Particles

Alpha particles consist of two protons and two neutrons, with an atomic mass of four. Alpha particles combine with electrons from the absorbed material and become helium atoms. Alpha particles have a positive charge of two units and react electrically with human tissue. Because of their large mass, they can travel only about 10 cm in air and are stopped by the outer layer of the skin. Alpha-emitters are considered to be only internal

Table 35.5 Units for Measuring Ionizing Radiation

property	SI
energy absorbed	gray (Gy) 1 J/kg 1 Gy = 100 rad
biological effect	sievert (Sv) Gy × quality factor 1 Sv = 100 rem

radiation hazards. They affect the bones, kidney, liver, lungs, and spleen.

Beta Particles

Beta particles are electrically charged particles ejected from the nuclei of radioactive atoms during disintegration. They have a negative charge of one unit and the mass of an electron. Beta particles can penetrate in human tissue to a depth of 20 to 130 mm and travel up to 9 m in air. Skin burns can result from an extremely high dose of beta radiation, but beta-emitters are primarily internal radiation hazards. Beta particles are more hazardous than alpha particles because they can penetrate deeper into tissue.

Neutrons

Neutron particles have no electrical charge and are released upon disintegration of certain radioactive materials. Their range in air and in human tissue depends on their kinetic energy, but the average depth of penetration in human tissue is 60 mm. Neutrons lose velocity when they are absorbed or deflected by the nuclei with which they collide. However, the nucleus is left with higher energy that is later released in the form of a proton, gamma ray, beta particle, or alpha particle. It is these secondary emissions from neutrons that produce damage in tissue and present a health hazard.

X-Rays

X-rays are produced by electron bombardment of target materials and are highly penetrating electromagnetic radiation. X-rays have a valuable scientific and commercial use in producing shadow pictures of objects. The energy of an X-ray is inversely proportional to its wavelength. X-rays of short wavelength are called *hard*, and they can penetrate several centimeters of steel. Long wavelength X-rays are called *soft*, and are less penetrating. The power of X-rays and gamma rays to penetrate matter is called *quality*. *Intensity* is the energy flux density.

Gamma Rays

Gamma rays, or gamma radiation, are a class of electromagnetic photons (radiation) emitted from the nuclei of radioactive atoms. They are highly penetrating and are an external radiation hazard. Gamma rays are emitted spontaneously from radioactive materials, and the energy emitted is specific to the radionuclide. Gamma rays present an external exposure problem because of their deep penetrating ability.

Radioactive Decay

Radioactive decay is measured in terms of *half-life*, the time to lose half of the activity of the original material. Decay activity can be calculated as follows.

$$A = A_o e^{-\lambda t} \qquad 35.8$$

The half-life can be calculated from the decay constant, $t_{1/2} = 0.693/\lambda$.

Radiation Effects on Humans

Ionizing radiation transfers energy to human tissue when it passes through the body. *Dose* refers to the amount of radiation that a body absorbs when exposed to ionizing radiation. The effects on the body from external radiation are quite different from the effects from internal radiation. Internal radiation is spread throughout the body to tissues and organs according to the chemical properties of the radiation. The effects of internal radiation depend on the energy and the residence time within the body. The principal effect of radiation on the body is destruction of or damage to cells. Damage may affect reproduction of cells or cause mutation of cells.

The effects of ionizing radiation on individuals include skin, lung, and other cancers; bone damage; cataracts; and a shortening of life. Effects on the population as a whole include possible damage to human reproductive elements, thereby affecting the genes of future generations.

Safety Factors

An environmental engineer should be aware of the basic safety factors for control of radiation exposure. These factors are time, distance, and shielding.

The exposure dose is directly related to the time exposed, so reducing the time of exposure will reduce the exposure dose. An individual's time of exposure can also be limited by spreading the exposure time among more workers.

Distance is another safety factor that can be changed to reduce the radiation dose. The exposure to external penetrating radiation decreases as the inverse of the square of the distance. By increasing the distance to a source from 2 m to 20 m, for example, the exposure would be reduced to 1% (2 m/20 m)^2.

Shielding involves placing a mass of material between a source and workers to reduce the exposure of the workers. The objective is to use a high-density material that will act as a barrier to X-ray or gamma-ray radiation. Lead and concrete are often used, with lead being the more effective material because of its greater density. For neutrons, different material is needed than for X-rays and gamma rays because neutrons produce secondary radiation from collisions with nuclei. Neutron shielding requires a light nucleus material. Typically water or graphite is used.

The shielding properties of materials are often compared using the *half-value thickness*, which is the thickness of the material required to reduce the radiation to half of the incident value. The half-value properties vary with the radiation source.

HEAT AND COLD STRESS

Thermal Stress

Heat and cold, or *thermal*, stress involves three zones of consideration relative to industrial hygiene. In the middle is the *comfort zone*, where workers feel comfortable in the work environment. On either side of the comfort zone is a *discomfort zone* where workers feel uncomfortable with the heat or cold, but a health risk is not present. Outside of each discomfort zone is a *health risk zone* where there is a significant risk of health disorders due to heat or cold. Industrial hygiene is primarily concerned with controlling worker exposure in the health risk zone.

The analysis of thermal stress involves taking a *heat balance* of the human body with the objective of determining whether the net heat storage is positive, negative, or zero. A simplified form of the heat balance is

$$Q_S = Q_M + Q_R + Q_C + Q_E \qquad 35.9$$

If Q_S is zero, heat gain is balanced by heat loss and the body is in equilibrium. If Q_S is positive, the body is gaining heat and if Q_S is negative, the body is losing heat.

The heat balance is affected by environmental and climatic conditions, work demands, and clothing. The metabolic rate, Q_M, is more significant for heat stress than for cold stress when compared with radiation and convection. The metabolic rate can affect heat gain by one to two orders of magnitude compared to radiation and convection, but it affects heat loss to about the same extent as radiation and convection.

Professional Publications, Inc.

Clothing affects the thermal balance through insulation, permeability, and ventilation. *Insulation* provides resistance to heat flow by radiation, convection, and conduction. *Permeability* affects the movement of water vapor and the amount of evaporative cooling. *Ventilation* influences evaporative and convective cooling.

Heat Stress

Heat stress can increase body temperature, heart rate, and sweating, which together are called *heat strain*.

The most serious heat disorder is *heatstroke*, because it involves a high risk of death or permanent damage. Fortunately, heatstroke is rare. Of lesser severity, *heat exhaustion* is the most commonly observed heat disorder for which treatment is sought. *Dehydration* is usually not noticed or reported, but without restoration of water loss, dehydration leads to heat exhaustion. The symptoms of these key heat stress disorders are as follows.

- heatstroke: chills, restlessness, irritability
- heat exhaustion: fatigue, weakness, blurred vision, dizziness, headache
- dehydration: no early symptoms, fatigue or weakness, dry mouth

Appropriate first aid and medical attention should be sought when any heat stress disorder is recognized.

Control of Heat Stress

Controls that are applicable to any heat stress situation are known as *general controls*. General controls include worker training, heat stress hygiene, and medical monitoring.

Specific controls are controls that are put in place for a particular job. They include engineering controls, administrative controls, and personal protection. *Engineering controls* include reducing the physical work demands to reduce the metabolic heat gain, reducing external heat gain from the air or surfaces, and enhancing external heat loss by increasing sweat evaporation and decreasing air temperature. *Administrative controls* include scheduling the work to allow worker acclimatization to occur, leveling work activity to reduce peak metabolic activity, and sharing or scheduling work so the heat exposure of individual workers is reduced. *Personal protection* entails using systems to circulate air or water through tubes or channels around the body, wearing ice garments, and wearing reflective clothing.

Cold Stress

The body reacts to cold stress by reducing blood circulation to the skin to insulate itself. The body also shivers to increase metabolism. These mechanisms are ineffective against extreme cold stress, so humans react by increasing clothing for more insulation, increasing body activity to increase metabolic heat gain, and finding a warmer location.

There are two main hazards from cold stress: hypothermia and tissue damage. *Hypothermia* depresses the central nervous system, causing sluggishness and slurred speech, and progresses to disorientation and unconsciousness. To avoid hypothermia, the minimum core body temperature must be above 36°C for prolonged exposure and above 35°C for occasional exposure of short duration.

Worker training, cold stress hygiene, and medical surveillance can control cold stress. Engineering controls, administrative controls, and personal protection measures can also be applied to control cold stress.

ERGONOMICS

Ergonomics is the study of human characteristics to determine how a work environment should be designed to make work activities safe and efficient. It includes both physiological and psychological effects on the worker, as well as health and safety and productivity aspects.

Work-Rest Cycles

Excessively heavy work should be broken by frequent short rest periods to reduce cumulative fatigue. The percentage of time a worker should rest can be estimated by the following equation.

$$t_{\text{rest}} = \frac{Q_{M,\text{max}} - Q_M}{Q_{M,\text{rest}} - Q_M} \times 100\% \qquad 35.10$$

Manual Handling of Loads

On many projects, loads must be handled manually because of the unavailability of mechanical equipment. The engineer needs to be aware of the hazards and limitations associated with manual handling of loads. Improper handling is the most common cause of injury and of the most severe injuries in the workplace.

Heavy loads can strain the body, particularly the lower back, due to the weight, size, or a lack of handles. Even light or small objects can cause risk of injury to the body if they are handled in a way that requires strain-inducing stretching, reaching, or lifting.

Cumulative Trauma Disorders

Because *cumulative trauma disorders* (CTD) can occur in almost any work situation, the environmental engineer needs to understand their causes and the appropriate preventative measures. CTDs result from repeated stresses that are not excessive individually but over time cause clinical disorders, injuries, and the inability to perform a job. High repetitiveness, or continuous use of the same body part results in fatigue followed by cumulative muscle strain. These cumulative injuries are often associated with tendons, tendon sheaths, and soft tissue. Moreover, the cumulative injuries can result in damage to nerves and restricted blood flow. CTDs are common in the hand, wrist, forearm, shoulder, neck, and back. Bone and the spinal vertebrae may also be damaged.

The manifestations of CTDs on soft tissues include stretching and straining of muscles, rough or torn tendons, inflammation of tendon sheaths, irritation and inflammation of bursa, and stretched (sprained) ligaments. Nerves can be affected by pressure from tendons or other soft tissue, resulting in loss of muscle control, sensations of numbness, tingling, or pain, and loss of response of nerves that control automatic functions such as body temperature and sweating. Blood vessels may be compressed, resulting in restricted blood flow and impaired control of tissues (muscles) dependent on that blood supply. Vibration, such as from operating vibrating tools, can cause the arteries in the fingers and hands to close down, resulting in numbness, tingling, and eventually loss of sensation and control.

Industrial hygienists have defined *high repetitiveness* as a cycle time of less than 30 s, or more than 50% of a cycle time spent performing the same fundamental motion. If the work activity requires the muscles to remain contracted at about 15% to 20% of their maximum capability, circulation can be restricted, which also contributes to CTDs. Also, severe deviation of the wrists, forearms, and other body parts can contribute to CTDs.

RECOGNITION OF CTDS

This section describes the characteristics of common CTDs and the activities that pose a risk for CTDs.

Carpal Tunnel Syndrome

Carpal tunnel syndrome (CTS) is the best known CTD. The American Industrial Hygiene Association has described CTS as an occupational illness of the hand and arm system. CTS results from rapid, repetitive finger and wrist movements.

The wrist has a "tunnel" created by the carpal bones on the outer side and ligaments, which are firmly attached to the bones, across the inner side. In the *carpal tunnel*, which is roughly oval in shape, are tendons and tendon sheaths of the fingers, several nerves, and arteries. If the wrist is bent up or down or flexed from side to side, the space in the carpal tunnel is reduced. Swelling of the tendons or tendon sheaths can place pressure on the nerves, blood vessels, and tendons. CTS occurs from compression of the nerves and blood vessels. Activities that can lead to this disorder include grinding, sanding, hammering, keyboarding, and assembly work.

Cubital Tunnel Syndrome

This disorder occurs from compression of the nerve in the forearm below the elbow and results in tingling, numbness, or pain in the fingers. Leaning over a workbench and resting the forearm on a hard surface or edge typically causes this disorder.

Epicondylitis

This disorder is also known as "tennis elbow" and "golfer's elbow." It results from irritation of the tendons of the elbow. It is caused by forceful wrist extensions, repeated straightening and bending of the elbow, and impacting throwing motions.

Ganglionitis

Ganglionitis is a swelling of a tendon sheath in the wrist. Examples of activities that can lead to this disorder include grinding, sanding, sawing, cutting, using pliers, and turning screws.

Neck Tension Syndrome

This disorder is characterized by an irritation of the muscles of the neck. It commonly occurs after repeated or sustained overhead work.

Pronator Syndrome

This disorder is compression of a nerve in the forearm. It results from rapid and forceful strenuous flexing of the elbow and wrist. Examples of activities that can lead to this disorder include buffing, grinding, polishing, and sanding.

Tendonitis

This is an inflammation of a tendon where its surface becomes thickened, bumpy, and irregular. Tendon fibers may become frayed or torn. This disorder can result from repetitious, forceful movements; contact with hard surfaces; and vibrations.

Shoulder tendonitis is irritation and swelling of the tendon or bursa of the shoulder. It is caused by continuous elevation of the arm.

Tenosynovitis

This disorder is characterized by a swelling of tendon sheaths that can lead to irritation of the tendon. It is known as *DeQuervain's syndrome* when it affects the thumb. Examples of activities that can lead to this disorder include grinding, polishing, sanding, sawing, cutting, and turning screws.

Trigger finger is a special case of tenosynovitis that results in the tendon of the trigger finger becoming nearly locked so that its forced movement is jerky. It comes from using hand tools with sharp edges pressing into the tissue of the finger or where the tip of the finger is flexed but the middle part is straight.

Thoracic Outlet Syndrome

This disorder is characterized by reduced blood flow to and from the arm due to compression of nerves and blood vessels between the collarbone and the ribs. It results in a numbing of the arm and constrains muscular activities.

Ulnar Artery Aneurysm

This disorder is characterized by a weakening of an artery in the wrist, causing a bubble that presses on the nerve. This often occurs from pounding or pushing with the heel of the hand, as in assembly work.

Ulnar Nerve Entrapment

This disorder involves pressure on a nerve in the wrist. It occurs from prolonged flexing of the wrist and repeated pressure on the palm. Examples of activities that can lead to this disorder include carpentry, brick laying, using pliers, and hammering.

White Finger

This disorder is also known as "dead finger," *Raynaud's syndrome*, or *vibration syndrome*. In this disorder, the finger turns cold and numb, tingles, and sensation and control may be lost. The cause is insufficient blood supply, which causes the finger to turn white. It results from closure of the arteries due to vibrations. Gripping vibrating tools, especially in the cold, is the common cause.

BIOLOGICAL HAZARDS

This section describes biological hazards that the environmental engineer needs to be familiar with in order to recognize when to obtain the advice of an industrial hygienist or biosafety specialist.

Biological Agents

Approximately 200 biological agents are known to produce infectious, allergenic, toxic, and carcinogenic reactions in workers. These agents and their reactions are as follows:

- Microorganisms (viruses, bacteria, fungi) and the toxins they produce cause infection, exposure, or allergic reactions.
- Arthropod (crustaceans, arachnids (spiders, scorpions, mites, and ticks), and insects) bites and stings cause skin inflammation, systemic intoxication, transmission of infectious agents, or allergic reactions.
- Allergens and toxins from higher plants cause dermatitis from skin contact and rhinitis (inflammation of the nasal mucus membranes) or asthma from inhalation.
- Protein allergens from vertebrate animals (urine, feces, hair, saliva, and dander) cause allergic reactions.

Also posing potential biohazards are lower plants other than fungi (e.g., lichens, liverworts, and ferns) and invertebrate animals other than arthropods (e.g., parasites, flatworms, and roundworms).

Microorganisms may be divided into prokaryotes and eukaryotes. *Prokaryotes* are organisms having DNA that is not physically separated from its cytoplasm (cell plasma that does not include the nucleus). They are small, simple, one-celled structures, less than 5 μm in diameter, with a primitive nuclear area consisting of one chromosome. Reproduction is normally by binary fission in which the parent cell divides into two daughter cells. All bacteria, both single-celled and multicellular, are prokaryotes, as are blue-green algae.

Eukaryotes are organisms having a nucleus that is separated from the cytoplasm by a membrane. Eukaryotes are larger cells (greater than 20 μm) than prokaryotes, with a more complex structure, and each cell contains a distinct membrane-bound nucleus with many chromosomes. They may be single-celled or multicellular, reproduction may be asexual or sexual, and complex life cycles may exist. This class of microorganisms includes fungi, algae (except blue-green), and protozoa.

Since prokaryotes and eukaryotes have all of the enzymes and biological elements to produce metabolic energy, they are considered organisms.

In contrast, a *virus* does not contain all of the elements needed to reproduce or sustain itself and must depend on its host for these functions. Viruses are nucleic acid molecules enclosed in a protein coat. A virus is inert outside of a host cell and must invade the host cell and use its enzymes and other elements for the virus's own reproduction. Viruses can infect very small organisms such as bacteria, as well as humans and animals. Viruses are 20 to 300 μm in diameter.

Smaller than the viruses by an order of magnitude are *prions*, small proteinaceous infectious particles. Prions have properties similar to viruses and cause degenerative disease in humans and animals.

Infection

The invasion of the body by pathogenic microorganisms and the reaction of the body to their presence and to the toxins they produce is called an *infection*. Infection may be *endogenous* where microorganisms that are normally present in the body (*indigenous*) at a particular site (such as *E. Coli* in the intestinal tract) reach another site (such as the urinary tract), causing infection there.

Infections from microorganisms not normally found on the body that gain entrance through the environment are called *exogenous infections*. The mechanisms of entry include inhalation, indirect or direct contact, penetration such as mosquito bites, and ingestion. Sometimes infectious agents can be transmitted from coworkers by inhalation, such as with measles or tuberculosis. Workers who do not show signs of the disease but are able to transmit the infection are known as *carriers*. The effect of the infection is dependent upon the virulence of the infectious agent, the route of infection, and the relative immunity of the worker.

The most common routes of exposure to infectious agents are through cuts, punctures, abrasions of the skin, inhalation of aerosols generated by accidents or work practices, contact between mucous membranes or contaminated material, and ingestion. In laboratory and medical settings, transmission of blood-borne pathogens through handling of blood products and human tissue is a serious concern.

Biohazardous Workplaces and Activities

Although engineers have long been concerned with waterborne diseases and their prevention in the design and operation of water supply and wastewater systems, pathogens may also be encountered in the workplace through air or direct contact. Workers engaged in a wide variety of activities may become exposed to one or more potential biohazards.

Microbiology and Public Health Laboratories

Workers in laboratories handling infectious agents have a potential risk of infection from accidental exposure or work practices.

Health Care Facilities

Health care facilities such as hospitals, doctor offices, blood banks, and outpatient clinics, present numerous opportunities for exposure to a wide variety of hazardous and toxic gases, liquids, and solids, as well as to infectious agents.

Biotechnology Facilities

Biotechnology is one of the newest technologies and involves a much greater scope and complexity than the historical use of microorganisms in the chemical and pharmaceutical industries. This technology now deals with DNA manipulation and the development of products for medicine, industry, and agriculture. The microorganisms used by the biotechnology industry often are genetically engineered bacteria, mold and yeast, and plant and animal cells. Allergies from worker exposure can be a major health issue.

Animal Facilities

Workers exposed to animals are at risk for animal-related allergies and infectious agents. Occupations include agricultural workers, veterinarians, workers in zoos and museums, taxidermists, and workers in animal-product processing plants.

Zoonotic diseases (diseases that affect both humans and animals) are the most common diseases reported by laboratory workers. Infection can occur from bites and scratches, puncture with contaminated needles, aerosols from animal respiration or excretions, and contact with infected tissue. Work acquired infections from nonhuman primates have occurred for years.

Some of the diseases of concern in animal facilities include Q fever, hantavirus, Ebola, Marburg viruses, and simian immunodeficiency viruses. An engineer could encounter potential infectious agents while dealing with solid and liquid wastes from animal processing facilities and animal agricultural activities.

Agriculture

Agricultural workers are readily exposed to infectious microorganisms through inhalation of aerosols, contact with broken skin or mucus membranes, and inoculation from injuries. Farmers and horticultural workers may be exposed to fungal diseases. Food and grain handlers may be exposed to parasitic diseases. Workers who process animal products may acquire bacterial skin diseases such as anthrax from contaminated hides, tularemia from skinning infected animals, and erysipelas from contaminated fish, shellfish, meat, or poultry. Infected turkeys, geese, and ducks can expose poultry workers to *psittacosis*, a bacterial infection. Workers handling grain may be exposed to *mycotoxins* from fungi and endotoxins from bacteria.

Utility Workers

Workers maintaining water systems may be exposed to Legionella pneumophila (Legionnaires' disease). Sewage collection and treatment workers may be exposed to enteric bacteria, hepatitis A virus, infectious bacteria, parasitic protozoa (giardia), and allergenic fungi. Solid waste handling and disposal facility workers may be exposed to blood-borne pathogens from infectious wastes.

Wood-Processing Facilities

Wood-processing workers may be exposed to bacterial endotoxins and allergenic fungi.

Mining

Miners may be exposed to zoonotic bacteria, mycobacteria, various fungi, and water and wastewater from mining activities.

Forestry

Forestry workers may be exposed to zoonotic diseases (rabies virus, Russian spring fever virus, Rocky Mountain spotted fever, Lyme disease, and tularemia) transmitted by ticks and fungi.

Blood-Borne Pathogens

The environmental engineer should have a general knowledge of blood-borne pathogens and be able to recognize when expert advice should be obtained on a project involving exposure to infectious wastes.

The risk from hepatitis B and human immunodeficiency virus (HIV) in health care and laboratory situations led OSHA to publish standards for occupational exposure to blood-borne pathogens. Some blood-borne pathogens are summarized as follows.

Human Immunodeficiency Virus (HIV)

HIV is the blood-borne virus that causes acquired immunodeficiency syndrome (AIDS). Contact with infected blood or other body fluids can transmit HIV. Transmission may occur from unprotected sexual intercourse, sharing of infected needles, accidental puncture wounds from contaminated needles or sharp objects, or transfusion with contaminated blood.

Symptoms of HIV include swelling of lymph nodes, pneumonia, intermittent fever, intestinal infections, weight loss, and tuberculosis. Death typically occurs from severe infection causing respiratory failure due to pneumonia.

Hepatitis

The hepatitis virus affects the liver. Symptoms of infection include jaundice, cirrhosis and liver failure, and liver cancer.

Hepatitis A can be contracted through contaminated food or water or by direct contact with blood or body fluids such as blood or saliva. Hepatitis B, known as *serum hepatitis*, may be transmitted through contact with infected blood, body fluids, and through blood transfusions. Hepatitis B is the most significant occupational infector of health care and laboratory workers. Hepatitis C is similar to hepatitis B, but can also be transmitted by shared needles, accidental puncture wounds, through blood transfusions, and unprotected sex.

Hepatitis D occurs when one of the other hepatitis viruses replicates. Individuals with chronic hepatitis D often develop cirrhosis of the liver. Chronic hepatitis may be present in carriers.

Syphilis

The bacterium responsible for the transmission of syphilis is called *treponema pallidum pallidum*. Syphilis is almost always transmitted through sexual contact, though it may be transmitted in utero through the placenta from mother to fetus. This is known as *congenital syphilis*. (Note: Treponema pallidum has 4 subspecies, so the extra "pallidum" indicates the virus that specifically causes syphilis.)

Toxoplasmosis

Toxoplasmosis is caused by a parasitic organism called *Toxoplasma gondii*, which may be transmitted by ingestion of contaminated meat, across the placenta, and through blood transfusions and organ transplants.

Rocky Mountain Spotted Fever

Ticks infected with the pathogen *Rickettsia rickettsii*, pass this disease from pets and other animals to humans. Symptoms and effects include headache, rash, fever, chills, nausea, vomiting, cardiac arrhythmia, and kidney dysfunction. Death may occur from renal failure and shock.

Bacteremia

Bacteremia is the presence of bacteria in the bloodstream, whether associated with active disease or not.

Bacteria- and Virus-Derived Toxins

An engineer needs to be aware of toxins derived from bacteria and viruses. The effect of some of these toxins varies from mild illness to debilitating illness or death.

Botulism

The organism *Clostridium botulinum*, produces the toxin that is responsible for botulism. There are four types of botulism: food-borne, infant, adult enteric (intestinal), and wound. *Food botulism* is associated with poorly preserved foods, and is the most widely recognized form. *Infant botulism* can occur in the second month after birth when the bacteria colonize the intestinal tract and produce the toxin. *Adult enteric botulism* is similar to infant botulism. *Wound botulism* occurs when the spores enter a wound through contaminated soil or needles. The toxin is absorbed in the bloodstream and blocks the release of a neurotransmitter. Severe cases can result in respiratory paralysis and death.

Lyme Disease

Lyme disease is transmitted to humans through bites of ticks infected with *Borrelia burgdorferi*.

Tetanus

Tetanus occurs from infection by the bacterium *Clostridium tetani*, which produces two exotoxins, tetanolysin and tetanospasmin. Routine immunizations will prevent the disease.

Toxic Shock Syndrome

Toxic shock syndrome (TSS) is caused by the bacterium *Staphylococcus aureus*, which produces a pyrogenic toxin.

Ebola (African Hemorrhagic Fever)

The Ebola and Marburg viruses produce an acute hemorrhagic fever in humans. Symptoms include headache, progressive fever, sore throat, and diarrhea.

Hantavirus

The hantavirus is found in rodents and shrews of the southwest and is spread by contact with their excreta.

Tuberculosis

Tuberculosis (TB) is a bacterial disease from *Mycobacterium tuberculosis*. Humans are the primary source of infection. TB affects a third of the world's population outside the United States. A drug-resistant strain is a serious problem worldwide, including in the United States. The risk of contracting active TB is increased among HIV-infected individuals.

Legionnaires' Disease

Legionnaires' Disease (legionellosis) is a type of pneumonia caused by inhaling the bacteria *Legionella pneumophilia*. Symptoms include fever, cough, headache, muscle aches, and abdominal pain. People usually recover in a few weeks and suffer no long-term consequences. *Legionellae* are common in nature and are associated with heat-transfer systems, warm-temperature water, and stagnant water. Sources of exposure include sprays from cooling towers or evaporative condensers and fine mists from showers and humidifiers. Proper design and operation of ventilation, humidification, and water-cooled heat-transfer equipment and other water systems equipment can reduce the risk. Good system maintenance includes regular cleaning and disinfection.

SAMPLE PROBLEMS

Problem 1

A 25 L sample of landfill gas was bubbled through 50 mL of a solution that has an 85% collection efficiency for methane. The sample was analyzed at 25°C and 750 mm Hg pressure. The laboratory reported the result of 25 μg/mL. What is the concentration of methane in the landfill gas in ppm?

Solution

The total mass of methane in the 25 L sample is

$$m = \frac{\left(\begin{array}{c}\text{concentration of}\\\text{sample in solution}\end{array}\right)\left(\begin{array}{c}\text{volume of}\\\text{solution}\end{array}\right)}{\text{collection efficiency}}$$

$$= \frac{\left(25\ \dfrac{\mu g}{mL}\right)(50\ mL)}{0.85}$$

$$= 1470.6\ \mu g$$

Correct the volume of gas per mole for temperature and pressure.

$$T_1 = 0°C + 273 = 273K \qquad p_1 = 760\ mm\ Hg$$
$$T_2 = 25°C + 273 = 298K \qquad p_2 = 750\ mm\ Hg$$

$$\frac{p_1 V_1}{T_1} = \frac{p_2 V_2}{T_2}$$

$$V_2 = V_1 \left(\frac{T_2}{T_1}\right)\left(\frac{p_1}{p_2}\right)$$

$$= \left(22.4\ \frac{L}{mol}\right)\left(\frac{298K}{273K}\right)\left(\frac{760\ mm\ Hg}{750\ mm\ Hg}\right)$$

$$= 24.8\ L/mol$$

The concentration of methane in ppm is

$$C_{ppm} = C_{m/V}\left(\frac{V_2}{MW}\right)$$

$$= \left(\frac{1470.6\ \mu g}{25\ L}\right)\left(\frac{24.8\ \dfrac{L}{mol}}{16\ \dfrac{g}{mol}}\right)$$

$$= 91.2\ \mu g/g \quad (91.2\ ppm)$$

Problem 2

An environmental engineer must prepare some wastewater samples for analysis using isopropyl alcohol (CH_3CH_3CHOH). Vapor levels over an 8 h workday will be 420 ppm (2 h), 350 ppm (2.5 h), and 390 ppm (3.5 h). The temperature of the lab is 30°C because the air conditioner is broken. Will the engineer's exposure level be in compliance if the threshold limit value (TLV) for isopropyl alcohol is 400 ppm at 25°C ?

Solution

At STP (0°C and 1 atm), 1 mol of any gas occupies 22.4 L. Calculate its volume at 25°C and 30°C.

$$T_1 = 0°C + 273 = 273K$$
$$T_2 = 25°C + 273 = 298K$$
$$T_3 = 30°C + 273 = 303K$$

At constant pressure (1 atm),

$$\frac{V_1}{T_1} = \frac{V_2}{T_2} = \frac{V_3}{T_3}$$

$$V_2 = V_1\left(\frac{T_2}{T_1}\right) = \left(22.4 \; \frac{\text{L}}{\text{mol}}\right)\left(\frac{298\text{K}}{273\text{K}}\right)$$

$$= 24.45 \; \text{L/mol}$$

$$V_3 = V_1\left(\frac{T_3}{T_1}\right) = \left(22.4 \; \frac{\text{L}}{\text{mol}}\right)\left(\frac{303\text{K}}{273\text{K}}\right)$$

$$= 24.86 \; \text{L/mol}$$

The molecular weight of isopropyl alcohol is

$$\text{MW} = (3)\left(12 \; \frac{\text{g}}{\text{mol}}\right) + (8)\left(1 \; \frac{\text{g}}{\text{mol}}\right) + (1)\left(16 \; \frac{\text{g}}{\text{mol}}\right)$$

$$= 60 \; \text{g/mol}$$

Calculate the concentration of isopropyl alcohol at 25°C for each exposure period. Note that as the temperature increases the volume increases, but the number of moles remains constant, so the concentration decreases.

$$C_2 = C_3\left(\frac{V_2}{V_3}\right) = (420 \text{ ppm})\left(\frac{24.45 \; \frac{\text{L}}{\text{mol}}}{24.86 \; \frac{\text{L}}{\text{mol}}}\right)$$

$$= 413 \text{ ppm}$$

Similarly,

$$350 \text{ ppm at } 30°\text{C} = 344 \text{ ppm at } 25°\text{C}$$

$$390 \text{ ppm at } 30°\text{C} = 384 \text{ ppm at } 25°\text{C}$$

Calculate TLV-TWA for the 8 h day.

$$\text{TLV-TWA} = \frac{\left(\begin{array}{c}(2 \text{ h})(413 \text{ ppm}) + (2.5 \text{ h})(344 \text{ ppm}) \\ +(3.5 \text{ h})(384 \text{ ppm})\end{array}\right)}{8 \text{ h}}$$

$$= 379 \text{ ppm}$$

The TLV-TWA of 400 ppm will not be exceeded.

FE-STYLE EXAM PROBLEMS

1. A concentration of 2.2 ppm of methane (CH_4) was measured at a landfill at an air temperature of 25°C. What is the concentration in mg/m^3?

(A) 0.14 mg/m^3
(B) 0.17 mg/m^3
(C) 1.4 mg/m^3
(D) 1.7 mg/m^3

Problems 2–4 refer to the following information.

During each shift, workers in a sewage treatment plant are exposed to noise from a sludge pump, a blower, and a ventilation fan as indicated.

source	sound pressure level, dB (reference: 20 μPa)	period of operation (h)
sludge pump	96	0800–1200
		1600–1700
blower	94	0800–1000
		1500–1700
ventilation fan	98	0800–1000
		1300–1500

All noise is at 1000 Hz and referenced to the dBA scale.

2. What is the sound pressure level the workers are exposed to during the period from 0800 to 1000 hours?

(A) 99 dB
(B) 100 dB
(C) 101 dB
(D) 103 dB

3. What is the time-weighted sound pressure level exposure for one shift?

(A) 95.2 dB
(B) 96.4 dB
(C) 97.8 dB
(D) 98.3 dB

4. Relative to OSHA noise exposure standards, which of the following statements is true for this situation?

(A) Standards are exceeded because the TWA exposure exceeds 85 dBA.
(B) Standards are not exceeded because none of the permissible noise levels are exceeded for any duration allowed per day.
(C) Standards are exceeded because the sum of the fractional allowable exposure duration for each sound pressure level exceeds 1.0 (100%).
(D) Standards are not exceeded because the TWA exposure does not exceed the permissible TWA exposures.

5. Which of the following diseases should be of special concern to the environmental engineer when designing, operating, or managing water supply projects?

I.   acquired immunodeficiency syndrome (AIDS)
II.   Rocky Mountain Spotted Fever
III. botulism
IV.   tuberculosis
V.   Legionnaires' disease
VI.   hepatitis A

(A) I, III, and VI
(B) II, IV, and VI
(C) V
(D) V and VI

SOLUTIONS TO FE-STYLE EXAM PROBLEMS

1. At STP (0°C and 1 atm), 1 mol of any gas occupies 22.4 L. Calculate its volume at 25°C.

$$T_1 = 0°C + 273 = 273K$$

$$T_2 = 25°C + 273 = 298K$$

$$V_2 = V_1 \left(\frac{T_2}{T_1} \right)$$
$$= \left(22.4 \ \frac{L}{mol} \right) \left(\frac{298K}{273K} \right)$$
$$= 24.45 \ L/mol$$

The molecular weight of methane is

$$MW = (1) \left(12 \ \frac{g}{mol} \right) + (4) \left(1 \ \frac{g}{mol} \right)$$
$$= 16 \ g/mol$$

Given a methane concentration of 2.2 ppm (2.2×10^{-6} liters of methane per liter of air), calculate the number of moles of methane per unit volume of air.

$$\frac{\text{moles methane}}{\text{volume air}} = \frac{\text{methane concentration}}{\text{methane volume}}$$
$$= \frac{2.2 \times 10^{-6} \ \frac{L}{L}}{24.45 \ \frac{L}{mol}}$$
$$= 9.0 \times 10^{-8} \ mol/L$$

Calculate the concentration in mg/m³ from the number of moles.

$$n = \frac{m}{MW}$$
$$\frac{m}{V_{air}} = \left(\frac{n}{V_{air}} \right) (MW)$$
$$= \left(9.0 \times 10^{-8} \ \frac{mol}{L} \right) \left(16 \ \frac{g}{mol} \right)$$
$$\times \left(10^3 \ \frac{mg}{g} \right) \left(10^3 \ \frac{L}{m^3} \right)$$
$$= 1.44 \ mg/m^3$$

Alternatively, the concentration of methane in mg/m³ can be calculated as

$$C_{m/V} = C_{ppm} \left(\frac{MW}{V_2} \right)$$
$$= \left(2.2 \times 10^{-6} \ \frac{L}{L} \right) \left(\frac{16 \ \frac{g}{mol}}{24.45 \ \frac{L}{mol}} \right)$$
$$\times \left(10^3 \ \frac{mg}{g} \right) \left(10^3 \ \frac{L}{m^3} \right)$$
$$= 1.44 \ mg/m^3$$

Answer is C.

2. Use Eq. 35.6.

$$L_{p,total} = 10 \log \sum_i^3 10^{L_i/10}$$
$$= 10 \log \left(10^{96/10} + 10^{94/10} + 10^{98/10} \right)$$
$$= 101.1 \ dB$$

Answer is C.

3. Calculate TWA exposure using Eq. 35.6.

period beginning hour	sludge pump (dB)	blower (dB)	ventilation fan (dB)	total exposure (dB)
0800	96	94	98	101.1
0900	96	94	98	101.1
1000	96			96
1100	96			96
1300			98	98
1400			98	98
1500		94		94
1600	96	94		98.1
total TWA				97.8

Answer is C.

4.

(A) False; the standards allow 85 dBA to be exceeded for specific durations at various sound pressure levels.

(B) False; the standards allow permissible noise levels to be exceeded for specific durations.

(C) From Eqs. 35.6 and 35.7 and Table 35.2,

total exposure (dB)	exposure duration, C (h)	permissible exposure duration, t (h)	C/t
101.1	2	1.5	2/1.5 = 1.33
96	2	3	2/3 = 0.67
98	2	2	2/2 = 1.00
94	1	4	1/4 = 0.25
98.1	1	2	1/2 = 0.50
Σ	8		3.75

Professional Publications, Inc.

Because the summation of 3.75 is greater than 1.0, the OSHA permissible noise exposure is exceeded. A hearing conservation program is required.

(D) False; the standards are based on an allowable duration at each exposure level, not on the TWA of the permissible exposures.

Answer is C.

5. Items I through IV are not concerns: AIDS is not a water-borne disease. Rocky Mountain spotted fever is passed to humans by ticks. Botulism is food borne or can enter the body through contaminated soil or needles. Tuberculosis is spread by inhalation of infectious droplets.

Legionnaires' disease, on the other hand, is associated with heat transfer systems, warm temperature water, and stagnant water. Hepatitis A can also be spread through contaminated water.

Answer is D.

$\mathit{36}$ Bioprocessing

Subjects

INTRODUCTION 36-1
ENVIRONMENTAL MICROBIOLOGY . . 36-2
    BOD Exertion 36-2
    Monod Kinetics 36-2
Half-Life of a Biologically Degraded
Contaminant, Asssuming a
    First-Order Rate Constant 36-2
ACTIVATED SLUDGE 36-2
FACULTATIVE POND 36-4
    BOD Loading 36-4
BIOTOWER 36-4
    Fixed-Film Equation without Recycle . 36-4
    Fixed-Film Equation with Recycle . . . 36-4
ANAEROBIC DIGESTER 36-4
    Standard Rate 36-4
    High Rate 36-4
AEROBIC DIGESTION 36-4
    Tank Volume 36-4

Nomenclature

A	surface area	m^2
A_{plan}	area of cross-section of a packed bed	m^2
F	fraction of influent BODs consisting of raw primary sewage	–
k	rate constant	d^{-1}
k_d	microbial death ratio, kinetic constant	
K_d	digester reaction rate coefficient	–
L	ultimate BOD (BOD remaining at time $t = \infty$)	mg/L
MLSS	mixed liquor suspended solids	mg/L
n	media characteristic coefficient	–
P	population	persons
P_v	volatile fraction of suspended solids	–
q	hydraulic loading	$m^3/m^2 \cdot min$
Q	volumetric flow rate	m^3/s
R	recycle ratio	–
S	BOD	mg/L
S	concentration	mg/L

S_0	initial BOD ultimate in mixing zone	mg/L
SVI	sludge volume index	mL/L
t	time	s
$t_{1/2}$	half-life	d
V	volume	m^3
V_1	raw sludge input	m^3/d
V_2	digested sludge accumulation	m^3/d
X	concentration	ppb or $\mu g/kg$
Y	yield coefficient	–
y_t	amount of BOD exerted at time t	mg/L

Symbols

θ	hydraulic residence time	d
θ_c	solids (cell) residence time or sludge age	d
μ	specific growth rate	1/s
ρ	density	kg/m^3

Subscripts

A	aeration basin
d	death, digester
e	effective, or effluent
i	influent
max	maximum
o	influent
r	reaction, residence
R	recycle
s	solute, sludge, or storage
t	thicken
T	at time T
v	volatile
w	aqueous phase, water, or waste sludge

INTRODUCTION

The term *bioprocessing* refers to systems that use living organisms, mostly microorganisms, to obtain desired results. One category of systems constitutes the large pharmaceutical industry, which uses bioprocessing to obtain desired products. The generic equations for these processes are provided in the stoichiometric section of Chapter 33. The second category of systems constitutes those that are used in wastewater processing. These commercial systems are used by local and regional government agencies. There is less variety among them than among those used for pharmaceutical purposes. The typical wastewater processing units are described in this chapter.

ENVIRONMENTAL MICROBIOLOGY

Microorganisms play an important role in the biological treatment of wastewater. They are useful for the removal of organic matter, colloids, and nitrogen and phosphorus compounds.

The important microorganisms in wastewater treatment are bacteria. Bacteria are single-celled organisms that survive within a narrow range of pH and temperature. Their size varies from 0.5 μm to 1 μm, and they generally reproduce by fission (division). (Fungi are aerobic organisms, multicellular in nature, that can survive in a low-pH environment. Protozoa are generally unicellular organisms, aerobic in nature, that consume bacteria.)

Bacteria placed in a vessel in a nutrient medium go through five growth phases; Fig. 33.3 shows the progression in graph form.

BOD Exertion

A laboratory analysis of organic matter in water often includes a biochemical oxygen demand (BOD) test. Most water quality laboratories have a ready supply of water that is saturated with oxygen, obtained by sparging air overnight through the water. To measure BOD, a sample of wastewater is diluted with oxygen-saturated water. A small amount of bacteria is fed to the sample. Oxygen concentrations are measured at the beginning of the test and also on a daily basis. The difference between the oxygen concentrations at the initial time and at time t gives a measure of the concentration of organic compounds in the water and is called the BOD exerted at time t. In this way, BOD provides a measure of the concentration of organic compounds in water without the complexity of analyzing the different compounds.

If the biological depletion of organic compounds in wastewater can be assumed to be an approximately first-order reaction and L corresponds to BOD remaining, then

$$-\frac{dL}{dt} = kL \qquad \text{36.1}$$

Integrating between the limits of $(t = 0,\ L = L)$ to $(t = t,\ L = L_t)$ gives

$$\frac{L_t}{L} = e^{-kt} \qquad \text{36.2}$$

The value $(L - L_t)$ is called y_t, the amount of BOD exerted at time t. Therefore,

$$y_t = L(1 - e^{-kt}) \qquad \text{36.3}$$

Monod Kinetics

Mathematically, in a batch reactor, the rate of growth of bacteria is given by

$$\frac{dX}{dt} = \mu X \qquad \text{36.4}$$

The value of μ is given by

$$\mu = \mu_{\max} \frac{S}{K_S + S} \qquad \text{36.5}$$

Half-Life of a Biologically Degraded Contaminant, Assuming a First-Order Rate Constant

The half-life of a reaction is the time taken to reduce the concentration (in this case, the organic compound) to half of its initial amount. The differential equation for a first-order reaction is shown as Eq. 36.1.

Integrating between the limits of $(t = 0,\ L = L)$ to $(t = t_{1/2},\ L = 0.5L)$ gives

$$0.5 = e^{kt_{1/2}} \qquad \text{36.6}$$

Solving for k gives

$$k = \frac{\ln 2}{t_{1/2}} = \frac{0.693}{t_{1/2}} \qquad \text{36.7}$$

ACTIVATED SLUDGE

The activated sludge process in its simplest form, as shown in the Fig. 36.1, consists of a mixing/aeration tank followed by a settler/clarifier. Mathematical analysis of the entire process is done by means of material balances on the different components and also by using the Monod kinetic model.

The results of the analysis for the reactor/aerator give a formula for the biomass concentration in the aerator, as well as the sludge flow rate and the solids residence time. The formulas for the organic loading rates are given. The biomass concentration in the aeration tank is

$$X_A = \frac{\theta_c Y (S_o - S_e)}{\theta (1 + k_d \theta_c)} \qquad \text{36.8}$$

The *yield coefficient* is

$$Y = \frac{\text{mass of biomass}}{\text{mass of BOD consumed}} \qquad \text{36.9}$$

The *hydraulic residence time* is

$$\theta = \frac{V}{Q} \qquad \text{36.10}$$

Figure 36.1 Activated Sludge Process

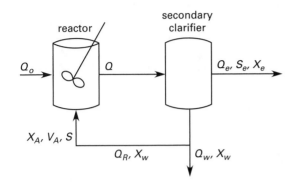

The *solids residence time* is

$$\theta_c = \frac{V_A X_A}{Q_w X_w + Q_e X_e} \qquad 36.11$$

The *sludge volume* per day is

$$Q_s = \frac{100M}{\rho_s(\% \text{ solids})} \qquad 36.12$$

$$\text{Solids loading rate} = \frac{QX}{A} \qquad 36.13$$

For an activated sludge secondary clarifier,

$$Q = Q_o + Q_R \qquad 36.14$$

$$\text{Organic loading rate (volumetric)} = \frac{Q_o S_o}{V} \qquad 36.15$$

$$\text{Organic loading rate (F:M)} = \frac{Q_o S_o}{V_A X_A} \qquad 36.16$$

$$\text{Organic loading rate (surface area)} = \frac{Q_o S_o}{A_M} \qquad 36.17$$

$$\text{SVI} = \frac{\text{sludge volume after settling}_{(\text{mL/L})}(1000)}{\text{MLSS}_{\text{mg/L}}} \qquad 36.18$$

The steady-state mass balance for a secondary clarifier is

$$(Q_o + Q_R)X_A = Q_e X_e + Q_R X_w + Q_w X_w \qquad 36.19$$

The *recycle ratio* is

$$R = \frac{Q_R}{Q_o} \qquad 36.20$$

The *recycle flow rate* is

$$Q_R = Q_o R \qquad 36.21$$

Design and operational parameters for activated-sludge treatment of municipal wastewater are given in Table 36.1.

FACULTATIVE POND

Facultative ponds, where the waste is kept in an open pond, are also called stabilization ponds. In these ponds, aerobic bacteria degrade the top layer, anaerobic bacteria degrade the bottom layer, and both aerobic and anaerobic bacteria degrade the middle layer.

Table 36.1 Design and Operational Parameters for Activated-Sludge Treatment of Municipal Wastewater

type of process	mean-cell residence time, θ_c (d)	food-to mass ratio (kg BOD$_5$/kg MLSS)	volumetric loading (V_L, kg BOD$_5$/m^3	hydraulic residence time in aeration basin, θ (h)	mixed liquor suspended solids, MLSS (mg/L)	recycle ratio (Q_r/Q)	flow regime*	BOD$_5$ removal efficiency (%)	air supplied m^3/kg BOD$_5$)
tapered aeration	5–15	0.2–0.4	0.3–0.6	4–8	1500–3000	0.25–0.5	PF	85–95	45–90
conventional	4–15	0.2–0.4	0.3–0.6	4–8	1500–3000	0.25–0.5	PF	85–95	45–90
step aeration	4–15	0.2–0.4	0.6–1.0	3–5	2000–3500	0.25–0.75	PF	85–95	45–90
completely mixed	4–15	0.2–0.4	0.8–2.0	3–5	3000–6000	0.25–1.0	CM	85–95	45–90
contact stabilization	4–15	0.2–0.6	1.0–1.2	–	–	0.25–1.0	–	–	45–90
contact basin	–	–	–	0.5–1.0	1000–3000	–	PF	80–90	–
stabilization basin	–	–	–	4–6	4000–10 000	–	PF	–	–
high-rate aeration	4–15	0.4–1.5	1.6–16	0.5–2.0	4000–10 000	1.0–5.0	CM	75–90	25–45
pure oxygen	8–20	0.2–1.0	1.6–4	1–3	6000–8000	0.25–0.5	CM	85–95	–
extended aeration	20–30	0.05–0.15	0.16–0.40	18–24	3000–6000	0.75–1.50	CM	75–90	90–125

BOD Loading

$$\text{mass} = \text{flow} \times \text{concentration} \qquad 36.22$$

A typical facultative pond system can process 35 lbm or less of $BOD_5/ac/d$. A system contains a minimum of three ponds, which are 3 ft to 8 ft deep. The minimum residence time is 90–120 days.

BIOTOWER

Also called a trickling filter, a biotower operates by having the wastewater fall through a packed bed or tower filled with permeable packing. The packing has both aerobic and anaerobic microorganisms attached to it. Material balance equations can be formulated using the flux from both the connective and diffusive processes. The rate of reaction is formulated by using the Monod kinetic model. The biotowers can be operated either with or without recycling.

Fixed-Film Equation without Recycle

$$\frac{S_e}{S_o} = e^{-kD/q^n} \qquad 36.23$$

Fixed-Film Equation with Recycle

$$\frac{S_e}{S_a} = \frac{e^{-kD/q^n}}{(1+R) - R(e^{-kD/q^n})} \qquad 36.24$$

$$S_a = \frac{S_o + RS_e}{1 + R} \qquad 36.25$$

Hydraulic loading with recycle is

$$q = \frac{Q_o + RQ_o}{A_{\text{plan}}} \qquad 36.26$$

The *treatability constant*, k, is given by Eq. 36.27, where the temperature, T, is in °C.

$$k_T = k_{20}(1.035)^{T-20} \qquad 36.27$$

As with activated sludge, the recycle ratio is

$$R = \frac{Q_o}{Q_R} \qquad 36.28$$

ANAEROBIC DIGESTER

The sludge from the primary settlers and the biological treatment processes can be treated to obtain methane, carbon dioxide, and other products. In the standard-rate digester, the digester is unmixed and not externally heated; in the high-rate digester, there is external heating and stirring of the contents.

Standard Rate

A standard rate digester must be sized to accommodate the raw sludge input, V_1, and the digested sludge accumulation, V_2, for the time for the sludge to digest and thicken (i.e., for the *residence time*, t_r) as well as to hold the accumulation for the period it is stored, t_s.

$$\text{Reactor volume} = \left(\frac{V_1 + V_2}{2}\right) t_r + V_2 t_s \qquad 36.29$$

High Rate

The first-stage reactor volume is selected to hold the raw sludge for as long as it takes for digestion to occur.

$$\text{Reactor volume} = V_1 t_r \qquad 36.30$$

The second-stage reactor volume is selected based on the time it takes for thickening to occur, t_t.

$$\text{Reactor volume} = \left(\frac{V_1 + V_2}{2}\right) t_t + V_2 t_s \qquad 36.31$$

AEROBIC DIGESTION

The aerobic digestion process is similar to the activated sludge process. Aeration is accomplished by means of diffusing equipment. The process can be either batch or continuous.

For a continuous process, the tank volume is

$$V = \frac{Q_i(X_i + FS_i)}{X_d \left(K_d P_v + \dfrac{1}{\theta_c}\right)} \qquad 36.32$$

SAMPLE PROBLEMS

Problem 1

Sludge is digested after being removed from a secondary treatment tank. During the digestion process, the gas that is produced is

    (A) carbon dioxide
    (B) methane
    (C) nitrogen oxide
    (D) hydrogen sulfide

Solution

The gas produced is primarily methane.

Answer is B.

Problem 2

Because of intermolecular forces, every kilogram of solids that enters a petroleum refinery's wastewater treatment system results in 10 kg of oily sludge. Oil that enters the wastewater treatment system and that is not entrained in oily sludge is recovered and processed into saleable product. After recovery and reprocessing, the recovered oil has an average value of $0.20/L and an average density of 0.8 g/mL. For this refinery, the sludge is composed of 10% solids, 40% oil, and 50% water (by mass). It is suggested that 50 kg/d of solids be prevented from entering the wastewater treatment system by sweeping refinery areas and instituting erosion control measures. If these measures were taken, how much would sales increase due to the increased recovery of oil solids?

(A) $4000/yr
(B) $12,000/yr
(C) $18,000/yr
(D) $27,000/yr

Solution

For every kilogram of solids that is prevented from going to wastewater treatment, 4 kg of oil are allowed to be recovered.

$$\text{sales} = \left(50 \; \frac{\text{kg solids}}{\text{d}}\right)\left(4 \; \frac{\text{kg oil}}{\text{kg solids}}\right)\left(0.20 \; \frac{\$}{\text{L oil}}\right)$$
$$\times \left(\frac{1 \text{ mL}}{0.8 \text{ g}}\right)\left(\frac{1 \text{ L}}{1000 \text{ mL}}\right)\left(365 \; \frac{\text{d}}{\text{yr}}\right)$$
$$\times \left(1000 \; \frac{\text{g}}{\text{kg}}\right)$$
$$= \$18,250 \quad (\$18\,000/\text{yr})$$

Answer is C.

FE-STYLE EXAM PROBLEMS

1. An aerobic digester has an inflow of 2.8 m^3/d. The influent 5 day BOD is 200 mg/L, while the influent suspended solids is 5 mg/L. The digester suspended solids concentration is 10 mg/L. The reaction rate constant is 0.50/d. The volatile fraction is 0.1. The solids retention time is 2 d. The fraction of influent 5 day BOD consisting of raw sewage is 0.5. Under these conditions, the volume of the aerobic digester in cubic meters is most nearly

(A) 5.6 m^3
(B) 28 m^3
(C) 34 m^3
(D) 53 m^3

2. A sample of wastewater is diluted by a factor of 10. The diluted wastewater has an initial dissolved oxygen concentration of 7.0 mg/L. After 5 days the dissolved oxygen concentration is 3.0 mg/L. The 5 day BOD of the initial undiluted wastewater is most nearly

(A) 3 mg/L
(B) 4 mg/L
(C) 7 mg/L
(D) 40 mg/L

3. A sample of wastewater has a kinetic rate constant of 0.1/d. The initial dissolved oxygen concentration is 8.0 mg/L. The concentration after 2 days without any additional oxygen being added is 6.0 mg/L. The ultimate BOD is most nearly

(A) 2.0 mg/L
(B) 9.0 mg/L
(C) 11 mg/L
(D) 21 mg/L

4. For a sample of wastewater, the BOD exerted in 5 days is 100 mg/L, and the reaction rate constant is 0.1 d^{-1}. The value of the ultimate BOD is most nearly

(A) 100 mg/L
(B) 200 mg/L
(C) 250 mg/L
(D) 280 mg/L

5. The half-life of a biologically degraded contaminant is 1 h. The kinetic constant of the degradation reaction is most nearly

(A) 17/d
(B) 24/d
(C) 73/d
(D) 96/d

6. The overall nitrification reaction when the ammonia ion is released to water is approximated by

$$\text{NH}_4^+ + 1.731\text{O}_2 + 1.962\text{HCO}_3^-$$
$$\rightarrow 0.038\text{C}_5\text{H}_7\text{NO}_2 + 0.962\text{NO}_3^-$$
$$+ 1.077\text{H}_2\text{O} + 1.769\text{H}_2\text{CO}_3$$

In this equation, $\text{C}_5\text{H}_7\text{NO}_2$ represents bacterial cells. If this is the sole removal process when 30 kg of ammonia is released from a livestock operation to a stream, the amount of oxygen consumed will most nearly be

(A) 2.9 kg O$_2$
(B) 52 kg O$_2$
(C) 76 kg O$_2$
(D) 92 kg O$_2$

SOLUTIONS TO FE-STYLE EXAM PROBLEMS

1. Use aerobic digester tank volume given by the formula in the NCEES Handbook.

$$
\begin{aligned}
V &= \frac{Q_i(X_i + FS)}{X_d\left(K_d P_v + \dfrac{1}{\theta_c}\right)} \\[2mm]
&= \frac{\left(2.8\ \dfrac{m^3}{d}\right)\left(5\ \dfrac{mg}{L} + (0.5)\left(200\ \dfrac{mg}{L}\right)\right)}{\left(10\ \dfrac{mg}{L}\right)\left(\left(0.5\ \dfrac{1}{d}\right)(0.1) + \dfrac{1}{2\ d}\right)} \\[2mm]
&= 53.45\ m^3 \quad (53\ m^3)
\end{aligned}
$$

Answer is D.

2. The 5 d BOD of the diluted wastewater is

$$
\begin{aligned}
\text{initial BOD} - \text{BOD}_5 &= 7.0\ \frac{mg}{L} - 3.0\ \frac{mg}{L} \\
&= 4\ mg/L
\end{aligned}
$$

Since the sample is diluted by a factor of 10, the original wastewater has a BOD_5 of

$$
\begin{aligned}
\text{BOD}_5 &= (\text{dilution factor})(\text{BOD of diluted water}) \\
&= (10)\left(4\ \frac{mg}{L}\right) \\
&= 40\ mg/L
\end{aligned}
$$

Answer is D.

3. The dissolved oxygen demand is

$$
y_t = \text{DO}_o - \text{DO}_t = L(1 - e^{-kt})
$$

Therefore,

$$
\begin{aligned}
y_{2\ d} &= \text{DO}_o - \text{DO}_{2\ d} \\
&= 8\ \frac{mg}{L} - 6\ \frac{mg}{L} \\
&= 2\ mg/L \\
2\ \frac{mg}{L} &= L\left(1 - e^{-\left(0.1\ \frac{1}{d}\right)(2\ d)}\right)
\end{aligned}
$$

Solve for L.

$$
\begin{aligned}
L &= \frac{2\ \dfrac{mg}{L}}{1 - e^{-0.2}} \\
&= 11.03\ mg/L \quad (11\ mg/L)
\end{aligned}
$$

Answer is C.

4. The dissolved oxygen demand is

$$
y_t = L\left(1 - e^{-kt}\right)
$$

Solve for the ultimate BOD, L.

$$
\begin{aligned}
L &= \frac{y_t}{1 - e^{-kt}} \\
&= \frac{100\ \dfrac{mg}{L}}{1 - e^{-\left(0.1\ \frac{1}{d}\right)(5\ d)}} \\
&= 254\ mg/L \quad (250\ mg/L)
\end{aligned}
$$

Answer is C.

5. The half-life is equal to

$$
\tau = \frac{1\ h}{24\ \dfrac{h}{d}} = \frac{1}{24}\ d
$$

The biological reaction is a Michelis-Menten type reaction which can be modeled as a first order reaction. For a first order reaction, the kinetic constant

$$
\begin{aligned}
k &= \frac{0.693}{\tau} = \frac{0.693}{\dfrac{1}{24}\ d} \\
&= 16.6/\text{day} \quad (17/\text{day})
\end{aligned}
$$

Answer is A.

6. The molecular weight of the ammonium ion, NH_4^+, is

$$
(1)\left(14\ \frac{g}{mol}\right) + (4)\left(1\ \frac{g}{mol}\right) = 18\ g/mol
$$

The molecular weight of oxygen is

$$
(2)\left(16\ \frac{g}{mol}\right) = 32\ g/mol
$$

From the chemical equation, for every mole of the ammonium ion, 1.731 moles of oxygen are consumed. Thus, the amount of oxygen consumed when 30 kg of the ammonium ion are released is

$$
\begin{aligned}
(30\ kg)&\left(\frac{1\ mol\ NH_4}{18\ g\ NH_4}\right)\left(1.731\ \frac{mol\ O_2}{mol\ NH_4}\right) \\
&\times \left(32\ \frac{g\ O_2}{mol\ O_2}\right) \\
&= 92.3\ kg\ O_2 \quad (92\ kg\ O_2)
\end{aligned}
$$

Answer is D.

Topic XII: Chemistry

Diagnostic Examination for Chemistry

Chapter 37   Atoms, Elements, and Compounds

Chapter 38   Chemical Reactions

Chapter 39   Solutions

CHEMISTRY

Hint: For the most current information about the exam, visit www.ppi2pass.com/fefaqs.html regularly.
Use the Exam Forum to compare notes with other FE examinees.

Diagnostic Examination

TOPIC XII: CHEMISTRY

TIME LIMIT: 45 MINUTES

1. What is a distinguishing characteristic of the halogens?

- (A) They are phosphorescent.
- (B) Next to the noble gases, they are the most chemically inactive group.
- (C) They readily accept an electron from another atom to form compounds.
- (D) They have a high electrical conductivity.

2. What is the valence (oxidation state) of carbon in sodium carbonate (Na_2CO_3)?

- (A) -4
- (B) -2
- (C) $+2$
- (D) $+4$

3. What mass of lead nitrate, $Pb(NO_3)_2$, must be dissolved in 1 L of water to produce a solution that contains 20 mg of lead ions? Assume 100% ionization.

- (A) 26 mg
- (B) 32 mg
- (C) 43 mg
- (D) 52 mg

4. What is the mass of 0.01 gram-moles of Na_2SO_4?

- (A) 0.71 g
- (B) 1.19 g
- (C) 1.42 g
- (D) 2.38 g

5. In an experiment, a compound was determined to contain 68.94% oxygen and 31.06% of an unknown element by weight. The molecular weight of this compound is 69.7 g/mol. What is this compound?

- (A) NO_2
- (B) F_2O_2
- (C) B_2O_3
- (D) SiO_4

6. Which of the following statements pertaining to acids and bases is incorrect?

- (A) Acids conduct electricity in aqueous solutions.
- (B) Acids turn blue litmus paper red.
- (C) Bases have a pH between 7 and 14.
- (D) Bases have a sour taste.

7. The reaction shown proceeds in a gaseous state. At equilibrium, the concentration of the components X, Y, and Z are measured to be 5.73×10^{-2} mol/L, 2.67×10^{-2} mol/L and 4.59×10^{-2} mol/L, respectively. What is the equilibrium constant for this reaction?

$$2X \rightleftharpoons Y + 2Z$$

- (A) 9.8×10^{-4} mol/L
- (B) 1.7×10^{-2} mol/L
- (C) 2.1×10^{-2} mol/L
- (D) 3.7×10^{-1} mol/L

8. Oxygen reacts stoichiometrically with methane to form 14 g of carbon monoxide. How many moles of methane are consumed?

$$2CH_4 + 3O_2 \longrightarrow 2CO + 4H_2O$$

- (A) 0.5 mol
- (B) 1 mol
- (C) 1.5 mol
- (D) 2 mol

9. What are the coefficients a, b, and c that are necessary to balance the following reaction?

$$a Pb(NO_3)_2 \longrightarrow b PbO + c NO_2 + \left(\frac{5}{7}\right) O_2$$

- (A) $\dfrac{5}{7}, \dfrac{5}{7}, \dfrac{5}{7}$
- (B) $\dfrac{10}{7}, \dfrac{10}{7}, \dfrac{20}{7}$
- (C) $\dfrac{10}{7}, \dfrac{15}{7}, \dfrac{20}{7}$
- (D) 1, 1, 2

10. The reaction shown occurs in a gaseous phase. Once equilibrium has been achieved in a particular reaction vessel, additional HI gas is injected directly into the reaction vessel. Compared to the initial conditions, which of the following statements is correct after the new equilibrium has been achieved?

$$H_2 + I_2 \rightleftharpoons 2HI$$

(A) The amount of H_2 will have decreased.
(B) The partial pressure of H_2 will have decreased.
(C) The amount of I_2 will have increased.
(D) The partial pressure of HI will have decreased.

11. What is the molarity of a solution obtained by dissolving 25 g of NaCl in enough water to produce 4 L of solution?

(A) 0.107 mol/L
(B) 0.365 mol/L
(C) 0.428 mol/L
(D) 6.25 mol/L

12. How much water must be added to 100 mL of a 0.75 molar solution of KCl to make a 0.04 molar solution?

(A) 0.188 L
(B) 1.78 L
(C) 1.88 L
(D) 1.98 L

Problems 13 and 14 refer to the following situation.

A current I passes through a solution of pure water. The electrolysis of water proceeds according to the reaction shown.

$$H_2O \longrightarrow H_2 + \left(\frac{1}{2}\right) O_2 \; \Delta H_f = 285.830 \text{ kJ/mol}$$

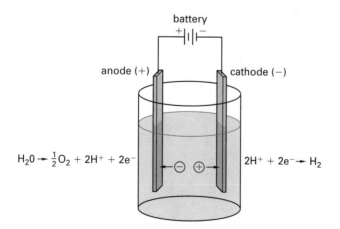

13. If the current, I, is 100 A, at what rate is oxygen produced?

(A) 8.29 mg/s
(B) 9.34 mg/s
(C) 16.7 mg/s
(D) 18.7 mg/s

14. Assuming all of the energy goes into the reaction, what electrical power is required to produce oxygen gas at a rate of 50 mg/s?

(A) 0.89 kW
(B) 1.5 kW
(C) 3.1 kW
(D) 9.2 kW

15. The decay of U-238 to Pb-206 can be used to estimate the age of inorganic matter. The half-life of U-238 is 4.5×10^9 years. In a particular rock sample, the ratio of the numbers of Pb-206 to U-238 atoms is 0.66. Assume all of the Pb-206 present is due to the decay of U-238. What is the age of the rock?

(A) 1.4×10^9 yr
(B) 3.3×10^9 yr
(C) 7.0×10^9 yr
(D) 9.3×10^9 yr

SOLUTIONS TO DIAGNOSTIC EXAMINATION TOPIC XI

1. The halogens need one electron to complete an electron shell. They readily accept this electron from other atoms to form compounds.

Answer is C.

2. Since sodium carbonate is a neutral compound, the sum of the oxidation numbers is zero. Oxygen has a valence of -2, and sodium has a valence of $+1$. The valence of carbon can be calculated as
$$(2)(+1) + x + (3)(-2) = 0$$
$$x - 4 = 0$$
$$x = +4$$

Answer is D.

3. The combining weights of each element in the molecule are
$$\text{Pb: } (1)\left(207.2 \; \frac{\text{g}}{\text{mol}}\right) = 207.2 \text{ g/mol}$$
$$\text{N: } (2)\left(14.01 \; \frac{\text{g}}{\text{mol}}\right) = 28.02 \text{ g/mol}$$
$$\text{O: } (6)\left(16.00 \; \frac{\text{g}}{\text{mol}}\right) = 96.00 \text{ g/mol}$$

The molecular weight of $Pb(NO_3)_2$ is

$$207.2 \; \frac{g}{mol} + 28.02 \; \frac{g}{mol} + 96.00 \; \frac{g}{mol} = 331.2 \; g/mol$$

Calculate the gravimetric fraction of Pb in $Pb(NO_3)_2$.

$$x_{Pb} = \frac{m_{Pb}}{m_{Pb(NO_3)_2}} = \frac{207.2 \; \frac{g}{mol}}{331.2 \; \frac{g}{mol}}$$

$$= 0.6256$$

The mass of $Pb(NO_3)_2$ required is

$$m_{Pb(NO_3)_2} = \frac{20 \; mg}{0.6256}$$

$$= 31.97 \; mg \quad (32 \; mg)$$

Answer is B.

4. The combining weights of each element are

$$Na: (2) \left(22.99 \; \frac{g}{mol} \right) = 45.98 \; g/mol$$

$$S: (1) \left(32.07 \; \frac{g}{mol} \right) = 32.07 \; g/mol$$

$$O: (4) \left(16.00 \; \frac{g}{mol} \right) = 64.00 \; g/mol$$

The molecular weight of Na_2SO_4 is

$$45.98 \; \frac{g}{mol} + 32.07 \; \frac{g}{mol} + 64.00 \; \frac{g}{mol} = 142.1 \; g/mol$$

The mass of Na_2SO_4 is

$$(0.01 \; mol) \left(142.1 \; \frac{g}{mol} \right) = 1.421 \; g \quad (1.42 \; g)$$

Answer is C.

5. This problem contains three unknowns: the number of oxygen atoms present (x), the unknown element's molecular weight (MW), and the number of atoms of the unknown element present (z).

The gravimetric fraction of oxygen is

$$\frac{x(MW)_O}{x(MW)_O + z(MW)} = 0.6894$$

The molecular weight of the compound is

$$x(MW)_O + z(MW) = 69.70 \; g/mol$$

There are three unknowns and only two equations. The number of equations is insufficient to solve for the three unknowns directly. However, the combination $z(MW)$

is the total mass of the unknown element in the compound, and the product $z(MW)$ can be determined. Solve the two equations for x and the product $z(MW)$.

$$\frac{x \left(16 \; \frac{g}{mol} \right)}{x \left(16 \; \frac{g}{mol} \right) + z(MW)} = 0.6894$$

$$\left(16 \; \frac{g}{mol} \right) x = \left(11.03 \; \frac{g}{mol} \right) x + 0.6894 z(MW)$$

$$\left(4.970 \; \frac{g}{mol} \right) x = 0.6894 z(MW)$$

$$x = \left(0.1387 \; \frac{mol}{g} \right) z(MW)$$

$$x \left(16 \; \frac{g}{mol} \right) + z(MW) = 69.70 \; g/mol$$

$$(0.1387 z(MW)) \left(16 \; \frac{g}{mol} \right)$$
$$+ z(MW) = 69.70 \; g/mol$$

$$3.22 z(MW) = 69.70 \; g/mol$$

$$z(MW) = 21.65 \; g/mol$$

$$x = \left(0.1387 \; \frac{mol}{g} \right) \left(21.65 \; \frac{g}{mol} \right)$$

$$= 3$$

Solve for z and MW by deductive reasoning. z must have (low) integer values (1, 2, 3, etc.). The atomic weight of the unknown element must be an integer divisor of 21.65 g/mol. Checking the periodic table, no element has a molecular weight of 21.65 g/mol. The element cannot, therefore, exist monatomically in the compound. Checking the first couple of quotients,

$$\frac{21.65 \; \frac{g}{mol}}{2} = 10.83 \; g/mol$$

$$\frac{21.65 \; \frac{g}{mol}}{3} = 7.22 \; g/mol$$

$$\frac{21.65 \; \frac{g}{mol}}{4} = 5.41 \; g/mol$$

Only the first quotient is reasonably close to the molecular weight of an element (boron): $MW_B = 10.81 \; g/mol$. A probable molecular formula for the compound is B_2O_3.

Check the valence of the components. Oxygen has a valence of -2, and boron has a valence of $+3$. These valences sum to zero in the compound, $(2)(+3) + (3)(-2) = 0$. This is a valid compound.

Answer is C.

6. Bases have a bitter taste, unlike acids which have a sour taste.

Answer is D.

7. The equilibrium constant for this reaction is

$$K_{eq} = \frac{[Y][Z]^2}{[X]^2}$$

$$= \frac{\left(2.67 \times 10^{-2}\ \frac{mol}{L}\right)\left(4.59 \times 10^{-2}\ \frac{mol}{L}\right)^2}{\left(5.73 \times 10^{-2}\ \frac{mol}{L}\right)^2}$$

$$= 1.71 \times 10^{-2}\ mol/L \quad (1.7 \times 10^{-2}\ mol/L)$$

Answer is B.

8. The molecular weight of carbon monoxide is

$$(MW)_{CO} = 12.01\ \frac{g}{mol} + 16.00\ \frac{g}{mol}$$

$$= 28.01\ g/mol$$

The number of moles of carbon monoxide produced is

$$\begin{array}{l} no.\ of\ moles\ CO \\ produced \end{array} = \frac{m_{CO}}{(MW)_{CO}}$$

$$= \frac{14\ g}{28.01\ \frac{g}{mol}}$$

$$= 0.5\ mol$$

The coefficients of the reactants and products can be interpreted as the number of moles. From the balanced reaction, each mole of methane consumed produces one mole of carbon monoxide.

Answer is A.

9. This equation can be balanced either by trial and error or by developing a series of equations to solve for the unknowns.

Write equations to balance each element.

$$O:\ 6a = b + 2c + \left(\frac{5}{7}\right)(2)$$

$$Pb:\ a = b$$

$$N:\ 2a = c$$

Solve these equations simultaneously.

$$6a = a + (2)(2a) + \left(\frac{5}{7}\right)(2)$$

$$a = 10/7$$

$$b = 10/7$$

$$c = 20/7$$

Answer is B.

10. According to Le Chatelier's principle, when a stress is applied to a system in equilibrium, the equilibrium shifts in such a way that tends to relieve the stress.

When the amount of HI is increased, the system will respond by decreasing it. This requires that the reverse reaction occurs, increasing the amount of H_2 and I_2 produced.

Two volumes of reactants (one each of hydrogen and iodine) produce two volumes of products; the new total pressure is not reduced by the reduction in HI. Therefore, all partial pressures increase from the original condition.

Answer is C.

11. The molecular weight of NaCl is

$$(MW)_{NaCl} = (MW)_{Na} + (MW)_{Cl}$$

$$= 22.99\ \frac{g}{mol} + 35.45\ \frac{g}{mol}$$

$$= 58.44\ g/mol$$

The number of moles of NaCl is

$$n_{NaCl} = \frac{m_{NaCl}}{(MW)_{NaCl}} = \frac{25\ g}{58.44\ \frac{g}{mol}}$$

$$= 0.4278\ mol$$

The molarity is the number of moles divided by the volume of the solution.

$$M = \frac{n}{V} = \frac{0.4278\ mol}{4\ L}$$

$$= 0.107\ mol/L$$

Answer is A.

12. Calculate the number of moles of solute present.

$$M = \frac{n}{V}$$

$$0.75 \, \frac{\text{mol}}{\text{L}} = \frac{n}{0.1 \, \text{L}}$$

$$n = \left(0.75 \, \frac{\text{mol}}{\text{L}}\right)(0.1 \, \text{L})$$

$$= 0.075 \, \text{mol}$$

Calculate the volume required for a 0.04 molar solution.

$$M = \frac{n}{V}$$

$$0.04 \, \frac{\text{mol}}{\text{L}} = \frac{0.075 \, \text{mol}}{V}$$

$$V = \frac{0.075 \, \text{mol}}{0.04 \, \frac{\text{mol}}{\text{L}}}$$

$$= 1.875 \, \text{L}$$

The amount of water that must be added is 1.875 L − 0.1 L = 1.775 L (1.78 L).

Answer is B.

13. The reaction that occurs at the cathode is

$$O^{-2} - 2e^- \longrightarrow \frac{1}{2}O_2$$

The molecular weight of oxygen gas is

$$(\text{MW})_{O_2} = (2)\left(16 \, \frac{\text{g}}{\text{mol}}\right) = 32 \, \text{g/mol}$$

From Faraday's law, one gram equivalent weight of water is dissociated at the anode for each faraday, or 96 485 C, of electricity passed through the solution. Four electrons are gained for each oxygen (O_2) molecule.

$$m_{\text{grams}} = \frac{It(\text{MW})}{(96\,485)(\text{change in oxidation state})}$$

$$\text{rate} = \frac{m_{\text{grams}}}{t}$$

$$= \frac{I(\text{MW})}{(96\,485)(\text{change in oxidation state})}$$

$$= \frac{(100 \, \text{A})\left(32 \, \frac{\text{g}}{\text{mol}}\right)}{\left(96\,485 \, \frac{\text{A·s}}{\text{mol}}\right)(4)}$$

$$= 8.29 \times 10^{-3} \, \text{g/s} \quad (8.29 \, \text{mg/s})$$

Answer is A.

14. The number of moles of oxygen gas (O_2) produced is

$$n_{O_2} = \frac{m}{\text{MW}} = \frac{50 \times 10^{-3} \, \frac{\text{g}}{\text{s}}}{32 \, \frac{\text{g}}{\text{mol}}}$$

$$= 0.0015625 \, \text{mol/s}$$

The coefficients in the reaction equation can be interpreted as the number of moles. Each mole of oxygen gas produced requires twice as many moles of water. Therefore, the number of moles of water dissociated per second is

$$n_{H_2O} = (2)\left(0.0015625 \, \frac{\text{mol}}{\text{s}}\right) = 0.003125 \, \text{mol/s}$$

The power required can be calculated from the enthalpy of reaction.

$$P = (\text{mole rate of reaction})\Delta H_f$$

$$= \left(0.003125 \, \frac{\text{mol}}{\text{s}}\right)\left(285.83 \, \frac{\text{kJ}}{\text{mol}}\right)\left(1000 \, \frac{\text{J}}{\text{kJ}}\right)$$

$$= 893.2 \, \text{W} \quad (0.89 \, \text{kW})$$

Answer is A.

15. Let N_o = number of original U-238 atoms, n = number of Pb-206 atoms, and N = number of remaining U-238 atoms.

$$\frac{n}{N} = \frac{N_o - N}{N} = 0.66$$

$$\frac{N_o}{N} - 1 = 0.66$$

$$\frac{N_o}{N} = 1.66$$

$$\frac{N}{N_o} = \frac{1}{1.66} = 0.602$$

$$N = N_o e^{-0.693t/\tau}$$

$$\frac{N}{N_o} = e^{-0.693t/\tau}$$

The fraction of remaining U-238 is 0.602.

$$0.602 = e^{-0.693t/4.5 \times 10^9 \, \text{yr}}$$

$$\ln(0.602) = \frac{-0.693t}{4.5 \times 10^9 \, \text{yr}}$$

$$t = 3.3 \times 10^9 \, \text{yr}$$

Answer is B.

37 Atoms, Elements, and Compounds

Subjects

ATOMIC STRUCTURE 37-1
THE PERIODIC TABLE 37-1
IONS AND ELECTRON AFFINITY 37-3
IONIC AND COVALENT BONDS 37-3
OXIDATION NUMBER 37-4
COMPOUNDS 37-4
MOLES 37-5
FORMULA AND MOLECULAR WEIGHT;
   EQUIVALENT WEIGHT 37-6
GRAVIMETRIC FRACTION 37-6
EMPIRICAL FORMULA DEVELOPMENT 37-6
RADIOACTIVE DECAY AND
   HALF-LIFE 37-6

Nomenclature

A	atomic weight	lbm/lbmol	kg/kmol
EW	equivalent weight	lbm/lbmol	kg/kmol
FW	formula weight	lbm/lbmol	kg/kmol
m	mass of element	lbm	kg
MW	molecular weight	lbm/lbmol	kg/kmol
N	number	–	–
N_A	Avogadro's number	–	L/mol
x	gravimetric fraction	–	–
Z	atomic number	–	–

Subscripts

i	element i
t	total

ATOMIC STRUCTURE

An *atom* is the smallest subdivision of an element that can take part in a chemical reaction. The atomic nucleus consists of neutrons and protons, which are both also known as *nucleons*. Protons have a positive charge and neutrons have no charge, but the masses of neutrons and protons are essentially the same, one *atomic mass unit* (amu). One amu is exactly 1/12 of the mass of an atom of carbon-12, approximately equal to 1.66×10^{-27} kg. The relative atomic weight or *atomic weight*, A, of an atom is approximately equal to the number of protons and neutrons in the nucleus. The *atomic number*, Z, of an atom is equal to the number of protons in the nucleus.

The atomic number determines the way an atom behaves chemically; thus, all atoms with the same atomic number are classified together as the same element. An *element* is a substance that cannot be decomposed into simpler substances during ordinary chemical reactions.

Although an element can have only a single atomic number, atoms of that element can have different atomic weights, and these are known as *isotopes*. The nuclei of isotopes differ from one another only in the number of neutrons. Isotopes behave the same way chemically for most purposes.

The atomic number and atomic weight of an element E are written in symbolic form as $_Z E^A$, E_Z^A, or $_Z^A E$. For example, carbon is the sixth element; radioactive carbon has an atomic mass of 14. Therefore, the symbol for carbon-14 is $^{14}_{6}C$. Because the atomic number and the chemical symbol give the same information, the atomic number can be omitted (e.g., C^{14} or C-14).

A *compound* is a combination of elements. A *molecule* is the smallest subdivision of an element or compound that can exist in a natural state.

THE PERIODIC TABLE

The *periodic table* (Table 37.1) is organized around the *periodic law*: Properties of the elements are periodic functions of their atomic numbers. Elements are arranged in order of increasing atomic numbers from left to right. The vertical columns are known as *groups*, numbered in Roman numerals. Each vertical group except 0 and VIII has A and B subgroups (*families*).

Adjacent elements in horizontal rows (i.e., in different groups) differ in both physical and chemical properties. However, elements in the same column (group) have similar properties. Graduations in properties, both physical and chemical, also occur in the periods (i.e., the horizontal rows).

There are several ways to categorize groups of elements in the periodic table. The biggest categorization of the elements is into metals and nonmetals.

Nonmetals (elements at the right end of the periodic chart) are elements 1, 2, 5–10, 14–18, 33–36, 52–54, and 85–86. The nonmetals include the *halogens* (Group

Table 37.1 The Periodic Table of Elements

The number of electrons in filled shells is shown in the column at the extreme left; the remaining electrons for each element are shown immediately below the symbol for each element. Atomic numbers are enclosed in brackets. Atomic weights (rounded, based on carbon-12) are shown above the symbols. Atomic weight values in parentheses are those of the isotopes of longest half-life for certain radioactive elements whose atomic weights cannot be precisely quoted without knowledge of origin of the element.

METALS — TRANSITION METALS — NONMETALS

Format of each cell: atomic weight, Symbol [atomic number], filled-shell electrons.

periods (shells)	I A	II A	III B	IV B	V B	VI B	VII B	VIII	VIII	VIII	I B	II B	III A	IV A	V A	VI A	VII A	O
1 (0)	1.0079 H [1] 1																	4.0026 He [2] 2
2 (2)	6.939 Li [3] 1	9.0122 Be [4] 2											10.81 B [5] 3	12.01115 C [6] 4	14.0067 N [7] 5	15.9994 O [8] 6	18.994 F [9] 7	20.183 Ne [10] 8
3 (2,8)	22.9898 Na [11] 1	24.312 Mg [12] 2											26.9815 Al [13] 3	28.086 Si [14] 4	30.9738 P [15] 5	32.064 S [16] 6	35.453 Cl [17] 7	39.948 Ar [18] 8
4 (2,8)	39.098 K [19] 8,1	40.08 Ca [20] 8,2	44.956 Sc [21] 9,2	47.90 Ti [22] 10,2	50.942 V [23] 11,2	51.996 Cr [24] 13,1	54.938 Mn [25] 13,2	55.847 Fe [26] 14,2	58.933 Co [27] 15,2	58.71 Ni [28] 16,2	63.546 Cu [29] 18,1	65.38 Zn [30] 18,2	69.72 Ga [31] 18,3	72.59 Ge [32] 18,4	74.922 As [33] 18,5	78.96 Se [34] 18,6	79.904 Br [35] 18,7	83.80 Kr [36] 18,8
5 (2,8,18)	85.47 Rb [37] 8,1	87.62 Sr [38] 8,2	88.905 Y [39] 9,2	91.22 Zr [40] 10,2	92.906 Nb [41] 12,1	95.94 Mo [42] 13,1	(98) Tc [43] 14,1	101.07 Ru [44] 15,1	102.905 Rh [45] 16,1	106.4 Pd [46] 18	107.868 Ag [47] 18,1	112.40 Cd [48] 18,2	114.82 In [49] 18,3	118.69 Sn [50] 18,4	121.75 Sb [51] 18,5	127.60 Te [52] 18,6	126.904 I [53] 18,7	131.30 Xe [54] 18,8
6 (2,8,18)	132.905 Cs [55] 18,8,1	137.34 Ba [56] 18,8,2	* [57–71]	178.49 Hf [72] 32,10,2	180.948 Ta [73] 32,11,2	183.85 W [74] 32,12,2	186.2 Re [75] 32,13,2	190.2 Os [76] 32,14,2	192.2 Ir [77] 32,15,2	195.09 Pt [78] 32,17,1	196.967 Au [79] 32,18,1	200.59 Hg [80] 32,18,2	204.37 Tl [81] 32,18,3	207.19 Pb [82] 32,18,4	208.980 Bi [83] 32,18,5	(210) Po [84] 32,18,6	(210) At [85] 32,18,7	(222) Rn [86] 32,18,8
7 (2,8,18,32)	(223) Fr [87] 18,8,1	226.025 Ra [88] 18,8,2	† [89–103]	Rf [104] 32,10,2	Ha [105] 32,11,2	[106] 32,12,2	[107]	[108]										

*LANTHANIDE SERIES

138.91 La [57] 18,9,2	140.12 Ce [58] 20,8,2	140.907 Pr [59] 21,8,2	144.24 Nd [60] 22,8,2	(147) Pm [61] 23,8,2	150.35 Sm [62] 24,8,2	151.96 Eu [63] 25,8,2	157.25 Gd [64] 25,9,2	158.924 Tb [65] 27,8,2	162.50 Dy [66] 28,8,2	164.930 Ho [67] 29,8,2	167.26 Er [68] 30,8,2	168.934 Tm [69] 31,8,2	173.04 Yb [70] 32,8,2	174.97 Lu [71] 32,9,2

† ACTINIDE SERIES

(227) Ac [89] 18,9,2	232.038 Th [90] 18,10,2	231.036 Pa [91] 20,9,2	238.03 U [92] 21,9,2	237.048 Np [93] 23,8,2	(242) Pu [94] 24,8,2	(243) Am [95] 25,8,2	(247) Cm [96] 25,9,2	(247) Bk [97] 26,9,2	(249) Cf [98] 28,8,2	(254) Es [99] 29,8,2	(253) Fm [100] 30,8,2	(256) Md [101] 31,8,2	(254) No [102] 32,8,2	(257) Lr [103] 32,9,2

Valence notes shown above group headers: $+1$ (I A), $+2$ (II A), -2 (VI A), -1 (VII A).

VIIA) and the *noble gases* (Group 0). Nonmetals are poor electrical conductors and have little or no metallic luster. Most are either gases or brittle solids under normal conditions; only bromine is liquid under ordinary conditions.

Metals are all of the remaining elements. The metals are further subdivided into the *alkali metals* (Group IA), the *alkaline earth metals* (Group IIA), *transition metals* (all B families and Group VIII), the *lanthanides* (also known as *lanthanons*, elements 57–71), and the *actinides* (also known as *actinons*, elements 89–103). Metals have low electron affinities, are reducing agents, form positive ions, and have positive oxidation numbers. They have high electrical conductivities, luster, generally high melting points, ductility, and malleability.

The electron-attracting power of an atom, which determines much of its chemical behavior, is called its *electronegativity* and is measured on an arbitrary scale of 0 to 4. Generally, the most electronegative elements are those at the right end of the periods. Elements with low electronegativities are the metals found at the beginning (i.e., left end) of the periods. Electronegativity generally decreases going down a group. In other words, the trend in any family is toward more metallic properties as the atomic weight increases. The electronegativities of some of the elements in the periodic table are listed in Table 37.2.

Table 37.2 Electronegativities of Some Elements [a]

									H 2.1								
Li 1.0	Be 1.5	B 2.0											C 2.5	N 3.0	O 3.5	F 4.0	
Na 0.9	Mg 1.2	Al 1.5	Ti 1.5	V 1.6	Cr 1.6	Mn 1.5	Fe 1.8	Co 1.8	Ni 1.8	Cu 1.9	Zn 1.6	Ga 1.6	Si 1.8	P 2.1	S 2.5	Cl 3.0	
K 0.8	Ca 1.0	Sc 1.3	Zr 1.4	Nb 1.6	Mo 1.8	Tc 1.9	Ru 2.2	Rh 2.2	Pd 2.2	Ag 1.9	Cd 1.7	In 1.7	Ge 1.8	As 2.0	Se 2.4	Br 2.8	
Rb 0.8	Sr 1.0	Y 1.2	Hf 1.3	Ta 1.5	W 1.7	Re 1.9	Os 2.2	Ir 2.2	Pt 2.2	Au 2.4	Hg 1.9	Tl 1.8	Sn 1.8	Sb 1.9	Te 2.1	I 2.5	
Cs 0.7	Ba 0.9	La–Lu 1.0–1.2	Th 1.3	Pa 1.5	U 1.7	Np–No 1.5–1.3							Pb 1.9	Bi 1.9	Po 2.0	At 2.2	
Fr 0.7	Ra 0.9	Ac 1.1															

College Chemistry: An Introductory Textbook of General Chemistry, 3/E, by Linus Pauling, copyright © 1964, W. H. Freeman and Company. Used with permission.

[a] The information in this table may not be provided in the actual exam.

IONS AND ELECTRON AFFINITY

The atomic number, Z, of chlorine is 17, which means there are 17 protons in the nucleus of a chlorine atom. There are also 17 electrons in various shells surrounding the nucleus.

Chlorine has only seven electrons in the outer shell. A stable shell requires eight electrons. In order to achieve this stable configuration, chlorine atoms tend to attract electrons from other atoms, a tendency known as *electron affinity*. The energy required to remove an electron from a neighboring atom is known as the *ionization energy*. The electrons attracted by chlorine atoms come from neighboring atoms with low ionization energies.

Chlorine, prior to taking a neighboring atom's electron, is electrically neutral. The fact that it needs one electron to complete its outer subshell does not mean that chlorine needs an electron to become neutral. On the contrary, the chlorine atom becomes negatively charged when it takes the electron. An atomic nucleus with a charge is known as an *ion*.

Negatively charged ions are known as *anions*. Anions lose electrons at the anode during electro-chemical reactions. Anions must lose electrons to become neutral. The loss of electrons is known as *oxidation*.

The charge on an anion is equal to the number of electrons taken from a neighboring atom. In the past, this charge has been known as the *valence*. (The term *charge* can usually be substituted for valence). Valence is equal to the number of electrons that must be gained for charge neutrality. For a chlorine ion, the valence is -1.

Sodium has one electron in its outer subshell; this electron has a low ionization energy and is very easily removed. If its outer electron is removed, sodium becomes positively charged. (For a sodium ion, the valence is $+1$.)

Positively charged ions are known as *cations*. Cations gain electrons at the cathode in electro-chemical reactions. The gaining of electrons is known as *reduction*. Cations must gain electrons to become neutral.

IONIC AND COVALENT BONDS

If a chlorine atom becomes an anion by attracting an electron from a sodium atom (which becomes a cation), the two ions will be attracted to each other by electrostatic force. The electrostatic attraction of the positive sodium to the negative chlorine effectively bonds the two ions together. This type of bonding, in which electrostatic attraction is predominant, is known as *ionic bonding*. In an ionic bond, one or more electrons are transferred from the valence shell of one atom to the valence shell of another. There is no sharing of electrons between atoms.

Ionic bonding occurs in compounds containing atoms with high electron affinities and atoms with low ionization energies. Specifically, the difference in electronegativities must be approximately 1.7 or greater for the bond to be classified as predominantly ionic.

Several common gases in their free states exist as diatomic molecules. Examples are hydrogen (H_2), oxygen (O_2), nitrogen (N_2), and chlorine (Cl_2). Since two atoms of the same element will have the same electronegativity and ionization energy, one atom cannot take electrons from the other. Therefore, the bond formed is not ionic.

The electrons in these diatomic molecules are shared equally in order to fill the outer shells. This type of bonding, in which sharing of electrons is the predominant characteristic, is known as *covalent bonding*. Covalent bonds are typical of bonds formed in organic compounds. Specifically, the difference in electronegativities must be less than approximately 1.7 for the bond to be classified as predominantly covalent.

If both atoms forming a covalent bond are the same element, the electrons will be shared equally. This is known as a *nonpolar covalent bond*. If the atoms are not both the same element, the electrons will not be shared equally, resulting in a *polar covalent bond*. For example, the bond between hydrogen and chlorine in HCl is partially covalent and partially ionic in nature. Thus, there is no sharp dividing line between ionic and covalent bonds for most compounds.

OXIDATION NUMBER

The *oxidation number (oxidation state)* is an electrical charge assigned by a set of prescribed rules. It is actually the charge, assuming all bonding is ionic. In a compound, the sum of the elemental oxidation numbers equals the net charge. For monoatomic ions, the oxidation number is equal to the charge.

In covalent compounds, all of the bonding electrons are assigned to the ion with the greater electronegativity. For example, nonmetals are more electronegative than metals. Carbon is more electronegative than hydrogen.

For atoms in an elementary free state, the oxidation number is zero. Hydrogen gas is a diatomic molecule, H_2. Thus, the oxidation number of the hydrogen molecule, H_2, is zero. The same is true for the atoms in O_2, N_2, Cl_2, and so on. Also, the sum of all the oxidation numbers of atoms in a neutral molecule is zero.

The *oxidation number* of an atom that forms a covalent bond is equal to the number of shared electron pairs. For example, each hydrogen atom has one electron. There are two electrons (i.e., a single shared electron pair) in each carbon-hydrogen bond in methane (CH_4). Therefore, the oxidation number of hydrogen is 1.

Fluorine is the most electronegative element, and it has an oxidation number of -1. Oxygen is second only to fluorine in electronegativity. Usually, the oxidation number of oxygen is -2, except in peroxides, where it is -1, and when combined with fluorine, where it is $+2$. Hydrogen is usually $+1$, except in hydrides, where it is -1.

The oxidation numbers of some common atoms and molecules are listed in Table 37.3.

COMPOUNDS

Combinations of elements are known as *compounds*. *Binary compounds* contain two elements; *ternary (tertiary) compounds* contain three elements. A *chemical formula* is a representation of the relative numbers of each element in the compound. For example, the formula $CaCl_2$ shows that there is one calcium atom and two chlorine atoms in one molecule of calcium chloride.

Generally, the numbers of atoms are reduced to their lowest terms. However, there are exceptions. For example, acetylene is C_2H_2, and hydrogen peroxide is H_2O_2.

For binary compounds with a metallic element, the positive metallic element is listed first. The chemical name ends in the suffix "-ide." For example, NaCl is sodium chloride. If the metal has two oxidation states, the suffix "-ous" is used for the lower state, and "-ic" is used for the higher state. Alternatively, the element name can be used with the oxidation number written in Roman numerals. For example,

$FeCl_2$: ferrous chloride, or iron (II) chloride
$FeCl_3$: ferric chloride, or iron (III) chloride

For binary compounds formed between two nonmetals, the more positive element is listed first. The number of atoms of each element is specified by the prefixes "di-" (2), "tri-" (3), "tetra-" (4), "penta-" (5), and so on. For example,

N_2O_5: dinitrogen pentoxide

Binary acids start with the prefix "hydro-," list the name of the nonmetallic element, and end with the suffix "-ic." For example,

HCl: hydrochloric acid

Ternary compounds generally consist of an element and a radical, with the positive part listed first in the formula. Ternary acids (also known as *oxy-acids*) usually contain hydrogen, a nonmetal, and oxygen, and can be grouped into families with different numbers of oxygen atoms. The most common acid in a family (i.e., the root acid) has the name of the nonmetal and the suffix "-ic." The acid with one more oxygen atom than the

root is given the prefix "per-" and the suffix "-ic." The acid containing one less oxygen atom than the root is given the ending "-ous." The acid containing two less oxygen atoms than the root is given the prefix "hypo-" and the suffix "-ous." For example,

HClO: hypochlorous acid
$HClO_2$:   chlorous acid
$HClO_3$:   chloric acid (the root)
$HClO_4$:   perchloric acid

Compounds form according to the *law of definite (constant) proportions*: A pure compound is always composed of the same elements combined in a definite proportion by mass. For example, common table salt is always NaCl. It is not sometimes NaCl and other times Na_2Cl or $NaCl_3$ (which do not exist).

Furthermore, compounds form according to the *law of (simple) multiple proportions*: When two elements combine to form more than one compound, the masses of the elements usually combine in ratios of the smallest possible integers.

In order to evaluate whether a compound formula is valid, it is necessary to know the oxidation numbers of the interacting atoms. Although some atoms have more than one possible oxidation number, most do not. The sum of the oxidation numbers must be zero if a neutral compound is to form. For example, H_2O is a valid compound because the two hydrogen atoms have a total positive oxidation number of $2 \times 1 = +2$. The oxygen ion has an oxidation number of -2. These oxidation numbers sum to zero.

On the other hand, $NaCO_3$ is not a valid compound formula. Sodium (Na) has an oxidation number of $+1$. However, the CO_3 radical has a charge number of -2. The correct sodium carbonate molecule is Na_2CO_3.

MOLES

The *mole* is a measure of the quantity of an element or compound. Specifically, a mole of an element (or compound) will have a mass equal to the element's atomic (or compound's molecular) weight.

The three main types of moles are based on mass measured in grams, kilograms, and pounds. Obviously, a gram-based mole of carbon (12.0 grams) is not the same quantity as a pound-based mole of carbon (12.0 pounds). Although "mol" is understood in SI countries to mean a gram-mole, the term *mole* is ambiguous, and the units mol (gmol), kmol (kgmol) or lbmol must be specified or the type of mole must be spelled out.

One gram-mole of any substance has a number of particles (atoms, molecules, ions, electrons, etc.) equal to

Table 37.3 Oxidation Numbers of Atoms and Charge Numbers of Radicals[a]

name	symbol	oxidation or charge number
acetate	$C_2H_3O_2$	-1
aluminum	Al	$+3$
ammonium	NH_4	$+1$
barium	Ba	$+2$
boron	B	$+3$
borate	BO_3	-3
bromine	Br	-1
calcium	Ca	$+2$
carbon	C	$+4, -4$
carbonate	CO_3	-2
chlorate	ClO_3	-1
chlorine	Cl	-1
chlorite	ClO_2	-1
chromate	CrO_4	-2
chromium	Cr	$+2, +3, +6$
copper	Cu	$+1, +2$
cyanide	CN	-1
dichromate	Cr_2O_7	-2
fluorine	F	-1
gold	Au	$+1, +3$
hydrogen	H	$+1$
hydroxide	OH	-1
hypochlorite	ClO	-1
iron	Fe	$+2, +3$
lead	Pb	$+2, +4$
lithium	Li	$+1$
magnesium	Mg	$+2$
mercury	Hg	$+1, +2$
nickel	Ni	$+2, +3$
nitrate	NO_3	-1
nitrite	NO_2	-1
nitrogen	N	$-3, +1, +2, +3, +4, +5$
oxygen	O	-2 [-1 in peroxides]
perchlorate	ClO_4	-1
permanganate	MnO_4	-1
phosphate	PO_4	-3
phosphorus	P	$-3, +3, +5$
potassium	K	$+1$
silicon	Si	$+4, -4$
silver	Ag	$+1$
sodium	Na	$+1$
sulfate	SO_4	-2
sulfite	SO_3	-2
sulfur	S	$-2, +4, +6$
tin	Sn	$+2, +4$
zinc	Zn	$+2$

[a] The information in this table may not be provided in the actual exam.

6.022×10^{23}, *Avogadro's number*, N_A. A pound-mole contains approximately 454 times the number of particles in a gram-mole.

"Molar" is used as an adjective when describing properties of a mole. For example, a molar volume is the volume of a mole.

FORMULA AND MOLECULAR WEIGHT; EQUIVALENT WEIGHT

The *formula weight*, FW, of a molecule (compound) is the sum of the atomic weights of all elements in the formula. The *molecular weight*, MW, is the sum of the atomic weights of all atoms in the molecule and is generally the same as the formula weight. The units of molecular weight are g/mol, kg/kmol, or lb/lbmol. However, units are sometimes omitted because weights are relative.

The *equivalent weight* (i.e., an *equivalent*), EW, is the amount of substance (in grams) that supplies one gram-mole (i.e., 6.022×10^{23}) of reacting units. For acid-base reactions, an acid equivalent supplies one gram-mole of H^+ ions. A base equivalent supplies one gram-mole of OH^- ions. In oxidation-reduction reactions, an equivalent of a substance gains or loses a gram-mole of electrons. Similarly, in electrolysis reactions, an equivalent weight is the weight of substance that either receives or donates one gram-mole of electrons at an electrode.

The equivalent weight can be calculated as the molecular weight divided by the change in oxidation number experienced in a chemical reaction. A substance can have several equivalent weights.

$$EW = \frac{MW}{\Delta \text{ oxidation number}} \qquad 37.1$$

GRAVIMETRIC FRACTION

The *gravimetric fraction*, x_i, of an element i in a compound is the fraction by weight of that element in the compound. The gravimetric fraction is found from an *ultimate analysis* (also known as a *gravimetric analysis*) of the compound.

$$\begin{aligned} x_i &= \frac{m_i}{m_1 + m_2 + \cdots + m_i + \cdots + m_n} \\ &= \frac{m_i}{m_t} \end{aligned} \qquad 37.2$$

The percentage composition is the gravimetric fraction converted to percentage.

$$\% \text{ composition} = x_i \times 100\% \qquad 37.3$$

If the gravimetric fractions are known for all elements in a compound, the *combining weights* of each element can be calculated from Eq. 37.2. (The term *weight* is used even though mass is the traditional unit of measurement.)

EMPIRICAL FORMULA DEVELOPMENT

It is relatively simple to determine the empirical formula of a compound from the atomic and combining weights of elements in the compound. The empirical formula gives the relative number of atoms (i.e., the formula weight is calculated from the empirical formula).

step 1: Divide the gravimetric fractions (or percentage compositions) by the atomic weight of each respective element.

step 2: Determine the smallest ratio from step 1.

step 3: Divide all of the ratios from step 1 by the smallest ratio.

step 4: Write the chemical formula using the results from step 3 as the numbers of atoms. Multiply through as required to obtain all integer numbers of atoms.

RADIOACTIVE DECAY AND HALF-LIFE

Most elements exist in different forms known as *isotopes*, differing in the number of neutrons each has. For example, hydrogen has three isotopes: regular hydrogen (H-1), *deuterium* (H-2), and *tritium* (H-3). Many times, two or more of the isotopes exist in nature simultaneously and will be intermixed in a naturally occurring sample. In chemical reactions, different isotopes typically react at different rates, but not in different manners. In nuclear reactions, however, different isotopes behave in decidedly different manners.

Some isotopes are unstable and will disintegrate spontaneously by a process known as *radioactive decay*. The instability is due to too many or too few neutrons in the nucleus. While the neutrons have no electrostatic effect, they contribute to the strong nuclear force needed to balance proton repulsion.

If a nucleus has too many neutrons, a neutron may spontaneously transform into a proton. An electron is also released to retain charge neutrality. This electron emission is known as $-\beta$ *decay*. If the nucleus has too few neutrons, a proton transforms into a neutron with a positron emission. This is known as $+\beta$ *decay*. α *decay* decreases the number of both protons and neutrons by two and may also result in a stable nucleus.

The disintegration of radioactive isotopes is described by a negative *exponential law*. An exponential law describes the behavior of a substance whose quantity changes at a rate proportional to the quantity present.

The rate of radioactive decay is specified by the *half-life*, $t_{1/2}$, which is essentially independent of the local environment (pressure, temperature, etc.). The half-life is equal to the time required for half of the original atoms to decay.

The number of atoms left at time t is calculated from Eq. 37.4. Time, t, is in the same units as the half-life.

$$N = N_0 e^{-0.693t/t_{1/2}} \qquad 37.4$$

SAMPLE PROBLEMS

Problem 1

For a given isotope of an element, the atomic number plus the atomic weight is 148, and their difference is 58. How many protons does an atom of the isotope contain?

(A) 45
(B) 58
(C) 90
(D) 148

Solution

The atomic number, Z, is equal to the number of protons in the nucleus. The atomic weight, A, is approximately equal to the number of protons and neutrons in the nucleus.

$A + Z = 148 = P$ protons $+ N$ neutrons $+ P$ protons
$A - Z = 58 = P$ protons $+ N$ neutrons $- P$ protons

From the second equation,

$$58 = N \text{ neutrons}$$

From the first equation,

$$148 = 2(P \text{ protons}) + 58$$
$$45 = P \text{ protons}$$

Answer is A.

Problem 2

The group of metals that includes lithium, sodium, potassium, rubidium, and cesium forms a closely related family known as the

(A) rare earth group.
(B) halogens.
(C) alkali metals.
(D) alkaline earth metals.

Solution

Lithium, sodium, potassium, rubidium, and cesium occupy the first column of the periodic chart, known as *Group IA* or the *alkali metals*.

Answer is C.

Problem 3

Which of the following compounds would be ionic, considering the electronegativities of the elements?

element	electronegativity
K	0.8
C	2.5
I	2.5
Cl	3.0
N	3.0
O	3.5
F	4.0

(A) CO
(B) NO
(C) I_2
(D) KCl

Solution

Consider the differences in electronegativities for each compound.

compound	electronegativities		difference
CO	C = 2.5	O = 3.5	1.0
NO	N = 3.0	O = 3.5	0.5
I_2	I = 2.5	I = 2.5	0
KCl	K = 0.8	Cl = 3.0	2.2

The difference in electronegativities must be greater than 1.7 for the bond to be considered ionic. Only KCl meets this requirement; the other compounds are considered to have covalent bonds.

Answer is D.

Problem 4

Which of the following chemical formulas is incorrect?

(A) $Ca(OH)_2$
(B) Na_2CO_3
(C) CaCl
(D) KOH

Solution

Examine the oxidation numbers of the molecular elements to check for neutral molecules.

molecule	compound or element	oxidation number	neutral?
$Ca(OH)_2$	Ca	$+2$	yes
	OH	$2 \times (-1)$	
Na_2CO_3	Na	$2 \times (+1)$	yes
	CO_3	-2	
CaCl	Ca	$+2$	no
	Cl	-1	
KOH	K	$+1$	yes
	OH	-1	

The answer is CaCl.

Answer is C.

Problem 5

What is the term for a quantity of a substance to which a chemical formula can be assigned and whose mass is equal to its formula weight?

(A) a molecule
(B) a mole
(C) an equivalent weight
(D) a one-normal solution

Solution

A mole of an element will have a mass equal to the element's molecular weight. The molecular weight is generally the same as the formula weight.

Answer is B.

Problem 6

A sample of an unknown compound is found to be 49.3% carbon, 9.6% hydrogen, 19.2% nitrogen, and 21.9% oxygen by weight. What is its molecular formula?

(A) C_4H_8NO
(B) C_4H_6NO
(C) $C_3H_6N_2O$
(D) C_3H_7NO

Solution

step 1:   Divide the percentage compositions by the atomic weights of each element.

$$C: \quad \frac{49.3 \text{ g}}{12.011 \frac{\text{g}}{\text{mol}}} = 4.1046 \text{ mol}$$

$$H: \quad \frac{9.6 \text{ g}}{1.0079 \frac{\text{g}}{\text{mol}}} = 9.5248 \text{ mol}$$

$$N: \quad \frac{19.2 \text{ g}}{14.007 \frac{\text{g}}{\text{mol}}} = 1.3707 \text{ mol}$$

$$O: \quad \frac{21.9 \text{ g}}{15.999 \frac{\text{g}}{\text{mol}}} = 1.3688 \text{ mol}$$

step 2:   Determine the smallest ratio from step 1.

$$\text{smallest ratio} = 1.3688$$

step 3:   Divide all ratios by the smallest ratio.

$$C: \quad \frac{4.1046}{1.3688} \approx 3$$

$$H: \quad \frac{9.5248}{1.3688} \approx 7$$

$$N: \quad \frac{1.3707}{1.3688} \approx 1$$

$$O: \quad \frac{1.3688}{1.3688} \approx 1$$

step 4:   Write the chemical formula using the results from step 3: C_3H_7NO.

Answer is D.

Problem 7

How many half-lives will it take for a substance to reduce to less than 1% of its original amount?

(A) 3
(B) 7
(C) 52
(D) 100

Solution

$$N = N_0 e^{-0.693t/t_{1/2}}$$

$$\frac{N}{N_0} = e^{-0.693t/t_{1/2}}$$

$$\ln\left(\frac{N}{N_0}\right) = (-0.693)\left(\frac{t}{t_{1/2}}\right)$$

$$\frac{t}{t_{1/2}} = \frac{\ln\left(\frac{N}{N_0}\right)}{-0.693} = \frac{\ln 0.01}{-0.693}$$

$$= 6.65 \quad (7)$$

Answer is B.

FE-STYLE EXAM PROBLEMS

1. What are the chemical formulas for the following compounds: aluminum nitrate, magnesium hydroxide, calcium oxide, and cupric carbonate?

(A) $Al(NO_3)_3$; $Mg(OH)_2$; CaO; $CuCO_3$
(B) Al_2NO_3; $Mg(OH)$; CaO_2; $CuCO_3$
(C) $AlNO_3$; $Mg(OH)_2$; CaO; $Cu(CO_3)_2$
(D) $AlNO_3$; $Mg(OH)$; Ca_2O_3; $CuCO_3$

2. Use Table 37.2 to arrange the following in order of increasing ionic character of their bonds: SO_2, H_2S, SF_2, OF_2.

(A) SO_2; H_2S; SF_2; OF_2
(B) H_2S; SF_2; SO_2; OF_2
(C) H_2S; OF_2; SO_2; SF_2
(D) SF_2; OF_2; SO_2; H_2S

3. What is the maximum possible positive oxidation number for the element Br?

(A) $+1$
(B) $+3$
(C) $+4$
(D) $+7$

4. In a laboratory experiment, a student analyzed a substance with 2.7626 g of lead, 0.00672 g of hydrogen, and 0.8534 g of oxygen. What is the empirical formula for the substance?

(A) $Pb_2O_4H_2$
(B) Pb_4O_2H
(C) Pb_4OH_2
(D) Pb_2O_8H

5. Vitamin C has the molecular formula of $C_6H_8O_6$. How many gram-moles are in 23 g of vitamin C?

(A) 0.13 mol
(B) 0.39 mol
(C) 3.08 mol
(D) 7.66 mol

For the following problems use the NCEES Handbook as your only reference.

6. What is the percentage (by mass) of hydrogen in glucose ($C_6H_{12}O_6$)?

(A) 6.7%
(B) 9.3%
(C) 17%
(D) 40%

7. A gaseous mixture consists of 2 kg of oxygen, 5 kg of nitrogen, and 3 kg of xenon. What is the mole fraction of the oxygen gas?

(A) 0.11
(B) 0.13
(C) 0.17
(D) 0.24

8. What is an isomer?

(A) a single atom
(B) a basic building block for large chemical chains
(C) different arrangements of the same atoms
(D) a substance containing a hydroxyl ion

9. While moving from left to right across the second row of the periodic table (i.e., from Li to Ne), the atomic radii tend to

(A) uniformly increase.
(B) uniformly decrease.
(C) remain the same.
(D) first increase, then decrease.

10. Which of the following elements does not exist as a diatomic molecule under normal (ambient) conditions?

(A) oxygen
(B) chlorine
(C) sulfur
(D) iodine

11. What is the gravimetric (i.e., mass) percentage of oxygen in K_2CrO_4?

(A) 33%
(B) 42%
(C) 57%
(D) 66%

12. Which of the following elements has the largest first ionization energy?

(A) Ba (barium)
(B) Cu (copper)
(C) Ne (neon)
(D) S (sulfur)

13. What is the half-life of a substance that decays to 25% of its original amount in 6 d?

(A) 0.08 d
(B) 3 d
(C) 8 d
(D) 12 d

14. A given sample of radioactive material has 80% of the original substance remaining after 10 years. How much will remain after 90 additional years?

- (A) 0.1%
- (B) 1.7%
- (C) 11%
- (D) 13%

15. The half-life of radioactive carbon is approximately 5700 years. If a sample is found to have 7000 atoms after 6000 years, how many atoms were present initially?

- (A) 13 800 atoms
- (B) 14 100 atoms
- (C) 14 300 atoms
- (D) 14 500 atoms

16. The diameter of a spherical mothball is observed to halve in 200 d. Approximately how long will it take for its remaining volume to become half of its volume at 200 d?

- (A) 67 d
- (B) 130 d
- (C) 160 d
- (D) 200 d

17. In the following reaction, which elements are the reducing and oxidizing agents?

$$2Mg(s) + O_2(g) \longrightarrow 2MgO(s)$$

- (A) Mg is the reducing agent; O_2 is the oxidizing agent.
- (B) O_2 is the reducing agent; Mg is the oxidizing agent.
- (C) Mg is the reducing agent; MgO is the oxidizing agent.
- (D) MgO is the reducing agent; Mg is the oxidizing agent.

18. What is the oxidation number for chromium (Cr) in the compound $BaCrO_4$?

- (A) +1
- (B) +2
- (C) +4
- (D) +6

19. Uranium-235 and uranium-238 have the same number of which of the following?

- (A) neutrons
- (B) protons
- (C) electrons
- (D) protons and electrons

20. Which of the following is a homogeneous mixture?

- (A) strawberry milkshake
- (B) curing concrete
- (C) sodium chloride
- (D) seawater

21. The activity of decay particles of a radioactive isotope depends on

- (A) temperature
- (B) pressure
- (C) density
- (D) mass

22. 10 000 liters of water are contaminated by a toxic compound with a concentration of 25.5 ppm. The concentration must be reduced to 500 ppb before the water can be discharged. What mass of compound must be removed in order to discharge the water?

- (A) 23 g
- (B) 25 g
- (C) 250 g
- (D) 375 g

23. The term "divalent" means

- (A) an ion's oxidation number is +2
- (B) an ion's oxidation number is −2
- (C) an ion's oxidation number can be +2 or −2
- (D) the ion can have two different oxidation numbers

24. Which of the following elements is the most electronegative?

- (A) Br
- (B) Cl
- (C) F
- (D) I

SOLUTIONS TO FE-STYLE EXAM PROBLEMS

1. Refer to the table of oxidation numbers.

aluminum nitrate: Al has an oxidation number of +3. NO_3 has an oxidation number of −1. The formula is $Al(NO_3)_3$.

magnesium hydroxide: Mg has an oxidation number of +2. OH has an oxidation number of −1. The formula is $Mg(OH)_2$.

calcium oxide: Ca has an oxidation number of +2. O has an oxidation number of −2. The formula is CaO.

cupric carbonate: Cu (cupric) has an oxidation number of +2. CO_3 has an oxidation number of −2. The formula is $CuCO_3$.

Answer is A.

2. The strength of an ionic bond comes from the difference in electronegativities of the bonding atoms.

compound	electronegativities		difference
SO_2	S = 2.5	O = 3.5	1.0
H_2S	H = 2.1	S = 2.5	0.4
SF_2	S = 2.5	F = 4.0	1.5
OF_2	O = 3.5	F = 4.0	0.5

The order of increasing ionic character is H_2S, OF_2, SO_2, and SF_2. None of these bonds is considered purely ionic because ionic bonds have differences in electronegativity greater than 1.7. Rather, these bonds would be considered covalent with partial ionic character.

Answer is C.

3. Bromine has an oxidation number of −1, which means that it normally accepts one electron to complete its outer shell of eight electrons. Alternatively, it could give up seven electrons to have a full outer shell.

Answer is D.

4. *step 1:* Find the gravimetric fractions of each element.

$$m_t = 2.7626 \text{ g} + 0.00672 \text{ g} + 0.8534 \text{ g}$$
$$= 3.62272 \text{ g}$$

$$x_{Pb} = \frac{m_{Pb}}{m_t} = \frac{2.7626 \text{ g}}{3.62272 \text{ g}}$$
$$= 0.76258$$

$$x_H = \frac{m_H}{m_t} = \frac{0.00672 \text{ g}}{3.62272 \text{ g}}$$
$$= 0.00185$$

$$x_O = \frac{m_O}{m_t} = \frac{0.8534 \text{ g}}{3.62272 \text{ g}}$$
$$= 0.23557$$

Divide the gravimetric fractions by the atomic weight of each element.

Pb: $\dfrac{0.76258 \text{ g}}{207.19 \dfrac{\text{g}}{\text{mol}}} = 3.6806 \times 10^{-3} \text{ mol}$

H: $\dfrac{0.00185 \text{ g}}{1.0079 \dfrac{\text{g}}{\text{mol}}} = 1.8355 \times 10^{-3} \text{ mol}$

O: $\dfrac{0.23557 \text{ g}}{15.999 \dfrac{\text{g}}{\text{mol}}} = 1.4724 \times 10^{-2} \text{ mol}$

step 2: Determine the smallest ratio from step 1.

smallest ratio = 1.8355×10^{-3} [by inspection]

step 3: Divide all of the ratios from step 1 by the smallest ratio.

Pb: $\dfrac{3.6806 \times 10^{-3}}{1.8355 \times 10^{-3}} = 2.005$

H: $\dfrac{1.8355 \times 10^{-3}}{1.8355 \times 10^{-3}} = 1.0$

O: $\dfrac{1.4724 \times 10^{-2}}{1.8355 \times 10^{-3}} = 8.02$

step 4: Write the chemical formula using results from step 3. (Recognize that there may be small errors present in the analysis that will give slight discrepancies.)

The formula is Pb_2O_8H.

Answer is D.

5. The approximate molecular weight of $C_6H_8O_6$ is

$$(6)\left(12 \frac{\text{g}}{\text{mol}}\right) + (8)\left(1 \frac{\text{g}}{\text{mol}}\right) + (6)\left(16 \frac{\text{g}}{\text{mol}}\right)$$
$$= 176 \text{ g/mol}$$

The number of moles is

$$n = \frac{m}{MW} = \frac{23 \text{ g}}{176 \dfrac{\text{g}}{\text{mol}}}$$
$$= 0.13 \text{ mol}$$

Answer is A.

6. The combining weights of each element are

C: $(6)\left(12 \dfrac{\text{g}}{\text{mol}}\right) = 72 \text{ g/mol}$

H: $(12)\left(1 \dfrac{\text{g}}{\text{mol}}\right) = 12 \text{ g/mol}$

O: $(6)\left(16 \dfrac{\text{g}}{\text{mol}}\right) = 96 \text{ g/mol}$

The molecular weight of glucose is

$$72 \, \frac{g}{mol} + 12 \, \frac{g}{mol} + 96 \, \frac{g}{mol} = 180 \text{ g/mol}$$

The mass fraction of hydrogen in glucose is

$$\frac{12 \, \frac{g}{mol}}{180 \, \frac{g}{mol}} = 0.0667 \quad (6.7\%)$$

Answer is A.

7. The approximate molecular weights of oxygen (O_2), nitrogen (N_2), and xenon (Xe) gases are 32 g/mol, 28 g/mol, and 131.3 g/mol, respectively. The units of kg/kmol can also be used.

Calculate the number of moles of each gas.

$$O_2: \quad \frac{2 \text{ kg}}{32 \, \frac{kg}{kmol}} = 0.0625 \text{ kmol}$$

$$N_2: \quad \frac{5 \text{ kg}}{28 \, \frac{kg}{kmol}} = 0.1786 \text{ kmol}$$

$$Xe: \quad \frac{3 \text{ kg}}{131.3 \, \frac{kg}{kmol}} = 0.0228 \text{ kmol}$$

The mole fraction of oxygen is

$$\frac{0.0625 \text{ kmol}}{0.0625 \text{ kmol} + 0.1786 \text{ kmol} + 0.0228 \text{ kmol}}$$
$$= 0.237 \quad (0.24)$$

Answer is D.

8. Isomers are different molecular arrangements of the same atoms.

Answer is C.

9. Elements with smaller radii are more stable than those with larger radii. Elements increase their stability toward the right side of the periodic table.

Answer is B.

10. There are seven common elements that exist as diatomic molecules under normal conditions. They are hydrogen, oxygen, nitrogen, chlorine, fluorine, bromine, and iodine.

Answer is C.

11. The combining weights are

$$K: (2) \left(39 \, \frac{g}{mol} \right) = 78 \text{ g/mol}$$

$$Cr: (1) \left(52 \, \frac{g}{mol} \right) = 52 \text{ g/mol}$$

$$O: (4) \left(16 \, \frac{g}{mol} \right) = 64 \text{ g/mol}$$

The gravimetric percentage is the same as the mass percentage.

$$\frac{64 \, \frac{g}{mol}}{78 \, \frac{g}{mol} + 52 \, \frac{g}{mol} + 64 \, \frac{g}{mol}} = 0.33 \quad (33\%)$$

Answer is A.

12. *Ionization energy* is the energy required to completely remove an electron from an atom. It is usually expressed in joules or joules per mole. When expressed per unit charge (i.e., in J/C, same as volts, V), it is known as *ionization potential*.

The first ionization energy is the energy required to remove an electron from the outermost shell. Ionization energy decreases as the number of electrons (shells) increases. Neon is not only a noble gas, but it has the fewest number of shells.

Answer is C.

13.
$$N = N_0 e^{-0.693t/t_{1/2}}$$
$$\frac{N}{N_0} = e^{-0.693t/t_{1/2}}$$
$$\ln \left(\frac{N}{N_0} \right) = \frac{-0.693t}{t_{1/2}}$$
$$t_{1/2} = \frac{-0.693t}{\ln \left(\frac{N}{N_0} \right)}$$
$$= \frac{(-0.693)(6 \text{ d})}{\ln 0.25}$$
$$= 3 \text{ d}$$

Answer is B.

14. First, calculate the half-life.

$$\frac{N}{N_0} = e^{-0.693t/t_{1/2}}$$
$$t_{1/2} = \frac{-0.693t}{\ln \left(\frac{N}{N_0} \right)} = \frac{(-0.693)(10 \text{ yr})}{\ln 0.8}$$
$$= 31.06 \text{ yr}$$

Now, use the half-life to calculate the surviving fraction.

$$\frac{N}{N_0} = e^{(-0.693)(100 \text{ yr})/31.06 \text{ yr}}$$

$$= 0.107 \quad (11\%)$$

Answer is C.

15.
$$N = N_0 e^{-0.693t/t_{1/2}}$$
$$7000 = N_0 e^{(-0.693)(6000 \text{ yr})/5700 \text{ yr}}$$

Take the natural logarithm of both sides.

$$\ln(7000) = \ln\left(N_0 e^{(-0.693)(6000 \text{ yr})/5700 \text{ yr}}\right)$$

$$\ln(7000) = \frac{-(0.693)(6000 \text{ yr})}{5700 \text{ yr}} \ln(N_0)$$

$$N_0 = 14\,518 \text{ atoms} \quad (14\,500 \text{ atoms})$$

Answer is D.

16. The volume of a sphere is

$$V = \left(\frac{1}{6}\right)\pi d^3$$

$$2V_2 = V_1$$

$$(2)\left(\frac{1}{6}\right)\pi d_2^3 = \left(\frac{1}{6}\right)\pi d_1^3$$

$$d_2^3 = \left(\frac{1}{2}\right)d_1^3$$

$$d_2 = \left(\frac{1}{2}\right)^{1/3} d_1$$

$$d = d_0 e^{-0.693t/t_{1/2}}$$

$$\left(\frac{1}{2}\right)^{1/3} d_1 = d_1 e^{-0.693t/200 \text{ d}}$$

$$\left(\frac{1}{2}\right)^{1/3} = e^{-0.693t/200 \text{ d}}$$

Take the natural log of both sides.

$$\left(\frac{1}{3}\right)\ln\left(\frac{1}{2}\right) = \frac{-0.693t}{200 \text{ d}}$$

$$t = \frac{\left(\frac{1}{3}\right)(200 \text{ d})\ln\left(\frac{1}{2}\right)}{-0.693}$$

$$= 66.7 \text{ d} \quad (67 \text{ d})$$

Answer is A.

17. The oxidation numbers of Mg and O_2 are both zero since they are both in their free elemental states. As MgO, the oxidation number of Mg is increased to $+2$ and O is decreased to -2. Mg is oxidized, and O is reduced. Therefore, Mg is the reducing agent and O is the oxidizing agent.

Answer is A.

18. Since barium (Ba) has an oxidation number of $+2$, the chromate ion (CrO_4) has an oxidation number of -2. Oxygen always has an oxidation number of -2, so the chromium must have an oxidation number of $+6$ in order to give the chromate ion a net oxidation number of -2.

Answer is D.

19. Uranium-235 and uranium-238 are both isotopes of uranium. Uranium has 92 protons and 92 electrons; the isotopes differ in the number of neutrons in the nucleus. Another notation for uranium-235 and uranium-238 is $^{235}_{92}U$ and $^{238}_{92}U$, respectively. Specifying the atomic number ($Z = 92$) is redundant because, by definition, uranium has 92 protons.

Answer is D.

20. A mixture of components consists of two or more materials that have been combined mechanically. In a homogeneous mixture, all components are well dispersed among the others. This describes a milkshake, which consists of water, milk solids, syrup, and strawberries blended together. Concrete is also a mechanical mixture, but its components can be seen by the naked eye and may separate by density. The components of sodium chloride and seawater are combined chemically, not mechanically.

Answer is A.

21. The activity is the number of frequency of decay (i.e., the number of particles decaying per unit time.) Activity is independent of temperature, pressure, and density, but it is proportional to the amount of isotope remaining.

Answer is D.

22. 25.5 ppm is the same as 25 500 ppb.

$$(10\,000 \text{ L}) \left(0.001 \ \frac{\text{m}^3}{\text{L}}\right) \left(1000 \ \frac{\text{kg}}{\text{m}^3}\right)$$
$$\times \left(\frac{25\,500 \text{ ppb} - 500 \text{ ppb}}{10^9 \text{ ppb}}\right) = 0.25 \text{ kg} \quad (250 \text{ g})$$

Answer is C.

23. The terms "divalent" and "bivalent" mean the same thing: a valence of 2.

Answer is C.

24. Electronegativity is the ability of an atom to attract an electron. According to periodic trends, electronegativity increases from left to right and bottom to top on the period table. Therefore, fluorine (F) is the most electronegative element, not only among the answer choices, but among all elements.

Answer is C.

38 Chemical Reactions

Subjects

CHEMICAL REACTIONS 38-1
OXIDATION-REDUCTION REACTIONS . 38-1
REVERSIBLE REACTIONS 38-2
LE CHÂTELIER'S PRINCIPLE 38-2
RATE AND ORDER OF REACTIONS . . . 38-2
EQUILIBRIUM CONSTANT 38-3
IDEAL GASES 38-3
AVOGADRO'S HYPOTHESIS 38-3
ACIDS AND BASES 38-3
ORGANIC CHEMISTRY 38-4

Nomenclature

k	reaction rate constant	–
K	equilibrium constant	–
m	mass	kg
MW	molecular weight	kg/kmol
n	number of moles	–
p	pressure	atm
r	rate of reaction	mol/L·s
R	specific gas constant	atm·m^3/kg·K
$\overline{R}$	universal gas constant	atm·L/mol·K
T	temperature	K
V	volume	m^3

Symbols

ν	specific volume	m^3/kg
ρ	density	kg/m^3

Subscripts

eq	equilibrium
p	partial pressures

CHEMICAL REACTIONS

During chemical reactions, bonds between atoms are broken and new bonds are formed. The starting substances are known as *reactants*; the ending substances are known as *products*. In a chemical reaction, reactants are either converted to simpler products or synthesized into more complex compounds.

The coefficients in front of element and compound symbols in chemical reaction equations are the numbers of molecules or moles taking part in the reaction. For gaseous reactants and products, the coefficients also represent the numbers of volumes. This is a direct result of *Avogadro's hypothesis* that equal numbers of molecules in the gas phase occupy equal volumes at the same conditions.

Because matter cannot be destroyed in a normal chemical reaction (i.e., mass is conserved), the numbers of each element must match on both sides of the equation. When the numbers of each element on both sides match, the equation is said to be *balanced*. The total atomic weights on both sides of the equation will be equal when the equation is balanced.

Balancing simple chemical equations is largely a matter of deductive trial and error. More complex reactions require the use of oxidation numbers.

OXIDATION-REDUCTION REACTIONS

Oxidation-reduction reactions (also known as *redox reactions*) involve the transfer of electrons from one element or compound to another. Specifically, one reactant is oxidized and the other reactant is reduced.

In *oxidation*, the substance's oxidation state increases, the substance loses electrons, and the substance becomes less negative. Oxidation occurs at the *anode* (positive terminal) in electrolytic reactions.

In *reduction*, the substance's oxidation state decreases, the substance gains electrons, and the substance becomes more negative. Reduction occurs at the *cathode* (negative terminal) in electrolytic reactions.

Whenever oxidation occurs in a chemical reaction, reduction must also occur. For example, consider the formation of sodium chloride from sodium and chlorine. This reaction is a combination of oxidation of sodium and reduction of chlorine. Notice that the electron released during oxidation is used up in the reduction reaction.

$$2Na + Cl_2 \longrightarrow 2NaCl$$
$$Na \longrightarrow Na^+ + e^-$$
$$Cl + e^- \longrightarrow Cl^-$$

The substance that causes oxidation to occur (chlorine in the preceding example) is called the *oxidizing agent* and is itself reduced (i.e., becomes more negative) in the process. The substance that causes reduction to occur

CHEMISTRY
Reactions

(sodium in the example) is called the *reducing agent* and is itself oxidized (i.e., becomes less negative) in the process.

The total number of electrons lost during oxidation must equal the total number of electrons gained during reduction. This is the main principle used in balancing redox reactions. Although there are several formal methods of applying this principle, balancing an oxidation-reduction equation remains somewhat intuitive and iterative.

The oxidation number change method of balancing redox reactions consists of the following steps.

step 1: Write an unbalanced equation that includes all reactants and products.

step 2: Assign oxidation numbers to each atom in the unbalanced equation.

step 3: Note which atoms change oxidation numbers, and calculate the amount of change for each atom. When more than one atom of an element that changes oxidation number is present in a formula, calculate the change in oxidation number for that atom per formula unit.

step 4: Balance the equation so that the number of electrons gained equals the number lost.

step 5: Balance (by inspection) the remainder of the chemical equation as required.

REVERSIBLE REACTIONS

Reversible reactions are capable of going in either direction and do so to varying degrees (depending on the concentrations and temperature) simultaneously. These reactions are characterized by the simultaneous presence of all reactants and all products. For example, the chemical equation for the exothermic formation of ammonia from nitrogen and hydrogen is

$$N_2 + 3H_2 \longleftrightarrow 2NH_3 + heat$$

At chemical equilibrium, reactants and products are both present. However, the concentrations of the reactants and products do not continue to change after equilibrium is reached.

LE CHÂTELIER'S PRINCIPLE

Le Châtelier's principle predicts the direction in which a reversible reaction at equilibrium will go when some condition (temperature, pressure, concentration, etc.) is stressed (i.e., changed). This principle states that when an equilibrium state is stressed by a change, a new equilibrium that reduces that stress is reached.

Consider the formation of ammonia from nitrogen and hydrogen. When the reaction proceeds in the forward direction, energy in the form of heat is released and the temperature increases. If the reaction proceeds in the reverse direction, heat is absorbed and the temperature decreases. If the system is stressed by increasing the temperature, the reaction will proceed in the reverse direction because that direction absorbs heat and reduces the temperature.

For reactions that involve gases, the reaction equation coefficients can be interpreted as volumes. In the nitrogen-hydrogen reaction, four volumes combine to form two volumes. If the equilibrium system is stressed by increasing the pressure, then the forward reaction will occur because this direction reduces the volume and pressure.

If the concentration of any substance is increased, the reaction proceeds in a direction away from the substance with the increase in concentration. For example, an increase in the concentration of the reactants shifts the equilibrium to the right, thus increasing the amount of products formed.

RATE AND ORDER OF REACTIONS

The time required for a reaction to proceed to equilibrium or completion depends on the rate of reaction. The *rate of reaction*, r, is the change in concentration per unit time, measured in moles/L·s.

$$r = \frac{\text{change in concentration}}{\text{time}} \qquad 38.1$$

Consider the following reversible reaction.

$$a\text{A} + b\text{B} \rightleftharpoons c\text{C} + d\text{D} \qquad 38.2$$

The *law of mass action* states that the speed of reaction is proportional to the equilibrium molar concentrations, [X] (i.e., the molarities), of the reactants. In Eqs. 38.3 and 38.4, the constants k_{forward} and k_{reverse} are the reaction rate constants needed to obtain the units of rate.

$$r_{\text{forward}} = k_{\text{forward}}[\text{A}]^a[\text{B}]^b \qquad 38.3$$

$$r_{\text{reverse}} = k_{\text{reverse}}[\text{C}]^c[\text{D}]^d \qquad 38.4$$

At equilibrium, the forward and reverse speeds of reaction are equal.

The rate of reaction for solutions is generally not affected by pressure, but is affected by the following factors.

- Types of substances in the reaction: Some substances are more reactive than others.
- Exposed surface area: The rate of reaction is proportional to the amount of contact between the reactants.
- Concentrations: The rate of reaction increases with increases in concentration.
- Temperature: The rate of reaction increases with increases in temperature.
- Catalysts: A *catalyst* is a substance that increases the reaction rate without being consumed in the reaction. If a catalyst is introduced, rates of reaction will increase (i.e., equilibrium will be reached more quickly), but the equilibrium will not be changed.

The *order of a reaction* is defined as the total number of reacting molecules in or before the slowest step in the mechanism, as determined experimentally. Consider the reversible reaction given by Eq. 38.2. The order of the forward reaction is $a + b$; the order of the reverse reaction is $c + d$.

$$a\text{A} + b\text{B} \rightleftharpoons c\text{C} + d\text{D}$$

EQUILIBRIUM CONSTANT

For reversible reactions, the *equilibrium constant*, K_{eq}, is equal to the ratio of the forward rate of reaction to the reverse rate of reaction. Except for catalysis, the equilibrium constant depends on the same factors affecting the reaction rate. For the reversible reaction given by Eq. 38.2, the equilibrium constant is given by the *law of mass action*.

$$K_{eq} = \frac{[\text{C}]^c[\text{D}]^d}{[\text{A}]^a[\text{B}]^b} = \frac{k_{\text{forward}}}{k_{\text{reverse}}} \qquad 38.5$$

If any of the reactants or products are in pure solid or pure liquid phases, their concentrations are omitted from the calculation of the equilibrium constant. For example, in weak aqueous solutions, the concentration of water, $[\text{H}_2\text{O}]$, is very large and essentially constant; therefore, that concentration is omitted.

For gaseous reactants and products, the concentrations (i.e., the numbers of atoms) will be proportional to the partial pressures. Therefore, an equilibrium constant can be calculated directly from the partial pressures

and is given the symbol K_p. For example, for the formation of ammonia gas from nitrogen and hydrogen, the equilibrium constant is

$$K_p = \frac{[p_{\text{NH}_3}]^2}{[p_{\text{N}_2}][p_{\text{H}_2}]^3} \qquad 38.6$$

K_{eq} and K_p are not numerically the same, but they are related by Eq. 38.7. Δn is the number of moles of products minus the number of moles of reactants.

$$K_p = K_{eq}(\overline{R}T)^{\Delta n} \qquad 38.7$$

IDEAL GASES

An *ideal gas* obeys the ideal gas laws (i.e., Eqs. 38.8 through 38.10). Under ideal gas conditions, the molecule size is insignificant compared with the distance between molecules, and molecules do not come into contact with each other. The density of an ideal gas can be calculated from Eq. 38.8 in which ρ is the density of the gas, ν is the specific volume, p is the absolute pressure, R is the specific gas constant, and T is the temperature.

$$\rho = \frac{1}{\nu} = \frac{p}{RT} \qquad 38.8$$

A general relationship that applies to any ideal gas experiencing any process is shown by Eq. 38.9.

$$\frac{p_1 V_1}{T_1} = \frac{p_2 V_2}{T_2} \qquad 38.9$$

AVOGADRO'S HYPOTHESIS

Avogadro's hypothesis (which is true) states that equal volumes of all gases at the same temperature and pressure contain equal numbers of gas molecules. Specifically, at *standard scientific conditions* (1.0 atm and 0°C), 1 gram-mole of any gas occupies 22.4 L.

Avogadro's law can be stated as the *equation of state* for ideal gases, Eq. 38.10. $\overline{R}$ is the *universal gas constant*, which has a value of 0.08206 atm·L/mol·K (or 8314 J/kmol·K) and can be used with any gas. The number of moles is n.

$$pV = n\overline{R}T \qquad 38.10$$

ACIDS AND BASES

An *acid* is any compound that dissociates in water into H^+ ions. (The combination of H^+ and water, H_3O^+, is known as the *hydronium ion*.) This is known as the *Arrhenius theory of acids*. Acids with one, two, and three ionizable hydrogen atoms are called *monoprotic*, *diprotic*, and *triprotic acids*, respectively.

The properties of acids are as follows.

- Acids conduct electricity in aqueous solutions.
- Acids have a sour taste.
- Acids turn blue litmus paper red.
- Acids have a pH between 0 and 7.
- Acids neutralize bases.
- Acids react with active metals to form hydrogen.

$$2H^+ + Zn \longrightarrow Zn^{++} + H_2$$

- Acids react with oxides and hydroxides of metals to form salts and water.

$$2H^+ + 2Cl^- + FeO \longrightarrow Fe^{++} + 2Cl^- + H_2O$$

- Acids react with salts of either weaker or more volatile acids (such as carbonates and sulfides) to form a new salt and a new acid.

$$2H^+ + 2Cl^- + CaCO_3 \longrightarrow H_2CO_3 + Ca^{++} + 2Cl^-$$

A *base* is any compound that dissociates in water into OH^- ions. This is known as the *Arrhenius theory of bases*. Bases with one, two, and three replaceable hydroxide ions are called *monohydroxic*, *dihydroxic*, and *trihydroxic* bases, respectively.

The properties of bases are as follows.

- Bases conduct electricity in aqueous solutions.
- Bases have a bitter taste.
- Bases turn red litmus paper blue.
- Bases have a pH between 7 and 14.
- Bases neutralize acids, forming salts and water.

A measure of the strength of an acid or base is the number of hydrogen or hydroxide ions in a liter of solution. Since these are very small numbers, a logarithmic scale is used.

$$pH = -\log_{10}[H^+] = \log_{10}\left(\frac{1}{[H^+]}\right) \quad 38.11$$

$$pOH = -\log_{10}[OH^-] = \log_{10}\left(\frac{1}{[OH^-]}\right) \quad 38.12$$

The quantities $[H^+]$ and $[OH^-]$ in square brackets are the ionic concentrations in moles of ions per liter. The number of moles can be calculated from Avogadro's law by dividing the actual number of ions per liter by 6.022×10^{23}.

A *neutral solution* has a pH of 7. Solutions with pH less than 7 are acidic; the smaller the pH, the more acidic the solution. Solutions with pH more than 7 are basic.

ORGANIC CHEMISTRY

Organic chemistry deals with the formation and reaction of compounds of carbon, many of which are produced by living organisms. Organic compounds typically have one or more of the following characteristics.

- Organic compounds are relatively insoluble in water.
- Organic compounds are soluble in organic solvents.
- Organic compounds are relatively nonionizing.
- Organic compounds are unstable at high temperatures.

Certain combinations of atoms occur repeatedly in organic compounds and remain intact during reactions. Such combinations are called *functional groups*. For example, the radical OH is known as a *hydroxyl group*. Table 38.1 contains the most important functional groups.

Table 38.1 Functional Groups of Organic Compounds

name	standard symbol	formula	number of single bonding sites
aldehyde		CHO	1
alkyl	[R]	C_nH_{2n+1}	1
alkoxy	[RO]	$C_nH_{2n+1}O$	1
amine (amino, $n = 2$)		NH_n	$3 - n \, [n = 0, 1, 2]$
aryl (benzene ring)	[Ar]	C_6H_5	1
carbinol		COH	3
carbonyl (keto)		CO	2
carboxyl		COOH	1
ester		COO	2
ether		O	2
halogen (halide)	[X]	Cl, Br, I, or F	1
hydroxyl		OH	1
nitrile		CN	1
nitro		NO_2	1

For convenience, organic compounds are categorized into families. Compounds within each family have similar structures based on similar combinations of groups. For example, all alcohols have the structure [R]–OH, where [R] is any alkyl group and OH is the hydroxyl group. Table 38.2 contains the most common organic families.

Families of compounds can be further subdivided into subfamilies. For example, the hydrocarbons are classified into *alkanes* (single carbon-carbon bond), *alkenes* (double carbon-carbon bond), and *alkynes* (triple carbon-carbon bond).

Table 38.2 Families of Organic Compounds

family	structure	example
acids		
carboxylic acids	[R]-COOH	acetic acid ($(CH_3)COOH$)
fatty acids	[Ar]-COOH	benzoic acid (C_6H_5COOH)
alcohols		
aliphatic	[R]-OH	methanol (CH_3OH)
aromatic	[Ar]-[R]-OH	benzyl alcohol ($C_6H_5CH_2OH$)
aldehydes	[R]-CHO	formaldehyde (HCHO)
alkyl halides	[R]-[X]	chloromethane (CH_3Cl)
amides	[R]-CO-NH$_n$	β-methylbutyramide ($C_4H_9CONH_2$)
amines	[R]$_{3-n}$-NH$_n$	methylamine (CH_3NH_2)
	[Ar]$_{3-n}$-NH$_n$	aniline ($C_6H_5NH_2$)
primary amines: $n=2$		
secondary amines: $n=1$		
tertiary amines: $n=0$		
amino acids	CH-[R]-(NH_2)COOH	glycine ($CH_2(NH_2)COOH$)
anhydrides	[R]-CO-O-CO-[R']	acetic anhydride ($(CH_3CO)_2O$)
aromatics	C_nH_{2n-6}	benzene (C_6H_6)
aryl halides	[Ar]-[X]	fluorobenzene (C_6H_5F)
carbohydrates	$C_x(H_2O)_y$	dextrose ($C_6H_{12}O_6$)
sugars		
polysaccharides		
esters	[R]-COO-[R']	methyl acetate (CH_3COOCH_3)
ethers	[R]-O-[R]	diethyl ether ($C_2H_5OC_2H_5$)
	[Ar]-O-[R]	methyl phenyl ether ($CH_3OC_6H_5$)
	[Ar]-O-[Ar]	diphenyl ether ($C_6H_5OC_6H_5$)
glycols	$C_nH_{2n}(OH)_2$	ethylene glycol ($C_2H_4(OH)_2$)
hydrocarbons		
alkanes[a]	C_nH_{2n+2}	octane (C_8H_{18})
saturated hydrocarbons		
cycloalkanes (cycloparaffins)		
	C_nH_{2n}	cyclohexane (C_6H_{12})
alkenes[b]	C_nH_{2n}	ethylene (C_2H_4)
unsaturated hydrocarbons		
cycloalkenes	C_nH_{2n-2}	cyclohexene (C_6H_{10})
alkynes	C_nH_{2n-2}	acetylene (C_2H_2)
unsaturated hydrocarbons		
ketones	[R]-[CO]-[R]	acetone ($(CH_3)_2CO$)
nitriles	[R]-CN	acetonitrile (CH_3CN)
phenols	[Ar]-OH	phenol (C_6H_5OH)

[a] Alkanes are also known as the *paraffin series* or *methane series*.

[b] Alkenes are also known as the *olefin series*.

SAMPLE PROBLEMS

Problem 1

Which of the following reactions are not balanced?

   I. $2Ca_3(PO_4)_2 + 6SiO_2 \longrightarrow 6CaSiO_3 + 2P_4O_{10}$
  II. $2LiH + B_2H_6 \longrightarrow 2LiBH_4$
 III. $N_2O_5 \longrightarrow 2NO_2 + O_2$
 IV. $HA + H_2O \longrightarrow H_3O^+ + A^-$

(A) I only
(B) IV only
(C) I and III
(D) II and III

Solution

The numbers of each element must match on both sides of the equation. The oxygens are unbalanced in both equations I and III. Equations II and IV have balanced elements.

Answer is C.

Problem 2

Nitroglycerin is made by combining glycerol, nitric acid, and sulfuric acid. What are the minimum coefficients needed to balance the equation of this reaction?

$$\_C_3H_8O_3 + \_HNO_3 + \_H_2SO_4$$
$$\longrightarrow \_C_3H_5N_3O_9 + \_H_2O + \_H_2SO_4$$

(A) 2, 6, 2, 2, 6, 2
(B) 1, 3, 3, 1, 3, 2
(C) 4, 2, 1, 1, 2, 4
(D) 1, 3, 1, 1, 3, 1

Solution

Choices (B) and (C) can be eliminated because the first and last coefficients on one side of the equation must be equal to the first and last coefficients, respectively, on the other side in order for carbon and sulfur to balance. Choices (A) and (D) are both balanced. The coefficients differ by a factor of two; the simplest form, choice (D), is the correct answer.

Answer is D.

Problem 3

Given the following reversible chemical reaction, assume all reactants and products are ideal gases.

$$N_2 + 3H_2 \rightleftharpoons 2NH_3 + heat$$

If the pressure in the reaction container is doubled, what would be the expected results? (Choose the best answer.)

(A) The amount of ammonia (NH_3) would double.
(B) There would be no change in the amount of ammonia (NH_3) present.
(C) More ammonia (NH_3) would be generated.
(D) The amount of ammonia (NH_3) would halve.

Solution

By Le Châtelier's principle, a reversible reaction will find a new equilibrium in response to an added stress, and the new equilibrium will reduce that stress. In this case, the stress is an increase in pressure. For gas reaction equations, the coefficients represent volumes as well as moles. In the given equation, four volumes of reactants produce two volumes of product. The reaction will proceed to a new equilibrium in which more ammonia is generated because this will reduce the pressure.

Answer is C.

Problem 4

What is the order of reaction with respect to reactant E and the overall order of the reaction described by the following rate law?

$$\text{rate} = k_2[\text{E}]^2$$

(A) second order with respect to E; fourth order overall
(B) second order with respect to E; second order overall
(C) first order with respect to E; second order overall
(D) first order with respect to E; fourth order overall

Solution

For a reaction of the form $a\text{A} + b\text{B} \longrightarrow c\text{C} + d\text{D}$, the rate of reaction is

$$r = k[\text{A}]^a[\text{B}]^b$$

For the given rate law, the reaction must be of the form

$$\text{E} + \text{E} \longrightarrow \text{product}$$

The order of a reaction is the total number of molecules reacting. In this case, the order of reaction is second with respect to reactant E and second overall (since there are no other reactants).

Answer is B.

Problem 5

Which of the following does a catalyst change?

(A) the concentration of product at equilibrium
(B) the equilibrium constant of a reaction
(C) the heat of reaction of a reaction
(D) the activation energy of a reaction

Solution

A catalyst is a substance that increases the rate of reaction without being consumed in the reaction. A catalyst lowers the activation energy. The energy at equilibrium is not affected by the catalyst, so the concentration of product and the equilibrium constant are unchanged.

Answer is D.

Problem 6

An unknown quantity of hydrogen gas has a volume of 2.5 L at STP (0°C and 1 atm). What is the mass of hydrogen?

(A) 0.073 g
(B) 0.19 g
(C) 0.22 g
(D) 0.51 g

Solution

Use the ideal gas law.

$$pV = n\overline{R}T$$
$$p = 1 \text{ atm}$$
$$V = 2.5 \text{ L}$$
$$\overline{R} = 0.08206 \text{ atm·L/mol·K}$$
$$T = 273\text{K}$$
$$n = \frac{pV}{\overline{R}T}$$
$$= \frac{(1 \text{ atm})(2.5 \text{ L})}{\left(0.08206 \, \frac{\text{atm·L}}{\text{mol·K}}\right)(273\text{K})}$$
$$= 0.1116 \text{ mol}$$
$$\text{MW of H}_2 = (2)(1.0079 \text{ g})$$
$$= 2.0158 \text{ g/mol}$$
$$m = n(\text{MW})$$
$$= (0.1116 \text{ mol})\left(2.0158 \, \frac{\text{g}}{\text{mol}}\right)$$
$$= 0.225 \text{ g} \quad (0.22 \text{ g})$$

Answer is C.

Problem 7

A compound in gas form has a mass of 0.377 g and occupies 191.6 mL at standard conditions ($0°C$ and 760 mm Hg). What is the formula of the compound?

(A) CH_4
(B) C_3H_8
(C) C_5H_{12}
(D) C_2H_6

Solution

By Avogadro's hypothesis, 1 gram-mole of any gas occupies 22.4 L. By a simple ratio analysis, the mass of 1 gram-mole of the compound is

$$m = \left(\frac{22.4 \text{ L}}{0.1916 \text{ L}}\right)(0.377 \text{ g})$$

$$= 44.08 \text{ g}$$

Calculate the molecular weights of the compounds listed.

compound	molecular weight
CH_4	$12.011 + (4)(1.0079) = 16.043$
C_3H_8	$(3)(12.011) + (8)(1.0079) = 44.096$
C_5H_{12}	$(5)(12.011) + (12)(1.0079) = 72.150$
C_2H_6	$(2)(12.011) + (6)(1.0079) = 30.069$

Answer is B.

Problem 8

An *alkyl radical* is best defined as

(A) an electron that is shared in a covalent bond.
(B) the remaining portion of an alkane after it loses a hydrogen atom.
(C) any functional group that substitutes for a hydrogen atom in an alkane.
(D) cancer-causing molecules found in foods.

Solution

Many hydrocarbons participate in chemical reactions in which they can be viewed as a charged radical (a group of atoms that combine as a unit) attached to a functional group. The chemical reactivity lies in the bond between the radical and the functional group.

An alkane is a hydrocarbon of the form C_nH_{2n+2}. When it loses a hydrogen atom, it becomes an alkyl radical of the form C_nH_{2n+1} with one bonding site.

Answer is B.

Problem 9

What is the following molecule?

$$CH_3 - \overset{\overset{\textstyle O}{\|}}{C} - O - C_2H_5$$

(A) an alcohol
(B) an aldehyde
(C) an amine
(D) an ester

Solution

An ester has the structure [R]-COO-[R'], where [R] and [R'] are alkyl functional groups of the formula C_nH_{2n+1}. Both CH_3 and C_2H_5 are alkyl groups. The molecule is an ester.

Answer is D.

FE-STYLE EXAM PROBLEMS

1. Balance the following reaction.

$$\_\,HBrO_3 + \_\,HBr \longrightarrow \_\,H_2O + \_\,Br_2$$

(A) $HBrO_3 + 4HBr \longrightarrow 3H_2O + Br_2$
(B) $2HBrO_3 + 4HBr \longrightarrow 3H_2O + 3Br_2$
(C) $3HBrO_3 + HBr \longrightarrow 2H_2O + 2Br_2$
(D) $HBrO_3 + 5HBr \longrightarrow 3H_2O + 3Br_2$

2. During a laboratory experiment at 1.0 atm and $25°C$, a student observed that oxygen gas was produced by decomposition of 15 g of sodium chlorate. What was the volume of oxygen?

(A) 1.27 L
(B) 3.85 L
(C) 5.17 L
(D) 6.54 L

3. Which of the following is (are) not a base when dissolved in water?

I. NH_3
II. sodium carbonate (Na_2CO_3)
III. sodium hydroxide ($NaOH$)
IV. C_6H_5COOH

(A) IV only
(B) I and III
(C) II and III
(D) II and IV

4. A solution is adjusted from pH 8 to pH 9. The relative concentration of the hydrogen [H^+] ion has changed by a factor of what?

(A) $\dfrac{1}{100}$

(B) $\dfrac{1}{10}$

(C) 5

(D) 10

5. What family of compounds is produced from the reaction between an alcohol and a carboxylic acid?

(A) amine
(B) ether
(C) ester
(D) ketone

For the following problems use the NCEES Handbook as your only reference.

6. Dimethyl hydrazine $(CH_3)_2NNH_2$ has been used as a fuel in space, with nitrogen tetraoxide (N_2O_4) as the oxidizer. The products of the reaction between these two in an engine are H_2O, CO_2, and N_2. What is the mass of nitrogen tetraoxide required to burn 50 kg of dimethyl hydrazine?

(A) 50 kg
(B) 100 kg
(C) 128 kg
(D) 153 kg

7. An unknown gas with a temperature of 25°C and a pressure of 740 mm Hg is collected in a sampling bag. The volume and mass of the gas are 24.0 L and 34.9 g, respectively. Which chemical formula could represent the gas?

(A) N_2
(B) Ar
(C) H_2S
(D) HCl

8. Hydrogen and chlorine gas combine in a 35 m³ reaction vessel to produce hydrogen chloride. The masses of hydrogen and chlorine are 4.5 kg and 160 kg, respectively. How much hydrogen chloride gas is produced?

(A) 21 kg
(B) 41 kg
(C) 82 kg
(D) 160 kg

9. The final temperature of the hydrogen and chlorine described in Prob. 8 is 30°C. What is the final pressure in the reaction vessel?

(A) 80 kPa
(B) 160 kPa
(C) 240 kPa
(D) 320 kPa

10. A transportation company specializes in the shipment of pressurized gaseous materials. An order is received for 100 L of a particular gas at STP (0°C and 1 atm). What minimum volume tank is necessary to transport the gas at 25°C and a maximum pressure of 8 atm?

(A) 10 L
(B) 12 L
(C) 14 L
(D) 16 L

11. Which of the following statements concerning reversible reactions is false?

(A) Both reactants and products are always present.
(B) Concentrations have no effect on the direction of the reaction.
(C) Temperature affects the direction of the reaction.
(D) Concentrations remain constant once equilibrium is reached.

12. It has been determined that 1.0 L of solution contain 52.7 g H_2SO_4, 240.8 g $KMnO_4$, 11.3 g K_2SO_4, and 5.5 g Mn_2O_7. The reaction equation is

$$H_2SO_4 + 2KMnO_4 \rightleftharpoons K_2SO_4 + Mn_2O_7 + H_2O$$

The molecular weights of the compounds are

$$H_2SO_4: 98 \text{ g/mol}$$
$$KMnO_4: 158 \text{ g/mol}$$
$$K_2SO_4: 174 \text{ g/mol}$$
$$Mn_2O_7: 222 \text{ g/mol}$$
$$H_2O: 18 \text{ g/mol}$$

What is the equilibrium constant for the reaction?

(A) 1.3×10^{-3}
(B) 2.6×10^{-3}
(C) 5.2×10^{-3}
(D) 6.9×10^{-3}

13. It is known that ozone (O_3) will decompose into oxygen (O_2) at a temperature of 100°C. One mole of ozone is sealed in a container at STP (0°C and 1 atm). What will be the pressure of the container once it is heated to 100°C?

(A) 1.4 kPa
(B) 2.1 kPa
(C) 37 kPa
(D) 210 kPa

14. There are 500 g of zinc sulfide (ZnS) in a load of zinc ore. The ZnS is roasted in excess air to form zinc oxide (ZnO) and sulfur dioxide (SO$_2$). How many grams of zinc can be subsequently recovered if 5% of the zinc is lost in the roasting process?

(A) 320 g
(B) 340 g
(C) 380 g
(D) 400 g

15. 10 g of solid PCl$_5$ is heated in a 0.5 m^3 container to 150°C, producing gaseous PCl$_3$ and Cl$_2$ gas according to the following decomposition reaction:

$$PCl_5 \text{ (s)} + \text{heat} \longrightarrow PCl_3 \text{ (g)} + Cl_2 \text{ (g)}$$

The molecular weights of the compounds are

$$PCl_5: 208.5 \text{ g/mol}$$
$$PCl_3: 137.5 \text{ g/mol}$$
$$Cl_2: 71 \text{ g/mol}$$

What is the increase in pressure in the container when 50% (by weight) of the PCl$_5$ is decomposed?

(A) 0.120 kPa
(B) 0.250 kPa
(C) 0.350 kPa
(D) 18 kPa

SOLUTIONS TO FE-STYLE EXAM PROBLEMS

1. To solve this problem quickly, look through the answer choices to see if any choices can be eliminated. Choice (A) has unbalanced hydrogens, while choices (B) and (C) have unbalanced oxygens. That leaves choice (D) as the only possible choice.

Balancing the equation from scratch requires some deductive trial and error. In the equation, the hydrogens and bromine are balanced but the oxygen is not. Multiply the H$_2$O by 3.

$$HBrO_3 + HBr \longrightarrow 3H_2O + Br_2$$

This leaves the left side short of four hydrogens. By trial and error, add four more HBr to the left side and two more Br$_2$ to the right side.

$$HBrO_3 + 5HBr \longrightarrow 3H_2O + 3Br_2$$

Answer is D.

2. The decomposition reaction is

$$2NaClO_3 \longrightarrow 2NaCl + 3O_2$$

The molecular weight of sodium chlorate is

$$22.990 \ \frac{g}{mol} + 35.453 \ \frac{g}{mol} + (3)\left(15.999 \ \frac{g}{mol}\right)$$
$$= 106.44 \text{ g/mol}$$

Calculate the moles of O$_2$ produced.

$$\frac{15 \text{ g NaClO}_3}{106.44 \ \frac{g}{mol}} = 0.14092 \text{ mol NaClO}_3$$

$$(0.14092 \text{ mol NaClO}_3)$$
$$\times \left(\frac{3 \text{ mol O}_2}{2 \text{ mol NaClO}_3}\right) = 0.21138 \text{ mol O}_2$$

Use the ideal gas law to calculate the volume.

$$pV = n\overline{R}T$$
$$\overline{R} = 0.08206 \text{ atm·L/mol·K}$$
$$T = 273 + 25\text{K} = 298\text{K}$$
$$p = 1 \text{ atm}$$
$$V = \frac{n\overline{R}T}{p}$$

$$= \frac{(0.21138 \text{ mol O}_2)\left(0.08206 \ \frac{\text{atm·L}}{\text{mol·K}}\right)(298\text{K})}{1 \text{ atm}}$$
$$= 5.17 \text{ L}$$

Answer is C.

3. According to the Arrhenius theory of bases, a base is any compound that dissociates in water into OH$^-$ ions.

$$NH_3 + H_2O \rightleftharpoons NH_4^+ + OH^-$$
$$Na_2CO_3 + H_2O \rightleftharpoons 2Na^+ + HCO_3^- + OH^-$$
$$NaOH + H_2O \rightleftharpoons Na^+ + OH^-$$
$$C_6H_5COOH + H_2O \rightleftharpoons C_6H_5COO^- + H^+$$

Answer is A.

4. The definition of pH is

$$pH = -\log_{10}[H^+] = \log_{10}\left(\frac{1}{[H^+]}\right)$$

For pH = 8 [H$^+$]=10^{-8}. For pH = 9, [H$^+$] = 10^{-9}. The change in [H$^+$] is by a factor of $10^{-9}/10^{-8}$, or 1/10.

Answer is B.

5. An alcohol has the structure [R]–OH. A carboxylic acid has the structure [R]–COOH. Together they react in a process referred to as *esterification*, in which a water molecule is removed (dehydration). The symbolic formula is

$$[R]\text{–}OH + [R']\text{–}COOH \longrightarrow [R]\text{–}COO\text{–}[R'] + H_2O$$

The resultant product is in the family called *esters*. This is the organic chemistry equivalent of reacting an acid and a base to obtain water and an inorganic salt.

Answer is C.

6. Balance the equation and calculate the combining weights.

$$\begin{array}{cccccc} (CH_3)_2NNH_2 & + & 2N_2O_4 & \longrightarrow & 4H_2O & + & 2CO_2 & + & 3N_2 \\ 60.1 & & (2)(92.0) & & (4)(18) & (2)(44) & (3)(28) \end{array}$$

By proportionality with the combining weights,

$$\frac{N_2O_4}{(CH_3)_2NNH_2} = \frac{(2)(92.0)}{60.1} = \frac{x}{50 \text{ kg}}$$

$$x = 153.1 \text{ kg}   (153 \text{ kg})$$

Answer is D.

7. Use the ideal gas law to convert the volume to standard conditions.

$$V_2 = \frac{p_1 V_1 T_2}{p_2 T_1}$$

$$= \frac{(740 \text{ mm Hg})(24.0 \text{ L})(0°C + 273)}{(760 \text{ mm Hg})(25°C + 273)}$$

$$= 21.4 \text{ L}$$

Avogadro's hypothesis states that 1 gram-mole of any ideal gas occupies 22.4 L at standard conditions (0°C and 760 mm Hg).

The molecular weight is

$$(MW)_{gas} = \frac{(34.9 \text{ g})(22.4 \text{ L})}{21.4 \text{ L}} = 36.5 \text{ g/mol}$$

The molecular weight of HCl is

$$(MW)_{HCl} = (1)\left(1 \frac{g}{mol}\right) + (1)\left(35.5 \frac{g}{mol}\right)$$

$$= 36.5 \text{ g/mol}$$

Answer is D.

8. The molecular weights of hydrogen and chlorine gas are approximately 2 and 71, respectively. Determine the number of gram-moles of each gas in the vessel.

$$H_2: \frac{4500 \text{ g}}{2 \frac{g}{mol}} = 2250 \text{ mol}$$

$$Cl_2: \frac{160\,000 \text{ g}}{71 \frac{g}{mol}} = 2254 \text{ mol}$$

The reaction equation is

$$H_2 + Cl_2 \longrightarrow 2HCl$$

For every mole of hydrogen, two moles of HCl gas are produced. Therefore, there are 4500 mol of HCl in the vessel.

$$(4500 \text{ mol})\left(36.5 \frac{g}{mol}\right) = 164\,250 \text{ g}   (160 \text{ kg})$$

Answer is D.

9. Use the ideal gas law to determine the pressure.

$$p_2 = \frac{n\overline{R}T}{V}$$

$$= \frac{(4500 \text{ mol})\left(8.314 \frac{J}{mol \cdot K}\right)(30°C + 273.15)}{35 \text{ m}^3}$$

$$= 324\,050 \text{ Pa}   (320 \text{ kPa})$$

Answer is D.

10. Use the ideal gas law.

$$V_2 = \frac{p_1 V_1 T_2}{p_2 T_1}$$

$$= \frac{(1 \text{ atm})(100 \text{ L})(25°C + 273.15)}{(8 \text{ atm})(0°C + 273.15)}$$

$$= 13.6 \text{ L}   (14 \text{ L tank minimum})$$

Answer is C.

11. According to Le Chatelier's principle, when a substance is added to an equilibrium, the reaction will proceed in the direction that reduces the quantity of the substance. Concentrations will affect the direction of the reaction.

Answer is B.

12. If any of the reactants or products are in pure solid or pure liquid phases, their concentrations should be omitted from the calculation of the equilibrium constant. Therefore, H_2O is omitted.

Since the masses given are per liter of solution, the molarities of each compound in solution are

$$[H_2SO_4]: \frac{52.7 \frac{g}{L}}{98 \frac{g}{mol}} = 0.538 \text{ M}$$

$$[KMnO_4]: \frac{240.8 \frac{g}{L}}{158 \frac{g}{mol}} = 1.524 \text{ M}$$

$$[K_2SO_4]: \frac{11.3 \frac{g}{L}}{174 \frac{g}{mol}} = 0.065 \text{ M}$$

$$[Mn_2O_7]: \frac{5.5}{222 \frac{g}{mol}} = 0.025 \text{ M}$$

The equilibrium constant is

$$K = \frac{[C]^c[D]^d}{[A]^a[B]^b}$$

$$= \frac{(0.065 \text{ M})(0.025 \text{ M})}{(0.538 \text{ M})(1.524 \text{ M})^2}$$

$$= 0.0013 \quad (1.3 \times 10^{-3})$$

Answer is A.

13. The decomposition reaction equation is

$$\text{heat energy} + 2O_3 \longrightarrow 3O_2$$

In reaction equations involving ideal gases, the coefficients can be interpreted as the number of molecules, the number of volumes, or the number of moles. In this case, 3 mol of oxygen are produced from 2 mol of ozone.

Use the ideal gas equation of state.

$$pV = n\overline{R}T$$

Since this is a constant-volume process,

$$V = \frac{n_1\overline{R}T_1}{p_1} = \frac{n_2\overline{R}T_2}{p_2}$$

$$p_2 = \frac{p_1 T_2 n_2}{T_1 n_1}$$

$$= \frac{(1 \text{ atm})(100°C + 273)(3 \text{ mol})}{(0°C + 273)(2 \text{ mol})}$$

$$= 2.05 \text{ atm}$$

$$= (2.05 \text{ atm})(101.3 \text{ kPa})$$

$$= 207.7 \text{ kPa} \quad (210 \text{ kPa})$$

Answer is D.

14. The fraction of Zn in ZnS is

$$\frac{65.4 \frac{g}{mol}}{65.4 \frac{g}{mol} + 32 \frac{g}{mol}} = 0.67$$

The amount of zinc that can be recovered is

$$(1 - 0.05)(0.67)(500 \text{ g}) = 318 \text{ g} \quad (320 \text{ g})$$

Answer is A.

15. 5 g of PCl_5 decompose. The mass of PCl_3 produced can be determined by calculating the ratio of PCl_3 to PCl_5 combining weights.

$$PCl_3/PCl_5: \frac{137.5 \frac{g}{mol}}{208.5 \frac{g}{mol}} = \frac{m}{5 \text{ g}}$$

$$m = 3.3 \text{ g of } PCl_3$$

The number of moles of PCl_3 is

$$n = \frac{m}{MW} = \frac{3.3 \text{ g}}{137.5 \frac{g}{mol}}$$

$$= 0.024 \text{ mol}$$

The reaction equation shows that equal numbers of moles, volumes, and molecules of both products are produced. 0.024 moles of Cl_2 will be produced. The total number of moles of gas produced will be

$$\Delta n = (2)(0.024 \text{ mol}) = 0.048 \text{ mol}$$

Use the ideal gas law.

$$\Delta pV = \Delta n\overline{R}T$$

$$\Delta p = \frac{\Delta n\overline{R}T}{V}$$

$$= \frac{(0.048 \text{ mol})\left(10^{-3} \frac{kmol}{mol}\right)\left(8314 \frac{J}{kmol\cdot K}\right)}{0.5 \text{ m}^3}$$

$$= \frac{\times (150°C + 273)}{0.5 \text{ m}^3}$$

$$= 337.6 \text{ Pa} \quad (0.350 \text{ kPa})$$

Answer is C.

39

Solutions

Subjects

UNITS OF CONCENTRATION 39-1
SOLUTIONS OF GASES IN LIQUIDS . . . 39-1
SOLUTIONS OF SOLIDS IN LIQUIDS . . . 39-1
SOLUBILITY PRODUCT 39-2
ENTHALPY OF FORMATION 39-2
ENTHALPY OF REACTION 39-2
HEAT OF SOLUTION 39-2
BOILING AND FREEZING POINTS 39-3
FARADAY'S LAWS OF ELECTROLYSIS . 39-3

Nomenclature

F	formality	FW/L
FW	formula weight	g
GEW	gram equivalent weight	g
h	Henry's law constant	atm
H	enthalpy	kcal/mol
I	current	A
K	constant	–
m	mass	kg
m	molality	mol/1000 g
M	molarity	mol/L
MW	molecular weight	kg/kmol
n	number of moles	–
N	normality	GEW/L
p	pressure	Pa
t	time	s
T	temperature	°C
x	mole fraction	–

Subscripts

b	boiling point
f	freezing point or formation
i	partial
r	reaction
sp	solubility product

UNITS OF CONCENTRATION

There are many units of concentration to express solution strengths.

F— formality: The number of gram formula weights (i.e., formula weights in grams) per liter of solution.

m—molality: The number of gram-moles of solute per 1000 grams of solvent. A "molal" (i.e., 1 m) solution contains 1 gram-mole per 1000 grams of solvent.

M—molarity: The number of gram-moles of solute per liter of solution. A "molar" (i.e., 1 M) solution contains 1 gram-mole per liter of solution. Molarity is related to normality: $N = M \times \Delta$ oxidation number.

N—normality: The number of gram equivalent weights of solute per liter of solution. A solution is "normal" (i.e., 1 N) if there is exactly one gram equivalent weight per liter. Molarity is related to normality: $N = M \times \Delta$ oxidation number.

x—mole fraction: The number of moles of solute divided by the number of moles of solvent and all solutes.

SOLUTIONS OF GASES IN LIQUIDS

Henry's law states that the amount (i.e., concentration, mass, weight, or mole fraction) of a slightly soluble gas dissolved in a liquid is proportional to the partial pressure of the gas as long as the gas and liquid are nonreacting. This law applies separately to each gas to which the liquid is exposed, as if each gas were present alone. The algebraic form of Henry's law is given by Eq. 39.1, in which h is the Henry's law constant in atmospheres.

$$p_i = hx_i \qquad 39.1$$

Generally, the solubility of gases in liquids decreases with increasing temperature.

SOLUTIONS OF SOLIDS IN LIQUIDS

When a solid is added to a liquid, the solid is known as the *solute* and the liquid is known as the *solvent*. If the dispersion of the solute throughout the solvent is at the molecular level, the mixture is known as a *solution*.

If the solute particles are larger than molecules, the mixture is known as a *suspension*.

In some solutions, the solvent and solute molecules bond loosely together. This loose bonding is known as *solvation*. If water is the solvent, the bonding process is also known as *aquation* or *hydration*.

The solubility of most solids in liquids increases with increasing temperature. Pressure has very little effect on the solubility of solids in liquids.

When the solvent has dissolved as much solute as it can, it is known as a *saturated solution*. Adding more solute to an already saturated solution will cause the excess solute to settle to the bottom of the container, a process known as *precipitation*. Other changes (in temperature, concentration, etc.) can be made to cause precipitation from saturated and unsaturated solutions.

SOLUBILITY PRODUCT

When an ionic solid is dissolved in a solvent, it dissociates. For example, consider the ionization of silver chloride in water.

$$AgCl(s) \rightleftharpoons Ag^+(aq) + Cl^-(aq)$$

If the equilibrium constant is calculated, the terms for pure solids and liquids (in this case, [AgCl] and [H_2O]) are omitted. Thus, the *solubility product*, K_{sp}, consists only of the ionic concentrations. The solubility product for slightly soluble solutes is essentially constant at a standard value.

$$K_{sp} = [Ag^+][Cl^-]$$

When the product of terms exceeds the standard value of the solubility product, solute will precipitate out until the product of the remaining ion concentrations attain the standard value. If the product is less than the standard value, the solution is not saturated.

The solubility products of nonhydrolyzing compounds are relatively easy to calculate. This encompasses chromates (CrO_4^{-2}), halides (F^-, Cl^-, Br^-, I^-), sulfates (SO_4^{-2}), and iodates (IO_3^-). However, compounds that hydrolyze must be evaluated differently.

ENTHALPY OF FORMATION

Enthalpy, H, is the potential energy that a substance possesses by virtue of its temperature, pressure, and phase. The *enthalpy of formation (heat of formation)*, ΔH_f, of a compound is the energy absorbed during the formation of one gram-mole of the compound from pure elements. The enthalpy of formation is assigned a value of zero for elements in their free states at 25°C and 1 atm. This is the so-called *standard state* for enthalpies of formation.

ENTHALPY OF REACTION

The *enthalpy of reaction (heat of reaction)*, ΔH_r, is the energy absorbed during a chemical reaction under constant volume conditions. It is found by summing the enthalpies of formation of all products and subtracting the sum of enthalpies of formation of all reactants. This is essentially a restatement of the energy conservation principle and is known as *Hess' law of energy summation*.

$$\Delta H_r = \sum \Delta H_{f,\text{products}} - \sum \Delta H_{f,\text{reactants}} \quad 39.2$$

Reactions that give off energy (i.e., have negative enthalpies of reaction) are known as *exothermic reactions*. Many (but not all) exothermic reactions begin spontaneously. On the other hand, endothermic reactions absorb energy and require heat or electrical energy to begin.

HEAT OF SOLUTION

The *heat of solution*, ΔH, is an amount of energy that is absorbed or released when a substance enters a solution. It can be calculated from the enthalpies of formation of the solution components. For example, the heat of solution associated with the formation of dilute hydrochloric acid from HCl gas and large amounts of water would be represented as follows.

$$HCl(g) \xrightarrow{H_2O} HCl(aq) + \Delta H$$

$$\Delta H = -17.21 \text{ kcal/mol}$$

If a heat of solution is negative (as it is for all aqueous solutions of gases), heat is given off when the solute dissolves in the solvent. This is an *exothermic reaction*. If the heat of solution is positive, heat is absorbed when the solute dissolves in the solvent. This is an *endothermic reaction*.

BOILING AND FREEZING POINTS

A liquid boils when its vapor pressure is equal to the surrounding pressure. Because the addition of a solute to a solvent decreases the vapor pressure (Raoult's law), the temperature of the solution must be increased to maintain the same vapor pressure. Thus, the boiling point (temperature), T_b, of a solution is higher than the boiling point of the pure solvent at the same pressure.

The *boiling point elevation* is given by Eq. 39.3. K_b is the *molal boiling point constant*, which is a property of the solvent only. The molal boiling point constant for water is 0.512 °C/m.

$$\Delta T_b = mK_b$$

$$= \frac{m_{\text{solute,in g}}K_b}{(\text{MW})m_{\text{solvent,in kg}}} \quad [\text{increase}] \quad 39.3$$

Similarly, the freezing (melting) point, T_f, will be lower for the solution than for the pure solvent. The freezing point depression depends on the *molal freezing point constant*, K_f, a property of the solvent only. The molal freezing point constant for water is 1.86°C/m.

$$\Delta T_f = -mK_f$$

$$= \frac{-m_{\text{solute,in g}}K_f}{(\text{MW})m_{\text{solvent,in kg}}} \quad [\text{decrease}] \quad 39.4$$

Equations 39.3 and 39.4 are for dilute, nonelectrolytic solutions and nonvolatile solutes.

FARADAY'S LAWS OF ELECTROLYSIS

An *electrolyte* is a substance that dissociates in solution to produce positive and negative ions. It can be an aqueous solution of a soluble salt, or it can be an ionic substance in molten form.

Electrolysis is the passage of an electric current through an electrolyte driven by an external voltage source. Electrolysis occurs when the positive terminal (the *anode*) and negative terminal (the *cathode*) of a voltage source are placed in an electrolyte. Negative ions (anions) will be attracted to the anode, where they are oxidized. Positive ions (cations) will be attracted to the cathode, where they will be reduced. The passage of ions constitutes the current.

Some reactions that do not proceed spontaneously can be forced to proceed by supplying electrical energy. Such reactions are called *electrolytic (electrochemical) reactions*.

Faraday's laws of electrolysis can be used to predict the duration and magnitude of a direct current needed to complete an electrolytic reaction.

law 1: The mass of a substance generated by electrolysis is proportional to the amount of electricity used.

law 2: For any constant amount of electricity, the mass of substance generated is proportional to its equivalent weight.

law 3: One *faraday* of electricity (96 485 C or 96 485 A·s) will produce one gram equivalent weight.

The number of grams of a substance produced at an electrode in an electrolytic reaction can be found from Eq. 39.5.

$$m_{\text{grams}} = \frac{It(\text{MW})}{(96\,485)(\text{change in oxidation state})}$$

$$= (\text{no. of faradays})(\text{GEW}) \quad 39.5$$

The number of gram-moles produced is

$$n = \frac{m}{\text{MW}}$$

$$= \frac{\text{no. of faradays}}{\text{change in oxidation state}}$$

$$= \frac{It}{(96\,485)(\text{change in oxidation state})} \quad 39.6$$

SAMPLE PROBLEMS

Problem 1

As the pressure of a gas increases, the solubility of that gas in a liquid

    (A) always increases.
    (B) always decreases.
    (C) is not changed.
    (D) cannot be determined.

Solution

By Henry's law, $x_i = Hp_i$, the fraction of a gas that is dissolved in a liquid is proportional to the partial pressure of the gas. As the pressure increases, so does its solubility in liquid.

Answer is A.

Problem 2

What would you have to know to determine whether a solvent poured into a beaker containing water will float on top of the water?

    (A) The solvent's specific gravity is more than one, and solubility is high.
    (B) The solvent's specific gravity is less than one, and solubility is high.
    (C) The solvent's specific gravity is less than one, regardless of solubility.
    (D) The solvent's specific gravity is less than one, and solubility is low.

Solution

For the solvent to float on water, it must be both lighter than water (specific gravity less than one) and without significant solubility in water. If the solvent has a high solubility, it would dissolve into the water and form a solution.

Answer is D.

Problem 3

The pH of a 0.001 M HCl solution is

(A) 1
(B) 3
(C) 5
(D) 7

Solution

Molarity (M) is the number of gram-moles of solute per liter of solution. To calculate pH, the ionic concentration of H^+ ions in moles per liter is needed. This is equal to the molarity for HCL.

$$
\begin{aligned}
\text{pH} &= \log_{10}\left(\frac{1}{[H^+]}\right) \\
&= \log_{10}\left(\frac{1}{0.001}\right) \\
&= 3
\end{aligned}
$$

Answer is B.

Problem 4

Two moles of sodium react with 2 mol of water to produce which of the following?

(A) 1 sodium hydroxide mol and 1 hydrogen mol
(B) 2 sodium hydroxide mol and 2 hydrogen mol
(C) 2 sodium hydroxide mol and 1 hydrogen mol
(D) 1 sodium hydroxide mol and 2 hydrogen mol

Solution

Sodium in water does not dissolve; it reacts. The chemical reaction is

$$2Na + 2H_2O \longrightarrow 2NaOH + H_2$$

The product is 2 mol of NaOH and 1 mol of H_2.

Answer is C.

Problem 5

Which of the following occurs when table salt (NaCl) is added to continuously heated boiling water?

(A) The water continues to boil.
(B) The water momentarily stops boiling.
(C) The water boils even more agitatedly.
(D) The temperature of the water decreases but boiling continues uninterrupted.

Solution

The boiling point of a solution is higher than the boiling point of the pure solvent at the same pressure. Addition of the solute will momentarily stop the boiling process until the elevated boiling point temperature is reached.

Answer is B.

FE-STYLE EXAM PROBLEMS

1. How many milliliters of 1 M NaOH solution will 25 mL of 2 M H_2SO_4 neutralize?

(A) 25 mL
(B) 50 mL
(C) 75 mL
(D) 100 mL

2. What is the normality of each of the following solutions?

I.   500 mL of 0.25 M H_2SO_4
II. 41.7 g of $K_2Cr_2O_7$ in 600 mL of solution (ionizes to Cr^{3+})
III. 0.135 gram-equivalents of H_2SO_4 in 400 mL of solution

(A) 2 N; 5.2 N; 1.66 N
(B) 6.25 N; 0.7 N; 0.338 N
(C) 0.5 N; 1.42 N; 0.338 N
(D) 2 N; 2.28 N; 0.56 N

3. 2.00 g of a substance dissolved in 250 g of water produces a boiling point elevation of 0.065°C. What is the molecular weight of the substance?

(A) 8
(B) 16
(C) 63
(D) 92

4. A current of 0.075 A passes through a solution of silver nitrate for 10 minutes. How much silver is deposited?

(A) 0.030 g
(B) 0.035 g
(C) 0.040 g
(D) 0.050 g

For the following problems use the NCEES Handbook as your only reference.

5. Water and SO_3 combine to sulfuric acid (H_2SO_4) according to the following reaction.

$$H_2O + SO_3 \longrightarrow H_2SO_4$$

How many grams of water must be added to 100 g of 20% oleum (20% SO_3 and 80% H_2SO_4 by weight) to produce a 95% solution (by weight) of sulfuric acid?

(A) 3.3 g
(B) 5.0 g
(C) 7.5 g
(D) 14 g

6. When a deliquescent substance is exposed to air, it

(A) oxidizes.
(B) crystallizes.
(C) loses water of hydration.
(D) becomes moist.

7. How many grams of copper will be deposited at an electrode if a current of 1.5 A is supplied for 2 hours to a $CuSO_4$ solution?

(A) 2.4 g
(B) 3.6 g
(C) 7.1 g
(D) 48 g

8. Enthalpy of formation is most closely defined as the

(A) potential energy of a substance.
(B) energy absorbed or released during a chemical reaction.
(C) sum of the enthalpy of reactions.
(D) energy absorbed during creation of 1 gram-mole of a compound from pure elements.

9. What is the enthalpy of reaction at 25°C for the combustion of ethane (C_2H_6)?

$$2C_2H_6 + 7O_2 \longrightarrow 4CO_2 + 6H_2O$$

$$\Delta H_f(C_2H_6) = -20.24 \text{ kcal/mol}$$
$$\Delta H_f(O_2) = 0.00 \text{ kcal/mol}$$
$$\Delta H_f(CO_2) = -94.05 \text{ kcal/mol}$$
$$\Delta H_f(H_2O) = -57.80 \text{ kcal/mol}$$

(A) −680 kcal/mol (exothermic)
(B) −340 kcal/mol (exothermic)
(C) 130 kcal/mol (endothermic)
(D) 340 kcal/mol (endothermic)

10. 6 g of a substance are dissolved in 1000 g of water. The solution freezes at −0.16°C. What is the molecular weight of the substance?

(A) 60 g/mol
(B) 70 g/mol
(C) 75 g/mol
(D) 100 g/mol

11. A wastewater treatment plant uses chlorine gas as a reactant. A tank is filled with 800 m^3 of 20°C water, and chlorine is added at a dosage of 125 g per cubic meter of water. (Assume all of the chlorine dissolves and none initially reacts chemically.) Henry's law constant for the chlorine is 15.2 atm. If the atmospheric pressure is 1.0 atm, what is the theoretical partial pressure of the chlorine gas at the tank surface immediately after the gas is added?

(A) 3.2×10^{-5} atm
(B) 4.8×10^{-4} atm
(C) 0.039 atm
(D) 0.11 atm

12. The solubility constant of strontium sulfate, $SrSO_4$, is 2.8×10^{-7}. How many grams of $SrSO_4$ must be dissolved in water to produce 1 L of saturated solution?

(A) 0.00005 g
(B) 0.0005 g
(C) 0.1 g
(D) 2 g

13. How many liters of $2M$ solution (i.e., a molarity of 2) can be produced from 184 g of ethyl alcohol (CH_3CH_2OH)?

(A) 1.5 L
(B) 2.0 L
(C) 2.5 L
(D) 5.0 L

14. How much energy is needed to convert ozone to oxygen?

$$\Delta H_f(O_2) = 0 \text{ kcal/mol}$$
$$\Delta H_f(O_3) = 34.0 \text{ kcal/mol}$$

(A) 0 kcal/mol
(B) 34 kcal/mol
(C) 68 kcal/mol
(D) 140 kcal/mol

SOLUTIONS TO FE-STYLE EXAM PROBLEMS

1. Balance the equation.

$$2NaOH + H_2SO_4 \longrightarrow Na_2SO_4 + 2H_2O$$

2 mol of NaOH neutralize 1 mol H_2SO_4.

The number of moles of H_2SO_4 is

$$n_{H_2SO_4} = \left(2 \; \frac{mol}{L}\right)(0.025 \; L)$$

$$= 0.05$$

The number of moles of NaOH needed is

$$n_{NaOH} = (0.05 \; mol \; H_2SO_4)\left(2 \; \frac{mol \; NaOH}{mol \; H_2SO_4}\right)$$

$$= 0.10$$

The volume of NaOH needed is

$$(0.10 \; mol \; NaOH)\left(\frac{1}{1 \; \frac{mol}{L}}\right) = 0.10 \; L \; NaOH$$

$$(100 \; mL \; NaOH)$$

Answer is D.

2. Normality is the number of gram equivalents of solute per liter of solution.

   I.   500 mL of 0.25 M H_2SO_4

$$N = M \times \Delta \; \text{oxidation number}$$

$$H_2SO_4 + H_2O \longrightarrow 2H^+ + SO_4^{--} + H_2O$$

$$\Delta \; \text{oxidation number} = 2$$

$$N = (0.25 \; M)(2)$$

$$= 0.5 \; N$$

  II.   41.7 g of $K_2Cr_2O_7$ in 600 mL of solution

$$K_2Cr_2O_7 + H_2O \longrightarrow 2Cr^{3+} + K_2O_7^{6-} + H_2O$$

$$\Delta \; \text{oxidation number} = 6$$

$$\begin{aligned} MW \; \text{of} \; K_2Cr_2O_7 &= (2)(39.1) + (2)(52.0) \\ &\quad + (7)(16) \\ &= 294.2 \; g/mol \end{aligned}$$

$$EW = \frac{294.2}{6} = 49.033 \; g/mol$$

$$N = \left(\frac{41.7 \; g}{49.033 \; \frac{g}{mol}}\right)\left(\frac{1}{0.6 \; L}\right) = 1.42 \; N$$

  III. 0.135 equivalents of H_2SO_4 in 400 mL of solution

$$N = \frac{0.135 \; \text{equivalents}}{0.4 \; L}$$

$$= 0.3375 \; N \quad (0.338 \; N)$$

Answer is C.

3. $$\Delta T_b = \frac{m_{\text{solute,in g}} K_b}{(MW) m_{\text{solvent,in kg}}}$$

$$MW = \frac{m_{\text{solute}} K_b}{\Delta T_b m_{\text{solvent}}} = \frac{(2.00 \; g)\left(0.512 \; \frac{°C}{m}\right)}{(0.065°C)(0.250 \; kg)}$$

$$= 63.0$$

Answer is C.

4. $$m_{\text{grams}} = \frac{It(MW)}{(96\,485)(\text{change in oxidation state})}$$

$$AgNO_3 \longrightarrow Ag^+ + NO_3^-$$

$$\text{change in oxidation state} = 1$$

Silver exists as single atoms, so the molecular weight is the atomic weight, 107.87 g/mol.

$$m_{\text{grams}} = \frac{(0.075 \; A)(10 \; min)\left(60 \; \frac{s}{min}\right)\left(107.87 \; \frac{g}{mol}\right)}{(96\,485 \; A\cdot s)(1)}$$

$$= 0.050 \; g$$

Answer is D.

5. The solution is already 80% sulfuric acid. To increase the concentration to 95% sulfuric acid will require adding water to combine with the SO_3. Thus, the final mass of the acid solution will be greater than 100 g, since water is being added.

The molecular weights of the reactants are

$$H_2O: (2)\left(1 \; \frac{g}{mol}\right) + 16 \; \frac{g}{mol} = 18 \; g/mol$$

$$SO_3: (1)\left(32.1 \; \frac{g}{mol}\right) + (3)\left(16 \; \frac{g}{mol}\right) = 80.1 \; g/mol$$

$$\begin{aligned} H_2SO_4: (2)\left(1 \; \frac{g}{mol}\right) &+ (1)\left(32.1 \; \frac{g}{mol}\right) \\ &+ (4)\left(16 \; \frac{g}{mol}\right) = 98.1 \; g/mol \end{aligned}$$

From the chemical reaction, the combining weights can be normalized to 1 g of water.

$$H_2O + SO_3 \longrightarrow H_2SO_4$$

18	80.1	98.1 (raw combining weights)
1	4.45	5.45 (normalized)
X	4.45X	5.45X (for unknown mass X)

For a 95% solution,

$$0.95 = \frac{m_{H_2SO_4}}{m_{H_2SO_4} + m_{SO_3}}$$

$$= \frac{80 \text{ g} + m_{\text{new } H_2SO_4}}{80 \text{ g} + m_{\text{new } H_2SO_4} + 20 \text{ g} - m_{\text{used } SO_3}}$$

Let X be the mass of the water added. The mass of SO_3 used is 4.45X. The mass of new H_2SO_4 produced is 5.45X.

$$0.95 = \frac{80 \text{ g} + 5.45X}{80 \text{ g} + 5.45X + 20 \text{ g} - 4.45X}$$

$$= \frac{80 \text{ g} + 5.45X}{100 \text{ g} + X}$$

$$X = 3.33 \text{ g} \quad (3.33 \text{ g})$$

Answer is A.

6. A deliquescent substance melts or dissolves by absorbing moisture from the air.

Answer is D.

7. The dissociation reaction for copper sulfate is

$$CuSO_4 \longrightarrow Cu^{+2} + SO_4^{-2}$$

The electrolytic reaction equation is

$$Cu^{+2} + 2e \longrightarrow Cu$$

The atomic weight of copper is 63.5. The change in oxidation number is 2 per atom of copper deposited. Use Faraday's law.

$$m = \frac{It(\text{MW})}{(96\,485)(\text{change in charge})}$$

$$= \frac{(1.5 \text{ A})(2 \text{ h})\left(3600 \frac{\text{s}}{\text{h}}\right)\left(63.5 \frac{\text{g}}{\text{mol}}\right)}{\left(96\,485 \frac{\text{A·s}}{\text{mol}}\right)(2)}$$

$$= 3.55 \text{ g} \quad (3.6 \text{ g})$$

Answer is B.

8. Enthalpy of formation, also known as the heat of formation, is the energy required to create one unit mass of a substance.

Answer is D.

9. The enthalpy of reaction is

$$H_r = \sum H_{f,\text{products}} - \sum H_{f,\text{reactants}}$$

$$= (4 \text{ mol})\left(-94.05 \frac{\text{kcal}}{\text{mol}}\right) + (6 \text{ mol})\left(-57.80 \frac{\text{kcal}}{\text{mol}}\right)$$

$$- (2 \text{ mol})\left(-20.24 \frac{\text{kcal}}{\text{mol}}\right) - (7 \text{ mol})\left(0.00 \frac{\text{kcal}}{\text{mol}}\right)$$

$$= -682.5 \text{ kcal} \quad (-680 \text{ kcal exothermic})$$

For 1 mol of ethane, the enthalpy of reaction is approximately -340 kcal.

Answer is B.

10. The addition of 1 mol of solute in 1000 g of water reduces the freezing temperature by 1.860°C. This is the molal freezing point constant, K_f, for water. Use Eq. 39.4.

$$\Delta T_f = -mK_f = \frac{-m_{\text{solute,g}}K_f}{(\text{MW})m_{\text{solvent,kg}}}$$

$$0°C - 0.16°C = \frac{-(6 \text{ g})\left(1.86 \frac{°C}{\text{kg·mol}}\right)}{(\text{MW})(1 \text{ kg})}$$

$$\text{MW} = 69.75 \text{ g/mol} \quad (70 \text{ g/mol})$$

Answer is B.

11. The mass of chlorine in the tank is

$$\left(0.125 \frac{\text{kg}}{\text{m}^3}\right)(800 \text{ m}^3) = 100 \text{ kg}$$

The molecular weight of chlorine gas is

$$(2)\left(35.5 \frac{\text{g}}{\text{mol}}\right) = 71 \text{ g/mol}$$

The number of moles of chlorine is

$$n_{Cl_2} = \frac{m}{\text{MW}} = \frac{100\,000 \text{ g}}{71 \frac{\text{g}}{\text{mol}}}$$

$$= 1408 \text{ mol}$$

The mass of water in the tank is

$$(800 \text{ m}^3)\left(1000 \frac{\text{kg}}{\text{m}^3}\right) = 800\,000 \text{ kg}$$

The number of moles of water in the tank is

$$n_{H_2O} = \frac{m}{MW}$$

$$= \frac{(800\,000 \text{ kg}) \left(1000 \frac{\text{g}}{\text{kg}}\right)}{18 \frac{\text{g}}{\text{mol}}}$$

$$= 4.44 \times 10^7 \text{ mol}$$

The mole fraction is

$$x = \frac{n_{Cl_2}}{n_{Cl_2} + n_{H_2O}}$$

$$= \frac{1408 \text{ mol}}{1408 \text{ mol} + 4.44 \times 10^7 \text{ mol}}$$

$$= 3.17 \times 10^{-5}$$

From Henry's law, the partial pressure is proportional to the mole fraction.

$$p_i = hx_i = (15.2 \text{ atm})(3.17 \times 10^{-5})$$

$$= 4.8 \times 10^{-4} \text{ atm}$$

Answer is B.

12. Disregarding the water (which is not in the equation for the solubility constant anyway), the dissociation reaction equation is

$$SrSO_4 \longrightarrow Sr^{+2} + SO_4^{-2}$$

Since the coefficient is 1 for all three compounds, the concentrations of Sr^{+2} and SO_4^{-2} will be the same at equilibrium. The number of moles of $SrSO_4$ will be the same as the number of moles of the products.

$$K_{sp} = 2.8 \times 10^{-7} = [Sr^{+2}][SO_4^{-2}]$$

$$[Sr^{+2}] = [SO_4^{-2}] = \sqrt{2.8 \times 10^{-7}} = 0.000529$$

Thus, concentration of the ions is 0.000529 mol/L.

The molecular weight of $SrSO_4$ is

$$(1)\left(87.6 \frac{\text{g}}{\text{mol}}\right) + (1)\left(32 \frac{\text{g}}{\text{mol}}\right) + (4)\left(16 \frac{\text{g}}{\text{mol}}\right)$$

$$= 183.6 \text{ g/mol}$$

The mass of $SrSO_4$ in 1 L of solution is

$$m = M(MW)$$

$$= \left(0.000529 \frac{\text{mol}}{\text{L}}\right)(1 \text{ L})\left(183.6 \frac{\text{g}}{\text{mol}}\right)$$

$$= 0.097 \text{ g} (0.1 \text{ g})$$

Answer is C.

13. The molecular weight of ethyl alcohol is

$$(2)\left(12 \frac{\text{g}}{\text{mol}}\right) + (6)\left(1 \frac{\text{g}}{\text{mol}}\right) + (1)\left(16 \frac{\text{g}}{\text{mol}}\right) = 46 \text{ g/mol}$$

The number of moles is

$$n = \frac{m}{MW} = \frac{184 \text{ g}}{46 \frac{\text{g}}{\text{mol}}}$$

$$= 4 \text{ mol}$$

The molarity is the number of moles per liter. Since the molarity is $M = n/V$,

$$V = \frac{n}{M} = \frac{4 \text{ mol}}{2 \frac{\text{mol}}{\text{L}}}$$

$$= 2 \text{ L}$$

Answer is B.

14. The thermochemical reaction is

$$2O_3 \longrightarrow 3O_2 + \Delta h$$

The enthalpy of reaction is

$$H_r = \sum H_{f,\text{products}} - \sum H_{f,\text{reactants}}$$

$$= (3 \text{ mol})\left(0 \frac{\text{kcal}}{\text{mol}}\right) - (2 \text{ mol})\left(34.0 \frac{\text{kcal}}{\text{mol}}\right)$$

$$= -68 \text{ kcal total} (-34 \text{ kcal per mole})$$

Since this is negative, the reaction is exothermic. The decomposition does not require any energy. Rather, it generates energy. No energy is needed.

Answer is A.

Topic XIII: Materials Science/ Structure of Matter

Diagnostic Examination for
    Materials Science/Structure of Matter
Chapter 40   Crystallography and Atomic Bonding
Chapter 41   Material Testing
Chapter 42   Metallurgy

MATERIALS

Hint: For the most current information about the exam, visit www.ppi2pass.com/fefaqs.html regularly. Use the Exam Forum to compare notes with other FE examinees.

Diagnostic Examination

TOPIC XIII: MATERIALS SCIENCE/STRUCTURE OF MATTER

TIME LIMIT: 45 MINUTES

---

1. How many atoms per cell are in the base-centered orthorhombic crystal shown?

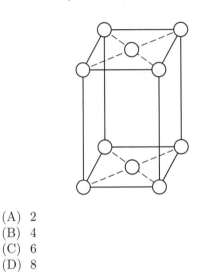

(A) 2
(B) 4
(C) 6
(D) 8

2. What is the packing factor for the body-centered tetragonal crystal with the dimensions shown? (Assume atoms are hard spheres and have a radius of r.)

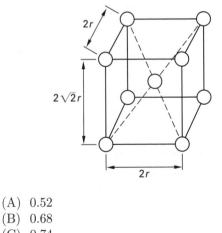

(A) 0.52
(B) 0.68
(C) 0.74
(D) 0.82

3. At a particular temperature, iron exhibits a body-centered cubic crystal structure with a cell dimension of 2.86 Å. What is the theoretical atomic radius of iron? (Assume atoms are hard spheres and have a radius of r.)

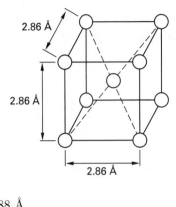

(A) 0.88 Å
(B) 0.95 Å
(C) 1.24 Å
(D) 1.43 Å

4. What are the Miller indices of the plane shown in its cubic crystallographic cell?

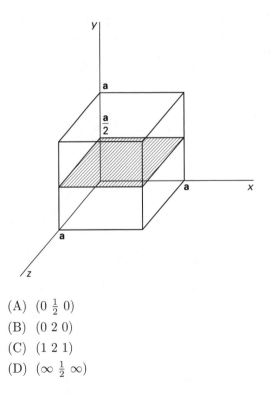

(A) $(0\ \frac{1}{2}\ 0)$
(B) $(0\ 2\ 0)$
(C) $(1\ 2\ 1)$
(D) $(\infty\ \frac{1}{2}\ \infty)$

5. Which of the following planes is a member of the family of equivalent planes containing $(1\ 0\ 1)$?

(A) $(0\ 1\ 0)$
(B) $(0\ \bar{1}\ 0)$
(C) $(\bar{1}\ 1\ \bar{1})$
(D) $(0\ \bar{1}\ \bar{1})$

6. The stress-strain curve for a nonlinear, perfectly elastic material is shown. A sample of the material is loaded until the stress reaches the value at point B. Then, the material is unloaded to zero stress. What is the approximate permanent set in the material?

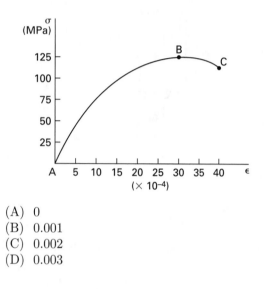

(A) 0
(B) 0.001
(C) 0.002
(D) 0.003

7. The stress-strain curve for an elastic-plastic material is shown. After loading and unloading, a test specimen of the material had a permanent set of 0.004. What was the maximum stress reached in the material during testing?

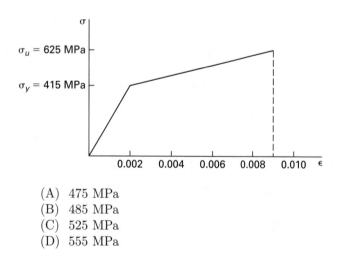

(A) 475 MPa
(B) 485 MPa
(C) 525 MPa
(D) 555 MPa

8. A cylindrical test specimen with a 15 mm diameter is tested axially in tension. A 0.20 mm elongation is recorded in a length of 200 mm when the load on the specimen is 36 kN. The material behaves elastically

during testing. What material could the specimen be made of?

(A) aluminum
(B) magnesium
(C) polystyrene
(D) steel

9. The stress-strain curve for an elastic-plastic material is as shown. What is the percent elongation of this material just before failure?

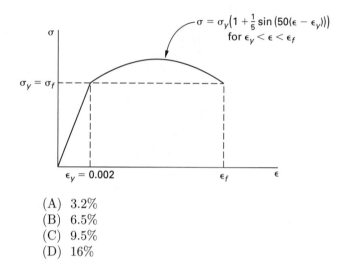

(A) 3.2%
(B) 6.5%
(C) 9.5%
(D) 16%

10. A 5 cm diameter bridge tension member experiences a loading of 275 kN every time a car crosses the bridge. The average traffic on the bridge is 50 car/h, and the bridge has a design life of 5 yr. The S-N curves for several materials are shown. From the standpoint of fatigue, what material(s) could the truss bar be constructed from if the bridge is to safely reach the end of its design life?

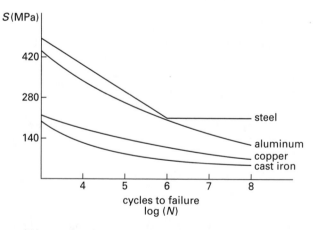

(A) steel only
(B) steel and aluminum
(C) steel, aluminum, and copper
(D) steel, aluminum, copper, and cast iron

11. All of the following metals will corrode if immersed in fresh water except

    (A) copper.
    (B) nickel.
    (C) chromium.
    (D) aluminum.

12. The movement of defects through a crystal by diffusion is described by which of the following?

    (A) Boyle's law
    (B) Fick's law
    (C) Dalton's law
    (D) Gibbs' rule

Problems 13 and 14 refer to the binary phase diagram shown.

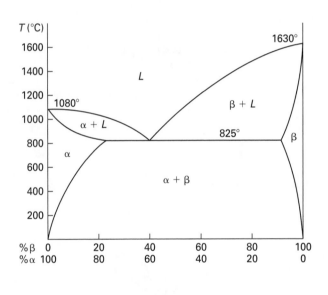

13. At what temperature will a sample composed of 20% α and 80% β begin to melt?

    (A) 825°C
    (B) 1050°C
    (C) 1500°C
    (D) 1630°C

14. A sample composed of 80% α and 20% β is heated to 900°C. What is the percentage of solid in the sample at this temperature?

    (A) 25%
    (B) 40%
    (C) 60%
    (D) 75%

15. The process of annealing can be used to achieve all of the following actions except

    (A) stress relief.
    (B) recrystallization.
    (C) grain growth.
    (D) toughness.

SOLUTIONS TO DIAGNOSTIC EXAMINATION TOPIC XII

1. Each corner atom is shared by eight cells, while each base atom is shared by two cells.

$$\left(\frac{1}{8}\right)(8) + \left(\frac{1}{2}\right)(2) = 2$$

Answer is A.

2. The packing factor is the volume of the atoms divided by the cell volume.

The number of atoms per cell is

$$\left(\frac{1}{8}\right)(8) + 1 = 2$$

The packing factor is

$$\mathrm{PF} = \frac{(2\ \mathrm{atoms})\left(\frac{4}{3}\right)\pi r^3}{(2r)(2r)(2\sqrt{2})r} = \frac{\pi}{3\sqrt{2}}$$
$$= 0.74$$

Answer is C.

3.

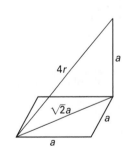

The atoms are touching along the cell diagonal. The length of the cell diagonal is four atomic radii.

$$(\sqrt{2}a)^2 + a^2 = (4r)^2$$
$$r^2 = \left(\frac{3}{16}\right)a^2$$

$$r = \left(\frac{\sqrt{3}}{4}\right) a$$

$$= \left(\frac{\sqrt{3}}{4}\right) (2.86 \text{ Å})$$

$$= 1.24 \text{ Å}$$

Answer is C.

4. The intercepts of the x-, y-, and z-axes are ∞, 1/2, and ∞, respectively. The Miller indices are the reciprocals of the intercepts, normalized to the lowest possible integer values. For this plane, the Miller indices are $(0\ 2\ 0)$.

Answer is B.

5. Due to the symmetry of the cubic structure, the choice of origin of a unit cell is arbitrary. Planes that belong to the same family of equivalent directions pass through the same sequence of atoms in the unit cell.

The $(1\ 0\ 1)$ plane is

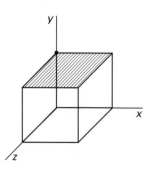

Choice (A): The $(0\ 1\ 0)$ plane is

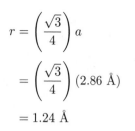

Choice (B): The $(0\ \bar{1}\ 0)$ plane is

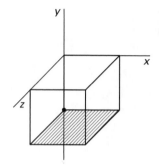

Choice (C): The $(\bar{1}\ 1\ \bar{1})$ plane is

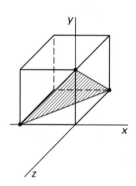

Choice (D): The $(0\ \bar{1}\ \bar{1})$ plane is

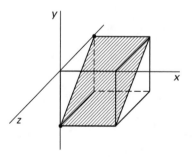

The plane that is crystallographically equivalent to $(1\ 0\ 1)$ is $(0\ \bar{1}\ \bar{1})$.

Answer is D.

6. A purely elastic material exhibits no permanent deformation upon unloading.

Answer is A.

7.

The modulus of elasticity is

$$E = \frac{\sigma_y}{\epsilon_y} = \frac{415 \text{ MPa}}{0.002}$$
$$= 207\,500 \text{ MPa}$$

The slope of the plastic portion of the curve is

$$m = \frac{\sigma_u - \sigma_y}{\epsilon_u - \epsilon_y} = \frac{625 \text{ MPa} - 415 \text{ MPa}}{0.009 - 0.002}$$
$$= 30\,000 \text{ MPa}$$

The equation of the plastic portion of the line is

$$\sigma = \sigma_y + m(\epsilon - 0.002)$$
$$= 415 \text{ MPa} + (30\,000 \text{ MPa})(\epsilon - 0.002)$$

The slope of the unloading curve is the same as the modulus of elasticity.

$$\sigma = E(\epsilon - 0.004)$$
$$= (207\,500 \text{ MPa})(\epsilon - 0.004)$$

These two equations can be solved simultaneously for σ. Solve for ϵ from the second equation and substitute into the first.

$$\epsilon = \left(\frac{1}{207\,500 \text{ MPa}}\right)\sigma + 0.004$$

$$\sigma = 415 \text{ MPa} + (30\,000 \text{ MPa})$$
$$\times \left(\left(\frac{1}{207\,500 \text{ MPa}}\right)\sigma + 0.004 - 0.002\right)$$

$$= 415 \text{ MPa} + 0.145\sigma + 60 \text{ MPa}$$

$$(0.855)\sigma = 475 \text{ MPa}$$

$$\sigma = 555 \text{ MPa}$$

Answer is D.

8. The stress in the specimen is

$$\sigma = \frac{P}{A} = \frac{36 \text{ kN}}{\left(\frac{1}{4}\right)\pi(0.015 \text{ m})^2}$$
$$= 203\,718 \text{ kPa} \quad (203.7 \text{ MPa})$$

The strain in the specimen is

$$\epsilon = \frac{\delta}{L} = \frac{0.2 \text{ mm}}{200 \text{ mm}}$$
$$= 0.001$$

The modulus of elasticity of the material is

$$E = \frac{\sigma}{\epsilon} = \frac{203.7 \text{ MPa}}{0.001}$$
$$= 203\,700 \text{ MPa} \quad (203.7 \text{ GPa})$$

The moduli of elasticity of aluminum, magnesium, polystyrene, and steel are approximately 70 GPa, 45 GPa, 2 GPa, and 205 GPa, respectively. This material is probably steel.

Answer is D.

9. From the diagram, at failure, $\sigma = \sigma_y$, and $\epsilon = \epsilon_f$.

$$\sigma_y = \sigma_y\left(1 + \frac{1}{5}\sin\left(50(\epsilon_f - \epsilon_y)\right)\right)$$
$$\sin\left(50(\epsilon_f - \epsilon_y)\right) = 0$$
$$(50)(\epsilon_f - \epsilon_y) = \pi$$
$$\epsilon_f = \frac{\pi}{50} + \epsilon_y$$
$$= \frac{\pi}{50} + 0.002$$
$$= 0.0648 \text{ rad}$$

$$\% \text{ elongation} = \epsilon_f \times 100\%$$
$$= 0.0648 \times 100\%$$
$$= 6.48\% \quad (6.5\%) \quad [\text{before failure}]$$

Answer is B.

10. The number of loading cycles during the life of the bridge is

$$N = \left(50 \frac{\text{cycle}}{\text{h}}\right)\left(24 \frac{\text{h}}{\text{d}}\right)\left(365 \frac{\text{d}}{\text{yr}}\right)(5 \text{ yr})$$
$$= 2.19 \times 10^6 \text{ cycles}$$
$$\log(N) = \log(2.19 \times 10^6 \text{ cycles})$$
$$= 6.34$$

The stress per cycle is

$$\sigma = \frac{P}{A} = \frac{275\,000 \text{ N}}{\frac{1}{4}\pi d^2}$$

$$= \frac{275\,000 \text{ N}}{\frac{1}{4}\pi(0.05 \text{ m})^2}$$

$$= 1.4 \times 10^8 \text{ Pa} \quad (140 \text{ MPa})$$

From the *S-N* curves, only steel and aluminum can be depended upon to resist the loading for the expected number of cycles without failing.

Answer is B.

11. From the table of oxidation potentials for corrosion reactions, copper has a positive voltage with respect to hydrogen, while the other metals have negative voltages. There is no driving potential for copper to go into solution, and it will not corrode in fresh water.

Answer is A.

12. The movement of defects through a crystal is described by Fick's law.

Answer is B.

13. The sample begins to melt once the solidus line is crossed. For this alloy, the solidus line is the horizontal line.

Answer is A.

14.

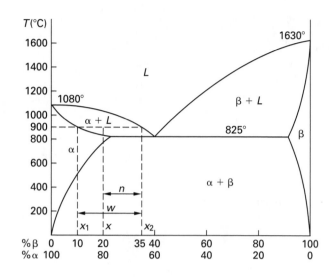

Use the lever rule.

$$\text{fraction solid} = \frac{x_2 - x}{x_2 - x_1} = \frac{n}{w}$$

$$= \frac{35\% - 20\%}{35\% - 10\%}$$

$$= 0.6 \quad (60\%)$$

Answer is C.

15. Annealing is used to relieve stress, recrystallize the grain, and increase grain growth. Annealing removes the dislocations caused by work hardening that makes a material tough.

Answer is D.

40 Crystallography and Atomic Bonding

Subjects

CRYSTALLOGRAPHY 40-1
    Common Metallic Crystalline
      Structures 40-1
    Number of Atoms in a Cell 40-2
    Packing Factor 40-2
    Coordination Number 40-3
    Miller Indices 40-3
ATOMIC BONDING 40-4

Nomenclature

a   lattice constant   m
r radius   m

CRYSTALLOGRAPHY

Common Metallic Crystalline Structures

The energy of a stable aggregation of atoms in a compound is lower than the energy of the individual atoms, and therefore, the aggregation is the more stable configuration. The atoms configure themselves spontaneously in this stable configuration, without requiring external energy.

Common table salt, NaCl, is a stable crystalline solid. The formation of table salt from positive sodium and negative chlorine ions proceeds spontaneously upon mixture. The ions form a three-dimensional cubic lattice of alternating sodium and chlorine ions. Each ion of one charge has six neighbors of the opposite charge. The strong electrostatic attraction of these six neighbors provides the force that causes the crystal to form.

If the ions were merely positive and negative point charges, the arrangement of charges would collapse into itself. However, the inner electron shells of both positive and negative ions provide the repulsive force to keep the ions apart. At the equilibrium position, this repulsion just balances the ionic attraction.

There are 14 different three-dimensional crystalline structures, known as *Bravais lattices*, as illustrated in Fig. 40.1. The smallest repeating unit of a Bravais lattice is known as a *cell* or *unit cell*.

Figure 40.1 Crystalline Lattice Structures

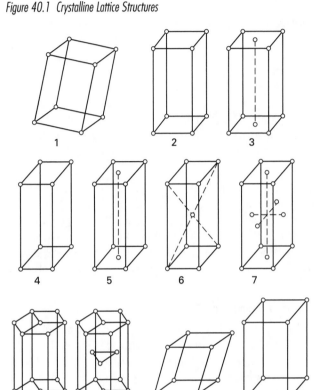

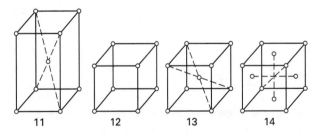

The 14 basic point-lattices are illustrated by a unit cell of each: (1) simple triclinic, (2) simple monoclinic, (3) base-centered monoclinic, (4) simple orthorhombic, (5) base-centered orthorhombic, (6) body-centered orthorhombic, (7) face-centered orthorhombic, (8) hexagonal, (9) rhombohedral, (10) simple tetragonal, (11) body-centered tetragonal, (12) simple cubic, (13) body-centered cubic, and (14) face-centered cubic.

MATERIALS
Cryst/Atom

There are seven different basic cell systems: the *cubic, tetragonal, orthorhombic, monoclinic, triclinic, hexagonal,* and *rhombohedral.* Characteristics of these systems are listed in Table 40.1.

Table 40.1 Characteristics of the Seven Different Crystalline Systems

system	sides	axial angles
cubic	$a_1 = a_2 = a_3$	all angles = 90°
tetragonal	$a_1 = a_2 \neq c$	all angles = 90°
orthorhombic	$a \neq b \neq c$	all angles = 90°
monoclinic	$a \neq b \neq c$	two angles = 90°; one angle $\neq$ 90°
triclinic	$a \neq b \neq c$	all angles different; none equals 90°
hexagonal	$a_1 = a_2 = a_3 \neq c$	angles = 90° and 120°
rhombohedral	$a_1 = a_2 = a_3$	all angles equal, but not 90°

Most common metallic crystals form in one of three cell systems: the *body-centered cubic* (BCC) cell, the *face-centered cubic* (FCC) cell, and the *hexagonal close-packed* (HCP) cell. Also, some of the simpler ceramic compounds (e.g., MgO, TiC, and $BaTiO_3$) are cubic. Table 40.2 lists common materials and their crystalline forms.

Number of Atoms in a Cell

When studying crystalline lattices, it is convenient to assume that the atoms have definite sizes and can be represented by hard spheres of radius r. The cell size, then, depends on the lattice type and the sizes of the touching spheres. On the basis of this representation, Fig. 40.2 illustrates three types of cubic structures and gives formulas for the center-to-center distances between the lattice atoms. The distances **a**, **b**, and **c** are known as *lattice constants*.

Some of the atoms in a unit cell are completely contained within the cell boundary (e.g., the center atom in a BCC structure). Other atoms are shared by adjacent cells (e.g., the corner atoms). Because of this sharing, the number of atoms attributable to a cell is not the number of whole atoms appearing in the lattice structures shown in Fig. 40.1. For example, there are nine atoms shown for the BCC structure. Although the center atom is completely enclosed, each of the eight corner atoms is shared by eight cells. Therefore, the number of atoms (also known as the number of *lattice points*) in a cell is $1 + (1/8)(8) = 2$.

Packing Factor

The *packing factor* is the volume of the atoms divided by the cell volume (i.e., a^3 for a cubic structure). These parameters are summarized in Table 40.3 for hard touching spheres of radius r.

Table 40.2 Crystalline Structures of Common Materials

body-centered cubic
  chromium
  iron, alpha (below 910°C)
  iron, delta (above 1390°C)
  lithium
  molybdenum
  potassium
  sodium
  tantalum
  titanium, beta (above 880°C)
  tungsten, alpha
face-centered cubic
  aluminum
  brass, alpha
  cobalt, beta
  copper
  gold
  iron, gamma (between 910°C and 1390°C)
  lead
  nickel
  platinum
  salts: NaCl, KCl, AgCl
  silver
hexagonal close-packed
  beryllium
  cadmium
  cobalt, alpha
  magnesium
  titanium, alpha (below 880°C)
  zinc

Figure 40.2(a) Cubic Lattice Dimensions Simple Cubic

distance between atoms	in terms of r	in terms of **a**
1 and 2	$2r$	**a**
1 and 4	$2\sqrt{2}r$	$\sqrt{2}$**a**
1 and 8	$2\sqrt{3}r$	$\sqrt{3}$**a**

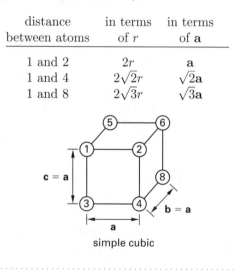

simple cubic

The low packing factor of the simple cubic and simple hexagonal structure indicates that these cells are wasteful of space. This is the primary reason simple cubic and hexagonal lattices seldom form naturally.

Figure 40.2(b)   Cubic Lattice Dimensions Body-Centered Cubic

distance between atoms	in terms of r	in terms of $\mathbf{a}$
1 and 2	$\dfrac{4r}{\sqrt{3}}$	$\mathbf{a}$
1 and 4	$\left(4\sqrt{\dfrac{2}{3}}\right)r$	$\sqrt{2}\mathbf{a}$
1 and 9	$2r$	$\left(\dfrac{\sqrt{3}}{2}\right)\mathbf{a}$
1 and 8	$4r$	$\sqrt{3}\mathbf{a}$

body-centered cubic

Figure 40.2(c)   Cubic Lattice Dimensions of a Face-Centered Cubic

distance between atoms	in terms of r	in terms of $\mathbf{a}$
1 and 2	$2\sqrt{2}r$	$\mathbf{a}$
1 and 10	$2r$	$\left(\dfrac{\sqrt{2}}{2}\right)\mathbf{a}$
1 and 4	$4r$	$\sqrt{2}\mathbf{a}$
1 and 8	$2\sqrt{6}r$	$\sqrt{3}\mathbf{a}$
1 and 11	$2\sqrt{3}r$	$\left(\sqrt{\dfrac{3}{2}}\right)\mathbf{a}$
10 and 11	$2r$	$\left(\dfrac{\sqrt{2}}{2}\right)\mathbf{a}$
9 and 11	$2\sqrt{2}r$	$\mathbf{a}$

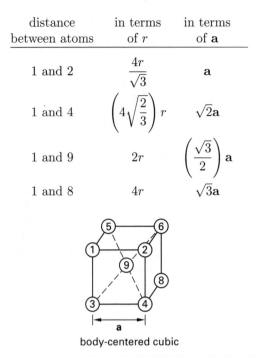

face-centered cubic

Table 40.3   Cell Packing Parameters (assuming hard touching spheres)

type of cell	number of atoms in a cell	packing factor	coordination number
simple cubic	1	0.52	6
body-centered cubic	2	0.68	8
face-centered cubic	4	0.74	12
simple hexagonal			
primitive cell	1	0.52	8
total structure	3	0.52	8
hexagonal close-packed			
primitive cell	2	0.74	12
total structure	6	0.74	12

Coordination Number

The *coordination number* of an atom in an ionic compound is the number of closest (touching) atoms. Since the atoms in an ionic solid are actually ions, another definition of coordination number is the number of anions surrounding each cation.

In reality, ions in a crystal are not all the same radius. The size of a cation that can fit in a site (known as *interstices* or *interstitial spaces*) between the anions is a function of the relative sizes of the ions.

Miller Indices

Crystallography uses a system known as *Miller indices* to specify planes in crystalline lattices. Planes are designated by numbers enclosed in parentheses, one for each coordinate direction in the unit cell.

Consider an orthorhombic cell that has cell dimensions $\mathbf{a}$, $\mathbf{b}$, and $\mathbf{c}$ in the x-, y-, and z-axes. The Miller indices are calculated as the *reciprocals* of the plane intercepts on these axes. If the plane does not intercept a cell axis, the intercept is infinity. By convention, there are no fractional intercepts, so the indices are multiplied by the least common denominator to clear all fractions. They are reduced to the smallest integers and written without commas, with overbars used to designate negative numbers.

For example, (0 1 1) indicates a plane that intersects the x-axis at infinity (i.e., it does not intersect), the y-axis at a unit distance $\mathbf{b}$ from the origin, and the z-axis at a unit distance $\mathbf{c}$ from the origin (Fig. 40.3(a)). (2 $\bar{1}$ 1) indicates a plane that intersects the x-axis at a distance of $\mathbf{a}/2$, the y-axis at a distance $-\mathbf{b}$, and the z-axis at a distance $\mathbf{c}$ from the origin (Fig. 40.3(b)).

Due to the symmetry of cubic systems, there are many *crystallographically equivalent* planes. Families of equivalent planes are written as $\{hkl\}$. For example, the

Figure 40.3 Miller Indices

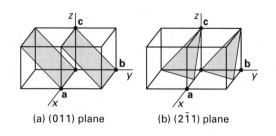

(a) (011) plane   (b) (2$\bar{1}$1) plane

{1 0 0} family contains the planes (0 1 0) and (0 0 1). Table 40.4 can be used to define equivalent cubic planes.

Sets of four numbers, $(hkil)$, are sometimes used to designate plane directions in hexagonal cells. h, k, and i are the respective reciprocals of the intercepts on three axes $\mathbf{a_1}$, $\mathbf{a_2}$, $\mathbf{a_3}$. l corresponds to the intercept on the longitudinal axis $\mathbf{c}$, the height of the hexagonal cell. With hexagonal systems, i is always equal to $-(h+k)$. This coordinate system is illustrated in Fig. 40.4.

Table 40.4 Cubic Cell Families of Equivalent Planes

family	equivalent directions		
{1 0 0}	(1 0 0),	(0 1 0),	(0 0 1),
	($\bar{1}$ 0 0),	(0 $\bar{1}$ 0),	(0 0 $\bar{1}$)
{1 1 0}	(1 1 0),	(1 0 1),	(0 1 1),
	($\bar{1}$ 1 0),	(1 $\bar{1}$ 0),	
	($\bar{1}$ 0 1),	(1 0 $\bar{1}$),	
	(0 $\bar{1}$ 1),	(0 1 $\bar{1}$)	
{1 1 1}	(1 1 1),		
	($\bar{1}$ 1 1),	(1 $\bar{1}$ 1),	(1 1 $\bar{1}$)
	($\bar{1}$ $\bar{1}$ 1),	($\bar{1}$ 1 $\bar{1}$),	(1 $\bar{1}$ $\bar{1}$)
	($\bar{1}$ $\bar{1}$ $\bar{1}$)		

Figure 40.4 Crystallographic Directions in Hexagonal Cells

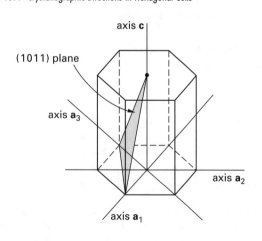

ATOMIC BONDING

There are three types of *primary bonds* between atoms in molecular structures: *ionic*, *covalent*, and *metallic*. Intermolecular bonds, known as *Van der Waals forces*, also exist. These are called *secondary bonds* because they are much weaker than the primary bonds.

Anions and cations are attracted to each other by electrostatic force. The electrostatic attraction of the positive cation to the negative anion effectively bonds the two ions together. This type of bonding, in which electrostatic attraction is predominant, is known as *ionic bonding*. One or more electrons are transferred from the valence shell of one atom to the valence shell of another. There is no sharing of electrons between atoms. Ionic bonding is characteristic of compounds of atoms with high electron affinities and atoms with low ionization energies (e.g., salts or metal oxides). The difference in electronegativities must be approximately 1.7 or greater for the bond to be classified as ionic.

Several common gases in their free states exist as *diatomic molecules*. Examples are hydrogen (H_2), oxygen (O_2), nitrogen (N_2), and chlorine (Cl_2). Since two atoms of the same element will have the same electronegativity and ionization energy, it is unlikely that one atom will take electrons from the other. Therefore, the bond formed is not ionic. Bonding in which the sharing of electrons is the predominant characteristic is known as *covalent bonding*. Covalent bonds are typical of bonds formed in organic and polymer compounds.

If the atoms are both the same element, the electrons will be shared equally and the bond will be purely covalent; but if the atoms are not both the same element, the electrons will not be shared equally and the bond will be partially covalent and partially ionic in nature. There is no sharp dividing line between ionic and covalent bonds for most compounds; if the difference in electronegativities is less than approximately 1.7, then the bond is classified as covalent.

Metallic bonding occurs when atoms contain electrons that are free to move from atom to atom. The sea of electrons is attracted to the positive ions in the metal structure, and this attraction bonds the atoms. Such bonding is nondirectional, as the electrons typically can move in three dimensions.

SAMPLE PROBLEMS

Problem 1

Which of the following elements does not have a face-centered cubic structure?

  (A) aluminum
  (B) copper
  (C) silver
  (D) sodium

Solution

Each of the materials except sodium has a face-centered cubic structure. Sodium has a body-centered cubic structure, but note that sodium chloride, NaCl, has a face-centered cubic structure.

Answer is D.

Problem 2

How many atoms are in the simple hexagonal structure shown?

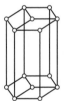

  (A) 1
  (B) 2
  (C) 3
  (D) 6

Solution

This is a simple hexagonal unit cell, not a hexagonal close-packed cell. The two center end atoms are each shared by two unit cells. The corner atoms are each shared by six unit cells.

$$\frac{\text{no. of atoms}}{\text{unit cell}} = \frac{2\ \text{atoms}}{2\ \text{cells}} + \frac{12\ \text{atoms}}{6\ \text{cells}}$$
$$= 1 + 2$$
$$= 3$$

Note that there are three primitive cells within the simple hexagonal structure. There is one atom in each primitive cell.

Answer is C.

Problem 3

What are the Miller indices of the plane shown?

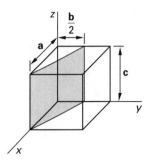

  (A) $\left(1\ \frac{1}{2}\ 0\right)$
  (B) $(2\ 1\ 0)$
  (C) $(1\ 2\ 0)$
  (D) $(1\ 2\ \infty)$

Solution

The intercepts along the x-, y-, and z-axes are 1, 1/2, and ∞. The reciprocals are $h = 1$, $k = 2$, and $l = 0$. The Miller indices are $(1\ 2\ 0)$.

Answer is C.

FE-STYLE EXAM PROBLEMS

1. What is the packing factor for the unit cell shown?

  (A) 0.35
  (B) 0.50
  (C) 0.52
  (D) 0.68

2. What is the coordination number of a face-centered cubic unit cell?

  (A) 6
  (B) 8
  (C) 10
  (D) 12

3. What are the Miller indices of the plane shown?

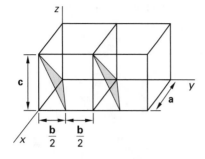

(A) $(\bar{1}\ 2\ 1)$

(B) $(0\ 2\ 1)$

(C) $(\bar{1}\ \frac{1}{2}\ \bar{1})$

(D) $(0\ \frac{1}{2}\ 1)$

4. Which of the following planes is a member of the $\{1\ 1\ 0\}$ family?

(A) $(1\ 0\ 0)$

(B) $(1\ 0\ \bar{1})$

(C) $(1\ 1\ \bar{1})$

(D) $(0\ 0\ 1)$

5. What type is the crystalline structure shown?

(A) simple cubic

(B) body-centered cubic

(C) face-centered cubic

(D) hexagonal close-packed

For the following problems use the NCEES Handbook as your only reference.

6. How many atoms are in a hexagonal close-packed cell?

(A) 2

(B) 4

(C) 6

(D) 8

7. Which of the following is the strongest type of bond?

(A) Van der Waals

(B) ionic

(C) covalent

(D) metallic

8. Which metals form crystals in a hexagonal close-packed structure?

I.   magnesium

II.   zinc

III. titanium

IV. iron

(A) I only

(B) I and II

(C) I and III

(D) I, II, and III

SOLUTIONS TO FE-STYLE EXAM PROBLEMS

1. The unit cell has a simple cubic structure. The packing factor is the volume of the atoms divided by the cell volume. In a simple cubic structure there is one atom per cell unit (each of the eight atoms shown shares with eight other unit cells). Assuming hard touching spheres of radius r, the length of the unit cell $\mathbf{a} = 2r$.

$$\begin{aligned}
\text{packing factor} &= \frac{(1\ \text{atom})\left(\dfrac{4\pi r^3}{3}\right)}{\mathbf{a}^3} \\[2mm]
&= \frac{4\pi r^3}{(3)(2r)^3} \\[2mm]
&= \frac{4\pi}{(3)(8)} \\[2mm]
&= 0.52
\end{aligned}$$

Answer is C.

2. Each face-centered atom has 12 closest neighbors, as the following illustration shows.

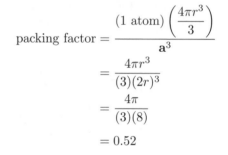

Answer is D.

3. This is a tricky question because the cell used to obtain the intercepts is not shown. Use the cell immediately behind the left-most cell shown.

The intercepts along the x-, y-, and z-axes are -1, $1/2$, and 1. The reciprocals are $h = -1$, $k = 2$, and $l = 1$. The Miller indices are written $(\bar{1}\ 2\ 1)$.

Answer is A.

4. Due to the symmetry of a cubic structure, the choice of origin in a unit cell may be arbitrary, and thus, families of planes contain planes that are equivalent.

The plane $(1\ 0\ \bar{1})$ is equivalent to plane $(1\ 1\ 0)$ and therefore, is in the same family. Verify this result by sketching the plane of the correct answer choice, and some of the incorrect answer choices, on a unit cell.

Answer is B.

5. The crystalline structure shown is the common body-centered cubic.

Answer is B.

6.

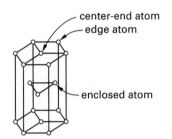

Each of the twelve edge atoms is shared by six cells (three in each layer). The two center atoms in the ends are shared by two complete cells (one in each layer). There are also three enclosed atoms.

$$\text{no. of atoms} = \frac{12}{6} + \frac{2}{2} + 3 = 6$$

Answer is C.

7. Covalent bonding is much stronger than ionic bonding (which is essentially electrostatic in nature). Electrons are easily displaced in metals by small voltages. Van der Waals forces in gases are so small as to be essentially negligible at normal temperatures.

Answer is C.

8. Hexagonal close-packed is the crystalline structure for magnesium, titanium, and zinc. Alpha- and delta-iron have a BCC structure, and gamma-iron has an FCC structure.

Answer is D.

41 Material Testing

Subjects

STRESS-STRAIN RELATIONSHIPS 41-1
    Engineering Stress and Strain 41-1
    Stress-Strain Curve 41-2
TESTING METHODS 41-3
    Standard Tensile Test 41-3
    Endurance Test 41-3
    Impact Test 41-4

Nomenclature

A	area	in^2	m^2
C_V	impact energy	ft-lbf	J
E	modulus of elasticity	lbf/in^2	MPa
F	force	lbf	N
L	length	in	m
N	number of cycles	–	–
S	strength	lbf/in^2	MPa

Symbols

γ	weight density	lbf/in^3	–
ϵ	engineering strain	in/in	m/m
σ	engineering stress	lbf/in^2	MPa

Subscripts

e	endurance
f	fracture, final
o	original
p	particular
u	ultimate
y	yield

STRESS-STRAIN RELATIONSHIPS

Engineering Stress and Strain

Figure 41.1 shows a *load-elongation curve* of *tensile test* data for a ductile ferrous material (e.g., low-carbon steel or other BCC transition metal). In this test, a prepared material sample (i.e., a *specimen*) is axially loaded in tension and the resulting elongation, ΔL, is measured as the load, F, increases.

When elongation is plotted against the applied load, the graph is applicable only to an object with the same length and area as the test specimen. To generalize the test results, the data are converted to stresses and

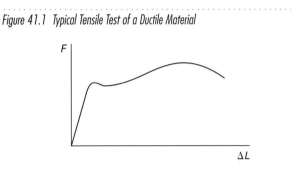

Figure 41.1 Typical Tensile Test of a Ductile Material

strains by use of Eqs. 41.1 and 41.2. *Engineering stress*, σ (usually called *stress*), is the load per unit original area. Typical engineering stress units are lbf/in^2 and MPa. *Engineering strain*, ϵ (usually called *strain*), is the elongation of the test specimen expressed as a percentage or decimal fraction of the original length. The units in/in and m/m are also used for strain.

$$\sigma = \frac{F}{A_o} \qquad 41.1$$

$$\epsilon = \frac{\Delta L}{L_o} \qquad 41.2$$

As the stress increases during a tensile test, the length of a specimen increases and the area decreases. Therefore, the engineering stress and strain are not *true stress and strain parameters*, which must be calculated from instantaneous values of length and area. Figure 41.2 illustrates engineering and true stresses and strains for a ferrous alloy. Although true stress and strain are more accurate, most engineering work has traditionally been

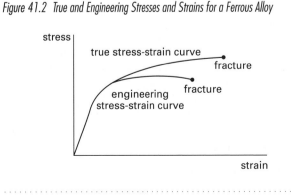

Figure 41.2 True and Engineering Stresses and Strains for a Ferrous Alloy

based on engineering stress and strain, which is justifiable for two reasons: (1) design using ductile materials is limited to the elastic region where engineering and true values differ little, and (2) the reduction in area of most parts at their service stresses is not known; only the original area is known.

Stress-Strain Curve

Segment OA in Fig. 41.3 is a straight line. The relationship between the stress and the strain in this linear region is given by *Hooke's law*, Eq. 41.3. The slope of the line segment OA is the *modulus of elasticity, E*, also known as *Young's modulus*. Table 41.1 lists approximate values of the modulus of elasticity for materials at room temperature. The modulus of elasticity will be lower at higher temperatures.

$$\sigma = E\epsilon \qquad 41.3$$

The stress at point A in Fig. 41.3 is known as the *proportionality limit* (i.e., the maximum stress for which the linear relationship is valid). Strain in the *proportional region* is called *proportional* (or *linear*) *strain*.

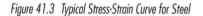

Figure 41.3 Typical Stress-Strain Curve for Steel

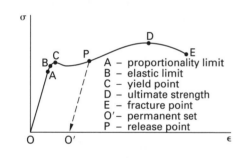

A – proportionality limit
B – elastic limit
C – yield point
D – ultimate strength
E – fracture point
O' – permanent set
P – release point

The *elastic limit*, point B in Fig. 41.3, is slightly higher than the proportionality limit. As long as the stress is kept below the elastic limit, there will be no *permanent set* (permanent deformation) when the stress is removed. Strain that disappears when the stress is removed is known as *elastic strain*, and the stress is said to be in the *elastic region*. When the applied stress is removed, the *recovery* is 100% and the material follows the original curve back to the origin.

If the applied stress exceeds the elastic limit, the recovery will be along a line parallel to the straight line portion of the curve, as shown in the line segment PO'. The strain that results (line OO') is *permanent set* (i.e., a permanent deformation). The terms *plastic strain* and *inelastic strain* are used to distinguish this behavior from the elastic strain.

For steel, the *yield point*, point C, is very close to the elastic limit. For all practical purposes, the *yield*

Table 41.1 Approximate Modulus of Elasticity of Representative Materials at Room Temperature

material	lbf/in^2	MPa
aluminum alloys	$10\text{–}11 \times 10^6$	$7\text{–}8 \times 10^4$
brass	$15\text{–}16 \times 10^6$	$10\text{–}11 \times 10^4$
cast iron	$15\text{–}22 \times 10^6$	$10\text{–}15 \times 10^4$
cast iron, ductile	$22\text{–}25 \times 10^6$	$15\text{–}17 \times 10^4$
cast iron, malleable	$26\text{–}27 \times 10^6$	$18\text{–}19 \times 10^4$
copper alloys	$17\text{–}18 \times 10^6$	$11\text{–}12 \times 10^4$
glass	$7\text{–}12 \times 10^6$	$5\text{–}8 \times 10^4$
magnesium alloys	6.5×10^6	4.5×10^4
molybdenum	47×10^6	32×10^4
nickel alloys	$26\text{–}30 \times 10^6$	$18\text{–}21 \times 10^4$
steel, hard[a]	30×10^6	21×10^4
steel, soft[a]	29×10^6	20×10^4
steel, stainless	$28\text{–}30 \times 10^6$	$19\text{–}21 \times 10^4$
titanium	$15\text{–}17 \times 10^6$	$10\text{–}11 \times 10^4$

(Multiply lbf/in^2 by 6.89×10^{-3} to obtain MPa.)

[a] common values given

strength or *yield stress, S_y*, can be taken as the stress that accompanies the beginning of plastic strain. Yield strengths are reported in lbf/in^2, kips/in^2, and MPa.

Most nonferrous materials, such as aluminum, magnesium, copper, and other FCC and HCP metals, do not have well-defined yield points. In such cases, the yield point is usually taken as the stress that will cause a 0.2% *parallel offset* (i.e., a plastic strain of 0.002), shown in Fig. 41.4. However, the yield strength can also be defined by other offset values, or by total strain characteristics.

Figure 41.4 Yield Strength of a Nonferrous Metal

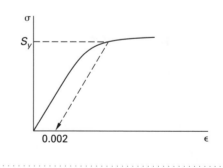

The *ultimate strength* or *tensile strength, S_u*, point D in Fig. 41.3, is the maximum stress the material can support without failure. This property is seldom used in the design of ductile material, since stresses near the ultimate strength are accompanied by large plastic strains.

The *breaking strength* or *fracture strength*, S_f, is the stress at which the material actually fails (point E in Fig. 41.3). For ductile materials, the breaking strength is less than the ultimate strength, due to the necking down in cross-sectional area that accompanies high plastic strains.

TESTING METHODS

Standard Tensile Test

Many useful material properties are derived from the results of a standard tensile test. As described previously, a tensile test is performed on a prepared material sample (i.e., a specimen) that is axially loaded in tension. The resulting elongation, ΔL, is measured as the load, F, increases.

The standard tensile test may be used to determine the modulus of elasticity, yield strength, ultimate tensile strength, and *ductility* of a specimen.

Ductility is the ability of a material to yield and deform prior to failure. The *percent elongation*, short for *percent elongation at failure*, is the total plastic strain at failure. (Percent elongation does not include the elastic strain, because after ultimate failure the material snaps back an amount equal to the elastic strain.)

$$\text{percent elongation} = \frac{L_f - L_o}{L_o} \times 100\%$$

$$= \epsilon_f \times 100\% \qquad 41.4$$

Highly ductile materials exhibit large percent elongations at failure. However, percent elongation is not the same as ductility. One typical definition of ductility is given by Eq. 41.5.

$$\text{ductility} = \frac{\text{ultimate failure strain}}{\text{yielding strain}} \qquad 41.5$$

Not all materials are ductile. Brittle materials, such as glass, cast iron, and ceramics, can support only small strains before they fail catastrophically without warning. As the stress is increased, the elongation is linear and Hooke's law can be used to predict the strain. Failure occurs within the linear region, and there is very little, if any, necking down. Since the failure occurs at a low strain, brittle materials are not ductile.

Endurance Test

A material can fail after repeated stress loadings even if the stress level never exceeds the ultimate strength, a condition known as *fatigue failure*.

The behavior of a material under repeated loadings is evaluated by an *endurance test* (or *fatigue test*). A specimen is loaded repeatedly to a specific stress amplitude, σ, and the number of applications of that stress required to cause failure, N, is counted. Rotating beam tests that load the specimen in bending (Fig. 41.5) are more common than alternating deflection and push-pull tests but are limited to round specimens. The *mean stress* is zero in rotating beam tests.

Figure 41.5 Rotating Beam Test

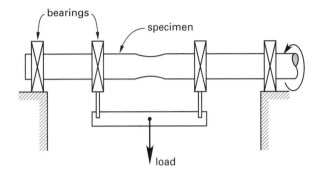

This procedure is repeated for different stresses using different specimens. The results of these tests are graphed on a semi-log plot, resulting in the *S-N curve* shown in Fig. 41.6.

Figure 41.6 Typical S-N Curve for Steel

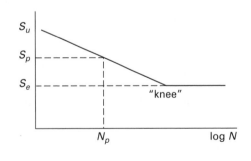

For a particular stress level, say S_p in Fig. 41.6, the number of cycles required to cause failure, N_p, is the *fatigue life*. S_p is the *fatigue strength* corresponding to N_p.

For steel subjected to fewer than approximately 10^3 loadings, the fatigue strength approximately equals the ultimate strength. (Although *low-cycle fatigue* theory has its own peculiarities, a part experiencing a small number of cycles can usually be designed or analyzed as for static loading.) The curve is linear between 10^3 and approximately 10^6 cycles if a logarithmic N-scale is used. Above 10^6 cycles, there is no further decrease in strength.

Therefore, below a certain stress level, called the *endurance limit, endurance stress*, or *fatigue limit, S'_e*, the material will withstand an almost infinite number of loadings without experiencing failure. This is characteristic of steel and titanium. If a dynamically loaded part is to have an infinite life, the stress must be kept below the endurance limit.

The yield strength is an irrelevant factor in cyclic loading. Fatigue failures are fracture failures, not yielding failures. They start with microscopic cracks at the material surface. Some of the cracks are present initially; others form when repeated cold working reduces the ductility in strain-hardened areas. These cracks grow minutely with each loading. Since cracks start at the location of surface defects, the endurance limit is increased by proper treatment of the surface. Such treatments include polishing, surface hardening, shot peening, and filleting joints.

The endurance limit is not a true property of the material since the other significant influences, particularly surface finish, are never eliminated. However, representative values of S'_e obtained from ground and polished specimens provide a baseline to which other factors can be applied to account for the effects of surface finish, temperature, stress concentration, notch sensitivity, size, environment, and desired reliability. These other influences are accounted for by reduction factors that are used to calculate a working endurance strength, S_e, for the material.

Impact Test

Toughness is a measure of a material's ability to yield and absorb highly localized and rapidly applied stress. A tough material will be able to withstand occasional high stresses without fracturing. Products subjected to sudden loading, such as chains, crane hooks, railroad couplings, and so on, should be tough. One measure of a material's toughness is the *modulus of toughness*, which is the *strain energy* or work per unit volume required to cause fracture. This is the total area under the stress-strain curve. Another measure is the *notch toughness*, which is evaluated by measuring the *impact energy* that causes a notched sample to fail.

In the *Charpy test* (Fig. 41.7), popular in the United States, a standardized beam specimen is given a 45-degree notch. The specimen is then centered on simple supports with the notch down. A falling pendulum striker hits the center of the specimen. This test is performed several times with different heights and different specimens until a sample fractures.

Figure 41.7 Charpy Test

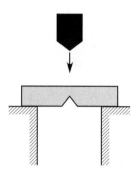

The kinetic energy expended at impact, equal to the initial potential energy, is calculated from the height. It is designated C_V and is expressed in either foot-pounds (ft-lbf) or joules (J). The energy required to cause failure is a measure of toughness. Note that without a notch, the specimen would experience uniaxial stress (tension and compression) at impact. The notch allows triaxial stresses to develop. Most materials become more brittle under triaxial stresses than under uniaxial stresses.

At 70°F (21°C), the energy required to cause failure ranges from 45 ft-lbf (60 J) for carbon steels to approximately 110 ft-lbf (150 J) for chromium-manganese steels. As temperature is reduced, however, the toughness decreases. In BCC metals, such as steel, at a low enough temperature the toughness decreases sharply. The transition from high-energy ductile failures to low-energy brittle failures begins at the *fracture transition plastic* (FTP) *temperature*.

Since the transition occurs over a wide temperature range, the *transition temperature* (also known as the *ductility transition temperature*) is taken as the temperature at which an impact of 15 ft-lbf (20 J) will cause failure. This occurs at approximately 30°F (−1°C) for low-carbon steel.

Table 41.2 Approximate Ductile Transition Temperatures

type of steel	ductile transition temperature, °F	°C
carbon steel	30°	−1°
high-strength, low-alloy steel	0° to 30°	−18° to −1°
heat-treated, high-strength carbon steel	−25°	−32°
heat-treated, construction alloy steel	−40° to −80°	−40° to −62°

The appearance of the fractured surface is also used to evaluate the transition temperature. The fracture can be fibrous (from shear fracture) or granular (from cleavage fracture), or a mixture of both. The fracture planes are studied and the percentages of ductile failure are plotted against temperature. The temperature at which the failure is 50% fibrous and 50% granular is known as *fracture appearance transition temperature*, FATT.

Not all materials have a ductile-brittle transition. Aluminum, copper, other FCC metals, and most HCP metals do not lose their toughness abruptly. Figure 41.8 illustrates the failure energy curves for several materials.

Figure 41.8 Failure Energy versus Temperature

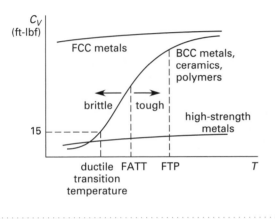

SAMPLE PROBLEMS

Problem 1

What is the ratio of stress to strain below the proportional limit called?

    (A) the modulus of rigidity
    (B) Hooke's constant
    (C) Poisson's ratio
    (D) Young's modulus

Solution

Young's modulus is defined by Hooke's law.

$$\sigma = E\epsilon$$

E is Young's modulus, or the modulus of elasticity, equal to the stress divided by strain within the proportional region of the stress-strain curve.

Answer is D.

Problem 2

What is the value of 40 MPa in the following illustration called?

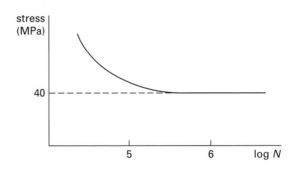

    I.   fatigue limit
    II. endurance limit
    III. proportional limit
    IV. yield stress

    (A) I only
    (B) I and II
    (C) II and IV
    (D) I, II, and IV

Solution

The diagram shows results of an endurance (or fatigue) test. The value of 40 MPa is called the endurance stress, endurance limit, or fatigue limit, and is equal to the maximum stress that can be be repeated indefinitely without causing the specimen to fail.

Answer is B.

Problem 3

What does the Charpy test determine?

    (A) endurance
    (B) yield strength
    (C) ductility
    (D) toughness

Solution

The Charpy test is an impact test to measure the toughness of the material—localized and rapidly applied stress.

Answer is D.

FE-STYLE EXAM PROBLEMS

Problems 1–5 refer to the following illustration.

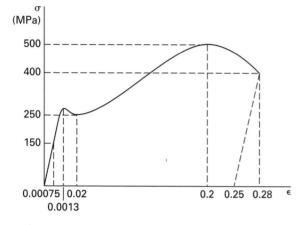

1. What test is represented by the diagram?

    (A) resilience test
    (B) rotating beam test
    (C) ductility test
    (D) tensile test

2. Which of the following is most likely the material that was tested to produce these results?

    (A) glass
    (B) concrete
    (C) low-carbon steel
    (D) aluminum

3. What is the modulus of elasticity?

    (A) 2×10^4 MPa
    (B) 8×10^4 MPa
    (C) 12.5×10^4 MPa
    (D) 20×10^4 MPa

4. What is the ductility?

    (A) 14
    (B) 19
    (C) 25
    (D) 215

5. What is the percent elongation at failure?

    (A) 14%
    (B) 19%
    (C) 25%
    (D) 28%

For the following problems use the NCEES Handbook as your only reference.

6. Which of the following material properties cannot be determined directly from a standard tensile test?

    (A) modulus of elasticity
    (B) yield strength
    (C) ultimate strength
    (D) hardness

7. The density of a particular metal is 2750 kg/m³. The modulus of elasticity for this metal is 210 GPa. A circular bar of this metal 3.5 m long and 160 cm² in cross-sectional area is suspended vertically from one end. What is the elongation of the bar due to its own mass?

    (A) 0.00055 mm
    (B) 0.00079 mm
    (C) 0.0016 mm
    (D) 0.0024 mm

8. Which of the following is the primary factor in determining if sheet metal can be bent and formed without experiencing stress fractures and other undesirable effects?

    (A) martensitic structure
    (B) surface hardness
    (C) modulus of elasticity
    (D) ductility

9. Which of the figures is (are) illustrative of a ductile metal after a tensile failure?

I.

II.

III.

(A) I only
(B) II only
(C) III only
(D) I and II

10. What test is used to determine the toughness of a material under shock loading?

(A) impact test
(B) hardness test
(C) fatigue test
(D) creep test

11. Which of the following cannot be used as a nonde-structive testing method for steel castings and forgings?

(A) radiography
(B) magnetic particle testing
(C) ultrasonic testing
(D) chemical analysis

SOLUTIONS TO FE-STYLE EXAM PROBLEMS

1. The diagram shows results from a tensile test. Both resilience and ductility may be calculated from the results, but the test is not known by those names. The rotating beam is a cyclic test and does not yield a mono-tonic stress-strain curve.

Answer is D.

2. The diagram shows results for a material with high ductility. Glass and concrete are relatively brittle materials. Aluminum may be ductile, but it does not have a well-defined yield point. Most likely, steel is the material tested.

Answer is C.

3. The modulus of elasticity (Young's modulus) is the slope of the stress-strain line in the proportional region.

$$E = \frac{\sigma}{\epsilon} = \frac{150 \text{ MPa}}{0.00075}$$

$$= 200\,000 \text{ MPa} \quad (20 \times 10^4 \text{ MPa})$$

Answer is D.

4. Ductility is the ratio of ultimate to yield strain.

$$\text{ductility} = \frac{\text{ultimate failure strain}}{\text{yielding strain}}$$

$$= \frac{0.28}{0.0013}$$

$$= 215$$

Answer is D.

5.
$$\text{percent elongation} = \epsilon_f \times 100\%$$
$$= 0.25 \times 100\%$$
$$= 25\%$$

The strain at failure used in the equation is found by extending a line from the failure point to the strain axis, parallel to the linear portion of the curve. Note that the percent elongation is an indicator of the ductility of a material, but it is not the same as the ductility, which was calculated in Prob. 4.

Answer is C.

6. The standard tensile test can determine the modulus of elasticity, yield strength, ultimate tensile strength, and ductility. Poisson's ratio can be determined from the lateral and axial strains. Although correlations between surface hardness and ultimate strength have been developed, hardness cannot be measured directly.

Answer is D.

7. The mass of the bar is

$$\text{mass} = \rho V = \rho AL$$

$$= \left(2750 \ \frac{\text{kg}}{\text{m}^3}\right) (160 \text{ cm}^2) \left(\frac{1 \text{ m}}{100 \text{ cm}}\right)^2 (3.5 \text{ m})$$

$$= 154 \text{ kg}$$

The total gravitational force is experienced by the metal at the suspension point. Farther down the rod, however, there is less volume contributing to the force, and the stress is reduced. The average force on the metal in the bar is half of the maximum value.

$$F_{\text{ave}} = \frac{1}{2} F_{\text{max}} = \frac{1}{2} mg$$

$$= \left(\frac{1}{2}\right) (154 \text{ kg}) \left(9.81 \ \frac{\text{m}}{\text{s}^2}\right)$$

$$= 755 \text{ N}$$

The elongation is

$$\Delta L = \epsilon L_o = \frac{\sigma}{E} L_o$$

$$= \frac{F}{AE} L_o$$

$$= \frac{(755\ \text{N})(3.5\ \text{m})}{(160\ \text{cm}^2)\left(\dfrac{1\ \text{m}}{100\ \text{cm}}\right)^2 (210 \times 10^9\ \text{Pa})}$$

$$= 7.86 \times 10^{-7}\ \text{m}\quad (0.00079\ \text{mm})$$

Answer is B.

8. Ductility is a measure of a material's ability to deform without failure.

Answer is D.

9. I is a necking failure, typical of very ductile materials. II is a cup-and-cone failure, common of moderately ductile materials. III is a shear failure, common of brittle materials. There is no reduction in the area at failure.

Answer is D.

10. Toughness is defined as a measure of a material's ability to absorb highly localized, rapidly applied (shock) loads. The impact test measures toughness.

Answer is A.

11. Chemical analysis requires a sample of the material to be taken. Tests that require taking samples of the material are not nondestructive tests. All of the other methods can be used with steel castings and forgings.

Answer is D.

42 Metallurgy

Subjects

CORROSION 42-1
DIFFUSION 42-3
BINARY PHASE DIAGRAMS 42-3
   Lever Rule 42-4
   Iron-Carbon Phase Diagram 42-5
   Gibbs' Phase Rule 42-5
THERMAL PROCESSING 42-6
HARDNESS AND HARDENABILITY . . . 42-7
ASTM GRAIN SIZE 42-7
COMPOSITE MATERIALS 42-8

Nomenclature

A	area	ft^2	m^2
c	specific heat	Btu/lbm-°F	kJ/kg·K
C	concentration	1/ft^3	1/m^3
C	number of components	–	–
D	diffusion coefficient	ft^2/sec	m^2/s
D_0	proportionality constant	ft^2/sec	m^2/s
E	modulus of elasticity	lbf/ft^2	Pa
f	volumetric fraction	–	–
F	degrees of freedom	–	–
F	force	lbf	N
J	defect flux	1/ft^2-sec	1/m^2·s
L	length	ft	m
n	grain size	–	–
N	number of grains	1/ft^2	1/m^2
P	number of phases	–	–
P_L	points per unit length	1/ft	1/m
Q	activation energy	Btu/lbmol	kJ/kmol
$\overline{R}$	universal gas constant	Btu/lbmol-°R	kJ/kmol·K
S_v	surface area per unit volume	1/ft	1/m
T	absolute temperature	–	K
T	temperature	°F	K
x	fraction by weight	–	–

Symbols

α	coefficient of linear thermal expansion	1/°F	1/K
ϵ	strain	ft/ft	m/m
ρ	density	lbm/ft^3	kg/m^3

Subscripts

C	composite
o	original

CORROSION

Corrosion is an undesirable degradation of a material resulting from a chemical or physical reaction with the environment. *Galvanic action* results from a difference in oxidation potentials of metallic ions. The greater the difference in oxidation potentials, the greater the galvanic corrosion will be. If two metals with different oxidation potentials are placed in an *electrolytic medium* (e.g., seawater), a *galvanic cell* (*voltaic cell*) will be created. The more electropositive metal will act as an anode and will corrode. The metal with the lower potential, being the cathode, will be unchanged.

A galvanic cell is a device that produces electrical current by way of an oxidation-reduction reaction—that is, chemical energy is converted into electrical energy. Galvanic cells typically have the following characteristics.

- The oxidizing agent is separate from the reducing agent.

- Each agent has its own electrolyte and metallic electrode, and the combination is known as a *half-cell*.

- Each agent can be in solid, liquid, or gaseous form, or can consist simply of the electrode.

- The ions can pass between the electrolytes of the two half-cells. The connection can be through a porous substance, salt bridge, another electrolyte, or other method.

The amount of current generated by a half-cell depends on the electrode material and the oxidation-reduction reaction taking place in the cell. The current-producing ability is known as the *oxidation potential*, *reduction potential*, or *half-cell potential*. *Standard oxidation potentials* have a zero reference voltage corresponding to the potential of a *standard hydrogen electrode*. Table 42.1 shows representative standard half-cell potentials for galvanic cell materials.

MATERIALS
Metallurgy

Table 42.1 Standard Oxidation Potentials for Corrosion Reactions

cathodic (protected) to anodic (corroded)

reaction	$\mathcal{E}^0$ (volts)[a]
$2F^- \longrightarrow F_2 + 2e^-$	-2.87
$O_2 + H_2O \longrightarrow O_3 + 2H^+ + 2e^-$	-2.07
$Ce^{+3} \longrightarrow Ce^{+4} + e^-$	-1.61
$2Cl^- \longrightarrow Cl_2 + 2e^-$	-1.36
$2H_2O \longrightarrow O_2 + 4H^+ + 4e^-$	-1.229
$Hg_2^{++} \longrightarrow 2Hg^{++} + 2e^-$	-0.920
$Ag \longrightarrow Ag^+ + e^-$	-0.799
$2Hg \longrightarrow Hg_2^{++} + 2e^-$	-0.789
$Fe^{++} \longrightarrow Fe^{+3} + e^-$	-0.771
$H_2O_2 \longrightarrow O_2 + 2H^+ + 2e^-$	-0.682
$2I^- \longrightarrow I_2 + 2e^-$	-0.536
$4(OH)^- \longrightarrow O_2 + 2H_2O + 4e^-$	-0.401
$Cu \longrightarrow Cu^{++} + 2e^-$	-0.337
$H_2 \longrightarrow 2H^+ + 2e^-$	0 (definition)
$Pb \longrightarrow Pb^{++} + 2e^-$	$+0.126$
$Sn \longrightarrow Sn^{++} + 2e^-$	$+0.136$
$Ni \longrightarrow Ni^{++} + 2e^-$	$+0.250$
$Co \longrightarrow Co^{++} + 2e^-$	$+0.277$
$Cd \longrightarrow Cd^{++} + 2e^-$	$+0.403$
$Fe \longrightarrow Fe^{++} + 2e^-$	$+0.440$
$Cr \longrightarrow Cr^{+3} + 3e^-$	$+0.74$
$Zn \longrightarrow Zn^{++} + 2e^-$	$+0.763$
$Mn \longrightarrow Mn^{++} + 2e^-$	$+1.18$
$Al \longrightarrow Al^{+3} + 3e^-$	$+1.66$
$Mg \longrightarrow Mg^{++} + 2e^-$	$+2.37$
$Na \longrightarrow Na^+ + e^-$	$+2.714$
$Li \longrightarrow Li^+ + e^-$	$+3.045$

[a] Reference: normal hydrogen electrode

To specify their tendency to corrode, metals are often classified according to their position in the galvanic series listed in Table 42.2. As expected, the metals in this series are in approximately the same order as their half-cell potentials listed in Table 42.1. However, alloys and proprietary metals are also included in the series.

Precautionary measures can be taken to inhibit or eliminate galvanic action when use of dissimilar metals is unavoidable.

- Use dissimilar metals that are close neighbors in the galvanic series.

- Use sacrificial anodes. In marine saltwater applications, sacrificial zinc plates can be used.

- Use protective coatings, oxides, platings, or inert spacers to reduce or eliminate the access of corrosive environments to the metals.

Table 42.2 The Galvanic Series in Seawater

(tendency to corrode increases with separation in the table)

cathodic to anodic
(less electropositive to more electropositive)

gold
graphite
titanium
Hastelloy C
Monel metal
stainless steels (passive)
silver
silver solder
Inconel
70/30 copper-nickel
90/10 copper-nickel
bronze (copper-tin)
copper
brass (copper-zinc)
nickel
tin
lead-tin alloys
lead
Hastelloy A
stainless steel (active)
   No. 316
   No. 404
   No. 430
   No. 410
cast iron
low-carbon steel
2024 aluminum alloy
cadmium
Alclad 3S
zinc
magnesium

It is not necessary that two dissimilar metals be in contact for corrosion by galvanic action to occur. Different regions within a metal may have different half-cell potentials. The difference in potential can be due to different phases within the metal (creating very small galvanic cells), heat treatment, cold working, and so on.

In addition to corrosion caused by galvanic action, there is also *stress corrosion*, *fretting corrosion*, and *cavitation*. Conditions within the crystalline structure can accentuate or retard corrosion. In one extreme type of intergranular corrosion, *exfoliation*, open endgrains separate into layers.

DIFFUSION

Real crystals possess a variety of imperfections and defects that affect *structure-sensitive properties*. Such properties include electrical conductivity, yield and ultimate strengths, creep strength, and semiconductor properties. Most imperfections can be categorized into *point*, *line*, and *planar* (*grain boundary*) imperfections. As shown in Fig. 42.1, *point defects* include vacant lattice sites, ion vacancies, substitutions of foreign atoms into lattice points or interstitial points, and occupation of interstitial points by atoms. *Line defects* consist of imperfections that are repeated consistently in many adjacent cells and thus have extension in a particular direction. *Grain boundary defects* are the interfaces between two or more crystals. This interface is almost always a mismatch in crystalline structures.

Figure 42.1   Point Defects

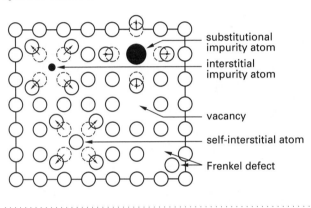

- substitutional impurity atom
- interstitial impurity atom
- vacancy
- self-interstitial atom
- Frenkel defect

All point defects can move individually and independently from one position to another through *diffusion*. The *activation energy* for such diffusion generally comes from heat and/or strain (i.e., bending or forming). In the absence of the activation energy, the defect will move very slowly, if at all.

Diffusion of defects is governed by *Fick's laws*. The first law predicts the number of defects that will move across a unit surface area per unit time. This number is known as the *defect flux*, J, and is proportional to the *defect concentration gradient* dC/dx in the direction of movement. The negative sign in Eq. 42.1 indicates that defects migrate to where the dislocation density is lower. *Fick's first law of diffusion* is

$$J = -D\frac{dC}{dx} \qquad 42.1$$

The *diffusion coefficient*, D (also known as the *diffusivity*), is dependent on the material, activation energy, and temperature. It is calculated from the *proportionality constant*, D_0, the activation energy, Q, the universal

gas constant, $\overline{R}$ (1.987 cal/gmol·K), and the absolute temperature, T. Equation 42.2 is typical of the method used to determine the diffusion coefficient.

$$D = D_0 e^{-Q/\overline{R}T} \qquad 42.2$$

BINARY PHASE DIAGRAMS

Most engineering materials are not pure elements but alloys of two or more elements. Alloys of two elements are known as *binary alloys*. Steel, for example, is an alloy of primarily iron and carbon. Usually one of the elements is present in a much smaller amount, and this element is known as the *alloying ingredient*. The primary ingredient is known as the *host ingredient*, *base metal*, or *parent ingredient*.

Sometimes, such as with alloys of copper and nickel, the alloying ingredient is 100% soluble in the parent ingredient. Nickel-copper alloy is said to be a *completely miscible alloy* or a *solid-solution alloy*.

The presence of the alloying ingredient changes the thermodynamic properties, notably the freezing (or melting) temperatures of both elements. Usually the freezing temperatures decrease as the percentage of alloying ingredient is increased. Because the freezing points of the two elements are not the same, one of them will start to solidify at a higher temperature than the other. Thus, for any given composition, the alloy might consist of all liquid, all solid, or a combination of solid and liquid, depending on the temperature.

A *phase* of a material at a specific temperature will have a specific composition and crystalline structure and distinct physical, electrical, and thermodynamic properties. (In metallurgy, the word "phase" refers to more than just solid, liquid, and gas phases.)

The regions of an *equilibrium diagram*, also known as a *phase diagram*, illustrate the various alloy phases. The phases are plotted against temperature and composition. The composition is usually a gravimetric fraction of the alloying ingredient. Only one ingredient's gravimetric fraction needs to be plotted for a binary alloy.

It is important to recognize that the equilibrium conditions do not occur instantaneously and that an equilibrium diagram is applicable only to the case of slow cooling.

Figure 42.2 is an equilibrium diagram for copper-nickel alloy. (Most equilibrium diagrams are much more complex.) The *liquidus line* is the boundary above which no solid can exist. The *solidus line* is the boundary below which no liquid can exist. The area between these two lines represents a mixture of solid and liquid phase materials.

Figure 42.2 Copper-Nickel Phase Diagram

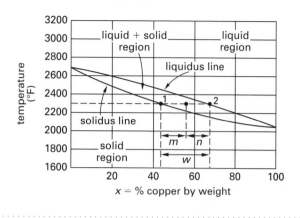

known as the *eutectic material*, and it will not solid-ify until the line BD (the *eutectic line*, *eutectic point*, or *eutectic temperature*)—the lowest point at which the eutectic material can exist in liquid form—is reached.

Since the two ingredients do not mix, reducing the temperature below the eutectic line results in crystals (layers or plates) of both pure ingredients forming. This is the microstructure of a solid eutectic alloy: alternating pure crystals of the two ingredients. Since two solid substances are produced from a single liquid substance, the process could be written in chemical reaction format as liquid → solid α + solid β. (Alternatively, upon heating, the reaction would be solid α + solid β → liquid.) For this reason, the phase change is called a *eutectic reaction*.

There are similar reactions involving other phases and states. Table 42.3 and Fig. 42.4 illustrate these.

Just as only a limited amount of salt can be dissolved in water, there are many instances where a limited amount of the alloying ingredient can be absorbed by the solid mixture. The elements of a binary alloy may be completely soluble in the liquid state but only partially soluble in the solid state.

When the alloying ingredient is present in amounts above the maximum solubility percentage, the alloying ingredient precipitates out. In aqueous solutions, the precipitate falls to the bottom of the container. In metallic alloys, the precipitate remains suspended as pure crystals dispersed throughout the primary metal.

In chemistry, a *mixture* is different from a *solution*. Salt in water forms a solution. Sugar crystals mixed with salt crystals form a mixture.

Figure 42.3 is typical of an equilibrium diagram for ingredients displaying a limited solubility.

Table 42.3 Types of Equilibrium Reactions

reaction name	type of reaction upon cooling
eutectic	liquid → solid α + solid β
peritectic	liquid + solid α → solid β
eutectoid	solid γ → solid α + solid β
peritectoid	solid α + solid γ → solid β

Figure 42.4 Typical Appearance of Equilibrium Diagram at Reaction Points

reaction name	phase reaction	phase diagram
eutectic	$L \rightarrow \alpha(s) + \beta(s)$ cooling	
peritectic	$L + \alpha(s) \rightarrow \beta(s)$ cooling	
eutectoid	$\gamma(s) \rightarrow \alpha(s) + \beta(s)$ cooling	
peritectoid	$\alpha(s) + \gamma(s) \rightarrow \beta(s)$ cooling	

Figure 42.3 Equilibrium Diagram of a Limited Solubility Alloy

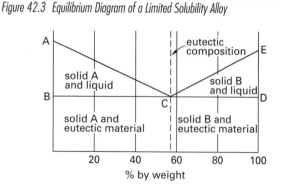

In Fig. 42.3, the components are perfectly miscible at point C only. This point is known as the *eutectic composition*. A *eutectic alloy* is an alloy having the composition of its eutectic point. The material in the region ABC consists of a mixture of solid component A crystals in a liquid of components A and B. This liquid is

Lever Rule

Within a liquid-solid region, the percentage of solid and liquid phases is a function of temperature and composition. Near the liquidus line, there is very little solid phase. Near the solidus line, there is very little liquid phase. The *lever rule* is an interpolation technique used to find the relative amounts of solid and liquid phase at any composition. These percentages are given in fraction (or percent) by weight.

Figure 42.2 shows an alloy with an average composition of 55% copper at 2300°F. (A horizontal line representing different conditions at a single temperature is known as a *tie line*.) The liquid composition is defined by point 2, and the solid composition is defined by point 1.

The fractions of solid and liquid phases depend on the distances m, n, and w (equal to $m + n$), which are measured using any convenient scale. (Although the distances can be measured in millimeters or tenths of an inch, it is more convenient to use the percentage alloying ingredient scale.) Then the fractions' solid and liquid can be calculated from Eqs. 42.3 and 42.4.

$$\text{fraction solid} = \frac{x_2 - x}{x_2 - x_1} = \frac{n}{w}$$
$$= 1 - \text{fraction liquid} \quad \textit{42.3}$$

$$\text{fraction liquid} = \frac{x - x_1}{x_2 - x_1} = \frac{m}{w}$$
$$= 1 - \text{fraction solid} \quad \textit{42.4}$$

The lever rule and method of determining the composition of the two components are applicable to any solution or mixture, liquid or solid, in which two phases are present.

Iron-Carbon Phase Diagram

The iron-carbon phase diagram (Fig. 42.5) is much more complex than idealized equilibrium diagrams due to the existence of many different phases. Each of these phases has a different microstructure and, therefore, different mechanical properties. By treating the steel in such a manner as to force the occurrence of particular phases, steel with desired wear and endurance properties can be produced.

Allotropes have the same composition but different atomic structures (microstructures), volumes, electrical resistance, and magnetic properties. *Allotropic changes* are reversible changes that occur at the *critical points* (i.e., *critical temperatures*).

Iron exists in three primary allotropic forms: alpha-iron, delta-iron, and gamma-iron. The changes are brought about by varying the temperature of the iron. Heating pure iron from room temperature changes its structure from BCC alpha-iron ($-460°F$ to $1670°F$), also known as *ferrite*, to FCC gamma-iron ($1670°F$ to $2552°F$), to BCC delta-iron (above $2552°F$).

Iron-carbon mixtures are categorized into *steel* (less than 2% carbon) and *cast iron* (more than 2% carbon) according to the amounts of carbon in the mixtures.

The most important eutectic reaction in the iron-carbon system is the formation of a solid mixture of austenite and cementite at approximately 2065°F (1129°C). *Austenite* is a solid solution of carbon in gamma-iron. It is nonmagnetic, decomposes on slow cooling, and does not normally exist below 1333°F (723°C), although it can be partially preserved by extremely rapid cooling.

Cementite (Fe_3C), also known as *carbide* or *iron carbide*, has approximately 6.67% carbon. Cementite is the hardest of all forms of iron, has low tensile strength, and is quite brittle.

The most important eutectoid reaction in the iron-carbon system is the formation of *pearlite* from the decomposition of austenite at approximately 1333°F (723°C). Pearlite is actually a mixture of two solid components, ferrite and cementite, with the common *lamellar (layered) appearance*.

Ferrite is essentially pure iron (less than 0.025% carbon) in BCC alpha-iron structure. It is magnetic and has properties complementary to cementite, since it has low hardness, high tensile strength, and high ductility.

Gibbs' Phase Rule

Gibbs' phase rule defines the relationship between the number of phases and elements in an equilibrium mixture. For such an equilibrium mixture to exist, the alloy must have been slowly cooled and thermodynamic equilibrium must have been achieved along the way. At equilibrium, and considering both temperature and pressure to be independent variables, Gibbs' phase rule is

$$P + F = C + 2 \quad \textit{42.5}$$

P is the number of phases existing simultaneously; F is the number of independent variables, known as *degrees of freedom*; and C is the number of elements in the alloy. Composition, temperature, and pressure are examples of degrees of freedom that can be varied.

For example, if water is to be stored in a condition where three phases (solid, liquid, gas) are present simultaneously, then $P = 3$, $C = 1$, and $F = 0$. That is, neither pressure nor temperature can be varied. This state corresponds to the *triple point* of water.

If pressure is constant, then the number of degrees of freedom is reduced by one, and Gibbs' phase rule can be rewritten as

$$P + F = C + 1|_{\text{constant pressure}} \quad \textit{42.6}$$

If Gibbs' rule predicts $F = 0$, then an alloy can exist with only one composition.

Figure 42.5 Iron-Carbon Diagram

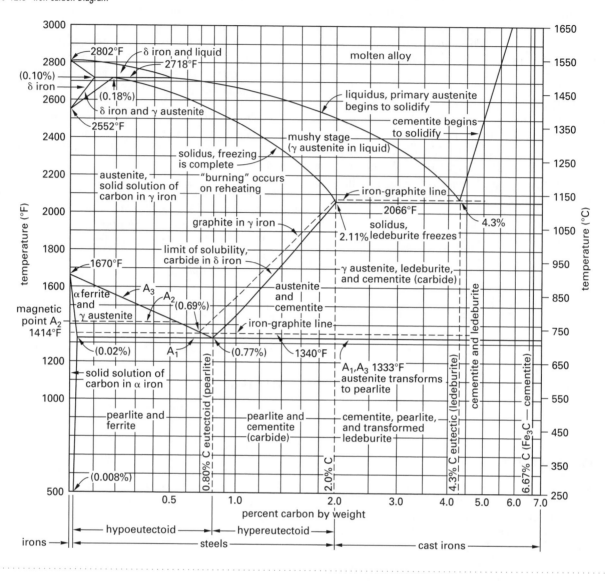

THERMAL PROCESSING

Thermal processing, including hot working, heat treating, and quenching, is used to obtain a part with desirable mechanical properties.

Cold and hot working are both forming processes (rolling, bending, forging, extruding, etc.). The term *hot working* implies that the forming process occurs above the *recrystallization temperature*. (The actual temperature depends on the rate of strain and the cooling period, if any.) *Cold working* (also known as *work hardening* and *strain hardening*) occurs below the recrystallization temperature.

Above the recrystallization temperature, almost all of the internal defects and imperfections caused by hot working are eliminated. In effect, hot working is a "self-healing" operation. Thus, a hot-worked part remains softer and has a greater ductility than a cold-worked part. Large strains are possible without strain hardening. Hot working is preferred when the part must go through a series of forming operations (passes or steps), or when large changes in size and shape are needed.

The hardness and toughness of a cold-worked part will be higher than that of a hot-worked part. Because the part's temperature during the cold working is uniform, the final microstructure will also be uniform. There are many times when these characteristics are desirable, and thus, hot working is not always the preferred forming method. In many cases, cold working will be the final operation after several steps of hot working.

Once a part has been worked, its temperature can be raised to slightly above the recrystallization temperature. This *heat treatment* operation is known as *annealing*, and is used to relieve stresses, increase grain size, and recrystallize the grains. Stress relief is also known as *recovery*.

Quenching is used to control the microstructure of steel by preventing the formation of equilibrium phases with undesirable characteristics. The usual desired result is hard steel, which resists plastic deformation. The quenching can be performed with gases (usually air), oil, water, or brine. Agitation or spraying of these fluids during the quenching process increases the severity of the quenching.

Time-temperature-transformation (TTT) *curves* are used to determine how fast an alloy should be cooled to obtain a desired microstructure. Although these curves show different phases, they are not equilibrium diagrams. On the contrary, they show the microstructures that are produced with controlled temperatures or when quenching interrupts the equilibrium process.

TTT curves are determined under ideal, isothermal conditions. However, the curves are more readily available than experimentally determined *controlled-cooling-transformation* (CCT) *curves*. Both curves are similar in shape, although the CCT curves are displaced downward and to the right from TTT curves.

Figure 42.6 shows a TTT diagram for a high-carbon (0.80% carbon) steel. Curve 1 represents extremely rapid quenching. The transformation begins at 420°F (216°C), and continues for 8 to 30 seconds, changing all of the austenite to martensite. The martensitic transformation does not depend on diffusion. Since martensite has almost no ductility, martensitic microstructures are used in applications such as springs and hardened tools where a high elastic modulus and low ductility are needed.

Curve 2 represents a slower quench that converts all of the austenite to fine pearlite.

A horizontal line below the critical temperature is a *tempering* process. If the temperature is decreased rapidly along curve 1 to 520°F (270°C) and is then held constant along cooling curve 3, bainite is produced. This is the principle of *austempering*. Bainite is not as hard as martensite, but it does have good impact strength and fairly high hardness. Performing the same procedure at 350°F to 400°F (180°C to 200°C) is *martempering*, which produces *tempered martensite*, a soft and tough steel.

Figure 42.6 TTT Diagram for High-Carbon Steel

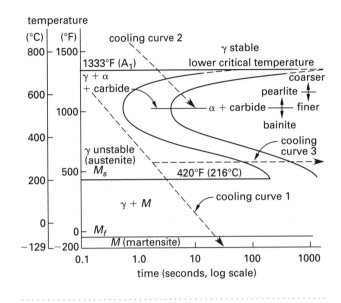

HARDNESS AND HARDENABILITY

Hardness is measure of resistance a material has to plastic deformation. Various *hardness tests* (e.g., Brinell, Rockwell, Meyer, Vickers, and Knoop) are used to determine hardness. These tests generally measure the depth of an impression made by a hardened penetrator set into the surface by a standard force.

Hardenability is a relative measure of the ease by which a material can be hardened. Some materials are easier to harden than others.

The hardness obtained also depends on the hardening method (e.g., cooling in air, other gases, water, or oil) and rate of cooling. Since the hardness obtained depends on the material, hardening data is often presented graphically. There are a variety of curve types used for this purpose.

ASTM GRAIN SIZE

One of the factors affecting hardness and hardenability is the average metal *grain size*. Grain size refers to the diameter of a three-dimensional spherical grain as determined from a two-dimensional micrograph of the metal.

The size of the grains formed depends on the number of nuclei formed during the solidification process. If there are many nuclei, as would occur when cooling is rapid, there will be many small grains. However, if cooling is slow, a smaller number of larger grains will be produced.

Thus, fast cooling produces fine-grained material, and slow cooling produces coarse-grained material.

At moderate temperatures and strain rates, fine-grained materials deform less (i.e., are harder and tougher) while coarse-grained materials deform more. For ease of cold-formed manufacturing, coarse-grained materials may be preferred. However, appearance and strength may suffer.

It is difficult to measure grain size because the grains are varied in size and shape and because a two-dimensional image does not reveal volume. Therefore, semi-empirical methods have been developed to automatically or semi-automatically correlate grain size with the number of intersections observed in samples. ASTM E112, "Determining Average Grain Size" (and the related ASTM E1382) describes a planimetric procedure for metallic and some nonmetallic materials that exist primarily in a single phase.

Data on grain size is obtained by counting the number of grains in any small two-dimensional area. The number of grains in a standard area (0.0645 mm^2) area can be extrapolated from the observations.

$$\frac{N_{standard}}{0.0645 \text{ mm}^2} = \frac{N_{actual}}{A_{actual}} \qquad 42.7$$

The grain "size," n (expressed as the nearest integer), is found from the standard number of grains, $N_{standard}$.

$$N_{standard} = 2^{n-1} \qquad 42.8$$

The grain-boundary surface area per unit volume, S_v, is taken as twice the number of points of intersection per unit length between the line and boundaries.

$$S_v = 2P_L \qquad 42.9$$

COMPOSITE MATERIALS

There are many types of modern composite material systems including dispersion-strengthened, particle-strengthened, and fiber-strengthened materials. High-performance composites are generally produced by dispersing large numbers of particles or whiskers of a strengthening component in a lightweight binder. (Steel-reinforced concrete and steel-plate-on-wood systems are also composite systems. However, these are designed and analyzed according to various building codes rather than to the theoretical methods presented in this section.)

Assuming a well-dispersed, well-bonded, and homogeneous mixture of components, the mechanical and thermal properties of a composite material can be predicted as volumetrically weighted fractions ($0 < f_i < 1.0$) of the properties of the individual components. For instance, the density, specific heat, and modulus of elasticity (Young's modulus) of a composite substance are given by Eqs. 42.10 through 42.12. Table 42.4 lists properties of common components used in producing composite materials.

$$\rho_C = \sum f_i \rho_i \qquad 42.10$$

$$c_C = \sum f_i c_i \qquad 42.11$$

$$E_C = \sum f_i E_i \qquad 42.12$$

Table 42.4 Properties of Components of Composite Materials

	density ρ (Mg/m^3)	modulus of elasticity E (GPa)	E/ρ (N·m/g)
binders/matrix			
polystyrene	1.05	2	2700
polyvinyl chloride	1.3	<4	3500
strengtheners			
alumina fiber	3.9	400	100 000
aluminum	2.7	70	26 000
aramide fiber	1.3	125	100 000
BeO fiber	3.0	400	130 000
beryllium fiber	1.9	300	160 000
boron fiber	2.3	400	170 000
carbon fiber	2.3	700	300 000
glass	2.5	70	28 000
magnesium	1.7	45	26 000
silicon carbide fiber	3.2	400	120 000
steel	7.8	205	26 000

Assuming perfect bonding, the strain in two adjacent components (e.g., strengthening whiskers and the supporting matrix) will be the same.

$$\epsilon_1 = \epsilon_2 \qquad 42.13$$

$$\left(\frac{\Delta L}{L_o}\right)_1 = \left(\frac{\Delta L}{L_o}\right)_2 \qquad 42.14$$

The strain can be due to an applied load (F/AE), changes in temperature ($\alpha \Delta T$), or a combination of the two.

$$(\epsilon_{thermal} + \epsilon_{mechanical})_1 = (\epsilon_{thermal} + \epsilon_{mechanical})_2$$

$$\left(\alpha \Delta T + \frac{F}{AE}\right)_1 = \left(\alpha \Delta T + \frac{F}{AE}\right)_2 \qquad 42.15$$

SAMPLE PROBLEMS

Problem 1

Why is aluminum more rust-resistant than steel?

- (A) The reaction rate with atmospheric oxygen is higher for steel.
- (B) The reaction rate with atmospheric oxygen is higher for aluminum.
- (C) Iron atoms are larger than aluminum atoms, and thus, the interstitial spaces are larger.
- (D) Iron has greater magnetic properties than aluminum.

Solution

Oxygen reacts faster with aluminum. In fact, it reacts so fast that it creates a film of aluminum oxide that acts as a protective coating.

Answer is B.

Problem 2

Which of the following metals do not have a face-centered cubic crystalline structure?

- I. aluminum
- II. gamma-iron
- III. delta-iron
- IV. lead

- (A) III only
- (B) II and III
- (C) III and IV
- (D) I, II, and IV

Solution

Aluminum, lead, and gamma-iron all have face-centered cubic structures. Delta-iron has a body-centered cubic structure.

Answer is A.

Problem 3

Which of the following will affect the hardenability of steel?

- I. composition of austenite
- II. composition of cementite
- III. austenite grain size
- IV. quenching medium
- V. carbon content

- (A) II only
- (B) I and V
- (C) III and V
- (D) I, II, III, and V

Solution

Carbon content and grain size are the primary factors affecting hardenability.

Answer is C.

FE-STYLE EXAM PROBLEMS

1. Which of the following characterize a hot-worked steel part in comparison with a cold-worked part?

- I. higher yield strength
- II. better surface finish
- III. greater hardness
- IV. greater toughness
- V. less ductility

- (A) I and V
- (B) II only
- (C) III and IV
- (D) none of the above

Problems 2–4 refer to the following phase diagram.

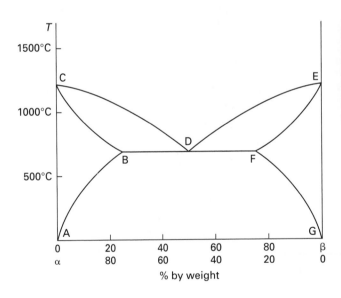

2. The region enclosed by points DEF can be described as which of the following?

- (A) a mixture of solid β component and liquid α component
- (B) a mixture of solid and liquid β component
- (C) a peritectic composition
- (D) a mixture of solid β component and the eutectic material

3. Which line(s) is (are) the liquidus?

(A) CBDFG
(B) CDE
(C) ABC and EFG
(D) CBFE

4. How much solid (as a percentage by weight) exists when the mixture is 30% α and 70% β and the temperature is 800°C?

(A) 0%
(B) 19%
(C) 30%
(D) 50%

5. Which of the following characteristics describes martensite?

I.   high ductility
II. formed by quenching austenite
III. high hardness

(A) I only
(B) I and II
(C) II and III
(D) I and III

For the following problems use the NCEES Handbook as your only reference.

6. The activation energy, Q, for aluminum in a copper solvent at 575°C is 1.6×10^8 J/kmol. What is the diffusion coefficient, D, if the constant of proportionality, D_0, is 7×10^{-6} m^2/s?

(A) 4.04×10^{-47} m^2/s
(B) 2.04×10^{-20} m^2/s
(C) 9.75×10^{-16} m^2/s
(D) 2.31×10^{-5} m^2/s

7. An iron alloy contains 2.5% carbon by weight. In what phase is the alloy at 900°C?

(A) liquid
(B) γ + liquid
(C) δ + carbide
(D) γ austenite and carbide

8. A mixture of ice and water is held at a constant temperature of 0°C. How many degrees of freedom does the mixture have?

(A) -1
(B) 0
(C) 1
(D) 2

9. A brass alloy is 40% zinc and 60% copper by weight. What is the approximate mole fraction of zinc?

(A) 5%
(B) 26%
(C) 39%
(D) 50%

10. The crystalline structure of metals can be modified by several processes. Plastic deformation of the crystalline structure resulting in misalignment of atoms, dislocations, and large stresses and strains in small regions are characteristic of which process?

(A) tempering
(B) cold forming
(C) twinning
(D) isostatic pressing

11. Which of the following figures is a cooling curve of a pure metal?

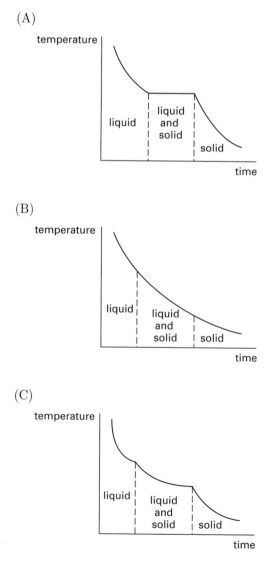

(D)

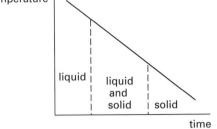

12. What is the hardest form of steel?

(A) pearlite
(B) ferrite
(C) bainite
(D) martensite

13. Which of the following processes can increase the deformation resistance of steel?

I. tempering
II. hot working
III. adding alloying elements
IV. hardening

(A) I and II
(B) I and IV
(C) II and III
(D) III and IV

14. Corrosion of iron can be inhibited with a more electropositive coating, while a less electropositive coating tends to accelerate corrosion. Which of the following coatings will contribute to corrosion of iron products?

(A) zinc
(B) gold
(C) aluminum
(D) magnesium

15. Which method of surface hardening does not depend on diffusion?

(A) induction hardening
(B) nitriding
(C) carburizing
(D) cyaniding

16. Steel is an alloy of which of the following elements?

(A) iron and silicon
(B) iron and carbon
(C) iron and chromium
(D) iron and molybdenum

17. Which of the following hardening methods depends on rapid diffusion through the surface?

(A) austenitizing
(B) carburizing
(C) martempering
(D) precipitation hardening

18. Austenite and martensite are

(A) both equilibrium solutions of carbon and iron
(B) an equilibrium solution and a nonequilibrium solution, respectively, of carbon and iron
(C) a nonequilibrium solution and an equilibrium solution, respectively, of carbon and iron
(D) both nonequilibrium solutions of carbon and iron

SOLUTIONS TO FE-STYLE EXAM PROBLEMS

1. The listed properties characterize a cold-worked steel part in comparison with a hot-worked part, rather than the opposite.

Answer is D.

2. The region describes a mixture of solid β component and the eutectic material, which is a liquid of components α and β.

Answer is D.

3. The liquidus line divides the diagram into two regions. Above the liquidus, the alloy is purely liquid, while below the liquidus the alloy may exist as solid phase or as a mixture of solid and liquid phases. The liquidus is line CDE.

Answer is B.

4.

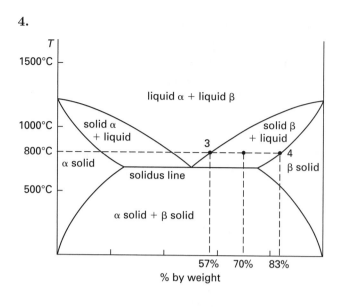

$$\text{fraction solid} = \frac{x - x_3}{x_4 - x_3}$$
$$= \frac{70\% - 57\%}{83\% - 57\%}$$
$$= 0.50 \quad (50\%)$$

Answer is D.

5. Martensite is a hard, strong, and brittle material formed by rapid cooling of austenite.

Answer is C.

6. The absolute temperature is

$$575°C + 273 = 848K$$

To apply the formula for the diffusion coefficient, the units in the exponent must cancel. Since the activation energy, Q, is given in units of joules per kmol, the universal gas constant $\overline{R}$ must also have those units.

$$D = D_0 e^{-Q/\overline{R}T}$$
$$= \left(7 \times 10^{-6} \, \frac{\text{m}^2}{\text{s}}\right) e^{-\left(1.6 \times 10^8 \, \frac{\text{J}}{\text{kmol}} \middle/ \left(8314 \, \frac{\text{J}}{\text{kmol·K}}\right)(848K)\right)}$$
$$= 9.75 \times 10^{-16} \, \text{m}^2/\text{s}$$

Answer is C.

7. Refer to the iron-carbon phase diagram. The point corresponding to 2.5% carbon and 900°C is below the liquid line and is in the γ austenite and carbide region.

Answer is D.

8. Since solid and liquid phases are present simultaneously, the number of phases, P, is 2. Only water is involved, so the number of compounds, C, is 1.

Gibbs' phase rule is applicable when both temperature and pressure can be varied. When the temperature is held constant, Gibbs' phase rule is

$$P + F = C + 1|_{\text{constant temperature}}$$
$$F = C + 1 - P$$
$$= 1 + 1 - 2$$
$$= 0$$

Answer is B.

9. Consider a sample of 100 g of brass. The sample will consist of 40 g of zinc and 60 g of copper.

From the periodic table, the atomic weight of zinc is 65.38 g/mol. The number of moles of zinc is

$$n_{\text{Zn}} = \frac{m}{\text{MW}} = \frac{40 \text{ g}}{65.38 \, \frac{\text{g}}{\text{mol}}}$$
$$= 0.612 \text{ mol}$$

From the periodic table, the atomic weight of copper is 65.546 g/mol. The number of moles of copper is

$$n_{\text{Cu}} = \frac{m}{\text{MW}} = \frac{60 \text{ g}}{63.546 \, \frac{\text{g}}{\text{mol}}}$$
$$= 0.944 \text{ mol}$$

The mole fraction of zinc is

$$x_{\text{Zn}} = \frac{n_{\text{Zn}}}{n_{\text{Zn}} + n_{\text{Cu}}}$$
$$= \frac{0.612 \text{ mol}}{0.612 \text{ mol} + 0.944 \text{ mol}}$$
$$= 0.393 \quad (39\%)$$

Answer is C.

10. Dislocations, defects, and large concentrations of stress and strain are characteristics of a metal that has been cold worked.

Answer is B.

11. The solidification of a molten metal is no different than the solidification of water into ice. During the phase change, the temperature remains constant as the heat of fusion is removed. The temperature remains constant during the phase change.

Answer is A.

12. Hard steel is obtained by rapid quenching. Martensite has a high hardness since it is rapidly quenched. Though martensite is hard, it has low ductility.

Answer is D.

13. Surface hardening processes will increase the deformation resistance of steel. Some alloying metals will also increase steel hardness. Tempering and hot working increase the ductility (deformation capability) of steel.

Answer is D.

14. Zinc, aluminum, and magnesium are all more electropositive (anodic) than iron and will corrode sacrificially to protect it. Gold is more cathodic and will be protected at the expense of the iron.

Answer is B.

15. Nitriding, carburizing, and cyaniding depend on the diffusion of nitrogen, carbon, and cyanide, respectively. Induction hardening uses high-frequency current to heat the metal and rapid quenching to capture the grain structure.

Answer is A.

16. Although additional elements (such as chromium and molybdenum) can be used to modify its properties, steel is an alloy of iron and carbon.

Answer is B.

17. Carburizing depends on the diffusion of carbon (the hardening element) through the surface of a heated piece of steel. Austenitizing, martempering, and precipitation hardening are grain-size modifications brought about by heat treatment.

Answer is B.

18. Austenite is an equilibrium substance consisting of carbon in gamma iron. It is found on the iron-carbon equilibrium diagram above 2065°F (1129°C). Martensite is not an equilibrium substance and cannot be found on the equilibrium diagram. Its grain structure is obtained during a rapid quenching process.

Answer is B.

Topic XIV: Electric Circuits

Diagnostic Examination for Electric Circuits

Chapter 43   Complex Numbers and Electrostatics

Chapter 44   Direct-Current Circuits

Chapter 45   Alternating-Current Circuits

Chapter 46   Rotating Machines

CIRCUITS

Hint: For the most current information about the exam, visit www.ppi2pass.com/fefaqs.html regularly. Use the Exam Forum to compare notes with other FE examinees.

Diagnostic Examination

TOPIC XIV: ELECTRIC CIRCUITS

TIME LIMIT: 45 MINUTES

1. The net force on the center charge is zero for the system of three colinear charges shown. What is the distance x?

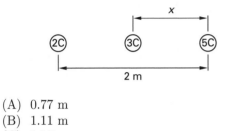

(A) 0.77 m
(B) 1.11 m
(C) 1.17 m
(D) 1.23 m

2. A heating element consists of two wires of different materials connected in series. At 20°C, they have resistances of 600 Ω and 300 Ω, and average temperature coefficients of 0.001 1/°C and 0.004 1/°C, respectively. What is most nearly the heating element's total resistance at 50°C?

(A) 900 Ω
(B) 950 Ω
(C) 980 Ω
(D) 990 Ω

3. A 5 Ω resistor is placed in series with a varying current. Most nearly how much energy is dissipated by the resistor over the 4 s time interval shown?

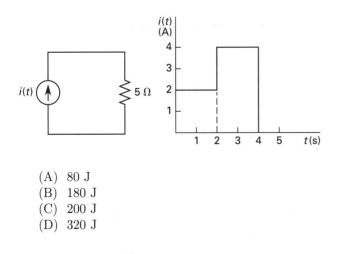

(A) 80 J
(B) 180 J
(C) 200 J
(D) 320 J

4. What is the total capacitance between terminals A and B?

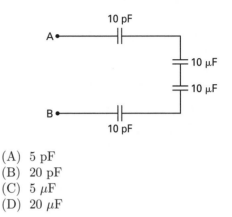

(A) 5 pF
(B) 20 pF
(C) 5 μF
(D) 20 μF

5. What is the voltage drop across the 7 Ω resistor?

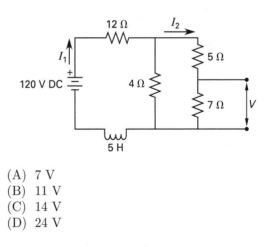

(A) 7 V
(B) 11 V
(C) 14 V
(D) 24 V

6. What is the voltage across the 10 Ω resistor in the circuit shown?

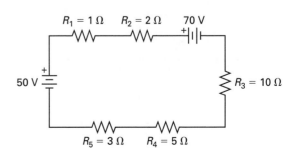

(A) 9.5 V
(B) 24 V
(C) 33 V
(D) 57 V

(A) $35\angle-90°$ A
(B) $35\angle90°$ A
(C) $50\angle-90°$ A
(D) $50\angle90°$ A

7. What are the Thevenin equivalent resistance and voltage between terminals A and B?

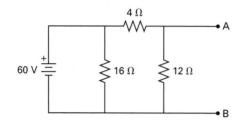

(A) $R_{Th} = 3\ \Omega$, $V_{Th} = 45$ V
(B) $R_{Th} = 7.5\ \Omega$, $V_{Th} = 7.5$ V
(C) $R_{Th} = 7.5\ \Omega$, $V_{Th} = 60$ V
(D) $R_{Th} = 12\ \Omega$, $V_{Th} = 5$ V

8. The capacitor is initially uncharged in the circuit shown. What will be the approximate current 3 ms after the switch is closed?

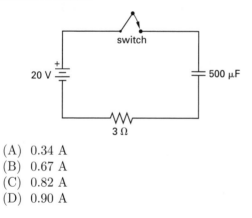

(A) 0.34 A
(B) 0.67 A
(C) 0.82 A
(D) 0.90 A

9. What is the correct phasor (polar) expression for the effective current in the graph shown?

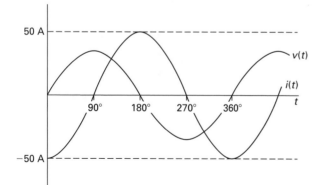

10. What is the equivalent impedance when the circuit shown is connected to a 120 V, 60 Hz source?

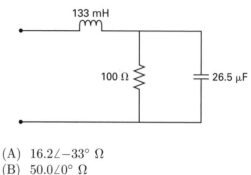

(A) $16.2\angle-33°\ \Omega$
(B) $50.0\angle0°\ \Omega$
(C) $70.7\angle-45°\ \Omega$
(D) $111\angle63°\ \Omega$

11. Which of the following expressions correctly relates power factor (p.f.) to real power (P), reactive power (Q), and complex power (S)?

(A) p.f. $= \dfrac{P}{Q}$

(B) p.f. $= \dfrac{Q}{P}$

(C) p.f. $= \dfrac{P}{S}$

(D) p.f. $= \dfrac{Q}{S}$

12. To replace a heat loss of 2 kW, a 20°C room is heated by a resistive element heater from a standard 120 V rms, 60 Hz power supply. Most nearly what must the element's resistance be to maintain the room at 20°C?

(A) 4.6 Ω
(B) 7.2 Ω
(C) 14 Ω
(D) 17 Ω

13. What is most nearly the resonant frequency of the circuit shown?

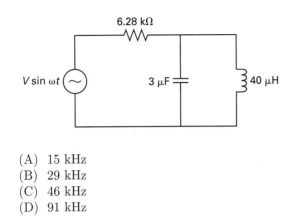

(A) 15 kHz
(B) 29 kHz
(C) 46 kHz
(D) 91 kHz

14. How much power is dissipated by the 1 kΩ resistor?

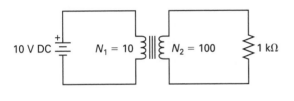

(A) 0 W
(B) 0.1 W
(C) 1 W
(D) 10 W

15. If $z_1 = 14\angle 31°$ and $z_2 = 30 + j40$, what are $z_1 z_2$ and z_1/z_2, respectively?

(A) $550\angle 62°$; $1.4\angle 67°$
(B) $700\angle 84°$; $0.28\angle -22°$
(C) $880\angle 35°$; $0.18\angle -120°$
(D) $900\angle 135°$; $-0.44\angle -93°$

SOLUTIONS TO DIAGNOSTIC EXAMINATION TOPIC XIV

1. The forces exerted by the two end charges on the middle charge are equal in magnitude.

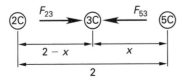

$$F_{23} = F_{53}$$

$$\frac{Q_2 Q_3}{4\pi\epsilon(2-x)^2} = \frac{Q_3 Q_5}{4\pi\epsilon x^2}$$

$$Q_2 Q_3 x^2 - Q_3 Q_5(2-x)^2 = 0$$

$$(2)(3)(x^2) - (3)(5)(2-x)^2 = 0$$

$$6x^2 - 60 + 60x - 15x^2 = 0$$

$$9x^2 - 60x + 60 = 0$$

$$
\begin{aligned}
x &= \frac{-b \pm \sqrt{b^2 - 4ac}}{2a} \quad \text{[quadratic formula]} \\
&= \frac{60 \pm \sqrt{(-60)^2 - (4)(9)(60)}}{(2)(9)} \\
&= 1.23,\ 5.44
\end{aligned}
$$

Answer is D.

2. Find the resistance of each wire.

For wire A,

$$
\begin{aligned}
R_A &= R_{0,A}\big(1 + \alpha(T - T_0)\big) \\
&= (600\ \Omega)\left(1 + \left(0.001\ \frac{1}{°C}\right)(50°C - 20°C)\right) \\
&= 618\ \Omega
\end{aligned}
$$

For wire B,

$$
\begin{aligned}
R_B &= R_{0,B}\big(1 + \alpha(T - T_0)\big) \\
&= (300\ \Omega)\left(1 + \left(0.004\ \frac{1}{°C}\right)(50°C - 20°C)\right) \\
&= 336\ \Omega
\end{aligned}
$$

$$
\begin{aligned}
R_{total} &= R_A + R_B \\
&= 618\ \Omega + 336\ \Omega \\
&= 954\ \Omega \quad (950\ \Omega)
\end{aligned}
$$

Answer is B.

3. Power is the energy dissipated per unit time. The current in this problem is not constant, so calculate the energy over two periods of time.

For $t = 0$ s to $t = 2$ s,

$$
\begin{aligned}
P &= I^2 R = (2\ A)^2(5\ \Omega) \\
&= 20\ W
\end{aligned}
$$

The energy dissipated is

$$E = Pt = (20 \text{ W})(2 \text{ s})$$
$$= 40 \text{ J}$$

For $t = 2$ s to $t = 4$ s,

$$P = I^2 R = (4 \text{ A})^2 (5 \text{ }\Omega)$$
$$= 80 \text{ W}$$
$$E = Pt = (80 \text{ W})(2 \text{ s})$$
$$= 160 \text{ J}$$
$$E_{\text{total}} = 40 \text{ J} + 160 \text{ J}$$
$$= 200 \text{ J}$$

Answer is C.

4. Capacitors connected in series combine like resistors in parallel.

$$C_{\text{eq}} = \cfrac{1}{\cfrac{1}{10 \times 10^{-12} \text{ F}} + \cfrac{1}{10 \times 10^{-6} \text{ F}} + \cfrac{1}{10 \times 10^{-6} \text{ F}} + \cfrac{1}{10 \times 10^{-12} \text{ F}}}$$

$$= 5 \times 10^{-12} \text{ F} \quad (5 \text{ pF})$$

Answer is A.

5. The inductor has zero resistance to DC in steady state, so it is not considered in the analysis. Determine the equivalent resistance of the two parallel legs.

$$R_{\text{parallel}} = \cfrac{1}{\cfrac{1}{4 \text{ }\Omega} + \cfrac{1}{5 \text{ }\Omega + 7 \text{ }\Omega}}$$
$$= 3 \text{ }\Omega$$

Determine the total equivalent resistance of the circuit. Because the voltage source is DC, the inductor contributes no resistance.

$$R_{\text{eq}} = 12 \text{ }\Omega + 3 \text{ }\Omega$$
$$= 15 \text{ }\Omega$$

Use Ohm's law to calculate the currents and voltages through individual elements.

$$I_1 = \frac{V}{R_{\text{eq}}} = \frac{120 \text{ V}}{15 \text{ }\Omega}$$
$$= 8 \text{ A}$$

The voltage across the parallel section is

$$V_{\text{parallel}} = I_1 R_{\text{parallel}} = (8 \text{ A})(3 \text{ }\Omega)$$
$$= 24 \text{ V}$$
$$I_2 = \frac{V_{\text{parallel}}}{R} = \frac{24 \text{ V}}{5 \text{ }\Omega + 7 \text{ }\Omega}$$
$$= 2 \text{ A}$$

The unknown voltage is calculated from Ohm's law.

$$V = I_2 R = (2 \text{ A})(7 \text{ }\Omega)$$
$$= 14 \text{ V}$$

Answer is C.

6. Follow the rules for simple series circuits. Find the equivalent resistance and equivalent voltage.

$$R_{\text{eq}} = R_1 + R_2 + R_3 + R_4 + R_5$$
$$= 1 \text{ }\Omega + 2 \text{ }\Omega + 10 \text{ }\Omega + 5 \text{ }\Omega + 3 \text{ }\Omega$$
$$= 21 \text{ }\Omega$$
$$V_{\text{eq}} = V_1 + V_2 = 70 \text{ V} - 50 \text{ V}$$
$$= 20 \text{ V}$$

Use Ohm's law to determine the current through the circuit.

$$I = \frac{V}{R} = \frac{V_{\text{eq}}}{R_{\text{eq}}}$$
$$= \frac{20 \text{ V}}{21 \text{ }\Omega}$$
$$= 0.952 \text{ A} \quad (9.5 \text{ V})$$

The voltage across any individual resistor is found from Ohm's law.

$$V = IR = (0.952 \text{ A})(10 \text{ }\Omega)$$
$$= 9.52 \text{ V} \quad (9.5 \text{ V})$$

Answer is A.

7. The Thevenin resistance is the resistance between the two terminals with the voltage source short-circuited. With the circuit shorted, the 16 Ω resistor is effectively bypassed, leaving the other two in a simple parallel arrangement.

$$R_{\text{Th}} = \cfrac{1}{\cfrac{1}{4 \text{ }\Omega} + \cfrac{1}{12 \text{ }\Omega}}$$
$$= 3 \text{ }\Omega$$

The Thevenin voltage is the voltage difference between the two terminals, which in this case is equal to the voltage drop across the 12 Ω resistor. Apply Ohm's law once to find the current $I_{12\,\Omega}$, and then again to find the voltage across the resistor.

$$I_{12\,\Omega} = \frac{V}{R} = \frac{60\text{ V}}{4\ \Omega + 12\ \Omega}$$
$$= 3.75\text{ A}$$

$$V_{\text{Th}} = I_{12\,\Omega}R = (3.75\text{ A})(12\ \Omega)$$
$$= 45\text{ V}$$

Answer is A.

8. For an RC circuit transient,

$$i(t) = \left(\frac{V - v_c(0)}{R}\right)e^{\frac{-t}{RC}}$$
$$= \left(\frac{20\text{ V} - 0\text{ V}}{3\ \Omega}\right)e^{\frac{-0.003\text{ s}}{(3\ \Omega)(500\times10^{-6}\text{ F})}}$$
$$= 0.902\text{ A}\quad(0.90\text{ A})$$

Answer is D.

9. The effective current is the rms (root-mean-square) current. The rms value is calculated from the peak value.

$$I_{\text{rms}} = \frac{I_{\text{max}}}{\sqrt{2}} = \frac{50\text{ A}}{\sqrt{2}}$$
$$= 35.35\text{ A}$$

The phase shift is determined using the voltage signal as a reference. The current reaches its peak after the voltage. Lagging currents have a negative phase shift.

$$\mathbf{I} = I_{\text{rms}}\angle\theta$$
$$= 35.35\angle-90°\text{ A}\quad(35\angle-90°\text{ A})$$

Answer is A.

10. The forcing frequency is

$$\omega = 2\pi f = 2\pi(60\text{ Hz})$$
$$= 377\text{ rad/s}$$

Find the impedances of the individual elements.

For the inductor,

$$\mathbf{Z}_L = j\omega L$$
$$= j\left(377\ \frac{\text{rad}}{\text{s}}\right)(133\times10^{-3}\text{ H})$$
$$= j50.1\ \Omega$$

For the resistor,

$$\mathbf{Z}_R = R = 100\ \Omega$$

For the capacitor,

$$\mathbf{Z}_C = \frac{-j}{\omega C}$$
$$= \frac{-j}{\left(377\ \dfrac{\text{rad}}{\text{s}}\right)(26.5\times10^{-6}\text{ F})}$$
$$= -j100\ \Omega$$

Combine the impedances using the same rules used for resistances in series and parallel.

$$\mathbf{Z}_{\text{eq}} = \frac{1}{\dfrac{1}{Z_R}+\dfrac{1}{Z_C}} + Z_L$$
$$= \frac{1}{\dfrac{1}{100\ \Omega}+\dfrac{j}{100\ \Omega}} + j50\ \Omega$$
$$= \frac{100\ \Omega}{1+j} + j50\ \Omega$$
$$= \left(\frac{100\ \Omega}{1+j}\right)\left(\frac{1-j}{1-j}\right) + j50\ \Omega$$
$$= \frac{100-j100}{2}\ \Omega + j50\ \Omega$$
$$= (50+j0)\ \Omega$$
$$= 50\angle0°\ \Omega$$

Answer is B.

11. Power factor is defined as $\cos\theta$, where θ is the angle between the real and complex power vectors in the complex power triangle.

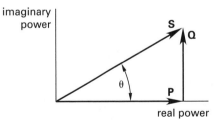

Standard trigonometric identities provide the relationship between θ and P, Q, and S.

$$\text{p.f.} = \cos\theta = \frac{\text{adjacent}}{\text{hypotenuse}}$$
$$= P/S$$

Answer is C.

12. The heater is a pure resistive element. The resistance can be solved directly by using the effective voltage.

$$P_{\text{ave}} = \frac{V_{\text{rms}}^2}{R}$$

$$R = \frac{V_{\text{rms}}^2}{P_{\text{ave}}} = \frac{(120 \text{ V})^2}{2000 \text{ W}}$$

$$= 7.2 \ \Omega$$

Answer is B.

13. The resonant frequency is

$$\omega_0 = 2\pi f_0 = \frac{1}{\sqrt{LC}}$$

$$f_0 = \frac{1}{2\pi\sqrt{LC}}$$

$$= \frac{1}{2\pi\sqrt{(40 \times 10^{-6} \text{ H})(3 \times 10^{-6} \text{ F})}}$$

$$= 14\,529 \text{ Hz} \quad (15 \text{ kHz})$$

Answer is A.

14. Transformers operate according to Faraday's law, which states that an induced voltage is generated in response to a time-dependent magnetic flux. Because the primary voltage in the circuit is DC, no voltage is induced in the secondary coil and no power is dissipated.

Answer is A.

15. First, convert z_2 to polar form.

$$|z_2| = c_2 = \sqrt{30^2 + 40^2}$$

$$= 50$$

$$\theta_2 = \tan^{-1}\frac{40}{30}$$

$$= 53.1°$$

To multiply numbers in polar form, find the product of their magnitudes and the sum of their angles.

$$z_1 z_2 = c_1 c_2 \angle \theta_1 + \theta_2$$

$$= (14)(50)\angle 31° + 53.1°$$

$$= 700\angle 84°$$

To find the quotient, divide the magnitudes, and subtract the angle of the denominator from the angle of the numerator.

$$\frac{z_1}{z_2} = \frac{c_1}{c_2}\angle \theta_1 - \theta_2$$

$$= \frac{14}{50}\angle 31 - 53.1°$$

$$= 0.28\angle -22°$$

Answer is B.

43 Complex Numbers and Electrostatics

Subjects

ALGEBRA OF COMPLEX NUMBERS . . 43-1
ELECTROSTATICS 43-2
   Electrostatic Fields 43-2
   Voltage 43-4
   Current 43-4
   Magnetic Fields 43-5
   Induced Voltage 43-6

Nomenclature

A	area	m^2
B	magnetic flux density	T
d	distance	m
e	induced voltage	V
E	electric field intensity	N/C or V/m
F	force	N
H	magnetic field strength	A/m
$i(t)$	time-varying current	A
I	constant current	A
L	length	m
N	number of turns	–
$q(t)$	time-varying charge	C
Q	constant charge	C
r	radius	m
s	distance	m
v	velocity	m/s
$v(t)$	time-varying voltage	V
V	constant voltage or potential difference	V
W	work	J

Symbols

ϵ	permittivity	F/m or $C^2/N{\cdot}m^2$
μ	permeability	H/m
θ	angle	rad
ρ	flux density	C/m, C/m^2, or C/m^3
ϕ	magnetic flux	Wb
ψ	electric flux	C

Subscripts

0	free space (vacuum)
encl	enclosed
H	magnetic field strength
L	per unit length
S	per unit area

ALGEBRA OF COMPLEX NUMBERS

A *complex number*, z, consists of the sum of real and imaginary numbers. *Real numbers* are rational and irrational numbers, and *imaginary numbers* are square roots of negative real numbers.

$$z = a + jb \qquad 43.1$$

In Eq. 43.1, a is the real component, b is the imaginary component, and j is the square root of negative 1. When expressed as a sum in the form of Eq. 43.1, the complex number is said to be in *rectangular* or *trigonometric form*. This is because the complex number can be plotted on the rectangular coordinate system known as the *complex plane*, as illustrated in Fig. 43.1. In the complex plane, the abscissa is the real component of the number and the ordinate is the imaginary component.

Figure 43.1 Rectangular Form of a Complex Number

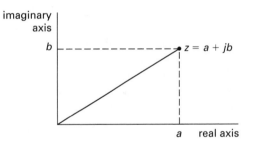

The complex number $z = a + jb$ can also be expressed in the *phasor form*, also called the *polar form*, as illustrated in Fig. 43.2. In Eq. 43.2, the quantity c is known as the *absolute value* or *modulus*, and θ is the *argument*.

$$z = c\angle\theta \qquad 43.2$$

Figure 43.2 Phasor Form of a Complex Number

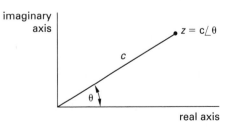

The rectangular form can be derived from c and θ.

$$a = c \cos \theta \qquad \qquad 43.3$$

$$b = c \sin \theta \qquad \qquad 43.4$$

$$z = a + jb$$

$$= c \cos \theta + jc \sin \theta$$

$$= c \left(\cos \theta + j \sin \theta \right) \qquad 43.5$$

Similarly, the phasor form can be derived from a and b.

$$c = \sqrt{a^2 + b^2} \qquad \qquad 43.6$$

$$\theta = \tan^{-1} \frac{b}{a} \qquad \qquad 43.7$$

$$z = c \angle \theta$$

$$= \sqrt{a^2 + b^2} \; \angle \tan^{-1} \frac{b}{a} \qquad 43.8$$

Equation 43.7 must be calculated carefully when using a standard calculator (a calculator that is not designed for complex number operations). When calculating the inverse tangent of Eq. 43.7 where a is a negative number, it is necessary to add 180° (or π rad) to the answer. A standard calculator does not distinguish between the phase angle of $1 + j$ and $-1 - j$. The inverse tangent of 1 divided by 1 and the inverse tangent of -1 divided by -1 are both 45° (or 0.5π rad) on a standard calculator. However, $-1 - j$ is in the third quadrant at 225° (or 1.5π rad). By tradition, complex numbers are usually shown with a phase angle that has a magnitude less than or equal to 180° (π rad), so 225° (1.5π rad) would be expressed as $-135°$ (-0.5π rad).

Most algebraic operations (addition, multiplication, exponentiation, etc.) work with complex numbers. When adding two complex numbers, real parts are added to real parts, and imaginary parts are added to imaginary parts. If $z_1 = a_1 + jb_1$, and $z_2 = a_2 + jb_2$, then

$$z_1 + z_2 = (a_1 + a_2) + j(b_1 + b_2) \qquad 43.9$$

$$z_1 - z_2 = (a_1 - a_2) + j(b_1 - b_2) \qquad 43.10$$

Multiplication of two complex numbers in rectangular form is accomplished by use of the algebraic distributive law. Wherever j^2 occurs in the resulting expression, it may be replaced by negative 1. Division of complex numbers in rectangular form requires use of the complex conjugate. The *complex conjugate* of the complex number $(a + jb)$ is $(a - jb)$. By multiplying the numerator and the denominator by the complex conjugate, the denominator will be converted to the real number $a^2 + b^2$. This technique is known as *rationalizing the denominator*.

Multiplication and division are often more convenient when the complex numbers are in polar form, as Eqs. 43.11 and 43.12 show. If $z_1 = c_1 \angle \theta_1$ and $z_2 = c_2 \angle \theta_2$, then

$$z_1 z_2 = c_1 c_2 \; \angle \theta_1 + \theta_2 \qquad 43.11$$

$$\frac{z_1}{z_2} = \left(\frac{c_1}{c_2} \right) \angle \theta_1 - \theta_2 \qquad 43.12$$

Complex numbers can also be expressed in exponential form. The relationship of the exponential form to the trigonometric form is given by *Euler's equations* (Eqs. 43.13 and 43.14).

$$e^{j\theta} = \cos \theta + j \sin \theta \qquad 43.13$$

$$e^{-j\theta} = \cos \theta - j \sin \theta \qquad 43.14$$

The trigonometric terms are related to the exponential form by Eqs. 43.15 and 43.16.

$$\cos \theta = \frac{e^{j\theta} + e^{-j\theta}}{2} \qquad 43.15$$

$$\sin \theta = \frac{e^{j\theta} - e^{-j\theta}}{2j} \qquad 43.16$$

The angle θ should always be in radians when complex numbers are expressed in exponential form, although θ in degrees is frequently observed in practice.

ELECTROSTATICS

Electric charge is a fundamental property of subatomic particles. The charge on an electron is negative one *electrostatic unit* (esu). The charge on a proton is positive one esu. A neutron has no charge. Charge is measured in the SI system in *coulombs* (C). One coulomb is approximately 6.24×10^{18} esu; the charge of one electron is -1.6×10^{-19} C.

Conservation of charge is a fundamental principle or law of physics. Electric charge can be distributed from one place to another under the influence of an electric field, but the algebraic sum of positive and negative charges in a system cannot change unless charged particles are added or removed.

Electrostatic Fields

An electric field, E, with units of newtons per coulomb or volts per meter (N/C, same as V/m) is generated in the vicinity of an electric charge. The imaginary lines of force, as illustrated in Fig. 43.3, are called the *electric flux*, ψ. The direction of the electric flux is the same as the force applied by the electric field to a positive charge introduced into the field. If the field is produced by a

Figure 43.3 Electric Field Around a Positive Charge

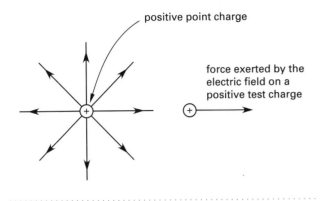

positive point charge

force exerted by the
electric field on a
positive test charge

positive charge, the force on another positive charge
placed nearby will try to separate the two charges, and
therefore, the line of force will leave the first positive
charge.

In general, the force on a test charge Q in an electric
field E is

$$\mathbf{F} = Q\mathbf{E} \qquad 43.17$$

The force experienced by point charge 2 (Q_2) in an
electric field E created by point charge 1 is given by
Coulomb's law, Eq. 43.18. Because charges with opposite signs attract, Eq. 43.18 is positive for repulsion
and negative for attraction. The unit vector $\mathbf{a}$ is defined
pointing from point charge 1 toward point charge 2. Although the unit vector $\mathbf{a}$ gives the direction explicitly,
the direction of force can usually be found by inspection
as the direction the object would move when released.
Vector addition (i.e., superposition) can be used with
systems of multiple point charges.

$$\mathbf{F}_2 = Q_2\mathbf{E}_1 = \frac{Q_1 Q_2}{4\pi\epsilon r^2}\mathbf{a} \qquad 43.18$$

The electric field is a vector quantity having both magnitude and direction. The orientations of the field and
flux lines always coincide (i.e., the direction of the electric field vector is always tangent to the flux lines).

The total electric flux generated by a point charge is
numerically equal to the charge.

$$\psi = Q \qquad 43.19$$

Electric flux does not pass equally well through all materials. It cannot pass through conductive metals at
all, and is canceled to various degrees by insulating media. The *permittivity* of a medium determines the flux
that passes through the medium. For free space or air,
$\epsilon = \epsilon_0 = 8.85 \times 10^{-12}$ F/m $= 8.85 \times 10^{-12}$ C^2/N·m^2.

Equation 43.20 is the electric field intensity in a medium
with permittivity ϵ at a distance r from a point charge

Q_1. The direction of the electric field is represented by
the unit vector $\mathbf{a}$.

$$\mathbf{E} = \frac{Q_1}{4\pi\epsilon r^2}\mathbf{a} \qquad 43.20$$

Not all electric fields are radial; the field direction depends on the shape and location of the charged bodies
producing the field. For a *line charge* with density ρ_L
(C/m) as shown in Fig. 43.4, the electric field is given
by Eq. 43.21. Flux density, ρ_S (C/m^2), is equal to the
number of flux lines crossing a unit area perpendicular
to the flux. In Eq. 43.21, ρ_L may be interpreted as the
flux density per unit width. The unit vector $\mathbf{a}$ is normal to the line of charge in the cylindrical coordinate
system.

$$\mathbf{E}_L = \frac{\rho_L}{2\pi\epsilon r}\mathbf{a} \qquad 43.21$$

Figure 43.4 Electric Field from a Line Charge

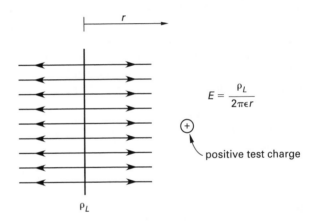

$$E = \frac{\rho_L}{2\pi\epsilon r}$$

positive test charge

For a *sheet charge* of density ρ_S (C/m^2) as shown in
Fig. 43.5, the electric field is given by Eq. 43.22. The
unit vector $\mathbf{a}$ is normal to the sheet charge.

$$\mathbf{E}_S = \frac{\rho_S}{2\epsilon}\mathbf{a} \qquad 43.22$$

Figure 43.5 Electric Field from a Sheet Charge

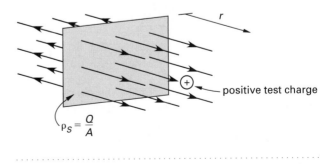

$\rho_S = \frac{Q}{A}$

positive test charge

Equation 43.20 has an inverse square relationship to r. Equation 43.21 has an inverse relationship to the separation distance, r, but Eq. 43.22 has no relationship to r. This is due to the assumptions that the line charge and the sheet charge are infinite in size, and r is small compared to the size of the line charge or sheet charge.

Gauss's law states that the electric flux passing out of a closed surface (i.e., the *Gaussian surface*) is equal to the total charge within the surface.

$$Q_{\text{encl}} = \psi \qquad 43.23$$

$$Q_{\text{encl}} = \oiint_S \epsilon \, \mathbf{E} \cdot d\mathbf{S} \qquad 43.24$$

The variable $d\mathbf{S}$ is a vector that represents an infinitesimal part of the closed surface, the direction of which is perpendicular to the surface. Imagine a cube with the charge outside and all the electric field lines coming from outside the cube perpendicular to one surface and exiting the other surface perpendicular. The charge within the cube is zero because the two vectors, $d\mathbf{S}$, are perpendicular to the surface and point in opposite directions. Alternatively, imagine a point charge at the center of a sphere. All the electric field lines are parallel to $d\mathbf{S}$ over the entire integral, so Q_{encl} is not zero and is equal to the charge at the center of the sphere.

The work, W, performed by moving a charge Q_1 radially from distance r_1 to r_2 in an electric field is given by Eq. 43.25.

$$W = -Q_1 \int_{r_1}^{r_2} \mathbf{E} \cdot d\mathbf{L} \qquad 43.25$$

The work, W, performed in moving a point charge Q_B in the radial direction from distance r_1 to r_2 within a field created by a point charge Q_A is given by Eq. 43.26. Work is positive if an external force is required to move the charges (e.g., to bring two repulsive charges together or move a charge against an electric field). Work is negative if the field does the work (allowing attracting charges to approach each other, or allowing repulsive charges to separate).

$$
\begin{aligned}
W &= -\int_{r_1}^{r_2} \mathbf{F} \cdot d\mathbf{r} \\
&= -\int_{r_1}^{r_2} \frac{Q_A Q_B}{4\pi\varepsilon r^2} \, dr \\
&= \left(\frac{Q_A Q_B}{4\pi\varepsilon}\right)\left(\frac{1}{r_2} - \frac{1}{r_1}\right) \qquad 43.26
\end{aligned}
$$

Work is performed only in moving the charges closer or further apart. Moving one point charge around the other in a constant-radius circle performs no work. In general, no work is performed in moving a charged object perpendicular to an electric field.

For a uniform field (as exists between two charged plates separated by a distance r), the work done in moving an object of charge Q a distance d parallel to the uniform field is given by Eq. 43.27.

$$
\begin{aligned}
W &= -\mathbf{F} \cdot \mathbf{d} = -EQd \\
&= \frac{-V_{\text{plates}} Q d}{r} \\
&= -Q\Delta V \qquad 43.27
\end{aligned}
$$

Voltage

Voltage is another way to describe the strength of an electric field, using a scalar quantity rather than a vector quantity. The *potential difference*, V, is the difference in electric potential between two points, defined as the work required to move one unit charge from one point to the other. This difference in potential is one volt if one joule of work is expended in moving one coulomb of charge from one point to the other.

The *electric potential gradient* in V/m is the change in potential per unit distance and is identical to the electric field strength in N/C. Thus, the electric field strength between two parallel plates with potential difference V and separated by a distance d is

$$E = \frac{V}{d} \qquad 43.28$$

Flux is directed from the positive to the negative plates.

Current

Current, $i(t)$, is the movement of charges. By convention, the current moves in a direction opposite to the flow of electrons (i.e., the current flows from the positive terminal to the negative terminal). Current is measured in amperes (A) and is the time rate change of charge (i.e., the current is equal to the number of coulombs of charge passing a point each second). If $q(t)$ is the instantaneous charge, then

$$i(t) = \frac{dq(t)}{dt} \qquad 43.29$$

If the rate of change in charge is constant, the current is

$$I = \frac{dQ}{dt} \qquad 43.30$$

The *current density*, ρ, is the density of charge (charged particles times their charge) moving per unit time through a volume. The *volume current density*, $\mathbf{J}$, (also called current density) is the vector current density with a magnitude equal to the scalar current density and direction of the velocity vector, $\mathbf{v}$. The velocity depends on the force moving the charge (voltage or electromotive force) and the force opposing the motion of the charge (impedance).

$$\mathbf{J} = \rho \mathbf{v} \qquad 43.31$$

The current in the direction perpendicular to a surface S can be determined by integrating the current density.

$$I = \int_S \mathbf{J} \cdot d\mathbf{S} \qquad 43.32$$

Magnetic Fields

A magnetic field can exist only with two opposite, equal poles called the *north pole* and *south pole*. This is unlike an electric field, which can be produced by a single charged object. Figure 43.6 illustrates two common permanent magnetic field configurations. It also illustrates the convention that the lines of magnetic flux are directed from the north pole (i.e., the *magnetic source*) to the south pole (i.e., the *magnetic sink*). The total amount of magnetic flux in a magnetic field is ϕ, measured in webers (Wb).

Figure 43.6 *Magnetic Fields from Permanent Magnets*

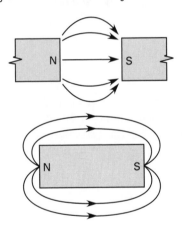

The magnetic flux density, $\mathbf{B}$, in teslas (T), equivalent to Wb/m^2, is one of two measures of the strength of a magnetic field. For this reason, it can be referred to as the *strength of the B-field*. ($\mathbf{B}$ should never be called the magnetic field strength as that name is reserved for the variable $\mathbf{H}$.) $\mathbf{B}$ is also known as the *magnetic induction*. The magnetic flux density is found by dividing the magnetic flux by an area perpendicular to it. Magnetic flux density is a vector quantity.

$$\mathbf{B} = \frac{\phi}{A} \mathbf{a} \qquad 43.34$$

The *magnetic field strength*, $\mathbf{H}$, with units of A/m, is derived from the magnetic flux density. The direction of the magnetic field, illustrated in Fig. 43.7, is given by the *right-hand rule*. In the case of a straight wire, the thumb indicates the current direction, and the fingers curl in the field direction; for a coil, the fingers indicate the current flow, and the thumb indicates the field direction.

$$\mathbf{H} = \frac{\mathbf{B}}{\mu} = \frac{I}{2\pi r} \mathbf{a} \quad \text{[straight wire]} \qquad 43.35$$

Figure 43.7 *Right-Hand Rule for the Magnetic Flux Direction in a Coil*

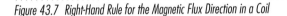
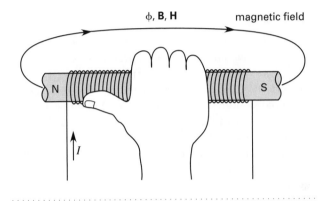

The magnetic flux density $\mathbf{B}$ is dependent on the *permeability* of the medium much like the electric flux density is dependent on permittivity. The permeability of free space (air or vacuum) is $\mu = \mu_0 = 4\pi \times 10^{-7}$ H/m.

The analogy to Coulomb's law, where a force is imposed on a stationary charge in an electric field, is that a magnetic field imposes a force on a moving charge. The force on a wire carrying a current I in a uniform magnetic field $\mathbf{B}$ is given by Eq. 43.36. $\mathbf{L}$ is the length vector of the conductor and points in the direction of the current. The force acts at right angles to the current and magnetic flux density directions.

$$\mathbf{F} = I\mathbf{L} \times \mathbf{B} \qquad 43.36$$

The energy stored in volume V within a magnetic field, $\mathbf{H}$, is given in Eq. 43.37.

$$W_{\mathbf{H}} = \frac{1}{2} \iiint_V \mathbf{B} \cdot \mathbf{H} dV \qquad 43.37$$

Assuming the $\mathbf{B}$ and $\mathbf{H}$ fields are in the same direction, Eq. 43.37 reduces to Eq. 43.38.

$$W_{\mathbf{H}} = \frac{1}{2} \iiint_V \mu |\mathbf{H}|^2 \, dV \qquad 43.38$$

Assuming the magnetic field is constant throughout the volume V, Eq. 43.38 reduces further to Eq. 43.39.

$$W_{\mathbf{H}} = \frac{\mu H^2 V}{2} = \frac{B^2 V}{2\mu} \qquad 43.39$$

The integral over any closed surface of magnetic flux density must be zero. Stated another way, magnetic flux lines must follow a closed path, and no matter how large or small the enclosing surface is, the path must be either entirely inside the surface or it must go out and back in. This law is referred to as the "no isolated magnetic charge" or "no magnetic monopoles" law (Eq. 43.33).

$$\phi = \oint \mathbf{B}{\cdot}d\mathbf{A} = 0 \qquad 43.40$$

Induced Voltage

Faraday's law of induction states that an induced voltage, v, also called the *electromotive force* or emf, will be generated in a circuit when there is a change in the magnetic flux. Figure 43.8 illustrates one of N series-connected conductors cutting across magnetic flux ϕ. The magnitude of the electromagnetic induction is given by *Faraday's law*, Eq. 43.41. The minus sign indicates the direction of the induced voltage, which is specified by *Lenz's law* to be opposite to the direction of the magnetic field.

$$v = \frac{-Nd\phi}{dt} = -NBL\frac{ds}{dt} \qquad 43.41$$

Figure 43.8 Conductor Moving in a Magnetic Field

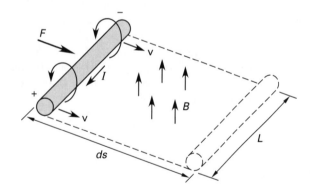

SAMPLE PROBLEMS

Problem 1

If $z_1 = 5\angle 25°$ and $z_2 = 3\angle 40°$, what is $z_1 + z_2$?

(A) $4.07\angle 81.72°$
(B) $5.81\angle 47.21°$
(C) $7.64\angle 32.57°$
(D) $7.94\angle 30.60°$

Solution

When adding complex numbers, real parts are added to real parts and imaginary parts are added to imaginary parts. This is accomplished by expressing z_1 and z_2 in rectangular form.

$$z_1 = (5)(\cos 25° + j \sin 25°)$$
$$z_2 = (3)(\cos 40° + j \sin 40°)$$
$$z_1 + z_2 = (5\cos 25° + 3\cos 40°)$$
$$+ j(5\sin 25° + 3\sin 40°)$$
$$= 6.83 + j4.04$$

Convert back to phasor form.

$$c = \sqrt{a^2 + b^2}$$
$$= \sqrt{(6.83)^2 + (4.04)^2}$$
$$= 7.94$$
$$\theta = \tan^{-1}\frac{b}{a} = \tan^{-1}\frac{4.04}{6.83}$$
$$= 30.6°$$
$$z_1 + z_2 = 7.94\angle 30.60°$$

Answer is D.

Problem 2

A 15 μC point charge is located on the y-axis at (0,0.25). A second charge of 10 μC is located on the x-axis at (0.25,0). If the two charges are separated by air, what is the force between them?

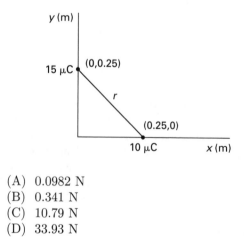

(A) 0.0982 N
(B) 0.341 N
(C) 10.79 N
(D) 33.93 N

Solution

The answer is given by Coulomb's law.

$$F = \frac{Q_1 Q_2}{4\pi\epsilon r^2}$$

For air, $\epsilon = 8.85 \times 10^{-12}$ F/m.

$$r = \sqrt{(0.25 \text{ m})^2 + (0.25 \text{ m})^2}$$

$$= (0.25)\left(\sqrt{2}\right) \text{m}$$

$$F = \frac{\left(15 \times 10^{-6} \text{ C}\right)\left(10 \times 10^{-6} \text{ C}\right)}{4\pi\left(8.85 \times 10^{-12} \dfrac{\text{F}}{\text{m}}\right)\left((0.25)\left(\sqrt{2}\right)\text{m}\right)^2}$$

$$= 10.79 \text{ N}$$

$$(\text{F/m} \equiv \text{C}^2/\text{N·m}^2)$$

Answer is C.

Problem 3

A thin metal plate with dimensions of 20 cm by 20 cm carries a total charge of 24 μC. What is the magnitude of the electric field 2.5 cm away from the center of the plate?

(A) 1.27×10^3 N/C
(B) 3.70×10^6 N/C
(C) 3.39×10^7 N/C
(D) 4.34×10^8 N/C

Solution

The separation distance is much smaller than the dimensions of the plate, so the plate can be considered infinite. The density of the sheet charge is the total charge divided by the plate area.

$$\rho_S = \frac{Q}{A}$$

$$= \frac{24 \times 10^{-6} \text{ C}}{(0.20 \text{ m})^2}$$

$$= 6 \times 10^{-4} \text{ C/m}^2$$

$$\mathbf{E}_S = \frac{\rho_S}{2\epsilon}\mathbf{a}$$

$$= \frac{6 \times 10^{-4} \dfrac{\text{C}}{\text{m}^2}}{(2)\left(8.85 \times 10^{-12} \dfrac{\text{F}}{\text{m}}\right)}$$

$$= 3.39 \times 10^7 \text{ N/C}$$

$$(\text{F/m} \equiv \text{C}^2/\text{N·m}^2)$$

Answer is C.

FE-STYLE EXAM PROBLEMS

1. What is the expression for the complex number $3 + j\,4$ in phasor form?

(A) $3\angle 36.87°$
(B) $5\angle 36.87°$
(C) $3\angle 53.13°$
(D) $5\angle 53.13°$

2. What is the work required to move a positive charge of 10 C for a distance of 5 m in the same direction as a uniform field of 50 V/m?

(A) -20 J
(B) -100 J
(C) -2500 J
(D) $-12\,500$ J

3. If $z_1 = 24.2\angle 32.3°$ and $z_2 = 16.2\angle 45.8°$, what are $z_1 z_2$ and z_1/z_2, respectively?

(A) $40.4\angle 78.1°$; $8.01\angle 13.5°$
(B) $392\angle 78.1°$; $1.49\angle -13.5°$
(C) $392\angle 39.1°$; $8.03\angle 1.4°$
(D) $241\angle 68.5°$; $134.8\angle 38.5°$

4. If the magnitude of the potential difference is generated by a single conductor passing through a magnetic field, which of the following statements is *false*?

(A) The potential difference depends on the speed with which the conductor cuts the magnetic field.
(B) The potential difference depends on the length of the conductor that cuts the magnetic field.
(C) The potential difference depends on the magnetic field density that is present.
(D) The potential difference depends on the diameter of the conductor that cuts the magnetic field.

5. A current of 10 A flows through a 1 mm diameter wire. What is the average number of electrons per second that pass through a cross section of the wire?

(A) 1.6×10^{18} electrons/s
(B) 6.2×10^{18} electrons/s
(C) 1.6×10^{19} electrons/s
(D) 6.3×10^{19} electrons/s

SOLUTIONS TO FE-STYLE EXAM PROBLEMS

1. The complex number is given in rectangular form: $z = a + jb$. Use trigonometry to convert to phasor form.

$$c = \sqrt{a^2 + b^2} = \sqrt{(3)^2 + (4)^2}$$
$$= 5$$

$$\theta = \tan^{-1}\frac{b}{a} = \tan^{-1}\frac{4}{3}$$
$$= 53.13°$$

$$z = c\angle\theta$$
$$= 5\angle 53.13°$$

Answer is D.

2. The work required to move a charge in a uniform field is

$$W = -Q_1\int_{r_1}^{r_2} E\,dL = -Q_1 Ed$$

$$= (-10\ \text{C})\left(50\ \frac{\text{V}}{\text{m}}\right)(5\ \text{m}) = -2500\ \text{C·V}$$

$$= -2500\ \text{J}$$

The positive charge moves in the direction of the field, thus no external work is required, and the charge returns potential energy to the field.

Answer is C.

3. To multiply numbers in polar form, find the product of their magnitudes and the sum of their angles.

$$z_1 z_2 = c_1 c_2 \angle \theta_1 + \theta_2$$
$$= (24.2)(16.2)\angle 32.3° + 45.8°$$
$$= 392\angle 78.1°$$

To find the quotient, divide the magnitudes and subtract the angles.

$$\frac{z_1}{z_2} = \frac{c_1}{c_2}\angle\theta_1 - \theta_2$$
$$= \frac{24.2}{16.2}\angle 32.3° - 45.8°$$
$$= 1.49\angle -13.5°$$

Answer is B.

4. The potential difference is the induced voltage described by Faraday's law.

$$e = \frac{-N\,d\phi}{dt}$$

The change in magnetic flux, $d\phi/dt$, will be influenced by the length of the conductor but not by the cross-sectional area or diameter of the conductor. For a single conductor, $N = 1$.

Answer is D.

5. Since an ampere is equivalent to the flow of 1 C of charge per second, a current of 10 A is equivalent to 10 C per second. One electron has a charge of approximately 1.6×10^{-19} C.

$$\dot{q} = \frac{I}{Q} = \frac{10\ \dfrac{\text{C}}{\text{s}}}{1.6 \times 10^{-19}\ \dfrac{\text{C}}{\text{electron}}}$$

$$= 6.25 \times 10^{19}\ \text{electrons/s} \quad (6.3 \times 10^{19}\ \text{electrons/s})$$

The wire diameter is irrelevant.

Answer is D.

44 Direct-Current Circuits

Subjects

DC CIRCUITS 44-1
   DC Voltage 44-1
   Resistivity 44-1
   Resistors in Series and Parallel 44-2
   Power in a Resistive Element 44-2
   Capacitors 44-2
   Inductors 44-3
   Capacitors and Inductors in
     Series and Parallel 44-3
DC CIRCUIT ANALYSIS 44-3
   Ohm's Law 44-3
   Kirchhoff's Laws 44-3
   Rules for Simple Resistive Circuits . . . 44-4
   Superposition Theorem 44-5
   Superposition Method 44-5
   Loop-Current Method 44-5
   Node-Voltage Method 44-5
   Source Equivalents 44-6
   Maximum Power Transfer 44-6
RC AND *RL* TRANSIENTS 44-7

Nomenclature

A	area	m^2
C	capacitance	F
d	distance	m
$i(t)$	time-varying current	A
I	constant current	A
L	inductance	H
L	length	m
P	power	W
$q(t)$	time-varying charge	C
Q	constant charge	C
R	resistance	Ω
t	time	s
T	temperature	°C
$v(t)$	time-varying voltage	V
V	constant voltage	V

Symbols

α	thermal coefficient of resistance	1/°C
ϵ	permittivity	F/m or $C^2/N{\cdot}m^2$
ρ	resistivity	$\Omega{\cdot}m$
ϕ	magnetic flux	Wb

Subscripts

0	initial
C	capacitive
eq	equivalent
L	inductive
N	Norton
oc	open circuit
sc	short circuit
Th	Thevenin

DC CIRCUITS

Electrical circuits contain active and passive elements. *Active elements* are elements that can generate electric energy, such as voltage and current sources. *Passive elements*, such as capacitors and inductors, absorb or store electric energy; other passive elements, such as resistors, dissipate electric energy.

An *ideal voltage source* supplies power at a constant voltage, regardless of the current drawn. An *ideal current source* supplies power at a constant current independent of the voltage across its terminals. However, real sources have internal resistances that, at higher currents, decrease the available voltage. Therefore, a real voltage source cannot maintain a constant voltage when currents become large. *Independent sources* deliver voltage and current at their rated values regardless of circuit parameters. *Dependent sources* deliver voltage and current at levels determined by voltages or currents elsewhere in the circuit.

The symbols for electrical circuit elements and sources are given in Table 44.1.

DC Voltage

Voltage, measured in volts (a combined unit equivalent to W/A, C/F, J/C, A/S, and Wb/s), is used to measure the voltage, also called *potential difference*, across terminals of circuit elements. Any device that provides electric energy is called a *seat of an electromotive force* (emf), and the electromotive force is measured in volts.

Resistivity

Resistance, R (measured in ohms, Ω), is the property of a circuit or circuit element to oppose current flow. A circuit with zero resistance is a *short circuit*, whereas an *open circuit* has infinite resistance.

Resistors are usually constructed from carbon compounds, ceramics, oxides, or coiled wire. *Resistance* depends on the *resistivity*, ρ (in $\Omega \cdot$m), which is a material property, and the length and cross-sectional area of the resistor. Resistors with larger surface areas have more free electrons available to carry charge, and therefore, have less resistance. Each of the free electrons has a limited ability to move, so the electromotive force must overcome the limited mobility for the entire length of the resistor. Therefore, the resistance increases with the length of the resistor.

$$R = \frac{\rho L}{A} \qquad 44.1$$

Resistivity depends on temperature. For most conductors, it increases with temperature. For most semiconductors, resistivity decreases with temperature. The variation of resistivity with temperature is specified by the *thermal coefficient of resistance*, α, with typical units of $1/^\circ$C. In Eqs. 44.2 and 44.3, R_0 and ρ_0 are the resistance and resistivity, respectively, at temperature T_0.

$$\rho = \rho_0(1 + \alpha(T - T_0)) \qquad 44.2$$

$$R = R_0(1 + \alpha(T - T_0)) \qquad 44.3$$

Resistors in Series and Parallel

Resistors connected in series share the same current and may be represented by an equivalent resistance equal to the sum of the individual resistances. For n resistors in series,

$$R_{\text{eq}} = R_1 + R_2 + \cdots + R_n \qquad 44.4$$

Resistors connected in parallel share the same voltage drop and may be represented by an equivalent resistance equal to the reciprocal of the sum of the reciprocals of the individual resistances. For n resistors in parallel,

$$R_{\text{eq}} = \cfrac{1}{\cfrac{1}{R_1} + \cfrac{1}{R_2} + \cdots + \cfrac{1}{R_n}} \qquad 44.5$$

The equivalent resistance of two resistors in parallel is described by Eq. 44.6.

$$R_{\text{eq}} = \frac{R_1 R_2}{R_1 + R_2} \qquad 44.6$$

Power in a Resistive Element

Energy represents a capacity to do work. *Power* is the time rate of energy performing work. *Work* is the result of power acting over a period of time, and the work

Table 44.1 Circuit Element Symbols

symbol	circuit element
$-\!\!\wedge\!\!\wedge\!\!\wedge\!\!-$ R	resistor
$-\!\!\vert\!\vdash\!-$ C	capacitor
$-\!\!\Box\!\!\frown\!\!\frown\!\!\frown\!-$ L	inductor
$-\!\!\bigcirc\!\!\pm\!-$ or $-\!\!\vert\!\vert\!\vert\!\vdash\!-$ V $\quad$ V	independent voltage source
$-\!\!\bigcirc\!\!\leftarrow\!-$ I	independent current source
$-\!\!\Diamond\!\!\pm\!-$ gV	dependent voltage source
$-\!\!\Diamond\!\!\leftarrow\!-$ gI	dependent current source

performed over that time is the integral of the power. In electric circuits, the energy is provided by voltage or current sources, and the work performed is the dissipation of heat in resistors. In DC circuits, the voltage and current do not change with time, except under transient conditions, so they can usually be written with uppercase "I" and "V".

The power dissipated across two terminals with resistance R and voltage drop V can be calculated from Eq. 44.7. This is known as *Joule's law*.

$$P = VI = \frac{V^2}{R} = I^2 R \qquad 44.7$$

Capacitors

A *capacitor* is a device that stores electric charge. A capacitor is constructed as two conducting surfaces separated by an insulator, such as oiled paper, mica, or air. A simple type of capacitor (i.e., the *parallel plate capacitor*) is constructed as two parallel plates. If the plates are connected across a voltage potential, charges of opposite polarity will build up on the plates and create an electric field between the plates. The amount of charge, Q, built up is proportional to the applied voltage. The constant of proportionality, C, is the *capacitance* in farads (F) and depends on the capacitor construction. Capacitance represents the ability to store charge; the greater the capacitance, the greater the charge stored.

$$Q = CV \quad [\text{constant } V] \qquad 44.8$$

$$q_C(t) = Cv_C(t) \quad [\text{varying } v(t)] \qquad 44.9$$

Equation 44.10 gives the capacitance of two parallel plates of equal area A separated by distance d. ϵ is the permittivity of the medium separating the plates.

$$C = \frac{\epsilon A}{d} \qquad 44.10$$

The current in a capacitor is the derivative of the voltage times the capacitance (Eq. 44.11). The voltage of a capacitor cannot change instantaneously, but the current can change instantaneously. The voltage will change as the capacitor integrates the current to produce a voltage that opposes the change (Eq. 44.12).

$$i_C(t) = C \frac{dv_C}{dt} \qquad 44.11$$

$$v_C(t) = v_C(0) + \frac{1}{C} \int_0^t i_C(\tau)d\tau \qquad 44.12$$

The total energy (in J) stored in a capacitor is

$$\text{energy} = \frac{CV^2}{2} = \frac{VQ}{2}$$

$$= \frac{Q^2}{2C} \qquad 44.13$$

In steady-state DC circuits, ideal capacitors have infinite resistance and, therefore, act like open circuits. Unless the voltage is varying with time, there is no current flow once a capacitor becomes charged. Transient behavior of capacitors is discussed later in this chapter.

Inductors

An *inductor* is basically a coil of wire. When connected across a voltage source, current begins to flow in the coil, establishing a magnetic field that opposes current changes. Usually, the wire is coiled around a core of magnetic material (high permeability) to increase the inductance. From Faraday's law, the induced voltage across the ends of the inductor is proportional to the change in flux linkage, which in turn is proportional to the current change. The constant of proportionality is the *inductance*, L, expressed in henries (H).

$$L = \frac{N\phi}{I} \qquad 44.14$$

The voltage across an inductor is the derivative of the current times the inductance (Eq. 44.16). The current of inductors cannot change instantaneously, but the voltage can change instantaneously. The current will change as the inductor integrates the voltage to produce a current that opposes the change (Eq. 44.16).

$$v_L(t) = L \frac{dI(t)}{dt} \qquad 44.15$$

$$i_L = i_L(0) + \frac{1}{L} \int_0^t v_L(\tau)d\tau \qquad 44.16$$

The total energy (in J) stored in an inductor carrying current I is

$$\text{energy} = \frac{LI^2}{2} \qquad 44.17$$

In steady-state DC circuits, ideal inductors have zero resistance and, therefore, act like a short circuit. Unless the current is varying with time, there is no voltage across the inductor. Transient behavior of inductors is discussed later in this chapter.

Capacitors and Inductors in Series and Parallel

The total capacitance of capacitors connected in series is

$$C_{\text{eq}} = \frac{1}{\dfrac{1}{C_1} + \dfrac{1}{C_2} + \cdots + \dfrac{1}{C_n}} \qquad 44.18$$

The total capacitance of capacitors connected in parallel is

$$C_{\text{eq}} = C_1 + C_2 + \cdots + C_n \qquad 44.19$$

The total inductance of inductors connected in series is

$$L_{\text{eq}} = L_1 + L_2 + \cdots + L_n \qquad 44.20$$

The total inductance of inductors connected in parallel is

$$L_{\text{eq}} = \frac{1}{\dfrac{1}{L_1} + \dfrac{1}{L_2} + \cdots + \dfrac{1}{L_n}} \qquad 44.21$$

DC CIRCUIT ANALYSIS

Most circuit problems involve solving for unknown parameters, such as the voltage or current across some element in the circuit. The methods that are used to find these parameters rely on combining elements in series and parallel, and applying *Ohm's law* or *Kirchhoff's laws* in some systematic manner.

Ohm's Law

The voltage drop, also known as the *IR drop*, across a circuit with resistance R is given by *Ohm's law*.

$$V = IR \qquad 44.22$$

Using Ohm's law implicitly assumes a *linear circuit* (i.e., one consisting of linear elements and linear sources). A *linear element* is a passive element whose performance can be represented by a linear voltage-current relationship. The output of a linear source is proportional to the first power of a voltage or current in the circuit. Many elements used in modern electronic devices do not obey Ohm's law.

Kirchhoff's Laws

Kirchhoff's current law (KCL) states that as much current flows out of a node (connection) as flows into it. Electrons must be conserved at any node in an electrical circuit. The direction of the current into or out of the node is arbitrary for each of the paths, so care should be taken to ensure the sign is correct.

$$\sum I_{\text{in}} = \sum I_{\text{out}} \qquad 44.23$$

Kirchhoff's voltage law (KVL) states that the algebraic sum of voltage drops around any closed path within a circuit is equal to the sum of the voltage rises. The forces from voltage sources in the circuit are opposed by equal and opposite reactions in the circuit elements, so the voltage sources must equal the voltage drops.

$$\sum V_{\text{rises}} = \sum V_{\text{drops}} \qquad 44.24$$

Rules for Simple Resistive Circuits

In a simple series (single-loop) circuit, such as the circuit shown in Fig. 44.1,

- the current is the same through all circuit elements.

$$I = I_{R1} = I_{R2} = I_{R3} \qquad 44.25$$

- the equivalent resistance is the sum of the individual resistances.

$$R_{\text{eq}} = R_1 + R_2 + R_3 \qquad 44.26$$

- the equivalent applied voltage is the algebraic sum of all voltage sources (polarity considered).

$$V_{\text{eq}} = V_1 + V_2 \qquad 44.27$$

- the sum of the voltage drops across all components is equal to the equivalent applied voltage (KVL).

$$V_{\text{eq}} = I R_{\text{eq}} \qquad 44.28$$

In a series circuit, the voltage across a resistor is the total circuit voltage times the resistance of that particular resistor divided by the total equivalent resistance. For example, the voltage across resistor R_1 of Fig. 44.1 is given in Eq. 44.29.

$$V_{R_1} = \left(\frac{R_1}{R_{\text{eq}}}\right) V_{\text{eq}}$$
$$= \left(\frac{R_1}{R_1 + R_2 + R_3}\right)(V_1 + V_2) \qquad 44.29$$

Figure 44.1 Simple Series Circuit

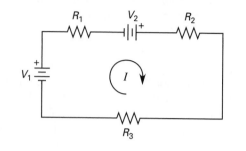

In a simple parallel circuit with only one active source, such as the circuit shown in Fig. 44.2,

- the voltage drop is the same across all legs.

$$V = V_{R1} = V_{R2} = V_{R3}$$
$$= I_1 R_1 = I_2 R_2 = I_3 R_3 \qquad 44.30$$

- the reciprocal of the equivalent resistance is the sum of the reciprocals of the individual resistances.

$$\frac{1}{R_{\text{eq}}} = \frac{1}{R_1} + \frac{1}{R_2} + \frac{1}{R_3} \qquad 44.31$$

- the total current is the sum of the leg currents (KCL).

$$I = I_1 + I_2 + I_3$$
$$= \frac{V}{R_1} + \frac{V}{R_2} + \frac{V}{R_3} \qquad 44.32$$

Figure 44.2 Simple Parallel Circuit

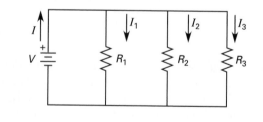

In a parallel circuit, the current through a resistor is the total circuit current times the total circuit resistance divided by the resistor's resistance. For example, the current through resistor R_1 of Fig. 44.2 is given in Eq. 44.33.

$$I_1 = \left(\frac{R_{\text{eq}}}{R_1}\right) I$$

$$= \left(\frac{\dfrac{1}{\dfrac{1}{R_1} + \dfrac{1}{R_2} + \dfrac{1}{R_3}}}{R_1}\right) I \qquad 44.33$$

The circuit analysis techniques in the following sections show how complicated linear circuits are simplified using circuit reduction and how they are analyzed as a system of n simultaneous equations and n unknowns. Circuit analysis can often be used to directly obtain the current or voltage at a component of interest or to reduce the number of simultaneous equations needed.

Use the following procedure to establish the current and voltage drops in a complicated resistive network. The circuit should be viewed from the perspective of the component of interest. Each step in the reduction should result in a circuit that is simpler as seen by the component of interest.

step 1: Combine series voltage and parallel current sources.

step 2: Combine series resistances to make combinations that more closely resemble a component in parallel with the component of interest.

step 3: Combine parallel resistances to make combinations that more closely resemble a component in series with the component of interest. Lines in the circuit represent zero resistance, and components connected by lines are connected to the same node. The lines can be moved to make parallel combinations more recognizable as long as the components remain connected to the node.

step 4: Repeat steps 2 through 4 as many times as needed.

This principle is only valid for linear circuits or nonlinear circuits that are operating in a linear range. The superposition theorem can be used to reduce a complicated circuit to multiple less-complicated circuits.

Superposition Theorem

The *superposition theorem* states that the response of (i.e., the voltage across or current through) a linear circuit element fed by two or more independent sources is equal to the response to each source taken individually with all other sources set to zero (i.e., voltage sources shorted and current sources opened).

Superposition Method

The *superposition method* analyzes the response of a component to each of the energy sources in a linear circuit separately and then sums the results. This requires circuit reduction analysis for each of the energy sources in the circuit. Superposition works equally well for finding unknown currents and unknown voltages. Superposition tends to be more efficient for less-complicated circuits or when there are more loops and nodes than power sources. The superposition method is not as commonly used as the methods discussed later in this chapter, because it tends to be inefficient for analyzing complicated circuits.

step 1: Choose one of the voltage or current sources, and short all other voltage sources and open all other current sources.

step 2: Make circuit reductions to simplify the circuit as seen by the component of interest.

step 3: Find the voltage or current at the component of interest.

step 4: Repeat steps 1, 2, and 3 for the other voltage and current sources.

step 5: Sum the voltages or currents (be careful to use the same direction conventions).

Loop-Current Method

The *loop-current method* (also known as the *mesh current method*) is a direct extension of Kirchhoff's voltage law and is particularly valuable in determining unknown currents in circuits with several loops and energy sources. It requires writing $n - 1$ simultaneous equations for an n-loop system.

step 1: Select $n-1$ loops (i.e., one less than the total number of loops).

step 2: Assume current directions for the chosen loops. (The choice of current direction is arbitrary, but some currents may end up being negative in step 4.) Show the direction with an arrow.

step 3: Write Kirchhoff's voltage law for each of the $n - 1$ chosen loops. A voltage source is positive when the assumed current direction is from the negative to the positive battery terminal. Voltage (IR) drops are always positive.

step 4: Solve the $n - 1$ equations (from step 3) for the unknown currents.

Node-Voltage Method

The *node-voltage method* is an extension of Kirchhoff's current law. Although currents can be determined with it, it is primarily used to find voltage potentials at various points (nodes) in the circuit. (A node is a point where three or more wires connect.)

step 1: Convert all current sources to voltage sources.

step 2: Choose one node as the voltage reference (i.e., 0 V) node. Usually, this will be the circuit ground—a node to which at least one negative battery terminal is connected.

step 3: Identify the unknown voltage potentials at all other nodes referred to the reference node.

step 4: Write Kirchhoff's current law for all unknown nodes. (This excludes the reference node.)

step 5: Write all currents in terms of voltage drops.

step 6: Write all voltage drops in terms of the node voltages.

Source Equivalents

Source equivalents are simplified models of two-terminal networks. They are used to represent a circuit when it is connected to a second circuit. Source equivalents simplify the analysis because the equivalent circuit is much simpler than the original.

Thevenin's theorem states that a linear, two-terminal network with dependent and independent sources can be represented by a *Thevenin equivalent* circuit consisting of a voltage source in series with a resistor, as illustrated in Fig. 44.3. The Thevenin equivalent voltage, or open-circuit voltage, V_{oc}, is the open-circuit voltage across terminals A and B. The Thevenin equivalent resistance, R_{eq}, is the resistance across terminals A and B when all independent sources are set to zero (i.e., short-circuiting voltage sources and open-circuiting current sources). The equivalent resistance can also be determined by measuring V_{oc} and the current with terminals A and B shorted together, I_{sc}, and using Eq. 44.35.

$$V_{oc} = V_A - V_B \qquad 44.34$$

$$R_{eq} = \frac{V_{oc}}{I_{sc}} \qquad 44.35$$

Norton's theorem states that a linear, two-terminal network with dependent or independent sources can be represented by an equivalent circuit consisting of a single current source and resistor in parallel, as shown in Fig. 44.4. The *Norton equivalent* current, I_{sc}, is the short-circuit current that flows through a shunt across terminals A and B. The Norton equivalent resistance, R_{eq}, is the resistance across terminals A and B when all

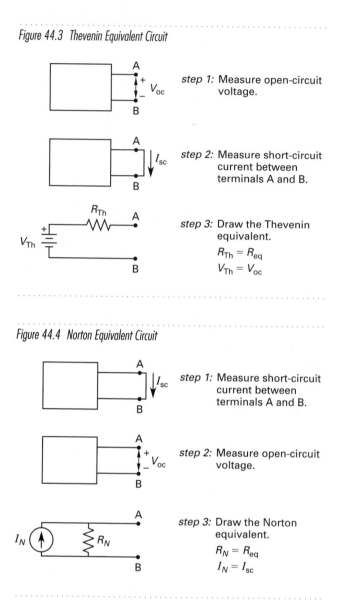

Figure 44.3 Thevenin Equivalent Circuit

step 1: Measure open-circuit voltage.

step 2: Measure short-circuit current between terminals A and B.

step 3: Draw the Thevenin equivalent.
$R_{Th} = R_{eq}$
$V_{Th} = V_{oc}$

Figure 44.4 Norton Equivalent Circuit

step 1: Measure short-circuit current between terminals A and B.

step 2: Measure open-circuit voltage.

step 3: Draw the Norton equivalent.
$R_N = R_{eq}$
$I_N = I_{sc}$

independent sources are set to zero (i.e., short-circuiting voltage sources and open-circuiting current sources). The Norton equivalent voltage, V_{oc}, is measured with terminals open.

$$V_{oc} = V_A - V_B \qquad 44.36$$

$$R_{eq} = \frac{V_{oc}}{I_{sc}} \qquad 44.37$$

Note that Norton's equivalent resistance is equal to Thevenin's equivalent resistance.

The Norton equivalent can be easily converted to a Thevenin equivalent and vice versa with Eqs. 44.38 through 44.40. The conversions from Norton to Thevenin or from Thevenin to Norton can aid in circuit analysis.

$$R_N = R_{Th} \qquad 44.38$$

$$V_{\text{Th}} = I_N R_N \qquad \text{44.39}$$

$$I_N = \frac{V_{\text{Th}}}{R_{\text{Th}}} \qquad \text{44.40}$$

Maximum Power Transfer

Electric circuits are often designed to transfer power from a source (e.g., generator, transmitter) to a load (e.g., motor, light, receiver). There are two basic types of power transfer circuits. In one type of system, the emphasis is on transmitting power with high efficiency. In this power system, large amounts of power must be transmitted in the most efficient way to the loads. In communication and instrumentation systems, small amounts of power are involved. The power at the transmitting end is small, and the main concern is that the maximum power reaches the load.

The *maximum power transfer* from a circuit will occur when the load resistance equals the Norton or Thevenin equivalent resistance of the source.

The method for determining the maximum power is counterintuitive, so a short version of the proof follows.

The power, P_L, delivered to the load, R_L, by a Thevenin equivalent voltage, V_{Th}, through a Thevenin equivalent resistance, R_{eq}, is a function of the load voltage.

$$P_L = V_{\text{Th}}^2 \left(\frac{R_L}{(R_{\text{Th}} + R_L)^2} \right) \qquad \text{44.41}$$

To find the maximum of this function, take the derivative with respect to R_L and set it equal to zero.

$$\frac{dP_L}{dR_L} = V_{\text{Th}}^2 \left((R_{\text{Th}} + R_L)^{-2} - 2R_L (R_{\text{Th}} + R_L)^{-3} \right)$$
$$= 0 \qquad \text{44.42}$$

Solving yields Eq. 44.43.

$$2R_L = R_{\text{Th}} + R_L \qquad \text{44.43}$$

This is equivalent to Eq. 44.44.

$$R_L = R_{\text{Th}} \qquad \text{44.44}$$

RC AND RL TRANSIENTS

When a charged capacitor is connected across a resistor, the voltage across the capacitor will gradually decrease and approach zero as energy is dissipated in the resistor. Similarly, when an inductor through which a steady current is flowing is suddenly connected across a resistor, the current will gradually decrease and approach zero. Both of these cases assume that any energy sources are disconnected at the time the resistor is connected. These gradual decreases are known as *transient behavior*. Transient behavior is also observed when a voltage or a current source is connected to a circuit with capacitors or inductors.

The *time constant*, τ, for a circuit is the time in seconds it takes for the current or voltage to reach e (i.e., the base of natural logarithms) times the difference between the steady-state value and the original value, or approximately 63.3% of its steady-state value. For a series-RL circuit, the time constant is L/R. For a series-RC circuit, the time constant is RC. In general, transient variables will have essentially reached their steady-state values after five time constants (99.3% of the steady-state value).

The following equations describe RC and RL transient response for source-free or energizing circuits. Time is assumed to begin when a switch is closed. Decay is a special case of the charging equations where $V = 0$ and either $v_C(0) \neq 0$ or $i_L(0) \neq 0$.

RC transient (Fig. 44.5):

$$v_C(t) = v_C(0)e^{-t/RC} + V \left(1 - e^{-t/RC} \right) \qquad \text{44.45}$$

$$i(t) = \left(\frac{V - v_C(0)}{R} \right) e^{-t/RC} \qquad \text{44.46}$$

$$v_R(t) = i(t)R$$
$$= (V - v_C(0))e^{-t/RC} \qquad \text{44.47}$$

$v_C(0)$ is the voltage across the terminals of the capacitor when the switch is closed.

Figure 44.5 RC Transient Circuit

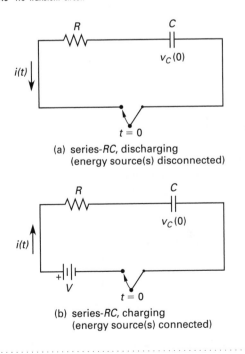

(a) series-*RC*, discharging
(energy source(s) disconnected)

(b) series-*RC*, charging
(energy source(s) connected)

RL transient (Fig. 44.6):

$$v_R(t) = i(t)R$$

$$= i(0)Re^{-Rt/L} + V\left(1 - e^{-Rt/L}\right) \quad \textit{44.48}$$

$$i(t) = i(0)e^{-Rt/L} + \frac{V}{R}\left(1 - e^{-Rt/L}\right) \quad \textit{44.49}$$

$$v_L(t) = L\frac{di}{dt}$$

$$= -i(0)Re^{-Rt/L} + Ve^{-Rt/L} \quad \textit{44.50}$$

$i(0)$ is the current through the inductor when the switch is closed.

Figure 44.6 RL Transient Circuit

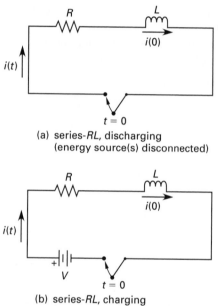

(a) series-*RL*, discharging
(energy source(s) disconnected)

(b) series-*RL*, charging
(energy source(s) connected)

SAMPLE PROBLEMS

Problem 1

What is the equivalent inductance of the circuit shown in the following illustration?

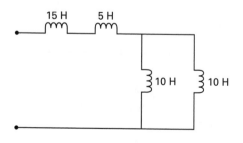

(A) 5 H
(B) 20 H
(C) 23.75 H
(D) 25 H

Solution

Inductors combine like resistors.

$$L_{eq} = 15\text{ H} + 5\text{ H} + \frac{(10\text{ H})(10\text{ H})}{10\text{ H} + 10\text{ H}}$$

$$= 25\text{ H}$$

Answer is D.

Problem 2

A solid copper conductor at $20°C$ has the following characteristics.

$$\text{resistivity} = 1.77 \times 10^{-8}\ \Omega\cdot\text{m}$$
$$\text{diameter} = 5\text{ mm}$$
$$\text{length} = 5000\text{ m}$$

What is the resistance of the conductor?

(A) $0.017\ \Omega$
(B) $4.5\ \Omega$
(C) $12\ \Omega$
(D) $18\ \Omega$

Solution

$$R = \frac{\rho L}{A}$$

$$= \frac{(1.77 \times 10^{-8}\ \Omega\cdot\text{m})(5000\text{ m})}{\frac{\pi}{4}\left((5\text{ mm})\left(\dfrac{1\text{ m}}{1000\text{ mm}}\right)\right)^2}$$

$$= 4.51\ \Omega \quad (4.5\ \Omega)$$

Answer is B.

Problem 3

What is the charge on the capacitor on plate A in the following illustration? The circuit is in steady state.

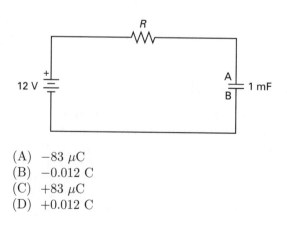

(A) $-83\ \mu\text{C}$
(B) -0.012 C
(C) $+83\ \mu\text{C}$
(D) $+0.012$ C

Solution

In steady state, all the voltage is across the capacitor.

$$Q = CV = (1 \times 10^{-3} \text{ F})(12 \text{ V})$$

$$= 0.012 \text{ C}$$

Answer is D.

Problem 4

What is the current I in the following illustration?

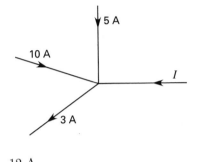

(A) -12 A
(B) 3 A
(C) 5 A
(D) 15 A

Solution

Kirchhoff's current law states that the sum of the currents entering a junction will be equal to the sum of the currents leaving the junction. To apply the law, use assumed directions for currents.

$$\sum I_{in} = I + 5 \text{ A} + 10 \text{ A}$$

$$\sum I_{out} = 3 \text{ A}$$

$$\sum I_{in} = \sum I_{out}$$

$$I = 3 \text{ A} - 5 \text{ A} - 10 \text{ A}$$

$$= -12 \text{ A}$$

Answer is A.

Problem 5

What is the voltage across the 5 Ω resistor in the center leg in the following illustration?

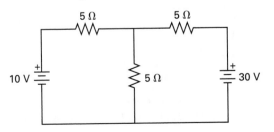

(A) 13.33 V
(B) 15.57 V
(C) 20.04 V
(D) 24.21 V

Solution

Use the loop-current method to solve for the voltage. Refer to the following figure. This is a three-loop network, so select $(3 - 1)$, or 2, loops. Current directions are arbitrary.

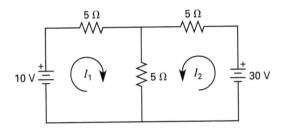

Write Kirchhoff's voltage law for each loop.

$$10 \text{ V} - I_1(5 \text{ }\Omega) - (I_1 + I_2)(5 \text{ }\Omega) = 0$$

$$30 \text{ V} - I_2(5 \text{ }\Omega) - (I_1 + I_2)(5 \text{ }\Omega) = 0$$

Solving for I_1 from the first equation,

$$(10 \text{ }\Omega)I_1 = 10 \text{ V} - (5 \text{ }\Omega)I_2$$

Substituting into the second equation,

$$30 \text{ V} - I_2(10 \text{ }\Omega) - \left(\frac{10 \text{ V} - (5 \text{ }\Omega)I_2}{10 \text{ }\Omega}\right)(5 \text{ }\Omega) = 0$$

$$I_2 = \frac{10}{3} \text{ A}$$

$$I_1 = -\frac{2}{3} \text{ A}$$

The voltage across the center resistor is

$$V = IR = \left(\frac{10 \text{ A}}{3} - \frac{2 \text{ A}}{3}\right)(5 \text{ }\Omega)$$

$$= 13.33 \text{ V}$$

Answer is A.

Problem 6

What are the Norton equivalent source and resistance values for the circuit shown?

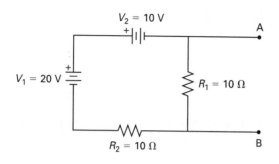

(A) $V_N = 5$ V; $R_N = 5$ Ω
(B) $V_N = 10$ V; $R_N = 20$ Ω
(C) $I_N = 1$ A; $R_N = 5$ Ω
(D) $I_N = 1$ A; $R_N = 10$ Ω

Solution

A Norton equivalent circuit contains a single-current source and a resistor in parallel, so the first two answers can be eliminated immediately.

To find the equivalent resistance, turn off all power sources (i.e., short circuit the voltage sources). The equivalent resistance across terminals A and B is

$$R_N = \frac{R_1 R_2}{R_1 + R_2} = \frac{(10 \ \Omega)(10 \ \Omega)}{10 \ \Omega + 10 \ \Omega}$$

$$= 5 \ \Omega$$

(This eliminates answer choices B and D.)

The Norton equivalent current is the short-circuit current through terminals A and B. The following illustration shows the circuit with a short circuit across terminals A and B.

Apply Kirchhoff's voltage law to the shorted circuit.

$$V_1 - V_2 - R_2 I_N = 0$$

$$20 \text{ V} - 10 \text{ V} = 10 I_N$$

$$I_N = 1 \text{ A}$$

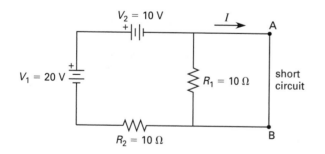

Notice that no current flows through R_1 because the terminals of R_1 are short-circuited.

Answer is C.

Problem 7

When a 20 V source is connected across terminals A and B, a current of 10 A is measured through R_1. What current would flow through R_1 if a 30 V source is connected across terminals A and B?

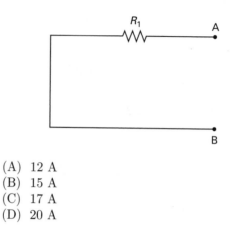

(A) 12 A
(B) 15 A
(C) 17 A
(D) 20 A

Solution

The applied voltage increased from 20 V to 30 V, or 1.5 times. By the linearity expressed in Ohm's law, the current is

$$I = (1.5)(10 \text{ A}) = 15 \text{ A}$$

Answer is B.

FE-STYLE EXAM PROBLEMS

1. Find the equivalent capacitance between terminals A and B.

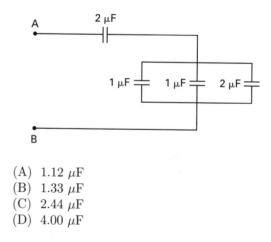

(A) 1.12 μF
(B) 1.33 μF
(C) 2.44 μF
(D) 4.00 μF

2. What are the voltage across and current through the 5 Ω resistor in the center leg? The circuit is at steady state.

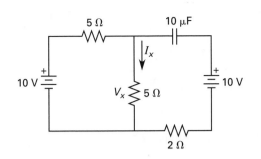

(A) 2 V; 1 A
(B) 2 V; 3 A
(C) 5 V; 1 A
(D) 5 V; 20 A

3. What is the Thevenin equivalent circuit between terminals A and B looking to the left for the following circuit?

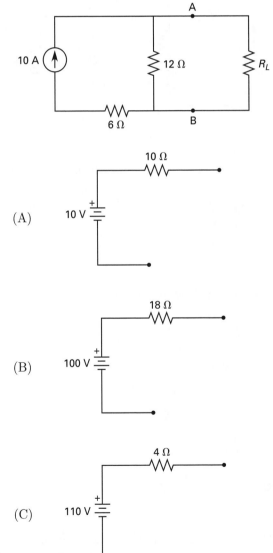

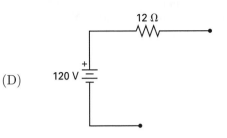

(D)

4. Size the resistor R_L to allow maximum power transfer through terminals A and B.

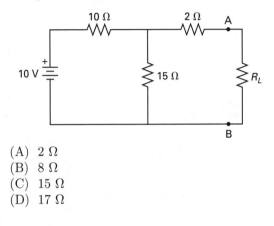

(A) 2 Ω
(B) 8 Ω
(C) 15 Ω
(D) 17 Ω

5. The initial voltage across the capacitor is 5 V. At $t = 0$, the switch is closed. What is the voltage across the capacitor 10 μs after the switch is closed?

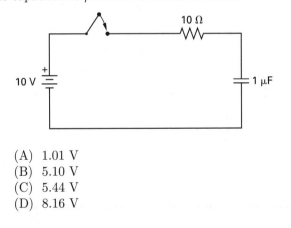

(A) 1.01 V
(B) 5.10 V
(C) 5.44 V
(D) 8.16 V

For the following problems use the NCEES Handbook as your only reference.

6. A power line is made of copper having a resistivity of 1.83×10^{-6} Ω·cm. The wire diameter is 2 cm. What is the resistance of 5 km of power line?

(A) 0.0032 Ω
(B) 0.29 Ω
(C) 0.67 Ω
(D) 1.8 Ω

7. A 10 kV power line 5 km long has a total resistance of 0.7 Ω. The current flowing is 10 A. What is the power dissipated in the line resistance?

(A) 7 W
(B) 14 W
(C) 70 W
(D) 14 kW

Problems 8–12 refer to the circuit shown and the data given.

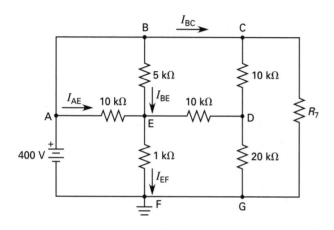

$$V_{BE} = 300 \text{ V}$$
$$V_{CD} = 200 \text{ V}$$
$$V_D = 200 \text{ V}$$
$$V_E = 100 \text{ V}$$
$$I_{BC} = 100 \text{ mA}$$

8. What current is flowing between nodes A and E?

(A) 0 mA
(B) 10 mA
(C) 30 mA
(D) 40 mA

9. What current is flowing between nodes B and E?

(A) 30 mA
(B) 60 mA
(C) 70 mA
(D) 80 mA

10. What is the current flowing between nodes E and F?

(A) 90 mA
(B) 100 mA
(C) 110 mA
(D) 120 mA

11. What is the output power of the battery?

(A) 9 W
(B) 12 W
(C) 16 W
(D) 76 W

12. What is resistance R_7?

(A) 4 kΩ
(B) 5 kΩ
(C) 10 kΩ
(D) 20 kΩ

Problems 13–15 refer to the following circuit. All circuit elements are ideal. The circuit is at steady state.

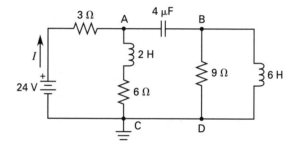

13. What current is flowing through the 3 Ω resistor?

(A) 2.00 A
(B) 2.67 A
(C) 3.64 A
(D) 8.00 A

14. What energy is stored in the capacitor?

(A) 14.3 μJ
(B) 128 μJ
(C) 512 μJ
(D) 1150 μJ

15. What energy is stored in the 2 H inductor?

(A) 7.1 J
(B) 13 J
(C) 14 J
(D) 29 J

16. The dielectric material in the capacitor shown has a permittivity of $24\epsilon_o$ (that is, the permittivity is 24 times that of a vacuum). What is the capacitance?

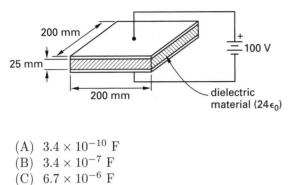

(A) 3.4×10^{-10} F
(B) 3.4×10^{-7} F
(C) 6.7×10^{-6} F
(D) 0.34 F

17. What energy is stored in the capacitor shown in Prob. 16?

(A) $0.17 \ \mu J$
(B) $0.48 \ \mu J$
(C) $1.7 \ \mu J$
(D) $2.4 \ \mu J$

18. Which of the following statements is true for steady-state conditions in a circuit powered by a dry cell?

(A) Voltage is dependent upon time.
(B) Inductors behave as short circuits.
(C) Capacitors behave as short circuits.
(D) Resistors behave as open circuits.

19. The switch in the circuit shown is closed at $t = 0$, after being open for a long time. What is the energy stored in the two inductors just prior to the switch being closed?

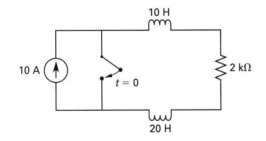

(A) 340 J
(B) 1500 J
(C) 1800 J
(D) 1900 J

20. What is the equivalent capacitance of the circuit shown?

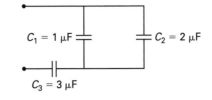

(A) $0.55 \ \mu F$
(B) $1.5 \ \mu F$
(C) $1.8 \ \mu F$
(D) $3.7 \ \mu F$

21. What is the total capacitance between terminals A and B?

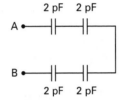

(A) 0.5 pF
(B) 2 pF
(C) 4 pF
(D) 6 pF

22. What is the equivalent capacitance seen by the battery for the circuit shown?

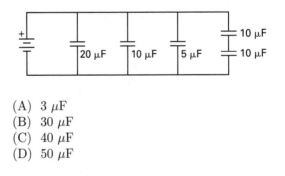

(A) $3 \ \mu F$
(B) $30 \ \mu F$
(C) $40 \ \mu F$
(D) $50 \ \mu F$

23. What is the internal resistance of a 9 V battery that delivers 100 A when its terminals are shorted? Assume the short circuit has negligible resistance.

(A) $0.09 \ \Omega$
(B) $1.0 \ \Omega$
(C) $11 \ \Omega$
(D) $90 \ \Omega$

24. In the circuit shown, all components are ideal. What is the magnitude of the current through the 20 Ω resistor? The circuit is at steady state.

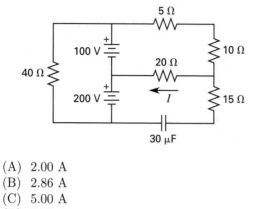

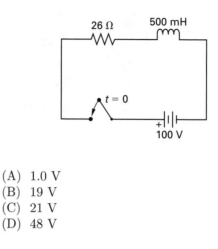

(A) 2.00 A
(B) 2.86 A
(C) 5.00 A
(D) 5.71 A

(A) 1.0 V
(B) 19 V
(C) 21 V
(D) 48 V

25. What is the current, I_B, through the 40 V battery?

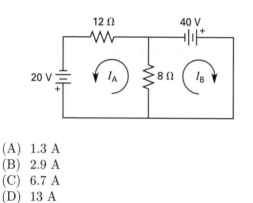

(A) 1.3 A
(B) 2.9 A
(C) 6.7 A
(D) 13 A

28. The switch in the circuit shown is closed at $t = 0$. How long will it take to charge the capacitor to 80% of the battery voltage?

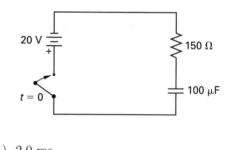

26. For the circuit shown, what equivalent resistance is seen by the battery?

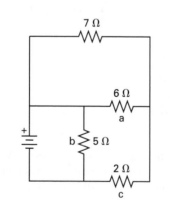

(A) 0.38 Ω
(B) 2.2 Ω
(C) 2.6 Ω
(D) 6.2 Ω

(A) 2.0 ms
(B) 10 ms
(C) 12 ms
(D) 24 ms

29. The initial voltage across the capacitor is 2.5 V. The switch is closed at $t = 0$. What is the current at $t = 0$ s?

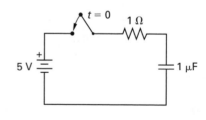

(A) 0.2 A
(B) 0.7 A
(C) 1.0 A
(D) 2.5 A

27. The switch in the circuit shown is closed at $t = 0$. What is the voltage across the inductor at $t = 30$ ms?

30. Which of the following statements is true?

(A) An ideal current source cannot be in parallel with a short circuit.

(B) An ideal voltage source can be in parallel with an open circuit.

(C) An ideal current source can be in series with an open circuit.

(D) An ideal voltage source cannot be in series with an ideal current source.

31. What is the average DC current through the inductor? The circuit is at steady state.

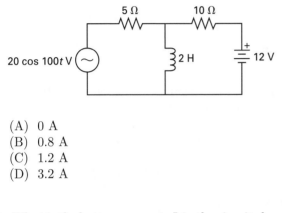

(A) 0 A
(B) 0.8 A
(C) 1.2 A
(D) 3.2 A

32. What is the battery current, I, in the circuit shown?

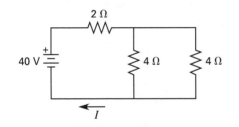

(A) 4.0 A
(B) 6.7 A
(C) 10 A
(D) 15 A

SOLUTIONS TO FE-STYLE EXAM PROBLEMS

1. The equivalent capacitance is

$$C_{eq} = \cfrac{1}{\cfrac{1}{2 \ \mu F} + \cfrac{1}{1 \ \mu F + 1 \ \mu F + 2 \ \mu F}}$$

$$= 1.33 \ \mu F$$

Answer is B.

2. A capacitor acts like an open circuit in DC circuits. Using Kirchhoff's voltage law, sum the voltage drops around the remaining loop.

$$10 \text{ V} - (5 \ \Omega)I_x - (5 \ \Omega)I_x = 0$$

$$I_x = \frac{10 \text{ V}}{10 \ \Omega} = 1 \text{ A}$$

Use Ohm's law to find the voltage.

$$V_x = I_x R = (1 \text{ A})(5 \ \Omega)$$

$$= 5 \text{ V}$$

The circuit loop through which the current runs is in a configuration known as a *voltage divider circuit*. Referring to part (a) of the following figure, the voltage could be more directly computed using the equation

$$V_2 = V \left(\frac{R_2}{R_1 + R_2} \right)$$

$$= (10 \text{ V}) \left(\frac{5 \ \Omega}{5 \ \Omega + 5 \ \Omega} \right)$$

$$= 5 \text{ V}$$

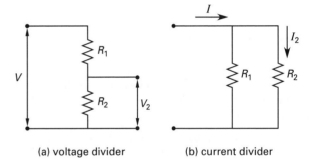

(a) voltage divider   (b) current divider

Another shortcut that is helpful for solving this type of problem is recognizing the *current divider circuit* shown in part (b). In this case, the current is

$$I_2 = I \left(\frac{R_1}{R_1 + R_2} \right)$$

$$= (2 \text{ A}) \left(\frac{5 \ \Omega}{5 \ \Omega + 5 \ \Omega} \right)$$

$$= 1 \text{ A}$$

Answer is C.

3. The Thevenin equivalent circuit consists of a voltage source in series with a resistor. The equivalent voltage is the open-circuit voltage between terminals A and B, with R_L disconnected. The voltage will be the same as that across the 12 Ω resistor.

$$V_{eq} = IR = (10 \text{ A})(12 \ \Omega)$$

$$= 120 \text{ V}$$

(This identifies the correct answer.)

The equivalent resistance is found by open-circuiting the current source. The equivalent resistance is the resistance across terminals A and B.

$$R_{eq} = 12 \ \Omega$$

Note: For a given circuit there is only one Thevenin equivalent.

Answer is D.

4. Maximum power transfer occurs when the load resistance equals the source resistance. To size R_L, find the Norton or Thevenin equivalent resistance for the circuit to the left of terminals A and B.

The Thevenin equivalent resistance is found by removing the voltage source and finding the equivalent resistance across terminals A and B, with R_L disconnected.

$$R_{eq} = 2 \ \Omega + \frac{(10 \ \Omega)(15 \ \Omega)}{10 \ \Omega + 15 \ \Omega} = 8 \ \Omega$$

Answer is B.

5. When the switch closes, the charge on the capacitor begins to increase from 5 V.

$$v_C(t) = v_C(0)e^{-t/RC} + V\left(1 - e^{-t/RC}\right)$$

$$\frac{-t}{RC} = \frac{-10 \times 10^{-6} \text{ s}}{(10 \ \Omega)(1 \times 10^{-6} \text{ F})} = -1$$

$$v_C(t) = 5e^{-1} + 10(1 - e^{-1})$$

$$= 8.16 \text{ V}$$

Answer is D.

6. The wire's resistance is directly proportional to its length and inversely proportional to its cross-sectional area.

$$R = \frac{\rho L}{A}$$

$$= \frac{(1.83 \times 10^{-6} \ \Omega \cdot \text{cm})(5 \text{ km})\left(10^5 \ \frac{\text{cm}}{\text{km}}\right)}{\left(\frac{\pi}{4}\right)(2 \text{ cm})^2}$$

$$= 0.29 \ \Omega$$

Answer is B.

7. Apply Joule's law.

$$P = I^2 R = (10 \text{ A})^2(0.7 \ \Omega)$$

$$= 70 \text{ W}$$

Answer is C.

8. V_A is at the battery potential of 400 V. V_E is given. From Ohm's law,

$$I = \frac{V_{AE}}{R} = \frac{V_A - V_E}{R}$$

$$= \frac{400 \text{ V} - 100 \text{ V}}{10 \times 10^3 \ \Omega}$$

$$= 0.03 \text{ A} \quad (30 \text{ mA})$$

Answer is C.

9. The voltage difference between nodes B and E is given. From Ohm's law,

$$I = \frac{V_{BE}}{R} = \frac{300 \text{ V}}{5 \times 10^3 \ \Omega}$$

$$= 0.06 \text{ A} \quad (60 \text{ mA})$$

Answer is B.

10. The voltage at node E is given. Node F is grounded (i.e., is at zero voltage). From Ohm's law,

$$I = \frac{V_{EF}}{R} = \frac{V_E - V_F}{R}$$

$$= \frac{100 \text{ V} - 0 \text{ V}}{1 \times 10^3 \ \Omega}$$

$$= 0.10 \text{ A} \quad (100 \text{ mA})$$

Answer is B.

11. Use Kirchhoff's current law to determine the current flowing through the battery. Current I_{BC} is given. All other currents have been determined in previous problems.

$$I_{total} = I_{AB} + I_{AE} = I_{BE} + I_{BC} + I_{AE}$$

$$= 60 \text{ mA} + 100 \text{ mA} + 30 \text{ mA}$$

$$= 190 \text{ mA}$$

$$P = I_{total}V$$

$$= (190 \text{ mA})\left(10^{-3} \ \frac{\text{A}}{\text{mA}}\right)(400 \text{ V})$$

$$= 76 \text{ W}$$

Answer is D.

12. First, find the current through the unknown resistor. Use Kirchhoff's current law. Current I_{BC} is given.

$$I_{CG} = I_{BC} - I_{CD} = I_{BC} - \frac{V_{CD}}{R}$$

$$= 100 \text{ mA} - \frac{(200 \text{ V}) \left(1000 \frac{\text{mA}}{\text{A}}\right)}{10 \times 10^3 \ \Omega}$$

$$= 100 \text{ mA} - 20 \text{ mA}$$

$$= 80 \text{ mA}$$

Next, find the voltage across the unknown resistor. By inspection, the entire battery voltage appears across R_7.

$$V_{CG} = 400 \text{ V}$$

Use Ohm's law.

$$R_7 = \frac{V_{CG}}{I_{CG}} = \frac{400 \text{ V}}{(80 \text{ mA}) \left(10^{-3} \frac{\text{A}}{\text{mA}}\right)}$$

$$= 5000 \ \Omega \quad (5 \text{ k}\Omega)$$

Answer is B.

13. A capacitor in a DC circuit has an infinite resistance. Therefore, there is no current flow through the capacitor between nodes A and B. An inductor in a DC circuit has no resistance. Therefore, the circuit simplifies to

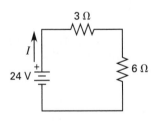

From Ohm's law,

$$I = \frac{V}{R} = \frac{24 \text{ V}}{3 \ \Omega + 6 \ \Omega}$$

$$= 2.67 \text{ A}$$

Answer is B.

14. The energy stored in the capacitor depends on the the capacitance and the voltage across the capacitor. Since there is no voltage drop across an ideal inductor, node B is grounded. The voltage at node A can be found from the current through leg AC. Since none of

the current found in Prob. 13 flows through the capacitor, $I_{AC} = 2.67$ A.

$$V_A = IR = (2.67 \text{ A})(6 \ \Omega)$$

$$= 16 \text{ V}$$

$$V_{AB} = V_A - V_B = 16 \text{ V} - 0 \text{ V}$$

$$= 16 \text{ V}$$

The energy stored in a capacitor is

$$\text{energy} = \tfrac{1}{2} C V_{AB}^2$$

$$= \left(\frac{1}{2}\right)(4 \times 10^{-6} \text{ F})(16 \text{ V})^2$$

$$= 512 \times 10^{-6} \text{ J} \quad (512 \ \mu\text{J})$$

Answer is C.

15. All of the current flowing passes through the 2 H inductor. From Prob. 13, $I = 2.67$ A. The energy stored in an inductor is

$$\text{energy} = \tfrac{1}{2} L I^2$$

$$= \left(\frac{1}{2}\right)(2 \text{ H})(2.67 \text{ A})^2$$

$$= 7.13 \text{ J} \quad (7.1 \text{ J})$$

Answer is A.

16. The permittivity of free space is

$$\epsilon_o = 8.85 \times 10^{-12} \text{ F/m}$$

The capacitance of parallel plates is

$$C = \frac{\epsilon A}{d} = \frac{24 \epsilon_0 A}{d}$$

$$= \frac{(24) \left(8.85 \times 10^{-12} \frac{\text{F}}{\text{m}}\right)(200 \times 10^{-3} \text{ m})^2}{25 \times 10^{-3} \text{ m}}$$

$$= 3.4 \times 10^{-10} \text{ F}$$

Answer is A.

17. From Prob. 16, the capacitance is 3.4×10^{-10} F.

The energy stored is

$$\text{energy} = \tfrac{1}{2} C V^2$$

$$= \left(\frac{1}{2}\right)(3.4 \times 10^{-10} \text{ F})(100 \text{ V})^2$$

$$= 1.7 \times 10^{-6} \text{ J} \quad (1.7 \ \mu\text{J})$$

Answer is C.

18. Inductors have no resistance in a DC circuit; therefore, they will act as short circuits. Capacitors have infinite resistance; thus, they act as open circuits. Capacitors placed across the dry cell will accumulate charge.

Answer is B.

19. The equivalent inductance is the sum of the individual inductances.

$$L_{eq} = L_1 + L_2$$
$$= 10 \text{ H} + 20 \text{ H}$$
$$= 30 \text{ H}$$

The current through the resistor is 10 A, as determined by the current source. The energy stored in the inductors is

$$\text{energy} = \tfrac{1}{2}LI^2 = \left(\frac{1}{2}\right)(30 \text{ H})(10 \text{ A})^2$$
$$= 1500 \text{ J}$$

Answer is B.

20. Capacitors 1 and 2 are parallel to each other, and that combination is in series with capacitor 3.

$$C_{eq} = \cfrac{1}{\cfrac{1}{C_1 + C_2} + \cfrac{1}{C_3}}$$
$$= \cfrac{1}{\cfrac{1}{1 \ \mu\text{F} + 2 \ \mu\text{F}} + \cfrac{1}{3 \ \mu\text{F}}}$$
$$= 1.5 \ \mu\text{F}$$

Answer is B.

21. The capacitors are connected in series.

$$C_{eq} = \cfrac{1}{\sum \cfrac{1}{C_i}}$$
$$= \cfrac{1}{\cfrac{1}{2 \text{ pF}} + \cfrac{1}{2 \text{ pF}} + \cfrac{1}{2 \text{ pF}} + \cfrac{1}{2 \text{ pF}}}$$
$$= 0.5 \text{ pF}$$

Answer is A.

22. Using the parallel and series capacitor relationships, the equivalent capacitance is

$$C_{eq} = 20 \ \mu\text{F} + 10 \ \mu\text{F} + 5 \ \mu\text{F} + \frac{(10 \ \mu\text{F})(10 \ \mu\text{F})}{10 \ \mu\text{F} + 10 \ \mu\text{F}}$$
$$= 40 \ \mu\text{F}$$

Answer is C.

23. From Ohm's law,

$$R = \frac{V}{I} = \frac{9 \text{ V}}{100 \text{ A}}$$
$$= 0.09 \ \Omega$$

Answer is A.

24. Since this is a DC circuit, the capacitor has infinite resistance. No current will flow through it.

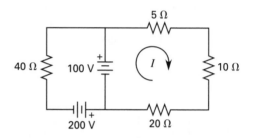

The only voltage source across the loop containing the 5 Ω, 10 Ω, and 20 Ω resistors is the 100 V battery. Write Kirchhoff's voltage law for the loop.

$$\sum V = \sum (IR) = I \sum R$$
$$100 \text{ V} = I(5 \ \Omega + 10 \ \Omega + 20 \ \Omega)$$
$$I = 2.86 \text{ A}$$

Answer is B.

25. Use the current-loop method to solve for the unknown current. Write Kirchhoff's voltage law for each loop.

$$\sum V = \sum (IR) = I \sum R$$
$$20 \text{ V} = (I_A + I_B)(8 \ \Omega) + (I_A)(12 \ \Omega)$$
$$40 \text{ V} = (I_A + I_B)(8 \ \Omega)$$

Solve for I_A from the second equation.

$$I_A = 5 \text{ A} - I_B$$

Substitute I_A into the first equation.

$$20 \text{ V} = (5 \text{ A} - I_B + I_B)(8 \ \Omega) + (5 \text{ A} - I_B)(12 \ \Omega)$$
$$I_B = 6.67 \text{ A} \quad (6.7 \text{ A})$$

Answer is C.

26. Redraw the circuit.

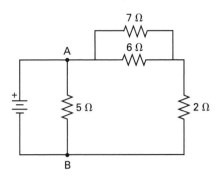

The equivalent resistance to the right of terminals A and B is

$$R_{AB} = \frac{(7\ \Omega)(6\ \Omega)}{7\ \Omega + 6\ \Omega} + 2\ \Omega$$

$$= 5.231\ \Omega$$

The total resistance seen by the battery is

$$R = \frac{(5.231\ \Omega)(5\ \Omega)}{5.231\ \Omega + 5\ \Omega}$$

$$= 2.556\ \Omega \quad (2.6\ \Omega)$$

Answer is C.

27. Use the RL transient equations. At $t = 0$ s, this is an open circuit, so $i(0) = 0$. 30 ms corresponds to 0.03 s.

$$v_L(t) = -i(0)Re^{-Rt/L} + Ve^{-Rt/L}$$

$$v_L(0.03\text{ s}) = 0 + (100\text{ V})e^{-(26\ \Omega)(0.03\text{ s})/0.5\text{ H}}$$

$$= 21\text{ V}$$

Answer is C.

28. Use the RC transient equations. Since the circuit is initially open, the initial voltage on the capacitor is zero.

$$v_C(t) = v_C(0)e^{-t/RC} + V(1 - e^{-t/RC})$$

$$0.8 = 1 - e^{-t/(150\ \Omega)(100\times 10^{-6}\text{ F})}$$

$$0.2 = e^{-t/0.015\text{ s}}$$

Take the natural log of both sides.

$$\ln 0.2 = \frac{-t}{0.015\text{ s}}$$

$$t = 0.024\text{ s} \quad (24\text{ ms})$$

Answer is D.

29. The transient current for an RC circuit is

$$i(t) = \left(\frac{V_{\text{bat}} - v_C(0)}{R}\right)e^{-t/RC}$$

$$i(0) = \left(\frac{5\text{ V} - 2.5\text{ V}}{1\ \Omega}\right)e^0$$

$$= 2.5\text{ A}$$

Answer is D.

30. If there is a complete circuit, ideal voltage and current sources deliver voltage and current at their rated values regardless of the circuit parameters. This eliminates the "cannot be in" alternatives. No current will flow in an open circuit, no matter how ideal the source is. This eliminates choice C. An ideal voltage source will still develop its full voltage potential even without a circuit.

Answer is B.

31. The average current from an AC source is zero, so short the AC voltage source. An inductor in a DC circuit has zero resistance. Because of that, all current bypasses the 5 Ω resistor. Use Ohm's law.

$$I = \frac{V}{R} = \frac{12\text{ V}}{10\ \Omega}$$

$$= 1.2\text{ A}$$

Answer is C.

32. Simplify the circuit by finding the equivalent resistance.

$$R_{\text{eq}} = 2\ \Omega + \frac{1}{\dfrac{1}{4\ \Omega} + \dfrac{1}{4\ \Omega}}$$

$$= 4\ \Omega$$

Use Ohm's law to find the current.

$$I = \frac{V}{R_{\text{eq}}} = \frac{40\text{ V}}{4\ \Omega}$$

$$= 10\text{ A}$$

Answer is C.

45

Alternating-Current Circuits

Subjects

AC CIRCUITS 45-1
    Alternating Waveforms 45-1
    Sine-Cosine Relations 45-1
    Phasor Transforms of Sinusoids 45-2
    Average Value 45-2
    Effective or rms Values 45-2
    Phase Angles 45-3
    Impedance 45-3
    Admittance 45-4
    Ohm's Law for AC Circuits 45-4
    Complex Power 45-5
    Resonance 45-5
    Transformers 45-6

Nomenclature

a	turns ratio	–
BW	bandwidth	Hz or rad/s
C	capacitance	H
f	frequency	Hz
$i(t)$	time-varying current	A
I	constant current	A
N	number of turns	–
p.f.	power factor	–
P	real power	W
Q	reactive power	VAR
Q	quality factor	–
R	resistance	Ω
S	complex power	VA
T	period	s
$v(t)$	time-varying voltage	V
V	constant voltage	V
x	time-varying general variable	–
X	constant general variable	–
X	reactance	Ω
Z	impedance	Ω

Symbols

θ	phase angle difference	rad
ϕ	phase angle	rad
ω	angular frequency	rad/s

Subscripts

0	at resonance
ave	average
C	capacitive
i	imaginary
L	inductive
max	maximum
P	primary
r	real
rms	effective or root-mean-square
S	secondary
t	total

AC CIRCUITS

Alternating Waveforms

The term *alternating waveform* describes any symmetrical waveform, including square, sawtooth, triangular, and sinusoidal waves, whose polarity varies regularly with time. However, the term AC (alternating current) almost always means that the current is produced from the application of a sinusoidal voltage.

Sinusoidal variables can be specified without loss of generality as either sines or cosines. If a sine waveform is used, Eq. 45.1 gives the instantaneous voltage as a function of time. $V_{\max}$ is the maximum value (also known as the *amplitude*) of the sinusoid. If $v(t)$ is not zero at $t = 0$, a *phase angle*, ϕ, must be used.

$$v(t) = V_{\max}\sin(\omega t + \phi) \qquad 45.1$$

Sine-Cosine Relations

The trigonometric relationships in Eqs. 45.2 and 45.3 will be useful for problems with alternating currents.

$$\cos(\omega t) = \sin\left(\omega t + \frac{\pi}{2}\right)$$
$$= -\sin\left(\omega t - \frac{\pi}{2}\right) \qquad 45.2$$
$$\sin(\omega t) = \cos\left(\omega t - \frac{\pi}{2}\right)$$
$$= -\cos\left(\omega t + \frac{\pi}{2}\right) \qquad 45.3$$

CIRCUITS
AC Circuits

Figure 45.1 illustrates the form of an AC voltage given by Eq. 45.1. The *period* of the waveform is T. (Because the horizontal axis corresponds to time and not to distance, the waveform does not have a wavelength.) The *frequency*, f, of the sinusoid is the reciprocal of the period in hertz (Hz). *Angular frequency*, ω, in radians per second (rad/s) can also be used.

$$f = \frac{1}{T} = \frac{\omega}{2\pi} \qquad 45.4$$

Figure 45.1   Sinusoidal Waveform with Phase Angle

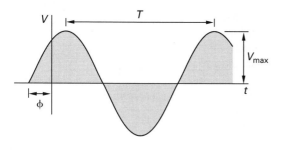

Phasor Transforms of Sinusoids

There are several equivalent methods of representing a sinusoidal waveform.

- *trigonometric*: $V_{\max}\cos(\omega t + \phi)$ 45.5
- *polar* or *phasor*: $V_{\text{eff}} \angle \phi$ 45.6
- *rectangular*:
  $$V_r + jV_i = V_m(\cos\theta + j\sin\theta) \qquad 45.7$$
- *exponential*: $V_m e^{j\phi}$ 45.8

Note that in polar, rectangular or exponential form, the frequency must be specified separately.

When given in polar or phasor form, the voltage is usually given as the effective value (discussed later in this chapter) and not the peak value. Always observe whether the values are peak or effective.

In trigonometric form, ω may be given in either rad/s or deg/s, but rad/s is more common. ϕ is usually given in degrees. This can result in a mismatch in units. (Unfortunately, this is common practice in electrical engineering.) In exponential form, ϕ should always be in radians, though some references incorrectly use degrees. In polar or phasor form and rectangular form, ϕ is usually in degrees.

Average Value

Equation 45.9 calculates the *average value* of any periodic variable (e.g., voltage or current).

$$X_{\text{ave}} = \left(\frac{1}{T}\right)\int_0^T x(t)\,dt \qquad 45.9$$

Waveforms that are symmetrical with respect to the horizontal time axis have an average value of zero. A full-wave rectified sinusoid is shown in Fig. 45.2(b); the average value of Eq. 45.9 for this waveform is Eq. 45.10.

$$X_{\text{ave}} = \frac{2X_{\max}}{\pi} \qquad \begin{bmatrix}\text{full-wave rectified}\\\text{sinusoid}\end{bmatrix} \qquad 45.10$$

The average value of the half-wave rectified sinusoid as shown in Fig. 45.2(c) is

$$X_{\text{ave}} = \frac{X_{\max}}{\pi} \qquad \begin{bmatrix}\text{half-wave rectified}\\\text{sinusoid}\end{bmatrix} \qquad 45.11$$

Figure 45.2   Average and Effective Values

$X_{\text{ave}} = 0$

$X_{\text{rms}} = \sqrt{\dfrac{1}{T}\displaystyle\int_0^T x^2(t)\,dt}$

(a) sinusoid

$X_{\text{ave}} = \dfrac{2X_{\max}}{\pi}$

$X_{\text{rms}} = \dfrac{X_{\max}}{\sqrt{2}}$

(b) full-wave rectified sinusoid

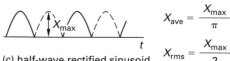

$X_{\text{ave}} = \dfrac{X_{\max}}{\pi}$

$X_{\text{rms}} = \dfrac{X_{\max}}{2}$

(c) half-wave rectified sinusoid

Effective or rms Values

Alternating waveforms are usually characterized by their *effective value*, also known as the *root-mean-square*, or rms value. A DC current of I produces the same heating effect as an AC current of I_{rms}.

The effective value of an alternating waveform is given by Eq. 45.12.

$$X_{\text{rms}} = \sqrt{\frac{1}{T}\int_0^T x^2(t)\,dt} \qquad 45.12$$

For a full-wave rectified sinusoidal waveform,

$$X_{\text{rms}} = \frac{X_{\max}}{\sqrt{2}} \qquad 45.13$$

For a half-wave rectified sinusoidal waveform,

$$X_{\text{rms}} = \frac{X_{\max}}{2} \qquad 45.14$$

In the United States, the value of the standard voltage used in households is 115–120 V; this is the effective value of the voltage. Therefore, the phasor form of the voltage is commonly depicted as

$$\mathbf{V} \equiv V_{\rm rms} \angle \theta = \left(\frac{V_{\rm max}}{\sqrt{2}} \right) \angle \theta \qquad \textit{45.15}$$

It should be assumed that values are effective unless otherwise specified.

Phase Angles

Ordinarily, the current and voltage sinusoids in an AC circuit do not peak at the same time. It is said that a *phase shift* exists between voltage and current, as illustrated in Fig. 45.3.

Figure 45.3 Leading Phase Angle Difference

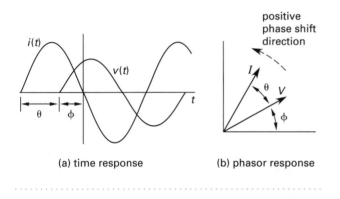

(a) time response   (b) phasor response

This shift is caused by the inductors and capacitors in the circuit. Capacitors and inductors each have a different effect on the phase angle. In a purely resistive circuit, no phase shift exists between voltage and current, and it is said that the current is in phase with the voltage.

It is common practice to use the voltage signal as a reference. In Fig. 45.3, the current *leads* the voltage. In a purely capacitive circuit, the current leads (is ahead of) the voltage by 90°; and, in a purely inductive circuit, the current *lags* behind the voltage by 90°. In a leading circuit, the phase angle difference is positive and the current reaches its peak before the voltage. In a lagging circuit, the phase angle difference is negative and the current reaches its peak after the voltage. For a leading circuit, the angle θ is a positive number in Eq. 45.17. For a lagging circuit, the angle θ is a negative number in Eq. 45.17.

$$v(t) = V_{\rm max} \sin(\omega t + \phi) \qquad \text{[reference]} \quad \textit{45.16}$$
$$i(t) = I_{\rm max} \sin(\omega t + \phi + \theta) \qquad \text{[leading]} \quad \textit{45.17}$$

Each AC *passive circuit element* (resistor, capacitor, or inductor) is assigned an angle, θ, known as its *impedance angle*, that corresponds to the phase angle shift produced when a sinusoidal voltage is applied across the element alone.

The choice of sine or cosine to represent AC waveforms is arbitrary, with the only distinction being that the relative phase angle, ϕ, differs by $\pi/2$ radians. The phasor form of complex values is shown relative to the cosine form in the trigonometric form, so it may be necessary to convert a sine representation into a cosine representation or vice versa.

$$\sin(\omega t) = -\cos\left(\omega t + \frac{\pi}{2}\right)$$
$$= \cos\left(\omega t - \frac{\pi}{2}\right) \qquad \textit{45.18}$$
$$\cos(\omega t) = \sin\left(\omega t + \frac{\pi}{2}\right)$$
$$= -\sin\left(\omega t - \frac{\pi}{2}\right) \qquad \textit{45.19}$$

Impedance

The term *impedance*, Z (with units of ohms), describes the combined effect circuit elements have on current magnitude and phase. Impedance is a complex quantity with a magnitude and an associated angle, and it is usually written in phasor form. However, it can also be written in rectangular form as the complex sum of its *resistive* (real part, R) and *reactive* (imaginary part, X) *components*, both having units of ohms. The resistive and reactive components combine trigonometrically in the *impedance triangle*, shown in Fig. 45.4. Note that resistance is always positive, while reactance may be either positive or negative.

In Fig. 45.4, the impedance is drawn in the complex plane with the real (resistive) part on the horizontal axis and the imaginary (reactive) part on the vertical axis. The impedance in Fig. 45.4 is lagging because the reactive part is positive imaginary. This designation derives from $I = V/Z$ where a positive angle for Z subtracts from the voltage phase angle.

$$\mathbf{Z} \equiv R \pm jX \qquad \textit{45.20}$$
$$R = Z \cos \theta \qquad \textit{45.21}$$
$$X = Z \sin \theta \qquad \textit{45.22}$$

Figure 45.4 Lagging Impedance Triangle

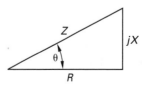

Equation 45.23 gives the impedance of an ideal resistor. An *ideal resistor* has neither inductance nor capacitance. The magnitude of the impedance is the resistance, R, and the phase angle difference is zero.

Therefore, current and voltage are in phase in a purely resistive circuit.

$$\mathbf{Z}_R = R\angle 0° = R + j0 = R \qquad 45.23$$

Equation 45.24 gives the impedance of an *ideal capacitor* with capacitance C. An ideal capacitor has neither resistance nor inductance. The magnitude of the impedance is the *capacitive reactance*, X_C, with units of ohms, and the phase angle difference is $-\pi/2$ ($-90°$). Therefore, current leads the voltage by $90°$ in a purely capacitive circuit. Some authors define X_C as a negative quantity such that Eq. 45.24 does not have a negative sign and Eq. 45.25 does have a negative sign. The nomenclature is not important if it is done consistently. The important thing to know is that the impedance of a capacitor is negative imaginary, regardless of how the reactance sign is defined.

$$\mathbf{Z}_C = X_C\angle -90° = -jX_C$$

$$= \frac{1}{j\omega C} \qquad 45.24$$

$$X_C = \frac{1}{\omega C} \qquad 45.25$$

Equation 45.26 gives the impedance of an ideal inductor with inductance L. An *ideal inductor* has no resistance or capacitance. The magnitude of the impedance is the *inductive reactance*, X_L, with units of ohms, and the phase angle difference is $\pi/2$ ($90°$). Therefore, current lags the voltage by $90°$ in a purely inductive circuit.

$$\mathbf{Z}_L = X_L\angle 90° = jX_L$$

$$= j\omega L \qquad 45.26$$

$$X_L = \omega L \qquad 45.27$$

Some circuits are shown with the capacitor and inductor impedance, rather than the capacitance or inductance, given in ohms. The reactances are at the circuit's operating frequency and can be used for circuit analysis for current dividers and voltage dividers, as with the DC circuits, although the analysis must be done with complex algebra.

Admittance

The reciprocal of impedance is the complex quantity *admittance*, $\mathbf{Y}$. Admittance is particularly useful in analyzing parallel circuits, since admittances of parallel circuit elements add together.

$$\mathbf{Y} = \frac{1}{\mathbf{Z}} = \frac{1}{Z}\angle -\theta \qquad 45.28$$

The reciprocal of the resistive part of impedance is *conductance*, G. The reciprocal of the reactive part of impedance is *susceptance*, B.

$$G = \frac{1}{R} \qquad 45.29$$

$$B = \frac{1}{X} \qquad 45.30$$

By multiplying the numerator and denominator by the complex conjugate, admittance can be written in terms of resistance and reactance, and vice versa.

$$\begin{aligned}\mathbf{Y} &= G + jB \\ &= \left(\frac{1}{R+jX}\right)\left(\frac{R-jX}{R-jX}\right) \\ &= \frac{R}{R^2+X^2} - j\left(\frac{X}{R^2+X^2}\right) \qquad 45.31\end{aligned}$$

$$\begin{aligned}\mathbf{Z} &= R + jX \\ &= \left(\frac{1}{G+jB}\right)\left(\frac{G-jX}{G-jX}\right) \\ &= \frac{G}{G^2+B^2} - j\left(\frac{B}{G^2+B^2}\right) \qquad 45.32\end{aligned}$$

Impedances are combined in the same way as resistances: impedances in series are added, while the reciprocals of impedances in parallel are added. For series circuits, the resistive and reactive parts of each impedance element are calculated separately and summed. For parallel circuits, the conductance and susceptance of each element are summed. The total impedance is found by a complex addition of the resistive (conductive) and reactive (susceptive) parts. It is convenient to perform the addition in rectangular form. Equations 45.33 and 45.34 represent the magnitude of the combined impedances for series and parallel circuits.

$$Z_{\text{eq}} = \sqrt{\left(\sum R\right)^2 + \left(\sum X_L - \sum X_C\right)^2}$$
$$\text{[series]} \qquad 45.33$$

$$Z_{\text{eq}} = \frac{1}{\sqrt{\left(\sum\left(\frac{1}{R}\right)\right)^2 + \left(\sum\left(\frac{1}{X_L}\right) - \sum\left(\frac{1}{X_C}\right)\right)^2}}$$
$$\text{[parallel]} \qquad 45.34$$

Ohm's Law for AC Circuits

Ohm's law for AC circuits with linear circuit elements is similar to Ohm's law for DC circuits.

$$\mathbf{V} = \mathbf{IZ} \qquad 45.35$$

It is important to recognize that V and I can both be either maximum values or effective values, but never a combination of the two. If the voltage source is specified by its effective value, then the current calculated from $I = V/Z$ will be an effective value.

The rules for solving AC circuit problems are the same as those for DC circuits, except that impedance is used instead of resistance, and phase angles must be considered.

Complex Power

The *complex power vector*, **S** (also called the *apparent power*), is the vector sum of the real (true, active) power vector, **P**, and the imaginary reactive power vector, **Q**. Its units are volt-amps (VA). The components of power combine as vectors in the *complex power triangle*, shown in Fig. 45.5.

$$\mathbf{S} \equiv P + jQ \qquad \text{45.36}$$

$$\mathbf{S} = \mathbf{I}^*\mathbf{V} \qquad \text{45.37}$$

The complex conjugate of the current, **I***, is used in the apparent power, resulting in a positive imaginary part and a positive power angle for a lagging current (which has a negative phase angle compared to the voltage), as shown in Fig. 45.5. For a leading current (which has a positive phase angle compared to the voltage), the power triangle has a negative imaginary part and a negative power angle.

Figure 45.5 Lagging Complex Power Triangle

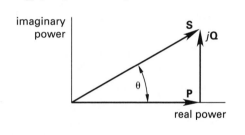

The *real power*, P, with units of watts (W), is defined as

$$P = \tfrac{1}{2}V_{\max}I_{\max}\cos\theta$$
$$= V_{\mathrm{rms}}I_{\mathrm{rms}}\cos\theta \qquad \text{45.38}$$

The *reactive power*, Q, in units of volt-amps reactive (VAR), is the imaginary part of **S**. The reactive power is given by Eq. 45.39.

$$Q = \tfrac{1}{2}V_{\max}I_{\max}\sin\theta \qquad \text{45.39}$$
$$= V_{\mathrm{rms}}I_{\mathrm{rms}}\sin\theta$$
$$= \frac{V_{\mathrm{rms}}^2}{X}$$

The part of Eqs. 45.38 and 45.39 with rms values is always true, while the part with max values is only true for sinusoidal signals.

The *power factor*, p.f. (usually given in percent), is $\cos\theta$. The angle θ is called the *power angle*, and is

the same as the overall impedance angle, or the angle between input voltage and current in the circuit. These are the voltage and current at the source (usually voltage source), which supplies the electric power to the circuit.

$$\text{p.f.} = \cos\theta \qquad \text{45.40}$$

The cosine is positive for both positive and negative angles. Therefore, the descriptions *lagging* (for an inductive circuit) and *leading* (for a capacitive circuit) must be used with the power factor.

For a purely resistive load, p.f. $= 1$, and the average real power is given by

$$P_{\mathrm{ave}} = V_{\mathrm{rms}}I_{\mathrm{rms}} = \frac{V_{\mathrm{rms}}^2}{R}$$
$$= I_{\mathrm{rms}}^2 R \qquad \text{45.41}$$

For a purely reactive load, p.f. $= 0$, and the average real power is given by

$$P_{\mathrm{ave}} = V_{\mathrm{rms}}I_{\mathrm{rms}}\cos 90^\circ = 0$$
$$= V_{\mathrm{rms}}I_{\mathrm{rms}}\cos(-90^\circ) = 0 \qquad \text{45.42}$$

Electric energy is stored in a capacitor or inductor during a fourth of a cycle and is returned to the circuit during the next fourth of the cycle. Only a resistance will actually dissipate energy.

The power factor of a circuit, and therefore the phase angle difference, can be changed by adding either inductance or capacitance. This is known as *power factor correction*.

Resonance

In a *resonant circuit*, input voltage and current are in phase, and therefore, the phase angle is zero. This is equivalent to saying that the circuit is purely resistive in its response to an AC voltage, although inductive and capacitive elements must be present for resonance to occur. At resonance, the power factor is equal to 1 and the reactance, X, is equal to zero, or $X_L = X_C$. The frequency at which the circuit becomes purely resistive, ω_0 or f_0, is the *resonant frequency*.

For both parallel and series circuits at the resonant frequency,

$$\omega_0 = 2\pi f_0 = \frac{1}{\sqrt{LC}} \qquad \text{45.43}$$

$$Z = R \qquad \text{45.44}$$

$$X_L = X_C \qquad \text{45.45}$$

$$\text{p.f.} = 1.0 \qquad \text{45.46}$$

$$\omega_0 L = \frac{1}{\omega_0 C} \qquad \text{45.47}$$

In a resonant *series-RLC circuit*, impedance is minimum, and the current and power dissipation are maximum. In a resonant *parallel-RLC circuit*, impedance is maximum, and the current and power dissipation are minimum.

For frequencies below the resonant frequency, a series-RLC circuit will be capacitive (leading) in nature. Above the resonant frequency, the circuit will be inductive (lagging) in nature.

For frequencies below the resonant frequency, a parallel-RLC circuit will be inductive (lagging) in nature; above the resonant frequency, the circuit will be capacitive (leading) in nature.

Circuits can become resonant in two ways. If the frequency of the applied voltage is fixed, the elements must be adjusted so that the capacitive reactance cancels the inductive reactance (i.e., $X_L - X_C = 0$). If the circuit elements are fixed, the frequency must be adjusted.

The behavior of a circuit at frequencies near the resonant frequency is illustrated in Fig. 45.6. The frequency difference between the *half-power points* is the *bandwidth*, BW, a measure of circuit selectivity. The smaller the bandwidth, the more selective the circuit.

$$\text{BW} = f_2 - f_1 = \frac{f_0}{Q} \quad \text{[in Hz]}$$
$$= \omega_2 - \omega_1 = \frac{\omega_0}{Q} \quad \text{[in rad/s]} \quad \textit{45.48}$$

The half-power points are so named because at those frequencies, the power dissipated in the resistor is half of the power dissipated at the resonant frequency.

$$Z_{\omega_1} = Z_{\omega_2} = \sqrt{2}R \quad \textit{45.49}$$
$$I_{\omega_1} = I_{\omega_2} = \frac{V}{Z_{\omega_1}} = \frac{V}{\sqrt{2}R}$$
$$= \frac{I_0}{\sqrt{2}} \quad \textit{45.50}$$
$$P_{\omega_1} = P_{\omega_2} = I^2R = \left(\frac{I_0}{\sqrt{2}}\right)^2 R$$
$$= \tfrac{1}{2}P_0 \quad \textit{45.51}$$

The *quality factor*, Q, for a circuit is a dimensionless ratio that compares, for each cycle, the reactive energy stored in an inductor to the resistive energy dissipated. Quality factor indicates the shape of the resonance curve. A circuit with a low Q has a broad and flat curve, while one with a high Q has a narrow and peaked curve.

The quality factor for a series-RLC circuit is

$$Q = \frac{\omega_0 L}{R} = \frac{1}{\omega_0 RC} \quad \textit{45.52}$$

Figure 45.6 Circuit Characteristics at Resonance

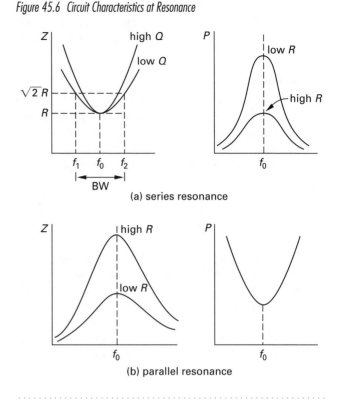

(a) series resonance

(b) parallel resonance

The quality factor for a parallel-RLC circuit is

$$Q = \omega_0 RC = \frac{R}{\omega_0 L} \quad \textit{45.53}$$

Assuming a fixed primary impedance, maximum power transfer in an AC circuit occurs when the source and load are complex conjugates (resistances are equal and their reactances are opposite). This is equivalent to having a resonant circuit as shown in Fig. 45.7.

Figure 45.7 Maximum Power Transfer at Resonance

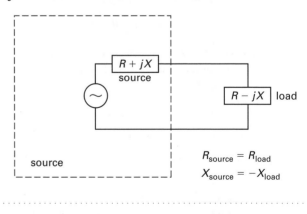

$$R_{source} = R_{load}$$
$$X_{source} = -X_{load}$$

Transformers

Transformers are used to change voltages, match impedances, and isolate circuits. They consist of coils of wire wound on a magnetically permeable core. The coils are grouped into primary and secondary windings. The winding connected to the source of electric energy is called the *primary*. The primary current produces a magnetic flux in the core, which induces a current in the secondary coil. Core and shell transformer designs are shown in Fig. 45.8.

Figure 45.8 Core and Shell Transformers

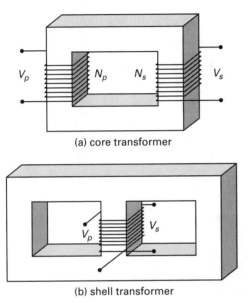

(a) core transformer

(b) shell transformer

The ratio of numbers of primary to secondary windings is the *turns ratio (ratio of transformation)*, a. If the turns ratio is greater than 1, the transformer decreases voltage and is a *step-down transformer*. If the turns ratio is less than 1, the transformer increases voltage and is a *step-up transformer*.

$$a = \frac{N_1}{N_2} = \frac{N_P}{N_S} \qquad 45.54$$

In a lossless (i.e., 100% efficient) transformer, the power absorbed by the primary winding equals the power generated by the secondary winding, so

$$I_P V_P = I_S V_S \qquad 45.55$$

$$a = \frac{N_1}{N_2} = \frac{V_P}{V_S} = \frac{I_S}{I_P} \qquad 45.56$$

A lossless transformer is called an *ideal transformer*; its windings are considered to have neither resistance nor reactance. An impedance of Z_S connected to the secondary of an ideal transformer is equivalent to an impedance of $a^2 Z_S$ connected to the source, as illustrated in Fig. 45.9. It is said that a secondary impedance of Z_S reflects as $a^2 Z_S$ on the primary side.

$$Z_{\text{eq}} = \frac{V_P}{I_P} = Z_P + a^2 Z_S \qquad 45.57$$

Figure 45.9 Equivalent Circuit with Secondary Impedance

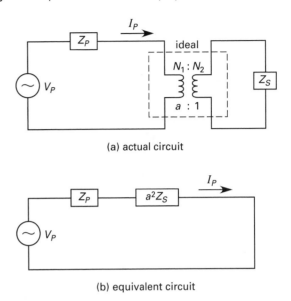

(a) actual circuit

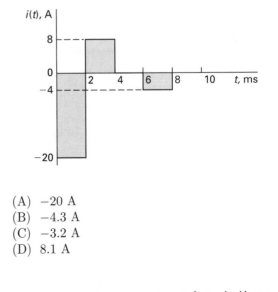

(b) equivalent circuit

Equation 45.57 shows how the impedance seen by the source changes when an impedance is connected to the secondary. This property is often used when impedances have to be matched for maximum power transfer.

SAMPLE PROBLEMS

Problem 1

The waveform shown repeats every 10 ms. What is the average value of the waveform?

(A) −20 A
(B) −4.3 A
(C) −3.2 A
(D) 8.1 A

Solution

The average value of a periodic waveform is

$$I_{\text{ave}} = \frac{1}{T} \int_0^T i(t)\,dt$$

$$= \left(\frac{1}{10 \text{ ms}}\right)((-20 \text{ A})(2 \text{ ms}) + (8 \text{ A})(2 \text{ ms})$$

$$+ (0 \text{ A})(2 \text{ ms}) + (-4 \text{ A})(2 \text{ ms})$$

$$+ (0 \text{ A})(2 \text{ ms}))$$

$$= \left(\frac{1}{10 \text{ ms}}\right)(-40 \text{ A·ms} + 16 \text{ A·ms} - 8 \text{ A·ms})$$

$$= -3.2 \text{ A}$$

Answer is C.

Problem 2

What is the current through the LC leg of the following circuit?

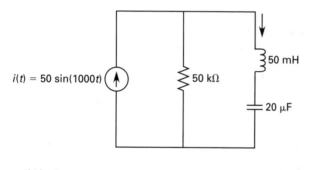

(A) 0

(B) $50\sin(1000t)$ A

(C) $50\sin\left(1000t + \dfrac{\pi}{4}\right)$ A

(D) $70.7\sin\left(1000t - \dfrac{3\pi}{4}\right)$ A

Solution

Use Ohm's law for AC circuits.

$$\mathbf{I} = \frac{\mathbf{V}}{\mathbf{Z}}$$

$$Z_{\text{leg}} = j\omega L + \frac{1}{j\omega C} = j\omega L - \frac{j}{\omega C}$$

$$= j\left(\left(1000 \ \frac{\text{rad}}{\text{s}}\right)(50 \times 10^{-3} \text{ H})\right.$$

$$\left. - \frac{1}{\left(1000 \ \frac{\text{rad}}{\text{s}}\right)(20 \times 10^{-6} \text{ F})}\right)$$

$$= 0$$

The current source is effectively short-circuited through the LC branch. Therefore, the current is equal to the current generated.

$$I_{LC} = 50\sin(1000t) \text{ A}$$

Answer is B.

Problems 3–5 refer to the following illustration.

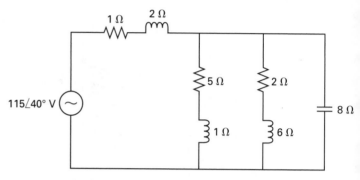

Problem 3

What is the average power dissipated by the circuit?

(A) 24 W

(B) 765 W

(C) 910 W

(D) 1970 W

Solution

$$P_{\text{ave}} = V_{\text{rms}} I_{\text{rms}} \cos\theta$$

The circuit impedance must be determined to find the impedance angle. First find the equivalent impedance of the parallel branch.

$$\frac{1}{Z_{\text{parallel}}} = \frac{1}{Z_{\text{leg1}}} + \frac{1}{Z_{\text{leg2}}} + \frac{1}{Z_{\text{leg3}}}$$

$$= \frac{1}{5 + j1} + \frac{1}{2 + j6} + \frac{1}{0 - j8}$$

Multiply each term by its complex conjugate.

$$\frac{1}{Z_{\text{parallel}}} = \frac{5 - j1}{(5 + j1)(5 - j1)} + \frac{2 - j6}{(2 + j6)(2 - j6)}$$

$$+ \frac{j8}{(-j8)(j8)}$$

$$= \frac{5 - j1}{26} + \frac{2 - j6}{40} + \frac{j8}{64}$$

$$= 0.24231 - j0.06346$$

$$\equiv 0.2505\angle{-14.68°}$$

$$Z_{\text{parallel}} = \frac{1}{0.2505\angle{-14.68°}} = 3.99\angle 14.68°$$

$$\equiv 3.86 + j1.01$$

Now add the resistor and inductor in series.

$$Z_{total} = Z_{series} + Z_{parallel}$$
$$= (1 + j2) + (3.86 + j1.01)$$
$$= 4.86 + j3.01$$
$$= 5.72\angle 31.77° \ \Omega$$
$$\theta = 31.77°$$

Use Ohm's law for AC circuits.

$$\mathbf{I} = \frac{\mathbf{V}}{\mathbf{Z}} = \frac{115\angle 40° \text{ V}}{5.72 \ \angle 31.77° \ \Omega}$$
$$= \frac{115}{5.72}\angle 40° - 31.77°$$
$$= 20.10\angle 8.23° \text{ A}$$

$$I_{rms} = 20.10 \text{ A}$$

$$P_{ave} = V_{rms}I_{rms}\cos\theta$$
$$= (115 \text{ V})(20.10 \text{ A})\cos 31.77°$$
$$= 1965 \text{ W} \quad (1970 \text{ W})$$

Answer is D.

Problem 4

What is the reactive power drawn by the circuit?

(A) 560 VAR
(B) 920 VAR
(C) 1220 VAR
(D) 1270 VAR

Solution

The equation for reactive power is

$$Q = V_{rms}I_{rms}\sin\theta$$

These values were calculated in Sample Prob. 3.

$$Q = (115 \text{ V})(20.1 \text{ A})\sin 31.77°$$
$$= 1217 \text{ VAR} \quad (1220 \text{ VAR})$$

Answer is C.

Problem 5

What are the average real and reactive powers taken by the capacitor?

(A) 0 W; −806 VAR
(B) 810 W; 0 VAR
(C) 50 W; 0 VAR
(D) 0 W; 1530 VAR

Solution

The real power taken by a capacitor is always zero. The reactive power taken by the capacitor is

$$Q = \frac{-V_C^2}{X_C}$$

V_C is the voltage across the capacitor, which is the same across each of the parallel branches. Use the voltage divider equation.

$$V_C = \left(\frac{Z_{parallel}}{Z_{series} + Z_{parallel}}\right)(V)$$
$$= \left(\frac{3.86 + j1.01}{(1 + j2) + (3.86 + j1.01)}\right)(115\angle 40°)$$
$$= \left(\frac{3.86 + j1.01}{4.86 + j3.01}\right)(115\angle 40°)$$
$$= \left(\frac{3.99\angle 14.7°}{5.72\angle 31.8°}\right)(115\angle 40°)$$
$$= (0.698\angle -17.1°)(115\angle 40°)$$
$$= 80.3\angle 22.9° \text{ V}$$
$$Q = \frac{-V_C^2}{X_C} = \frac{-(80.3 \text{ V})^2}{8 \ \Omega}$$
$$= -806 \text{ VAR}$$

Answer is A.

FE-STYLE EXAM PROBLEMS

1. The reactances of a 10 mH inductor and a 0.2 μF capacitor are equal when the frequency is

(A) 3.56 kHz
(B) 7.12 kHz
(C) 14 kHz
(D) 21 kHz

2. Find the effective value of the voltage for the repeating waveform.

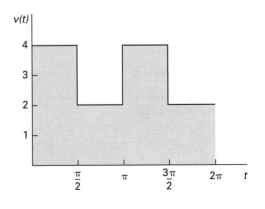

(A) 2.45
(B) 2.75
(C) 3.0
(D) 3.16

3. For the circuit shown, calculate the resonant frequency (in rad/s) and the quality factor.

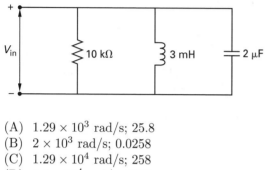

(A) 1.29×10^3 rad/s; 25.8
(B) 2×10^3 rad/s; 0.0258
(C) 1.29×10^4 rad/s; 258
(D) 1.5×10^4 rad/s; 2580

4. For the circuit shown, determine the value of the capacitor C_x that makes the power factor 100%.

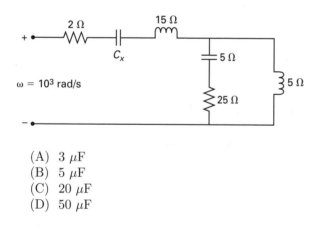

(A) 3 μF
(B) 5 μF
(C) 20 μF
(D) 50 μF

5. What is the average power dissipated by an electric heater with resistance of 50 Ω drawing a current of $20\sin(30t)$ A?

(A) 0 kW
(B) 10 kW
(C) 14.14 kW
(D) 20 kW

For the following problems use the NCEES Handbook as your only reference.

6. A 13.2 kV circuit has a 10 000 kVA load with a 0.85 lagging power factor. How much capacitive reactive power (in kVAR) is needed to correct the power factor to 0.97 lagging?

(A) −2500 kVAR
(B) −3138 kVAR
(C) −4753 kVAR
(D) −5156 kVAR

7. What is the turns ratio $(N_1 : N_2)$ for maximum power transfer in the following circuit?

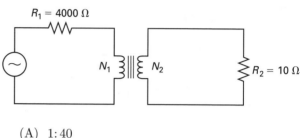

(A) 1:40
(B) 1:20
(C) 20:1
(D) 40:1

8. What measurements are required to determine the phase angle of a single-phase circuit?

(A) the power in watts consumed by the circuit
(B) the frequency, capacitance, and inductance
(C) the power in watts, voltage, and current
(D) the resistance, current, and voltage

9. The electric field (in V/m) of a particular plane wave propagating in a dielectric medium is given by the following expression.

$$E(t,z) = \mathbf{a_x} \cos\left(10^8 t - \frac{z}{3}\right) - \mathbf{a_y} \sin\left(10^8 t - \frac{z}{3}\right)$$

What is the wave frequency?

(A) 6.28 MHz
(B) 7.00 MHz
(C) 16.0 MHz
(D) 160 MHz

10. What is the resonant frequency for the circuit shown?

(A) 0 rad/s
(B) 0.2 rad/s
(C) 9 rad/s
(D) 200 rad/s

11. An AC circuit consisting of a capacitor, an inductor, and a resistor in series is resonant when the

(A) in-phase current equals the out-of-phase current.
(B) maximum current passes through the circuit.
(C) power dissipation is minimum.
(D) power factor is zero.

12. Which of the following statements is true for an ideal transformer with primary and secondary windings designated 1 and 2, respectively? (I is current, V is applied voltage, Z is impedance, and N is the number of turns.)

(A) $I_1 = I_2$

(B) $Z_2 = \dfrac{V_1}{I_1}$

(C) $\dfrac{I_1}{I_2} = \dfrac{N_2}{N_1}$

(D) $I_1 \times V_1 \times N_1 = I_2 \times V_2 \times N_2$

Problems 13–15 are based on the following illustration of an ideal transformer.

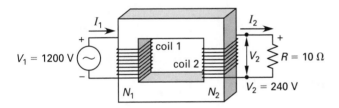

13. If coil 1 has 500 turns and V_2 is 240 V, how many turns does coil 2 have?

(A) 100
(B) 200
(C) 500
(D) 1000

14. If the turns ratio, N_1/N_2, is 5 and V_2 is 240 V, what is the current flowing through coil 1?

(A) 1.9 A
(B) 4.8 A
(C) 24 A
(D) 120 A

15. If the turns ratio is 5, and if the current through coil 1 is 4.8 A, what is the current flowing through coil 2?

(A) 12 A
(B) 24 A
(C) 29 A
(D) 120 A

16. In the circuit shown, what capacitance is needed to achieve a power factor of 1.0?

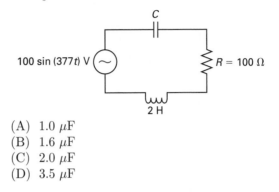

(A) 1.0 μF
(B) 1.6 μF
(C) 2.0 μF
(D) 3.5 μF

17. What is the resonant frequency for the circuit shown?

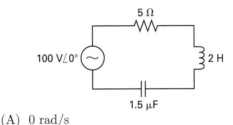

(A) 0 rad/s
(B) 92 rad/s
(C) 260 rad/s
(D) 580 rad/s

SOLUTIONS TO FE-STYLE EXAM PROBLEMS

1. Set the reactances equal to one another and solve for the frequency.

$$X_L = X_C$$

$$\omega L = \frac{1}{\omega C}$$

$$2\pi f L = \frac{1}{2\pi f C}$$

$$f^2 = \frac{1}{4\pi^2 LC}$$

$$f = \frac{1}{2\pi\sqrt{LC}} \quad \text{[same as resonant frequency]}$$

$$= \frac{1}{2\pi}\left(\frac{1}{\sqrt{(10\times 10^{-3}\ \text{H})(0.2\times 10^{-6}\ \text{F})}}\right)$$

$$= 3559\ \text{Hz} \quad (3.56\ \text{kHz})$$

Answer is A.

2. For an alternating waveform,

$$V_{\text{rms}} = \sqrt{\frac{1}{T} \int_0^T v^2(t)dt}$$

$$= \sqrt{\frac{1}{T} \left(\int_0^{\frac{T}{2}} (4)^2 dt + \int_{\frac{T}{2}}^T (2)^2 dt \right)}$$

$$= \sqrt{\frac{1}{T} \left(16t \Big|_0^{\frac{T}{2}} + 4t \Big|_{\frac{T}{2}}^T \right)}$$

$$= \sqrt{\frac{1}{T} \left(\frac{16T}{2} + 4T - \frac{4T}{2} \right)}$$

$$= \sqrt{\frac{1}{T} (8T + 4T - 2T)}$$

$$= \sqrt{\frac{1}{T} (10T)} = \sqrt{10}$$

$$= 3.16$$

Answer is D.

3. The resonant frequency is

$$\omega_0 = \frac{1}{\sqrt{LC}}$$

$$= \frac{1}{\sqrt{(3 \times 10^{-3} \text{ H})(2 \times 10^{-6} \text{ F})}}$$

$$= 1.29 \times 10^4 \text{ rad/s}$$

For a parallel-RLC circuit,

$$Q = \frac{R}{\omega_0 L}$$

$$= \frac{10 \times 10^3 \text{ } \Omega}{\left(1.29 \times 10^4 \text{ } \frac{\text{rad}}{\text{s}} \right) (3 \times 10^{-3} \text{ H})}$$

$$= 258.4 \quad (258)$$

Answer is C.

4. Determine the equivalent impedance of the circuit.

$$\frac{1}{Z_{\text{parallel}}} = \frac{1}{Z_{\text{leg1}}} + \frac{1}{Z_{\text{leg2}}}$$

$$= \frac{1}{25 - j5} + \frac{1}{j5}$$

Multiply each term by its complex conjugate.

$$\frac{1}{Z_{\text{parallel}}} = \frac{25 + j5}{(25 - j5)(25 + j5)} - \frac{j5}{(j5)(-j5)}$$

$$= \frac{25 + j5}{650} - \frac{j5}{25}$$

$$= 0.03846 - j0.1923$$

$$\equiv 0.1961 \angle -78.69°$$

$$Z_{\text{parallel}} = \frac{1}{0.1961 \angle -78.69°}$$

$$= 5.10 \angle 78.69°$$

$$\equiv 1.000 + j5.000$$

$$Z_{\text{total}} = Z_{\text{series}} + Z_{\text{parallel}}$$

$$= 2 - jX_C + j15 + 1 + j5$$

$$= 3 + j20 + (-jX_C)$$

A power factor of 100% means the load is purely resistive and there is no reactive component to the impedance.

$$j20 - jX_C = 0$$

$$X_C = 20 \text{ } \Omega = \frac{1}{\omega C}$$

$$C = \frac{1}{\omega X_C} = \frac{1}{\left(10^3 \text{ } \frac{\text{rad}}{\text{s}} \right) (20 \text{ } \Omega)}$$

$$= 5 \times 10^{-5} \text{ F} \quad (50 \text{ } \mu\text{F})$$

Answer is D.

5. Average power is usually calculated in terms of rms values, but the relationship can be converted to use maximum values, as I_{max} is given in the problem statement. (The form of the sine wave is $I_{\text{max}} \sin(\omega t + \phi) = 20 \sin 30t$.)

$$P_{\text{ave}} = I_{\text{rms}}^2 R = \frac{I_{\text{max}}^2 R}{2}$$

$$= \frac{(20 \text{ A})^2 (50 \text{ } \Omega)}{2}$$

$$= 10\,000 \text{ W} \quad (10 \text{ kW})$$

Answer is B.

6. The complex power vector is given in phasor form as $10\,000$ kVA$\angle \cos^{-1} 0.85 = 10\,000$ kVA$\angle 31.79°$. Convert this to rectangular form to identify the real and reactive components.

$$P = S \cos \theta = (10\,000)(\cos 31.79°)$$

$$= 8500 \text{ kW}$$

$$Q = S \sin\theta = (10\,000)(\sin 31.79°)$$
$$= 5268 \text{ kVAR}$$

$$\mathbf{S} = P + jQ$$
$$= 8500 + j5268$$

To change the power factor, add capacitive kVAR, which affects the reactive component only.

$$\text{desired p.f.} = 0.97$$
$$\cos\theta = 0.97$$
$$\theta = \arccos 0.97$$
$$= 14.07°$$

The power triangles for the current and desired circuits are shown in the following illustration. The new reactive component will be

$$Q = P \tan\theta$$
$$= 8500 \tan 14.07°$$
$$= 2130 \text{ kVAR}$$

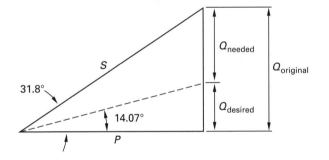

The capacitive kVAR needed is the difference between the original and desired reactive powers.

$$Q_{\text{needed}} = Q_{\text{desired}} - Q_{\text{original}}$$
$$= 2130 \text{ kVAR} - 5268 \text{ kVAR}$$
$$= -3138 \text{ kVAR}$$

Answer is B.

7. Maximum power transfer occurs in an ideal transformer in which primary and secondary powers are equal.

$$P_P = P_S$$
$$I_P^2 R_P = I_S^2 R_S$$
$$\left(\frac{I_S}{I_P}\right)^2 = \frac{R_P}{R_S}$$

$$a^2 = \frac{R_P}{R_S}$$

$$a = \sqrt{\frac{R_P}{R_S}} = \sqrt{\frac{4000 \text{ }\Omega}{10 \text{ }\Omega}}$$
$$= 20$$

$$N_1 : N_2 = 20 : 1$$

Answer is C.

8. The equation for power is

$$P = \tfrac{1}{2} V_{\max} I_{\max} \cos\theta$$
$$= V_{\text{rms}} I_{\text{rms}} \cos\theta$$

Therefore, to find the phase angle, θ, the power, voltage, and current must be shown.

Answer is C.

9. The standard expression for a sinusoidal wave is

$$E = K_1 \cos(\omega t + \phi)\mathbf{a_x} + K_2 \sin(\omega t + \phi)\mathbf{a_y}$$

From the given expression, $\omega = 10^8$ rad/s.

Even without knowing the standard expression of the sinusoidal wave, it is evident that the angular frequency is the time-dependant part of the sine and cosine functions. Therefore, the number that multiplies the variable t is the angular frequency, ω.

$$f = \frac{\omega}{2\pi} = \frac{10^8 \frac{\text{rad}}{\text{s}}}{2\pi}$$
$$= 15.9 \times 10^6 \text{ Hz} \quad (16 \text{ MHz})$$

Answer is C.

10. For a series-RLC circuit, the resonant frequency is

$$\omega_0 = \frac{1}{\sqrt{LC}} = \frac{1}{\sqrt{(3 \text{ H})(8 \times 10^{-6} \text{ F})}}$$
$$= 204 \text{ rad/s} \quad (200 \text{ rad/s})$$

Answer is D.

11. In a resonant series-RLC circuit, impedance is a minimum and current is a maximum. The magnitude of the current is the magnitude of the voltage divided by the magnitude of the impedance. The magnitude of the impedance is a minimum when the imaginary part cancels (resonance).

Answer is B.

---

12. An ideal transformer is lossless. All the power in the primary is transferred to the secondary winding.

$$I_1 V_1 = I_2 V_2$$

The turns ratio, a, is defined as

$$a = \frac{N_1}{N_2} = \frac{I_2}{I_1} = \frac{V_1}{V_2}$$

Inverting this equation yields

$$\frac{1}{a} = \frac{N_2}{N_1} = \frac{I_1}{I_2} = \frac{V_2}{V_1}$$

Answer is C.

---

13. The turns ratio, a, is defined as

$$a = \frac{N_1}{N_2} = \frac{V_1}{V_2}$$

$$N_2 = \frac{N_1 V_2}{V_1} = \frac{(500 \text{ turns})(240 \text{ V})}{1200 \text{ V}}$$

$$= 100 \text{ turns}$$

Answer is A.

---

14. The current flowing through coil 2 is

$$I_2 = \frac{V_2}{R_2} = \frac{240 \text{ V}}{10 \ \Omega}$$

$$= 24 \text{ A}$$

The current flowing through coil 1 is

$$I_1 = \frac{I_2}{a} = \frac{24 \text{ A}}{5}$$

$$= 4.8 \text{ A}$$

Answer is B.

---

15. The current flowing through coil 2 is

$$I_2 = a I_1$$

$$= (5)(4.8 \text{ A})$$

$$= 24 \text{ A}$$

Answer is B.

---

16. A circuit has a power factor of 1.0 when it is resonant. The input frequency is given as 377 rad/s. Size the capacitor so that the circuit resonates at the input frequency.

$$C = \frac{1}{\omega^2 L} = \frac{1}{\left(377 \ \dfrac{\text{rad}}{\text{s}}\right)^2 (2 \text{ H})}$$

$$= 3.52 \times 10^{-6} \text{ F} \quad (3.5 \ \mu\text{F})$$

Answer is D.

---

17. For a series-RLC circuit,

$$\omega_0 = \frac{1}{\sqrt{LC}} = \frac{1}{\sqrt{(2 \text{ H})(1.5 \times 10^{-6} \text{ F})}}$$

$$= 577 \text{ rad/s} \quad (580 \text{ rad/s})$$

Answer is D.

46 Rotating Machines

Subjects

AC MACHINES 46-1
    Synchronous Machines 46-2
    Induction Motors 46-2
DC MACHINES 46-3
    DC Generators 46-3
    DC Motors 46-4

Nomenclature

B	magnetic flux density	N/A·m
E	induced voltage	V
f	frequency	Hz
I	current	A
K_a	armature constant	m²/rpm
K_f	field constant	H
L	inductance	H
n	speed	rpm
p	number of poles	–
P	power	W
R	resistance	Ω
s	slip	–
T	torque	N·m
V	voltage	V

Symbols

ϕ	magnetic flux	Wb
ω	angular velocity (electrical)	rad/s
Ω	rotational velocity (mechanical)	rad/s

Subscripts

a	armature
e	electrical
g	generated
h	heat
m	mechanical
s	synchronous

AC MACHINES

Rotating machines are broadly categorized as AC and DC machines. Both categories include machines that use power (i.e., motors) and those that generate power (alternators and generators). Most AC machines can be constructed in either single-phase or polyphase (usually three-phase) configurations. The rotating part of the machine is called the *rotor*. The stationary part of the machine is called the *stator*. (In AC machines, the rotor is the field and the stator is the armature if the machine is a generator, and vice versa if the machine is a motor. The terms "armature" and "field" are not commonly used in relation to AC machines.)

For simplicity, Fig. 46.1 shows only one of the poles for each phase. But there is actually a winding on the opposite side from each winding depicted with the voltage in the opposite polarity, such that the magnetic fields produced by the windings have opposite polarities (north and south). When V_a is a maximum value, the field in the a-phase windings will be in the up direction, while the b-phase and c-phase fields will be equal and negative so their net field will be up too. When V_c reaches the maximum negative voltage, the field in the c-phase windings will be in the negative c-direction (60° counterclockwise), while the a-phase and b-phase fields will be equal and positive so their net field will be in the negative c-direction too. When V_b is a maximum value, the field in the b-phase windings will be in the positive b-direction (120° counterclockwise), while the a-phase and c-phase fields will be equal and negative so their net field will be in the positive b-direction too. Therefore, the field makes a complete rotation around the stator for one cycle of three-phase voltage.

The number of rotations around the stator depends on the number of poles. The motor shown in Fig. 46.1 is a two-pole machine (two poles in each phase). If the motor were a four-pole machine there would be twice as many windings around the stator and it would take 720° of electrical phase to complete 360° of mechanical rotation. The rotating magnetic field principle applies to three-phase synchronous motors and induction motors.

The rotating magnetic field rotates with a speed known as the *synchronous speed*, n_s; the electrical frequency, f, of the generated potential is given by Eq. 46.1. The actual rotational speed, n, is known as the *mechanical frequency*. Care must be taken to distinguish between the armature speed, n (in rpm), the angular armature speed, Ω, (in rad/s), and the linear and angular voltage frequencies, f and ω (in Hz and rad/s, respectively).

$$n_s = \frac{120f}{p} = \frac{60\Omega}{2\pi}$$
$$= \frac{60\omega}{\pi p} \quad \text{[synchronous speed]} \qquad 46.1$$

Figure 46.1 Rotating Magnetic Field

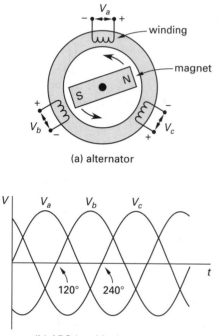

(a) alternator

(b) ABC (positive) sequence

Synchronous Machines

A synchronous machine has a permanent magnet or, more typically, a DC electromagnet that produces a constant magnetic field in the rotor. For the electromagnet, a DC current is provided to windings on the rotor and is transferred by stationary brushes making contact with *slip rings* or *collector rings* on the rotating shaft. The term "slip ring" implies that the conductor is a continuous type and slips under the brush. The term "collector ring" implies that the ring collects current from more than one winding (a characteristic that is usually true). Synchronous machines derive their name from the fact the magnetic fields and electric currents in the stator and rotor are at the same frequency at all times, as related by Eq. 46.1.

The rotor of a synchronous generator is rotated by a mechanical force. As the rotor rotates, both the current and the magnetic field remain constant, but the angle the magnetic field makes with each of the windings changes as the rotor rotates. For each of the stator windings, the flux lines from the rotor will be perpendicular to the loops at some angles and parallel to them at other angles. The current induced in the loops is greatest when the flux lines are perpendicular to the loops, and is zero when the flux lines are parallel to the loops. In this way, a rotating constant magnetic field induces alternating electrical currents in the stator.

The induced voltage, E, is commonly called the *electromotive force* (emf). In an elementary alternator, emf is the desired end result to the load. In a motor, emf is also produced but is referred to as *back-emf* (*counter-emf*), since it opposes the input current.

A *synchronous motor* is essentially a synchronous generator operating in reverse. There is no difference in the construction of the machine. Alternating current is supplied to the stationary stator windings. A rotating magnetic field is produced when three-phase power is applied to the windings. Direct current is applied to the windings in the rotor through brushes and slip rings, as in the synchronous generators. The rotor current interacts with the stator field, causing the rotor to turn. Since the stator field frequency is fixed, the motor runs only at a single *synchronous speed*. (See Eq. 46.1.)

Induction Motors

An *induction motor* is essentially a constant-speed device that receives power through induction, without using brushes or slip rings. A motor can be considered to be a rotating transformer secondary (the *rotor*) with a stationary primary (the *stator*). The stator field rotates at the synchronous speed given by Eq. 46.1. Stator construction for an induction motor is the same as that of a synchronous motor. An emf is induced as the stator field moves past the rotor conductors. Since the rotor windings have reactance, the rotor field lags the induced emf.

Induction motors are commonly used for most industrial applications. They are more common than synchronous motors since they are more rugged, requires less maintenance, and are less expensive. Induction motors consume the majority of all electric power generated and come in sizes from a fraction of a horsepower to many thousand horsepower. They are particularly useful in applications that require smooth starting under a load, such as hoists, mixers, or conveyors. Synchronous motors are much better than induction motors in power-generation applications.

Induction generators have few significant commercial uses, although some applications, such as windmills that generate power via induction motors, are gaining popularity. Induction generators are similar in concept to induction motors and are not discussed further here.

For the rotor of an induction motor to have a magnetic field, there must always be a variation in the magnetic flux as a function of time. To have a variation in magnetic flux, the rotor must turn slower than the synchronous speed so that the magnetic fields of the stator and rotor are never in the same direction. The difference in speed is small, but essential. *Percent slip*, s, typically 2% to 5%, is the percentage difference in speed between the rotor and stator field. Slip can be expressed as either a decimal or a fraction (e.g., 0.05 slip or 5% slip). Slip in rpm is the difference between actual and synchronous speeds.

$$s = \frac{n_s - n}{n_s} = \frac{\Omega_s - \Omega}{\Omega_s} \qquad 46.2$$

$$\text{percent slip} = s(100\%) \qquad 46.3$$

Induction motor slip is normally 2% to 5% except for motors designed with a variable rotor resistance that can be increased for speed control.

DC MACHINES

DC machines have a constant magnetic field in the stator (called the *field*). The magnetic field of the rotor (called the *armature*) responds to the stator field. The armature will respond by inducing current if the machine is a generator or by applying torque if the machine is a motor.

In most DC machines, the magnetic field is provided by electromagnets. (In a few small DC machines, it is supplied by permanent magnets.) The stator in a DC machine is composed of windings that produce the machine's field. Field poles are located on the stator and project inward. The poles alternate north and south around the machine and are separated by $360°/p$ around the machine. Each pole has a narrow ferromagnetic (e.g., iron) core around which the winding is wrapped. Each coil may consist of two or more separate windings. The magnetic flux produced by the field is linearly related to the field current, I_f, by a constant, K_f, that depends on the construction of the machine, as long as the current is low enough that the magnetization can be ignored.

$$\phi = K_f I_f \qquad 46.4$$

A simplified explanation of magnetization is that the current in the coils creates a magnetic flux in the core that causes currents to circulate in the tiny magnetic domains in the core and reinforces the core's magnetic flux. As the current in the coil increases, the currents in the magnetic domains increase approximately linearly until the currents in the magnetic domains cannot respond fully, and part of the energy in the flux is lost as heat. At this stage, the magnetic material is said to be *saturating* and the response is no longer linear. As

the current in the coil continues to increase, it reaches a point where the currents in the magnetic domains are going as fast as they can, and any additional energy in the flux from the coil is dissipated as heat. At this stage, the magnetic material is said to be *completely saturated*. The armature circuit and the field circuit are actually loops of wire around magnetic materials, but they are each modeled as a resistor in series with an inductor. The variables R_a and L_a are used for the armature resistance and inductance. The variables R_f and L_f are used for the field resistance and inductance. The inductance is usually ignored in DC machines, except for transient conditions.

DC Generators

A *DC generator* is a device that produces DC potential. The actual voltage induced is sinusoidal (i.e., AC). However, brushes on *split-ring commutators* make the connection to the rotating armature and rectify AC potential. The DC generator is, in fact, not DC. Rather, it is rectified AC, because for energy to be created by magnetism, something must be changing. The armature in a simple DC generator consists of a single coil with several turns (loops) of wire. The two ends of the coil terminate at a *commutator*. The commutator consists of a single ring split into two halves known as *segments*. (This arrangement is shown in Fig. 46.2. The field windings and rotor core are omitted for clarity.) The brushes slide on the commutator and make contact with the adjacent segment every half-rotation of the coil. As shown in Fig. 46.2, the coil is nearly perpendicular to the magnetic flux density, $\mathbf{B}$. The magnetic flux through the coil will be at its maximum when the coil is perpendicular to the magnetic flux density. The magnetic flux through the coil will be zero when the coil is parallel to the magnetic flux density. Therefore, the rate of change in the magnetic flux depends on the speed of rotation. The coil rotates and the magnetic flux density, $\mathbf{B}$, maintains its orientation.

Figure 46.2 Commutator Action

As shown in Fig 46.2, the current flows from the positive brush to the negative brush. As the coil continues to rotate to be parallel to the magnetic flux density, the commutator rotates such that the brushes are at the gaps between the commutator segments. As the coil rotates further, the positive brush will touch the commutator segment that the negative brush is touching in the figure, and vice versa. The current will be in the same direction because the magnetic flux density is in the same relative direction as the brushes. Figure 46.3 shows the full-wave rectified signal that results from this simple arrangement, which is obviously not suitable as a DC generator (unless a filtering circuit was used).

Figure 46.4 shows a simplified DC armature with two coils at 90° to each other. (The field windings and rotor core are omitted for clarity). Each of the coils will produce a full-wave rectified emf, as in Fig. 46.3. Since the coils are connected in series (in the modern closed-coil winding arrangement), the emf induced is the sum of the emfs induced in the individual coils, as shown in Fig. 46.5. The voltage induced in each coil of a DC generator with multiple coils is still sinusoidal, but the terminal output is nearly constant, not a (rectified) sinusoid. The slight variations in the voltage are known as *ripple*. The more coils there are, the smoother the DC voltage. The output may be filtered or passed through a DC voltage regulator to reduce or eliminate the ripple.

As shown in Fig. 46.6, DC generators can be modeled as an emf, E_g, which represents the part of the voltage induced in the armature circuit by the magnetic flux as it rotates in the field, and a resistor, R_a, which represents the part of the voltage that is dissipated as heat in the armature windings.

As shown in Fig. 46.7, the model for DC generators can be simplified by ignoring the resistance of the armature windings.

Since the rate of change of the magnetic flux in the armature depends on the speed of rotation, and because the armature is an inductor, the current depends on the magnetic flux. The voltage of an inductor is directly dependent upon the rate of change of the current; therefore, the voltage is directly dependent upon the rotational speed. Also, the greater the magnetic flux, the greater the current induced; therefore, the voltage is directly dependent upon the magnetic flux. The proportionality constant that relates the speed and magnetic flux to the generated voltage is K_a and depends upon the construction of the machine. For the generator model in Fig. 46.6, this results in

$$E_g = K_a n\phi \qquad 46.5$$

The terminal voltage of the generator model in Fig. 46.6 is

$$V_a = E_g + I_a R_a = K_a n\phi + I_a R_a \qquad 46.6$$

The terminal voltage of the generator model in Fig. 46.7 is

$$V_a = K_a n\phi \qquad 46.7$$

Figure 46.3 Rectified DC Voltage Induced in a Single Coil

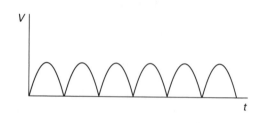

Figure 46.4 Two-Coil, Four-Segment Closed-Coil Armature

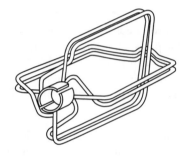

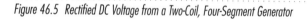

Figure 46.5 Rectified DC Voltage from a Two-Coil, Four-Segment Generator

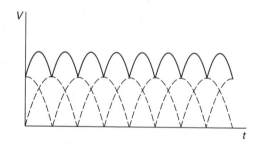

Figure 46.6 DC Machine Equivalent Circuit

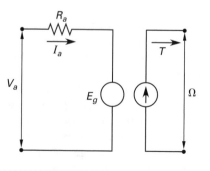

Figure 46.7 Simplified DC Machine Equivalent Circuit

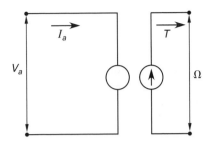

DC Motors

A *DC motor* is very similar to a DC generator, only the direction of the current changes. The voltage applied to the terminals of the motor, V_a, with a current, I_a, results in the supply of electrical power. Some of that power is dissipated as heat in the armature winding's resistance, and some is converted into mechanical power to turn the armature. The power for the motor model shown in Fig. 46.6 is determined from

$$P_e = P_h + P_m = I_a^2 R_a + I_a E_g \qquad 46.8$$

Ignoring the power dissipated as heat for the motor model shown in Fig. 46.7, the power is determined from

$$P_m = V_a I_a \qquad 46.9$$

The coil in Fig. 46.2 is nearly perpendicular to the magnetic field. If the load were replaced with a voltage source that caused the current to flow in the opposite direction from I in Fig. 46.2, the magnetic field of the armature would be pointing in almost the opposite direction from B. The armature would rotate in the opposite direction from Ω in Fig. 46.2, to align with B. If the load were replaced with a voltage source that caused the current to flow in the same direction as I in Fig. 46.2, the magnetic field of the armature would be pointing in almost the same direction as B. The armature would rotate in the same direction as Ω in Fig. 46.2, to align with B. The direction of rotation depends on the polarity of the voltage. As the armature continues to rotate in the case where current is in the same direction as I to the point where B is perpendicular to the loop of wire, the armature continues rotation due to its momentum, and the brushes make contact with the opposite commutator segments. This reverses the direction of the current and the direction of the magnetic field of the armature so it moves in the opposite direction from B and continues rotating in the same direction, to align with B.

The torque of the DC motor is linearly related to the strength of the field and the magnetic field of the armature. The stronger each of these fields is, the stronger

the force would need to be to bring them into line (which never happens, due to the commutator). The magnetic field of the armature is linearly related to the current of the armature (ignoring magnetization). The proportionality constant that relates the armature current and magnetic flux to the torque voltage is the same K_a as used in Eq. 46.7 and depends upon the construction of the machine.

$$T_m = \frac{60}{2\pi} K_a \phi I_a \qquad 46.10$$

The mechanical power can also be expressed in terms of torque, T, and the mechanical rotational velocity, Ω.

$$P_m = T\Omega \qquad 46.11$$

SAMPLE PROBLEMS

Problem 1

A two-pole AC motor operates on a three-phase, 60 Hz, 240 V_{rms} line-to-line supply. What is its synchronous speed?

    (A) 1000 rpm
    (B) 1800 rpm
    (C) 2400 rpm
    (D) 3600 rpm

Solution

$$n_s = \frac{120f}{p}$$
$$= \frac{\left(120 \;\frac{\text{rpm}}{\text{Hz}}\right)(60 \text{ Hz})}{2}$$
$$= 3600 \text{ rpm}$$

Answer is D.

Problem 2

A DC generator is operating at 1200 rpm with a magnetic flux of 0.02 Wb. The constant of the machine is $K_a = 1$ V/rpm-Wb. What is most nearly the output voltage?

    (A) 12 V
    (B) 20 V
    (C) 24 V
    (D) 48 V

Solution

$$E_g = K_a n \phi$$
$$= \left(1 \;\frac{\text{V}}{\text{rpm·Wb}}\right)(1200 \text{ rpm})(0.02 \text{ Wb})$$
$$= 24 \text{ V}$$

Answer is C.

Problem 3

A DC motor armature is drawing on 100 A while generating a magnetic flux of 0.02 Wb. The constant of the machine is $K_a = 1$ V/rpm·Wb. What is most nearly the output torque?

(A) 19 N·m
(B) 24 N·m
(C) 32 N·m
(D) 51 N·m

Solution

$$T_m = \frac{60}{2\pi} K_a \phi I_a$$
$$= \left(\frac{60}{2\pi}\right)\left(1\ \frac{\text{V}}{\text{rpm·Wb}}\right)(0.02\ \text{Wb})(100\ \text{A})$$
$$= 19\ \text{N·m}$$

Answer is A.

FE-STYLE EXAM PROBLEMS

1. An AC synchronous motor is operating at 1200 rpm on a supply of 60 Hz. The supply frequency is changed to 50 Hz. What is most nearly the operating mechanical rotational velocity when the motor reaches steady-state?

(A) 18 rad/s
(B) 65 rad/s
(C) 100 rad/s
(D) 380 rad/s

2. A synchronous motor is operating under a mechanical load with a unity power factor. The mechanical load is gradually increased. The load remains within the capability of the motor. The motor speed will

(A) decrease and the motor will draw more power
(B) decrease and the motor will draw less power
(C) stay the same and the motor will draw more power
(D) stay the same and the motor will draw less power

3. A three-phase synchronous generator is operating at 1200 rpm with an output frequency of 60 Hz. What is the number of poles in the generator?

(A) 2
(B) 4
(C) 6
(D) 8

4. A synchronous motor and an induction motor are operating on the same three-phase supply voltage. Which statement is true about the rotating speeds of the motors?

(A) The synchronous motor will rotate slower than the induction motor.
(B) The synchronous motor will rotate faster than the induction motor.
(C) The synchronous motor will rotate at the same speed as the induction motor.
(D) The difference in speeds depends on the mechanical loads on the two machines.

5. A two-pole induction motor operates on a three-phase, 60 Hz, 240 V_{rms} line-to-line supply. The motor speed is 3420 rpm. What is most nearly the slip?

(A) 5%
(B) 7%
(C) 11%
(D) 15%

6. A four-pole induction motor operates on a three-phase, 60 Hz, 240 V_{rms} line-to-line supply. The slip is 2%. What is most nearly the operating speed?

(A) 1240 rpm
(B) 1660 rpm
(C) 1760 rpm
(D) 1800 rpm

7. A DC generator is producing 24 V while operating at 1200 rpm with a magnetic flux of 0.02 Wb. The same generator is operated at 1000 rpm with a magnetic flux of 0.05 Wb. What is most nearly the new voltage the generator produces? Disregard armature resistance.

(A) 20 V
(B) 50 V
(C) 60 V
(D) 100 V

8. A DC generator is producing 200 V at 1800 rpm. The speed is decreased to 1000 rpm. What is most nearly the new operating voltage? Ignore armature resistance.

(A) 100 V
(B) 110 V
(C) 200 V
(D) 360 V

9. The polarity of the voltage to a DC motor is reversed. Which statement is true?

    (A) The rotor will not turn, because the commutator rectifies voltage.
    (B) The rotor will reverse direction.
    (C) The rotor will rotate in the same direction.
    (D) The motor will burn up.

10. A DC generator produces 100 V under an operating condition where the field current is 2 A. The field current is increased to 3 A. What is most nearly the new operating voltage? Ignore armature resistance.

    (A) 150 V
    (B) 200 V
    (C) 250 V
    (D) 300 V

11. A DC generator produces a current of 10 A into a resistive load when operating at 2000 rpm. The resistance of the load is halved while the speed is reduced to 1600 rpm. What is most nearly the new current produced when the generator reaches steady-state conditions? Ignore armature resistance.

    (A) 12 A
    (B) 16 A
    (C) 25 A
    (D) 50 A

12. The mechanical load to a DC motor is increased for a brief time from 100% of rated capacity to 102% of rated capacity. When the motor is at 102% of rated capacity it will

    (A) stall
    (B) be destroyed
    (C) operate at rated torque
    (D) operate at a higher torque than rated torque

13. Which of the following could serve as a polyphase inductor?

    (A) an induction motor
    (B) heating coils
    (C) a synchronous capacitor
    (D) a plating tank

SOLUTIONS TO FE-STYLE EXAM PROBLEMS

1. The synchronous speed is linearly related to the supply frequency.

$$\frac{n_{s2}}{n_{s1}} = \frac{f_2}{f_1}$$

$$n_{s2} = \frac{f_2}{f_1} n_{s1}$$

$$= \left(\frac{50 \text{ Hz}}{60 \text{ Hz}}\right)(1200 \text{ rpm})$$

$$= 1000 \text{ rpm}$$

The units of mechanical rotational velocity are rad/s, so the angular velocity is

$$\Omega_2 = \frac{n_{s2}\left(2\pi \dfrac{\text{rad}}{\text{rev}}\right)}{60 \dfrac{\text{sec}}{\text{min}}}$$

$$= \frac{\left(1000 \dfrac{\text{rev}}{\text{min}}\right)\left(2\pi \dfrac{\text{rad}}{\text{rev}}\right)}{60 \dfrac{\text{sec}}{\text{min}}}$$

$$= 104.7 \text{ rad/s} \quad (100 \text{ rad/s})$$

Answer is C.

2. Synchronous motors rotate at the same synchronous speed as long as the mechanical load is within the operating capability of the motor. Under the new operating condition, the motor must produce more torque to maintain this speed. So the motor will draw more power.

Answer is C.

3. Solve the synchronous speed equation for the number of poles.

$$p = \frac{120f}{n_s}$$

$$= \frac{\left(120 \dfrac{\text{rpm}}{\text{Hz}}\right)(60 \text{ Hz})}{1200 \text{ rpm}}$$

$$= 6$$

Answer is C.

4. A synchronous motor will always rotate at the synchronous speed. The induction motor speed will always have a slip relative to the synchronous speed, to maintain the magnetic field in its rotor, so it must rotate at a speed that is slightly slower than the synchronous speed. Therefore, the synchronous motor will rotate faster than the induction motor.

Answer is B.

5.
$$n_s = \frac{120f}{p}$$
$$= \frac{\left(120 \, \frac{\text{rpm}}{\text{Hz}}\right)(60 \text{ Hz})}{2}$$
$$= 3600 \text{ rpm}$$

$$\% \text{ slip} = \frac{n_s - n}{n_s} \times 100\%$$
$$= \left(\frac{3600 \text{ rpm} - 3420 \text{ rpm}}{3600 \text{ rpm}}\right) \times 100\%$$
$$= 5\%$$

Answer is A.

6.
$$n_s = \frac{120f}{p}$$
$$= \frac{\left(120 \, \frac{\text{rpm}}{\text{Hz}}\right)(60 \text{ Hz})}{4}$$
$$= 1800 \text{ rpm}$$

Solving the slip equation for the operating speed yields

$$n = n_s (1 - s)$$
$$= (1800 \text{ rpm})(1 - 0.02)$$
$$= 1764 \text{ rpm} \quad (1760 \text{ rpm})$$

Answer is C.

7. Solving the DC generator voltage equation for the constant of the generator yields the generator constant for the original operating conditions.

$$K_a = \frac{V_a}{n\phi}$$
$$= \frac{24 \text{ V}}{(1200 \text{ rpm})(0.02 \text{ Wb})}$$
$$= 1 \, \frac{\text{V}}{\text{rpm·Wb}}$$

Under the new operating conditions, the voltage produced is

$$V_{a,\text{new}} = K_a n_{\text{new}} \phi_{\text{new}}$$
$$= \left(1 \, \frac{\text{V}}{\text{rpm·Wb}}\right)(1000 \text{ rpm})(0.05 \text{ Wb})$$
$$= 50 \text{ V}$$

Answer is B.

8. The output voltage is approximately proportional to the speed (ignoring armature resistance).

$$\frac{V_2}{V_1} = \frac{n_2}{n_1}$$
$$V_2 = \left(\frac{n_2}{n_1}\right) V_1$$
$$= \left(\frac{1000 \text{ rpm}}{1800 \text{ rpm}}\right)(200 \text{ V})$$
$$= 111 \text{ V} \quad (110 \text{ V})$$

Answer is B.

9. The current through the armature windings will be reversed when the voltage polarity is reversed. The direction of the magnetic flux is reversed, and the torque on the rotor is reversed. Therefore, the direction of rotation is reversed.

Answer is B.

10. The output voltage is approximately proportional to the magnetic flux (ignoring armature resistance). The magnetic flux is proportional to the field current. Therefore, the output voltage is approximately proportional to the field current.

$$\frac{V_2}{V_1} = \frac{I_{f2}}{I_{f1}}$$
$$V_2 = \left(\frac{I_{f2}}{I_{f1}}\right) V_1$$
$$= \left(\frac{3 \text{ A}}{2 \text{ A}}\right)(100 \text{ A})$$
$$= 150 \text{ V}$$

Answer is A.

11. The voltage is approximately proportional to the speed (ignoring armature resistance).

$$\frac{V_2}{V_1} = \frac{n_2}{n_1}$$
$$= \frac{1600 \text{ rpm}}{2000 \text{ rpm}}$$
$$= 0.80$$

The current is proportional to the voltage by Ohm's law. Assuming no change in resistance,

$$I_2 = \left(\frac{V_2}{V_1}\right) I_1$$
$$= (0.80)(10 \text{ A})$$
$$= 8.0 \text{ A}$$

However, the resistance was halved, and the current is inversely proportional to the resistance.

$$\frac{I_3}{I_2} = \frac{R_2}{R_3}$$

$$I_3 = \left(\frac{R_2}{R_3}\right) I_2$$

$$= \left(\frac{2}{1}\right) (8.0 \text{ A})$$

$$= 16 \text{ A}$$

Answer is B.

12. The rated value for mechanical load can be exceeded by a small percentage for a brief time without stalling or damage to the motor. The rated value is established to keep the operation of the motor linear and to avoid excessive heating. During the time the motor is loaded to 102% of the rated capacity, it will produce approximately 102% of the rated torque.

Answer is D.

13. Heating coils are essentially purely resistive and do not have a large inductive effect. A synchronous capacitor is a synchronous motor that runs unloaded and draws a leading current, often used for power factor correction. A plating tank is not an inductive load. The stator of a three-phase induction motor consists of three sets of primary windings connected in either a wye or delta configuration. The induction motor is essentially a three-phase inductor.

Answer is A.

Topic XV: Computers, Measurement, and Controls

Diagnostic Examination for Computers, Measurement, and Controls

Chapter 47   Computer Hardware
Chapter 48   Computer Software
Chapter 49   Measurement
Chapter 50   Controls

COMPUTERS

Diagnostic Examination

TOPIC XV: COMPUTERS, MEASUREMENT, AND CONTROLS

TIME LIMIT: 45 MINUTES

---

1. Which of the following flowcharts does *not* represent a complete program?

(A)

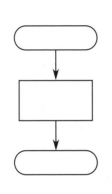

(B)

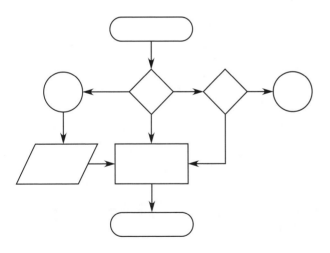

(C)

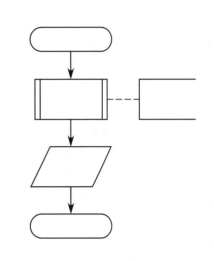

(D)

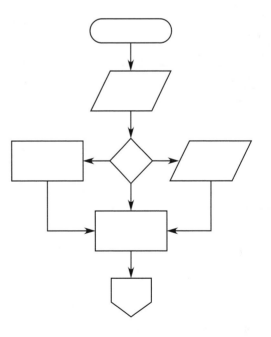

2. The flowchart shown represents the summer training schedule for a college athlete. What is the regularly scheduled workout on Wednesday morning?

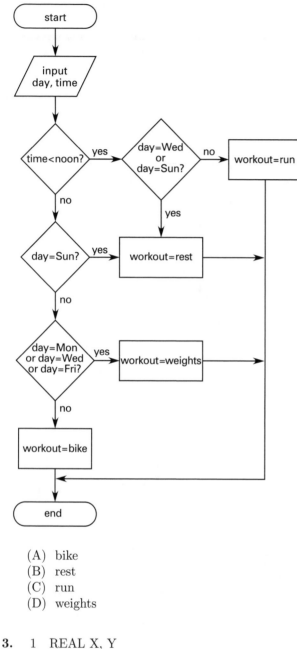

(A) bike
(B) rest
(C) run
(D) weights

3.
```
1   REAL X, Y
2   X = 3
3   Y = COS(X)
4   PRINT Y
```

In the structured programming fragment shown, line 2 is most accurately described as a(n)

(A) assignment
(B) command
(C) declaration
(D) function

4. What sum is calculated by the following pseudocode program segment?

```
INPUT N, X
S = 1
T = 1
FOR K = 1 TO N
T = (−1)*T*X^2/(2*K*(2*K − 1))
S = S+T
NEXT K
```

(A) $S = 1 - \dfrac{x^2}{2} - \dfrac{x^2}{12} - \dfrac{x^2}{30} - \cdots - \dfrac{x^2}{2n(2n-1)}$

(B) $S = 1 - \dfrac{x^2}{2} - \dfrac{x^4}{12} - \dfrac{x^6}{30} - \cdots - \dfrac{x^{2n}}{2n(2n-1)}$

(C) $S = 1 - \dfrac{x^2}{2!} + \dfrac{x^4}{4!} - \dfrac{x^6}{6!} + \cdots + (-1)^n \left(\dfrac{x^{2n}}{(2n)!} \right)$

(D) $S = 1 - \dfrac{x^3}{2!} + \dfrac{x^4}{4!} - \dfrac{x^5}{6!} + \cdots + (-1)^n \left(\dfrac{x^{n+2}}{(2n)!} \right)$

5. A program contains the following structured programming segment.

```
INPUT X
Y = 0
Z = X + 1
FOR K = 1 TO Z
Y = Y + Z
NEXT K
Y = Y − (2*Z − 1)
```

What is the final value of Y in terms of X?

(A) $X^2 - 1$
(B) X^2
(C) $X^2 + 1$
(D) $X^2 + 2X + 1$

6. In a typical spreadsheet program, what cell is directly below cell AB4?

(A) AB5
(B) AC4
(C) AC5
(D) BC4

7. The cells in a spreadsheet are initialized as shown. The formula B1 + $A$1*A2 is entered into cell B2 and then copied into B3 and B4. What value is displayed in B4?

	A	B		
1	3	111		
2	4			
3	5			
4	6			
5				

(A) 123
(B) 147
(C) 156
(D) 173

8. An operating system is being developed for use in the control unit of an industrial machine. For efficiency, it is decided that any command to the machine will be represented by a combination of four characters, and that only eight distinct characters will be supported. What is the minimum number of bits required to represent one command?

(A) 5
(B) 7
(C) 12
(D) 16

9. Data are being collected automatically from an experiment at a rate of 14.4 kbps. How long will it take to completely fill a diskette whose capacity is 1.44 MB?

(A) 1.67 min
(B) 10.0 min
(C) 13.3 min
(D) 14.0 min

10. Before being used, programs must generally be "loaded into memory." What storage component does this "memory" refer to?

(A) cache
(B) EPROM
(C) RAM
(D) ROM

11. A computer signal that stops the execution of the current instruction is called which of the following?

(A) keystroke
(B) dispatch table
(C) interrupt
(D) switch

12. The computer hardware acronym "ALU" means which of the following?

(A) Arithmetic and Logic Unit
(B) Arithmetic Logic User
(C) Analog and Logic Unit
(D) Address Last Unit

13. Which of the following best defines a micro-operation?

(A) the complex operations that are performed by the CPU during one micro-second
(B) an elementary operation performed on information stored in one or more registers during one clock cycle
(C) an operation that takes the smallest amount of time to execute
(D) an elementary operation performed on information stored in one or more registers during one revolution of the disk drive

14. The American Standard Code for Information Interchange (ASCII) has a standard set of English characters (letters, numbers, and punctuation) for data transfer. There are several larger extended sets of ASCII that are dependent on the operating system. How many bits represent the standard and extended sets, respectively?

(A) 7, 7
(B) 7, 8
(C) 8, 12
(D) 8, 16

15. A 32MB RAM uses 32-bit words. How many bus lines are required to pass data for bidirectional communication between RAM and the CPU? Do not count clock, sync, or other protocol lines.

(A) 8
(B) 16
(C) 32
(D) 64

SOLUTIONS TO DIAGNOSTIC EXAMINATION TOPIC XV

1. A flowchart must begin and end with a terminal symbol. The symbol at the bottom of choice (D) is the "off-page" symbol, which indicates that the flowchart continues on the next page. This is not a complete program.

Answer is D.

2. The day and time being considered are Wednesday morning. At the first decision follow the "yes" branch, which leads to the question "Is the day Wednesday or Sunday?" Again the answer is "yes," so the scheduled workout is actually a resting day.

Answer is B.

3. Line 2 is an assignment. A command directs the computer to take some action, such as PRINT. A declaration states what type of data a variable will contain (like REAL) and reserves space for it in memory. A function performs some specific operation (like finding the COSine of a number) and returns a value to the program.

Answer is A.

4. The values of S and T in each loop depend on their values from the preceding loop. Make a table listing their values through the first few loops, until the pattern becomes obvious.

loop	T	S
0	1	1
1	$\dfrac{-x^2}{2}$	$1 - \dfrac{x^2}{2}$
2	$\dfrac{x^4}{24}$	$1 - \dfrac{x^2}{2} + \dfrac{x^4}{24}$
3	$\dfrac{-x^6}{720}$	$1 - \dfrac{x^2}{2} + \dfrac{x^4}{24} - \dfrac{x^6}{720}$
⋮	⋮	⋮
n	$(-1)^n \left(\dfrac{x^{2n}}{(2n)!} \right)$	$1 - \dfrac{x^2}{2} + \dfrac{x^4}{24} - \dfrac{x^6}{720}$
		$+ \cdots + (-1)^n \left(\dfrac{x^{2n}}{(2n)!} \right)$

Answer is C.

5. Y is initialized to zero and then has Z added to it Z times.

$$Y = 0 + (Z)(Z)$$
$$= Z^2$$

After the FOR ... NEXT loop, $(2Z - 1)$ is taken away from the previous sum.

$$Y = Z^2 - (2Z - 1)$$
$$Z = X + 1$$

$$Y = (X^2 + 2X + 1) - \big(2(X + 1) - 1\big)$$
$$= X^2$$

Answer is B.

6. Spreadsheets generally label a cell by giving its column and row, in that order. Cell AB4 is in column AB, row 4. The cell directly below AB4 is in column AB, row 5, designated as AB5.

Answer is A.

7. When the formula is copied into cells B3 and B4, the relative references will be updated. The resulting spreadsheet (with formulas displayed) should look like this:

	A	B
1	3	111
2	4	B1 + $A$1*A2
3	5	B2 + $A$1*A3
4	6	B3 + $A$1*A4
5		

$$B4 = B3 + (A1)(A4)$$
$$= B2 + (A1)(A3) + (A1)(A4)$$
$$= B1 + (A1)(A2) + (A1)(A3) + (A1)(A4)$$
$$= 111 + (3)(4) + (3)(5) + (3)(6)$$
$$= 156$$

Answer is C.

8. Since $2^3 = 8$, three bits are sufficient to represent 8 characters. Since there are 4 characters per command, a string a minimum of $4 \times 3 = 12$ bits is needed.

Answer is C.

9. A megabyte is 2^{20} bytes; kbps is the abbreviation for 10^3 bits per second.

$$t = \frac{(1.44 \text{ MB}) \left(2^{20} \, \dfrac{\text{bytes}}{\text{MB}} \right) \left(8 \, \dfrac{\text{bits}}{\text{byte}} \right)}{\left(14.4 \times 10^3 \, \dfrac{\text{bits}}{\text{s}} \right) \left(60 \, \dfrac{\text{s}}{\text{min}} \right)}$$

$$= 13.98 \text{ min} \quad (14.0 \text{ min})$$

Answer is D.

10. RAM is typically used to store the operational portion of executing programs. Occasionally, when a program is too large to fit entirely in RAM, sections of it will be swapped back and forth with the hard drive. This generally slows program execution and is avoided whenever possible. The hard drive's intended purpose is long-term storage of data. ROM is permanent storage, used for instructions vital to the computer's operation. Cache memory is used to store small amounts of data that the processor needs to access frequently, and is not generally large enough to store entire programs.

Answer is C.

11. An interrupt is a signal that informs a program that an event has occurred. When a program receives an interrupt signal, it takes a specified action, which transfers control to another memory location, subroutine, or program. Depending on the nature of the interrupt and the current execution, the next execution may be the same as before the interrupt, but the interrupt still stops the current instruction.

A dispatch table is a table of interrupt vectors (pointers to routines that tell the program what to do for each interrupt). A keystroke is an input function and may or may not cause an interrupt.

Depending on the context, there are several definitions of "switch" in computer applications, none of which means that a switch stops the execution of the current instruction.

Answer is C.

12. "ALU" stands for arithmetic and logic unit.

Answer is A.

13. A micro-operation is an action that is taken during one clock cycle.

Answer is B.

14. The standard set of ASCII characters is encoded by 7 bits for 128 combinations representing the English characters. In the extended set, 8 bits encode 256 characters. The first 128 combinations represent the standard set of English characters. The second 128 combinations represent foreign letters, graphic symbols, or whatever characters are needed for the particular system.

Answer is B.

15. All of the bits in a word are passed in parallel to the CPU. There is one data line per data bit. Bidirectional communication between RAM and the CPU does not occur simultaneously.

Answer is C.

COMPUTERS

47 Computer Hardware

Subjects

COMPUTER ARCHITECTURE 47-1
MICROPROCESSORS 47-1
CONTROL OF COMPUTER
   OPERATION 47-2
COMPUTER MEMORY 47-2
PARITY 47-3
INPUT/OUTPUT DEVICES 47-3
RANDOM SECONDARY STORAGE
   DEVICES 47-4
SEQUENTIAL SECONDARY STORAGE
   DEVICES 47-5
REAL-TIME AND BATCH
   PROCESSING 47-5
MULTITASKING AND
   TIME-SHARING 47-5
BACKGROUND AND FOREGROUND
   PROCESSING 47-6
TELEPROCESSING 47-6
DISTRIBUTED SYSTEMS AND
   LOCAL-AREA NETWORKS 47-6

COMPUTER ARCHITECTURE

All digital computers, from giant supercomputers to the smallest microcomputers, contain three main components—a central processing unit (CPU), main memory, and external (peripheral) devices. Figure 47.1 illustrates a typical integration of these components.

Figure 47.1 Simplified Computer Architecture

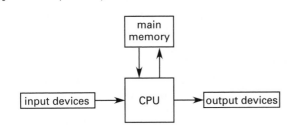

MICROPROCESSORS

A *microprocessor* is a central processing unit (CPU) on a single chip. With *large-scale integration* (LSI), most microprocessors are contained on one chip, although other chips in the set can be used for memory and input/output control. The most popular microprocessor families for personal computers are produced by Intel (P6, Pentium, and 80X86 families) and Motorola (680X0 family), although work-alike clones of these chips may be produced by other companies.

Microprocessor CPUs consist of an arithmetic and logic unit, several accumulators, one or more registers, stacks, and a control unit. The *control unit* fetches and decodes the incoming instructions and generates the signals necessary for the arithmetic and logic unit to perform the intended function. The *arithmetic and logic unit* (ALU) executes commands and manipulates data.

Accumulators hold data and instructions for further manipulation in the ALU. Registers are used for temporary storage of instructions or data. The *program counter* (PC) is a special register that always points to (contains) the address of the next instruction to be executed. Another special register is the *instruction register* (IR), which holds the current instruction during its execution. *Stacks* provide temporary data storage in sequential order—usually on a last-in, first-out (LIFO) basis. Because their operation is analogous to spring-loaded tray holders in cafeterias, the name *pushdown stack* is also used.

Microprocessors communicate with support chips and peripherals through connections in a *bus* or *channel*, which is logically subdivided into three different functions. (The term *bus* refers to the physical path—that is, wires or circuit board traces, along which the signal travels. The term *channel* refers to the logical path.) The *address bus* directs memory and input/output device transfers. The *data bus* carries the actual data and is the busiest bus. The *control bus* communicates control and status information. The number of lines in the address bus determines the amount of random access memory (RAM) that can be directly addressed. When there are n address lines in the bus, 2^n words of memory can be addressed.

Microprocessors can be designed to operate on 4-bit, 8-bit, 16-bit, 32-bit, and 64-bit words, although microprocessors with 4- and 8-bit words are now used primarily in process control applications. Using appropriate control sequencing circuitry, microprocessors can be combined, and the resulting larger unit is known as a *bit-slice microprocessor*, where each processor handles a *bit field* (i.e., *slice*) of the operand. For example, two 64-bit microprocessors might be combined into a 128-bit-slice microprocessor. This method of obtaining greater computing power was popular prior to the introduction of 16-bit and larger microprocessors.

All microprocessors use a crystal-controlled *clock* to control instruction and data movements. The *clock rate* is specified in microprocessor cycles per second (e.g., 2 GHz). Ideally, the clock rate is the number of instructions the microprocessor can execute per second. However, one or more cycles may be required for complex instructions (*macrocommands*). For example, executing a complex instruction may require one or more cycles each to fetch the instruction from memory, decode the instruction to see what to do, execute the instruction, and store (write) the result. (In some cases, other chips, for example, *memory management units*, can perform some of these tasks.) Since operations on floating point numbers are macrocommands, the speed of a microprocessor can also be specified in *flops*, the number of floating point operations it can perform per second. Similar units of processing speed are *mips* (millions of instructions per second).

Most microprocessors are rich in complex executable commands (i.e., the *command set*) and are known as *complex instruction-set computer* (CISC) microprocessors. In order to increase the operating speed, however, *reduced instruction-set computer* (RISC) microprocessors are limited to performing simple, standardized format instructions but are otherwise fully featured.

Some microprocessors can emulate the operation of other microprocessors. For example, an 80486 chip can operate in virtual 8086 mode. *Emulation mode* is also referred to as *virtual mode*.

CONTROL OF COMPUTER OPERATION

The user interface and basic operation of a computer are controlled by the *operating system* (OS), also known as the *monitor program*. The operating system is a program that controls the computer at its most basic level and provides the environment for application programs. The operating system manages the memory, schedules processing operations, accesses peripheral devices, communicates with the user/operator, and (in multitasking environments) resolves conflicting requirements for resources. Since one of the functions of the operating system is to coordinate use of the peripheral devices (disk drives, keyboards, etc.). The term *basic input/output system* (BIOS) is also used.

All or part of the operating system can be stored in read-only memory (ROM). In early computers, a small part of executable code that was used to initiate data transfers and logical operations when the computer was first started was known as a *bootstrap loader*. Although modern start-up operations are more sophisticated, the phrase "booting the computer" is still used today.

During program operation, peripherals and other parts of the computer signal the operating system through interrupts. An *interrupt* is a signal that stops the execution of the current instruction (or, in some cases, the current program) and transfers control to another memory location, subroutine, or program. Other interrupts signal error conditions, such as division by zero, overflow and underflow, and syntax errors. The operating system intercepts, decodes, and acts on these interrupts.

COMPUTER MEMORY

Computer memory consists of many equally sized storage locations, each of which has an associated address. The contents of a storage location may change, but the address does not.

The total number of storage locations in a computer can be measured in various ways. A *bit* (binary digit) is the smallest changeable data unit. Bits can only have values of 1 or 0. Bits are combined into *nibbles* (4 bits), *bytes* (8 bits, the smallest number of bits that can represent one alphanumeric character), *half-words* (8 and 16 bits), *words* (8, 16, and 32 bits) and *doublewords* (16, 32, and 64 bits). The distinction between doublewords, words, and half-words depends on the computer. Sixteen bits would be a word in a 16-bit computer but would be a half-word in a 32-bit computer. Furthermore, *double-precision (double-length) words* double the number of bytes normally used. The abbreviations kB (*kilobytes*) and kW (*kilowords*) used by some manufacturers do not help much to clarify the ambiguity.

The number of memory storage locations is always a multiple of two. The abbreviations k, M, and G are used to designate the quantities $(2)^{10}$ (1024), $(2)^{20}$ (1,048,576), and $(2)^{30}$ (1,073,741,824), respectively. For example, a 6 MB memory would contain $6 \times (2)^{20}$ bytes. (k and M do not mean one thousand and one million exactly.)

Most of the memory locations are used for user programs and data. However, portions of the memory may be used for video memory, I/O cache memory, the BIOS, and other purposes. *Video memory* (known as VRAM) contains the text displayed on the screen of a terminal. Since the screen is refreshed many times per second, the screen information must be repeatedly read from video memory. *Cache memory* holds the most recently read and frequently read data in memory, making subsequent retrieval of that data much faster than reading from a tape or disk drive, or even from main memory. (A high-speed supercomputer may require 200 to 500 nanoseconds to access main memory but only 20 to 50 nanoseconds to access cache memory.) *OS memory* contains the BIOS that is read in when the computer is first started. *Scratchpad memory* is high-speed memory used to store a small amount of data temporarily so that the data can be retrieved quickly.

Modern memory hardware is semiconductor based. (The term *core*, derived from the ferrite cores used in early computers, is seldom used today.) Memory is designated as RAM (random access memory), ROM (read-only memory), PROM (programmable read-only memory), and EPROM (erasable programmable read-only memory). While data in RAM is easily changed, data in ROM cannot be altered. PROMs are initially blank but once filled, they cannot be changed. EPROMs are initially blank but can be filled, erased, and refilled repeatedly. (Most EPROMs can be erased by exposing them to ultraviolet light.) EEPROM (electrically erasable PROM) is a class of PROM that can be erased with an electrical charge. EEPROM is similar to flash memory (also called flash EEPROM). Flash memory allows data to be written or erased in blocks, while EEPROM requires data to be written or erased one byte at a time. Therefore, flash memory can be written and erased faster. The term *firmware* is used to describe programs stored in ROMs and EPROMs.

The contents of a *volatile memory* are lost when the power is turned off. RAM is usually volatile, while ROM, PROM, and EPROM are *nonvolatile*. With *static memory*, data does not need to be refreshed and remains as long as the power stays on. With *dynamic memory*, data must be continually refreshed. Static and dynamic RAM (i.e., DRAM) are both volatile.

Synchronous DRAM (SDRAM) is a type of DRAM that can run at much higher clock speeds than conventional memory. SDRAM is capable of running at 133 MHz, about three times faster than conventional FPM (fast page mode) RAM, and about twice as fast as EDO (extended data out) DRAM and BEDO (burst extended data out) DRAM. SDRAM is replacing EDO DRAM in many newer computers.

Rambus DRAM (RDRAM) is a type of DRAM developed by Rambus, Inc. RDRAM transfers data at up to 1600 MHz. RDRAM has been used in place of VRAM in some graphics accelerator boards. RDRAM has been superceded by *double data rate* (DDR) SDRAM which is capable of speeds of 200 MHz, 400 MHz, and 800 MHz in DDR, DDR2, and DDR3 configurations, respectively.

Virtual memory (storage) (VS) is a technique by which programs and data larger than main memory can be accessed by the computer. (Virtual memory is not synonymous with *virtual machine*, described in "Multitasking and Time-Sharing.") In virtual memory systems, some of the disk space is used as an extension of the semiconductor memory. A large application or program is divided into modules of equal size called *pages*. Each page is switched into (and out of) RAM from (and back to) disk storage as needed, a process known as *paging*. This interchange is largely transparent to the user. Of course, access to data stores on a disk drive is much slower than semiconductor memory access. *Thrashing* is a deadlock situation that occurs when a program references a different page for almost every instruction, and there is not even enough real memory to hold most of the virtual memory.

Most memory locations are filled and managed by the CPU. However, *direct memory access* (DMA) is a powerful I/O (input/output) technique that allows peripherals (e.g., tape and disk drives) to transfer data directly into and out of memory without affecting the CPU. Although special DMA hardware is required, DMA does not require explicit program instructions, making data transfer faster.

PARITY

Parity is a technique used to ensure that the bits within a memory byte are correct. For every eight data bits, there is a ninth bit—the parity bit—that serves as a *check bit*. The nine bits together constitute a *frame*. In *odd-parity recording*, the parity bit will be set so there is an odd number of one-bits in the frame. In *even-parity recording*, the parity bit will be set so that there is an even number of one-bits in the frame. When the data are read, the nine bits are checked to ensure valid data (although this does not detect whether two of the bits in the frame are incorrect).

INPUT/OUTPUT DEVICES

Devices that feed data to, or receive data from, the computer are known as *input/output* (I/O) *devices*. Terminals, light pens, digitizers, printers and plotters, and tape and disk drives are common peripherals. (CRT, the abbreviation for *cathode ray tube*, is often used to mean a terminal.) Point-of-sale (POS) devices, bar code readers, and magnetic ink character recognition (MICR) and optical character recognition (OCR) readers are less common devices.

Peripherals are connected to their computer through multiline cables. With a *parallel interface* (used in a

**COMPUTERS
Hardware**

parallel device), there are as many separate lines in the cable as there are bits (typically seven, eight, or nine) in the code representing a character. An additional line is used as the *strobe signal* to carry a timing signal. With a *serial interface* (used in a *serial device*), all bits pass one at a time along a single line in the cable. The *transmission speed (baud rate)* in bps is the number of bits that pass through the data line per second. The name *baud rate* is derived from the use of the *Baudot code*. One *baud* is one modulation per second. If there is a one-to-one correspondence between modulations and bits, one baud unit is the same as one bit per second (bps). In general, the unit bps should always be used.

The USB (universal serial bus) is an external bus standard that supports data transfer rates of 12 Mbps. A single USB port can be used to connect up to 127 peripheral devices. USB also supports plug-and-play installation (the ability to plug in a device and operate it, without setting configuration elements) and hot plugging (the ability to add and remove devices while the computer is operating and the operating system automatically recognizes the change).

Peripherals such as terminals and printers typically do not have large memories. They only need memories large enough to store the information before the data are displayed or printed. The small memories are known as *buffers*. The peripheral can send the status (i.e., full, empty, off-line, etc.) of its buffer to the computer in several different ways. This is known as *flow control* or *handshaking*.

If the computer and peripheral are configured so that each can send and receive data, the peripheral can send a single character (e.g., the XOFF *character* for transmission off) to the computer when its buffer is full. Simililarly, a different character (e.g., the XON *character* for transmission on) can be sent when the peripheral is ready for more data. The computer must monitor the incoming data line for these characters. This is known as *software flow control* or *software handshaking*.

If there are enough separate lines between the computer and the peripheral, one or more of them can be used for *hardware handshaking*. (Serial terminals and printers usually require only two or three lines—data in, data out, and ground. Most computer cables contain more lines than this, and one of the extra lines can be used for DTR, *data terminal ready*, or CTS, *clear to send*, handshaking.) In this method of flow control, the peripheral keeps the voltage on one of the lines high (or low) when it is able to accept more data. The computer monitors the voltage on this line.

Most peripheral devices are connected to the computer by a dedicated channel (cable). However, a pair of multiplexers *(statistical multiplexers or concentrators)* can be used to carry data for several peripherals along a single cable known as the *composite link*. There are two methods of achieving multiplexed transmission: *frequency division multiplexing* (FDM) and *time division multiplexing* (TDM). With FDM, the available transmission band is divided into narrower bands, each used for a separate channel. In TDM, the connecting channel is operated at a much higher clock rate (proportional to the capacity of the multiplexer), and each peripheral shares equally in the available cycles. Figure 47.2 illustrates multiplexed peripheral connections.

The Electronics Industries Association (EIA) RS-232 standard was developed in an attempt to standardize the connectors and pin uses in serial device cables.

Figure 47.2 Multiplexed Peripherals

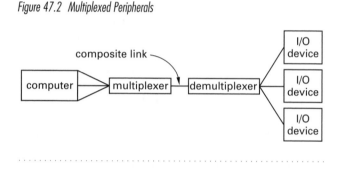

RANDOM SECONDARY STORAGE DEVICES

Random access (direct access) storage devices, also known as *mass storage devices*, include magnetic and optical disk drive units. They are random access because individual records can be accessed without having to read through the entire file.

Magnetic disk drives (*hard drives*) are composed of several *platters*, each with one or more read/write heads. The platers turn at high speeds (e.g., thousands of rpm). Data on a surface are organized into tracks, sectors, and cylinders. *Tracks* are the concentric storage areas. *Sectors* are pie-shaped subdivisions of each track. A *cylinder* consists of the same numbered track on all drive platters. Some platters and *disk packs* are removable, but most hard drives are fixed (i.e., nonremovable).

Depending on the media, *optical disk drives* can be *read only* (R/O) or *read/write* (R/W) in nature. WORM drives (write once—read many, or write once—read mostly) can be written by the user, while others such as CD-ROM (compact disk read-only memory) can only be read.

In addition to *storage capacity* (usually specified in megabytes—MB) there are several parameters that describe the performance of a disk drive. *Areal density* is

Figure 47.3 Tracks, Sectors, and Cylinders

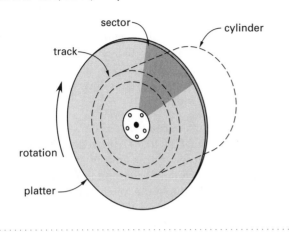

a measure for the number of data bits stored per square inch of disk surface. It is calculated by multiplying the number of bits per track by the number of tracks per (radial) inch. The *average seek time* is the average time it takes to move a head from one location to a new location. The *track-to-track seek time* is the time required to move a head from one track to an adjacent track. *Latency* or *rotational delay* is the time it takes for a disk to spin a particular sector under the head for reading. On the average, latency is one-half of the time to spin a full revolution. The *average access time* is the time to move to a new sector and read the data. Access time is the sum of average latency and average seek time.

Floppy disks (diskettes) are suitable for low-capacity random-access storage. Although their capacities are comparatively low (typically less than one or two million characters for magnetic media), they are used to transfer programs and data between computers (primarily microcomputers). The capacity of a diskette depends on its size, the recording density, number of tracks, and number of sides.

SEQUENTIAL SECONDARY STORAGE DEVICES

Tape units are *sequential access devices* because a computer cannot access information stored at the end of a tape without first reading or passing by the information stored at the front of the tape. Some tapes use *indexed sequential format* in which a directory of files on the tape is placed at the start of the tape. The tape can be rapidly wound to (near) the start of the target file without having to read everything in between.

There are a number of tape formats in use. One of the few standardized commercial formats is the *nine-track* format. (Other formats that have received some acceptance include *quarter inch cartridge*, QIC, and *digital audio tape*, DAT.) The tape is divided into nine tracks running the length of the tape. The width of the tape is

divided into frames (characters). Eight tracks are used to record the data in either ASCII or EBCDIC format. The remaining track is used to record the parity bit. 1600 bpi (bits or frames per inch) is still a common standard *recording density* for sharing data, although densities of 9600 bpi and above are in use.

Frames are grouped into fixed-length *blocks* separated by *interblock gaps* (IBG). Reflective spots for photo-electric detection are used to indicate the beginnings and ends of magnetic tapes. These spots are called *load point* and *end-of-file* (EOF) *markers*, respectively.

Magnetic tape is often used to back up hard disks. A *streaming tape* operates in a continuously running—or streaming—mode, with data being written or read while the tape is running.

REAL-TIME AND BATCH PROCESSING

Programs run on a computer in one of two main ways: batch mode and real-time mode. In some data processing environments, programs are held (either by the operator or by the operating system) and eventually grouped into efficient categories requiring the same peripherals and resources. This is called *batch mode processing* since all jobs of a particular type are batched together for subsequent processing. There is usually no interaction between the user and the computer once batch processing begins. In *real-time (interactive) processing*, a program runs when it is submitted, often with user interaction during processing.

MULTITASKING AND TIME-SHARING

If a computer's main memory is large enough and the CPU is fast enough, it is possible to allocate the main memory among several users running applications simultaneously. This is the concept of a *virtual machine* (VM)—each user appears to have his or her own computer. This is also known as *multitasking* and *multiprogramming* since multiple tasks can be performed simultaneously. A *multi-user system* is similar to a multitasking system in that several users can use the computer simultaneously, although that term also means that all users are using the same program.

Time-sharing (swapping) is a technique where each user takes turns (under the control of the operating system) using the entire computer main memory for a certain length of time (usually less than a second). At the end of that time period, all of the active memory is written to a private area, and the memory for the next user is loaded. The swapping occurs so frequently that all users are able to accomplish useful work on a real-time basis.

BACKGROUND AND FOREGROUND PROCESSING

A program running in real time is an example of *foreground processing*. There are times, however, when it is convenient to start a long program running while the same computer is used for a second program. The first program continues to run unseen in the background. *Background processing* can be accomplished by segmenting the main computer memory (i.e., establishing virtual machines or by time-sharing (i.e., allowing the background application to have all cycles not used by the foreground application).

TELEPROCESSING

Teleprocessing is the access of a computer from a remote station, usually over a telephone line (although fiber optic, coaxial, and microwave links can also be used). Since these media transmit analog signals, a *modem (modulator-demodulator)* is used to convert to/from the digital signals required by a computer.

With *simplex communication*, transmission is only in one direction. With *half-duplex communication*, data can be transmitted in both directions, but only in one direction at a time. With *duplex* or *full duplex communication*, data can be transmitted in both directions simultaneously. 28,800 to 33,600 bps is the maximum practical transmission speed over voice-grade lines without data compression. Speeds up to 56,600 bps can be achieved with data compression. Much higher rates, however, are possible over dedicated data lines and wide-band lines.

In *asynchronous* or *start-stop transmission*, each character is preceded and followed by special signals (i.e., *start* and *stop bits*). Thus, every 8-bit character is actually transmitted as 10 bits, and the character transmission rate is one-tenth of the transmission speed in bits per second. (For historical reasons, a second stop bit is used when data are sent at ten characters per second. This is referred to as 110 bps.) With asynchronous transmission, it is possible to distinguish the beginning and end of each character from the bit stream itself.

Synchronous equipment transmits a block of data continuously without pause and requires a built-in clock to maintain synchronization. Synchronous transmission is preceded and interwoven with special clock-synchronizing characters, and the separation of a bit stream into individual characters is done by counting bits from the start of the previous character. Since start and stop bits are not used, synchronous communication is approximately 20 percent faster than asynchronous communication.

There are three classes of communication lines—narrow-band, voice-grade, and wide-band—depending on the bandwidth (i.e., range of frequencies) available for signaling. *Narrow-band* may only support a single channel of communication, as the bandwidth is too narrow for modulation. *Wide-band channels* support the highest transfer rates, since the bandwidth can be divided into individual channels. *Voice-grade lines*, supporting frequencies between 300 and 3000 to 3300 Hz, are midrange in bandwidth.

Errors in transmission can easily occur over voice grade lines at the rate of 1 in 10,000. In general, methods of ensuring the accuracy of transmitted and received data are known as *communications protocols* and *transmission standards*. A simple way of checking the transmission is to have the receiver send each block of data back to the sender. This process is known as *loop checking* or *echo checking*. If the characters in a block do not match, they are re-sent. While accurate, this method requires sending each block of data twice.

Another method of checking the accuracy of transmitted data is for both the receiver and sender to calculate a *check digit* or *block check character* derived from each block of *characters* sent. (A common transmission block size is 128 characters.) With *cycling redundancy checking* (CRC), the block check character is the remainder after dividing all the serialized bits in a transmission block by a predetermined binary number. Then, the block check character is sent and compared after each block of data.

DISTRIBUTED SYSTEMS AND LOCAL-AREA NETWORKS

Distributed data processing systems assume many configurations. In the traditional situation, a centrally located main computer interacts with, and is fed by, smaller computers in other locations. In a second configuration, many identical computers are linked together in order to share storage and printing resources. This latter case is known as a *local-area network* (LAN). Local-area networks typically communicate at speeds between 200 kbps and 50 Mbps.

SAMPLE PROBLEMS

Problem 1

Which of the following best defines a buffer?

- (A) a region where extra information goes once the main memory is full
- (B) a temporary storage region used to compensate for signal time differences
- (C) the same thing as main memory
- (D) a permanent memory region where start-up information is stored

Solution

A buffer is a temporary storage region that holds data until it is used.

Answer is B.

Problem 2

Which of the following best defines a bit?

(A) a basic unit of a computer used to encode a single character of text

(B) the smallest portion of computer memory that can represent a distinct computer address

(C) a binary digit that represents one of two possible values or states

(D) computer memory that can represent one of two possible values or states

Solution

A binary digit (i.e., a bit) can be in one of two possible states.

Answer is C.

Problem 3

Which of the following best defines a nibble?

(A) half a byte of memory

(B) a read-only operation (without any change in memory)

(C) a CPU operation that is performed in half a clock cycle

(D) a CPU operation that is performed in one clock cycle

Solution

A nibble is a block of four contiguous bits (i.e., half a byte).

Answer is A.

FE-STYLE EXAM PROBLEMS

1. Which of the following best defines a byte?

(A) eight bits typically used to encode a single character of text

(B) the smallest number of bits that can be used in arithmetic operations

(C) the basic unit of information accessed by a computer

(D) a binary digit that represents one of two equally accessible values or states

2. Which of the following best defines a word?

(A) the equivalent of four bytes; the basic unit of data transfer

(B) the largest number of bytes that can be used in arithmetic operations

(C) eight contiguous bits in computer memory

(D) the smallest number of bytes that can be used in arithmetic operations

3. Which data transfer path(s) is (are) used the most in typical computer operations?

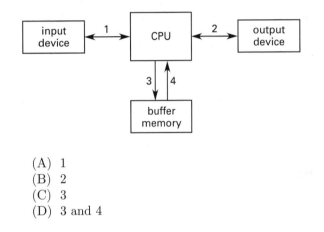

(A) 1

(B) 2

(C) 3

(D) 3 and 4

4. A 256K-word memory uses 16-bit words. How many parallel data lines are required to pass data to the CPU for processing? Do not count clock, sync, or other protocol lines.

(A) 2

(B) 8

(C) 9

(D) 16

5. Which of the following best describes a direct memory access process?

(A) the address of the address of the operand is specified in the instruction

(B) the address of the operand is specified in the instruction

(C) the operand itself is specified as part of the instruction

(D) the operand address is the same as the register and base addresses

For the following problems use the NCEES Handbook as your only reference.

6. How does a CPU know whether it is executing instructions from a commercial database management program or from a program executing from an on-line programmer?

(A) The micro-operations used are different for the two programs.

(B) One of the programs uses memory references while the other uses register references.

(C) One of the programs uses compiled code while the other uses interpreted code.

(D) The CPU does not know where instructions originate.

7. A simple controller board has two thousand 8-bit memory locations and two 8-bit registers. How many different states can this board be in?

(A) 2002

(B) $(2)^8$

(C) $(2)^{2002}$

(D) $(2)^{16,016}$

8. How long will it ideally take to transmit a 400k (byte) text file using a 28.8k modem in synchronous simplex mode?

(A) 1 min

(B) 2 min

(C) 11 min

(D) 14 min

9. How many 8-bit bytes are in 2 MB of memory?

(A) 16,000

(B) 16,256

(C) 2,000,000

(D) 2,097,152

10. Which of the following types of memory is lost when the power is removed?

(A) RAM

(B) ROM

(C) PROM

(D) EPROM

11. A hard disk drive with three 5 cm (diameter) platters turns at 3000 rpm. What is the average latency for the drive?

(A) 0.003 s

(B) 0.01 s

(C) 0.02 s

(D) 0.04 s

12. Which of the following terms is *not* a synonym for the others?

(A) bps

(B) baud

(C) bits per second

(D) data rate

13. A 12.7 cm diameter disk drive platter spins at 5600 rpm. The average access time is 11 ms. What is the average rotational delay?

(A) 5.17 ms

(B) 5.36 ms

(C) 5.50 ms

(D) 5.64 ms

14. A manufacturer advertises a 5 GB (gigabyte) drive. How many bits are contained in 5 GB of hard disk storage?

(A) 4.29×10^9

(B) 5.00×10^9

(C) 40.0×10^9

(D) 42.9×10^9

15. Flash memory is classified as which type of memory?

(A) ROM

(B) WORM

(C) RAM

(D) EPROM

16. The acronym RDRAM refers to

(A) Rambus Dynamic Random Access Memory

(B) Rambus Direct Random Access Memory

(C) Redundant Dual Random Access Memory

(D) Random Duplex Random Access Memory

17. What information normally would *not* be transmitted on a microprocessor's control bus?

(A) contents of the instruction cache

(B) contents of data registers

(C) I/O commands

(D) keyboard entries

SOLUTIONS TO FE-STYLE EXAM PROBLEMS

1. A byte is a group of eight bits. A byte can store a single character or command.

Answer is A.

2. Computer words are the smallest memory units that can be manipulated in arithmetic operations. The number of bytes in a word depends on the computer architecture.

Answer is D.

3. Input and output devices (e.g., keyboards and printers) are used far less frequently than memory-transfer operations. Data enters and is read (i.e., or is replaced) in buffer memory at approximately the same rates.

Answer is D.

4. All of the bits in a word are passed in parallel to the CPU. There is one data line per data bit.

Answer is D.

5. In direct memory access, the instruction must specify the address of the desired memory for recall (the operand).

Answer is B.

6. A CPU executes micro-operations regardless of their origins.

Answer is D.

7. The number of bits in the memory and registers is

$$(8)(2000 + 2) = 16{,}016$$

Each bit can take on two different values. The number of different states is

$$(2)^{16{,}016}$$

Answer is D.

8. Each kB contains $(2)^{10} = 1024$ bytes. The number of bits to be transmitted is

$$(400 \text{ kbytes}) \left(1024 \; \frac{\text{bytes}}{\text{kbyte}} \right) \left(8 \; \frac{\text{bits}}{\text{byte}} \right) = 3{,}276{,}800 \text{ bits}$$

The modem transmits at 28,800 bits/sec. (Notice that "k" here refers to exactly 1000, in contrast to 1024 in kB. In synchronous transmission there are start and stop bits to add to the transmission time.) In simplex mode, transmission is in one direction only. The time required is

$$t = \frac{3{,}276{,}800 \text{ bits}}{\left(28{,}800 \; \frac{\text{bits}}{\text{s}} \right) \left(60 \; \frac{\text{s}}{\text{min}} \right)}$$

$$= 1.90 \text{ min} \quad (2 \text{ min})$$

Answer is B.

9. MB stands for "megabyte," roughly a million bytes. The actual number of bytes in a megabyte is

$$(2)^{20} = 1{,}048{,}576$$
$$(2)(1{,}048{,}576) = 2{,}097{,}152$$

Answer is D.

10. ROM (read-only memory), PROM (programmable read-only memory), EPROM (erasable, programmable read-only memory), and WORM (write-once, read many) devices retain their information when the power is removed.

Answer is A.

11. Latency is the rotational delay (i.e., the time it takes for the information needed to appear under the read-write head.) On the average, latency is one-half of the time to turn a full revolution.

$$\text{latency} = \left(\frac{1}{2} \right) \left(\frac{60 \; \frac{\text{s}}{\text{min}}}{3000 \; \frac{\text{rev}}{\text{min}}} \right) = 0.01 \text{ s}$$

Answer is B.

12. bps, bits per second, and baud are all synonymous. The data rate depends on the encoding scheme. For example, with ASCII encoding, each character will require 8 or 9 (or more) bits to encode.

Answer is D.

13. Rotational delay is the same as latency, half of the period of rotation.

$$\text{latency} = \left(\frac{1}{2} \right) \left(60 \; \frac{\text{s}}{\text{min}} \right) \left(\frac{1000 \; \frac{\text{ms}}{\text{s}}}{5600 \; \frac{\text{rev}}{\text{min}}} \right)$$

$$= 5.357 \text{ ms} \quad (5.36 \text{ ms})$$

Answer is B.

14. A computer "gigabyte" contains $(2)^{30}$ bytes. $(2)^{30}$ is the first power of 2 that exceeds 10^9. Each byte contains 8 bits.

$$(5 \text{ GB}) \left((2)^{30} \frac{\text{bytes}}{\text{GB}} \right) \left(8 \frac{\text{bits}}{\text{bytes}} \right) = 42.9 \times 10^9 \text{ bits}$$

Answer is D.

15. Flash memory is a type of EEPROM (electrically erasable programmable read-only memory), which makes it classified as EPROM.

Answer is D.

16. RDRAM computer dynamic memory chips utilizing the Rambus signaling level (RSL) technology to achieve higher random access speeds are used in many microcomputers with clock speeds in excess of 400 MHz.

Answer is A.

17. In a microprocessor, the bus unit can be generalized into a control bus (for carrying instructions) and a data bus (for moving values taken from data registers and caches). A keyboard can be the original source of both data and instructions. However, contents of data registers would be transmitted via the data bus only.

Answer is B.

48 Computer Software

Subjects

CHARACTER CODING 48-1
PROGRAM DESIGN 48-1
FLOWCHARTING SYMBOLS 48-1
LOW-LEVEL LANGUAGES 48-2
HIGH-LEVEL LANGUAGES 48-2
RELATIVE COMPUTATIONAL SPEED . . 48-2
STRUCTURE, DATA TYPING, AND
  PORTABILITY 48-3
STRUCTURED PROGRAMMING 48-3
SPREADSHEETS 48-4
FIELDS, RECORDS, AND
  FILE TYPES 48-4
FILE INDEXING 48-5
SORTING 48-5
SEARCHING 48-5
HASHING 48-5
DATABASE STRUCTURES 48-6
HIERARCHICAL AND RELATIONAL
  DATA STRUCTURES 48-6
ARTIFICIAL INTELLIGENCE 48-7

CHARACTER CODING

Alphanumeric data refers to characters that can be displayed or printed, including numerals and symbols ($, %, &, etc.) but excluding *control characters* (tab, carriage return, form feed, etc.). Since computers can handle binary numbers only, all symbolic data must be represented by binary codes. *Coding* refers to the manner in which alphanumeric data and control characters are represented by sequences of bits.

The *American Standard Code for Information Interchange*, ASCII, is a seven-bit code permitting 128 (2^7) different combinations. It is commonly used in microcomputers, although use of the high order (eighth) bit is not standardized. ASCII-coded magnetic tape and disk files are used to transfer data and documents between computers of all sizes that would otherwise be unable to share data structures.

The *Extended Binary Coded Decimal Interchange Code*, EBCDIC (pronounced eb'-sih-dik), is in widespread use in supercomputers. It uses eight bits (a byte) for each character, allowing a maximum of 256 (2^8) different characters.

Since strings of binary digits (bits) are difficult to read, the *hexadecimal* (or "packed") format is used to simplify working with EBCDIC data. Each byte is converted into two strings of four bits each. The two strings are then converted to hexadecimal. Since $(1111)_2 = (15)_{10} = (F)_{16}$, the largest possible EBCDIC character is coded FF in hexadecimal.

Example 48.1

A number is represented as 11110111 in EBCDIC. What is this in hexadecimal format?

Solution

The first four bits are 1111, which is $(15)_{10}$ or $(F)_{16}$. The last four bits are 0111, which is $(7)_{10}$ or $(7)_{16}$. The hexadecimal representation is F7.

PROGRAM DESIGN

A *program* is a sequence of computer instructions that performs some function. The program is designed to implement an *algorithm*, which is a procedure consisting of a finite set of well-defined steps. Each step in the algorithm usually is implemented by one or more instructions (e.g., READ, GOTO, OPEN, etc.) entered by the programmer. These original "human-readable" instructions are known as *source code statements*.

Except in rare cases, a computer will not understand source code statements. Therefore, the source code is translated into machine-readable object code and absolute memory locations. Eventually, an executable program is produced.

If the executable program is kept on disk or tape, it is normally referred to as *software*. If the program is placed in ROM or EPROM, it is referred to as *firmware*. The computer mechanism itself is known as the *hardware*.

FLOWCHARTING SYMBOLS

A *flowchart* is a step-by-step drawing representing a specific procedure or algorithm. Figure 48.1 illustrates

COMPUTERS
Software

the most common flowcharting symbols. The terminal symbol begins and ends a flowchart. The input/output symbol defines an I/O operation, including those to and from keyboard, printer, memory, and permanent data storage. The processing symbol and predefined process symbol refer to calculations or data manipulation. The decision symbol indicates a point where a decision must be made or two items are compared. The connector symbol indicates that the flowchart continues elsewhere. The off-page symbol indicates that the flowchart continues on the following page. Comments can be added in an annotation symbol.

Figure 48.1 Flowcharting Symbols

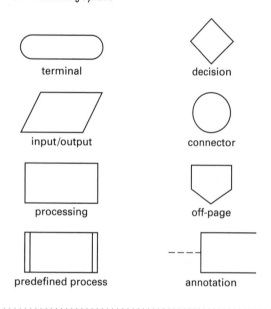

LOW-LEVEL LANGUAGES

Programs are written in specific languages, of which there are two general types: low-level and high-level. Low-level languages include machine language and assembly language.

Machine language instructions are intrinsically compatible with and understood by the computer's CPU. They are the CPU's native language. An instruction normally consists of two parts: the operation to be performed (*op-code*) and the operand expressed as a storage location. Each instruction ultimately must be expressed as a series of bits, a form known as *intrinsic machine code*. However, octal and hexadecimal coding are more convenient. In either case, coding a machine language program is tedious and seldom done by hand.

Assembly language is more sophisticated (i.e., is more symbolic) than machine language. Mnemonic codes are used to specify the operations. The operands are

Table 48.1 Comparison of Typical ADD Commands

language	instruction
intrinsic machine code	1111 0001
machine language	1A
assembly language	AR
FORTRAN	+

referred to by variable names rather than by the addresses. Blocks of code that are to be repeated verbatim at multiple locations in the program are known as *macros (macro instructions)*. Macros are written only once and are referred to by a symbolic name in the source code.

Assembly language code is translated into machine language by an *assembler* (*macro-assembler* if macros are supported). After assembly, portions of other programs or function libraries may be combined by a *linker*. In order to run, the program must be placed in the computer's memory by a *loader*. Assembly language programs are preferred for highly efficient programs. However, the coding inconvenience outweighs this advantage for most applications.

HIGH-LEVEL LANGUAGES

High-level languages are easier to use than low-level languages because the instructions resemble English. High-level statements are translated into machine language by either an interpreter or a compiler. Table 48.1 shows a comparison of a typical ADD command in different computer languages. A *compiler* performs the checking and conversion functions on all instructions only when the compiler is invoked. A true stand-alone executable program is created. An *interpreter*, however, checks the instructions and converts them line by line into machine code during execution but produces no stand-alone program capable of being used without the interpreter. (Some interpreters check syntax as each statement is entered by the programmer. Some languages and implementations of other languages blur the distinction between interpreters and compilers. Terms such as *pseudo-compiler* and *incremental compiler* are used in these cases.)

RELATIVE COMPUTATIONAL SPEED

Certain languages are more efficient (i.e., execute faster) than others. (Efficiency can also, but seldom does, refer to the size of the program.) While it is impossible to be specific, and exceptions abound, assembly language programs are fastest, followed in order of decreasing speed by compiled, pseudo-compiled, and interpreted programs.

Similarly, certain program structures are more efficient than others. For example, when performing a repetitive operation, the most efficient structure will be a single equation, followed in order of decreasing speed by a stand-alone loop and a loop within a subroutine. Incrementing the loop variables and managing the exit and entry points is known as *overhead* and takes time during execution.

STRUCTURE, DATA TYPING, AND PORTABILITY

A language is said to be *structured* if subroutines and other procedures each have one specific entry point and one specific return point. (Contrast this with BASIC, which permits (1) a GOSUB to a specific subroutine with a return from anywhere within the subroutine and (2) unlimited GOTO statements to anywhere in the main program.) A language has *strong data types* if integer and real numbers cannot be combined in arithmetic statements.

A *portable language* can be implemented on different machines. Most portable languages are either sufficiently rigidly defined (as in the cases of ADA and C) to eliminate variants and extensions, or (as in the case of Pascal) are compiled into an intermediate, machine-independent form. This so-called *pseudocode (p-code)* is neither source nor object code. The language is said to have been "ported to a new machine" when an interpreter is written that converts p-code to the appropriate machine code and supplies specific drivers for input, output, printers, and disk use. (Some companies have produced Pascal engines that run p-code directly.)

STRUCTURED PROGRAMMING

Structured programming (also known as *top-down programming*, *procedure-oriented programming*, and *GOTO-less programming*) divides a procedure or algorithm into parts known as subprograms, subroutines, modules, blocks, or procedures. (The format and readability of the source code—improved by indenting nested structures, for example—do not define structured programming.) Internal subprograms are written by the programmer; external subprograms are supplied in a library by another source. Ideally, the mainline program will consist entirely of a series of calls (references) to these subprograms. Liberal use is made of FOR/NEXT, DO/WHILE, and DO/UNTIL commands. Labels and GOTO commands are avoided as much as possible.

Very efficient programs can be constructed in languages that support *recursive calls* (i.e., permit a subprogram to call itself). Some languages permit recursion; others do not.

Variables whose values are accessible strictly within the subprogram are *local variables*. *Global variables* can be referred to by the main program and all other subprograms.

Calculations are performed in a specific order in an instruction, with the contents of parentheses done first. The symbols used for mathematical operations in programming are

$+$	add
$-$	subtract
$*$	multiply
$/$	divide

Raising one expression to the power of another expression depends on the language used. Examples of how X^B might be expressed are

$$X**B$$
$$X^{\hat{}}B$$

Following are brief descriptions of some commonly used structured programming functions.

IF THEN statements: In an IF <condition> THEN <action> statement, the condition must be satisfied, or the action is not executed and the program moves on to the next operation. Sometimes an IF THEN statement will include an ELSE statement in the format of IF <condition> THEN <action 1> ELSE <action 2>. If the condition is satisfied, then action 1 is executed. If the condition is not satisfied, action 2 is executed.

DO/WHILE loops: A set of instructions between the DO WHILE <condition> and the ENDWHILE lines of code is repeated as long as the condition remains true. The number of times the instructions are executed depends on when the condition is no longer true. The variable or variables that control the condition must eventually be changed by the operations, or the WHILE loop will continue forever.

DO/UNTIL loops: A set of instructions between the DO UNTIL <condition> and the ENDUNTIL lines of code is repeated as long as the condition remains false. The number of times the instructions are executed depends on when the condition is no longer false. The variable or variables that control the condition must eventually be changed by the operations, or the UNTIL loop will continue forever.

FOR loops: A set of instructions between the FOR <counter range> and the NEXT <counter> lines of code is repeated for a fixed number of loops that depends on the counter range. The counter is a variable that can be used in operations in the loop, but the value of the counter is not changed by anything in the loop besides the NEXT <counter> statement.

GOTO: A GOTO operation moves the program to a number designator elsewhere on the program. The GOTO statement has fallen from favor and is avoided whenever possible in structured programming.

SPREADSHEETS

Spreadsheet application programs (often referred to as *spreadsheets*) are computer programs that provide a table of values arranged in rows and columns and that permit each value to have a predefined relationship to the other values. If one value is changed, the program will also change other related values.

In a spreadsheet, the items are arranged in rows and columns. The rows are typically assigned with numbers (1, 2, 3, ...) along the vertical axis, and the columns are assigned with letters (A, B, C, ...), as shown in Fig. 48.2.

Figure 48.2 Typical Spreadsheet Cell Assignments

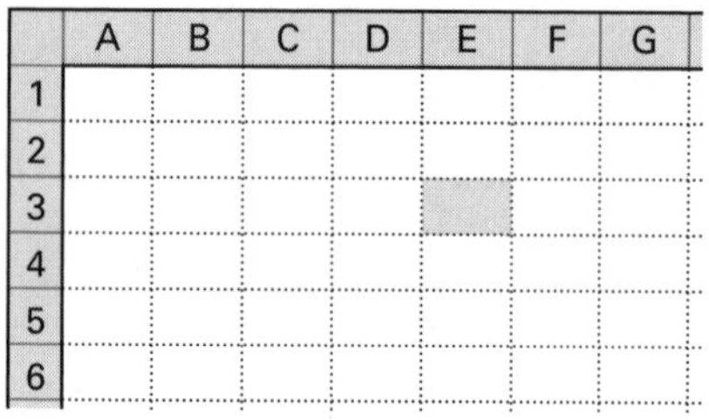

A *cell* is a particular element of the table identified by an address that is dependent on the row and column assignments of the cell. For example, the address of the shaded cell in Fig. 48.2 is E3. A cell may contain a number, a formula relating its value to another cell or cells, or a label (usually descriptive text).

When the contents of one cell are used for a calculation in another cell, the address of the cell being used must be referenced so the program knows what number to use.

An absolute cell reference identifies a particular cell and will have a "$" before both the row and column designators. For example, $A$1 identifies the cell in the first column and first row, $A$3 identifies the cell in the first column and third row, and $C$1 identifies the cell in the third column and first row, regardless of the cell

the reference is located in. If the absolute cell reference is copied and pasted into another cell, it continues referring to the exact same cell.

An absolute column, relative row cell reference has an absolute column reference (indicated with a "$") and a relative row reference. For example, the cell reference $A1 depends on what row it is entered in; if it is entered into a cell in the third row, it really refers to a cell that is in the first column in the current row. If this reference is copied and pasted into a cell in the third row, the reference will become $A3. Similarly, a reference of $A3 in the second row refers to a cell in the first column one row below the current row. If this reference is copied and pasted into the third row, it becomes $A4.

A relative column, absolute row cell reference has a "$" on the row designator, and the column reference depends on the column it is entered in. For example, a cell reference of B$4 in the fourth column (column D) refers to a cell two columns to the left in the fourth row. If this reference is copied and pasted into the sixth column (column F), it becomes D$4.

A cell reference that does not include a "$" is entirely dependent on the cell in which is it located. For example, a cell reference to B4 in the cell C2 refers to a cell that is one column to the left and two rows below. If this reference is copied and pasted into cell D3, it becomes A6.

The syntax for calculations with rows, columns, or blocks of cells can differ from one brand of spreadsheet to another.

Cells can be called out in square or rectangular blocks, usually for a SUM function. The difference between the row and column designations in the call will define the block. For example, SUM(A1:A3) says to sum the cells A1, A2, and A3; SUM(D3:D5) says to sum the cells D3, D4, and D5; and SUM(B2:C4) says to sum cells B2, B3, B4, C2, C3, and C4.

FIELDS, RECORDS, AND FILE TYPES

A collection of *fields* is known as a *record*. For example, name, age, and address might be fields in a personnel record. Groups of records are stored in a *file*.

A *sequential file* structure (typical of data on magnetic tape) contains consecutive records and must be read starting at the beginning. An *indexed sequential file* is one for which a separate index file (see "File Indexing") is maintained to help locate records.

With a *random (direct access) file structure*, any record can be accessed without starting at the beginning of the file.

FILE INDEXING

It is usually inefficient to place the records of an entire file in order. (A good example is a mailing list with thousands of names. It is more efficient to keep the names in the order of entry than to sort the list each time names are added or deleted.) Indexing is a means of specifying the order of the records without actually changing the order of those records.

An index (key or keyword) file is analogous to the index at the end of this book. It is an ordered list of items with references to the complete record. One field in the data record is selected as the key field (record index). More than one field can be indexed. However, each field will require its own index file. The sorted keys are usually kept in a file separate from the data file. One of the standard search techniques is used to find a specific key.

SORTING

Sorting routines place data in ascending or descending numerical or alphabetical order.

With the method of *successive minima*, a list is searched sequentially until the smallest element is found and brought to the top of the list. That element is then skipped, and the remaining elements are searched for the smallest element, which, when found, is placed after the previous minimum, and so on. A total of $n(n-1)/2$ comparisons will be required. When n is large, $n^2/2$ is sometimes given as the number of comparisons.

In a *bubble sort*, each element in the list is compared with the element immediately following it. If the first element is larger, the positions of the two elements are reversed (swapped). In effect, the smaller element "bubbles" to the top of the list. The comparisons continue to be made until the bottom of the list is reached. If no swaps are made in a pass, the list is sorted. A total of approximately $n^2/2$ comparisons are needed, on the average, to sort a list in this manner. This is the same as for the successive minima approach. However, swapping occurs more frequently in the bubble sort, slowing it down.

In an *insertion sort*, the elements are ordered by rewriting them in the proper sequence. After the proper position of an element is found, all elements below that position are bumped down one place in the sequence. The resulting vacancy is filled by the inserted element. At worst, approximately $n^2/2$ comparisons will be required. On the average, there will be approximately $n^2/4$ comparisons.

Disregarding the number of swaps, the number of comparisons required by the successive minima, bubble, and insertion sorts is on the order of n^2. When n is large, these methods are too slow. The *quicksort* is more complex but reduces the average number of comparisons (with random data) to approximately $n \times \log(n)/\log(2)$, generally considered as being on the order of $n \times \log(n)$. (However, the quicksort falters, in speed, when the elements are in near-perfect order.) The maximum number of comparisons for a *heap sort* is $n \times \log(n)/\log(2)$, but it is likely that even fewer comparisons will be needed.

SEARCHING

If a group of records (i.e., a list) is randomly organized, a particular element in the list can be found only by a linear search (sequential search). At best, only one comparison and, at worst, n comparisons will be required to find something (an event known as a *hit*) in a list of n elements. The average is $n/2$ comparisons, described as being on the order of n. (Note that the term *probing* is synonomous with *searching*.)

If the records are in ascending or descending order, a binary search will be superior. (A binary search is unrelated to a binary tree. A binary tree structure, see "Database Structures," greatly reduces search time but does not use a sorted list.) The search begins by looking at the middle element in the list. If the middle element is the sought-for element, the search is over. If not, half the list can be disregarded in further searching since elements in that portion will be either too large or too small. The middle element in the remaining part of the list is investigated, and the procedure continues until a hit occurs or the list is exhausted. The number of required comparisons in a list of n elements will be $\log(n)/\log(2)$ (i.e., on the order of $\log(n)$).

HASHING

An index file is not needed if the record number (i.e., the storage location for a read or write operation) can be calculated directly from the key, a technique known as *hashing*. The procedure by which a numeric or nonnumeric key (e.g., a last name) is converted into a record number is called the *hashing function* or *hashing algorithm*. Most hashing algorithms use a remaindering modulus—the remainder after dividing the key by the number of records, n, in the list. Excellent results are obtained if n is a prime number; poor results occur if n is a power of 2. (Finding a record in this manner requires it to have been written in a location determined by the same hashing routine.)

Not all hashed record numbers will be correct. A *collision* occurs when an attempt is made to use a record number that is already in use. Chaining, linear probing, and double hashing are techniques used to resolve such collisions.

DATABASE STRUCTURES

Databases can be implemented as indexed files, linked lists, and tree structures; in all three cases, the records are written and remain in the order of entry.

An indexed file such as that shown in Fig. 48.3 keeps the data in one file and maintains separate index files (usually in sorted order) for each key field. The index file must be recreated each time records are added to the field. A *flat file* has only one key field by which records can be located. Searching techniques (see "Searching") are used to locate a particular record. In a *linked list* (*threaded list*), each record has an associated *pointer* (usually a record number or memory address) to the next record in key sequence. Only two pointers are changed when a record is added or deleted. Generally, a linear search following the links through the records is used. Figure 48.4(a) shows an example of a linked list structure.

Figure 48.3 Key and Data Files

key file

key	ref.
ADAMS	3
JONES	2
SMITH	1
THOMAS	4

data file

record	last name	first name	age
1	SMITH	JOHN	27
2	JONES	WANDA	39
3	ADAMS	HENRY	58
4	THOMAS	SUSAN	18

Pointers are also used in *tree structures*. Each record has one or more pointers to other records that meet certain criteria. In a binary tree structure, each record has two pointers—usually to records that are lower and higher, respectively, in key sequence. In general, records in a tree structure are referred to as *nodes*. The first

record in a file is called the *root node*. A particular node will have one node above it (the *parent* or *ancestor*) and one or more nodes below it (the *daughters* or *offspring*). Records are found in a tree by starting at the root node and moving sequentially according to the tree structure. The number of comparisons required to find a particular element is $1 + (\log(n)/\log(2))$, which is on the order of $\log(n)$. Figure 48.4(b) shows an example of a binary tree structure.

Figure 48.4 Database Structures

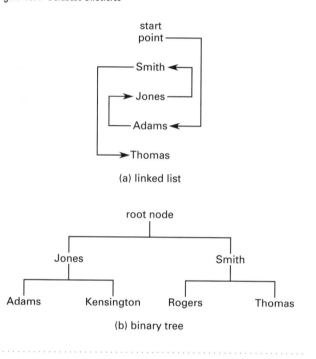

(a) linked list

(b) binary tree

HIERARCHICAL AND RELATIONAL DATA STRUCTURES

A *hierarchical database* contains records in an organized, structured format. Records are organized according to one or more of the indexing schemes. However, each field within a record is not normally accessible. Figure 48.5 shows an example of a hierarchical structure.

Figure 48.5 A Hierarchical Personnel File

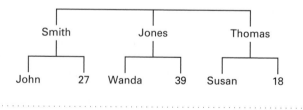

A *relational database* stores all information in the equivalent of a matrix. Nothing else (no index files, pointers, etc.) is needed to find, read, edit, or select information. Any information can be accessed directly by referring to the field name or field value. Figure 48.6 shows an example of relational structure.

Figure 48.6 A Relational Personnel File

rec. no.	last	first	age	
1	Smith	John	27	
2	Jones	Wanda	39	
3	Thomas	Susan	18	

ARTIFICIAL INTELLIGENCE

Artificial intelligence (AI) in a machine implies that the machine is capable of absorbing and organizing new data, learning new concepts, reasoning logically, and responding to inquiries. AI is implemented in a category of programs known as *expert systems* that "learn" rules from sets of events that are entered whenever they occur. (The manner in which the entry is made depends on the particular system.) Once the rules are learned, an expert system can participate in a dialogue to give advice, make predictions and diagnoses, or draw conclusions.

SAMPLE PROBLEMS

Problem 1

In a spreadsheet, the formula \$A\$3 + \$B3 + D2 is entered into cell C2. The contents of cell C2 are copied and pasted into cell D5. The formula in cell D5 is

(A) \$A\$3 + C\$2 + C4
(B) \$B\$6 + \$C4 + C4
(C) \$A\$3 + \$B6 + E5
(D) \$A\$3 + \$B2 + B2

Solution

The first absolute cell reference is unchanged by the paste operation and remains \$A\$3.

The second cell reference will have the row reference increased by three and become \$B6.

The third cell reference will have the column reference increased by one and the row reference increased by three and will become E5.

Answer is C.

Problem 2

A computer structured programming segment contains the following program segment. What is the value of X after the segment is executed?

```
X = 0
DO FOR T = -2 TO 1
    X = X + T
NEXT T
```

(A) −3
(B) −2
(C) 0
(D) 4

Solution

The FOR statement of the loop is repeated four times, when T = −2, T = −1, T = 0, and T = 1.

So, in order, the operations are

$$\begin{aligned} X &= X + T = 0 - 2 = -2 \\ X &= X + T = -2 + -1 = -3 \\ X &= X + T = -3 + 0 = -3 \\ X &= X + T = -3 + 1 = -2 \end{aligned}$$

Answer is B.

Problem 3

A computer structured programming segment contains the following program segment. What is the value of G after the segment is executed?

```
Set G = 1 and X = 0
DO WHILE G ≤ 5
    G = G * X + 1
    X = G
ENDWHILE
```

(A) 5
(B) 26
(C) 63
(D) The loop never ends.

Solution

The first execution of the WHILE loop results in

$$\begin{aligned} G &= (1)(0) + 1 = 1 \\ X &= 1 \end{aligned}$$

The second execution of the WHILE loop results in

$$\begin{aligned} G &= (1)(1) + 1 = 2 \\ X &= 2 \end{aligned}$$

COMPUTERS Software

The third execution of the WHILE loop results in

$$G = (2)(2) + 1 = 5$$
$$X = 5$$

The WHILE condition is still satisfied, so the instruction is executed a fourth time.

$$G = (5)(5) + 1 = 26$$
$$X = 26$$

Answer is B.

Problem 4

What operation is typically represented by the following program flowchart symbol?

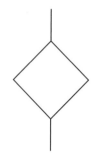

(A) input-output
(B) processing
(C) storage
(D) branching

Solution

Branching, comparison, and decision operations are typically represented by the diamond symbol.

Answer is D.

Problem 5

Structured programming is to be used to determine whether examinees pass a test. A passing score is 70 or more out of a possible 100. Which of the following IF statements would set the variable PASSED to 1 (true) when the variable SCORE is passing and set the variable PASSED to 0 (false) when the variable SCORE is not passing?

(A) IF SCORE > 70 PASSED = 1 ELSE PASSED = 0
(B) IF SCORE > 69 PASSED = 1 ELSE PASSED = 0
(C) IF SCORE < 69 PASSED = 1
(D) IF SCORE < 69 PASSED = 0 ELSE PASSED = 1

Solution

Answer A will not give the correct response when SCORE = 70. Answer C will not set PASSED to 0. Answer D will not give the correct response when SCORE = 69. Answer B will set PASSED to 1 for SCORE = 70 to 100 and will set PASSED to 0 for SCORE = 0 to 69.

Answer is B.

FE-STYLE EXAM PROBLEMS

1. Which of the following terms is best defined as a formula or set of steps for solving a particular problem?

(A) program
(B) software
(C) firmware
(D) algorithm

2. Which of the following is the computer language that is executed within a computer's central processing unit?

(A) DOS
(B) high-level language
(C) assembly language
(D) machine language

3. Which of the following best defines a compiler?

(A) hardware that is used to translate high-level language to machine code
(B) software that collects and stores executable commands in a program
(C) software that is used to translate high-level language into machine code
(D) hardware that collects and stores executable commands in a program

4. Which of the following program flowchart symbols is typically used to indicate the end of a process or program?

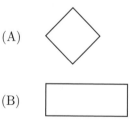

(C)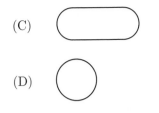

(D)

5. The cells in a computer spreadsheet program are as shown.

	A	B	C	D
1	4	−1	3	0
2	3	A3+C1	−3	6
3	0.5	m	33	2
4	Smith	1	B2*B3	C4
5				

Instructions in macro-commands are scanned from left to right. What is the value of n in the following macro-commands?

$$m = 5$$
$$p = m*2 + 6$$
$$n = D4 - 3*p**0.5$$

- (A) 4
- (B) 5.5
- (C) 10.5
- (D) 1041

For the following problems use the NCEES Handbook as your only reference.

6. How many times will the second line be executed?

```
            M = 42
LOOPSTART   M = M − 1
            P = INTEGER PART OF (M/2)
            IF P > 15, THEN GO TO
                LOOPSTART, OTHERWISE
                GO TO END
END         PRINT "DONE"
```

- (A) 8
- (B) 9
- (C) 10
- (D) 11

7. What flowchart element would be used to represent an IF...THEN statement?

(A)

(B)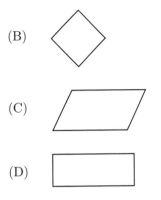

(C)

(D)

8. In the program segment shown, how many times will the line labeled "START" execute?

```
        I = 1
        J = 1
START   J = J + I
        I = J^2
        If J<100 THEN GO TO START
        ELSE GO TO FINISH
FINISH  PRINT J
```

- (A) 3
- (B) 4
- (C) 5
- (D) 6

9. Based on the following program segment, what is the ANSWER?

```
I = 1
J = 2
K = 3
L = 4
ANSWER = C
IF (I>J) OR (K<L) THEN ANSWER = A
IF (I<J) OR (K>L) THEN ANSWER = B
IF (I>J) OR (K>L) THEN I = 5
IF (I<J) AND (K<L) THEN ANSWER = D
```

- (A) A
- (B) B
- (C) C
- (D) D

10. How many cells are in the range C5...Z30?

- (A) 575
- (B) 598
- (C) 600
- (D) 624

11. The Taylor Series approximation for the exponential function is $e^x = \Sigma(x^n/n!)$. A spreadsheet set up like the one shown is being used to calculate e^x. The cells in row 2 will be totalled to arrive at the approximation. Which formula, typed into C2 and then copied to the rest of row 2, will provide the correct total?

	A	B	C	D	E
1	0	1	2	3	4
2	1	x			
3					
4					
5					

(A) B2^B1/FACTORIAL(B1)
(B) B2 * $B$2/B1
(C) B2 * $B$2/C1
(D) B2^C1/(C1 * B1 * A1)

12. A portion of a spreadsheet is as shown. The value of cell C1 is set to (A1 + B1)/2. This formula is copied into the range of cells, C2:C4. The value of cell C5 is set to SUM(C1:C4) * 0.05. What is the number in cell C5?

	A	B	C
1	1	5	
2	2	6	
3	3	7	
4	4	8	
5			

(A) 0.40
(B) 0.45
(C) 0.60
(D) 0.90

13. For the program segment shown, the input value of X is 5. What is the output value of T?

INPUT X
N = 0
T = 0
   DO WHILE N < X
   $T = T + X^N$
   N = N + 1
   END DO
OUTPUT T

(A) 156
(B) 629
(C) 781
(D) 3906

Problems 14 and 15 relate to the following portion of a spreadsheet.

	A	B	C	D
1	10	11	12	13
2	1	A2^2	B2*A$1	
3	2	A3^2	B3*B$1	
4	3	A4^2	B4*C$1	
5	4	A5^2	B5*D$1	

14. What will be the top-to-bottom values in column B?

(A) 11,1,2,3,4
(B) 11,1,3,6,10
(C) 11,1,4,9,16
(D) 11,1,5,12,22

15. What will be the top-to-bottom values in column C?

(A) 12,1,4,9,16
(B) 12,10,44,108,208
(C) 12,19,84,207,408
(D) 12,19,100,250,500

16. In programming, a recursive function is one that

(A) calls previously used functions
(B) generates functional code to replace symbolic code
(C) calls itself
(D) compiles itself in real time

17. A computer structured programming segment contains the following program segment. What is the value of Y after the segment is executed?

Y = 4
B = 4
Y = 3 * B − 6
IF Y > B THEN Y = B − 2
IF Y < B THEN Y = Y + 2
IF Y = B THEN Y = B + 2

(A) 2
(B) 6
(C) 8
(D) 12

18. The numbers −3, 4, 0, −5, −1, 5, ... are in a file to be read and processed by the structured program shown. The number after the program is executed is most nearly

```
Set I = 1 and Y = 0
WHILE I ≤ 3
        Read a value from the file and set X equal
            to that value
        If X < 0 GOTO 1
        ELSE Y = Y + X*X
1       Increment I by 1
ENDWHILE
    Z = Y/I
```

(A) 1.0
(B) 2.7
(C) 4.0
(D) 26

19. A spreadsheet evaluating the performance of a resistive component in a direct-current circuit is shown. What value should be displayed in cell G2?

	E	F	G
1	amps	ohms	watts
2	5.0	2.0	

(A) 10.0
(B) 12.5
(C) 20.0
(D) 50.0

20. The effect of using recursive functions in a program is generally to

(A) use less code and less memory
(B) use less code and more memory
(C) use more code and less memory
(D) use more code and more memory

21. When can an 8-bit system correctly access more than 128 different integers?

(A) when the integers are in the range of [−256, 0]
(B) when the integers are in the range of [−128, 0]
(C) when the integers are in the range of [−128, 128]
(D) when the integers are in the range of [0, 512]

22. What computer operating system (OS) is required to view a document saved in HTML (hypertext markup language) format?

I.   Apple Macintosh OS
II. MS DOS
III. Windows
IV. Unix

(A) I or III
(B) I, II, or III
(C) I, III, or IV
(D) I, II, III, or IV

23. The following flowchart represents an algorithm. Which of the given structured programming segments correctly translates the algorithm?

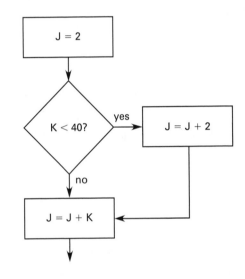

(A) J = 2
     IF K < 40 THEN J = J + 2
     J = J + K

(B) J = 2
     IF K < 40 THEN J = J + 2
     ELSE J = J + K

(C) J = 2
     DO WHILE K < 40
     J = J + 2
     ENDWHILE
     J = J + K

(D) J = 2
     DO UNTIL K < 40
     J = J + 2
     ENDUNTIL
     J = J + K

24. The structured programming segment shown implements which of the following equations? The variable N is an integer greater than 0.

$$A = X$$
$$DO\ UNTIL\ N = 0$$
$$Y = A * X$$
$$A = Y$$
$$N = N - 1$$
$$END\ UNTIL$$

(A) $Y = X!$
(B) $Y = X^{N+1}$
(C) $Y = X^{N-1}$
(D) $Y = X^N$

25. The spreadsheet shown is to be used to calculate the power dissipated by circuits with R_1 in series with the parallel combination of R_2 and R_3. Which of the given formulas can be entered in cell E2 and then copied and pasted into the rest of the cells in the E column to accomplish this task?

	A	B	C	D	E
1	V (volts)	R1 (ohms)	R2 (ohms)	R3 (ohms)	P (watts)
2	5	3	6	14	
3	5	5	7	18	
4	5	7	8	19	
5	5	9	9	21	
6	5	11	10	22	

(A) (A2 * A2)/ ((C$3 * D$4)/(C$3 + D$4))
(B) ($A2 * A2)/(B2 + (($C2 * $D2)/(C2 + D2)))
(C) $A2/(B2 + ((C3 * D4)/($C3 + $D4)))
(D) ($A$2 * A2)/
    (B2 + ((C$2 * D$2)/(C$2 + D$2)))

26. A portion of a typical spreadsheet is shown. The contents of column B are copied and pasted into columns C and D. What number will be represented in cell D4?

	A	B	C	D
1	5	A$1		
2	7	$B1		
3	−3	B2 + A2		
4	0	A$3*B3		

(A) 47
(B) 60
(C) 100
(D) 150

27. A typical spreadsheet for economic evaluation of alternatives uses cell F4 to store the percentage value of inflation rate. The percentage rate is assumed to be constant throughout the lifetime of the study. What variable should be used to access that value throughout the model?

(A) F4
(B) $F4
(C) %F4
(D) $F$4

SOLUTIONS TO FE-STYLE EXAM PROBLEMS

1. A program is a sequence of instructions that implements a formula or set of steps but is not a formula or set of steps. Software and firmware are programs on media. An algorithm is a formula or set of steps for solving a particular problem that is often implemented as a program.

Answer is D.

2. The central processing unit executes a version of the program that has been compiled into the machine language and that is in the form of operations and operands specific to the machine's coding.

Answer is D.

3. A compiler is a program (i.e., software) that converts programs written in higher-level languages to lower-level languages that the computer can understand.

Answer is C.

4. Termination (end of the program) is commonly represented by the symbol in choice (C).

Answer is C.

5. Evaluate the macro. Use the following typical precedence rules:

- exponentiation before multiplication and division
- multiplication and division before addition and subtraction
- operations inside parentheses before operations outside

$$m = 5$$
$$p = m * 2 + 6 = (5 * 2) + 6 = 16$$

$$n = D4 - 3*p**0.5$$
$$= C4 - 3*p**0.5$$
$$= B2*B3 - 3*p**0.5$$
$$= (A3 + C1)*m - 3*p**0.5$$
$$= ((0.5 + 3)*5) - (3)*(16)**(0.5)$$
$$= ((3.5)*5) - (3)*(16)**(0.5)$$
$$= ((3.5)*5) - 12$$
$$= 17.5 - 12$$
$$= 5.5$$

Answer is B.

6. The values of the variables for each iteration are

iteration	M	P
1	41	20
2	40	20
3	39	19
4	38	19
5	37	18
6	36	18
7	35	17
8	34	17
9	33	16
10	32	16
11	31	15

When P reaches 15, P is no longer greater than 15. Line 2 is executed 11 times.

Answer is D.

7. An IF...THEN statement alters the flow of a program based on some criterion that can be evaluated as true or false. The diamond-shaped symbol is used to represent a decision.

Answer is B.

8. Construct a table listing the values of I and J during each loop. Notice that J is adjusted before I, so the "new" value of J must be used to calculate I.

loop	I	J
(initially) 0	1	1
1	4	2
2	36	6
3	1764	42
4	3,261,636	1806

After four loops J is greater than 100, so the program moves on.

Answer is B.

9. The code must be followed all the way from beginning to end. After the first IF, ANSWER is A; after the second it's B. The third IF evaluates to "false," so the ANSWER remains B, but the fourth statement is true.

Answer is D.

10. A range from m to n, inclusive, has $m - n + 1$ elements, not $m - n$.

$$\text{no. of rows} = (30 - 5) + 1$$
$$= 26$$
$$\text{no. of columns} = (Z - C) + 1$$
$$= (26 - 3) + 1$$
$$= 24$$
$$\text{no. of cells} = (\text{no. of rows})(\text{no. of columns})$$
$$= (26)(24)$$
$$= 624$$

Answer is D.

11. The elements in this series, T_n, could be represented using exponents and factorial functions. In this case, absolute cell references must be used for x, but a relative reference works well for n.

$$T_n = \frac{x^n}{n!} = \frac{\$B\$2^\wedge C1}{\text{FACTORIAL}(C1)}$$

Since this is not one of the available answers, develop a recursive definition for the series.

$$T_{n+1} = \frac{x^{n+1}}{(n+1)!} = \frac{(x^n)(x)}{(n!)(n+1)}$$
$$= T_n \left(\frac{x}{n+1} \right)$$
$$= B2*\$B\$2/C1$$

Answer is C.

12.

	A	B	C
1	1	5	3
2	2	6	4
3	3	7	5
4	4	8	6
5			0.9

The formula in cell C1, (A1 + B1)/2, is a relational formula; the formula (AN + BN)/2 is copied into cell CN.

Completing the spreadsheet, the value in cell C5 is

$$(3 + 4 + 5 + 6) \times 0.05 = 0.90$$

Answer is D.

13. The program segment defines the series

$$\sum_{0}^{N=X-1} X^N$$

The value of T at the conclusion of the loop can be computed as

N	T	
0	1	
1	$1 + 5$	$= 6$
2	$1 + 5 + 25$	$= 31$
3	$1 + 5 + 25 + 125$	$= 156$
4	$1 + 5 + 25 + 125 + 625$	$= 781$

Answer is C.

14. Except for the first entry (which is 11), column B calculates the square of the values in column A. Thus, the entries are 11, $(1)^2$, $(2)^2$, $(3)^2$, and $(4)^2$.

Answer is C.

15. Except for the first entry (which is 12), column C is found by taking the numbers from column B and then multiplying by the entries in row 1. For example, B2A$1 means to multiply the entry in B2, which was calculated to be 1 in the previous problem, by the number entered in cell A1, which is 10. This product is 10.

The entries are 12, 1×10, 4×11, 9×12, 16×13, or 12, 10, 44, 108, 208.

Answer is B.

16. A recursive function calls itself.

Answer is C.

17. The first operation changes the value of Y.

$$Y = 3 \times 4 - 6 = 6$$

The first IF statement is satisfied, so the operation is performed.

$$Y = 4 - 2 = 2$$

However, the program execution does not end here.

The value of Y is then less than B, so the second IF statement is executed. This statement is satisfied, so the operation is performed.

$$Y = 2 + 2 = 4$$

The value of Y is then equal to B, so the third IF statement is executed. This statement is satisfied, so the operation is performed.

$$Y = 4 + 2 = 6$$

Answer is B.

18. First, the while loop will repeat for I = 1, 2, and 3.

When I = 1, X = −3, so the Y = Y + X * X instruction is not executed.

When I = 2, X = 4, so Y = Y + X * X = 0 + 4 * 4 = 16.

When I = 3, X = 0, so Y = Y + X * X = 16 + 0 * 0 = 16.

I is incremented to I = 4.

The while loop is exited, and the last instruction is executed: Z = Y/I = 16/4 = 4.

Answer is C.

19. In a resistive direct-current circuit, the power is

$$P = I^2 R = (5.0 \text{ A})^2 (2.0 \ \Omega) = 50.0 \text{ W}$$

Answer is D.

20. A recursive function calls itself. Since the function does not need to be coded in multiple places, less code is used. Each subsequent call of the function must be carried out in a different location, so more memory is used.

Answer is B.

21. An 8-bit system can represent $(2)^8 = 256$ different distinct integers. Normally, the 8th bit is used for the sign and only 7 bits are used for magnitude, resulting in a range of $[-127, 128]$ or $[-128, 127]$ (counting zero as one of the integers). If all of the integers are known or assumed to have the same sign, a range of 256 integers is available.

Answer is A.

22. HTML may be viewed on any computer with a compatible browser.

Answer is D.

23. The decision block indicates an IF statement because both outcomes progress toward the end. If the flowchart looped back for one of the outcomes, the chart might indicate a DO WHILE or DO UNTIL loop. The flowchart indicates that the operation J = J + K will occur regardless of the outcome of the IF statement; therefore, there is no ELSE as shown in answer B.

Answer is A.

24. The DO loop will be executed N times. After the first execution, $Y = X^2$. After each subsequent execution of the loop, Y is multiplied by X. Therefore, the segment calculates $Y = X^{N+1}$

Answer is B.

25. The power is the voltage squared divided by the resistance.

$$P = \frac{V^2}{R}$$

The total resistance is R_1 plus the parallel combination R_2 and R_3.

$$R = R_1 + \frac{R_2 R_3}{R_2 + R_3}$$

The formula that will satisfy this relationship in cell D2 with all cell references relative is

$$(A2 * A2)/(B2 + ((C2 * D2)/(C2 + D2)))$$

Answers A and C are eliminated because they do not match this formula.

When the formula is copied and pasted into the other cells in the column, it does not matter if the column references are relative or absolute; however, all cell references must have relative row references for the copy operation to yield the correct answer. $ must not appear before a number.

Answer is B.

26. The "$" symbol represents the absolute cell references; all other references will be dependent on the location. Only the column references will change.

	A	B	C	D
1	5	A$1	B$1	C$1
2	7	$B1	$B1	$B1
3	−3	B2 + A2	C2 + B2	D2 + C2
4	0	A$3*B3	B$3*C3	C$3*D3

The formula for cell D4 is

$$\begin{aligned}
C3 * D3 &= (C2 + B2) * (D2 + C2) \\
&= (B1 + B1) * (B1 + B1) \\
&= (A1 + A1) * (A1 + A1) \\
&= (5 + 5)(5 + 5) \\
&= 100
\end{aligned}$$

Answer is C.

27. The dollar sign ($) is used in spreadsheets to "fix" the column and/or row designator following it when other columns or rows are permitted to vary.

Answer is D.

49 Measurement

Subjects

TRANSDUCERS 49-1
SENSITIVITY 49-1
LINEARITY 49-1
ACCURACY 49-2
PRECISION 49-2
STABILITY 49-2
RESISTANCE TEMPERATURE
   DETECTORS 49-2
THERMISTORS 49-2
STRAIN GAGES 49-2
WHEATSTONE BRIDGES 49-3
SAMPLING 49-4
ANALONG-TO-DIGITAL
   CONVERSION 49-5
MEASUREMENT UNCERTAINTY 49-5

Nomenclature

F	force	N
GF	gage factor	–
I	current	A
k	constant	various
L	shaft length	m
R	resistance	Ω
t	time	s
T	temperature	K
V	voltage	V
w	measurement error	–

Symbols

α	temperature coefficient	$1/°R$	$1/K$
β	temperature coefficient	$1/°R^2$	$1/K^2$
Δ	change	–	–
ϵ	strain	–	–
ϵ_v	resolution	–	–

Subscripts

0	reference
g	gage
I	interest
N	Nyquist
o	original or out
s	sample
t	total
T	at temperature T
v	variable

TRANSDUCERS

A *transducer* is any device used to covert a physical phenomenon into an electrical signal. For example, a microphone is a transducer that transforms sound energy into electrical signals that can be recorded, amplified, or transmitted. Other transducers sense phenomena like temperature, pressure, physical movement, or light intensity.

The signal from a transducer may be used with a measurement device such as an ohmmeter or a voltmeter. The signal may be used to provide feedback to the circuit, to control the physical phenomenon in some way. For example, a thermostat is a transducer that is sensitive to temperature. When the temperature dips below the temperature set on the thermostat, a circuit turns on a space heater until the temperature of the room equals the temperature setting.

SENSITIVITY

Sensitivity is the ratio of the change in electrical signal magnitude to the change in magnitude of the physical phenomena parameter being measured. If a transducer exhibits a significant change in an electrical characteristic (voltage, current, resistance, capacitance, or inductance) in response to a change in a parameter of a physical phenomenon, then it is sensitive to that parameter. The greater the change in the electrical signal for the same change in the parameter, the greater the sensitivity of the transducer. It is desirable to use transducers that are sensitive to only one parameter and insensitive to all others. For example, an odometer in a car should respond only to the rotation of the car wheels, and the measurement should not depend on other factors like wind speed, temperature, or humidity.

LINEARITY

The *linearity* of a transducer is the degree to which the sensitivity is in direct proportion to the parameter being measured. Transducers are usually designed and selected to closely approximate linear sensitivity over the range of measurements of the parameter. This is not always practical, so a second-order term and even a third-order term may be included.

ACCURACY

A measurement is said to be *accurate* if it is substantially unaffected by (i.e., is insensitive to) all variation outside of the measurer's control.

For example, suppose a rifle is aimed at a point on a distant target and several shots are fired. The target point represents the "true value" of a measurement—the value that should be obtained. The impact points represent the measured values—what is obtained. The distance from the centroid of the points of impact to the target point is a measure of the alignment accuracy between the barrel and the sights. This difference between the true and measured values is known as the measurement *bias*.

PRECISION

Precision is not synonymous with *accuracy*. Precision is concerned with the repeatability of the measured results. If a measurement is repeated with identical results, the experiment is said to be precise. The average distance of each impact from the centroid of the impact group is a measure of precision. Thus, it is possible to take highly precise measurements and still have a large bias.

Most measurement techniques (e.g., taking multiple measurements and refining the measurement methods or procedures) that are intended to improve accuracy actually increase the precision.

Sometimes, the term *reliability* is used with regard to the precision of a measurement. A *reliable measurement* is the same as a *precise estimate*.

STABILITY

Stability and *insensitivity* are synonymous terms. (Conversely, *instability* and *sensitivity* are synonymous.) A stable measurement is insensitive to minor changes in the measurement process.

RESISTANCE TEMPERATURE DETECTORS

Resistance temperature detectors (RTDs), also known as *resistance thermometers*, make use of changes in their resistance to determine changes in temperature. A fine wire is wrapped around a form and protected with glass or a ceramic coating. Nickel and copper are commonly used for industrial RTDs. Platinum is used when precision resistance thermometry is required. RTDs are connected through resistance bridges to compensate for lead resistance.

Resistance in most conductors increases with temperature. The resistance of RDTs has greater sensitivity and more linear response to temperature than that of standard resistors. The resistance at a given temperature can be calculated from the *coefficients of thermal resistance*, α and β. (Higher-order terms—third, fourth, etc.—are used when extreme accuracy is required.) The variation of resistance with temperature is nonlinear, though β is small and is often insignificant over short temperature ranges. Therefore, a linear relationship is often assumed and only α is used. In Eq. 49.1, R_0 is the resistance at the reference temperature, T_0, usually 100 Ω at 32°F (0°C) second-order approximation. In commercial RTDs, α is referred to by the literal term *alpha-value*.

$$R_T \approx R_0 \left(1 + \alpha \Delta T + \beta \Delta T\right) \qquad 49.1$$
$$\Delta T = T - T_0 \qquad 49.2$$

The first-order approximation in Eq. 49.3 is sufficient in many practical applications.

$$R_T \approx R_0 \left(1 + \alpha \left(T - T_0\right)\right) \qquad 49.3$$

THERMISTORS

Thermistors are temperature-sensitive semiconductors constructed from oxides of manganese, nickel, and cobalt and from sulfides of iron, aluminum, and copper. Thermistor materials are encapsulated in glass or ceramic materials to prevent penetration of moisture. Unlike RTDs, the resistance of thermistors decreases as the temperature increases.

Thermistor temperature-resistance characteristics are exponential. Depending on the brand, material, and construction, β typically varies between 3400 K and 3900 K.

$$R = R_o e^k \qquad 49.4$$
$$k = \beta \left(\frac{1}{T} - \frac{1}{T_o}\right) \quad [T \text{ in K}] \qquad 49.5$$

STRAIN GAGES

A *bonded strain gage* is a metallic resistance device that is cemented to the surface of the unstressed member. It is constructed by bonding the folded wire to the surface of the member. The gage consists of a metallic conductor (known as the *grid*) on a backing (known as the *substrate*). The grids of strain gages were originally of the folded-wire variety. For example, nichrome wire with a total resistance under 1000 Ω was commonly used.

Modern strain gages are generally of the foil type manufactured by printed circuit techniques. Semiconductor gages are also used when extreme sensitivity (i.e., gage factors in excess of 100) is required. However, semiconductor gages are extremely temperature-sensitive.

The substrate and grid experience the same strain as the surface of the member. The resistance of the gage changes as the member is stressed due to changes in conductor cross section and intrinsic changes in resistivity with strain. Temperature effects must be compensated by the circuitry or by using a second unstrained gage as part of the bridge measurement system. (See "Wheatstone Bridges.")

When simultaneous strain measurements in two or more directions are needed, it is convenient to use a commercial *rosette strain gage*. A rosette consists of two or more *grids* properly oriented for application as a single unit.

Figure 49.1 Strain Gage

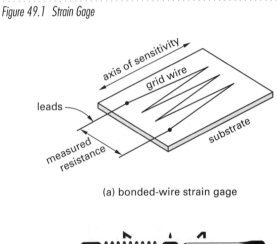

(a) bonded-wire strain gage

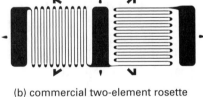

(b) commercial two-element rosette

The *gage factor (strain sensitivity factor)*, GF, is the ratio of the fractional change in resistance to the fractional change in length (strain) along the detecting axis of the gage. The gage factor is a function of the gage material. It can be calculated from the grid material's properties and configuration. The higher the gage factor, the greater the sensitivity of the gage. From a practical standpoint, however, the gage factor and gage resistance are provided by the gage manufacturer. Only the change in resistance is measured.

$$GF = \frac{\frac{\Delta R}{R}}{\frac{\Delta L}{L}} = \frac{\frac{\Delta R}{R}}{\epsilon} \qquad 49.6$$

Table 49.2 Approximate Gage Factors

material	GF
constantan	2.0
iron, soft	4.2
isoelastic	3.5
manganin	0.47
monel	1.9
nichrome	2.0
nickel	$-12^{(a)}$
platinum	4.8
platinum-iridium	5.1

[a]Value depends on amount of preprocessing and cold working.

Constantan and isoelastic wires and metal foil with gage factors of approximately 2 and initial resistances of less than 1000 Ω (typically 120 Ω, 350 Ω, 600 Ω, and 700 Ω) are commonly used. In practice, the gage factor and initial gage resistance, R_g, are specified by the manufacturer of the gage. Once the strain sensitivity factor is known, the strain, ϵ, can be determined from the change resistance. Strain is often reported in units of μin/in (μm/m) and is given the name *microstrain*.

$$\epsilon = \frac{\Delta R}{(GF)R} \qquad 49.7$$

WHEATSTONE BRIDGES

The *Wheatstone bridge* shown in Fig. 49.2 is one type of *resistance bridge*. The bridge can be used to determine the unknown resistance of a resistance transducer (e.g., thermistor or resistance-type strain gage), say R_1 in Fig. 49.2. The potentiometer is adjusted (i.e., the bridge is "balanced") until no current flows through the meter or until there is no voltage across the meter, hence the name *null indicator* or alternatively, *zero-indicating bridge* or *null-indicating bridge*. The unknown resistance can also be determined from the amount of voltage unbalance shown by the meter reading, in which case, the bridge is known as a *deflection bridge* rather than a null-indicating bridge. When the bridge is balanced and no current flows through the meter leg, Eqs. 49.8 through 49.11 are applicable.

$$I_1 = I_2 \quad \text{[balanced]} \qquad 49.8$$

$$I_4 = I_3 \quad \text{[balanced]} \qquad 49.9$$

$$V_4 + V_3 = V_1 + V_2 \quad \text{[balanced]} \qquad 49.10$$

$$\frac{R_4}{R_1} = \frac{R_3}{R_2} \quad \text{[balanced]} \qquad 49.11$$

Figure 49.2 Series-Balanced Wheatstone Bridge

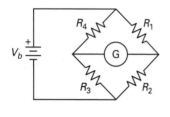

Any one of the four resistances can be the unknown, up to three of the remaining resistances can be fixed or adjustable, and the battery and meter can be connected to either of two diagonal corners. Therefore, it is sometimes confusing to apply Eq. 49.11 literally. However, the following bridge law statement can be used to help formulate the proper relationship: *When a series Wheatstone bridge is null-balanced, the ratio of resistance of any two adjacent arms equals the ratio of resistance of the remaining two arms, taken in the same sense.* In this statement, "taken in the same sense" means that both ratios must be formed reading either left to right, right to left, top to bottom, or bottom to top.

A special case of the Wheatstone bridge circuit is the quarter bridge circuit shown in Fig. 49.3. The quarter bridge circuit has three identical resistors and one resistor that differs slightly in resistance from the other three. The difference, ΔR, can be positive or negative. This different resistor is the transducer.

$$R_1 = R_2 = R_3 = R \qquad 49.12$$

$$R_4 = R + \Delta R \qquad 49.13$$

$$|\Delta R| << R \qquad 49.14$$

Figure 49.3 Wheatstone Quarter Bridge

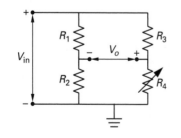

Most Wheatstone bridge circuits are difficult to analyze if they are not balanced, but the Wheatstone quarter bridge has a simple approximation, useful for many instrumentation applications, that is given in Eq. 49.15.

$$V_o \approx \left(\frac{\Delta R}{4R} \right) V_{\text{in}} \qquad 49.15$$

SAMPLING

As part of the *analog-to-digital conversion* process, continuous-time signals are sampled to a discrete-time system. The analog signal is sampled at regular time intervals Δt. The sampling rate or frequency is given by Eq. 49.16.

$$f_s = \frac{1}{\Delta t} \qquad 49.16$$

Shannon's sampling theorem states that a time-continuous signal is completely determined by (i.e., can be reconstructed from) its values at an infinite sequence of equally spaced times, if the frequency of sampling times is greater than twice the highest frequency component, known as the *Nyquist* frequency, f_I, of the signal.[1] Of course, it is not possible to take an infinite number of samples, but the sampling theorem is sufficiently accurate for most practical applications.

The signal may also contain frequencies that are higher than twice the sampling rate, a situation that may be acceptable if the frequencies are not of interest. For example, sampling of an audio signal does not need to represent frequencies that are beyond the range that the human ear can hear.

The *Nyquist rate*, f_N, for sampling a signal for analog-to-digital is two times the highest frequency of interest, the Nyquist frequency, f_I. If the signal is a pure sinusoidal signal at the highest frequency, then sampling at greater than the Nyquist frequency will ensure at least one sample in every positive half-cycle and every negative half-cycle. If the signal also includes lower frequency components, then sampling at more than twice the Nyquist frequency will ensure at least one sample each time the highest frequency component causes the total signal to increase or decrease.

$$f_N = 2f_I \qquad 49.17$$

To be able to reproduce the signal, the sampling frequency must be greater than the Nyquist rate.

$$f_s > f_N \quad \text{[reproducible sampling]} \qquad 49.18$$

---

[1]Engineers are inconsistent in their use of the terms "Nyquist frequency" and "Nyquist rate." "Nyquist frequency" may be used by some authorities as the sampling rate.

If sampling is done at a lower rate than twice the Nyquist rate, then the higher frequencies in the measured signal are not accurately represented and will distort the lower frequencies' content in the sampled data. Frequencies greater than the sampling frequencies and at integer multiples of the sampling frequency appear as lower frequencies and are known as *alias frequencies*.

ANALOG-TO-DIGITAL CONVERSION

The resolution of analog-to-digital (A/D) conversion is an important factor. The resolution determines the accuracy that is possible for the measurement. The digital number that represents the analog sample does not represent the actual value, but rather it indicates that the actual value is somewhere within a range, and that range is the resolution.

An analog measurement in the range from a high voltage, V_H, to a low voltage, V_L, that is measured by a digital system with n bits has a voltage resolution given by Eq. 49.19.

$$\varepsilon_V = \frac{V_H - V_L}{2^n} \qquad 49.19$$

The digital number, N, that represents the analog value has a range from 0 to $2^n - 1$. This means that the maximum value the digital number can represent is one resolution (a resolution is defined by Eq. 49.19) less than the analog value can obtain, $V_H - \epsilon_V$. Thus if all the bits are "1" then the analog value is somewhere between V_H and $V_H - \epsilon_V$. To calculate the analog value from the digital value, use Eq. 49.20.

$$V = \epsilon_V N + V_L \qquad 49.20$$

MEASUREMENT UNCERTAINTY

Measurement uncertainty of a function $R = f(x_1, x_2, x_3, ..., x_n)$ whose values have uncertainties $x_1 \pm w_1, x_2 \pm w_2, x_3 \pm w_3$, and so on is given by Eq. 49.21, the Kline-McClintock equation.

$$w_R = \sqrt{\left(w_1 \frac{\partial f}{\partial x_1}\right)^2 + \left(w_2 \frac{\partial f}{\partial x_2}\right)^2 + \cdots + \left(w_n \frac{\partial f}{\partial x_n}\right)^2} \qquad 49.21$$

The *Kline-McClintock equation* is a method for estimating the uncertainty in a function that depends on more than one measurement. Generally, the measurements will not be at the most extreme of the inaccuracy (which is known as a *worst-case stack-up* of the inaccuracy). The Kline-McClintock method is closer to

the real inaccuracy than averaging the inaccuracies, in most cases. If the function R is the sum of the measurements (i.e., $R = x_1 + x_2 + x_3 + ... + x_n$), then the Kline-McClintock method reduces to Eq. 49.22. This is called the *root sum square* (RSS) value.

$$w_R = \sqrt{w_1^2 + w_2^2 + \cdots + w_n^2} \qquad 49.22$$

If the function R is a sum of the measurements multiplied by constants (i.e., $R = a_1 x_1 + a_2 x_2 + a_3 x_3 + ... + a_n x_n$), then the Kline-McClintock method reduces to Eq. 49.23. This is called a *weighted RSS* value.

$$w_R = \sqrt{a_1^2 w_1^2 + a_2^2 w_2^2 + \cdots + a_n^2 w_n^2} \qquad 49.23$$

FE-STYLE EXAM PROBLEMS

1. A Wheatstone bridge is used to measure an unknown resistance, R_x, as shown. At the null point, the variable resistor, R_v, has a resistance of 2100 Ω. What is the unknown resistance?

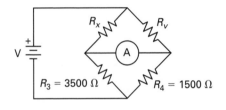

(A) 900 Ω
(B) 2100 Ω
(C) 2500 Ω
(D) 4900 Ω

2. What can be used to construct a quarter-bridge measurement circuit?

    (A) one strain gage, one variable resistor, and two fixed resistors
    (B) one strain gage, one variable resistor, and one reference resistor
    (C) two strain gages and one variable resistor
    (D) two strain gages, one variable resistor, and one reference resistor

3. Which of the following is *not* a transducer?

    (A) computer mouse
    (B) microphone
    (C) electrical outlet
    (D) light switch

4. An analog-to-digital conversion process has a resolution of approximately 1.52588×10^{-4} V. The voltage range is 0 V to 10 V. What is the number of bits?

(A) 4
(B) 8
(C) 16
(D) 32

5. A calculation is made by combining three measurements, x_1, x_2, and x_3, using the equation shown. The uncertainties of the measurements are ± 0.03, ± 0.05, ± 0.07, respectively. What is most nearly the estimated uncertainty of the calculation?

$$f = 3x_1 - 5x_2 + 7x_3$$

(A) 0.34
(B) 0.56
(C) 0.67
(D) 0.79

6. In the circuit shown, $V_{\text{in}} = 10.00$ V, $V_o = 0.0125$ V, and $R_1 = R_2 = R_3 = R = 10.00$ kΩ. What is most nearly the resistance of R_4?

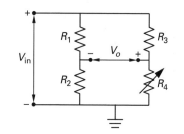

(A) 10,000 Ω
(B) 10,050 Ω
(C) 10,500 Ω
(D) 15,000 Ω

7. It is desired to choose a resistance temperature detector (RTD) transducer for a measurement. The actual value of the temperature is not important because the initial temperature is known by other means, but the change in temperature as the test item is heated is important. Which statement is true about the selection of the RTD?

(A) The precision is important, and the accuracy is not important.
(B) The accuracy is important, and the precision is not important.
(C) Neither the accuracy nor the precision is important.
(D) Both accuracy and precision are important.

8. A resistance temperature detector (RTD) transducer is to be chosen such that the analog-to-digital conversion will have a resolution of 0.001°C. The conversion uses 16 bits. The RTD will be in a circuit with a lower voltage of −0.150 V at the lowest temperature extreme and an upper voltage of 0.300 V at the highest temperature extreme. The RTD will have an output of 0.000 V at 0.000°C. What is most nearly the sensitivity of the RTD, and what is most nearly the temperature range that the analog-to-digital conversion can represent?

(A) 96.0 °C/V; −43.9°C to 98.6°C
(B) 146 °C/V; −21.9°C to 43.7°C
(C) 160 °C/V; −11.7°C to 38.8°C
(D) 222 °C/V; 1.90°C to 78.7°C

9. A strain gage is to be used in measuring the strain on a test specimen. A strain gauge with an initial resistance of 120 Ω exhibits a decrease of 0.12 Ω. The gage factor is 2.00. The initial length of the gage was 1.000 cm. What is most nearly the final length of the strain gage?

(A) 0.9995 cm
(B) 1.0000 cm
(C) 1.0005 cm
(D) 1.0050 cm

SOLUTIONS TO FE-STYLE EXAM PROBLEMS

1. At the null point,

$$\frac{R_x}{R_3} = \frac{R_v}{R_4}$$

$$R_x = \frac{R_3 R_v}{R_4} = \frac{(3500 \ \Omega)(2100 \ \Omega)}{1500 \ \Omega}$$

$$= 4900 \ \Omega$$

Answer is D.

2. A Wheatstone-type measurement bridge has a total of four resistive elements. In a full-bridge arrangement, all four elements are strain gages. In a half-bridge arrangement, only two strain gages are used. A fixed (reference) resistor and an adjustment (variable) resistor constitute the rest of the bridge. In a quarter bridge arrangement, only one strain gage is used. Two fixed (reference ratio) resistors and an adjustment (variable) resistor constitute the rest of the bridge.

Answer is A.

3. A computer mouse is transducer that transforms hand movement into electrical signals that control a cursor. A microphone is a transducer that transforms sound energy into electrical signals. A light switch is a transducer that transforms switch movement into electrical signals that turn a light on or off. An electrical outlet does not convert energy from one form to another; therefore, it is not a transducer.

Answer is C.

4. Solving the resolution equation for 2 raised to the number of bits yields

$$2^n = \frac{V_H - V_L}{\epsilon_V}$$
$$= \frac{10 - 0 \text{ V}}{1.52588 \times 10^{-4} \text{ V}}$$
$$= 65{,}536$$

This equation can be solved with some calculators directly, or it can be solved by taking the log base 2 of both sides and using the logarithm identity.

$$\log_b x = \frac{\log_a x}{\log_b b}$$

A more efficient method is to substitute the possible answers from the problem statement.

$$(2)^{16} = 65{,}536$$

Answer is C.

5. The Kline-McClintock equation estimates the uncertainty of the calculation.

$$w_R = \sqrt{\left(w_1 \frac{\partial f}{\partial x_1}\right)^2 + \left(w_2 \frac{\partial f}{\partial x_2}\right)^2 + \cdots + \left(w_n \frac{\partial f}{\partial x_n}\right)^2}$$

$$\frac{\partial f}{\partial x_1} = 3$$

$$\frac{\partial f}{\partial x_2} = -5$$

$$\frac{\partial f}{\partial x_3} = 7$$

$$w_R = \sqrt{\left(w_1 \frac{\partial f}{\partial x_1}\right)^2 + \left(w_2 \frac{\partial f}{\partial x_2}\right)^2 + \left(w_3 \frac{\partial f}{\partial x_3}\right)^2}$$
$$= \sqrt{((0.03)(3))^2 + ((0.05)(-5))^2 + ((0.07)(7))^2}$$
$$= 0.5574 \quad (0.56)$$

Answer is B.

6. The circuit is a Wheatstone bridge. If R_4 is close to 10 kΩ, then the quarter bridge approximation can be used.

$$V_o \approx \left(\frac{\Delta R}{4R}\right) V_{\text{in}}$$

$$\Delta R \approx \left(\frac{V_o}{V_{\text{in}}}\right) 4R$$
$$= \left(\frac{0.0125 \text{ V}}{10.00 \text{ V}}\right)(4)\left(10 \times 10^3 \text{ }\Omega\right)$$
$$= 50 \text{ }\Omega$$

$$R_4 = R + \Delta R$$
$$= 10{,}000 \text{ }\Omega + 50 \text{ }\Omega$$
$$= 10{,}050 \text{ }\Omega$$

The quarter bridge approximation is valid because

$$R_4 = 10{,}050 \text{ }\Omega$$
$$\approx 10{,}000 \text{ }\Omega$$
$$= R$$

Answer is B.

7. Precision is a measure of the repeatability of results, which is important for this measurement. The accuracy can have a significant bias, and the RTD would be acceptable for this measurement, so the accuracy is not important.

Answer is A.

8. The resolution of the circuit, in volts, is

$$\epsilon_V = \frac{V_H - V_L}{2^n}$$
$$= \frac{0.30 \text{ V} - (-0.150 \text{ V})}{(2)^{16}}$$
$$= 6.8664 \times 10^{-6} \text{ V}$$

This voltage is to represent 0.001°C with the RTD; therefore, the sensitivity of the RTD should be

$$\frac{0.001°C}{6.8664 \times 10^{-6} \text{ V}} = 145.64 \text{ °C/V} \quad (146 \text{ °C/V})$$

The temperature range that this sensor can represent is derived from the number of bits and the value that represents 0°C. The voltage range is

$$(0.001°C)((2)^{16}) = 65.54°C$$

The value that represents 0°C is 0 V, and one third of the voltage range is below zero while two thirds of the voltage range is above zero. The lower and upper limits the sensor can represent are

$$\frac{-65.54°C}{3} = -21.85°C \quad (-21.9°C)$$

$$\frac{(2)(65.54°C)}{3} = 43.69°C \quad (43.7°C)$$

Answer is B.

9. The gage factor can be expressed in terms of length and resistance.

$$GF = \frac{\frac{\Delta R}{R}}{\frac{\Delta L}{L}}$$

Solve for the change in length.

$$\Delta L = \frac{\frac{\Delta R}{R}}{\frac{GF}{L}} = \frac{\frac{0.120\,\Omega}{120\,\Omega}}{\frac{2.00}{1.000\text{ cm}}}$$

$$= 5.00 \times 10^{-4}\text{ cm}$$

The resistance decreased, so length of the strain gage also decreased.

$$L_{\text{final}} = L - \Delta L$$

$$= 1.000\text{ cm} - 0.0005\text{ cm}$$

$$= 0.9995\text{ cm}$$

Answer is A.

$\mathit{50}$ Controls

Subjects

FEEDBACK THEORY	50-1
BLOCK DIAGRAM ALGEBRA	50-2
PREDICTING SYSTEM TIME RESPONSE	50-2
INITIAL AND FINAL VALUES	50-2
SPECIAL CASES OF STEADY-STATE RESPONSE	50-3
POLES AND ZEROS	50-3
PREDICTING SYSTEM TIME RESPONSE FROM RESPONSE POLE-ZERO DIAGRAMS	50-3
FREQUENCY RESPONSE	50-4
GAIN CHARACTERISTIC	50-4
PHASE CHARACTERISTIC	50-5
STABILITY	50-5
BODE PLOTS	50-5
ROOT-LOCUS DIAGRAMS	50-6
ROUTH CRITERION	50-6
APPLICATION TO CONTROL SYSTEMS	50-6
STATE MODEL REPRESENTATION	50-7

Nomenclature

BW	bandwidth
$e(t)$	error
$E(s)$	error, $\mathcal{L}(e(t))$
$F(s)$	forcing function, $\mathcal{L}(f(t))$
$G(s)$	forward transfer function
$H(s)$	reverse transfer function
j	$\sqrt{-1}$
K	gain
L	line length
$p(t)$	arbitrary function
$P(s)$	arbitrary function, $\mathcal{L}(p(t))$
Q	quality factor
$R(s)$	response function, $\mathcal{L}(r(t))$
t	time
$T(s)$	transfer function, $\mathcal{L}(t(t))$
u	input variable
V	voltage
x	state variable
y	output variable

Symbols

ϵ	a small number
ω	natural frequency

Subscripts

f	forced or feedback
i	in
n	natural
o	out

FEEDBACK THEORY

The output signal is returned as input in a feedback loop (feedback system). A basic feedback system consists of two black box units (a *dynamic unit* and a *feedback unit*), a pick-off point (take-off point), and a summing point (*comparator* or *summer*). The summing point is assumed to perform positive addition unless a minus sign is present. The incoming signal, V_i, is combined with the feedback signal, V_f, to give the *error (error signal)*, e. Whether addition or subtraction is used in Eq. 50.1 depends on whether the summing point is additive (i.e., a positive feedback system) or subtractive (i.e., a negative feedback system), respectively. $E(s)$ is the *error transfer function (error gain)*.

$$E(s) = \mathcal{L}(e(t)) = V_i(s) \pm V_f(s)$$
$$= V_i(s) \pm H(s)V_o(s) \qquad \mathit{50.1}$$

The ratio $E(s)/V_i(s)$ is the *error ratio (actuating signal ratio)*.

$$\frac{E(s)}{V_i(s)} = \frac{1}{1 + G(s)H(s)} \quad \text{[negative feedback]} \quad \mathit{50.2}$$

$$\frac{E(s)}{V_i(s)} = \frac{1}{1 - G(s)H(s)} \quad \text{[positive feedback]} \quad \mathit{50.3}$$

Since the dynamic and feedback units are black boxes, each has an associated transfer function. The transfer function of the dynamic unit is known as the *forward transfer function (direct transfer function)*, $G(s)$. In most feedback systems—amplifier circuits in particular—the magnitude of the forward transfer function is

Figure 50.1 Feedback System

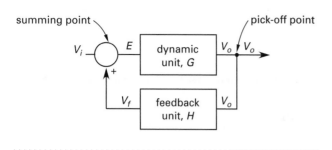

known as the *forward gain* or *direct gain*. $G(s)$ can be a scalar if the dynamic unit merely scales the error. However, $G(s)$ is normally a complex operator that changes both the magnitude and the phase of the error.

$$V_o(s) = G(s)E(s) \qquad 50.4$$

The pick-off point transmits the output signal, V_o, from the dynamic unit back to the feedback element. The output of the dynamic unit is not reduced by the pick-off point. The transfer function of the feedback unit is the *reverse transfer function* (*feedback transfer function*, *feedback gain*, etc.), $H(s)$, which can be a simple magnitude-changing scalar or a phase-shifting function.

$$V_f(s) = H(s)V_o(s) \qquad 50.5$$

The ratio $V_f(s)/V_i(s)$ is the *feedback ratio (primary feedback ratio)*.

$$\frac{V_f(s)}{V_i(s)} = \frac{G(s)H(s)}{1+G(s)H(s)} \quad \text{[negative feedback]} \quad 50.6$$

$$\frac{V_f(s)}{V_i(s)} = \frac{G(s)H(s)}{1-G(s)H(s)} \quad \text{[positive feedback]} \quad 50.7$$

The *loop transfer function* (*loop gain, open-loop gain*, or *open-loop transfer function*) is the gain after going around the loop one time, $\pm G(s)H(s)$.

The *overall transfer function* (*closed-loop transfer function, control ratio, system function, closed-loop gain*, etc.), $G_{\text{loop}}(s)$, is the overall transfer function of the feedback system. The quantity $1 + G(s)H(s) = 0$ is the *characteristic equation*. The *order of the system* is the largest exponent of s in the characteristic equation. (This corresponds to the highest-order derivative in the system equation.)

$$G_{\text{loop}}(s) = \frac{V_o(s)}{V_i(s)} = \frac{G(s)}{1+G(s)H(s)} \quad \text{[negative feedback]}$$
$$50.8$$

$$G_{\text{loop}}(s) = \frac{V_o(s)}{V_i(s)} = \frac{G(s)}{1-G(s)H(s)} \quad \text{[positive feedback]}$$
$$50.9$$

With positive feedback and $G(s)H(s)$ less than 1.0, G_{loop} will be larger than $G(s)$. This increase in gain is a characteristic of positive feedback systems. As $G(s)H(s)$ approaches 1.0, the closed-loop transfer function increases without bound, usually an undesirable effect.

In a negative feedback system, the denominator of Eq. 50.8 will be greater than 1.0. Although the closed-loop transfer function will be less than $G(s)$, there may be other desirable effects. Generally, a system with negative feedback will be less sensitive to variations in temperature, circuit component values, input signal frequency, and signal noise. Other benefits include distortion reduction, increased stability, and impedance matching. (For circuits to be directly connected in series without affecting their performance, all input impedances must be infinite and all output impedances must be zero.)

BLOCK DIAGRAM ALGEBRA

The functions represented by several interconnected black boxes (*cascaded blocks*) can be simplified into a single block operation. Some of the most important simplification rules of block diagram algebra are shown in Fig. 50.2. Case 3 represents the standard feedback model.

PREDICTING SYSTEM TIME RESPONSE

The transfer function is derived without knowledge of the input and is insufficient to predict the time response of the system. The system time response will depend on the form of the input function. Since the transfer function is expressed in the s-domain, the forcing and response functions must be also.

$$R(s) = T(s)F(s) \qquad 50.10$$

The time-based response function, $r(t)$, is found by performing the inverse Laplace transform.

$$r(t) = \mathcal{L}^{-1}(R(s)) \qquad 50.11$$

INITIAL AND FINAL VALUES

The initial and final (steady-state) values of any function, $P(s)$, can be found from the *initial* and *final value theorems*, respectively, providing the limits exist. Equations 50.12 and 50.13 are particularly valuable in determining the steady-state response (substitute $R(s)$ for $P(s)$) and the steady-state error (substitute $E(s)$ for $P(s)$).

$$\lim_{t\to 0^+} p(t) = \lim_{s\to\infty}(sP(s)) \quad \text{[initial value]} \quad 50.12$$

$$\lim_{t\to\infty} p(t) = \lim_{s\to 0}(sP(s)) \quad \text{[final value]} \quad 50.13$$

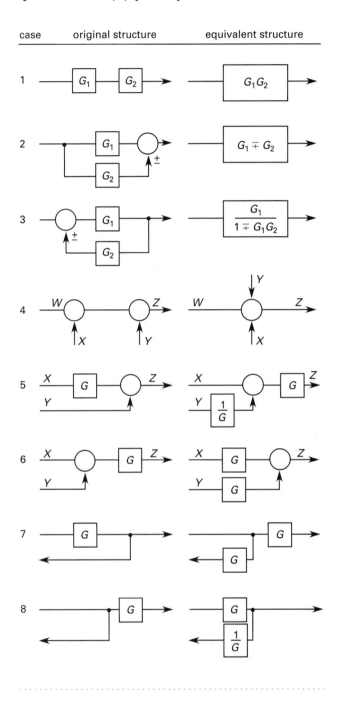

Figure 50.2 Rules of Simplifying Block Diagrams

SPECIAL CASES OF STEADY-STATE RESPONSE

In addition to determining the steady-state response from the final value theorem (see "Initial and Final Values"), the steady-state response to a specific input can be easily derived from the transfer function, $T(s)$, in a few specialized cases. For example, the steady-state response function for a system acted upon by an impulse is simply the transfer function. That is, a pulse has no long-term effect on a system.

$$R(\infty) = T(s) \quad \text{[pulse input]} \qquad 50.14$$

The steady-state response for a *step input* (often referred to as a *DC input*) is obtained by substituting 0 for s everywhere in the transfer function. (If the step has magnitude h, the steady-state response is multiplied by h.)

$$R(\infty) = T(0) \quad \text{[unit step input]} \qquad 50.15$$

The steady-state response for a sinusoidal input is obtained by substituting $j\omega_f$ for s everywhere in the transfer function, $T(s)$. The output will have the same frequency as the input. It is particularly convenient to perform sinusoidal calculations using phasor notation.

$$R(\infty) = T(j\omega_f) \qquad 50.16$$

POLES AND ZEROS

A *pole* is a value of s that makes a function, $P(s)$, infinite. Specifically, a pole makes the denominator of $P(s)$ zero. (Pole values are the system *eigenvalues*.) A *zero* of the function makes the numerator of $P(s)$ (and hence $P(s)$ itself) zero. Poles and zeros need not be real or unique; they can be imaginary and repeated within a function.

A *pole-zero diagram* is a plot of poles and zeros in the *s-plane*—a rectangular coordinate system with real and imaginary axes. Zeros are represented by $\bigcirc$'s; poles are represented as $\times$'s. Poles off the real axis always occur in conjugate pairs known as *pole pairs*.

Sometimes it is necessary to derive the function $P(s)$ from its pole-zero diagram. This will be only partially successful since repeating identical poles and zeros are not usually indicated on the diagram. Also, scale factors (scalar constants) are not shown.

PREDICTING SYSTEM TIME RESPONSE FROM RESPONSE POLE-ZERO DIAGRAMS

A response pole-zero diagram based on $R(s)$ can be used to predict how the system responds to a specific input. (Note that this pole-zero diagram must be based on the product $T(s)F(s)$ since that is how $R(s)$ is calculated. Plotting the product $T(s)F(s)$ is equivalent to plotting $T(s)$ and $F(s)$ separately on the same diagram.)

The system will experience an *exponential decay* when a single pole falls on the real axis. A pole with a value of $-r$, corresponding to the linear term $(s + r)$, will decay at the rate of e^{-rt}. The quantity $1/r$ is the decay *time constant*, the time for the response to achieve approximately 63% of its steady-state value. Thus, the farther left the point is located from the vertical imaginary axis, the faster the motion will die out.

Undamped sinusoidal oscillation will occur if a pole pair falls on the imaginary axis. A conjugate pole pair with the value of $\pm j\omega$ indicates oscillation with a natural frequency of ω rad/s.

Pole pairs to the left of the imaginary axis represent *decaying sinusoidal* response. The closer the poles are to the real (horizontal) axis, the slower will be the oscillations. The closer the poles are to the imaginary (vertical) axis, the slower will be the decay. The *natural frequency*, ω, of undamped oscillation can be determined from a *conjugate pole pair* having values of $r \pm \omega_f$.

$$\omega = \sqrt{r^2 + \omega_f^2} \qquad 50.17$$

Figure 50.3 Types of Response Determined by Pole Location

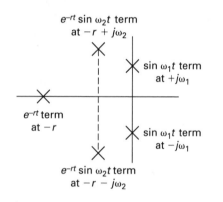

The magnitude and phase shift can be determined for any input frequency from the pole-zero diagram with the following procedure: Locate the angular frequency, ω_f, on the imaginary axis. Draw a line from each pole (i.e., a pole-line) and from zero (i.e., a zero-line) of $T(s)$ to this point. The angle of each of these lines is the angle between it and the horizontal real axis. The overall magnitude is the product of the lengths of the zero-lines divided by the product of the lengths of the pole-lines. (The scale factor must also be included because it is not shown on the pole-zero diagram.) The phase is the sum of the pole-angles less the sum of the zero-angles.

$$|R| = \frac{K \prod_z |L_z|}{\prod_p |L_p|} = \frac{K \prod_z \text{length}}{\prod_p \text{length}} \qquad 50.18$$

$$\underline{/R} = \sum_p \alpha - \sum_z \beta \qquad 50.19$$

Figure 50.4 Calculating Magnitude and Phase from a Pole-Zero Diagram

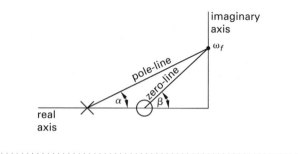

FREQUENCY RESPONSE

The gain and phase angle frequency response of a system will change as the forcing frequency is varied. The *frequency response* is the variation in these parameters, always with a sinusoidal input. *Gain* and *phase characteristics* are plots of the steady-state gain and phase angle responses with a sinusoidal input versus frequency. While a linear frequency scale can be used, frequency response is almost always presented against a logarithmic frequency scale.

The steady-state gain response is expressed in decibels, while the steady-state phase angle response is expressed in degrees. The gain is calculated from Eq. 50.20 where $|T(s)|$ is the absolute value of the steady-state response.

$$\text{gain} = 20 \log |T(j\omega)| \quad \text{[in dB]} \qquad 50.20$$

A doubling of $|T(j\omega)|$ is referred to as an *octave* and corresponds to a 6.02 dB increase. A tenfold increase in $|T(j\omega)|$ is a *decade* and corresponds to a 20 dB increase.

$$\text{no. of octaves} = \frac{\text{gain}_2 - \text{gain}_1 \quad \text{[in dB]}}{6.02}$$
$$= 3.32 \times \text{no. of decades} \qquad 50.21$$

$$\text{no. of decades} = \frac{\text{gain}_2 - \text{gain}_1 \quad \text{[in dB]}}{20}$$
$$= 0.301 \times \text{no. of octaves} \qquad 50.22$$

GAIN CHARACTERISTIC

The *gain characteristic* (M-curve for magnitude) is a plot of the gain as ω_f is varied. It is possible to make a rough sketch of the gain characteristic by calculating the gain at a few points (pole frequencies, $\omega = 0$, $\omega = \infty$, etc.). The curve will usually be asymptotic to several lines. The frequencies at which these asymptotes intersect are *corner frequencies*. The peak gain, M_p, coincides with the natural (resonant) frequency of the system. The gain characteristic peaks when the forcing frequency equals the natural frequency. It is also said that this peak corresponds to the resonant frequency. Strictly speaking, this is true, although the gain may not actually be resonant (i.e., may not be infinite). Large peak gains indicate lowered stability and large overshoots. The *gain crossover point*, if any, is the frequency at which log(gain) = 0.

The *half-power points (cut-off frequencies)* are the frequencies for which the gain is 0.707 (i.e., $\sqrt{2}/2$ times the peak value.) This is equivalent to saying the gain is 3 dB less than the peak gain. The *cut-off rate* is the slope of the gain characteristic in dB/octave at a half-power point. The frequency difference between the half-power points is the *bandwidth*, BW. The *closed-loop bandwidth* is the frequency range over which the closed-loop gain falls 3 dB below its value at $\omega = 0$. (The term "bandwidth" often means closed-loop bandwidth.) The *quality factor*, Q, is

$$Q = \frac{\omega_n}{\text{BW}} \qquad\qquad 50.23$$

Figure 50.5 Bandwidth

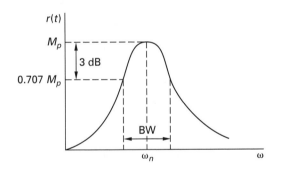

Since a low or negative gain (compared to higher parts of the curve) effectively represents attenuation, the gain characteristic can be used to distinguish between low- and high-pass filters. A low-pass filter will have a large gain at low frequencies and a small gain at high frequencies. Conversely, a high-pass filter will have a high gain at high frequencies and a low gain at low frequencies.

PHASE CHARACTERISTIC

The phase angle response will also change as the forcing frequency is varied. The *phase characteristic (α curve)* is a plot of the phase angle as ω_f is varied.

STABILITY

A stable system will remain at rest unless disturbed by external influence and will return to a rest position once the disturbance is removed. A pole with a value of $-r$ on the real axis corresponds to an exponential response of e^{-rt}. Since e^{-rt} is a decaying signal, the system is stable. Similarly, a pole of $+r$ on the real axis corresponds to an exponential response of e^{rt}. Since e^{rt} increases without limit, the system is unstable.

Since any pole to the right of the imaginary axis corresponds to a positive exponential, a *stable system* will have poles only in the left half of the s-plane. If there is an isolated pole on the imaginary axis, the response is stable. However, a conjugate pole pair on the imaginary axis corresponds to a sinusoid that does not decay with time. Such a system is considered to be unstable.

Passive systems (i.e., the homogeneous case) are not acted upon by a forcing function and are always stable. In the absence of an energy source, exponential growth cannot occur. *Active systems* contain one or more energy sources and may be stable or unstable.

There are several *frequency response (domain) analysis* techniques for determining the stability of a system, including Bode plot, root-locus diagram, Routh stability criterion, Hurwitz test, and Nichols chart. The term *frequency response* almost always means the steady-state response to a sinusoidal input.

The value of the denominator of $T(s)$ is the primary factor affecting stability. When the denominator approaches zero, the system increases without bound. In the typical feedback loop, the denominator is $1 \pm GH$, which can be zero only if $|GH| = 1$. It is logical, then, that most of the methods for investigating stability (e.g., Bode plots, root-locus, Nyquist analysis, and the Nichols chart) investigate the value of the open-loop transfer function, GH. Since $\log(1) = 0$, the requirement for stability is that $\log(GH)$ must not equal 0 dB.

A negative feedback system will also become unstable if it changes to a positive feedback system, which can occur when the feedback signal is changed in phase more than 180°. Therefore, another requirement for stability is that the phase angle change must not exceed 180°.

BODE PLOTS

Bode plots are gain and phase characteristics for the open-loop $G(s)H(s)$ transfer function that are used to determine the *relative stability* of a system. The gain characteristic is a plot of $20 \log(|G(s)H(s)|)$ versus ω for a sinusoidal input. (Bode plots, though similar in appearance to the gain and phase frequency response charts, are used to evaluate stability and do not describe the closed-loop system response.)

The *gain margin* is the number of decibels that the open-loop transfer function, $G(s)H(s)$, is below 0 dB at the *phase crossover frequency* (i.e., where the phase angle is -180 degrees). (If the gain happens to be plotted on a linear scale, the gain margin is the reciprocal of the gain at the phase crossover point.) The gain margin must be positive for a stable system, and the larger it is, the more stable the system will be.

The *phase margin* is the number of degrees the phase angle is above -180 degrees at the *gain crossover point* (i.e., where the logarithmic gain is 0 dB or the actual gain is 1).

In most cases, large positive gain and phase margins will ensure a stable system. However, the margins could have been measured at other than the crossover frequencies. Therefore, a Nyquist stability plot is needed to verify the absolute stability of a system.

Figure 50.6   Gain and Phase Margin Bode Plots

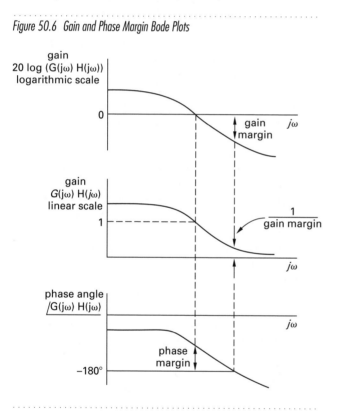

ROOT-LOCUS DIAGRAMS

A *root-locus diagram* is a pole-zero diagram showing how the poles of $G(s)H(s)$ move when one of the system parameters (e.g., the gain factor) in the transfer function is varied. The diagram gets its name from the need to find the roots of the denominator (i.e., the poles). The locus of points defined by the various poles is a line or curve that can be used to predict *points of instability* or other critical operating points. A point of

instability is reached when the line crosses the imaginary axis into the right-hand side of the pole-zero diagram.

A root-locus curve may not be contiguous, and multiple curves will exist for different sets of roots. Sometimes the curve splits into two branches. In other cases, the curve leaves the real axis at *breakaway points* and continues on with constant or varying slopes approaching asymptotes. One branch of the curve will start at each open-loop pole and end at an open-loop zero.

ROUTH CRITERION

The *Routh criterion* uses the coefficients of the polynomial characteristic equation. A table (the *Routh table*) of these coefficients is formed. The Routh-Hurwitz criterion states that the number of sign changes in the first column of the table equals the number of positive (unstable) roots. Therefore, the system will be stable if all entries in the first column have the same sign.

The table is organized in the following manner.

$$
\begin{array}{cccc}
a_0 & a_2 & a_4 & a_6 \cdots \\
a_1 & a_3 & a_5 & a_7 \cdots \\
b_1 & b_2 & b_3 & b_4 \cdots \\
c_1 & c_2 & c_3 & c_4 \cdots \\
\vdots & \vdots & \vdots & \vdots
\end{array}
$$

The remaining coefficients are calculated in the following pattern until all values are zero.

$$b_1 = \frac{a_1 a_2 - a_0 a_3}{a_1} \qquad 50.24$$

$$b_2 = \frac{a_1 a_4 - a_0 a_5}{a_1} \qquad 50.25$$

$$b_3 = \frac{a_1 a_6 - a_0 a_7}{a_1} \qquad 50.26$$

$$c_1 = \frac{b_1 a_3 - a_1 b_2}{b_1} \qquad 50.27$$

Special methods are used if there is a zero in the first column but nowhere else in that row. One of the methods is to substitute a small number, represented by ϵ or δ, for the zero and calculate the remaining coefficients as usual.

APPLICATION TO CONTROL SYSTEMS

A control system monitors a process and makes adjustments to maintain performance within certain acceptable limits. Feedback is implicitly a part of all control systems. The *controller (control element)* is the part of the control system that establishes the acceptable limits of performance, usually by setting its own reference inputs. The controller transfer function for a proportional controller is a constant: $G_1(s) = K$. The *plant (controlled system)* is the part of the system that responds to the controller. Both of these are in the forward loop. The input signal, $R(s)$, in Fig. 50.7 is known in a control system as the *command* or *reference value*. Figure 50.7 is known as a *control logic diagram* or *control logic block diagram*.

Figure 50.7 *Typical Feedback Control System*

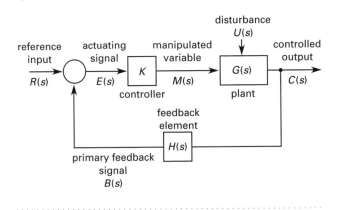

A *servomechanism* is a special type of control system in which the controlled variable is mechanical position, velocity, or acceleration. In many servomechanisms, $H(s) = 1$ (i.e., unity feedback) and it is desired to keep the output equal to the reference input (i.e., maintain a zero error function). If the input, $R(s)$, is constant, the descriptive terms *regulator* and *regulating system* are used.

STATE MODEL REPRESENTATION

While the classical methods of designing and analyzing control systems are adequate for most situations, state model representations are preferred for more complex cases, particularly those with multiple inputs and outputs or when behavior is nonlinear or varies with time. This evaluation method almost always is carried out on a digital or analog computer.

The state variables completely define the dynamic state (position, voltage, pressure, etc.), $x_i(t)$, of the system at time t. (In simple problems, the number of state variables corresponds to the number of *degrees of freedom*, n, of the system.) The n state variables are written in matrix form as a state vector, $\mathbf{X}$.

$$\mathbf{X} = \begin{pmatrix} x_1 \\ x_2 \\ x_3 \\ \vdots \\ x_n \end{pmatrix} \qquad 50.28$$

It is a characteristic of state models that the state vector is acted upon by a first-degree derivative operator, d/dt, to produce a differential term, $\mathbf{X}'$, of order 1,

$$\mathbf{X}' = \frac{d\mathbf{X}}{dt} \qquad 50.29$$

Equations 50.30 and 50.31 illustrate the general form of a state model representation: $\mathbf{U}$ is an r-dimensional (i.e., an $r \times 1$ matrix) *control vector*; $\mathbf{Y}$ is an m-dimensional (i.e., an $m \times 1$ matrix) *output vector*; $\mathbf{A}$ is an $n \times n$ *system matrix*; $\mathbf{B}$ is an $n \times r$ *control matrix*; and $\mathbf{C}$ is an $m \times n$ *output matrix*. The actual unknowns are the x_i's. The y_i's, which may not be needed in all problems, are only linear combinations of the x_i's. (For example, the x's might represent spring end positions; the y's might represent stresses in the spring. Then, $y = k\Delta x$.) Equation 50.30 is the *state equation*, and Eq. 50.31 is the *response equation*.

$$\mathbf{X}' = \mathbf{AX} + \mathbf{BU} \qquad 50.30$$
$$\mathbf{Y} = \mathbf{CX} \qquad 50.31$$

A conventional block diagram can be modified to show the multiplicity of signals in a state model, as shown in Fig. 50.8. (The block $\mathbf{I}/s$ is a diagonal identity matrix with elements of $1/s$. This effectively is an integration operator.) The actual physical system does not need to be a feedback system. The form of Eqs. 50.30 and 50.31 is the sole reason that a feedback diagram is appropriate.

Figure 50.8 *State Variable Diagram*

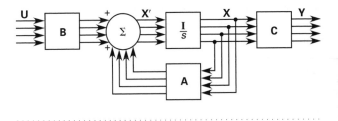

A state variable model permits only first-degree derivatives, so additional x_i state variables are used for higher-order terms (e.g., acceleration).

System controllability exists if all of the system states can be controlled by the inputs, $\mathbf{U}$. In state model language, system controllability means that an arbitrary initial state can be steered to an arbitrary target state in a finite amount of time. *System observability* exists if the initial system states can be predicted from knowing the inputs, **U**, and observing the outputs, **Y**. (*Kalman's theorem* based on matrix rank is used to determine system controllability and observability.)

SAMPLE PROBLEMS

Problem 1

Which of the following is not a direct benefit of using negative feedback?

   (A) improvement of a circuit's input and output impedances
   (B) reduction in temperature sensitivity
   (C) reduction in frequency sensitivity
   (D) reduction in power loss

Solution

Negative feedback does not reduce power losses.

Answer is D.

Problem 2

How is the sensitivity of a feedback system defined?

   (A) as the ratio of the percentage change in the loop transfer function to the percentage change in the forward transfer function
   (B) as the ratio of the output signal to the input signal
   (C) as the ratio of the forward transfer function to the reverse transfer function
   (D) as the ratio of change in the forward transfer function to the percentage change in the loop transfer function

Solution

The sensitivity of a feedback system is the percentage change in the loop transfer function divided by the percentage change in the forward transfer function.

Answer is A.

Problem 3

An 8-bit programmable module will be used to control temperature. If the temperature sensor's range is 0 to 300°F and the controller's range is 0 to 5 VDC, what is the resolution capacity of the controller?

   (A) 0.02 V/step
   (B) 1.17 °F/step
   (C) 2.34 °F/step
   (D) 60 °F/V

Solution

Since $2^8 = 256$, the controller can be in one of 256 distinct states (steps). The temperature monitored can range from 0°F to 300°F, so the resolution capacity is

$$\frac{300°\text{F} - 0°\text{F}}{256 \text{ steps}} = 1.17 \; °\text{F/step}$$

Answer is B.

Problem 4

A positioning servomechanism is represented by the block diagram shown. What is the magnitude of the overall feedback gain of the system?

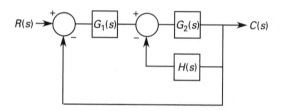

   (A) 0
   (B) 1
   (C) $H(s)$
   (D) $\dfrac{G_1(s)G_2}{H(s)}$

Solution

The output feeds directly back to the comparator. This is a unity feedback gain (feedback gain of 1).

Answer is B.

Problem 5

For the block diagram shown in Prob. 4, what is the overall transfer function?

   (A) $\dfrac{G_1(s)G_2(s)}{H(s)}$

   (B) $\dfrac{G_1(s) + G_2(s)}{1 - G_2(s)H(s)}$

   (C) $\dfrac{G_1(s)G_2(s)}{1 + G_2(s)H(s)}$

   (D) $\dfrac{\dfrac{G_1(s)G_2(s)}{1 + G_2(s)H(s)}}{1 + \dfrac{G_1(s)G_2(s)}{1 + G_2(s)H(s)}}$

Solution

The blocks can be rearranged and combined as shown.

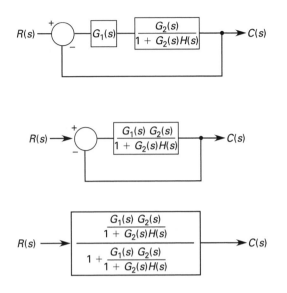

Answer is D.

Topic XVI: Engineering Economics

Diagnostic Examination for Engineering Economics

Chapter 51   Cash Flow and Equivalence
Chapter 52   Depreciation and Special Topics
Chapter 53   Comparison of Alternatives

ECONOMICS

Hint: For the most current information about the exam, visit www.ppi2pass.com/fefaqs.html regularly.
Use the Exam Forum to compare notes with other FE examinees.

Diagnostic Examination

TOPIC XVI: ENGINEERING ECONOMICS

TIME LIMIT: 45 MINUTES

1. Approximately how many years will it take to double an investment at a 6% effective annual rate?

- (A) 10 years
- (B) 12 years
- (C) 15 years
- (D) 17 years

2. An individual contributes $200 per month to a 401(k) retirement account. The account earns interest at a nominal annual interest rate of 8%, with interest being credited monthly. What is the value of the account after 35 years?

- (A) $368,000
- (B) $414,000
- (C) $447,000
- (D) $459,000

3. A graduating high school student decides to take a year off and work to save money for college. The student plans to invest all money earned in a savings account earning 6% interest, compounded quarterly. The student hopes to have $5000 by the time school starts in 12 months. How much money will the student have to save each month?

- (A) $396/month
- (B) $405/month
- (C) $407/month
- (D) $411/month

4. A gold mine is projected to produce $20,000 during its first year of operation, $19,000 the second year, $18,000 the third year, and so on. If the mine is expected to produce for a total of 10 years, and the effective annual interest rate is 6%, what is its present worth?

- (A) $118,000
- (B) $125,000
- (C) $150,000
- (D) $177,000

5. $5000 is put into an empty savings account with a nominal interest rate of 5%. No other contributions are made to the account. With monthly compounding, how much interest will have been earned after five years?

- (A) $1250
- (B) $1380
- (C) $1410
- (D) $1420

6. An engineer deposits $10,000 in a savings account on the day her child is born. She deposits an additional $1000 on every birthday after that. The account has a 5% nominal interest rate, compounded continuously. How much money will be in the account the day after the child's 21st birthday?

- (A) $36,200
- (B) $41,300
- (C) $64,800
- (D) $84,300

7. A machine costs $10,000 and can be depreciated over a period of four years, after which its salvage value will be $2000. What is the straight-line depreciation in year 3?

- (A) $2000
- (B) $2500
- (C) $4000
- (D) $6000

8. A groundwater treatment system is needed to remediate a solvent-contaminated aquifer. The system costs $2,500,000. It is expected to operate a total of 130,000 hours over a period of 10 years and then have a $250,000 salvage value. During its first year in service, it is operated for 6500 hours. What is its depreciation in the first year using the MACRS method?

- (A) $113,000
- (B) $125,000
- (C) $225,000
- (D) $250,000

9. A machine initially costing $25,000 will have a salvage value of $6000 after five years. Using MACRS depreciation, what will its book value be after the third year?

(A) $5470
(B) $7200
(C) $10,000
(D) $13,600

10. Given the following cash flow diagram and an 8% effective annual interest rate, what is the equivalent annual expense over the five-year period?

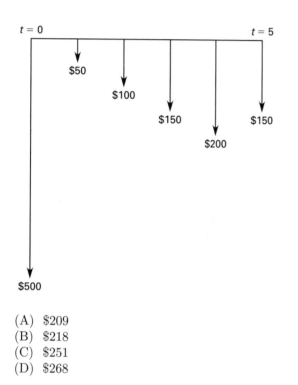

$t = 0$ .... $t = 5$

$50

$100

$150 $150

$200

$500

(A) $209
(B) $218
(C) $251
(D) $268

11. The construction of a volleyball court for the employees of a highly successful mid-sized publishing company in California is expected to cost $1200 and have annual maintenance costs of $300. At an effective annual interest rate of 5%, what is the project's capitalized cost?

(A) $1500
(B) $2700
(C) $7200
(D) $18,000

12. A warehouse building was purchased 10 years ago for $250,000. Since then, the effective annual interest rate has been 8%, inflation has been steady at 2.5%, and the building has had no deterioration or decrease in utility. What should the warehouse sell for today?

(A) $427,000
(B) $540,000
(C) $678,000
(D) $691,000

13. A delivery company is expanding its fleet by five vans at a total cost of $75,000. Operating and maintenance costs for the new vehicles are projected to be $20,000/year for the next eight years. After eight years, the vans will be sold for a total of $10,000. Annual revenues are expected to increase by $40,000 with the expanded fleet. What is the company's rate of return on the purchase?

(A) 19.7%
(B) 20.8%
(C) 21.7%
(D) 23.2%

14. A company is considering replacing its air conditioner. Management has narrowed the choices to two alternatives that offer comparable performance and considerable savings over their present system. The effective annual interest rate is 8%. What is the benefit-cost ratio of the better alternative?

	I	II
initial cost	$7000	$9000
annual savings	$1500	$1900
salvage value	$500	−$1250
life	15 years	15 years

(A) 1.73
(B) 1.76
(C) 1.84
(D) 1.88

15. A gourmet ice cream store has fixed expenses (rent, utilities, etc.) of $50,000/year. Its two full-time employees each earn $25,000 per year. There is also a part-time employee who makes $14,000 plus $6000 in overtime if sales reach $120,000 in a year. The ice cream costs $4/L to produce and sells for $7/L. What is the minimum number of liters the store must sell to break even?

(A) 38 000 L
(B) 39 000 L
(C) 40 000 L
(D) 41 000 L

SOLUTIONS TO DIAGNOSTIC EXAMINATION TOPIC XVI

1. Determine the number of years for the compound amount factor to equal 2.

$$F = 2P = P(F/P, i\%, n)$$
$$2 = (F/P, 6\%, n)$$
$$= (1 + i)^n$$
$$= (1 + 0.06)^n$$

$$\ln 2 = \ln 1.06^n$$
$$= n(\ln 1.06)$$

$$n = \frac{\ln 2}{\ln 1.06}$$
$$= 11.9 \quad (12 \text{ years})$$

Alternatively, use the 6% factor table. n is approximately 12 years.

Answer is B.

2. The effective rate per month is

$$i = \frac{r}{m} = \frac{0.08}{12} = 0.00667$$

Use the uniform series compound amount factor.

$$F = A(F/A, i\%, n)$$

Because compounding is monthly, n is the number of months.

$$n = (35 \text{ years})\left(12 \ \frac{\text{months}}{\text{year}}\right)$$
$$= 420 \text{ months}$$

$$F = A\left(\frac{(1+i)^n - 1}{i}\right)$$

$$= (\$200)\left(\frac{(1+0.00667)^{420} - 1}{0.00667}\right)$$

$$= \$459,227 \quad (\$459,000)$$

Answer is D.

3. The effective rate per quarter is

$$i = \frac{r}{m} = \frac{0.06}{4} = 0.015$$

There are four compounding periods during the year.

$$n = 4$$

Use the sinking fund factor.

$$A = F(A/F, i\%, n)$$
$$= (\$5000)(A/F, 1.5\%, 4)$$

$$= (\$5000)\left(\frac{0.015}{(1+0.015)^4 - 1}\right)$$

$$= \$1222$$

$$\text{monthly savings} = \frac{1222 \ \dfrac{\$}{\text{quarter}}}{3 \ \dfrac{\text{months}}{\text{quarter}}}$$

$$= \$407/\text{month}$$

Answer is C.

4. This cash flow is equivalent to a $20,000 annual series with a −$1000/year gradient. Use the tables of factors.

$$P = (\$20,000)(P/A, 6\%, 10) - (\$1000)(P/G, 6\%, 10)$$
$$= (\$20,000)(7.3601) - (\$1000)(29.6023)$$
$$= \$117,600 \quad (\$118,0000)$$

Answer is A.

5. The effective annual interest rate is

$$i_e = \left(1 + \frac{r}{m}\right)^m - 1$$

$$= \left(1 + \frac{0.05}{12}\right)^{12} - 1$$

$$= 0.05116$$

The total future value is

$$F = P(F/P, i\%, n) = P(1+i)^n$$
$$= (\$5000)(1 + 0.05116)^5$$
$$= \$6417$$

The interest available is

$$\text{interest} = F - P = \$6417 - \$5000$$
$$= \$1417 \quad (\$1420)$$

(This problem can also be solved by calculating the effective interest rate per period and compounding for 60 months.)

Answer is D.

6. The uniform series compound amount factor does not include a contribution at $t = 0$. Therefore, calculate the future value as the sum of a single payment and an annual series.

$$F = P(F/P, r\%, n) + A(F/A, r\%, n)$$

$$= P(e^{rn}) + A\left(\frac{e^{rn} - 1}{e^r - 1}\right)$$

$$= (\$10,000)e^{(0.05)(21)} + (\$1000)\left(\frac{e^{(0.05)(21)} - 1}{e^{(0.05)} - 1}\right)$$

$$= \$64,808 \quad (\$64,800)$$

Answer is C.

7. With the straight-line method, depreciation is the same in each year.

$$D_3 = D = \frac{C - S_n}{n}$$

$$= \frac{\$10,000 - \$2000}{4 \text{ years}}$$

$$= \$2000/\text{year} \quad (\$2000)$$

Answer is A.

8. MACRS depreciation depends only on the original cost, not on the salvage cost or hours of operation.

$$D_j = C(\text{factor})$$

$$D_1 = (\$2{,}500{,}000)(0.10)$$

$$= \$250{,}000$$

Answer is D.

9. Book value is the initial cost less the accumulated depreciation. Use the MACRS factors for a five-year recovery period.

$$BV = C - \sum_{j=1}^{t} D_j$$

$$= C - \sum_{j=1}^{t} \left(C(\text{factor}_j) \right)$$

$$= C \left(1 - \sum_{j=1}^{3} \text{factor}_j \right)$$

$$= (\$25{,}000)\left(1 - (0.20 + 0.32 + 0.192)\right)$$

$$= \$7200$$

Answer is B.

10. First, find the present worth of all of the cash flows.

$$P = \$500 + (\$50)(P/A, 8\%, 5) + (\$50)(P/G, 8\%, 4)$$
$$\quad + (\$100)(P/F, 8\%, 5)$$
$$= \$500 + (\$50)(3.9927) + (\$50)(4.6501)$$
$$\quad + (\$100)(0.6806)$$
$$= \$1000$$

Next, find the effective uniform annual expense (cost).

$$\text{EUAC} = (\$1000)(A/P, 8\%, 5)$$
$$= (\$1000)(0.2505)$$
$$= \$251$$

Answer is C.

11. Find the capitalized cost of the annual maintenance and add the initial construction cost to it.

$$P = C + \frac{A}{i} = \$1200 + \frac{\$300}{0.05}$$

$$= \$7200$$

Answer is C.

12. Ideally, the current price should be the future worth (from 10 years ago) adjusted for inflation. Use the inflation-adjusted interest rate, d, together with the single payment compound amount factor.

$$d = i + f + if$$
$$= 0.08 + 0.025 + (0.08)(0.025)$$
$$= 0.107$$

$$F = P(F/P, d\%, n)$$
$$= (\$250{,}000)(1 + 0.107)^{10}$$
$$= \$690{,}902 \quad (\$691{,}000)$$

Answer is D.

13. Rate of return is the effective annual interest rate that would make the investment's present worth zero.

$$P = 0 = -(\$75{,}000)$$
$$\quad + (\$40{,}000 - \$20{,}000)(P/A, i\%, 8)$$
$$\quad + (\$10{,}000)(P/F, i\%, 8)$$

$$\$75{,}000 = (\$20{,}000)\left(\frac{(1+i)^8 - 1}{i(1+i)^8}\right)$$
$$\quad + (\$10{,}000)(1+i)^{-8}$$

By trial and error, $i = 0.217$ (21.7%).

Answer is C.

14. Compute the present worth of the benefits and costs for each alternative. Salvage value should be counted as a decrease in cost, not as a benefit.

For alternative I,

$$B = (\$1500)(P/A, 8\%, 15)$$
$$= (\$1500)(8.5595)$$
$$= \$12{,}839$$

$$C = \$7000 - (\$500)(P/F, 8\%, 15)$$
$$= \$7000 - (\$500)(0.3152)$$
$$= \$6842$$

$$\frac{B}{C} = \frac{\$12{,}839}{\$6842} = 1.88$$

For alternative II,

$$B = (\$1900)(P/A, 8\%, 15)$$
$$= (\$1900)(8.5595)$$
$$= \$16{,}263$$

$$C = \$9000 + (\$1250)(P/F, 8\%, 15)$$
$$= \$9000 + (\$1250)(0.3152)$$
$$= \$9394$$

$$\frac{B}{C} = \frac{\$16{,}263}{\$9394} = 1.73$$

The alternatives cannot be compared to one another based simply on their ratios. Instead, perform an incremental analysis.

$$\frac{B_2 - B_1}{C_2 - C_1} = \frac{\$16{,}263 - \$12{,}839}{\$9394 - \$6842}$$
$$= 1.34$$

Because the incremental analysis ratio is greater than one, alternative II is superior.

Answer is A.

15. Calculate the costs and revenues assuming sales of $120,000 are exceeded.

$$\text{costs} = \$50{,}000 + (2)(\$25{,}000) + \$14{,}000$$
$$+ \$6000 + \left(4\,\frac{\$}{\text{L}}\right)Q$$
$$\text{revenues} = \left(7\,\frac{\$}{\text{L}}\right)Q$$

At the break-even point, costs equal revenues.

$$\text{revenues} = \text{costs}$$
$$\left(7\,\frac{\$}{\text{L}}\right)Q = \$120{,}000 + \left(4\,\frac{\$}{\text{L}}\right)Q$$
$$Q = 40\,000\ \text{L}$$

Check the assumption that sales exceed $120,000.

$$\left(7\,\frac{\$}{\text{L}}\right)(40\,000\ \text{L}) = \$280{,}000 \quad [\text{ok}]$$

Answer is C.

51 Cash Flow and Equivalence

Subjects

CASH FLOW 51-1
TIME VALUE OF MONEY 51-2
DISCOUNT FACTORS AND
    EQUIVALENCE 51-2
    Single Payment Equivalence 51-3
    Uniform Series Equivalence 51-3
    Gradient Equivalence 51-5
FUNCTIONAL NOTATION 51-5
NONANNUAL COMPOUNDING 51-5
CONTINUOUS COMPOUNDING 51-6

Nomenclature

A annual amount or annual value
C initial cost, or present worth
          (present value) of all costs
F future worth or future value
G uniform gradient amount
i interest rate per period
m number of compounding periods per year
n number of compounding periods
P present worth (present value)
r nominal rate per year (rate per annum)

Subscripts

0 initial
e annual effective rate
j at time j
n at time n

CASH FLOW

The sums of money recorded as receipts or disbursements in a project's financial records are called *cash flows*. Examples of cash flows are deposits to a bank, dividend interest payments, loan payments, operating and maintenance costs, and trade-in salvage on equipment. Whether the cash flow is considered to be a receipt or disbursement depends on the project under consideration. For example, interest paid on a sum in a bank account will be considered a disbursement to the bank and a receipt to the holder of the account.

Because of the time value of money, the timing of cash flows over the life of a project is an important factor. Although they are not always necessary in simple problems (and they are often unwieldy in very complex problems), *cash flow diagrams* can be drawn to help visualize and simplify problems that have diverse receipts and disbursements.

The following conventions are used to standardize cash flow diagrams.

- The horizontal (time) axis is marked off in equal increments, one per period, up to the duration of the project.

- *Receipts* are represented by arrows directed upward. *Disbursements* are represented by arrows directed downward. The arrow length is approximately proportional to the magnitude of the cash flow.

- Two or more transfers in the same period are placed end to end, and these may be combined.

- Expenses incurred before $t = 0$ are called *sunk costs*. Sunk costs are not relevant to the problem unless they have tax consequences in an after-tax analysis.

For example, consider a mechanical device that will cost $20,000 when purchased. Maintenance will cost $1000 each year. The device will generate revenues of $5000 each year for five years, after which the salvage value is expected to be $7000. The cash flow diagram is shown in Fig. 51.1(a), and a simplified version is shown in Fig. 51.1(b).

In order to evaluate a real-world project, it is necessary to present the project's cash flows in terms of standard cash flows that can be handled by engineering economic analysis techniques. The standard cash flows are single payment cash flow, uniform series cash flow, and gradient series cash flow.

A *single payment cash flow* can occur at the beginning of the time line (designated as $t = 0$), at the end of the time line (designated as $t = n$), or at any time in between.

The *uniform series cash flow*, illustrated in Fig. 51.2, consists of a series of equal transactions starting at $t = 1$ and ending at $t = n$. The symbol A (representing an

Figure 51.1 Cash Flow Diagrams

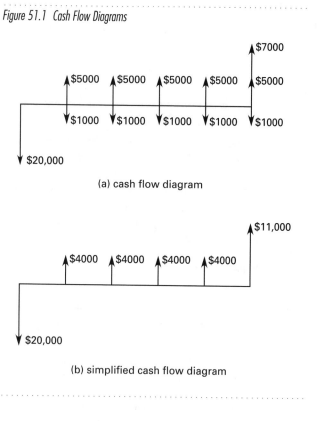

(a) cash flow diagram

(b) simplified cash flow diagram

Figure 51.2 Uniform Series

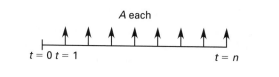

annual amount) is typically given to the magnitude of each individual cash flow.

Notice that the cash flows do not begin at the beginning of a year (i.e., the year 1 cash flow is at $t = 1$, not $t = 0$). This convention has been established to accommodate the timing of annual maintenance and other cash flows for which the *year-end convention* is applicable. The year-end convention assumes that all receipts and disbursements take place at the end of the year in which they occur. The exceptions to the year-end convention are *initial project cost* (purchase cost), *trade-in allowance*, and other cash flows that are associated with the inception of the project at $t = 0$.

The *gradient series cash flow*, illustrated in Fig. 51.3, starts with a cash flow (typically given the symbol G) at $t = 2$ and increases by G each year until $t = n$, at which time the final cash flow is $(n-1)G$. The value of the gradient at $t = 1$ is zero.

Figure 51.3 Gradient Series

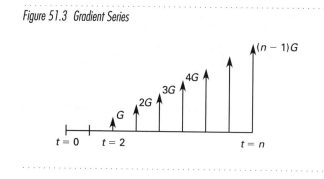

TIME VALUE OF MONEY

Consider $100 placed in a bank account that pays 5% effective annual interest at the end of each year. After the first year, the account will have grown to $105. After the second year, the account will have grown to $110.25.

The fact that $100 today grows to $105 in one year at 5% annual interest is an example of the *time value of money* principle. This principle states that funds placed in a secure investment will increase in value in a way that depends on the elapsed time and the interest rate.

The interest rate that is used in calculations is known as the *effective interest rate*. If compounding is once a year, it is known as the *effective annual interest rate*. However, effective quarterly, monthly, or daily interest rates are also used.

DISCOUNT FACTORS AND EQUIVALENCE

Assume that you will have no need for money during the next two years, and any money you receive will immediately go into your account and earn a 5% effective annual interest rate. Which of the following options would be more desirable to you?

option a: receive $100 now

option b: receive $105 in one year

option c: receive $110.25 in two years

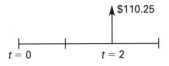

None of the options is superior under the assumptions given. If you choose the first option, you will immediately place $100 into a 5% account, and in two years the account will have grown to $110.25. In fact, the account will contain $110.25 at the end of two years regardless of which option you choose. Therefore, these alternatives are said to be *equivalent*.

The three options are equivalent only for money earning 5% effective annual interest rate. If a higher interest rate can be obtained, then the first option will yield the most money after two years. Thus, equivalence depends on the interest rate, and an alternative that is acceptable to one decision maker may be unacceptable to another who invests at a higher rate. The procedure for determining the equivalent amount is known as *discounting*.

Single Payment Equivalence

The equivalent future amount, F, at $t = n$, of any *present amount*, P, at $t = 0$ is called the *future worth* and can be calculated from Eq. 51.1. In this equation, and for all the other discounting formulas, the interest rate used must be the effective rate per period. The basis of the rate (annually, monthly, etc.) must agree with the type of period used to count n. Thus, it would be incorrect to use an effective annual interest rate if n was the number of compounding periods in months.

$$F = P(1 + i)^n \qquad 51.1$$

The factor $(1 + i)^n$ is known as the *single payment compound amount factor*.

Similarly, the equivalence of any future amount to any present amount is called the *present worth* and can be calculated from Eq. 51.2.

$$P = F(1 + i)^{-n} = \frac{F}{(1 + i)^n} \qquad 51.2$$

The factor $(1 + i)^{-n}$ is known as the *single payment present worth factor*.

Rather than actually writing the formula for the compound amount factor (which converts a present amount to a future amount), it is common convention to substitute the standard functional notation of $(F/P, i\%, n)$. This notation is interpreted as, "Find F, given P, using an interest rate of $i\%$ over n years." Thus, the future value in n periods of a present amount would be symbolically written as

$$F = P(F/P, i\%, n) \qquad 51.3$$

Similarly, the present worth factor has a functional notation of $(P/F, i\%, n)$. The present worth of a future amount n periods from now would be symbolically written as

$$P = F(P/F, i\%, n) \qquad 51.4$$

The discounting factors are listed in Table 51.1 in symbolic and formula form. Normally, it will not be necessary to calculate factors from these formulas. Values of these cash flow (discounting) factors are tabulated in the tables at the end of this chapter for various combinations of i and n. For intermediate values, computing the factors from the formulas may be necessary, or linear interpolation can be used as an approximation.

Uniform Series Equivalence

A cash flow that repeats at the end of each year for n years without change in amount is known as an *annual amount* and is given the symbol A. (This is shown in Fig. 51.2.) Although the equivalent value for each of the n annual amounts could be calculated and then summed, it is more expedient to use one of the uniform series factors. For example, it is possible to convert from an annual amount to a future amount by using the *(F/A) uniform series compound amount factor*.

$$F = A(F/A, i\%, n) \qquad 51.5$$

Example 51.1

Suppose you deposited $200 at the end of every year for seven years in an account that earned 6% annual effective interest. At the end of seven years, how much would the account be worth?

Solution

$$\begin{aligned} F &= (\$200)(F/A, 6\%, 7) \\ &= (\$200)\left(\frac{(1 + 0.06)^7 - 1}{0.06}\right) \\ &= (\$200)(8.3938) \\ &= \$1678.76 \end{aligned}$$

(The value of 8.3938 could easily have been obtained directly from Table 51.2 at the end of this chapter.)

Table 51.1 Discount Factors for Discrete Compounding

factor name	converts	symbol	formula
single payment compound amount	P to F	$(F/P, i\%, n)$	$(1+i)^n$
single payment present worth	F to P	$(P/F, i\%, n)$	$(1+i)^{-n}$
uniform series sinking fund	F to A	$(A/F, i\%, n)$	$\dfrac{i}{(1+i)^n - 1}$
capital recovery	P to A	$(A/P, i\%, n)$	$\dfrac{i(1+i)^n}{(1+i)^n - 1}$
uniform series compound amount	A to F	$(F/A, i\%, n)$	$\dfrac{(1+i)^n - 1}{i}$
uniform series present worth	A to P	$(P/A, i\%, n)$	$\dfrac{(1+i)^n - 1}{i(1+i)^n}$
uniform gradient present worth	G to P	$(P/G, i\%, n)$	$\dfrac{(1+i)^n - 1}{i^2(1+i)^n} - \dfrac{n}{i(1+i)^n}$
uniform gradient future worth	G to F	$(F/G, i\%, n)$	$\dfrac{(1+i)^n - 1}{i^2} - \dfrac{n}{i}$
uniform gradient uniform series	G to A	$(A/G, i\%, n)$	$\dfrac{1}{i} - \dfrac{n}{(1+i)^n - 1}$

A *sinking fund* is a fund or account into which annual deposits of A are made in order to accumulate F at $t = n$ in the future. Because the annual deposit is calculated as $A = F(A/F, i\%, n)$, the (A/F) factor is known as the *sinking fund factor*.

Example 51.2

Suppose you want exactly $1600 in the previous investment account at the end of the seventh year. By using the sinking fund factor, you could calculate the necessary annual amount you would need to deposit.

Solution

$$A = F(A/F, 6\%, 7) = (\$1600)\left(\frac{0.06}{(1+0.06)^7 - 1}\right)$$
$$= (\$1600)(0.1191)$$
$$= \$190.56$$

An *annuity* is a series of equal payments, A, made over a period of time. Usually, it is necessary to "buy into"

an investment (a bond, an insurance policy, etc.) in order to fund the annuity. In the case of an annuity that starts at the end of the first year and continues for n years, the purchase price, P, would be

$$P = A(P/A, i\%, n) \qquad 51.6$$

Example 51.3

Suppose you will retire in exactly one year and want an account that will pay you $20,000 a year for the next 15 years. (The fund will be depleted at the end of the fifteenth year.) Assuming a 6% annual effective interest rate, what is the amount you would need to deposit now?

Solution

$$P = A(P/A, 6\%, 15)$$
$$= (\$20,000)\left(\frac{(1+0.06)^{15} - 1}{(0.06)(1+0.06)^{15}}\right)$$
$$= (\$20,000)(9.7122)$$
$$= \$194,244$$

Gradient Equivalence

If the cash flow has the proper form (i.e., Fig. 51.3), its present worth can be determined by using the *uniform gradient factor*, $(P/G, i\%, n)$. The uniform gradient factor finds the present worth of a uniformly increasing cash flow. By definition of a uniform gradient, the cash flow starts in year 2, not in year 1.

There are three common difficulties associated with the form of the uniform gradient. The first difficulty is that the first cash flow starts at $t = 2$. This convention recognizes that annual costs, if they increase uniformly, begin with some value at $t = 1$ (due to the year-end convention), but do not begin to increase until $t = 2$. The tabulated values of (P/G) have been calculated to find the present worth of only the increasing part of the annual expense. The present worth of the base expense incurred at $t = 1$ must be found separately with the (P/A) factor.

The second difficulty is that, even though the $(P/G, i\%, n)$ factor is used, there are only $n - 1$ actual cash flows. n must be interpreted as the *period number* in which the last gradient cash flow occurs, not the number of gradient cash flows.

Finally, the sign convention used with gradient cash flows may seem confusing. If an expense increases each year, the gradient will be negative, since it is an expense. If a revenue increases each year, the gradient will be positive. In most cases, the sign of the gradient depends on whether the cash flow is an expense or a revenue.

Example 51.4

A bonus package pays an employee $1000 at the end of the first year, $1500 at the end of the second year, and so on, for the first nine years of employment. What is the present worth of the bonus package at 6% interest?

Solution

$$P = (\$1000)(P/A, 6\%, 9) + (\$500)(P/G, 6\%, 9)$$
$$= (\$1000)\left(\frac{(1 + 0.06)^9 - 1}{(0.06)(1 + 0.06)^9}\right)$$
$$+ (\$500)\left(\frac{(1 + 0.06)^9 - 1}{(0.06)^2(1 + 0.06)^9}\right.$$
$$\left. - \frac{9}{(0.06)(1 + 0.06)^9}\right)$$
$$= (\$1000)(6.8017) + (\$500)(24.5768)$$
$$= \$19,090$$

FUNCTIONAL NOTATION

There are several ways of remembering what the functional notation means. One method of remembering which factor should be used is to think of the factors as *conditional probabilities*. The conditional probability of event A given that event B has occurred is written as $P\{A|B\}$, where the given event comes after the vertical bar. In the standard notational form of discounting factors, the given amount is similarly placed after the slash. What you want, A, comes before the slash. (F/P) would be a factor to find F given P.

Another method of remembering the notation is to interpret the factors algebraically. Thus, the (F/P) factor could be thought of as the fraction F/P. The numerical values of the discounting factors are consistent with this algebraic manipulation. Thus, the (F/A) factor could be calculated as $(F/P)(P/A)$. This consistent relationship can be used to calculate other factors that might be occasionally needed, such as (F/G) or (G/P). For instance, the annual cash flow that would be equivalent to a uniform gradient may be found from

$$A = G(P/G, i\%, n)(A/P, i\%, n) \qquad 51.7$$

NONANNUAL COMPOUNDING

If $100 is invested at 5%, it will grow to $105 in one year. If only the original principal accrues interest, the interest is known as *simple interest* and the account will grow to $110 in the second year, $115 in the third year, and so on. Simple interest is rarely encountered in engineering economic analyses.

More often, both the principal and the interest earned accrue interest, and this is known as *compound interest*. If the account is compounded yearly, then during the second year, 5% interest continues to be accrued, but on $105, not $100, so the value at year end will be $110.25. The value after the third year will be $115.76, and so on.

The interest rate used in the discount factor formulas is the *interest rate per period*, i (called the *yield* by banks). If the interest period is one year (i.e., the interest is compounded yearly), then the interest rate per period, i, is equal to the *annual effective interest rate*, i_e. The annual effective interest rate is the rate that would yield the same accrued interest at the end of the year if the account were compounded yearly.

The term *nominal interest rate*, r (*rate per annum*), is encountered when compounding is more than once per year. The nominal rate does not include the effect of compounding and is not the same as the annual effective interest rate.

ECONOMICS
Equivalence

The effective interest rate can be calculated from the nominal rate if the number of compounding periods per year is known. If there are m compounding periods during the year (two for semiannual compounding, four for quarterly compounding, twelve for monthly compounding, etc.), the *effective interest rate per period*, i, is r/m. The effective annual interest rate, i_e, can be calculated from the interest rate per period by using Eq. 51.9.

$$i = \frac{r}{m} \qquad\qquad 51.8$$

$$i_e = (1 + i)^m - 1$$
$$= \left(1 + \frac{r}{m}\right)^m - 1 \qquad 51.9$$

Sometimes, only the effective rate per period (e.g., per month) is known. However, compounding for m periods at an effective interest rate per period is not affected by the definition or length of the period. For example, compounding for 365 periods (days) at an interest rate of 0.03808% is the same as compounding for 12 periods (months) at an interest rate of 1.164%, or once at an effective annual interest rate of 14.9%. In each case, the interest rate per period is different, but the effective annual interest rate is the same. If only the daily effective rate were given, the discount factor formulas could be used with $i = 0.03808\%$ and $n = 365$ to represent each yearly cash flow. Equation 51.9 could be used to calculate $i_e = 14.9\%$ to use with $n = 1$ for each yearly cash flow.

Since they do not account for the effect of compounding, nominal rates cannot be compared unless the method of compounding is specified. The only practical use for a nominal rate is for calculating the effective rate.

The following rules may be used to determine what type of interest rate is given in a problem.

- Unless specifically qualified in the problem, the interest rate given is an annual rate. If the compounding period is not specified, the interest rate is the annual effective interest rate, i_e.

- If the compounding is annual, the rate given is the effective rate, i_e. If compounding is not annual, the rate given is the nominal rate, r.

CONTINUOUS COMPOUNDING

Discount factors for continuous compounding are different from those for discrete compounding. The discounting factors can be calculated directly from the nominal interest rate, r, and number of years, n, without having to find the effective interest rate per period.

$$(F/P, r\%, n) = e^{rn} \qquad\qquad 51.10$$
$$(P/F, r\%, n) = e^{-rn} \qquad\qquad 51.11$$

$$(A/F, r\%, n) = \frac{e^r - 1}{e^{rn} - 1} \qquad 51.12$$

$$(F/A, r\%, n) = \frac{e^{rn} - 1}{e^r - 1} \qquad 51.13$$

$$(A/P, r\%, n) = \frac{e^r - 1}{1 - e^{-rn}} \qquad 51.14$$

$$(P/A, r\%, n) = \frac{1 - e^{-rn}}{e^r - 1} \qquad 51.15$$

The effective annual interest rate determined on a daily compounding basis will not be significantly different than if continuous compounding is assumed.

SAMPLE PROBLEMS

Problem 1

If a credit union pays 4.125% interest compounded quarterly, what is the effective annual interest rate?

   (A) 4.189%
   (B) 8.250%
   (C) 12.89%
   (D) 17.55%

Solution

4.125% is the nominal annual rate, r.

$$i_e = (1 + i)^m - 1$$
$$= \left(1 + \frac{0.04125}{4}\right)^4 - 1$$
$$= 0.04189 \quad (4.189\%)$$

Answer is A.

Problem 2

The national debt is approximately $4 trillion. What is the required payment per year to completely pay off the debt in 20 years, assuming an interest rate of 6%?

   (A) $315 billion
   (B) $325 billion
   (C) $350 billion
   (D) $415 billion

Solution

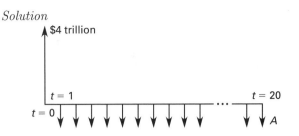

Use the capital recovery discount factor from the tables.

$$(A/P, 6\%, 20) = 0.0872$$
$$A = P(A/P, 6\%, 20)$$
$$= (\$4{,}000{,}000{,}000{,}000)(0.0872)$$
$$= \$348{,}800{,}000{,}000 \quad (\$350 \text{ billion})$$

Answer is C.

Problem 3

The president of a growing engineering firm wishes to give each of 50 employees a holiday bonus. How much is needed to invest monthly for a year at 12% nominal interest rate, compounded monthly, so that each employee will receive a \$1000 bonus?

(A) \$2070
(B) \$3840
(C) \$3940
(D) \$4170

Solution

The total holiday bonus is

$$(50)(\$1000) = \$50{,}000$$

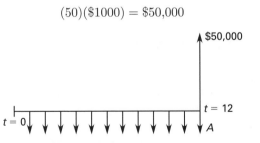

Use the uniform series sinking fund discount factor. The interest period is one month, there are 12 compounding periods, and the effective interest rate per interest period is $12\%/12 = 1\%$.

$$(A/F, 1\%, 12) = \frac{0.01}{(1 + 0.01)^{12} - 1} = 0.0788$$
$$A = F(A/F, 1\%, 12)$$
$$= (\$50{,}000)(0.0788)$$
$$= \$3940$$

Answer is C.

Problem 4

If the nominal interest rate is 3%, how much is \$5000 worth in 10 years in a continuously compounded account?

(A) \$3180
(B) \$4490
(C) \$5420
(D) \$6750

Solution

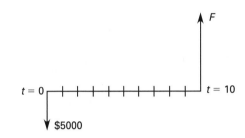

Use the single payment compound amount factor for continuous compounding.

$$n = 10$$
$$r = 3\%$$
$$(F/P, r\%, n) = e^{rn}$$
$$(F/P, 3\%, 10) = e^{(0.03)(10)} = 1.34986$$
$$F = P(F/P, 3\%, 10)$$
$$= (\$5000)(1.34986)$$
$$= \$6749$$

Answer is D.

Problem 5

An engineering graduate plans to buy a home. She has been advised that her monthly house and property tax payment should not exceed 35% of her disposable monthly income. After researching the market, she determines she can obtain a 30-year home loan for 6.95% annual interest per year, compounded monthly. Her monthly property tax payment will be approximately \$150. What is the maximum amount she can pay for a house if her disposable monthly income is \$2000?

(A) \$80,000
(B) \$83,100
(C) \$85,200
(D) \$90,500

Solution

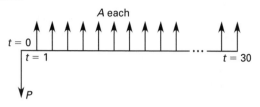

The amount available for monthly house payments, A, is

$$(\$2000)(0.35) - \$150 = \$550$$

Use the uniform series present worth discount factor. The effective rate per period is

$$i = \frac{0.0695}{12 \text{ months}} = 0.00579 \text{ per month}$$

$$n = (30 \text{ years})\left(12 \frac{\text{months}}{\text{year}}\right) = 360 \text{ months}$$

There are no tables for this interest rate.

$$(P/A, 0.579\%, 360) = \frac{(1+i)^n - 1}{i(1+i)^n}$$

$$= \frac{(1 + 0.00579)^{360} - 1}{(0.00579)(1 + 0.00579)^{360}}$$

$$= 151.10$$

$$P = A(P/A, 0.579\%, 360)$$

$$= (\$550)(151.10)$$

$$= \$83,105$$

Answer is B.

Problem 6

The designer of the penstock for a small hydroelectric cogeneration station has the option of using steel pipe, which costs $150,000 installed and requires $5000 yearly for painting and leak-checking maintenance, or DSR4.3 (heavy-duty plastic) pipe, which costs $180,000 installed and requires $1200 yearly for leak-checking maintenance. Both options have an expected life of 25 years. If the interest rate is 8%, which choice has the lower present equivalent cost and how much lower is it?

    (A) DSR4.3 costs less by $10,600.
    (B) Steel pipe costs less by $10,600.
    (C) DSR4.3 costs less by $65,000.
    (D) Steel pipe costs less by $65,000.

Solution

The problem requires a comparison of the uniform series present worth of each alternative.

For steel pipe,

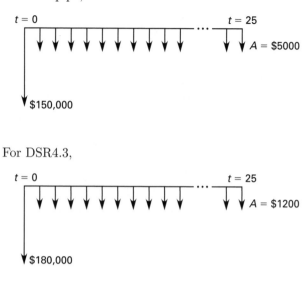

For DSR4.3,

$$P(\text{steel pipe}) = \$150,000 + A(P/A, 8\%, 25)$$
$$= \$150,000 + (\$5000)(10.6748)$$
$$= \$203,374$$
$$P(\text{DSR4.3}) = \$180,000 + A(P/A, 8\%, 25)$$
$$= \$180,000 + (\$1200)(10.6748)$$
$$= \$192,810$$

Using DSR4.3 is less expensive by

$$\$203,374 - \$192,810 = \$10,564 \quad (\$10,600)$$

Answer is A.

FE-STYLE EXAM PROBLEMS

1. If the interest rate on an account is 11.5% compounded yearly, approximately how many years will it take to triple the amount?

    (A) 8 years
    (B) 9 years
    (C) 10 years
    (D) 11 years

2. Fifteen years ago $1000 was deposited in a bank account, and today it is worth $2370. The bank pays interest semi-annually. What was the nominal annual interest rate paid on this account?

    (A) 2.9%
    (B) 4.4%
    (C) 5.0%
    (D) 5.8%

3. Mr. Jones plans to deposit $500 at the end of each month for 10 years at 12% annual interest, compounded monthly. The amount that will be available in two years is

(A) $13,000
(B) $13,500
(C) $14,000
(D) $14,500

4. The purchase price of a car is $25,000. Mr. Smith makes a down payment of $5000 and borrows the balance from a bank at 6% interest for five years. Calculate the nearest value of the required monthly payments to pay off the loan.

(A) $350
(B) $400
(C) $450
(D) $500

5. A piece of machinery can be bought for $10,000 cash or for $2000 down and payments of $750 per year for 15 years. What is the annual interest rate for the time payments?

(A) 1.51%
(B) 4.61%
(C) 7.71%
(D) 12.0%

For the following problems use the NCEES Handbook as your only reference.

6. You have borrowed $5000 and must pay it off in five equal annual payments. Your annual interest rate is 10%. How much interest will you pay in the first two years?

(A) $855
(B) $868
(C) $875
(D) $918

7. A company puts $25,000 down and will pay $5000 every year for the life of a machine (10 years). If the salvage value is zero and the interest rate is 10% compounded annually, what is the present value of the machine?

(A) $55,700
(B) $61,400
(C) $75,500
(D) $82,500

8. You borrow $3500 for one year from a friend at an interest rate of 1.5% per month instead of taking a loan from a bank at a rate of 18% per year. Compare how much money you will save or lose on the transaction.

(A) You will pay $55 more than if you borrowed from the bank.
(B) You will pay $630 more than if you borrowed from the bank.
(C) You will pay $685 more than if you borrowed from the bank.
(D) You will save $55 by borrowing from your friend.

9. If you invest $25,000 at 8% interest compounded annually, approximately how much money will be in the account at the end of 10 years?

(A) $31,000
(B) $46,000
(C) $54,000
(D) $75,000

10. A college student borrows $10,000 today at 10% interest compounded annually. Four years later, the student makes the first repayment of $3000. Approximately how much money will the student still owe on the loan after the first payment?

(A) $7700
(B) $8300
(C) $11,000
(D) $11,700

11. A 40-year-old consulting engineer wants to set up a retirement fund to be used starting at age 65. $20,000 is invested now at 6% compounded annually. Approximately how much money will be in the fund at retirement?

(A) $84,000
(B) $86,000
(C) $88,000
(D) $92,000

12. The maintenance cost for a car this year is expected to be $500. The cost will increase $50 each year for the subsequent 9 years. The interest is 8% compounded annually. What is the approximate present worth of maintenance for the car over the full 10 years?

(A) $4300
(B) $4700
(C) $5300
(D) $5500

13. A house is expected to have a maintenance cost of $1000 the first year. It is believed that the maintenance cost will increase $500 per year. The interest rate is 6% compounded annually. Over a 10-year period, what will be the approximate effective annual maintenance cost?

(A) $1900
(B) $3000
(C) $3500
(D) $3800

14. You deposited $10,000 in a savings account five years ago. The account has earned 5.25% interest compounded continuously since then. How much money is in the account today?

(A) $12,800
(B) $12,900
(C) $13,000
(D) $13,600

15. A young engineer wants to surprise her husband with a European vacation for their tenth anniversary, which is five years away. She determines that the trip will cost $5000. Assuming an interest rate of 5.50% compounded daily, approximately how much money does she need to deposit today for the trip?

(A) $3790
(B) $3800
(C) $3880
(D) $3930

16. A young woman plans to retire in 30 years. She intends to contribute the same amount of money each year to her retirement fund. The fund earns 10% compounded annually. She would like to withdraw $100,000 each year for 20 years, starting 1 year after the last contribution is made. Approximately how much money should she contribute to her retirement fund each year?

(A) $490
(B) $570
(C) $5200
(D) $11,000

17. A deposit of $1000 is made in a bank account that pays 8% interest compounded annually. Approximately how much money will be in the account after 10 years?

(A) $1890
(B) $2000
(C) $2160
(D) $2240

18. A deposit of $1000 is made in a bank account that pays 24% interest per year compounded quarterly. Approximately how much money will be in the account after 10 years?

(A) $7000
(B) $7200
(C) $8600
(D) $10,000

19. A machine costs $20,000 today and has an estimated scrap cash value of $2000 after eight years. Inflation is 8% per year. The effective annual interest rate earned on money invested is 8%. How much money needs to be set aside each year to replace the machine with an identical model eight years from now?

(A) $2970
(B) $3000
(C) $3290
(D) $3510

20. At what rate of annual interest will an investment quadruple itself in 12 years?

(A) 10.1%
(B) 11.2%
(C) 12.2%
(D) 13.1%

SOLUTIONS TO FE-STYLE EXAM PROBLEMS

1. The future amount will be three times the present amount when the (F/P) factor is equal to 3.

$$(F/P, i\%, n) = (1 + i)^n$$
$$(1 + 0.115)^n = 3$$
$$n \log 1.115 = \log 3$$
$$n = \frac{\log 3}{\log 1.115}$$
$$= 10.09 \text{ years} \quad (10 \text{ years})$$

Answer is C.

2.

$P = \$1000$

$n = (15 \text{ years})(2 \text{ compounding periods per year})$

$\quad = 30 \text{ compounding periods}$

$F = P(F/P, i\%, n)$

$\$2370 = (\$1000)(F/P, i\%, 30)$

$2.37 = (F/P, i\%, 30)$

Use the formula for single payment compounding. If the table values were available, i could be determined by using linear interpolation. The effective rate per period is

$$2.37 = (1 + i)^{30}$$
$$i = 0.02918 \quad (2.918\%)$$

The nominal annual interest rate is twice this amount, or 5.8%.

Answer is D.

3.

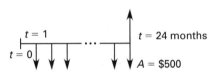

Use the uniform series compound amount discount factor.

$F = A(F/A, i\%, n)$

$A = \$500$

$i = \dfrac{12\%}{12 \text{ compounding periods per year}} = 1\%$

$n = (2 \text{ years})(12 \text{ compounding periods per year})$

$\quad = 24 \text{ compounding periods}$

$F = (\$500)(F/A, 1\%, 24) = (\$500)\left(\dfrac{(1 + 0.01)^{24} - 1}{0.01}\right)$

$\quad = (\$500)(26.9735)$

$\quad = \$13{,}487 \quad (\$13{,}500)$

Answer is B.

4.

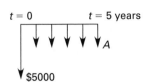

Use the capital recovery discount factor.

$A = P(A/P, i\%, n)$

$P = \$25{,}000 - \$5000 = \$20{,}000$

$i = \dfrac{6\%}{12 \text{ compounding periods per year}} = 0.5\%$

$n = (5 \text{ years})(12 \text{ months per year}) = 60$

$A = (\$20{,}000)(A/P, 0.5\%, 60)$

$\quad = (\$20{,}000)\left(\dfrac{(0.005)(1 + 0.005)^{60}}{(1 + 0.005)^{60} - 1}\right)$

$\quad = (\$20{,}000)(0.0193)$

$\quad = \$386 \quad (\$400)$

Answer is B.

5.

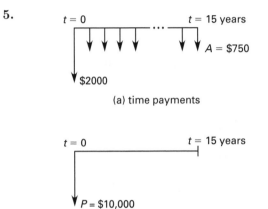

(a) time payments

(b) single payment

Use the uniform series present worth discount factor.

$P = \$10{,}000 - \$2000 = \$8000$

$A = \$750$

$n = 15$

$P = A(P/A, i\%, n)$

$\$8000 = (\$750)(P/A, i\%, 15)$

$10.67 = (P/A, i\%, 15)$

From the factor tables, $i\%$ is below 6%. The formula for uniform series present worth can be used to determine the interest rate more closely.

$$(P/A, i\%, 15) = \dfrac{(1 + i)^n - 1}{i(1 + i)^n}$$

$$10.67 = \dfrac{(1 + i)^{15} - 1}{i(1 + i)^{15}}$$

By trial and error,

$$i = 0.0461 \quad (4.61\%)$$

Answer is B.

6.

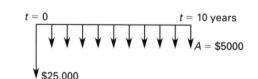

$A = (\$5000)(A/P, 10\%, 5)$

Find the amount you will pay each year.

$$P = \$5000$$
$$i = 10\%$$
$$n = 5$$
$$(A/P, 10\%, 5) = 0.2638$$
$$A = P(A/P, i\%, n)$$
$$= (\$5000)(0.2638)$$
$$= \$1319$$

The interest paid at the end of the first year is

$$(\$5000)(0.10) = \$500$$

The principal left after the first year is

$$\$5000 - (\$1319 - \$500) = \$4181$$

The interest paid at the end of the second year is

$$(\$4181)(0.10) = \$418$$

The total interest paid at the end of two years is

$$\$500 + \$418 = \$918$$

Answer is D.

7.

$t = 0$ $t = 10$ years

$A = \$5000$

$\$25,000$

Use the uniform series present worth factor.

$$A = \$5000$$
$$i = 10\%$$
$$n = 10 \text{ years}$$
$$(P/A, 10\%, 10) = 6.1446$$
$$P = A(P/A, i\%, n) + \$25,000$$
$$= (\$5000)(6.1446) + \$25,000$$
$$= \$55,723 \quad (\$55,700)$$

Answer is A.

8. The amount that you will pay your friend is

$$F = P(F/P, i\%, n)$$
$$= (\$3500)(F/P, 1.5\%, 12)$$
$$= (\$3500)(1.015)^{12}$$
$$= (\$3500)(1.1956)$$
$$= \$4185$$

The amount that you would have paid the bank is

$$F = P(F/P, i\%, n)$$
$$= (\$3500)(F/P, 18\%, 1)$$
$$= (\$3500)(1.18)$$
$$= \$4130$$

The difference is

$$\$4185 - \$4130 = \$55$$

You have paid your friend \$55 more than you would have paid the bank.

Answer is A.

9. The future worth of \$25,000 10 years from now is

$$F = P(F/P, 8\%, 10) = (\$25,000)(2.1589)$$
$$= \$53,973 \quad (\$54,000)$$

Answer is C.

10. The amount owed at the end of four years is

$$F = P(F/P, 10\%, 4)$$
$$= (\$10,000)(1.4641)$$
$$= \$14,641$$

The amount owed after the \$3000 payment is

$$\$14,641 - \$3000 = \$11,641 \quad (\$11,700)$$

Answer is D.

11. Determine the future worth of \$20,000 25 years from now.

$$F = P(F/P, 6\%, 25)$$
$$= (\$20,000)(4.2919)$$
$$= \$85,838 \quad (\$86,000)$$

Answer is B.

12. To find the present worth of maintenance, use both the uniform gradient and the uniform series factors. Notice that both factors are evaluated for 10 years.

$$P = A(P/A, 8\%, 10) + G(P/G, 8\%, 10)$$
$$= (\$500)(6.7101) + (\$50)(25.9768)$$
$$= \$4654 \quad (\$4700)$$

Answer is B.

13. Use the uniform gradient uniform series equivalent factor to determine the effective annual cost.

$$A = A_1 + G(A/G, 6\%, 10)$$
$$= \$1000 + (\$500)(4.0220)$$
$$= \$3011 \quad (\$3000)$$

Answer is B.

14. Use the future worth continuous compounding factor.

$$F = Pe^{rn}$$
$$= \$10{,}000 e^{(0.0525)(5)}$$
$$= \$13{,}002 \quad (\$13{,}000)$$

Answer is C.

15. Daily compounding is essentially equivalent to continuous compounding. Use the present worth equivalent factor for continuous compounding interest.

$$P = Fe^{-rn}$$
$$= \$5000 e^{-(0.055)(5)}$$
$$= \$3798 \quad (\$3800)$$

Answer is B.

16. Use the uniform series present worth factor to determine the amount of money needed in the account at retirement.

$$P = A(P/A, 10\%, 20)$$
$$= (\$100{,}000)(8.5136)$$
$$= \$851{,}360$$

Use the sinking fund factor to determine the annual contribution.

$$A = F(A/F, 10\%, 30)$$
$$= (\$851{,}360)(0.0061)$$
$$= \$5193 \quad (\$5200)$$

It does not matter if the annual contribution is made at the beginning or end of each year. The withdrawals start one year after the last contribution. Only the time between contributions and withdrawals is relevant.

Answer is C.

17.
$$F = P(F/P, 8\%, 10)$$
$$= (\$1000)(2.1589)$$
$$= \$2159 \quad (\$2160)$$

Answer is C.

18. $r = 24\%$ [nominal rate per year]

$$\phi = \frac{i}{m} = \frac{24\%}{4}$$
$$= 6\% \quad \text{[effective rate per quarter]}$$

$$n = (10 \text{ years})\left(4 \ \frac{\text{quarters}}{\text{year}}\right)$$
$$= 40 \text{ quarters}$$

$$F = P(F/P, 6\%, 40)$$
$$= (\$1000)(10.2857)$$
$$= \$10{,}286 \quad (\$10{,}000)$$

This problem did not require calculating the effective annual interest rate per year because 6% tables were available.

Answer is D.

19. Use the single payment factor to determine the cost of the machine in eight years.

$$F = P(F/P, 8\%, 8)$$
$$= (\$20{,}000)(1.8509)$$
$$= \$37{,}018$$

$2000 of this cost will be offset by the scrap value. Use the sinking fund factor to determine the required annual amount to set aside.

$$A = F(A/F, 8\%, 8)$$
$$= (\$37{,}018 - \$2000)(0.0940)$$
$$= \$3292 \quad (\$3290)$$

Answer is C.

20.

$$F = P(1 + i)^n$$

$$F/P = 4 = (1 + i)^{12}$$

$$\ln 4 = \ln\left((1 + i)^{12}\right) = 12 \ln(1 + i)$$

$$\frac{\ln 4}{12} = 0.1155 = \ln(1 + i)$$

$$e^{0.1155} = 1.122 = 1 + i$$

$$i = 0.122 \quad (12.2\%)$$

Answer is C.

Table 51.2 Factor Table $i = 6.00\%$

n	P/F	P/A	P/G	F/P	F/A	A/P	A/F	A/G
1	0.9434	0.9434	0.0000	1.0600	1.0000	1.0600	1.0000	0.0000
2	0.8900	1.8334	0.8900	1.1236	2.0600	0.5454	0.4854	0.4854
3	0.8396	2.6730	2.5692	1.1910	3.1836	0.3741	0.3141	0.9612
4	0.7921	3.4651	4.9455	1.2625	4.3746	0.2886	0.2286	1.4272
5	0.7473	4.2124	7.9345	1.3382	5.6371	0.2374	0.1774	1.8836
6	0.7050	4.9173	11.4594	1.4185	6.9753	0.2034	0.1434	2.3304
7	0.6651	5.5824	15.4497	1.5036	8.3938	0.1791	0.1191	2.7676
8	0.6274	6.2098	19.8416	1.5938	9.8975	0.1610	0.1010	3.1952
9	0.5919	6.8017	24.5768	1.6895	11.4913	0.1470	0.0870	3.6133
10	0.5584	7.3601	29.6023	1.7908	13.1808	0.1359	0.0759	4.0220
11	0.5268	7.8869	34.8702	1.8983	14.9716	0.1268	0.0668	4.4213
12	0.4970	8.3838	40.3369	2.0122	16.8699	0.1193	0.0593	4.8113
13	0.4688	8.8527	45.9629	2.1239	18.8821	0.1130	0.0530	5.1920
14	0.4423	9.2950	51.7128	2.2609	21.0151	0.1076	0.0476	5.5635
15	0.4173	9.7122	57.5546	2.3966	23.2760	0.1030	0.0430	5.9260
16	0.3936	10.1059	63.4592	2.5404	25.6725	0.0990	0.0390	6.2794
17	0.3714	10.4773	69.4011	2.6928	28.2129	0.0954	0.0354	6.6240
18	0.3505	10.8276	75.3569	2.8543	30.9057	0.0924	0.0324	6.9597
19	0.3305	11.1581	81.3062	3.0256	33.7600	0.0896	0.0296	7.2867
20	0.3118	11.4699	87.2304	3.2071	36.7856	0.0872	0.0272	7.6051
21	0.2942	11.7641	93.1136	3.3996	39.9927	0.0850	0.0250	7.9151
22	0.2775	12.0416	98.9412	3.6035	43.3923	0.0830	0.0230	8.2166
23	0.2618	12.3034	104.7007	3.8197	46.9958	0.0813	0.0213	8.5099
24	0.2470	12.5504	110.3812	4.0489	50.8156	0.0797	0.0197	8.7951
25	0.2330	12.7834	115.9732	4.2919	54.8645	0.0782	0.0182	9.0722
30	0.1741	13.7648	142.3588	5.7435	79.0582	0.0726	0.0126	10.3422
40	0.0972	15.0463	185.9568	10.2857	154.7620	0.0665	0.0065	12.3590
50	0.0543	15.7619	217.4574	18.4202	290.3359	0.0634	0.0034	13.7964
75	0.0126	16.4558	258.4527	79.0569	1300.9487	0.0608	0.0008	15.7058
100	0.0029	16.6175	272.0471	339.3021	5638.3681	0.0602	0.0002	16.3711

Table 51.3 Factor Table $i = 8.00\%$

n	P/F	P/A	P/G	F/P	F/A	A/P	A/F	A/G
1	0.9259	0.9259	0.0000	1.0800	1.0000	1.0800	1.0000	0.0000
2	0.8573	1.7833	0.8573	1.1664	2.0800	0.5608	0.4808	0.4808
3	0.7938	2.5771	2.4450	1.2597	3.2464	0.3880	0.3080	0.9487
4	0.7350	3.3121	4.6501	1.3605	4.5061	0.3019	0.2219	1.4040
5	0.6806	3.9927	7.3724	1.4693	5.8666	0.2505	0.1705	1.8465
6	0.6302	4.6229	10.5233	1.5869	7.3359	0.2163	0.1363	2.2763
7	0.5835	5.2064	14.0242	1.7138	8.9228	0.1921	0.1121	2.6937
8	0.5403	5.7466	17.8061	1.8509	10.6366	0.1740	0.0940	3.0985
9	0.5002	6.2469	21.8081	1.9990	12.4876	0.1601	0.0801	3.4910
10	0.4632	6.7101	25.9768	2.1589	14.4866	0.1490	0.0690	3.8713
11	0.4289	7.1390	30.2657	2.3316	16.6455	0.1401	0.0601	4.2395
12	0.3971	7.5361	34.6339	2.5182	18.9771	0.1327	0.0527	4.5957
13	0.3677	7.9038	39.0463	2.7196	21.4953	0.1265	0.0465	4.9402
14	0.3405	8.2442	43.4723	2.9372	24.2149	0.1213	0.0413	5.2731
15	0.3152	8.5595	47.8857	3.1722	27.1521	0.1168	0.0368	5.5945
16	0.2919	8.8514	52.2640	3.4259	30.3243	0.1130	0.0330	5.9046
17	0.2703	9.1216	56.5883	3.7000	33.7502	0.1096	0.0296	6.2037
18	0.2502	9.3719	60.8426	3.9960	37.4502	0.1067	0.0267	6.4920
19	0.2317	9.6036	65.0134	4.3157	41.4463	0.1041	0.0241	6.7697
20	0.2145	9.8181	69.0898	4.6610	45.7620	0.1019	0.0219	7.0369
21	0.1987	10.0168	73.0629	5.0338	50.4229	0.0998	0.0198	7.2940
22	0.1839	10.2007	76.9257	5.4365	55.4568	0.0980	0.0180	7.5412
23	0.1703	10.3711	80.6726	5.8715	60.8933	0.0964	0.0164	7.7786
24	0.1577	10.5288	84.2997	6.3412	66.7648	0.0950	0.0150	8.0066
25	0.1460	10.6748	87.8041	6.8485	73.1059	0.0937	0.0137	8.2254
30	0.0994	11.2578	103.4558	10.0627	113.2832	0.0888	0.0088	9.1897
40	0.0460	11.9246	126.0422	21.7245	259.0565	0.0839	0.0039	10.5699
50	0.0213	12.2335	139.5928	46.9016	573.7702	0.0817	0.0017	11.4107
75	0.0031	12.4611	152.8448	321.2045	4002.5566	0.0802	0.0002	12.2658
100	0.0005	12.4943	155.6107	2199.7613	27,484.5157	0.0800	–	12.4545

Table 51.4 Factor Table $i = 10.00\%$

n	P/F	P/A	P/G	F/P	F/A	A/P	A/F	A/G
1	0.9091	0.9091	0.0000	1.1000	1.0000	1.1000	1.0000	0.0000
2	0.8264	1.7355	0.8264	1.2100	2.1000	0.5762	0.4762	0.4762
3	0.7513	2.4869	2.3291	1.3310	3.3100	0.4021	0.3021	0.9366
4	0.6830	3.1699	4.3781	1.4641	4.6410	0.3155	0.2155	1.3812
5	0.6209	3.7908	6.8618	1.6105	6.1051	0.2638	0.1638	1.8101
6	0.5645	4.3553	9.6842	1.7716	7.7156	0.2296	0.1296	2.2236
7	0.5132	4.8684	12.7631	1.9487	9.4872	0.2054	0.1054	2.6216
8	0.4665	5.3349	16.0287	2.1436	11.4359	0.1874	0.0874	3.0045
9	0.4241	5.7590	19.4215	2.3579	13.5735	0.1736	0.0736	3.3724
10	0.3855	6.1446	22.8913	2.5937	15.9374	0.1627	0.0627	3.7255
11	0.3505	6.4951	26.3962	2.8531	18.5312	0.1540	0.0540	4.0641
12	0.3186	6.8137	29.9012	3.1384	21.3843	0.1468	0.0468	4.3884
13	0.2897	7.1034	33.3772	3.4523	24.5227	0.1408	0.0408	4.6988
14	0.2633	7.3667	36.8005	3.7975	27.9750	0.1357	0.0357	4.9955
15	0.2394	7.6061	40.1520	4.1772	31.7725	0.1315	0.0315	5.2789
16	0.2176	7.8237	43.4164	4.5950	35.9497	0.1278	0.0278	5.5493
17	0.1978	8.0216	46.5819	5.5045	40.5447	0.1247	0.0247	5.8071
18	0.1799	8.2014	49.6395	5.5599	45.5992	0.1219	0.0219	6.0526
19	0.1635	8.3649	52.5827	6.1159	51.1591	0.1195	0.0195	6.2861
20	0.1486	8.5136	55.4069	6.7275	57.2750	0.1175	0.0175	6.5081
21	0.1351	8.6487	58.1095	7.4002	64.0025	0.1156	0.0156	6.7189
22	0.1228	8.7715	60.6893	8.1403	71.4027	0.1140	0.0140	6.9189
23	0.1117	8.8832	63.1462	8.9543	79.5430	0.1126	0.0126	7.1085
24	0.1015	8.9847	65.4813	9.8497	88.4973	0.1113	0.0113	7.2881
25	0.0923	9.0770	67.6964	10.8347	98.3471	0.1102	0.0102	7.4580
30	0.0573	9.4269	77.0766	17.4494	164.4940	0.1061	0.0061	8.1762
40	0.0221	9.7791	88.9525	45.2593	442.5926	0.1023	0.0023	9.0962
50	0.0085	9.9148	94.8889	117.3909	1163.9085	0.1009	0.0009	9.5704
75	0.0008	9.9921	99.3317	1271.8954	12,708.9537	0.1001	0.0001	9.9410
100	0.0001	9.9993	99.9202	13,780.6123	137,796.1234	0.1000	–	9.9927

52 Depreciation and Special Topics

Subjects

DEPRECIATION 52-1
    Straight-Line Depreciation 52-1
    Accelerated Cost Recovery System
      (ACRS) 52-1
BOOK VALUE 52-2
EQUIVALENT UNIFORM ANNUAL
    COST 52-2
CAPITALIZED COST 52-2
BONDS 52-3
INFLATION 52-3
PROBABILISTIC PROBLEMS 52-3

Nomenclature

A	annual amount or annual value
BV	book value
C	initial cost, or present worth
	(present value) of all costs
d	inflation-adjusted interest rate
D_j	depreciation in year j
EUAC	equivalent uniform annual cost
f	constant inflation rate
F	future worth or future value
n	service life
p	probability
P	present worth or present value
S_n	expected salvage value in year n

Subscripts

j	at time j
n	at time n

DEPRECIATION

Depreciation is an artificial expense that spreads the purchase price of an asset or other property over a number of years. Generally, tax regulations do not allow the cost of an asset to be treated as a deductible expense in the year of purchase. Rather, portions of the expense must be allocated to each of the years of the asset's depreciation period. The amount that is allocated each year is called the *depreciation*.

The inclusion of depreciation in engineering economic analysis problems will increase the after-tax present worth (profitability) of an asset. The larger the depreciation, the greater the profitability will be. Therefore, individuals and companies that are eligible to utilize depreciation desire to maximize and accelerate the depreciation available to them.

The *depreciation basis* of an asset is the part of the asset's purchase price that is spread over the *depreciation period*, also known as the *service life*. The depreciation basis may or may not be equal to the purchase price.

A common depreciation basis is the difference between the purchase price and the expected salvage value at the end of the depreciation period.

$$\text{depreciation basis} = C - S_n \qquad 52.1$$

There are several methods of calculating the year-by-year depreciation of an asset. Equation 52.1 is not universally compatible with all depreciation methods. If a depreciation basis does not consider the salvage value, it is known as an *unadjusted basis*.

Straight-Line Depreciation

With the *straight-line method*, depreciation is the same each year. The depreciation basis $(C - S_n)$ is allocated uniformly to all of the n years in the depreciation period. Each year, the depreciation will be

$$D_j = \frac{C - S_n}{n} \qquad 52.2$$

Accelerated Cost Recovery System (ACRS)

In the United States, property placed into service in 1981 and thereafter must use the *Accelerated Cost Recovery System* (ACRS) and property placed into service after 1986 must use *Modified Accelerated Cost Recovery System* (MACRS) or other statutory method. Other methods, such as the straight-line method, cannot be used except in special cases. Property placed into service in 1980 or before must continue to be depreciated according to the method originally chosen. ACRS cannot be used.

Under ACRS and MACRS, the cost recovery amount in the jth year of an asset's cost recovery period is calculated by multiplying the initial cost by a factor. The initial cost used is not reduced by the asset's salvage value.

$$D_j = C \times \text{factor} \qquad 52.3$$

ECONOMICS
Depreciation

The factor used depends on the asset's cost recovery period. Such factors are subject to continuing legislation changes. Representative depreciation factors are shown in Table 52.1.

Table 52.1 Representative MACRS Depreciation Factors

	recovery period (years)			
	3	5	7	10
year j	recovery rate (percent)			
1	33.3	20.0	14.3	10.0
2	44.5	32.0	24.5	18.0
3	14.8	19.2	17.5	14.4
4	7.4	11.5	12.5	11.5
5		11.5	8.9	9.2
6		5.8	8.9	7.4
7			8.9	6.6
8			4.5	6.6
9				6.5
10				6.5
11				3.3

BOOK VALUE

The difference between the original purchase price and the accumulated depreciation is known as *book value*. At the end of each year, the book value (which is initially equal to the purchase price) is reduced by the depreciation in that year.

It is important to distinguish the difference between beginning-of-year book value and end-of-year book value. In Eq. 52.4, the book value BV_j means the book value at the end of the jth year after j years of depreciation have been subtracted from the original purchase price.

$$BV_j = \text{initial cost} - \text{accumulated depreciation}$$

$$= C - \sum_{j=1}^{t} D_j \qquad 52.4$$

The ratios of book value to initial costs for an asset depreciated using both the straight-line and the MACRS methods are illustrated in Fig. 52.1.

EQUIVALENT UNIFORM ANNUAL COST

Alternatives with different lifetimes will generally be compared by way of *equivalent uniform annual cost*, or EUAC. An EUAC is the annual amount that is equivalent to all of the cash flows in the alternative. The

Figure 52.1 Book Value with Straight-Line and MACRS Methods

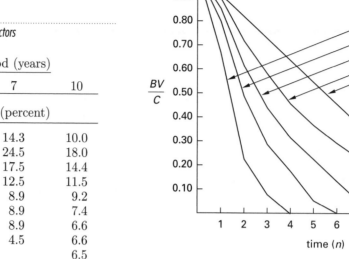

EUAC differs in sign from all of the other cash flows. Costs and expenses expressed as EUACs, which would normally be considered negative, are considered positive. Conversely, benefits and returns are considered negative. The term *cost* in the designation EUAC serves to make clear the meaning of a positive number.

CAPITALIZED COST

The present worth of a project with an infinite life is known as the *capitalized cost*. Capitalized cost is the amount of money at $t = 0$ needed to perpetually support the project on the earned interest only. Capitalized cost is a positive number when expenses exceed income.

Normally, it would be difficult to work with an infinite stream of cash flows since most discount factor tables do not list factors for periods in excess of 100 years. However, the (A/P) discount factor approaches the interest rate as n becomes large. Since the (P/A) and (A/P) factors are reciprocals of each other, it is possible to divide an infinite series of annual cash flows by the interest rate in order to calculate the present worth of the infinite series.

$$\text{capitalized cost} = P = \frac{A}{i} \qquad \left[\begin{array}{c}\text{infinite}\\\text{series}\end{array}\right] \qquad 52.5$$

Equation 52.5 can be used when the annual costs are equal in every year. If the operating and maintenance costs occur irregularly instead of annually, or if the costs vary from year to year, it will be necessary to somehow determine a cash flow of equal annual amounts that is equivalent to the stream of original costs (i.e., to determine the EUAC).

BONDS

A *bond* is a method of obtaining long-term financing commonly used by governments, states, municipalities, and very large corporations. The bond represents a contract to pay the bondholder specified amounts of money at specific times. The holder purchases the bond in exchange for these payments of interest and principal. Typical municipal bonds call for quarterly or semiannual interest payments and a payment of the *face value of the bond* on the *date of maturity* (end of the bond period). Because of the practice of discounting in the bond market, a bond's face value and its purchase price will generally not coincide.

The *bond value* is the present worth of the bond, considering all interest payments that are paid out, plus the face value of the bond when it matures.

The *bond yield* is the bondholder's actual rate of return from the bond, considering the purchase price, interest payments, and face value payment (or value realized if the bond is sold before it matures). By convention, bond yield is specified as a nominal rate (rate per annum), not as an effective rate per year. The bond yield should be determined by finding the effective rate of return per payment period (e.g., per semiannual interest payment) as a conventional rate of return problem. Then the nominal annual rate can be found by multiplying the effective rate per period by the number of payments per year.

INFLATION

Economic studies must be performed in terms of constant-value dollars. There are several methods used to accomplish this when inflation is present. One alternative is to replace the effective annual interest rate, i, with a value adjusted for *inflation*. This adjusted value, d, is

$$d = i + f + if \qquad 52.6$$

In Eq. 52.6, f is a constant *inflation rate* per year. The inflation-adjusted interest rate should be used to compute present worth values.

PROBABILISTIC PROBLEMS

If an alternative's cash flows are specified by an implicit or explicit probability distribution rather than being known exactly, the problem is *probabilistic*. Probabilistic problems typically possess the following characteristics.

- There is a chance of loss that must be minimized (or, more rarely, a chance of gain that must be maximized) by selection of one of the alternatives.

- There are multiple alternatives. Each alternative offers a different degree of protection from the loss. Usually, the alternatives with the greatest protection will be the most expensive.

- The outcome is independent of the alternative selected.

Probabilistic problems are typically solved using annual costs and expected values. An *expected value* is similar to an average value, since it is calculated as the mean of the discrete values. If cost 1 has a probability of occurrence of p_1, cost 2 has a probability of occurrence of p_2, and so on, the expected value is

$$\mathcal{E} = (p_1)(\text{cost } 1) + (p_2)(\text{cost } 2) + \cdots$$

SAMPLE PROBLEMS

Problem 1

A machine has an initial cost of \$50,000 and a salvage value of \$10,000 after 10 years. What is the straight-line depreciation rate as a percentage of the initial cost?

    (A) 4%
    (B) 8%
    (C) 10%
    (D) 12%

Solution
The straight-line depreciation per year is

$$D_j = \frac{C - S_n}{n} = \frac{\$50{,}000 - \$10{,}000}{10 \text{ years}}$$

$$= \$4000 \text{ per year}$$

The depreciation rate is

$$\frac{\$4000}{\$50{,}000} = 0.08 \quad (8\%)$$

Answer is B.

Problem 2

Referring to the machine described in Prob. 1, what is the book value after five years using straight-line depreciation?

    (A) \$12,500
    (B) \$16,400
    (C) \$22,300
    (D) \$30,000

Solution

Using straight-line depreciation, the depreciation every year is the same.

$$BV_5 = C - \sum_{t=1}^{5} D_j$$
$$= \$50,000 - (5 \text{ years})(\$4000 \text{ per year})$$
$$= \$30,000$$

Answer is D.

Problem 3

Referring to the machine described in Prob. 1, what is the book value after five years using the MACRS method of depreciation?

(A) $12,500
(B) $16,400
(C) $18,500
(D) $21,900

Solution

$$BV = C - \sum_{j=1}^{5} D_j$$

To compute the depreciation in the first five years, use the MACRS factors for a 10-year recovery period.

year	factor (%)	$D_j = (\text{factor})C$
1	10.0	$(0.10)(\$50,000) = \$5000$
2	18.0	$(0.18)(\$50,000) = \$9000$
3	14.4	$(0.144)(\$50,000) = \$7200$
4	11.5	$(0.115)(\$50,000) = \$5750$
5	9.2	$(0.092)(\$50,000) = \underline{\$4600}$
		$\sum D_j = \$31,550$

$$BV = \$50,000 - \$31,550 = \$18,450 \quad (\$18,500)$$

Answer is C.

Problem 4

A machine that costs $20,000 has a 10-year life and a $2000 salvage value. If straight-line depreciation is used, what is the book value of the machine at the end of the second year?

(A) $14,000
(B) $14,400
(C) $15,600
(D) $16,400

Solution

Use straight-line depreciation.

$$D_j = \frac{C - S_n}{n}$$
$$= \frac{\$20,000 - \$2000}{10 \text{ years}}$$
$$= \$1800 \text{ per year}$$

$$BV_2 = C - \sum_{j=1}^{2} D_j$$
$$= \$20,000 - (2 \text{ years})(\$1800 \text{ per year})$$
$$= \$16,400$$

Answer is D.

Problem 5

A $1000 face-value bond pays dividends of $110 at the end of each year. If the bond matures in 20 years, what is the approximate bond value at an interest rate of 12% per year, compounded annually?

(A) $890
(B) $930
(C) $1000
(D) $1820

Solution

The bond value is the present value of the sum of annual interest payments and the present worth of the future face value of the bond.

$$P = (\$110)(P/A, 12\%, 20) + (\$1000)(P/F, 12\%, 20)$$
$$= (\$110)\left(\frac{(1.12)^{20} - 1}{(0.12)(1.12)^{20}}\right) + (\$1000)(1 + 0.12)^{-20}$$
$$= (\$110)(7.4694) + (\$1000)(0.1037)$$
$$= \$925 \quad (\$930)$$

Answer is B.

FE-STYLE EXAM PROBLEMS

Problems 1–7 refer to the following situation.

A company is considering buying one of the following two computers.

	computer A	computer B
initial cost	$3900	$5500
salvage value	$1800	$3100
useful life	10 years	13 years
annual		$275 (year 1 to 8)
maintenance	$390	$425 (year 9 to 13)
interest rate	6%	6%

1. What is the equivalent uniform annual cost of computer A?

(A) $740
(B) $780
(C) $820
(D) $850

2. What is the equivalent uniform annual cost of computer B?

(A) $770
(B) $780
(C) $850
(D) $940

3. If computer A was to be purchased and kept forever without any change in the annual maintenance costs, what would be the present worth of all expenditures?

(A) $3970
(B) $7840
(C) $10,000
(D) $10,400

4. What is the annual straight-line depreciation for computer A?

(A) $210/year
(B) $225/year
(C) $262/year
(D) $420/year

5. What is the total straight-line depreciation value of computer A after the fifth year?

(A) $1000
(B) $1050
(C) $1125
(D) $1250

6. What is the book value of computer B after the second year, using the MACRS method of depreciation and a 10-year recovery period?

(A) $3360
(B) $3780
(C) $3960
(D) $4120

7. What is the present worth of the costs for computer A?

(A) $5330
(B) $5770
(C) $6670
(D) $6770

For the following problems use the NCEES Handbook as your only reference.

8. An investment proposal calls for a $100,000 payment now and a second $100,000 payment 10 years from now. The investment is for a project with a perpetual life. The effective annual interest rate is 6%. What is the approximate capitalized cost?

(A) $156,000
(B) $160,000
(C) $200,000
(D) $267,000

9. Depreciation allowance is best defined as

(A) the value that a buyer will give a machine's owner at the end of the machine's useful life.
(B) the amount awarded to industries involved in removing natural limited resources from the earth.
(C) the amount used to recover the cost of an asset so that a replacement can be purchased.
(D) a factor whose use is regulated by federal law.

10. Twenty thousand dollars is invested today. If the annual inflation rate is 6% and the effective annual return on investment is 10%, what will be the approximate future value of the investment, adjusted for inflation, in five years?

(A) $26,800
(B) $32,200
(C) $42,000
(D) $43,100

11. An investment currently costs $28,000. If the current inflation rate is 6% and the effective annual return on investment is 10%, approximately how long will it take for the investment's future value to reach $40,000?

(A) 1.8 years
(B) 2.3 years
(C) 2.6 years
(D) 3.4 years

Problems 12–16 refer to the following information.

An oil company is planning to install a new pipeline to connect storage tanks to a processing plant 1500 m away. The connection will be needed for the foreseeable future. Both 80 mm and 120 mm pipes are being considered.

	80 mm pipe	120 mm pipe
initial cost	$1500	$2500
service life	12 years	12 years
salvage value	$200	$300
annual maintenance	$400	$300
pump cost/hour	$2.50	$1.40
pump operation	600 hours/year	600 hours/year

For this analysis, the company will use an annual interest rate of 8%. Annual maintenance and pumping costs may be considered to be paid in their entireties at the end of the years in which their costs are incurred.

12. Disregarding the initial and replacement pipe costs, what is the approximate capitalized cost of the maintenance and pumping costs for the 80 mm pipe?

(A) $15,100
(B) $20,100
(C) $23,800
(D) $27,300

13. What is the approximate equivalent uniform annual cost of the 80 mm pipe, considering all costs and expenses?

(A) $1710
(B) $1800
(C) $1900
(D) $2100

14. What is the approximate equivalent uniform annual cost of the 120 mm pipe, considering all costs and expenses?

(A) $1250
(B) $1290
(C) $1380
(D) $1460

15. What is the approximate depreciation allowance for the 120 mm pipe in the first year? Use MACRS depreciation assuming a 10-year life.

(A) $193
(B) $210
(C) $230
(D) $250

16. If the annual effective rates for inflation and interest have been 5% and 9%, respectively, what was the uninflated present worth of the 120 mm pipe three years ago?

(A) $1590
(B) $1670
(C) $1710
(D) $1780

17. Permanent mineral rights on a parcel of land are purchased for an initial lump-sum payment of $100,000. Profits from mining activities are $12,000 each year, and these profits are expected to continue indefinitely. What approximate interest rate is being earned on the initial investment?

(A) 8.33%
(B) 9.00%
(C) 10.0%
(D) 12.0%

18. Flood damage in a typical year is given according to the following table.

value of flood damage	probability
$0	0.75
$10,000	0.20
$20,000	0.04
$30,000	0.01

If the effective annual interest rate is 6%, what is the most likely present worth of flood damage over the next 10-year period?

(A) $3100
(B) $9600
(C) $16,000
(D) $23,000

SOLUTIONS TO FE-STYLE EXAM PROBLEMS

1. The equivalent uniform annual cost is the annual amount that is equivalent to all of the cash flows in the alternative. For computer A,

$$
\begin{aligned}
(\text{EUAC})_A &= (\$3900)(A/P, 6\%, 10) + \$390 \\
&\quad - (\$1800)(A/F, 6\%, 10) \\
&= (\$3900)(0.1359) + \$390 - (\$1800)(0.0759) \\
&= \$783 \quad (\$780)
\end{aligned}
$$

Answer is B.

2. For computer B,

$$(\text{EUAC})_\text{B} = (\$5500)(A/P, 6\%, 13) + \$275$$
$$+ (\$425 - \$275)(P/A, 6\%, 5)$$
$$\times (P/F, 6\%, 8)(A/P, 6\%, 13)$$
$$- (\$3100)(A/F, 6\%, 13)$$
$$= (\$5500)(0.1130) + \$275$$
$$+ (\$150)(4.2124)(0.6274)(0.1130)$$
$$- (\$3100)(0.0530)$$
$$= \$777 \quad (\$780)$$

Alternate Solution

$$(\text{EUAC})_\text{B} = (\$5500)(A/P, 6\%, 13) + \$425$$
$$- (\$425 - \$275)(P/A, 6\%, 8)$$
$$\times (A/P, 6\%, 13) - (\$3100)$$
$$\times (A/F, 6\%, 13)$$
$$= (\$5500)(0.1130) + \$425$$
$$- (\$150)(6.2098)(0.1130)$$
$$- (\$3100)(0.0530)$$
$$= \$777 \quad (\$780)$$

Answer is B.

3.   capitalized cost $= $ initial cost $+ \dfrac{\text{annual cost}}{i}$

$$= \$3900 + \dfrac{\$390}{0.06}$$
$$= \$10,400$$

Answer is D.

4. For straight-line depreciation,

$$D_j = \frac{C - S_n}{n} = \frac{\$3900 - \$1800}{10 \text{ years}}$$
$$= \$210/\text{year}$$

Answer is A.

5. The total depreciation after five years is

$$\sum D = (5 \text{ years})(\$210 \text{ per year})$$
$$= \$1050$$

Answer is B.

6. Subtract the depreciation of the first two years from the original cost.

$$\text{BV} = C - \sum_{j=1}^{2} D_j$$

year	factor (%)	D_j
1	10.0	$(0.10)(\$5500) = \$550$
2	18.0	$(0.18)(\$5500) = \underline{\$990}$

$$\sum D_j = \$1540$$
$$\text{BV} = \$5500 - \$1540 = \$3960$$

Answer is C.

7. Bring all costs and benefits into the present.

$$P = \$3900 + (\$390)(P/A, 6\%, 10)$$
$$- (\$1800)(P/F, 6\%, 10)$$
$$= \$3900 + (\$390)(7.3601) - (\$1800)(0.5584)$$
$$= \$5765 \quad (\$5770)$$

Answer is B.

8. Capitalized cost is the present worth of a project with an infinite life.

$$\text{capitalized cost} = \text{initial cost} + F(P/F, 6\%, 10)$$
$$= \$100,000 + (\$100,000)(0.5584)$$
$$= \$155,840 \quad (\$156,000)$$

Answer is A.

9. Depreciation rates are closely regulated by federal law. Although depreciation does in fact result in a reduction in income taxes, it never is able to fully recover the original cost of the asset.

Answer is D.

10. The interest rate adjusted for inflation is

$$d = i + f + if$$
$$= 0.10 + 0.06 + (0.10)(0.06)$$
$$= 0.166$$

Use the single payment factor to determine the future worth of the investment.

$$F = P(1 + d)^n$$
$$= (\$20,000)(1 + 0.166)^5$$
$$= \$43,105 \quad (\$43,100)$$

Answer is D.

11. The interest rate adjusted for inflation is

$$d = i + f + if$$
$$= 0.10 + 0.06 + (0.10)(0.06)$$
$$= 0.166$$

Use the present worth factor to determine the number of years.

$$P = F(1+i)^{-n}$$
$$\$28{,}000 = (\$40{,}000)(1+0.166)^{-n}$$
$$0.7 = (1+0.166)^{-n}$$

Solve by taking the log of both sides.

$$\log(0.7) = -n \; \log(1.166)$$
$$-0.1549 = -n(0.0667)$$
$$n = 2.32 \text{ years} \quad (2.3 \text{ years})$$

Answer is B.

12. The capitalized cost is the present worth of an infinite investment. The (P/A) factor for an infinite number of years is the reciprocal of the interest rate.

$$\text{capitalized cost} = (\text{annual cost})(P/A, i\%, \infty)$$

$$= \frac{\text{annual cost}}{i}$$

$$= \frac{400 \; \dfrac{\$}{\text{year}} + \left(2.50 \; \dfrac{\$}{\text{hour}}\right) \times \left(600 \; \dfrac{\text{hours}}{\text{year}}\right)}{0.08}$$

$$= \$23{,}750 \quad (\$23{,}800)$$

Answer is C.

13.
$$(\text{EUAC})_{80} = (\$1500)(A/P, 8\%, 12) + \$400$$
$$+ \left(2.50 \; \frac{\$}{\text{hour}}\right)(600 \text{ hours})$$
$$- (\$200)(A/F, 8\%, 12)$$
$$= (\$1500)(0.1327) + \$400 + \$1500$$
$$- (\$200)(0.0527)$$
$$= \$2089 \quad (\$2100)$$

Answer is D.

14.
$$(\text{EUAC})_{120} = (\$2500)(A/P, 8\%, 12) + \$300$$
$$+ \left(1.40 \; \frac{\$}{\text{hour}}\right)(600 \text{ hours})$$
$$- (\$300)(A/F, 8\%, 12)$$
$$= (\$2500)(0.1327) + \$300 + \$840$$
$$- (\$300)(0.0527)$$
$$= \$1456 \quad (\$1460)$$

Answer is D.

15. Use the MACRS factor for 10-year depreciation.

$$D_{10} = (\text{cost})(\text{MACRS factor})$$
$$= (\$2500)(0.10)$$
$$= \$250$$

Answer is D.

16. Determine the adjusted interest rate.

$$d = i + f + if$$
$$= 0.09 + 0.05 + (0.09)(0.05)$$
$$= 0.1445$$

Use the present worth equation.

$$P = F(1+d)^{-n}$$
$$= (\$2500)(1+0.1445)^{-3}$$
$$= \$1668 \quad (\$1670)$$

Answer is B.

17. Use the capitalized cost equation.

$$i = \frac{A}{P} = \frac{\$12{,}000}{\$100{,}000}$$
$$= 0.12 \quad (12\%)$$

Answer is D.

18. The expected value of flood damage in any given year is

$$\mathcal{E} = (0.75)(\$0) + (0.20)(\$10{,}000)$$
$$+ (0.04)(\$20{,}000) + (0.01)(\$30{,}000)$$
$$= \$3100$$

The present worth of 10 years of expected flood damage is

$$P = (\$3100)(P/A, 6\%, 10)$$
$$= (\$3100)(7.3601)$$
$$= \$22{,}816 \quad (\$23{,}000)$$

Answer is D.

53 Comparison of Alternatives

Subjects

ALTERNATIVE COMPARISONS 53-1
PRESENT WORTH ANALYSIS 53-1
ANNUAL COST ANALYSIS 53-1
RATE OF RETURN ANALYSIS 53-1
BENEFIT-COST ANALYSIS 53-2
BREAK-EVEN ANALYSIS 53-2

Nomenclature

A	annual amount or annual value
B	present worth of all benefits
C	initial cost, or present worth of all costs
EUAC	equivalent uniform annual cost
F	future worth or future value
i	effective interest rate per period
MARR	minimum attractive rate of return
n	number of years
P	present worth or present value
PBP	pay-back period
ROR	rate of return

ALTERNATIVE COMPARISONS

In the real world, the majority of engineering economic analysis problems are alternative comparisons. In these problems, two or more mutually exclusive investments compete for limited funds. A variety of methods exists for selecting the superior alternative from a group of proposals. Each method has its own merits and applications.

PRESENT WORTH ANALYSIS

When two or more alternatives are capable of performing the same functions, the economically superior alternative will have the largest present worth. The *present worth method* is restricted to evaluating alternatives that are mutually exclusive and that have the same lives. This method is suitable for ranking the desirability of alternatives.

ANNUAL COST ANALYSIS

Alternatives that accomplish the same purpose but that have unequal lives must be compared by the *annual cost method*. The annual cost method assumes that each alternative will be replaced by an identical twin at the end of its useful life (i.e., infinite renewal). This method, which may also be used to rank alternatives according to their desirability, is also called the *annual return method* or *capital recovery method*.

The alternatives must be mutually exclusive and repeatedly renewed up to the duration of the longest-lived alternative. The calculated annual cost is known as the *equivalent uniform annual cost* (EUAC) or *equivalent annual cost* (EAC). Cost is a positive number when expenses exceed income.

RATE OF RETURN ANALYSIS

An intuitive definition of the *rate of return* (ROR) is the effective annual interest rate at which an investment accrues income. That is, the rate of return of an investment is the interest rate that would yield identical profits if all money was invested at that rate. Although this definition is correct, it does not provide a method of determining the rate of return.

The present worth of a \$100 investment invested at 5% is zero when $i = 5\%$ is used to determine equivalence. Therefore, a working definition of rate of return would be the effective annual interest rate that makes the present worth of the investment zero. Alternatively, rate of return could be defined as the effective annual interest rate that makes the benefits and costs equal.

A company may not know what effective interest rate, i, to use in engineering economic analysis. In such a case, the company can establish a minimum level of economic performance that it would like to realize on all investments. This criterion is known as the *minimum attractive rate of return*, or MARR.

Once a rate of return for an investment is known, it can be compared with the minimum attractive rate of return. If the rate of return is equal to or exceeds the minimum attractive rate of return, the investment is qualified (i.e., the alternative is viable). This is the basis for the rate of return method of alternative viability analysis.

ECONOMICS
Alternatives

If rate of return is used to select among two or more investments, an *incremental analysis* must be performed. An incremental analysis begins by ranking the alternatives in order of increasing initial investment. Then, the cash flows for the investment with the lower initial cost are subtracted from the cash flows for the higher-priced alternative on a year-by-year basis. This produces, in effect, a third alternative representing the costs and benefits of the added investment. The added expense of the higher-priced investment is not warranted unless the rate of return of this third alternative exceeds the minimum attractive rate of return as well. The alternative with the higher initial investment is superior if the incremental rate of return exceeds the minimum attractive rate of return.

Finding the rate of return can be a long, iterative process, requiring either interpolation or trial and error. Sometimes, the actual numerical value of rate of return is not needed; it is sufficient to know whether or not the rate of return exceeds the minimum attractive rate of return. This comparative analysis can be accomplished without calculating the rate of return simply by finding the present worth of the investment using the minimum attractive rate of return as the effective interest rate (i.e., $i = $ MARR). If the present worth is zero or positive, the investment is qualified. If the present worth is negative, the rate of return is less than the minimum attractive rate of return and the additional investment is not warranted.

The present worth, annual cost, and rate of return methods of comparing alternatives yield equivalent results, but they are distinctly different approaches. The present worth and annual cost methods may use either effective interest rates or the minimum attractive rate of return to rank alternatives or compare them to the MARR. If the incremental rate of return of pairs of alternatives are compared with the MARR, the analysis is considered a rate of return analysis.

BENEFIT-COST ANALYSIS

The *benefit-cost ratio method* is often used in municipal project evaluations where benefits and costs accrue to different segments of the community. With this method, the present worth of all benefits (irrespective of the beneficiaries) is divided by the present worth of all costs. (Equivalent uniform annual costs can be used in place of present worths.) The project is considered acceptable if the ratio equals or exceeds 1.0 (i.e., $B/C \geq 1.0$). This will be true whenever $B - C \geq 0$.

When the benefit-cost ratio method is used, disbursements by the initiators or sponsors are *costs*. Disbursements by the users of the project are known as *disbenefits*. It is often difficult to determine whether a cash flow is a cost or a disbenefit (whether to place it in the denominator or numerator of the benefit-cost ratio calculation).

Regardless of where the cash flow is placed, an acceptable project will always have a benefit-cost ratio greater than or equal to 1.0, although the actual numerical result will depend on the placement. For this reason, the benefit-cost ratio alone should not be used to rank competing projects.

If ranking is to be done by the benefit-cost ratio method, an incremental analysis is required, as it is for the rate-of-return method. The incremental analysis is accomplished by calculating the ratio of differences in benefits to differences in costs for each possible pair of alternatives. If the ratio exceeds 1.0, alternative 2 is superior to alternative 1. Otherwise, alternative 1 is superior.

$$\frac{B_2 - B_1}{C_2 - C_1} \geq 1 \qquad\qquad 53.1$$

BREAK-EVEN ANALYSIS

Break-even analysis is a method of determining when the value of one alternative becomes equal to the value of another. It is commonly used to determine when costs exactly equal revenue. If the manufactured quantity is less than the *break-even quantity*, a loss is incurred. If the manufactured quantity is greater than the break-even quantity, a profit is made.

An alternative form of the break-even problem is to find the number of units per period for which two alternatives have the same total costs. Fixed costs are spread over a period longer than one year using the EUAC concept. One of the alternatives will have a lower cost if production is less than the break-even point. The other will have a lower cost if production is greater than the break-even point.

The *pay-back period*, PBP, is defined as the length of time, n, usually in years, for the cumulative net annual profit to equal the initial investment. It is tempting to introduce equivalence into pay-back period calculations, but the convention is not to.

$$C - (\text{PBP})(\text{net annual profit}) = 0 \qquad 53.2$$

SAMPLE PROBLEMS

Problem 1

A company purchases a piece of construction equipment for rental purposes. The expected income is $3100 annually for its useful life of 15 years. Expenses are estimated to be $355 annually. If the purchase price is $25,000 and there is no salvage value, what is the prospective rate of return, neglecting taxes?

(A) 5.2%
(B) 6.4%
(C) 6.8%
(D) 7.0%

Solution

The rate of return can be viewed as the effective annual interest rate that makes the present worth of the investment equal to zero.

$$P = 0 = -\$25,000$$
$$+ (\$3100)(P/A, i\%, 15)$$
$$- (\$355)(P/A, i\%, 15)$$
$$\$25,000 = (\$2745)(P/A, i\%, 15)$$
$$9.1075 = (P/A, i\%, 15)$$

Use linear interpolation as an approximation.

$$(P/A, 6\%, 15) = 9.7122$$
$$(P/A, 8\%, 15) = 8.5595$$
$$\frac{9.7122 - 9.1075}{9.7122 - 8.5595} = 0.5245$$
$$i\% = 6\% + (0.5245)(8\% - 6\%)$$
$$= 7.049\% \quad (7.0\%)$$

Answer is D.

Problems 2–4 refer to the following situation.

An industrial firm uses an economic analysis to determine which of two different machines to purchase. Each machine is capable of performing the same task in a given amount of time. Assume the minimum attractive rate of return is 8%.

Use the following data in this analysis.

	machine X	machine Y
initial cost	$6000	$12,000
estimated life	7 years	13 years
salvage value	none	$4000
annual maintenance cost	$150	$175

Problem 2

What is the approximate equivalent uniform annual cost of machine X?

(A) $1000
(B) $1120
(C) $1190
(D) $1300

Solution

$$(EUAC)_X = (\$6000)(A/P, 8\%, 7) + \$150$$
$$= (\$6000)(0.1921) + \$150$$
$$= \$1302.60 \quad (\$1300)$$

Answer is D.

Problem 3

What is the equivalent uniform annual cost of machine Y?

(A) $1160
(B) $1300
(C) $1490
(D) $1510

Solution

$$(EUAC)_Y = (\$12,000)(A/P, 8\%, 13) + \$175$$
$$- (\$4000)(A/F, 8\%, 13)$$
$$= (\$12,000)(0.1265) + \$175$$
$$- (\$4000)(0.0465)$$
$$= \$1507$$

Answer is D.

Problem 4

Which, if either, of the two machines should the firm choose and why?

(A) machine X because $(EUAC)_X < (EUAC)_Y$
(B) machine X because $(EUAC)_X > (EUAC)_Y$
(C) machine Y because $(EUAC)_X < (EUAC)_Y$
(D) machine Y because $(EUAC)_X > (EUAC)_Y$

Solution

$$(EUAC)_X < (EUAC)_Y$$

Machine X represents the superior alternative, based on a comparison of EUACs.

Answer is A.

Problem 5

Going Broke County is using a 10% annual interest rate to decide if it should buy snowplow A or snowplow B.

	snowplow A	snowplow B
initial cost	$300,000	$400,000
life	10 years	10 years
annual operations and maintenance	$45,000	$35,000
annual benefits	$150,000	$200,000
salvage value	$0	$10,000

What are the benefit-cost ratios for snowplows A and B, respectively, and which snowplow should Going Broke County buy?

(A) 2.2, 1.8; snowplow A
(B) 2.6, 2.1; snowplow A
(C) 1.4, 1.8; snowplow B
(D) 1.6, 2.0; snowplow B

Solution

The benefit-cost method requires the cash flows to be converted to present worths.

For snowplow A,

$$C = \$300,000 + (\$45,000)(P/A, 10\%, 10)$$
$$= \$300,000 + (\$45,000)(6.1446)$$
$$= \$576,507$$
$$B = (\$150,000)(P/A, 10\%, 10)$$
$$= (\$150,000)(6.1446)$$
$$= \$921,690$$
$$\frac{B}{C} = \frac{\$921,690}{\$576,507}$$
$$= 1.60$$

For snowplow B,

$$C = \$400,000 + (\$35,000)(P/A, 10\%, 10)$$
$$\quad - (\$10,000)(P/F, 10\%, 10)$$
$$= \$400,000 + (\$35,000)(6.1446) - (\$10,000)(0.3855)$$
$$= \$611,206$$
$$B = (\$200,000)(P/A, 10\%, 10)$$
$$= (\$200,000)(6.1446)$$
$$= \$1,228,920$$
$$\frac{B}{C} = \frac{\$1,228,920}{\$611,206}$$
$$= 2.01$$

To rank the projects using the benefit-cost ratio method, use an incremental analysis.

$$\frac{B_2 - B_1}{C_2 - C_1} \geq 1 \quad \text{[for choosing alternative 2]}$$

$$\frac{B_2 - B_1}{C_2 - C_1} = \frac{\$1,228,920 - \$921,690}{\$611,206 - \$576,507}$$
$$= 8.85 > 1$$

The additional investment is warranted. Alternative 2 is superior; choose snowplow B.

Answer is D.

Problem 6

A company produces a gear that is commonly used by several lawnmower manufacturing companies. The base cost of operation (rent, utilities, etc.) is $750,000 per year. The cost of manufacturing is $1.35 per gear. If these gears are sold at $7.35 each, how many must be sold each year to break even?

(A) 65,000 per year
(B) 90,000 per year
(C) 100,000 per year
(D) 125,000 per year

Solution

The break-even point for this problem is the point at which costs equal revenues.

$$\text{costs} = \$750,000 + (\$1.35)(\text{no. of gears})$$

$$\text{revenues} = (\$7.35)(\text{no. of gears})$$

$$\$750,000 + (\$1.35)(\text{no. of gears}) = (\$7.35)(\text{no. of gears})$$

$$\text{no. of gears} = \frac{\$750,000}{\$7.35 - \$1.35}$$
$$= 125,000$$

Answer is D.

FE-STYLE EXAM PROBLEMS

1. Calculate the rate of return for an investment with the following characteristics.

initial cost	$20,000
project life	10 years
salvage value	$5000
annual receipts	$7500
annual disbursements	$3000

(A) 19.6%
(B) 20.6%
(C) 22.9%
(D) 24.5%

2. Grinding mills M and N are being considered for a 12-year service in a chemical plant. The minimum attractive rate of return is 10%. What are the equivalent uniform annual costs of mills M and N, respectively, and which is the more economic choice?

	mill M	mill N
initial cost	$7800	$14,400
salvage value	$0	$2700
annual operating cost	$1745	$1200
annual repair cost	$960	$540

(A) $3840, $3620; mill N
(B) $3850, $3730; mill N
(C) $4330, $3960; mill N
(D) $3960, $5000; mill M

3. You want to purchase one of the following milling machines.

	machine A	machine B
initial cost	$20,000	$30,000
life	10 years	10 years
salvage value	$2000	$5000
annual receipts	$9000	$12,000
annual disbursements	$3500	$4500

What are the approximate rates of return for machines A and B, respectively?

(A) 22.5%, 28.2%
(B) 23.9%, 27.0%
(C) 24.8%, 22.1%
(D) 25.0%, 26.8%

4. Consider the two machines described in Prob. 3. If machine A is the preferred economic choice, what is the lowest value that the minimum attractive rate of return can be?

(A) 10%
(B) 17%
(C) 22%
(D) 25%

5. The annual maintenance cost of a machine shop is $10,000. The cost of making a forging is $2.00, and the selling price is $3.00. How many forgings should be produced each year in order to break even?

(A) 5000
(B) 10,000
(C) 13,000
(D) 17,000

For the following problems use the NCEES Handbook as your only reference.

Problems 6 and 7 refer to the following situation.

A company plans to manufacture a product and sell it for $3.00 per unit. Equipment to manufacture the product will cost $250,000 and will have a net salvage value of $12,000 at the end of its estimated economic life of 15 years. The equipment can manufacture up to 2,000,000 units per year. Direct labor costs are $0.25 per unit, direct material costs are $0.85 per unit, variable administrative and selling expenses are $0.25 per unit, and fixed overhead costs are $200,000, not including depreciation.

6. If capital investments and return on the investment are excluded, what is the number of units that the company must manufacture and sell in order to break even with all other costs?

(A) 86,900
(B) 94,900
(C) 121,200
(D) 131,000

7. If straight-line depreciation is used, what is the number of units that the company must manufacture and sell to yield a before-tax profit of 20%?

(A) 187,700
(B) 203,000
(C) 225,300
(D) 270,000

8. Which of the following five situations are examples of making mutually exclusive decisions?

I.   The maintenance department has requested a new air compressor and either a larger paint booth or an additional air compressor.
II. The machine shop needs new inspection and locating equipment.
III. The steno pool needs either a new, faster word processor or an additional office assistant.
IV. The budget committee must decide among building an employees' convenience store, an on-site cafeteria, an enclosed pool, or an in-house exercise room.
V.   The newly elected union representative must resign due to a conflict of interest.

(A) I, II, and IV
(B) I, II, and V
(C) I, III, and IV
(D) II, III, and V

9. A particular gate valve can be repaired, replaced, or left alone. It will cost $12,500 to repair the valve and $25,000 to replace it. The cost due to a failure of the valve seat is $13,000; for a failure of the stem, $21,000; and for a failure of the body, $35,000. All amounts are the present values of all expected future costs. The probabilities of failure of the valve are known.

| | valve component | | |
course of action	seat	stem	body
repair valve	50%	41%	21%
replace valve	35%	27%	9%
no action	65%	53%	42%

What plan of action should be chosen based on a present worth economic basis?

- (A) Repair the valve.
- (B) Replace the valve.
- (C) Either repair or replace the valve.
- (D) Do nothing.

10. Instead of paying $10,000 in annual rent for office space at the beginning of each year for the next 10 years, an engineering firm has decided to take out a 10-year, $100,000 loan for a new building at 6% interest. The firm will invest $10,000 of the rent saved and earn 18% annual interest on that amount. What will be the difference between the firm's annual revenue and expenses?

- (A) The firm will need $3300 extra.
- (B) The firm will need $1800 extra.
- (C) The firm will break even.
- (D) The firm will have $1600 left over.

Problems 11–13 refer to the following information.

An oil company is planning to install a new pipeline to connect storage tanks to a processing plant 1500 m away. Both 120 mm and 180 mm pipes are being considered.

	120 mm pipe	180 mm pipe
initial cost	$2500	$3500
service life	12 years	12 years
salvage value	$300	$400
annual maintenance	$300	$200
pump cost/hour	$1.40	$1.00
pump operation	600 hours/year	600 hours/year

For this analysis, the company will use an annual interest rate of 10%. Annual maintenance and pumping costs may be considered to be paid in their entireties at the end of the years in which their costs are incurred.

11. What is the approximate present worth of the 120 mm pipe over the first 12 years of operation?

- (A) $9200
- (B) $10,200
- (C) $11,900
- (D) $12,100

12. What is the present worth of the 180 mm pipe over the first 12 years of operation if operating costs increase by $0.75 (to $1.75 per hour) beginning in year 7?

- (A) $8790
- (B) $9010
- (C) $9380
- (D) $9930

13. If the annual benefit for the 180 mm pipe is $2000, what is the benefit-cost ratio?

- (A) 1.10
- (B) 1.35
- (C) 1.49
- (D) 1.54

14. A machine has an initial cost of $40,000 and an annual maintenance cost of $5000. Its useful life is 10 years. The annual benefit from purchasing the machine is $18,000. The effective annual interest rate is 10%. What is the machine's benefit-cost ratio?

- (A) 1.51
- (B) 1.56
- (C) 1.73
- (D) 2.24

SOLUTIONS TO FE-STYLE EXAM PROBLEMS

1.

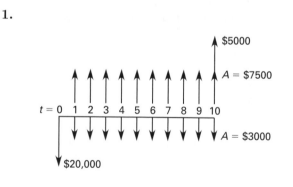

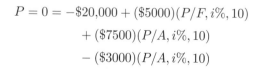

$$P = 0 = -\$20{,}000 + (\$5000)(P/F, i\%, 10)$$
$$+ (\$7500)(P/A, i\%, 10)$$
$$- (\$3000)(P/A, i\%, 10)$$

$$\$20{,}000 = (\$5000)(P/F, i\%, 10)$$
$$+ (\$4500)(P/A, i\%, 10)$$
$$= (\$5000)(1 + i)^{-10}$$
$$+ (\$4500)\left(\frac{(1 + i)^{10} - 1}{i(1 + i)^{10}}\right)$$

By trial and error, $i = 19.6\%$.

The answer may also be obtained by linear interpolation of values from the discount factor tables, but this will not give an exact answer.

Answer is A.

2. For mill M,

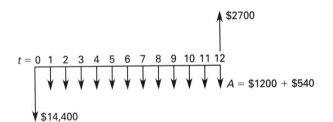

$$(\text{EUAC})_M = (\$7800)(A/P, 10\%, 12) + \$1745 + \$960$$
$$= (\$7800)(0.1468) + \$1745 + \$960$$
$$= \$3850$$

For mill N,

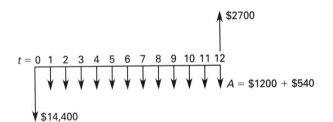

$$(\text{EUAC})_N = (\$14{,}400)(A/P, 10\%, 12)$$
$$- (\$2700)(A/F, 10\%, 12)$$
$$+ \$1200 + \$540$$
$$= (\$14{,}400)(0.1468) - (\$2700)(0.0468)$$
$$+ \$1200 + \$540$$
$$= \$3728 \quad (\$3730)$$

$$\$3730 < \$3850$$

Choose mill N.

Answer is B.

3. For machine A,

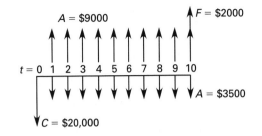

$$P = 0 = -\$20{,}000 + (\$2000)(P/F, i\%, 10)$$
$$+ (\$9000)(P/A, i\%, 10)$$
$$- (\$3500)(P/A, i\%, 10)$$
$$\$20{,}000 = (\$2000)(P/F, i\%, 10)$$
$$+ (\$5500)(P/A, i\%, 10)$$
$$= (\$2000)(1 + i)^{-10}$$
$$+ (\$5500)\left(\frac{(1 + i)^{10} - 1}{i(1 + i)^{10}}\right)$$

By trial and error, $i = 24.8\%$.

For machine B,

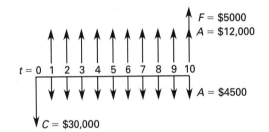

$$P = 0$$
$$= -\$30{,}000 + (\$5000)(P/F, i\%, 10)$$
$$+ (\$12{,}000)(P/A, i\%, 10)$$
$$- (\$4500)(P/A, i\%, 10)$$
$$\$30{,}000 = (\$5000)(P/F, i\%, 10)$$
$$+ (\$7500)(P/A, i\%, 10)$$
$$= (\$5000)(1 + i)^{-10}$$
$$+ (\$7500)\left(\frac{(1 + i)^{10} - 1}{i(1 + i)^{10}}\right)$$

By trial and error, $i = 22.1\%$.

Answer is C.

4. To compare alternatives using rate of return, it is not appropriate to simply compare the rates of return. An incremental analysis must be performed.

Machine B initially costs more than machine A. Subtract the cash flows of machine A from those of machine B to obtain a third alternative representing the costs and benefits of the added investment.

machine B: $-\$30,000 + (\$5000)(P/F, i\%, 10) + (\$12,000)(P/A, i\%, 10) - (\$4500)(P/A, i\%, 10)$

machine A: $-\$20,000 + (\$2000)(P/F, i\%, 10) + \ \ (\$9000)(P/A, i\%, 10) - (\$3500)(P/A, i\%, 10)$

$$-\$10,000 + (\$3000)(P/F, i\%, 10) + \ \ (\$3000)(P/A, i\%, 10) - (\$1000)(P/A, i\%, 10)$$

Set the cash flows equal to zero and find the rate of return.

$$- \$10,000 + (\$3000)(1 + i)^{-10}$$
$$+ (\$2000)\left(\frac{(1+i)^{10} - 1}{i(1+i)^{10}}\right) = 0$$

By trial and error, $i = 16.9\%$ (17%).

For machine B to be worth the extra initial investment, the incremental rate of return must exceed the minimum attractive rate of return. Thus, if machine A is the preferred alternative, MARR > 17%.

Answer is B.

5. At the break-even point, costs equal revenues.

$$\text{costs} = \$10,000 + (\$2.00)(\text{no. of forgings})$$
$$\text{revenues} = (\$3.00)(\text{no. of forgings})$$

$$\$10,000 + (\$2.00)(\text{no. of forgings})$$
$$= (\$3.00)(\text{no. of forgings})$$

$$\text{no. of forgings} = \frac{\$10,000}{\$3.00 - \$2.00} = 10,000$$

Answer is B.

6. $\quad \text{costs} = \$200,000 + (\$0.25)(\text{no. of units})$
$$+ (\$0.85)(\text{no. of units})$$
$$+ (\$0.25)(\text{no. of units})$$
$$= \$200,000 + (\$1.35)(\text{no. of units})$$

$$\text{revenues} = (\$3.00)(\text{no. of units})$$

$$\$200,000 + (\$1.35)(\text{no. of units}) = (\$3.00)(\text{no. of units})$$

$$\text{no. of units} = \frac{\$200,000}{\$3.00 - \$1.35}$$
$$= 121,212 \quad (121,000)$$

Answer is C.

7. $\quad D_j = \dfrac{\$250,000 - \$12,000}{15 \text{ years}}$
$$= \$15,867 \text{ per year}$$

$$\text{costs} = \$15,867 + \$200,000 + (\$1.35)(\text{no. of units})$$

$$\text{revenues} = (\$3.00)(\text{no. of units})$$

For a before-tax profit of 20% of costs,

$$(\$3.00)(\text{no. of units}) = (1.2)(\$215,867$$
$$+ (\$1.35)(\text{no. of units}))$$
$$\text{no. of units} = \frac{(1.2)(\$215,867)}{\$3.00 - (1.2)(\$1.35)}$$
$$= 187,710 \quad (187,700)$$

Answer is A.

8. Alternatives are mutually exclusive when selecting one precludes the others. Situations I, III, and IV require a decision that will eliminate one part or another of the alternative. There is no alternative from which to choose in situations II and V.

Answer is C.

9. Determine the expected cost of each option. The expected cost of an event is the product of the probability and cost for that event.

$$\text{expected cost} = \text{cost of option}$$
$$+ p\{\text{seat failure}\}(\text{cost of seat failure})$$
$$+ p\{\text{stem failure}\}(\text{cost of stem failure})$$
$$+ p\{\text{body failure}\}(\text{cost of body failure})$$

The cost to do nothing is

$$\$0 + (0.65)(\$13,000) + (0.53)(\$21,000)$$
$$+ (0.42)(\$35,000) = \$34,280$$

The cost of repair is

$$\$12,500 + (0.50)(\$13,000) + (0.41)(\$21,000)$$
$$+ (0.21)(\$35,000) = \$34,960$$

The cost to replace is

$$\$25,000 + (0.35)(\$13,000) + (0.27)(\$21,000)$$
$$+ (0.09)(\$35,000) = \$38,370$$

The least expensive option is to do nothing.

Answer is D.

10. The annual loan payment will be

$$P(A/P, 6\%, 10) = (\$100,000)(0.1359)$$
$$= \$13,590$$

The annual return from the investment will be

$$P(A/P, 18\%, 1) = (\$10,000)(1.1800)$$
$$= \$11,800$$

The difference between the loan payment and the return on the investment is

$$\$13,590 - \$11,800 = \$1790 \quad (\$1800)$$

Answer is B.

11.
$$P = \$2500 + (\$300)(P/A, 10\%, 12)$$
$$+ \left(1.40 \ \frac{\$}{\text{hour}}\right)\left(600 \ \frac{\text{hours}}{\text{year}}\right)$$
$$\times (P/A, 10\%, 12)$$
$$- (\$300)(P/F, 10\%, 12)$$
$$= \$2500 + (\$300 + \$840)(6.8137)$$
$$- (\$300)(0.3186)$$
$$= \$10,172 \quad (\$10,200)$$

Answer is B.

12. $P = \$3500 + (\$200)(P/A, 10\%, 12)$
$$+ \left(1.00 \ \frac{\$}{\text{hour}}\right)(600 \ \text{hours})(P/A, 10\%, 6)$$
$$+ \left(1.75 \ \frac{\$}{\text{hour}}\right)(600 \ \text{hours})(P/A, 10\%, 12)$$
$$- \left(1.75 \ \frac{\$}{\text{hour}}\right)(600 \ \text{hours})(P/A, 10\%, 6)$$
$$- (\$400)(P/F, 10\%, 12)$$
$$= \$3500 + (\$200)(6.8137)$$
$$+ (\$600)(4.3553)$$
$$+ (\$1050)(6.8137)$$
$$- (\$1050)(4.3553)$$
$$- (\$400)(0.3186)$$
$$= \$9930$$

Answer is D.

13. The effective uniform annual cost of the 180 mm pipe is

$$\text{EUAC} = (\$3500)(A/P, 10\%, 12) + \$200$$
$$+ \left(1.00 \ \frac{\$}{\text{hour}}\right)(600 \ \text{hours})$$
$$- (\$400)(A/F, 10\%, 12)$$
$$= (\$3500)(0.1468) + \$200 + \$600$$
$$- (\$400)(0.0468)$$
$$= \$1295$$

The benefit-cost ratio of annual amounts is

$$B/C = \frac{\$2000}{\$1295} = 1.54$$

Answer is D.

14. The effective uniform annual cost for the machine is

$$\text{EUAC} = (\$40,000)(A/P, 10\%, 10) + \$5000$$
$$= (\$40,000)(0.1627) + \$5000$$
$$= \$11,508$$

$$B/C = \frac{\$18,000}{\$11,508}$$
$$= 1.56$$

Answer is B.

Topic XVII: Ethics

Diagnostic Examination for Ethics

Chapter 54 Ethics

Hint: For the most current information about the exam, visit www.ppi2pass.com/fefaqs.html regularly.
Use the Exam Forum to compare notes with other FE examinees.

Diagnostic Examination

TOPIC XVII: ETHICS

TIME LIMIT: 45 MINUTES

1. Which of the following establishes ethical behavior among professionals?

(A) taxation regulations
(B) rules of play for professional sports
(C) Hippocratic oath
(D) legal contracts

2. Which of the following methods of advertising is most likely to violate an ethical standard for engineering design firms?

(A) radio or television advertising
(B) yellow-page (phone book) advertising
(C) distribution of company calendars to clients
(D) company brochures using self-laudatory language

3. Whistle blowing is best described as calling public attention to

(A) your own previous unethical behavior.
(B) unethical behavior of employees under your control.
(C) secret illegal behavior by your employer.
(D) unethical or illegal behavior in a government agency you are monitoring as a private individual.

4. Complete the sentence: "Obeying direct and specific orders from a person with higher authority

(A) will never result in ethical behavior."
(B) always results in ethical behavior."
(C) is a defense against unethical behavior."
(D) may place you in an ethical dilemma."

5. Why is competitive bidding often considered to be counterproductive?

(A) Competitive bidding may reduce the client's cost but will increase the design professional's cost.
(B) Competition among different design firms is unprofessional, undignified, and unethical.

(C) The best design is not always the least expensive.
(D) A design professional who reduces the final cost in order to be competitive may eliminate important design steps or features.

6. Complete the sentence: "It is generally considered unethical to moonlight as a consulting engineer while you are working for your primary employer because

(A) you shouldn't be competing with your primary employer for clients."
(B) you can't do a good job for your primary employer if you come to work in the morning tired."
(C) you might be tempted to use proprietary information from your current employer."
(D) "double dipping" (i.e., drawing two paychecks) is unfair to the other engineers in the company."

7. Which of the following could you accept within most codes of ethical behavior?

(A) a trip to the Bahamas from a vendor to learn about that vendor's products
(B) a pen-and-pencil set sent by a blueprint reproduction company
(C) a smoked Thanksgiving turkey sent by a previous client in gratitude for a previous successful job
(D) a monetary incentive sent from a vendor in a country where such incentives are legal and common

8. Complete the sentence: "A professional engineer who took the licensing examination in mechanical engineering

(A) may not design in electrical engineering."
(B) may design in electrical engineering if she feels competent."

(C) may design in electrical engineering if she feels competent and the electrical portion of the design is insignificant and incidental to the overall job."

(D) may design in electrical engineering if another engineer checks the electrical engineering work."

9. Engineering is considered by many to be one of the learned "professions" rather than a more fundamental "occupation." Which of the following is not a characteristic of a profession?

(A) A profession satisfies an indispensable and beneficial social need.

(B) A profession is based on knowledge and skills not common to the general public.

(C) A profession depends to a great extent on the personal judgment of its members.

(D) A profession utilizes entrance exams as a means of limiting its membership to the "elite."

10. An environmental engineer with five years of experience reads a story in the daily paper about a proposal being presented to the city council to construct a new sewage treatment plant near protected wetlands. Based on professional experience and the facts presented in the newspaper, the engineer suspects the plant would be extremely harmful to the local ecosystem. Which of the following would be an acceptable course of action?

(A) The engineer should contact appropriate agencies to get more data on the project before making a judgment.

(B) The engineer should write an article for the paper's editorial page urging the council not to pass the project.

(C) The engineer should circulate a petition through the community condemning the project, and present the petition to the council.

(D) The engineer should do nothing because he doesn't have enough experience in the industry to express a public opinion on the matter.

11. An engineer is consulting for a construction company that has been receiving bad publicity in the local papers about its waste-handling practices. Knowing that this criticism is based on public misperceptions and the paper's thirst for controversial stories, the engineer would like to write an article to be printed in the paper's editorial page. What statement best describes the engineer's ethical obligations?

(A) The engineer's relationship with the company makes it unethical for him to take any public action on its behalf.

(B) The engineer should request that a local representative of the engineering registration board review the data and write the article in order that an impartial point of view be presented.

(C) As long as the article is objective and truthful, and presents all relevant information including the engineer's professional credentials, ethical obligations have been satisfied.

(D) The article must be objective and truthful, present all relevant information including the engineer's professional credentials, and disclose all details of the engineer's affiliation with the company.

12. A survey crew is hired by the general contractor on a large government project to verify pertinent data on the owner-supplied plans. While performing their functions, the survey crew is approached by a subcontractor who wants them to perform some work for him on the same project. He states that he will pay for this additional work and notes that it will be easy for the survey crew to perform both services at the same time. What should the survey crew do?

(A) The survey crew should accept this additional work as long as they have the equipment and capacity to perform both services adequately.

(B) The survey crew should accept this additional work as long as the circumstances are fully disclosed and agreed to by all interested parties.

(C) The survey crew should not accept this additional work as it will be a conflict of interest.

(D) The survey crew should not accept compensation for any additional work because they cannot bill two parties for work performed on the same job.

13. An engineering firm is hired by a developer to prepare plans for a shopping mall. Prior to the final bid date, several contractors who have received bid documents and plans contact the engineering firm with requests for information relating to the project. What can the engineering firm do?

(A) The firm can supply requested information to the contractors as long as it does so fairly and evenly. It cannot favor or discriminate against any contractor.

(B) The firm should supply information to only those contractors that it feels could safely and economically perform the construction services.

(C) The firm cannot reveal facts, data or information relating to the project that might prejudice a contractor against submitting a bid on the project.

(D) The firm cannot reveal facts, data or information relating to the project without the consent of the client as authorized or required by law.

14. Without your knowledge, an old classmate applies to the company you work for. Knowing that you recently graduated from the same school, the director of engineering shows you the application and resume your friend submitted and asks your opinion. It turns out that your friend has exaggerated his participation in campus organizations, even claiming to have been an officer in an engineering society that you are sure he was never in. On the other hand, you remember him as being a highly intelligent student and believe that he could really help the company. How should you handle the situation?

(A) Your should remove yourself from the ethical dilemma by claiming that you don't remember enough about the applicant to make an informed decision.

(B) You should follow your instincts and recommend the applicant. Almost everyone stretches the truth a little in their resumes, and the thing you're really being asked to evaluate is his usefulness to the company. If you mention the resume padding, the company is liable to lose a good prospect.

(C) You should recommend the applicant, but qualify your recommendation by pointing out that you think he may have exaggerated some details on his resume.

(D) You should point out the inconsistencies in the applicant's resume and recommend against hiring him.

15. While working to revise the design of the suspension for a popular car, an engineer discovers a flaw in the design currently being produced. Based on a statistical analysis, the company determines that although this mistake is likely to cause a small increase in the number of fatalities seen each year, it would be prohibitively expensive to do a recall to replace the part. Accordingly, the company decides not to issue a recall notice. What should the engineer do?

(A) The engineer should go along with the company's decision. The company has researched its options and chosen the most economic alternative.

(B) The engineer should send an anonymous tip to the media, suggesting that they alert the public and begin an investigation of the company's business practices.

(C) The engineer should notify the National Transportation Safety Board, providing enough details for them to initiate a formal inquiry.

(D) The engineer should resign from the company. Because of standard nondisclosure agreements, it would be unethical as well as illegal to disclose any information about this situation. In addition, the engineer should not associate with a company that is engaging in such behavior.

SOLUTIONS TO DIAGNOSTIC EXAMINATION TOPIC XVII

1. Ethical restrictions do not carry the force of law. Ethics are not enforced by statutes, referees, umpires, or state troopers. The Hippocratic oath taken by graduating medical students is the earliest known example of a code of ethics for professionals.

Answer is C.

2. Although the standards for what is considered to be ethical advertising are considerably more lenient than they have been in the past, all advertising that is self-laudatory is derogatory to the dignity of the profession and is unethical. Scantily clad women at trade shows and "we're the best" language in company brochures are examples of poor-taste advertising.

Answer is D.

3. "Whistle blowing" is calling public attention to illegal actions taken in the past or being taken currently by your employer. Whistle blowing jeopardizes your own good standing with your employer.

Answer is C.

4. Actions ordered by a military, governmental, or commercial authority may be unethical, immoral, or illegal. Blind obedience to authority is not a valid defense of improper behavior.

Answer is D.

5. A design professional knows the steps necessary to satisfy the client. Each bidding professional will have a different design, will use different methods and suppliers, and will work at a different pace. Some professionals, due to the nature of their design and their own capabilities, will be more expensive. To arbitrarily reduce a bid to "beat" someone else's bid (for a different design or method) may require cutting corners later in the design or construction process.

Answer is D.

6. Choice (D) is irrelevant. Choices (A) and (C) are true, but are not the reason moonlighting is discouraged by most ethical codes.

Answer is B.

7. Technically, any gift, no matter what the value and no matter whether the gift is legal or illegal, is unethical if it influences you or biases you toward the giver. Choices (A), (B), and (D) are all intended to gain favor for future work. Choice (C) is a method of saying "thank you" for past work.

Answer is C.

8. Although the laws vary from state to state, engineers are usually licensed generically. Engineers are licensed as "professional engineers." The scope of their work is limited only by their competence. In the states where the license is in a particular engineering discipline, an engineer may "touch upon" another discipline when the work is insignificant and/or incidental.

Answer is C.

9. Being as dependent as they are on the dissemination of knowledge, professions must be dedicated to both the training and evaluation of potential members of the field. Entrance exams are intended not to limit the membership of the profession, but instead to ensure that all applicants meet a minimum requirement of knowledge and skills.

Answer is D.

10. The engineer certainly has more experience and knowledge in the field than the general public or even the council members who will have to vote on the issue. Therefore, the engineer is qualified to express his opinion if he wishes to do so. Before the engineer takes any public position, however, the engineer is obligated to make sure that all the available information has been collected.

Answer is A.

11. It is ethical for the engineer to issue a public statement concerning a company he works for, provided he makes that relationship clear and provided the statement is truthful and objective.

Answer is D.

12. It is legal and ethical for the survey crew to work for more than one party on the same project as long as the circumstances are fully disclosed and agreed to by all interested parties, and as long as neither of the parties are adversely affected.

Answer is B.

13. It is normal for engineers and architects to clarify the bid documents. However, some information may be proprietary to the developer. The engineering firm should only reveal information that has already been publicly disseminated or approved for release with the consent of the client.

Answer is D.

14. Engineers are ethically obligated to prevent the misrepresentation of their associates' qualifications. You must make your employer aware of the incorrect facts on the resume. On the other hand, if you really believe that the applicant would make a good employee, you should make that recommendation as well. Unless you are making the hiring decision, ethics requires only that you be truthful. If you believe the applicant has merit, you should state so. It is the company's decision to remove or not remove the applicant from consideration because of this transgression.

Answer is C.

15. The engineer's highest obligation is to the public's safety. In most instances, it would be unethical to take some public action on a matter without providing the company with the opportunity to resolve the situation internally. In this case, however, it appears as though the company's senior officers have already reviewed the case and made a decision. The engineer must alert the proper authorities, the NTSB, and provide them with any assistance necessary to investigate the case. To contact the media, although it might accomplish the same goal, would fail to fulfill the engineer's obligation to notify the authorities.

Answer is C.

54 Ethics

Subjects

CREEDS, CODES, CANONS, STATUTES, AND RULES	54-1
PURPOSE OF A CODE OF ETHICS	54-1
ETHICAL PRIORITIES	54-2
DEALING WITH CLIENTS AND EMPLOYERS	54-2
DEALING WITH SUPPLIERS	54-3
DEALING WITH OTHER ENGINEERS	54-3
DEALING WITH (AND AFFECTING) THE PUBLIC	54-3
COMPETITIVE BIDDING	54-4

CREEDS, CODES, CANONS, STATUTES, AND RULES

It is generally conceded that an individual acting on his or her own cannot be counted on to always act in a proper and moral manner. Creeds, statutes, rules, and codes all attempt to complete the guidance needed for an engineer to do "...the correct thing."

A *creed* is a statement or oath, often religious in nature, taken or assented to by an individual in ceremonies. For example, the *Engineers' Creed* adopted by the National Society of Professional Engineers is[1]

> I pledge...
>
> ... to give the utmost of performance;
>
> ... to participate in none but honest enterprise;
>
> ... to live and work according to the laws of man and the highest standards of professional conduct;
>
> ... to place service before profit, the honor and standing of the profession before personal advantage, and the public welfare above all other considerations.
>
> In humility and with need for Divine Guidance, I make this pledge.

A canon is a system of nonstatutory, nonmandatory *codes* of personal conduct. A *canon* is a fundamental

---

[1]The *Faith of an Engineer* adopted by the Accreditation Board for Engineering and Technology (ABET) is a similar but more detailed creed.

belief that usually encompasses several rules. For example, the code of ethics of the American Society of Civil Engineers (ASCE) contains the following seven canons.

1. Engineers shall hold paramount the safety, health, and welfare of the public in the performance of their professional duties.

2. Engineers shall perform services only in areas of their competence.

3. Engineers shall issue public statements only in an objective and truthful manner.

4. Engineers shall act in professional matters for each employer or client as faithful agents or trustees and shall avoid conflicts of interest.

5. Engineers shall build their professional reputation on the merit of their service and shall not compete unfairly with others.

6. Engineers shall act in such a manner as to uphold and enhance the honor, integrity, and dignity of the engineering profession.

7. Engineers shall continue their professional development throughout their careers and shall provide opportunities for the professional development of those engineers under their supervision.

A *rule* is a guide (principle, standard, or norm) for conduct and action in a certain situation. A *statutory rule* is enacted by the legislative branch of state or federal government and carries the weight of law. Some U.S. engineering registration boards have statutory *rules of professional conduct*.

PURPOSE OF A CODE OF ETHICS

Many different sets of *codes of ethics (canons of ethics, rules of professional conduct*, etc.) have been produced by various engineering societies, registration boards, and other organizations.[2] The purpose of these ethical

---

[2]All of the major engineering technical and professional societies in the United States (ASCE, IEEE, ASME, AIChE, NSPE, etc.) and throughout the world have adopted codes of ethics. Most U.S. societies have endorsed the *Code of Ethics of Engineers* developed by the Accreditation Board for Engineering and Technology (ABET), formerly the Engineers' Council for Professional

guidelines is to guide the conduct and decision making of engineers. Most codes are primarily educational. Nevertheless, from time to time they have been used by the societies and regulatory agencies as the basis for disciplinary actions.

Fundamental to ethical codes is the requirement that engineers render faithful, honest, professional service. In providing such service, engineers must represent the interests of their employers or clients and, at the same time, protect public health, safety, and welfare.

There is an important distinction between what is legal and what is ethical. Many legal actions can be violations of codes of ethical or professional behavior. For example, an engineer's contract with a client may give the engineer the right to assign the engineer's responsibilities, but doing so without informing the client would be unethical.

Ethical guidelines can be categorized on the basis of who is affected by the engineer's actions—the client, vendors and suppliers, other engineers, or the public at large. (Some authorities also include ethical guidelines for dealing with the employees of an engineer. However, these guidelines are no different for an engineering employer than they are for a supermarket, automobile assembly line, or airline employer. Ethics is not a unique issue when it comes to employees.)

ETHICAL PRIORITIES

There are frequently conflicting demands on engineers. While it is impossible to use a single decision-making process to solve every ethical dilemma, it is clear that ethical considerations will force engineers to subjugate their own self-interests. Specifically, the ethics of engineers dealing with others need to be considered in the following order from highest to lowest priority.

- society and the public
- the law
- the engineering profession
- the engineer's client
- the engineer's firm
- other involved engineers
- the engineer personally

Development (ECPD). The National Council of Examiners for Engineering and Surveying (NCEES) has developed its *Model Rules* as a guide for state registration boards in developing guidelines for FE professional engineers in those states.

DEALING WITH CLIENTS AND EMPLOYERS

The most common ethical guidelines affecting engineers' interactions with their employer (the *client*) can be summarized as follows.[3]

- Engineers should not accept assignments for which they do not have the skill, knowledge, or time to complete.

- Engineers must recognize their own limitations. They should use associates and other experts when the design requirements exceed their abilities.

- The client's interests must be protected. The extent of this protection exceeds normal business relationships and transcends the legal requirements of the engineer-client contract.

- Engineers must not be bound by what the client wants in instances where such desires would be unsuccessful, dishonest, unethical, unhealthy, or unsafe.

- Confidential client information remains the property of the client and must be kept confidential.

- Engineers must avoid conflicts of interest and should inform the client of any business connections or interests that might influence their judgment. Engineers should also avoid the *appearance* of a conflict of interest when such an appearance would be detrimental to the profession, their client, or themselves.

- The engineers' sole source of income for a particular project should be the fee paid by their client. Engineers should not accept compensation in any form from more than one party for the same services.

- If the client rejects the engineer's recommendations, the engineer should fully explain the consequences to the client.

- Engineers must freely and openly admit to the client any errors made.

All courts of law have required an engineer to perform in a manner consistent with normal professional standards. This is not the same as saying an engineer's work must be error-free. If an engineer completes a design, has the design and calculations checked by another competent engineer, and an error is subsequently shown to have been made, the engineer may be held responsible, but will probably not be considered negligent.

---

[3]These general guidelines contain references to contractors, plans, specifications, and contract documents. This language is common, though not unique, to the situation of an engineer supplying design services to an owner-developer or architect. However, most of the ethical guidelines are general enough to apply to engineers in industry as well.

DEALING WITH SUPPLIERS

Engineers routinely deal with manufacturers, contractors, and vendors (*suppliers*). In this regard, engineers have great responsibility and influence. Such a relationship requires that engineers deal justly with both clients and suppliers.

An engineer will often have an interest in maintaining good relationships with suppliers since this often leads to future work. Nevertheless, relationships with suppliers must remain highly ethical. Suppliers should not be encouraged to feel that they have any special favors coming to them because of a long-standing relationship with the engineer.

The ethical responsibilities relating to suppliers are listed as follows.

- The engineer must not accept or solicit gifts or other valuable considerations from a supplier during, prior to, or after any job. An engineer should not accept discounts, allowances, commissions, or any other indirect compensation from suppliers, contractors, or other engineers in connection with any work or recommendations.

- The engineer must enforce the plans and specifications (i.e., the *contract documents*) but must also interpret the contract documents fairly.

- Plans and specifications developed by the engineer on behalf of the client must be complete, definite, and specific.

- Suppliers should not be required to spend time or furnish materials that are not called for in the plans and contract documents.

- The engineer should not unduly delay the performance of suppliers.

DEALING WITH OTHER ENGINEERS

Engineers should try to protect the engineering profession as a whole, to strengthen it, and to enhance its public stature. The following ethical guidelines apply.

- An engineer should not attempt to maliciously injure the professional reputation, business practice, or employment position of another engineer. However, if there is proof that another engineer has acted unethically or illegally, the engineer should advise the proper authority.

- An engineer should not review someone else's work while the other engineer is still employed unless the other engineer is made aware of the review.

- An engineer should not try to replace another engineer once the other engineer has received employment.

- An engineer should not use the advantages of a salaried position to compete unfairly (i.e., moonlight) with other engineers who have to charge more for the same consulting services.

- Subject to legal and proprietary restraints, an engineer should freely report, publish, and distribute information that would be useful to other engineers.

DEALING WITH (AND AFFECTING) THE PUBLIC

In regard to the social consequences of engineering, the relationship between an engineer and the public is essentially straightforward. Responsibilities to the public demand that the engineer place service to humankind above personal gain. Furthermore, proper ethical behavior requires that an engineer avoid association with projects that are contrary to public health and welfare or that are of questionable legal character.

- Engineers must consider the safety, health, and welfare of the public in all work performed.

- Engineers must uphold the honor and dignity of their profession by refraining from self-laudatory advertising, by explaining (when required) their work to the public, and by expressing opinions only in areas of their knowledge.

- When engineers issue a public statement, they must clearly indicate if the statement is being made on anyone's behalf (i.e., if anyone is benefitting from their position).

- Engineers must keep their skills at a state-of-the-art level.

- Engineers should develop public knowledge and appreciation of the engineering profession and its achievements.

- Engineers must notify the proper authorities when decisions adversely affecting public safety and welfare are made (a practice known as *whistle-blowing*).

COMPETITIVE BIDDING

The ethical guidelines for dealing with other engineers presented here and in more detailed codes of ethics no longer include a prohibition on *competitive bidding*. Until 1971, most codes of ethics for engineers considered competitive bidding detrimental to public welfare, since cost cutting normally results in a lower quality design.

However, in a 1971 case against the National Society of Professional Engineers that went all the way to the U.S. Supreme Court, the prohibition against competitive bidding was determined to be a violation of the Sherman Antitrust Act (i.e., it was an unreasonable restraint of trade).

The opinion of the Supreme Court does not *require* competitive bidding—it merely forbids a prohibition against competitive bidding in NSPE's code of ethics. The following points must be considered.

- Engineers and design firms may individually continue to refuse to bid competitively on engineering services.

- Clients are not required to seek competitive bids for design services.

- Federal, state, and local statutes governing the procedures for procuring engineering design services, even those statutes that prohibit competitive bidding, are not affected.

- Any prohibitions against competitive bidding in individual state engineering registration laws remain unaffected.

- Engineers and their societies may actively and aggressively lobby for legislation that would prohibit competitive bidding for design services by public agencies.

(See the Introduction to NCEES' "Model Rules of Professional Conduct" at the end of this chapter.)

SAMPLE PROBLEMS

Problem 1

"Ethics" is best defined as

I.   a philosophical concept dealing with moral conduct.
II. a set of standards establishing right and wrong actions.
III. rules that describe your duty to society and to your fellow professionals.
IV. guidelines that help you make decisions.

(A) I only
(B) II only
(C) I and III
(D) I, II, III, and IV

Solution

Ethics describes moral conduct; right and wrong actions; and duty to society, your employer, your fellow professionals, and yourself. These rules help you make practical decisions.

Answer is D.

Problem 2

What is the direct result of ethical behavior?

I.   Your reputation will be enhanced.
II. You will be rewarded economically.
III. You will feel good about yourself.

(A) I only
(B) II only
(C) III only
(D) neither I, II, nor III

Solution

Ethical behavior promises nothing, not even a "warm fuzzy" feeling for having done "what is right." Many ethical decisions are bittersweet.

Answer is D.

Problem 3

Ethical behavior is invariant with respect to

I.   time.
II. location.
III. culture.

(A) I only
(B) II only
(C) III only
(D) neither I, II, nor III

Solution

Ethics are not universal. They depend on your culture. Ethics of a culture change over time.

Answer is D.

Problem 4

Complete the sentence: "State registration boards, boards of ethical review, oversight committees, and internal audit departments are used in industry and government because

I.   illegal and unethical actions must be punished."
II. people must be shown that the rules will be enforced."
III. there is something to be learned from all errors in judgment."

(A) I only
(B) II only
(C) III only
(D) I, II, and III

Solution

It is said that "eternal vigilance is the price we pay..." Professionals can make mistakes. They may act out of ignorance, inexperience, or greed. Regardless of the motive, in order to protect the public, the system must react to keep the action from being repeated by the same individual or by others.

Answer is D.

Problem 5

Complete the sentence: "If you check the calculations for a licensed (registered) friend who has gone into a consulting engineering business for himself/herself,

(A) you should be paid for your work."
(B) your friend's client should be told of your involvement."
(C) you do not need to be licensed or registered yourself."
(D) your friend assumes all the liability for your work."

Solution

Whether you are paid or not is between you and your friend. Both you and your friend need to be licensed, and both of you can be held liable for the work. The client has a right to know who worked on the design, and your friend has an ethical obligation to notify the client.

Answer is B.

Problem 6

Which of the following can override your ethical requirement to perform a thorough analysis and check of the work for your client?

(A) time constraints
(B) budgetary constraints
(C) legal constraints, including subpoenas and judicial orders
(D) other ethical obligations

Solution

Ethical obligations to society can take precedence over ethical obligations to a particular person.

Answer is D.

Problem 7

What does it mean when a design professional accepts a punishment for an unethical act from his technical society "with prejudice?"

(A) The professional's race, creed, and national origin were considered in deciding on the punishment.
(B) The professional's race, creed, and national origin were not considered in deciding on the punishment.
(C) Even after the sentence is served or punishment is completed, there may be further actions taken.
(D) The design professional is held in bad report for the period of prejudice.

Solution

When a punishment is meted out "with prejudice," there may still be further ramifications. For example, a professional whose membership is revoked with prejudice may need to have a future hearing in order to rejoin the society at a future date.

Answer is C.

Problem 8

Which of the following principles is not embodied in codes of ethics for engineering consultants?

I.   Consulting engineers will place service to humankind above personal gain.
II. Consulting engineers will serve clients faithfully, honestly, and professionally.
III. Consulting engineers will be fair and will act with integrity and courtesy.
IV. Consulting engineers will encourage the development of the engineering and consulting profession.

(A) I
(B) II
(C) III
(D) neither I, II, III, nor IV

Solution

Codes of ethics for consulting engineers contain provisions pertaining to all four principles.

Answer is D.

Problem 9

While supervising a construction project in a developing country, an engineer discovers that his client's project manager is treating laborers in an unsafe and inhumane (but for that country, legal) manner. When he protests, the engineer is told by company executives that the company has no choice in the matter if it wishes to remain competitive in the region, and he should just accept this as the way things are. What would ethics require the engineer to do?

- (A) Take no action—the company is acting in a perfectly legal manner.
- (B) Withdraw from the project, returning any fees he may already have received.
- (C) Report the company to the proper authorities for its human rights abuses.
- (D) Assist the laborers in organizing a strike to obtain better working conditions.

Solution

The company hasn't broken any laws, so there is no one to report them to, but it is using unethical business practices. The engineer should at the least withdraw from the project as a form of protest and sever any business relations with the company. He could go so far as to assist the workers in protesting, but this might actually be illegal in the country in question. In any case, such activism would be a personal choice on the part of the engineer, not something he is obligated to do under a code of ethics.

Answer is B.

Problem 10

An engineering professor with a professional engineering license and 20 years of experience in engineering education is asked to consult on a building design. Can the professor accept this request?

- (A) Yes, but she should review and comment on only those portions of the project in which she is qualified by education and experience.
- (B) Yes, a professor is a subject matter expert and as such should be fully competent to review the design.
- (C) Yes, as a licensed professional engineer, the professor has demonstrated her competence in engineering and may review the design.
- (D) No, there is a tremendous difference between working in academia and having professional experience. The review should be conducted by a practicing engineer.

Solution

It is perfectly legal and ethical for the professor to consult on the building design. She should, however, review and comment on only those portions of the design that deal with matters in which she is technically competent; the fact that she is a professor with 20 years of experience does not necessarily mean that she is fully knowledgeable of all current design procedures and practices.

Answer is A.

Problem 11

Two engineers submitted sealed bids to a prospective client for a design project. The client told engineer A how much engineer B had bid and invited engineer A to beat that amount. Engineer A really wants the project and honestly believes he can do a better job than engineer B. What should he do?

- (A) He should submit another quote, but only if he can perform the work adequately at the reduced price.
- (B) He should withdraw from consideration for the project.
- (C) He should remain in consideration for the project, but not change his bid.
- (D) He should bargain with the client for the cost of the work.

Solution

It would be unfair and unethical for engineer A to submit another bid. Depending upon the regulating agency, it may also be illegal. He does not, however, have to remove himself from consideration for the project.

Answer is C.

Problem 12

A local engineering professor acts as technical advisor for the city council in a town. A few weeks before the council is scheduled to award a large construction contract, the professor is approached by one of the competing companies and offered a consulting position. Under what circumstances would it be ethical to accept the job?

- (A) Both the company and the council must know about and approve of the arrangement.
- (B) The professor should arrange not to begin work until after the council's vote.
- (C) The professor may accept the job if the advisory position to the council is on a volunteer basis.
- (D) The professor must not participate in any discussions concerning the project for which the company is competing.

Solution

The professor's association with the company is bound to influence his advice to the board, even if the job doesn't start until later. Regardless of whether the board understands and approves of this situation, the engineer is obligated to withdraw from any of the council's discussions concerning the project.

Answer is D.

FE-STYLE EXAM PROBLEMS

1. What is the best description of "going along with the crowd" and "doing what the Romans do?"

    (A) ethical behavior

    (B) legal behavior

    (C) moral behavior

    (D) mob action

2. Which of the following is not a code of ethics?

    (A) the ten commandments

    (B) the ancient code of Hammurabi

    (C) the Rosetta stone

    (D) the scouting creed

3. Ethics requires you to take into consideration the effects of your behavior on which group(s) of people?

    I.   your employer

    II. the nonprofessionals in society

    III. other professionals

    (A) II only

    (B) I and II

    (C) II and III

    (D) I, II, and III

4. Complete the sentence: "Guidelines of ethical behavior among engineers are needed because

    (A) engineers are analytical and they don't always think in terms of right or wrong."

    (B) all people, including engineers, are inherently unethical."

    (C) rules of ethics are easily forgotten."

    (D) it is easy for engineers to take advantage of clients."

5. Which of the following terms is not related to ethics?

    (A) integrity

    (B) honesty

    (C) morality

    (D) profitability

For the following problems use the NCEES Handbook as your only reference.

6. What actions can be taken by a state-regulating agency against a design professional who violates one or more of its rules of conduct?

    I.   the professional's license may be revoked or suspended

    II. notice of the violation may be published in the local newspaper

    III. the professional may be asked to make restitution

    IV. the professional may be required to complete a course in ethics

    (A) I and II

    (B) I and III

    (C) I and IV

    (D) I, II, III, and IV

7. Which of the following activities is not commonly required by codes of ethics for engineers?

    (A) acting as a faithful agent or trustee for the client

    (B) accepting payment for services only from the client

    (C) submitting competitive bids to the client

    (D) spelling out all known conditions of bids and proposals

8. Which organizations typically do not have codes of ethics for engineers?

    (A) technical societies (e.g., ASCE, ASME, IEEE)

    (B) national professional societies (e.g., the National Society of Professional Engineers)

    (C) state professional societies (e.g., the Michigan Society of Professional Engineers)

    (D) companies that write, administer, and grade licensing exams

9. Plan stamping is best defined as

    (A) the legal action of signing off on a project you didn't design but are taking full responsibility for.

    (B) the legal action of signing off on a project you didn't design or check but didn't accept money for.

    (C) the illegal action of signing off on a project you didn't design but did check.

    (D) the illegal action of signing off on a project you didn't design or check.

10. Complete the sentence: "The U.S. Department of Justice's successful action in the 1970s against engineering codes of ethics that formally prohibited competitive bidding was based on the premise that

(A) competitive bidding allowed minority firms to participate."

(B) competitive bidding was required by many government contracts."

(C) the prohibitions violated antitrust statutes."

(D) engineering societies did not have the authority to prohibit competitive bidding."

11. An engineering firm specializes in designing wood-framed houses. Which organizations would it be ethical for the owner of that firm to belong to?

I.   the local Rotary Club
II. the local Chamber of Commerce
III. a national timber-research foundation
IV. a pro-logging lobbying group

(A) I and II
(B) III and IV
(C) III only
(D) I, II, III, and IV

12. An engineer working for a big design firm has decided to start a consulting business, but it will be a few months before she leaves. How should she handle the impending change?

(A) The engineer should discuss her plans with her current employer.

(B) The engineer may approach the firm's other employees while still working for the firm.

(C) The engineer should immediately quit.

(D) The engineer should return all of the pens, pencils, pads of paper, and other equipment she has brought home over the years.

13. During the day, an engineer works for a scientific research laboratory doing government research. During the night, the engineer uses some of the lab's equipment to perform testing services for other consulting engineers. Why is this action probably unethical?

(A) The laboratory has not given its permission for the equipment use.

(B) The government contract prohibits misuse and misappropriation of the equipment.

(C) The equipment may wear out or be broken by the engineer and the replacement cost will be borne by the government contract.

(D) The engineer's fees to the consulting engineers can undercut local testing services' fees because the engineer has a lower overhead.

14. Which of the following methods of charging for professional services is unethical?

(A) lump sum at the start of the job
(B) per diem, billed monthly in advance
(C) per hour, billed at the end of each week
(D) retainer, plus per hour billed at the end of each week

15. An engineer spends all of his free time (outside of work) gambling illegally. Is this a violation of ethical standards?

(A) No, the engineer is entitled to a life outside of work.

(B) No, the engineer's employer, his clients, and the public are not affected.

(C) No, not as long as the engineer stays debt-free from the gambling activities.

(D) Yes, the engineer should associate only with reputable persons and organizations.

16. During routine inspections, a field engineer discovers that one of the company's pipelines is leaking hazardous chemicals into the environment. The engineer recommends that the line be shut down so that seals can be replaced and the pipe can be inspected more closely. His supervisor commends him on his thoroughness, and says the report will be passed on to the company's maintenance division. The engineer moves on to his next job, assuming things will be taken care of in a timely manner. While working in the area again several months later, the engineer notices that the problem hasn't been corrected and is in fact getting worse. What should the engineer do?

(A) Give the matter some more time. In a large corporate environment, it is understandable that some things take longer than people would like them to.

(B) Ask the supervisor to investigate what action has been taken on the matter.

(C) Personally speak to the director of maintenance and insist that this project be given high priority.

(D) Report the company to the EPA for allowing the situation to worsen without taking any preventative measures.

17. A senior licensed professional engineer with 30 years of experience in geotechnical engineering is placed in charge of a multidisciplinary design team consisting of a structural group, a geotechnical group, and an environmental group. In this role, she is responsible for supervising and coordinating the efforts of the groups when working on large interconnected projects. In order to facilitate coordination, designs are prepared by the groups under the direct supervision of the group leader, and then they are submitted to her for review and approval. This arrangement is ethical as long as

(A) she signs and seals each design segment only after being fully briefed by the appropriate group leader.

(B) she signs and seals only those design segments pertaining to geotechnical engineering.

(C) each design segment is signed and sealed by the licensed group leader responsible for its preparation.

(D) she signs and seals each design segment only after it has been reviewed by an independent consulting engineer who specializes in the field in which it pertains.

18. A relatively new engineering firm is considering running an advertisement for their services in the local newspaper. An ad agency has supplied them with four concepts. Of the four types of ads, which one(s) would be acceptable from the standpoint of professional ethics?

I. an advertisement contrasting their successes over the past year with their nearest competitors' failures

II. an advertisement offering a free television to anyone who hires them for work valued at over $10,000

III. an advertisement offering to beat the price of any other engineering firm for the same services

IV. an advertisement that tastefully depicts their logo against the backdrop of the Golden Gate Bridge

(A) I and III

(B) I, III and IV

(C) II, III and IV

(D) neither I, II, III, nor IV

19. An engineer works at a large firm for several years, during which he participates in the development of a new production technique. After leaving the company to start a consulting business, a competitor of his original employer asks for help with a similar problem, and the engineer is sure that the only solution is to use the process developed by his previous employer. Can the engineer ethically accept the job?

(A) No. This would constitute accepting payment from more than one party for the same project.

(B) No. The engineer would have to use information obtained while working for his original employer.

(C) Yes. Because he is no longer employed by the original company, any nondisclosure agreements are invalid.

(D) Yes. It is understood that consulting engineers often work for competing clients, and that some knowledge transfer is inevitable.

20. An engineering firm receives much of its revenue from community construction projects. Which of the following activities would it be ethical for the firm to participate in?

(A) Contribute to the campaigns of local politicians.

(B) Donate money to the city council to help finance the building of a new city park.

(C) Encourage employees to volunteer in community organizations.

(D) Rent billboards to increase the company's name recognition.

21. A building designed by an engineer-architect team leaks during heavy rains. The building's owner may not be able to recover the cost of repairs because of

(A) an act-of-God clause in the owner's building insurance policy.

(B) a hold-harmless clause in the contract between the engineer and the architect.

(C) the statute of limitations.

(D) insolvency of the construction bonding company.

22. Rebecca works for a company in its laboratory division. She is required to keep accurate records of the results of her experimental work. What is the first thing Rebecca should log into her lab notebook?

(A) her name

(B) her position

(C) the date and time

(D) her supervisor's name

23. You are an engineer in charge of receiving bids for an upcoming project. One of the contractors bidding the job is your former employer. The former employer laid you off in a move to cut costs. Which of the following should you do?

I.   say nothing
II. inform your present employer of the situation
III. remain objective when reviewing the bids

(A) I and II
(B) I and III
(C) II only
(D) II and III

24. To whom/what is a registered engineer's foremost responsibility?

(A) client
(B) employer
(C) state and federal laws
(D) public welfare

25. If one is aware that a registered engineer willfully violates a state's rule of professional conduct, one should

(A) do nothing.
(B) report the violation to the state's engineering registration board.
(C) report the violation to the employer.
(D) report the violation to the parties it affects.

26. Which of the following acts would normally not be permitted by an engineering code of conduct?

(A) revealing facts, data, or information obtained in a professional capacity without the consent of the client or employer
(B) approving and sealing only those documents and surveys that conform to accepted engineering standards
(C) providing the state board information of a violation of its engineering code of conduct by a registered engineer
(D) undertaking assignments requiring assistance to complete some elements outside your area of expertise

27. Which would be considered an ethical gift for an engineer to accept from a supplier?

(A) paid trip to an industry trade show
(B) ticket for a trip to Europe
(C) coffee mug with supplier's corporate logo
(D) a $100 check

28. In dealing with suppliers, an engineer may

(A) unduly delay vendor performance if the client agrees
(B) spend personal time outside of the contract to ensure adequate performance
(C) prepare plans containing ambiguous design-build references as cost-saving measures
(D) enforce plans and specifications to the letter, without regard to fairness

29. You are a city engineer in charge of receiving bids on behalf of the city council. A contractor's bid arrives with two tickets to a professional football game. The bid is the lowest received. What should you do?

(A) Return the tickets and accept the bid.
(B) Return the tickets and reject the bid.
(C) Discard the tickets and accept the bid.
(D) Discard the tickets and reject the bid.

30. Mr. S.W. Frank, PE, is the owner and founder of Frank and Sons Engineering, Inc., a state corporation founded in compliance with state laws. None of Mr. Frank's sons or employees are professional engineers. Mr. Frank, now 94 years old and with failing eyesight, comes into the office daily and asks questions about the projects. When his sons complete their reports and plans, he applies his engineer's seal (stamp) to them.

(A) This mode of operation is permitted as long as the company's name contains Mr. Frank's name.
(B) This mode of operation is permitted as long as Mr. Frank's sons work for him.
(C) This mode of operation is permitted as long as Mr. Frank continues to ask questions about the projects.
(D) This mode of operation is not permitted.

31. Seventeen years ago, Susan designed a corrugated steel culvert for a rural road. Her work was accepted and paid for by the county engineering department. Last winter, the culvert collapsed as a loaded logging truck passed over. Although there were no injuries, there was damage to the truck and roadway, and the county tried unsuccessfully to collect on Susan's company's bond. The judge denied the claim on the basis that the work was done too long ago. This defense is known as

(A) privity of contract
(B) duplicity of liability
(C) statute of limitations
(D) caveat emptor

SOLUTIONS TO FE-STYLE EXAM PROBLEMS

1. Group actions may be illegal, immoral, and unethical. The best you can say is that crowd actions are the action of the group. A mob is a group or crowd that acts with a singular purpose.

Answer is D.

2. All choices except (C) describe how you should behave. It is not necessary for a code of ethics to be written. The Rosetta stone described events; it did not describe behavior.

Answer is C.

3. Ethical behavior places restrictions on behavior that affect you, your employer, other engineers, your clients, and society as a whole.

Answer is D.

4. Untrained members of society are at the mercy of the professionals (e.g., doctors, lawyers, engineers) they employ. Even a cab driver can take advantage of a new tourist who doesn't know the shortest route between two points. In many cases, the unsuspecting public needs protection from unscrupulous professionals, engineers included, who act in their own interest.

Answer is D.

5. Ethical actions may or may not be profitable.

Answer is D.

6. All four punishments are commonly used by state engineering licensing boards.

Answer is D.

7. Codes of ethics do not require engineers to submit competitive bids. In fact, competitive bids were prohibited in most engineering codes of ethics for many years.

Answer is C.

8. Answer is D.

9. It is legal to stamp (i.e., sign off on) plans that you personally designed and/or checked. It is illegal to stamp plans that you didn't personally design or check, regardless of whether you got paid. It is legal to work as a "plan checker" consultant.

Answer is D.

10. The U.S. Department of Justice's successful challenge was based on antitrust statutes. Prohibiting competitive bidding was judged to inhibit free competition among design firms.

Answer is C.

11. The owner should not belong to organizations that would cast the profession in a bad light. Though self-serving, none of the organizations listed are illegal or committed to immoral actions. The owner would be able to inform his client of any connections, interests, or affiliations that might influence his judgment. Since the firm specializes in timber construction already, it is unlikely that any of these organizations would cause the owner to favor timber.

Answer is D.

12. There is nothing wrong with wanting to go into business for oneself. The ethical violation occurs when one of the parties does not know what is going on. Even if the engineer acts ethically, takes nothing, and talks to no one about her plans, there will still be the appearance of impropriety if she leaves later. The engineer should discuss her plans with her current employer. That way, there will be minimal disruption to the firm's activities. The engineer shouldn't quit unless her employer demands it. (Those pencils, pens, and pads of paper probably shouldn't have been brought home in the first place.)

Answer is A.

13. Choices (A), (B), (C), and (D) may all be valid. However, the rationale for specific ethical prohibitions on using your employer's equipment for a second job is economic. When you don't have to pay for the equipment, you don't have to recover its purchase price in your fees for services.

Answer is D.

14. A lump sum at the start of the job is probably not wise, but it could be an ethical agreement. Billings for actual costs are ethical. It is common to charge a retainer at the start of a contract and to charge a percentage add-on for materials purchased for the benefit of the client. It is probably not ethical to bill for time you intend to spend in the future.

Answer is B.

15. If the gambling activities were legitimate and legal, this wouldn't be a question, since legal activities are by definition, ethical. The gambling is illegal. Engineers should do nothing that brings them and their profession into disrepute. It is impossible to separate people from their professions. When engineers participate in disreputable activities, it casts the entire profession in a bad light.

Answer is D.

16. While it is true that corporate bureaucracy tends to slow things down, several months is too lengthy a period for an environmental issue. On the other hand, it is by no means clear that the company is ignoring the situation. There could have been some action taken that the engineer is unaware of, or extenuating circumstances that are delaying the repair. To go outside the company or even over the head of his supervisor would be premature without more information. The engineer should ask his supervisor to look into the issue, and should only take further measures if he is dissatisfied with the response.

Answer is B.

17. According to the NCEES model code,

Licensees may accept assignments for coordination of an entire project, provided that each design segment is signed and sealed by the registrant responsible for preparation of that design segment.

Answer is C.

18. None of the ads is acceptable from the standpoint of professional ethics. Concepts I and II are explicitly prohibited by the NCEES model code. Concept III demeans the profession of engineering by placing the emphasis on price as opposed to the quality of services. Concept IV is a misrepresentation; the picture of the Golden Gate Bridge in the background might lead some potential clients to believe that the engineering firm in question had some role in the design or construction of that project.

Answer is D.

19. Nondisclosure agreements are not limited to the term of employment. Consulting engineers are expected to maintain the confidentiality of their clients' proprietary knowledge just as regular employees are, and often they have to sign nondisclosure agreements with each client. Using knowledge gained while working for one company to help another would not be considered working on the same project, but would be a breach of confidentiality.

Answer is B.

20. Contributing to local politics, either to individual campaigns or in the form of a gift to the city, would be seen as an attempt to gain political favor. The renting of billboards, while not as well-defined an issue, implies the sort of self-laudatory advertising that ethical professionals prefer to avoid. Encouraging the company's employees to volunteer their own time to the community is acceptable because the company is unlikely to get any specific benefit from it.

Answer is C.

21. Insurance for building owners routinely covers water damage from rain leakage. A hold-harmless agreement between the engineer and the architect would not affect the client-engineer relationship. Insolvency of the original bonding company would be covered by back-up or federal insurance. A statute of limitations would prohibit the owner from collecting if the building was completed more than a certain number of years before.

Answer is C.

22. An important responsibility in a laboratory environment is to record the date and time that an experiment is conducted.

Answer is C.

23. Registrants should remain objective at all times and should notify their employers of conflicts of interest or situations that could influence the registrants' ability to make objective decisions.

Answer is D.

24. The purpose of engineering registration is to protect the public. This includes protection from harm due to conduct as well as competence. No individual or organization may legitimately direct a registered engineer to harm the public.

Answer is D.

25. A violation should be reported to the organization that has promulgated the rule.

Answer is B.

26. An engineer may not ethically reveal facts that have been confided.

Answer is A.

27. Choices A, B, and D are gifts with relatively high monetary value and therefore are unethical. An engineer should only accept items that are of insignificant value.

Answer is C.

28. An engineer not only may, but is required to, ensure performance consistent with plans and specifications. If a job is intentionally or unintentionally underbid, the engineer will have to use personal time to complete the project.

Answer is B.

29. Registrants should not accept gifts from parties expecting special consideration, so the tickets cannot be kept. They also should not be merely discarded, for several reasons. Inasmuch as the motive of the contractor is not known with certainty, in the absence of other bidding rules, the bid may be accepted.

Answer is A.

30. This problem deals with "plan stamping." Although all professional engineers can stamp (seal) plans, the term "plan stamping" is specifically used to describe stamping plans that someone else has developed without verifying the validity of the plans personally. Plan stamping is in violation of all state engineering acts. In this case, it is unlikely that Mr. Frank's questions are sufficient to determine the validity of the plan he stamps.

Answer is D.

31. Most states have statutes of limitations. Unless a crime or fraudulent act has been committed, defects appearing after a certain amount of time are not actionable.

Answer is C.

Introduction to NCEES' "Model Rules of Professional Conduct"[1]

Engineering is considered to be a "profession" rather than an "occupation" because of serveral important characteristics shared with other recognized learned professions, law, medicine, and theology: special knowledge, special privileges, and special responsibilities. Professions are based on a large knowledge base requiring extensive training. Professional skills are important to the well-being of society. Professions are self-regulating, in that they control the training and evaluation processes that admit new persons to the field. Professionals have autonomy in the workplace; they are expected to utilize their independent judgment in carrying out their professional responsibilities. Finally, professions are regulated by ethical standards.

The expertise possessed by engineers is vitally important to public welfare. In order to serve the public effectively, engineers must maintain a high level of technical competence. However, a high level of technical expertise without adherence to ethical guidelines is as much a threat to public welfare as is professional incompetence. Therefore, engineers must also be guided by ethical principles.

The ethical principles governing the engineering profession are embodied in codes of ethics. Such codes have been adopted by state boards of registration, professional engineering societies, and even by some private industries. An example of one such code is the NCEES Rules of Professional Conduct, found in Section 240 of *Model Rules* and presented here. As part of his/her responsibility to the public, an engineer is responsible for knowing and abiding by the code. Additional rules of conduct are also included in *Model Rules*.

The three major sections of the model rules address (1) Licensee's Obligation to Society, (2) Licensee's Obligation to Employers and Clients, and (3) Licensee's Obligation to Other Licensees. The principles amplified in these sections are important guides to appropriate behavior of professional engineers.

Application of the code in many situations is not controversial. However, there may be situations in which applying the code may raise more difficult issues. In particular, there may be circumstances in which terminology in the code is not clearly defined, or in which two sections of the code may be in conflict. For example, what constitutes "valuable consideration" or "adequate" knowledge may be interpreted differently by qualified professionals. These types of questions are called *conceptual issues*, in which definitions of terms may be in dispute. In other situations, *factual issues* may also affect ethical dilemmas. Many decisions regarding engineering design may be based upon interpretation of disputed or incomplete information. In addition, *tradeoffs* revolving around competing issues of risk *vs* benefit, or safety *vs* economics may require judgments that are not fully addressed simply by application of the code.

No code can give immediate and mechanical answers to all ethical and professional problems that an engineer may face. Creative problem solving is often called for in ethics, just as it is in other areas of engineering.

Excerpt from NCEES' "Model Rules"

240.15 Rules of Professional Conduct

A. Licensee's Obligation to Society

1. Licensees, in the performance of their services for clients, employers, and customers, shall be cognizant that their first and foremost responsibility is to the public welfare.

2. Licensees shall approve and seal only those design documents and surveys that conform to accepted engineering and surveying standards and safeguard the life, health, property, and welfare of the public.

3. Licensees shall notify their employer or client and such other authority as may be appropriate when their professional judgment is overruled under circumstances where the life, health, property, or welfare of the public is endangered.

4. Licensees shall be objective and truthful in professional reports, statements, or testimony. They shall include all relevant and pertinent information in such reports, statements, or testimony.

(continued)

---

[1]Reproduced with permission from *Fundamentals of Engineering Supplied-Reference Handbook*, 7th Ed., pg. 99, copyright © by the National Council of Examiners for Engineering and Surveying® (www.ncees.org).

(continued)

5. Licensees shall express a professional opinion publicly only when it is founded upon an adequate knowledge of the facts and a competent evaluation of the subject matter.

6. Licensees shall issue no statements, criticisms, or arguments on technical matters which are inspired or paid for by interested parties, unless they explicitly identify the interested parties on whose behalf they are speaking and reveal any interest they have in the matters.

7. Licensees shall not permit the use of their name or firm name by, nor associate in the business ventures with, any person or firm which is engaging in fraudulent or dishonest business or professional practices.

8. Licensees having knowledge of possible violations of any of these Rules of Professional Conduct shall provide the board with the information and assistance necessary to make the final determination of such violation. *(Section 150, Disciplinary Action, NCEES Model Law)*

B. Licensee's Obligation to Employer and Clients

1. Licensees shall undertake assignments only when qualified by education or experience in the specific technical fields of engineering or surveying involved.

2. Licensees shall not affix their signatures or seals to any plans or documents dealing with subject matter in which they lack competence, nor to any such plan or document not prepared under their direct control and personal supervision.

3. Licensees may accept assignments for coordination of an entire project, provided that each design segment is signed and sealed by the licensee responsible for preparation of that design segment.

4. Licensees shall not reveal facts, data, or information obtained in a professional capacity without the prior consent of the client or employer except as authorized or required by law. Licensees shall not solicit or accept gratuities, directly or indirectly, from contractors, their agents, or other parties in connection with work for employers or clients.

5. Licensees shall make full prior disclosures to their employers or clients of potential conflicts of interest or other circumstances which could influence or appear to influence their judgment or the quality of their service.

6. Licensees shall not accept compensation, financial or otherwise, from more than one party for services pertaining to the same project, unless the circumstances are fully disclosed and agreed to by all interested parties.

7. Licensees shall not solicit or accept a professional contract from a governmental body on which a principal or officer of their organization serves as a member. Conversely, licensees serving as members, advisors, or employees of a government body or department, who are the principals or employees of a private concern, shall not participate in decisions with respect to professional services offered or provided by said concern to the governmental body which they serve. *(Section 150, Disciplinary Action, NCEES Model Law)*

C. Licensee's Obligation to Other Licensees

1. Licensees shall not falsify or permit misrepresentation of their, or their associates', academic or professional qualifications. They shall not misrepresent or exaggerate their degree of responsibility in prior assignments nor the complexity of said assignments. Presentations incident to the solicitation of employment or business shall not misrepresent pertinent facts concerning employers, employees, associates, joint ventures, or past accomplishments.

2. Licensees shall not offer, give, solicit, or receive, either directly or indirectly, any commission, or gift, or other valuable consideration in order to secure work, and shall not make any political contribution with the intent to influence the award of a contract by public authority.

3. Licensees shall not attempt to injure, maliciously or falsely, directly or indirectly, the professional reputation, prospects, practice, or employment of other licensees, nor indiscriminately criticize other licensees' work. *(Section 150, Disciplinary Action, NCEES Model Law)*

Reprinted with permission from *Model Rules*, copyright © 2008 by the NCEES. As adapted with permission from C. E. Harris, M. S. Pritchard, and M. J. Rabins, *Engineering Ethics: Concepts and Cases*, copyright © 1995 by Wadsworth Publishing Company, pp. 27, 28. Revised editorially to match the third edition, December 1997.

ETHICS

Sample Examination
Morning Section

To make taking this sample exam as realistic as possible,
use the NCEES Handbook as your only reference.

1.	Ⓐ Ⓑ Ⓒ Ⓓ	31.	Ⓐ Ⓑ Ⓒ Ⓓ	61.	Ⓐ Ⓑ Ⓒ Ⓓ	91.	Ⓐ Ⓑ Ⓒ Ⓓ
2.	Ⓐ Ⓑ Ⓒ Ⓓ	32.	Ⓐ Ⓑ Ⓒ Ⓓ	62.	Ⓐ Ⓑ Ⓒ Ⓓ	92.	Ⓐ Ⓑ Ⓒ Ⓓ
3.	Ⓐ Ⓑ Ⓒ Ⓓ	33.	Ⓐ Ⓑ Ⓒ Ⓓ	63.	Ⓐ Ⓑ Ⓒ Ⓓ	93.	Ⓐ Ⓑ Ⓒ Ⓓ
4.	Ⓐ Ⓑ Ⓒ Ⓓ	34.	Ⓐ Ⓑ Ⓒ Ⓓ	64.	Ⓐ Ⓑ Ⓒ Ⓓ	94.	Ⓐ Ⓑ Ⓒ Ⓓ
5.	Ⓐ Ⓑ Ⓒ Ⓓ	35.	Ⓐ Ⓑ Ⓒ Ⓓ	65.	Ⓐ Ⓑ Ⓒ Ⓓ	95.	Ⓐ Ⓑ Ⓒ Ⓓ
6.	Ⓐ Ⓑ Ⓒ Ⓓ	36.	Ⓐ Ⓑ Ⓒ Ⓓ	66.	Ⓐ Ⓑ Ⓒ Ⓓ	96.	Ⓐ Ⓑ Ⓒ Ⓓ
7.	Ⓐ Ⓑ Ⓒ Ⓓ	37.	Ⓐ Ⓑ Ⓒ Ⓓ	67.	Ⓐ Ⓑ Ⓒ Ⓓ	97.	Ⓐ Ⓑ Ⓒ Ⓓ
8.	Ⓐ Ⓑ Ⓒ Ⓓ	38.	Ⓐ Ⓑ Ⓒ Ⓓ	68.	Ⓐ Ⓑ Ⓒ Ⓓ	98.	Ⓐ Ⓑ Ⓒ Ⓓ
9.	Ⓐ Ⓑ Ⓒ Ⓓ	39.	Ⓐ Ⓑ Ⓒ Ⓓ	69.	Ⓐ Ⓑ Ⓒ Ⓓ	99.	Ⓐ Ⓑ Ⓒ Ⓓ
10.	Ⓐ Ⓑ Ⓒ Ⓓ	40.	Ⓐ Ⓑ Ⓒ Ⓓ	70.	Ⓐ Ⓑ Ⓒ Ⓓ	100.	Ⓐ Ⓑ Ⓒ Ⓓ
11.	Ⓐ Ⓑ Ⓒ Ⓓ	41.	Ⓐ Ⓑ Ⓒ Ⓓ	71.	Ⓐ Ⓑ Ⓒ Ⓓ	101.	Ⓐ Ⓑ Ⓒ Ⓓ
12.	Ⓐ Ⓑ Ⓒ Ⓓ	42.	Ⓐ Ⓑ Ⓒ Ⓓ	72.	Ⓐ Ⓑ Ⓒ Ⓓ	102.	Ⓐ Ⓑ Ⓒ Ⓓ
13.	Ⓐ Ⓑ Ⓒ Ⓓ	43.	Ⓐ Ⓑ Ⓒ Ⓓ	73.	Ⓐ Ⓑ Ⓒ Ⓓ	103.	Ⓐ Ⓑ Ⓒ Ⓓ
14.	Ⓐ Ⓑ Ⓒ Ⓓ	44.	Ⓐ Ⓑ Ⓒ Ⓓ	74.	Ⓐ Ⓑ Ⓒ Ⓓ	104.	Ⓐ Ⓑ Ⓒ Ⓓ
15.	Ⓐ Ⓑ Ⓒ Ⓓ	45.	Ⓐ Ⓑ Ⓒ Ⓓ	75.	Ⓐ Ⓑ Ⓒ Ⓓ	105.	Ⓐ Ⓑ Ⓒ Ⓓ
16.	Ⓐ Ⓑ Ⓒ Ⓓ	46.	Ⓐ Ⓑ Ⓒ Ⓓ	76.	Ⓐ Ⓑ Ⓒ Ⓓ	106.	Ⓐ Ⓑ Ⓒ Ⓓ
17.	Ⓐ Ⓑ Ⓒ Ⓓ	47.	Ⓐ Ⓑ Ⓒ Ⓓ	77.	Ⓐ Ⓑ Ⓒ Ⓓ	107.	Ⓐ Ⓑ Ⓒ Ⓓ
18.	Ⓐ Ⓑ Ⓒ Ⓓ	48.	Ⓐ Ⓑ Ⓒ Ⓓ	78.	Ⓐ Ⓑ Ⓒ Ⓓ	108.	Ⓐ Ⓑ Ⓒ Ⓓ
19.	Ⓐ Ⓑ Ⓒ Ⓓ	49.	Ⓐ Ⓑ Ⓒ Ⓓ	79.	Ⓐ Ⓑ Ⓒ Ⓓ	109.	Ⓐ Ⓑ Ⓒ Ⓓ
20.	Ⓐ Ⓑ Ⓒ Ⓓ	50.	Ⓐ Ⓑ Ⓒ Ⓓ	80.	Ⓐ Ⓑ Ⓒ Ⓓ	110.	Ⓐ Ⓑ Ⓒ Ⓓ
21.	Ⓐ Ⓑ Ⓒ Ⓓ	51.	Ⓐ Ⓑ Ⓒ Ⓓ	81.	Ⓐ Ⓑ Ⓒ Ⓓ	111.	Ⓐ Ⓑ Ⓒ Ⓓ
22.	Ⓐ Ⓑ Ⓒ Ⓓ	52.	Ⓐ Ⓑ Ⓒ Ⓓ	82.	Ⓐ Ⓑ Ⓒ Ⓓ	112.	Ⓐ Ⓑ Ⓒ Ⓓ
23.	Ⓐ Ⓑ Ⓒ Ⓓ	53.	Ⓐ Ⓑ Ⓒ Ⓓ	83.	Ⓐ Ⓑ Ⓒ Ⓓ	113.	Ⓐ Ⓑ Ⓒ Ⓓ
24.	Ⓐ Ⓑ Ⓒ Ⓓ	54.	Ⓐ Ⓑ Ⓒ Ⓓ	84.	Ⓐ Ⓑ Ⓒ Ⓓ	114.	Ⓐ Ⓑ Ⓒ Ⓓ
25.	Ⓐ Ⓑ Ⓒ Ⓓ	55.	Ⓐ Ⓑ Ⓒ Ⓓ	85.	Ⓐ Ⓑ Ⓒ Ⓓ	115.	Ⓐ Ⓑ Ⓒ Ⓓ
26.	Ⓐ Ⓑ Ⓒ Ⓓ	56.	Ⓐ Ⓑ Ⓒ Ⓓ	86.	Ⓐ Ⓑ Ⓒ Ⓓ	116.	Ⓐ Ⓑ Ⓒ Ⓓ
27.	Ⓐ Ⓑ Ⓒ Ⓓ	57.	Ⓐ Ⓑ Ⓒ Ⓓ	87.	Ⓐ Ⓑ Ⓒ Ⓓ	117.	Ⓐ Ⓑ Ⓒ Ⓓ
28.	Ⓐ Ⓑ Ⓒ Ⓓ	58.	Ⓐ Ⓑ Ⓒ Ⓓ	88.	Ⓐ Ⓑ Ⓒ Ⓓ	118.	Ⓐ Ⓑ Ⓒ Ⓓ
29.	Ⓐ Ⓑ Ⓒ Ⓓ	59.	Ⓐ Ⓑ Ⓒ Ⓓ	89.	Ⓐ Ⓑ Ⓒ Ⓓ	119.	Ⓐ Ⓑ Ⓒ Ⓓ
30.	Ⓐ Ⓑ Ⓒ Ⓓ	60.	Ⓐ Ⓑ Ⓒ Ⓓ	90.	Ⓐ Ⓑ Ⓒ Ⓓ	120.	Ⓐ Ⓑ Ⓒ Ⓓ

Problems for Sample Examination

MORNING SECTION

1. What is the integral of the following equation?

$$15x^4 - 8x^3 + \frac{1}{x} + 7$$

(A) $3x^5 - 2x^4 + \ln |x| + 7x + C$

(B) $60x^3 - 24x^2 - \frac{1}{x^2} + C$

(C) $60x^4 - 24x^3 - \frac{1}{x^2} + 7x + C$

(D) $\frac{15}{4}x^5 - \frac{8}{3}x^4 - \frac{1}{x^2} + 7x + C$

2. What is the general solution for the following differential equation?

$$2\left(\frac{d^2y}{dx^2}\right) - 4\left(\frac{dy}{dx}\right) + 4y = 0$$

(A) $C_1 \cos x + C_2 \sin x$
(B) $C_1 e^x + C_2 e^{-x}$
(C) $e^{-x}(C_1 \cos x - C_2 \sin x)$
(D) $e^x(C_1 \cos x + C_2 \sin x)$

3. What is the distance between points P $(1, -3, 5)$ and Q $(-3, 4, -2)$?

(A) $\sqrt{10}$
(B) $\sqrt{14}$
(C) 8
(D) $\sqrt{114}$

4. If the rectangular coordinates of a point are $(-3, -5.2)$, what are the polar coordinates (r, θ) of the point?

(A) $(-6, -120°)$
(B) $(6, -120°)$
(C) $(6, 120°)$
(D) $(6, -150°)$

5. What are the three roots of the equation $(x - 1)^3 = 0$?

(A) $1, 1, 1$
(B) $-1, -1, -1$
(C) $-1, 1, 1$
(D) $1, 1, 0$

6. Determine the (x, y) coordinates of the center of the circle defined by the following equation.

$$x^2 - 8x + y^2 - 10y + 25 = 0$$

(A) $(3,2)$
(B) $(3,4)$
(C) $(4,5)$
(D) $(5,4)$

7. What is the partial derivative with respect to x of the following function?

$$z = e^{xy}$$

(A) e^{xy}

(B) $\frac{e^{xy}}{x}$

(C) $\frac{e^{xy}}{y}$

(D) ye^{xy}

8. What is the determinant of the following matrix?

$$\begin{bmatrix} 3 & 2 & 7 \\ 3 & 7 & 1 \\ 1 & 5 & 2 \end{bmatrix}$$

(A) -113
(B) -7
(C) 12
(D) 73

9. What is the value of the following limit?

$$\lim_{x \to 0}\left(\frac{\sin 5x}{x}\right)$$

(A) 0
(B) 0.087
(C) 1
(D) 5

Problems 10 and 11 are based on the following vector equations.

$$V_1 = 3i + 2j + k$$
$$V_2 = 2i + 4j + 6k$$
$$V_3 = 2i + 3j + 2k$$

10. What is most nearly the angle between V_1 and V_2?

(A) 35.5°
(B) 44.4°
(C) 45.6°
(D) 88.5°

11. What is the cross product of V_2 and V_3?

(A) $-10i + 8j - 2k$
(B) $26i + 16j + 14k$
(C) $-2i + 8j - 10k$
(D) $4i + 12j + 12k$

12. For the ellipse shown, what is most nearly the distance z?

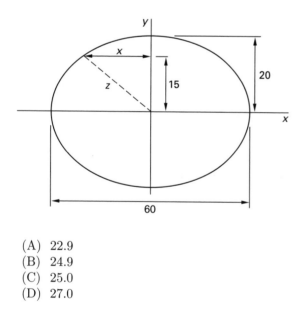

(A) 22.9
(B) 24.9
(C) 25.0
(D) 27.0

13. Solve the system of simultaneous equations for x, y, and z.

$$x - y + z = 0$$
$$x - y - z = -3$$
$$-x - y + z = 5$$

(A) $\left(-\dfrac{5}{2}, -4, -\dfrac{3}{2}\right)$

(B) $\left(\dfrac{5}{2}, 4, \dfrac{3}{2}\right)$

(C) $(0, -1, 4)$

(D) $\left(-\dfrac{5}{2}, -1, \dfrac{3}{2}\right)$

14. Which of the following statements is true for a polynomial of degree 3 with constant real coefficients?

(A) It always has three distinct real roots.
(B) It has either two distinct imaginary roots and one real root, or three distinct real roots.
(C) It can have zero, one, two, or three distinct roots, all of which are real.
(D) It has none of the above.

15. What is most nearly the length of the longest side of the triangle defined by the following lines?

$$y = 2x - 2$$
$$y = -2x + 3$$
$$y = 12x - 7$$

(A) 1.20
(B) 1.68
(C) 1.73
(D) 2.58

16. Matrix **C** is the product of the matrix multiplication **AB**. What is the size of matrix **B**?

$$A = \begin{vmatrix} 4 & 7 & 0 \\ 2 & 0 & 2 \\ 3 & 1 & 3 \\ 1 & 1 & 1 \end{vmatrix} \qquad C = \begin{vmatrix} 1 & 2 & 0 & 4 \\ 2 & 1 & 0 & 7 \\ 3 & 2 & 1 & 6 \\ 5 & 5 & 4 & 2 \end{vmatrix}$$

(A) 3×3
(B) 3×4
(C) 4×3
(D) 4×4

17. An ancient square pyramid is 60 m tall and 200 m on each side at the base. The density of the building stone is 1300 kg/m³. If the pyramid is solid throughout, what is the pyramid's approximate mass?

(A) 0.8 Mg
(B) 2.4 Mg
(C) 1.0 Gg
(D) 1.0 Tg

18. Use logarithmic identities to evaluate the numerical value for the following expression.

$$\log_3 \left(\frac{3}{2} \right) + \log_3 12 - \log_3 2$$

(A) 0.95
(B) 1.33
(C) 2.00
(D) 2.20

19. In the following progression, what is the value of the term a?

$$125, \ 50, \ a, \ 8, \ 3.2$$

(A) 16
(B) 20
(C) 25
(D) 30

20. The scores for a final exam are distributed according to the table shown. What are the approximate mode and arithmetic mean of the scores?

score	frequency
38	5
45	2
69	7
76	10
82	12
90	8
91	19
95	15

(A) mode 79, mean 73
(B) mode 82, mean 79
(C) mode 85, mean 85
(D) mode 91, mean 82

21. A cloth bag contains six cards numbered 1 through 6. Two cards are drawn without replacement. What is the probability that the sum of the numbers on the two drawn cards is 7?

(A) $\dfrac{1}{12}$

(B) $\dfrac{7}{36}$

(C) $\dfrac{1}{5}$

(D) $\dfrac{7}{18}$

22. Four dice are thrown. The probability of rolling at least one six is most nearly

(A) 0.50
(B) 0.52
(C) 0.63
(D) 0.67

23. What is the sample standard deviation for the data set $\{4, 6, 9, 9\}$?

(A) 2.1
(B) 2.4
(C) 4.2
(D) 6.0

24. On average, a piece of machinery jams three times a week. Assuming a Poisson distribution for the frequency of jams, the probability that the machine will jam exactly five times in a given week is most nearly

(A) 0.0033
(B) 0.10
(C) 0.33
(D) 0.40

25. The least squares method is used to plot a straight line through the data points $(1, 6)$, $(2, 7)$, $(3, 11)$, and $(5, 13)$. The slope of the line is most nearly

(A) 0.867
(B) 1.73
(C) 1.86
(D) 2.03

26. A hypothesis test is useful in determining, to a certain degree of probability, whether a

(A) straight line is a good fit for a set of points
(B) sample size is large enough to represent a population
(C) mathematical model can correctly predict results
(D) set of data has come from a particular distribution

27. The density of gold is 19.32 g/cm^3. Most nearly, how many atoms are there in 1 cm^3 of gold?

(A) 5.9×10^{19} atoms
(B) 5.9×10^{22} atoms
(C) 1.7×10^{24} atoms
(D) 1.7×10^{29} atoms

28. Most nearly, how many grams of sodium hydroxide will be formed when 200 g of sodium react completely with water?

(A) 170 g
(B) 200 g
(C) 230 g
(D) 350 g

29. Most nearly, how much 2 M sulfuric acid is required to neutralize 50 mL of 5 M sodium hydroxide?

(A) 20 mL
(B) 53 mL
(C) 63 mL
(D) 70 mL

30. 2 mol of aluminum react with hydrochloric acid to form which of the following?

(A) 2 mol of hydrogen gas
(B) 6 mol of hydrogen gas
(C) 1 mol of aluminum chloride
(D) 3 mol of hydrogen gas

31. Which of the following is a member of the halogen family?

(A) sodium
(B) fluorine
(C) hydrogen chloride
(D) phosphorus

32. What is the empirical formula for a compound containing 31.9% potassium, 29.0% chlorine, and 39.2% oxygen by weight?

(A) $KClO_3$
(B) $KClO$
(C) $K_2Cl_2O_3$
(D) KCl_2O

33. What are most nearly the molar H^+ and OH^- concentrations of a solution with a pH of 1.3?

(A) $[H^+] = 2.7 \times 10^{-1}$ mol/L
  $[OH^-] = 3.7 \times 10^{-14}$ mol/L
(B) $[H^+] = 5.0 \times 10^{-2}$ mol/L
  $[OH^-] = 3.7 \times 10^{-14}$ mol/L
(C) $[H^+] = 1.3 \times 10^{-7}$ mol/L
  $[OH^-] = 0$ mol/L
(D) $[H^+] = 5.0 \times 10^{-2}$ mol/L
  $[OH^-] = 2.0 \times 10^{-13}$ mol/L

34. Substances A and B in liquid form have vapor pressures at 85°C of 800 mm Hg and 300 mm Hg, respectively. What will be the molar composition of a mixture of these substances that boils at 85°C under 1 atm of pressure?

(A) 23% A; 77% B
(B) 54% A; 46% B
(C) 73% A; 27% B
(D) 92% A; 8% B

35. Most nearly what volume of 3 M HCl is required to neutralize 25 mL of 5 M NaOH?

(A) 15 mL
(B) 42 mL
(C) 45 mL
(D) 50 mL

36. Most nearly how much of a 1 kg sample of U-238 will be present after 500 million years? The half-life of U-238 is 4.5×10^9 years.

(A) 110 g
(B) 930 g
(C) 940 g
(D) 990 g

Problems 37–39 are based on the following information.

The following section of pseudocode approximates the integral of some function, $F(x)$, over the interval from a to b. Assume that a separate subroutine has already been written to calculate the value of $F(x)$.

```
1    INPUT a, b, n
2    d = (b − a)/n
3    S = F(a)/2 + F(b)/2
4    FOR k FROM 1 TO (n − 1)
5    S = S + F(a + k*d)
6    NEXT k
7    S = S*d
```

37. Which numerical method does this code represent?

(A) Euler's method
(B) Newton's method
(C) Simpson's rule
(D) trapezoidal rule

38. Using the pseudocode's algorithm with $n = 4$, what would be the approximation of $\int_0^\pi \sin x \, dx$?

(A) 1.4
(B) 1.6
(C) 1.9
(D) 2.0

39. Lines 3 and 4 are changed as shown. Which numerical method does the code now represent?

```
1    INPUT a, b, n
2    d = (b − a)/n
3    S = 0
4    FOR k FROM 0 TO (n − 1)
5    S = S + F(a + k*d)
6    NEXT k
7    S = S*d
```

(A) Euler's method
(B) Newton's method
(C) Simpson's rule
(D) trapezoidal rule

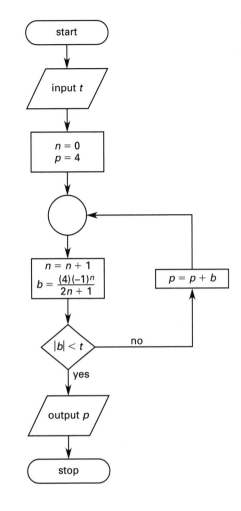

40. How many individual spreadsheet cells are contained in the range C3:J12?

(A) 8
(B) 63
(C) 72
(D) 80

(A) 2.90
(B) 3.34
(C) 4.00
(D) 4.44

41. Which of the following types of memory is lost when a computer's power is removed?

(A) RAM
(B) ROM
(C) PROM
(D) EPROM

42. The flowchart for a computer program is as shown. Variable t has an initial value of 0.5. What is the value of variable p at the conclusion of the routine?

43. A computer program segment written in a generic language is shown. For which input values X will the ouput value of T be equal to X?

```
INPUT X
IF X < 5
        T = X*X/(16 − X)
ELSE T = X*5/(10 − X)
ENDIF
OUTPUT T
```

(A) 0
(B) 0, 5
(C) 0, 5, 8
(D) none

44. A 256K-word memory uses 16 bit-words. How many parallel data lines are required to pass data to the CPU for processing?

    (A) 2
    (B) 8
    (C) 9
    (D) 16

45. What are the Miller indices of the plane shown?

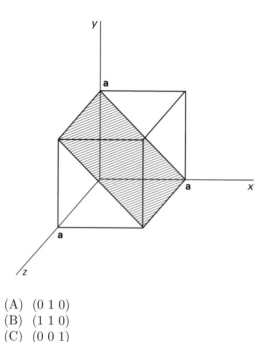

    (A) (0 1 0)
    (B) (1 1 0)
    (C) (0 0 1)
    (D) (1 1 ∞)

46. The spreadsheet shown is to be used to calculate the current in R_1, where R_1, R_2, and R_3 are resistors in parallel and V is the voltage applied across all of them. Which of the formulas below can be entered in cell E6 and copied and pasted into the rest of the cells in the E column to accomplish this task?

	A	B	C	D	E
1	V (volts)	R_1 (ohms)	R_2 (ohms)	R_3 (ohms)	I (amps)
2	5	3	6	14	
3	5	5	7	18	
4	5	7	8	19	
5	5	9	9	21	
6	5	11	10	22	

    (A) $A6 * (((\$C6 * D6)/(\$C6+D6))/((1/\$C6 +(1/D6)))$
    (B) $A6 * (((C6 * D\$6)/(\$C6+D6))/((1/\$B6 +(1/C\$6)+(1/D\$6)))$

    (C) $A6/\$B6$
    (D) $A6 * (\$B6/((1/B6)+(1/\$C6)+(1/D6))))$

47. A new junior engineer in a design company notices a detail in a design that she feels has the potential to be dangerous to the end users. Her superiors explain that this detail was incorporated in the design by the company to save manufacturing time. Furthermore, they assure her that although the analysis is technically correct, this "shortcut" has been used for several years and the company has never been accused of any wrongdoing. What should the engineer do?

    (A) Go along with the advice of the more senior engineers. They have more experience in the field and are most likely right. Besides, if some harm does come from the design, they will take the blame, not her.
    (B) Ask one of her college professors, an expert in the field, to look at the plans and make a recommendation.
    (C) Bring the issue to the attention of the company's upper nontechnical management.
    (D) Report the company's violation to the state board and any other appropriate regulatory agencies.

48. On his own time, a professional engineer writes and self-publishes a technical book. The engineer designs his own cover and writes his own ad copy. The cover and promotional literature both brazenly claim that the book is "the best available," the "most concise," the "best buy," and the "most authoritative." Both the cover and the ad copy repeatedly slam the book's other major competitor. A dissatisfied reader who relied on the cover copy when buying the book complains to the engineer's professional society that such advertising methods violate the society's code of ethics. What should the society do?

    (A) Do nothing. Codes of ethics are intended to guide providers of consulting services, not authors.
    (B) Do nothing. Such "puffery" is a common marketing tool, and nobody believes it anyway.
    (C) Request that the engineer provide supporting data, comparisons, and studies to support the claims before taking further steps.
    (D) Immediately charge the engineer with making outlandish, unsupported, and misleading claims. File an injunction and/or lawsuit prohibiting the engineer from using such wording in the future.

49. What is the relationship between ethical requirements and legal requirements?

(A) They place restrictions on the same actions.

(B) They cover none of the same actions.

(C) Ethical requirements add restrictions not required by the law.

(D) Legal actions are ethical actions.

50. Complete the sentence: "Engineers are to uphold the health, safety, and public

(A) trust."

(B) welfare."

(C) confidence."

(D) good."

51. Ideally, what is the precedence (from highest importance to lowest importance) of your ethical responsibility?

I.   yourself

II. your client

III. society as a whole

IV. your profession

(A) I-II-IV-III

(B) IV-II-III-I

(C) II-IV-III-I

(D) III-II-IV-I

52. The National Council of Examiners for Engineering and Surveying *Model Rules* contains guidelines for licensees' rules of professional conduct. This outlines a licensee's obligation to which of the following entities?

I.   society

II. employer and clients

III. licensing board

IV. other licensees

(A) I and II only

(B) I and IV only

(C) I, II, and IV

(D) I, II, III, and IV

53. Complete the sentence: "Information that is proprietary to a client

(A) may not be used in the work performed for the client."

(B) may be divulged to third parties."

(C) may be shared with other members of your design firm."

(D) should be kept absolutely confidential."

54. An engineering firm is hired by a state agency to supervise the construction of a large wastewater treatment plant. The design of the plant is prepared by the state agency. A clause in the contract states that any money saved as a result of design improvements suggested by the contractor shall be split evenly between the state and the contractor. Knowing that they are thoroughly familiar with the plans, the contractor wants to hire the engineering firm performing the inspection services to review the plans for improvements. Should the engineering firm accept this employment?

(A) Yes, without hesitation. It is a win-win-win situation; the engineering firm makes money, the contractor makes money, and the state saves money. Furthermore, the design is ultimately improved.

(B) Yes, as long as the contractor and the state evenly split the engineering fees.

(C) No, accepting work from the contractor while acting as a representative of the owner would be a conflict of interest.

(D) No, money should be paid to the contractor only for original ideas. It is unethical for any engineering firm to assist the contractor in devising improvements.

55. After making a presentation for an international project, an engineer is told by a foreign official that his company will be awarded the contract, but only if it hires the official's brother as an advisor to the project. The engineer sees this as a form of extortion and informs his boss. His boss tells him that, while it might be illegal in the United States, it is a customary and legal business practice in the foreign country. The boss impresses upon the engineer the importance of getting the project, but leaves the details up to the engineer. What should the engineer do?

(A) He should hire the official's brother, but insist that he perform some useful function for his salary.

(B) He should check with other companies doing business in the country in question, and if they routinely hire relatives of government officials to secure work, then he should do so too.

(C) He should withdraw his company from consideration for the project.

(D) He should inform the government official that his company will not hire the official's brother as a precondition for being awarded the contract, but invite the brother to submit an application for employment with the company.

56. An asset is purchased for $100,000. The useful life is estimated at 7 years. The salvage value is estimated as $15,000. If the asset is subsequently sold after 3 years, most nearly what will be its remaining book value if straight-line depreciation is used?

(A) $12,100
(B) $36,400
(C) $57,100
(D) $63,600

57. What is most nearly the annual effective interest rate of money invested at 5% per annum and compounded quarterly?

(A) 1.3%
(B) 5.0%
(C) 5.1%
(D) 20%

58. A machine has an initial cost of $7600 and an annual maintenance cost of $380. The life of the machine is 12 years, and the salvage value is $2000. Assuming an effective annual interest rate of 6%, the equivalent uniform annual cost for the machine is most nearly

(A) $1200
(B) $1400
(C) $1600
(D) $1800

59. A tractor manufacturer signs a long-term contract with a farm consortium to provide a new tractor every 5 years. The cost of each tractor is $24,000. What is most nearly the capitalized cost of the contract if interest is 6% over the life of the contract?

(A) $950
(B) $5700
(C) $80,000
(D) $95,000

60. A bank advertises an interest rate of 4.6% per annum with continuous compounding. What is most nearly the effective annual interest rate?

(A) 4.62%
(B) 4.71%
(C) 4.89%
(D) 4.94%

61. A lump sum of money is to be invested now at 8% interest, so that $800 can be withdrawn at the end of each year for 8 years. The minimum amount that must be invested is most nearly

(A) $4400
(B) $4600
(C) $5100
(D) $5900

62. A machine has an initial cost of $14,000, a life of 15 years, and a salvage value of $1200. The straight line depreciation value of this machine is most nearly

(A) $780
(B) $800
(C) $830
(D) $850

63. A factory that manufactures chairs costs $180,000 per year to operate, including rent, depreciation charges on equipment, and salaries. Each chair costs $31 to make, and sells for $59. What is the factory's break-even sales volume?

(A) 5300 chairs/yr
(B) 5600 chairs/yr
(C) 6400 chairs/yr
(D) 6900 chairs/yr

64. A machine has an initial cost of $18,000, and operating costs of $2500 each year. The salvage value decreases by $3000 each year. The machine is now three years old. Assuming an effective annual interest rate of 12%, the cost of owning and operating the machine for 1 more year is most nearly

(A) $5500
(B) $6100
(C) $6900
(D) $7100

65. The area shown is 60 cm^2, and the area moment of inertia about the x-axis is 3870.3 cm^4. What is most nearly the area moment of inertia about the x'-axis?

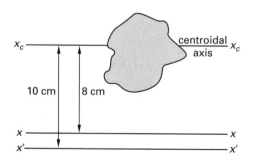

(A) 270 cm^4
(B) 2200 cm^4
(C) 4100 cm^4
(D) 6000 cm^4

66. What is most nearly force F if point O remains stationary?

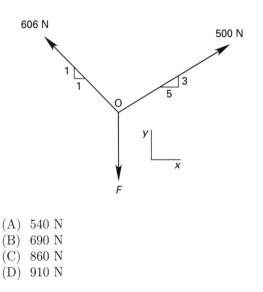

(A) 540 N
(B) 690 N
(C) 860 N
(D) 910 N

67. What is most nearly the force reaction at point O?

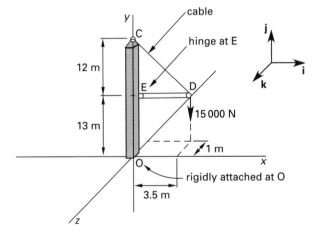

(A) $R_O = (-4375 \text{ N})\mathbf{i} + (15\,000 \text{ N})\mathbf{j} + (343 \text{ N})\mathbf{k}$
(B) $R_O = (15\,000 \text{ N})\mathbf{j}$
(C) $R_O = (-4375 \text{ N})\mathbf{i} - (15\,000 \text{ N})\mathbf{j} - (562 \text{ N})\mathbf{k}$
(D) $R_O = (-4550 \text{ N})\mathbf{i} + (15\,675 \text{ N})\mathbf{j}$

68. A homogeneous block with a mass of 10 kg with dimensions $D \times H$, $H \geq 4D$, rests on a level surface with coefficient of friction $\mu > 0.2$. What is most nearly the maximum value of h for which no tipping of the block can occur?

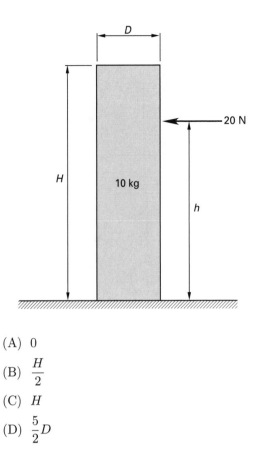

(A) 0

(B) $\dfrac{H}{2}$

(C) H

(D) $\dfrac{5}{2}D$

69. A 40 kg box is to be lifted out of a hole as shown. Two forces, F_1 and F_2, are applied at different angles. Neglecting friction, what are most nearly the minimum forces F_1 and F_2 that will lift the box out of the hole without binding?

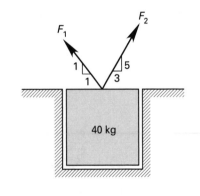

(A) $F_1 = 190$ N; $F_2 = 350$ N
(B) $F_1 = 210$ N; $F_2 = 290$ N
(C) $F_1 = 240$ N; $F_2 = 400$ N
(D) $F_1 = 360$ N; $F_2 = 150$ N

70. What tension T must be applied to the rope to keep the system shown in equilibrium?

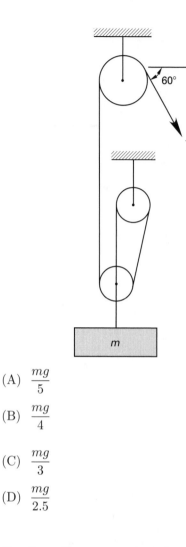

(A) $\dfrac{mg}{5}$

(B) $\dfrac{mg}{4}$

(C) $\dfrac{mg}{3}$

(D) $\dfrac{mg}{2.5}$

71. For the pulley-mass system shown, the torque, M, applied to the pulley B, needed to initiate motion is given by which of the following? Rollers C and D are frictionless.

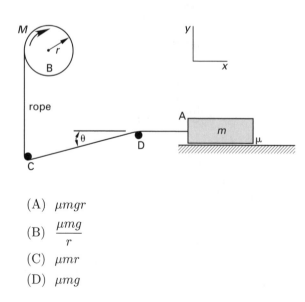

(A) μmgr

(B) $\dfrac{\mu mg}{r}$

(C) μmr

(D) μmg

72. The centripetal force acting on a particle traveling at constant velocity in a circular path is given by which of the following?

(A) $m\omega^2 r$

(B) $\dfrac{1}{2}m\omega^2 r$

(C) $\dfrac{1}{2}m\omega^2 r^2$

(D) $m\omega\text{v}$

73. A block is pushed up a plane inclined at 25° to the horizontal. If the block has a mass of 5 kg and the coefficient of friction between the plane and the block is 0.25, most nearly how much force is required to accelerate the block up the plane at the rate of 5 m/s²?

(A) 13 N
(B) 57 N
(C) 180 N
(D) 240 N

74. A 2 kg clay ball moving at a rate of 40 m/s collides with a 5 kg ball of clay moving in the same direction at a rate of 10 m/s. What is the final velocity of both balls if they stick together after colliding?

(A) 11.8 m/s
(B) 12.4 m/s
(C) 15.3 m/s
(D) 18.6 m/s

75. A 1000 kg car enters an unbanked curve of 1 km radius with a velocity of 80 km/h. What is the minimum coefficient of friction between the road and the car that will allow the car to travel through the turn without braking?

(A) 0.05
(B) 0.25
(C) 0.35
(D) 0.65

76. Two masses are connected to a spring as shown. Both masses are motionless. The string connecting the masses is suddenly cut. Assuming that the spring never goes "solid," what is most nearly the amplitude of oscillation for m_1?

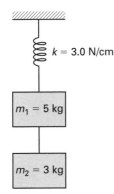

$k = 3.0$ N/cm

$m_1 = 5$ kg

$m_2 = 3$ kg

(A) 9.81 cm
(B) 16.4 cm
(C) 19.6 cm
(D) 26.2 cm

77. Which of the following properties cannot be derived from a tensile test?

(A) modulus of elasticity
(B) modulus of resilience
(C) endurance limit
(D) ultimate tensile strength

Problems 78 and 79 are based on the following illustration and statements.

A steel rod is fixed at each end to two rigid end pieces. The end pieces are separated by two aluminum bars. The structure is unstressed at 20°C but is slowly heated to a higher temperature.

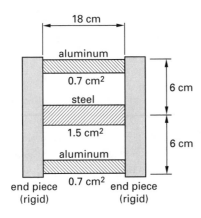

18 cm

aluminum
0.7 cm²
steel
1.5 cm²
aluminum
0.7 cm²

6 cm

6 cm

end piece (rigid)   end piece (rigid)

$$\alpha_{\text{aluminum}} = 23 \times 10^{-6} \ 1/°C$$
$$\alpha_{\text{steel}} = 11.7 \times 10^{-6} \ 1/°C$$
$$E_{\text{aluminum}} = 6.9 \times 10^{10} \ \text{Pa}$$
$$E_{\text{steel}} = 2.1 \times 10^{11} \ \text{Pa}$$

78. At what temperature will the strains in the steel and aluminum bars be equal? Neglect buckling.

(A) 22°C
(B) 27°C
(C) 32°C
(D) at all temperatures

79. What is most nearly the stress in the steel when the stress in the aluminum is 4000 Pa?

(A) 1.3 kPa
(B) 1.8 kPa
(C) 3.7 kPa
(D) 12 kPa

80. A steel bracket is acted upon by a force, F, of 1000 N at point A. For steel, the modulus of elasticity is 2.1×10^{11} Pa and the density is 7860 kg/m³. Neglecting any bending moment, what is most nearly the maximum shear stress at the fixed end of the column?

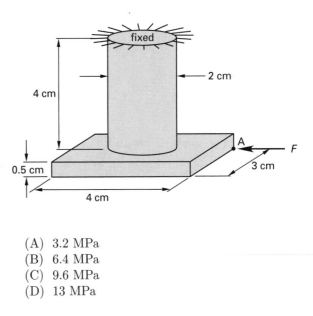

fixed

2 cm

4 cm

0.5 cm

A

F

3 cm

4 cm

(A) 3.2 MPa
(B) 6.4 MPa
(C) 9.6 MPa
(D) 13 MPa

Problems 81–83 are based on the following illustration and statements.

A force of 250 N is applied at the tip of the beam. The beam also has a distributed load, w, acting on it as shown.

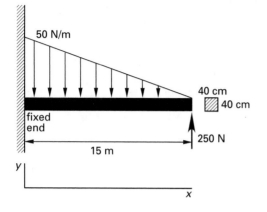

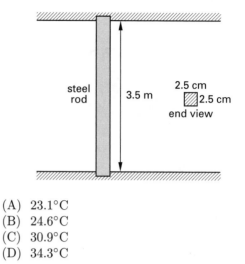

(A) 23.1°C
(B) 24.6°C
(C) 30.9°C
(D) 34.3°C

81. What is most nearly the maximum shear?

(A) 100 N
(B) 130 N
(C) 200 N
(D) 250 N

85. How many atoms are contained in a complete hexagonal close-packed cell?

(A) 3
(B) 5
(C) 6
(D) 14

82. What is most nearly the maximum bending moment?

(A) 1900 N·m
(B) 2000 N·m
(C) 2800 N·m
(D) 5200 N·m

86. A steel rod is stressed in tension slightly beyond its yield point so that it suffers permanent deformation without rupture. The rod is then unloaded so that there is no residual stress. If the rod is again stressed in tension, which of the following statements is true?

(A) The yield point is unchanged.
(B) The ductility of the steel is increased.
(C) The ultimate strength is decreased.
(D) A new yield point is established, which is higher than the original one.

83. The beam is made of hot-rolled AISI 1020 steel (modulus of elasticity = 210 GPa). What is most nearly the deflection of the beam 7.5 m from the wall due to the tip load alone? (Neglect buckling and the beam's own weight.)

(A) 0.0024 mm
(B) 0.012 mm
(C) 0.18 mm
(D) 0.2 mm

87. What is the difference between annealing and recrystallization?

(A) Annealing is a process in which a metal is heated to the critical point and allowed to cool gradually, while recrystallization occurs naturally in nonmetals, such as silicon.
(B) Recrystallization requires a long period at a high temperature, while annealing involves quenching.
(C) Recrystallization occurs above the Curie point, while annealing occurs below the eutectoid temperature.
(D) Recrystallization is the same as annealing except at lower temperatures.

84. The steel rod shown (modulus of elasticity = 210 GPa) is mounted between two rigid walls at 20°C. The coefficient of linear thermal expansion is $11.7 \times 10^{-6} \ 1/°C$, the moment of inertia is $32.55 \times 10^{-9} \ m^4$, the cross-sectional area is $6.25 \times 10^{-4} \ m^2$, and the yield stress is 250 MPa. Assuming the walls are rigid, what is most nearly the temperature at which the rod will buckle?

88. Poisson's ratio is the ratio of which of the following?

(A) the unit lateral deformation to the unit longi-
tudinal deformation
(B) the unit stress to the unit strain
(C) the elastic limit to the proportional limit
(D) the shear strain to the compressive strain

89. Zinc galvanizing is used to protect a steel (iron)
fence. Which of the following statements is true?

(A) The zinc atom has a greater tendency to lose
electrons and form positive ions than the iron
atom.
(B) The zinc atom has a greater tendency to lose
electrons and form negative ions than the iron
atom.
(C) The zinc ion has a greater tendency to gain
electrons and form ions than the iron atom.
(D) Zinc is lower than iron in the galvanic series.

90. A force of 315 N is carried by a wire 60 m long with
a diameter of 1.75 mm. The wire's modulus of elasticity
is 200 GN/m^2. Neglect the mass of the wire. What is
most nearly the unit elongation?

(A) 4.32×10^{-3} m/m
(B) 5.41×10^{-2} m/m
(C) 6.55×10^{-4} m/m
(D) 7.89×10^{-4} m/m

91. What is the definition of packing factor?

(A) the coordination number divided by the volume of a unit cell
(B) the number of atoms in a unit cell multiplied by
the atomic weight and divided by Avogadro's
number
(C) the volume of atoms divided by the cell volume
(D) the number of unit cells in a lattice sheet

92. Carbon monoxide and oxygen gas react at 4000K
and 1 atm pressure in a closed, flexible container to produce carbon dioxide. The standard enthalpy of formation for this reaction is -283 kJ/mol CO_2. At equilibrium, the mole fractions of the reactants and products
are 0.08 carbon monoxide, 0.03 oxygen, and 0.89 carbon
dioxide, respectively.

After equilibrium has been attained, the temperature of
the system is suddenly decreased from 4000K to 1000K.
Which of the following statements is correct?

(A) The equilibrium constant will increase.
(B) The concentration of carbon monoxide will in-
crease.
(C) The mole fraction of carbon dioxide will de-
crease.
(D) The reaction will stop.

93. The tank shown is filled with water. The gate is
mounted on frictionless bearings on the hinged edge.
What is most nearly the resultant force (per meter of
wall width) required to keep the gate closed?

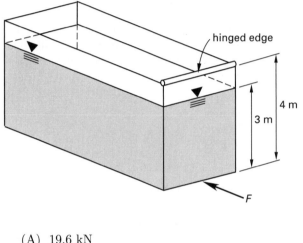

(A) 19.6 kN
(B) 27.6 kN
(C) 29.4 kN
(D) 33.1 kN

94. How does the Reynolds number of a 1/15 scale
model relate to the Reynolds number of a dimensionally
similar prototype?

(A) They are both equal.
(B) The Reynolds number of the model is 15 times
lower.
(C) The Reynolds number of the prototype is 15
times lower.
(D) The Reynolds number of the model is a function of the velocity ratio.

95. A pressure vessel with an internal pressure, p, is
connected to the simple U-tube open to the atmosphere
as shown. A 10 cm deflection of mercury is observed.
The density of mercury is 13.6 Mg/m^3. What is most
nearly the pressure, p, if the atmospheric pressure is
101 kPa?

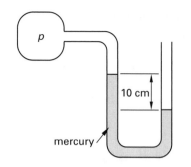

(A) 1.0 kPa
(B) 13 kPa vacuum
(C) 39 kPa vacuum
(D) 78 kPa

96. Fluid flows at 5 m/s in a 5 cm diameter pipe section. The section is connected to a 10 cm diameter section. At most nearly what velocity does the fluid flow in the 10 cm section?

(A) 1.00 m/s
(B) 1.25 m/s
(C) 2.50 m/s
(D) 10.0 m/s

97. A venturi meter installed horizontally is used to measure the flow of water in a pipe. The area ratio, A_2/A_1, of the meter is 0.5, and the velocity through the throat of the meter is 3 m/s. What is most nearly the pressure differential across the venturi meter?

(A) 1.5 kPa
(B) 2.3 kPa
(C) 3.5 kPa
(D) 6.8 kPa

98. The 8 cm by 12 cm rectangular flume shown is filled to three-quarters of its height. What is most nearly the hydraulic radius of the flow?

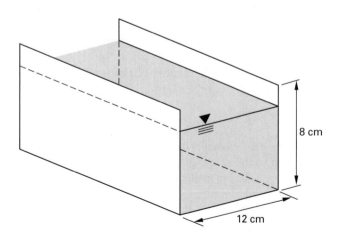

(A) 1.5 cm
(B) 2.5 cm
(C) 3.0 cm
(D) 5.0 cm

99. Water is flowing at the rate of 0.01 m³/s in the vertical pipe system shown. What is most nearly the pressure at point 2?

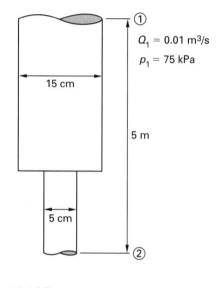

(A) 49.4 kPa
(B) 111 kPa
(C) 375 kPa
(D) 675 kPa

100. A waterfall has a total drop of 15 m and a flow rate of 2 m³/s. Most nearly what theoretical power is available?

(A) 294 kW
(B) 375 kW
(C) 439 kW
(D) 750 kW

101. To what height will 20°C ethyl alcohol (surface tension 0.0227 N/m, contact angle 0°, specific heat 0.68, kcal/kg-°C, density 0.79 g/cm³) rise in a 0.25 mm internal diameter glass capillary tube? The density of the alcohol is 790 kg/m³.

(A) 0.012 m
(B) 0.047 m
(C) 0.12 m
(D) 0.47 m

102. What is most nearly the equivalent capacitance of the following circuit?

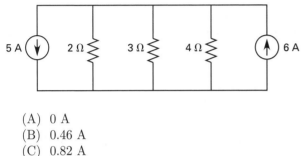

A •—||—||—||—||—||—• B
50 μF 50 μF 50 μF 50 μF 50 μF

(A) 10 μF
(B) 100 μF
(C) 250 μF
(D) 10 F

103. What is most nearly the current in the 2 Ω resistor?

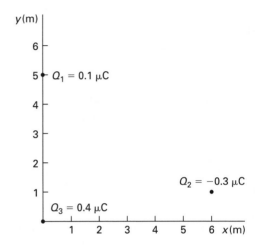

5 A ↓   2 Ω   3 Ω   4 Ω   ↑ 6 A

(A) 0 A
(B) 0.46 A
(C) 0.82 A
(D) 1.1 A

104. Kirchhoff's laws state which of the following?

(A) Voltages across parallel branches are equal, and currents through series branches are equal.
(B) A voltage source in series with a resistor may be replaced by a current source in parallel with a resistor, and vice versa.
(C) The sum of the currents in branches comprising any closed loop in a circuit is zero, and the sum of all voltages in branches connected to a common node is zero.
(D) The sum of the currents flowing into a node is zero, and the sum of the voltages in the branches of any closed loop of a circuit is zero.

105. What is most nearly the equivalent resistance between terminals A and B? All resistors are 10 Ω.

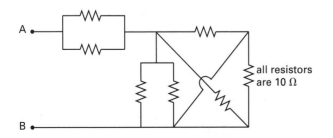

all resistors are 10 Ω

(A) 7.5 Ω
(B) 7.9 Ω
(C) 8.3 Ω
(D) 10 Ω

106. Point charges Q_1, Q_2, and Q_3 are arranged as shown in the following illustration.

y(m)

6

5 • $Q_1 = 0.1\ \mu C$

4

3

2

1                                $Q_2 = -0.3\ \mu C$
                                        •
  $Q_3 = 0.4\ \mu C$

      1   2   3   4   5   6 x(m)

The magnitude of the force on Q_3 due to Q_1 and Q_2 is most nearly

(A) 3.0×10^{-5} N
(B) 2.3×10^{-5} N
(C) 9.8×10^{-4} N
(D) 5.1×10^{-2} N

107. What is a passive element?

(A) an element with reactance but no resistance
(B) an element that plays no role in a circuit
(C) an element whose characteristics are controlled by the characteristics of another element
(D) an element that only stores, releases, or dissipates power

108. A large, hollow metallic sphere surrounds a smaller metallic sphere of radius r with a charge of Q. What is the radial field intensity, a distance R from the surface of the smaller sphere and between the spheres, proportional to?

(A) $\dfrac{Q}{R^2}$

(B) $\dfrac{Q}{r^2}$

(C) $\dfrac{Q}{(R+r)^2}$

(D) $\dfrac{Q}{(R-r)^2}$

109. Most nearly what are the maximum amplitude and frequency of the following sinusoidal voltage?

$$V(t) = 100 \ \cos \ (20\pi t + 45°)$$

(A) 50 V; 45 Hz
(B) 100 V; 10 Hz
(C) 100 V; 2π Hz
(D) 200 V; 20 Hz

110. What most nearly is the work performed in moving an electron through a potential of 1 million volts?

(A) -1.6×10^{-19} J
(B) -1.6×10^{-13} J
(C) 0 J
(D) 1.6×10^{-13} J

111. Which of the following is a correct definition for power factor in an alternating current circuit?

(A) the ratio of apparent power to actual power
(B) the ratio of resistance to impedance
(C) the sine of the phase angle between the voltage and current of the load
(D) the ratio of impedance to resistance

112. Which of the following is the best definition of enthalpy?

(A) the ratio of heat added to temperature change induced in a substance
(B) the total amount of useful energy in a system
(C) the heat required to cause a complete conversion between two phases at a constant temperature
(D) the amount of energy in a system that is no longer available to do useful work

113. What does the triple point on a pressure-temperature diagram indicate?

(A) the point at which a solid can be converted to a vapor without passing through the liquid phase
(B) the point at which three phases can coexist
(C) the point above which different phases are indistinguishable
(D) the point at which a liquid is vaporized by the addition of heat

114. Given the following properties of saturated steam at 9 MPa and 303.4°C, what is most nearly the specific volume at 85% quality?

$$v_f = 1.4178 \ \text{cm}^3/\text{g}$$
$$v_g = 20.48 \ \text{cm}^3/\text{g}$$

(A) 0.00126 m³/kg
(B) 0.00428 m³/kg
(C) 0.01620 m³/kg
(D) 0.01762 m³/kg

115. A turbine with an isentropic efficiency of 85% receives steam with a pressure of 1.4 MPa and an enthalpy of 3900 kJ/kg. At the output, the pressure is 0.1 MPa and the enthalpy is 2680 kJ/kg. What is most nearly the turbine's work output?

(A) 1040 kJ/kg
(B) 1220 kJ/kg
(C) 1510 kJ/kg
(D) 1700 kJ/kg

116. A Carnot engine operates on steam between 65°C and 425°C. What is most nearly the ideal efficiency?

(A) 19%
(B) 48%
(C) 52%
(D) 81%

117. 2.8 m³ of a perfect gas at 310°C is compressed in an isobaric process to 1.4 m³. What is most nearly the final temperature?

(A) 19°C
(B) 54°C
(C) 110°C
(D) 150°C

118. A reduction in the relative humidity in a test chamber will result from which of the following?

(A) an increase in the air temperature
(B) a drop in the air temperature
(C) a decrease in the partial pressure of water vapor
(D) a reduction in the water vapor pressure in relation to the chamber temperature

119. An oxygen tank has a total volume capacity of 1 m³. The gas constant for oxygen is $R = 259.8$ J/kg·K. If the tank and oxygen are at 30°C and the absolute pressure in the tank is 100 kPa, what is most nearly the mass of oxygen inside the tank?

(A) 1.1 kg
(B) 1.3 kg
(C) 12 kg
(D) 13 kg

120. 550 g of air at 200 kPa and 375°C is cooled at constant volume to 160°C. The resulting pressure is most nearly

(A) 30 kPa
(B) 85 kPa
(C) 130 kPa
(D) 185 kPa

Solutions for Sample Examination

MORNING SECTION

1. A	31. B	61. B	91. C
2. D	32. A	62. D	92. A
3. D	33. D	63. C	93. D
4. B	34. D	64. C	94. A
5. A	35. B	65. D	95. B
6. C	36. B	66. B	96. B
7. D	37. D	67. B	97. C
8. D	38. C	68. D	98. C
9. D	39. A	69. B	99. B
10. B	40. D	70. C	100. A
11. A	41. A	71. A	101. B
12. B	42. A	72. A	102. A
13. D	43. B	73. B	103. B
14. D	44. D	74. D	104. D
15. D	45. B	75. A	105. A
16. B	46. C	76. A	106. A
17. D	47. C	77. C	107. D
18. C	48. D	78. D	108. C
19. B	49. C	79. C	109. B
20. D	50. B	80. D	110. D
21. C	51. D	81. D	111. B
22. B	52. C	82. B	112. B
23. B	53. D	83. D	113. B
24. B	54. C	84. D	114. D
25. C	55. D	85. C	115. B
26. D	56. D	86. D	116. C
27. B	57. C	87. D	117. A
28. D	58. A	88. A	118. C
29. C	59. D	89. A	119. B
30. D	60. B	90. C	120. C

1. $\displaystyle\int \left(15x^4 - 8x^3 + \frac{1}{x} + 7\right) dx$

$$= \frac{15x^5}{5} - \frac{8x^4}{4} + \ln\ |x| + 7x + C$$

$$= 3x^5 - 2x^4 + \ln\ |x| + 7x + C$$

Answer is A.

2. This is a second-order, homogeneous, linear differential equation. Start by putting it in general form.

$$y'' + 2ay' + by = 0$$
$$2y'' - 4y' + 4y = 0$$
$$y'' - 2y' + 2y = 0$$
$$a = -1$$
$$a^2 = 1$$
$$b = 2$$

Because $a^2 < b$, this equation represents an underdamped condition.

$$y = e^{\alpha x}(C_1\ \cos\beta x + C_2\ \sin\beta x)$$

$$\alpha = -a = 1$$

$$\beta = \sqrt{b - a^2}$$
$$= \sqrt{2 - 1}$$
$$= 1$$

$$y = e^x(C_1\ \cos x + C_2\ \sin x)$$

Answer is D.

3. $\text{PQ} = \sqrt{(-3-1)^2 + \left(4 - (-3)\right)^2 + (-2-5)^2}$
$$= \sqrt{114}$$

Answer is D.

4. $\qquad r = \sqrt{x^2 + y^2}$
$$= \sqrt{(-3)^2 + (-5.2)^2}$$
$$= 6$$
$$\theta = \tan^{-1}\left(\frac{y}{x}\right)$$
$$= \tan^{-1}\left(\frac{-5.2}{-3}\right)$$
$$= -120° \quad \text{[a clockwise angle]}$$

Answer is B.

5. $\qquad (x-1)^3 = (x-1)(x-1)(x-1) = 0$

The roots are $x = 1, 1, 1$.

Answer is A.

6. $x^2 + y^2 - 8x - 10y + 25 = 0$ is the general form for the circle.

Convert to center-radius form by simultaneously completing the square for both x and y.

$$\left(x - \frac{8}{2}\right)^2 + \left(y - \frac{10}{2}\right)^2 = -25 + \left(\frac{-8}{2}\right)^2 + \left(\frac{-10}{2}\right)^2$$
$$(x-4)^2 + (y-5)^2 = -25 + (-4)^2 + (-5)^2$$
$$= 16$$

The circle is located at center (4,5) with radius 4.

Answer is C.

7. $\qquad \dfrac{\partial z}{\partial x} = \dfrac{\partial e^{xy}}{\partial x} = e^{xy}\ \dfrac{\partial(xy)}{\partial x} = ye^{xy}$

Answer is D.

8. $\begin{vmatrix} 3 & 2 & 7 \\ 3 & 7 & 1 \\ 1 & 5 & 2 \end{vmatrix} = (3)((7)(2) - (1)(5))$

$$- (2)((3)(2) - (1)(1))$$
$$+ (7)((3)(5) - (7)(1))$$
$$= 73$$

Answer is D.

9. Use L'Hôpital's rule.

$$\lim_{x\to 0}\left(\frac{\sin 5x}{x}\right) = \lim_{x\to 0}\left(\frac{\frac{d}{dx}(\sin 5x)}{\frac{d}{dx}(x)}\right)$$
$$= \lim_{x\to 0}\left(\frac{5\ \cos 5x}{1}\right)$$
$$= 5$$

Answer is D.

10. $\theta = \cos^{-1}\left(\dfrac{\mathbf{V}_1 \cdot \mathbf{V}_2}{|\mathbf{V}_1|\,|\mathbf{V}_2|}\right)$

$= \cos^{-1}\left(\dfrac{(3)(2) + (2)(4) + (1)(6)}{\left(\sqrt{(3)^2 + (2)^2 + (1)^2}\right) \times \left(\sqrt{(2)^2 + (4)^2 + (6)^2}\right)}\right)$

$= \cos^{-1}\left(\dfrac{20}{(\sqrt{14})(\sqrt{56})}\right)$

$= 44.4°$

Answer is B.

11. $\mathbf{V}_2 \times \mathbf{V}_3 = \begin{vmatrix} \mathbf{i} & \mathbf{j} & \mathbf{k} \\ 2 & 4 & 6 \\ 2 & 3 & 2 \end{vmatrix}$

$= ((4)(2) - (6)(3))\mathbf{i}$
$\quad - ((2)(2) - (6)(2))\mathbf{j}$
$\quad + ((2)(3) - (4)(2))\mathbf{k}$

$= -10\mathbf{i} + 8\mathbf{j} - 2\mathbf{k}$

Answer is A.

12. The standard form for an ellipse is

$$\frac{(x - h)^2}{a^2} + \frac{(y - k)^2}{b^2} = 1$$

$(h,k) = (0,0)$ [center]

$a = $ semimajor distance $= \dfrac{60}{2} = 30$

$b = $ semiminor distance $= 20$

$$\frac{x^2}{900} + \frac{y^2}{400} = 1$$

$x = \sqrt{\left(1 - \dfrac{(15)^2}{400}\right)(900)} = 19.84$

$z = \sqrt{x^2 + y^2}$
$\quad = \sqrt{(19.84)^2 + (15)^2}$
$\quad = 24.87 \quad (24.9)$

Answer is B.

13. Add the second and third equations together. The x and z terms cancel.

$$-2y = 2$$
$$y = -1$$

Add the first and second equations together and substitute $y = -1$. The z terms cancel.

$$2x - 2y = -3$$
$$2x - (2)(-1) = -3$$
$$x = -\frac{5}{2}$$

It is not necessary to solve for z, as only choice (D) satisfies the known values of x and y.

Answer is D.

14. The polynomial has the form

$$Ax^3 + Bx^2 + Cx + D = 0$$

The possible root combinations are (a) one real; (b) two complex and one real; (c) two real; and (d) three real.

Answer is D.

15. Determine the coordinates of the vertices by setting all possible pairs of equations equal.

point A: $2x - 2 = -2x + 3$

$x = \dfrac{5}{4}$

$y = (2)\left(\dfrac{5}{4}\right) - 2 = \dfrac{1}{2}$

point B: $-2x + 3 = 12x - 7$

$x = \dfrac{5}{7}$

$y = (-2)\left(\dfrac{5}{7}\right) + 3 = \dfrac{11}{7}$

point C: $12x - 7 = 2x - 2$

$x = \dfrac{1}{2}$

$y = (12)\left(\dfrac{1}{2}\right) - 7 = -1$

$|AB| = \sqrt{\left(\dfrac{5}{4} - \dfrac{5}{7}\right)^2 + \left(\dfrac{1}{2} - \dfrac{11}{7}\right)^2} = 1.198$

$|BC| = \sqrt{\left(\dfrac{5}{7} - \dfrac{1}{2}\right)^2 + \left(\dfrac{11}{7} - (-1)\right)^2} = 2.580$

$|CA| = \sqrt{\left(\dfrac{1}{2} - \dfrac{5}{4}\right)^2 + \left(-1 - \dfrac{1}{2}\right)^2} = 1.677$

Answer is D.

16. For the matrix product $\mathbf{AB}$ to be possible, matrix $\mathbf{A}$ must have the same number of columns as matrix $\mathbf{B}$ has rows. If $\mathbf{A}$ is an $m \times n$ matrix and $\mathbf{B}$ is an $n \times s$ matrix, $\mathbf{AB}$ is an $m \times s$ matrix.

$$(m \times n)(n \times s) = (m \times s)$$
$$(4 \times 3)(3 \times s) = (4 \times 4)$$
$$s = 4$$

Answer is B.

17. The volume of a pyramid is

$$V = \frac{1}{3}A_{\text{base}}h$$
$$= \frac{1}{3}s^2 h$$
$$= \left(\frac{1}{3}\right)(200 \text{ m})^2(60 \text{ m})$$
$$= 800\,000 \text{ m}^3$$

$$m = \rho V$$
$$= \left(1300 \, \frac{\text{kg}}{\text{m}^3}\right)\left(1000 \, \frac{\text{g}}{\text{kg}}\right)(800\,000 \text{ m}^3)$$
$$= 1.04 \times 10^{12} \text{ g} \quad (1.0 \text{ Tg})$$

Answer is D.

18. Use the logarithmic identities.

$$\log x + \log y = \log xy$$
$$\log x - \log y = \log \frac{x}{y}$$

$$\log_3 \frac{3}{2} + \log_3 12 - \log_3 2 = \log_3 \frac{\left(\frac{3}{2}\right)(12)}{2}$$
$$= \log_3 9$$
$$= 2.00$$

Answer is C.

19. Check to see if the progression is either an arithmetic or geometric sequence. The difference between consecutive terms is not constant, so this sequence is not arithmetic. The ratio of consecutive terms is constant, so the sequence appears to be geometric.

$$r = \frac{125}{50} = \frac{8}{3.2} = 2.5$$
$$a = \frac{50}{2.5} = 20$$

Answer is B.

20. The mode is the observation that occurs most frequently. From the data, this is 91. To find the arithmetic mean, divide the sum of the scores by the total number of scores.

$$X\{\text{mean}\} = \left(\frac{1}{n}\right)(X_1 + X_2 + X_3 + \ldots + X_n)$$
$$= \left(\frac{1}{78}\right)((38)(5) + (45)(2) + (69)(7)$$
$$+ (76)(10) + (82)(12) + (90)(8)$$
$$+ (91)(19) + (95)(15))$$
$$= 81.81 \quad (82)$$

Answer is D.

21. Suppose a 1 is drawn first. To total seven, the second card must be the 6. The probability that the 1 will be drawn first (out of six cards) and the 6 will be drawn second (out of the five remaining cards) is

$$P(1,6) = \left(\frac{1}{6}\right)\left(\frac{1}{5}\right) = \frac{1}{30}$$

No matter which card is chosen first, there is exactly one card out of the five remaining cards that will bring the sum to seven. Since there are six different starting possibilities, the total probability is

$$(6)\left(\frac{1}{30}\right) = \frac{1}{5}$$

Answer is C.

22. First calculate the probability of rolling no sixes, and then subtract that from 1 to get the probability of rolling at least one six.

$$p\{\text{four dice, no sixes}\} = p\{\text{one die, no six}\}^4$$
$$= \left(\frac{5}{6}\right)^4$$
$$= 0.482$$
$$p\{\text{four dice, at least one six}\} = 1 - p\{\text{four dice, no sixes}\}$$
$$= 1 - 0.482$$
$$= 0.518 \quad (0.52)$$

Answer is B.

23. The mean is

$$\bar{x} = \frac{\sum x_i}{n} = \frac{4 + 6 + 9 + 9}{4}$$
$$= 7$$

The sample standard deviation is

$$\sigma = \sqrt{\frac{\sum (x_i - \bar{x})^2}{n-1}}$$
$$= \sqrt{\frac{(4-7)^2 + (6-7)^2 + (9-7)^2 + (9-7)^2}{4-1}}$$
$$= 2.449 \quad (2.4)$$

Answer is B.

24. The mean of the Poisson distribution, λ, is 3 jams. The probability of exactly 5 jams is

$$p\{5\} = \frac{e^{-\lambda}\lambda^x}{x!} = \frac{e^{-3}(3)^5}{5!}$$
$$= 0.101 \quad (0.10)$$

Answer is B.

25. First calculate the following values.

$$\sum x_i = 1 + 2 + 3 + 5 = 11$$
$$\sum y_i = 6 + 7 + 11 + 13 = 37$$
$$\sum x_i^2 = (1)^2 + (2)^2 + (3)^2 + (5)^2 = 39$$
$$\sum x_i y_i = (1)(6) + (2)(7) + (3)(11) + (5)(13) = 118$$

The slope can be found by

$$b = \frac{\sum x_i y_i - \frac{(\sum x_i)(\sum y_i)}{n}}{\sum x_i^2 - \frac{(\sum x_i)^2}{n}}$$
$$= \frac{118 - \frac{(11)(37)}{4}}{39 - \frac{(11)^2}{4}}$$
$$= 1.857 \quad (1.86)$$

Answer is C.

26. A hypothesis test can be used to determine whether a set of data came from a population whose mean and standard deviation are known. For example, such a test can tell, within a certain level of confidence, whether a set of recent observations is consistent with historical data, indicating whether the conditions leading to those observations are likely to have changed.

Answer is D.

27. no. of atoms = no. of moles × N_A

$$= \frac{m}{MW} \times N_A = \frac{\rho V N_A}{MW}$$
$$= \frac{\left(\begin{array}{c} \left(19.32 \frac{g}{cm^3}\right)(1 \text{ cm}^3) \\ \times \left(6.022 \times 10^{23} \frac{\text{atoms}}{\text{mol}}\right) \end{array}\right)}{196.97 \frac{g}{\text{mol}}}$$
$$= 5.91 \times 10^{22} \text{ atoms}$$
$$\quad (5.9 \times 10^{22} \text{ atoms})$$

Answer is B.

28. Sodium hydroxide is NaOH. The molecular weight of NaOH is

$$22.990 + 15.999 + 1.0079 = 39.997$$

The atomic weight of Na is 22.990.

$$\text{no. of grams NaOH} = (200 \text{ g})\left(\frac{39.997}{22.990}\right)$$
$$= 347.95 \text{ g} \quad (350 \text{ g})$$

Answer is D.

29. $$H_2SO_4 + 2NaOH \rightleftharpoons 2H_2O + Na_2SO_4$$

The balanced equation shows that 1 mol of H_2SO_4 will neutralize 2 mol of NaOH. Since molarity, by definition, is the number of moles per liter,

$$\text{no. of moles NaOH} = (0.05 \text{ L})\left(5 \frac{\text{mol}}{\text{L}}\right)$$
$$= 0.25 \text{ mol}$$

To neutralize 0.25 gmol of NaOH, 0.125 gmol of H_2SO_4 are necessary.

$$\text{no. of liters } H_2SO_4 \text{ needed} = \frac{0.125 \text{ mol}}{2 \frac{\text{mol}}{\text{L}}} = 0.0625 \text{ L}$$

Approximately 63 mL H_2SO_4 are required.

Answer is C.

30. Aluminum has an oxidation number of +3, and chlorine has an oxidation number of −1. Therefore,

3 mol of HCl are required per mole of aluminum to balance the number of electrons gained and lost. For 2 mol of aluminum,

$$2Al + 6HCl \rightarrow 2AlCl_3 + 3H_2 \text{ gas}$$

3 mol of hydrogen gas are produced.

Answer is D.

31. Fluorine is a member of the halogen family.

Answer is B.

32. The atomic weights are K = 39.098, Cl = 35.453, and O = 15.999. Divide the percentage composition for each by the atomic weight.

$$\text{K:} \frac{31.9}{39.098} = 0.816$$

$$\text{Cl:} \frac{29.0}{35.453} = 0.818$$

$$\text{O:} \frac{39.2}{15.999} = 2.45$$

Divide each ratio by the smallest ratio.

$$\text{K:} \frac{0.816}{0.816} = 1.0$$

$$\text{Cl:} \frac{0.818}{0.816} = 1.00245$$

$$\text{O:} \frac{2.45}{0.816} = 3.002$$

The chemical formula is $KClO_3$.

Answer is A.

33.
$$\text{pH} = -\log_{10} [\text{H}^+]$$
$$[\text{H}^+] = \text{antilog}_{10}(-1.3)$$
$$= 0.0501 \text{ mol/L} \quad (5.0 \times 10^{-2} \text{ mol/L})$$
$$\text{pH} + \text{pOH} = 14$$
$$\text{pOH} = 14 - 1.3 = 12.7$$
$$[\text{OH}^-] = \text{antilog}_{10}(-12.7)$$
$$= 2.0 \times 10^{-13} \text{ mol/L}$$

Answer is D.

34. The temperature given is irrelevant. The mixture boils when the atmospheric pressure and vapor pressure are the same. One atmospheric pressure = 760 mm Hg.

From Dalton's law, the total pressure is the sum of the partial pressures.

$$p = \sum p_i$$
$$= x_A p_A + x_B p_B$$
$$= x_A p_A + (1 - x_A) \, p_B$$
$$760 \text{ mm Hg} = x_A (800 \text{ mm Hg})$$
$$+ (1 - x_A)(300 \text{ mm Hg})$$
$$500 x_A = 760 - 300$$
$$x_A = 0.92 \quad (92\%)$$
$$x_B = 1 - 0.92$$
$$= 0.08 \quad (8\%)$$

Answer is D.

35. $\quad$ no. of moles NaOH $= \left(5 \, \dfrac{\text{mol}}{\text{L}} \right) (0.025 \text{ L})$
$$= 0.125 \text{ mol}$$
$$\text{HCl} + \text{NaOH} \rightleftharpoons \text{H}_2\text{O} + \text{NaCl}$$

From the reversible reaction equation, 1 mol of HCl plus 1 mol NaOH produces 1 mol of H_2O and 1 mol of NaCl.

$$0.125 \text{ mol HCl} = \left(3 \, \frac{\text{mol}}{\text{L}} \right) V_{\text{HCl}}$$
$$V_{\text{HCl}} = \frac{0.125 \text{ mol}}{3 \, \dfrac{\text{mol}}{\text{L}}}$$
$$= 0.0417 \text{ L} \quad (42 \text{ mL})$$

Answer is B.

36. The radioactivity decay equation can be written in terms of mass as well as number of atoms.

$$N = N_o e^{-0.693t/\tau}$$
$$m = m_o e^{-0.693t/\tau}$$
$$= (1 \text{ kg}) e^{(-0.693)(500 \times 10^6 \text{ yr})/(4.5 \times 10^9 \text{ yr})}$$
$$= 0.926 \text{ kg} \quad (930 \text{ g})$$

Answer is B.

37. The pseudocode compiles a sum, S, which is used to approximate the integral. Starting with line 3,

$$S = \frac{F(a)}{2} + \frac{F(b)}{2}$$

The FOR ... NEXT loop (lines 4, 5, and 6) is effectively a summation.

$$S = S + \sum_{1}^{n-1} F(a + kd)$$

$$= \frac{F(a)}{2} + \frac{F(b)}{2} + \sum_{1}^{n-1} F(a + kd)$$

Finally, line 7 multiplies the previous sum by d.

$$S = d\left(\frac{F(a)}{2} + \frac{F(b)}{2} + \sum_{1}^{n-1} F(a + kd)\right)$$

$$= \left(\frac{d}{2}\right)\left(F(a) + F(b) + 2\sum_{1}^{n-1} F(a + kd)\right)$$

This is an implementation of the trapezoidal rule.

Answer is D.

38. Use the algorithm provided to approximate the integral.

$$d = \frac{b - a}{2} = \frac{\pi - 0}{4}$$

$$= \pi/4$$

$$S = d\left(\frac{F(a)}{2} + \frac{F(b)}{2} + F(a + d) + F(a + 2d)\right.$$

$$\left. + F(a + 3d)\right)$$

$$= \left(\frac{\pi}{4}\right)\left(\begin{array}{c}\dfrac{F(0)}{2} + \dfrac{F(\pi)}{2} + F\left(\dfrac{\pi}{4}\right) \\ + F\left(\dfrac{\pi}{2}\right) + F\left(\dfrac{3\pi}{4}\right)\end{array}\right)$$

$$= \left(\frac{\pi}{4}\right)\left(0 + 0 + \frac{\sqrt{2}}{2} + 1 + \frac{\sqrt{2}}{2}\right)$$

$$= 1.896 \quad (1.9)$$

Answer is C.

39. Line 3 merely initializes the sum to zero, so the method doesn't actually start until the FOR ... NEXT loop.

$$S = \sum_{k=0}^{n-1} F(a + kd)$$

The sum is multiplied by d.

$$S = d\left(\sum_{0}^{n-1} F(a + kd)\right)$$

This is Euler's method.

Answer is A.

40. The number of letters from C to J inclusive is 8. The number of numbers from 3 to 12 inclusive is 10. The range of cells C3:J12 defines an 8×10 grid. There are $8 \times 10 = 80$ cells in this grid.

Answer is D.

41. ROM (read-only memory), PROM (programmable read-only memory), EPROM (erasable, programmable read-only memory), and WORM (write-once, read many) devices retain their information when the power is removed.

Answer is A.

42. Construct a table of the values of the variables for each iteration.

iteration	n	p	b	$|b| < t$?
1	1	4.00	$\dfrac{-4}{3}$	no
2	2	2.67	$\dfrac{4}{5}$	no
3	3	3.47	$\dfrac{-4}{7}$	no
4	4	2.90	$\dfrac{4}{9}$	yes

Answer is A.

43. Set the X equal to the expression for T in the first equation.

$$X = T$$

$$X = \frac{X^2}{16 - X}$$

$$16X - X^2 = X^2$$

$$X^2 - 8X = 0$$

This equation is satisfied for $X = 0$ and $X = 8$. Since X must be less than 5 in order to implement this first equation, an input value of $X = 8$ will never occur. Only $X = 0$ is valid.

Setting T = X in the second equation,

$$X = \frac{5X}{10 - X}$$
$$10X - X^2 = 5X$$
$$X^2 - 5X = 0$$

This equation is satisfied by X = 0 and X = 5. However, X must be greater than or equal to 5 for this second equation to be implemented. Within the range of applicability, the condition T = X is satisfied for an input value of X = 5.

T will have values only when X is 0 and 5.

Answer is B.

44. All of the bits in a word are passed in parallel to the CPU. There is one data line per data bit.

Answer is D.

45. The (x, y, z) intercepts are $(1, 1, \infty)$. The reciprocals are 1, 1, 0. The indices are (1 1 0).

Answer is B.

46. The current in R_1 is the voltage across it divided by its resistance.

$$I_{R_1} = \frac{V}{R_1}$$

The formula that will satisfy this relationship in cell D6 with all cell references relative is

A6/B6

When the formula is copied and pasted into the other cells in column E, it does not matter if the column references are relative or absolute. However, all cell references must have relative row references for the copy operation to yield the correct answer. $A6/$B6 is functionally equivalent to A6/B6 in this case.

Answer is C.

47. As verified by the superiors, the company is using dangerous manufacturing practices. The fact that they have never been caught or implicated in any accidents is irrelevant. To ask someone outside the company to review the design without the company's permission would itself be unethical and probably a breach of the engineer's employment contract. However, to "blow the whistle" is probably too extreme at this point. Given

an opportunity, the company's management will probably make an effort to correct the problem. If not, then notification of the authorities is the next alternative.

Answer is C.

48. Exaggerating one's capabilities in order to "get the business" is a violation of probably all ethical codes. Speaking disparagingly of the work of other engineers is similarly a violation. However, brand name comparisons and "no one does it better" claims are usually legal as long as they are factual, no matter how much in poor taste they are. Since a complaint was filed, the society should follow up on it to determine if the claims have any factual basis. This eliminates choices (A) and (B). While the engineer's competitor probably should consider choice (D), the professional society should gather more information before taking any action.

Answer is C.

49. Codes of ethics are guidelines of behavior. Ethics place more restrictions on your behavior than does the law. An action may be unethical (e.g., moonlighting) and still be legal. Some ethical actions (e.g., a reporter who refuses a judge's order to specify a secret source of information) may be illegal.

Answer is C.

50. The phrase "health, safety, and public welfare" appears in many codes and canons.

Answer is B.

51. Clearly, the safety and welfare of society as a whole are your highest responsibilities, and you (yourself) are the lowest. The honor, dignity, and reputation of the profession requires a professional to maintain the trust of the client. Therefore, the needs of the profession are served when the client is served first.

Answer is D.

52. The Rules of Professional Conduct as described by the NCEES *Model Rules* include sections on a licensee's obligation to society, employer and clients, and other licensees. Though NCEES governs the rules of conduct and can take disciplinary action, the *Model Rules* do not elaborate a licensee's obligation to the licensing board.

Answer is C.

Solutions for Sample Examination Morning Section   SEAM-27

SAMPLE EXAM
Morning Solutions

53. Confidential, proprietary, and trade secret information may be used and divulged only with the permission of the client. The permission does not need to be in writing, though it is common practice to obtain such permission in writing. Otherwise, the information must be kept to yourself.

Answer is D.

54. Working for both the owner and the contractor on the same project would be a conflict of interest, especially if it pertains to work that the engineer will later have to supervise and evaluate. Accepting the contractor's offer might appear to influence future decisions and judgments on the part of the engineer.

Answer is C.

55. Hiring the official's brother as a precondition for being awarded the contract is a form of extortion. Depending on the circumstances, however, it may be legal to do so according to U.S. law. (The Foreign Corrupt Practices Act of 1977 allows American companies to pay extortion in some cases.) This practice, however, is not approved by the NCEES model code:

> Registrants shall not offer, give, solicit, or receive, either directly or indirectly, any commission or gift, or other valuable consideration in order to secure work.

Answer is D.

56. The depreciation each year will be

$$D = \frac{C - S_n}{n} = \frac{\$100{,}000 - \$15{,}000}{7}$$
$$= \$12{,}143 \text{ per year}$$

$$(BV)_3 = C - 3D = \$100{,}000 - (3)(\$12{,}143)$$
$$= \$63{,}571 \quad (\$63{,}600)$$

Answer is D.

57. $i = \left(1 + \dfrac{r}{m}\right)^m - 1 = \left(1 + \dfrac{0.05}{4}\right)^4 - 1$
$$= 0.05095 \quad (5.1\%)$$

Answer is C.

58. The equivalent uniform annual cost (EUAC) is the sum of the equivalent annual values of its initial and maintenance costs, minus the equivalent annual value of its salvage value.

$$\begin{aligned}
\text{EUAC} &= (\$7600)(A/P, 6\%, 12) + \$380 \\
&\quad - (\$2000)(A/F, 6\%, 12) \\
&= (\$7600)(0.1193) + \$380 - (\$2000)(0.0593) \\
&= \$1168.08 \quad (\$1200)
\end{aligned}$$

Answer is A.

59. The effective uniform annual cost is

$$\begin{aligned}
\text{EUAC} &= C(A/P, 6\%, 5) \\
&= (\$24{,}000)(0.2374) \\
&= \$5698
\end{aligned}$$

The capitalized cost is

$$\text{capitalized cost} = \frac{\text{EUAC}}{i} = \frac{\$5698}{0.06}$$
$$= \$94{,}967 \quad (\$95{,}000)$$

Answer is D.

60. $i = \left(1 + \dfrac{r}{k}\right)^k - 1$

With continuous compounding, k can be any large number. Use $k = 1000$.

$$i = \left(1 + \frac{0.046}{1000}\right)^{1000} - 1$$
$$= 0.04707 \quad (4.71\%)$$

Answer is B.

61. Use the cash flow factor $(P/A, i\%, n)$, where the interest rate, i, is 8%, and the number of periods, n, is 8. From a table of factors, this value is 5.7466.

$$\begin{aligned}
P &= A\,(P/A,\ 8\%,\ 8) \\
&= (\$800)(5.7466) \\
&= \$4597.28 \quad (\$4600)
\end{aligned}$$

Answer is B.

62. The straight line depreciation value is equal to the initial cost minus the salvage value, divided by the number of years.

$$D = \frac{C - S_n}{n} = \frac{\$14{,}000 - \$1200}{15}$$
$$= \$853.33 \quad (\$850)$$

Answer is D.

Professional Publications, Inc.

63. The breakeven point occurs when costs (which are negative) and sales (which are positive) total zero. Letting x be the number of chairs per year at breakeven,

$$\text{annual operating cost}$$
$$+ \text{annual manufacturing cost}$$
$$+ \text{ annual sales}$$
$$= \$0$$
$$- \$180,000 - (\$31)\left(x_{\text{chairs/yr}}\right)$$
$$+ (\$59)\left(x_{\text{chairs/yr}}\right)$$
$$= \$0$$

$$\$28\left(x_{\text{chairs/yr}}\right) = \$180,000$$
$$x = 6429 \text{ chairs/yr} \quad (6400 \text{ chairs/yr})$$

Answer is C.

64. The cost of owning and operating the machine one more year is equal to the operating costs plus the lost salvage value. After three years, the machine's salvage value is $9000, and after another year it will be $6000. However, the value of having the cash one year earlier must also be considered.

$$F = (\$9000)(F/P, 12\%, 1)$$
$$= (\$9000)(1.1500)$$
$$= \$10,350$$

The cost of one more year of ownership and operation is thus

$$C = \text{operating cost} + \text{lost salvage value}$$
$$= \$2500 + +\$3000 + (0.15)(\$9000)$$
$$= \$6850 \quad (\$6900)$$

Answer is C.

65.
$$I_x = I_c + Ad^2$$
$$= I_c + A(8 \text{ cm})^2$$
$$I_{x'} = I_c + A(10 \text{ cm})^2$$
$$= I_x - A(8 \text{ cm})^2 + A(10 \text{ cm})^2$$
$$= 3870.3 \text{ cm}^4 - (60 \text{ cm}^2)(8 \text{ cm})^2$$
$$+ (60 \text{ cm}^2)(10 \text{ cm})^2$$
$$= 6030.3 \text{ cm}^4 \quad (6000 \text{ cm}^4)$$

Answer is D.

66. For equlibrium, the y-components must balance.

$$\sum F_y = 0: (606 \text{ N})\left(\frac{1}{\sqrt{(1)^2 + (1)^2}}\right)$$
$$+ (500 \text{ N})\left(\frac{3}{\sqrt{(5)^2 + (3)^2}}\right) - F$$
$$= 0$$
$$F = 685.75 \text{ N} \quad (690 \text{ N})$$

Answer is B.

67. The only force acting on the structure is the 15 000 N downward force. The only reaction at point O is an upward force of 15 000 N.

$$R_O = (15\,000 \text{ N})\mathbf{j}$$

Answer is B.

68. The block will tend to rotate about the lower left corner. Take the sum of moments about this point to find the maximum value of h.

$$\sum M = 0: -\left(\frac{D}{2}\right)(10 \text{ kg})\left(9.81 \frac{\text{m}}{\text{s}^2}\right) + h(20 \text{ N})$$
$$= 0$$
$$h = \frac{\left(\frac{D}{2}\right)(10 \text{ kg})\left(9.81 \frac{\text{m}}{\text{s}^2}\right)}{20 \text{ N}}$$
$$= 2.45D \quad \left(\frac{5}{2}D\right)$$

Answer is D.

69.

$$\sum F_x = 0$$
$$= -F_1\left(\frac{1}{\sqrt{(1)^2 + (1)^2}}\right) + F_2\left(\frac{3}{\sqrt{(3)^2 + (5)^2}}\right)$$
$$F_1 = \left(\frac{3\sqrt{2}}{\sqrt{34}}\right)F_2$$
$$\sum F_y = 0$$
$$= F_1\left(\frac{1}{\sqrt{2}}\right) + F_2\left(\frac{5}{\sqrt{34}}\right) - (40 \text{ kg})\left(9.81 \frac{\text{m}}{\text{s}^2}\right)$$
$$= \left(\frac{3\sqrt{2}}{\sqrt{34}}\right)\left(\frac{1}{\sqrt{2}}\right)F_2 + F_2\left(\frac{5}{\sqrt{34}}\right)$$
$$- (40 \text{ kg})\left(9.81 \frac{\text{m}}{\text{s}^2}\right)$$

$$F_2 = \frac{(40 \text{ kg}) \left(9.81 \dfrac{\text{m}}{\text{s}^2}\right) \sqrt{34}}{8}$$

$$= 286.0 \text{ N} \quad (290 \text{ N})$$

$$F_1 = \left(\frac{3\sqrt{2}}{\sqrt{34}}\right) (286.0 \text{ N})$$

$$= 208.1 \text{ N} \quad (210 \text{ N})$$

Answer is B.

70. The tension in the pulley rope is constant. From the free body of the lowest pulley,

$$3T = mg$$

$$T = \frac{mg}{3}$$

Answer is C.

71.

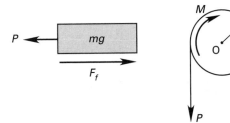

The free body of the mass is

$$\sum F_x = 0 = -P + F_f$$

$$P = F_f = \mu N = \mu mg$$

The free body of the pulley is

$$\sum M_O = 0 = M - Pr$$

$$M = Pr = \mu mgr$$

Answer is A.

72.
$$F_c = ma_n = \frac{mv^2}{r} = \frac{m(\omega r)^2}{r}$$

$$= m\omega^2 r$$

Answer is A.

73. The condition of dynamic equilibrium is

$$ma = F - mg \sin\theta - \mu N$$

$$F = ma + mg \sin\theta + \mu mg \cos\theta$$

$$= (5 \text{ kg}) \left(5 \dfrac{\text{m}}{\text{s}^2}\right) + (5 \text{ kg}) \left(9.81 \dfrac{\text{m}}{\text{s}^2}\right) \sin 25°$$

$$+ (0.25)(5 \text{ kg}) \left(9.81 \dfrac{\text{m}}{\text{s}^2}\right) \cos 25°$$

$$= 56.8 \text{ N} \quad (57 \text{ N})$$

Answer is B.

74.
$$e = 0 = \frac{v_1' - v_2'}{v_2 - v_1}$$

$$v_2' = v_1' = v'$$

From the conservation of momentum,

$$m_1 v_1 + m_2 v_2 = (m_1 + m_2) v'$$

$$v' = \frac{(2 \text{ kg}) \left(40 \dfrac{\text{m}}{\text{s}}\right) + (5 \text{ kg}) \left(10 \dfrac{\text{m}}{\text{s}}\right)}{2 \text{ kg} + 5 \text{ kg}}$$

$$= 18.6 \text{ m/s}$$

Answer is D.

75. The frictional force must counteract the centrifugal force.

$$F_f = F_c$$

$$\mu N = \frac{mv^2}{r}$$

$$\mu mg = \frac{mv^2}{r}$$

$$\mu = \frac{v^2}{gr}$$

$$= \frac{\left(\left(80 \dfrac{\text{km}}{\text{h}}\right) \left(1000 \dfrac{\text{m}}{\text{km}}\right) \left(\dfrac{1 \text{ h}}{3600 \text{ s}}\right)\right)^2}{\left(9.81 \dfrac{\text{m}}{\text{s}^2}\right) (1000 \text{ m})}$$

$$= 0.05$$

Answer is A.

76. Determine the extension of the spring prior to the string being cut.

$$F = k\delta = W = mg$$

$$k\delta = mg$$

$$\delta_o = \frac{mg}{k}$$

$$= \frac{(5 \text{ kg} + 3 \text{ kg}) \left(9.81 \dfrac{\text{m}}{\text{s}^2}\right)}{300 \dfrac{\text{N}}{\text{m}}}$$

$$= 0.2616 \text{ m} \quad (26.16 \text{ cm})$$

Determine the static deflection of m_1 alone. This is the equilibrium position.

$$\delta_{st,1} = \frac{m_1 g}{k}$$

$$= \frac{(5 \text{ kg}) \left(9.81 \dfrac{\text{m}}{\text{s}^2}\right)}{300 \dfrac{\text{N}}{\text{m}}}$$

$$= 0.1635 \text{ m} \quad (16.35 \text{ cm})$$

When the string is cut, at $t = 0$, the initial displacement of m_1 from its equilibrium position is

$$x_o = \delta_o - \delta_{st,1}$$
$$= 26.16 \text{ cm} - 16.35 \text{ cm}$$
$$= 9.81 \text{ cm}$$

The mass will oscillate about its static equilibrium position with a maximum displacement equal to its initial displacement. The amplitude is equal to the initial displacement.

$$A = x_o = 9.81 \text{ cm}$$

Answer is A.

77. The endurance limit is measured by a fatigue test and cannot be found from a tensile test. The other three quantities can be obtained by a tensile test.

Answer is C.

78. Since the rods and bars are restrained by the rigid end pieces, the elongation (i.e., strain) is the same for all three members at all temperatures.

Answer is D.

79. The total compressive force in the two aluminum bars is equal to the tensile force in the steel bar.

$$2F_{\text{aluminum}} = F_{\text{steel}}$$
$$2A_{\text{aluminum}}\sigma_{\text{aluminum}} = A_{\text{steel}}\sigma_{\text{steel}}$$
$$\sigma_{\text{steel}} = \frac{2A_{\text{aluminum}}\sigma_{\text{aluminum}}}{A_{\text{steel}}}$$
$$= \frac{(2)(0.7 \text{ cm}^2)(4000 \text{ Pa})}{1.5 \text{ cm}^2}$$
$$= 3733 \text{ Pa} \quad (3.7 \text{ kPa})$$

Answer is C.

80.

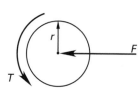

The force, F, can be replaced by a force-couple system acting through the axis of the column. Then the torsional and direct shear stresses can be combined.

$$T = (1000 \text{ N})(0.015 \text{ m})$$
$$= 15 \text{ N·m}$$

$$\tau = \frac{Tr}{J} + \frac{F}{A}$$
$$= \frac{Tr}{\dfrac{\pi r^4}{2}} + \frac{F}{\pi r^2}$$
$$= \frac{(15 \text{ N·m})(0.01 \text{ m})}{\left(\dfrac{\pi}{2}\right)(0.01 \text{ m})^4} + \frac{1000 \text{ N}}{\pi(0.01 \text{ m})^2}$$
$$= 1.273 \times 10^7 \text{ Pa} \quad (13 \text{ MPa})$$

Answer is D.

81. The resultant of the distributed load is

$$R = \left(\frac{1}{2}\right)\left(50 \ \frac{\text{N}}{\text{m}}\right)(15 \text{ m}) = 375 \text{ N}$$

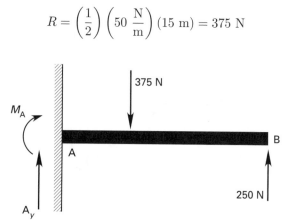

The resultant acts through the centroid of the loading distribution, 5 m from the fixed end.

At point A,

$$\sum M_\text{A} = 0 = (375 \text{ N})(5 \text{ m}) - (250 \text{ N})(15 \text{ m}) + M_\text{A}$$
$$M_\text{A} = 1875 \text{ N·m}$$
$$\sum F_y = 0 = 250 \text{ N} - 375 \text{ N} + A_y$$
$$A_y = 125 \text{ N}$$

The shear is $+125$ N at A, and it decreases parabolically to -250 N at B. The maximum shear is 250 N (sign is not relevant).

Answer is D.

82. The moment is a maximum where the shear is equal to zero. The slope of the triangular loading curve is $50 \text{ N}/15 \text{ m} = 10/3 \text{ N/m}$.

$$V = 125 - \int \left(50 - \frac{10}{3}x\right) dx = 0$$
$$\frac{5}{3}x^2 - 50x + 125 = 0$$

$$x = \frac{-(-50) \pm \sqrt{(-50)^2 - (4)\left(\frac{5}{3}\right)(125)}}{(2)\left(\frac{5}{3}\right)}$$

$$= 2.753 \text{ m or } 27.25 \text{ m} \quad \begin{bmatrix} 27.25 \text{ m is beyond} \\ \text{the beam end} \end{bmatrix}$$

$$M = 1875 + \int \left(125 - 50x + \frac{5}{3}x^2\right) dx$$

$$= 1875 + 125x - 25x^2 + \frac{5x^3}{9}$$

$$M_{2.753} = 1875 + (125)(2.753) - (25)(2.753)^2$$

$$+ \left(\frac{5}{9}\right)(2.753)^3$$

$$= 2041 \text{ N·m} \quad (2000 \text{ N·m})$$

Answer is B.

83. The deflection is

$$y_x = \left(\frac{Px^2}{6EI}\right)(-x + 3a)$$

$$y_{7.5} = \frac{(250 \text{ N})(7.5 \text{ m})^2}{(6)(2.1 \times 10^{11} \text{ Pa})\left(\frac{(0.4 \text{ m})(0.4 \text{ m})^3}{12}\right)}$$

$$\times \left(-7.5 \text{ m} + (3)(15 \text{ m})\right)$$

$$= 0.000196 \text{ m} \quad (0.2 \text{ mm}) \quad \text{[upward]}$$

Answer is D.

84. For a column with fixed ends, the effective length is

$$l' = kl = (0.5)(3.5 \text{ m})$$

$$= 1.75 \text{ m}$$

$$r = \text{radius of gyration} = \sqrt{\frac{I}{A}}$$

$$= \sqrt{\frac{\left(\frac{1}{12}\right)(0.025 \text{ m})(0.025 \text{ m})^3}{(0.025 \text{ m})(0.025 \text{ m})}}$$

$$= 0.007217 \text{ m}$$

$$\sigma_{cr} = \frac{\pi^2 E}{\left(\frac{l'}{r}\right)^2}$$

$$= \frac{\pi^2 \left(210 \times 10^9 \ \frac{\text{N}}{\text{m}^2}\right)}{\left(\frac{1.75 \text{ m}}{0.007217 \text{ m}}\right)^2}$$

$$= 3.525 \times 10^7 \text{ Pa}$$

$$\sigma = E\epsilon_{\text{th}} = E\alpha(T_2 - T_1)$$

$$T_2 = \frac{\sigma}{E\alpha} + T_1$$

$$= \frac{3.525 \times 10^7 \text{ Pa}}{(210 \times 10^9 \text{ Pa})\left(11.7 \times 10^{-6} \ \frac{1}{°\text{C}}\right)} + 20°\text{C}$$

$$= 34.3°\text{C}$$

Answer is D.

85. The number of atoms in a cell is the number of atoms appearing in that cell, each divided by the number of cells sharing those atoms. For HCP structures,

$$\text{no. of atoms} = \frac{\text{corner}}{\text{atoms}} + \frac{\text{end}}{\text{atoms}} + \frac{\text{interior}}{\text{atoms}}$$

$$= \frac{12 \text{ atoms}}{6 \text{ cells}} + \frac{2 \text{ atoms}}{2 \text{ cells}} + \frac{3 \text{ atoms}}{1 \text{ cell}}$$

$$= 6 \text{ atoms}$$

Answer is C.

86.

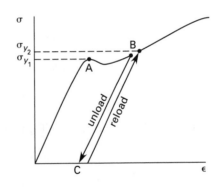

The stress-strain curve for the steel rod is shown in the figure. The rod is loaded first past the yield point, point A, to point B. The rod is then unloaded to point C. When tension is again applied, point B is the new yield point.

Answer is D.

87. Recrystallization is the same as annealing, except at a lower temperature.

Answer is D.

88. Poisson's ratio is the ratio of the lateral strain to the axial strain.

Answer is A.

89. Zinc has a higher standard oxidation potential than iron in the galvanic series, indicating that it has a greater tendency than iron to lose electrons and form positive ions. The zinc will corrode more than the iron.

Answer is A.

90.
$$\delta = \frac{PL}{AE} = \frac{(315 \text{ N})(60 \text{ m})}{\left(\frac{\pi}{4}\right)(0.00175 \text{ m})^2 \left(200 \times 10^9 \ \frac{\text{N}}{\text{m}^2}\right)}$$
$$= 0.03929 \text{ m}$$

$$\epsilon = \frac{\delta}{L} = \frac{0.03929 \text{ m}}{60 \text{ m}}$$
$$= 6.55 \times 10^{-4} \text{ m/m}$$

Answer is C.

91. The packing factor is the volume of atoms in the cell divided by the cell volume.

Answer is C.

92. According to Le Châtelier's principle, when a stress is applied to a system in equilibrium, the equilibrium shifts in such a way as to relieve the stress.

In this case, the stress is a decrease in temperature. When the temperature is decreased, the system will respond by increasing the rate at which CO_2 is produced since this is an exothermic reaction (as evidenced by the negative enthalpy of formation). The concentration of carbon dioxide and the equilibrium constant will increase as a result.

Answer is A.

93. The average pressure of water on the gate is

$$\overline{p} = \frac{1}{2}\rho g(h_1 + h_2)$$
$$= \left(\frac{1}{2}\right)\left(1000 \ \frac{\text{kg}}{\text{m}^3}\right)\left(9.81 \ \frac{\text{m}}{\text{s}^2}\right)(0 + 3 \text{ m})$$
$$= 14\,715 \text{ Pa}$$

The resultant water force per meter of width is

$$R = \overline{p}A$$
$$= (14\,715 \text{ Pa})(3 \text{ m})(1 \text{ m})$$
$$= 44\,145 \text{ N}$$

The resultant acts at the centroid of the pressure distribution.

$$h_R = \frac{2}{3}h = \left(\frac{2}{3}\right)(3 \text{ m}) + 1 \text{ m}$$
$$= 3 \text{ m} \quad \text{[from the hinge]}$$

Sum moments about the horizontal hinge at the top.

$$\sum M_{\text{hinge}} = 0$$
$$= (44\,145 \text{ N})(3 \text{ m}) - F(4 \text{ m})$$
$$F = 33\,109 \text{ N} \quad (33.1 \text{ kN})$$

Answer is D.

94. For dynamic similarity between a model and its prototype, the ratios of all types of forces must be equal. The ratio of the inertial force to the viscous force is expressed as the Reynolds number.

Answer is A.

95.
$$p_{\text{gage}} = p_{\text{absolute}} - p_{\text{atm}} = -\rho g h$$
$$= -\left(1.36 \times 10^4 \ \frac{\text{kg}}{\text{m}^3}\right)\left(9.81 \ \frac{\text{m}}{\text{s}^2}\right)(0.1 \text{ m})$$
$$= -13\,342 \text{ Pa} \quad (-13.3 \text{ kPa})$$
$$= 13.3 \text{ kPa vacuum} \quad (13 \text{ kPa vacuum})$$

Answer is B.

96. Q is constant.
$$v_1 A_1 = v_2 A_2$$
$$v_2 = \frac{v_1 A_1}{A_2} = \frac{v_1 D_1^2}{D_2^2}$$
$$= \frac{\left(5 \ \frac{\text{m}}{\text{s}}\right)(0.05 \text{ m})^2}{(0.10 \text{ m})^2}$$
$$= 1.25 \text{ m/s}$$

Answer is B.

97. For a venturi meter,
$$v_2 = \frac{Q}{A_2}$$
$$= \left(\frac{C_v}{\sqrt{1 - \left(\frac{A_2}{A_1}\right)^2}}\right)\sqrt{2\left(\frac{p_1}{\rho} + z_1 - \frac{p_2}{\rho} - z_2\right)}$$

Since the venturi is horizontal, $z_1 = z_2$.

$$p_1 - p_2 = \frac{v_2^2 \left(1 - \left(\dfrac{A_2}{A_1}\right)^2\right) \rho}{2C_v^2}$$

$$= \frac{\left(3 \ \dfrac{m}{s}\right)^2 (1 - (0.5)^2)\left(1000 \ \dfrac{kg}{m^3}\right)}{(2)(0.98)^2}$$

$$= 3514 \text{ Pa} \quad (3.5 \text{ kPa})$$

Answer is C.

98.
$$R_H = \frac{\text{cross-sectional area}}{\text{wetted perimeter}}$$

$$= \frac{(12 \text{ cm})(8 \text{ cm})\left(\dfrac{3}{4}\right)}{(2)\left(\dfrac{3}{4}\right)(8 \text{ cm}) + 12 \text{ cm}}$$

$$= 3.0 \text{ cm}$$

Answer is C.

99.
$$v_1 = \frac{Q_1}{A_1} = \frac{0.01 \ \dfrac{m^3}{s}}{\left(\dfrac{\pi}{4}\right)(0.15 \text{ m})^2}$$

$$= 0.566 \text{ m/s}$$

$$A_1 v_1 = A_2 v_2$$

$$v_2 = v_1 \left(\frac{A_1}{A_2}\right) = v_1 \left(\frac{D_1}{D_2}\right)^2$$

$$= \left(0.566 \ \frac{m}{s}\right)\left(\frac{15 \text{ cm}}{5 \text{ cm}}\right)^2$$

$$= 5.09 \text{ m/s}$$

$$\frac{p_2}{\rho g} + \frac{v_2^2}{2g} + z_2 = \frac{p_1}{\rho g} + \frac{v_1^2}{2g} + z_1$$

$$p_2 = \rho g\left(\left(\frac{1}{2g}\right)(v_1^2 - v_2^2) + \frac{p_1}{\rho g} + (z_1 - z_2)\right)$$

$$= \left(1000 \ \frac{kg}{m^3}\right)\left(9.81 \ \frac{m}{s^2}\right)$$

$$\times \left(\frac{\left(0.566 \ \dfrac{m}{s}\right)^2 - \left(5.09 \ \dfrac{m}{s}\right)^2}{(2)\left(9.81 \ \dfrac{m}{s^2}\right)}\right.$$

$$\left. + \frac{75\,000 \ \dfrac{N}{m^2}}{\left(1000 \ \dfrac{kg}{m^3}\right)\left(9.81 \ \dfrac{m}{s^2}\right)} + (5 \text{ m} - 0 \text{ m})\right)$$

$$= 111\,256 \text{ Pa} \quad (111 \text{ kPa})$$

Answer is B.

100. $P = Q\gamma h = Q\rho g h$

$$= \left(2 \ \frac{m^3}{s}\right)\left(1000 \ \frac{kg}{m^3}\right)\left(9.81 \ \frac{m}{s^2}\right)(15 \text{ m})$$

$$= 294\,300 \text{ W} \quad (294 \text{ kW})$$

Answer is A.

101. $\sigma = 0.0227 \text{ N/m}$

$\beta = 0°$

$$\rho = \frac{\left(0.790 \ \dfrac{g}{cm^3}\right)\left(10^6 \ \dfrac{cm^3}{m^3}\right)}{\left(1000 \ \dfrac{g}{kg}\right)}$$

$$= 790 \ \frac{kg}{m^3}$$

$$h = \frac{4\sigma \cos \beta}{\rho d_{\text{tube}} g}$$

$$= \frac{(4)\left(0.0227 \ \dfrac{N}{m}\right)(1.0)\left(1000 \ \dfrac{mm}{m}\right)}{\left(790 \ \dfrac{kg}{m^3}\right)(0.25)\left(9.81 \ \dfrac{m}{s^2}\right)}$$

$$= 0.0469 \text{ m} \quad (0.047 \text{ m})$$

Answer is B.

102.
$$C_{\text{eq}} = \frac{1}{\sum \dfrac{1}{C_i}} = \frac{1}{(5)\left(\dfrac{1}{50 \ \mu F}\right)}$$

$$= 10 \ \mu F$$

Answer is A.

103. Draw an equivalent circuit.

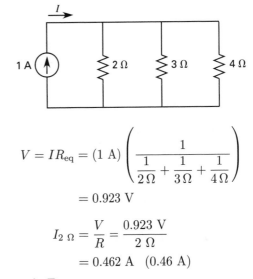

$$V = I R_{\text{eq}} = (1 \text{ A})\left(\frac{1}{\dfrac{1}{2 \ \Omega} + \dfrac{1}{3 \ \Omega} + \dfrac{1}{4 \ \Omega}}\right)$$

$$= 0.923 \text{ V}$$

$$I_{2 \ \Omega} = \frac{V}{R} = \frac{0.923 \text{ V}}{2 \ \Omega}$$

$$= 0.462 \text{ A} \quad (0.46 \text{ A})$$

Answer is B.

104. Kirchhoff's current law states that as much current flows out of a node as flows into it. Kirchhoff's voltage law states that the sum of the voltage drop around any closed loop is equal to the sum of the applied voltages.

Answer is D.

105. Untwist the network.

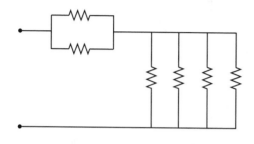

Both terminals of the right-most resistor are connected to the same point, and that resistor is omitted.

The circuit consists of two parallel resistors in series with four parallel resistors.

$$R_{eq} = \left(\frac{1}{\frac{1}{10\ \Omega} + \frac{1}{10\ \Omega}} \right) + \left(\frac{1}{(4)\left(\frac{1}{10\ \Omega}\right)} \right)$$

$$= 5\ \Omega + 2.5\ \Omega$$

$$= 7.5\ \Omega$$

Answer is A.

106. The force between Q_1 and Q_3 is repulsive and is entirely in the negative y-direction.

The distance between Q_1 and Q_3 is 5 m.

$$F_{3-1} = \frac{Q_1 Q_3}{4\pi\varepsilon r^2} = \frac{(0.1 \times 10^{-6}\ \text{C})(0.4 \times 10^{-6}\ \text{C})}{4\pi\left(8.85 \times 10^{-12}\ \frac{\text{F}}{\text{m}}\right)(5\ \text{m})^2}$$

$$= 1.44 \times 10^{-5}\ \text{N}$$

The distance between Q_2 and Q_3 is

$$\sqrt{(1\ \text{m})^2 + (6\ \text{m})^2} = 6.08\ \text{m}$$

The force between Q_2 and Q_3 is attractive and is partly in the positive y-direction and partly in the positive x-direction.

$$F_{3-2} = \frac{Q_2 Q_3}{4\pi\varepsilon r^2} = \frac{(-0.3 \times 10^{-6}\ \text{C})\,(0.4 \times 10^{-6}\ \text{C})}{4\pi\left(8.85 \times 10^{-12}\ \frac{\text{F}}{\text{m}}\right)(6.08\ \text{m})^2}$$

$$= -2.92 \times 10^{-5}\ \text{N}$$

The portion of the force between Q_2 and Q_3 in the positive y-direction is

$$(-2.92 \times 10^{-5}\ \text{N})\left(\frac{1\ \text{m}}{6.08\ \text{m}}\right) = -4.8 \times 10^{-6}\ \text{N}$$

The portion of the force between Q_2 and Q_3 in the positive x-direction is

$$(-2.92 \times 10^{-5}\ \text{N})\left(\frac{6\ \text{m}}{6.08\ \text{m}}\right) = -2.88 \times 10^{-5}\ \text{N}$$

The total force is the vector sum of the two forces.

$$F_{total} = \sqrt{\begin{array}{c}\left(1.44 \times 10^{-5}\ \text{N} - 4.8 \times 10^{-6}\ \text{N}\right)^2 \\ + \left(-2.88 \times 10^{-5}\ \text{N}\right)^2\end{array}}$$

$$= 3.04 \times 10^{-5}\ \text{N} \quad (3.0 \times 10^{-5}\ \text{N})$$

Answer is A.

107. Passive elements absorb and/or release energy.

Answer is D.

108. All of the charge can be assumed to be located at the smaller sphere's center. The field intensity is proportional to the square of the distance from the smaller sphere's center.

$$E = \frac{Q}{(R+r)^2} \quad \left(E \propto \frac{Q}{(R+r)^2}\right)$$

Answer is C.

109.
$$v(t) = V_{max}\cos(\omega t + \theta)$$
$$V_{max} = 100\ \text{V}$$
$$\omega = 2\pi f$$
$$f = \frac{\omega}{2\pi} = \frac{20\pi\ \frac{\text{rad}}{\text{s}}}{2\pi\ \frac{\text{rad}}{\text{cycle}}}$$
$$= 10\ \text{Hz}$$

Answer is B.

110.
$$W = -Q\Delta V$$
$$= -\left(-1.6 \times 10^{-19} \text{ C}\right)\left(1.0 \times 10^{6} \text{ V}\right)$$
$$= 1.6 \times 10^{-13} \text{ J}$$

Answer is D.

111. Impedance is the complex sum of resistance and reactance. Considering the impedance triangle, the impedance can be written as

$$Z = \frac{R}{\cos\theta}$$

The quantity $\cos\theta$ is called the *power factor*. θ is the impedance angle—the angle between voltage and current.

$$\cos\theta = \frac{R}{Z}$$

Answer is B.

112. Enthalpy represents the total useful energy of a system.

Answer is B.

113. The triple point of a substance is a unique state at which solid, liquid, and gaseous phases can coexist.

Answer is B.

114.
$$v = v_f + xv_{fg} = v_f + x(v_g - v_f)$$

$$= 1.4178 \ \frac{\text{cm}^3}{\text{g}} + (0.85)\left(20.48 \ \frac{\text{cm}^3}{\text{g}} - 1.4178 \ \frac{\text{cm}^3}{\text{g}}\right)$$

$$= 17.62 \text{ cm}^3/\text{g}$$

Convert to m^3/kg.

$$v = \frac{\left(17.62 \ \dfrac{\text{cm}^3}{\text{g}}\right)\left(1000 \ \dfrac{\text{g}}{\text{kg}}\right)}{\left(100 \ \dfrac{\text{cm}}{\text{m}}\right)^3} = 0.01762 \text{ m}^3/\text{kg}$$

Answer is D.

115.
$$W_{\text{out}} = h_1 - h_2$$
$$= 3900 \ \frac{\text{kJ}}{\text{kg}} - 2680 \ \frac{\text{kJ}}{\text{kg}}$$
$$= 1220 \text{ kJ/kg}$$

Notice that the actual (not ideal) exit enthalpy was given.

Answer is B.

116.
$$\eta_{\text{Carnot}} = \frac{T_H - T_L}{T_H}$$
$$\eta = \frac{425°\text{C} - 65°\text{C}}{425°\text{C} + 273}$$
$$= 0.516 \quad (52\%)$$

Answer is C.

117.
$$\frac{p_1 V_1}{T_1} = \frac{p_2 V_2}{T_2}$$

In an isobaric process, $p_1 = p_2$.

$$T_2 = \frac{T_1 V_2}{V_1} = (310°\text{C} + 273)\left(\frac{1.4 \text{ m}^3}{2.8 \text{ m}^3}\right)$$
$$= 291.5\text{K}$$
$$T_2 = 291.5\text{K} - 273$$
$$= 18.5°\text{C} \quad (19°\text{C})$$

Answer is A.

118. $\phi = \dfrac{p_v}{p_g} = \dfrac{\text{partial pressure of water vapor}}{\text{saturation pressure}}$

A decrease in the partial pressure of water vapor would reduce the humidity.

Answer is C.

119. $pV = mRT$

$$m = \frac{pV}{RT} = \frac{(100 \text{ kPa})\left(1000 \ \dfrac{\text{Pa}}{\text{kPa}}\right)(1 \text{ m}^3)}{\left(259.8 \ \dfrac{\text{J}}{\text{kg·K}}\right)(30\text{K} + 273)}$$

$$= 1.27 \text{ kg} \quad (1.3 \text{ kg})$$

Answer is B.

120.
$$\frac{p_1 V_1}{T_1} = \frac{p_2 V_2}{T_2}$$

As volume is constant,

$$\frac{p_1}{T_1} = \frac{p_2}{T_2}$$
$$p_2 = \frac{p_1 T_2}{T_1} = \frac{(200 \text{ kPa})(160°\text{C} + 273°)}{(375°\text{C} + 273°)}$$
$$= 133.6 \text{ kPa} \quad (130 \text{ kPa})$$

Answer is C.

Sample Examination
Afternoon Section

To make taking this sample exam as realistic as possible,
use the NCEES Handbook as your only reference.

1. Ⓐ Ⓑ Ⓒ Ⓓ   16. Ⓐ Ⓑ Ⓒ Ⓓ   31. Ⓐ Ⓑ Ⓒ Ⓓ   46. Ⓐ Ⓑ Ⓒ Ⓓ

2. Ⓐ Ⓑ Ⓒ Ⓓ   17. Ⓐ Ⓑ Ⓒ Ⓓ   32. Ⓐ Ⓑ Ⓒ Ⓓ   47. Ⓐ Ⓑ Ⓒ Ⓓ

3. Ⓐ Ⓑ Ⓒ Ⓓ   18. Ⓐ Ⓑ Ⓒ Ⓓ   33. Ⓐ Ⓑ Ⓒ Ⓓ   48. Ⓐ Ⓑ Ⓒ Ⓓ

4. Ⓐ Ⓑ Ⓒ Ⓓ   19. Ⓐ Ⓑ Ⓒ Ⓓ   34. Ⓐ Ⓑ Ⓒ Ⓓ   49. Ⓐ Ⓑ Ⓒ Ⓓ

5. Ⓐ Ⓑ Ⓒ Ⓓ   20. Ⓐ Ⓑ Ⓒ Ⓓ   35. Ⓐ Ⓑ Ⓒ Ⓓ   50. Ⓐ Ⓑ Ⓒ Ⓓ

6. Ⓐ Ⓑ Ⓒ Ⓓ   21. Ⓐ Ⓑ Ⓒ Ⓓ   36. Ⓐ Ⓑ Ⓒ Ⓓ   51. Ⓐ Ⓑ Ⓒ Ⓓ

7. Ⓐ Ⓑ Ⓒ Ⓓ   22. Ⓐ Ⓑ Ⓒ Ⓓ   37. Ⓐ Ⓑ Ⓒ Ⓓ   52. Ⓐ Ⓑ Ⓒ Ⓓ

8. Ⓐ Ⓑ Ⓒ Ⓓ   23. Ⓐ Ⓑ Ⓒ Ⓓ   38. Ⓐ Ⓑ Ⓒ Ⓓ   53. Ⓐ Ⓑ Ⓒ Ⓓ

9. Ⓐ Ⓑ Ⓒ Ⓓ   24. Ⓐ Ⓑ Ⓒ Ⓓ   39. Ⓐ Ⓑ Ⓒ Ⓓ   54. Ⓐ Ⓑ Ⓒ Ⓓ

10. Ⓐ Ⓑ Ⓒ Ⓓ   25. Ⓐ Ⓑ Ⓒ Ⓓ   40. Ⓐ Ⓑ Ⓒ Ⓓ   55. Ⓐ Ⓑ Ⓒ Ⓓ

11. Ⓐ Ⓑ Ⓒ Ⓓ   26. Ⓐ Ⓑ Ⓒ Ⓓ   41. Ⓐ Ⓑ Ⓒ Ⓓ   56. Ⓐ Ⓑ Ⓒ Ⓓ

12. Ⓐ Ⓑ Ⓒ Ⓓ   27. Ⓐ Ⓑ Ⓒ Ⓓ   42. Ⓐ Ⓑ Ⓒ Ⓓ   57. Ⓐ Ⓑ Ⓒ Ⓓ

13. Ⓐ Ⓑ Ⓒ Ⓓ   28. Ⓐ Ⓑ Ⓒ Ⓓ   43. Ⓐ Ⓑ Ⓒ Ⓓ   58. Ⓐ Ⓑ Ⓒ Ⓓ

14. Ⓐ Ⓑ Ⓒ Ⓓ   29. Ⓐ Ⓑ Ⓒ Ⓓ   44. Ⓐ Ⓑ Ⓒ Ⓓ   59. Ⓐ Ⓑ Ⓒ Ⓓ

15. Ⓐ Ⓑ Ⓒ Ⓓ   30. Ⓐ Ⓑ Ⓒ Ⓓ   45. Ⓐ Ⓑ Ⓒ Ⓓ   60. Ⓐ Ⓑ Ⓒ Ⓓ

Problems for Sample Examination

AFTERNOON SECTION

1. What is the Cartesian equation describing the path of a particle whose position is given by the following equations?

$$x(t) = \sqrt{2} \, \cos\left(\frac{\sqrt{2}}{2} t\right)$$

$$y(t) = \frac{1}{\sqrt{2}} \, \sin\left(\frac{\sqrt{2}}{2} t\right)$$

(A) $\dfrac{x^2}{2} + 2y^2 = 1$

(B) $y = \dfrac{x}{2} \, \tan\left(\dfrac{\sqrt{2}}{2} t\right)$

(C) $2x^2 + \dfrac{y^2}{2} = 1$

(D) $x^2 + y^2 = \dfrac{3}{2} \, \cos^2\left(\dfrac{\sqrt{2}}{2} t\right)$

Problems 2–4 are based on the following equation.
$$f(x) = x^3 + 7x^2 - 5x + 6$$

2. What is most nearly the x-coordinate of the centroid of the area bounded by $y = 0$, $f(x)$, $x = 0$, and $x = 20$?

(A) 7.62
(B) 9.43
(C) 14.3
(D) 15.7

3. What is most nearly the moment of inertia about the y-axis of the area bounded by $y = 0$, $f(x)$, $x = 0$, and $x = 20$?

(A) 6.25×10^5
(B) 8.23×10^6
(C) 9.89×10^6
(D) 1.50×10^7

4. What are most nearly the x- and y-coordinates of the points where the curve intersects the function $f(x) = x^3 + 6x^2 + 1$?

(A) $(0.52, 24.3)$ and $(5.46, 227)$
(B) $(1.38, 15.1)$ and $(3.62, 127)$
(C) $(5.25, 63.4)$ and $(12.1, 2740)$
(D) $(6.48, 540)$ and $(7.56, 800)$

Problems 5 and 6 are based on the following matrices.

$$\mathbf{A} = \begin{bmatrix} 3 & 0 & 1 \\ 5 & -4 & 2 \\ -1 & 2 & 1 \end{bmatrix} \qquad \mathbf{B} = \begin{bmatrix} 7 \\ 1 \\ 2 \end{bmatrix}$$

5. What is the determinant of $\mathbf{A}$?

(A) -18
(B) -6
(C) 1
(D) 12

6. What is the matrix product of $\mathbf{A}$ and $\mathbf{B}$?

(A) $\begin{bmatrix} 7 & 12 & 4 \end{bmatrix}$

(B) $\begin{bmatrix} 23 & 35 & -3 \end{bmatrix}$

(C) $\begin{bmatrix} 7 \\ 12 \\ 4 \end{bmatrix}$

(D) $\begin{bmatrix} 23 \\ 35 \\ -3 \end{bmatrix}$

Problems 7 and 8 are based on the following test results.

no. of occurrences	value
5	0.89
10	0.942
45	0.955
20	0.996
13	1.092
10	1.15
7	1.24

7. What is most nearly the sample standard deviation?

(A) 0.091
(B) 0.098
(C) 0.198
(D) 0.320

8. What is most nearly the sample variance?

(A) 0.0084
(B) 0.039
(C) 0.103
(D) 0.127

9. The least squares method is used to plot a straight line through the data points $(5, -5)$, $(3, -2)$, $(2, 3)$, and $(-1, 7)$. The correlation coefficient is most nearly

(A) -0.80
(B) -0.88
(C) -0.92
(D) -0.97

10. Testing has shown that, on the average, 3% of the bearings produced at a factory are defective. 12 bearings are chosen at random. The probability that exactly two of them are defective is most nearly

(A) 0.037
(B) 0.044
(C) 0.059
(D) 0.066

11. The best curve of the form $y = a + b\sqrt{x}$ is fitted to the (x, y) points $(2.0, 8.8)$, $(3.1, 9.5)$, $(5.5, 10.8)$, and $(6.8, 11.3)$. The value of b is most nearly

(A) -0.015
(B) 0.48
(C) 0.52
(D) 2.1

12. A sample of wastewater is diluted 10 times. The diluted solution has an ultimate biochemical oxygen demand, BOD, L_d, of 30 mg/L and a reaction rate constant, K, of 0.20/d. The values of ultimate BOD and the reaction rate constant of the original wastewater are most nearly

(A) $L = 3.0$ mg/L, $K = 2.00$/d
(B) $L = 30$ mg/L, $K = 0.20$/d
(C) $L = 300$ mg/L, $K = 2.0$/d
(D) $L = 300$ mg/L, $K = 0.20$/d

13. The concentration of carbon monoxide, CO, in a landfill is 0.90 ppm at 25°C and 1 atm pressure. The CO concentration in mg/m^3 is most nearly

(A) 0.82 mg/m^3
(B) 0.90 mg/m^3
(C) 1.00 mg/m^3
(D) 1.03 mg/m^3

14. An analysis of breathing zone air in a workroom shows that butane and pentane reach measured average 8-hour concentrations (MAC-TWAs) of 810 ppm and 590 ppm, respectively. The TLV-TWA (threshold limit values – time-weighted averages) for butane and pentane are 800 and 600 ppm, respectively. To bring the workroom air to safe standards, the engineer should

(A) reduce the 8-hour butane average to 800 ppm and leave the pentane at 590 ppm
(B) reduce the butane to 800 ppm and allow the pentane to increase to 600 ppm
(C) reduce the butane to 400 ppm and the pentane to 400 ppm
(D) reduce the butane to 200 ppm and the pentane to 200 ppm

Problems 15–17 are based on the following information.

A trucking company considers buying either a new diesel engine or a second-hand gasoline engine for a route that is to last 10 years. The company estimates 5000 km per year will be driven. The minimum attractive rate of return for the company is 6%.

	diesel	gasoline
initial cost	$10,000	$5000
salvage value after 10 years	$1000	0
annual fuel cost	10¢/km	15¢/km
annual repair cost	$150	$350

15. For both engines to be equivalent investments, what distance would have to be driven each year?

(A) 2000 km/yr
(B) 4500 km/yr
(C) 5000 km/yr
(D) 8100 km/yr

16. The diesel engine is purchased by the company, and an employee offers to buy it for the book value at the end of 5 years. Straight-line depreciation is used. If the employee's account earns 5%, most nearly how much should the employee put into a savings account now in order to have enough money to buy the diesel engine in 5 years?

(A) $4300
(B) $5000
(C) $5500
(D) $7800

17. What is most nearly the capitalized cost of the gasoline engine?

    (A) $18,300
    (B) $23,300
    (C) $26,700
    (D) $29,700

18. Warehouse A with a life of 10 years can be constructed now for $100,000, with no repair costs and a salvage value of $10,000. Alternatively, warehouse B with a life of 12 years can be constructed for $70,000 now with a salvage value of $5000, but requires $18,000 worth of repairs every 3 years. Both have equal usefulness and are needed indefinitely. Assuming the cost of money is 6%, which warehouse is a better deal and by most nearly how much per year?

    (A) A, by $140 per year
    (B) A, by $190 per year
    (C) B, by $190 per year
    (D) A, by $880 per year

19. A sum of $110,000 will be needed for factory renovation in 7 years. To generate this amount, a sinking fund is established, into which four equal payments will be made, one at the end of each of the first 4 years. After the fourth year, no further payments will be made. If an interest rate of 9% can be expected, the amount that must be paid into the fund each year is most nearly

    (A) $18,600
    (B) $19,500
    (C) $21,200
    (D) $25,200

20. A woman invests $1500 at the start of each year for 9 years. The invested money earns 6% interest. At the end of the ninth year and after nine payments, the value of her investment is most nearly

    (A) $16,800
    (B) $18,300
    (C) $19,800
    (D) $20,800

21. A cube-shaped space truss is loaded as shown. What is most nearly the force in member AH?

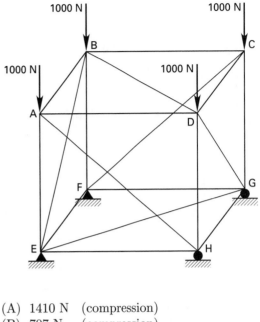

    (A) 1410 N   (compression)
    (B) 707 N   (compression)
    (C) 354 N   (compression)
    (D) 0 N

22. A 1 kg block rests on the centerline of a rough ramp as shown. The pulleys at points A and B are frictionless, and the lengths of line segments AC and BC are 4.5 m each. Most nearly what frictional force between the block and ramp is necessary to maintain equilibrium?

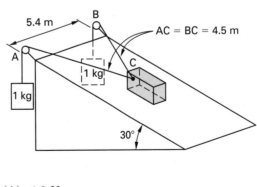

    (A) 4.9 N
    (B) 8.2 N
    (C) 9.1 N
    (D) 11 N

23. For the pinned arch and loading shown, what is most nearly the horizontal reaction at point A?

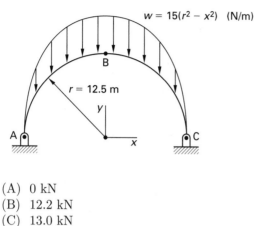

$w = 15(r^2 - x^2)$ (N/m)

$r = 12.5$ m

(A) 0 kN
(B) 12.2 kN
(C) 13.0 kN
(D) 19.5 kN

24. A 100 kg cylindrical roller is balanced on top of two 50 kg half-cylinders as shown. Friction between the cylinders is negligible. The radii of all of the cylinders are the same. The minimum coefficient of friction between the half-cylinders and the horizontal surface necessary to maintain equilibrium is most nearly

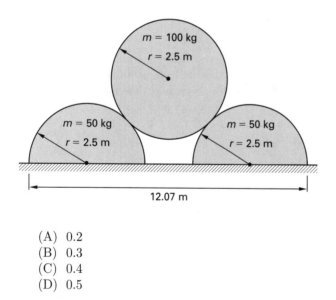

$m = 100$ kg
$r = 2.5$ m

$m = 50$ kg
$r = 2.5$ m

$m = 50$ kg
$r = 2.5$ m

12.07 m

(A) 0.2
(B) 0.3
(C) 0.4
(D) 0.5

25. A 5 m long uniform pipe with a total mass of 100 kg is hinged at the bottom and held in position by a rope as shown. The coefficient of friction between the rope and the fixed drum at point C is 0.2. The minimum number of revolutions that the rope must make around the drum is most nearly

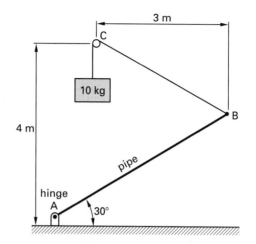

3 m

C

10 kg

4 m

B

pipe

hinge
A

30°

(A) 0.3 rev
(B) 1.3 rev
(C) 2.3 rev
(D) 3.3 rev

26. A nonlinear girder supports a cantilever load at its tip of 10 kN. The girder's depth changes parabolically (i.e., $x = f(y^2)$) from 3 m at the built-in end to 1 m at the tip. The girder is uniformly 1 m wide at all points. The average density of the girder is 2400 kg/m^3. The moment at the support is most nearly

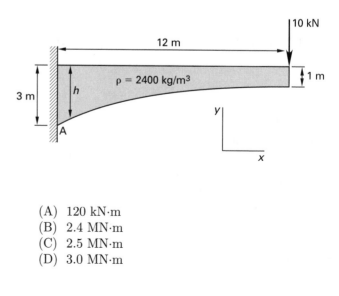

10 kN

12 m

$\rho = 2400$ kg/m^3

1 m

3 m

h

A

y

x

(A) 120 kN·m
(B) 2.4 MN·m
(C) 2.5 MN·m
(D) 3.0 MN·m

27. What is most nearly the moment at point O?

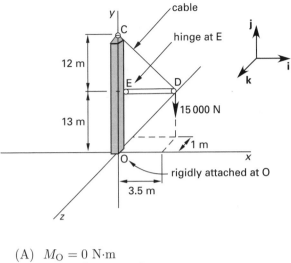

(A) $M_O = 0$ N·m
(B) $M_O = (5250$ N·m$)\mathbf{k}$
(C) $M_O = (46\,250$ N·m$)\mathbf{i} + (25\,703$ N·m$)\mathbf{k}$
(D) $M_O = (15\,000$ N·m$)\mathbf{i} + (52\,500$ N·m$)\mathbf{k}$

28. A 2 kg homogeneous triangular plate rests against a frictionless wall as shown. Determine the minimum force, P, necessary to prevent the triangle from sliding down the wall.

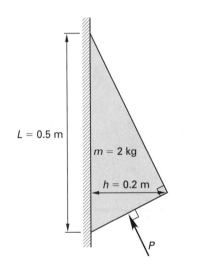

(A) 17 N
(B) 22 N
(C) 27 N
(D) 36 N

29. The aluminum beam (modulus of elasticity = 69 GPa) shown is simply supported at points A and B. The weight of the beam can be neglected.

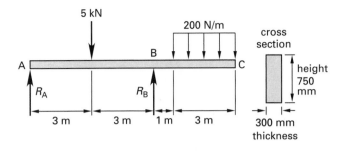

Which of the following illustrations best represents the shear diagram for the beam?

(A)

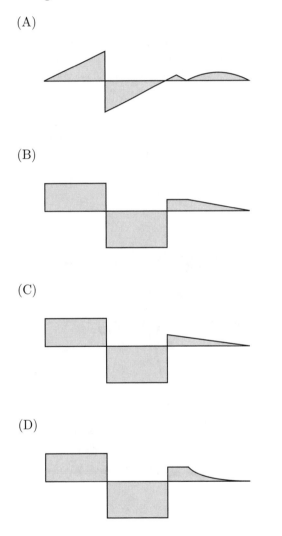

(B)

(C)

(D)

30. A pure-aluminum bar with rectangular cross section has the dimensions shown. In this particular form, aluminum metal atoms have an atomic radius of 1.43 Å, and the solid takes on a face-centered cubic crystal structure. Assuming the atoms are hard-packed spheres, approximately how many aluminum atoms are contained in the bar?

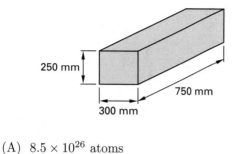

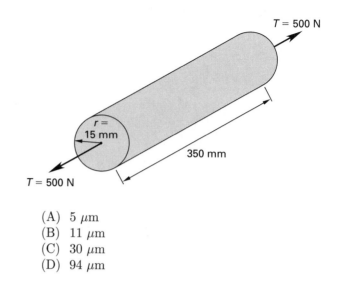

(A) 8.5×10^{26} atoms
(B) 3.4×10^{27} atoms
(C) 4.6×10^{27} atoms
(D) 6.2×10^{27} atoms

31. The stress-strain curve for an anisotropic elastic-plastic material is as shown. A 2 cm diameter, initially unstressed cylindrical specimen of the material is loaded gradually in tension to 32 MN. The specimen experiences the following loading cycle: $0 \longrightarrow 32$ MN $\longrightarrow 0 \longrightarrow 32.8$ MN $\longrightarrow 0 \longrightarrow 32$ MN $\longrightarrow 0$. Most nearly how much permanent set remains in the specimen at the conclusion of the loading cycle?

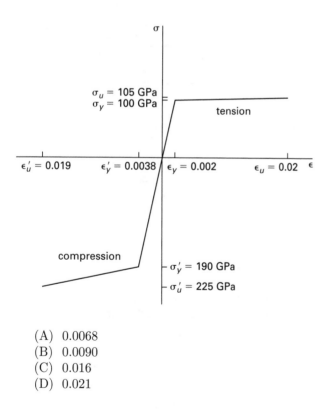

(A) 0.0068
(B) 0.0090
(C) 0.016
(D) 0.021

32. The fiberglass cylindrical test specimen shown is composed of 70% glass and 30% polyester by volume. The moduli of elasticity of glass and polyester are 70 GPa and 2.75 GPa, respectively. The elongation of the specimen as a result of the applied loading is most nearly

(A) $5 \ \mu m$
(B) $11 \ \mu m$
(C) $30 \ \mu m$
(D) $94 \ \mu m$

33. The following illustration shows the compressive strength of concrete for various water-cement ratios (W/C).

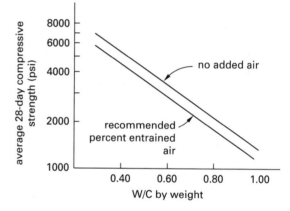

For non-air-entrained concrete with a water-cement ratio of 0.20, the average 28-day compressive strength is most nearly

(A) 4500 psi
(B) 5500 psi
(C) 6000 psi
(D) 7000 psi

34. Which of the following is not a common additive used to modify the properties of polymers?

(A) aldehydes
(B) halogenated paraffins
(C) hydroxyls
(D) phenols

35. A 4 m long cylindrical metal bar is suspended vertically from one end. The metal has a density of 2500 kg/m³ and modulus of elasticity of 210 GPa. The bar's cross-sectional area is 150 cm^2. What is the elongation of the bar due to its own mass?

(A) 1.87×10^{-10} m
(B) 9.34×10^{-9} m
(C) 1.87×10^{-8} m
(D) 9.34×10^{-7} m

Problems 36–39 are based on the following information and illustration.

Water at 10°C is pumped through smooth steel pipes from tank 1 to tank 2. The discharge rate is 0.1 m^3/min. The kinematic viscosity of water at 10°C is 1.31×10^{-6} m^2/s.

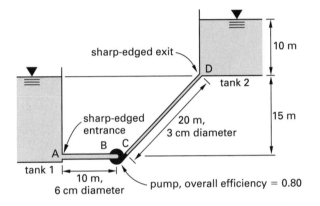

36. What is most nearly the ratio of the water velocity in the 3 cm pipe to the velocity in the 6 cm pipe?

(A) 1:1
(B) 2:1
(C) 4:1
(D) 8:1

37. What is most nearly the total friction head loss in the pipes between points A and D?

(A) 4.7 m
(B) 5.3 m
(C) 9.6 m
(D) 11 m

38. If at some operating point, the friction head loss in the pipes between points A and D is 10 m, most nearly what head does the pump add?

(A) 15 m
(B) 20 m
(C) 25 m
(D) 30 m

39. If the pump adds 15 m of head while pumping 0.1 m^3/min of 10°C water, what is most nearly the pump's brake power?

(A) 32 W
(B) 250 W
(C) 310 W
(D) 19 kW

40. Determine the approximate fluid flow friction factor for a Reynolds number, Re, of 400,000 and a relative roughness, ε/D, of 0.004.

(A) 0.015
(B) 0.017
(C) 0.020
(D) 0.028

41. A rectangular channel on a 0.002 slope is constructed of finished concrete with a roughness coefficient of 0.012. The channel is 2.4 m wide. Water flows at a depth of 1.5 m. The flow rate is most nearly

(A) 6.9 m^3/s
(B) 7.4 m^3/s
(C) 10 m^3/s
(D) 15 m^3/s

Problems 42 and 43 are based on the following information.

A pipe draws water from the bottom of a reservoir and discharges it freely at point C, 30 m below the surface. The flow is frictionless.

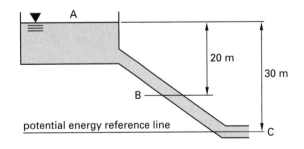

42. The total specific energy at an elevation 20 m below the water surface (i.e., at point B) is most nearly

(A) 98 J/kg
(B) 200 J/kg
(C) 290 J/kg
(D) 490 J/kg

43. The velocity at point C is most nearly

(A) 14 m/s
(B) 20 m/s
(C) 24 m/s
(D) 31 m/s

44. Water at 10°C is pumped through 300 m of steel pipe (inside diameter of 84.45 mm) with a friction factor of 0.0195 at a velocity of 2.3 m/s. The friction loss is most nearly

(A) 2 m
(B) 8.6 m
(C) 19 m
(D) 24 m

Problems 45–48 are based on the following circuit.

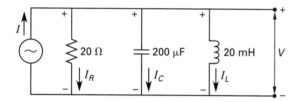

45. What is most nearly the circuit's resonant frequency?

(A) 0.5 rad/s
(B) 2 rad/s
(C) 20 rad/s
(D) 500 rad/s

46. What is most nearly the impedance seen by the current source at resonance?

(A) 0 Ω
(B) 10 Ω
(C) 20 Ω
(D) 40 Ω

47. What is most nearly the power consumed at resonance if $I_{max} = 2$ A?

(A) 0 W
(B) 10 W
(C) 20 W
(D) 40 W

48. If the effective (rms) source current is 150 mA and the angular frequency is 1000 rad/s, the effective (rms) voltage at the circuit terminals is most nearly

(A) 0.6 V
(B) 1 V
(C) 6.3 V
(D) 7.4 V

49. A DC motor is producing a torque of 300 N·m. The current of the armature is halved. What is the approximate new torque? Ignore armature resistance.

(A) 150 N·m
(B) 200 N·m
(C) 300 N·m
(D) 350 N·m

50. In the circuit shown, $V_0 = 0.0500$ V, $R_1 = R_2 = R_3 = R = 10.00$ kΩ, $R_4 = 10.25$ kΩ.

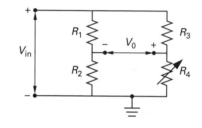

The voltage V_{in} is most nearly

(A) 2 V
(B) 4 V
(C) 5 V
(D) 8 V

51. A circuit is driven by a 60 Hz voltage source. What is most nearly the admittance seen by the voltage source?

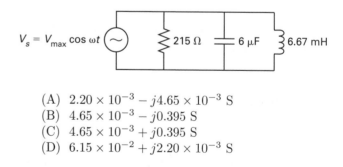

(A) $2.20 \times 10^{-3} - j4.65 \times 10^{-3}$ S
(B) $4.65 \times 10^{-3} - j0.395$ S
(C) $4.65 \times 10^{-3} + j0.395$ S
(D) $6.15 \times 10^{-2} + j2.20 \times 10^{-3}$ S

Problems 52–57 refer to the following information and illustration.

A simple Rankine cycle operates between 20°C and 100°C and uses water as the working fluid. The turbine has an isentropic efficiency of 80%, and the pump has an isentropic efficiency of 65%. The steam leaving the boiler and the water leaving the condenser are both saturated.

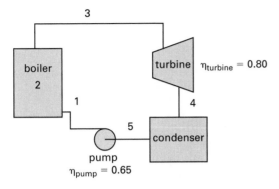

The properties of steam at the two temperatures are as follows.

For a saturation temperature of 20°C,

pressure:	2.339 kPa
specific volume:	$v_f = 0.001002 \text{ m}^3/\text{kg}$
	$v_g = 57.79 \text{ m}^3/\text{kg}$
enthalpy:	$h_f = 83.95 \text{ kJ/kg}$
	$h_{fg} = 2454.1 \text{ kJ/kg}$
	$h_g = 2538.1 \text{ kJ/kg}$
entropy:	$s_f = 0.2966 \text{ kJ/kg·K}$
	$s_g = 8.6672 \text{ kJ/kg·K}$

For a saturation temperature of 100°C,

pressure:	101.35 kPa
specific volume:	$v_f = 0.001044 \text{ m}^3/\text{kg}$
	$v_g = 1.6729 \text{ m}^3/\text{kg}$
enthalpy:	$h_f = 419.04 \text{ kJ/kg}$
	$h_{fg} = 2257.0 \text{ kJ/kg}$
	$h_g = 2676.1 \text{ kJ/kg}$
entropy:	$s_f = 1.3069 \text{ kJ/kg·K}$
	$s_g = 7.3549 \text{ kJ/kg·K}$

52. What is most nearly the turbine work output per kilogram of steam circulated?

(A) 150 kJ/kg
(B) 240 kJ/kg
(C) 420 kJ/kg
(D) 510 kJ/kg

53. Most nearly what work is done by the pump per kilogram of steam circulated?

(A) 0.153 kJ/kg
(B) 2.54 kJ/kg
(C) 6.24 kJ/kg
(D) 9.48 kJ/kg

54. Most nearly what energy is put into each kilogram of steam in the boiler?

(A) 230 kJ/kg
(B) 510 kJ/kg
(C) 1280 kJ/kg
(D) 2590 kJ/kg

55. What is most nearly the energy removed from each kilogram of steam in the condenser?

(A) 84 kJ/kg
(B) 420 kJ/kg
(C) 1500 kJ/kg
(D) 2100 kJ/kg

56. What is most nearly the thermal efficiency of the cycle?

(A) 16%
(B) 20%
(C) 77%
(D) 80%

57. What is most nearly the maximum theoretical efficiency of a power cycle operating between these temperatures?

(A) 21%
(B) 32%
(C) 60%
(D) 80%

58. What is most nearly the theoretical power required for the compression of 25 m³ of water per minute from 100 kPa to 850 kPa?

(A) 26 kW
(B) 30 kW
(C) 150 kW
(D) 310 kW

59. If the ratio of mass of vapor to liquid in a mixture is 0.8, the quality of the mixture is most nearly

(A) 0.20
(B) 0.25
(C) 0.44
(D) 0.80

60. The latent heat of fusion for ice is 333 kJ/kg. The power required to drive a refrigeration system with a coefficient of performance of 5 to remove the heat equivalent of 150 kg of ice per day is most nearly

    (A) 82 W
    (B) 115 W
    (C) 140 W
    (D) 165 W

Solutions for Sample Examination

AFTERNOON SECTION

1. **A** (B) (C) (D)
2. (A) (B) (C) **D**
3. (A) (B) (C) **D**
4. (A) **B** (C) (D)
5. **A** (B) (C) (D)
6. (A) (B) (C) **D**
7. **A** (B) (C) (D)
8. **A** (B) (C) (D)
9. (A) (B) (C) **D**
10. (A) **B** (C) (D)
11. (A) (B) (C) **D**
12. (A) (B) (C) **D**
13. (A) (B) (C) **D**
14. (A) (B) (C) **D**
15. (A) (B) (C) **D**

16. **A** (B) (C) (D)
17. (A) (B) (C) **D**
18. (A) (B) (C) **D**
19. **A** (B) (C) (D)
20. (A) **B** (C) (D)
21. (A) (B) (C) **D**
22. (A) (B) (C) **D**
23. (A) **B** (C) (D)
24. (A) (B) (C) **D**
25. (A) **B** (C) (D)
26. (A) (B) **C** (D)
27. (A) (B) (C) **D**
28. (A) **B** (C) (D)
29. (A) **B** (C) (D)
30. (A) **B** (C) (D)

31. (A) (B) **C** (D)
32. **A** (B) (C) (D)
33. (A) (B) (C) **D**
34. (A) (B) **C** (D)
35. (A) (B) (C) **D**
36. (A) (B) **C** (D)
37. (A) **B** (C) (D)
38. (A) **B** (C) (D)
39. (A) (B) **C** (D)
40. (A) (B) (C) **D**
41. (A) (B) **C** (D)
42. (A) (B) **C** (D)
43. (A) (B) **C** (D)
44. (A) (B) **C** (D)
45. (A) (B) (C) **D**

46. (A) (B) **C** (D)
47. (A) (B) (C) **D**
48. (A) **B** (C) (D)
49. **A** (B) (C) (D)
50. (A) (B) (C) **D**
51. (A) **B** (C) (D)
52. (A) (B) **C** (D)
53. **A** (B) (C) (D)
54. (A) (B) (C) **D**
55. (A) (B) (C) **D**
56. **A** (B) (C) (D)
57. **A** (B) (C) (D)
58. (A) (B) (C) **D**
59. (A) (B) **C** (D)
60. (A) **B** (C) (D)

1.
$$\frac{x}{\sqrt{2}} = \cos\left(\frac{\sqrt{2}}{2}\,t\right)$$

$$\frac{x^2}{2} = \cos^2\left(\frac{\sqrt{2}}{2}\,t\right)$$

$$\sqrt{2}\,y = \sin\left(\frac{\sqrt{2}}{2}\,t\right)$$

$$2y^2 = \sin^2\left(\frac{\sqrt{2}}{2}\,t\right)$$

Use the trigonometric relationship.

$$\sin^2\theta + \cos^2\theta = 1$$

$$\frac{x^2}{2} + 2y^2 = 1$$

Answer is A.

2.
$$x_c = \frac{\int x\,dA}{A} = \frac{\int xf(x)\,dx}{A}$$

$$\int xf(x)\,dx = \int_0^{20}(x^4 + 7x^3 - 5x^2 + 6x)\,dx$$

$$= \frac{x^5}{5} + \frac{7x^4}{4} - \frac{5x^3}{3} + \frac{6x^2}{2}\,\Big|_0^{20}$$

$$= 907\,867$$

$$A = 57\,787$$

$$x_c = \frac{907\,867}{57\,787}$$

$$= 15.71 \quad (15.7)$$

Answer is D.

3.
$$I_y = \int x^2\,dA = \int x^2 f(x)\,dx$$

$$= \int (x^5 + 7x^4 - 5x^3 + 6x^2)\,dx$$

$$= \frac{x^6}{6} + \frac{7x^5}{5} - \frac{5x^4}{4} + \frac{6x^3}{3}\,\Big|_0^{20}$$

$$= 1.5 \times 10^7$$

Answer is D.

4.
$$x^3 + 7x^2 - 5x + 6 = x^3 + 6x^2 + 1$$
$$x^2 - 5x + 5 = 0$$

Use the quadratic equation.

$$x_1, x_2 = \frac{-(-5) \pm \sqrt{(-5)^2 - (4)(1)(5)}}{(2)(1)}$$

$$= \frac{5 \pm \sqrt{5}}{2}$$

$$= 3.618, 1.382$$

$$f(x_1) = (3.618)^3 + (6)(3.618)^2 + 1$$
$$= 126.9$$

$$f(x_2) = (1.382)^3 + (6)(1.382)^2 + 1$$
$$= 15.1$$

The two intersection points are at

$$(1.38, 15.1) \text{ and } (3.62, 127)$$

Answer is B.

5.
$$\begin{vmatrix} 3 & 0 & 1 \\ 5 & -4 & 2 \\ -1 & 2 & 1 \end{vmatrix}$$

$$= 3\begin{vmatrix} -4 & 2 \\ 2 & 1 \end{vmatrix} - 0\begin{vmatrix} 5 & 2 \\ -1 & 1 \end{vmatrix} + 1\begin{vmatrix} 5 & -4 \\ -1 & 2 \end{vmatrix}$$

$$= (3)((-4)(1) - (2)(2)) - 0$$
$$\quad + (1)((5)(2) - (-4)(-1))$$

$$= -18$$

Answer is A.

6. $\mathbf{A \cdot B} = \begin{bmatrix} (3)(7) & + & (0)(1) & + & (1)(2) \\ (5)(7) & + & (-4)(1) & + & (2)(2) \\ (-1)(7) & + & (2)(1) & + & (1)(2) \end{bmatrix}$

$$= \begin{bmatrix} 23 \\ 35 \\ -3 \end{bmatrix}$$

Answer is D.

7.
$$s = \sqrt{\frac{\sum(x_i - \bar{x})^2}{n - 1}}$$

x_i	$(x_i - \bar{x})^2$	n	$\sum(x_i - \bar{x})^2$
0.89	0.0144	5	0.0720
0.942	0.00462	10	0.0462
0.955	0.00303	45	0.13613
0.996	0.000196	20	0.00392
1.092	0.00672	13	0.08741
1.15	0.0196	10	0.1960
1.24	0.0529	7	0.3703
	total:	110	0.912

$$s = \sqrt{\frac{0.912}{110 - 1}}$$
$$= 0.0915 \quad (0.091)$$

Answer is A.

8. The sample variance is

$$s^2 = (0.0915)^2$$
$$= 0.00837 \quad (0.0084)$$

Answer is A.

9. First calculate the following values.

$$\sum x_i = 5 + 3 + 2 + (-1)$$
$$= 9$$
$$\sum y_i = (-5) + (-2) + 3 + 7$$
$$= 3$$
$$\sum x_i^2 = (5)^2 + (3)^2 + (2)^2 + (-1)^2$$
$$= 39$$
$$\sum y_i^2 = (-5)^2 + (-2)^2 + (3)^2 + (7)^2$$
$$= 87$$
$$\sum x_i y_i = (5)(-5) + (3)(-2) + (2)(3) + (-1)(7)$$
$$= -32$$

The correlation coefficient is

$$r = \frac{\sum x_i y_i - \frac{(\sum x_i)(\sum y_i)}{n}}{\sqrt{\left(\sum x_i^2 - \frac{(\sum x_i)^2}{n}\right)\left(\sum y_i^2 - \frac{(\sum y_i)^2}{n}\right)}}$$

$$= \frac{-32 - \frac{(9)(3)}{4}}{\sqrt{\left(39 - \frac{(9)^2}{4}\right)\left(87 - \frac{(3)^2}{4}\right)}}$$

$$= -0.972 \quad (-0.97)$$

Answer is D.

10. Use the binomial distribution.

$$p = 0.03$$
$$q = 1 - p = 0.97$$

$$p\{2\} = \left(\frac{n!}{x!\,(n-x)!}\right) p^x q^{n-x}$$

$$= \left(\frac{12!}{(2!)\,(12-2)!}\right)(0.03)^2\,(0.97)^{12-2}$$

$$= 0.0438 \quad (0.044)$$

Answer is B.

11. Obtain a linear relationship by letting z equal $\sqrt{x}$. Then plot a straight line of the form $y = a + bz$ through the points $(1.414, 8.8)$, $(1.761, 9.5)$, $(2.345, 10.8)$, and $(2.608, 11.3)$.

Calculate the following values.

$$\sum z_i = 1.414 + 1.761 + 2.345 + 2.608$$
$$= 8.128$$
$$\sum y_i = 8.8 + 9.5 + 10.8 + 11.3$$
$$= 40.4$$
$$\sum z_i^2 = (1.414)^2 + (1.761)^2 + (2.345)^2 + (2.608)^2$$
$$= 17.40$$
$$\sum z_i y_i = (1.414)(8.8) + (1.761)(9.5) + (2.345)(10.8)$$
$$\quad\quad + (2.608)(11.3)$$
$$= 83.97$$

The slope of the line $y = a + bz$ is

$$b = \frac{\sum z_i y_i - \frac{(\sum z_i)(\sum y_i)}{n}}{\sum z_i^2 - \frac{(\sum z_i)^2}{n}}$$

$$= \frac{83.97 - \frac{(8.128)(40.4)}{4}}{17.40 - \frac{(8.128)^2}{4}}$$

$$= 2.124 \quad (2.1)$$

Answer is D.

12. The ultimate BOD of the original solution is 10 times that of the diluted solution.

$$L = 10 L_d$$
$$= (10)\left(30\ \frac{\text{mg}}{\text{L}}\right)$$
$$= 300\ \text{mg/L}$$

The value of the reaction rate constant does not change with dilution.

$$K = 0.20/\text{d}$$

Answer is D.

13.
$$\mathrm{MW_{CO}} = 12\ \frac{\mathrm{g}}{\mathrm{mol}} + 16\ \frac{\mathrm{g}}{\mathrm{mol}}$$
$$= 28\ \mathrm{g/mol}$$
$$T = 25^\circ\mathrm{C} + 273 = 298\mathrm{K}$$
$$p = 1.0\ \mathrm{atm}$$

The concentration in $\mathrm{mg/m^3}$ is

$$C_{\mathrm{CO,mg/m^3}} = \left(\frac{C_{\mathrm{CO,ppm}}(\mathrm{MW})}{22.414\ \dfrac{\mathrm{L}}{\mathrm{mol}}}\right)\left(\frac{273\mathrm{K}}{T}\right)p$$

$$= (0.9\ \mathrm{ppm})\left(28\ \frac{\mathrm{g}}{\mathrm{mol}}\right)\left(\frac{1\ \mathrm{mol}}{22.414\ \mathrm{L}}\right)$$

$$\times \left(\frac{273\mathrm{K}}{298\mathrm{K}}\right)(1.0\ \mathrm{atm})$$

$$= \frac{\left(1000\ \dfrac{\mathrm{mg}}{\mathrm{g}}\right)\left(1000\ \dfrac{\mathrm{L}}{\mathrm{m^3}}\right)}{10^6}$$

$$= 1.03\ \mathrm{mg/m^3}$$

Answer is D.

14. To be considered safe, the workroom air must meet two conditions. The average 8-hour concentrations, E, of both butane and pentane must be below their respective TLV-TWAs and the sum of the MAC-TWA to TLV-TWA ratios must be less than 1.00.

$$\frac{E_{\mathrm{butane}}}{(\mathrm{TLV\text{-}TWA})_{\mathrm{butane}}} + \frac{E_{\mathrm{pentane}}}{(\mathrm{TLV\text{-}TWA})_{\mathrm{pentane}}} < 1.00$$

In all options, the 8-hour averages for butane and pentane are at or below their TLV-TWAs.

Sum the MAC-TWA/TLV-TWA ratios for the two gases for all of the options.

A. $\left(\dfrac{800\ \mathrm{ppm}}{800\ \mathrm{ppm}}\right) + \left(\dfrac{590\ \mathrm{ppm}}{600\ \mathrm{ppm}}\right) = 1.99$

B. $\left(\dfrac{800\ \mathrm{ppm}}{800\ \mathrm{ppm}}\right) + \left(\dfrac{600\ \mathrm{ppm}}{600\ \mathrm{ppm}}\right) = 2.00$

C. $\left(\dfrac{400\ \mathrm{ppm}}{800\ \mathrm{ppm}}\right) + \left(\dfrac{400\ \mathrm{ppm}}{600\ \mathrm{ppm}}\right) = 1.17$

D. $\left(\dfrac{200\ \mathrm{ppm}}{800\ \mathrm{ppm}}\right) + \left(\dfrac{200\ \mathrm{ppm}}{600\ \mathrm{ppm}}\right) = 0.53$

Options A, B, and C all have sums greater than 1.00. The sum for option D is less than 1.00.

Answer is D.

15. At the break-even point,

$$(\mathrm{EUAC})_{\mathrm{gas}} = (\mathrm{EUAC})_{\mathrm{diesel}}$$

Let x = number of kilometers driven each year.

$$\$0.15x + (\$5000)(0.1359) + (\$350)$$
$$= \$0.10x + \$150$$
$$+ (\$10{,}000)(0.1359)$$
$$- (\$1000)(0.0759)$$
$$\$0.15x + \$1029.50 = \$0.10x + \$1433.10$$
$$\$0.05x = \$403.60$$
$$x = 8072\ \mathrm{km/yr}\quad(8100\ \mathrm{km/yr})$$

Answer is D.

16.
$$D_j = \frac{C - S_n}{n}$$
$$= \frac{\$10{,}000 - \$1000}{10\ \mathrm{yr}}$$
$$= \$900/\mathrm{yr}$$

After five years,

$$\sum D_j = (5\ \mathrm{yr})\left(900\ \frac{\$}{\mathrm{yr}}\right)$$
$$= \$4500$$
$$BV = \$10{,}000 - \$4500$$
$$= \$5500$$

The amount to be put into savings is

$$P = (\$5500)(P/F, 5\%, 5)$$
$$= (\$5500)(1 + 0.05)^{-5}$$
$$= (\$5500)(0.7835)$$
$$= \$4309\quad(\$4300)$$

Answer is A.

17. First, find the EUAC of the gasoline engine.

$$(\mathrm{EUAC})_{\mathrm{gas}} = \left(0.15\ \frac{\$}{\mathrm{km}}\right)(5000\ \mathrm{km})$$
$$+ (\$5000)(A/P, 6\%, 10) + \$350$$
$$= \$750 + (\$5000)(0.1359) + \$350$$
$$= \$1780$$

Next, find the capitalized cost.

$$\text{capitalized cost} = \frac{\mathrm{EUAC}}{i} = \frac{\$1780}{0.06}$$
$$= \$29{,}666\quad(\$29{,}700)$$

Answer is D.

18. For warehouse A,

$$\begin{aligned}
\text{EUAC} &= (\$100{,}000)(A/P, 6\%, 10) \\
&\quad - (\$10{,}000)(A/F, 6\%, 10) \\
&= (\$100{,}000)(0.1359) \\
&\quad - (\$10{,}000)(0.0759) \\
&= \$12{,}831
\end{aligned}$$

For warehouse B,

$$\begin{aligned}
\text{EUAC} &= (\$70{,}000)(A/P, 6\%, 12) \\
&\quad + ((\$18{,}000)(P/F, 6\%, 3))(A/P, 6\%, 12) \\
&\quad + ((\$18{,}000)(P/F, 6\%, 6))(A/P, 6\%, 12) \\
&\quad + ((\$18{,}000)(P/F, 6\%, 9))(A/P, 6\%, 12) \\
&\quad + ((\$18{,}000)(P/F, 6\%, 12))(A/P, 6\%, 12) \\
&\quad - (\$5000)(A/F, 6\%, 12) \\
&= (0.1193)((\$70{,}000) \\
&\quad + (\$18{,}000)(0.8396 + 0.705 + 0.5919 \\
&\quad + 0.4970)) - (\$5000)(0.0593) \\
&= \$13{,}710
\end{aligned}$$

$$\$13{,}710 - \$12{,}831 = \$879 \text{ per year} \quad (\$880/\text{yr})$$

Warehouse A is a better deal.

Answer is D.

19. The irregular cash flow can be divided into two standard cash flows. The first four years form a uniform annual payment cash flow. An equivalent single payment at the end of the fourth year can be calculated, and then the last three years can be treated as a single payment cash flow.

Work backward from the known final sum. First consider only the second cash flow. Calculate what single payment at the end of four years will give a future value of $110,000 at the end of seven years. Use the cash flow factor $(P/F, i\%, n)$, where the interest rate, i, is 9%, and the number of periods, n, is 3.

$$\begin{aligned}
(P/F, 9\%, 3) &= (1+i)^{-n} = (1 + 0.09)^{-3} \\
&= 0.7722 \\
P_{4 \text{ yr}} &= F(P/F, 9\%, 3) \\
&= (\$110{,}000)(0.7722) \\
&= \$84{,}942
\end{aligned}$$

This amount, $84,942, then becomes the future value for the uniform annual payments. In order to have $84,942 at the end of four years, the amount that must be paid into the fund at the end of each year is

$$\begin{aligned}
(A/F, 9\%, 4) &= \frac{i}{(1+i)^n - 1} = \frac{0.09}{(1 + 0.09)^4 - 1} \\
&= 0.2187 \\
A = F(A/F, 9\%, 4) &= (\$84{,}942)(0.2187) \\
&= \$18{,}576.82 \quad (\$18{,}600)
\end{aligned}$$

Answer is A.

20. Standard cash flow factors for uniform annual payments assume that payments are made at the ends of years, not the beginnings. The cash flow pattern in this situation can be seen as a standard uniform series cash flow, plus a single payment at the beginning of the first year, minus a single payment at the end of the ninth year (or beginning of the tenth). The future value of the investment after nine years is therefore

$$\begin{aligned}
F &= A(F/A, 6\%, 9) + P_{1 \text{ yr}}(F/P, 6\%, 9) - P_{10 \text{ yr}} \\
&= (\$1500)(11.4913) + (\$1500)(1.6895) - \$1500 \\
&= \$18{,}271.20 \quad (\$18{,}300)
\end{aligned}$$

Answer is B.

21.

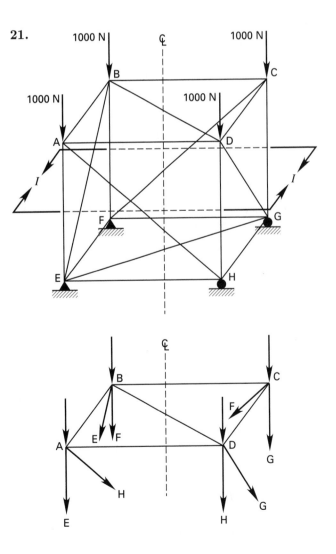

By symmetry, the force in the four diagonal members AH, BE, CF, and DG must be the same. Cut the truss along plane I-I and sum moments about the centerline. Force in members AE, DH, CG, and BF are parallel to the centerline and do not contribute to the moment.

$$\text{AH}\cos 45° + \text{BE}\cos 45°$$
$$+\text{CF}\cos 45° + \text{DG}\cos 45° = 0$$
$$4\text{AH}\cos 45° = 0$$
$$\text{AH} = 0$$

Answer is D.

22. At equilibrium, the tension in cables AC and BC will be

$$F_{\text{AC}} = F_{\text{BC}} = mg$$
$$= (1\text{ kg})\left(9.81\ \frac{\text{m}}{\text{s}^2}\right)$$
$$= 9.81\text{ N}$$

The dimension triangle that is similar to the cable force triangle has a leg of length

$$\sqrt{(4.5\text{ m})^2 - \left(\frac{5.4\text{ m}}{2}\right)^2} = 3.6\text{ m}$$

By similar triangles, the total component of forces AC and BC up the plane is

$$F_{\text{up}} = \frac{(2\text{ cables})\left(9.81\ \frac{\text{N}}{\text{cable}}\right)(3.6\text{ m})}{4.5\text{ m}}$$
$$= 15.7\text{ N}$$

The component of gravitational force on the block directed down the plane is

$$F_{\text{down}} = mg\sin\theta = (1\text{ kg})\left(9.81\ \frac{\text{m}}{\text{s}^2}\right)\sin 30°$$
$$= 4.91\text{ N}$$

The required frictional force is

$$F_f = F_{\text{up}} - F_{\text{down}}$$
$$15.7\text{ N} - 4.91\text{ N} = 10.8\text{ N}\quad(11\text{ N})$$

Answer is D.

23. The maximum loading occurs at the center of the arch where $x = 0$.

$$w = \left(15\ \frac{\text{N}}{\text{m}^3}\right)\left((12.5\text{ m})^2 - (0)^2\right)$$
$$= 2344\text{ N/m}$$

The area of a full parabola with width $2a$ and height b is $4ab/3$. By symmetry, the vertical reactions at points A and C are equal to one-half of the applied loading.

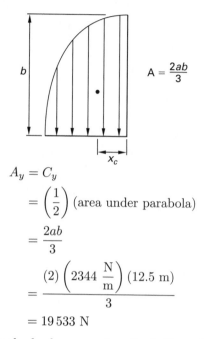

$$A_y = C_y$$
$$= \left(\frac{1}{2}\right)(\text{area under parabola})$$
$$= \frac{2ab}{3}$$
$$= \frac{(2)\left(2344\ \frac{\text{N}}{\text{m}}\right)(12.5\text{ m})}{3}$$
$$= 19\,533\text{ N}$$

For the free body shown, replace the half-parabola loading with a concentrated force of magnitude equal to the area under the half-parabola and applied at its centroid. The centroid of a semi-parabolic shape with width a and height b is located at $3a/8$ from the major axis.

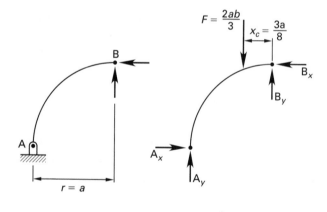

$$\frac{3a}{8} = \frac{(3)(12.5\text{ m})}{8} = 4.6875\text{ m}$$

Take moments about point B.

$$\sum M_{\text{B}} = F\left(\frac{3a}{8}\right) - A_y(12.5\text{ m}) + A_x(12.5\text{ m}) = 0$$
$$(19\,533\text{ N})(4.6875\text{ m}) - (19\,533\text{ N})(12.5\text{ m})$$
$$+ A_x(12.5\text{ m}) = 0$$
$$A_x = 12\,208\text{ N}\quad(12.2\text{ kN})$$

Answer is B.

24. The forces between the cylinders are directed between the centers of the cylindrical sections. The distance between the two centers of the half-cylinders is

$$12.07 \text{ m} - (2)(2.5 \text{ m}) = 7.07 \text{ m}$$

The angle between the horizontal and the line of force between the centers is

$$\arccos\left(\frac{\frac{7.07 \text{ m}}{2}}{5 \text{ m}}\right) = 45°$$

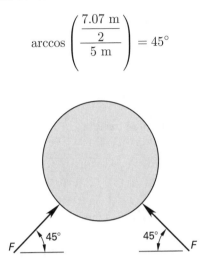

The top cylinder is supported vertically by each half-cylinder equally. Since the angle is 45°, the horizontal and vertical components of force are the same. The horizontal component of force on each half-cylinder due to the top cylinder is

$$F_x = F_y = mg$$
$$= \frac{(100 \text{ kg})\left(9.81 \frac{\text{m}}{\text{s}^2}\right)}{2}$$
$$= 490 \text{ N}$$

The normal (vertical) force on the horizontal surface from each half-cylinder is

$$N = mg + F_y$$
$$= \left(50 \text{ kg} + \frac{100 \text{ kg}}{2}\right)\left(9.81 \frac{\text{m}}{\text{s}^2}\right)$$
$$= 981 \text{ N}$$

At equilibrium, the frictional force will equal the separation force.

$$F_x - F_f = 0$$
$$F_x - \mu N = 0$$
$$490 \text{ N} - \mu(981 \text{ N}) = 0$$
$$\mu = 0.5$$

Answer is D.

25.

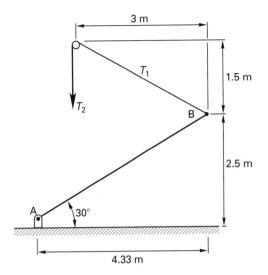

The horizontal extent of the pipe is

$$(5 \text{ m}) \cos 30° = 4.33 \text{ m}$$

The height of the tip of the pipe from the plane of the hinge is

$$(5 \text{ m}) \sin 30° = 2.5 \text{ m}$$

The rope rises another $4 \text{ m} - 2.5 \text{ m} = 1.5 \text{ m}$.

The length of the inclined portion of the rope is

$$\sqrt{(3 \text{ m})^2 + (1.5 \text{ m})^2} = 3.35 \text{ m}$$

The components of force in the rope supporting the pipe end are

$$T_{1y} = \left(\frac{1.5 \text{ m}}{3.35 \text{ m}}\right)T_1 = 0.448T_1$$
$$T_{1x} = \left(\frac{3.0 \text{ m}}{3.35 \text{ m}}\right)T_1 = 0.896T_1$$

For the purpose of taking moments, the mass of the pipe can be considered to be concentrated at the midpoint. To determine the tension in the rope supporting the pipe end, take moments about the hinge (point A).

$$\sum M_A = (100 \text{ kg})\left(9.81 \frac{\text{m}}{\text{s}^2}\right)\left(\frac{5 \text{ m}}{2}\right)\cos 30°$$
$$- (0.448T_1)(4.33 \text{ m}) - (0.896T_1)(2.5 \text{ m})$$
$$= 0$$
$$T_1 = 508 \text{ N}$$

The tension in the rope supporting the mass is

$$T_2 = mg$$
$$= (10 \text{ kg})\left(9.81 \frac{\text{m}}{\text{s}^2}\right)$$
$$= 98.1 \text{ N}$$

The angle of wrap is

$$\theta = \frac{\ln\left(\dfrac{T_1}{T_2}\right)}{\mu} = \frac{\ln\left(\dfrac{508\text{ N}}{98.1\text{ N}}\right)}{0.2}$$

$$= 8.22\text{ rad}$$

The number of revolutions of the rope is

$$\frac{8.22\text{ rad}}{2\pi\ \dfrac{\text{rad}}{\text{rev}}} = 1.31\text{ rev}\quad(1.3\text{ rev})$$

Answer is B.

26.

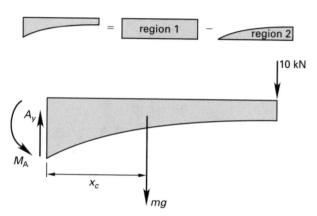

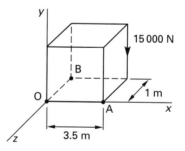

For calculating the weight of the girder and the location of its centroid, the girder can be divided into the regions shown.

For region 1 (rectangular),

$$A_1 = (3\text{ m})(12\text{ m})$$

$$= 36\text{ m}^2$$

$$x_{c1} = \frac{12\text{ m}}{2}$$

$$= 6\text{ m}$$

For region 2 (parabolic), the depth of the parabolic section at the girder tip is

$$b = 3\text{ m} - 1\text{ m}$$

$$= 2\text{ m}$$

$$A_2 = \frac{2ab}{3} = \frac{(2)(12\text{ m})(2\text{ m})}{3}$$

$$= 16\text{ m}^2$$

$$x_{c2} = \frac{3a}{5} = \frac{(3)(12\text{ m})}{5}$$

$$= 7.2\text{ m}$$

The x-coordinate of the girder centroid is

$$x_c = \frac{\sum_{c,i} A_i}{\sum A_i}$$

$$= \frac{(6\text{ m})(36\text{ m}^2) - (7.2\text{ m})(16\text{ m}^2)}{36\text{ m}^2 - 16\text{ m}^2}$$

$$= 5.04\text{ m}$$

The volume of the girder is

$$V = (A_1 - A_2)t$$

$$= (36\text{ m}^2 - 16\text{ m}^2)(1\text{ m})$$

$$= 20\text{ m}^3$$

The total vertical force (weight) of the girder is

$$W = \rho g V$$

$$= \left(2400\ \frac{\text{kg}}{\text{m}^3}\right)\left(9.81\ \frac{\text{m}}{\text{s}^2}\right)(20\text{ m}^3)$$

$$= 470\,880\text{ N}\quad(470.88\text{ kN})$$

The resultant weight of the girder can be applied at the centroid of the area. Taking moments about point A,

$$\sum M_A = M_A - (470.88\text{ kN})(5.04\text{ m}) - (10\text{ kN})(12\text{ m})$$

$$= 0$$

$$M = 2493\text{ kN}\cdot\text{m}\quad(2.5\text{ MN}\cdot\text{m})$$

Answer is C.

27. The original condition is

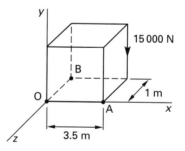

The force could be moved to position A if moment M_x was added.

$$\mathbf{M}_x = \mathbf{r} \times \mathbf{F}$$

$$= (-1\mathbf{k}\text{ m}) \times (-15\,000\mathbf{j}\text{ m})$$

$$= 15\,000\mathbf{j} \times \mathbf{k}\text{ N}\cdot\text{m}$$

$$= -15\,000\mathbf{i}\text{ N}\cdot\text{m}$$

(Use the right-hand rule to verify the sign of $\mathbf{M}_x$.)

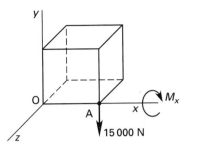

The force could be moved to position B if moment M_z was added.

$$\mathbf{M}_z = \mathbf{r} \times \mathbf{F}$$
$$= (3.5\mathbf{i} \text{ m})(-15\,000\mathbf{j} \text{ N})$$
$$= -52\,500\mathbf{i} \times \mathbf{j} \text{ N·m}$$
$$= -52\,500\mathbf{k} \text{ N·m}$$

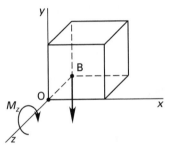

There is no moment about the y-axis.

$$M_O = (-15\,000 \text{ N·m})\mathbf{i} + 0\mathbf{j} + (-52\,500 \text{ N·m})\mathbf{k}$$

The sign of the terms will depend on the sign convention used for positive moments. Signs given here are derived from the coordinate axes.

Answer is D.

28.

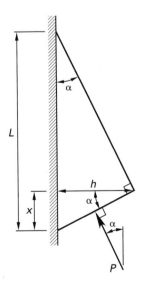

$$\tan \alpha = \frac{h}{L - x} = \frac{x}{h}$$
$$x(L - x) = h^2$$
$$x^2 - xL + h^2 = 0$$
$$x = \frac{L \pm \sqrt{L^2 - 4h^2}}{2}$$
$$= \frac{0.5 \text{ m} \pm \sqrt{(0.5 \text{ m})^2 - (4)(0.2 \text{ m})^2}}{2}$$
$$= 0.1 \text{ m}, \ 0.4 \text{ m} \quad [\text{discard } 0.4 \text{ m}]$$
$$\tan \alpha = \frac{x}{h}$$
$$\alpha = \tan^{-1}\left(\frac{x}{h}\right) = \tan^{-1}\left(\frac{0.1 \text{ m}}{0.2 \text{ m}}\right)$$
$$= 26.56°$$
$$\sum F_y = 0: P \cos \alpha - W = 0$$
$$P = \frac{W}{\cos \alpha} = \frac{mg}{\cos \alpha}$$
$$= \frac{(2 \text{ kg})\left(9.81 \frac{\text{m}}{\text{s}^2}\right)}{\cos 26.56°}$$
$$= 21.93 \text{ N} \quad (22 \text{ N})$$

Answer is B.

29. Without calculating any values, the conventions of drawing shear diagrams dictate that (1) from point A to 3 m to the right of point A, the shear is a positive constant value equal to R_A; (2) the shear is reduced by 5000 N and remains constant to point B; (3) the shear increases by R_B at point B and remains constant to 1 m to the right of point B; and (4) the shear decreases linearly by 200 N/m to point C.

The only choice that matches this shape is (B).

Answer is B.

30. The volume of each aluminum atom is

$$V = \frac{4}{3}\pi r^3$$
$$= \frac{4}{3}\pi(1.43 \times 10^{-10} \text{ m})^3$$
$$= 1.22 \times 10^{-29} \text{ m}^3$$

For a face-centered cubic crystal, there are 4 atoms per cell and the packing factor is 0.74.

The cell volume is

$$V_{\text{cell}} = \frac{\left(\dfrac{\text{no. of atoms}}{\text{cell}}\right)(\text{atomic volume})}{\text{packing factor}}$$

$$= \frac{(4)(1.22 \times 10^{-29} \text{ m}^3)}{0.74}$$

$$= 6.62 \times 10^{-29} \text{ m}^3$$

The number of cells in the bar is

$$\text{no. of cells} = \frac{V_{\text{bar}}}{V_{\text{cell}}}$$

$$= \frac{(0.25 \text{ m})(0.3 \text{ m})(0.75 \text{ m})}{6.62 \times 10^{-29} \text{ m}^3}$$

$$= 8.5 \times 10^{26} \text{ cells}$$

The number of atoms is

$$\text{no. of atoms} = (\text{no. of atoms per cell})(\text{no. of cells})$$

$$= \left(4 \, \frac{\text{atoms}}{\text{cell}}\right)(8.5 \times 10^{26} \text{ cells})$$

$$= 3.4 \times 10^{27} \text{ atoms}$$

Answer is B.

31.

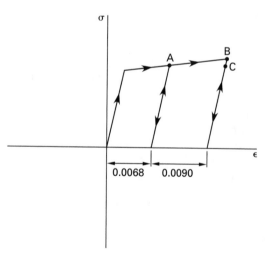

The area of the specimen is

$$A = \pi r^2 = \pi (0.01 \text{ m})^2$$

$$= 0.000314 \text{ m}^2$$

Calculate the stresses induced by the loads. For the 32 MN force, for example,

$$\sigma = \frac{P}{A} = \frac{32 \times 10^6 \text{ N}}{0.000314 \text{ m}^2}$$

$$= 1.01 \times 10^{11} \text{ Pa} \quad (101.9 \text{ GPa})$$

The other stresses are calculated similarly. The stress cycle is

$$0 \longrightarrow 101.9 \text{ GPa} \longrightarrow 0 \longrightarrow 104.4 \text{ GPa} \longrightarrow$$
$$0 \longrightarrow 101.9 \text{ GPa} \longrightarrow 0$$

The modulus of elasticity (i.e., the slope of the elastic portion of the stress-strain curve) is

$$E_1 = \frac{\sigma_y}{\epsilon_y} = \frac{100 \text{ GPa}}{0.002}$$

$$= 50\,000 \text{ GPa}$$

The plastic modulus (i.e., the slope of the plastic portion of the stress-strain curve) is

$$E_2 = \frac{\sigma_u - \sigma_y}{\epsilon_u - \epsilon_y}$$

$$= \frac{105 \text{ GPa} - 100 \text{ GPa}}{0.02 - 0.002}$$

$$= 277.8 \text{ GPa}$$

The permanent set in the material due to the first loading of the cycle is

$$\epsilon_A = \frac{\sigma - \sigma_y}{E_2}$$

$$= \frac{101.9 \text{ GPa} - 100 \text{ GPa}}{277.8 \text{ GPa}}$$

$$= 0.0068$$

The new yield stress in tension is equal to the highest stress experienced in the plastic region: 101.9 GPa.

The permanent set in the material due to the second loading of the cycle is

$$\epsilon_B = \frac{\sigma - \sigma_{y'}}{E_2'}$$

$$= \frac{104.4 \text{ GPa} - 101.9 \text{ GPa}}{277.8 \text{ GPa}}$$

$$= 0.009$$

The new yield stress in tension is equal to 104.4 GPa.

Since the stress in the sample due to the third loading of the cycle is less than the new yield stress of the material, there will be no additional permanent set in the material as a result of this loading.

The total permanent set is

$$0.0068 + 0.0090 = 0.0158 \quad (0.016)$$

Answer is C.

32. The modulus of elasticity of the composite is

$$E_c = f_g E_g + f_p E_p$$
$$= (0.70)(70 \text{ GPa}) + (0.30)(2.75 \text{ GPa})$$
$$= 49.825 \text{ GPa}$$

The elongation of the specimen is

$$\delta = \frac{PL}{AE} = \frac{PL}{\pi r^2 E_c}$$
$$= \frac{(500 \text{ N})(0.35 \text{ m})}{\pi (0.015 \text{ m})^2 (49.825 \times 10^9 \text{ Pa})}$$
$$= 4.97 \times 10^{-6} \text{ m} \quad (5 \text{ } \mu\text{m})$$

Answer is A.

33. The water-cement (W/C) ratio is the primary factor affecting the strength of concrete. For both air-entrained and non-air-entrained concrete, strength decreases with an increase in the W/C ratio. From the illustration, non-air-entrained (no added air) concrete with a W/C ratio of 0.20 has a compressive strength of 7000 psi.

Answer is D.

34. Aldehydes, halogenated paraffins, and phenols are all common polymer additives. Hydroxyls are not commonly used as a polymer additive.

Answer is C.

35. The mass of the bar is

$$\text{mass} = \rho V = \rho A L$$
$$= \left(2500 \text{ } \frac{\text{kg}}{\text{m}^3} \right) (150 \text{ cm}^2) \left(\frac{1 \text{ m}}{100 \text{ cm}} \right)^2 (4 \text{ m})$$
$$= 150 \text{ kg}$$

The total gravitational force is experienced by the metal at the suspension point. Farther down the rod there is less volume contributing to the force, and the stress is reduced. The average force on the metal in the bar is half of the maximum value.

$$F_{\text{ave}} = \left(\frac{1}{2} \right) F_{\text{max}} = \left(\frac{1}{2} \right) mg$$
$$= \left(\frac{1}{2} \right) (150 \text{ kg}) \left(9.81 \text{ } \frac{\text{m}}{\text{s}^2} \right)$$
$$= 735.75 \text{ N}$$

The elongation is

$$\Delta L = \varepsilon L_o = \left(\frac{\sigma}{E} \right) L_o$$
$$= \left(\frac{F}{AE} \right) L_o$$
$$= \frac{(735.75 \text{ N})(4 \text{ m})}{\left(\begin{array}{c} (150 \text{ cm}^2) \left(\dfrac{1 \text{ m}}{100 \text{ cm}} \right)^2 (210 \text{ GPa}) \\ \times \left(1.0 \times 10^9 \text{ } \dfrac{\text{Pa}}{\text{GPa}} \right) \end{array} \right)}$$
$$= 9.34 \times 10^{-7} \text{ m}$$

Answer is D.

36.
$$v_{3 \text{ cm}} = \frac{Q}{A_{3 \text{ cm}}}$$
$$v_{6 \text{ cm}} = \frac{Q}{A_{6 \text{ cm}}}$$
$$\frac{v_{3 \text{ cm}}}{v_{6 \text{ cm}}} = \frac{Q A_{6 \text{ cm}}}{Q A_{3 \text{ cm}}} = \frac{D_{6 \text{ cm}}^2}{D_{3 \text{ cm}}^2}$$
$$= \frac{(6 \text{ cm})^2}{(3 \text{ cm})^2} = 4$$

Answer is C.

37.
$$v_{3 \text{ cm}} = \frac{Q}{A_{3 \text{ cm}}}$$
$$= \frac{0.1 \text{ } \dfrac{\text{m}^3}{\text{min}}}{\left(\dfrac{\pi}{4} \right) (0.03 \text{ m})^2 \left(60 \text{ } \dfrac{\text{s}}{\text{min}} \right)}$$
$$= 2.358 \text{ m/s}$$

$$v_{6 \text{ cm}} = \frac{v_{3 \text{ cm}}}{4} = \frac{2.358 \text{ } \dfrac{\text{m}}{\text{s}}}{4}$$
$$= 0.5895 \text{ m/s}$$

$$\text{Re} = \frac{vD}{\nu}$$

$$\text{Re}_{3 \text{ cm}} = \frac{\left(2.358 \text{ } \dfrac{\text{m}}{\text{s}} \right) (0.03 \text{ m})}{1.31 \times 10^{-6} \text{ } \dfrac{\text{m}^2}{\text{s}}}$$
$$= 54\,000$$

$$\text{Re}_{6 \text{ cm}} = \frac{\left(0.5895 \text{ } \dfrac{\text{m}}{\text{s}} \right) (0.06 \text{ m})}{1.31 \times 10^{-6} \text{ } \dfrac{\text{m}^2}{\text{s}}}$$
$$= 27\,000$$

For smooth steel pipes, $\epsilon = 0.046$ mm. The relative roughnesses are

$$\frac{\epsilon}{D_{3\;cm}} = \frac{4.6 \times 10^{-5}\;m}{0.03\;m}$$
$$= 0.0015$$

$$\frac{\epsilon}{D_{6\;cm}} = \frac{4.6 \times 10^{-5}\;m}{0.06\;m}$$
$$= 0.0008$$

From the Moody chart,

$$f_{3\;cm} = 0.025$$

$$f_{6\;cm} = 0.026$$

$$h_{f,3\;cm} = \frac{v^2 L f}{2gD}$$
$$= \frac{\left(2.358\;\frac{m}{s}\right)^2 (20\;m)(0.025)}{(2)\left(9.81\;\frac{m}{s^2}\right)(0.03\;m)}$$
$$= 4.72\;m$$

$$h_{f,6\;cm} = \frac{\left(0.5895\;\frac{m}{s}\right)^2 (10\;m)(0.026)}{(2)\left(9.81\;\frac{m}{s^2}\right)(0.06\;m)}$$
$$= 0.0768\;m$$

For minor losses,

$$C_{3\;cm,sharp\;entrance} = 0.5 = C_{6\;cm,sharp\;entrance}$$

$$C_{3\;cm,sharp\;exit} = 1.0 = C_{6\;cm,sharp\;exit}$$

$$h_{f,total} = h_{f,3\;cm} + h_{f,6\;cm}$$
$$+ (C_{entrance,3\;cm} + C_{exit,3\;cm}) h_{v,3\;cm}$$
$$+ (C_{entrance,6\;cm} + C_{exit,6\;cm}) h_{v,6\;cm}$$
$$= 4.72\;m + 0.0768\;m$$
$$+ (0.5 + 1.0)\left(\frac{\left(2.358\;\frac{m}{s}\right)^2}{(2)\left(9.81\;\frac{m}{s^2}\right)}\right)$$
$$+ (0.5 + 1.0)\left(\frac{\left(0.5895\;\frac{m}{s}\right)^2}{(2)\left(9.81\;\frac{m}{s^2}\right)}\right)$$
$$= 5.25\;m\quad(5.3\;m)$$

Answer is B.

38. If the static pressures on either side of the pump were known, the pump head could be calculated from the change in pressure and velocity heads across the pump. However, the static pressures are not known. Therefore, it is necessary to use two locations for which all terms (static, velocity, and potential heads) are known. In this problem, the tank surfaces are the most convenient locations. The velocities and gage pressures are both zero at these points.

$$h_{pump} = \Delta h_v + \Delta h_p + \Delta h_g + h_{f,total}$$
$$= 0\;m + 0\;m + (25\;m - 15\;m) + 10\;m$$
$$= 20\;m$$

Answer is B.

39. $$W = \frac{Q\gamma h}{\eta} = \frac{Q\rho g h}{\eta}$$
$$= \frac{\left(0.1\;\frac{m^3}{min}\right)\left(1000\;\frac{kg}{m^3}\right)\left(9.81\;\frac{m}{s^2}\right)(15\;m)}{(0.8)\left(60\;\frac{s}{min}\right)}$$
$$= 306.6\;W\quad(310\;W)$$

Answer is C.

40. Using the Moody diagram, the friction factor is approximately 0.028.

Answer is D.

41. The hydraulic radius is

$$R = \frac{A}{P} = \frac{(2.4\;m)(1.5\;m)}{1.5\;m + 2.4\;m + 1.5\;m}$$
$$= 0.67\;m$$

The Manning coefficient is determined by

$$C = \left(\frac{1.00}{n}\right) R^{1/6}$$
$$= \left(\frac{1.00}{0.012}\right)(0.67\;m)^{1/6}$$
$$= 77.9$$

The discharge is

$$Q = vA = C\sqrt{RS}\,A$$
$$= \left(77.9\;\frac{\sqrt{m}}{s}\right)\left(\sqrt{(0.67\;m)(0.002)}\right)$$
$$\times (1.5\;m)(2.4\;m)$$
$$= 10.3\;m^3/s\quad(10\;m^3/s)$$

Answer is C.

42. At point A, the velocity and gage pressure are both zero. Therefore, the total energy consists only of potential energy. Point C is chosen as the reference ($z = 0$) elevation.

$$E_A = z_A g = (30 \text{ m}) \left(9.81 \, \frac{\text{m}}{\text{s}^2}\right)$$
$$= 294.3 \text{ m}^2/\text{s}^2 \quad (290 \text{ J/kg})$$

At point B, the fluid is moving and possesses kinetic energy. The fluid is also under hydrostatic pressure and possesses pressure energy. These energy forms have come at the expense of potential energy. (This is a direct result of the Bernoulli equation.) Also, the flow is frictionless. Thus, there is no net change in the total energy between points A and B.

$$E_B = E_A$$
$$= 294.3 \text{ m}^2/\text{s}^2 \quad (290 \text{ J/kg})$$

Answer is C.

43. At point C, the gage pressure and pressure energy are again zero, since the discharge is at atmospheric pressure. The potential energy is zero, since $z = 0$. The total energy of the system has been converted to kinetic energy.

$$E_t = E_p + E_v + E_z$$
$$= \frac{p}{\rho} + \frac{v^2}{2} + zg$$
$$294.3 \, \frac{\text{m}^2}{\text{s}^2} = 0 + \frac{v^2}{2} + 0$$
$$v = 24.3 \text{ m/s} \quad (24 \text{ m/s})$$

Answer is C.

44. Use the Darcy-Weisbach equation.

$$h_f = \frac{fLv^2}{2Dg}$$
$$= \frac{(0.0195)(300 \text{ m}) \left(2.3 \, \frac{\text{m}}{\text{s}}\right)^2}{(2)(84.45 \text{ mm}) \left(\dfrac{1 \text{ m}}{1000 \text{ mm}}\right) \left(9.81 \, \frac{\text{m}}{\text{s}^2}\right)}$$
$$= 18.7 \text{ m} \quad (19 \text{ m})$$

Answer is C.

45. For a parallel-RLC circuit at resonance,

$$\omega_0 = \sqrt{\frac{1}{LC}}$$
$$= \sqrt{\frac{1}{(20 \times 10^{-3} \text{ H})(200 \times 10^{-6} \text{ F})}}$$
$$= 500 \text{ rad/s}$$

Answer is D.

46. At resonance, $Z = R = 20 \, \Omega$.

Answer is C.

47. At resonance, the circuit is purely resistive and the power factor $= 1$.

$$P = \frac{1}{2} V_{\max} I_{\max} \cos\theta = \frac{1}{2} I_{\max}^2 R \cos\theta$$
$$= \left(\frac{1}{2}\right)(2 \text{ A})^2 (20 \, \Omega)(1)$$
$$= 40 \text{ W}$$

Answer is D.

48. For a parallel circuit,

$$Z = \frac{1}{\sum Y} = \frac{1}{\sqrt{\left(\dfrac{1}{R}\right)^2 + \left(\dfrac{1}{X_L} - \dfrac{1}{X_C}\right)^2}}$$
$$= \frac{1}{\sqrt{\left(\dfrac{1}{R}\right)^2 + \left(\dfrac{1}{\omega L} - \omega C\right)^2}}$$
$$= \frac{1}{\sqrt{\left(\dfrac{1}{20 \, \Omega}\right)^2 + \left(\dfrac{\dfrac{1}{\left(1000 \, \frac{\text{rad}}{\text{s}}\right)(20 \times 10^{-3} \text{ H})}}{- \left(1000 \, \frac{\text{rad}}{\text{s}}\right)(200 \times 10^{-6} \text{ F})}\right)^2}}$$
$$= 6.325 \, \Omega$$
$$V = IZ = (150 \times 10^{-3} \text{ A})(6.325 \, \Omega)$$
$$= 0.95 \text{ V} \quad (1 \text{ V})$$

Answer is B.

49. The torque is approximately proportional to the armature current (ignoring armature resistance).

$$\frac{T_{m2}}{T_{m1}} = \frac{I_{a2}}{I_{a1}}$$
$$T_{m2} = \frac{I_{a2}}{I_{a1}} T_{m1} = \left(\frac{1}{2}\right)(300 \text{ N·m})$$
$$= 150 \text{ N·m}$$

Answer is A.

50. The circuit is a Wheatstone bridge. If R_4 is close to 10 kΩ then the quarter bridge approximation can be used.

$$V_o \approx \frac{\Delta R}{4R} V_{\text{in}}$$

$$V_{\text{in}} \approx \frac{4R}{\Delta R} V_o$$

$$= \left(\frac{(4)(10.00 \times 10^3 \ \Omega)}{10.25 \times 10^3 \ \Omega - 10.00 \times 10^3 \ \Omega} \right) (0.0500 \text{ V})$$

$$= 8.00 \text{ V}$$

Answer is D.

51.
$$\omega = 2\pi f = \left(2\pi \ \frac{\text{rad}}{\text{cycle}} \right) \left(60 \ \frac{\text{cycle}}{\text{s}} \right)$$

$$= 377 \ \frac{\text{rad}}{\text{s}}$$

The admittance is the complex sum of the admittances of the parallel components.

$$Y = \frac{1}{R} + j\omega C - \frac{j}{\omega L}$$

$$= \frac{1}{215 \ \Omega} + j \left(377 \ \frac{\text{rad}}{\text{s}} \right) (6 \times 10^{-6} \text{ F})$$

$$- \frac{j}{\left(377 \ \frac{\text{rad}}{\text{s}} \right) (6.67 \times 10^{-3} \text{ H})}$$

$$= 4.65 \times 10^{-3} - j0.395 \text{ S}$$

Answer is B.

52.

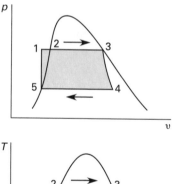

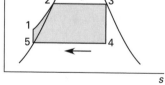

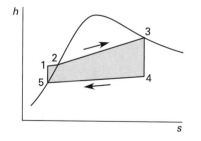

At state 3,

$$T_3 = 100°\text{C} \quad \text{[saturated]}$$

$$h_3 = h_g = 2676.1 \text{ kJ/kg}$$

$$s_3 = s_g = 7.3549 \text{ kJ/kg·K}$$

At state 4,

$$s_4 = s_3 = 7.3549 \text{ kJ/kg·K}$$

$$x = \frac{s - s_f}{s_{fg}} = \frac{s - s_f}{s_g - s_f}$$

$$= \frac{7.3549 \ \frac{\text{kJ}}{\text{kg·K}} - 0.2966 \ \frac{\text{kJ}}{\text{kg·K}}}{8.6672 \ \frac{\text{kJ}}{\text{kg·K}} - 0.2966 \ \frac{\text{kJ}}{\text{kg·K}}}$$

$$= 0.843$$

$$h_4 = h_f + x h_{fg}$$

$$= 83.96 \ \frac{\text{kJ}}{\text{kg}} + (0.843) \left(2454.1 \ \frac{\text{kJ}}{\text{kg}} \right)$$

$$= 2153 \text{ kJ/kg}$$

$$W_{\text{turbine}} = (h_3 - h_4)\eta_{\text{turbine}}$$

$$= \left(2676.1 \ \frac{\text{kJ}}{\text{kg}} - 2153 \ \frac{\text{kJ}}{\text{kg}} \right) (0.80)$$

$$= 418.5 \text{ kJ/kg} \quad (420 \text{ kJ/kg})$$

Answer is C.

53. At state 5,

$$T_5 = 20°\text{C}$$

$$h_5 = h_{f,4} = 83.95 \text{ kJ/kg} \quad \text{[saturated]}$$

$$v_5 = v_1 = 0.001002 \text{ m}^3/\text{kg}$$

$$p_5 = p_4 = 2.339 \text{ kPa}$$

For state 1,

$$p_1 = p_2 = 101.35 \text{ kPa}$$

$$W_{\text{pump}} = \frac{v_5(p_1 - p_5)}{\eta_{\text{pump}}}$$

$$= \frac{\left(0.001002 \ \frac{\text{m}^3}{\text{kg}} \right) (101.35 \text{ kPa} - 2.339 \text{ kPa})}{0.65}$$

$$= 0.153 \text{ kJ/kg}$$

Answer is A.

54. For state 1,

$$h_1 = W_{\text{pump,in}} + h_5 = 0.153 \ \frac{\text{kJ}}{\text{kg}} + 83.95 \ \frac{\text{kJ}}{\text{kg}}$$

$$= 84.103 \text{ kJ/kg}$$

$$Q_{\text{in}} = h_3 - h_1 = 2676.1 \,\frac{\text{kJ}}{\text{kg}} - 84.103 \,\frac{\text{kJ}}{\text{kg}}$$
$$= 2592 \text{ kJ/kg} \quad (2590 \text{ kJ/kg})$$

Answer is D.

55.
$$Q_{\text{out}} = h_4 - h_5$$
$$= 2153 \,\frac{\text{kJ}}{\text{kg}} - 83.95 \,\frac{\text{kJ}}{\text{kg}}$$
$$= 2069 \text{ kJ/kg} \quad (2100 \text{ kJ/kg})$$

Answer is D.

56.
$$\eta_{\text{th}} = \frac{W_{\text{turbine}} - W_{\text{pump}}}{Q_{\text{in}}}$$
$$= \frac{419 \,\frac{\text{kJ}}{\text{kg}} - 0.153 \,\frac{\text{kJ}}{\text{kg}}}{2592 \,\frac{\text{kJ}}{\text{kg}}}$$
$$= 0.162 \quad (16\%)$$

Answer is A.

57. The most efficient power cycle is the Carnot cycle.

$$\eta_{\text{Carnot}} = \frac{T_H - T_L}{T_H}$$
$$= \frac{(100°\text{C} + 273) - (20°\text{C} + 273)}{100°\text{C} + 273}$$
$$= 0.214 \quad (21\%)$$

Answer is A.

58. For an incompressible liquid,

$$W = \int p \, dV$$
$$\approx \Delta p V$$
$$P = \dot{W} = \Delta p \dot{V}$$
$$= \frac{(850 \text{ kPa} - 100 \text{ kPa}) \left(25 \,\dfrac{\text{m}^3}{\text{min}} \right)}{60 \,\dfrac{\text{s}}{\text{min}}}$$
$$= 312.5 \text{ kW} \quad (310 \text{ kW})$$

Answer is D.

59.
$$x = \frac{m_g}{m_g + m_f}$$
$$\frac{m_g}{m_f} = 0.8$$
$$m_g = 0.8 m_f$$
$$x = \frac{0.8 m_f}{0.8 m_f + m_f}$$
$$= \frac{0.8}{1.8} = 0.44$$

Answer is C.

60.
$$Q = \frac{\left(150 \,\dfrac{\text{kg}}{\text{d}} \right) \left(333 \,\dfrac{\text{kJ}}{\text{kg}} \right) \left(1000 \,\dfrac{\text{J}}{\text{kJ}} \right)}{\left(24 \,\dfrac{\text{h}}{\text{d}} \right) \left(60 \,\dfrac{\text{min}}{\text{h}} \right) \left(60 \,\dfrac{\text{s}}{\text{min}} \right)}$$
$$= 578 \text{ W}$$
$$\text{COP} = \frac{Q}{W}$$
$$W = \frac{Q}{\text{COP}}$$
$$= \frac{578 \text{ W}}{5}$$
$$= 115.6 \text{ W} \quad (115 \text{ W})$$

Answer is B.

Index

A

Absolute
  column, 48-4
  dynamic viscosity, 22-3
  pressure, 22-2
  temperature, 26-4
  value, 43-1
  viscosity, 22-3
  zero temperature, 26-4
Absorbed dose, radiation, 35-10
Absorption, dermal, 34-2
Absorptivity, 31-2
AC
  current, 45-1
  machine, 46-1
Acceleration
  angular, 16-2
  constant, 14-2
  normal, 14-3, 16-3
  of gravity, 1-2
  tangential, 14-3
Access time, 47-5
Accumulator, 47-1
Accuracy, 49-2
ACGIH, 34-7, 35-6
Acid, 38-3
  amino, 38-4 (tbl)
  Arrhenius theory of, 38-3
  binary, 37-4
  carboxylic, 38-5 (tbl)
  deoxyribonucleic, 33-2
  diprotic, 38-4
  fatty, 33-10, 38-5 (tbl)
  monoprotic, 38-4
  organic, 38-5 (tbl)
  properties of, 38-4
  ribonucleic, 33-2
  strength of, 38-4
  ternary, 37-4
  triprotic, 38-4
  volatile, 33-10
Acoustic trauma, 35-9
Acquired immunodeficiency syndrome,
    33-11
ACRS, 52-1
Actinide, 37-3
Actinon, 37-3
Action
  galvanic, 42-1
  line of, 5-5
  mass, law of, 38-2, 38-3
Activated sludge, 36-2, 36-3 (tbl)
Activation energy, 42-3
Active element, electrical, 44-1
Actuating signal ratio, 50-1
Additive effect, law of, 34-8
Address bus, 47-1

Adenosine triphosphate, 33-9
Adenovirus, 33-6
Adiabatic
  process, 25-2, 27-1
  tip, 31-3
Adjacent side, 4-4
Adjoint, classical, 5-4
Admittance, 45-4
Advantage
  mechanical, 12-1
  pulley, 12-1
Aerobe, obligate, 33-6
Aerobic
  decomposition, 33-9
  digestion, 36-4
Affinity, electron, 37-3
Agent
  oxidizing, 38-1
  reducing, 38-2
AI, 48-7
AIDS, 33-11, 35-16
Air
  atmospheric, 29-3
  atmospheric, enthalpy of, 29-5
  composition of, 30-2
  conditioner, 28-4
  excess, 30-2
  -fuel ratio, 30-2
  ideal (combustion), 30-1
  percent excess, 30-2
  percent theoretical, 30-2
  properties, 25-2, 26-8
  saturated, 29-3
  -standard cycle, 28-4
  stoichiometric, 30-1, 30-2
  unsaturated, 29-3
Alcohols
  aliphatic, 38-5 (tbl)
  aromatic, 38-5 (tbl)
Aldehyde, 38-4 (tbl), 38-5 (tbl)
Algae, 33-8
Algal bloom, 33-8
Algebra of complex numbers, 43-1
Algorithm, 48-1
Aliphatic alcohol, 38-5 (tbl)
Alkali metal, 37-3
Alkaline earth metal, 37-3
Alkane, 38-4, 38-5 (tbl)
Alkene, 38-4
Alkoxy, 38-4 (tbl)
Alkyl, 38-4 (tbl)
Alkyl halide, 38-5 (tbl)
Alkyne, 38-4, 38-5 (tbl)
Allergen, 35-14
Allotrope, 42-5
Allotropic change, 42-5
Allowable stress, 18-5

Allowance, trade-in, 51-2
Alloy
  binary, 42-3
  eutectic, 42-4
  miscible, 42-3
  solid-solution, 42-3
Alloying ingredient, 42-3
Alpha
  decay, 37-6
  -iron, 42-5
  particle, 35-10
Alphanumeric data, 48-1
Alternating
  current, 45-1
  waveform, 45-1
Alternative comparison, 53-1
ALU, 47-1
Alveoli, 34-2, 35-6
Amagat-Leduc's rule, 29-2
Amagat's law, 29-2
Amide, 38-5 (tbl)
Amine, 38-4 (tbl), 38-5 (tbl)
Amino acid, 38-4 (tbl), 38-5 (tbl)
Amplitude, sinusoidal, 45-1
Amu, 37-1
Anaerobe, obligate, 33-6
Anaerobic
  decomposition, 33-10
  digester, 36-4
Analog-digital conversion, 49-4, 49-5
Analogy
  Chilton and Colburn, 32-3
  transport phenomena, 32-3
Analysis (*see also* Method)
  benefit-cost, 53-2
  break-even, 53-2
  dimensional, 25-4
  Fourier, 8-2
  gravimetric, 37-6
  incremental, 53-2
  numerical, 9-1
  present worth, 53-1
  rate of return, 53-1
  ultimate, 37-6
Ancestor node, 48-6
Angle
  cone-generating, 4-2
  of contact, 22-4
  of twist, 19-3
  of wrap, 12-2
  phase, 45-3
  pitch, 12-2
  power, 45-5
  thread friction, 12-2
Angular
  acceleration, 16-2
  distance, 16-2

frequency, 15-4, 45-2
momentum, 16-2
motion, 14-3
velocity, 16-2
Anhydride, 38-5 (tbl)
Anion, 37-3
Annealing, 42-7
Annual
    cost, equivalent uniform, 52-2, 53-1
    cost method, 53-1
    effective interest rate, 51-5
    return method, 53-1
Annuity, 51-4
Annum, rate per, 51-5
Anode, 38-1, 39-3, 42-1
Anoxic
    condition, 33-6
    decomposition, 33-10
    denitrification, 33-6
Antiderivative, 7-3
Apparatus, Golgi, 33-2
Approach, velocity of, 25-2
Approximation, Euler's, 8-5
Aquation, 39-2
Archimedes' principle, 23-5
Area
    centroid of, 13-1, 13-9 (tbl)
    first moment of, 13-1
    logarithmic mean, 31-2
    moment of inertia, 7-4
    second moment of, 13-2
Areal density, 47-4
Argument, 43-1
Arithmetic
    and logic unit, 47-1
    mean, 6-2
    progression, 5-7
Arm, moment, 10-2
Armature, AC machine, 46-3
Aromatic, 38-5 (tbl)
Aromatic alcohol, 38-5 (tbl)
Arrhenius theory
    of acids, 38-3
    of bases, 38-4
Arthropod, 35-14
Artificial intelligence, 48-7
Aryl, 38-4 (tbl)
Aryl halide, 38-5 (tbl)
Asbestos, 35-6
Asbestosis, 35-6
ASCII, 48-1
Asphyxiant, 35-3
Assembler, 48-2
Assembly language, 48-2
ASTM grain size, 42-8
Asynchronous transmission, 47-6
Atmospheric
    air, 29-3
    pressure, standard, 22-2, 26-4 (tbl)
Atom, 37-1
Atomic
    bond, 40-4
    mass unit, 37-1
    number, 37-1
    weight, 37-1
Atoms in a cell, 40-2, 40-3 (tbl)

ATP, 33-9
Attractive rate of return, 53-1
Audible sound, 35-7
Audiometry, 35-10
Austempering, 42-7
Austenite, 42-5
Autotroph, 33-4, 33-6
Availability, 28-6
Average (*see also* Mean)
    pressure, fluid, 23-3
    value, sinusoid, 45-2
Averaging duration, 34-7
Avogadro's
    hypothesis, 38-1, 38-3
    law, 26-7
    number, 37-6
Axes, transformation (stress), 18-3
Axial
    loading, 21-1
    strain, 18-1
Axis
    conjugate, hyperbola, 4-3
    ellipse, 4-3
    neutral, 20-2, 20-3
    parallel, theorem of, 7-5, 16-1
    transfer, theorem of, 7-5
    transverse, hyperbola, 4-3

B

Back-emf, 46-2
Background processing, 47-6
Bacteremia, 35-16
Bacteria, 33-7 (tbl), 33-8
    bioprocessing, 36-2
    iron, 33-6
    nonphotosynthetic, 33-6
Balance, heat, human body, 35-11
Bandwidth, 45-6, 50-5
    closed-loop, 50-5
Banked curve, 16-3
Barometer, 23-2
Barometric pressure, 22-2
Base, 38-4
    Arrhenius theory of, 38-4
    logarithm, 5-1
    metal, 42-3
    unit, 1-4
Basic input/output system, 47-2
Basis
    depreciation, 52-1
    unadjusted, 52-1
Batch mode, 47-5
Baud rate, 47-4
Baudot code, 47-4
BCC, 40-2
BCF, 34-7
Beam, 20-1
    boundary condition, 20-4
    -column, 21-1
    composite, 20-4
    deflection, 20-4
    shear stress distribution, 20-3
    steel, 20-4
    stress in, 20-2
Behavior, transient, 44-7
Bell-shaped curve, 6-3
Belt friction, 12-2

Bending stress, 20-2
Bend, pipe, 24-7
Benefit-cost analysis, 53-2
Benign tumor, 34-6
Benzene ring, 38-4 (tbl)
Bernoulli
    equation, 24-2, 24-3
    equation (thermodynamic), 27-3
Beryllium, exposure to, 35-7
Beta
    decay, 37-6
    particle, radiation, 35-10
B-field, 43-5
Bias, 49-2
Biased estimator, 6-2
Biaxial loading, 18-2
Bidding, competitive, 54-3
Bimetallic element, 19-2
Binary
    acid, 37-4
    alloy, 42-3
    compound, 37-4
    fission, 33-8
    phase diagram, 42-3
    tree, 48-5
Binder, 42-8
Binomial
    coefficient, 6-1, 6-3
    distribution, 6-3
Biochemical oxygen demand, 33-9, 36-2
Biocide, 33-8
Bioconcentration factor, 34-7
Biological
    hazard, 35-2, 35-14
    reaction, 33-11
Biology, cellular, 33-1
Biomass composition, 33-11
Bioprocessing, 36-1
BIOS, 47-2
Biotower, 36-4
Bisection method, 9-1
Bit, 47-2
Bit-slice microprocessor, 47-2
Black
    body, 31-2
    lung disease, 35-7
Blade
    force on, 24-8
    velocity, tangential, 24-9
Block, 47-5
    and tackle, 12-1
    diagram algebra, 50-2
Blood
    -borne pathogen, 35-16
    cell, 34-3
BOD, 33-9
    exerted, 33-10
    exertion, 36-2
    loading, 36-4
    test, 36-2
Bode plot, 50-5
Body
    black, 31-2
    -centered cubic cell, 40-2
    rigid, 14-1
Boiler, 27-4

Boiling point, 26-4
  constant, 39-3
  elevation, 39-3
Bond
  atomic, 40-4
  covalent, 37-4, 40-4
  financial, 52-3
  ionic, 37-3, 40-4
  metallic, 40-4
  nonpolar covalent, 37-4
  polar covalent, 37-4
  primary, 40-4
  secondary, 40-4
  value, 52-3
  yield, 52-3
Bonded
  exponential growth curve, 33-4
  strain gage, 49-2
Book value, 52-2
Bootstrap loader, 47-2
Botulism, 35-16
Boundary
  condition, beam, 20-4
  system, 27-2
Box, closed, 19-3
Boyle's law, 25-2, 27-2
Braced column, 21-2
Brake pump power, 25-2
Bravais lattice, 40-1
Breakaway point, 50-6
Break-even
  analysis, 53-2
  quantity, 53-2
Breaking strength, 41-3
Breakthrough time, 35-5
Bridge
  truss, 11-1
  Wheatstone, 49-3
Brinnel test, 42-7
British thermal unit, 26-5
Brittle material, 18-5, 41-3
Btu, 26-5
Bubble sort, 48-5
Buckingham π-theorem, 25-5
Buckling
  column, 21-2
  load, 21-2
Buffer, 47-4
Bulk velocity, 24-4
Buoyancy
  center of, 23-4
  theorem, 23-4
Buoyant force, 23-5
Bus, 47-1
Byte, 47-2

C
Cable, ideal, 12-1
Cache memory, 47-2
Calculus, 7-1
  fundamental theorem of, 7-3
  integral, 7-3
Cancer risk, excessive lifetime, 34-6
Canon, 54-1
Capacitance, 44-2, 44-3
Capacitive reactance, 45-4

Capacitor, 44-2
  ideal, 45-4
Capacity
  carrying, 33-4
  heat, 26-5, 26-6 (tbl), 26-7 (tbl)
Capillary action, 22-4
Capitalized cost, 52-2
Capital recovery method, 53-1
Carbide, 42-5
Carbinol, 38-4 (tbl)
Carbohydrate, 38-5 (tbl)
Carbonyl, 38-4 (tbl)
Carboxylic acid, 38-5 (tbl)
Carcinogen, 34-5, 34-6
  potency factor, 34-6
Cardiotoxicity, 34-3
Carnot cycle
  power, 28-2
  refrigeration, 28-5
Carpal Tunnel Syndrome, 35-13
Carrying capacity, 33-4
Cartesian
  triad, 5-5
  unit vector, 5-5
Cash flow, 51-1
  diagram, 51-1
  gradient series, 51-2
  single payment, 51-1
  uniform series, 51-1
CASRN, 34-6
Cast iron, 42-5
Catabolic process, 33-9
Catalyst, 38-3
Cathode, 38-1, 39-3, 42-1
  ray tube, 47-3
Cation, 37-3
Cauchy number, 25-5
Cavitation, 42-2
CCT curve, 42-7
Ceiling threshold limit value, 34-8
Cell
  body-centered cubic, 40-2
  crystalline, 40-1
  cubic, 40-1
  face-centered, 40-2
  galvanic, 42-1
  half-, 42-1
  hexagonal, 40-1
  hexagonal close-packed, 40-2
  membrane, 33-1
  monoclinic, 40-1
  number of atoms in, 40-2, 40-3 (tbl)
  orthorhombic, 40-1
  red blood, 34-3
  rhombohedral, 40-1
  spreadsheet, 48-4
  structure, 33-1
  structural, 11-1
  tetragonal, 40-1
  transport, 33-3
  triclinic, 40-1
  unit, 40-1
  voltaic, 42-1
Cellular biology, 33-1
Cementite, 42-5

Center
  instant, 16-2
  of buoyancy, 23-4
  of gravity, 13-2
  of mass, 13-2
  of pressure, 23-4
  of rotation, instantaneous, 16-2
  parabola, 4-2
  -radius form, 4-4
Central
  limit theorem, 6-4
  nervous system, 34-4
  processing unit, 47-1
  tendency, 6-2
Centrifugal force, 16-3
Centripetal force, 16-3
Centroid, 7-4, 13-1, 13-14 (tbl)
  of a line, 13-1
  of an area, 13-1, 13-14 (tbl)
  of a volume, 13-1
Centroidal
  mass moment of inertia, 16-1, 16-14 (tbl)
  moment of inertia, 7-5, 13-2, 13-14 (tbl)
CFR, 35-3
cgs system, 1-4
Change, allotropic, 42-5
Channel, 47-1
Characteristic
  dimension, 24-6
  equation, 8-1, 50-2
Characteristics
  of metals, 37-3
  of nonmetals, 37-3
Charge
  conservation, 43-2
  line, 43-3
  sheet, 43-3
  valence, 37-3
Charles' law, 27-2
Charpy test, 41-4
Chart
  Moody friction factor, 24-5
  psychrometric, 29-3, 29-4 (fig)
Check
  bit, 47-3
  digit, 47-6
Chemical
  abstract service registry number, 34-6
  asphyxiant, 35-3
  formula, 37-4
  hazard, 35-2
  reaction, 38-1
Chemistry
  inorganic, 38-1
  organic, 38-4
Chemotroph, 33-6
Chiller, 28-4
Chilton and Colburn analogy, 32-3
Chloroplast, 33-2
Chord (truss), 11-1
Chronic daily intake, 34-6
Circle
  degenerate, 4-4
  general form, 4-4
  line tangent to, 4-4
  Mohr's, 18-4

standard form, 4-4
  unit, 4-5
Circuit
  element, linear, 44-3
  linear, 44-3
Circular
  motion, 14-3
  sector, mensuration, 4-7
  segment, mensuration, 4-7
Circumferential stress, 19-2, 19-3
CISC, 47-2
Clapeyron equation, 29-5
Class of noise exposure, 35-9
Classical adjoint, 5-4
Clausius
  inequality, 28-5
  statement, 28-6
Clear to send, 47-4
Client, ethics 54-2
Clock rate, 47-2
Closed
  box, 19-3
  -cup flash point, 35-5
  feedwater heater, 27-4
  system, 27-2
Closed-loop
  bandwidth, 50-5
  gain, 50-2
  transfer function, 50-2
Close-packed, hexagonal, 40-2
Coal dust, exposure to, 35-7
Code
  ethical, 54-1
  of Federal Regulations, 35-3
Coding, data, 48-1
Coefficient
  binomial, 6-1, 6-3
  constant, 8-1
  diffusion, 42-3
  drag, 24-10
  end-restraint, 21-2, 21-3 (tbl)
  flow (meter), 25-4
  Fourier, 8-2
  Hazen-Williams, roughness, 24-10
  heat transfer, film, 32-2
  loss, 24-6
  Manning's roughness, 24-10
  mass transfer, 32-3
  matrix, 5-5
  of contraction, 25-3
  of discharge, 25-4
  of friction, 12-2, 15-2
  of performance, 28-5
  of resistance, thermal, 44-2, 49-2
  of restitution, 17-3
  of thermal expansion, 19-1, 19-2 (tbl)
  of variation, 6-3
  of velocity, 25-3
  partition, 33-3
  undetermined, method of, 7-4, 8-2
  yield, 33-11
Cofactor
  expansion by, 5-4
  matrix, 5-4
Coherent system of units, 1-2 (ftn)

Cold
  stress, 35-11, 35-12
  working, 42-6
Coliform, 33-9
Collector ring, 46-2
Collision, 17-2
Column, 21-1
  braced, 21-2
  effective length, 21-2
  long, 21-1
Combination, 6-1
Combined
  sound sources, 35-8
  stress, 18-3
Combining weight, 37-6
Combustible liquid, 35-5
Combustion, 30-1
  complete, 30-2
  incomplete, 30-2
  reaction, 30-1
Comfort zone, temperature, 35-11
Command
  set, 47-1
  value, 50-7
Common
  difference, 5-7
  logarithm, 5-1
  ratio, 5-7
Communication protocol, 47-6
Commutator, 46-3
Comparator, 50-1
Comparison, alternative, 53-1
Competitive bidding, 54-3
Compiler, 48-2
Complementary
  equation, 8-2
  probability, 6-1
Complete combustion, 30-2
Complex
  conjugate, 5-2, 43-2, 45-4
  Goldi, 33-2
  instruction-set computing, 47-2
  number, 5-1, 43-1
  number, algebra, 43-2
  plane, 43-1
  power, 45-5
  power vector, 45-5
Component
  normal, 15-3
  radial, 14-2, 15-3
  tangential, 14-3, 15-3
  transverse, 14-2
Composite
  beam, 20-4
  link, 47-4
  material, 42-8
Composition
  air, 30-2
  biomass, 33-11
  eutectic, 42-4
Compound, 37-1, 37-4
  binary, 37-4
  continuous, 51-6
  interest, 51-5
  non-annual, 51-5
  organic, 38-4

probability, 6-1
  ternary, 37-4
Compounds, naming, 37-5
Compression ratio, 28-4
Compressor, 27-4
Computer
  architecture, 47-1
  memory, 47-2
Concentration
  gradient, 32-3
  gradient, defect, 42-3
  ionic, 38-4
  legal, 34-5
  units of, 39-1
Concentrator, 47-4
Concentric loading, 21-1
Concurrent forces, 10-3
Condenser, 27-4
Condition
  beam boundary, 20-4
  standard scientific, 38-3
Conditional probability, 6-1
Conductance, 45-4
Conduction, 31-1
Cone
  -generating angle, 4-2
  right circular, mensuration, 4-7
Confidence
  interval, 6-4
  level, 6-4
  limit, 6-4
Configuration factor, 31-3
Congenital syphilis, 35-16
Conic section, 4-2
Conjugate
  axis, hyperbola, 4-3
  complex, 5-2, 45-4
Conservation
  energy, principle, 17-2
  of charge, 43-2
  of energy (fluid), 24-3
  of mass (fluid), 24-1
  of momentum, law, 15-1
Consistency index, 22-3, 24-4
Consistent
  deformations, 20-4
  units, 1-2
Constant
  acceleration, 14-2
  boiling point, 39-3
  coefficient, 8-1
  entropy process, 27-1
  equilibrium, 38-3
  freezing point, 39-3
  gas, 38-3
  gravitational, 1-2
  Henry's law, 29-2
  Joule's, 26-5
  lattice, 40-2
  logistic growth rate, 33-4
  of integration, 7-3
  pressure process, 27-1–27-3
  pressure, specific heat at, 25-2
  proportions, law of, 37-5
  specific gas, 25-2, 26-7
  spring, 17-2

Stefan-Boltzmann, 31-3
temperature process, 27-1, 27-3
torsional spring, 19-3
time, 44-7
universal gas, 25-2, 26-7 (tbl), 38-3
volume process, 27-1, 27-2
volume, specific heat at, 25-2
Constants, fundamental, 2-1
Contact, angle of, 22-4
Container label, 35-3
Continuity equation, 24-1
Continuous
compounding, 51-6
distribution, 6-2
noise, 35-9
random variable, 6-3
Continuum, 22-1
Contracta, vena, 25-3
Contraction
coefficient, 25-3
pipe, 24-6
Contraflexure point, 7-1
Control
character, 48-1
element, 50-7
logic block diagram, 50-7
logic diagram, 50-7
mass, 27-2
matrix, 50-7
ratio, 50-2
surface, 27-2
system, 50-6
unit, 47-1
vector, 50-7
volume, 27-2
Controllability, 50-7
Controlled
-cooling-transformation curve, 42-7
system, 50-7
Controller, 50-7
Controls, 50-1
Convection, 31-1, 31-2
forced, 31-1
natural, 31-1
Newton's law of, 31-2
Convention
sign, moment, 20-2
sign, shear, 20-1
sign, thermodynamic, 27-2
year-end, 51-2
Convergence, interval of, 5-7
Convergent sequence, 5-7
Conversion
factors, 3-1, 3-2
of energy, 17-2
Coordinates
polar, 5-2
rectangular, 14-1
Coordination number, 40-3
COP, 28-5
Core, 47-3
column, 21-1
Cornea, 34-3
Corner frequency, 50-3

Corrosion, 42-1
fretting, 42-2
stress, 42-2
Cosecant, 4-5
Cosine, 4-5, 10-1
-sine relationships, 45-1
Cosines, law of, 4-6
Cost
capitalized, 52-2
equivalent annual, 53-1
equivalent uniform annual, 52-2, 53-1
project, 51-2
recovery system, accelerated, 52-1
sunk, 51-1
Cotangent, 4-5
Coulomb, 43-2
friction, 12-2
Coulomb's law, 43-2
Counter-emf, 46-2
Couple, 10-2
Coupling moment, 10-3
Covalent bond, 37-4, 40-4
CPF, 34-6
CPU, 47-1
Cramer's rule, 5-5
CRC, 47-6
Creed, 54-1
Critical
isobar, 26-3
load (column), 21-2
point, 7-1, 26-2 (fig), 26-3, 42-5
Reynolds number, 24-4
slenderness ratio, 21-2
temperature, 42-5
values, F-distribution, 6-14 (tbl)
zone, 24-4
Critically damped, 8-2
Crossover
frequency, phase, 50-6
gain, 50-6
Cross product, 5-6
CRT, 47-3
Cryogenic liquid, 35-3
Crystalline cell, 40-1
Crystallography, 40-1
CTD, 35-12
CTS, 35-13, 47-4
Cubic
body-centered, 40-2
cell, 40-1
face-centered, 40-2
Cubital Tunnel Syndrome, 35-13
Cumulative trauma disorder, 35-12
Curl, 7-5
Current, 43-4
density, 43-4
source, ideal, 44-1
Curvature, 7-2, 14-3, 20-4
Curve
banked, 16-3
bell-shaped, 6-3
controlled-cooling-transformation, 42-7
load-elongation, 41-1
S-N, 41-3
time-temperature-transformation, 42-7
Curvilinear motion, 14-2

Cut-off frequency, 50-5
Cycle
air-standard, 28-4
Carnot (power), 28-2
Carnot (refrigeration), 28-5
four-stroke, 28-4
Otto, 28-3
power, 28-1, 28-2
Rankine, 28-2, 28-3
refrigeration, 28-4
Cycling redundancy checking, 47-6
Cycloalkene, 38-5 (tbl)
Cylinder, 16-14, 47-4
hollow, 16-14
right circular, mensuration, 4-7
Cytoplasm, 33-2
Cytosol, 33-2

D

Daily intake, chronic, 34-6
Dalton's law, 29-2
Darcy
equation, 24-4
friction factor, 24-4
Data bus, 47-1
Data terminal ready, 47-4
Database, 48-6
Daughter node, 48-6
DC
circuit, 44-1
generator, 46-3
machine, 46-1, 46-3
motor, 46-5
voltage, 46-3
DDT, 34-3
Dead finger, 35-14
Death growth phase, 33-4
Decay
alpha, 37-6
beta, 37-6
radioactive, 35-11, 37-6
Declining growth phase, 33-4
Decomposition, 33-9, 33-10
DEDO, 47-3
Defect
concentration gradient, 42-3
grain, 42-3
grain boundary, 42-3
planar, 42-3
point, 42-3
Definite integral, 7-3
Definition proportions, law of, 37-5
Deflect flux, 42-3
Deflection
beam, 20-4
bridge, 49-3
Deflector, force on, 24-8
Deformation, 18-2
Deformations, consistent, 20-4
Degenerate circle, 4-4
Degree
of freedom, 6-4, 29-5, 42-5, 50-7
of indeterminacy, 10-4
of reduction, 33-11
Dehydration, 35-12
Del operator, 7-5
Delta-iron, 42-5

de Moivre's formula, 5-2
Demand, oxygen, biochemical, 33-9
Denominator, rationalizing, 5-2
Density, 22-1, 22-2
    current, 43-4
    function, probability, 6-3
    ideal gas, 38-3
    mass, 22-1
    recording, 47-5
Denitrification, 33-10
Denitrifier, 33-6
Deoxyribonucleic acid, 33-2
Dependent source, electrical, 44-1
Depreciation, 52-1
Derivative, 7-1, 7-2, 7-14 (tbl)
Derived unit, 1-4
Dermal absorption, 34-2
Dermis, 34-2
Destruction, nerve, 34-4
Determinacy, 10-4
Determinant, matrix, 5-3
Determinate
    system, 10-4
    truss, 11-2
Deuterium, 37-6
Deviation, standard, 6-2
Dew-point temperature, 29-3
Diagram
    binary phase, 42-3
    cash flow, 51-1
    equilibrium, 26-2, 42-3
    free-body, 10-4
    moment, 20-2
    phase, 26-2, 42-3
    shear, 20-2
    Stanton, 24-5
Diameter
    equivalent, 24-6
    hydraulic, 24-6
Diatomic
    gas, 37-4
    molecule, 37-4, 40-4
Difference
    common, 5-7
    equation, 8-4
    of means, 6-5
Differential
    equation, 8-1, 8-2
    manometer, 23-2, 25-3
    term, 7-3
Diffuser, 27-4
Diffusion, 42-3
    coefficient, 42-3
    facilitated, 33-3
    molecular, 32-3
    passive, 33-3
Diffusivity, 42-3
Digester, biological, 36-4
Digital audio tape, 47-5
Dihydroxic base, 38-4
Dilatant fluid, 22-3, 24-4
Dilution ventilation, 35-7
Dimensional analysis, 25-4
Dimension, characteristic, 24-6
Dimensions, primary, 25-4
Diprotic acid, 38-4

Direct
    access, 47-4
    central impact, 17-3
    contact heater, 27-5
    current, 44-1
    gain, 50-2
    human exposure, 34-6
    memory access, 47-3
    transfer function, 50-1
Direction cosine, 10-1
Directrix, parabola, 4-2
Disbursement (cash), 51-1
Discharge
    coefficient of, 25-4
    free fluid, 25-4
Discomfort zone, temperature, 35-11
Discount factor, 51-2, 51-4 (tbl)
Discrete
    numerical event, 6-2
    random variable, 6-3
Discriminant, 4-2
Disease
    waterborne, 33-11
    zoonotic, 35-15
Disk pack, 47-4
Dispersion, 6-2, 6-3
Distance
    angular, 16-2
    between points, 4-1, 4-4
    semimajor, ellipse, 4-3
    semiminor, ellipse, 4-3
Distortion energy theory, 18-5
Distributed data processing system, 47-6
Distribution
    binomial, 6-3
    continuous, 6-2
    fluid velocity, 24-4
    function, probability, 6-3
    Gaussian, 6-3
    normal, 6-3
    shear stress, beam, 20-3
    Student's t-, 6-4, 6-9 (tbl)
    t-, 6-4, 6-5, 6-9 (tbl)
Divergence, 7-5
Divergent sequence, 5-7
DMA, 47-3
DNA, 33-2
Dome, vapor, 26-2
Dose
    infective, 33-11
    lethal, 34-5
    radiation, 35-11
    -response curve, 34-4
    -response relationship, 34-4
    safe human, 34-5
Dot product, 5-5
Double
    -angle formula, 4-5
    root, 4-2
DO/UNTIL, 48-3
Double
    -length word, 47-2
    -precision, 47-2
Doubleword, 47-2
DO-WHILE, 48-3
Drag, 24-10

Drag coefficient, 24-10
DRAM, 47-3
Dry-bulb temperature, 29-3
DTR, 47-4
Ductility, 41-3
Ductility transition temperature, 41-4
Du Nouy apparatus, 22-4
Duplex communication, 47-6
Duration
    average, 34-7
    exposure, 34-6
Dust, radioactive, 35-7
Dynamic
    friction, 12-2, 15-2
    memory, 47-3
    similarity, 25-5
    unit, 50-1
    viscosity, absolute, 22-3
Dynamics, 14-1
Dynamics, fluid, 24-1

E
E. coli, 33-3, 33-6, 33-9
EBCDIC, 48-1
Ebola, 35-17
Eccentricity, 4-2, 4-3, 21-1
Eccentric loading, 21-1
Echo checking, 47-6
Effect
    Frank, 34-5
    systemic, 34-3
Effective
    annual interest rate, 51-5
    interest rate, 51-2
    interest rate per period, 51-6
    length (column), 21-2
    stress, 18-5
    value, sinusoid, 45-2
Efficiency
    fin, 31-3
    isentropic, 27-4
    nozzle, 27-4
    pump, 25-2
    radiator, 31-3
    thermal, 28-2
Efflux, Torricelli's speed of, 24-8
Eigenvalue, 50-3
Elastic
    impact, 17-3
    limit, 41-2
    potential energy, 17-2
    region, 41-2
    section modulus, 20-3
    strain, 41-2
    strain energy, 18-2
Elasticity, modulus of, 18-1, 41-2
Electric
    flux, 43-2, 43-3
    potential gradient, 43-4
Electrochemical reaction, 39-3
Electrolysis, 39-3
Electrolyte, 39-3
Electrolytic
    medium, 42-1
    reaction, 39-3
Electromotive force, 43-6, 46-2
Electron affinity, 37-3

Electronegativity, 37-3
Electrostatic unit, 43-2
Element, 37-1
   bimetallic, 19-2
Elevation
   boiling point, 39-3
   head, 24-2
Ellipse, 4-2, 4-3
   eccentricity, 4-3
   general form, 4-3
   latus rectum, 4-3
   mensuration, 4-7
   semimajor distance, 4-3
   semiminor distance, 4-3
   standard form, 4-3
Elongation, percent at failure, 41-3
EMF, 43-6
Emissivity, 31-2
Emulation mode, 47-2
End
   -of-file, 47-5
   post, 11-1
   -restrain coefficient, 21-2, 21-3 (tbl)
Endocytosis, 33-3
Endogenous infection, 35-15
Endoplasmic reticulum, 33-2
Endothermic reaction, 39-2
Endurance
   limit, 41-4
   stress, 41-4
   test, 41-3
Energy
   activation, 42-3
   conservation of, fluid, 24-3
   conservation principle, 17-2
   conversion, 17-2
   distortion, 18-5
   electrical, 44-2
   equation, steady flow, 27-3
   flow, 26-5
   fluid, 24-2
   free, 29-5
   grade line, 24-2
   impact (toughness), 41-4
   internal, 26-4
   ionization, 37-3
   kinetic, 17-1
   kinetic (fluid), 24-2
   line, fluid, 24-2
   of a mass, 17-1
   potential, 17-2
   potential (fluid), 24-2
   pressure, 24-2
   specific pressure, 24-2
   storage, magnetic, volume, 43-5
   strain, 18-2, 41-4
   transfer, 32-2
   -work theorem, 17-2
Engine, 28-2
Engineering
   strain, 41-1
   stress, 41-1
Engineers, ethics, 54-3
English
   engineering system, 1-2
   gravitational system, 1-4

Enterovirus, 33-6
Enthalpy, 26-5, 39-2
   atmospheric air, 29-5
   chemical, 39-2
   of formation, 39-2
   of reaction, 39-2
Entropy, 26-5
   change, 28-5
   production, 26-5, 28-5
Environment, 27-2
Enzyme, 33-6
Epicondylitis, 35-13
Epidermis, 34-2
EPROM, 47-3
Equation
   Bernoulli, 24-2, 24-3
   Bernoulli (thermodynamic), 27-3
   characteristic, 8-1
   Clapeyron, 29-5
   complementary, 8-2
   continuity, 24-1
   Darcy, 24-4
   difference, 8-4
   differential, 8-1, 8-2
   extended Bernoulli, 24-3
   field, 24-2
   Hagen-Poiseuille, 24-6
   Hazen-Williams, 24-10
   Henderson-Hasselbach, 33-3
   Kline-McClintock, 49-5
   logistic, 33-4
   Manning's, 24-10
   of state, 25-2, 26-7, 38-3
   quadratic, 4-1
   reduced, 8-2
   steady flow energy, 27-3
Equations, simultaneous linear, 5-4
Equilibrium
   constant, 38-3
   diagram, 26-2, 42-3
   static, 10-3
   thermal, 26-4
Equipment, personal protective, 35-5
Equivalence, 51-3
   gradient, 51-5
   single payment, 51-3
   uniform series, 51-3
Equivalent
   annual cost, 53-1
   chemical, 37-6
   diameter, 24-6
   exposure, 34-8
   uniform annual cost, 52-2, 53-1
   weight, 37-6
Ergonomic hazard, 35-2
Ergonomics, 35-12
Error, 50-1
   ratio, 50-1
   signal, 50-1
   transfer function, 50-1
Escherichia coli, 33-3
Ester, 38-4 (tbl), 38-5 (tbl)
Estimator
   biased, 6-2
   unbiased, 6-2
Ether, 38-4 (tbl), 38-5 (tbl)

Ethical priority, 54-2
Ethics, 54-1
EUAC, 52-2
Eukaryote, 33-1, 35-14
Euler load, 21-2
Euler's
   approximation, 8-5
   equation, 43-2
   formula, 21-2
Eutectic
   alloy, 42-4
   composition, 42-4
   reaction, 42-4 (tbl)
   temperature, 42-4
Eutectoid reaction, 42-4 (tbl)
Evaporator, 27-4
Even parity, 47-3
Event, 6-1, 6-2
Excess air, 30-2
Excessive lifetime cancer risk, 34-6
Exchanger, heat, 27-4
Exfoliation, 42-2
Exocytosis, 33-3
Exogenous infection, 35-15
Exothermic reaction, 39-2
Expansion
   by cofactors, 5-4
   pipe, 24-6
   thermal coefficient of, 19-1, 19-2 (tbl)
Expected
   exposure dose, 34-5
   value, 6-3, 52-3
Expert system, 48-7
Explosive limit
   lower, 35-5
   upper, 35-5
Exponential
   decay, 50-3
   decay law, 37-7
   form
      sinusoidal, 45-2
      trigonometric terms, 43-2
   growth phase, 33-4
Exposure
   dose, 34-5
   duration, 34-6
   level, short-term, 34-7
   limit, permissible, 34-8, 35-3
   pathway, 34-2
Extended Bernoulli equation, 24-3
Extended surface, 31-3
Extensive property, 26-3
External force, 10-1
Extraction heater, 27-4
Extrema, 7-1
Extreme
   fiber, 20-2
   point, 7-1
Eye, 34-3, 34-4

F
Face
   -centered cubic cell, 40-2
   value (of a bond), 52-3
Facilitated diffusion, 33-3
Factor
   bioconcentration, 34-7

configuration, 31-3
Darcy friction, 24-4
discount, 51-2, 51-4 (tbl)
potency, 34-6
slope, 34-6
Factory, yield, 33-11
Facultative
bacteria, 33-6
heterotroph, 33-8
pond, 36-3
Fanning friction, 32-2
j-, 32-3
Moody friction, 32-2
packing, 40-3
shape, 31-3
single payment compound amount, 51-3
single payment present worth, 51-3
sinking fund, 51-4
stress concentration, 18-5
uniform gradient, 51-5
uniform series compound amount, 51-3
Factors, conversion, 3-1, 3-2
Failure
fatigue, 41-3
theory, 18-5
Family
elemental, 37-1
organic compound, 38-4
Fanning friction factor, 32-2
Faraday's
law, 43-6
laws of electrolysis, 39-3
Faraday (unit), 39-3
Fatigue
failure, 41-3
life, 41-4
limit, 41-4
strength, 41-4
test, 41-3
Fatty
acid, 33-10, 38-5 (tbl)
tissue, 34-2
FCC, 40-2
FCC metals, 41-2, 41-5
F-distribution, critical values, 6-14 (tbl)
FDM, 47-4
Fecal coliform, 33-9
Feedback, 50-1
ratio, 50-2
transfer function, 50-2
unit, 50-1
Feedwater heater, 27-4
Fermentation, 33-10
Ferrite, 42-5
Fiber
airborne, 35-6
extreme, 20-2
-strengthened material, 42-8
Fick's law, 32-3, 42-3
Field
AC machine, 46-3
data, 48-4
equation, 24-2
magnetic, 43-5

File
indexing, 48-5
data, 48-4
Film coefficient, heat transfer, 32-2
Fin, 31-3
Final
value, 50-2
value theorem, 8-3, 8-5
Finite
difference equation, 4-5
series, 5-7
Firmware, 47-3, 48-1
First
area moment, 13-1
derivative, 7-1
law of motion, Newton's, 15-1
law of thermodynamics, 27-2
law of thermodynamics, open system, 27-3
moment, 20-3
moment of the area, 7-4, 13-1
-order differential equation, 8-1
Fish, bioconcentration factors, 34-7 (tbl)
Fission, binary, 33-8
Fixed
blade, force on, 24-8
-film bioprocessing, 36-4
-point iteration, 9-2
pulley, 12-1
Flagella, 33-2, 33-8
Flagellated protozoa, 33-8
Flammable
limit, lower, 35-5
limit, upper, 35-5
liquid, 35-5
Flange (structural steel), 13-3
Flash point, 35-5
Flat file, 48-6
Flatworm, 33-8
Flexural stress, 20-2
Floppy disk, 47-5
Flow
cash, 51-1
coefficient (meter), 25-4
control, 47-4
energy, 26-5
gravity, 24-6
laminar, 24-3
open channel, 24-6, 24-11
pressure, 24-6
rate, recycle, 36-3
shear, 19-3
turbulent, 24-3, 24-4
work, 26-5, 27-2
Flowchart, 48-1
Fluid, 22-1
dynamics, 24-1
energy, 24-2
friction, 12-2
ideal, 22-2
non-Newtonian, 22-3
power law, 22-3, 24-3
statics, 23-1
stress, 22-2, 22-3
surface tension, 22-3
velocity distribution, 24-4
Fluke, 33-8

Flux
defect, 42-3
electric, 43-2, 43-3
molar, 32-3
Focus, parabola, 4-2
Food
botulism, 35-16
poisoning, 35-16
FOR loop, 48-3
Force
buoyant, 23-5
centrifugal, 16-3
centripetal, 16-3
concurrent, 10-3
-couple system, 10-3
external, 10-1
frictional, 15-2
hydrostatic, 23-3
internal, 10-1, 11-1
normal, 12-2, 15-2
on a pipe bend, 24-7
Van der Waals, 40-4
Forced
convection, 31-1
vibration, 15-3
Foreground processing, 47-6
Forcing function, 8-2
Form
center-radius, 4-4
general, circle, 4-4
general, conic section, 4-2
general, ellipse, 4-3
general, hyperbola, 4-3
general, parabola, 4-2
general, sphere, 4-4
general, straight line, 4-1
point-slope, 4-1
polar, 5-2
rectangular, 5-2
rectangular coordinator (position), 14-2
slope-intercept, 4-1
standard, circle, 4-4
standard, ellipse, 4-3
standard, hyperbola, 4-3
standard, parabola, 4-2
standard, sphere, 4-4
standard, straight line, 4-1
trigonometric, 5-2
unit vector (position), 14-2
vector (position), 14-2
Formality, 39-1
Formation
enthalpy of, 39-2
Gibbs' function, 29-5
heat of, 39-2
Formula
chemical, 37-4
de Moivre's, 5-2
development, 37-6
double-angle, 4-5
Euler's, 21-2
half-angle, 4-6
Taylor's, 5-8
two-angle, 4-5
weight, 37-6

Forward
  gain, 50-2
  transfer function, 50-??
Fourier
  analysis, 8-2
  coefficient, 8-2
  series, 8-2
Fourier's
  law, 31-1
  theorem, 8-2
Four-stroke cycle, 28-4
Fourth-power law, 31-3
Fraction
  gravimetric, 29-1, 37-6
  mass, 29-1
  mole, 29-1, 39-1
  partial, 7-4
Fracture
  appearance transition temperature, 41-5
  strength, 41-3
  transition temperature, 41-4
Frame, 11-1, 47-3
Frank effect level, 34-5
Free
  -body diagram, 10-4
  discharge (fluid), 25-4
  energy, 29-5
  moment, 10-3
  pulley, 12-1
  surface, 24-6
  vibration, 15-3
  vibration, torsional, 16-3
Freedom
  degrees, 6-4, 29-5, 42-5
  degree of, 50-7
Freezing point constant, 39-3
Frequency, 35-7, 45-2
  angular, 15-4, 45-2
  corner, 50-5
  cut-off, 50-5
  division multiplexer, 47-4
  fundamental, 8-3
  mechanical, 46-1
  natural, 8-3
  natural, vibration, 15-4
  resonance, 45-5
  response, 50-5
Fretting corrosion, 42-2
Friction, 12-2, 15-2
  angle, thread, 12-2
  belt, 12-2
  coefficient of, 12-2, 15-2
  Coulomb, 12-2
  dynamic, 12-2, 15-2
  factor chart, Moody, 24-5
  factor, Darcy, 24-4
  factor, Fanning, 32-2
  factor, Moody, 32-2
  fluid, 12-2
  head loss, 24-3
  static, 12-2, 15-2
Frictional force, 15-2
Frictionless surface, 10-5
Froude number, 25-5
Full-duplex communication, 47-6

Fume, 35-6
  welding, 35-7
Function
  forcing, 8-2
  Gibbs', 26-5, 29-5
  Helmholtz, 26-5, 29-5
  probability density, 6-3
  probability distribution, 6-3
Functional group, organic, 38-4
Function of formation, Gibbs', 29-5
Fundamental
  constants, 2-1
  frequency, 8-3
  theorem of calculus, 7-3
Fund, sinking, 51-4
Fungi, 33-8
Future worth, 51-3

G
Gage
  factor, 49-3
  pressure, 22-2
Gain
  characteristic, 50-3
  crossover, 50-6
  crossover point, 50-4
  direct, 50-2
  forward, 50-2
  margin, 50-6
Galvanic
  action, 42-1
  cell, 42-1
  series, 42-1, 42-2 (tbl)
Gamma
  -iron, 42-5
  ray, radiation, 35-11
Ganglionitis, 35-13
Gas
  constant, 38-3
  constant, specific, 25-2, 26-7
  constant, universal, 25-2, 26-7 (tbl), 38-3
  diatomic, 37-4
  ideal, 26-2, 26-6, 38-3
  inert (see Gas, noble)
  mixture, 26-2, 29-2
  noble, 37-3
  perfect, 25-2 (see also Gas, ideal)
  real, 26-2
Gases in liquid, 39-1
Gauss' law, 43-4
Gaussian
  distribution, 6-3
  surface, 43-4
General
  form, circle, 4-4
  form, conic section, 4-2
  form, ellipse, 4-3
  form, hyperbola, 4-3
  form, parabola, 4-2
  form, sphere, 4-4
  form, straight line, 4-1
  strain, 18-4
  term, sequence, 5-7
  triangle, 4-6
Generator, 46-3
Geometric
  mean, 6-2

progression, 5-7
  similarity, 25-5
Gibbs'
  function, 26-5, 29-5
  function of formation, 29-5
  phase rule, 29-5, 42-5
  rule, 29-2
  theorem, 29-2
Global variable, 48-3
Gloves, 35-5
Glycol, 38-5 (tbl)
Golfer's elbow, 35-13
Golgi
  apparatus, 33-2
  complex, 33-2
GOTO statement, 48-4
GOTO-less programming, 48-3
Grade line, hydraulic, 24-2
Gradient
  concentration, 32-3
  defect concentration, 42-3
  equivalence, 51-5
  function, 7-5
  of a scalar function, 7-5
  series cash flow, 51-2
  velocity, 22-3
Grain
  boundary defect, 42-3
  size, 42-8
  unit of mass, 29-3
Gram-mole, 37-5
Gravimetric
  analysis, 37-6
  fraction, 29-1, 37-6
Gravitational
  constant, 1-2
  head, 24-2
  potential energy, 17-2
Gravity
  acceleration, 1-2
  center of, 13-2
  flow, 24-6
  head, 24-2
  specific, 22-2
Gray, unit, 35-10
Group
  elemental, 37-1
  pi-, 25-5
Growth, organismal, 33-3, 33-4
Growth rate constant, 33-4
Gyration, 13-3, 13-14 (tbl), 16-2, 16-14 (tbl)

H
Hagen-Poiseuille equation, 24-6
Half
  -angle formula, 4-6
  -cell, 42-1, 42-2 (tbl)
  -duplex communication, 47-6
  -life, 35-11, 37-7
  -life, biological contaminant, 36-2
  -power point, 45-6, 50-5
  -value thickness, 35-11
  -word, 47-2
Halide, 38-4 (tbl), 38-5 (tbl)
Halogen, 37-3, 38-4 (tbl)
Handshaking, 47-4
Hantavirus, 35-17

Hard drive, 47-4
Hard X-ray, 35-10
Hardenability, 42-7
Hardening, 42-6
Hardness, 42-7 (*see also by type*)
Hardness test, 42-7
Harmonic series, 8-2
Hardware, 48-1
   flow control, 47-4
Hashing, 48-5
Hazard, 35-3
   biological, 35-2
   chemical, 35-2
   ergonomic, 35-2
   physical, 35-2
   ratio, 34-5
   ration, 34-8
Hazen-Williams equation, 24-10
HCP, 40-2
HCP metals, 41-2, 41-5
Head
   elevation, 24-2
   gravitational, 24-2
   gravity, 24-2
   loss, 24-3, 24-6
   potential, 24-2
   pressure, 24-2
   velocity, 24-2
Health risk zone, temperature, 35-11
Heap sort, 48-5
Hearing
   conservation program, 35-7
   loss, 35-9
   protection, 35-9
Heart, 34-4
Heat
   balance, human body, 35-11
   capacity, 26-5, 26-6 (tbl), 26-7 (tbl)
   engine, 28-2
   exchanger, 27-4
   exhaustion, 35-12
   graphical (thermodynamic), 28-7
   of formation, 39-2
   of reaction, 39-2
   of solution, 39-2
   pump, 28-4
   specific, 26-5, 26-6 (tbl), 26-7 (tbl)
   strain, 35-12
   stress, 35-11, 35-12
   transfer, 31-1
   transfer film coefficient, 32-2
   transfer, radiant, 31-3
   treatment, 42-6
Heater
   direct contact, 27-4
   extraction, 27-4
   feedwater, 27-4
   mixing, 27-4
Heatstroke, 35-12
Helmholtz function, 26-5, 29-5
Hematoxicity, 34-3
Henderson-Hasselbach equation, 33-3
Henry's
   law, 29-2, 39-1
   law constant, 29-2
Hepatitis, 33-6, 35-16

Hepatoxicity, 34-4
Hess' law of energy summation, 39-2
Heterotroph, 33-6
Hexadecimal, 48-1
Hexagonal
   cell, 40-1
   close-packed cell, 40-2
Hierarchical database, 48-6
High rate digester, 36-4
HIV, 33-11, 35-16
Hollow
   cylinder, 16-14
   shell, 19-3
Homogeneous
   differential equation, 8-1
   system of units, 1-2 (ftn)
Hooke's law, 18-1, 41-2
Hoop stress, 19-2, 19-3
Horsepower, 25-1
Host ingredient, 42-3
Hot working, 42-6
HR, 34-5
Human immunodeficiency virus, 33-11
Humidity
   ratio, 29-3
   relative, 29-3
   specific, 29-3
Hydration, 39-2
Hydraulic
   diameter, 24-6
   grade line, 24-2
   horsepower, 25-1
   power, 25-1
   radius, 24-6
   residence time, 36-2
Hydrocarbon, 38-4, 38-5 (tbl)
Hydrogen isotope, 37-6
Hydronium ion, 38-3
Hydrostatic
   force, 23-3
   paradox, 23-1
   pressure, 23-1, 23-3
   resultant, 23-3, 23-4
Hydroxyl, 38-4 (tbl)
Hygiene, industrial, 35-2
Hyperbola, 4-2, 4-3
   conjugate axis, 4-3
   focus, 4-3
   general form, 4-3
   standard form, 4-3
   transverse axis, 4-3
Hypotenuse, 4-4
Hypothermia, 35-12
Hypothesis
   Avogadro's, 38-1, 38-3
   test, 6-5

I
IC, 16-2
Ice point, 26-4
Ideal
   air (combustion), 30-1
   cable, 12-1
   capacitor, 45-4
   current source, 44-1
   fluid, 22-2
   gas, 26-2, 26-6, 38-3

gas, density, 38-3
gas mixture, 29-1
inductor, 45-4
radiator, 31-2
resistor, 45-3
transformer, 45-7
voltage source, 44-1
Identities
   trigonometric, 4-5
   vector, 5-6
Identity, logarithm, 5-1
Identity matrix, 5-3
IDLH, 35-4 (tbl)
IF-THEN, 48-3
Imaginary number, 5-1, 43-1
Immunotoxicity, 34-4
Impacts, 17-2
   direct central, 17-3
   elastic, 17-3
   energy (toughness), 41-4
   inelastic, 17-3
   noise, 35-9
   plastic, 17-3
   test, 41-4
Impedance
   electrical, 45-3
   triangle, 45-3
Imperfection (*see* Defect)
Impulse, 24-7
   linear, 17-2
   -momentum principle, 17-2, 24-6
   turbine, 24-9
Incomplete combustion, 30-2
Incremental, analysis, 53-2
Indefinite integral, 7-3, 7-14 (tbl)
Independent source, electrical, 44-1
Indeterminacy, 10-4
Indeterminate system, 10-4
Index
   consistency, 22-3, 24-4
   Miller, 40-3
Indexed
   sequential file, 48-4
   sequential format, 47-5
Indicator organism, 33-9
Indigenous infection, 35-15
Induced voltage, 43-6
Inductance, 44-3
Induction
   motor, 46-2
   magnetic, 43-5
Inductive reactance, 45-4
Inductor, 44-3
   ideal, 45-4
Industrial Hygiene, 35-2
Inelastic
   impact, 17-3
   strain, 41-2
Inequality of Clausius, 28-5
Inert gas (*see* Noble gas)
Inertia
   centroidal moment of, 7-4, 13-2, 13-14 (tbl)
   mass moment of, 16-1, 16-14 (tbl)
   moment of, 13-2, 13-14 (tbl)
   moment of, area, 7-4

polar moment of, 13-2, 13-14 (tbl), 19-3
   product of, 13-3, 16-14 (tbl)
Infection, 35-15
Infective dose, 33-11
Infinite series, 5-7
Inflation, 52-3
Inflection point, 7-1
Ingredient
   alloying, 42-3
   host, 42-3
   parent, 42-3
Ingestion, 34-3
Inhalation, 34-2
Initial
   value, 50-2
   value problem, 8-2
   value theorem, 8-3
   value theorem, z-transform, 8-5
Initiation stage, cancer, 34-6
Input/output device, 47-3
Insensitivity, 49-2
Insertion sort, 48-5
Instability, point of, 50-6
Instantaneous center of rotation, 16-2
Instant center, 16-2
Instruction register, 47-1
Insulated tip, 31-3
Insulation, 35-12
Integral, 7-3, 7-14 (tbl)
   calculus, 7-3
   definite, 7-3
   indefinite, 7-3, 7-14 (tbl)
Integrand, 7-3
Integrated Risk Information System, 34-6
Integration, 7-3
   by parts, 7-3
   by substitution, 7-3
   constant of, 7-3
   variable of, 7-3
Intensity
   radiation, 35-10
   sound, 35-8
Intensive property, 26-3
Interblock gap, 47-5
Intercept, 4-1
Interest
   compound, 51-5
   rate, effective, 51-2
   rate, effective annual, 51-5
   rate, nominal, 51-5
   rate per period, 51-5
   rate per period, effective, 51-6
   simple, 51-5
Interference, nerve, 34-4
Intermittent noise, 35-9
Internal
   energy, 26-4
   force, 10-1, 11-1
   work, 18-2
Interpreter, 48-2
Interrupt, 47-2
Intersection lines, 4-1
Interstice, 40-3
Interstitial space, 40-3

Interval
   confidence, 6-4
   of convergence, 5-7
Intrinsic machine code, 48-2
Inverse Laplace transform, 8-3
Inverse matrix, 5-4
Ion, 37-3, 38-3
Ionic
   bond, 37-3, 40-4
   concentration, 38-4
Ionization energy, 37-3
Ionizing radiation, 35-10
IRIS, 34-6
Iron
   alpha-, 42-5
   bacteria, 33-6
   carbide, 42-5
   -carbon phase diagram, 42-5 (fig)
   cast, 42-5
   delta-, 42-5
   gamma, 42-5
Irrational number, 5-1
Irreversibility, process, 28-6
Isentropic
   efficiency, 27-4
   process, 25-2, 26-8, 27-1, 27-3
Isobar, 26-3
Isobaric process, 27-1
Isochoric process, 27-1
Isolated system, 27-2
Isometric process, 27-1
Isotherm, 26-3
Isothermal process, 27-1, 27-3
Isotope, 37-1, 37-6
Iteration, fixed-point, 9-2

J
Jet
   propulsion, 24-8
   velocity (fluid), 24-8
j-factor, 32-3
Joint, 6-1, 11-1
Joints, method of, 11-2
Joule's
   constant, 26-5
   law, 44-2

K
Kalman's theorem, 50-7
KB, 47-2
KCL, 44-4
Kelvin-Planck statement, 28-6
Kelvin temperature, 26-4
Kern, 21-1
Kernel, 21-1
Ketone, 38-5 (tbl)
Kidney, 34-3, 34-4
Kilobyte, 47-2
Kiloword, 47-2
Kinematic
   similarity, 25-5
   viscosity, 22-3
Kinematics, 14-1
Kinetic
   energy, 17-1
   energy (fluid), 24-2
Kinetics, 15-1, 15-3

Kingdom, 33-4
Kirchhoff
   current law, 44-4
   laws, 44-3
   voltage law, 44-4
Kline-McClintock equation, 49-5
Knoop hardness, 42-7
KVL, 44-4
kW, 47-2

L
Label, container, 35-3
Lag growth phase, 33-3
Lagging
   circuit, 45-5
   current, 45-3
   impedance triangle, 45-3
Lamellar appearance, 42-5
Lame's solution, 19-3
Laminar flow, 24-3, 24-4
LAN, 47-6
Lanthanide, 37-3
Lanthanon, 37-3
Laplace transform, 8-3, 8-4 (tbl)
Laplacian, 7-5
Large-scale integration, 47-1
Latency, 47-5
Lattice
   Bravais, 40-1
   constant, 40-2
   point, 40-3
Latus rectum
   ellipse, 4-3
   parabola 4-2
Law
   Amagat's, 29-2
   Avogadro's, 26-7
   Boyle's, 25-2, 27-2
   Charles', 27-2
   conservation of momentum, 15-1
   Dalton's, 29-2
   exponential decay, 37-7
   Faraday's, 43-6
   Fick's, 32-3, 42-3
   Fourier's, 31-1
   fourth-power, 31-3
   Gauss', 43-4
   Henry's, 29-2, 39-1
   Hess', 39-2
   Hooke's, 18-1, 41-2
   Joule's, 44-2
   Kirchhoff, 44-3, 44-4
   Lenz', 43-6
   Newton's first, 15-1
   Newton's second, 15-2
   of additive effects, 34-8
   of compound probability, 6-1
   of constant proportions, 37-5
   of convection, Newton's, 31-2
   of cosines, 4-6
   of definite proportions, 37-5
   of joint probability, 6-1
   of mass action, 38-2, 38-3
   of mass conservation (fluid), 24-1
   of multiple proportions, 37-5
   of simple proportions, 37-5
   of sines, 4-6

of thermodynamics, first, 27-2, 27-3
of thermodynamics, second, 28-6
of thermodynamics, zeroth, 26-4
of total probability, 6-1
of viscosity, Newton's, 22-3
Pascal's, 23-1
perfect gas, 25-2
periodic, 37-1
power (fluid), 22-3, 24-3
Raoult's, 29-2, 39-3
Stefan-Boltzmann, 31-3
LC-50, 34-5
LD-50, 34-5
Lead, 12-2
   exposure to, 35-6
Leading
   circuit, 45-5
   current, 45-3
Le Châtelier's principle, 38-2
Legal concentration, 34-5
Legionnaires' disease, 35-17
LEL, 35-4 (tbl), 35-5
Length, effective column, 21-2
Lenz' law, 43-6
Lethal dose, 34-5
Leukemia, 34-4
Level
   confidence, 6-4
   sound power, 35-8
   sound pressure, 35-8
Lever rule (alloys), 42-5
LFL, 35-5
L'Hôpital's rule, 7-2, 7-3
Life
   fatigue, 41-4
   half-, 37-7
   service, 52-1
Lifting, 35-12
Limit, 7-2
   confidence, 6-4
   elastic, 41-2
   endurance, 41-4
   explosive, upper, 35-5
   fatigue, 41-4
   flammable, upper, 35-5
   lower explosive, 35-5
   lower flammable, 35-5
   proportionality, 41-2
   value, threshold, 35-3
Line
   centroid of, 13-1
   charge, 43-3
   defect, 42-3
   energy, fluid, 24-2
   energy grade, 24-2
   hydraulic grade, 24-2
   liquidus, 42-3
   of action (vector), 5-5
   saturated liquid, 26-3
   saturated vapor, 26-3
   solidus, 42-3
   straight, 4-1
Linear
   circuit, 44-3
   differential equation, 8-1
   equations, simultaneous, 5-4

expansion, thermal, 19-1
impulse, 17-2
kinetic energy, 17-1
momentum, 15-1
strain, 18-1, 41-2
system, 14-1
Linearity (Laplace transforms), 8-3, 49-1
Lines
   intersecting, 4-1
   perpendicular, 4-1
Linked list, 48-6
Liquid
   saturated, 26-2
   subcooled, 26-2, 26-3
Liquidus line, 42-3
Liquid-vapor mixture, 26-2, 26-3, 26-6
Lithotroph, 33-6
Liver, 34-3, 34-4
Load
   buckling (column), 21-2
   critical (column), 21-2
   -elongation curve, 41-1
   Euler, 21-2
Loader, 48-2
Loading
   axial, 21-1
   biaxial, 18-2
   BOD, 36-4
   concentric, 21-1
   eccentric, 21-1
   rate, solids, 36-3
   triaxial, 18-2
   uniaxial, 18-2
Local
   -area network, 47-6
   variable, 48-3
LOEL, 34-5
Log (see Logarithm)
Logarithm, 5-1
Logarithmic
   growth phase, 33-4
   mean area, 31-2
Long
   column, 21-1
   stress, 19-2
Logistic
   equation, 33-4
   growth rate constant, 33-4
Longitudinal
   fin, 31-3
   strain, 18-1
   stress, 19-2
Loop
   checking, 47-6
   -current method, 44-5
   gain, 50-2
   transfer, 50-2
Loss
   coefficient, 24-6
   head, minor, 24-6
   minor, 24-6
Loudness, 35-8
Low-cycle fatigue theory, 41-4
Lower
   explosive limit, 35-5
   flammable limit, 35-5

Lowest
   observed adverse effect level, 34-5
   observed effect level, 34-5
LSI, 47-1
Lyme disease, 35-16
Lysosome, 33-2

M
Mach number, 25-2, 26-8
Machine
   DC, 46-3
   language, 48-2
   synchronous, 46-2
Maclaurin series, 5-8
Macro, 48-2
   -assembler, 48-2
   -commands, 47-2
   -fouling, 33-8
MACRS, 52-1
Magnetic
   field, 43-5
   field strength, 43-5
   induction, 43-5
   permeability, 43-5
   sink, 43-5
   source, 43-5
Main memory, 47-1
Major axis, ellipse, 4-3
Malignant tumor, 34-6
Manning's
   equation, 24-10
   roughness coefficient, 24-10
Manometer, 23-2, 25-3
Margin
   gain, 50-6
   phase, 50-6
Martempering, 42-7
Martensite, tempered, 42-7
Mass, 1-1, 26-4
   action, law of, 38-2, 38-3
   center of, 13-2
   conservation of (fluid), 24-1
   control, 27-2
   density, 22-1
   flow rate per unit area, 32-3
   fraction, 29-1
   moment of inertia, 16-1, 16-14 (tbl)
   -spring system, 15-3
   storage device, 47-4
   transfer, 32-2
   transfer coefficient, 32-3
   unit, atomic, 37-1
   velocity, 32-3
Material
   brittle, 41-3
   composite, 42-8
   Safety Data Sheet, 35-3
   testing, 41-1
Matrices, 5-2
Matrix
   coefficient, 5-5
   composite material, 42-8
   determinant, 5-3
   identity, 5-3
   inverse, 5-4
   nonsingular, 5-4
   operations, 5-3

order, 5-3
singular, 5-4
transpose, 5-3
triangular, 5-3
variable, 5-5
Maturity date, 52-3
Maxima, 7-1
Maximum
normal stress theory, 18-5
point, 7-1
power transfer, 44-7
shear stress theory, 18-5
Mean
area, logarithmic, 31-2
arithmetic, 6-2
geometric, 6-2
population, 6-2
sample, 6-2
stress, fatigue, 41-3
weighted, 6-2
Means
difference of, 6-5
sum of, 6-5
Measurement, 49-1
Mechanical
advantage, 12-1
frequency, 46-1
similarity, 25-5
Median, 6-2
Medium electrolytic, 42-1
Megabyte, 47-4
Member
redundant, 10-4, 11-2
three-force, 10-3
two-force, 10-3
zero-force, 11-2
Membrane
cell, 33-1
pumping, 33-3
Memory
cache, 47-2
computer, 47-2
dynamic, 47-3
nonvolatile, 47-3
scratchpad, 47-3
video, 47-2
Meniscus, 22-4
Mensuration, 4-6
Mercaptan compound, 33-10
Mesh current method, 44-5
Mesophile, 33-6 (tbl)
Mesothelioma, 35-6
Metabolism, 33-9
Metal, 37-3
alkali, 37-3
alkaline earth, 37-3
base, 42-3
fume fever, 35-7
transition, 37-3
Metallic bond, 40-4
Metallurgy, 42-1
Metals
FCC, 41-2, 41-5
HCP, 41-2, 41-5
Metastasis, 34-6

Meter
orifice, 25-3
venturi, 25-3
Methane series, 38-5 (tbl)
Method
annual cost, 53-1
annual return, 53-1
bisection, 9-1
capital recovery, 53-1
Newton's, roots, 9-2
of consistent deformations, 20-4
of joints, 11-2
of loss coefficients, 24-6
of sections, 11-2
of undetermined coefficients, 7-4, 8-2
parallelogram, 5-5
polygon, 5-5
present worth, 53-1
superposition, 44-5
Metric system, 1-4
Meyer hardness, 42-7
Microbe, 33-4
organisms, 33-7 (tbl)
Microorganisms, 33-4, 35-14
bioprocessing, 36-2
Microprocessor, 47-1
Microstrain, 49-3
Miller index, 40-3
Minima, 7-1
Minimum
attractive rate of return, 53-1
point, 7-1
Minor
axis, ellipse, 4-3
loss, 24-6
of a matrix element, 5-4
Miscible alloy, 42-3
Mist, airborne, 35-6
Mitochondria, 33-2
Mixing heater, 27-4
Mixture
alloy, 42-4
gas, 26-2
ideal gas, 29-1
liquid-vapor, 26-2, 26-3, 26-6
properties, gas, 29-2
vapor-gas, 26-2
mks system, 1-4
Mneumonic code, 48-2
Mode, 6-2
Model
AC machine, 46-4, 46-5
similar, 25-5
Modem, 47-6
Modified accelerated cost recovery
system, 52-1
Modular ratio, 20-4
Modulator-demodulator, 47-6
Modulus, 43-1
of elasticity, 18-1, 41-2
of rigidity, 18-2
of toughness, 41-4
section, 20-3
shear, 18-2
Young's, 18-1, 41-2
Mohr's circle, 18-4

Molal
boiling point constant, 39-3
freezing point constant, 39-3
Molality, 39-1
Molar, 37-6
flux, 32-3
properties, gas mixture, 29-2
Molarity, 39-1
Mole, 37-5
Molecular
diffusion, 32-3
formula, biomass, 33-11
weight, 37-6
Molecule, 37-1, 37-4, 40-4
Mole fraction, 29-1, 39-1
Mollusk, 33-8
Moment, 10-2, 20-1
and shear relationship, 20-2
arm, 10-2
coupling, 10-3
diagram, 20-2
first, 20-3
free, 10-3
of inertia, 7-4, 13-2, 13-14 (tbl)
of inertia, centroidal, 7-4, 13-2, 13-14 (tbl)
of inertia, mass, 16-1, 16-14 (tbl)
of inertia, polar, 13-2, 13-14 (tbl), 19-3
of the area, first, 7-4, 13-1
of the area, second, 13-2
sign convention, 20-2
statical, 13-1, 20-3
Momentum, 15-1
angular, 16-2
conservation, law, 15-1
fluid, 24-7
-impulse principle, 17-2, 24-6
linear, 15-1
rate of, 24-7
transfer, 32-1, 32-2
Money, time value of, 51-2
Monitor program, 47-2
Monochromatic radiation, 31-2
Monoclinic cell, 40-1
Monod kinetics, 36-2
Monohydroxic base, 38-4
Monoprotic acid, 38-4
Moody friction factor, 32-2
Moody friction factor chart, 24-5
Most probable number index, 33-9
Motion
angular, 14-3
circulation, 14-3
curvilinear, 14-2
plane circular, 14-3
plane, rigid body, 16-2
projectile, 14-3
rectilinear, 14-1
rotational particle, 14-3
simple harmonic, 15-3
Motor, synchronous, 46-2
Moving blade, force on, 24-8
MPN, 33-9
MSDS, 35-3
Multipath pipeline, 24-9
Multiple proportions, law of, 37-5
Multiplexer, 47-4

Multitasking, 47-5
Multi-user, 47-5
Mussell, Zebra, 33-8
Mycotoxin, 35-15

N

Naming compounds, 37-4
Narrow-band line, 47-6
National Institute for Occupational Safety
    and Health, 35-2
Natural
    convection, 31-1
    frequency, 8-3
    frequency of vibration, 15-4
    logarithm, 5-1
    vibration, 15-3
NCRP, 35-10
Neck tension, syndrome, 35-13
Negative feedback, 50-1
NEL, 34-5
Neoplasm, 34-6
Nephrotoxicity, 34-4
Nernst theorem, 26-5
Nervous system, 34-4
Neurotoxicity, 34-4
Neutral
    axis, 20-2, 20-3
    solution, 38-4
Neutron, 37-1
    radiation, 35-10
Newton's
    first law of motion, 15-1
    law of convection, 31-2
    law of viscosity, 22-3
    method, roots, 9-2
    second law of motion, 15-1
Nibble, 47-2
Nine-track tape, 47-5
NIOSH, 34-8, 35-2
Nitrate nitrogen, 33-6
Nitrile, 38-4 (tbl), 38-5 (tbl)
Nitro, 38-4 (tbl)
No observed
    adverse effect level, 34-5
    effect level, 34-5
    effect, 34-5
NOAEL, 34-5
Noble gas, 37-3
Node, 48-6
    -voltage method, 44-6
NOEL, 34-5
Noise, 35-7, 35-9
    control, 35-9
    exposure factors, 35-9
    exposure, permissible, 35-8 (tbl)
    -induced hearing loss, 35-9
    reduction rating, 35-9
Nominal interest rate, 51-5
Nonannual compounding, 51-5
Noncarcinogen, 34-5
Nonhomogeneous differential equation, 8-2
Non-ionizing radiation, 35-10
Non-Newtonian
    fluid, 22-3
    Reynolds number, 24-4
Nonphotosynthetic bacteria, 33-6
Nonpolar covalent bond, 37-4

Nonpotable water, 33-4
Nonsingular matrix, 5-4
Nonvolatile memory, 47-3
Normal
    acceleration, 14-3, 16-3
    component, 15-3
    distribution, 6-3
    force, 12-2, 15-2
    stress, 18-1
    stress (in a fluid), 22-2
    stress theory, maximum, 18-5
    table, unit, 6-4, 6-12 (tbl)
Normality, 39-1
Norton's
    equivalent, 44-6
    theorem, 44-6
Norwalk virus, 33-6
Notch toughness, 41-4
Nozzle, 27-4
Nozzle efficiency, 27-4
NRR, 35-9
Nuclear radiation, 35-10
Nucleon, 37-1
Null
    -indicating bridge, 49-3
    indicator, 49-3
    vector, 5-6
Number
    atomic, 37-1
    Avogadro's, 37-6
    Cauchy, 25-5
    complex, 5-1
    coordination, 40-3
    critical Reynolds, 24-4
    Froude, 25-5
    imaginary, 5-1
    irrational, 5-1
    Mach, 25-2, 26-8
    of atoms in a cell, 40-2, 40-3 (tbl)
    oxidation, 37-4
    Prandtl, 32-2
    rational, 5-1
    real, 5-1
    Reynolds, 24-3, 25-5
    Schmidt, 32-3
    Sherwood, 32-3
    Stanton, 32-3
    Weber, 25-5
Numerical
    analysis, 9-1
    event, 6-2
Nyquist rate, 49-4

O

Obligate
    aerobe, 33-6
    anaerobe, 33-6
Observability, system, 50-7
Observed effect level, 34-5
Occupational Safety and Health Act, 34-8,
    35-2
Odd parity, 47-3
Offset, parallel, 41-2
Offspring node, 48-6
Ohm's law, 44-3
    AC, 45-4
Olefin series, 38-5 (tbl)

One-tail confidence limit, 6-4
Open
    channel flow, 24-6, 24-11
    feedwater heater, 27-4
    -loop gain, 50-2
    manometer, 23-2
    system, 27-2, 27-3
    system, steady-flow, 27-2
Operating system, 47-2
Operations
    matrix, 5-3
    vector, 5-5
Opposite side, 4-4
Optical drive, 47-4
Order
    differential equation, 8-1
    of a matrix, 5-3
    of a reaction, 38-3
    of difference equation, 4-5
    of the system, 50-2
Organelle, 33-1
Organic
    acid, 38-5 (tbl)
    chemistry, 38-4
    compound, 38-4
    compound family, 38-4
    functional group, 38-4
    loading rate, 36-3
Organism
    indicator, 33-9
    microbe, 33-7 (tbl)
Organismal growth, 33-3
Organotroph, 33-6
Orifice
    meter, 25-3
    plate, 25-3
    submerged, 25-4
Orthorhombic cell, 40-1
OS, 47-2
Oscillation, period of, 15-4
OSHA, 34-8, 35-2
    noise, 35-9
Otto cycle, 28-3
Output
    matrix, 50-7
    vector, 50-7
Overall transfer function, 50-2
Overdamped, 8-1
Overhead, computer, 48-3
Oxidation, 37-3, 38-1
    number, 37-4, 37-5 (tbl)
    potential, 42-1, 42-2 (tbl)
    -reduction reaction, 38-1
    state, 37-4
Oxidizing agent, 38-1
Oxygen demand, biochemical, 33-9

P

Packing factor, 40-3
Panel (truss), 11-1
Parabola, 4-2
    center, 4-2
    directrix, 4-2
    focus, 4-2
    latus rectum, 4-2
    mensuration, 4-6
Paraboloid of revolution, 4-7

Paradox, hydrostatic, 23-1
Paraffin series, 38-5 (tbl)
Parallel
    axis theorem, 7-5, 13-2, 16-1
    capacitors, 44-3
    device, 47-4
    interface, 47-3
    offset, 41-2
    pipeline, 24-9
    plate capacitor, 44-2
    resistance, 44-2
    RLC, 45-6
Parallelogram, 4-7
Parallelogram method, 5-5
Parasite, aquatic, 33-8
Parent
    ingredient, 42-3
    node, 48-6
Parity, 47-3
Parseval
    equality, 8-3
    relation, 8-3
Partial
    derivative, 7-2
    fraction, 7-4
    pressure, 29-2
    pressure ratio, 29-2
    volume, 29-2
Particle
    kinetics, 15-3
    motion, rotational, 14-3
Particular solution, 8-2
Particulate, 35-6
Partition coefficient, 33-3
Parts, integration by, 7-3
Pascal's law, 23-1
Passing element, electrical, 44-1
Passive
    circuit element, 45-3
    diffusion, 33-3
    system, 50-5
Path of projectile, 14-3
Pathogen, 33-4, 33-5 (tbl)
    blood-borne, 35-16
Pathway, exposure, 34-2
Pay-back period, 53-2
P-code, 48-3
Pearlite, 42-5
Pedestal, 21-2
PEL, 34-8
    noise, 35-9
Pendulum, torsional, 16-3
Percent
    elongation, at failure 41-3
    excess air, 30-2
    slip, 46-3
    theoretical air, 30-2
Perfect gas, 26-2 (see also Ideal gas)
Performance, coefficient of, 28-5
Perimeter, wetted, 24-6
Period
    depreciation, 52-1
    of oscillation, 15-4
    pay-back, 53-2
    transient, 31-1
    waveform, 8-3, 45-2

Periodic
    law, 37-1
    table, 37-1, 37-2
Peripheral
    devices, 47-1
    nervous system, 34-4
Peritectic reaction, 42-4 (tbl)
Peritectoid reaction, 42-4 (tbl)
Permanent set, 41-2
Permeability
    clothing, 35-12
    magnetic, 43-5
    of free-space, 43-5
Permeation rate, 35-5
Permissible
    exposure level, noise, 35-9
    exposure limit, 34-8, 35-3
Permittivity, 43-3
Permutation, 6-1
Perpendicular
    axis theorem, 13-2
    lines, 4-1
Persistence, 33-11
Personal protective equipment, 35-5
pH, 38-4
Phase
    alloy, 42-3
    angle, 45-3
    angle, sinusoidal, 45-1
    characteristic, 50-5
    crossover frequency, 50-6
    diagram, 26-2, 42-3
    diagram, binary, 42-3
    growth, organismal, 33-3, 33-4
    margin, 50-6
    rule, Gibbs', 29-5, 42-5
    shift, 45-3
    thermodynamic, 26-1
Phasor form, 43-1, 43-2
Phenol, 38-5 (tbl)
Photoautotroph, 33-8
Photosynthesis, 33-2
Phototroph, 33-6
Physical hazard, 35-2
Pier, 21-2
Pi-group, 25-5
Pin joint, 11-1
Pinned support, 10-5
Pipe
    bend, 24-7
    fitting, 24-6
Pipeline
    multipath, 24-9
    parallel, 24-9
Pitch
    angle, 12-2
    screw, 12-2
Pitot tube, 25-2
Planar defect, 42-3
Plane
    circular motion, 14-3
    motion, rigid body, 16-2
    stress, 18-3
    surface, submerged, 23-3
Plant, 50-7
Plasma, 34-3

Plastic
    impact, 17-3
    strain, 41-2
Plate, orifice, 25-3
Platelet, 34-3
Platter, 47-4
pOH, 38-4
Point
    boiling, 26-4
    contraflexure, 7-1
    critical, 7-1, 26-2 (fig), 26-3, 42-5
    defect, 42-3
    extreme, 7-1
    ice, 26-4
    inflection, 7-1
    lattice, 40-3
    maximum, 7-1
    minimum, 7-1
    of instability, 50-6
    -slope form, line, 4-1
    stagnation, 25-2
    triple, 26-3, 26-4
    triple, water, 42-6
    yield, 41-2
Points, distance between, 4-1, 4-4
Poisson's ratio, 18-1
Polar
    coordinates, 5-2
    covalent bond, 37-4
    form, 5-2, 43-1, 43-2
    form, sinusoidal, 45-2
    moment of inertia, 13-2, 13-14 (tbl), 19-3
Pole
    magnetic, 43-5
    stability, 50-3
    -zero diagram, 50-3
Polygon
    mensuration, 4-7
    method, 5-5
Polysaccaride, 38-5 (tbl)
Polytropic process, 27-1, 27-3
Pond, facultative, 36-3
Population, 6-2
Portable language, 48-3
Position
    forms of, 14-2
    vector, 10-2
Positive feedback, 50-1
Potency factor, 34-6
Potential
    difference, 43-4, 44-1
    energy, 17-2
    energy, elastic, 17-2
    energy (fluid), 24-2
    energy, gravitational, 17-2
    gradient, electric, 43-4
    half-cell, 42-1, 42-2 (tbl)
    head, 24-2
    oxidation, 42-1, 42-2 (tbl)
    reduction, 42-1, 42-2 (tbl)
Power
    angle, 45-5
    brake pump, 25-2
    cycle, 28-1, 28-2
    dissipation, motor, 46-5
    electrical, 44-2

factor, 45-5
factor correction, 45-5
hydraulic, 25-1
law (fluid), 22-3, 24-3
level, sound, 35-8
pump, 25-1
screw, 12-2
series, 5-7
shaft, 27-3
reactive, 45-5
real, 45-5
resistive element, 44-2
transfer, maximum, 44-7
turbine (fluid), 24-9
water, 25-1
Prandtl number, 32-2
Precipitation, 39-2
Precision, 49-2
Present worth, 51-3
analysis, 53-1
method, 53-1
Pressure, 22-2, 26-4
absolute, 22-2
average fluid, 23-3
barometric, 22-2
center of, 23-4
drop in a pipe, 24-3
energy, 24-2
flow, 24-6
fluid, 23-1
gage, 22-2
head, 24-2
hydrostatic, 23-1, 23-3
partial, 29-2
ratio, partial, 29-2
saturation, 29-3
sound, 35-8
stagnation, 25-3
standard atmospheric, 22-2, 26-4 (tbl)
vapor, 35-3
vessel, 19-2
vessel, thick-walled, 19-3
Primary
bond, 40-4
dimensions, 25-4
winding, 45-7
Principal
Archimedes', 23-5
energy conservation, 17-2
impulse-momentum, 17-2, 24-6
Le Châtelier's, 38-2
stress, 18-3
stress (tanks), 19-2
Prion, 35-14
Priority, ethical, 54-2
Probabilistic economics problems, 52-3
Probable risk, 34-7
Probability, 6-1
complementary, 6-1
compound, law of, 6-1
conditional, 6-1
density function, 6-3
distribution function, 6-3
joint, law of, 6-1
total, law of, 6-1
Probing, 48-5

Problem, initial value, 8-2
Procedure-oriented programming, 48-3
Process
activated sludge, 36-2
adiabatic, 25-2, 27-1
constant entropy, 27-1
constant pressure, 27-1–27-3
constant temperature, 27-1, 27-3
constant volume, 27-1, 27-2
irreversibility, 28-6
isentropic, 25-2, 26-8, 27-1, 27-3
isobaric, 27-1
isochoric, 27-1
isometric, 27-1
isothermal, 27-1, 27-3
polytropic, 27-1, 27-3
quasiequilibrium, 27-1
quasistatic, 27-1
reversible, 27-2
thermodynamic, 27-1
throttling, 27-1, 27-4
Processing, thermal, 42-6
Product, 38-1
cross, 5-6
dot, 5-5
of inertia, 13-3, 16-14 (tbl)
scalar, 5-5
solubility, 39-2
vector, 5-6
Production, entropy, 26-5
Professional conduct, rule, 54-1
Program
computer, 48-1
counter, 47-1
Progression, 5-7
arithmetic, 5-7
geometric, 5-7
state, cancer, 34-6
Project cost, 51-2
Projectile
motion, 14-3
path, 14-3
Prokaryote, 33-1, 35-14
Promotion stage, cancer, 34-6
Pronator syndrome, 35-13
Properties
gas mixture, 29-2
of acids, 38-4
of air, 25-2
of bases, 38-4
of water, 22-4
structure-sensitive, 42-3
Property
extensive, 26-3
intensive, 26-3
Proportional
region, 41-2
strain, 41-2
Proportionality limit, 41-2
Proportions
law of constant, 37-5
law of definite, 37-5
law of multiple, 37-5
law of simple, 37-5
Propulsion, jet, 24-8
Protection, hearing, 35-9

Protective equipment, personal, 35-5
Protein allergen, 35-14
Protista, 33-4
Proton, 37-1
Prototype, similar, 25-5
Protozoa, 33-8
flagellated, 33-8
Pseudocode, 48-3
Pseudoplastic fluid, 22-3, 24-4
Psittacosis, 35-15
Psychrometric chart, 29-3, 29-4 (fig)
Psychrometrics, 29-3
Psychrophile, 33-6 (tbl)
Public, ethics, 54-3
Pulley, 12-1
advantage, 12-1
fixed, 12-1
free, 12-1
Pulmonary toxicity, 34-3
Pump, 27-4
efficiency, 25-2
heat, 28-5
power, 25-1
power, brake, 25-2
Pumping, membrane, 33-3
Pushdown stack, 47-1
p-V work, 26-5, 27-2
Pyrex, 19-2

Q
Quadratic equation, 4-1
Quality, 26-6
electrical, 45-6
factor, 50-5
radiation, 35-10
Quantity, break even, 53-2
Quasiequilibrium, 27-2
Quasiequilibrium process, 27-1
Quasistatic process, 27-1
Quenching, 42-7
Quicksort, 48-5
Quotient, respiratory, 33-11

R
Radial component, 14-2, 15-3
Radiant heat transfer, 31-3
Radiation, 31-1, 31-2
effect on humans, 35-11
exposure to, 35-10
monochromatic, 31-2
spectral, 31-2
Radiator
efficiency, 31-3
ideal, 31-2
Radical, 37-5 (tbl)
Radioactive
decay, 35-11, 37-6
dust, 35-7
Radius
hydraulic, 24-6
of curvature, 7-2, 14-3, 20-4
of gyration, 13-3, 13-14 (tbl), 16-2, 16-14 (tbl)
RAM, 47-3
Random
access device, 47-4

file, 48-4
variable, 6-3
Rankine
cycle, 28-2, 28-3
temperature, 26-4
Raoult's law, 29-2, 39-3
Rate
annual effective interest, 51-5
inflation, 52-3
interest, effective, 51-2
nominal interest, 51-5
of momentum, 24-7
of reaction, 38-2
of refrigeration, 28-5
of return analysis, 53-1
of return, minimum attractive, 53-1
of shear formation, 22-3
of strain (fluid), 22-3
per annum, 51-5
permeation, 35-5
shear, 22-3
Ratio
air-fuel, 30-2
benefit-cost, 53-2
common, 5-7
compression, 28-4
critical slenderness, 21-2
hazard, 34-8
humidity, 29-3
modular, 20-4
of specific heats, 25-2, 26-8
of transformation, 45-7
partial pressure, 29-2
Poisson's, 18-1
recycle, 36-3
slenderness, 21-2
turns, 45-7
vapor-hazard number, 35-5
Rationalizing the denominator, 5-2, 43-2
Rational number, 5-1
Raynaud's syndrome, 35-14
RC transient, 44-7
RDRAM, 47-3
Reactance, capacitive, 45-4
Reactant, 38-1
Reaction, 10-4, 10-6
biologic, 33-11
chemical, 38-1
combustion, 30-1
electrochemical, 39-3
electrolytic, 39-3
endothermic, 39-2
enthalpy of, 39-2
exothermic, 39-2
heat of, 39-2
order of, 38-3
oxidation-reduction, 38-1
rate of, 38-2
redox, 38-1
reversible, 38-2
speed of, 38-2
Reactive
part, 45-3
power, 45-5
Reactivity, 35-3
Read-only, 47-4

Read-write, 47-4
Real
gas, 26-2
number, 5-1, 43-1
part, 45-3
power, 45-5
-time processing, 47-5
Receipt (cash), 51-1
Recommended Exposure Limits, 35-2
Recording density, 47-5
Recovery, 41-2, 42-7
Recrystallization temperature, 42-6
Rectangular
coordinate form (position), 14-2
coordinates, 14-1
fin, 31-3
form, 43-1, 43-2
form, complex number, 5-1, 5-2
sinusoidal, 45-2
Rectilinear motion, 14-1
Rectum
latus, ellipse, 4-3
latus, parabola, 4-2
Recursive call, 48-3
Recycle
flow rate, 36-3
ratio, 36-3
Red blood cell, 34-3
Redox reaction, 38-1
Reduced
equation, 8-2
Reducing agent, 38-2
Reduction, 37-3, 38-1
degree of, 33-11
potential, 42-1, 42-2 (tbl)
Redundant
member, 10-4, 11-2
support, 10-4
Reference
concentration, 34-5
dose, 34-5
value, 50-7
Reflectivity, 31-2
Refrigeration
cycle, 28-4
rate, 28-5
Region
elastic, 41-2
proportional, 41-2
transition, 24-4
Register, instruction, 47-1
Relation, Parseval's, 8-3
Relative
column, 48-4
dispersion, 6-3
humidity, 29-3
roughness, 24-5
stability, 50-5
velocity difference, 24-8
Reliability, 49-2
Relief, stress, 42-7
Reovirus, 33-6
Repetitive stress, 35-13
Reproductive toxicity, 34-4
Reservoir, heat, 28-1

Residence time, 36-4
hydraulic, 36-2, 36-3
Resistance, 44-2
temperature detector, 49-2
thermal, 31-2
thermal coefficient of, 44-2
Resistive
circuit, 44-4
part, 45-3
Resistivity, 44-1, 44-2
Resistor, 44-2
ideal, 45-3
Resonance, 8-2, 45-5
Resonant
circuit, 45-5
frequency, 45-5
Respirator, 35-7
Respiratory quotient, 33-11
Respiratory tract, 34-2
Response, 50-2
equation, 50-7
versus dose, 34-4
steady-state, 50-3
Restitution, coefficient of, 17-3
Resultant, 10-1
hydrostatic, 23-3, 23-4
vector, 5-5
Reticulum, endoplasmic, 33-2
Return
minimum attractive rate, 53-1
rate of, 53-1
Reverse transfer function, 50-2
Reversible
process, 27-2
reaction, 38-2
work, 27-2
Revolution, paraboloid of, 4-8
Reynolds number, 24-3–25-5
RfC, 34-5
RfD, 34-5
Rhombohedral cell, 40-1
Ribonucleic acid, 33-2
Right
circular cone, mensuration, 4-7
circular cylinder, mensuration, 4-7
-hand rule, 5-6, 10-2, 43-5
triangle, 4-4
Rigid
body, 14-1
body plane motion, 16-2
truss, 11-1
Rigidity, modulus of, 18-2
Ring
benzene, 38-4 (tbl)
slender, 16-14
Ripple, electrical, 46-4
Risk, probable, 34-7
RL transient, 44-7
rms value, 6-2, 49-5
RNA, 33-2
Roadway banking, 16-3
Rockwell hardness, 42-7
Rocky Mountain spotted fever, 35-16
Rod, slender, 16-14
Roller support, 10-5
Rolling wheel, 16-3

Root
    double, 4-2
    node, 48-6
    -locus diagram, 50-6
    -mean-square, 35-8, 49-5
    -mean-square value, 6-2, 45-2
    of quadratic, 4-1
ROR, 53-1
Rosette train gage, 49-3
Rotating
    beam test, 41-3
    machine, 46-1
Rotational
    delay, 47-5
    kinetic energy, 17-1
    particle motion, 14-3
Rotation, instantaneous center of, 16-2
Rotavirus, 33-6
Rotifer, 33-8
Rotor, 46-1
Roughness
    relative, 24-5
    specific, 24-5
Roughness coefficient
    Manning's, 24-10
    Hazen-Williams, 24-10
Routh
    criterion, 50-6
    table, 50-6
RS-232, 47-4
RTD, 49-2
Rule
    Amagat-Leduc, 29-2
    Cramer's, 5-5
    Gibbs' phase, 29-2, 29-5, 42-5
    lever, 42-5
    L'Hôpital's, 7-2, 7-3
    professional conduct, 54-1
    right-hand, 5-6, 10-2, 43-5
Rules
    of Professional Conduct, NCEES, 54-14
    SI system, 1-6
Runaway speed, 24-9

S
Safe human dose, 34-5
Safety solvent, 35-5
Sample
    mean, 6-2
    standard deviation, 6-2
    variance, 6-2
Sampling, 49-4
Saturated
    air, 29-3
    hydrocarbon, 38-5 (tbl)
    liquid, 26-2
    liquid line, 26-3
    solution, 39-2
    vapor, 26-2
    vapor line, 26-3
Saturation
    magnetic, 46-3
    pressure, 29-3
Scalar, 5-5
Schmidt number, 32-3
Scientific condition, standard, 38-3
Scratchpad memory, 47-3

Screw, 12-2
S-curve, 33-4
SDRAM, 47-3
Searching, 48-5
Seat of EMF, 44-1
Secant, 4-5
Second
    law of motion, Newton's, 15-1
    law of thermodynamics, 28-6
    moment of the area, 13-2
    -order differential equation, 8-1
Secondary
    bond, 40-4
    winding, 45-7
Section
    conic, 4-2
    modulus, 20-3
Sections, method of, 11-2
Sector, 47-4
    circular, mensuration, 4-7
Seek time, 47-5
Segment
    circular, mensuration, 4-7
    commutator, 46-3
Semimajor distance, ellipse, 4-3
Sensitivity, 49-1
Sequence, 5-6
    convergent, 5-6
    divergent, 5-6
Sequential
    access device, 47-5
    file, 48-4
Serial
    device, 47-4
    interface, 47-4
Series, 5-7
    capacitors, 44-3
    finite, 5-7
    Fourier, 8-2
    galvanic, 42-1, 42-2 (tbl)
    harmonic, 8-2
    infinite, 5-7
    Maclaurin, 5-8
    methane, 38-5 (tbl)
    olefin, 38-5 (tbl)
    paraffin, 38-5 (tbl)
    power, 5-7
    resistance, 44-2
    RLC, 45-6
    Taylor's, 5-8
Serum hepatitis, 35-16
Service life, 52-1
Servomechanism, 50-7
Set, permanent, 41-2
Shaft, 19-3
    power, 27-3
    work, 27-3
Shannon's sampling theorem, 49-4
Shape factor, 31-3
Shear, 20-1
    and moment relationship, 20-2
    diagram, 20-2
    flow, 19-3
    formation, rate of, 22-3
    modulus, 18-2
    rate (fluid), 22-3

sign convention, 20-1
strain, 18-1
stress, 18-1
stress distribution, beam, 20-3
stress (in a fluid), 22-3
stress theory, maximum, 18-5
Sheave, 12-1
Sheet charge, 43-3
Shell, 19-3
    hollow, 19-3
    thin-walled, 19-3
Sherwood number, 32-3
Shielding, radiation, 35-11
Short-term exposure level, 34-7
Shot peening, 41-4
Side
    adjacent, 4-4
    opposite, 4-4
Sievert, unit, 35-10
Sigmoidal shape, 33-4
Sign convention
    moment, 20-2
    shear, 20-1
    thermodynamic, 27-2
Silica, 35-6
Similarity, 25-5
    dynamic, 25-5
    geometric, 25-5
    kinematic, 25-5
    mechanical, 25-5
Similitude, 25-5
Simple
    asphyxiant, 35-3
    harmonic motion, 15-3
    interest, 51-5
    proportions, law of 37-5
    support, 10-5
Simplex communication, 47-6
Simultaneous linear equations, 5-4
Sine, 4-5
    -cosine relationships, 45-1
    law, 4-6
Single payment
    cash flow, 51-1
    compound amount factor, 51-3
    equivalence, 51-3
    present worth factor, 51-3
Singular matrix, 5-4
Sink, reservoir, 28-1
Sinking fund, 51-4
Sinusoidal waveform, 45-1, 45-2
Skin, 34-2
SI system, 1-6
Slender
    ring, 16-14
    rod, 16-14
Slenderness ratio, 21-2
Sliding plate viscometer, 22-3
Slip
    ac machine, 46-3
    ring, 46-2
Slope, 4-1
    factor, 34-6
    -intercept form, line 4-1
Sludge
    activated, 36-2, 36-3 (tbl)

volume, 36-3
Slug, 1-4
Small intestine, 34-3
S-N curve, 41-3
Soft X-ray, 35-10
Software, 48-1
  flow control, 47-4
  handshaking, 47-4
Solid
  phase, 26-2
  -solution alloy, 42-3
Solids
  in liquids, solutions of, 39-1
  loading rate, 36-3
  residence time, 36-3
Solidus line, 42-3
Solubility product, 39-2
Solute, 39-2
Solution
  heat of, 39-2
  neutral, 38-4
  particular, 8-2
  saturated, 39-2
Solutions
  of gases in liquids, 39-1
  of solids in liquids, 39-1
Solvent, 39-2
  safety, 35-5
Sorting, 48-5
Sound, 35-7
  audible, 35-7
  intensity, 35-8
  power level, 35-8
  pressure, 35-8
  pressure level, 35-8
  sources, combining, 35-8
  speed of, 25-2, 35-8 (tbl)
Source
  code, 48-1
  equivalent, 44-6
  electrical, dependent, 44-1
  electrical, independent, 44-1
  reservoir, 28-1
Sources, sound, combining, 35-8
Space, interstitial, 40-3
Specific
  gas constant, 25-2, 26-7
  gravity, 22-2
  heat, 26-5, 26-6 (tbl), 26-7 (tbl)
  heat at constant pressure, 25-2
  heat at constant volume, 25-2
  heats, ratio of, 25-2, 26-8
  humidity, 29-3
  kinetic energy (fluid), 24-2
  potential energy, 24-2
  pressure energy, 24-2
  roughness, 24-5
  volume, 22-1, 26-4
  weight, 1-3, 22-1
Specimen, test, 41-1
Spectral radiation, 31-2
Speed
  of efflux, Torricelli's, 24-8
  of reaction, 38-2
  of sound, 25-2, 35-8
  runaway, 24-9

synchronous, 46-1, 46-2
Sphere, 4-4
  general form, 4-4
  mensuration, 4-7
  solid, 16-14
  standard form, 4-4
Spherical tank, 19-2
s-plane, 50-3
Split-ring, 46-3
Spring
  constant, 17-2
  constant, torsional, 19-3
  -mass system, 15-3
  stiffness, 17-2
Spreadsheet, 48-4
Square screw threads, 12-2
Stability, 49-2, 50-5
  relative, 50-5
Stable system, 50-5
Stack, 47-1
Stack-up, 49-5
Stagnation
  point, 25-2
  pressure, 25-3
Standard
  atmospheric pressure, 22-2, 26-4 (tbl)
  deviation, 6-2
  form, circle, 4-4
  form, ellipse, 4-3
  form, hyperbola, 4-3
  form, parabola, 4-2
  form, sphere, 4-4
  form, straight line, 4-1
  oxidation potential, 42-1, 42-2 (tbl)
  rate digester, 36-4
  scientific condition, 38-3
  state, enthalpy, 39-2
Stanton
  diagram, 24-5
  number, 32-3
Start bit, 47-6
State
  equation of, 25-2, 26-7, 38-3
  oxidation, 37-4
  standard, enthalpy, 39-2
  thermodynamic, 26-1
Statement
  Clausius, 28-6
  Kelvin-Planck, 28-6
Static, friction, 12-2, 15-2
Statically
  determinate, 10-4
  determinate truss, 11-2
  indeterminate, 10-4
Statical moment, 13-1, 20-3
Statics, 10-1, 23-1
Stationary growth phase, 33-4
Stator, 46-1
Statute, 54-1
Statutory rule, 54-1
Steady
  -flow energy equation, 27-3
  -flow open system, 27-2
  -flow system, 27-3
  -state response, 50-3
Steel, 42-5

Steel beam, 20-4
Stefan-Boltzmann
  constant, 31-3
  law, 31-3
Stenothermophiles, 33-6 (tbl)
Step-down transformer, 45-7
Step-up transformer, 45-7
Stiffness
  spring, 17-2
  torsional, 19-3
Stoichiometric air, 30-1, 30-2
Stoichiometry, biologic, 33-11
Stop bit, 47-6
Storage capacity, 47-4
STP, 38-5
Straight
  fin, 31-3
  line, 4-1
Strain, 18-1, 41-1
  axial, 18-1
  elastic, 41-2
  energy, 18-2, 41-4
  engineering, 41-1
  gage, 49-2
  hardening, 42-6
  heat, 35-12
  inelastic, 41-2
  linear, 18-1, 41-2
  longitudinal, 18-1
  plastic, 41-2
  proportional, 41-2
  rate of (fluid), 22-3
  sensitivity factory, 49-3
  shear, 18-1
  thermal, 19-1
  three-dimensional, 18-4
  true, 41-1
Streaming tape, 47-5
Strength, 41-3
  breaking, 41-3
  fatigue, 41-4
  magnetic field, 43-5
  of acid, 38-4
  of base, 38-4
  of the B-field, 43-5
  tensile, 41-2
  ultimate, 41-2
  yield, 41-2
Stress, 18-1, 41-1
  allowable, 18-5
  beam, 20-2
  bending, 20-2
  circumferential, 19-2, 19-3
  cold, 35-11, 35-12
  combined, 18-3
  concentration factor, 18-5
  corrosion, 42-2
  effective, 18-5
  endurance, 41-4
  engineering, 41-1
  flexural, 20-2
  heat, 35-11, 35-12
  hoop, 19-2, 19-3
  in a fluid, 22-2, 22-3
  long, 19-2
  longitudinal, 19-2

mean fatigue, 41-3
  normal, 18-1
  plane, 18-3
  principal, 18-3
  principal (tanks), 19-2
  relief, 42-7
  shaft, 19-3
  shear, 18-1
  shear, distribution, beam, 20-3
  tangential, 19-2, 19-3
  torsional, 19-3
  true, 41-1
  von Mises, 18-5
  yield, 41-2
Strobe signal, 47-3
Strong data type, 48-3
Structural cell, 11-1
Structure, cell, 33-1
Structured programming, 48-3
Structure-sensitive properties, 42-3
Student's t-distribution, 6-4, 6-9 (tbl)
Subcooled liquid, 26-2, 26-3
Subcutaneous fatty tissue, 34-2
Successive minima, 48-5
Submerged
  orifice, 25-4
  plane surface, 23-3
Substitution
  integration by, 7-3
  trigonometric, 7-3
Sugars, 38-5 (tbl)
Summer, 50-1
Sum of means, 6-5
Sunk cost, 51-1
Superheated vapor, 26-2, 26-3
Superheating, 28-3
Superposition, 20-4
  method, 44-5
  Laplace transforms, 8-3
  theorem, 44-5
Supplementary unit, 1-4
Supplier, ethics, 54-3
Support
  pinned, 10-5
  reaction, 10-5
  redundant, 10-4
  roller, 10-5
  simple, 10-5
Surface
  control, 27-2
  extended, 31-3
  free, 24-6
  frictionless, 10-5
  tension, 22-3
  treatment, 41-4
Surroundings, 27-2
Susceptance, 45-4
Suspension, 39-2
Symbol, flowchart, 48-2
Synchronous
  machine, 46-2
  memory, 47-3
  motor, 46-2
  speed, 46-1, 46-2
  transmission, 47-6
Syphilis, 35-16

System
  accelerated cost recovery, 52-1
  boundary, 27-2
  closed, 27-2
  controllability, 50-7
  English engineering, 1-2
  English gravitational, 1-4
  function, 50-2
  isolated, 27-2
  linear, 14-1
  metric, 1-4
  observability, 50-7
  open, 27-2, 27-3
  response, 50-2
  spring-mass, 15-3
  steady-flow, 27-3
  steady-flow open, 27-2
  steady-state, 27-3
  thermodynamic, 27-2
Systemic effect, 34-3

T

Table, periodic, 37-1, 37-2 (tbl)
Tangent, 4-5, 4-4
Tangential
  acceleration, 14-3
  blade velocity, 24-9
  component, 14-3, 15-3
  stress, 19-2, 19-3
  stress (in a fluid), 22-2
  velocity, 14-3
Tank
  spherical, 19-2
  thin-walled, 19-2
Tape, streaming, 47-5
Tapeworm, 33-8
Taylor's
  formula, 5-8
  series, 5-8
t-distribution, 6-4, 6-5, 6-9 (tbl)
TDM, 47-4
Teleprocessing, 47-6
Temperature, 26-4
  absolute, 26-4
  critical, 42-5
  dew-point, 29-3
  dry-bulb, 29-3
  ductility transition, 41-4
  eutectic, 42-4
  fracture appearance transition, 41-5
  fracture transition, 41-4
  Kelvin, 26-4
  Rankine, 26-4
  range, bacterial growth, 33-6
  recrystallization, 42-6
  wet-bulb, 29-3
Tempered martensite, 42-7
Tempering, 42-7
Tendency, central, 6-2
Tendonitis, 35-13
Tennis elbow, 35-13
Tenosynovitis, 35-13
Tensile
  strength, 41-2
  test, 41-1, 41-3
Tension, surface, 22-3
Tensor, 5-5

Term, differential, 7-3
Ternary
  acid, 37-4
  compound, 37-4
Test
  Charpy, 41-4
  endurance, 41-3
  fatigue, 41-3
  for inflection point, 7-2
  for maximum, 7-2
  for minimum, 7-2
  hardness, 42-7
  hypothesis, 6-5
  impact, 41-4
  rotating beam, 41-3
  tensile, 41-1, 41-3
Testing, material, 41-1
Tetanus, 35-17
Tetragonal cell, 40-1
Theorem
  Buckingham π-, 25-5
  buoyancy, 23-4
  central limit, 6-4
  final value, 8-3
  final value, z-transform, 8-5
  Fourier's, 8-2
  fundamental of calculus, 7-3
  Gibbs', 29-2
  initial value, 8-3
  initial value, z-transform, 8-5
  Kalman, 50-7
  Nernst, 26-5
  parallel axis, 7-5, 13-2, 16-1
  perpendicular axis, 13-2
  Shannon's sampling, 49-4
  superposition, 44-5
  transfer axis, 7-5, 13-2
  work-energy, 17-2
Theoretical air, percent, 30-2
Theory
  Arrhenius, of acids, 38-3
  Arrhenius, of bases, 38-4
  distortion energy, 18-5
  failure, 18-5
  low-cycle fatigue, 41-4
  maximum normal stress, 18-5
  maximum shear stress, 18-5
Thermal
  coefficient of resistance, 44-2
  efficiency, 28-2
  equilibrium, 26-4
  expansion, coefficient of, 19-1, 19-2 (tbl)
  processing, 42-6
  resistance, 31-1
  strain, 19-1
  stress, 35-11
Thermistor, 49-2
Thermodynamic
  process, 27-1
  sign convention, 27-2
  state, 26-1
  system, 27-2
Thermodynamics
  first law of, 27-3
  second law of, 28-6
  zeroth law of, 26-4

Thermophile, 33-6 (tbl)
Thevenin's
  equivalent, 44-6
  theorem, 44-6
Thickness, half-value, 35-11
Thick-walled pressure vessel, 19-2, 19-3
Thin-walled
  shell, 19-3
  tank, 19-2
Thread
  friction angle, 12-2
  screw, 12-2
Threaded list, 48-6
Three-force member, 10-4
Threshold limit value, 34-7, 35-3
  ceiling, 34-8
Throttling
  process, 27-1, 27-4
  valve, 27-4
Ticks, 35-16
Time
  breakthrough, 35-5
  constant, 44-7
  division multiplexing, 47-4
  -sharing, 47-5
  -temperature-transformation curve, 42-7
  value of money, 51-2
  -weighted average, 34-8
Tip
  adiabatic, 31-3
  insulated, 31-3
TLV, 34-7, 35-3, 35-4 (tbl), 35-6
  -C, 34-8
  noise, 35-9 (tbl)
  -STEL, 34-7
Ton (refrigeration), 28-5
Top-down programming, 48-3
Toracic outlet Syndrome, 35-14
Torque, 10-2
  DC motor, 46-5
Torricelli's speed of efflux, 24-8
Torsion, 19-3
Torsional
  free vibration, 16-3
  pendulum, 16-3
  spring constant, 19-3
  stiffness, 19-3
  stress, 19-3
Total
  coliform, 33-9
  probability, law of, 6-1
Toughness, 41-4
  modulus of, 41-4
  notch, 41-4
Toxic shock syndrome, 35-17
Toxicant, 34-3
Toxicity, 35-3
  pulmonary, 34-3
Toxicology, 34-1
Toxin, 35-14
Toxoplasmosis, 35-16
Track, hard drive, 47-4
Trade-in allowance, 51-2
Transducer, 49-1
Transfer
  axis theorem, 7-5, 13-2

energy, 32-1, 32-2
  function, 50-1
  heat, 31-1
  mass, 32-2
  momentum, 32-1, 32-2
  reverse, 50-2
Transform
  Laplace, 8-3, 8-4 (tbl)
  z-, 8-5
Transformation
  method, 20-4
  of axes (stress), 18-3
  ratio, 45-7
Transformer
  electrical, 45-6
  ideal, 45-7
Transient
  electrical, 44-7
  period, 31-1
  region, 24-4
  temperature, ductility, 41-4
  temperature, fracture, 41-4
  temperature, fracture appearance, 41-5
Transition metals, 37-3
Transmission
  speed, 47-4
  standard, 47-6
Transmissivity, 31-2
Transport
  cell, 33-3
  phenomena, 32-1
  phenomena analogy, 32-3
Transpose matrix, 5-3
Transverse
  axis, hyperbola, 4-3
  component, 14-2
Treatability constant, 36-4
Treatment
  heat, 42-6
  surface, 41-4
Tree structure, 48-6
Trestle, 11-1
Triad, Cartesian, 5-5
Triangle
  general, 4-6
  law of sines, 4-6
  right, 4-4
Triangular matrix, 5-3
Triaxial loading, 18-2
Triclinic cell, 40-1
Trigger finger, 35-14
Trigonometric
  form, 43-1, 43-2
  form, complex number, 5-1, 5-2
  identities, 4-5
  sinusoidal, 45-2
  substitution, 7-3
Trihydroxic base, 38-4
Triple point, 26-3, 26-4, 42-6
Triprotic acid, 38-4
Tritium, 37-6
True
  strain, 41-1
  stress, 41-1
Truss, 11-1
  bridge, 11-1

rigid, 11-1
  statically determinate, 11-2
  types of, 11-1
TTT curve, 42-7
Tube, pitot, 25-2
Tuberculosis, 35-17
Tumor, 34-6
Turbine, 27-4
  impulse, 24-9
  power (fluid), 24-9
Turbulent flow, 24-3, 24-4
Turns ratio, 45-7
TWA, 34-8
Twist, angle of, 19-3
Two-angle formula, 4-5
Two-force member, 10-3
Two-tail confidence limit, 6-5

U
UEL, 35-4 (tbl), 35-5
UFL, 35-5
Ulnar
  artery Aneurysm, 35-14
  nerve entrapment, 35-14
Ultimate
  analysis, 37-6
  strength, 41-2
Unadjusted basis, 52-1
Unbiased estimator, 6-2
Uncertainty, measurement, 49-5
Underdamped, 8-2
Undetermined coefficients, method of, 7-4,
  8-2
Uniaxial loading, 18-2
Uniform
  annual cost, equivalent, 52-2
  gradient factor, 51-5
  series cash flow, 51-1
  series compound amount factor, 51-3
  series equivalence, 51-3
Unit
  atomic mass, 37-1
  base, 1-4
  British thermal, 26-5
  cell, 40-1
  circle, 4-5
  derived, 1-4
  normal table, 6-4, 6-8 (tbl)
  operation, 32-1
  supplementary, 1-4
  vector, 5-5, 10-1
  vector, Cartesian, 5-5
  vector form (position), 14-2
  weight, 1-3
Units, 1-1
  consistent system of, 1-2
  of concentration, 39-1
  radiation, 35-10 (tbl)
Universal
  gas constant, 25-2, 26-7 (tbl), 38-3
  serial bus, 47-4
Universe (thermodynamic), 27-2
Unsaturated
  air, 29-3
  hydrocarbon, 38-5 (tbl)

Upper
  explosive limit, 35-5
  flammable limit, 35-5
USB, 47-4

V
Vacuole, 33-2
Vacuum, 22-2
Valence, 37-3, 37-5 (tbl)
Value, 52-2
  bond, 52-3
  expected, 6-3, 52-3
  face (of a bond), 52-3
  future, 51-3
  initial, problem, 8-2
  present, 51-3
  time, 51-2
Valve, throttling, 27-4
Van der Waals force, 40-4
Vapor
  dome, 26-2
  -gas mixture, 26-2
  -hazard ratio number, 35-5
  pressure, 35-3
  saturated, 26-2
  superheated, 26-2, 26-3
Variable
  continuous random, 6-3
  matrix, 5-5
  of integration, 7-3
  random, discrete, 6-3
Variance, 6-2
Variation, coefficient, 6-3
Vector, 5-5
  Cartesian unit, 5-5
  complex power, 45-5
  form (position), 14-2
  identities, 5-6
  null, 5-6
  operations, 5-5
  position, 10-2
  product, 5-6
  resultant, 5-5
  unit, 5-5, 10-1
Vector del operator, 7-5
Velocity, 16-2
  bulk, 24-4
  coefficient of, 25-3
  difference, relative, 24-8
  distribution, fluid, 24-4
  fluid jet, 24-8
  gradient, 22-3
  head, 24-2
  mass, 32-3
  noise, 35-8
  of approach, 25-2
  tangential, 14-3
  tangential blade, 24-9
Vena contracta, 25-3
Ventilation, 35-5, 35-7, 35-12
  dilution, 35-7
Venturi meter, 25-3
Vertex, 4-2

Vessel
  pressure, 19-2
  pressure, thick-walled, 19-3
Vibration
  forced, 15-3
  free, 15-3
  natural, 15-3
  natural frequency, 15-4
  syndrome, 35-14
  torsional, free, 16-3
Vickers hardness, 42-7
Video memory, 47-2
Villi, 34-3
Virtual
  machine, 47-5
  memory, 47-3
  memory storage, 47-3
  mode, 47-2
Virus, 33-6, 35-14
Viscometer, sliding plate, 22-3
Viscosity, 22-3
  absolute, 22-3
  kinematic, 22-3
  Newton's law of, 22-3
Voice-grade line, 47-6
Volatile
  acid, 3-10
  memory, 47-3
Voltage, 43-4
  induced, 43-6
  source, ideal, 44-1
Voltaic cell, 42-1
Volume
  centroid of, 13-1
  control, 27-2
  current density, 43-4
  partial, 29-2
  sludge, 36-3
  specific, 22-1, 26-4
von Mises stress, 18-5
VRAM, 47-3

W
Water
  boiling point constant, 39-3
  freezing point constant, 39-3
  horsepower, 25-1
  nonpotable, 33-4
  power, 25-1
  properties, 22-4
  triple point, 42-6
Waterborne disease, 33-11
Weber number, 25-5
Web (structural steel), 13-3
Weight, 1-1, 1-3, 13-2, 15-2
  and proportion problem, biologic, 33-11
  atomic, 37-1
  combining, 37-6
  density, 1-3
  equivalent, 37-6
  formula, 37-6
  molecular, 37-6
  of evidence, 34-6
  specific, 1-3, 22-1

Weighted
  mean, 6-2
  root-mean-square, 49-5
Welding fumes, 35-7
Wet-bulb temperature, 29-3
Wetted perimeter, 24-6
Wheatstone bridge, 49-3
Wheel, rolling, 16-3
Whistle-blowing, 54-3
White finger, 35-14
Wide-band line, 47-6
Work, 17-1
  charge, 43-4
  electrical, 44-2
  -energy theorem, 17-2
  flow, 26-5, 27-2
  graphical (thermodynamic), 28-6
  hardening, 42-6
  internal, 18-2
  p-V, 27-2
  -rest cycle, 35-12
  reversible, 27-2
  shaft, 27-3
Working
  cold, 42-6
  hot, 42-6
Worm, 33-8
Worst-case stack up, 49-5
Worth
  future, 51-3
  present, 51-3
Wrap angle, 12-2

X
XOFF, 47-4
XON, 47-4
X-ray, radiation, 35-10

Y
Year-end convention, 51-2
Yield
  biomass, 36-222
  bond, 52-3
  coefficient, 33-11
  factor, 33-11
  financial, 51-5
  point, 41-2
  strength, 41-2
  stress, 41-2
Young's modulus, 18-1, 41-2

Z
Zebra mussel, 33-8
Zero
  -force member, 11-2
  -indicating bridge, 49-3
  stability, 50-3
Zeroth law of thermodynamics, 26-4
Zone, critical, 24-4
Zoonotic disease, 35-15
z-transform, 8-4